T0328618

ECOFRIENDLY PEST MANAGEMENT FOR FOOD SECURITY

ECOFRIENDLY PEST MANAGEMENT FOR FOOD SECURITY

Edited by

OMKAR

Centre of Excellence in Biocontrol of Insect Pests
Ladybird Research Laboratory, Department of Zoology
University of Lucknow, Lucknow, India

AMSTERDAM • BOSTON • HEIDELBERG • LONDON
NEW YORK • OXFORD • PARIS • SAN DIEGO
SAN FRANCISCO • SINGAPORE • SYDNEY • TOKYO

Academic Press is an imprint of Elsevier

ISBN: 978-0-12-803265-7

British Library Cataloguing-in-Publication Data
A catalogue record for this book is available from the British Library

Library of Congress Cataloging-in-Publication Data
A catalog record for this book is available from the Library of Congress

For information on all Academic Press publications
visit our website at www.elsevier.com

Working together
to grow libraries in
developing countries

www.elsevier.com • www.bookaid.org

Publisher: Nikki Levy
Acquisition Editor: Nancy Maragioglio
Editorial Project Manager: Billie Jean Fernandez
Production Project Manager: Julie-Ann Stansfield
Designer: Maria Inês Cruz

Typeset by TNQ Books and Journals
www.tnq.co.in

Printed and bound in the United States of America

Contents

List of Contributors

Dunston P. Ambrose Entomology Research Unit, St. Xavier's College (Autonomous), Palayamkottai, Tamil Nadu, India

N. Bakthavatsalam ICAR-National Bureau of Agricultural Insect Resources, Bangalore, India

Chandish R. Ballal ICAR-National Bureau of Agricultural Insect Resources, Bangalore, India

Ajoy Kr. Choudhary Department of Botany and Biotechnology, TNB College, Bhagalpur, India

N. Dhandapani Department of Entomology, Tamil Nadu Agricultural University, Coimbatore, Tamil Nadu, India

Yaghoub Fathipour Department of Entomology, Faculty of Agriculture, Tarbiat Modares University, Tehran, Iran

S.K. Jalali Division of Molecular Entomology, National Bureau of Agricultural Insect Resources, Bengaluru, India

M. Kalyanasundaram Department of Agricultural Entomology, Agricultural College and Research Institute, Madurai, India

I. Merlin Kamala Department of Agricultural Entomology, Agricultural College and Research Institute, Madurai, India

P. Karuppuchamy Agricultural Research Station, Tamil Nadu Agricultural University, Bhavanisagar, Tamil Nadu, India

G. Keshavareddy Department of Entomology, University of Agriculture Sciences, GKVK, Bangalore, India

Opender Koul Insect Biopesticide Research Centre, Jalandhar, India

A.R.V. Kumar Department of Entomology, University of Agriculture Sciences, GKVK, Bangalore, India

Bhupendra Kumar Centre of Excellence in Biocontrol of Insect Pests, Ladybird Research Laboratory, Department of Zoology, University of Lucknow, Lucknow, India

A. Ganesh Kumar Entomology Research Unit, St. Xavier's College (Autonomous), Palayamkottai, Tamil Nadu, India

Priyanka Kumari Department of Botany and Biotechnology, TNB College, Bhagalpur, India

B.L. Lakshmi Priority Setting, Monitoring and Evaluation Cell, National Bureau of Agricultural Insect Resources, Bengaluru, India

Bahador Maleknia Department of Entomology, Faculty of Agriculture, Tarbiat Modares University, Tehran, Iran

Geetanjali Mishra Centre of Excellence in Biocontrol of Insect Pests, Ladybird Research Laboratory, Department of Zoology, University of Lucknow, Lucknow, India

Shikha Mishra CSIR-Central Drug Research Institute, Lucknow, India

Prashanth Mohanraj Division of Insect Systematics, National Bureau of Agricultural Insect Resources, Bengaluru, India

M. Muthulakshmi Department of Nematology, Tamil Nadu Agricultural University, Coimbatore, Tamil Nadu, India

Omkar Centre of Excellence in Biocontrol of Insect Pests, Ladybird Research Laboratory, Department of Zoology, University of Lucknow, Lucknow, India

Ahmad Pervez Department of Zoology, Radhey Hari Government Post Graduate College, Kashipur, India

Vivek Prasad Molecular Plant Virology Laboratory, Department of Botany, University of Lucknow, Lucknow, India

T.P. Rajendran   National Institute of Biotic Stress Management, Baronda, Raipur, India

Rashmi Roychoudhury   Department of Botany, University of Lucknow, Lucknow, India

Pallavi Sarkar   Department of Entomology, Tamil Nadu Agricultural University, Coimbatore, Tamil Nadu, India

K. Shankarganesh   Division of Entomology, Indian Agricultural Research Institute, New Delhi, India

Manish Shukla   Plant Production Research Centre, Directorate General of Agriculture and Livestock Research, Muscat, The Sultanate of Oman

Devendra Singh   Division of Agricultural Chemicals, Indian Agricultural Research Institute, New Delhi, India

Garima Singh   Department of Zoology, Rajasthan University, Jaipur, India

Rachana Singh   Amity Institute of Biotechnology, Amity University Uttar Pradesh, Lucknow, India

Rajendra Singh   Department of Zoology, Deendayal Upadhyay Gorakhpur University, Gorakhpur, India

Kaushal K. Sinha   Department of Botany, TM Bhagalpur University, Bhagalpur, India

Shalini Srivastava   Molecular Plant Virology Laboratory, Department of Botany, University of Lucknow, Lucknow, India

S. Subramanian   Department of Nematology, Tamil Nadu Agricultural University, Coimbatore, Tamil Nadu, India; Division of Entomology, Indian Agricultural Research Institute, New Delhi, India

Rajesh K. Tiwari   Amity Institute of Biotechnology, Amity University Uttar Pradesh, Lucknow, India

Arun K. Tripathi   CSIR-Central Institute of Medicinal and Aromatic Plants, Lucknow, India

Mala Trivedi   Amity Institute of Biotechnology, Amity University Uttar Pradesh, Lucknow, India

Sheela Venugopal   Agricultural Research Station, Tamil Nadu Agricultural University, Bhavanisagar, Tamil Nadu, India

Kazutaka Yamada   Tokushima Prefectural Museum, Tokushima, Japan

Preface

The human population across the world is expanding at an alarming exponential pace, with that of India at a staggering 1.27 billion and of my state Uttar Pradesh at more than 200 million. Paradoxically, as the need to feed higher amounts of quality food becomes increasingly urgent, agricultural lands are shrinking at an even more rapid pace, owing to accelerated growth of industries as well as a pressing need for housing. The use of chemical fertilizers for enhancing agricultural output, while reaping immediate benefits, has in the long term "killed the golden goose"—the quality of agricultural land has deteriorated immensely.

There are also numerous insect species as well as crop pathogens that cause losses to crop production either on account of their infestation and/or by causing and spreading diseases in crop plants. To overcome these problems, various synthetic chemicals have been used in agroecosystems, which have not only killed the beneficial insect species and caused development of resistance in pest species against them but also led to the introduction of new pest species, deterioration of the environment, i.e., ambient air and drinking water quality, as well as affecting human and animal health.

Providing adequate amounts of quality food seems to have become an increasingly elusive proposition, unless we aim to radically and dramatically change our approach to the above issues. A return to the lap of nature by smartly adopting ecofriendly management of agricultural land, pests, and vectors seems to be the need of the hour.

As a student of zoology and entomology, pest management has been an area of great fascination to me. My PhD research that revealed the adverse effects of pesticides to nontarget species, beneficial for aquaculture, further whetted my interest in ecofriendly approaches. In view of past experience, I selected Ladybird beetles, already established biocontrol agents, as a research model for investigations on reproductive strategies, age and aging, various aspects of ecology, prey–predator interactions, cannibalism, intraguild predation, and the role of chemicals in these phenomena.

While I have been fortunate to have received adequate funds from different state and central agencies for advancing my research work, it was the generous grant under the program of the Centre of Excellence by the Department of Higher Education, Government of Uttar Pradesh, that dramatically increased my excitement. Consequently, my team and I organized a National Symposium on Modern Approaches to Insect Pest Management, whose selected presentations were published under the title Modern Approaches to Insect Pest Management followed by a catalog on Ladybird Beetles of Uttar Pradesh under the aegis of the Centre of Excellence Program. In the follow-up of the same, I conceived the idea of publishing a book entitled *Ecofriendly Pest Management for Food Security* having contributions from renowned experts for international readership. The present book starts with the introduction of insects and pests, followed by biocontrol of pests, aphids and their biocontrol, role of parasitoids, predators, pathogens including *Bacillus thuringiensis*, semiochemicals, hormones as insecticides, biotechnological approaches, to GMOs and food security. I am confident that this book will not only provide interesting resource material for students, teachers, and researchers of this field but will also be quite useful to those involved in the policy planning.

I am grateful to the book's contributors for sparing valuable time from their busy schedules to write their chapters, as well as for positively accepting my criticism and sometimes harsh comments (and also for modifying their respective chapters as per suggestions). I am especially thankful to my past research team, including Drs R. B. Bind, Shefali Srivastava, Barish Emeline James, Ahmad Pervez, Geetanjali Mishra, Kalpana Singh, A. K. Gupta, Satyendra K. Singh, Rajesh Kumar, Shuchi Pathak, Priyanka Saxena, Shruti Rastogi, Pooja Pandey, Jyotsna Sahu, Uzma Afaq, Gyanendra Kumar, Mahadev Bista, Bhupendra Kumar, Neha Singh and Mohd. Shahid for being my strength and

to my present team, Dr Geetanjali Mishra (Assistant Professor, Grade III), Ms Garima Pandey, Ms Ankita Dubey, Mr Shashwat Singh, Mr Desh Deepak Choudhary, Ms Arshi Siddiqui, and Ms Swati Saxena for their unstinting support and help while I was working on this project. The generous financial support from the Department of Higher Education, Government of Uttar Pradesh, Lucknow under the Centre of Excellence in Biocontrol of Insect Pests is gratefully acknowledged. I am also thankful to my wife, Ms Kusum Upadhyay, for her sacrifice by sparing me for this work. Last, but not least, I also express my thanks to the Academic Press Division of Elsevier, Inc., USA, especially Ms Nancy Maragioglio, Ms Billie Jean Fernandez, and Ms Julie-Ann Stansfield for taking keen interest in this project and publishing this book in time, thus turning my dream into reality.

Omkar
July 2015

1

Insects and Pests

T.P. Rajendran[1], Devendra Singh[2]

[1]National Institute of Biotic Stress Management, Baronda, Raipur, India; [2]Division of
Agricultural Chemicals, Indian Agricultural Research Institute, New Delhi, India

1. INTRODUCTION

Insects have been recorded on this planet for 480 million years, since the early Ordovi-
cian era (Rohdendorf and Rasnitsyn, 1980; Rasnitsyn and Quicke, 2002). This conclusion was
confirmed on the basis of molecular data of genome sequences (Misof et al., 2014; Caterino
et al., 2000). This was approximately the time when plants also originated on Earth. The
chronology of events of coevolution witnessed the insects selecting various flora as their
primary food resource; the plants also provide food to other herbivores of the food chain of
our universe. Herbivory as a concept of exploitation of food resources is seen at its best in
the class Insecta under the phylum Arthropoda. In the geological upheavals due to weather
conditions and factors that determined adaptations of hexapods and flora on which they
were dependent for food and shelter, the evolutionary radiation did bring about a plethora
of variability of insects in their potency to exploit plant and animal resources. Phytophagous
insects became more predominant because of the availability of various flora. However, overgrazing of flora is controlled by regulating the herbivory of insects through defense chemicals
in target feeding tissues. Insects also adapted to the changing food resource and learned to
adapt to the chemical ecology over many millennia. The ability of phytophagous insects to
detoxify phytochemicals in host plants enables them to succeed with unchallenged survivorship. In the coevolution process, the changes in phytochemical profiles and the genetic ability
of phytophagous insects to survive these changes made them finally rule the roost as highly
successful herbivores on the primary producers—plants in the ecosystem. Wide adaptations
over several million years to all of the ecologies of the planet make insects ubiquitous in
their presence in almost all natural and manmade habitats. Their numbers and impacting
damage to various commodities in agricultural crop fields, in storage, and in the health and
well-being of animals and human beings have caused human beings to declare them strong
competitors of our civilization. Anthropogenic ecology called agriculture has modified the
behavior and bionomics of many naturally occurring insect genera and species, making them
disproportionate in numbers in the given nutritional host crop profile when compared with

other ecosystems. The key natural mortality factors of these insects are lower because of agricultural practices for crop production. The challenge on carrying capacity of these insects makes them survive on an r-/K-selection basis (Southwood, 1975, 1978). Their survivorship depends on the quality of host tissues and pressure from natural mortality factors (Andrewartha and Birch, 1971). The spatial structure of insect population dynamics is related to the food source and favorable weather conditions (Hassell et al., 1991).

2. PHYTOPHAGOUS INSECTS

The herbivores are a specialized group with specific adaptations to live on various plant species that have been evolutionarily adapted for food and shelter to complete their biology. Their metabolic needs are met by exploiting phytochemicals for energy, nutrition, and other metabolic needs. In turn, they also spend energy to detoxify many toxic phytochemicals that get into their body through the food they take from plants. Polyphagy, oligophagy, and monophagy in insects have been defined by Cates (1980) and Bernays and Chapman (1994) in the context of resource exploitation in various host plants. Specialist feeders are oligophagous or monophagous types that have higher sensitivity for host selection (Bernays and Funk, 1999). The phytochemistry profile of host plants does determine the preferential choice between females and males of heteropteran insects, and their neural sensitivity shall decide the diet breadth and evolution of host plant association (Bernays, 2001). A specific increase in damage due to sap-sucking pests in crops has been noteworthy in this millennium to suggest that the manmade agroecologies have destabilized natural biodiversity (Blumler et al., 1991) so as to affect their key natural mortality factors. Indeed, several research reports of this millennium suggest that the transient ecologies as in agriculture provide adequate evolutionary challenge for insects to sustain adaptations into genetic variations that can be fixed into the speciation process.

2.1 Agriculture for Commodity Production

The crop production concept for farming of crop commodities is supposed to have originated on Earth approximately 12,000 years ago. The domestication of useful flora from the wild into domesticated crop plants led to the development of agriculture, including cropping and livestock management, dating back 9500–12,000 years from the present time, spread across different continents of the planet. The Indian subcontinent that is beyond the present political barriers of nations around India took to practicing agriculture between 7000 and 9500 years ago. Agriculture as a manmade ecosystem, agroecology, became alienated from natural ecosystems as in forests. The invasion into natural habitats for homing plants as crops and farming them for profitable sustenance of human life made agriculture a specialized human endeavor (Sanderson et al., 2002). Food production became the harbinger of civilizations. Over several centuries, as the human population competitively grew to a large size, the competition with co-living organisms for sharing all natural elements and resources became the order of the day. Agriculture also forged in animal husbandry along with crop husbandry. Livestock and fisheries became more than livelihood assurance to the nutritional security of communities. In the current millennium in which world trade order is decisive to make nations prospect agriculture for higher economic gain, the trade of agricultural commodities has literally steered the policy and practice of crop production in the modern world in accordance with the trade value and volume. This has resulted in specialized monocropping instead of the earlier philosophy of mixed cropping. The challenge thrown at humanity today is in optimizing the efficiency of utilization of natural resources and other agricultural inputs.

In the context of the production of food, fiber, fodder, and feed, the methodology developed appreciated the role of several pest species that depredated these crops. The evolutionary development of herbivory on these crop plant species was coevolved from their wild relatives as well as from these natural selections of crop plant strains that bore commodities with desirable traits. The early phase of agriculture was free from major herbivory. The severity of insect damage due to increased pestilence on crop plants is now known to be due to various crop husbandry measures that tend to make crops more nutritious over their naturally occurring counterparts. It is also important to realize that extensive seasonal monocropping provided manmade opportunity for insects to more extensively exploit the agricultural resources. Thus, the insects evolved as pests in large scale because of anthropogenic agroecology, which drives the need for satiation of human needs through farming-system-based agriculture. Gould (1991) explained the evolutionary potential of crop pests in agricultural systems. The evolution of pest insects due to cropping systems based on their polyphagy or oligophagy progressed over many centuries. Complex adaptations have led to specializations in herbivory to make phytophagous insects develop specific host plant selection. This evolutionary relationship makes the insects acquire shelter and safety from their potential natural enemies by using a phytochemical-based nonrecognition mechanism (Paschold et al., 2006).

2.2 Pest Incidence in Crops

Pest is the broad term given to noxious insect species that damage crops, animals, humans, stored commodities, timber, and many such products that come to be used by man. In the broadest term, insects using all of these items of commodities as biological resources for their survivorship have been called pests. The competition between insects and human beings for exploitation of the same natural resource has led to the origin of the term *pestilence*. Insect species increased their population on host crops according to the favorable conditions and caused economic damage to crop plants. The economic measures of the damage in terms of injury to plant parts and the threshold level of pest number and damage were used to define the timing of suitable intervention for pest management (Southwood and Norton, 1973; Pedigo et al., 1986).

The prominent crops and their insect pests are compiled in Table 1. This list has the insects that regularly damage crops grown in Indian agricultural farms such as cereals (rice, wheat, maize, sorghum, etc.), pulses (pigeonpea, chickpea, green gram, black gram, etc.), oilseeds (rapeseed mustard, sesame, groundnut, etc.), vegetables, fruits, and other crops of economic significance. This is not exhaustive, but it is indicative of the insect numbers that have adapted to various crops. Looking at the distribution of similar genera and species of insects in crop plants, one can find a pattern of flora taxa in the case of oligophagy whereas in polyphagous pests, because of their wide adaptability, they can make use of plants from diverse families for food resource.

TABLE 1 Insects Pests of Major Indian Crops (Insect Numbers in Parentheses)

Crop	Pests
Rice (57)	*Acrida exaltata* (Walker), *Ampittia dioscorides* (Fabricius), *Anomala dimidiata* Hope, *Bothrogonia albidicans* (Walker), *Callitettix versicolor* (Fabricius), *Chilo partellus* (Swinhoe), *Chilo polychrysa* (Meyrick), *Cletus punctiger* (Dallas), *Cnaphalocrocis medinalis* Guenée, *Cnaphalocrocis trapezalis* (Guenée), *Cofana spectra* (Distant), *Cofana unimaculata* (Signoret), *Corcyra cephalonica* (Stainton), *Creatonotos gangis* (Linnaeus), *Cryptocephalus schestedti* Fabricius, *Dicladispa armigera* (Olivier), *Diostrombus carnosa* (Westwood), *Euproctis similis* (Moore), *Helcystogramma arotraea* (Meyrick), *Hieroglyphus banian* (Fabricius), *Hispa ramosa* group, *Hispa stygia* (Chapuis), *Lenodora vittata* (Walker), *Leptispa pygmaea* Baly, *Leptocorisa acuta* (Thunberg), *Leptocorisa oratorius* (Fabricius), *Lyclene* sp., *Melanitis leda* Linnaeus, *Menida versicolor* (Gmelin), *Mocis frugalis* (Fabricius), *Myllocerus dentifer* (Fabricius), *Mythimna loreyi* (Duponchel), *Mythimna separata* (Walker), *Nephotettix malayanus* Ishihara & Kawase, *Nephotettix nigropictus* (Stal), *Nephotettix parvus* Ishihara & Kawase, *Nezara viridula* (Linnaeus), *Nilaparvata lugens* (Stal), *Nisia nervosa* (Motschulsky), *Oedaleus senegalensis* (Krauss), *Oxya hyla* Serville, *Parapoynx fluctuosalis* (Zeller), *Parapoynx stagnalis* (Zeller), *Psalis pennatula* (Fabricius), *Psalydolytta rouxi* (Castelnau), *Pyrilla perpusilla* (Walker), *Recilia dorsalis* (Motschulsky), *Schistophleps bipuncta* Hampson, *Scirpophaga incertulas* (Walker), *Scirpophaga innotata* (Walker), *Scotinophara* sp.1, *Scotinophara* sp.2, *Sesamia inferens* (Walker), *Sitotroga cerealella* (Olivier), *Sogatella furcifera* (Horvath), *Trilophidia annulata* (Thunberg), *Tropidocephala serendiba* (Melichar)
Maize (50)	*Agonoscelis nubilis* (Fabricius), *Agrotis* sp., *Amsacta albistriga* (Walker), *A. dimidiata* Hope, *Atherigona soccata* (Rondani), *Atractomorpha crenulata* (Fabricius), *C. versicolor* (Fabricius), *Cerococcus indicus* (Maskell), *Chaetocnema* sp., *Chiloloba acuta* (Wiedemann), *C. partellus* (Swinhoe), *Chrysodeixis chalcites* (Esper), *Clinteria klugi* (Hope), *C. trapezalis* (Guenée), *Cyrtacanthacris tatarica* (L.), *D. carnosa* (Westwood), *Dolycoris indicus* Stal, *Eumeta crameri* (Westwood), *Hieroglyphus nigrorepletus* Bolívar, *Hysteroneura setariae* (Thomas), *M. leda* Linnaeus, *M. frugalis* (Fabricius), *Monolepta signata* Olivier, *Myllocerus discolor* Boheman, *Myllocerus undecimpustulatus* Faust, *Myllocerus viridanus* Fabricius, *M. loreyi* (Duponchel), *M. separata* (Walker), *N. viridula* (Linnaeus), *O. senegalensis* (Krauss), *Olene mendosa* Hubner, *Oxycetonia versicolor* (Fabricius), *Patanga succincta* (Johannson), *Peregrinus maidis* (Ashmead), *Pericallia ricini* (Fabricius), *Piezodorus hybneri* (Gmelin), *Protaetia alboguttata* Vigors, *Protaetia cinerea* (Kraatz), *Protaetia squamipennis* Burmeister, *Proutista moesta* (Westwood), *P. rouxi* (Castelnau), *P. perpusilla* (Walker), *Rhopalosiphum maidis* (Fitch), *Schistocerca gregaria* (Forskål), *S. inferens* (Walker), *Sitophilus oryzae* (Linnaeus), *S. cerealella* (Olivier), *S. furcifera* (Horvath), *Spodoptera exigua* (Huebner), *Spodoptera litura* (Fabricius)
Sorghum (30)	*A. exaltata* (Walker), *A. nubilis* (Fabricius), *Amsacta lactinea* (Cramer), *Archips micaceana* (Walker), *A. crenulata* (Fabricius), *B. albidicans* (Walker), *C. versicolor* (Fabricius), *Chaetocnema* sp., *C. partellus* (Swinhoe), *C. punctiger* (Dallas), *C. gangis* (Linnaeus), *D. carnosa* (Westwood), *E. similis* (Moore), *Eyprepocnemis alacris* (Serville), *Helicoverpa armigera* (Hübner), *H. setariae* (Thomas), *Luperomorpha vittata* Duvivier, *M. frugalis* (Fabricius), *M. signata* Olivier, *M. discolor* Boheman, *M. viridanus* Fabricius, *M. separata* (Walker), *N. viridula* (Linnaeus), *O. senegalensis* (Krauss), *O. hyla* Serville, *P. rouxi* (Castelnau), *S. inferens* (Walker), *Somena scintillans* (Walker), *S. litura* (Fabricius), *Tetraneura nigriabdominalis* (Sasaki)
Sugarcane (39)	*Abdastartus atrus* (Motschulsky), *Aceria sacchari* Wang, *Aleurolobus barodensis* (Maskell), *A. dioscorides* (Fabricius), *Antonina graminis* (Maskell), *Ceratovacuna lanigera* Zehntner, *Chilo infuscatellus* Snellen, *C. partellus* (Swinhoe), *Chilo sacchariphagus indicus* (Kapur), *C. punctiger* (Dallas), *C. trapezalis* (Guenée), *C. spectra* (Distant), *Cofana subvirescens* (Stål), *Colemania sphenarioides* Bolívar, *C. schestedti* Fabricius, *H. ramosa* group, *H. stygia* (Chapuis), *Holotrichia serrata* (Fabricius), *H. setariae* (Thomas), *Icerya pilosa* Green, *Kiritshenkella sacchari* (Green), *Lepidiota mansueta* (Burmeister), *Melanaphis sacchari* (Zehntner), *Melanaspis glomerata* (Green), *M. frugalis* (Fabricius), *M. separata* (Walker), *Neomaskellia bergii* (Signoret), *Odonaspis* sp., *Oryctes rhinoceros* (Linnaeus), *Poophilus costalis* (Walker), *P. moesta* (Westwood), *P. pennatula* (Fabricius), *P. perpusilla* (Walker), *Saccharicoccus sacchari* (Cockerell), *S. gregaria* (Forskål), *Scirpophaga excerptalis* (Walker), *S. inferens* (Walker), *T. serendiba* (Melichar), *Varta rubrofasciata* Distant

TABLE 1 Insects Pests of Major Indian Crops (Insect Numbers in Parentheses)—cont'd

Crop	Pests
Pigeonpea (66)	*Amata passalis* (Fabricius), *A. albistriga* (Walker), *Anoplocnemis phasiana* (Fabricius), *Aphis craccivora* Koch, *Atteva fabriciella* (Swederus), *Aularches miliaris* (Linnaeus), *Callosobruchus* sp., *Chrysocoris stolli* Wolff, *Clavigralla gibbosa* (Spinola), *Clavigralla scutellaris* (Westwood), *C. punctiger* (Dallas), *Coccidohystrix insolita* (Green), *Coptosoma variegata* (Herrich-Schäffer), *D. indicus* Stal, *Drepanococcus cajani* (Maskell), *Drepanococcus chiton* (Green), *Episomus lacerta* (Fabricius), *Etiella zinckenella* (Treitschke), *Euchrysops cnejus* (Fabricius), *Eurybrachys* sp., *Eurystylus* sp., *Exelastis atomosa* (Walsingham), *Ferrisia virgata* (Cockerell), *Glyphodes bivitralis* Guenée, *Halyomorpha picus* (Fabricius), *H. armigera* (Hübner), *Icerya purchasi* Maskell, *Indozacladus theresiae* (Dalla Torre), *Lampides boeticus* (Linnaeus), *Lobesia aeolopa* Meyrick, *Maruca vitrata* (Fabricius), *Megacopta cribraria* (F.), *Melanagromyza obtusa* (Malloch), *M. leda* Linnaeus, *Menida formosa* (Westwood), *M. versicolor* (Gmelin), *Merilia lunulata* (Fabricius), *Metacanthus pulchellus* Dallas, *Mocis undata* (Fabricius), *Mylabris pustulata* Thunberg, *Myllocerus dorsatus* (Fabricius), *M. undecimpustulatus* Faust, *Nanaguna breviuscula* Walker, *Neostauropus alternus* Walker, *N. viridula* (Linnaeus), *Oecophylla smaragdina* (Fabricius), *O. mendosa* Hubner, *Omiodes indicata* Fabricius, *Ophiomyia phaseoli* (Tryon), *Orgyia postica* (Walker), *O. versicolor* (Fabricius), *Pammene critica* (Meyrick), *P. hybneri* (Gmelin), *Plautia crossota* (Dallas), *Poppiocapsidea biseratense* (Distant), *Riptortus linearis* (Linnaeus), *Sagra femorata* (Drury), *S. gregaria* (Forskål), *Scutellera perplexa* (Westwood), *S. scintillans* (Walker), *Sphenarches caffer* (Zeller), *Sphenoptera perroteti* Guerin-Meneville, *Spilosoma obliqua* Walker, *Sternuchopsis collaris* (Pascoe), *Tanaostigmodes cajaninae* La Salle, *Tanymecus indicus* Faust
Chickpea (3)	*E. zinckenella* (Treitschke), *H. armigera* (Hübner), *M. obtusa* (Malloch)
Green & black grams (20)	*Agrius convolvuli* (Linnaeus), *Alcidodes fabricii* (Fabricius), *Anomis flava* (Fabricius), *A. craccivora* Koch, *Apion* sp., *C. scutellaris* (Westwood), *Chionaema peregrina* (Walker), *Epilachna ocellata* Redtenbacher, *E. zinckenella* (Treitschke), *E. cnejus* (Fabricius), *H. armigera* (Hübner), *M. vitrata* (Fabricius), *M. obtusa* (Malloch), *M. frugalis* (Fabricius), *M. undata* (Fabricius), *Obereopsis brevis* (Swedenbord), *O. indicata* Fabricius, *P. pennatula* (Fabricius), *P. rouxi* (Castelnau), *S. collaris* (Pascoe)
Horse gram (5)	*A. fabricii* (Fabricius), *Apion* sp., *B. albidicans* (Walker), *Chauliops choprai* sweet and Schaeffer, *E. zinckenella* (Treitschke)
Cowpea (24)	*A. passalis* (Fabricius), *A. phasiana* (Fabricius), *A. craccivora* Koch, *Callosobruchus maculatus* (Fabricius), *E. lacerta* (Fabricius), *E. zinckenella* (Treitschke), *E. cnejus* (Fabricius), *H. picus* (Fabricius), *L. boeticus* (Linnaeus), *M. vitrata* (Fabricius), *M. cribraria* (F.), *M. obtusa* (Malloch), *M. versicolor* (Gmelin), *M. undata* (Fabricius), *M. signata* Olivier, *M. dorsatus* (Fabricius), *M. undecimpustulatus* Faust, *Neptis hylas* Linnaeus, *O. indicata* Fabricius, *O. phaseoli* (Tryon), *P. hybneri* (Gmelin), *P. crossota* (Dallas), *R. linearis* (Linnaeus), *S. scintillans* (Walker)
Soybean (14)	*Agrotis* sp., *Aproaerema modicella* (Deventer), *C. choprai* sweet and Schaeffer, *C. chalcites* (Esper), *E. zinckenella* (Treitschke), *M. vitrata* (Fabricius), *M. cribraria* (F.), *M. undata* (Fabricius), *O. brevis* (Swedenbord), *O. indicata* Fabricius, *P. hybneri* (Gmelin), *Platypria hystrix* (Fabricius), *S. litura* (Fabricius), *Thysanoplusia orichalcea* (Fabricius)
Field bean (17)	*Acherontia styx* (Westwood), *Adisura atkinsoni* (Moore), *Alcidodes liae* Alonso-Zarazaga, *A. phasiana* (Fabricius), *Coridius janus* (Fabricius), *E. lacerta* (Fabricius), *E. zinckenella* (Treitschk), *E. cnejus* (Fabricius), *L. boeticus* (Linnaeus), *M. vitrata* (Fabricius), *M. cribraria* (F.), *M. dorsatus* (Fabricius), *P. hystrix* (Fabricius), *P. crossota* (Dallas), *S. femorata* (Drury), *S. scintillans* (Walker), *S. caffer* (Zeller)
Sesame (6)	*A. styx* (Westwood), *Eysarcoris* sp., *Nesidiocoris tenuis* (Reuter), *Orosius albicinctus* Distant, *P. ricini* (Fabricius), *P. hystrix* (Fabricius)

(Continued)

TABLE 1 Insects Pests of Major Indian Crops (Insect Numbers in Parentheses)—cont'd

Crop	Pests
Castor (39)	*Acanthodelta janata* (Linnaeus), *Agrotis* sp., *A. albistriga* (Walker), *A. lactinea* (Cramer), *Ariadne ariadne* (Linnaeus), *Ariadne merione* (Cramer), *Artaxa guttata* Walker, *Artaxa* sp. (*vitellina*-group), *Asota ficus* (Fabricius), *A. crenulata* (Fabricius), *A. miliaris* (Linnaeus), *Belippa* sp., *Berta chrysolineata* Walker, *Biston suppressaria* (Guenée), *Dichocrocis punctiferalis* (Guenée), *Dysgonia algira* (Linnaeus), *E. crameri* (Westwood), *Eupterote undata* Blanchard, *Euwallacea fornicatus* (Eichhoff), *Hasora chromus* (Cramer), *H. armigera* (Hübner), *Hyposidra talaca* Walker, *Liriomyza trifolii* (Burgess), *L. aeolopa* Meyrick, *N. viridula* (Linnaeus), *O. mendosa* Hubner, *O. postica* (Walker), *Retithrips syriacus* (Mayet), *Parasa lepida* (Cramer), *P. ricini* (Fabricius), *Pinnaspis strachani* (Cooley), *Pseudococcus longispinus* (Targioni-Tozetti), *S. scintillans* (Walker), *Spatulifimbria castaneiceps* Hampson, *S. obliqua* Walker, *S. exigua* (Huebner), *S. litura* (Fabricius), *Tetranychus urticae* Koch, *Trypanophora semihyalina* Kollar
Groundnut (22)	*A. styx* (Westwood), *A. convolvuli* (Linnaeus), *Agrotis* sp., *A. albistriga* (Walker), *A. craccivora* Koch, *A. modicella* (Deventer), *A. crenulata* (Fabricius), *Caryedon serratus* (Fabricius), *C. chalcites* (Esper), *C. gangis* (Linnaeus), *Dudua aprobola* (Meyrick), *Franliniella schultzei* (Trybom), *H. armigera* (Hübner), *M. pustulata* Thunberg, *M. viridanus* Fabricius, *O. indicata* Fabricius, *O. versicolor* (Fabricius), *S. perroteti* Guerin-Meneville, *S. obliqua* Walker, *S. exigua* (Huebner), *S. litura* (Fabricius), *T. indicus* Faust
Sunflower (14)	*A. convolvuli* (Linnaeus), *A. lactinea* (Cramer), *Blosyrus inequalis* Boheman, *C. chalcites* (Esper), *D. indicus* Stal, *H. armigera* (Hübner), *Macherota* sp., *M. undecimpustulatus* Faust, *O. brevis* (Swedenbord), *P. ricini* (Fabricius), *Pseudaulacaspis cockerelli* (Cooley), *S. obliqua* Walker, *S. exigua* (Huebner), *T. orichalcea* (Fabricius)
Safflower (3)	*Dioxyna sororcula* (Wiedemann), *Prospalta capensis* (Guenée), *T. indicus* Faust
Cotton (50)	*Agrotis* sp., *Amorphoidea arcuata* Motschulsky, *Amrasca biguttula biguttula* (Ishida), *A. albistriga* (Walker), *A. flava* (Fabricius), *Anomis sabulifera* (Guenée), *Aphis gossypii* Glover, *Bemisia tabaci* (Gennadius), *C. indicus* (Maskell), *C. chalcites* (Esper), *Coccus hesperidum* Linnaeus, *C. tatarica* (L.), *Dysdercus koenigii* Fabricius, *Dysdercus* sp., *Earias vittella* (Fabricius), *Eurybrachys* sp., *F. virgata* (Cockerell), *F. schultzei* (Trybom), *H. picus* (Fabricius), *H. armigera* (Hübner), *Hemiberlesia lataniae* (Signoret), *Hermolaus typicus* Distant, *Maconellicoccus hirsutus* (Green), *Macrocheraia grandis* (Gray), *Megapulvinaria maxima* (Green), *M. formosa* (Westwood), *M. versicolor* (Gmelin), *M. undata* (Fabricius), *M. signata* Olivier, *M. pustulata* Thunberg, *M. dorsatus* (Fabricius), *Myllocerus subfasciatus* Guerin-Meneville, *M. undecimpustulatus* Faust, *N. viridula* (Linnaeus), *Nipaecoccus viridis* (Newstead), *Odontopus varicornis* (Distant), *O. mendosa* Hubner, *Oxycarenus hyalinipennis* (Costa), *Pectinophora gossypiella* (Saunders), *Pempherulus affinis* (Faust), *P. ricini* (Fabricius), *Phenacoccus madeirensis* Green, *Phenacoccus solenopsis* Tinsley, *P. hybneri* (Gmelin), *P. crossota* (Dallas), *P. biseratense* (Distant), *Rastrococcus iceryoides* (Green), *S. gregaria* (Forskål), *S. litura* (Fabricius), *Syllepte derogata* (Fabricius), *Xanthodes albago* (Fabricius), *Xanthodes transversa* Guenée
Sunhemp (9)	*Asota caricae* Fabricius, *E. zinckenella* (Treitschke), *E. cnejus* (Fabricius), *L. boeticus* (Linnaeus), *Longitarsus belgaumensis* Jacoby, *Mangina argus* (Kollar), *O. mendosa* Hubner, *O. indicata* Fabricius, *Utetheisa pulchella* (Linnaeus)
Brinjal (41)	*Acanthocoris scabrator* (Fabricius), *A. styx* (Westwood), *A. biguttula biguttula* (Ishida), *A. phasiana* (Fabricius), *A. gossypii* Glover, *A. crenulata* (Fabricius), *Autoba olivacea* (Walker), *Bactrocera latifrons* (Hendel), *B. tabaci* (Gennadius), *Ceratoneura indi* Girault, *C. indicus* (Maskell), *C. chalcites* (Esper), *Cletomorpha hastata* (Fabricius), *C. insolita* (Green), *C. janus* (Fabricius), *D. chiton* (Green), *E. ocellata* Redtenbacher, *Epilachna vigintioctopunctata* (Fabricius), *Goethella asulcata* Girault, *Herpetogramma bipunctalis* (Fabricius), *Insignorthezia insignis* (Browne), *Leucinodes orbonalis* (Guenée), *Lygus* sp., *M. pulchellus* Dallas, *M. subfasciatus* Guerin-Meneville, *N. tenuis* (Reuter), *N. viridula* (Linnaeus), *O. mendosa* Hubner, *O. versicolor* (Fabricius), *P. ricini* (Fabricius), *P. madeirensis* Green, *P. solenopsis* Tinsley, *Phthorimaea operculella* (Zeller), *P. hybneri* (Gmelin), *P. strachani* (Cooley), *P. alboguttata* Vigors, *P. cinerea* (Kraatz), *P. squamipennis* Burmeister, *S. litura* (Fabricius), *Urentius hystricellus* (Richter), *X. transversa* Guenée

TABLE 1 Insects Pests of Major Indian Crops (Insect Numbers in Parentheses)—cont'd

Crop	Pests
Okra (22)	*A. biguttula biguttula* (Ishida), *A. flava* (Fabricius), *A. sabulifera* (Guenée), *A. gossypii* Glover, *B. albidicans* (Walker), *D. carnosa* (Westwood), *D. koenigii* Fabricius, *E. vittella* (Fabricius), *E. ocellata* Redtenbacher, *Eurybrachys* sp., *H. armigera* (Hübner), *M. grandis* (Gray), *M. pustulata* Thunberg, *M. viridanus* Fabricius, *N. viridis* (Newstead), *P. gossypiella* (Saunders), *P. affinis* (Faust), *P. solenopsis* Tinsley, *S. exigua* (Huebner), *S. derogata* (Fabricius), *X. albago* (Fabricius), *X. transversa* Guenée
Cluster bean (5)	*A. liae* Alonso-Zarazaga, *B. inequalis* Boheman, *Cylindralcides bubo* (Fabricius), *M. cribraria* (F.), *S. gregaria* (Forskål)
Sweet potato (14)	*A. convolvuli* (Linnaeus), *Amata cyssea* Stoll, *A. lactinea* (Cramer), *Aspidimorpha miliaris* (Fabricius), *B. tabaci* (Gennadius), *Cassida circumdata* Herbst, *Chiridopsis bipunctata* (Linnaeus), *C. gangis* (Linnaeus), *Cylas formicarius* (Linnaeus), *Junonia orithya* Linnaeus, *P. ricini* (Fabricius), *Physomerus grossipes* (Fabricius), *Planococcus citri* (Risso), *S. obliqua* Walker
Potato (23)	*A. dimidiata* Hope, *A. gossypii* Glover, *Aphis spiraecola* Patch, *Atmetonychus peregrinus* (Olivier), *A. crenulata* (Fabricius), *C. chalcites* (Esper), *E. ocellata* Redtenbacher, *E. vigintioctopunctata* (Fabricius), *H. armigera* (Hübner), *Melolontha indica* Hope, *M. undata* (Fabricius), *M. dorsatus* (Fabricius), *M. subfasciatus* Guerin-Meneville, *N. viridula* (Linnaeus), *O. mendosa* Hubner, *O. albicinctus* Distant, *P. madeirensis* Green, *P. solenopsis* Tinsley, *P. operculella* (Zeller), *P. hybneri* (Gmelin), *S. exigua* (Huebner), *T. urticae* Koch., *T. orichalcea* (Fabricius)
Cucurbits (26)	*A. albistriga* (Walker), *Anadevidia peponis* (Fabricius), *A. gossypii* Glover, *Apomecyna saltator* (Fabricius), *Aulacophora* sp., *Aulacophora cincta* (Fabricius), *Aulacophora foveicollis* (Lucas), *Bactrocera cucurbitae* (Coquillett), *Bactrocera diversa* (Coquillett), *B. latifrons* (Hendel), *Ceroplastes ceriferus* (Fabricius), *C. janus* (Fabricius), *Diaphania indica* (Saunders), *E. ocellata* Redtenbacher, *Eucalymnatus tessellatus* (Signoret), *Henosepilachna elaterii* (Rossi), *L. trifolii* (Burgess), *M. pulchellus* Dallas, *N. tenuis* (Reuter), *N. viridula* (Linnaeus), *P. ricini* (Fabricius), *P. strachani* (Cooley), *P. citri* (Risso), *S. scintillans* (Walker), *S. caffer* (Zeller), *S. obliqua* Walker
Amaranthus (7)	*C. hastata* (Fabricius), *C. punctiger* (Dallas), *D. indicus* Stal, *Hypolixus truncatulus* (Fabricius), *J. orithya* Linnaeus, *N. viridula* (Linnaeus), *Spoladea recurvalis* (Fabricius)
Onion (2)	*A. crenulata* (Fabricius), *C. chalcites* (Esper)
Chillies (6)	*B. tabaci* (Gennadius), *C. hesperidum* Linnaeus, *D. carnosa* (Westwood), *G. asulcata* Girault, *H. armigera* (Hübner), *P. hybneri* (Gmelin)
Colocasia (3)	*M. signata* Olivier, *S. litura* (Fabricius), *Stephanitis typicus* Distant
Curryleaf (9)	*Aleurocanthus woglumi* Ashby, *Aonidiella aurantii* (Maskell), *Aonidiella orientalis* (Newstead), *Diaphorina citri* Kuwayama, *Papilio demoleus* (Linnaeus), *Papilio polytes* Linnaeus, *P. strachani* (Cooley), *Psorosticha zizyphi* (Stainton), *Silana farinosa* (Boheman)
Tomato (23)	*Aleurodicus dispersus* Russell, *A. gossypii* Glover, *B. diversa* (Coquillett), *B. latifrons* (Hendel), *B. tabaci* (Gennadius), *C. chalcites* (Esper), *E. ocellata* Redtenbacher, *E. vigintioctopunctata* (Fabricius), *F. virgata* (Cockerell), *Frankliniella occidentalis* (Pergande), *F. schultzei* (Trybom), *H. armigera* (Hübner), *Icerya aegyptiaca* (Douglas), *L. trifolii* (Burgess), *M. pulchellus* Dallas, *N. tenuis* (Reuter), *P. madeirensis* Green, *P. solenopsis* Tinsley, *P. operculella* (Zeller), *P. hybneri* (Gmelin), *S. litura* (Fabricius), *T. urticae* Koch, *Tuta absoluta* (Meyrick)
Crucifers (16)	*Agrotis* sp., *A. spiraecola* Patch, *Athalia proxima* (Klug), *Bagrada hilaris* (Burmeister), *Brevicoryne brassicae* (Linnaeus), *Chromatomyia horticola* (Goureau), *Crocidolomia pavonana* (Fabricius), *E. ocellata* Redtenbacher, *Eurydema* sp., *Hellula undalis* (Fabricius), *Lipaphis erysimi* (Kaltenbach), *N. viridula* (Linnaeus), *Pieris brassicae* (Linnaeus), *Pieris canidia* Linnaeus, *Plutella xylostella* (L.), *S. litura* (Fabricius)

(Continued)

TABLE 1   Insects Pests of Major Indian Crops (Insect Numbers in Parentheses)—cont'd

Crop	Pests
Mango (85)	*A. janata* (Linnaeus), *Acrocercops syngramma* Meyrick, *Amplypterus panopus* (Cramer), *Amrasca splendens* Ghauri, *Amritodus atkinsoni* Lethierry, *Amritodus brevistylus* Viraktamath, *Anacridium flavescens* (Fabricius), *A. aurantii* (Maskell), *A. orientalis* (Newstead), *Aphis odinae* (van der Goot), *Apoderus tranquebaricus* Fabricius, *A. micaceana* (Walker), *Aspidiotus destructor* Signoret, *A. miliaris* (Linnaeus), *Bactrocera caryeae* (Kapoor), *Bactrocera correcta* (Bezzi), *Bactrocera dorsalis* Hendel, *Bactrocera zonata* (Saunders), *Batocera parryi* Hope, *Batocera rufomaculata* (De Geer), *Ceroplastes floridensis* Comstock, *Ceroplastes rubens* Maskell, *Ceroplastes stellifer* (Westwood), *Cerosterna scabratrix* (Fabricius), *Chlumetia transversa* (Walker), *Chrysomphalus aonidum* (Linnaeus), *C. hesperidum* Linnaeus, *Coptops aedificator* (Fabricius), *Cricula trifenestrata* (Helfer), *Deanolis sublimbalis* (Snellen), *Deporaus marginatus* (Pascoe), *D. punctiferalis* (Guenée), *Drosicha mangiferae* Green, *D. aprobola* (Meyrick), *E. tessellatus* (Signoret), *Eumeta variegata* (Snellen), *Euthalia aconthea* (Cramer), *E. fornicatus* (Eichhoff), *Gastropacha pardale* (Walker), *Gatesclarkeana* sp., *Gonodontis clelia* (Cramer), *Hemaspidoproctus cinereus* (Green), *H. lataniae* (Signoret), *Heterorrhina elegans* (Fabricius), *Howardia biclavis* (Comstock), *I. aegyptiaca* (Douglas), *Icerya seychellarum* (Westwood), *Idioscopus clypealis* (Lethierry), *Idioscopus nagpurensis* (Pruthi), *Idioscopus nitidulus* (Walker), *Indarbela tetraonis* Moore, *Labioproctus poleii* (Green), *Lymantria marginata* Walker, *Lypesthes kanarensis* (Jacoby), *M. maxima* (Green), *M. discolor* Boheman, *M. undecimpustulatus* Faust, *N. breviuscula* Walker, *N. viridis* (Newstead), *O. smaragdina* (Fabricius), *O. mendosa* Hubner, *Orthaga exvinacea* (Hampson), *P. lepida* (Cramer), *Paratachardina lobata* (Chamberlin), *Pauropsylla* sp., *Penicillaria jocosatrix* Guenée, *Perina nuda* (Fabricius), *P. strachani* (Cooley), *P. citri* (Risso), *Prococcus acutissimus* (Green), *Pseudaonidia trilobitiformis* (Green), *P. cockerelli* (Cooley), *P. longispinus* (Targioni-Tozetti), *Pulvinaria psidii* Maskell, *Rastrococcus* sp., *R. iceryoides* (Green), *Rastrococcus mangiferae* (Green), *Rathinda amor* (Fabricius), *Rhynchaenus mangiferae* Marshall, *Siphanta* sp., *S. scintillans* (Walker), *S. castaneiceps* Hampson, *Sternochetus mangiferae* (Fabricius), *Strepsicrates rhothia* (Meyrick), *Thalassodes falsaria* Prout
Citrus spp. (53)	*A. janata* (Linnaeus), *A. woglumi* Ashby, *A. aurantii* (Maskell), *A. orientalis* (Newstead), *A. odinae* (van der Goot), *A. spiraecola* Patch, *B. caryeae* (Kapoor), *B. correcta* (Bezzi), *B. diversa* (Coquillett), *B. dorsalis* Hendel, *Cappaea taprobanensis* (Dallas), *C. ceriferus* (Fabricius), *C. rubens* Maskell, *C. stellifer* (Westwood), *Chelidonium cinctum* Guérin-Méneville, *Chondracris rosea* (De Geer), *C. aonidum* (Linnaeus), *C. hesperidum* Linnaeus, *Coccus viridis* (Green), *C. aedificator* (Fabricius), *Dialeurodes citri* (Ashmead), *D. citri* Kuwayama, *D. mangiferae* Green, *E. tessellatus* (Signoret), *Eudocima homaena* (Hubner), *Eudocima materna* (Linnaeus), *E. variegata* (Snellen), *H. cinereus* (Green), *I. aegyptiaca* (Douglas), *I. purchasi* Maskell, *I. seychellarum* (Westwood), *I. tetraonis* Moore, *L. poleii* (Green), *Oberea lateapicalis* Pic, *O. smaragdina* (Fabricius), *O. versicolor* (Fabricius), *P. demoleus* (Linnaeus), *P. polytes* Linnaeus, *P. lobata* (Chamberlin), *Phyllocnistis citrella* Stainton, *Phyllocoptruta oleivora* (Ashmead), *P. strachani* (Cooley), *P. citri* (Risso), *P. zizyphi* (Stainton), *P. psidii* Maskell, *R. iceryoides* (Green), *S. gregaria* (Forskål), *S. farinosa* (Boheman), *Siphoninus phillyreae* (Haliday), *S. scintillans* (Walker), *Stromatium barbatum* (Fabricius), *Virachola isocrates* (Fabricius), *Zeuzera coffeae* Nietner
Guava (32)	*A. dispersus* Russell, *A. miliaris* (Linnaeus), *B. caryeae* (Kapoor), *B. correcta* (Bezzi), *B. diversa* (Coquillett), *B. dorsalis* Hendel, *B. zonata* (Saunders), *C. indicus* (Maskell), *C. viridis* (Green), *D. punctiferalis* (Guenée), *D. cajani* (Maskell), *E. tessellatus* (Signoret), *E. fornicatus* (Eichhoff), *Helopeltis antonii* Signoret, *Helopeltis bradyi* Waterhouse, *H. cinereus* (Green), *F. virgata* (Cockerell), *I. aegyptiaca* (Douglas), *I. purchasi* Maskell, *I. seychellarum* (Westwood), *I. tetraonis* Moore, *L. poleii* (Green), *M. hirsutus* (Green), *Metanastria hyrtaca* (Cramer), *M. discolor* Boheman, *N. viridis* (Newstead), *Peltotrachelus cognatus* Marshall, *P. psidii* Maskell, *Rastrococcus* sp., *R. iceryoides* (Green), *S. litura* (Fabricius), *V. isocrates* (Fabricius)

TABLE 1 Insects Pests of Major Indian Crops (Insect Numbers in Parentheses)—cont'd

Crop	Pests
Pomegranate (23)	*A. janata* (Linnaeus), *C. scabratrix* (Fabricius), *D. punctiferalis* (Guenée), *D. algira* (Linnaeus), *E. crameri* (Westwood), *E. fornicatus* (Eichhoff), *H. picus* (Fabricius), *H. cinereus* (Green), *H. lataniae* (Signoret), *I. tetraonis* Moore, *L. poleii* (Green), *N. viridis* (Newstead), *P. lepida* (Cramer), *P. cognatus* Marshall, *P. solenopsis* Tinsley, *P. strachani* (Cooley), *P. citri* (Risso), *P. psidii* Maskell, *Rhipiphorothrips cruentatus* Hood, *S. phillyreae* (Haliday), *Trabala vishnou* (Lefèbvre), *V. isocrates* (Fabricius), *Z. coffeae* Nietner
Grapes (11)	*A. aurantii* (Maskell), *A. orientalis* (Newstead), *H. lataniae* (Signoret), *Hippotion celerio* (Linnaeus), *I. aegyptiaca* (Douglas), *M. hirsutus* (Green), *N. viridis* (Newstead), *P. citri* (Risso), *R. cruentatus* Hood, *Scelodonta strigicollis* (Motschulsky), *Theretra alecto* (Linnaeus)
Banana (21)	*A. destructor* Signoret, *A. miliaris* (Linnaeus), *B. diversa* (Coquillett), *B. dorsalis* Hendel, *C. rubens* Maskell, *Ceroplastes rusci* (Linnaeus), *Cosmopolites sordidus* (Germar), *Darna* sp., *Erionota torus* Evans, *Hishimonus phycitis* (Distant), *I. aegyptiaca* (Douglas), *Ischnaspis longirostris* (Signoret), *Odoiporus longicollis* Olivier, *Pamendanga* sp., *P. lepida* (Cramer), *P. ricini* (Fabricius), *Prodromus clypeatus* Distant, *P. cockerelli* (Cooley), *S. litura* (Fabricius), *S. typicus* Distant, *Tirathaba* sp. nr. *rufivena* (Walker)
Jack (12)	*A. odinae* (van der Goot), *A. miliaris* (Linnaeus), *C. rubens* Maskell, *Cosmocarta relata* Distant, *Glyphodes caesalis* Walker, *I. aegyptiaca* (Douglas), *I. seychellarum* (Westwood), *N. viridis* (Newstead), *Ochyromera artocarpi* Marshall, *O. smaragdina* (Fabricius), *Olenecamptus bilobus* (Fabricius), *P. nuda* (Fabricius)
Sapota (21)	*A. woglumi* Ashby, *Anarsia* sp., *B. caryeae* (Kapoor), *B. correcta* (Bezzi), *B. diversa* (Coquillett), *B. dorsalis* Hendel, *Banisia myrsusalis* (Walker), *C. stellifer* (Westwood), *Cryptophlebia ombrodelta* (Lower), *Dereodus mastos* Herbst, *Gatesclarkeana* sp., *I. aegyptiaca* (Douglas), *M. hyrtaca* (Cramer), *M. discolor* Boheman, *O. mendosa* Hubner, *P. lobata* (Chamberlin), *P. cognatus* Marshall, *Phycita erythrolophia* Hampson, *P. acutissimus* (Green), *Trymalitis margarias* Meyrick, *V. isocrates* (Fabricius)
Ber (18)	*A. janata* (Linnaeus), *A. guttata* Walker, *Carpomya vesuviana* Costa, *Castalius rosimon* (Fabricius), *C. ceriferus* (Fabricius), *Cryptocephalus sexsignatus* (Fabricius), *D. cajani* (Maskell), *H. cinereus* (Green), *M. discolor* Boheman, *Neptis jumbah* Moore, *O. mendosa* Hubner, *O. postica* (Walker), *Otinotus oneratus* (Walker), *Pinnaspis aspidistrae* (Signoret), *S. scintillans* (Walker), *Streblote siva* (Lefèbvre), *Thiacidas postica* Walker, *T. semihyalina* Kollar
Custard apple (11)	*A. orientalis* (Newstead), *B. zonata* (Saunders), *C. hesperidum* Linnaeus, *F. virgata* (Cockerell), *H. biclavis* (Comstock), *I. aegyptiaca* (Douglas), *N. viridis* (Newstead), *P. strachani* (Cooley), *P. citri* (Risso), *Pseudococcus jackbeardsleyi* Gimpel and Miller, *R. iceryoides* (Green)
Jamun (17)	*A. styx* (Westwood), *A. tranquebaricus* Fabricius, *B. diversa* (Coquillett), *Carea angulata* (Fabricius), *Carea chlorostigma* Hampson, *C. stellifer* (Westwood), *Curculio c-album* Fabricius, *Homona coffearia* (Nietner), *Megatrioza hirsuta* (Crawford), *M. hyrtaca* (Cramer), *M. discolor* Boheman, *P. longispinus* (Targioni-Tozzetti), *P. zizyphi* (Stainton), *Singhiella bicolor* (Singh), *Tambila gravelyi* Distant, *T. vishnou* (Lefèbvre), *Trioza jambolanae* Crawford
Cashew (32)	*A. syngramma* Meyrick, *A. albistriga* (Walker), *Anigraea cinctipalpis* (Walker), *A. odinae* (van der Goot), *A. tranquebaricus* Fabricius, *C. floridensis* Comstock, *C. trifenestrata* (Helfer), *D. aprobola* (Meyrick), *D. koenigii* Fabricius, *E. variegata* (Snellen), *Eurystylus* sp., *E. aconthea* (Cramer), *H. antonii* Signoret, *H. bradyi* Waterhouse, *I. tetraonis* Moore, *M. hyrtaca* (Cramer), *Neoplocaederus ferrugineus* (L.), *Oligonychus coffeae* Nietner, *O. exvinacea* (Hampson), *Pachypeltis maesarum* Kirkaldy, *P. cognatus* Marshall, *P. jocosatrix* Guenée, *Pleuroptya balteata* (Fabricius), *P. acutissimus* (Green), *P. trilobitiformis* (Green), *P. cockerelli* (Cooley), *P. longispinus* (Targioni-Tozzetti), *R. cruentatus* Hood, *S. bipuncta* Hampson, *S. obliqua* Walker, *S. rhothia* (Meyrick), *Thylacoptila paurosema* Meyrick

(Continued)

TABLE 1   Insects Pests of Major Indian Crops (Insect Numbers in Parentheses)—cont'd

Crop	Pests
Amla (6)	*Caloptilia acidula* (Meyrick), *N. viridis* (Newstead), *Rhodoneura emblicalis* (Moore), *Schoutedenia emblica* (Patel & Kulkarni), *S. perplexa* (Westwood), *V. isocrates* (Fabricius)
Papaya (13)	*A. dispersus* Russell, *A. orientalis* (Newstead), *A. destructor* Signoret, *B. correcta* (Bezzi), *B. dorsalis* Hendel, *D. chiton* (Green), *E. tessellatus* (Signoret), *F. virgata* (Cockerell), *H. lataniae* (Signoret), *N. viridis* (Newstead), *P. trilobitiformis* (Green), *P. jackbeardsleyi* Gimpel and Miller, *Parococcus marginatus* Williams and Granara de Willink
Apple (12)	*Aeolesthes holosericea* (Fabricius), *A. dimidiata* Hope, *A. spiraecola* Patch, *Apriona germari* Hope, *Eriosoma lanigerum* (Hausmann), *I. aegyptiaca* (Douglas), *Meristata quadrifasciata* (Hope), *M. subfasciatus* Guerin-Meneville, *P. cognatus* Marshall, *Tanymecus circumdatus* (Wiedemann), *Tessaratoma javanica* (Thunberg), *T. paurosema* Meyrick
Peach (4)	*B. correcta* (Bezzi), *B. dorsalis* Hendel, *B. zonata* (Saunders), *P. cognatus* Marshall
Pear (5)	*B. dorsalis* (Hendel), *D. punctiferalis* (Guenée), *E. lanigerum* (Hausmann), *P. cognatus* Marshall, *T. javanica* (Thunberg)
Turmeric (4)	*D. punctiferalis* (Guenée), *Lasioderma serricorne* (F.), *S. typicus* Distant, *Udaspes folus* (Cramer)
Coriander, aniseed (2)	*A. nubilis* (Fabricius), *Systole albipennis* Walker
Tamarind (14)	*A. orientalis* (Newstead), *C. serratus* (Fabricius), *C. ombrodelta* (Lower), *D. mangiferae* Green, *D. aprobola* (Meyrick), *E. crameri* (Westwood), *Exechesops* sp., *H. picus* (Fabricius), *N. alternus* Walker, *N. viridis* (Newstead), *O. oneratus* (Walker), *P. grossipes* (Fabricius), *P. strachani* (Cooley), *V. isocrates* (Fabricius)
Pepper (11)	*A. spiraecola* Patch, *C. trifenestrata* (Helfer), *F. virgata* (Cockerell), *H. antonii* Signoret, *H. lataniae* (Signoret), *Idioscopus decoratus* Viraktamath, *Lanka ramakrishnai* Prathapan & Viraktamath, *Lepidosaphes piperis* (Green), *Marsipococcus marsupialis* (Green), *P. aspidistrae* (Signoret), *P. strachani* (Cooley)
Cardamom (3)	*D. punctiferalis* (Guenée), *Eupterote* sp., *S. typicus* Distant
Tobacco (14)	*Agrotis* sp., *A. spiraecola* Patch, *A. crenulata* (Fabricius), *B. tabaci* (Gennadius), *C. chalcites* (Esper), *E. vigintioctopunctata* (Fabricius), *F. schultzei* (Trybom), *H. armigera* (Hübner), *L. serricorne* (F.), *M. pulchellus* Dallas, *N. tenuis* (Reuter), *P. operculella* (Zeller), *S. exigua* (Huebner), *S. litura* (Fabricius)
Coconut (30)	*Aleurocanthus arecae* David & Manjunatha, *A. orientalis* (Newstead), *A. destructor* Signoret, *A. miliaris* (Linnaeus), *Callispa keram* Shameem and Prathapan, *Cerataphis brasiliensis* (Hempel), *Ceroplastes* sp. (*rusci*-group), *C. rubens* Maskell, *C. rusci* (Linnaeus), *C. stellifer* (Westwood), *C. aonidum* (Linnaeus), *Darna* sp., *E. torus* Evans, *Diocalandra frumenti* (Fabricius), *Elymnias caudata* Butler, *Elymnias hypermnestra* (Linnaeus), *Gangara thyrsis* (Fabricius), *I. longirostris* (Signoret), *L. pygmaea* Baly, *Opisina arenosella* Walker, *O. rhinoceros* (Linnaeus), *P. lepida* (Cramer), *Phalacra* sp., *P. acutissimus* (Green), *P. moesta* (Westwood), *P. trilobitiformis* (Green), *P. cockerelli* (Cooley), *Rhynchophorus ferrugineus* Olivier, *S. typicus* Distant, *Tirathaba* sp. nr. *rufivena* (Walker)
Arecanut (13)	*A. arecae* David & Manjunatha, *A. woglumi* Ashby, *Araecerus fasciculatus* (De Geer), *A. miliaris* (Linnaeus), *B. suppressaria* (Guenée), *C. stolli* Wolff, *E. caudata* Butler, *E. hypermnestra* (Linnaeus), *Leucopholis burmeisteri* Brenske, *Mircarvalhoia arecae* (Miller & China), *P. strachani* (Cooley), *P. acutissimus* (Green), *Tirathaba* sp. nr. *rufivena* (Walker)

TABLE 1 Insects Pests of Major Indian Crops (Insect Numbers in Parentheses)—cont'd

Crop	Pests
Coffee (31)	*A. woglumi* Ashby, *A. lactinea* (Cramer), *Antestiopsis cruciata* (Fabricius), *A. fasciculatus* (De Geer), *A. micaceana* (Walker), *A. miliaris* (Linnaeus), *B. dorsalis* Hendel, *Cephonodes hylas* (Linnaeus), *C. ceriferus* (Fabricius), *Clinteria imperialis truncata* Arrow, *C. hesperidum* Linnaeus, *C. viridis* (Green), *C. gangis* (Linnaeus), *Darna* sp., *E. crameri* (Westwood), *E. fornicatus* (Eichhoff), *H. coffearia* (Nietner), *Hypothenemus hampei* (Ferrari), *I. insignis* (Browne), *I. longirostris* (Signoret), *N. alternus* Walker, *O. mendosa* Hubner, *O. coffeae* Nietner, *Parasa* sp., *P. lepida* (Cramer), *P. citri* (Risso), *P. psidii* Maskell, *S. scintillans* (Walker), *S. barbatum* (Fabricius), *Xylotrechus quadripes* Chevrolat, *Z. coffeae* Nietner
Tea (29)	*A. orientalis* (Newstead), *A. caricae* Fabricius, *A. crenulata* (Fabricius), *Attacus atlas* Linnaeus, *A. miliaris* (Linnaeus), *B. suppressaria* (Guenée), *C. chlorostigma* Hampson, *C. trifenestrata* (Helfer), *Darna* sp., *D. chiton* (Green), *E. crameri* (Westwood), *E. fornicatus* (Eichhoff), *H. antonii* Signoret, *H. bradyi* Waterhouse, *Helopeltis theivora* Waterhouse, *H. lataniae* (Signoret), *H. coffearia* (Nietner), *H. talaca* Walker, *O. mendosa* Hubner, *O. coffeae* Nietner, *O. postica* (Walker), *P. lepida* (Cramer), *P. aspidistrae* (Signoret), *P. pennatula* (Fabricius), *R. iceryoides* (Green), *S. scintillans* (Walker), *S. castaneiceps* Hampson, *T. alecto* (Linnaeus), *Z. coffeae* Nietner
Rose (20)	*A. janata* (Linnaeus), *A. aurantii* (Maskell), *A. guttata* Walker, *C. ceriferus* (Fabricius), *Cohicaleyrodes caerulescens* (Singh), *D. aprobola* (Meyrick), *E. aconthea* (Cramer), *H. armigera* (Hübner), *H. elegans* (Fabricius), *N. alternus* Walker, *N. viridis* (Newstead), *O. versicolor* (Fabricius), *P. lepida* (Cramer), *P. cinerea* (Kraatz), *S. scintillans* (Walker), *Streblote siva* (Lefèbvre), *T. urticae* Koch, *T. vishnou* (Lefèbvre), *T. semihyalina* Kollar, *Wahlgreniella nervata* (Gillette)
Oleander (4)	*Aphis nerii* Boyer de Fonscolombe, *Daphnis nerii* (Linnaeus), *Euploea core* (Cramer), *P. strachani* (Cooley)
Hibiscus (25)	*A. flava* (Fabricius), *A. gossypii* Glover, *C. indicus* (Maskell), *C. hesperidum* Linnaeus, *Danaus chrysippus* (Linnaeus), *E. vittella* (Fabricius), *Eurybrachys* sp., *H. lataniae* (Signoret), *H. biclavis* (Comstock), *I. aegyptiaca* (Douglas), *Indomias cretaceus* (Faust), *M. hirsutus* (Green), *M. signata* Olivier, *N. jumbah* Moore, *O. brevis* (Swedenbord), *O. hyalinipennis* (Costa), *P. gossypiella* (Saunders), *P. madeirensis* Green, *P. solenopsis* Tinsley, *P. aspidistrae* (Signoret), *P. trilobitiformis* (Green), *P. jackbeardsleyi* Gimpel and Miller, *S. scintillans* (Walker), *S. caffer* (Zeller), *X. albago* (Fabricius)
Champaka (2)	*Graphium agamemnon* (Linnaeus), *Graphium doson* (Felder)
Mulberry (11)	*Acanthophorus serraticornis* (Olivier), *A. germari* Hope, *A. miliaris* (Linnaeus), *Glyphodes pulverulentalis* Hampson, *L. vittata* Duvivier, *M. hirsutus* (Green), *M. maxima* (Green), *M. discolor* Boheman, *M. viridanus* Fabricius, *P. cognatus* Marshall, *Pseudodendrothrips mori* (Niwa)
Jatropha (7)	*C. stolli* Wolff, *Pempelia morosalis* (Saalmuller), *P. aspidistrae* (Signoret), *R. syriacus* (Mayet), *R. cruentatus* Hood, *S. perplexa* (Westwood), *Stomphastis thraustica* (Meyrick)
Bamboo (12)	*A. woglumi* Ashby, *Asterolecanium* sp., *Chaetococcus bambusae* (Maskell), *Matapa aria* (Moore), *M. leda* Linnaeus, *N. bergii* (Signoret), *Noorda blitealis* Walker, *Pseudoregma montana* (van der Goot), *Purohita* sp., *S. barbatum* (Fabricius), *Symplana viridinervis* Kirby, *Udonga montana* (Distant)
Sandal (15)	*A. cyssea* Stoll, *A. passalis* (Fabricius), *Ascotis imparata* (Walker), *A. fabriciella* (Swederus), *Calodia kirkaldyi* Nielson, *Cardiococcus bivalvata* (Green), *E. crameri* (Westwood), *Hotea curculionoides* (Herrich-Schäffer), *Hotea nigrorufa* Walker, *H. talaca* Walker, *Nyctemera lacticinia* (Cramer), *Purpuricenus sanguinolentus* (Olivier), *Tajuria cippus* (Fabricius), *Teratodes monticollis* (Gray), *T. vishnou* (Lefèbvre)
Casuarina (9)	*C. scabratrix* (Fabricius), *E. crameri* (Westwood), *E. variegata* (Snellen), *H. lataniae* (Signoret), *H. phycitis* (Distant), *I. aegyptiaca* (Douglas), *I. purchasi* Maskell, *I. tetraonis* Moore, *P. longispinus* (Targioni-Tozzetti)

(Continued)

TABLE 1 Insects Pests of Major Indian Crops (Insect Numbers in Parentheses)—cont'd

Crop	Pests
Erythrina (2)	*Cyclopelta siccifolia* Westwood, *Terastia egialealis* (Walker)
Neem (3)	*A. imparata* (Walker), *H. antonii* Signoret, *H. serrata* (Fabricius)
Pongamia (13)	*C. bivalvata* (Green), *C. stolli* Wolff, *Clanis phalaris* (Cramer), *Curetis thetis* (Drury), *C. siccifolia* Westwood, *H. chromus* (Cramer), *I. tetraonis* Moore, *Jamides celeno* (Cramer), *Maruca amboinalis* Felder, *N. jumbah* Moore, *N. viridis* (Newstead), *P. lobata* (Chamberlin), *S. perplexa* (Westwood)
Ficus spp. (22)	*A. janata* (Linnaeus), *A. atkinsoni* Lethierry, *A. caricae* Fabricius, *A. ficus* (Fabricius), *Badamia exclamationis* (Fabricius), *B. rufomaculata* (De Geer), *C. ceriferus* (Fabricius), *C. stellifer* (Westwood), *Cirrhochrista fumipalpis* Felder & Rogenhofer, *E. core* (Cramer), *G. bivitralis* Guenée, *Greenidea* sp., *Gynaikothrips uzeli* (Zimmermann), *H. talaca* Walker, *I. aegyptiaca* (Douglas), *Ocinara varians* (Walker), *Pauropsylla depressa* Crawford, *P. nuda* (Fabricius), *Phycodes radiata* (Ochsenheimer), *S. scintillans* (Walker), *S. barbatum* (Fabricius), *Zanchiophylus hyaloviridis* Duwal, Yasunaga & Lee
Teak (27)	*A. styx* (Westwood), *A. lactinea* (Cramer), *Aristobia approximator* (Thomson), *A. imparata* (Walker), *A. caricae* Fabricius, *A. miliaris* (Linnaeus), *C. hylas* (Linnaeus), *Cyphicerus emarginatus* Faust, *D. punctiferalis* (Guenée), *E. undata* Blanchard, *E. fornicatus* (Eichhoff), *G. clelia* (Cramer), *Hyblaea puera* Cramer, *H. talaca* Walker, *I. aegyptiaca* (Douglas), *O. mendosa* Hubner, *O. postica* (Walker), *Pagyda salvalis* Walker, *Paliga machoeralis* (Walker), *Parotis vertumnalis* Guenée, *Pontanus puerilis* (Drake & Poor), *P. pennatula* (Fabricius), *Psiloptera* sp., *S. barbatum* (Fabricius), *T. monticollis* (Gray), *T. alecto* (Linnaeus), *Z. coffeae* Nietner
Calotropis (10)	*A. albistriga* (Walker), *C. janus* (Fabricius), *Dacus* (*Leptoxyda*) *persicus* Hendel, *D. chrysippus* (Linnaeus), *Eurybrachys* sp., *Paramecops farinosus* (Wiedemann), *Phygasia silacea* (Illiger), *Poekilocerus pictus* (Fabricius), *P. cockerelli* (Cooley), *S. obliqua* Walker
Storage pests (16)	*Acanthoscelides obtectus* (Say), *A. fasciculatus* (De Geer), *Callosobruchus* sp., *C. maculatus* (Fabricius), *C. serratus* (Fabricius), *C. cephalonica* (Stainton), *Cryptolestes pusillus* (Schäenherr), *L. serricorne* (F.), *Oryzaephilus mercator* (Fauvel), *Oryzaephilus surinamensis* (Linnaeus), *Rhyzopertha dominica* (Fabricius), *Sinoxylon* sp., *S. oryzae* (Linnaeus), *S. cerealella* (Olivier), *Stegobium paniceum* (Linnaeus), *Tribolium castaneum* (Herbst)

Depredation of crop plants in agricultural farms leads to noxious pestilence (Pedigo, 1996a). These insects may be tissue borers, chewers, cutters, rappers, sap suckers, etc. Dependence of insects on agricultural crops as a life resource has become an adaptation that they took to along with human interest to domesticate wild plants to cultivate them to reap profitable harvest. Insects use plants and animals as food resources based on their feeding biology. Using them as a resource for food and shelter to complete insect life cycles would bring in the strategy of co-living with all of those organisms including humans. Thus, agriculture redefined pest incidence in terms of herbivory and ectoparasitism on animals. Strictly speaking, parasitism is broadly the terminology for herbivory and ectoparasitism in the domesticated animals of farms. Insects that have taken to pestilence cut across most of the insect orders. The genera and species that were specialized to exploit agricultural resources, such as crop plants and domesticated animals, were those that did so in the wild natural habitat (Southwood, 1975). The ecological specialization that the agroecologies offered to insects (Hassell et al., 1991) made them build up as communities with r-selection organisms, albeit the fact that the theory of r-/K-selections (Southwood, 1975). This theory has had typical aberrance (Parry, 1981) in the evolutionary biology of insects.

Crop production is undertaken according to facilitating conditions for optimal crop growth. The trend of seasonal occurrence of these insect pests in accordance with the crop species and cropping patterns has aligned the insect pests to the seasons across the Indian geography. In general, the Indian scenario shows that insects prefer to be much more active when monsoon rains bring about better metabolic activity in plants. In agricultural ecologies, the insects have the choice of pasturage based on the crops that are offered in cropping systems of agricultural farms. To utilize the annual crops, the insects adjust their life cycle through the year in different hosts, available in farms and in the wild. Several insects as pests may exploit the ecosystem by oversummering or overwintering in farms and reappear when favorable conditions arise after rains when new crops growth begins. Many insects, such as aphids, plant and leaf hoppers, whiteflies and several moths and butterflies, move across large geographic tracts, either through involuntary wind currents or through wafting by strong gales and storms. Colonization of insects on crops happens based on the cues for perception of canopy color, canopy odor from phytochemical(s) consortia. Multiple crop hosts make the survivorship of polyphagous insects better in every season. The survivorship of oligophagous insect pests is more challenging because of a narrow host band and higher pressure from natural enemies and other key mortality factors (Andrewartha and Birch, 1971, 1984). Looking at the evolutionary trends for adaptation for survivorship, insects have gained good advantage over many other similar invertebrates because of their cuticular body, spiracles, and active limbs with extensive sensillary support to perfectly gauge the chemical environment in their niches and habitats. The holometabolous or hemimetabolous life cycle and parthenogenetic reproduction, including vivipary, have provided better opportunity to successfully tide over an adverse environment. In accordance with the habitats, their fast adapting metabolic corrections could sustain insects to the best of opportunism. Although we exploit insects for crop pollination, apiculture, lac culture, sericulture etc., the recent deployment of the biocontrol agent insects in farm lands has fortified the toolbox for efficient integrated pest management in crops of various seasons.

Numerical outbreak in relation to crop biology and health as well as a short seasonal period made these insects different from their wild counterparts who have their resources assured through the annum. Biological specialization of insects invading and damaging crop plants was predominant. Agricultural biodiversity that is predominantly manmade, became different from natural biodiversity. Selection of crop species for high yield and for other commodity traits resulted in their extensive monocropping in large geographies and resulted in the selection of certain insects that also became selected entities of the new agroecology. Insects of the wilderness became pests of crops when the crops were tended and husbanded by specific packages to drive for the best genetic yield that was derived in immense measure through modern agricultural practices. Calling for reasons for increased herbivory by endemic species, key pests, primary pests, or invasive species developing as new pests of crops and assessing crop loss due to multiple pest damage are the systemic entomological issues in contemporary agriculture. They needed to adapt for the load of phytochemicals that the plants fed along with their tissues to sustain efficient insect metabolism.

2.3 Plant Defense Systems—Regulation of Herbivory by Plants

In addition to metabolic compensation processes, plants defend themselves by using phytochemical communication mechanisms. Scientific evidence to elucidate this arose in the early part of the last century in monocropping systems of agriculture. Kessler and Baldwin (2001) gave insight into the herbivore-induced plant volatile emissions in nature that are taken as cues by natural enemies of the herbivores. It appears that the affected plant calls for the services of natural enemies to reach the plants which are depredated upon (Paschold et al., 2006). Low and Merida (2001) explained the plant defense through the production of reactive oxygen species to signal varied defense responses to different stresses. The phytochemicals, both volatiles and tissue-rich ones, become cues for plant-defending insects to move in to take on the herbivores that become pests in crops. Thus, insects as defenders of crop plants to reduce overgrazing by pest insects are a wonderful food chain support that nature has built in to sustain herbivory without annihilation of target plant species.

2.4 Insects in Commodity Storage

It is difficult to secure commodities from those insect species that home and damage the commodities under storage for human use. Many entomologists such as Lefroy (1906), Fletcher (1916), Fletcher and Ghosh (1919), Turner (1994), and Cotton (1956) have enumerated various insect species (Table 2), loss caused by them to commodities, and measures to mitigate the damage. The biology of these insects is very specialized because they must live in a niche with humidity ranging between 6% and 12% with low gas exchange. Postharvest storage of agricultural commodities is the key to secure wanton loss of the commodities, being deprived for human use. Suitable storage structures with modified environment in silos of varying capacity could prevent the buildup of store-grain pests that spoil the commodities extensively to the tune of 2–10% across the country in various years according to favorable weather and storage ecologies. Securing commodities from insects has been worked upon from the time storage was contemplated for using during leaner availability of the year. Different storage structures prevent insect entry after the commodities are cleaned and stored. Highly toxic chemical fumigants, such as methyl bromide, ethyl bromide, phosgene, carbon dioxide, sulfuryl sulfide, and several others, are in use to secure commodities from destruction due to storage insects. However, the risk from poisonous gases to human health and saving of grains with such gases is debatable. Traditional methods, such as mixing with repellent tree leaves, physical aberration-based mortality by missing with the commodities, any abrasive substance such as sand, fly ash, wood ash, and many such products, shall be of immense use to reduce losses in storage.

TABLE 2   List of Insects That Damage Agricultural Commodities in Storage

Commodity/seeds	Name of insect (16 species)
Cereals, pulses, oilseeds, spices, vegetables, condiments, coffee, tobacco	*Acanthoscelides obtectus* (Say), *Araecerus fasciculatus* (De Geer), *Callosobruchus* sp., *Callosobruchus maculatus* (Fabricius), *Caryedon serratus* (Fabricius), *Corcyra cephalonica* (Stainton), *Cryptolestes pusillus* (Schäenherr), *Lasioderma serricorne* (F.), *Oryzaephilus mercator* (Fauvel), *Oryzaephilus surinamensis* (Linnaeus), *Rhyzopertha dominica* (Fabricius), *Sinoxylon* sp., *S. oryzae* (Linnaeus), *Sitotroga cerealella* (Olivier), *Stegobium paniceum* (Linnaeus), *Tribolium castaneum* (Herbst)

2.5 Insect Ectoparasites on Farm Animals

Milk, meat, eggs, and animal products such as hair, wool, leather, and many such products come from livestock, piggery, and poultry. Agriculture also contains the animal component for ectoparasites, which are a specialized group of insects that are blood sucking (i.e., hematophagous or chewing on skin and keratin of hair in their food habits). They live on all animals to complete their nutritional needs. Fleas, flies, grubs, caterpillars, beetles, bugs, etc., are specialized to feed on animal skin, keratin, and blood. Their chewing/sucking mouthparts enable them to seek their nutrition from the respective host animal. The animal-parasitic insects are a large group of highly evolved hematophagous insects that are ectoparasitic in habit and have evolved with the animal kingdom (Hopla et al., 1994; Grimaldi and Engel, 2005). Blood-sucking dipteran insects are exclusively females whereas their males feed on plants. Table 3 provides the various insect orders with families and number of species in parentheses that are animal ectoparasites. Some of them are capable of transmitting disease pathogens. These obligate insect parasites have adapted to their hosts through coevolution. They cause anemia, dermatitis, and irritability. Some of them act as vectors of diseases (e.g., transmission of swine pox in pigs by *Haematopinus suis*). Skin-infecting fungus, *Trichophyton verrucosum*, in cattle is transmitted by 994 species of cattle lice. Dog louse is known to transmit endoparasitic tapeworms. There are several such vector associations on scientific record in veterinary science.

3. HOW DO INSECTS INFLUENCE FARM ECONOMY?

Farm economy is directly affected when the crops suffer more than 30% crop loss in a season (Pimental, 2002) if they are not appropriately protected. Crop damage due to extensive herbivory as with that of army worms, leaf feeders, plant hoppers, leaf hoppers, etc., has been recurrent and unbridled. Their population explosion in crops makes it highly damaging in large tracts of agricultural farms. The economic injury level (EIL) of these pests and consequent crop loss is at times extremely high. In recent years, the severe damage caused by invasive pests introduced from other countries has been high. The invasion of coconut eryiophid mites was the first of those kinds to invade coconut palms in southern states reported in the country in the late 1980s. The sugarcane woolly aphid (*Ceratovacuna lanigera*) invaded the crop from eastern India and spread to the rest of India during 2002–2005, causing huge economic loss; in terms of recovered sugar, the estimate was Rs 600 crores (Rabindra et al., 2004; Joshi et al., 2010). Cotton mealy bug (2008–2010) was the next to be tackled in cotton and a wide variety of crops. Papaya mealy bug (*Paracoccus marginatus*) invasion (2010–2013) (Muniappan et al., 2006) caused a crop loss estimated to be Rs 1200 crores (Shylesha et al., 2012). The present invasion in the country of the South American tomato leaf miner, *Tuta absoluta* (Meyrick) (Lepidoptera: Gelechiidae), is causing damage in protected and open cultivation of this crop. Prevention of crop loss in tomato and solanaceous crops such as potato is in progress. The establishment of natural enemies is evolving in nature and the pest could be confidently contained. The point to recognize is that many invasive insects that are introduced through various channels become a challenge for pest management, and the country has now geared up to effectively tackle such events.

TABLE 3   Insect Ectoparasites and Their Hosts

Ectoparasitic insect order and family/ subfamily/(known species)	Host	Tissue
Dermaptera		
Hemimeridae (11)	Rodents	Skin serration/secretions
Anexenia (5)	Bats	
Siphonaptera (2500 spp.)	Mammals, birds	Blood
Phthiraptera	Mammals, birds	Chewing skin/keratin
Amblycera (900)	Mammals, birds	Chewing skin/keratin
Ischnocera (1800)	Mammals, birds	Chewing skin/keratin
Rhynchophyirina (3)	Mammals, birds	Chewing skin/keratin
Anopleura (500)	Mammals, birds	Chewing skin/keratin
Hemiptera		
Reduviidae Triatominae (155)	Mammals, birds	Blood
Lygaeidae—Cleradinin (50)	Mammals, birds	Blood
Cimicidae (90)	Mammals, birds	Blood
Polyctenidae (32)	Mammals, birds	Blood
Diptera		
Psychodidiae Sycorinae (100) Phlebotominae (700)	Vertebrates	Blood
Culicidae (3000)	Vertebrates	Blood
Corethrellidae	Frogs	Blood
Chironomidae (10)	Vertebrates	Blood
Ceratopogonidae (1000)	Vertebrates	Blood
Simuliidae (1500)	Birds/mammals	Blood
Rhagionidae (50)	Birds/mammals	Blood
Tabanidae (3500)	Mammals	Blood
Heleomyzidae Chiropteromyza (1) Mormotomyidae (1)	Bats	Sponging/tissue not known
Carnidae—Carnus (4)	Birds	Blood
Some Muscidae (50)	Mammals/birds	Blood
Clossinidae (25)	Mammals/birds	Blood
Pupipara (400)	Mammals/birds	Blood

TABLE 3   Insect Ectoparasites and Their Hosts—cont'd

Ectoparasitic insect order and family/subfamily/(known species)	Host	Tissue
Calliphoridae (50)	Birds	Blood
Mystacinobidae	Bats	Sponging/tissue not known
Lepidoptera		
Noctuidae—Calyptra	Mammal	Blood
Pyralidae—Bradypophila	Sloths	Blood
Coleoptera		
Leiodidae—Catopidius (1) Leptinidae (6)	Rabbits/hares	Chewing skin/keratin
Platypsyllidae (2)	Beavers	Chewing skin/keratin
Staphylinidae—Amblyopinini (60)	Rodents	Chewing skin/keratin
Sanguiridae (3)—Uroxys, Trichillum	Sloths	Chewing skin/keratin

Adapted from Grimaldi and Engel (2005).

Farming for profitability is the axiom for developing the manmade ecology called agriculture. The term *profitable* did mean adequate farm outturn for satiation farmers' needs. Tissue damage in the form of cutting, chewing, boring, sap sucking, rasping and the like result in physiological and metabolic drag in crop plants that have been bred for high yield through more efficient usage of nutrients, light, water, and all required natural resources. This ultimately affects the crop yield, which in economic terms is measured in terms of EIL and economic threshold (ET). EIL and ET (Higley and Wintersteen, 1992; Pedigo and Higley, 1992; Higley and Peterson, 1996; Higley and Pedigo, 1996) became concepts in modern insect pest management and were further applied to pestilence due to other species such as mites, nematodes, etc., in crops (Pedigo et al., 1986; Pedigo, 1996b; Peterson and Higley, 2000; Southwood and Norton, 1973; Poston et al., 1983; Riley, 2004). The definable relationship of the given insect species to the host crops and animals in terms of damage to their well-being drives the quantification of yield loss. Pedigo and Higley (1992) provided definition and perspective to the damage potential of insect pests. Varying applicability of these definitions in annual and perennial cropping patterns as well as in those farms that followed specific cropping systems, as guided by profitability and agroecology's sustainability, was defined over the last century. Insects as pests did guide biological solutions for human efforts to suppress their anomalous damage to crops. These concepts became fruitful in light of seeking food security for humanity. Sustained suppression at critical phase of crops through integrated pest management (IPM) made it possible to develop reliable toolboxes with effective expert systems to predict and proactively apply IPM to reduce the impact of insects as pests in agriculture. The global loss of crop commodities in the farm fields is up to 40% due to biotic stresses (insects, pathogens, nematodes, larger animals, etc.) and 20% during storage of commodities, mainly due to insects (Pimental, 2002).

3.1 Compensation Systems of Host Plants against Intense Pasturage

Many processes have been defined in attaining the compensation potential of plants for thwarting extremities of depredation. Metabolic adjustments to repair tissue damage, phytochemical support to toxify the herbivore, palatability alterations of tissues in different phenology of crop growth, and involvement of protective tissues (e.g., pubescence, glands, and tissue anatomy) have been defined to support the biological process of plants to control overgrazing by insects. Antibiosis is efficiently deployed by plants to reduce impedance of insect colonization. Moderation of herbivory enables the plants to efficiently survive. Crop plants also exhibit similar processes. The crop compensation potential does reduce the potential economic loss. This genetic trait is exploited in the development of crop varieties with insect tolerance.

3.2 Efficient Pest Management Service

Efficient pest management service can considerably reduce avoidable crop loss (Pedigo et al., 1986; Pimental, 2002; Higley and Pedigo, 1996; Riley, 2004). India adopted the toolbox of IPM in the 1990s and worked for the promotion of judicious integration of all of the potential and relevant tools, such as biological agents, clean cultivation, tolerant varieties of crops, application of pesticides that have the least environmental impact, etc. (Rajendran and Basu, 1999; Rajendran, 2005; Ramamurthy et al., 2009; Rajendran, 2013). The country could implement pest scouting during seasons in major crops and could plan on the steering of IPM to contain any emerging threat in large geographic farm areas in every state. The network of institutions that have been geared up in government and nongovernment institutions including farmers' groups are effectively able to take up pest management through such supervised management. Export-oriented crops, such as Basmati rice, grapes, pomegranate, vegetables, etc., have special supervision for managing their biotic stresses (Rajendran, 2009).

4. BENEVOLENT INSECTS

Insects have been utilized in human civilization from time immemorial. The featured knowledge in this regard has been in cross-pollination that is engineered by various species of flora in nature. However, other uses affected have been in sericulture, lac production, apiculture, and in recent times in biocontrol for IPM in crops.

4.1 Silk, Lac, and Honeybees

There has been excellent industrial setup for sericulture in modern times, and exploitation of silk moths occurred in our civilization for centuries before Christ and onward (Barber, 1992). India is one of the major silk-production countries in the world. India continues to be the second largest producer of silk in the world. Among the four varieties of silk produced, as in 2012–2013, Mulberry accounts for 79% (18,715 MT), Tasar (tropical and temperate) 7.3% (1729 MT), Eri 13.2% (3116 MT), and Muga 0.5% (119 MT) of the total raw silk production of 23,679 MT. The economic value in terms of livelihood assurance and business outturn in the country is invaluable (Anonymous, 2014). As a natural fiber, silk in apparel and fashion along with many other industrial applications have affected the business economics and livelihood assurance of the countries where industrial silk production occurs.

Knowledge regarding the lac insect (*Kerria lacca*) in India is very ancient, and today India is leading producer of lac, which has several applications in various products that favor modern man's consumption trend. India and Thailand are the major producers, producing on average 1700 tonnes of lac each year, followed by China. Its industrial application in food processing, health, cosmetics, paints and varnishes, and a host of others is immense, and as a natural product its value is very much appreciated.

Honeybees have fascinated human civilization and substantially influenced anthropology. The sweet material collected and stored in honeycombs by gathering and converting nectar from plants has been a subject of study that has enabled crop pollination and harvesting of bee products, including honey, in today's commercial beekeeping. Propolis is a major product of beekeeping that has been extensively used for geriatric and pediatric nutrient supplements. Wild honey from the forests is considered more nutritious than apiculture-originated honey. Honey from *Trigona* spp. is highly valued because of a special medicinal property and is 10 times more costly than Apis hive honey. Beehive products, such as propolis, bee wax, bee pollen, etc., have high market value and beekeepers get good return from them. In India, migratory apiculture is livelihood and employment assurance. The yeoman service beekeeping industry and undertaking the assurance of quality and quantity of crop commodity production is enormous. The Indian Council of Agricultural Research under the Ministry of Agriculture focuses special attention on lac culture and apiculture for research on improved strains, improved culture techniques, and improved product processing from these two groups of insects. The Indian Institute of Natural Resins and Gums, Ranchi (Jharkhand), and the All India Coordinated Research Project on honeybees and crop pollinators have time-tested contributions in the promotion of these two important rural enterprises.

4.2 Natural Enemies of Pests

Insects are highly adaptive and have undergone diverse adaptations across the planet. Their roles as pests are to be rated against the roles they take in balancing the pest community population in agriculture and in nature. Widely adaptive predatory and parasitic insects have earned human awe. In crop husbandry system, insects take the role of both depredatory pests as well as that of natural enemies of the crop pests, enabling biological/natural control of crop pests. In fact, insect science in the last century flourished more through scientific analysis of trophic relationships between prey–predator and host–parasite interactions. Global research efforts as well as those in the National Agricultural Research and Education System (NARES) under the Indian Council of Agricultural Research (ICAR) provided impetus and thrust to this mode of pest management in crops during the last few decades of the twentieth century (Rajendran and Ballal, 2008).

Commercial exploitation of some of the natural enemies for classical biocontrol and for inundative biocontrol over crop seasons has been highly successful in India. A list of these insects is given in Table 2. In fact, several entrepreneurs have made good livelihood of this scientific knowledge and utilized the technical know how of mass production/manufacture of these natural enemies (parasitoids and predators) for use by farmers to contain pests in wide variety of crops. Biocontrol as a basic element of IPM could be an axis on which to focus to sustain principles of IPM in Indian farms. In recent years, the ICAR National Institute of Agricultural Insect Resources is the beacon of service to farmers and to manufacturing entrepreneurs who make up the supply chain to farmers of several predatory and parasitoid insects. The organizational support system such as that of the Crop Research Institute under the Indian Council of Agricultural Research, Ministry of Agriculture, Government of India drove this biological reality into commercial enterprises of factories producing insects and releasing in crop fields as biological pesticides. Although the microbial pest control tools robbed the term *biological control*, the genesis of this term was for exploitation of the insect behavior of hunting insect fauna as their biological resource for sustenance. Extensive application of the utilization of natural enemies globally occurred in crop pest management. Many large insectaries were created in the 1970s in undivided Russia and other Eurasian countries (Russel, 2004). Similar manufacturers were developed and patronized in Maharashtra, Tamil Nadu, Andhra Pradesh, and Karnataka and subsequently in most states that cater to specialized crop production systems such as sugarcane, cotton, vegetables, pulse crops, and the like. During the 1990s institutions such as the Central Institute of Cotton Research have spearheaded skill development among farm families and educated youth to establish similar insectaries on a factory scale, and they indeed have been their livelihood assurance units even today. The symbiosis between cultivators and these factories built up a natural process of regaining biological balance of agroecologies, which had been smothered by injudicious pesticide application.

4.3 Crop Pollination Service

Pollination of flowers as an evolutionary assurance for genetic improvement is the natural principle that plants adopted to sustain their generations without genetic breakdown. Floral biology and anatomy are tailored to suit the selection of pollinator species and enable effective pollination. A major component of the class of natural pollinators is insects. Their feeding habits, anatomy of mouthparts, and behavior has been tailored to the pollination needs of the flora that they visit. The natural flora has designed their flowering anatomy and biology to suit the favor that they call for from insect pollinators. Pollination service in nature has been undertaken in nature by many thousands of insect species. Although Hymenopteran insects have specialized in this job and are chartered by flower biology of flora, insects from orders such as Diptera, Coleoptera, Lepidoptera, and Neuroptera as well as several unknown orders of insects are recognized.

In the perspectives of growth in agriculture across the globe at large as well as that in India, sheer dependence on deployment of such insect fauna called natural enemies in crop field management could sustain crop yields by intelligent integration with synthetic pesticide application at timed upsurge of pestilence in crops. Insects have been benevolent to human communities that practice agriculture by providing the biggest agricultural input of cross-pollination to yield the best quality fruits, be it in horticultural crops or in field crops such as pigeonpea, sesame, etc. Deprived of the various insect pollinators, such as minute thrips to large beetles, butterflies, moths, and many others, the agricultural productivity of cross-pollinated crops could be in jeopardy. The dipteran pollinators can produce the cocoa pods containing bold seeds from which cocoa is processed, thereby commonly connoted as "no chocolate without flies.". The pollination service by insects is the basis of sustaining crop productivity and quality of commodity. Crop production in temperate regions of the Earth is fully dependent on bee pollination alongside encouraged and enabled natural pollination.

4.4 Other Uses of Insects—Insects for Food

The use of sarcophagous maggots to clean up wounds and prevent gangrene has been in vogue for many decades. From ancient Indian medical records, there has been mention about the use of maggots of various insect species to cure infections in wounds and enable faster healing along with medicines.

Several insects are useful as food material. Rich sources of protein and minerals are known to be present in various insects. Grubs, caterpillars, adult morphs, and pupae are consumed by various human communities. Many insects are used to feed pigs and poultry. The wastes arising out of sericulture and apiculture are used in animal feeding in many countries. There are no official records on the quantities that are used as insect-originated feed and human food; however, it is common knowledge that many human communities in various nations have in their diets seasonal ingredients from the insect world.

5. INSECTS AS VECTORS OF CROPS DISEASES

A large class of insects are called vectors that are phytophagous and in turn transmit various plant viruses that cause various diseases. Vector insects are sap-sucking insects that take in viral particles along with phloem sap and transmit them while feeding on healthy plants. The movement of viruses between plants through sap-feeding insects has evolved to attain seamless movement within and between plants. In the case of plants, many practices are integrated to reduce the vector load at the critical time of increase in pathogen load. Plant viruses have been managed by effective reduction of the vector population in vegetable crops and potatoes. Special techniques are developed and put to practice by which the crop escapes pathogen load and survives disease incidence.

Research into the role of insects and vectors of pathogens has resulted in specialized vector control programs in animal and human health management. Specific vector management efforts are in place by health ministries of the union and states to reduce overwhelming disease inoculum load. Protozoa, bacteria, and viruses move through vector insects.

6. INSECTS FOR CIVILIZATION CHANGE TO HUMANITY

Human civilization has learned major lessons in social life, homing, and designing of internal weather ambiance of homes from insects. Honeycomb geometry has fascinated mathematicians and architects alike. Precision in the design from such structures (e.g., knowledge of material science from paper wasp nests) has enhanced the structural and chemical knowledge. The natural structures that have been built by insects for nesting have several properties that are widely studied for use in modern human civilization.

7. CONCLUSIONS

Insects are fascinating organisms that have successfully colonized our planet, have challenged man by being competitors for food from plants, and have taught several lessons in ecology. Agriculture is a specialized ecosystem in which insects as pests damage and destroy crops at different phenological stages, resulting in reduced profitability. Their presence in crops stimulates human efforts to ward them off using toxic chemicals called pesticides. These chemical entities have many negative effects, including adverse effects on several non-target organisms in agroecologies. Survivorship of insects with complex adaptations to man-made ecology such as agriculture arose out of extensive adaptations. Insect science became rich with knowledge assembly from systematic research. The role of insects as vectors of plant, animal, and human diseases is recognized. Plant disease vectors such as aphids, thrips, whitefly, and sap suckers are used by viruses, phytoplasma, and other such organisms to move among plants through feeding of insects. The pests as a nuisance value or economically threatening organisms in cropping are a concern for farmers. Their presence in crops sends distress signals to farmers. Integrated pest management was designed with location-specific ingredients to effectively mitigate crop damage. Storage of commodities has special challenges of carrying field infestation in crop grains and the risk of using chemical fumigants.

Beneficial insects such as pollinators enhance crop production of commodities. The enhancement of quality and quantity of cross-pollinated fruits is well recognized. Other useful products, such as silk, lac, honey, and such other natural products, originate from scientifically culturing those insects. The enhanced income and livelihood sustenance of the communities who take up sericulture, apiculture, and lac culture are understood, and governmental support for their patronage is offered.

Acknowledgments

The authors acknowledge the assistance from Dr Ajanta Bira, Dr Pratibha Menon, and Dr Prasad Burange in providing many inputs that enabled this chapter to be prepared with good material support. Public resource material available from various government sites has been used to provide suitable and relevant picturization.

References

Andrewartha, H.G., Birch, L.C., 1971. Introduction to Animal Populations. University of Chicago Press, Chicago; Methuen & Co., London. ISBN:0-226-02029-0.

Andrewartha, H.G., Birch, L.C., 1984. The Ecological Web: More on the Distribution and Abundance of Animals. University of Chicago Press, Chicago. ISBN:0-226-02033-9.

Anonymous, 2014. Note on the Performance of Indian Silk Industry & Functioning of Central Silk Board. Central Silk Board, Bengaluru, p. 28.

Barber, E.J.W., 1992. Prehistoric Textiles: The Development of Cloth in the Neolithic and Bronze Ages with Special Reference to the Aegean (reprint, illustrated ed.). Princeton University Press. pp. 31. ISBN:978-0-691-00224-8.

Bernays, E.A., 2001. Neural limitations in phytophagous insects: implications for diet breadth and evolution of host affiliation. Annual Review of Entomology 46, 703–727.

Bernays, E.A., Chapman, R.F., 1994. Host Plant Selection by Phytophagous Insects (Contemporary Topics in Entomology). Springer, New York.

Bernays, E.A., Funk, D.J., 1999. Specialists make faster decisions than generalists: experiments with aphids. Proceedings of the Royal Society of London B: Biological Sciences 266, 151–156.

Blumler, M.A., Byrne, R., Belfer-Cohen, A., Bird, R.M., Bohrer, V.L., Byrd, B.F., Dunnell, R.C., Hillman, G., Moore, A.M.T., Olszewski, D.I., Redding, R.W., Riley, T.J., 1991. The ecological genetics of domestication and the origins of agriculture. Current Anthropology 32 (1), 23–54.

Caterino, M.S., Cho, S., Sperling, F.A.H., 2000. The current state of insect molecular systematics: a thriving Tower of Babel. Annual Review of Entomology 45, 1–54.

Cates, R.G., 1980. Feeding patterns of monophagous, oligophagous and polyphagous insect herbivores: the effect of resource abundance and plant chemistry. Oecologia 46, 22–31.

Cotton, R.T., 1956. Pests of Stored Grains and Grain Products. Burgess Publishing Co., Minneapolis, Minnesota.

Fletcher, T.B., 1916. Rep. Imp. Ent. Res. Inst. and Coll. Pusa, pp. 58–77.

Fletcher, T.B., Ghosh, C.C., 1919. Stored grain pests. In: 3rd Ent. Meeting, Pusa (Calcutta). Rep. Proc. II, pp. 712–761.

Gould, F., 1991. The evolutionary potential of crop pests. American Scientist 79 (6), 496–507.

Grimaldi, D., Engel, M.S., 2005. Evolution of Ectoparasites and Blood Feeders of Vertebrates. Cambridge Univ. Press. 489 p.

Hassell, M.P., Comins, H.N., May, R.M., 1991. Spatial structure and chaos in insect population dynamics. Nature 353, 255–258.

Higley, L.G., Pedigo, L.P., 1996. The EIL concept. In: Higley, L.G., Pedigo, L.P. (Eds.), Economic Thresholds for Integrated Pest Management. University of Nebraska, Lincoln, Nebraska. 327 pp.

Higley, L.G., Peterson, R.K.D., 1996. The biological basis of the EIL. pp. 22–40. In: Pedig, L.P., Higley, L.G. (Eds.), Economic Thresholds for Integrated Pest Management. University of Nebraska Press, Lincoln, Nebraska. 327 pp.

Higley, L.G., Wintersteen, W.K., 1992. A novel approach to environmental risk assessment of pesticides as a basis for incorporating environmental costs into economic injury levels. American Entomologist 38, 34–49.

Hopla, C.E., Durden, L.A., Keirans, J.E., 1994. Ectoparasites and classification. Revue Scientifique et Technique Office International des Epizooties 13 (4), 985–1017.

Joshi, S., Rabindra, R.J., Rajendran, T.P., 2010. Biological control of aphids. Journal of Biological Control 24 (2), 1–10.

Kessler, A., Baldwin, I.T., 2001. Defensive function of herbivore-induced plant volatile emissions in nature. Science 291, 2141–2144.

Lefroy, H.M., 1906. Indian Insect Pests. pp. 251–260.

Low, P.S., Merida, J.R., 2001. The oxidative burst in plant defence: function and signal transduction. Physiologia Plantarum 96, 533–542.

Misof, B., Liu, S., Meusemann, K., Peters, R.S., Donath, A., Mayer, C., Frandsen, P.B., Ware, J., Flouri, T., Beutel, R.G., Niehuis, O., Petersen, M., Izquierdo-Carrasco, F., Wappler, T., Rust, J., Aberer, A.J., Aspock, U., Aspock, H., Bartel, D., Blanke, A., Berger, S., Bohm, A., Buckley, T.R., Calcott, B., Chen, J., Friedrich, F., Fukui, M., Fujita, M., Greve, C., Grobe, P., Gu, S., Huang, Y., Jermiin, L.S., Kawahara, A.Y., Krogmann, L., Kubiak, M., Lanfear, R., Letsch, H., Li, Y., Li, Z., Li, J., Lu, H., Machida, R., Mashimo, Y., Kapli, P., McKenna, D.D., Meng, G., Nakagaki, Y., Navarrete-Heredia, J.L., Ott, M., Ou, Y., Pass, G., Podsiadlowski, L., Pohl, H., von Reumont, B.M., Schutte, K., Sekiya, K., Shimizu, S., Slipinski, A., Stamatakis, A., Song, W., Su, X., Szucsich, N.U., Tan, M., Tan, X., Tang, M., Tang, J., Timelthaler, G., Tomizuka, S., Trautwein, M., Tong, X., Uchifune, T., Walzl, M.G., Wiegmann, B.M., Wilbrandt, J., Wipfler, B., Wong, T.K.F., Wu, Q., Wu, G., Xie, Y., Yang, S., Yang, Q., Yeates, D.K., Yoshizawa, K., Zhang, Q., Zhang, R., Zhang, W., Zhang, Y., Zhao, J., Zhou, C., Zhou, L., Ziesmann, T., Zou, S., Li, Y., Xu, X., Zhang, Y., Yang, H., Wang, J., Wang, J., Kjer, K.M., Zhou, X., 2014. Phylogenomics resolves the timing and pattern of insect evolution. Science 346 (6210), 763.

Muniappan, R.D., Meyerdirk, E., Sengebau, F.M., Berringer, D.D., Reddy, G.V.P., 2006. Classical biological control of the papaya mealybug, *Paracoccus marginatus* (Hemiptera: Pseudococcidae) in the Republic of Palau. Florida Entomologist 89 (2), 212–217.

Parry, G.D., 1981. The meanings of r- and K-selection. Oecologia 48 (2), 260–264.

Paschold, A., Halitschke, R., Baldwin, I.T., 2006. Using 'mute' plants to translate volatile signals. The Plant Journal 45, 275–291.

Pedigo, L.P., 1996a. Entomology and Pest Management, second ed. Prentice-Hall Pub., Englewood Cliffs, NJ. 679 pp.

Pedigo, L.P., 1996b. General models of economic thresholds. pp. 41–57. In: Pedigo, L.P., Higley, L.G. (Eds.), Economic Thresholds for Integrated Pest Management. University of Nebraska Press, Lincoln, Nebraska. 327 pp.

Pedigo, L.P., Higley, L.G., 1992. A new perspective of the economic injury level concept and environmental quality. American Entomologist 38, 12–21.

Pedigo, L.P., Hutchins, S.H., Higley, L.G., 1986. Economic injury levels in theory and practice. Annual Review of Entomology 31, 341.

Peterson, R.K.D., Higley, L.G. (Eds.), 2000. Biotic Stress and Yield Loss. CRC Press, Boca Raton, Florida, USA, p. 258.

Pimental, D. (Ed.), 2002. Encyclopaedia of Pest Management. Marcel Decker, Inc., New York, USA, p. 933.

Poston, F.L., Pedigo, L.P., Welch, S.M., 1983. Economic injury levels: reality and practicality. Bulletin of the Entomological Society of America 29, 49–53.

Rajendran, T.P., 2005. Plant protection quagmire – cotton as a case study. pp. 90–112. In: Ramamurthy, V.V., Singh, V.S., Gupta, G.P., Paul, A.V.N. (Eds.), Gleanings in Entomology – in Commemoration of 100 Years of Service. Entomology Division, IARI, New Delhi, p. 317.

Rajendran, T.P., 2009. Integrated pest management – policy directions in the context of climate change. pp. 8–13. In: Ramamurthy, V.V., Gupta, G.P., Puri, S.N. (Eds.), Proceedings of the National Symposium – Integrated Strategies to Combat Emerging Pests in the Current Scenario of Climate Change. Entomological Society of India, New Delhi, p. 412.

Rajendran, T.P., 2013. 50 years after Silent Spring: looking backwards and forwards. pp. 313–320. In: Shetty, P.K., Ayyappan, S., Swaminathan, M.S. (Eds.), Climate Change and Sustainable Food Security. National Institute of Advanced Studies, IISC, Bangalore. 324 p.

Rajendran, T.P., Ballal, C., 2008. Biopesticides in the management of Indian agricultural pests. pp. 18–41. In: Kaul, O., Dhaliwal, G.S., Kaul, V.K. (Eds.), Sustainable Crop Protection – Biopesticide Strategies. Kalyani Publishers, New Delhi. 326 p.

Rajendran, T.P., Basu, A.K., 1999. Integrated pest management in cotton – history, present and future perspectives. pp. 205–232. In: Sundaram, V., Basu, A.K., Narayanan, S.S., Krishna Iyer, K.R., Rajendran, T.P. (Eds.), Handbook of Cotton. Indian Society for Cotton Improvement, Mumbai. 600 p.

Rabindra, R.J., Joshi, S., Mohanraj, P., 2004. Biological Control of Sugarcane Woolly Aphid – A Success Story. http://www.nbair.res.in/achievements/sugarcane.pdf.

Ramamurthy, V.V., Gupta, G.P., Puri, S.N., 2009. Proceedings of the National Symposium – Integrated Strategies to Combat Emerging Pests in the Current Scenario of Climate Change. Entomological Society of India, New Delhi, p. 412.

Rasnitsyn, A.P., Quicke, D.L.J., 2002. History of Insects. Kluwer Academic Publishers. ISBN:1-4020-0026-X.

Riley, D.G., 2004. Economic injury level (EIL) and economic threshold (ET) concepts in pest management. pp. 744–748. In: Capinera, J.L. (Ed.), Encyclopedia of Entomology. Kluwer Academic Publishers, Dordrecht.

Rohdendorf, B.B., Rasnitsyn, A.P. (Eds.), 1980. Historical Development of the Class Insecta (Trans. Paleontol. Inst. Acad. Sci. USSR), vol. 175. Nauka Press, Moscow, 270 pp.

Russel, D., 2004. Integrated pest management for insects of cotton for less developed countries. pp. 141–173. In: Horowitz, R., Ishayaa, I. (Eds.), Insect Pest Management. Springer-Verlag Berlin Heidelberg. 344 p.

Sanderson, E.W., Jaileh, M., Levy, M.A., Redford, K.H., Wannebo, A.V., Woolmer, G., 2002. The human footprint and the last of the wild. BioScience 52, 891–904.

Shylesha, A.N., Joshi, S., Rabindra, R.J., Bhumannawar, B.S., 2012. Classical Biological Control of Papaya Mealybug. http://www.nbair.res.in/achievements/papaya.pdf.

Southwood, T.R.E., 1975. The dynamics of insect populations. In: Pimentel, D. (Ed.), Insects, Science and Society. Academic Press, London, pp. 151–199.

Southwood, T.R.E., 1978. Ecological Methods. Chapman and Hall, New York, 524 pp.

Southwood, T.R.E., Norton, G.A., 1973. Economic aspects of pest management strategies and decisions. Memoirs of the Ecological Society of Australia 1, 168–184.

Turner, B.D., 1994. *Liposcelis bostrychophila* (Psocoptera: Liposcelididae), a stored food pest in the UK. International Journal of Pest Management 40, 179–190.

Biocontrol of Insect Pests

Omkar, Bhupendra Kumar

Centre of Excellence in Biocontrol of Insect Pests, Ladybird Research Laboratory, Department of Zoology, University of Lucknow, Lucknow, India

1. INTRODUCTION

Crop production must increase significantly in the near future to meet the needs of the fast increasing human population. One of the ways to emphatically increase the food availability is to reduce the crop damage caused by insect pests and to improve the management of pests. Despite exhaustive measures of crop protection, approximately 10% of agricultural yields are destroyed globally by animal pests before harvest (Oerke, 2006). This is despite a 15- to 20-fold increase in the usage of chemical pesticides in the past 40 years (Oerke, 2006).

In the 1990s, insecticides were recognized as the tools primarily responsible for saving the crops from pests leading to increased agricultural output. However, with their increased liberal usage, many damning facts about insecticides began to accrue. In the present day, although insecticides are still recognized as a source of rapid suppression of pest populations, direct exposure to them is also known to cause severe heart diseases, stomach ulcers, neurological and reproductive disorders, liver damage, cancer, and even death of human beings (Hoppin et al., 2006; Roldan-Tapia et al., 2006; Remor et al., 2009; Pathak et al., 2011, 2013; Fareed et al., 2013). Insecticide application may also lead to complete wipeout of insect pests instead of managing their populations below the economic threshold level, thereby ultimately disrupting food chains and food webs. Excessive and injudicious prophylactic use of insecticides can result in management failure through pest resurgence, secondary pest problems, or the development of heritable resistance. Worldwide, more than 500 species of arthropod pests have become resistant to one or more insecticides (Hajek, 2004), whereas there are close to 200 species of herbicide-resistant weeds (Heap, 2010). In addition, the developmental cost and time of each insecticide are very high (Sparks, 2013). Approximately 140,000 insecticidal compounds need to be screened to find 1 successful compound, and once that is identified, it can take more than $250 million and 8–12 years for the insecticide to be developed and registered (Sparks, 2013).

In 1992 Agenda 21 of the Earth Summit proposed the need for corrective measures to attain sustainable agriculture and environmental safety (Singh, 2004). One of such means that was identified was control using natural enemies because they are responsible for an estimated 50–90% of the pest control occurring in crop fields (Pimentel, 2005). This form of control using natural enemies or biological control (i.e., biocontrol) carries great pest management potential.

The scientific basis of biological forms of pest control is very complex. The possibilities for their development widen with an increase in scientific knowledge and can be exploited in accordance with economic and social needs. Biocontrol of pests by the natural enemies has taken place since the origin of crop plants. Individual examples of the use of natural enemies to control pests have existed for centuries, but biocontrol emerged as a scientific method only late in the nineteenth century. As the field of entomology developed in the nineteenth century, the awareness grew of the importance of predators, parasitoids, and pathogens in the limitation of insect numbers, and suggestions were made for the practical use of such natural enemies (Huffaker et al., 1976; Huffaker, 2012; Naranjo et al., 2015).

Biocontrol strongly reduces the exposure of crops to toxic pesticides; this results in lack of residues on the marketed products. Furthermore, by limiting or delaying pesticide applications and contributing to pest suppression, biocontrol can also postpone the onset and cost of pest resistance (Holt and Hochberg, 1997; Liu et al., 2014). With the use of biocontrol methods, no premature abortion of flowers and fruit takes place. Release of natural enemies usually occurs shortly after the planting period when the grower has sufficient time to check for successful development of natural enemies; thereafter, the system is reliable for months with only occasional checks. With biocontrol there is no safety period between application and harvesting the crop; therefore, harvesting can be done at any moment, which is particularly important with strongly fluctuating market prices. Biocontrol methods result in contribution to protection or even improvement of biodiversity, and there is low risk of food, water, and environmental pollution. As a result, biocontrol is generally appreciated by the general public more than the chemical/other control methods and result in a quicker sale of crops produced under biocontrol, a better price for these crops, or both.

Yet another notable feature of biocontrol has been its freedom from harmful side effects, especially when compared with chemical control. The safety with which biocontrol has been conducted has depended on several factors. The biotic agents used in biocontrol are selected from the naturally occurring, self-balancing population systems, in which many of the natural enemies, such as entomophagous and/or phytophagous insects, often show a high degree of host specificity. This allows them to be used with considerable confidence that undesirable complications will not arise (Table 1). In addition, the preliberation research now undertaken is being practiced at deeper levels with respect to the characteristics of the natural enemies and the pest and its habitat and with the increasing expertise that arises from greater ecological understanding and scientific background (Huffaker et al., 1976; Huffaker, 2012).

The advantages of biocontrol are numerous and include a high level of control at low cost; self-perpetuation at little and/or no cost; and absence of harmful effects on humans and their cultivated plants, domesticated animals, wildlife, and other beneficial organisms on the land or in the sea. The ability of biocontrol agents to reproduce rapidly and to search out their hosts and survive at relatively low host densities makes outstanding advantages possible (Waterhouse and Sands, 2001; Mason and Huber, 2002; Neuenschwander et al., 2003). The development of host resistance to introduced bioagents, with the consequence of jeopardizing a whole program, is virtually unknown, although host resistance to insect parasitoids and other types of parasites is common.

TABLE 1 Comparison of Data on Performance of Chemical and Biocontrol (van Lenteren, 1997, 2008)

Properties	Chemical control	Biocontrol
Success ratio	1:200,000	1:10
Developmental costs	~US $150 million	~US $2 million
Developmental time	~10 years	~10 years
Benefit/cost ratio	2:1	20:1
Risks of resistance	Very large	Very small
Specificity	Very small	Very large
Harmful side effects	Numerous	Nil/few

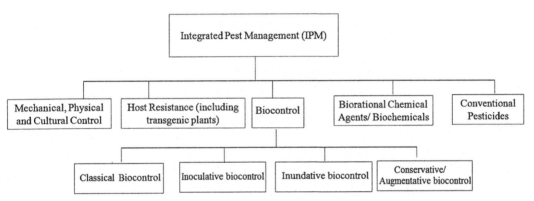

FIGURE 1 Relationship between biocontrol and other strategies for pest management (see Eilenberg et al., 2001).

During the past two decades, steady progress has been made worldwide to safeguard the crops from pests through the use and manipulation of biocontrol agents (i.e., natural enemies). Nevertheless, there is an increasing demand to search for more natural enemies (viz., predators, parasitoids, and pathogens) and assess their efficacy against various agricultural pests. Several biocontrol agents are undergoing scientific examinations, and the field trials on the efficacy of many of them have been recently introduced in various countries. The information concerning progress in this discipline is very important for scientists and the farmers because it would place greater emphasis on large-scale demonstrations of biocontrol methods as a part of an integrated pest management (IPM) strategy (Figure 1).

Despite a growing demand for biocontrol agents, major populations of the world, especially those of developing nations, are still unaware about the ways by which they are being exploited in IPM. Although numerous efforts have been made in the past by scientists and researchers in this direction, synthetic pesticides are still widely used in agriculture. In light of earlier scientific work, this chapter is an attempt to develop an ecofriendly approach in the minds of the people when dealing with agricultural pests. After reading the chapter, it is expected that the readers will be aware of (1) meanings and methods of biocontrol; (2) biocontrol agents, including definitions, types, successes, and/or disappointments; (3) applications and limitations of biocontrol methods; and (4) the future of biocontrol methods within the IPM system.

2. BIOCONTROL: MEANING AND METHODS

The term *biological control* (biocontrol) was first introduced by Smith (1919) for the "top-down" action of natural enemies/biocontrol agents (viz., predators, parasitoids and pathogens) in maintaining the pest population density at a lower level than what may have occurred in their absence. According to DeBach (1974) and Crump et al. (1999), "Biocontrol is the use of living organisms/natural enemies to suppress the population density or impact of a specific pest organism, making it less abundant or less damaging than it would otherwise be." Although this definition is an ecological approach to biocontrol, in an applied sense biocontrol includes the reduction of pests by manipulating the population of biocontrol agents (Hodek et al., 2012).

Biocontrol is an age-old practice. Approximately 4000 years ago, Egyptians used domestic cats as tools for rodent control (Wilson and Huffaker, 1976). Centuries ago, the farmers of China and other Asian countries maintained colonies of predatory ants to reduce pest populations on citrus trees (Chander, 1999). In 1602, Aldrovandi from Italy observed the reduction of the common cabbage butterfly, *Pieris rapae* Linnaeus, by a parasitoid, *Apanteles glomeratus* (L.). By 1762 the first attempt of biocontrol was made by introducing the mynah bird, *Acridotheres tristis* Linnaeus, from India to the island of Mauritius for the control of red locust, *Nomadacris septemfasciata* Serville. Darwin, in 1800, suggested the use of natural enemies to control insect pests. In 1863, mealybugs from North India were introduced in South India to control cactus (Wilson and Huffaker, 1976) and was the first ever instance of biocontrol of weeds. The small Indian mongoose (*Herpestes auropunctatus*) was introduced in 1872 primarily for rat control from the vicinity of Kolkata to Jamaica, subsequently from Jamaica to other islands in the West Indies (occasionally to control snakes and rats) and to Hawaii, and independently from Asia to several other islands around the world (Simberloff, 2012).

Although several stories exist regarding the successful utilization of biocontrol agents, the impetus to biocontrol arose from a famous and very successful campaign in the United States in 1889. *Icerya purchasi* Maskell, native to Australia, made its way to California on acacia plants around 1868, and in about 10 years it was threatening the citrus orchards of Southern California (Ebeling, 1959). Charles V. Riley (Chief of the Division of Entomology, U.S. Department of Agriculture) sent Albert Koebele to Australia in 1888 to collect natural enemies of the scale insect because the pest was of Australian origin. Koebele found two natural enemies, viz. a dipterous parasite, *Cryptochaetum iceryae* (Williston) and the vedalia beetle, *Rodolia cardinalis* (Mulsant), attacking this scale on citrus, but he selected the vedalia beetle for augmentation and subsequent release. Within a year, *R. cardinalis* reduced the population of *I. purchasi* below the economic threshold level. It was the spectacular success of *R. cardinalis* that subsequently initiated biocontrol agent introductions (Figure 2).

In 1955, rosy wolf snail, *Euglandina rosea*, was introduced widely, first from Florida to Hawaii, and then to various islands primarily in the Pacific to control the previously introduced giant African snail, *Lissachatina fulica*. In Florida, South American alligator weed (*Alternanthera phyloxeroides*) has been successfully controlled by the alligator weed flea beetle,

  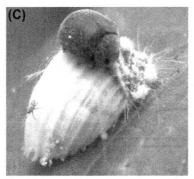

FIGURE 2 (A) Albert Koebele, who was sent to Australia to identify the natural enemy viz. vedalia beetle, *Rodolia cardinalis* (Mulsant) of *Icerya purchasi* Maskell that was threatening the citrus orchards of Southern California. (B) Larvae and (C) adults of *R. cardinalis* attacking *I. purchasi*.

TABLE 2 Pertinent Dates in the Early History of Biocontrol (Huffaker et al., 1976)

Year	Natural insect enemies	Pest species
1200	Ladybird beetles (Coleoptera)	Aphids and scale insects
1602	*Apanteles glomeratus* (Hymenoptera)	*Pieris rapae*
1734	Aphidivorous fly (Diptera)	Aphids
1763	*Calosoma sycophanta* (Coleoptera)	Caterpillars
1776	*Reduvius personatus* (Hemiptera)	*Cimex lectularius*
1800	Ichneumonids (Hymenoptera)	Cabbage caterpillars
1844	Staphylinids and carabids (Coleoptera)	Garden pests
1870	*Aphytis mytilaspidis* (Hymenoptera)	Scale insects
1882	*Trichogramma* sp. (Hymenoptera)	*Nematus ribesii*
1883	*A. glomeratus* (Hymenoptera)	*P. rapae*

Agasicles hygrophila (Center et al., 1997). On the island of St. Helena, the tropical American scale insect (*Orthezia insignis*), a threat to endemic gumwood tree (*Commidendrum robustum*), was controlled by the South American ladybird beetle, *Hyperaspis pantherina* (Booth et al., 2001). The Brazilian weevil, *Cyrtobagous salviniae*, has dramatically suppressed the floating fern giant salvinia (*Salvinia molesta*), an invasive weed of water bodies in many tropical and subtropical countries (Room et al., 1984; Julien et al., 2002).

Biocontrol is a rapidly growing area that brings together scientists from many disciplinary backgrounds, including ecologists, entomologists, weed scientists, plant pathologists, insect pathologists, and microbiologists. It is generally applied to control (1) invertebrate pests using predators, parasitoids, and pathogens; (2) weeds using herbivores and pathogens; and (3) plant pathogens using antagonistic microorganisms and induced plant resistance. In addition, the application of biocontrol to veterinary and human medicine research and practice is now being explored (Eilenberg et al., 2001; Matthew et al., 2010, Table 2).

It is noteworthy that the definitions of biocontrol stress the point that "living organisms" are used (i.e., including insect viruses) whereas genes or gene fragments, metabolites from insect or weed pathogens, or competitors to plant pathogens, used without the organisms producing them, are excluded (Eilenberg et al., 2001).

The scope of biocontrol has expanded over the decades, and some biology-based non-chemical forms of control are also included in this term by many authors. The inclusion of any of the following such methods is not generally approved within the term biocontrol:

1. Development of strains of crops that are resistant to, or tolerant of, pests or diseases.
2. Modification of cultural practices in a way that avoids or reduces infestation, such as change of planting date, avoidance of continuity of the crop in successive seasons on the same land, or plowing, pruning, flooding, etc.
3. The release of sterile males, which has proven effective against screwworms and fruit flies.
4. Techniques such as genetic, pheromonal, and other actual or potential forms of pest control that arise from new scientific knowledge.

Doutt (1972) has said that such an expansion of meaning "has the damaging effect of obscuring the unique functional and ecological basis of biocontrol" (Huffaker, 2012).

The ecological and applied approaches of biocontrol may be achieved through the following methods: (1) classical biocontrol (including new association); (2) inoculative biocontrol; (3) inundative biocontrol; and (4) conservative or augmentative biocontrol.

2.1 Classical Biocontrol

Classical biocontrol may be defined as, "The intentional introduction of an exotic, usually co-evolved, biocontrol agent for permanent establishment and long-term pest control" (Greathead, 1994; FAO, 1996; Coombs and Hall, 1998). The basis of classical biocontrol is the "enemy release hypothesis." According to this hypothesis, the exotic species become pests in new environments by escaping the influence of those natural enemies that suppress their populations in their native range. Thus, classical biocontrol reestablishes the top-down control by reintroducing the natural enemies of the pest into its new range (Van Driesche and Bellows, 1995; Crawley, 1997; Keane and Crawley, 2002; Hajek, 2004; Naranjo et al., 2015).

Thus, classical biocontrol primarily describes the releases of insect predators, parasitoids, and pathogens to control other insect pests and insect herbivores to control weeds. This form of biocontrol is appropriate when insects that spread or are introduced (usually accidentally) to areas outside of their natural range become pests mainly because of the absence of their natural enemies. The same strategy has also been called "importation" by Nordlund (1996) and "introduction of natural enemies" by Van Driesche and Bellows (1995). The primary objective of classical biocontrol is the permanent establishment of a biocontrol agent for self-sustained long-term control and depends on finding an appropriate biocontrol agent that is not native to the area where the pest needs to be controlled (Eilenberg et al., 2001). However, classical biocontrol requires the introduction of an "exotic" organism; it also provides an unparalleled situation to study the biology of introduced populations (see Marsico et al., 2011).

This field emerged after the stupendous success of the vedalia beetle, R. (*Vedalia*) *cardinalis* Mulsant (Coleoptera: Coccinellidae) against the cottony cushion scale, *I. purchasi* Maskell

(Homoptera: Mgarodidae) in California in the late 1800s. Likewise, ash whitefly (*Siphoninus phillyreae*) that had spread to 28 of the United States, including California, Arizona, and New Mexico in 1988, was under complete control within 2 years of classical biocontrol introductions in 1990.

As far as classical biocontrol of crop pests on the Indian subcontinent is concerned, perhaps the first intentionally introduced agent was the ladybird predator, *Cryptolaemus montrouzieri*, which was introduced in June 1898. Although, the predator did not control soft green scale, *Coccus viridis*, which was its specific target, in July 1951, it controlled various mealybugs infesting fruit crops, coffee, ornamental plants, etc., in south India. The predator is now effective in suppressing mealybug infestations on citrus, guava, grapes, mulberry, coffee, mango, pomegranate, custard apple, ber, etc., and green shield scale on sapota, mango, guava, brinjal, and crotons in Karnataka (Singh, 2004).

Similarly, the woolly aphid, *Eriosoma lanigerum* (native of Eastern United States), was accidentally introduced to India from England. It soon spread to all of the apple-growing areas of the country and started causing severe damage. For the control of woolly aphid, exotic aphelinid parasitoid, *Aphelinus mali* (native of North America), was introduced from the United Kingdom to Saharanpur (Uttar Pradesh, India). However, the parasitoid failed to establish itself because of the intense activity of a ladybird beetle, *Coccinella septempunctata*, which fed indiscriminately on the parasitized as well as unparasitized woolly aphids. *Coccinella septempunctata* has provided satisfactory control of the pest, but the parasitoid has also established itself in all apple-growing areas of the country, being more effective in valleys rather than on mountain slopes (Singh, 2004). During the period of activity of other predators, the population of this exotic parasitoid has diminished (Figure 3).

The spiraling whitefly, *Aleurodicus dispersus*, a native of the Caribbean region and Central America, was introduced into India in 1995. First reported from Kerala, it soon spread to all of the southern states, causing serious damage to several plants. It started attacking more than 253 plant species belonging to 176 genera and 60 families. However, serious damage was caused to avocado, banana, cassava, guava, papaya, and mango in addition to several ornamental and avenue trees. As a result, exotic aphelinid parasitoids, *Encarsia guadeloupae* and *Encarsia meritoria* (Origin: Caribbean region/Central America), were collected from Minicoy Island of Lakshadweep and brought to the mainland, where the parasitoid established well,

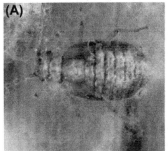 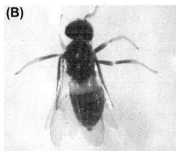 

FIGURE 3 (A) The wooly aphid *Eriosoma lanigerum* that was accidentally introduced to India from England. (B) Aphelinid parasitoid, *Aphelinus mali*, was introduced from the United Kingdom to Saharanpur (Uttar Pradesh, India) to control the aphid pest. (C) However, *Coccinella septempunctata* has also provided satisfactory control of the aphid pest.

causing a perceptible reduction in the pest population. Although the parasitism levels due to both parasitoids vary from 29% to 70% and exceed 90% during some parts of the year, the former is performing better than the latter and has established itself in Kerala, Karnataka, and several parts of Andhra Pradesh, where it was previously absent (Singh, 2004).

Despite having many beneficial aspects, classical biocontrol is currently not being encouraged because negative environmental effects may arise through ill-considered introductions of exotic natural enemies. Many introduced agents have failed to control pests; for example, more than 60 predators and parasitoids have been introduced into the northeastern part of North America with little effect on the target gypsy moth, *Lymantria dispar* (Lymantriidae). Some introductions have strengthened the pest problems whereas others have become pests themselves. Exotic introductions generally are irreversible, and nontarget species can suffer worse consequences from efficient natural enemies than from chemical insecticides, which are unlikely to cause total extinctions of native insect species. There are documented cases of introduced biocontrol agents wiping out native invertebrates. Several endemic Hawaiian insects (target and nontarget) have become extinct largely as a result of biocontrol introductions (Howarth, 1983). The endemic snail fauna of Polynesia has been almost completely replaced by accidentally and deliberately introduced alien species (Gullan and Cranston, 2005, 2009).

Although initial classical biocontrol was predominantly focused on control of introduced pests with the goal of reestablishing host/natural enemy associations that keep pests in check in their areas of origin, the strategy has now also been applied against native pests (e.g., see Carruthers and Onsager, 1993). Under such circumstances, the term *neoclassical biocontrol* is used when an exotic natural enemy is introduced against a native pest. However, the introduction (importation, augmentation, and release) of exotic natural enemies against exotic pests with which they did not coevolve is termed *new-association biocontrol* (novel association biocontrol; Hokkanen and Pimentel, 1989). The former term is a subcategory within classical biocontrol whereas the latter is considered as the standard type of classical biocontrol, not needing a specialized name (Eilenberg et al., 2001).

Such new associations are supposed to be very effective at controlling pests because the pest has not coevolved with the introduced enemies. Unfortunately, the exotic species that are most likely to be effective biocontrol agents because of their ability to utilize new hosts are also those most likely to be a threat to nontarget species. An example of the possible dangers of neoclassical control is provided by the work of Jeffrey Lockwood, who campaigned against the introduction of a parasitic wasp and an entomophagous fungus from Australia as control agents of native rangeland grasshoppers in the western United States. The potential adverse environmental effects of such introductions include the suppression or extinction of many nontarget grasshopper species with probable concomitant losses of biodiversity and existing weed control and disruptions to food chains and plant community structure. The inability to predict the ecological outcomes of neoclassical introductions means that they are highly risky, especially in systems where the exotic agent is free to expand its range over large geographical areas.

Polyphagous agents have the greatest potential to harm nontarget organisms. Natural enemies with broad host ranges would not provide the desirable biocontrol, because they may attack important nontarget organisms in the new environment and become exotic pests in their own right (Harris, 1990; Howarth, 1991; Louda et al., 1997; Follett and Duan, 2000;

Wajnberg et al., 2001). As a result, classical biocontrol programs generally highlight host specificity in selecting agents for introduction to avoid these undesirable nontarget effects.

Moreover, native species in tropical and subtropical environments may be especially vulnerable to exotic introductions because in comparison with temperate areas biotic interactions can be more important than abiotic factors in regulating their populations. Unfortunately, the countries and states that may have the most to lose from inappropriate introductions are exactly those with the most careless quarantine restrictions and few or no protocols for the release of alien/exotic organisms.

Thus, classical biocontrol has tremendous promises, but those who attempt to introduce biocontrol agent(s) beyond their normal range need to undertake serious responsibilities. Before introductions can be regarded as safe, it must be conclusively proven that the biocontrol agent(s) will not harm the new environment, including the native flora and fauna, human health and aesthetic values, and/or local industry.

2.2 Inoculative Biocontrol

According to Crump et al. (1999), inoculative biocontrol may be defined as "The intentional release of a living organism as a biocontrol agent with the expectation that it will multiply and control the pest for an extended period, but not permanently." It appears that if an exotic organism is released with the aim of long-term control without additional releases, it is classical biocontrol; however, if the releases only result in temporary control and additional releases are needed, it is inoculative biocontrol (Eilenberg et al., 2001). In glasshouses, the early release of predators and parasitoids, often with alternative food sources, is inoculative biocontrol.

Examples of inoculative biocontrol are (1) the releases of *Encarsia formosa* Gahan (Hymenoptera: Aphelinidae) and other natural enemies for pests control, now commonly practiced in glasshouses (Eilenberg et al., 2000; van Lenteren, 2000); (2) experimental control of the two-spotted spider mite, *Tetranychus urticae* (Koch), with early-season releases of *Phytoseiulus persimilis* Athias-Henriot (Hussey and Bravenboer, 1971); (3) experimental control of brown soft scale, *Coccus hesperidum* L., with *Microterys flavus* (Howard) (Hart, 1972); and (4) experimental control of the sugarcane stem borer *Diatraea saccharalis* (F.) in many Central and South American countries with the parasitoids *Lixophaga diatraea* (Townsend), *Paratheresia claripalpis* (Wulp), *Metagonistylum minense* Townsend, and *Trichogramma* spp. (Bennet, 1971).

2.3 Inundative Biocontrol

This type of biocontrol involves release of many biocontrol agents for immediate reduction of a damaging or near-damaging pest population in isolated pockets (glasshouses) and the biocontrol agents are mass multiplied in conservatories or greenhouses. According to Van Driesche and Bellows (1995), inundative biocontrol is "the use of living organisms to control pests when control is achieved exclusively by the released organisms themselves." Although inoculative and inundative biocontrol appear similar, their practical approach and ecological implications distinguish them from each other.

Inundatively released biocontrol agents must normally contact and kill a sufficiently high proportion of the pest population or reduce the damage level to give economic control before being dispersed or inactivated. The most common example of inundative biocontrol is the use of living *Bacillus thuringiensis* spores for insect control, in which the bacterium is released in high numbers per unit area with the aim of quickly killing a sufficiently large number of target insects. Over time, *B. thuringiensis* spores decrease in number and there is no expectation of long-term pest control. It is also noteworthy to mention here that the agents used for inundative releases, especially microorganisms, are commonly called "biopesticides" (Eilenberg et al., 2001).

In contrast to the inundative releases, if an inoculative release is planned, then sufficient pest numbers or other means for the growth of biocontrol agent must be maintained after the initial release to support a second or third generation of the released agent, and attention must be focused on ensuring that conditions enable this multiplication to take place (Eilenberg et al., 2001).

For inoculative and inundative releases, in addition to the cost of production and release of biocontrol agents, their control potential per individual must also be considered. However, the cost of rearing is far more important than the control potential, such that a species that is only half as effective as another may be used if it can be reared for less than half of the cost (Knipling, 1966). Moreover, consistent production of a biocontrol agent is needed more than maximizing the production at the risk of a wide difference in quantity and quality of the insects, particularly in periodic mass releases in which a short time interval exists between the recognition of the infestation and the need for the treatment (Huffaker et al., 1976).

Selection for strains with certain desirable characteristics, such as resistance to insecticides and/or high or low temperature tolerance, could prove valuable (Lingren and Ridgway, 1967; Hoyt and Caltagirone, 1971). Other biological parameters such as (1) the stage of the host attacked in relation to the damaging stage, (2) the host plant preferences, and (3) the behavioral patterns may also affect the efficiency of biocontrol agents for inundative or inoculative releases (Huffaker et al., 1976).

2.4 Conservative or Augmentative Biocontrol

According to Debach (1974), conservative/augmentative biocontrol is the "modification of the environment or existing practices to protect and enhance specific natural enemies or other organisms to reduce the effect of pests." Thus, conservative/augmentative biocontrol involves changes to the environment (potentially including factors outside of a given field) that may either reduce those factors that check the growth of biocontrol agents (i.e., predators, parasitoids, or pathogens), such as using less frequent, lower rate or less toxic pesticides, and/or provide much needed resources to these agents to increase their populations without much human interference (such as providing them alternative food such as pollen, nectar, alternative prey, and/or shelter that includes unsprayed refuge habitat, crop covers, mulches) (Naranjo et al., 2015).

Importantly, conservative biocontrol is distinguished from other strategies in that natural enemies are not released (Eilenberg et al., 2001). As a conservative biocontrol approach, an estimated 50–70% of strawberry acreage in California uses the beneficial mite *P. persimilis* against the pest, two-spotted spider mite *T. urticae*. The use of this beneficial mite grew rapidly in 1987 when the use of the pesticide Plictran declined from the U.S. market as a result of the federal regulations.

Because conservative biocontrol is a combination of protecting biocontrol agents and providing resources so that they can be more effective, conservative practices include limited and selective use of pesticides but also active processes such as providing refuges adjacent to crops or within crops, facilitating transfer of natural enemies between crops, or even directly provisioning food or shelter for natural enemies (Van Driesche and Bellows, 1995).

Van Driesche and Bellows (1995) called the practice of "conservative biocontrol" the "natural enemy conservation." With a more divergent viewpoint, Nordlund (1996) proposed that "conservation" includes "actions taken to protect or maintain existing populations of biocontrol agents." Under a category of "augmentation," Nordlund (1996) further lists "environmental manipulation" as "alterations of the environment designed to protect or increase existing populations of biocontrol agents." For many crop systems, environmental manipulation can greatly enhance the effect of natural enemies in reducing pest populations. This typically involves altering the habitat available to insect predators and parasitoids to improve conditions for their growth and reproduction by the provision or maintenance of shelter (including overwintering sites), alternative foods, and/or oviposition sites.

Habitat management is a conservative biological pest control strategy focused on manipulating the environment to enhance natural enemy populations (Landis et al., 2000; Paredes et al., 2013a,b; Naranjo et al., 2015). A common conservative biocontrol practice in perennial crops is the establishment and management of a ground cover (Boller et al., 2004), which often consists of an intertree herbaceous vegetation strip, although it can extend as a continuous covering across the crop. Ground cover can be formed by single species or a polyculture, with species typically belonging to Poaceae and Fabaceae and other herbs (Song et al., 2010; Beizhou et al., 2011), and often comprises self-regenerating vegetation (Silva et al., 2010; Aguilar-Fenollosa et al., 2011). Ground cover can improve the diversity of vegetation, especially if replacing bare ground, and it creates new habitat structure (Landis et al., 2000). As a result, the abundance (Silva et al., 2010) and diversity of natural enemies increase (Beizhou et al., 2011; Ditner et al., 2013), which might promote higher rates of predation and parasitism of insect pests by natural enemies. However, the identity of ground cover species can benefit some pests, so "selective plants" have been recommended such that only natural enemies derive benefit.

These secondary plants fall into several categories, including companion, repellent, barrier, indicator, trap, insectary, and banker, and they may increase the efficiency and sustainability of biocontrol of pests by natural enemies (Parolin et al., 2012). Some plants (e.g., companion plants) may directly increase crop plant productivity (e.g., by fixing nitrogen). They can also have negative effects on pests (e.g., by emitting repellent chemical substances). This has indirect positive effects on the crops. Other plants (barrier, indicator, and trap plants) are directly and negatively affected by the pest that attacks them. In this way, pests are less concentrated on the crops, the outcome of which is the indirect beneficial influence of secondary plants on the crops. Other secondary plants (e.g., banker plants and insectary plants) have indirect positive effects on crop productivity. The plants influence the presence of natural enemies by facilitating their establishment. The natural enemies then have a negative effect on the pest organisms that indirectly benefits the crop plants (which are less affected by the pests and consequently are more productive).

Moreover, growth of perennial vegetation close to the focal crop may also provide an alternative strategy for conservative biocontrol. Habitat stability is naturally higher in perennial systems (e.g., forests, orchards, and ornamental gardens) than in annual or seasonal crops (especially monocultures) because of differences in the duration of the crop (Paredes et al., 2015).

Likewise, in landscapes containing large amounts of natural or seminatural habitat, natural enemies are often more diverse and abundant than in structurally simple, intensely cultivated landscapes. This may be due to the higher availability of food (e.g., nectar, pollen, and alternative prey), overwintering sites, and refuges from disturbance in these landscapes (Landis et al., 2000; Naranjo et al., 2015). Thus, the complex landscapes might overcome the problems faced by the farmers while using ladybird beetles in conservative/augmentative biocontrol programs. For example, in the United States, the results of research on the convergent ladybird beetles, *Hippodamia convergens*, have consistently been disappointing because most of them fly away within 24 h after release (Huffaker et al., 1976; Huffaker, 2012). In other words, conservative biological pest control is expected to be much higher in complex than in simple landscapes (Bianchi et al., 2006; Tscharntke et al., 2007; Attwood et al., 2008; Straub et al., 2008; Letourneau et al., 2009; Veres et al., 2013) and may efficiently replace control by pesticides (Martin et al., 2013; Rusch et al., 2013).

For instance, it has been found that fecundity, body size, and condition of polyphagous carabid beetles are positively correlated with landscape heterogeneity, suggesting lower food availability in homogeneous landscapes (Bommarco, 1998; Ostman et al., 2001). Likewise, parasitoid fecundity and longevity have been found to be enhanced when supplied with nectar, suggesting higher biocontrol in complex landscapes supporting more abundant and diverse floral resources (Olson and Wackers, 2007). Finally, because most predator species require seminatural habitats for overwintering, fields located in complex landscapes are assumed to support higher predator colonization rates and conservative biocontrol potential (Corbett and Rosenheim, 1996).

Although conservative biocontrol has tremendous potential to suppress insect pests, its economic issues have been largely ignored (Naranjo et al., 2015). Conservative biocontrol in the form of habitat management (locally and regionally) is an active area of investigation. More comprehensive study is needed to demonstrate not only that habitat management is effective for enhancing natural enemies and suppressing pests, but also that these manipulations are realistic and profitable for the grower and advantageous for public and private investment. In addition education about conservative biocontrol and then detailed surveys of stakeholder-recognized value may be a traceable and fruitful method to identify and convey the economic value of this ecosystem service (Naranjo et al., 2015, Figure 4).

FIGURE 4 Habitat management is a conservative biological pest control strategy that focuses on manipulation of the environment to enhance natural enemy populations (A, B, and C).

3. BIOCONTROL AGENTS

The potential of biocontrol is much higher in tropical than in temperate countries because of high arthropod diversity and year-round activity of natural enemies. Numerous factors influence the effectiveness of biocontrol agents: their high adaptability, host specificity (monophagous), voracious feeding capacity, synchronized development with pest species, more spatial and temporal dispersal, multiplication capacity, short life cycle with multiple generations, unending host search capacity, being nonpalatable for predators, and resistance to pathogens and parasites. Moreover, systemic studies on the ecology and behavior of the biocontrol agents, and their relationships with their prey or hyperparasites/pathogens/predators (tritrophic relationship), would better enable us to understand the biological pest suppression mechanisms (Rao and Tanweer, 2011, Figure 5).

3.1 Predators

Because of the obvious act of predation, predators were mentioned for pest control long ago in many independent sources (Needham, 1956, 1986; and numerous authors in Smith et al., 1973). However, biocontrol was first applied when Egyptian people started keeping cats to protect stored grain from damage by the rodents some 4000 years ago. Likewise, nests of an ant, *Oecophylla smaragdina*, were sold near Canton (China) in the third century to use for control of citrus pests such as *Tesseratoma papillossa* (Chi Han, approximately 300 AD, Nan Fang Tshao Mu Chuang, records of the plants and trees of the southern regions). The earliest graphic record of an insect also concerns a predator, the hornet *Vespa orientalis*, which was depicted as a hieroglyph representing the Kingdom of Lower Egypt by King Menes approximately 3100 BC (Harpez, 1973) and can still be seen today on wall paintings in many of the ancient temples and tombs in the Nile Valley. Therefore, it appears that predators such as mongooses, owls, toads, ants, etc., have been rigorously used as biocontrol agents from the very beginning of human civilization (van Lenteren, 2005, 2008).

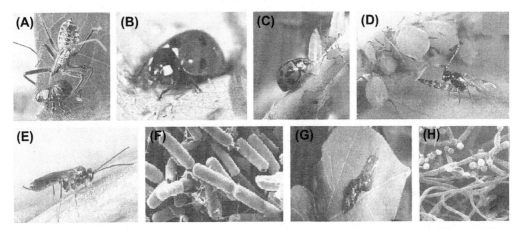

FIGURE 5   Predators (A, spider; B and C, ladybird beetles), parasitoids (D and E), and pathogens (F, bacteria; G, viral-infected insect pest; H, fungi) for control insect pests.

As far as entomophagous insect predators are concerned, most of them are arachnids, particularly spiders (order Araneae) and mites (Acarina, also called Acari); beetles (Coleoptera: Coccinellidae and Carabidae); or lacewings (Neuroptera: Chrysopidae and Hemerobiidae). Predatory mites not only regulate the populations of phytophagous mites, but some of them parasitize immature and adult insects or even feed on insect eggs and are potentially useful control agents for certain scale insects, grasshoppers, and stored-product pests. Spiders are diverse and efficient predators with a much greater effect on insect populations than mites, particularly in tropical ecosystems.

Predatory beetles and lacewings have also been used successfully in the biocontrol of agricultural pests (Dixon, 2000; Omkar and Pervez, 2000). Most of these predatory species are polyphagous/generalist predators and feed on numerous prey species. Entomophagous insect predators are generally larger than their prey. They may feed on several or all stages (from egg to adult) of their prey and each predator usually consumes several individual prey organisms during its life, with the predaceous habit often characterizing immature and adult instars. However, some highly efficient predatory coccinellids sometimes eliminate their food organisms so effectively that their own populations die out, with subsequent uncontrolled resurgence of the pest. In such cases, limited biocontrol of the pest's natural enemies may be necessary.

Nowadays natural enemies are commercially available, with the most commonly sold arthropods being various species of predatory insects (especially ladybird beetles such as *C. montrouzieri* and *H. convergens*, and mirid (*Macrolophus*) and anthocorid (*Orius* spp.) bugs) and predatory mites (*Amblyseius* and *Hypoaspis* spp.).

3.2 Parasitoids

The other major type of entomophagous insect is parasitic as a larva and free-living as an adult. The larva develops either as an endoparasite within its insect host or externally as an ectoparasite. In both the cases the host is either consumed and/or killed by the time the fully fed larva pupates in (if endoparasite) or near the remains of (if ectoparasite) the host. Such insects, called parasitoids, are holometabolous insects and most are wasps (Hymenoptera: especially superfamilies Chalcidoidea, Ichneumonoidea, and Platygasteroidea) or flies (Diptera: especially the Tachinidae).

Although the true parasites live at the expense of their hosts without actually causing the death of the host, the parasitoids always kill their host after spending the larval period as a true parasite and their adult period as free-living. They can be egg, larval, pupal, or adult parasitoids. These natural enemies are similar in size and only have a slightly longer developmental time than their hosts (Srivastava, 2004). They are specialized in their choice of host, and only females search for a host. Parasitoids often are parasitized themselves by secondary parasitoids, called hyperparasitoids, which may reduce the effectiveness of the primary parasitoid in controlling the primary host—the pest insect (van Lenteren, 2005, 2008).

Insect parasitism was understood much later than predation because of the complex biological relationships between parasitoids and their hosts. It was known in China for a long time in the form of parasitic tachinid flies of silkworms (*Bombyx mori* L.). Although the discovery of insect parasitism in Africa; North, Central, and South America; Asia (except China); Australia; and New Zealand took place after 1700 (van Lenteren, 2005), the discovery of insect parasitism in the eleventh century in China and in the seventeenth century in Europe has led to the highly successful and environmentally safe use of hundreds of species of parasitoids in biocontrol (Gurr and Wratten, 2000; van Lenteren, 2003; van Lenteren et al., 2006). However, the first Chinese record with a correct description of the life cycle of a hymenopteran parasitoid dates from 1704 whereas in Europe, the life cycles of insect parasitoids were studied by Antoni van Leeuwenhoek, John Ray, and Antonio Vallisnieri around the year 1700.

Some common parasitoids are *A. glomeratus* (on *Pieris brassicae*); *Campoletis chloridae* (on *Helicoverpa armigera*); *Epiricania melanolenca* (on *Pyrilla perpusilla*); and *Trichogramma minutum*, *Trichogramma chelonis*, and *Trichogramma japonica* (on *P. perpusilla*). In Europe, more than 80 species of natural enemies are available commercially, and among them various species of parasitoid wasps include *Aphidius*, *Encarsia*, *Leptomastix*, and *Trichogramma* spp.

3.3 Pathogens

Although the research publications of the sixteenth, seventeenth, and eighteenth centuries have emphasized the diseases of silkworms, De Réaumur (1726) was the first to illustrate a fungus, *Cordyceps*, infecting a noctuid larva. Moreover, Agostino Bassi (1836) was the first to experimentally demonstrate that a microorganism, *Beauvaria bassiana*, causes the muscardine disease of silkworms. He further published the idea to use microorganisms for insect pest control in 1836. Later in 1874, Pasteur suggested the use of microorganisms against the grape phylloxera in France. In 1879 Metchnikoff concluded that the fungus, *Metarrhizium anisopliae*, when mass produced and properly introduced in the field, results in effective biocontrol (van Lenteren, 2005, 2008).

Success of entomopathogens depends on virulence, easy application and production, low cost, good storage properties, safe and aesthetic accessibility by the farmers, and efficacy to reduce pest problems below subeconomic levels (Jaiswal et al., 2008; Prasad and Mallikarjuna Rao, 2009; Srivastava et al., 2009). Bacteria, fungi, oomycetes, viruses, and protozoa are all used for the biocontrol of pestiferous insects, plant pathogens, and weeds (Kloepper et al., 2004; Spadaro and Gullino, 2004; Whipps et al., 2008; Chouhdary and Johri, 2009; Berg, 2009).

3.3.1 *Bacteria*

The entomogenous bacteria are categorized as spore formers and non-spore formers, crystalliferous and noncrystalliferous, and obligate and facultative pathogens. To date, different genus of bacteria such as *Bacillus*, *Clostridium*, *Brevifaciens*, *Pseudomonas*, *Aerobacter*, *Cloaca*, *Proteus*, *Serratia*, etc., are identified as entamopathogenic forms, but two genera from order *Eubacteriales* (i.e., *Bacillus* (*Bacillaceae*) and *Serratiae* (*Enterobacteriaceae*)) are frequently used for insect control programs (Table 3).

Bacillus thuringiensis (Bt) is a rod-shaped, spore-forming, gram-positive, and aerobic bacteria that produces a protein crystal (the Bt δ-endotoxin) during bacterial spore formation that is capable of causing lysis of gut cells when consumed by susceptible insects (Gill et al., 1992). The δ-endotoxin is host specific and can cause host death within 48 h (Siegel, 2001). It does not harm vertebrates and is safe to people, beneficial organisms, and the environment (Lacey and Siegel, 2000). Microbial Bt biopesticides consist of bacterial spores and δ-endotoxin crystals mass produced in fermentation tanks and formulated as a sprayable product. Bt sprays are a growing tactic for pest management on fruit and vegetable crops where their high level of selectivity and safety are considered desirable and where resistance to synthetic insecticides

TABLE 3   Some Bt-Based Commercial Products (see Rao and Tanweer, 2011)

Bt variety	Trade name	Target pest	Crop
Bt *var. kurstaki*	Halt	Diamondback moth	Cabbage
	Biolep	Lepidopterous caterpillar	Cotton
	Bioasap	Cotton bollworm	Cotton
	Delfin	Cotton bollworm	Cotton
		Cotton bollworm	Castor
	Dipel	Tobacco caterpillar	Cotton
		Cotton bollworm	Tomato
		Fruit borer	Brinjal
Bt *var. galleriae*	Spicturin	American bollworm	Cotton
		Tobacco caterpillar	Chillies
		Diamondback moth	Cabbage
			Cauliflower
Bt *var. aizawai*	Xentari	Lepidopterous caterpillar	Vegetables
Bt *var. morrisoni*	Foil	Coleopteran beetles	Vegetables
	Novodor	Coleopteran beetles	Vegetables

Bacillus thuringiensis (Bt) var. *kurstaki*, Bt var. *israeliensis*, Bt var. *sandiego*, Bt var. *tenebrionis* are used for the control of various insect pests (Kumar et al., 2004; Pawar, 2004).

is a problem (Van Driesche et al., 2008). Bt sprays have also been used on broad-acre crops such as maize, soybean, and cotton, but in recent years these have been superseded by transgenic Bt crop varieties.

3.3.2 *Virus*

Thus far, viruses belonging to Baculoviridae, Reaviridae, Iridoviridae, Poxviridae, Parovoviridae, Picnaviridae, and Rhabdoviridae have been identified as entomopathogenic organisms (Zhang et al., 2007). However, nuclear polyhedrosis virus (NPV), granulosis viruses (GV) belonging to Baculoviridae, and the cytoplasmic polyhedrosis virus of the Reaviridae family are commonly used to control various insect pests (Demir et al., 2008; Jiang et al., 2008). Furthermore, using these bioagents along with compatible chemical pesticides is much more effective than using either of the two controls alone (Shanmugam et al., 2006; Raymond et al., 2006; Polanczyk et al., 2006; Bale et al., 2008).

The virus is ingested by the insect with plant material. Upon ingestion in the midgut, the proteinaceous inundative is subjected to digestive enzymes at high pH and is rapidly solubilized, thereby releasing highly infectious virions. Once the viral particles are liberated, the nucleocapsid envelope fuses with the microvillar membrane of the gut wall cells. The nucleocapsids are released to enter the nucleus where viral DNA can replicate and produce secondary infections that invade other tissues, such as the fat body and the hemolymph.

TABLE 4 Commercially Available Viral Insecticides (see Rao and Tanweer, 2011)

Host insect	Crop	Commercial name
Autographa californica	Cabbage, cotton	VPN 80
Heliothis virescens	Cotton	Elcar
Helicoverpa armigera	Cotton, tomato	Virin-HS
Helicoverpa zea	Cotton	Eclar, Gemstar
Lymantria dispar	Forests	Gypcheck, dispavirus
Spodoptera exigua	Ornamentals, vegetables garden pea	SPOD–X
Agrotis segetum, Cydia pomonella	GV apple, pears	Agrovir carpovirusine CYD–X
Phthorimoea operculellea	Potato	PTM baculovirus

In Brazil, the nucleopolyhedrovirus of the soybean caterpillar, *Anticarsia gemmatalis* was used on up to 4 million ha (~35%) of the soybean crop in the mid-1990s (Moscardi, 1999). In the United States and Europe, the *Cydia pomonella granulovirus* (CpGV) is used as an inundative biopesticide against codling moth on apples. In Washington State, the United States's biggest apple producer, it is used on 13% of the apple crop (National Agricultural Statistics Service, 2008, Table 4).

The symptoms of virus infection appear when the insect is near death. The infected insect stops feeding and becomes sluggish and pale in color. It may swell slightly, and then becomes limp and flaccid. The integument becomes fragile and ruptures easily to emit blood with disintegrated tissues and turbid body fluid. The viral pesticides are host specific, are highly virulent, and have fast action. However, the time duration required by a viral pesticide to kill its host is long, and the host insects continue to feed on crop plants during this period. Moreover, viral pesticide persists on the crop plants for a long time after its application (Srivastava, 2004).

3.3.3 Fungi

Mycobiocontrol is the use of fungi in biological processes to lower the insect density with the aim of reducing disease-producing activity and consequently crop damage (Sandhu et al., 2012). All groups of insects may be affected, and over 700 species of fungi have been recorded as pathogens. Members of Deuteromycetes, Zygomycetes, Oomycetes, Chytridiomycetes, and Trichomycetes are generally used as entomopathogenic fungal pathogens (Purwar and Sachan, 2006; Rabindra and Ramanujam, 2007). However, *Beauveria, Metarhizium* sp., *Nomuraea* spp., *Verticellium* spp., and *Hirsutella* sp. exhibit insecticidal activities against numerous insect pests belonging to orders Coleoptera, Lepidoptera, Orthoptera, and Diptera (Ferron, 1978; Roy and Cottrell, 2008; Makaka, 2008).

Some of these fungi have restricted host ranges (e.g., *Aschersonia aleyrodes* infects only scale insects and whiteflies) whereas other fungal species have a wide host range, with individual isolates being more specific to target pests. Some species are facultative generalist pathogens, such as *Aspergillus* and *Fusarium*. However, most species are obligate pathogens, often specific and rarely found (e.g., many species of *Cordyceps*). Entomopathogens, such as *M. anisopliae* and *B. bassiana*, are well characterized in respect to pathogenicity to several insects and have been used as mycobiocontrol agents for the biocontrol of agriculture pests worldwide.

Fungi of the genus *Trichoderma* are soil borne, green-spored ascomycetes that are ubiquitous in nature. *Trichoderma* spp. are characterized by rapid growth, mostly bright green conidia and a repetitively branched conidiophore structure. As opportunistic plant symbionts and effective mycoparasites, numerous species of this genus have the potential to become commercial biopesticides (Saba et al., 2012). This biocontrol agent has no harmful effects on humans, wildlife, or other beneficial organisms. It is safe and effective in natural and controlled environments and does not accumulate in the food chain. The extensive studies on diverse physiological traits are available and still progressing for *Trichoderma* and make these fungi versatile model organisms for research on industrial fermentations and natural phenomena (Saba et al., 2012).

The effective usage of entomopathogenic fungi in field conditions requires saturated or near saturated humidity in the atmosphere or water film for spore generation and viability (Parker et al., 2000; El-Husseini, 2006; Shah et al., 2009). Entomopathogenic fungi are important because they are virulent, infect by contact, and persist in the environment for a long period of time. These can be mass produced in liquid or solid media. Most of the entomopathogenic fungi are facultative parasites that exist as saprotrophs; therefore, they can be grown apart from living hosts. Few groups are obligate parasites, which must be reared in living hosts.

Introduction of fungal pathogens into the pest population initiates epizootic (an epidemic outbreak of disease) and reduces the damage caused by the pest. The initiation of artificial epizootics has been accomplished for long-term control, especially in areas where high humidity prevails. There are several defense mechanisms in insects that prevent the penetration and growth of fungi. The most common is the melanization of the cuticle at the infection site. Entomopathogenic fungi can display either a very broad host spectrum (e.g., *M. anisopliae* and *B. bassiana*) or they have a very narrow host range (e.g., *Aschersonia* spp.; Silverio et al., 2009).

The infective unit is usually a spore or conidium that attaches to the insect epicuticle. The conidium then germinates into a short germ tube, which extends a small swelling called the appresoria. The fungi penetrate the cuticle either directly by germ tubes or indirectly by infection pegs from appresoria. The hyphae then penetrate the layers of integument, ramifying first in the cuticle and then reaching the hemocoel and internal organs. The fungal mycelium continues to involve insect tissue until the insect is filled with fungus and becomes firm in touch. After a certain stage, the host insect dies and then the fungus saprophytically grows. The conidiospores erupt through the cuticle and produce spores on the outside of the insect; thus, they also infect the nearly healthy insects (Sandhu et al., 2012, Table 5).

The early symptoms of fungal infection are loss of appetite, followed by decreased irritability, general or partial paralysis, discolored patches on integument, and increased acidity of the blood. The death of the insect takes place within 24 h to 1 week because of blockage of the circulation and gut and toxins produced by the fungi.

The advantages of using fungi as mycobiocontrol agents are as follows: (1) high degree of specificity for pest control; (2) absence of effects on mammals and thus the reduction of the hazards normally encountered with insecticide applications, such as pollution of the environment; (3) lack of problems caused to insect resistance and prolonged pest control; (4) high potential for further development by biotechnological research; and (5) high persistence in

TABLE 5 Some Fungal Pathogens Available as Commercial Products for IPM (see Rao and Tanweer, 2011)

Fungal pathogen	Trade name	Target pest
Beauveria bassiana	Congaurd	Bark feeding insect
	Mycotrol	Hairy caterpillar
	Naturalis L	Leaf roller
	Botani gard	Horn caterpillar
Metarhizium anisopliae	Green muscle	Jassids
	Metathripol	Thrips
	Greenguard	Whiteflies
	Biogreen	Mealybugs
	Bioblast	Chafer beetle
Verticillium lecani	Mycotol	Jassids, thrips
	Verticillium–50	Chafer beetle
	Bio-Catch	Mealybugs, scales

the environment provides long-term effects of entomopathogenic fungi on pest suppression (Sandhu et al., 2012).

However, there are also several constraints on the use of fungi as insecticides: (1) Two to three weeks are required to kill the insects whereas chemical insecticides may need only 2–3h. (2) Application needs to coincide with high relative humidity, low pest numbers, and a fungicide-free period. (3) Because of the high specificity, additional control agents are needed for other pests. (4) Their production is relatively expensive and the short shelf life of spores necessitates cold storage. (5) The persistence and efficacy of entomopathogenic fungi in the host population varies among different insects species; thus, insect-specific application techniques need to be optimized to retain long-term effects. (6) A potential risk to immunosuppressive people (Sandhu et al., 2012).

3.3.4 Protozoans

Protozoa consists of seven distinct phyla, and members from four phyla—*Ciliophora, Sarcomastigophora, Apicomplexa*, and *Microspora*—have been found to be parasites of invertebrates. The infection in insects occurs when resistant spores are ingested. In the gut of insect hosts, the spores extrude polar tubes that penetrate into the gut epithelial cells through which infective sporoplasm then passes. The infective agent multiplies in the cytoplasm of the host cell and eventually produces more spores, which invade other cells. After a certain level of tissue infection, the insect host may suffer from loss of vigor, reduction in number and duration of matings, decrease in fecundity, and change in behavior before the onset of death, which may occur after many days or even weeks.

Because of the large production cost, protozoan pesticides become commercially uneconomical and hence not feasible. In addition, low pathogenicity and difficulty in large-scale production makes protozoans unattractive as biopesticides (Srivastava, 2004).

TABLE 6 Some EPNs and Their Target Pests
(see Rao and Tanweer, 2011)

Name of the nematode	Name of the insect pest
Heterorhabditis indica	*Helicoverpa armigera* *Spodoptera litura* *Galleria mellonella*
H. indica *Steinernema* spp.	*G. mellonella* *S. litura* *Spodoptera exigua* *Plutella xylostella* *Henedecasis duplifacialis*
Steinernema riobrave	*Tribolium castaneum* *Plodia interpunctella*

3.3.5 Nematodes

The first entamopathogenic nematode (EPN) used for insect control was *Deladinus siricidicola*. Later, many EPNs were identified, but those belonging to three families—*Steinernematidae*, *Heterorhabditidae*, and *Mermithidae*—are most potent in killing the insect hosts (Waterhouse and Norris, 1987a,b; Georgis et al., 2006; Sepulveda-Cano et al., 2008). The first two families have terrestrial insect hosts that are associated with symbiotic bacteria and they kill their insect hosts via septicemia. The third family has aquatic nematodes that kill their host upon exit through the cuticle. The infective stage is third instar juveniles, which are ingested with food by the host insect. They molt in the gut, penetrate the foregut, and invade the hemocoel, where mating takes place. The males die and females grow to become gravid worms. Eggs hatch within the female and juveniles consume the mother's tissue before escaping into the hosts' hemocoel (Table 6).

The infective stages of nematodes survive only under the conditions of high humidity, and in nature infections occur usually in damp habitats. Thus, high humidity works as a limiting factor for their success in cryptic and soil habitats (Srivastava, 2004).

4. LIMITATIONS OF BIOCONTROL

1. **Bioinvasion and indigenous species**: Certain biocontrol agents, either through intentional or accidental introductions, have acquired the attention as invasive species with significant effects on biodiversity and human economies beyond their contributions to biocontrol (Table 7).

   Furthermore, many prominent species are unsuited for foreign introduction although they may be important agents of biocontrol in their native ecosystems. Once established, they may simply replace the mortality of pests provided by local species, cause nontarget effects on other herbivores, or even displace autochthonous species from particular habitats (Hodek et al., 2012). For example, the native ladybird species *Stethorus punctum* was replaced by the Palearctic *Stethorus pusillus* in Ontario around 1940 after its inadvertent establishment in Canada (Putman, 1955). *Harmonia axyridis* has gained the

TABLE 7 Nontarget Effects of Released Exotic Biocontrol Agents on Native Insect Species (see van Lenteren et al., 2006)

Suspected effect	Introduced natural enemy	Affected native species	Location, time
Possible global extinction	*Bessa remota* (aldrich)	*Heteropan dolens* Druce (nontarget)	Fiji, 1925
Reduction in distribution range	*Cotesia glomerata* (L.)	*Pieris napi oleraceae* Harris	Massachusetts, late 1800s
Reduction in population levels	*Compsilura concinnata* (Meigen)	Several species of native Lepidoptera (Saturniidae)	North America, 1906 onward
Partial displacement of native natural enemies	*Coccinella septempunctata* (L.)	*Coccinella novemnotata* Herbst; *Coccinella transversoguttata* Fald.; *Adalia bipunctata* (L.)	North America, 1973 onward
Interbreeding with native natural enemies	*Aphelinus varipes* Foerster, species complex	*A. varipes* Foerster, species complex	North America, 1987 onward
Lowering effectiveness of an introduced natural enemy for weed control	*Cryptolaemus montrouzieri* Mulsant	*Dactylopius opuntiae* (Cockerell)	Mauritius, late 1940s

status of a pest in North America, something that might have been anticipated based on the behavior of this species in its native China (Wang et al., 2011).

2. **Problems in establishment/colonization**: The rate at which biocontrol agents disperse from release sites has important implications for their establishment and spread. Low rates of dispersal can yield spread that is too slow and may necessitate redistribution efforts for importation biocontrol and a high density of release sites for augmentation. Low dispersal rates may also lead to inbreeding at the site of release. On the other hand, high rates of dispersal can lead to Allee's effects (i.e., some species may rely on large groups to deter predators or to provide the necessary stimulation for breeding activities) at the leading edge of the invasion front, potentially reducing the likelihood of establishment. Given these disadvantages associated with low and high dispersal rates, intermediate rates of dispersal are likely to maximize the probability of establishment and appropriate spread for biocontrol agents released in the context of either importation or augmentative biocontrol (Goldilock's hypothesis; Heimpel and Asplen, 2011).

3. **Asynchrony of life cycles**: For numerous natural enemies, there is an asynchrony in life cycles in between them and their hosts (or prey), and this asynchrony constitutes a major limitation. This asynchrony results from (1) the differential responses to climate, (2) the intrinsic limitations, and/or (3) the interactive association between the enemy and the host. For example, the tachinid parasitoid *Hyperecteina aldrichi* Mesnil is not able to provide reliable control of its host, the Japanese beetle *Popillia japonica* Newman, in the eastern United States because of the asynchrony in life cycles of the parasitoid and its host (Clausen, 1956).

4. **Agroclimatic conditions**: In classical biocontrol, great care is usually taken specifically to exclude the natural hyperparasitoids of primary parasitoids and the parasitoids and

specialized predators of other introduced exotic natural enemies. More commonly, exotic parasitoids that are imported free of their natural hyperparasitoids are utilized by indigenous hyperparasitoids in the new habitat, with varying detrimental effects on the biocontrol system. Little can be done to solve this latter problem, except to test the host-switching abilities of some indigenous hyperparasitoids before introductions of the natural enemies. Of course, the same problem applies to introduced predators, which may become subject to parasitization and predation by indigenous insects in the new area. Such hazards of classical biocontrol systems result from the complexities of food webs, which can be unpredictable and difficult to test in advance of introductions.

5. **Cultural practices**: The manner or timing of certain cultural or agronomic practices used in growing crop plants often interferes with the natural enemies and constrains biocontrol programs. For example, the parasitic and predatory fauna associated with cotton are seriously reduced when the plants are destroyed and the fields are plowed in winter to control the pink bollworm after the cotton is harvested (Clausen, 1956). Likewise, in Louisiana (United States), where sugarcane is grown as an annual crop and is harvested during the autumn season, the habitat and hosts of several parasites (e.g., *Lixophaga*, *Paratheresia*, and *Metagonistylum*) of the sugarcane borer *D. saccharalis* are eliminated. Thus, the abundance of these natural enemies of the sugarcane borer is reduced each year (Clausen, 1956).

6. **Pesticides**: The harmful effect of pesticides on the natural enemies is globally well known. In the mid-1940s, the use of dichlorodiphenyltrichloroethane in the northwestern United States for control of the codling moth led to the local extinction of the very effective parasitoid, *A. mali*, with subsequent occurrences of outbreaks of woolly apple aphid, *E. lanigerum* (Clausen, 1956). In a similar manner, Safavi (1968) described that in Iran when apple pests were treated with pesticides during winter, the pesticides destroyed *Trissolcus* spp. adults, the egg parasitoids of the cereal pest *Eurygaster integriceps*. The adult wasps used the trunks of apple trees as overwintering sites.

7. **Cannibalism and intraguild predation**: A guild is a group of species in a community that share similar resources (food or space) regardless of differences in tactics of resource acquisition and in taxonomic position (Polis et al., 1989). Intraspecific (cannibalism) and/or interspecific (intraguild predation) competitions may occur within a guild of predators and/or parasitoids, particularly for generalist feeders and facultative hyperparasitoids, and may inhibit the efficacy of biocontrol agents in suppressing the populations of insect pest species. Such interactions generally affect the pest control antagonistically by reducing the activity of predators and decreasing the mortality of pests (Mills, 2006; Omkar et al., 2002, 2014; Kumar et al., 2014, 2015, Figure 6).
When different parasitoids species were introduced to control the California red scale, *Aonidiella aurantii*, in Southern California, a series of competitive interferences were produced between the parasitoids. Likewise, when *Aphytis lingnanensis* was introduced in 1947 it soon displaced the earlier established *Aphytis chrysomphali* (Mercet) (DeBach, 1965, 1969).

8. **The "lottery approach"**: This is a multiple release strategy in classical biocontrol that promotes the use of multiple host-specific biocontrol agents for each target pest (Hokkanen and Pimentel, 1984; Myers, 1985; McEvoy and Coombs, 2000). The result of multiple release strategies/lottery approach is that the exotic organisms intentionally

FIGURE 6 Predator-predator interactions: (A) predatory interaction between the larva and adult of *Coccinella septempunctata* (L.) for the common prey, (B) intraguild predation by an adult *Podisus maculiventris* on a larva of *Coleomegilla maculata* lengi in a soybean field (Hodek et al., 2012), and (C) intraguild predation by a *Thanatus* sp. spider on an adult of *Harmonia axyridis* (Hodek et al., 2012).

introduced for classical biocontrol exceed the number of exotic pests targeted for control (Hokkanen and Pimentel, 1984; Myers, 1985; McEvoy and Coombs, 1999). The lottery approach has been challenged on the grounds that (1) it is risky to introduce more biocontrol agents than are necessary to achieve effective control and (2) multiple biocontrol agents can just as well negatively affect the outcome of biocontrol as result in additive or synergistic interactions as intended (Myers, 1985; Myers et al., 1989; McEvoy and Coombs, 2000; Pearson and Callaway, 2003).

Until now, the assumption that host-specific biocontrol agents are safe has helped to continue multiple release approaches such as the lottery approach despite these attacks. However, there can be serious indirect nontarget effects.

a. **Ecological replacement**: The biocontrol agent is host specific and strongly suppresses the target pest/weed, thereby releasing suppressed natives, but this also weakens dependencies that have developed between the pest/weed and other native species, thereby negatively affecting these nontarget species.

b. **Compensatory response**: The biocontrol agent is host specific and the overall interaction between the biocontrol agent and the pest/weed is top-down, but the target pest is only weakly affected because it displaces the negative effects onto nontarget species through compensatory responses.

c. **Food-web interaction**: The biocontrol agent is host specific, but the overall interaction between the biocontrol agent and the pest is strongly bottom-up; therefore, the biocontrol agent becomes superabundant and then serves to promote other natural enemies in the system. These natural enemies then translate this promotion into significant negative interactions with other nontarget species (Pearson and Callaway, 2005).

9. **Maintenance of stock in hope of pest arrival**: Recent studies on augmentation biocontrol have indicated that the predators (especially the coccinellids) are rarely utilized in such programs (Collier and van Steenwyck, 2004; Powell and Pell, 2007) because of two limiting factors: the cost of their production and the education required for their effective application by end users. Thus, many demonstration projects have shown potential efficacy but have failed to evolve into tactics that are attractive or economically viable pest control alternatives (Baker et al., 2003). Ladybird augmentation programs

may sometimes be useful in developing countries where labor costs associated with rearing and distribution are low, but they are unlikely to prove a viable method of pest suppression on the large-scale commercial farms of the developed world, especially in low-value row crops (Hodek et al., 2012).

5. LEGISLATION AND REGULATION OF BIOCONTROL AGENTS

Today, more than 150 natural enemy species are available in the markets of Europe and the United States for biocontrol of pests, and there are approximately 85 commercial producers of natural enemies for augmentative forms of biocontrol (van Lenteren, 2008). In addition, there are hundreds of state- or farmer-funded production units that may sell natural enemies (van Lenteren and Bueno, 2003; van Lenteren, 2000, 2008). Recent findings have shown that the natural enemies commercially available in Europe are much more in number than the United States because of the much larger greenhouse industry in Europe, and the commercial biocontrol suppliers in Europe are of larger size than their counterparts in the United States (van Lenteren, 2008).

Different international legal frameworks control the introductions of exotic species from their native ranges to new environments (Fasham and Trumper, 2001; Sheppard et al., 2003). The purpose behind such introductions is to prevent the entry, release, and/or control of organisms that are harmful either to animal or plant health or to biodiversity (CBD, 2001; Shine et al., 2000; Genovesi and Shine, 2003). The two core instruments of relevant international legislation with respect to the introduction of exotic organisms are the International Plant Protection Convention and the Convention on Biological Diversity (see van Lenteren, 2008).

Most of the countries with experience in classical biocontrol (e.g., Australia, Canada, New Zealand, South Africa, the United Kingdom, and the United States) have legislation and procedures to control imports and analyze the risks of introducing non-native biocontrol agents (Sheppard et al., 2003). Those that do not have such legislations or procedures (most developing countries) rely on the Food and Agricultural Organization Code of Conduct (FAO, 1996; adopted as ISPM3 by IPPC, 2005) that addresses the application of control measures before the import and export and introduced procedures of an internationally acceptable level (van Lenteren and Loomans, 2006; Loomans and van Lenteren, 2005; van Lenteren et al., 2006; van Lenteren, 2008).

6. STEPS IN ESTABLISHING A BIOCONTROL PROGRAM FOR INSECT PESTS

1. **Identification of the pest species:** Identifying the insect pest is generally the first step to determine whether action is required to protect the crops against pest destruction or not. This includes recognizing that the damage is occurring, determining the specific cause of the damage, and accurate taxonomic identification of the pest insect. Thus, once the pest species is taxonomically identified, it not only provides access to the available literature on the pest, but it can also explore the possibilities of utilizing the natural enemies of the related pest species (Coppel and Mertins, 1977; Fisher and Bellows, 2005).

2. **Origin, geographic distribution and ecological requirements of the pest species:** Once the exact identity of the pest species is determined, it is easy to gather the information from the existing literature on how to proceed further for its biological suppression. First, the previously known geographic distribution of the pest and its probable origin must be determined. Thereafter, examining the previous distribution and the ecological conditions of the pest under which it survives provides some idea of the pest's potential range in the area and the possible natural enemies in its homeland (Coppel and Mertins, 1977; Fisher and Bellows, 2005).

3. **Field study of target insect pest and its natural enemy:** Once the literature survey is complete, it is understood affirmatively where to go in the world in search of the target pest species, what its pest status and ecology are there, and what its recorded natural enemies are. Field studies may be needed in the native habitat for the following:

    a. Confirmation of the presence of the beneficial species/natural enemies listed in the literature and familiarize the foreign collector with their recognition,

    b. Allowing one to know with great certainty the identities of natural enemies introduced, not only to ensure that one knows what to look for when assessing establishment after introduction, but also to make sure that one recognizes and screens every potentially useful bioagent, even morphologically similar sibling species (Rosen and DeBach, 1973),

    c. Selection of the most suitable collecting areas for obtaining large quantities of natural enemies and for conveniently moving them from the field to the point of shipment, and

    d. Filling in the gaps in the knowledge thus increased. In addition, the identification and screening processes through the field studies avoid wasting time on the beneficial species, which may already be present in the intended area for introduction, and to avoid the introduction of potentially inimical species such as the hyperparasitoids (Coppel and Mertins, 1977).

    If the natural enemies are already present in the proposed area for introduction, then conservative/augmentative biocontrol will be needed and no exotic beneficial species are to be imported. Otherwise, natural enemies are to be introduced from outside, and in such a situation classical biocontrol will work.

4. **Selection of the natural enemies:**

    a. The natural enemies should be monophagous so that they may remain host-specific and do not kill other insects.

    b. They should be active at low host density at their native home, which will indicate their searching ability.

    c. They should be of correct genetic strain because one strain of the same species may be more effective than the other.

    d. They should have a wide range of climatic adaptability, which includes temperature, photoperiod, and humidity.

    e. Their life cycle should be synchronized with their host prey species.

    f. They should not be harmful to other nontarget species and should show the fewest interactions with other natural enemies or the individuals of the same species.

    g. They may have a certain tolerance level to insecticides/pesticides (Coppel and Mertins, 1977; Fisher and Bellows, 2005).

5. Preparations in the country for arrival before foreign exploration:
   a. A laboratory with satisfactory breeding facilities should be arranged in the country for arrival. Such facility is also needed for the multiplication of indigenous natural enemies before augmentative field releases.
   b. Before that, the permission of the government to import a living organism is also needed.
   c. The custom authorities in the country of arrival should be informed with the nature of the parcels that are going to arrive.

6. Collection and shipment of the natural enemy:
   a. The main objective of collecting natural enemies for importation is usually to get as many as possible in the briefest time possible.
   b. Collection sites in the exporting country need not be remote, exotic, or inaccessible, and they should possibly be well situated to the central point of export.
   c. In the early history of biocontrol, transporting natural enemies/bioagents frequently involved shipboard transits of several weeks duration. Since there was either little or no refrigeration to suspend activity of the natural enemies; so to provide food and water to sustain life of natural enemies during the long journey, living host/prey insects on potted plants were used to be transported. Moreover, the entomologist collector often accompanied his charges on the journey to ensure the adequate care of the natural enemies.
   d. The use of modern air transportation has greatly benefitted biological insect pest suppression.
   e. The container used for shipment ought to be strong enough to survive rough handling and should be constructed in such a manner so as to prevent the escape of any insect that may be active during the journey hours. The containers are usually constructed of wood and are often provided with some air ventilation, controlled humidity, and sometimes a means of customs inspection.
   f. Honey or honey mixtures are the most common nutrient and moisture sources for the natural enemies that are required if the trip lasts for more than 2–3 days.
   g. Sometimes, it is necessary to separate individuals within the shipment to prevent cannibalism or injury due to their aggressive or protective behavior.
   h. Cardboard mailing tubes or aerated plastic containers are the most suitable for airmail shipments under moderate conditions, but vacuum flasks or foam-insulated containers are also necessary to maintain the contents of shipments at lowered temperatures if they are potentially subject to exposure to lethal heat levels.
   i. Containers should be suitably labeled to caution against temperature extremes or other adverse conditions.
   j. Although insects can be shipped at any stage of development, the resting stage is considered the easiest to handle. Predators are generally sent as fully fed adults, whereas the parasitoids are shipped in host material or as cocoons (Hymenoptera), puparia (Diptera), or adults.
   k. An information sheet should be a part of each shipment indicating at least the identity and source of the enclosed material; the number of predators, parasitoids, or parasitized hosts shipped; and the dates of collection and shipment.
   l. Finally, it is better to send material in a series of small shipments rather than in single large one so as to minimize the possibility of losing them all because of an unusual delay or other chance adversity (Coppel and Mertins, 1977; Fisher and Bellows, 2005).

7. **Quarantine:**
   a. The process of quarantine is generally done to prevent the premature escape of the imported insects and to prevent the contamination of entomophage cultures by native species.
   b. Preventing the premature escape of the imported insects not only ensures that the greatest numbers of the imported beneficial species are retained for release where they are needed, but it also allows adequate screening (rearing the imported species for one or more generations under controlled conditions) of the shipment before release to eliminate hyperparasitoids, unwanted contaminating species, and diseases that might reduce the effectiveness of the introduction.
   c. Thorough knowledge of the identity and range of useful habits of the insects is required before confidently deciding that the species under culture is the one to be released and that it is free of contamination.
   d. Facilities used for quarantine are specially designed for this function and include insect-tight construction with triple-glazed windows, double-sealed air lock doors, closed-air ventilation systems, and special systems of illumination, which discourages the passage of insects through exit doors.
   e. Most of the laboratories also include specialized rearing facilities for propagation of beneficial organisms and fumigation and incineration capability for treating shipping containers and rearing cages.
   f. Admittance to the laboratories is usually restricted to authorized personnel only, and special laboratory clothing is frequently required while working in such laboratories (Coppel and Mertins, 1977).

8. **Augmentation of the natural enemies:**
   a. It includes all activities designed to increase numbers of natural enemies.
   b. This may be achieved by releasing additional numbers of a natural enemy into a system or modifying the environment in a way to promote greater numbers.
   c. The periodic release may be inoculative and/or inundative.
   d. Inoculative release may be made as infrequently as once a year to reestablish a species of natural enemy, which is periodically killed out in an area by unfavorable conditions during part of the year but operates effectively the rest of the year.
   e. Inundative releases involve mass culture and release of natural enemies to directly suppress the pest population as in the case of conventional insecticides. These are most economical against pests that have only one or at the most a few discrete generations every year (Huffaker et al., 1976).

9. **Colonization (release and recovery):**
   After release, the natural enemies establish themselves in a new environment and this is the most critical step in biocontrol.
   a. The site chosen for the release of natural enemies and colonization should be thoroughly protected from use of pesticides and cultural practices.
   b. The natural enemy should be stored at low temperature to curtail activity but permit some feeding while suitable weather for release is being awaited.
   c. The timing of release should coincide with the availability of the host stages suitable for release.
   d. The weather should be favorable and preferably timed in the morning.

e. The release stage should be egg or adult.

f. Insects about to be released should be fed and mated before release.

g. Release of insects should be done in two ways: open field release and confined release.

h. The number of releases depends on the natural enemy's host searching ability.

i. Recovery attempts are undertaken soon after initial release just to ascertain if released insects are surviving and without relevance whether or not control is being provided (Coppel and Mertins, 1977; Fisher and Bellows, 2005).

10. **Evaluation of effectiveness of the natural enemies:**
   A minimum period of 3 years is to be given before assessing the results. Evaluation is done by the following:

   a. *Addition method*: This involves comparing the host population of the pest before and after release.

   b. *Exclusion/subtraction method*: This involves the elimination and subsequent exclusion of resident natural enemies from several plots that can then be compared with a like number of otherwise comparable plots where the natural enemies are not disturbed.

   c. *Interference method*: It involves reducing the efficiency of natural enemies in one group of plots, as contrasted to another group having natural enemies undisturbed. This method further includes the biological check method (interference by ants), the insecticide check method (interference by insecticides), the hand removal method, and the trap method (Huffaker et al., 1976; Huffaker, 2012).

7. MISTAKES AND MISUNDERSTANDINGS ABOUT BIOCONTROL

1. Does biocontrol create new pests?
   Use of biocontrol against one specific pest is said to encourage new pests because the spraying of broad-spectrum pesticides is terminated. The criticism roused when the new pests that were unintentionally imported in Europe in the year 1975 (e.g., *Spodoptera exigua, Liriomyza trifolii, Liriomyza huidobrensis, Frankliniella occidentalis, Bemisia tabaci*) created serious problems in glasshouses under biological and chemical control. They threatened the biocontrol of other pests because their natural enemies could not be identified quickly, but the pests were already resistant to most pesticides before they were imported into Europe. Several of these pests are now so hard to chemically control that biocontrol appears to be the only viable option (see van Lenteren, 1992, 2008).

2. Is biocontrol unreliable?
   The idea that biocontrol is less reliable than chemical control has emerged because of strong pressure to market natural enemies that were not fully tested for efficacy. The criticism also arose because some nonprofessional producers of natural enemies did not check the biocontrol efficacy of the agents against the target pests before their launch in the market. However, the philosophy of most biocontrol workers is to support the use of only those natural enemies that have proven to be effective under practical conditions and within the total pest and disease program for a certain crop (see van Lenteren, 1992, 2008).

3. Is biocontrol research expensive?

   This is not true because the cost-benefit analyses show that biocontrol research is more cost effective than chemical control (cost-benefit ratios of 30:1 for biocontrol and 5:1 for chemical control (Tisdell, 1990; van Driesche and Bellows, 1995; Neuenschwander, 2001)). It is also often thought that finding a natural enemy is more expensive than identifying a new chemical agent. However, the costs for developing a natural enemy are on average US $2 million, and those for developing a pesticide are on average US $180 million (see van Lenteren, 1992, 2008).

4. Is commercial biocontrol application expensive for the farmers?

   Ramakers estimated the costs (agent and labor) for chemical and biological pest controls in 1980 and found that chemical control of whitefly was twice as expensive as biocontrol with the parasitoid *E. formosa* (van Lenteren, 1992, 2008). Likewise, van Lenteren (1990) and Wardlow and O'Neill (1992) found that chemical control of *T. urticae* Koch and pests of tomato/cucumber were more expensive than when controlled with predatory mites and other biocontrol agents, respectively.

5. Does the practical use of biocontrol develop very slowly?

   This is also not true. More than 5000 introductions of approximately 2000 species of exotic arthropod agents for control of arthropod pests in 196 countries or islands have been made during the past 120 years, and more than 150 species of natural enemies (parasitoids, predators, and pathogens) are currently commercially available (van Lenteren et al., 2006). It is important to mention that the identification and mass production of natural enemies has been so successful during the past 40 years that there are currently more species of natural enemies available in Northwest Europe (more than 150 species) than there are registered active ingredients for use in insecticides (<100; see van Lenteren, 1992, 2008).

8. FUTURE OF BIOCONTROL IN IPM

Despite being an effective approach to solving most agricultural problems, the past 50 years of research in IPM have been frustrating with regard to the very limited support from the public and the governing bodies. Although biocontrol is a key ecosystem service and underlying pillar of IPM, and it is likely to provide one of the highest returns on investment available in IPM, its economic value is hardly evaluated (Naranjo et al., 2015). Therefore, assessing the economic value of biocontrol will not only broaden its utility in crop protection, but it will also raise its standard by encouraging all stakeholders of agriculture, including the governing bodies who make funding decisions, to work jointly and improve the functioning of the biocontrol of insect pests in agriculture.

Nowadays, the outlook of the general public and the government has changed toward the use of biocontrol approaches for pest control because the methods are environmentally friendly and cost effective. These changes in the opinion of the public and the governing bodies will have a stimulating effect on further development and implementation of IPM systems in the agricultural sector. In the future, the IPM system will hopefully be converted into an integrated crop management program enwrapping all agricultural practices, especially the biocontrol methods, leading to a healthy growth of the plant; higher yields with a low cost of production; and being ecologically viable and farmer friendly by emphasizing insect resistance management, genetically modified crops, push–pull strategies, IT technologies, etc. (Rao and Tanweer, 2011). Moreover, co-operative farming, contact farming, and involvement of government and nongovernmental organizations in a pest management system would definitely result in sustainable agriculture and effective pest management ensuring food security (Rao and Tanweer, 2011).

9. CONCLUSIONS

During the past two decades, steady progress has been made worldwide to safeguard crops from pests through the use and manipulation of bioagents. The advantages of biocontrol are numerous and the biocontrol of pests by biocontrol agents has been taking place since the beginning of the evolutionary process of crop plants. Although the importance of biocontrol agents for pest control has been appreciated for more than 1000 years, only recently have scientists and practitioners attempted to assign economic value to biocontrol. The classical biocontrol attempts to manage primarily invasive pests through the introduction of exotic biocontrol agents; and is a publicly funded enterprise with social and private benefits (Naranjo et al., 2015). However, the innoculative and inundative biocontrol rely on one-time or repeated supplemental introductions of biocontrol agents to suppress pests, are market-driven efforts, and benefit the investors. Although the conservative biocontrol involves effecting positive changes to the environment or in control tactics to favor the abundance and activity of biocontrol agents and has social benefits, the benefits are generally unnoticed, unvalued, and are more difficult to quantify (Naranjo et al., 2015).

The biocontrol agents are not only the alternative to insecticides, but they are environmentally friendly. Although the biocontrol agents eliminate the selection pressure due to insecticides, they have certain limitations. For example, when they are used to control the pests of cotton, their perpetuation is hampered by the annual habitat of cotton, and the complexity of cotton pests makes insecticide use a must to affect the sustenance of biocontrol agents. Moreover, the slow action of biocontrol agents and their susceptibility to environmental factors make them less suitable over the insecticides. The role of biocontrol agents in biocontrol programs has also diminished as various unexpected ecological effects of exotic species have come to light and certain large, dominant natural enemies of pests have gained recognition as invasive species. The short life span and the storage and quality control of biocontrol agents also pose problems (Hodek et al., 2012).

In addition, the effect of augmentation of biocontrol agents varies among locations and over years in the same location. Consequently, regulatory authorities are likely to permit only highly specialized, noninvasive species for use in biocontrol programs in the future. Thus, the mass production of biocontrol agents and their application methods need refinement.

In this regards, the future appears brighter for improving conservation and enhancing the efficacy of biocontrol agents in open systems. More studies are needed in the future to improve the holistic understanding of the ecological roles of biocontrol agents and their ability to harmonize other beneficial species so that the information would be significant for the development of novel approaches to habitat management that could improve the efficiency of established biocontrol agents in particular agroecosystems and enhance their ability to track the economically important prey species in time and space.

Acknowledgments

The authors thank the Department of Higher Education, Government of Uttar Pradesh, India, for providing financial assistance in the form of Center of Excellence in Biocontrol of Insect Pests.

References

Aguilar-Fenollosa, E., Pascual-Ruiz, S., Hurtado, M.A., Jacas, J.A., 2011. Efficacy and economics of ground cover management as a conservation biological control strategy against *Tetranychus urticae* in clementine mandarin orchards. Crops Protection 30, 1328–1333.

Aldrovandi, U., 1602. De Animalibus Insectis Libri Septem. Bologna (in Latin).

Attwood, S.J., Maron, M., House, A.P.N., Zammit, C., 2008. Do arthropod assemblages display globally consistent responses to intensified agricultural land use and management. Global Ecology and Biogeography 17, 585–599.

Baker, S.C., Elek, J.A., Bashford, R., Paterson, S.C., Madden, J., Battaglia, M., 2003. Inundative release of coccinellid beetles into eucalypt plantations for biological control of chrysomelid leaf beetles. Agricultural and Forest Entomology 5, 97–106.

Bale, J.S., van Lenteren, J.C., Bigler, F., 2008. Biological control and sustainable food production. Philosophical Transactions of the Royal Society B 363, 761–776.

Bassi, A., 1836. Del mal del segno e di altre malattie dei bachi da seta – Parte seconda, Practica. Tipografia Orcesi, Lodi, p. 58.

Beizhou, S., Zhang, J., Jinghui, H., Hongying, W., Yun, K., Yuncong, Y., 2011. Temporal dynamics of the arthropod community in pear orchard intercropped with aromatic plants. Pest Managment Science 67, 1107–1114.

Bennet, F.D., 1971. Some recent successes in the field of biological control in the West Indies. Revista Peruana de Entomologia 14 (2), 369–373.

Berg, G., 2009. Plant–microbe interactions promoting plant growth and health: perspectives for controlled use of microorganisms in agriculture. Applied Microbiology and Biotechnology 84, 11–18.

Bianchi, F.J., Booij, C.J., Tscharntke, T., 2006. Sustainable pest regulation in agricultural landscapes: a review on landscape composition, biodiversity and natural pest control. Proceedings of Biological Sciences 273 (1595), 1715–1727.

Boller, E.F., Häni, F., Poehling, H.M., 2004. Ecological Infrastructures: Ideabook on Functional Biodiversity at the Farm Level. Landwirtschaftliche Beratungszentrale Lindau, Lindau, Suisse.

Bommarco, R., 1998. Reproduction and energy reserves of a predatory carabid beetle relative to agroecosystem complexity. Ecological Applications 8, 846–853.

Booth, R.G., Cross, A.E., Fowler, S.V., Shaw, R.H., 2001. Case study 5.24. Biological control of an insect to save an endemic tree on St. Helena. In: Wittenberg, R., Cock, M.J.W. (Eds.), Invasive Alien Species: A Toolkit of Best Prevention and Management Practices. CAB International, Wallingford, p. 192.

Carruthers, R.I., Onsager, J.A., 1993. Perspective on the use of exotic natural enemies for biological control of pest grasshoppers (Orthoptera: Acrididae). Environmental Entomology 22, 885–903.

Center, T.D., Frank, J.H., Dray Jr., F.A., 1997. Biological control. In: Simberloff, D., Schmitz, D.C., Brown, T.C. (Eds.), Strangers in Paradise. Impact and Management of Non-indigenous Species in Florida. Island Press, Washington, DC, pp. 245–263.

Chander, S., 1999. Biological control of insect pests in agriculture. Everyman's Science 34 (3), 127–131.

Choudhary, D.K., Johri, B.N., 2009. Interactions of *Bacillus* spp. and plants—with special reference to induced systemic resistance (ISR). Microbiology Research 164, 493–513.

Clausen, C.P., 1956. Biological Control of Insect Pests in the Continental United States. U.S. Department of Agriculture, p. 151.

Collier, T., van Steenwyck, R., 2004. A critical evaluation of augmentative biological control. Biological Control 31, 245–256.

Convention on Biological Diversity, CBD, 2001. Review of the Efficiency and Efficacy of Existing Legal Instruments Applicable to Invasive Alien Species. CBD Technical Series no. 2. SCBD, Montreal, p. 42.

Coombs, J., Hall, K.E., 1998. Dictionary of Biological Control and Integrated Pest Management. CPL Press, Newbury, UK, p. 196.

Coppel, H.C., Mertins, J.W., 1977. Biological Insect Pest Suppression. Springer-Verlag.

Corbett, A., Rosenheim, J.A., 1996. Impact of a natural enemy overwintering refuge and its interaction with the surrounding landscape. Ecological Entomology 21, 155–164.

Crawley, M.J., 1997. Plant Ecology. Blackwell Scientific, London.

Crump, N.S., Cother, E.J., Ash, G.J., 1999. Clarifying the nomenclature in microbial weed control. Biocontrol Science and Technology 9, 89–97.

DeBach, P., 1965. Weather and the success of parasites in population regulation. The Canadian Entomologist 97, 848–863.

DeBach, P., 1969. Uniparental, sibling and semi-species in relation to taxonomy and biological control. Israel Journal of Entomology 4, 11–28.

DeBach, P., 1974. Biological Control by Natural Enemies. Cambridge University Press, Cambridge, p. 323.

Demir, I., Nalcacoglu, R., Demirbag, Z., 2008. The significance of insect viruses in biotechnology. Tarim Bilimleri Dergisi 14, 193–201.

De Réaumur, R.A., 1726. Remarques sur la plante appelle'e a la Chine Hia Tsao Tom Tschom ou plante ver. Mémoires de l' Académie Royale des Sciences, Paris, pp. 302–306.

Ditner, N., Oliver, B., Beck, J., Brick, T., Nagel, P., Luka, H., 2013. Effects of experimentally planting non-crop flowers into cabbage fields on the abundance and diversity of predators. Biodiversity and Conservation 22, 1049–1061.

Dixon, A.F.G., 2000. Insect Predator–Prey Dynamics, Ladybird Beetles and Biological Control. Cambridge University Press, Cambridge, UK, p. 257.

Doutt, R.L., 1972. Biological control: parasites and predators. In: Pest Control Strategies for the Future. National Academy of Science, National Research Council, Washington, DC, pp. 228–297.

Ebeling, W., 1959. Avocado Pests: Subtropical Fruit Pests. University of California Press, Division of Agricultural Sciences, Berkeley, CA, pp. 285–332.

Eilenberg, J., Enkegaard, A., Vestergaard, S., Jensen, B., 2000. Biocontrol of pests on plant crops in Denmark: present status and future potential. Biocontrol Science and Technology 10, 703–716.

Eilenberg, J., Hajek, A., Lomer, C., 2001. Suggestions for unifying the terminology in biological control. BioControl 46, 387–400.

El-Husseini, M.M., 2006. Microbial control of insect pests: is it an effective and environmentally safe alternative? Arab Journal for Plant Protection 24, 162–169.

Food and Agricultural Organization, FAO, 1996. International Standards for Phytosanitary Measures. Part 1. Import Regulations. Code of Conduct for the Import and Release of Exotic Biological Control Agents. Publ. No. 3. FAO, Rome, p. 21.

Fareed, M., Pathak, M.K., Bihari, V., Kamal, R., Srivastava, A.K., Kesavachandran, C.N., 2013. Adverse respiratory health and hematological alterations among agricultural workers occupationally exposed to organophosphate pesticides: a cross-sectional study in North India. PLoS One 8 (7), e69755. http://dx.doi.org/10.1371/journal.pone.0069755.

Fasham, M., Trumper, K., 2001. Review on non-natives species legislation and guidance. Ecoscope.

Ferron, P., 1978. Biological control of insect pests by entomogenous fungi. Annual Review of Entomology 16, 259–263.

Fisher, T., Bellows, T., 2005. Hand Book of Biological Control: Principles and Applications of Biological Control. San Diego, California. , p. 1046.

Follett, P.A., Duan, J.J., 2000. Nontarget Effects of Biological Control. Kluwer Academic Publishers, Boston, MA.

Genovesi, P., Shine, C., 2003. European strategy on invasive alien species. Final version convention on the conservation of European wildlife and natural habitats. In: Council of Europe, 23rd Meeting Standing Committee, Strasbourg, 1–5 December 2003.

Georgis, R., Koppenhofer, A.M., Lacey, L.A., Belair, G., Duncan, L.W., Grewal, P.S., Samish, M., Tan, L., Torr, P., Tol, R.W., Van, H.M., 2006. Successes and failures in the use of parasitic nematodes for pest control. Biological Control 38, 103–123.

Gill, S.S., Cowles, E.A., Pietrantonio, P.V., 1992. The mode of action of *Bacillus thuringiensis* endotoxins. Annual Review of Entomology 37, 615–636.

Greathead, D.J., 1994. History of biological control. Antenna 18 (4), 187–199.

Gullan, P.J., Cranston, P.S., 2009. The Insects: An Outline of Entomology. John Wiley and Sons, p. 528.

Gullan, P.J., Cranston, P.S., 2005. The Insects: An Outline of Entomology. Blackwell Publishing, Oxford, p. 505.

Gurr, G., Wratten, S., 2000. Measures of Success in Biological Control. Kluwer Academic Publishers, Dordrecht, p. 429.

Hajek, A.E., 2004. Natural Enemies: An Introduction to Biological Control. Cambridge University Press, Cambridge, UK, p. 378.

Harpez, I., 1973. Early entomology in the Middle East. In: Smith, R.F., Mittler, T.E., Smith, C.N. (Eds.), History of Entomology. Annual Reviews Inc., Palo Alto, California, pp. 21–36.

Harris, P., 1990. Environmental impact of introduced biological control agents. In: Mackauer, Ehler, L.E., Roland, J. (Eds.), Critical Issues in Biological Control. Intercept, Andover, Hants, UK, pp. 289–300.

Hart, W.G., 1972. Compensatory releases of *Microterys flavus* as a biological control agent against brown soft scale. Environmental Entomology 1 (4), 414–419.

Heap, I., 2010. International Survey of Herbicide Resistant Weeds. http://www.weedscience.com.

Heimpel, G.E., Asplen, M.K., 2011. A 'Goldilocks' hypothesis for dispersal of biological control agents. BioControl 56 (4), 441–450.

Hodek, I., Van Emden, H.F., Honek, A., 2012. Ecology and Behavior of the Ladybird Beetles (Coccinellidae). Blackwell Publishing Limited, UK, p. 600.

Hokkanen, H., Pimentel, D., 1984. New approach for selecting biological control agents. The Canadian Entomologist 116, 1109–1121.

Hokkanen, M.T., Pimentel, D., 1989. New associations in biological control: theory and practice. The Canadian Entomologist 121, 829–840.

Holt, R.D., Hochberg, M.E., 1997. When is biological control evolutionarily stable (or is it)? Ecology 78, 1673–1683.

Hoppin, J.A., Umbach, D.A., London, S.J., Lynch, C.F., Alavanja, M.C.R., Sandler, D.P., 2006. Pesticides and adult respiratory outcomes in the agricultural health study. Annals of New York Academy of Sciences 1076, 343–354.

Howarth, F.G., 1991. Environmental impacts of classical biological control. Annual Review of Entomology 36, 485–509.

Howarth, F.G., 1983. Classical biocontrol: panacea or Pandora's box. Proceedings of the Hawaiian Entomological Society 24 (2), 239–244.

Hoyt, S.C., Caltagirone, L.E., 1971. The developing programs of integrated control of pests of apples in Washington and peaches in California. Biological Control 395–421.

Huffaker, C.B., Simmonds, F.J., Laing, J.E., 1976. The theoretical and empirical basis of biological control. In: Huffaker, C.B., Messenger, P.S. (Eds.), Theory and Practice of Biological Control. Academic Press, New York, pp. 41–78.

Huffaker, C.B., 2012. Theory and Practice of Biological Control. Elsevier, p. 810.

Hussey, N.T., Bravenboer, L., 1971. Control of pests in glasshouse culture by the introduction of natural enemies. Biological Control 195–216.

International Plant Protection Convention, IPPC, 2005. Revision of ISPM No. 3: Guidelines for the Export, Shipment, Import and Release of Biological Control Agents and Beneficial Organisms. Draft for country consultation 2004. International Plant Protection Convention (IPPC). Available at: https://www.ippc.int/IPP/En/default.jsp.

Jaiswal, A.K., Bhattacharya, A., Kumar, S., Singh, J.P., 2008. Evaluation of *Bacillus thuringiensis* Berliner subsp. kurstaki for management of lepidopteran pests of lac insect. Entomon 33, 65–69.

Jiang, H., Ya, H., Hu, J., Zhang, L., Min, J., Yang, Y.H., 2008. Advances in application of recombinant insect viruses as biopesticides. Acta Entomologica Sinica 51, 322–327.

Julien, M.H., Center, T.D., Tipping, P.W., 2002. Floating fern (*Salvinia*). In: van Driesche, R., Blossey, B., Hoddle, M., Lyon, S., Reardon, R. (Eds.), Biological Control of Invasive Plants in the Eastern United States (FHTET-2002–2004). United States Department of Agriculture Forest Service. Forest Health Technology Enterprise Team, Morgantown, pp. 17–32.

Keane, R.M., Crawley, M.J., 2002. Exotic plant invasions and the enemy release hypothesis. Trends in Ecology & Evolution 17 (4), 164–170.

Kloepper, J.W., Ryu, C.M., Zhang, S., 2004. Induced systemic resistance and promotion of plant growth by Bacillus spp. Phytopathology 94 (11), 1259–1266.

Knipling, E.F., 1966. Some basic principles in insect population suppression. Bulletin of the Entomological Society of America 12 (1), 7–15.

Kumar, B., Mishra, G., Omkar, 2014. Functional response and predatory interactions within conspecific and heterospecific guilds of two congeneric species (Coleoptera: Coccinellidae). European Journal of Entomology 111 (2), 257–265.

Kumar, B., Mishra, G., Omkar, 2015. Prey species modify predatory interactions between ladybird beetles: a case study using two sympatric *Coccinella* species. Journal of Asia-Pacific Entomology 18 (2), 109–116.

Kumar, S., Vasuda, G., Khan, M.A., 2004. Biological control: a potential weapon in agriculture. Indian Farming 39–44.

Lacey, L.A., Siegel, J.P., 2000. Safety and ecotoxicology of entomopathogenic bacteria. In: Charles, J.F., Delecluse, A., Nielsen-LeRoux, C. (Eds.), Entomopathogenic Bacteria: from Laboratory to Field Application. Kluwer Academic Press, Dordrecht, The Netherlands, pp. 253–273.

Landis, D.A., Wratten, S.D., Gurr, G.M., 2000. Habitat management to conserve natural enemies of arthropod pests in agriculture. Annual Review of Entomology 45, 175–201.

Letourneau, D.K., Jedlicka, J.A., Bothwell, S.G., Moreno, C.R., 2009. Effects of natural enemy biodiversity on the suppression of arthropod herbivores in terrestrial ecosystems. Annual Review of Ecology Evolution and Systematics 40, 573–592.

Lingren, P.D., Ridgway, R.L., 1967. Toxicity of five insecticides to several insect predators. Journal of Economic Entomology 60 (6), 1639–1641.

Liu, X., Chen, M., Collins, H.L., Onstad, D.W., Roush, R.T., Zhang, Q., Earle, E.D., Shelton, A.M., 2014. Natural enemies delay insect resistance to Bt crops. PLoS One 9 (3), e90366.

Loomans, A.J.M., van Lenteren, J.C., 2005. Tools for environmental risk assessment of invertebrate biological control agents: a full and quick scan method. In: ISBCA Davos Proceedings.

Louda, S.M., Kendall, D., Connor, J., Simberloff, D., 1997. Ecological effects of an insect introduced for the biological control of weeds. Science 277, 1088–1090.

Makaka, C., 2008. The efficacy of two isolates of *Metarhizium anisopliae* (Metschin) Sorokin (Deuteromycotina: Hyphomycetes) against the adults of the black maize beetle *Heteronychus licas* Klug (Coleoptera: Scarabidae) under laboratory conditions. African Journal of Agricultural Research 3, 259–265.

Marsico, T.D., Wallace, L.E., Ervin, G.N., Brooks, C.P., McClure, J.E., Welch, M.E., 2011. Geographic patterns of genetic diversity from the native range of *Cactoblastis cactorum* (Berg) support the documented history of invasion and multiple introductions for invasive populations. Biological Invasions 13 (4), 857–868.

Martin, E.A., Reineking, B., Seo, B., Steffan-Dewenter, I., 2013. Natural enemy interactions constrain pest control in complex agricultural landscapes. PNAS 110 (14), 5534–5539.

Mason, P.G., Huber, J.T., 2002. Biological Control Programmes in Canada, 1981–2000. CABI, Wallingford.

Matthew, J.W.C., van Lenteren, J.C., Brodeur, J., Barratt, B.I.P., Bigler, F., Bolckmans, K., Consoli, F.L., Haas, F., Mason, P.G., Parra, J.R.P., 2010. Do new access and benefit sharing procedures under the convention on biological diversity threaten the future of biological control? BioControl 55 (2), 199–218.

McEvoy, P.B., Coombs, E.M., 1999. Biological control of plant invaders: regional patterns, Weld experiments, and structured population models. Ecological Applications 9, 387–401.

McEvoy, P.B., Coombs, E.M., 2000. Why things bite back: unintended consequences of biological weed control. In: Follett, P.A., Duan, J.J. (Eds.), Nontarget Effects of Biological Control. Kluwer Academic Publishers, Boston, MA, pp. 167–194.

Mills, N., 2006. Interspecific competition among natural enemies and single versus multiple introductions in biological control. In: Trophic and Guild in Biological Interactions Control. Springer, The Netherlands, pp. 191–220.

Moscardi, F., 1999. Assessment of the application of baculoviruses for control of Lepidoptera. Annual Review of Entomology 44, 257–289.

Myers, J.H., 1985. How many insects are necessary for successful biological control of weeds. In: Delfosse, E.S. (Ed.), Proceedings of the VI International Symposium on Biological Control of Weeds, August 19–25, 1984, Vancouver, BC, Agriculture Canada, Ottawa, pp. 77–82.

Myers, J.H., Higgins, C., Kovacs, E., 1989. How many insect species are necessary for the biological control of insects. Environmental Entomology 18, 541–547.

Naranjo, S.E., Ellsworth, P.C., Frisvold, G.B., 2015. Economic value of biological control in integrated pest management of managed plant systems. Annual Review of Entomology 60, 621–645.

National Agricultural Statistics Service, NASS, 2008. Agricultural Chemical Usage 2007: Field Crops Summary. See http://usda.mannlib.cornell.edu/usda/current/AgriChem-UsFruits/AgriChemUsFruits-05-21-2008.pdf (accessed 24.03.10.).

Needham, J., 1956. Science and Civilisation in China. II. History of Scientific Thought. Cambridge University Press, Cambridge.

Needham, J., 1986. Science and Civilisation in China. VI. I. Botany. Cambridge University Press, Cambridge.

Neuenschwander, P., 2001. Biological control of the cassava mealybug in Africa: a review. Biological Control 21, 214–229.

Neuenschwander, P., Borgemeister, C., Langewald, J., 2003. Biological Control in IPM Systems in Africa. CABI Publishing, Wallingford, UK, p. 414.

Nordlund, D.A., 1996. Biological control, integrated pest management and conceptual models. Biocontrol News and Information 17, 35N–44N.

Oerke, E.C., 2006. Crop losses to pests. Journal of Agricultural Science 144 (1), 31–43.

Olson, D.M., Wackers, F.L., 2007. Management of field margins to maximize multiple ecological services. Journal of Applied Ecology 44 (1), 13–21.

Omkar, Pervez, A., 2000. Biodiversity of predaceous coccinellids (Coleoptera: Coccinellidae) in India: a review. Journal of Aphidology Gorakhpur 14 (1&2), 41–66.

Omkar, Mishra, G., Pervez, A., 2002. Intraguild predation by ladybeetles: an ultimate survival strategy or an aid to advanced aphid biocontrol? In: Pandey, K.C., Agrawal, N., Omkar. (Eds.), Professor S.B. Singh Commemoration Volume. Zoological Society of India, Lucknow, India, pp. 77–90.

Omkar, Mishra, G., Kumar, B., Singh, N., Pandey, G., 2014. Risks associated with tandem release of large and small ladybirds in heterospecific aphidophagous guilds. The Canadian Entomologist 146 (1), 52–66.

Ostman, O., Ekbom, B., Bengtsson, J., 2001. Landscape heterogeneity and farming practice influence biological control. Basic and Applied Ecology 2, 365–371.

Paredes, D., Cayuela, L., Campos, M., 2013b. Synergistic effects of ground cover and adjacent vegetation on natural enemies of olive insect pests. Agriculture Ecosystem and Environment 173, 72–80.

Paredes, D., Cayuela, L., Gurr, G.M., Campos, M., 2013a. Effect of non-crop vegetation types on conservation biological control of pests in olive groves. Peer J 1, e116. http://dx.doi.org/10.7717/peerj.116. PMID: 23904994.

Paredes, D., Cayuela, L., Gurr, G.M., Campos, M., 2015. Is ground cover vegetation an effective biological control enhancement strategy against olive pests? PLoS One 10 (2), e0117265. http://dx.doi.org/10.1371/journal.pone.0117265.

Parker, B.L., Skinner, M., Brownbridge, M., El-Bouhssini, M., 2000. Control of insect pests with entomopathogenic fungi. Arab Journal of Plant Protection 18, 133–138.

Parolin, P., Bresch, C., Desneux, N., Brun, R., Bout, A., Boll, R., Poncet, C., 2012. Secondary plants used in biological control: a review. International Journal of Pest Management 58 (2), 91–100.

Pathak, M.K., Fareed, M., Bihari, V., Mathur, N., Srivastava, A.K., Kuddus, M., Nair, K.C., 2011. Cholinesterase levels and morbidity in pesticide sprayers in North India. Occupational Medicine 61 (7), 512–514.

Pathak, M.K., Fareed, M., Srivastava, A.K., Pangtey, B.S., Bihari, V., Kuddus, M., Kesavachandran, C., 2013. Seasonal variations in cholinesterase activity, nerve conduction velocity and lung function among sprayers exposed to mixture of pesticides. Environmental Science and Pollution Research 20 (10), 7296–7300.

Pawar, A.D., 2004. Biological control of crop pests and weeds and integrated pest management. Plant Protection Buletin 56, 1–5.

Pearson, D.E., Callaway, R.M., 2005. Indirect nontarget effects of host-specific biological control agents: implications for biological control. Biological Control 35, 288–298.

Pearson, D.E., Callaway, R.M., 2003. Indirect effects of host-specific biological control agents. Trends in Ecology and Evolution 18, 456–461.

Pimentel, D., 2005. Environmental and economic costs of the application of pesticides primarily in the United States. Environment, Development and Sustainability 7 (2), 229–252.

Polanczyk, R.A., Pratissoli, D., Vianna, U.R., Oliveira, R.G., Dos, S., Andrade, G.S., 2006. Interaction between natural enemies: *Trichogramma* and *Bacillus thuringiensis* in pest control. Acta Scientiarum Agronomy 28, 233–239.

Polis, G.A., Myers, C.A., Holt, R.D., 1989. The ecology and evolution of intraguild predation: potential competitors that eat each other. Annual Review of Ecology and Systematics 20, 297–330.

Powell, W., Pell, J.K., 2007. Biological control. In: van Emden, H.F., Harrington, R. (Eds.), Aphids as Crop Pests. CABI, Wallingford, pp. 469–513.

Prasad, N.V.V.S.D., Mallikarjuna Rao, Hariprasad Rao, N., 2009. Performance of Bt. cotton and non Bt. cotton hybrids against pest complex under unprotected conditions. Journal of Biopesticides 2, 107–110.

Purwar, J.P., Sachan, G.C., 2006. Insect pest management through entomogenous fungi: a review. Journal of Applied Bioscience 32, 1–26.

Putman, W.L., 1955. Bionomics of *Stethorus punctillum* Weise in Ontario. The Canadian Entomologist 87, 527–579.

Rabindra, R.J., Ramanujam, B., 2007. Microbial control of sucking pests using entamopathogenic fungi. Journal of Biological Control 21, 21–28.

Rao, P.N., Tanweer, A., 2011. Concepts and components of integrated pest management. Pests and Pathogens: Management Strategies 543.

Raymond, B., Sayyed, A.H., Wright, D.J., 2006. The compatibility of a nuclear poly hedrosis virus control with resistance management for *Bacillus thuringiensis*: co-infection and cross-resistance studies with the diamondback moth, *Plutella xylostella*. Journal of Invertebrates Pathology 93, 114–120.

Remor, A.P., Totti, C.C., Moteira, D.A., Dutra, G.P., Heuser, V.D., Boeira, J.M., 2009. Occupational exposure of farm workers to pesticides: biochemical parameters and evaluation of genotoxicity. Environment International 35, 273–278.

Roldan-Tapia, L., Nieto-Escamez, F.A., Del Aguilla, E.M., Laynez, F., Parron, T., Sanchez-Santed, F., 2006. Neurophysiologial sequela from acute poisoning and long term exposure to carbamate and organophosphate pesticides. Neurotoxicology and Teratology 28, 694–703.

Room, P.M., Forno, I.W., Taylor, M.F.J., 1984. Establishment in Australia of two insects for biological control of the floating weed Salvinia molesta. Bulletin of Entomological Research 74, 505–516.

Rosen, D., DeBach, P., 1973. Systematics, morphology, and biological control. Entomophaga 18, 215–222.

Roy, H.E., Cottrell, T.E., 2008. Forgotten natural enemies: interactions between coccinellids and insect-parasitic fungi. European Journal of Entomology 105, 391–398.

Rusch, A., Bommarco, R., Jansson, M., Smith, H.G., Ekbom, B., 2013. Flow and stability of natural pest control services depend on complexity and crop rotation at the landscape scale. Journal of Applied Ecology 50, 345–354.

Saba, H., Vibhash, D., Manisha, M., Prashant, K.S., Farhan, H., Tauseef, A., 2012. Trichoderma-a promising plant growth stimulator and biocontrol agent. Mycosphere 3 (4), 524–531.

Safavi, M., 1968. Etude biologique et ecologique des hymenopteres parasites des oeufs des pinaises des cereals. Entomophaga 13, 381–495 (French with English summary).

Sandhu, S.S., Sharma, A.K., Beniwal, V., Goel, G., Batra, P., Kumar, A., Jaglan, S., Sharma, A.K., Malhotra, S., 2012. Myco-biocontrol of insect pests: factors involved, mechanism, and regulation. Journal of Pathogens. http://dx.doi.org/10.1155/2012/126819. Article ID 126819 (10 pages).

Sepulveda-Cano, P.A., Lopez-Nunez, J.C., Soto-Giraldo, A., 2008. Effect of two entomopathogenic nematodes on Cosmopolites sordidus (Coleoptera: Dryophthoridae). Revista Colombiana de Entomologia 34, 62–67.

Shah, F.A., Ansari, M.A., Watkins, J., Phelps, Z., Cross, J., Butt, T.M., 2009. Influence of commercial fungicides on the germination, growth and virulence of four species of entomopathogenic fungi. Biocontrol Science and Technology 19, 743–753.

Shanmugam, P.S., Balagurunathan, R., Sathiah, N., 2006. Bio-intensive integrated pest management for Bt. cotton. International Journal of Zoological Research 2, 116–122.

Sheppard, A.W., Hill, R., DeClerck-Floate, R.A., McClay, A., Olckers, T., Quimby Jr., P.C., Zimmermann, H.G., 2003. A global review of risk-benefit-cost analysis for the introduction of classical biological control against weeds: a crisis in the making? Biocontrol News and Information 24 (4), 91N–108N.

Shine, C., Williams, N., Gündling, L., 2000. A Guide to Designing Legal and Institutional Frameworks on Alien Invasive Species. IUCN, Gland, Switzerland Cambridge and Bonn, p. 138.

Siegel, J.P., 2001. The mammalian safety of Bacillus thuringiensis-based insecticides. Journal of Invertebrate Pathology 77 (1), 13–21.

Silva, E.B., Franco, J.C., Vasconcelos, T., Branco, M., 2010. Effect of ground cover vegetation on the abundance and diversity of beneficial arthropods in citrus orchards. Bulletin of Entomological Research 100, 489–499.

Silverio, F.O., de Alvarenga, E.S., Moreno, S.C., Picanco, M.C., 2009. Synthesis and insecticidal activity of new pyrethroids. Pest Management Science 65, 900–905.

Simberloff, D., 2012. Risks of biological control for conservation purposes. BioControl 57, 263–276.

Singh, S.P., 2004. Some Success Stories in Classical Biological Control of Agricultural Pests in India. Asia-Pacific Association of Agricultural Research Institutions, Bangkok, Thailand, p. 73.

Smith, H.S., 1919. On some phases of insect control by the biological method. Journal of Economic Entomology 12, 288–292.

Smith, R.F., Mittler, T.E., Smith, C.N., 1973. History of Entomology. Annual Reviews Inc., Palo Alto, California.

Song, B.Z., Wu, H.Y., Kong, Y., Zhang, J., Du, Y.L., Hu, J.H., Yao, Y.C., 2010. Effects of intercropping with aromatic plants on the diversity and structure of an arthropod community in a pear orchard. BioControl 55, 741–751.

Spadaro, D., Gullino, M.L., 2004. State of the art and future prospects of the biological control of postharvest diseases. International Journal of Food Microbiology 91, 185–194.

Sparks, T.C., 2013. Insecticide discovery: an evaluation and analysis. Pesticides, Biochemistry and Physiology 107, 8–17.

Srivastava, C.N., Maurya, P., Sharma, P., Mohan, L., 2009. A review on futuristic domain approach for efficient Bacillus thuringiensis (Bt.) applications. Journal of Entomological Research 33, 15–31.

Srivastava, K.P., 2004. A Textbook of Applied Entomology (Methods of Insect Pest Control), vol. I. Kalayani Publishers, New Delhi.

Straub, C.S., Finke, D.L., Snyder, W.E., 2008. Are the conservation of natural enemy biodiversity and biological control compatible goals? Biological Control 45 (2), 225–237.

Tisdell, C.A., 1990. Economic impact of biological control of weeds and insects. In: Mackauer, M., Ehler, L.E., Roland, J. (Eds.), Critical Issues in Biological Control. Intercept, Andover, pp. 301–316.

Tscharntke, T., Bommarco, R., Clough, Y., Crist, T.O., Kleijn, D., Rand, T.A., Tylianakis, J.M., Van Nouhuys, S., Vidal, S., 2007. Conservation biological control and enemy diversity on a landscape scale. Biological Control 43, 294–309.

Van Driesche, R., Hoddle, M., Center, T., 2008. Control of Pests and Weeds by Natural Enemies: An Introduction to Biological Control. Blackwell Publishing, Oxford, UK.

Van Driesche, R.G., Bellows, T.S., 1995. Biological Control. Chapman and Hall, New York, 539 pp.

van Lenteren, J.C., 1990. Implementation and commercialization of biological control in West Europe. In: Int. Symp. on Biological Control Implementation, McAllen, Texas, April 4–6, 1989, NAPPO Bulletin 6, pp. 50–70.

van Lenteren, J.C., 1992. Biological control in protected crops: where do we go? Pesticide Science 36, 321–327.

van Lenteren, J.C., 1997. From *Homo economicus* to *Homo ecologicus*: towards environmentally safe pest control. In: Rosen, D., Tel-Or, E., Hadar, Y., Chen, Y. (Eds.), Modern Agriculture and the Environment. Kluwer Acadamic Publishers, Dordrecht, pp. 17–31.

van Lenteren, J.C., 2000. A greenhouse without pesticides: fact of fantasy? Crop Protection 19, 375–384.

van Lenteren, J.C., 2003. Quality Control and Production of Biological Control Agents: Theory and Testing Procedures. CABI Publishing, Wallingford, UK.

van Lenteren, J.C., 2005. Early entomology and the discovery of insect parasitoids. Biological Control 32, 2–7.

van Lenteren, J.C., 2008. IOBC Internet Book of Biological Control, Version 5. www.IOBC-Global.org.

van Lenteren, J.C., Bale, J., Bigler, F., Hokkanen, H.M.T., Loomans, A.J.M., 2006. Assessing risks of releasing exotic biological control agents of arthropod pests. Annual Review of Entomology 51, 609–634.

van Lenteren, J.C., Bueno, V.H.P., 2003. Augmentative biological control of arthropods in Latin America. BioControl 48, 123–139.

van Lenteren, J.C., Loomans, A.J.M., 2006. Environmental risk assessment: methods for comprehensive evaluation and quick scan. In: Bigler, F., Babendreier, D., Kuhlmann, U. (Eds.), Environmental Impact of Invertebrates in Biological Control of Arthropods: Methods and Risk Assessment. CAB Int., Wallingford, UK.

Veres, A., Petit, S., Conord, C., Lavigne, C., 2013. Does landscape composition affect pest abundance and their control by natural enemies? A review. Agriculture, Ecosystems and Environment 166, 110–117.

Wajnberg, E., Scott, J.K., Quimby, P.C., 2001. Evaluating Indirect Ecological Effects of Biological Control. CAB Int., Wallingford, UK.

Wang, S., Michaud, J.P., Tan, X.L., Zhang, F., Guo, X.J., 2011. The aggregation behavior of *Harmonia axyridis* (Coleoptera: coccinellidae) in Northeast China. BioControl 56 (2), 193–206.

Wardlow, L.R., O'Neill, T.M., 1992. Management strategies for controlling pests and diseases in glasshouse crops. Pesticide Science 36 (4), 341–347.

Waterhouse, D.F., Sands, D.P.A., 2001. Classical Biological Control of Arthropods in Australia. Australian Centre for International Agricultural Research, Canberra (ACIAR monograph no. 77).

Waterhouse, D.F., Norris, K.R., 1987a. Biological Control Pacific Prospects Australian Centre for International Agricultural Research. Inkata Press, Melbourne, p. 453.

Waterhouse, D.F., Norris, K.R., 1987b. Know your crop. It's pest problem and control. Pesticides 7, 77–81.

Whipps, J.M., Sreenivasaprasad, S., Muthumeenakshi, S., Rogers, C.W., Challen, M.P., 2008. Use of *Coniothyrium minitans* as a biocontrol agent and some molecular aspects of sclerotial mycoparasitism. European Journal of Plant Pathology 121, 323–330.

Wilson, F., Huffaker, C.B., 1976. The philosophy, scope and importance of biological control. In: Huffaker, C.B., Messenger, P.S. (Eds.), Theory and Practice of Biological Control. Academic Press, p. 788.

Zhang, Y., Qu, L.J., Wang, Y.Z., 2007. Using virus to restore and construct table forest ecosystem for pest insects control. Chinese Forest Science Technology 6, 53–61.

Aphids and Their Biocontrol

Rajendra Singh[1], Garima Singh[2]

[1]Department of Zoology, Deendayal Upadhyay Gorakhpur University, Gorakhpur, India;
[2]Department of Zoology, Rajasthan University, Jaipur, India

1. INTRODUCTION

The suppression of pest populations in agricultural crops has always been a complex eco-logical, and bioecological problem with the arsenal of chemical control agents reinforced by synthetic compounds with a broad spectrum of effects. In recent years, the recommenda-tion of conventional pesticides has decreased due to concerns about their residues in food and water and the adverse effects on nontarget organisms. In managed natural ecosystems, plants, herbivore pests, and natural enemies of pests have evolved a set of interactions influ-encing the population levels of one or all of them. However, in the agroecosystem, these naturally-evolved mechanisms are ignored while devising plant protection strategies. This led to an exclusively pesticide-based pest management strategy in the previous millennium which has thrown up problems. It is therefore necessary, in the current millennium, that pest management strategies must be biointensive. This has generated a need for awareness of several naturally-occurring phenomena. An insect population attains pest status by invasion and ecological and socioeconomic changes.

Advances in transportation technology over the past century have increased global inter-action among the nations and encouraged trade among them. However, increased trade resulted in an increased risk of new pest invasion (Sailer, 1983; Meyerdirk, 1992). Ecologi-cal changes are another major cause of pest outbreaks. The history of agriculture has been the history of constant ecological changes. Various agrotechnical practices, e.g., monoculture, selection of high-yielding plant cultivars, elimination of the competitors or natural enemies, etc., created conditions that favor certain insect species and has thus induced a manyfold increase in their population. Actually, these factors disrupt the interactions between herbivo-rous insects and their natural enemies (entomophages) that contribute to the regulation of insect population (DeBach, 1964, 1974; Huffaker, 1971; Huffaker and Messenger, 1976; Clausen, 1978; Rosen, 1995; van Driesche and Bellows, 1996; Dent, 2000). Whenever this interaction is disrupted, the population of the herbivorous insects increases tremendously, attaining pest status because they become free from the constraints imposed by the entomophages. The use of natural enemies (parasitoids, predators, and pathogens) in pest management is mainly concerned with redressing the imbalance that has occurred through this dissociation, either by their reintroduction into the system or by trying to recreate conditions where the above association can occur. Thus, the problem of insect control is ecological and not chemical, as they have been in use since 1940s (Dent, 2000).

The opportunities and needs for effective biocontrol in this century are greater than ever. At present, pest control approaches are aimed to maximize productivity and emphasize efficiency and the long-term sustainability of agroecosystems (Williams and Leppla, 1992). Because of the uncertainty over the durability and public acceptance of genetically modified (GM) pest-resistant crops, and the possible effects of GM crops on natural enemies of pests (Singh, 2006), biocontrol is considered a cornerstone of many integrated pest management (IPM) programs (Kogan, 1998).

Biocontrol programs are generally less pursued in developing countries, where there exists a need to thoroughly explore and evaluate their native natural enemies as promising biocontrol agents. In India, native natural enemies have little prospect of being exterminated by synthetic poisons due to lack of pesticides, unavailable pesticides, or poor quality pesticides commonly reported by the media. It is, therefore, obvious that such prevailing situations in India, like other third world countries, will lead to a more conducive environment for the implementation of biological control programs designed for native pests, or even for introduction programs involving the utilization of exotic natural enemies (Napompeth, 1987). The current revival of interest in biocontrol is also driven by a change from pest control approaches that aim to maximize productivity to approaches that emphasize efficiency and the long-term sustainability of agroecosystems. The philosophy of IPM is based on the management of entire pest populations, not just localized populations, where a single control technique is employed (Metcalf and Luckmann, 1994). In IPM, emphasis is placed on the use of combinations of the methods aimed at providing cheap but long-term reliability with minimal harmful side effects (Kogan, 1986). The philosophy and methodology of modern IPM programs is thus compatible with the philosophy and methodology of biocontrol. Indeed, biocontrol has been a central core around which IPM has been commonly developed. The reason for this is that natural enemies constitute major natural control factors that can be manipulated.

Parasitoids, predators, and pathogens are components of biocontrol of insect pests (Memmott et al., 2000). In biocontrol, parasitoids are favored over predators because they are host-specific, better-adapted and synchronized in interrelationships, and have a lower food requirement per individual thereby maintaining a balance with their host species at lower host densities, and their larvae do not need to search for food (van Lenteren, 1986; Sigsgaard and Hansen, 2000). Parasitoids are used more frequently than predators in biocontrol programs and comprise about 80% of all biocontrol (Hokkanen, 1985). More than 300 agricultural and urban insect pests in more than 100 countries are now being managed by biocontrol agents. Although the above figure of success is very low (about 10,000 insect species are recognized as pests), it does not underestimate the efficacy of biocontrol. The research and development efforts on biocontrol are very meager compared with synthetic pesticides. Sailer (1976) estimated the total cost of research into biocontrol in the USA during 1928–1972 (55 years) was less than US$20 million, compared with US$110 million on chemical pesticides in 1973 alone. The basic reason for this relative neglect was that biocontrol was widely perceived as

TABLE 1 Comparison of Aspects Related to the Development and Application of Chemical and Biological Control

Particulars	Chemical control	Biological control
Number of "ingredients" tested	>3.5 million	3000
Success ratio	1:200,000	1:20
Developmental costs	US$180 million	US$2 million
Developmental time	10 years	10 years
Benefit per unit of money invested	2.5–5	30
Risk of resistance	Large	Nil/small
Specificity	Low	High
Harmful side effects	Many	Nil/few

Updated to 2004. Adapted from van Lenteren (1997a,b).

unreliable compared with chemical control but the fact that most pests developed resistance against pesticides, causing failure of this control measure was not thought out.

Cost-to-benefit analyses suggest that research on biocontrol is more cost-effective than chemical control (30:1 and 5:1, respectively; Tisdell, 1990; van Driesche and Bellows, 1995). Despite this advantage, the main reason that biocontrol is not used on a larger scale is due to problems associated with the production and distribution of natural enemies; particularly, the limited shelf-life (days, or at most weeks) of most natural enemy species and the impossibility of patenting a naturally-occurring, unmodified species (Bale et al., 2008).

Table 1 clearly demonstrates that more than 1150-fold more chemical compounds have been tested against insect pests than natural enemies. Though there is no scarcity of such species in nature, they need to be explored for biocontrol. Even the success rate of finding a suitable natural enemy is much higher (20,000 times) than for a synthetic pesticide. The development costs for chemical pesticides excluding the future ecological cost are much higher (90 times) than biocontrol. In addition, chemical control has inherent disadvantages as pests develop resistance against them. Most synthetic pesticides kill other nontarget animals including their natural enemies, whereas natural enemies used in biocontrol are usually highly host-specific, killing only one or a few related species of prey/host. Biocontrol programs have no side effects, unlike chemicals (van Lenteren et al., 2006). No doubt, biocontrol has an enormous undeveloped potential that needs to be exploited through improved procedures.

2. APHIDS AS INSECT PESTS

Aphids (Homoptera: Aphididae) are small sap-sucking insects infesting both aerial and subaerial parts of a variety of plant species. They are cosmopolitan but are most abundant in temperate climates. They are unique on account of their peculiar mode of reproduction, development, and polymorphism. They may reproduce either by parthenogenesis, zygogenesis, or pedogenesis. They may either be oviparous or viviparous. The sexes may be unequally

FIGURE 1　A parthenogenetic aphid colony consisting of individuals of different morphs (alate and apterous) and age/structure along with predatory larva.

represented (males are frequently rare) in certain generations. Parthenogenetic reproduction allows rapid increase in numbers and results in populations consisting of clones. Some species reproduce both parthenogenetically and sexually (holocyclic species), whereas only a few reproduce solely through parthenogenesis (anholocyclic species) (Dixon, 1985; Singh and Ghosh, 2002) (Figure 1).

2.1 Economic Importance of Aphids

In suitable conditions, the number of aphids rapidly rises above economic threshold levels. All parts of the plants, including the roots, are attacked. Some of them directly damage the plants by sucking their nutrients, causing curling and twisting of tender shoots and general devitalization of plants, especially those of agricultural as well as horticultural importance. Very young seedlings may die after being attacked by them. The inflorescence may fail to open fully when the part of the plant is heavily infested. Sometimes fruits fail to develop normally and may also show various malformations, like twisting of pods, impaired developments of seeds, etc. Subaerial infestations by aphids also cause yellowing of foliage and stunted growth. In gall-making aphids, direct injury is caused by making different types of leaf and stem galls and these galls subsequently serve as temporary abodes for those aphids. These symptoms are observed on perennial forest trees. As well as these direct effects, aphids have also some indirect effects. The physical presence of a large number of aphids can be a cause for concern, and the honeydew they excrete promotes the growth of black sooty molds, which in turn reduce the crop's photosynthesis as well as its aesthetic value. Dust, dirt, and skins shed in molting adhere to the viscous substance, making plants unsightly.

Out of an estimated world fauna of over 4700 species of aphids, about 787 species belonging to 211 genera are known to be from India. Over 1200 species of plants belonging to nearly 700 genera and 175 families are infested by these aphids in India (Ghosh, 2000). In India, fewer than 100 species are pests of crops of economic importance. A list of major aphid pests in India and abroad is given in Table 2. Blackman and Eastop (2006) described the taxonomy, biology, food plant relations, and economic importance of few aphid pests. Polymorphism, cyclic

TABLE 2 List of Aphid Pests of Agricultural Importance in India and Abroad

Aphid species	Plants of economic importance
Acyrthosiphon kondoi	Alfalfa
Acyrthosiphon pisum	Peas, alfalfa, lucern, gram, clover
Amphorophora rubi	Raspberry
Aphis craccivora	Bean, groundnut, pigeon pea, citrus, gram
Aphis fabae	Citrus, tobacco, rose, potato, bamboo
Aphis gossypii	Cotton, groundnut, pigeon pea, brinjal, cucurbits, some brassicas
Aphis helianthi	Sunflower
Aphis nasturtii	Groundnut, sweet potato, some cucurbits
Aphis pomi	Apple, pear, many other fruits
Aphis spiraecola	Citrus, spiraea
Brachycaudus helichrysi	Pigeon pea, potato, tomato, tobacco, some brassicas
Brevicoryne brassicae	Cabbage, cauliflower, rapeseed mustard, radish
Ceratovacuna lanigera	Sugarcane
Chromaphis juglandicola	Walnut
Diuraphis noxia	Wheat, barley, other cereal crops
Eriosoma lanigerum	Apple, pear
Hyalopterus amygdali	Peaches
Hysteroneura setariae	Paddy, sugarcane, wheat, maize
Lipaphis pseudobrassicae	Mustard, rapeseed, cabbage, radish, cauliflower
Macrosiphum euphorbiae	Potato, tomato, tobacco
Melanaphis sacchari	Millets, sugarcane, maize, sorghum
Metopolophium dirhodum	Cereal crops
Myzus nicotianae	Tobacco
Myzus persicae	Peas, beans, pears, apricot, peaches, potato, tobacco, brassicas
Pemphigus bursarius	Lettuce
Pomaphis mali	Apple, pear
Rhopalosiphum maidis	Maize, millets, sugarcane, wheat, barley
Rhopalosiphum padi	Wheat, maize
Schizaphis graminum	Wheat, millets, maize, barley
Sipha flava	Sugarcane, wheat, sorghum
Sitobion avenae	Millets, sugarcane, barley, wheat

(Continued)

TABLE 2   List of Aphid Pests of Agricultural Importance in India and Abroad—cont'd

Aphid species	Plants of economic importance
Therioaphis trifolii	Alfalfa, yellow clover
Tinocallis plantani	Elm
Toxoptera aurantii	Tea, citrus, and tomato
Uroleucon compositae	Safflower
Viteus vitifoliae	Grape vines

parthenogenesis, and heteroecy have enabled aphids to exploit the host plant to a greater extent than any other insect group. Polyphagia may reach as high as that of *Aphis gossypii*, which can develop on 569 plant species that belong to 103 families in India (Singh et al., 2014a). Most of them are serious pests on vegetables, pulses, cereals, and oil crops. A by-product of their probing and feeding behavior is also the transmission of hundreds of plant viruses. Nearly 200 species of aphids are vectors for over 200 plant viruses (Hogenhout et al., 2008). Most of the nonpersistent viruses (e.g., beet yellow stunt virus) and persistent viruses (e.g., potato leaf roll virus) are also transmitted by them (Ghosh, 1980). Aphid-transmitted viruses belong to 19 of the 70 recognized virus genera and comprise approximately 275 virus species (Nault, 1997). A list of virus pathogens transmitted by aphid species and regarded as vectors of plant viruses in India has been provided by Ghosh (1974). The green peach aphid, *Myzus persicae*, alone transmits more than 100 plant viruses (Eastop, 1958).

2.2 Life History of Aphids

About 85% of the species described from India are parthenogenetic virginoparous for most of the year but are capable of sexual reproduction with production of eggs. They develop in parthenogenetic females without fertilization. Even embryos inside parthenogenetic females may contain embryos, i.e., a mother can have developing embryos in her ovarioles which in turn also contain embryos, the future granddaughters. Thus, there is a telescopic generation due to parthenogenesis and viviparity in aphids (Minks and Harrewijn, 1987). This results in reduced postnatal development periods and generation time. There is more or less regular cyclic or anholocyclic alternation of parthenogenetic, oviparous, and viviparous generations associated with polymorphism, changes of food plants, and mode of life. Several generations often succeed each other, in which the males are extremely rare or are totally absent. Individuals of the same generation often differ considerably from one another. Some have fully-developed wings others have atrophied wings or are apterous.

The life history of aphids is highly fascinating though very complicated. The aphids have several biological peculiarities such as prolific breeding, polyphagy, advanced degree of polymorphism, anholocyclic and/or holocyclic reproduction, host alternation, and high potential for rapid evolutionary changes because of parthenogenesis and polyvoltinism (Behura, 1994). Some aphids are anholocyclic (continuously parthenogenetic), while others living in temperate climates are holocyclic (sexual generation alternates with parthenogenetic reproduction).

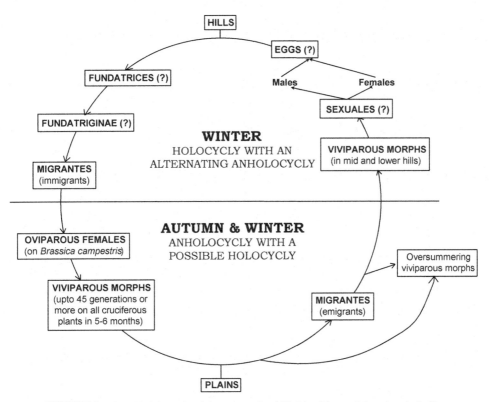

FIGURE 2 Possible life cycle of the mustard aphid, *Lipaphis pseudobrassicae*, in India.

In a year's time, numerous generations may succeed one another, for even at moderate mean temperatures nymphs, that molt four times at most, complete their development in little more than 10 days and some of the aphid species have more than 10 generations in one year (Iversen and Harding, 2007).

A generalized life cycle pattern is illustrated in Figure 2. Sexual females, like asexual females, have two sex chromosomes, i.e., XX. Males have only one sex chromosome, i.e., OX. In theory this means males could produce sperm with either no sex chromosomes, i.e., O, or one sex chromosome, i.e., X. However, in reality sperm with an O sex chromosome degenerates very rapidly and never contributes to an embryo. This means that all offspring of a sexual mating must have XX as their sex chromosomes, because females always contribute an X chromosome, and therefore, all aphids resulting from sexual matings are female. Eggs are laid during the Autumn as the overwintering stage in many temperate forms, and as explained above, gives rise to females whether they are the result of sexual mating or not. In other species a special overwintering form develops in the Autumn called a "hiemalis", while in some species the adults are the overwintering stage. Ova within a viviparously reproducing female start to develop immediately after ovulation—this occurs long before birth. This means that an embryo can exist inside another larger and more mature embryo. In fact, a newly-born summer aphid can contain within herself not only the developing embryos of her daughters but also those of her granddaughters, which are developing within her daughters. Parthenogenesis, combined with this "telescoping of generations", gives aphids an exceedingly rapid turnover of generations meaning that they can build up immense populations very quickly.

Although low temperature, short day length, and physical condition of host plants are regarded as important factors governing the production of sexuales, the discovery of apterous oviparous females of the mustard aphid, *Lipaphis pseudobrassicae*, on mustard in the arid and semiarid region of Jaipur (Rajasthan, India; Ghosh and Rajendran, 1988) suggests that some other factors are also operative in the phenomenon.

Aphids are remarkable on account of their peculiar mode of development and the polymorphism exhibited in different generations of the same species. Females may exhibit up to eight discrete phenotypes which are genetically identical individuals. They differ in morphology, physiology, numbers, timing of production, progeny sizes, developmental periods, longevity, host preferences, and the ability to locate and utilize alternative host plants. The associated phenomena concerning reproduction are: (1) parthenogenesis; (2) oviparity and viviparity; and (3) the occurrence of generations in which the sexes are very unequally represented—males are frequently rare. With regard to structure, the phenomena are: (1) the production of totally different types of individual of the same sex either in the same or different generations; (2) the production of individuals with perfect and also atrophied mouthparts; and (3) the production of individuals of the same sex but differing as to the gonads. Associated with habits are: (1) host alternation, involving migration to totally different plant hosts; (2) different modes of life of the same species on the same host; and (3) different habits of individuals of the same generation. In extreme cases almost all the above phenomena may occur associated with the annual cycle of an individual species (Singh and Ghosh, 2002).

3. BIOINTENSIVE MANAGEMENT

The development of resistant varieties of crops is the most practicable approach to tackle aphid problems. However, development of biotypes is generally a major factor limiting the use of resistant varieties. Biotypes can be defined as "populations within an arthropod species that differ in their ability to utilize a particular trait in a particular plant genotype" (Smith, 2005). Almost 50% of the known 120 biotypes registered in 36 arthropod species, belonging to 17 families and 6 orders, belong to the aphids. Hence, biotype management becomes a significant area of research. A biparental, sexually-reproducing, monocyclic insect is less likely to develop insect biotypes because offspring will be genetically heterogeneous compared with uniparental, parthenogenetically reproducing, polycyclic insects such as aphids (Dent, 2000). Because of this, almost 50% of the known insect biotypes registered are aphids (Smith, 2005). Hence, biotype management becomes a significant area of research (Michel et al., 2011). Bakhetia and Chander (1997) reviewed the management strategies of aphids in relation to host plant resistance. Conservation of natural resources (air, soil, and water) is essential and hence IPM strategies should aim at developing the approaches which do not lead to degradation and/or depletion of these gifts from nature (Fiedler et al., 2008). Thus, it is high time that we laid more emphasis on evolving ecofriendly nonchemical approaches of pest management. Though use of synthetic aphidicides has provided effective control of almost all aphid pests, their undesirable side effects (pest outbreaks, resurgence, resistance development, harmful residues, etc.) limit their continued use. The use of cultural control Sachan (1997), biocontrol agents (parasitoids and predators), and host plant resistance have many advantages (van Emden, 2007).

In the current and worldwide situations pertaining to crop protection, avocations are made and emphasis is given to alternative strategies as opposed to past traditional systems dominated by the use of pesticides. The word alternative seems to be misnomer for biocontrol of insect pests as this has long been in existence but largely overlooked and neglected. It is more appropriate to refer to biocontrol strategies as being underutilized rather than the alternative, judging from the fact that many successful cases of pest control achieved through biocontrol predated the era of agrochemicals. Having been long forgotten and dominated by the use of agrochemicals, biocontrol is now identified and considered as an alternative method of insect control, together with other nonpesticidal or biologically-based control measures (Bale et al., 2008).

4. BIOCONTROL AGENTS OF APHIDS

Predators and parasitoids are the most common bioagents used to control other insects. Predators such as ladybird beetles or dragonflies devour prey insects and usually consume many prey in their lifetimes. Parasitoids are a type of parasite that often do not prey on other insects as adults, but instead deposit their eggs inside (internal parasitoid) or on (ectoparasitoid) a host insect. After hatching, the immature form, i.e., the larvae, consumes the living host from within. Certain parasitic wasps and flies are good examples of parasitoids. Often, there is a difference between the two: predators are mostly generalist in their feeding habits, i.e., they attack a wide range of insect species, while parasitoids are specialist and are host-specific, as they attack a narrow range of insect hosts. They may be monophagous (parasitizing one host species) or oligophagous (parasitizing a few closely-related species). There are few parasitoids that are polyphagous parasitizing distantly-related insect hosts (Singh and Singh, 2015).

4.1 Aphid Predators

Aphid predators belong to four orders of insects: Coleoptera (families Coccinellidae and Carabidae), Diptera (families Chamaemyiidae, Syrphidae, and Cecidomyiidae), Hemiptera (families Anthocoridae and Geocoridae), and Neuroptera (family Chrysopidae) (Völkl et al., 2007).

Ladybird beetles are most common aphid predators encountered throughout the world. The common genera predaceous on aphids are: *Adalia, Adonia, Brumoides, Coccinella, Cheilomenes, Exochomus, Hippodamia, Oenopia, Micraspis, Scymnus,* etc. Aphidophagous coccinellids have a long history of importation in classical biocontrol with only few recognized successes (Hagen and van den Bosch, 1968; Hodek, 1973; Hagen, 1974; Majerus, 1994; Hodek and Honek, 1996). In 1874, *Coccinella undecimpunctata,* was imported to New Zealand where it became established as an important predator of aphids and mealybugs in various fruit and forage crops (Dumbleton, 1936). It was also established in various regions of North America

(Wheeler and Hoebeke, 1981) although its introduction was thought to be accidental. Dixon (2000) judged only one to be substantially successful after 155 tallied intentional introductions of coccinellid species worldwide that specifically targeted aphids. The most common species used in the biocontrol of aphids are *Coccinella septempunctata, C. undecimpunctata, Menochilus sexmaculatus, Harmonia axyridis, Adalia bipunctata, Brumoides suturalis, Synonycha grandis, Coelophora biplagiata, Propylea japonica, Hippodamia convergens*, etc. Because most ladybirds are polyphagous, they usually disrupt existing biocontrol by introduced ladybirds as well as reduce the potential of indigenous ladybirds. They often feed on parasitized aphids (mummies) (Snyder and Ives, 2001; Ferguson and Stiling, 1996; Kaneko, 2004) and may affect efficacy of existing aphid parasitoids. Consequently, their role in suppressing a particular pest must be considered within the larger context of the entire guild of natural enemies that contribute to pest mortality. Obrycki and Kring (1998) and Michaud (2012) reviewed the status of predatory ladybirds in biocontrol.

The carabid beetles are also important aphid predators mostly belonging to the genera *Agonum, Notiobia, Crossonychus, Feroniomorpha*, and *Metius* (Zaviezo et al., 2004).

Among aphidophagous Diptera, members of Chamaemyiidae such as *Leucopis glyphinivora* and *Neoleucopis obscura* are potential biocontrol agents against aphidine and adelgidine aphids, respectively (Ghadiri et al., 2003). The aphid midge, *Aphidoletes aphidimyza*, is a cecidomyiid fly whose larvae are effective predators of aphids, an important component of biocontrol for greenhouse crops, and is commercially available (e.g., APHIDEND®, Koppert B.V., The Netherlands). The larvae of syrphids are important aphidophagous insects (Agarwala et al., 1984). Several species of syrphids have been evaluated as biocontrol agents against aphids and few species are commercially available, e.g., *Episyrphus balteatus* (SYRPHIDEND®, Koppert B.V., The Netherlands). Joshi and Ballal (2013) reviewed the status of syrphid flies in biocontrol of aphids.

Among Hemiptera, members of the families Anthocoridae, Nabidae, Miridae, and Geochoridae predominate among other aphidophagous bugs (Lucas, 2005; Alhmedi et al., 2007).

Green lacewings, particularly members of the genera *Chrysopa, Chrysoperla*, and *Mallada* (Chrysopidae), and brown lacewing (Hemerobiidae) are major biocontrol agents of aphids among Neuroptera and have been used against aphids in several parts of the world (Abd-Rabou, 2008; Hayashi and Nomura, 2011; Pappas et al., 2011).

4.2 Aphid Parasitoids

Most aphid parasitoids belong to Hymenoptera (Braconidae, Figure 3(A), and Aphelinidae, Figure 3(B)), and a few to Diptera (Cecidomyiidae). Aphid parasitoids are themselves parasitized by other insects (hyperparasitoids, Figure 3(C)) and preyed by predators belonging to different taxa of insects. The parasitized aphids are also infected by several species of fungi. All these components (hyperparasitoids, predators, fungi) have detrimental effects on the biocontrol of released or naturally existing parasitoids. Boivin et al. (2012) reviewed the status of aphid parasitoids in biocontrol.

All the hymenopteran parasitoids belong to the families Braconidae (subfamily Aphidiinae) and Aphelinidae. Aphidiinae is a monophyletic group that parasitizes only aphids (Kambhampati et al., 2000). They are solitary endoparasitoids, i.e., only one individual completes its development inside the host in spite of superparasitism (a female parasitoid lays

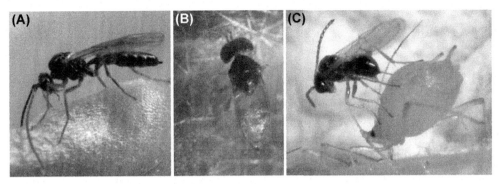

FIGURE 3   (A) Braconid parasitoid, (B) aphelinid parasitoid, and (C) hyperparasitoid of aphids.

more than one egg in hosts that support the development of only one egg) and multiparasitism (two or more female parasitoids lay eggs in hosts that support the development of only one egg). Aphid parasitoids are also koinobiont, as the parasitoid larva develops in a living host that continues to feed and grow.

The world fauna of Aphidiinae comprises over 400 species and 50 genera (Mackauer and Starý, 1967; Starý, 1987) and are kept under four tribes: Aclitini, Aphidiini, Ephedrini, and Praini (Smith and Kambhampati, 2000). They are cosmopolitan in distribution so that one species, *Diaeretiella rapae*, parasitizes about 98 species of the aphids infesting more than 180 plant species belonging to 43 plant families distributed in 87 countries throughout the world (Singh and Singh, 2015). Most of the aphidiine parasitoids used in biocontrol belong to the genera: *Aphidius*, *Binodoxys*, *Diaeretiella*, *Ephedrus*, *Praon*, and *Trioxys* (Wei et al., 2005; Vollhardt et al., 2008; Boivin et al., 2012).

The family Aphelinidae contains over 1000 species in 50 genera (Viggiani, 1984) and parasitize not only aphids but also other members of Homoptera (whiteflies, scale insects, etc.) (Starý, 1987). They are also cosmopolitan in distribution but most of them are distributed in Palearctic and Australian regions (Viggiani, 1984). The genera that parasitize aphids are *Aphelinus*, *Marietta*, *Protaphelinus*, and *Mesidiopsis* (Viggiani, 1984; Wei et al., 2005). All the species of *Aphelinus* are solitary koinobiont endoparasitoids of aphids (van Lenteren et al., 1997b).

The family Cecidomyiidae (Diptera) has six species of the genus *Endaphis* and is known to parasitize aphids (Muratori et al., 2009). These flies lay eggs on the leaves near an aphid colony. After hatching, the larvae search for hosts while crawling on the leaf, and then penetrate the aphid between the legs and thorax and develop as koinobiont endoparasitoids. They emerge from the aphid anus and fall to the ground to pupate in the soil (Muratori et al., 2009). Unlike aphidiine and aphelinid parasitoids, aphids parasitized by these flies do not form mummies.

4.3 Biology of Aphidiine Parasitoids

The biology of aphid parasitoids has been investigated for several aphidiine species (Starý, 1987). In order to plan controlled release experiments in greenhouses or in the fields, reliable ecological data are necessary (Singh and Agarwala, 1992; Singh, 2001). To understanding the

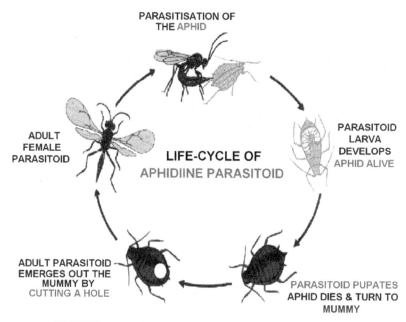

PARASITISATION OF
THE APHID

ADULT
FEMALE
PARASITOID

LIFE-CYCLE OF
APHIDIINE PARASITOID

PARASITOID
LARVA
DEVELOPS
APHID ALIVE

ADULT PARASITOID
EMERGES OUT THE
MUMMY BY
CUTTING A HOLE

PARASITOID PUPATES
APHID DIES & TURN TO
MUMMY

FIGURE 4   The life cycle of a generalized aphidiine parasitoid.

ecology of an organism, it is important to know many aspects of its developmental biology, such as the length of its life stages, species response to temperature, its intrinsic pattern of survivorship and fecundity, density responses, factors affecting progeny sex ratio, host alternation, etc. Constructing a life table is the most common method of measuring these parameters. Laboratory age-specific life tables follow the fate of a cohort of individuals from birth to death and are used to measure intrinsic attributes of the species, such as oviposition pattern, fecundity, longevity, and rate of natural increase under nonlimiting conditions (Tripathi and Singh, 1990; Mishra and Singh, 1990).

The life cycle of a generalized aphidiine parasitoid is illustrated in Figure 4. Aphidiine parasitoids are solitary endoparasitoids of aphids. Soon after emergence, the female is usually able to oviposit after insemination. Virgin females respond to courtship by a conspecific male through signaling receptivity and adopting the copulation posture (Mishra and Singh, 1993a). Usually, females are monandrous but males are always polygynous and mate with as many as 22 females (Tripathi and Singh, 1990; Mishra and Singh, 1993a; Pandey et al., 1996). After copulation, the females search for suitable hosts for oviposition. Oviposition behavior typically involves antennal tapping of the aphid host, then the female stands on erect legs, bending her abdomen forwards beneath the thorax and between her legs (Figure 5(A)). The duration of oviposition ranges from about 1 s to about 1 min (Singh and Sinha, 1982a).

During oviposition, the females insert their ovipositor into the host aphids. An egg is not necessarily laid during each insertion. Females have the ability to distinguish between parasitized and healthy aphids. Superparasitism and multiple parasitism are common in the

FIGURE 5   A female parasitoid parasitizing an aphid (A), parasitized aphids as "mummies" (B), isolated mummies with emergence hole (C), adult parasitoids after making a hole emerging out the mummy (D).

laboratory as well as in the field under certain circumstances (Singh and Sinha, 1982b,c, 1983; Mishra and Singh, 1993b). Oviposition behavior is quite consistent throughout the group, although there are many minor deviations from the general procedure (Singh and Agarwala, 1992; Rehman and Powell, 2010).

Aphidiine parasitoids have three to four larval instars (Singh and Agarwala, 1992). Before completing development, the last instar larva spins a cocoon inside or under the empty aphid exoskeleton. The cuticle of the host aphid hardens, stretches, and dries before the parasitoid larva turns into a pupa (Quicke, 1997). At this stage, the aphid exoskeleton becomes a "mummy" (Figure 5(B)). The prepupal, pupal, and adult stages develop within the mummy. The adults emerge through a circular hole (Figure 5(C)). The newly-emerged adults (Figure 5(D)) need a short time to mature. Males often emerge slightly earlier than females and mating occurs soon after emergence which lasts for several seconds.

Host searching behavior involves several steps and their sequential operation progressively decreases the searching space and increases the chances of finding suitable hosts for oviposition (Singh and Sinha, 1982a). Finding a host starts with the selection of a suitable habitat, with the food plants of the host aphids playing an important role because the parasitoids seem to be attracted to them (Du et al., 1996; Royer and Boivin, 1999; Rehman and Powell, 2010). The attractiveness of the host aphids to the parasitoids is apparently due to the perception of their kairomones, which seem to be present in the honeydew (Bouchard and Cloutier, 1984). Host and parasitoid population densities are also important in conditioning the searching activity of the parasitoid. Contact kairomones present in the host cuticle and secretion of siphunculi are involved in host recognition (Powell et al., 1998). Visual cues also play a role in host recognition, color being an important short-range cue (Singh and Sinha, 1982a; Powell et al., 1998; Rehman and Powell, 2010). If the aphid colony is present nearby, the females start parasitizing aphids on the same patch or the same host plant (Weisser and Völkl, 1997), or she may disperse in search of a new aphid colony. In most species, the females search for hosts within a few meters (Weisser and Völkl, 1997; Langhof et al., 2005). However, long-distance dispersal occurs, especially when hosts are rare, and some species may disperse over 100 km in a year (Cameron et al., 1981).

Aphidiinae are typically arrhenotokous, with mated females producing fertilized (diploid) eggs that develop into females and virgin females producing unfertilized (haploid) eggs that develop into males. The developmental rate is influenced by temperature but usually takes 2 weeks. Adults are active on warm sunny days, especially in the late morning and afternoon, showing a positive phototactic response. Longevity is variable; minimum survival occurs without water and food. The sex ratio in the field typically favors females but is variable under different environmental factors (Singh et al., 2014b).

Host stage preference may vary among species within the Aphidiinae. They show clear preferences not only toward specific aphid species, but within a species they prefer some instars over others (Singh and Sinha, 1982a; Singh and Agarwala, 1992; Barrette et al., 2009; Rehman and Powell, 2010). This preference is based on the rate of resource acquisition by the female parasitoid and as a result, not all aphid instars are exploited equally within a patch (Barrette et al., 2010). Patch residence time and the level of exploitation of an aphid colony is influenced by the size and quality of the colony (van Steenis et al., 1996; Pierre et al., 2003), the distance between colonies (Tentelier et al., 2006), and the presence of competitors (Singh and Agarwala, 1992; Le Lann et al., 2011).

The reproductive capacity of the aphid parasitoids is also variable; from less than 50 to more than 850 eggs/embryo/mummies/offspring per female depending upon food plants, host density, temperature etc. (Table 3). The adult parasitoids survive for 1–4 weeks depending upon nutrition and temperature.

The adult parasitoids disperse by flight or by walking, or immatures can disperse within their live aphid hosts or within mummies. Differences in short-distance dispersal can be found between parasitized live aphids and healthy aphids (Starý, 1987). Prior to mummification, parasitized aphids typically leave their colony and move to microhabitats that are microclimatically favorable for the parasitoids. Long-distance dispersal of adult parasitoids and parasitized alate aphids occurs by flight (Starý, 1987).

Some parasitoids are restricted to a single host species (monophagy), two or more species of the same aphid genus, two or more genera of the same subfamily, two or more genera of two or more subfamilies of the same family, or to two or more aphid families (polyphagy) (Moreno-Mari et al., 1999; Stilmant et al., 2008; Singh and Singh, 2015). The choice of the best aphid parasitoid to use in a biocontrol program should rely on information about the level of specialization of the parasitoid.

Although parasitism ultimately results in mummification and death, parasitized aphids can develop for some time. If the first or second instars of aphids are parasitized, they do not develop to adults, while aphids parasitized in third instar reach the adult stage; producing few or no offspring prior to mummification. Aphids parasitized in fourth instar

TABLE 3 The Mean Fecundity, Female Longevity (Days), and Intrinsic Rate of Natural Increase (r_m) of Some Species of Aphid Parasitoids (Braconidae: Aphidiidae).

Species (Parasitoid/host)	Temp. °C	Fecundity	Longevity	r_m	References
Aphidius colemani / Aphis gossypii	20	48.9b	15.4	0.37	van Steenis and El-Khawass (1955)
	25	50.2b	14.4	0.48	van Steenis and El-Khawass (1955)
	30	49.5b	14.2	0.47	van Steenis and El-Khawass (1955)
Aphidius gifuensis/Myzus persicae	20	532.0	17.6	–	Fukui and Takada (1988)
Aphidius matricariae/D. noxia	20	66.0c	23.8	–	Reed et al. (1992)
A. matricariae/M. persicae	21	313	7.0	0.16	Giri et al. (1982)
	25	–	–		Shijko (1989)
	20	308a	13.7	–	Shalaby and Rabasse (1979)
Aphidius nigripes / Macrosiphum euphorbiae	20	374b	20.5	–	Cloutier et al. (1981)
Aphidius rhopalosiphi / S. avenae	18	212c	13.1	–	Shirota et al. (1983)
Aphidius rosae/Macrosiphum sp.	–	180–200	–	–	Sorokina (1970)
Aphidius smithi/Acyrthosiphon pisum	20.5	870.0	8.0	0.36	Mackauer (1983)
	20.5	547.0a	11.3	0.40	Mackauer (1983)
	20.5	258.0a	12.3	0.34	Mackauer and Kambhampati (1988)
Aphidius sonchi/H. lactucae	22	236.0	4.9	0.34	Liu (1985)
Aphidius urticae/H. humilis	–	257.0	18–20	–	Dransfield (1979)
Aphidius uzbekistanicus/M. dirhodum	–	478.0	–	–	Dransfield (1979)
Diaeretiella rapae/Brachycorynella asparagi	20	17–25	–	–	Hayakawa et al. (1990)
D. rapae/B. brassicae	20	506.0	–	–	Broussal (1966)
D. rapae/D. noxia	20	88c	19.7	0.26	Reed et al. (1992)
D. rapae/M. persicae	20	289.0	14.7	–	Fukui and Takada (1988)
	20	235.0	15.5	–	Hsieh and Allen (1986)
Dyscritulus planiceps/ Drepanosiphum platanoides	–	67d	14.0	–	Hamilton (1974)
Ephedrus cerasicola/M. persicae	23	1193a	13.4	0.39	Cohen and Mackauer (1987)
	21	961.0	18.0	0.29	Hågvar and Hofsvang (1990b)
Ephedrus persicae/Dysaphis reamuri	–	340–370	–	–	Shirota et al. (1983)

(Continued)

TABLE 3   The Mean Fecundity, Female Longevity (Days), and Intrinsic Rate of Natural Increase (r_m) of Some Species of Aphid Parasitoids (Braconidae: Aphidiidae).—cont'd

Species (Parasitoid/host)	Temp. °C	Fecundity	Longevity	r_m	References
Ephedrus plagiator/Macrosiphum avenae	21	60b	15.0	–	Jackson et al. (1974)
Ephedrus plagiator/R. maidis	21	183b	19.0	–	Jackson et al. (1974)
Ephedrus plagiator/R. padi	21	162b	23.0	–	Jackson et al. (1974)
Ephedrus plagiator/Schizaphis graminum	21	255b	22.0	–	Jackson et al. (1974)
Lysiphlebus delhiensis/R. maidis	22	–	7.9	0.27	Mishra and Singh (1990)
	20	271.0	8.8	0.31	Mishra and Singh (1991)
Lysiphlebia japonica/Toxoptera citricida	20	99.0	7.3	–	Takanashi (1990)
Monoctonus pseudoplatani/D. platanoides	–	236.0	19.0	–	Hamilton (1974)
Praon palitans/Therioaphis trifolii	21.1	578.0	17.2	0.24	Force and Messenger (1964)
	21	300.0	22.0	–	Force and Messenger (1964)
	10	155d	27.0	–	Schlinger and Hall (1960)
Trioxys complanatus/T. trifolii	21.1	844.0	14.3	0.38	Force and Messenger (1964)
Trioxys crisii/Drepanosiphum platanoides	—	252d	13.0	–	Hamilton (1974)
Binodoxys indicus/Aphis craccivora	20	240.0	7.0	0.31	Singh and Srivastava (1989)
	25	143.0	6.1	–	Singh and Srivastava (1989)
B. indicus/A. gossypii	25	165.0	8.3	0.39	Bhatt and Singh (1991a)
B. indicus/R. maidis	20	–	6.0	0.26	Tripathi and Singh (1990)
	20	185.0	5.5	0.26	Tripathi and Singh (1991)
Trioxys utilis/T. trifolii	21.1	450.0	–	–	Flint (1980)
	10	152d	27.0	–	Schlinger and Hall (1961)

Numerals followed by alphabets in the fecundity column denote: a as an Egg, b as Mummies, c as Offspring, d as an Egg in an ovary.

or as adults do reproduce but to a limited extent. Parasitized aphids consume more food but assimilate it less efficiently. They also gain more weight than unparasitized aphids. Parasitized aphids also produce more honeydew (Cloutier and Mackauer, 1980; Hoy and Nuygen, 2000).

For detailed biology of aphid parasitoids, reviews of Starý (1987), Hågvar and Hofsvang (1991), Singh and Agarwala (1992), Hoy and Nuygen (2000), Rehman and Powell (2010), and Boivin et al. (2012) should be consulted.

4.4 Biocontrol in Fields

Natural enemies may not be an important component in the long-term regulation of aphid populations, but there is no doubt they can be the deciding factor in preventing pest levels of aphids developing on a particular cropped area within a growing season (Powell and Pell, 2007). The interest in biocontrol of aphids has increased since the 1990s. At present, commercial use of natural enemies to control aphids in the protected crop is widespread; however, progress has been slower in outdoor crops. The major constraints in augmentative control in open fields are associated with mass rearing and release methods of the parasitoids. However, the advancement of technology in both mass rearing and field release are addressing these constraints, and a growing number of natural enemy species is becoming commercially available (van Driesche and Bellows, 1996; Nordlund et al., 2001; Copping, 2004) (Table 4).

Biocontrol of aphids in the fields has been successfully achieved in several parts of the world because their parasitoids have great potential in the managing their populations in spite of certain limitations (Hughes et al., 1994; Singh and Rao, 1995). More than 100 biocontrol programs have been monitored against at least 30 species of aphids and about 50% of them proved successful. These programs include the introduction of about 25 species of parasitoids. The parasitoids become established in 34 out of 57 attempts. The following paragraphs detail aphid parasitoids that have successfully controlled the aphid populations (Singh, 2001).

The introduction of *Aphelinus mali* in France to control the woolly aphid, *Eriosoma lanigerum*, in apples was probably the first attempt of biocontrol of aphids (Howard, 1929). After several attempts, the species established, but the control of the aphid remained variable depending on the regions and climatic conditions. *Aphelinus mali* was then introduced in several European countries, Australia, New Zealand (Howard, 1929), and in India (Rahman and Khan, 1941).

In California, the introduced parasitoids *Praon palitans* and *Trioxys utilis* against the two spotted alfalfa aphid (van den Bosch, 1956) and *Trioxys pallidus* against the walnut aphid, *Chromaphis juglandicola*, gave substantial control (Frazer and van den Bosch, 1973; van den Bosch et al., 1979). The Indian species *Aphidius smithi* quickly established in Mexico after was importing it from India (Clancy, 1967). It also established in Canada and USA in the fields of alfalfa (Mackauer, 1971). Similarly, *Aphidius eadyi* successfully controlled the pea aphid, *Acyrthosiphon pisum*, in New Zealand and *Trioxys complanatus* and *Trioxys tenuicaudus* suppressed the population of alfalfa aphid, *Therioaphis trifolii*, and elm aphid, *Tinocallis plantani*, respectively, in the USA (Hughes, 1989). *Aphidius ervi* was also established in North America against alfalfa aphid, *Acyrthosiphon kondoi* and *A. pisum*, in California (Mackauer and Kambhampati, 1986; Gonzalez et al., 1992). Chambers et al. (1986) reported low populations of the aphid *Sitobion avenae* on cereals caused by the action of their natural enemies in England. *Lysiphlebus testaceipes* readily established against black citrus aphid *Toxoptera aurantii* in southern France (imported from Cuba in 1973–1974) (Starý et al., 1988). Similarly, *T. complanatus* and *A. ervi* have been the major mortality factors against *T. trifolii* and *A. kondoi*, respectively, in Australia (Hughes et al., 1987). Messing and Aliniazi (1989) reported successful establishment of a European biotype of *T. pallidus* in Oregon, USA against the filbert aphid, *Myzocallis coryli*, a major pest in hazelnuts. Völkl et al. (1990) evaluated *L. testaceipes* and *Aphidius colemani* against the banana aphid *Pentalonia nigronervosa* in the South Pacific.

TABLE 4 Biological Control of Aphids Using Aphid Parasitoids

Parasitoid species	Aphid species	Crop	Country	References
Aphelinus certus	*Aphis glycines*	Soybean	USA	Heimpel et al. (2010)
Aphelinus mali	*Eriosoma lanigerum*	Apple	India	Rahman and Khan (1941)
			France	Howard (1929)
Aphidius colemani	*Aphis gossypii*	?	Netherlands	van Steenis (1995)
		Flower crops	USA	Heinz (1998)
Aphidius eadyi	*Acyrthosiphon pisum*	Peas, lucerne	New Zealand	Hughes (1989)
Aphidius ervi	*Acyrthosiphon kondoi*	Lucerne	Australia	Hughes et al. (1987)
			USA	Mackauer and Kambhampati (1986)
	Sitobion avenae	Wheat	Denmark	Sigsgaard and Hansen (2000)
Aphidius gifuensis	*Myzus persicae*	Tobacco	China	Yang et al. (2009)
Aphidius matricariae	*A. gossypii*	Cucumber	Netherlands	van Lenteren and Woets (1988)
	Aphis nasturtii	?	Bulgaria	Loginova et al. (1987)
	Dysaphis plantaginea	Apple orchards	Canada	Boivin et al. (2012)
	Macrosiphum euphorbiae	Cucumber	Netherlands	van Lenteren and Woets (1988)
	M. persicae	Cucumber	Netherlands	van Lenteren and Woets (1988)
		Eggplant	France	Rabasse et al. (1983)
		Chrysanthemum	UK	Wyatt (1985)
		Sweet pepper	UK	Buxton et al. (1990)
			Netherlands	Ramakers (1989) and van Schelt et al. (1990)
		Sweet pepper	Russia	Popov et al. (1987) and Shijko (1989)
		?	Bulgaria	Loginova et al. (1987)
		Tomato	Canada	Gilkeson (1990)
Aphidius rhopalosiphi	*S. avenae*	Cereals	Netherlands	Levie et al. (1999)
			Denmark	Sigsgaard and Hansen (2000)
	S. avenae	Cereals	Belgium	Levie et al. (2005)
Aphidius rosae	*Macrosiphum rosae*	Rose	Australia	Jörg Kitt (1996)
Aphidius salicis	*Cavariella aegopodii*	Carrot	Australia	Waterhouse and Sands (2001)
Aphidius smithi	*Therioaphis trifolii maculata*	Alfalfa	Mexico, Canada, USA	Clancy (1967)
	A. pisum	Pea	USA	Halfhill and Featherstone (1973)

TABLE 4 Biological Control of Aphids Using Aphid Parasitoids—cont'd

Parasitoid species	Aphid species	Crop	Country	References
Aphidius sonchi	*Hyperomyzus lactucae*	Lettuce	Australia	
Aphidius spp.	*M. persicae*	Vegetables	Germany	Albert (1990)
Binodoxys indicus	*Aphis craccivora*	Pigeon pea	India	Singh and Agarwala (1992)
	A. gossypii	Cucurbits	India	Singh and Rao (1995)
Diaeretiella rapae	*Diuraphis noxia*	Wheat	USA	Brewer and Elliott (2004)
Ephedrus cerasicola	*M. persicae*	Sweet pepper	Norway	Mackauer and Campbell (1972), Clausen (1978), and Hågvar and Hofsvang (1990a)
Ephedrus persicae	*Dysaphis plantaginea*	Apple orchards	Canada	Boivin et al. (2012)
Lipolexis oregmae	*Toxoptera citricida*	Citrus	USA	Hoy and Nguyen (2000) and Persad et al. (2007)
Lysiphlebus testaceipes	*Toxoptera aurantii*	Citrus	France	Starý et al. (1988)
Multiple spp.	*D. noxia*	Wheat	USA	Kindler and Springer (1989)
	A. gossypii	?	Netherlands	van Steenis (1992)
Pauesia anatolica	*Cinara cedri*	*Cedrus atlantica*	Turkey	Michelena et al. (2005)
Pauesia cedrobii	*Cinara laportei*	Pines	France	Fabré and Rabasse (1987)
Pauesia cinaravora	*Cinara cronartii*	Pines	South Africa	Kfir et al. (2003)
Pauesia juniperorum	*Cinara cupressivora*	Pines	Kenya, Uganda, Malawi	Kfir et al. (2003) and Day et al. (2003)
Pauesia momicola	*Cinara todocola*	Pines	Japan	Yamaguchi and Takai (1977)
Praon palitans	*Chromaphis juglandicola*	Walnut	USA	Frazer and van den Bosch (1973)
Trioxys complanatus	*T. trifolii maculata*	Lucern	Australia	Hughes et al. (1987)
			USA	Mackauer and Kambhampati (1986)
Trioxys curvicaudus	*Eucallipterus tiliae*	Linden, lime	USA	Hughes (1989)
Trioxys pallidus	*C. juglandicola*	Walnut	USA	Frazer and van den Bosch (1973)
	Myzocallis coryli	Hazelnuts	USA	Messing and Aliniazee (1989)
Trioxys tenuicaudus	*Tinocallis plantani*	Elm	USA	Hughes (1989)
Trioxys utilis	*C. juglandicola*	Walnut	USA	Frazer and van den Bosch (1973)
Xenostigmus bifasciatus	*Cinara atlantica*	Pines	South Africa	Reis-Filho et al. (2004) and Oliveira (2006)

Hughes et al. (1992) and van Steenis (1992) demonstrated considerable control of the population of *A. gossypii* by introducing its parasitoids in the Netherlands. Similarly, Starý (1993a) recovered several parasitoids of cereal aphids in Chile after their introduction.

Active biocontrol attempts have been made by the introduction of *D. rapae* against the Russian wheat aphid *Diuraphis noxia* with partial success (Gonzalez et al., 1989, 1990, 1992; Starý, 1999; Brewer and Elliott, 2004). Singh and Agarwala (1992) and Singh and Rao (1995) demonstrated successful control of *Aphis craccivora* on pigeon pea and *A. gossypii* on cucurbits by introducing the indigenous parasitoid *Binodoxys indicus*. The classic biocontrol program against the Russian wheat aphid *D. noxia* involved introduction of 12 parasitoid species along with a few species of predators from 1988–1996 and several of these species are now established in Idaho. *Aphidius rosae* was observed to control the population growth of the rose aphid *Macrosiphum rosae* into Australia (Jörg Kitt, 1996). Similarly, introduction of *A. colemani* on flower crops infested with *A. gossypii* in Texas, USA gave effective aphid suppression that was not only economical but also yielded a high quality crop (Heinz, 1997, 1998; Thompson et al., 1999). Sigsgaard and Hansen (2000) observed about 40% parasitism of cereal aphids in the winter wheat field caused by *A. ervi* Haliday and *Aphidius rhopalosiphi* De Stefani-Perez. *Lipolexis oregmae* was imported from Guam, evaluated in quarantine, mass reared, and released into citrus groves in Florida during 2000, 2001, and 2002 in a classical biocontrol program against the brown citrus aphid, *Toxoptera citricida* (Persad et al., 2007). Using the molecular assay with brown citrus aphids and other aphid species collected from citrus, weeds, and vegetables near release sites they observed parasitism of several aphid species such as black citrus aphids, cowpea aphids, spirea aphids, and melon aphids, as well as in the brown citrus aphid indicating that *L. oregmae* was established and widely distributed in the area.

When several species of cereal aphids particularly *Schizaphis graminum* and *S. avenae* invaded South America in the 1960s, farmers sprayed insecticides several times per season (Zuniga, 1990). A number of aphid parasitoids were introduced and established successfully leading to complete control of the two main aphid species within 10 years (Starý et al., 1993). The benefit is estimated at about US$52 million annually in reduced pesticide applications in Argentina, Brazil, and Chile for a project with a total cost of less than US$350000 (Zuniga, 1990).

Levie et al. (2005) showed that the release of 20,000 *A. rhopalosiphi* per hectare in wheat crops, twice at 1-week intervals allowed the control of the aphid, *S. avenae*. In China, mass release of *Aphidius gifuensis* was used to control *M. persicae* in tobacco crops (Yang et al., 2009). In apple orchards, the inundative release of two parasitoid species, *Ephedrus persicae* and *Aphidius matricariae*, controlled the population of rosy apple aphid, *Dysaphis plantaginea* (Boivin et al., 2012). *Aphidius colemani* and *A. ervi* were evaluated by Enkegaard et al. (2013) against the shallot aphid, *Myzus ascalonicus*, on strawberries and found suitable for release for its control. *Ephedrus cerasicola* has also proved to be a promising agent in this project. The soybean aphid, *Aphis glycines*, Asian in origin invaded the USA and its native aphelinid parasitoid *Aphelinus certus* was also accidentally introduced along with the aphid (Heimpel et al., 2010; Frewin et al., 2010) with high levels of parasitism in Quebec and Ontario, but the parasitoid is not common elsewhere. A previously-introduced strain of the European parasitoid *Aphelinus atriplicis* was released against soybean aphid in 2002. This strain was first released in the western United States against the Russian wheat aphid (Hopper et al., 1998; Prokrym et al., 1998; Heraty et al., 2007) and was found to attack the soybean aphid as well (Wu et al., 2004). Although this strain of *A. atriplicis* was released in nine Minnesota sites in 2002 (Heimpel et al., 2004), it has not been recovered since then. Another Asian aphidiine parasitoid, *Binodoxys communis* was released in the United States beginning in 2007; unfortunately, it did not establish (Ragsdale et al., 2011). Hopper and Diers (2014) studied the impact of resistant cultivars of soybean on the biological parameters of *Aphelinus glycinis* and demonstrated that although the rate of parasitism decreased on resistant plants, the aphids were parasitized. In an ongoing program, the Chinese strain of this parasitoid was released against soybean aphid at three sites in Minnesota, USA (Anonymus, 2014).

Pauesia cinaravora, a native Nearctic aphid parasitoid, was introduced from North America against *Cinara cronartii* on pines (*Pinus*) in South Africa and has become established and dispersed as a specific and effective parasitoid of the target aphid in the area (Kfir and Kirsten, 1991; Kfir et al., 1985, 2003; Marsh, 1991). *Pauesia juniperorum*, a native European species (Starý, 1966) has been found useful biocontrol agent against an exotic species, *Cinara cupressivora*, on ornamental and forest Cupressaceae in Kenya, Uganda, and Malawi (Day et al., 2003; Kfir et al., 2003), but only became established in Malawi (Anonymus, 1997). *Cinara pinivora* and *Cinara atlantica* were detected as pests on pine in Brazil in the end of the 1990s, but no parasitoids were found in the local fauna (Penteado et al., 2000). In 2002 and 2003, the parasitoid species, *Xenostigmus bifasciatus*, was introduced from the USA to the states of Paraná, Santa Catarina, and São Pauloand and became established in Brazil, dispersing 80 km per year from the release point (Reis-Filho et al., 2004; Oliveira, 2006). *Pauesia momicola* was introduced against *Cinara todocola* in Todo fir plantations in Japan where it rapidly dispersed to almost all infested trees and brought about significant reductions of aphid populations (Yamaguchi and Takai, 1977). *Pauesia anatolica* sourcing from *Cinara cedri/Cedrus libani* was imported from Anatolia, Turkey to the Golan area, Israel where it has become established on a conspecific host but not on *Cedrus atlantica* (Michelena et al., 2005). *Pauesia cedrobii* was introduced against *Cinara laportei* and became established in the southeastern France (Fabré and Rabasse, 1987).

Before 1962, the carrot aphid, *Cavariella aegopodii*, damaged the crops by transmitting the carrot motley dwarf virus complex. The parasitoid *Aphidius salicis* was introduced from California in 1962, and became established. However, the impact of the aphid parasitoid occurred concomitantly with the introduction of carrot varieties resistant to the virus and aphid (Waterhouse and Sands, 2001). Likewise, the lettuce aphid *Hyperomyzus lactucae* caused more than 90% loss in 1980 by transmitting the lettuce necrotic yellows rhabdovirus. The parasitoid *Aphidius sonchi* was introduced several times, and in 1984 the parasitoid settled (Waterhouse and Sands, 2001).

Waterhouse (1998) summarized the attempts of biocontrol of *A. craccivora* and *A. gossypii* using their parasitoids in several countries such as Australia, China, Columbia, Cuba, East Asia, France, India, Iraq, Israel, Italy, Japan, Korea, Malaysia, Netherlands, Pakistan, Philippines, USA, Russia, Vietnam, etc.

4.5 Control in Glasshouses

Glasshouse crop cultivation is a striking example of recent development in the field of biocontrol. Around 50 years ago, even specialists had serious doubt about the success of biocontrol in the glasshouses because this method of crop raising is economically vulnerable.

Despite of the complexity of this technology and the degree of specialized knowledge required, biocontrol became common practice in greenhouse vegetables in Northern Europe and in North America during the 1990s. Parr and Scopes (1970) described the problems associated with biocontrol of glasshouse pests. According to them, biocontrol gives more predictable control lasting several weeks to months despite being cheaper and ecofriendly. Paprikas, tomatoes, lettuces, chrysanthemums, and other ornamental pot plants are cultivated in glasshouses mostly in Europe. All these plants severely suffer with *M. persicae*. Successful biocontrol of *M. persicae* was achieved by introducing *A. matricariae* (Hussey and Scopes, 1985) and *E. cerasicola* (Hofsvang and Hågvar, 1980; Hågvar and Hofsvang, 1990a,b). Hofsvang and Hågvar (1980) introduced four mummies of the parasitoid per plant twice with 10 days interval in an 8 m² glasshouse with 24 paprika plants. The parasitoid kept the aphid population below economic injury level for more than 4 months. They suggested that if the crop is highly infested with the aphid, the introduction of mummies of the parasitoid at a parasitoid-to-host ratio of 1:10 is sufficient enough to check the aphid population. Introduction methods and amounts of *E. cerasicola* in small paprika glasshouses were studied by Hofsvang and Hågvar (1979, 1980). Three aphid parasitoids, *A. colemani* (against *M. persicae, A. gossypii* on melon), *A. matricariae* (against *M. persicae* on potato), and *A. ervi* (against *Macrosiphum euphorbiae* on potato and *Aulacorthum solani* on foxglove) were successfully released in glasshouses (Gill, 2012). Similar success of biocontrol of *M. persicae* in glasshouses were reported by Ramakers (1989), Popov et al. (1987), van Lenteren and Woets (1988), Shijko (1989), Gilkeson (1990), van Schelt et al. (1990) through introduction of *A. matricariae*, the most widely-used aphid parasitoid species in glasshouses in Europe.

Biocontrol through inundative or inoculative releases is applied in greenhouses where it gives the best results (Eilenberg et al., 2000, 2001; van Lenteren, 2000). During 2006, more than 37,000 ha of greenhouses were under biocontrol programs (Parrella, 2008). However, the augmentative use of parasitoids for aphid biocontrol requires the release of thousands of individuals. For instance, under greenhouse cultivation, quantities of parasitoids released for aphid control range from 2500 to 10,000 individuals per ha (van Lenteren et al., 1997a; van Lenteren, 2003a).

4.6 Mass Propagation

Mass rearing of the parasitoids in the laboratory on alternative host complex (insect as well as plant host) is a challenge for biocontrol workers. Unless natural enemies are made as readily available as chemical pesticides, biocontrol is likely to be treated as a subject of mere academic interest. It should be our endeavor to translate biocontrol precepts to practice.

Maintenance conditions, size, and genetic composition of the source colony and parasitoid diet are the main factors that determine the success or failure of the rearing program. This holds both in the case of small-scale production for research purpose and of mass production for release. Of many factors that influence the outcome of a rearing program, the nutritional requirement of the parasitoid is the most important (House, 1977). Schlinger and Hall (1960, 1961) mass reared *P. palitans* and *T. utilis* on their natural host. Similarly, Starý (1970a,b) developed methods for mass rearing, storage, and release of *A. smithi*. Featherstone and Halfhill (1966), Halfill (1967) and Halfhill and Featherstone (1967) manufactured portable cages for mass rearing of pea aphid parasitoids. Simpson et al. (1975) described a technique for the mass production of *D. rapae* and *Praon* sp. with limited insectary facilities. Using green peach aphids feeding on Chinese cabbage, two technicians working each day were able to produce 16,000 parasitoids per week. Mackauer and Kambhampati (1988) reviewed the status of mass rearing of aphid parasitoids. Singh et al. (1996, 2000) evaluated honeydew as food supplement for *Lysiphlebia mirzai* Shuja-Uddin, a cereal aphid parasitoid and recommended for its use in the mass culture of the parasitoid. In mass propagation, the quality of parasitoids may be improved by selective hybridization between the population reared on different host plant–host aphid combinations (Tripathi and Singh, 1997) or using semiochemicals (Srivastava and Singh, 1988a,b; Zhua et al., 2006).

A worldwide survey indicated that more than 170 species of natural enemies are produced by 85 commercial producers of natural enemies: 25 in Europe, 20 in North America, 6 in Australia and New Zealand, 5 in South Africa, about 15 in Asia, and about 15 in Latin America (van Lenteren, 2008; Cock et al., 2010). Currently, it is estimated that around 600 companies in the USA and 200 in Europe are involved in the production, shipment, and distribution of natural enemies. The global value of the insect biocontrol agent market was estimated at US$50 million in 2000, and this emerging biotechnology shows a continuous 15–20% annual growth (van Lenteren, 2008). Ten species of parasitoids are currently sold to control aphid pest species (Table 5) and thus require mass rearing techniques (Table 5).

Koppert Biological Systems, The Netherlands mass rear certain aphid parasitoids such as *Aphelinus abdominalis*, *A. colemani*, *A. ervi*, *A. matricariae*, *E. cerasicola*, and *Praon volucre*

TABLE 5 List of Aphid Parasitoid Species that are Presently Marketed Worldwide

Parasitoid species	Target	Culture
Aphelinus abdominalis	*Aulacorthum solani*	Tomato, sweet pepper, eggplant, french bean, gerbera, rose, chrysanthemum, strawberry
	Macrosiphum euphorbiae	
	Macrosiphum rosae	
	Myzus persicae	
	Rhodobium porosum	
Aphidius colemani	*Aphis gossypii*	Sweet pepper, cucumber, melon, eggplant, rose, chrysanthemum, strawberry
	Aphis craccivora	
	Aphis ruborum	
	M. persicae	
Aphidius ervi	*M. euphorbiae*	Sweet pepper, cucumber, eggplant, gerbera, rose, chrysanthemum, strawberry, French bean
	M. rosae	
	A. solani	
	M. persicae	
	R. porosum	
Aphidius gifuensis	*M. persicae*	Tobacco

(Continued)

TABLE 5   List of Aphid Parasitoid Species that are Presently Marketed Worldwide —cont'd

Parasitoid species	Target	Culture
Aphidius matricariae	*M. persicae*	Strawberry, sweet pepper, tobacco
	A. craccivora	
	Aphis fabae	
	A. gossypii	
	Aphis nasturii	
	A. ruborum	
Aphidius urticae	*A. solani*	Strawberry
Ephedrus cerasicola	*A. solani*	Strawberry, sweet pepper
	M. persicae	
Lysiphlebus fabarum	*A. gossypii*	Melon, cucumber
Lysiphlebus testaceipes	*A. gossypii*	Melon, cucumber
Praon volucre	*Acyrthosiphon malvae*	Strawberry
	A. craccivora	
	A. fabae	
	A. gossypii	
	Aphis nasturtii	
	M. euphorbiae	
	M. rosae	
	M. persicae	

Data from Takada (1992), Yano (2010), van Lenteren (2008), Wei et al. (2003).

and supply 240 mummies per unit and advise dose of 1 unit/200 m² (Figure 6). Most of the parasitoids are commercially reared in the USA, UK, Belgium, Germany, The Netherlands, Canada, Italy, and Thailand. The species most reared are *A. colemani* (13 suppliers) followed by *A. abdominalis* (9 suppliers), *A. ervi* (7 suppliers), and *A. matricariae* (4 suppliers).

Mass production of insect parasitoids faces a major constraint as the product must be sold at a cost affordable to the growers. Rearing a parasitoid also requires the multiplication of the host together with the host's food plant. The maintenance of three trophic levels is very tedious, expensive, and demands large greenhouse or growth chamber facilities and considerable manpower. Parra (2010) estimated that the manpower amounts 70–80% of the total production costs, however, it varies from country to country. The production cost of *A. gifuensis* is less than US$0.06 per 1000 mummies (Wei et al., 2003). Due to the importance of manpower cost, most mass production programs are limited in countries offering very low salaries.

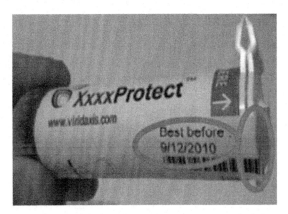

FIGURE 6 Supply unit of aphid parasitoids designed by the Koppert Biological System.

Low cost production of the parasitoids needs techniques that involve synthetic diet of the host aphid without altering the potential of their parasitoids. Greany et al. (1984) recognized the potential to produce parasitoids using artificial media or alternate hosts. Artificial diets for aphid (host to rear parasitoid) production eliminate the need to grow host plants. An artificial diet was formulated for propagation of *M. persicae* (Hamilton, 1935) and a complete synthetic diet was developed (Mittler and Dadd, 1962). However, these diets resulted in low fecundity as well as low survival of the aphids and were thus unsuitable for aphid mass production. van Emden and Kifle (2002) reared *A. colemani* on a completely chemically-defined diet (holidic diet). Though the fecundity of the parasitoid was unchanged, their size was reduced and development time was increased. Reduced size of the adult parasitoid is known to reduce the rate of parasitism (Cloutier et al., 2000). The second option is to rear the parasitoid on an artificial diet that mimics the nutritional value of the aphid host. Fewer than 20 species of natural enemies (both parasitoids and predators) are reared on a large scale on artificial diet, none of aphid parasitoid (van Lenteren, 2008).

Lysiphlebus fabarum was the first aphid parasitoid reared on a completely artificial medium by Rotundo et al. (1988). Use of small bag of parafilm membranes containing host hemolymph and microcapsule of polymer containing oviposition elicitor (Hance et al., 2001) were designed to rear parasitoids but still none of the aphid parasitoids were reared on a commercial scale. In addition, there are several other constraints that are limiting factors for mass culture of healthy parasitoids, such as temperature (Biswas and Singh, 1997), humidity, food availability (Singh et al., 2001), host-to-parasitoid ratio (Tripathi and Singh, 1991), genetic polymorphism, dispersal ability of the parasitoid, selection of proper host (Singh and Sinha, 1982a), progeny sex ratio and associated factors (Grenier and De Clercq, 2003; Singh et al., 2014b), pathogen infections (Bjørnson and Schütte, 2003), and superparasitism (Singh and Sinha, 1982b,c).

As the use of biocontrol agents has been increasing worldwide to control insect pests, there is a need for strict quality control in the production and use of these natural enemies (van Lenteren, 2003b).

4.7 Cold Storage of Parasitoids

Conservation and tolerance of low temperature by parasitoids has a practical importance in biocontrol because: (1) sufficient numbers of the parasitoid must be secured by accumulating mummies for release in infested plants (this implies that a small laboratory can propagate large number of parasitoids), and (2) unfavorable weather conditions may prevent timely releases. Cold storage is also a simple method for keeping the parasitoids alive when they are of no use (Hofsvang and Hågvar, 1977). The mummies as well as the adults can be kept at low temperatures. The mummies have been found to be more suitable for long-term storage (Starý, 1970a,b; Singh and Srivastava, 1988). Several workers have studied the effect of cold storage on the emergence of adult aphidiid parasitoids and were successful in conserving mummies for over 1 month, for example, *L. testaceipes* (Archer et al., 1973), *A. matricariae* (Scopes et al., 1973; Polgar, 1986), *E. cerasicola*, *A. colemani* (Hofsvang and Hågvar, 1977), *Aphidius uzbekistanicus* (Rabasse and Ibrahim, 1987), *B. indicus* (Singh and Srivastava, 1988), and *A. ervi* (Frère et al., 2011).

Colinet and Hance (2010) examined interspecific variation in the ability to tolerate cold storage of five species of aphidiine parasitoids: *A. colemani*, *A. ervi*, *A. matricariae*, *E. cerasicola*, and *P. volucre*, and observed that *A. matricariae* and *A. ervi* were most cold tolerant, *P. volucre* and *E. cerasicola* had an intermediate cold sensitivity, while *A. colemani* was most cold sensitive.

However, cold storage of parasitoids leads to a decrease in their viability, in particular in their survival, mobility, and sex ratio (Langer and Hance, 2000; Sigsgaard, 2000; Colinet et al., 2006; Ismail et al., 2010). *Aphidius ervi* mummies (host aphid: *A. pisum*) survive storage for long periods at temperatures close to 0 °C (Michel, 2007). Frère et al. (2011) demonstrated that the mummies of *A. ervi* (host *S. avenae* and *A. pisum*) can be kept for a maximum of 2 weeks at 7 °C without emergence of adults and for 7 weeks at 2 °C without emergence or mortality. Moreover, storage of the mummies at 7 or 2 °C does not affect fertility. The parasitoid pupae in *A. pisum* mummies suffered a higher mortality and took longer to complete their development. Lins et al. (2013) demonstrated that the percentage adult emergence, mummy body mass, flight capacity, and number of eggs in the ovarioles of *P. volucre* females decreased with increasing the period of storage, while the longevity of females was only slightly affected. Fat content of mummies, percentage of parasitized aphids, and survival of progeny to emergence decreased with increase in the period of storage. Storage of *P. volucre* prepupae for up to 5 days at 5 °C did not affect any of the abovementioned parameters. Ismail et al. (2013) demonstrated that resistant individuals of *A. ervi* occurred at 0 °C (constant temperature) that produced individuals without any decrease in the measured fitness traits.

Tolerance to cold storage is a very plastic trait influenced by a wide range of endogenous (biotic) and exogenous (abiotic) factors experienced before, during, or after cold exposure. In fact, every hierarchical level from interspecies to interindividuals shows a high plasticity in the response to cold exposure (Boivin and Colinet, 2011). Boivin and Colinet (2011) reviewed cold storage of insect parasitoids considering the genotypic-based plasticity in cold storage tolerance and the complex network of endogenous and exogenous factors affecting the phenotypic plasticity in cold storage tolerance of the parasitoids; and also examined the wealth of fitness-related traits affected by cold storage in the parasitoids. They also provided a comprehensive list of documented factors that must be taken into account when designing cold storage protocols.

4.8 Enhancement of Biotic Potential

Several factors influence the biotic potential of parasitoids at all trophic levels. It includes the provision of supplementary resources such as alternative hosts (both food plant resources as well as aphid hosts), provision of adult food, agricultural practices, climatic variations, infochemicals, selective hybridizations, biotechnological approaches, etc. (Coppel and Mertins, 1977; van Driesche and Bellows, 1996).

4.8.1 *Manipulation of Host Habitat*

Powell (1986), van Emden (1986, 1990) and Landis et al. (2000) reviewed habitat management to enhance the activity of natural enemies of insect pests with several references of aphid parasitoids. Agroecosystems are among the most difficult environments for efficient operation of biocontrol agents. This is because they usually lack adequate resources for the effective performance of the natural enemies and many of the cultural practices used in annual cropping are damaging to natural enemies (Powell, 1986).

In several studies, emphasis were given to increase the diversity within the agroecosystem by introducing multiple cropping, intercropping, strip harvesting, selective retention of weeds within the crop, or conservation of wild plants at field margins (Altieri and Letourneau, 1982; Powell, 1986). Increasing diversity within crops is predicted to provide a greater number of opportunities for natural enemies to survive in agricultural systems and also tends to increase natural enemy abundance and diversity, providing a system more resilient to pest population increase (Rodriguez-Saona et al., 2012). Vegetational diversity also provides support for insect biocontrol at local and landscape levels (Thies et al., 2003; Roschewitz et al., 2005; Bianchi et al., 2006; Gardiner et al., 2009). The plants serve as a reservoir of the alternative host species, and flowering plants are important sources for food as the adult parasitoids do not necessarily feed only on honeydews but also on pollen, nectar, and other sugary plant secretions.

In addition to pollen, a honeydew or sugar solution was found to increase the longevity and fecundity of some species of parasitoids (Singh et al., 1996; Lee et al., 2004; Hogervorst et al., 2007; Wäckers et al., 2008; Vollhardt et al., 2010). Therefore, intercropping of such plants or tropical application of honeydew not only attract the parasitoids but also increased their retention time which is directly related with the rate of parasitism (Bouchard and Cloutier, 1984; Srivastava and Singh, 1988a,b; Budenberg, 1990). However, the ability of the parasitoid to switch between host resources is crucial (Starý, 1993b). Studies demonstrated that several parasitoids can easily be transferred on hosts feeding on a variety of food plants or cultivars (Tripathi and Singh, 1997) or between the aphid species (Powell and Wright, 1988; Singh et al., 2000).

In the past, studies on the biocontrol of insect pests have paid little attention to the properties of the hosts and supporting plants, while emphasizing the ability of natural enemies to search for herbivore insects. Fritz (1992) demonstrated that the genetic variation among crop cultivars could directly influence the nutritional status of the host, profoundly affecting normal parasitoid development and exerting long-term effects on its fecundity, longevity, etc. Plant architecture, plant organ, plant structural refuges, and plant surface structure influence the herbivore distribution and alter the capacity of a natural enemy to exploit potential hosts/prey (Boethel and Eikenbary, 1986; Brodeur and McNeil, 1992). Marcovitch (1935) was the first to report that the glandular hairs on petunia could trap the braconid *L. testaceipes.*

Several authors have demonstrated a host plant effect on aphid parasitoids (Bhatt and Singh, 1989, 1991a,b,c,d,e; Reed et al., 1992; van Steenis and El-Khawass, 1995). The suitability of the host plant for the herbivore might be unrelated to its suitability for the parasitoid. This could be for several reasons. Both Powell and Zhi-Li (1983) and Reed et al. (1992) have observed that food plant quality affects the developmental period and other life table parameters of an aphid host and this can directly influence the behavior of the parasitoid. However, host quality is not a static property, but varies during the course of an interaction, often in relation to the amount and the quality of the food consumed by the host itself. For instance, the cabbage aphid, *Brevicoryne brassicae*, is highly susceptible to differences in plant quality, so that members of the same cohort often grow at very different rates. These differences were reflected in the developmental rate of its parasitoid, *D. rapae* (Mackauer and Kambhampati, 1984). Reed et al. (1991) recorded a shorter developmental period for *D. rapae* on resistant wheat cultivar than susceptible wheat. Starks et al. (1972) showed that resistant cultivars of barley and sorghum synergize the activity of *L. testaceipes* in reducing plant damage by decreasing the population of the aphid *S. graminum*. However, resistant cultivars exhibiting antibiosis are more likely to have negative third trophic level effects than the cultivars having other means of resistance (Salto et al., 1983). Campbell et al. (1990) found that the resistant cultivars of barley exhibiting antibiosis significantly affect the reproduction and developmental rates and also the ultimate size of *L. testaceipes* parasitizing *S. graminum* compared with the susceptible barley cultivars. Biswas and Singh (1997, 1998) reported synergism between host plant resistance and biocontrol between resistant corn cultivar and the action of *Lysiphlebus delhiensis* against *Melanaphis sacchari*.

4.8.2 Genetic Improvements

Attempts at genetic improvement of parasitoids have been made extensively, but relatively few laboratory-selected parasitoids have been field-tested. Genetic manipulation will remain a controversial technique in biocontrol until we quantify the likelihood of achieving successful laboratory selection responses and document the fitness and efficacy of the selected parasitoids under field conditions. Advances in genetic engineering have allowed insects from a range of orders to be engineered with a range of traits, and thus genetic engineering of parasitoids and predators is becoming a real possibility (Atkinson et al., 2001). However, ecological (genetic and organismic biodiversity) and ethical issues like modification of plants will exist for parasitoids too. By maneuvering the genetics of parasitoids, the dynamics of tritrophic interactions will also be affected, questioning the stability of the interaction. It is rather difficult to genetically manipulate parasitoids with the confidence that it will maximize only the ecological benefits and minimize the ecological risks.

Hoy et al. (1989, 1990) and Hoy and Cave (1991) developed a strain of *T. pallidus*, a walnut aphid parasitoid in California, resistant to Guthion and azinphos-methyl, respectively. The biological potential of the parasitoids may also be enhanced either by selective hybridization or through mutagenesis, recombinant DNA technology, etc.

Poppy and Powell (2004) discussed the potential of genetic manipulation of parasitoids to improve the efficiency of biocontrol, highlighting methodologies to directly manipulate the genetics of the parasitoid itself or to exploit our growing knowledge of the influence of the plant on parasitoid foraging behavior and genetically manipulate the plant to improve parasitoid efficiency.

4.8.3 *Manipulation of Behavior by Semiochemicals*

Lewis et al. (1982) and Chiri and Legner (1983), and more recently Heuskin et al. (2012), conducted a number of experiments demonstrating the potential of kairomones in the manipulation of the behavior of the parasitoids for pest management. Kairomones increase the rate of parasitism by three ways: (1) by stimulating the female parasitoids (Srivastava and Singh, 1988a), (2) by retaining the female parasitoids on the treated host patch (Srivastava and Singh, 1988b), and (3) by improving egg distribution among the hosts (Singh and Srivastava, 1989). By application of kairomones in the fields infested with aphid pests at low density, the female parasitoid can be retained for longer period on the treated plants. The retention and activation increases the chance for host contact and results in an increased extent of host mortality. Also, the parasitoid can be attracted toward the infestation site by applying the kairomones, e.g., the female *P. volucre* responds the sex pheromones of the aphid host (functioning as kairomones) and thus may be attracted in the fields by putting traps containing such lures (Hardie et al., 1994; Glinwood et al., 1998).

5. CONSTRAINTS IN BIOCONTROL

5.1 Hyperparasitism

The term hyperparasitism (Fiske, 1910) and superparasitism (Salt, 1934) are both misnomers, as the prefix "hyper" means over and thus implies excess and "super" means the development of one parasitoid's progeny through or on that of another. These terms do not accurately describe the adaptation of one parasitoid for development on another (hyperparasitism) and more progeny are deposited in a host than can survive (superparasitism). Instead, it should be vice versa. Hyperparasitoids are secondary parasitoid of a primary parasitoid and constitute a fourth trophic level (Fiske, 1910). Aphid hyperparasitoids independently evolved in seven insect families (Chalcidoidea: Encyrtidae: *Aphidencyrtus, Litomastix, Prionomitus, Syrphophagus, Tassonia*; Eulophidae: *Prospeltella, Tetrastichus*; Pteromalidae: *Asaphes, Corona, Pachyneuron*; Signiphoridae: *Chartocerus*; Ceraphronoidea; Ceraphronidae: *Aphanogmus, Ceraphron*; Megaspilidae: *Dendrocerus*; Cynipoidea; Figitidae: *Alloxysta, Lytoxysta, Phaenoglyphis*) which represent three superfamilies of the order Hymenoptera (Sullivan, 1988; Singh and Tripathi, 1991).

There is disagreement over the importance of hyperparasitism in biocontrol (Valentine, 1974; Ehler, 1979; Bennett, 1981; Mackauer and Völkl, 1993; Sullivan and Völkl, 1999; Brodeur and Rosenheim, 2000; Schooler et al., 2011; Hofsvang et al., 2014). The hyperparasitoids have traditionally been considered harmful to the beneficial primary parasitoids of phytophagous insect pests; it has been the policy in biocontrol programs to exclude them whenever the parasitoids are to be introduced into an area (Hagen and van den Bosch, 1968; Sullivan, 1987, 1988). The percentage hyperparasitism of some aphid parasitoids are summarized by Singh and Sinha (1980) and range from 3% of *A. craccivora* (parasitized by *B. indicus*) to 88% of *B. brassicae* (parasitized by *D. rapae*). Hofsvang et al. (2014) observed about 90% hyperparasitism of mummies of *A. pisum* (parasitized by *L. testaceipes*). Hassell and Waage (1984) reviewed the multispecies interactions between primary parasitoids and hyperparasitoids and noted that there have been relatively few attempts to assess the ecological impact of obligate hyperparasitoids (Chua, 1979; Gutierrez and van den Bosch, 1970; Kfir et al., 1976; Rabasse and Brunei, 1977; Singh and Sinha, 1980; Singh and Srivastava, 1990) and to analyze this mathematically (Beddington and Hammond, 1977; May and Hassell, 1981; Holt and Hochberg, 1998).

Schooler et al. (2011) performed glasshouse experiments using cages containing 48 plants to address whether the hyperparasitoid *Asaphes suspensus* can potentially eliminate the population of primary parasitoid *A. ervi*, a biocontrol agent of the pea aphid, *A. pisum*. They observed that although *A. suspensus* has a low intrinsic rate of increase, only one-half of *A. ervi* and one-third that of pea aphid was capable of eliminating the *A. ervi* population within its seven generations. In contrast, in the absence of the hyperparasitoid, the parasitoid eliminates the pea aphid population. Field surveys found that the hyperparasitoid does not eliminate entire natural populations of the parasitoid in lucerne crops, probably due to the high frequency of disturbance that favors high intrinsic rates of increase and short generation times. Nonetheless, the ability of *A. suspensus* to eliminate *A. ervi* in cages despite its low intrinsic rate of increase underscores the potential for hyperparasitism to disrupt biocontrol. Small populations are expected to be particularly susceptible to hyperparasitism, such as when releases of a new biocontrol agent are made.

The debate continues about the actual harm and the possible positive beneficial role of hyperparasitoids in maintaining a balance between populations of insect species in the ecosystem, because multispecies complexity might help to effect community stability. Hence, aphid hyperparasitoids provide a microcosm for research on this fascinating and practical ecological puzzle (Sullivan, 1987).

5.2 Intraguild Predation

There is competition between predators and parasitoids for aphids. Organisms that share a common prey/host are known as a guild. When the predator consumes aphid parasitized by a parasitoid, a dynamic phenomenon known as intraguild predation takes place, where one organism feeds on another within the same predatory guild (Brodeur and Rosenheim, 2000). Through intense intraguild predation, predators can reduce the impact of other guild members such as the parasitoids that are released to control the aphid population, and trigger or worsen prey outbreaks indirectly and potentially allow for an increase in the amount of aphid damage to the plants (Rosenheim et al., 1993; Snyder and Ives, 2001, 2003). Intraguild predation decreases the pressure of top-down interactions and thus the lower trophic level organism may no longer be under control (Finke and Denno, 2004). It has been suggested that generalist predators do not distinguish between healthy and parasitized aphids and thus function as facultative predators of parasitoids (Phoofolo et al., 2007). However, few studies have shown that ladybirds and hoverflies are able to discriminate healthy and parasitized aphids. The parasitized aphids serve as inferior prey for ladybirds (Lebusa, 2004; Bilu and Coll, 2009). Meyling et al. (2004) observed that the females of an anthocorid bug (*Anthocoris nemorum*) during feeding was unable to discriminate between mummies of *M. persicae* parasitized by *A. colemani*. If the parasitized aphids serve as inferior prey for the predators in the same guild, it demonstrates an indirect parasitoid defense mechanism that under natural conditions reduces competition and prevents ladybirds from eliminating the local parasitoid population. This concept is new to the study of intraguild interactions and may help to explain guild dynamics in agricultural systems in future (Bilu and Coll, 2007). Additional field work is required to reach conclusions about short and long-term effects of hyperparasitism and intraguild predation on parasitoid population dynamics and the consequences on aphid population suppression.

6. FUTURE OF BIOCONTROL OF APHIDS

The main source of information on biology, ecology, behavior, trophic interaction of aphid parasitoids, etc. is laboratory studies, and a few field studies. At present, the fundamental biological studies of the aphid parasitoids are carried out by the ecologists and biocontrol workers. The following questions need exploration and answers (Boivin et al., 2012):

1. What is the fate of aphid parasitoids once they have been introduced as biocontrol agents in inoculative or augmentative releases?
2. To what extent and how rapidly do they spread in the environment?
3. After their introduction, what is the level of intra and interspecific competition?
4. Do introduced parasitoids have the capacity to include the target aphid species in their host range?
5. If yes, at what rates and under which conditions do they evolve toward new aphid host associations?

These questions are relevant to biocontrol, but studying the dispersal over long distances and the host selection behavior of parasitoids over time remains challenging under field conditions (Casas, 2000). The potential for mass rearing aphid parasitoids is bright because new technologies are being developed to produce aphids and parasitoids using artificial media. However, there is a need to consider the trade-off between producing and storing large quantities of parasitoids at low costs and the overall quality of the individuals that are released in crops. Measuring the fitness of mass-produced individuals remains a challenge, and such a measure is likely to change depending on the crop where the parasitoids are released and the aphid species present. Quantifying these parameters should increase the reliability of biocontrol programs and contribute to its acceptance by the growers (Boivin et al., 2012).

7. CONCLUSION

Biocontrol is generally regarded as most effective in stable ecosystems (e.g., forests, orchards, range lands). However, the studies conducted and reports available demonstrate that it can still be effective in agroecosystems where the environment is continually disrupted and diversity is minimal. It implies, therefore, that although the biocontrol is not the answer to everything, and is not only more than a trial that succeeds by chance, it can be employed and can minimize the application of poisonous chemicals. Biocontrol is most successful if a professional pest manager advises growers and assists with monitoring. An extension agent from a government-supported research station can perform this role, or a cooperative farming could employ a specialist. The biocontrol approach may not be applicable to every crop, however, thousands of growers around the world already exists who employ such measures with great success. With more research, better supply and distribution of bioagents, and the employment of trained pest managers to work with growers, biocontrol of aphids can be quite successful.

References

Abd-Rabou, S., 2008. Evaluation of the green lacewing, *Chrysoperla carnea* (Stephens) (Neuroptera; Chrysopidae) against aphids on different crops. Journal of Biological Control 22 (2), 299–310.

Agarwala, B.K., Laska, P., Raychaudhuri, D.N., 1984. Prey records of aphidophagous syrphid flies from India (Diptera: Syrphidae). Acta Entomologica Bohemoslovaka 81, 15–21.

Albert, R., 1990. Experiences with biological control measures in glasshouses in southwest Germany. SROP/WPRS Bulletin XIII/5, 1–5.

Alhmedi, A., Haubruge, E., Bodson, B., Francis, F., 2007. Aphidophagous guilds on nettle (*Urtica dioica*) strips close to fields of green pea, rape and wheat. Insect Science 14, 419–424.

Altieri, M.A., Letourneau, D.K., 1982. Vegetation management and biological control in agroecosystems. Crop Protection 1, 405–430.

Anonymous, 1997. Annual Report 1997. International Institute of Biological Control. CAB International, Wallingord, UK.

Anonymus, 2014. http://www.ncsrp.com/pdf_doc/Soybean_Aphid_Tilman_SAR2014.pdf.

Archer, T.L., Murray, C.L., Eikenbary, R.D., Starks, K.J., Morrison, R.D., 1973. Cold storage of *Lysiphlebus testaceipes* mummies. Environmental Entomology 2, 1104–1108.

Atkinson, P.W., Pinkerton, A.C., O'Brochta, D.A., 2001. Genetic transformation system in insects. Annual Review of Entomology 46, 317–348.

Bakhetia, D.R.C., Chander, H., 1997. Aphids and their management with particular reference to host plant resistance. Journal of Aphidology 11, 1–19.

Bale, J.S., van Lenteren, J.C., Bigler, F., 2008. Biological control and sustainable food production. Philosophical Transactions of the Royal Society of London, B Biological Sciences 363 (1492), 761–776.

Barrette, M., Wu, M., Brodeur, J., Giraldeau, L.A., Boivin, G., 2009. Testing competing measures of profitability for mobile resources. Oecologia 158, 757–764.

Barrette, M., Boivin, G., Brodeur, J., Giraldeau, L.A., 2010. Travel time affects optimal diets in depleting patches. Behavioral Ecology and Sociobiology 64, 593–598.

Beddington, J.R., Hammond, P.S., 1977. On the dynamics of host-parasite, hyperparasite interactions. Journal of Animal Ecology 46, 811–821.

Behura, B.K., 1994. The mystery of aphid life history. Journal of Aphidology 8, 1–18.

Bellows Jr., T.S., Fisher, T.W. (Eds.), 1999. Handbook of Biological Control: Principles and Applications. Academic Press, San Diego, CA, p. 1046.

Bennett, F.D., 1981. Hyperparasitism in the practice of biological control. In: The Role of Hyperparasitism in Biological Control: A Symposium. Division of Agricultural Science, University of California, Berkeley, pp. 43–49.

Bhatt, N., Singh, R., 1989. Bionomics of an aphidiid parasitoid *Trioxys indicus* Subba Rao and Sharma 30. Effect of host plants on reproductive and developmental factors. Biological Agriculture and Horticulture 6, 149–157.

Bhatt, N., Singh, R., 1991a. Bionomics of an aphidiid parasitoid *Trioxys indicus* Subba Rao and Sharma 36. Impact of male on the fecundity and sex allocation of the parasitoid through its host *Aphis gossypii* Glover on some cucurbit food plants. Zeitschrift fur Angewnadte Zoologie 78, 3–10.

Bhatt, N., Singh, R., 1991b. Bionomics of an aphidiid parasitoid *Trioxys indicus* Subba Rao and Sharma 32. Effect of food plants on the functional response, area of discovery and sex ratio of F_1 offspring at varying densities of its host *Aphis gossypii* Glover. Journal of Applied Entomology 111, 263–269.

Bhatt, N., Singh, R., 1991c. Bionomics of an aphidiid parasitoid *Trioxys indicus* Subba Rao and Sharma 33. Impact of food plants on the behaviour and sex allocation by the female parasitoid at her varying densities. Biological Agriculture and Horticulture 7, 247–259.

Bhatt, N., Singh, R., 1991d. Bionomics of an aphidiid parasitoid *Trioxys indicus* Subba Rao and Sharma 35. Influence of food plants on life-table statistics of the parasitoid through its host *Aphis gossypii* Glover. Insect Science and Its Application 12, 385–389.

Bhatt, N., Singh, R., 1991e. Impact of male parasitoid *Trioxys (Binodoxys) indicus* Subba Rao and Sharma on the reproductive strategies of female on different food plants. Indian Journal of Ecology 18, 36–40.

Bianchi, F.J.J.A., Booij, C.J.H., Tscharntke, T., 2006. Sustainable pest regulation in agricultural landscapes: a review on landscape composition, biodiversity and natural pest control. Proceedings of the Royal Society of London B 273, 1715–1727.

Bilu, E., Coll, M., 2007. The importance of intraguild interactions to the combined effect of a parasitoid and a predator on aphid population suppression. BioControl 52, 753–763.

Bilu, E., Coll, M., 2009. Parasitized aphids are inferior prey for a coccinellid predator: implications for intraguild predation. Environmental Entomology 38 (1), 153–158.

Biswas, S., Singh, R., 1997. Interrelationships between host density, temperature, offspring sex ratio and intrinsic rate of natural increase of the aphid parasitoid *Lysiphlebia mirzai* reared on resistant and susceptible cultivars (Hymenoptera: Braconidae: Aphidiinae). Entomologia Generalis 22, 239–249.

Biswas, S., Singh, R., 1998. Interaction between host plant resistance and the biocontrol of a cereal aphid: a laboratory study. Biological Agriculture and Horticulture 16, 25–36.

Bjørnson, S., Schütte, C., 2003. Pathogens of mass-produced natural enemies and pollinators. In: van Lenteren, J.C. (Ed.), Quality Control and Production of Biological Control Agent. CAB International, Wallingford, UK, pp. 133–165.

Blackman, R.L., Eastop, V.F., 2006. Aphids on the World's Herbaceous Plants and Shrubs. Host List and Keys, vol. 1. John Wiley & Sons, Chichester, UK, p. 1460.

Boethel, D.G., Eikenbary, R.D. (Eds.), 1986. Interaction of Plant Resistance and Parasitoids and Predators of Insects. John Wiley and Sons, New York, p. 224.

Boivin, G., Colinet, H., 2011. Insect parasitoids cold storage: a comprehensive review of factors of variability and consequences. Biological Control 58 (2), 83–95.

Boivin, G., Hance, T., Brodeur, J., 2012. Aphid parasitoids in biological control. Canadian Journal of Plant Science 92 (1), 1–12.

Bouchard, Y., Cloutier, C., 1984. Honeydew as a source of host-searching kairomones for the aphid parasitoid *Aphidius nigripes* (Hymenoptera: Aphidiidae). Canadian Journal of Zoology 62, 1513–1520.

Brewer, M.J., Elliott, N.C., 2004. Biological control of cereal aphids in North America and mediating effects of host plant and habitat manipulations. Annual Review of Entomology 49, 219–242.

Brodeur, J., McNeil, J.N., 1992. Host behaviour modification by the endoparasitoid *Aphidius nigripes*: a strategy to reduce hyperparasitism. Ecological Entomology 17, 97–104.

Brodeur, J., Rosenheim, J.A., 2000. Intraguild predation in aphid parasitoids: a review. Entomologia Experimentalis et Applicata 97, 93–108.

Broussal, G., 1966. (thèsis). Etude d'un complexe parasitaire: les hyménoptères parasites et hyperparasites de *Brevicoryne brassicae* (Homoptère: Aphididae), vol. 3. Faculty of Science, Reims, pp. 1–187.

Budenberg, W., 1990. Honeydew as a contact kairomone for aphid parasitoids. Entomologia Experimentalis et Applicata 55, 139–148.

Buxton, J.H., Jacobsen, R., Saynor, M., Storer, R., Wardlow, L., 1990. An integrated pest management programme for peppers, three years trials experience. SROP/WPRS Bulletin XIII/5, 45–50.

van den Bosch, R., Hom, R., Matteson, P., Frazer, B.D., Messenger, P.S., 1979. Biological control of the walnut aphid in California: Impact of the parasite *Trioxys pallidus*. Hilgardia 47, 1–13.

van den Bosch, R., 1956. Parasites of the alfalfa aphid. California Agriculture 10, 7–15.

Cameron, P.J., Walker, G.P., Allan, D.J., 1981. Establishment and dispersal of the introduced parasite *Aphidius eadyi* (Hymenoptera: Aphidiidae) in the North Island of New Zealand, and its initial effect on pea aphid. New Zealand Journal of Zoology 8, 105–112.

Campbell, R.K., Salto, C.E., Sumner, L.C., Eikenbary, R.D., 1990. Tritrophic interactions between grains, aphids and a parasitoid. In: Szentesi, A., Jermy, T. (Eds.), Insects-Plant. Symp. Biol. Hungary, vol. 39. Academizi Kiado, Budapest, pp. 393–401.

Casas, J., 2000. Host location and selection in the field. In: Hochberg, M.E., Ives, A.R. (Eds.), Parasitoid Population Biology. Princeton University Press, Princeton, NJ, pp. 17–26.

Chambers, R.J., Sunderland, K.D., Stacey, D.L., Wyatt, I.J., 1986. Control of cereal aphids in winter wheat by natural enemies, aphid-specific predators, parasitoids and pathogenic fungi. Annals of Applied Biology 108, 219–231.

Chiri, A.A., Legner, E.F., 1983. Field application of host searching kairomones to enhance parasitisation of the pink bollworm (Lepidoptera: Gelechiidae). Journal of Economic Entomology 76, 254–255.

Chua, T.H., 1979. A comparison of searching efficiency of a parasite and a hyperparasite. Researches on Population Ecology 20, 179–187.

Clancy, D.W., 1967. Discovery of the pea aphid parasite, *Aphidius smithi* in central Mexico. Journal of Economic Entomology 60, 17–43.

Clausen, C.P. (Ed.), 1978. Aphididae. In: Introduced Parasites and Predators of Arthropod Pest and Weeds, a World Review. Agricultural Handbook of USDA, 480, Washington D.C., pp. 35–55.

Cloutier, C., Mackauer, M., 1980. The effect of superparasitism by *Aphidius smithi* (Hymenoptera: Aphidiidae) on the food budget of the pea aphid, *Acyrthosiphon pisum* (Homoptera: Aphididae). Canadian Journal of Zoology 58 (2), 241–244.

Cloutier, C., McNeil, J.N., Regniere, J., 1981. Fecundity, longevity, and sex ratio of *Aphidius nigripes* (Hymenoptera: Aphidiidae) parasitizing different stages of its host, *Macrosiphum euphorbiae* (Homoptera: Aphididae). Canadian Entomologist 113, 193–198.

Cloutier, C., Duperron, J., Tertuliano, M., McNeil, J.N., 2000. Host instar, body size and fitness in the koinobiotic parasitoid *Aphidius nigripes*. Entomologia Experimentalis et Applicata 97 (1), 29–40.

Cock, M.J.W., van Lenteren, J., Brodeur, J., Barratt, B., Bigler, F., Bolckmans, K., Consoli, F.L., Haas, F., Mason, P.G., Parra, J.R.P., 2010. Do new access and benefit sharing procedures under the convention on biological diversity threaten the future of biological control? Biocontrol 55, 199–218.

Cohen, M.B., Mackauer, M., 1987. Intrinsic rate of increase and temperature coefficients of the aphid parasite *Ephedrus californicus* Baker (Hymenoptera: Aphidiidae). Canadian Entomologists 119, 231–237.

Colinet, H., Hance, T., 2010. Interspecific variation in the response to low temperature storage in different aphid parasitoids. Annals of Applied Biology 156 (1), 147–156.

Colinet, H., Hance, T., Vernon, P., 2006. Water relations, fat reserves, survival, and longevity of a cold-exposed parasitic wasp, *Aphidius colemani* (Hymenoptera: Aphidiinae). Environmental Entomology 35, 228–236.

Coppel, H.C., Mertins, J.M., 1977. Biological Insect Pest Suppression. Springer Verlag, Berlin, Heidelberg, New York, p. 314.

Copping, L.G., 2004. The Manual of Biocontrol Agents: A World Compendium. British Crop Protection Council, p. 752.

Day, R., Kairo, M.T.K., Abraham, Y.J., Kfir, R., Murphy, S.T., Mutitu, K.E., Chilima, C.Z., 2003. Homopteran pests of conifers. In: Neuenschwander, P., Borgemeister, E. (Eds.), Biological Control in IPM Systems in Africa. CAB International, Wallingdford, pp. 101–112.

DeBach, P. (Ed.), 1964. Biological Control of Insect Pests and Weeds. Chapman and Hall, London, UK, p. 844.

DeBach, P., 1974. Biological Control by Natural Enemies. Cambridge University Press, p. 323.

Dent, D., 2000. Insect Pest Management. Cambridge University Press, p. 410.

Dixon, A.F.G., 1985. Aphid Ecology. Blackie and Sons Ltd, Biosphobriggs, Glasgow, p. 157.

Dixon, A.F.G., 2000. Insect Predator–Prey Dynamics, Ladybird Beetles and Biological Control. Cambridge University Press, Cambridge, UK, p. 257.

Dransfield, R.D., 1979. Aspects of host—parasitoid interactions of two aphid parasitoids, *Aphidius urticae* (Haliday) and *Aphidius uzbeckistanicus* (Luzhetski) (Hymenoptera, Aphidiidae). Ecological Entomology 4 (4), 307–316.

Du, Y.J., Poppy, G.M., Powell, W., 1996. Relative importance of semiochemicals from first and second trophic levels in host foraging behavior of *Aphidius ervi*. Journal of Chemical Ecology 22, 1591–1605.

Dumbleton, L.D., 1936. The biological control of fruit pests in New Zealand. New Zealand Journal of Science and Technology 18, 288–295.

van Driesche, R.G., Bellows, T.S., 1995. Biological Control. Chapman & Hall, New York, NY.

van Driesche, R.G., Bellows Jr., T.S., 1996. Biological Control. Chapman & Hall, London, p. 56.

Eastop, V.F., 1958. A Study of Aphididae (Homoptera) of East Africa. H.M.S.O, London, p. 126.

Ehler, L.E., 1979. Utility of facultative secondary parasites in biological control. Environment 8, 829–832.

Eilenberg, J., Enkegaard, A., Vestergaard, S., Jensen, B., 2000. Biocontrol of pests on plant crops in Denmark: present status and future potential. Biocontrol Science and Technology 10, 703–716.

Eilenberg, J., Hajek, A., Lomer, C., 2001. Suggestions for unifying the terminology in biological control. BioControl 46, 387–400.

Enkegaard, A., Sigsgaard, L., Kristensen, K., 2013. Shallot aphids, *Myzus ascalonicus*, in strawberry: biocontrol potential of three predators and three parasitoids. Journal of Insect Science 13, 83.

van Emdem, H.F., Kifle, A.T., 2002. Performance of the parasitoid *Aphidius colemani* when reared on *Myzus persicae* on a fully defined artificial diet. Biocontrol 47, 607–616.

van Emden, H.F., 1986. The interaction of plant resistance and natural enemies: effects on population of sucking insects. In: Boethel, D.J., Eikenbery, R.D. (Eds.), Interactions of Plant Resistance and Parasitoids and Predators of Insects. John Wiley and Sons, New York, pp. 39–57.

van Emden, H.F., 1990. Plant diversity and natural enemy efficiency in agroecosystems. In: Mackauer, M., Ehler, L.E., Roland, J. (Eds.), Critical Issues in Biological Control. Intercept, Andover, Hants, pp. 63–80.

van Emden, H.F., 2007. Host-plant resistance. In: van Emden, H.F., Harrington, R. (Eds.), Aphids as Crop Pests. CAB International, pp. 447–468.

Fabré, J.P., Rabasse, J.M., 1987. Introduction dans le sud-est de la France d´un parasite: *Pauesia cedrobii* (Hym.: Aphidiidae) du puceron: *Cedrobium laportei* (Hom.: Lachnidae) du cedre de l´Atlas: *Cedrus atlantica*. Entomophaga 32, 127–141.

Featherstone, P.E., Halfhill, J.E., 1966. A portable field cage for mass culturing aphid parasites. USDA, Research Surveys, ARS 8, 33–113.

Ferguson, K.I., Stiling, P., 1996. Non-additive effects of multiple natural enemies on aphid populations. Oecologia 108, 375–379.

Fiedler, A.K., Landis, D.A., Wratten, D., 2008. Maximizing ecosystems services from conservation biocontrol: the role of habitat management. Biological Control 45, 254–271.

Finke, D.L., Denno, R.F., 2004. Predator diversity dampens trophic cascades. Nature 429, 407–410.

Fiske, W.E., 1910. Superparasitism: an important factor in the natural control of insects. Journal of Economic Entomology 3, 88–97.

Flint, M.L., 1980. Climatic ecotypes in *Trioxys complanatus*, a parasite of the spotted alfalfa aphid. Environmental Entomology 9, 501–507.

Force, D.C., Messenger, P.S., 1964. Duration of development, generation time, and longevity of three hymenopterous parasites of *Therioaphis maculata*, reared at various constant temperatures. The Annals of Entomological Society of America 57, 405–413.

Frazer, B.D., van den Bosch, R., 1973. Biological control of the walnut aphid in California. The interrelationship of the aphid and its parasites. Environmental Entomology 2, 561–568.

Frère, I., Balthazar, C., Sabri, A., Hance, T., 2011. Improvement in the cold storage of *Aphidius ervi* (Hymenoptera: Aphidiinae). European Journal of Environmental Sciences 1 (1), 33–40.

Frewin, A.J., Xue, Y.E., Welsman, J.A., Broadbent, A.B., Schaafsma, A.W., Hallett, R.H., 2010. Development and parasitism by *Aphelinus certus* (Hymenoptera: Aphelinidae), a parasitoid of *Aphis glycines* (Hemiptera: Aphididae). Environmental Entomology 39, 1570–1578.

Fritz, R.S., 1992. Natural enemy impact varies with host plant genotype. In: Menken, S.B.J., Visser, J.H., Harrewijn, P. (Eds.), Proc. 8th Int. Sym. Insect-plant Relationships. Kluwer Acad. Publ, Dordrecht, pp. 367–369.

Fukui, M., Takada, H., 1988. Fecundity, oviposition period and longevity of *Diaeretiella rapae* M'Intosh and *Aphidius gifuensis* Ashmead (Hymenoptera: Aphidiidae), two parasitoids of *Myzus persicae*. Japanese Journal of Applied Entomology and Zoology 32, 331–333.

Gardiner, M.M., Landis, D.L., Gratton, C., DiFonzo, C.D., O'Neal, M., Chacon, J.M., Wayo, M.T., Schmidt, N.P., Mueller, E.E., Heimpel, G.E., 2009. Landscape diversity enhances biological control of an introduced crop pest in the North-Central USA. Ecological Applications 19, 143–154.

Ghadiri, R.S., Hatami, B., Asadi, G., 2003. Biology of *Leucopis glyphynivora* Tanas. (Dip.: Chamaemyiidae) and its efficiency in biological control of *Aphis fabae* Scop. JWSS – Isfahan University of Technology 6 (4), 195–207.

Ghosh, L.K., Rajendran, T.P., 1988. First record of an aphid sexual (Homoptera: Aphididae) from Rajasthan, India. Journal of Aphidology 2, 51–58.

Ghosh, A.K., 1974. A list of aphids (Homoptera: Aphididae) from India and adjacent countries. Journal of Bombay Natural History Society 71, 201–225.

Ghosh, A.K., 1980. The Fauna of India and Adjacent Countries. (Homoptera: Aphidiodea, Subfamily: Chaitophorinae), Part I. Zoological Survey of India, Calcutta, p. 124.

Ghosh, L.K., 2000. Contribution to the Indian Aphididae (Homoptera) (D.Sc. thesis). D.D.U.Gorakhpur University, Gorakhpur, India.

Gilkeson, L.A., 1990. Biological control of aphids in glasshouse sweet peppers and tomatoes. SROP/WPRS Bulletin 13 (5), 64–70.

Gill, S., 2012. Biological Control of Aphids Using Banker Plants. Central Maryland Research and Education Center, The University of Maryland, College of Agriculture and Natural Resources.

Giri, M.K., Pass, B.C., Yeargan, K.V., Parr, J.C., 1982. Behavior, net reproduction, longevity, and mummy-stage survival of *Aphidius matricariae* (Hym.: Aphididae). Entomophaga 27, 147–153.

Glinwood, R.T., Powell, W., Tripathi, C.P.M., 1998. Increased parasitisation of aphids on trap plants alongside vials releasing synthetic aphid sex pheromone and effective range of the pheromone. Biocontrol Science and Technology 8, 607–614.

Gonzalez, D., Gilstrap, F., Starý, P., Mckinnon, L., 1989. Foreign exploration for Russian wheat aphid natural enemies: a summary of co-operative efforts by State University Agricultural Experiment Stations, USDA/APHIS, USDA/ARS, CIBC, ICARDA, and ITGC. In: Proceedings of Russian Wheat Aphid Conference, Albuquerque, New Mexico, pp. 113–128.

Gonzalez, D., Gilstrap, F., Zhang, G., Zhang, J., Zareh, N., Whang, R., Dijkstra, E., McKinnon, L., Starý, P., Wooley, J., 1990. Foreign exploration of natural enemies of Russian wheat aphid in China, Iran, Turkey, and the Netherlands. In: Proceedings of Fourth Russian Wheat Aphid Workshop, Bozeman, Montana, pp. 154–165.

Gonzalez, D., Gilstrap, F., McKinnon, L., Zhang, J., Zareh, N., Zhang, G., Starý, P., Wooley, J., Wang, R., 1992. Foreign exploration for natural enemies of Russian wheat aphid in Iran, and in the Kunlun, Tian Shan, and Altai Mountain Valleys of the People's Republic of China. In: Proceedings of Russian Wheat Aphid Conference, Ft.Worth, Texas, pp. 197–208.

Greany, P.D., Vinson, S.B., Lewis, W.J., 1984. Insect parasitoids: finding new opportunities for biological control. BioScience 34, 690–696.

Grenier, S., De Clerq, P., 2003. Comparison of artificially v/s naturally reared natural enemies and their potentials in biological control. In: van Lenteren, J.C. (Ed.), Quality Control and Production of Biological Control Agent. CAB International, Wallingford, UK, pp. 115–131.

Gutierrez, A.P., van Den Bosch, R., 1970. Studies on host selection and host specificity of the aphid hyperparasite, *Charips victrix* (Hymenoptera: Cynipidae). 1. Review of hyperparasitism and the field ecology of *Charips victrix*. Annals Entomological Society of America 63, 1345–1354.

Hagen, K.S., van den Bosch, R., 1968. Impact of pathogens, parasite and predators of aphids. Annual Review of Entomology 13, 325–384.

Hagen, K.S., 1974. The significance of predaceous Coccinellidae in biological and integrated control of insects. Entomophaga 7, 25–44.

Hågvar, E.B., Hofsvang, T., 1990a. The aphid parasitoid *Ephedrus cerasicola*, a possible candidate for biological control in glasshouses. SROP/WPRS Bulletin XIII (5), 87–90.

Hågvar, E.B., Hofsvang, T., 1990b. Fecundity and intrinsic rate of increase of aphid parasitoid *Ephedrus cerasicola*. (Hymenoptera: Aphidiidae). Journal of Applied Entomology 109, 262–267.

Hågvar, E.B., Hofsvang, T., 1991. Aphid parasitoids (Hymenoptera: Aphidiidae) – biology, host selection and use in biological control. Biocontrol News and Information 12, 13–41.

Halfhill, J.E., Featherstone, P.E., 1967. Propagation of braconid parasites of the pea aphid. Journal of Economic Entomology 60, 1756.

Halfhill, J.E., Featherstone, P.E., 1973. Inundative release of *Aphidius smithi* against *Acyrthosiphon pisum*. Environmental Entomology 2, 469–472.

Halfhill, J.E., 1967. Mass propagation of pea aphids. Journal of Economic Entomology 60, 298–299.

Hamilton, M.A., 1935. Further experiments on the artificial feeding of *Myzus persicae* (Sulz.). Annals of Applied Biology 22, 243–258.

Hamilton, P.A., 1974. The biology of *Monoctonus pseudoplatani*, *Trioxys cirsii* and *Dyscritulus planiceps*, with notes on their effectiveness as parasites of sycamore aphid, *Drepanosiphum platanoides*. Annales de la Société Entomologique de France 10, 821–840.

Hance, T., Debatty-Mestdagh, M., Cambier, V., Boegen, C., Muratori, F., Lebbe, O., Dos Santos Goncalves, A.M., 2001. Hydrogel beads or capsules as artificial media for insect oviposition and rearing of endoparasitoid. Biological Control 58 (3), 167–173.

Hardie, J., Hick, A.J., Holler, C., Mann, J., Merritt, L., Nottingham, S.F., Powell, W., Wadhams, L.J., Witthinrich, J., Wright, A.F., 1994. The responses of *Praon* spp. parasitoids to aphid sex pheromone components in the field. Entomologia Experimentalis et Applicata 71, 95–99.

Hassell, M.P., Waage, J.K., 1984. Host parasitoid population interactions. Annual Review of Entomology 29, 89–114.

Hayakawa, D.L., Grafius, E., Stehr, F.W., 1990. Effects of temperature on longevity, reproduction, and development of the asparagus aphid and the parasitoid *Diaeretiella rapae*. Environmental Entomology 19, 890–897.

Hayashi, M., Nomura, M., 2011. Larvae of the green lacewing *Mallada desjardins* (Neuroptera: Chrysopidae) protect themselves against aphid-tending ants by carrying dead aphids on their backs. Applied Entomology and Zoology 46 (3), 407–413.

Heimpel, G.E., Ragsdale, D.W., Venette, R., Hopper, K.H., O'Neil, R.J., Rutledge, C.E., Wu, Z., 2004. Prospects for importation biological control of the soybean aphid: anticipating potential costs and benefits. Annals of Entomological Society 97, 249–258.

Heimpel, G.E., Frelich, L.E., Landis, D.A., Hopper, K.R., Hoelmer, K.A., Sezen, Z., Asplen, M.K., Wu, K.M., 2010. European buckthorn and Asian soybean aphid as components of an extensive invasional meltdown in North America. Biological Invasions 12, 2913–2931.

Heinz, K.M., 1997. Biological aphid control. Grower Notes 2, 1–3.

Heinz, K.M., 1998. Dispersal and dispersion of aphids and selected natural enemies in spatially subdivided greenhouse environments. Environmental Entomology 27, 1029–1038.

Heraty, J.M., Woolley, J.B., Hopper, K.R., Hawks, D.L., Kim, J.W., Buffington, M., 2007. Molecular phylogenetics and reproductive incompatibility in a complex of cryptic species of aphid parasitoids. Molecular Phylogenetics and Evolution 45, 480–493.

Heuskin, S., Lorge, S., Godin, B., Leroy, P.D., Frère, I., Verheggen, F., 2012. Optimisation of a semiochemical slow-release alginate formulation attractive towards *Aphidius ervi* Haliday parasitoids. Pest Management Science. 68 (1), 127–136. http://www.researchgate.net/journal/1526-4998_Pest_Management_Science.

Hodek, I., Honek, A., 1996. Ecology of Coccinellidae. Kluwer, Dordrecht, p. 480.

Hodek, I., 1973. Biology of Coccinellidae. Academia, Prague and Dr. W. Junk, The Hague, p. 260.

Hofsvang, T., Hågvar, E.B., 1977. Cold storage tolerance and supercooling points of mummies of *Ephedrus cerasicola* Starý and *Aphidius colemani* Viereck (Hymenoptera: Aphidiidae). Norwegian Journal of Entomology 24, 1–6.

Hofsvang, T., Hågvar, E.B., 1979. Different introduction methods of *Ephedrus cerasicola* Starý to control *Myzus persicae* (Sulzer) in small paprika glass houses. Journal of Applied Entomology 88, 16–23.

Hofsvang, T., Hågvar, E.B., 1980. Use of mummies of *Ephedrus cerasicola* Starý to control *Myzus persicae* (Sulzer) in small glass houses. Journal of Applied Entomology 90, 220–226.

Hofsvang, T., Godonou, I., Tepa-Yotto, G.T., Sæthre, M.G., 2014. The native hyperparasitoid complex of the invasive aphid parasitoid *Lysiphlebus testaceipes* (Hymenoptera: Braconidae: Aphidiinae) in Benin, West Africa. International Journal of Tropical Insect Science 34 (1), 9–13.

Hogenhout, S.A., Ammar, E.D., Whitfield, A.E., Redinbaugh, M.G., 2008. Insect vector interactions with persistently transmitted viruses. Annual Review of Entomology 46, 327–359.

Hogervorst, P.A., Wäckers, F.L., Romeis, J., 2007. Effects of honeydew sugar composition on the longevity of Aphidius ervi. Entomologia Experimentalis et Applicata 122, 223–232.

Hokkanen, H.M.T., 1985. Successes in Biological control. CRC in Plant Science 3, 35–72.

Holt, R.D., Hochberg, M.E., 1998. The co-existence of competing parasites. Part II. Hyperparasitism and food chain dynamics. Journal of Theoretical Biology 193, 485–495.

Hopper, K.R., Diers, B.W., 2014. Parasitism of soybean aphid by *Aphelinus* species on soybean susceptible versus resistant to the aphid. Biological Control 76, 101–106.

Hopper, K.R., Coutinot, D., Chen, K., Mercadier, G., Halbert, S.E., Kazmer, D.J., Miller, R.H., Pike, K.S., Tanigoshi, L.K., 1998. Exploration for natural enemies to control *Diuraphis noxia* in the United States. In: Quissenbery, S.S., Peairs, F.B. (Eds.), A Response Model for an Introduced Pest – the Russian Wheat Aphid. Thomas Say Publications in Entomology, Proceedings of Entomological Society of America, Lanham, MD, pp. 166–182.

House, R.L., 1977. In: Ridgway, R.L., Vinson, S.B. (Eds.), Biological Control by Augmentation of Natural Enemies. Plenum Press, New York, pp. 183–218.

Howard, L.O., 1929. *Aphelinus mali* and its travel. Annals of Entomological Society of America 22, 341–368.

Hoy, M.A., Cave, F.E., 1991. Genetic improvement of a parasitoid: response of *Trioxys pallidus* to laboratory selection with azinphosmethyl. Biocontrol Science and Technology 1, 31–41.

Hoy, M.A., Nuygen, R., 2000. Classical biological control of brown citrus aphid, release of *Lipolexis scutellaris*. Citrus Industry 81, 24–26.

Hoy, M.A., Cave, F.E., Beede, R.H., Grant, J., Krueger, W.H., Olson, W.H., Spollen, K.M., Barnett, W.W., Hendricks, L.C., 1989. Guthion-resistant walnut aphid parasite. California Agriculture 43, 21–23.

Hoy, M.A., Cave, F.E., Beede, R.H., Grant, J., Krueger, W.H., Olson, W.H., Spollen, K.M., Barnett, W.W., Hendricks, L.C., 1990. Release, dispersal and recovery of a laboratory selected strain of the walnut aphid parasite *Trioxys pallidus* (Hymenoptera: Aphidiidae). Journal of Economic Entomology 83, 89–96.

Hsieh, c-Y., Allen, W.W., 1986. Effects of insecticide on emergence, survival, longevity, and fecundity of the parasitoid *Diaeretiella rapae* (Hymenoptera: Aphidiidae) from mummified *Myzus persicae* (Homoptera: Aphidiidae). Journal of Economic Entomology 79, 1599–1602.

Huffaker, C.B., Messenger, P.S. (Eds.), 1976. Theory and Practice of Biological Control. Academic Press, Inc, New York, London, p. 788.

Huffaker, C.B. (Ed.), 1971. Biological Control. Plennum Press, New York.

Hughes, R.D., Woolcock, L.T., Roberts, J.A., Hughes, M.A., 1987. Biological control of the spotted alfalfa aphid, *Therioaphis trifolii*, *T. maculata* on lucern crops in Australia, by the introduced parasitic hymenopteran *Trioxys complanatus*. Applied Ecology 24, 515–537.

Hughes, R.D., Woolcock, L.T., Hughes, M.A., 1992. Laboratory evaluation of parasitic hymenoptera used in attempts to biologically control aphid pests of crops in Australia. Entomologia Experimentalis et Applicata 63, 177–185.

Hughes, R.D., Hughes, M.A., Aeschlimann, J.P., Woolcock, L.T., Carvar, M., 1994. An attempt to anticipate biological control of *Diuraphis noxia* (Homoptera: Aphididae). Entomophaga 39, 211–223.

Hughes, R.D., 1989. Biological control in the open field. In: Minks, A.K., Herrewijn, P. (Eds.), World Crop Pests. Aphids. Their biology, natural enemies and control, vol. C. Elsevier, Amsterdam, pp. 167–198.

Hussey, N.W., Scopes, N., 1985. Biological Pest Control: The Glasshouse Experience. Blanferd Press, Pooe, Dorset.

Ismail, M., Vernon, P., Hance, T., Van Baaren, J., 2010. Physiological cost of cold exposure on the parasitoid *Aphidius ervi*, without selection pressure and under constant or fluctuating temperatures. Bio Control 55, 729–740.

Ismail, M., Baaren, J.V., Hance, T., Pierre, J.S., Vernon, P., 2013. Stress intensity and fitness in the parasitoid *Aphidius ervi* (Hymenoptera: Braconidae): temperature below the development threshold combined with a fluctuating thermal regime is a must. Ecological Entomology 38 (4), 355–363.

Iversen, T., Harding, S., 2007. Life table parameters affecting the population development of the woolly beech aphid, *Phyllaphis fagi*. Entomologia Experimentalis et Applicata 123, 109–117.

Jackson, H.B., Rogers, C.E., Eikenbary, R.D., Starks, K.J., McNew, R.W., 1974. Biology of *Ephedrus plagiator* on different aphid hosts and at various temperatures. Environmental Entomology 3, 618–620.

Jörg Kitt, 1996. The Introduction of *Aphidius Rosae* (Braconidae, Aphidiinae) as a Biological Control Agent of the Rose Aphid *Macrosiphum rosae* (Hemiptera, Aphididae) into Australia (Ph.D. thesis). The University of Adelaide, Australia.

Joshi, S., Ballal, C.R., 2013. Syrphid predators for biological control of aphids. Journal of Biological Control 27 (3), 151–170.

Kambhampati, S., Völkl, W., Mackauer, M., 2000. Phylogenetic relationships among genera of Aphidiinae (Hymenoptera: Braconidae) based on DNA sequence of the mitochondrial 16S rRNA gene. Systematic Entomology 25, 437–445.

Kaneko, S., 2004. Positive impacts of aphid-attending ants on the number of emerging adults of aphid primary parasitoids and hyperparasitoids through exclusion of intraguild predators. Japanese Journal of Entomology 7, 173–183.

Kfir, R., Kirsten, F., 1991. Seasonal abundance of *Cinara cronartii* (Homoptera: Aphididae) and the effect of an introduced parasite, *Pauesia* sp. (Hymenoptera: Aphidiidae). Journal of Economic Entomology 84, 76–82.

Kfir, R., Podolor, H., Rosen, D., 1976. The area of discovery and searching strategy of a primary parasite and two hyperparasites. Ecological Entomology 1, 287–295.

Kfir, R., Kirsten, F., van Rensburg, N.J., 1985. *Pauesia* sp. (Hymenoptera: Aphidiidae): a parasite introduced into South Africa for biological control of the black pine aphid, *Cinara cronartii* (Homoptera: Aphididae). Environmental Entomology 14, 597–601.

Kfir, R., van Rensburg, N.J., Kirsten, F., 2003. Biological control of the black pine aphid *Cinara cronartii* (Homoptera: Aphididae) in South Africa. African Entomology 11, 117–121.

Kindler, S.D., Springer, T.L., 1989. Alternate hosts of Russian wheat aphid (Homoptera: Aphididae). Journal of Economic Entomology 82, 1358–1362.

Kogan, M. (Ed.), 1986. Ecological Theory and Integrated Pest Management Practice. A Wiley Interscience Publ, p. 362.

Kogan, M., 1998. Integrated pest management: historical perspective and contemporary developments. Annual Review of Entomology 43, 243–270.

Landis, D.A., Wratten, S.D., Gurr, G.M., 2000. Habitat management to conserve natural enemies of arthropod pests in Agriculture. Annual Review of Entomology 45, 175–201.

Langer, A., Hance, T., 2000. Overwintering strategies and cold hardiness of two aphid parasitoid species (Hymenoptera: Braconidae: Aphidiinae). Journal of Insect Physiology 46, 671–676.

Langhof, M., Meyhôfer, R., Poehling, H.M., Gathmann, A., 2005. Measuring the field dispersal of *Aphidius colemani* (Hymenoptera: Braconidae). Agriculture Ecosystem and Environment 107, 137–143.

Le Lann, C., Outreman, Y., Van Alphen, J.J.M., Van Baaren, J., 2011. First in, last out: asymmetric competition influences patch exploitation of a parasitoid. Behavioural Ecology 22, 101–107.

Lebusa, M.M., 2004. Suitability of Greenbugs (*Schizaphis graminum*) Parasitized by *Lysiphlebus Testaceipes* as a Food Source for Predatory Coccinellidae: *Coccinella septempunctata* and *Hippodamia convergens* (Masters thesis). Oklahoma State University.

Lee, J.C., Heimpel, G.E., Leibee, G.L., 2004. Comparing floral nectar and aphid honeydew diets on the longevity and nutrient levels of a parasitoid wasp. Entomologia Experimentalis et Applicata 111, 189–199.

Levie, A., Marques, N., Dogot, P., Hance, T., 1999. Control capacities of *Aphidius rhopalosiphi* Hymenoptera: Aphidiinae) on cereal aphids in field cages: preliminary results. Mededelingen: Faculteit landbouwkundige en toegepaste biologische wetenschappen, Universiteit Gent 64, 17–24.

Levie, A., Legrand, M.A., Dogot, P., Pels, C., Baret, P.V., Hance, T., 2005. Mass releases of *Aphidius rhopalosiphi* (Hymenoptera: Aphidiinae), and strip management to control of wheat aphids. Agriculture Ecosystem and Environment 105, 17–21.

Lewis, W.J., Nordlund, D.A., Gueldner, R.C., 1982. Semiochemicals influencing behaviour of entomophagous – role and strategies for their employment in pest control. Le Mediateurs Chimiques, INRA 225–242.

Lins Jr., J.C., Bueno, B.H.P., Sidney, L.A., Silva, D.B., Sampaio, M.V., Pereira, J.M., Nomelini, Q.S.S., van Lenteren, J.C., 2013. Cold storage affects mortality, body mass, lifespan, reproduction and flight capacity of *Praon volucre* (Hymenoptera: Braconidae). European Journal of Entomology 110 (2), 263–270.

Liu, S.-S., 1985. Development, adult size and fecundity of *Aphidius sonchi* reared in two instars of its aphid host, *Hyperomyzus lactucae*. Entomologia Experimentalis et Applicata 37 (1), 41–48.

Loginova, E., Atanassov, N., Georgiev, G., 1987. Biological control of pests and diseases in glasshouses in Bulgaria-today and in the future. SROP/WPRS Bulletin X/2, 101–105.

Lucas, E., 2005. Intraguild predation among aphidophagous predators. European Journal of Entomology 102 (3), 351–364.

van Lenteren, J.C., Woets, J., 1988. Biological and integrated pest control in greenhouses. Annual Review of Entomology 33, 239–269.

van Lenteren, J.C., Roskam, M.M., Timmer, R., 1997a. Commercial mass production and pricing of organisms for biological control of pests in Europe. Biological Control 10, 143–149.

van Lenteren, J.C., Drost, Y.C., Van Roermund, H.J.W., Posthuma-Doodeman, C.J.A.M., 1997b. Aphelinid parasitoids as sustainable biological control agents in greenhouses. Journal of Applied Entomology 121, 476–485.

van Lenteren, J.C., Bale, J.S., Bigler, F., Hokkanen, H.M.T., Loomans, A.J.M., 2006. Assessing risks of releasing exotic biological control agents of arthropod pests. Annual Review of Entomology 51, 609–634.

van Lenteren, J.C., 1986. Evaluation, mass production quality control and release of entomophagous insects. In: Frang, J.M. (Ed.), Biological Plant and Health Protection. Fisher Verlag, Stutt, pp. 31–56.

van Lenteren, J.C., 2000. Measures of success in biological control of arthropods by augmentation of natural enemies. In: Gurr, G., Wratten, S. (Eds.), Measures of Success in Biological Control. Kluwer Academic, Dordrecht, The Netherlands, pp. 77–103.

van Lenteren, J.C., 2003a. Environmental risk assessment of exotic natural enemies used in inundative biological control. BioControl 48, 3–38.

van Lenteren, J.C., 2003b. Need for quality control of mass produced biological control agents. In: van Lenteren, J.C. (Ed.), Quality Control and Production of Biological Control Agent. CAB International, Wallingford, UK, pp. 1–18.

van Lenteren, J.C., January 2008. IOBC Internet Book of Biological Control, v. 5. http://www.iobcglobal.org/publications_iobc_internet_book_of_biological_control.html.

Mackauer, M., Campbell, A., 1972. The establishment of three exotic aphid parasites (Hymenoptera: Aphididae) in British Columbia. Journal of Entomological Society of British Columbia 69, 54–88.

Mackauer, M., Kambhampati, S., 1984. Reproduction and longevity of the cabbage aphid, *Brevicoryne brassicae* (Homoptera: Aphididae) parasitised by *Diaeretiella rapae* (Hymenoptera: Aphidiidae). The Canadian Entomologist 116, 1605–1610.

Mackauer, M., Kambhampati, S., 1986. Structural changes in the parasite guild attacking the pea aphid in North America. In: Hodek, I. (Ed.), Ecology of Aphidophaga. Acadmica, Prague, pp. 347–356.

Mackauer, M., Kambhampati, S., 1988. Sampling and rearing of aphid parasites. In: Minks, A.K., Harrewijn, P. (Eds.), Aphids, Their Biology, Natural Enemies and Control, vol. B. Elsevier Science Publ., Amsterdam, Netherlands, pp. 205–217.

Mackauer, M., Starý, P., 1967. World Aphidiidae (Hymenoptera: Ichneumonoidea). Le Francois, Paris.

Mackauer, M., Völkl, W., 1993. Regulation of aphid populations by aphidiid wasps: does parasitoid foraging behaviour or hyperparasitism limit impact? Oecologia 94, 339–350.

Mackauer, M., 1971. *Acyrthosiphon pisum* (Harris), pea aphid (Homoptera: Aphididae). In: Biological Control Programmers Against Insects and Weeds in Canada (1959–1968), 4. Technical Communication, pp. 3–10.

Mackauer, M., 1983. Quantitative assessment of *Aphidius smithi* (Hymenoptera: Aphidiidae): fecundity, intrinsic rate of increase, and functional response. The Canadian Entomologist 115 (4), 399–415.

Majerus, M.E.N., 1994. Ladybirds. Harper Collins, London, p. 367.

Marcovitch, S., 1935. Experimental evidence on the value of strip farming as a method for the natural control of injurious insects with special reference to plant lice. Journal of Economic Entomology 28, 62–70.

Marsh, P.M., 1991. A new species of *Pauesia* (Hymenoptera: Braconidae, Aphidiinae) from Georgia and introduced into South Africa against the black pine aphid (Homoptera: Aphididae). Journal of Entomological Science 26, 81–84.

May, R.M., Hassell, M.P., 1981. The dynamics of multi-parasitoid-host interactions. American Nature 117, 234–261.

Memmott, J., Martinez, N.D., Cohen, J.E., 2000. Predators, parasitoids and pathogen, species richness, trophic generality and body size in a natural food web. Journal of Animal Ecology 69, 1–15.

Messing, R.H., Aliniazi, M.T., 1989. Introduction and establishment of *Trioxys pallidus* (Hymenoptera: Aphidiidae) in Oregon, USA for control of filbert aphid *Myzocallis coryli* (Homoptera: Aphididae). Entomophaga 34, 153–163.

Metcalf, R.L., Luckmann, W.H. (Eds.), 1994. Introduction to Insect Pest Management, third ed. John Wiley and Sons, New York, p. 672.

Meyerdirk, D.E., 1992. International opportunities for classical biological control. In: Kauffman, W.C., Nichols, J.E. (Eds.), Selection Criteria and Ecological Consequences of Importing Natural Enemies. Proceedings of Thomas Say Publications, Entomological Society of America, pp. 7–14.

Meyling, N.V., Enkegaard, A., Brødsgaard, H., 2004. Intraguild predation by *Anthocoris nemorum* (Heteroptera: Anthocoridae) on the aphid parasitoid *Aphidius colemani* (Hymenoptera: Braconidae). Biocontrol Science and Technology 14 (6), 627–630.

Michaud, J.P., 2012. Coccinellids in biological control. In: Hodek, I., van Emden, H.F., Honek, A. (Eds.), Ecology and Behaviour of the Ladybird Beetles (Coccinellidae). Blackwell Publishing Ltd, pp. 448–519.

Michel, C., 2007. Etude générale de l'impact du froid sur la survie et la morphologie du système reproducteur d'*Aphidius ervi* (Hymenoptera: Aphidiinae). Mémoire de fin d'étude, Université Catholique de Louvain, Louvain-la-Neuve.

Michel, A.P., Mittapalli, O., Rouf Mian, M.A., 2011. Evolution of soybean aphid biotypes: understanding and managing virulence to host-plant resistance. In: Sudaric, A. (Ed.), Soybean-Molecular Aspects of Breeding, pp. 355–372.

Michelena, J.M., Assael, F., Mendel, Z., 2005. Description of *Pauesia (Pauesia) anatolica* (Hymenoptera: Braconidae, Aphidiinae) sp.nov., a parasitoid of the cedar aphid *Cinara cedri*). Phytoparasitica 33 (5), 499–505.

Minks, A.K., Harrewijn, P. (Eds.), 1987. Aphids: Their Biology, Natural Enemies and Control, vol. A. Elsevier Science Publisher B.V.Amsterdam, p. 450.

Mishra, S., Singh, R., 1990. Life-time performance characteristics of an aphidiid parasitoid, *Lysiphlebus delhiensis* (Hymenoptera: Aphidiidae). Entomologia Generalis 15, 173–179.

Mishra, S., Singh, R., 1991. Effect of host densities on the demographic statistics of an aphid parasitoid *Lysiphlebus delhiensis* (Subba Rao and Sharma). Biological Agriculture and Horticulture 7, 281–302.

Mishra, S., Singh, R., 1993a. Mating behaviour of *Lysiphlebus delhiensis* (Subba Rao and Sharma), an aphidiid parasitoid of corn aphid *Rhopalosiphum maidis* (Fitch). Zeitschrift für Angewandte Zoologie 79, 319–328.

Mishra, S., Singh, R., 1993b. Factors affecting supernumerary egg deposition by the parasitoid *Lysiphlebus delhiensis* (Subba Rao and Sharma) (Hymenoptera: Aphidiidae) in its host *Rhopalosiphum maidis* (Fitch). Biological Agriculture and Horticulture 10, 39–45.

Mittler, D.E., Dadd, R.H., 1962. Artificial feeding and rearing of the aphid *Myzus persicae* (Sulzer), on a completely defined synthetic diet. Nature 195, 404.

Moreno-Mari, J., Dominguez-Romero, M., Oltra-Moscardo, M.T., Jimenez-Peydro, R., 1999. Phytophagous and natural enemies associated with aromatic plants in glasshouse. Annales de la Société Entomologique de France 35, 521–524.

Muratori, F.B., Gagne, R.J., Messing, R.H., 2009. Ecological traits of a new aphid parasitoid, *Endaphis fugitiva* (Diptera: Cecidomyiidae), and its potential for biological control of the banana aphid, *Pentalonia nigronervosa* (Hemiptera: Aphididae). Biological Control 50, 185–193.

Napompeth, B., 1987. Biological control and integrated pest control in the tropics – an overview. Accademia Nazionale delle Scienze 50, 415–428.

Nault, L.R., 1997. Arthropod transmission of plant viruses: a new synthesis. Annals Entomological Society of America 90 (5), 521–541.

Nordlund, D.A., Cohen, A.C., Smith, R.A., 2001. Mass-rearing, release techniques and augmentation. In: McEwen, P.K., New, T.R., Wliittington, A.E. (Eds.), Lacewings in the Crop Environment. Cambridge University Press, Cambridge, pp. 303–319.

Obrycki, J.J.1, Kring, T.J., 1998. Predaceous Coccinellidae in biological control. Annual Review of Entomology 43, 295–321.

Oliveira, S., 2006. Fatores biológicos e comportamentais do parasitóide *Xenostigmus bifasciatus* Ashmead 1891 (Hymenoptera, Braconidae) visando a otimização de criação massal em laboratório e índice de parasitismo em casa-de-vegetação (Dissertation). UFPR, Curitiba, p. 78.

Pandey, S., Upadhyay, B.S., Srivastava, P.N., Singh, R., 1996. Progeny sex ratio of *Lysiphlebia mirzai* Shuja-Uddin (Hym.: Braconidae) inseminated by a multiple mated male. Journal of Aphidology 10, 43–48.

Pappas, M.L., Broufas, G.D., Koveos, D.S., 2011. Chrysopid predators and their role in biological control. Journal of Entomology 8, 301–326.

Parr, W.J., Scopes, N.E.A., 1970. Problems associated with biological control of glasshouse pests. N.A.A.S.Quarterly Review 89, 113–121.

Parra, J.R.P., 2010. Mass rearing of egg parasitoid for biological control programs. In: Consoli, F.L., Parra, J.R.P., Zucchi, R.A. (Eds.), Progress in Biological Control – Egg Parasitoid in Agroecosystems with Emphasis on Trichogramma. Springer, Dordrecht, the Netherlands, pp. 267–292.

Parrella, M.P., 2008. Biological control in protected culture: will it continue to expand? Phytoparasitica 36, 3–6.

Penteado, S.R.C., Trentini, R.F., Iede, E.T., Reis Filho, W., 2000. Pulgões do Pinus: nova praga florestal. Anais do 1° Simpósio do Cone Sul sobre manejo de pragas e doenças do Pinus. Série Técnica IPEF 13, 97–102.

Persad, A.B., Hoy, M.A., Guyen, N., 2007. Establishment of *Lipolexis oregmae* (Hymenoptera: Aphidiidae) in a classical biological control program directed against the brown citrus aphid (Homoptera: Aphididae) in Florida. Florida Entomologist 90 (1), 204–213.

Phoofolo, M.W., Giles, K.L., Elliott, N.C., 2007. Quantitative evaluation of suitability of the greenbug, *Schizaphis graminum*, and the bird cherry-oat aphid, *Rhopalosiphum padi*, as prey for *Hippodamia convergens* (Coleoptera: Coccinellidae). Biological Control 41, 25–32.

Pierre, J.S., van Baaren, J., Boivin, G., 2003. Patch leaving decisions rules in parasitoids: do they use sequential decisional sampling? Behaviour Ecology and Sociobiology 54, 147–155.

Polgar, L., 1986. Effect of cold storage on the emergence, sex ratio and fecundity of *Aphidius matricariae*. In: Hodek, I. (Ed.), Ecology of Aphidophaga. Prague Acad, pp. 255–260.

Popov, N.A., Belousov, Y.V., Zabudskaya, I.A., Khudyakova, O.A., Shevtscenko, V.B., Shijko, E.S., 1987. Biological control of glasshouse pests in the south of the USSR. SROP/WPRS Bulletin X/2, 155–157.

Poppy, G.M., Powell, W., 2004. Genetic manipulation of natural enemies: can we improve biological control by manipulating the parasitoid and/or the plant? In: Ehler, L.E., Sforza, R., Mateille, T. (Eds.), Genetics, Evolution and Biological Control. CAB International, pp. 219–233.

Powell, W., Pell, J.K., 2007. Biological control. In: van Emden, H.F., Harrington, R. (Eds.), Aphids as Crop Pests. CAB International, Cambridge, Massachusetts, USA, pp. 469–513.

Powell, W., Wright, A.F., 1988. The abilities of the aphid parasitoid *Aphidius ervi* Haliday and *A. rhopalosiphi* De Stefani Perez (Hymenoptera: Braconidae) to transfer between different known host species and the implication for the use of alternative host in pest control strategies. Bulletin of Entomological Research 78, 683–693.

Powell, W., Zhi-Li, Z., 1983. The reactions of two cereal aphid parasitoids, *Aphidius uzbekistanicus* and *A.ervi* to host aphids and their food plants. Physiological Entomology 8, 439–443.

Powell, W., Pennacchio, F., Poppy, G.M., Tremblay, E., 1998. Strategies involved in the location of hosts by the parasitoid *Aphidius ervi* Haliday (Hymenoptera: Braconidae: Aphidiinae). Biological Control 11, 104–112.

Powell, W., 1986. Enhancing parasitoid activity in crops. In: Waage, J., Greathead, D. (Eds.), Insect Parasitoids. Academic press, London, pp. 319–340.

Prokrym, D.R., Pike, K.S., Nelson, D.J., 1998. Biological control of *Diuraphis noxia* (Homoptera: Aphididae): Implementation and evaluation of natural enemies. In: Quisenberry, S.S., Peairs, F. (Eds.), Response Model for an Introduced Pest-the Russian Wheat Aphid. Thomas Say Publications, Entomological Society of America, Lanham, MD, pp. 183–208.

Quicke, D.L.J., 1997. Parasitic Wasps. Chapman & Hall, New York, NY, p. 470.

Rabasse, J.M., Brunei, E., 1977. *Cavariella aegopodii* Scop. (Homoptera: Aphididae) en culture de carottes dans l'Ouest de la France. II. Régulation naturelle par aphidiides (Hymenoptera) et Entomophthorales. Annales de Zoologie Ecologie Animale 9, 481–496.

Rabasse, J.M., Ibrahim, A.M.A., 1987. Conservation of *Aphidius uzbekistanicus* Luz. (Hym., Aphidiidae), parasite on *Sitobion avenae* F. (Hom., Aphididae). SROP/WPRS Bulletin X/1, 54–56.

Rabasse, J.M., Lafont, J.P., Delpuech, I., Silvie, P., 1983. Progress in aphid control in protected crops. SROP/WPRS Bulletin VI/3, 151–162.

Ragsdale, D.W., Landis, D.A., Brodeur, J., Heimpel, G.E., Desneux, N., 2011. Ecology and management of soybean aphid in North America. Annu Rev Entomol 56, 375–399.

Rahman, K.A., Khan, A.W., 1941. Observations on *Aphelinus mali* Hald. in the Punjab. Indian Journal of Agricultural Science 11, 265–278.

Ramakers, P.L.J., 1989. Biological control in green house. In: Mink, A.K., Harriwijn, P. (Eds.), World Crop Pest: Aphids Their Biology, Natural Enemies and Control, vol.C. Elsevier, Amsterdam, pp. 199–208.

Reed, D.K., Webster, J.A., Jones, B.G., Burd, J.D., 1991. Tritrophic relationships of Russian wheat aphid (Homoptera: Aphididae), a hymenopterous parasitoid *(Diaeretiella rapae* M'Intosh), and resistant and susceptible small grains. Biological Control 1, 35–41.

Reed, D.K., Kinder, S.D., Springer, T.L., 1992. Interactions of Russian Wheat aphid, a hymenopterous parasitoid and resistant and susceptible slender wheat grasses. Entomologia Experimentalis et Applicata 64, 239–240.

Rehman, A., Powell, W., 2010. Host selection behaviour of aphid parasitoids (Aphidiidae: Hymenoptera). Journal of Plant Breeding and Crop Science 2 (10), 299–311.

Reis-Filho, W., Penteado, S.R.C., Iede, E.T., 2004. Controle biológico de pulgão-gigante-do-pinus, *Cinara atlantica* (Hemiptera: Aphididae), pelo parasitóide, *Xenostigmus bifasciatus* (Hymenoptera: Braconidae). Comunicado Técnico 122, Colombo, Embrapa Floresta, p. 3.

Rodriguez-Saona, C., Blaauw, B.R., Isaacs, R., 2012. Manipulation of natural enemies in agroecosystems: habitat and semiochemicals for sustainable insect pest control. In: Larramendy, M., Soloneski, S. (Eds.), Integrated Pest Management and Pest Control – Current and Future Tactics. Intech, Rijeka, Croatia, pp. 89–126.

Roschewitz, I., Hucker, M., Tscharntke, T., Thies, C., 2005. The influence of landscape context and farming practices on parasitism of cereal aphids. Agriculture Ecosystems and Environment 108, 218–227.

Rosen, D., 1995. Readings in Biological Pest Control. Intercept Ltd, Andover, p. 530.

Rosenheim, J.A., Wilhoit, L.R., Armer, C.A., 1993. Influence of intraguild predation among generalist insect predators on the suppression of an herbivore population. Oecologia 96, 439–449.

Rotundo, G., Cavalloro, R., Tremblay, E., 1988. In vitro rearing of *Lysiphlebus fabarum* (Hym.: Braconidae). Entomophaga 33, 261–267.

Royer, L., Boivin, G., 1999. Infochemicals mediating the foraging behaviour of *Aleochara bilineata* Gyllenhal adults: sources of attractants. Entomologia Experimentalis et Applicata 90, 199–205.

Sachan, G.C., 1997. Cultural control of aphids: a review and bibliography. Journal of Aphidology 11, 37–47.

Sailer, R.I., 1976. Future role of biological control in management of pests. Proceedings Tall Timber Conference on Ecology of Animals: Control by Habitat Management 6, 195.

Sailer, R.I., 1983. History of insect introductions. In: Wilson, C.L., Graham, C.L. (Eds.), Exotic Plant Pest and North American Agriculture. Academic Press, New York, pp. 15–39.

Salt, G., 1934. Experimental studies in insect parasitism II. Superparasitism. Proceeding of Royal Entomological Society, Series B 114, 455–476.

Salto, D.H., Eikenbary, R.D., Starks, K.J., 1983. Compatibility of *Lysiphlebus testaceipes* (Hym.: Braconidae) with green bug (Hem.: Aphididae) biotypes "C" and "E" reared on susceptible and resistant oat varieties. Environmental Entomology 12, 603–604.

Schlinger, E.I., Hall, J.C., 1960. The biology, behaviour, and morphology of *Praon palitans* Muesebeck, an internal parasite of the spotted alfalfa aphid, *Therioaphis maculata* (Buckton) (Hymenoptera: Braconidae: Aphidiinae). Annals Entomological Society of America 53, 144–160.

Schlinger, E.I., Hall, J.C., 1961. The biology, behaviour, and morphology of *Trioxys (Trioxys) utilis*, an internal parasite of the spotted alfalfa aphid, *Therioaphis maculata* (Hymenoptera: Braconidae: Aphidiidae). Annals Entomological Society of America 54, 34–45.

Schooler, S.S., De Barroa, P., Ivesb, A.R., 2011. The potential for hyperparasitism to compromise biological control: why don't hyperparasitoids drive their primary parasitoid hosts extinct? Biological Control 58 (3), 167–173.

Scopes, N.E.A., Biggerstaff, S.M., Goodaal, D.E., 1973. Cool storage of some parasites used for pest control in glasshouses. Plant Pathology 22, 189–193.

Shalaby, F.F., Rabasse, J.M., 1979. On the biology of *Aphidius matricariae* Hal. (Hymenoptera: Aphidiidae), a parasite of *Myzus persicae* (Sulz.) (Homoptera: Aphididae). Annals of Agricultural Science, Moshtohor 11, 75–96.

Shijko, E.S., 1989. Rearing and application of the peach aphid parasite, *Aphidius matricariae* (Hymenoptera: Aphidiidae). Acta Entomologica Fennica 53, 53–56.

Shirota, Y., Carters, N., Rabbinge, R., Ankersmit, G.W., 1983. Biology of *Aphidius rhopalosiphi*, a parasitoid of cereal aphids. Entomologia Experimentalis et Applicata 34 (1), 27–34.

Sigsgaard, L., Hansen, L.M., 2000. Perspective for biological control of cereal aphids with parasitoids. Jile, Denmark; Danmarks J. Forsk., DJF Rapport, Markburg. 24, 125–130.

Sigsgaard, L., 2000. The temperature-dependent duration of development and parasitism of three cereal aphid parasitoids, *Aphidius ervi A. rhopalosiphi*, and *Praon volucre*. Entomologia Experimentalis et Applicata 95, 173–184.

Simpson, B.A., Shands, W.A., Simpson, G.W., 1975. Mass rearing of the parasites *Praon* sp. and *Diaretiella rapae*. Annals Entomological Society of America 68, 257–260.

Singh, R., Agarwala, B.K., 1992. Biology, ecology, and control efficiency of the aphid parasitoid *Trioxys indicus* Subba Rao and Sharma (Hymenoptera: Aphidiidae): a review and bibliography. Biological Agriculture & Horticulture 8, 271–298.

Singh, R., Ghosh, S., 2002. The glimpses of Indian aphids (Insecta: Hemiptera, Aphididae). Proceedings of the National Academy of Sciences, Allahabad 72b (3&4), 215–234.

Singh, R., Rao, S.N., 1995. Biological control of *Aphis gossypii* Glover (Homoptera: Aphididae) on cucurbits by *Trioxys indicus* Subba rao and Sharma (Hymenoptera: Aphidiidae). Biological Agriculture and Horticulture 12, 227–236.

Singh, R., Singh, G., 2015. Systematics, distribution and host range of *Diaeretiella rapae* (McIntosh) (Hymenoptera: Braconidae, Aphidiinae). International Journal of Research Studies in Biosciences 3 (1), 1–36.

Singh, R., Sinha, T.B., 1980. Bionomics of *Trioxys (Binodoxys) indicus* Subba Rao & Sharma, an aphidiid parasitoid of *Aphis craccivora* Koch.5.The extent of hyperparasitism. Zeitschrift für Angewandte Entomologie 90, 141–146.

Singh, R., Sinha, T.B., 1982a. Bionomics of *Trioxys (Binodoxys) indicus* Subba Rao & Sharma, an aphidiid parasitoid of *Aphis craccivora* Koch. XIII. Host selection by the parasitoid. Zeitschrift für Angewandte Entomologie 93, 64–75.

Singh, R., Sinha, T.B., 1982b. Bionomics of *Trioxys (Binodoxys) indicus*, an aphidiid parasitoid of *Aphis craccivora* Koch. 10. Superparasitism caused by confinement with the host. Entomologia Experimentalis et Applicata 32, 227–231.

Singh, R., Sinha, T.B., 1982c. Factors responsible for the superparasitic ability of the parasitoid wasp *Trioxys indicus* (Hymenoptera: Aphidiidae). Entomologia Generalis 7, 295–300.

Singh, R., Sinha, T.B., 1983. The development of ability of host discrimination in the parasitoid wasp *Trioxys indicus* (Hymenoptera: Aphidiidae). Entomologia Generalis 8, 225–234.

Singh, R., Srivastava, M., 1988. Effect of cold storage of mummies of *Aphis craccivora* Koch subjected to different pre-storage temperature on percent emergence of *Trioxys indicus* Subba Rao and Sharma. Insect Science and its Applications 9, 655–657.

Singh, R., Srivastava, M., 1989. Bionomics of *Trioxys indicus* (Hymenoptera: Aphidiidae), a parasitoid of *Aphis craccivora* (Hemiptera: Aphididae). 31.Effect of host haemolymph on the numerical response of the parasitoid. Entomophaga 34, 139–144.

Singh, R., Srivastava, P.N., 1990. Host specificity and seasonal distribution of *Alloxysta pleuralis* (Cameron), a cynipoid aphid hyperparasitoid. Ecological Entomology 15, 215–224.

Singh, R., Tripathi, R.N., 1991. Records of aphid hyperparasitoids in India. Bioved 1, 141–150.

Singh, R., Biswas, S., Pandey, S., 1996. Dietary role of honeydew on the life-table parameters of a cereal aphid parasitoid, *Lysiphlebia mirzai* Shuja-Uddin (Hymenoptera: Braconidae). Journal of Applied Zoological Research 7, 102–103.

Singh, R., Singh, K., Upadhyay, B.S., 2000. Honeydew as a food source for an aphid parasitoid *Lipolexis scutellaris* Mackauer (Hymenoptera: Braconidae.). Journal of Advanced Zoology 21, 76–83.

Singh, R., Singh, A., Pandey, S., 2001. Effect of host-patch size and parasitoid density on the progeny sex ratio of an aphid parasitoid *Binodoxys (=Trioxys) indicus* (Subba Rao & Sharma) (Hymenoptera: Braconidae, Aphidiinae). Journal of Advanced Zoology 22 (2), 100–108.

Singh, G., Singh, N.P., Singh, R., 2014a. Food plants of a major agricultural pest *Aphis gossypii* Glover (Homoptera: Aphididae) from India : an updated checklist. International Journal of Life Sciences Biotechnology & Pharma Research 3 (2), 1–26.

Singh, G., Singh, N.P., Singh, R., 2014b. Maternal manipulation of progeny sex ratio in parasitic wasps with reference to Aphidiinae (Hymenoptera: Braconidae): a review. Indo-American Journal of Life Sciences and Biotechnology 2 (1), 1–23.

Singh, R., 2001. Biological control of the aphids by utilising parasitoids. In: Upadhyay, et al. (Ed.), Biocontrol Potential and its Exploitation in Sustainable Agriculture, vol. 2. Kulwer Academic/Plenum Publishers, USA, pp. 57–73.

Singh, R., 2006. Transgenic crop technology for pest management: an ecological assessment. Journal of Aphidology 20 (2), 1–18.

Smith, P.T., Kambhampati, S., 2000. Evolutionary transitions in aphidiinae (Hymenoptera: Braconidae). In: Austin, A.D., Dowton, M. (Eds.), Hymenoptera: Evolution, Biodiversity and Biological Control. CSIRO Publishing, Collingwood, Australia, pp. 106–113.

Smith, C.M., 2005. Plant Resistance to Arthropods. Springer, Dordrecht, the Netherlands, p. 426.

Snyder, W.E., Ives, A.R., 2001. Generalist predators disrupt biological control by specialist parasitoid. Ecology 82, 705–716.

Snyder, W.E., Ives, A.R., 2003. Interactions between specialist and generalist natural enemies: parasitoids, predators, and pea aphid biocontrol. Ecology 84, 91–107.

Sorokina, A.P., 1970. Structure and development of the reproductive organs and potential fecundity in the female of some aphid parasites (Hymenoptera: Aphidiidae). Entomological Review, Washington 49, 27–31.

Srivastava, M., Singh, R., 1988a. Bionomics of *Trioxys indicus*, an aphidiid parasitoid of *Aphis craccivora*. 26. Impact of host extract on the ovipostion response of the parasitoid. Biological Agriculture and Horticulture 5, 169–176.

Srivastava, M., Singh, R., 1988b. Bionomics of *Trioxys (Binodoxys) indicus* Subba Rao and Sharma, an aphidiid parasitoid of *Aphis craccivora* Koch. 29. Impact of host extract on reproductive behaviour of the parasitoid at different host densities. Journal of Applied Entomology 106, 408–412.

Starks, K.J., Muniappan, R., Eikenbary, R.D., 1972. Interaction between plant resistance and parasitism against green bug on barley and sorghum. Annals Entomological Society of America 65, 650–655.

Starý, P., Lyon, J.P., Lecant, F., 1988. Biocontrol of aphids by the introduced *Lysiphlebus testaceipes* (Cresson) (Hymenoptera: Aphidiidae) in Mediterranean France. Journal of Applied Entomology 105, 74–87.

Starý, P., Gerdingh, M., Norambuenea, H., Remaudière, G., 1993. Environmental research on aphid parasitoid biocontrol agents in Chile (Hym., Aphidiidae; Hom., Aphidoidea). Journal of Applied Research 115, 292–306.

Starý, P., 1966. Aphid Parasites of Czechoslovakia. Academia, Prague co-edit. Dr.W.Junk, The Hague, pp. 1–242.

Starý, P., 1970a. Biology of Aphid Parasites (Hymenoptera: Aphidiidae) with Respect to Integrated Control. Dr. W. Junk, B.V, The Hague, p. 643.

Starý, P., 1970b. Methods of mass rearing, collection and release of *Aphidius smithi* (Hymenoptera: Aphidiidae) in Czechoslovakia. Acta Entomologica Bohemslovaca 67, 339–346.

Starý, P., 1987. Aphidiidae. In: Minks, A.K., Harrewijn, P. (Eds.), Aphids: their Biology, Natural Enemies and Control, vol. 2B. Elsevier, New York, NY, pp. 171–184.

Starý, P., 1993a. The fate of released parasitoid (Hymenoptera: Braconidae, Aphidiinae) for biological control of aphids in Chile. Bulletin of Entomological Research 83, 633–639.

Starý, P., 1993b. Alternative host and parasitoid is first method in aphid pest management in glasshouses. Journal of Applied Entomology 116, 187–191.

Starý, P., 1999. Parasitoids and biocontrol of Russian wheat aphid, *Diuraphis noxia* (Kurdj.) expanding in central Europe. Journal of Applied Entomology 123, 273–279.

Stilmant, D., van Bellinghen, C., Hance, T., Boivin, G., 2008. Host specialization in habitat specialists and generalists. Oecologia 156, 905–912.

Sullivan, D.J., Völkl, W., 1999. Hyperparasitism: multitrophic ecology and behavior. Annual Review of Entomology 44, 291–315.

Sullivan, D.J., 1987. Insect hyperparasitism. Annual Review of Entomology 32, 49–70.

Sullivan, D.J., 1988. Hyperparasites. In: Minks, A.K., Harrewijn, P. (Eds.), Aphids: their Biology, Natural Enemies and Control, vol. 2B. Elsevier, New York, NY, pp. 189–203.

van Schelt, J., Douma, J.B., Ravensberg, W.J., 1990. Recent developments in the control of aphids in sweet pepper and cucumbers. SROP/WPRS, Bulletin XIII/5, 190–193.

van Steenis, M.J., EL-Khawass, K.A.M.H., 1995. Life history of *Aphis gossypii* on cucumber: influence of temperature, host plant and parasitism. Entomologia Experimentalis et Applicata 76, 121–131.

van Steenis, M.J., El-Khawass, K.A.M.H., Hemerik, L., Van Lenteren, J.C., 1996. Time allocation of the parasitoid *Aphidius colemani* (Hymenoptera: Aphidiidae) foraging for *Aphis gossypii* (Homoptera: Aphidae) on cucumber leaves. Journal of Insect Behaviour 9, 283–295.

van Steenis, M.J., 1992. Biological control of the cotton aphid, *Aphis gossypii* Glover (Homoptera: Aphididae): pre introduction evaluation of natural enemies. Journal of Applied Entomology 114, 362–380.

van Steenis, M.J., 1995. Evaluation of four aphidiine parasitoids for biological control of *Aphis gossypii*. Entomologia Experimentalis et Applicata 75, 151–157.

Takada, H., 1992. Transmission by aphid vectors of virus diseases of papaya and banana. A. Aphid parasitoids as biological control agents of vector aphids of papaya ring spot virus and banana bunchy top virus. Food & Fertilizer Technology Group ASPAC. Technical Bulletin 132, 1–11.

Takanashi, M., 1990. Development and reproductive ability of *Lysiphlebus japonicus* Ashmead (Hymenoptera: Aphidiidae) parasitising the citrus brown aphid, *Toxoptera citricidus* (Kirkaldy) (Homoptera: Aphidiidae). Japanese Journal of Applied Entomology and Zoology 34 (3), 237–243.

Tentelier, C., Desouhant, E., Fauvergue, X., 2006. Habitat assessment by parasitoids: mechanisms for patch use behavior. Behaviour & Ecology 17, 515–521.

Thies, C., Steffan-Dewenter, I., Tscharntke, T., 2003. Effects of landscape context on herbivory and parasitism at different spatial scales. Oikos 101, 18–25.

Thompson, S., Krauter, P.C., Heinz, K.M., 1999. Battling bugs biologically. Greenhouse Grower 17, 26–34.

Tisdell, C.A., 1990. Economic impact of biological control of weeds and insects. In: Mackauer, M., Ehler, L.E., Roland, J. (Eds.), Critical Issues in Biological Control. Intercept, Andover, UK, pp. 301–316.

Tripathi, R.N., Singh, R., 1990. Fecundity, reproductive rate, intrinsic rate of increase of an aphidiid parasitoid *Lysiphlebia mirzai* Shuja-Uddin. Entomophaga 35, 601–610.

Tripathi, R.N., Singh, R., 1991. Progeny and sex allocation by an aphidiid parasitoid *Lysiphlebia mirzai* (Hymenoptera: Aphidiidae) at different parasitoid-host ratios. Entomologia Generalis 16, 23–30.

Tripathi, S.K., Singh, R., 1997. Effect of host-food plant transfer on the life-table of *Lysiphlebia mirzai* Shuja-Uddin, an aphid parasitoid of cereal aphids. Malayasian Journal of Applied Biology 26, 45–55.

Valentine, E.W., 1974. Differential host relations of the sexes in some New Zealand parasitic Hymenoptera. N. A. Entomology 3, 6–8.

Viggiani, G., 1984. Bionomics of the Aphelinidae. Annual Review of Entomology 29, 257–276.

Völkl, W., Stechman, D.H., Starý, P., 1990. Suitability of five species of Aphidiidae (Hymenoptera) for the biological control off the banana aphid *Pentalonia nigronervosa* Coq. (Homoptera: Aphididae) in South Pacific. Tropical Pest Management 36, 249–257.

Völkl, W., Mackauer, M., Pell, J., Brodeur, J., 2007. Predators, parasitoids and pathogens. In: van Emden, H., Harrington, R. (Eds.), Aphids as Crop Pests. CAB International, Oxford, UK, pp. 187–233.

Vollhardt, I.M.G., Bianchi, F.J.J.A., Wäckers, F.L., Thies, C., Tscharntke, T., 2010. Spatial distribution of flower v/s honeydew resources in cereal fields may affect aphid parasitism. Biological Control 53, 204–213.

Vollhardt, I.M.G., Tscharntke, T., Wäckers, F.L., Bianchi, F.J.J.A., Thies, C., 2008. Diversity of cereal aphid parasitoids in simple and complex landscapes. Agriculture Ecosystem and Environment 126, 289–292.

Wäckers, F.L., Van Rijn, P.C., Heimpel, G.E., 2008. Honeydew as a food source for natural enemies: making the best of a bad meal? Biological Control 45, 176–184.

Waterhouse, D.F., Sands, D.P.A., 2001. Classical Biological Control of Arthropods in Australia, vol. 77. ACIAR Monograph, Cambera, Australia, p. 560.

Waterhouse, D.F., 1998. Biological control of insect pests: Southeast asian prospects. ACIAR Monograph No. 51 548.

Wei, J., Li, T., Kuang, R., Wang, Y., Yin, T., Wu, X., Zou, L., Zhao, W., Cao, J., Deng, J., 2003. Mass rearing of *Aphidius gifuensis* (Hymenoptera: Aphidiidae) for biological control of *Myzus persicae* (Homoptera: Aphididae). Biocontrol Science and Technology 13, 87–97.

Wei, J.N., Bai, B.B., Yin, T.S., Wang, Y., Yang, Y., Zhao, L.H., Kuang, R.P., Xiang, R.J., 2005. Development and use of parasitoids (Hymenoptera: Aphidiidae & Aphelinidae) for biological control of aphids in China. Biocontrol Science and Technology 15, 533–551.

Weisser, W.W., Völkl, W., 1997. Dispersal in the aphid parasitoid, *Lysiphlebus cardui* (Marshall) (Hym.: Aphidiidae). Journal of Applied Entomology 21, 23–28.

Wheeler, A.G., Hoebeke, E.R., 1981. A Revised Distribution of *Coccinella Undecimpunctata* L. In Eastern and Western North America (Coleoptera: Coccinellidae).

Williams, D.W., Leppla, N.C., 1992. The future of augmentation of beneficial arthropods. In: Kauffman, W.C., Nechols, J.E. (Eds.), Selection Criteria and Ecological Consequences of Importing Natural Enemies. Thomas Say Publications in Entomology, Entomological Society of America, Maryland, pp. 87–102.

Wu, Z.S., Hopper, K.R., O'Neil, R.J., Voegtlin, D.J., Prokrym, D.R., Heimpel, G.E., 2004. Reproductive compatibility and genetic variation between two strains of *Aphelinus albipodus* (Hymenoptera: Aphelinidae), a parasitoid of the soybean aphid, *Aphis glycines* (Homoptera: Aphididae). Biological Control 31, 311–319.

Wyatt, I.J., 1985. Aphid control by parasites. In: Hussey, N.W., Scopes, N. (Eds.), Biological Pest Control. The Glasshouse Experience. Blanford Press, Poole, Dorset, pp. 134–137.

Yamaguchi, H., Takai, M., 1977. An Integrated Control System for the Todo-fir Aphid, *Cinara Todocola* Inouye in Young Todo-fir Plantations. Bulletin of the Government Experiment Station, Tokyo, vol. 295, pp. 61–96.

Yang, S., Yang, S.Y.Y., Zhang, C.P., Wei, J., Kuang, R.P., 2009. Population dynamics of *Myzus persicae* on tobacco in Yunnan province, China, before and after augmentative releases of *Aphidius gifuensis*. Biocontrol Science and Technology 19, 219–228.

Yano, E., 2010. Ecological considerations for biological control of aphids in protected culture. Population Ecology 48, 335–339.

Zaviezo, T., Grez, A.A., Donoso, D., 2004. Dina´mica temporal de coleo´pteros asociados a alfalfa. Ciencia e Investigación Agraria 31, 29–38.

Zhua, J., Zhang, A., Parkc, K., Bakera, T., Lange, B., Jurenkac, R., Obryckic, J.J., Gravesg, w.r., Picketth, J.A., Smileyh, D., Chauhand, K.R., Klund, J.A., 2006. Sex pheromone of the soybean aphid, *Aphis glycines* Matsumura, and its potential use in semiochemical-based control. Environmental Entomology 35 (2), 249–257.

Zúniga, E., 1990. Biological control of cereal aphids in the southern cone of South America. In: Burnett, P.A. (Ed.), World Perspectives on Barley Yellow Dwarf. CIMMYT, Mexico, DF (Mexico), pp. 362–367.

4

Parasitoids

M. *Kalyanasundaram*, I. *Merlin Kamala*

Department of Agricultural Entomology, Agricultural College
and Research Institute, Madurai, India

1. INTRODUCTION

Natural biocontrol by predators, parasites/parasitoids, and pathogens of agricultural pests has occurred since the beginning of evolutionary process of crop cultivation. This age-old sustainable practice of controlling crop pests has gained impetus, as conventional pesticides cannot achieve the desired level of insect pest control due to the development of resistance to them, and deleterious effects like environmental pollution, pest resurgence, and reduction of natural enemies in the agroecosystem (Waage, 1989). Biocontrol involving the introduction, augmentation, and conservation of natural enemies has been accepted as an effective, environmentally safe, technically appropriate, economically viable, and socially acceptable method of pest management. Parasitoids are an important biological tool used widely in agriculture for the suppression of various pest species (DeBach, 1964; Hagen et al., 1971; Huffaker and Messenger, 1976). The potential of biocontrol by parasitoids has largely remained untapped because it has been underused, underexploited, underestimated, and often untried and, therefore, unproven (Hokkanen, 1993). In fact, the use of biocontrol agents including parasitoids should be the primary concern in any pest management program. As the world is moving toward organic products due to environmental concerns, the use of parasitoids in managing agricultural pests has been extensively increased.

The term "parasitoid" embraces an exceedingly large number of insect species (Gauld, 1986) that kill their hosts and are able to complete their development on a single host (Vinson, 1976). The term parasitoid is derived from the more general term "parasite." Parasites are organisms living in (endoparasites) or on (ectoparasites) another organism, termed the host. Most insects parasitic upon other insects are protelean parasites, *i.e.*, they are parasitic during their immature stages and the adults are mostly free-living. In the process of parasitization, they may slowly weaken and eventually kill larger host insects. Parasitoids are often called parasites, but the term parasitoid is more technically accepted. Generally, the parasitoids kill their hosts but in some circumstances the host may complete their full life cycle before death.

Ecofriendly Pest Management for Food Security
http://dx.doi.org/10.1016/B978-0-12-803265-7.00004-X

Parasitoids are the most common type of natural enemy introduced for biocontrol of insect pests. They are employed in insect pest management for centuries (Orr and Suh, 1998). Over the last century there has been a dramatic increase in their use as well in an understanding of how they can be manipulated for effective and safe use in insect pest management systems (Orr and Suh, 1998).

2. HISTORICAL PERSPECTIVE

In 1602, Aldrovandi witnessed parasitoids emerging from butterfly larvae and misunderstood the process as a transformation in which the parasitoids were another stage of the larvae produced by metamorphosis. Insect parasitism was first recognized by Antonio Vallisnieri (1661–1730) who noted the unique association between the parasitic wasp, *Apanteles glomeratus* (Linnaeus), and the cabbage butterfly, *Pieris rapae* (Linnaeus), in Italy (DeBach, 1974). The first person to publish a correct interpretation of insect parasitism was the British physician Martin Lister reported that ichneumon wasps emerging from caterpillars were the result of eggs laid in the caterpillars by adult female ichneumons (Van Driesche and Bellows, 1996a). In 1700, Antonie van Leeuwenhoek, the Dutch microscopist correctly interpreted parasitism of aphids by *Aphidius* spp. In 1800, Darwin noticed the destruction of cabbage butterfly caterpillar by the small ichneumon fly, which deposits its eggs in their backs. In 1827, the gathering and storing of parasitized caterpillars in order to harvest parasitoid adults for later release was proposed by Hartig, in Germany (Howard, 1930).

The first suggestion to transport parasitoids between countries for pest control was made in 1855 by the American entomologist, Asa Fitch (1809–1899), who proposed importing the parasitoids of the wheat midge, *Sitodiplosis mosellana* (Gehin) from Europe, however, no importations were made (Hagen and Franz, 1973). In 1883, the first parasitoid species, *A. glomeratus*, was successfully transported between continents from England to the USA and established for the suppression of the cabbage butterfly (Bajpai, 2010).

In 1904, the US Horticultural Commission initiated a project for the biocontrol of the apple codling moth, *Cydia pomonella* (L.), by importing the parasitoid, *Ephialtes caudatus* (Ratzeburg) from Spain (Doutt, 1964). In 1910, scientists from the University of Kansas collected and distributed large number of dead aphids parasitized by *Aphidius testaceipes* (Cresson) to the north region of the United States for the management of the wheat green bug, *Schizaphis graminum* (Rondani).

3. CHARACTERISTICS OF INSECT PARASITOIDS

Insect parasitoids are smaller than their host and are specialized in choice of host. Only the female searches for hosts and usually destroy their hosts during development. Different parasitoid species attack at different life stages of the host. Eggs are usually laid in, on, or near the host. The immature stages remain on or in the host and almost always kill the host (Hoffmann and Frodsham, 1993). Adult parasitoids are free-living, mobile, and may be predaceous. With respect to population dynamics, parasitoids are similar to predatory insects.

Parasitoids attack either by penetrating the body wall and laying eggs inside the host or attaching eggs to the outer body surface. The immature parasitoid develops on or within the host, consumes all or most of the host's body fluid, and pupates either within or external to the host (Hoy, 1994). The adult parasitoids emerge from pupae and start the next generation afresh by actively searching for a host to oviposit. The life cycle of the pest and the parasitoid can coincide, or that of the pest may be altered by the parasitoid to accommodate its development. Most adult parasitoids require food such as nectar, pollen, or honey dew and many feed on host's body fluids, whereas others require free water as adults (DeBach and Rosen, 1991; Altieri and Nicholls, 1998). Parasitism is found in at least five insect orders, *i.e.*, Hymenoptera, Diptera, Coleoptera, Lepidoptera, and Neuroptera (Eggleton, 1992), but the most parasitoids that have been used in biocontrol are from the insect orders Hymenoptera and, to a lesser degree, Diptera (Van Driesche and Bellows, 1996b). Hymenopteran parasitoids account for nearly 78% of the estimated number of species and consequently have served as models of choice for nearly all research on insect parasitoids (Godfray, 1994; Hawkins and Sheehan, 1994; Waage and Greathead, 1986). The superfamilies Ichneumonoidea and Chalcidoidea are among the largest assemblages within parasitoids. Members of these superfamilies are parasitoids of economic importance and have been used in many biocontrol programs (LaSalle, 1993). There are several unique factors that are responsible for the success of hymenopterans as parasitoids, they are the only holometabolous parasitoids retaining a primitive lepismatid form of ovipositor and associated accessory glands. This modified ovipositor is used to gain direct access to concealed host, secrete venom to paralyze the prey, lay eggs, secrete feeding tubes, and locate prey via sensitive neural receptors. Only hymenopteran parasitoids are haplodiploids. Such a sex-determining system gives females control over the sex ratio of the progeny (Charnov, 1982). The most frequently used groups in the order Hymenoptera are Braconidae and Ichneumonidae in the superfamily Ichneumonoidea, and the Eulophidae, Pteromalidae, Encyrtidae, and Aphelinidae in the superfamily Chalcidoidea. About 66% of all successful biocontrol programs have involved parasitoids that belong to the order Hymenoptera.

Dipterans include an estimated 16,000 described species of parasitoids, about 20% of the known species with this lifestyle (Eggleton, 1992). Within Diptera, several families are parasitic but are less amenable in biocontrol programs. These families include Sciomyzidae, Bombyliidae (Yeates, 1994), Calliphoridae (Rognes, 1991), Phoridae (Disney, 1994), Sarchophagidae (Pape, 1987), Rhinophoridae (Pape, 1986), and Tachinidae (Belshaw, 1993; Hertig, 1960; Mellini, 1990; Pape, 1992). The family Tachinidae is the most frequently employed group as parasitoids (Greathead, 1986). The dipteran parasitoids in Tachinidae appear to be less host-specific. Some species of tachinid flies have been found to attack more than 100 different host insects. The family Tachinidae, which is exclusively parasitic, has many species that have been manipulated for biocontrol for a range of pest taxa. Examples of dipterans as parasitoids include the parasitoid, *Lydella thompsoni*, that was used against European corn borer, *Ostrinia nubilalis*; *Lixophaga diatreae* used in the Caribbean against the sugar cane borer, *Diatraea saccharalis*; and *Cyzenis albicans* introduced to Canada against the winter moth *Operophtera brumata* (Arnaud, 1978; McAlpine, 1981). Although parasitoids have also been recorded in the insect orders Strepsiptera and Coleoptera, parasitism is not common in them (Greathead, 1986). Table 1 depicts the common parasitoids found in agroecosystem.

TABLE 1   Common Parasitoids Found in Agroecosystems

Order	Family	Host	Internal/external
Diptera	Tachinidae	Beetles, butterflies, and moths	Internal
	Nemestrinidae	Locusts, beetles	Internal
	Phoridae	Ant, caterpillar, termites, flies, others	Internal
	Cryptochaetidae	Scale insects	Internal
Hymenoptera	Chalcididae	Flies and butterflies (larvae and pupae)	Internal or external
	Encyrtidae	Various insect eggs, larvae, or pupa	Internal or external
	Eulophidae	Various insect eggs, larvae, or pupa	Internal or external
	Aphelinidae	Whiteflies, scales, mealybugs, aphids	Internal or external
	Trichogrammatidae	Moth eggs	Internal
	Myrmaridae	True bugs, flies, beetles, leafhopper eggs	Internal
	Scelionidae	Insect eggs of true bugs and moths	Internal
	Ichneumonidae	Larvae or pupae of beetles, caterpillars, and wasps	Internal or external
	Braconidae	Larvae of beetles, caterpillars, flies, and saw flies	Internal (mostly)
	Pteromalidae	Larvae and pupae of lepidopteran and Coleopteran insects	Internal

Modified from Hagler (2000).

3.1 Classification of Parasitoids

A wide group of parasitoids lay eggs on or inside the body of insect hosts, which are then used as food by the developing larvae. The most important parasitoid groups that parasitize the eggs of the host insects are trichogrammatids, ichneumonids, scleonids, evanids, chalcids, and tachinids (Quicke, 1997). The trichogrammatids parasitize the eggs of several insect species and have been used extensively in biocontrol programs (Flint and Dreistadt, 1998). The braconids, bethylids, and ichneumonids parasitize mainly the larvae of moths and butterflies (Hrcek et al., 2013). The chalcids, braconids, and encyrtids parasitize the eggs and larvae of insects. Certain ichneumonids, eulopids, and chalcids parasitize both larval and pupal stages of insects. An epricania, *Epiricania melanoleuca* (Fletcher), parasitizes *Pyrilla perpusilla* (Walker) on sugar cane, and is an adult parasitoid belonging to the order Lepidoptera (Kumarasinghe and Wratten, 1996).

Parasitoids may be classified depending upon whether they are external feeders (ectoparasitoids) or internal parasitoids (endoparasitoids) (Doutt et al., 1976). Ectoparasitic species are generally found in situations where the host is protected, *i.e.*, leaf miners or scale insects.

In these circumstances, the adult parasitoid may simply deposit an egg inside a feeding tunnel (leaf miner) or near the host without making a direct contact with the host. Endoparasitism is more common in hosts that are not protected, such as aphids or caterpillars, where the adult parasitoids directly deposit the egg inside the host with their ovipositors (Doutt et al., 1976).

Parasitoids may have only one generation (univoltine) or two or more generations (multivoltine) (Van Driesche and Bellows, 1996b). Life cycles are generally short, ranging from 10 days to 4 weeks or so in summer but correspondingly longer in cold weather. However, some require a year or more if they have hosts having only a single generation per year. In general, they all have great potential rates of increase.

Parasitoids may be either solitary, when only one larva develops per individual host, or gregarious, when more than one larva develops upon a single host. However, this is not always clear cut, as many species may develop facultatively as solitary upon a small host and gregarious upon a larger one (DeBach and Rosen, 1991). Adult parasitoids seek out their hosts using a variety of cues such as visual, olfactory, and tactile, from the target host and its habitat. Parasitoids may be either idiobiont, *i.e.*, the host's development is arrested or terminated upon parasitization (e.g., egg parasitoids) or koinobionts, *i.e.*, the host continues to develop following parasitization (e.g., larval–pupal parasitoids) (Doutt et al., 1976).

Some species of parasitoids are polyembryonic, where a single egg gives rise to many embryos. Other parasitoids display superparasitism, where more than one adult of the same species attacks the host, and multiparasitism, where more than one species attacks the same host. Both super and multiparasitism are generally regarded as undesirable situations because much reproductive capacity is wasted.

When one parasitoid is attacked by another parasitoid the phenomenon is called hyperparasitism. Hyperparasitoids are secondary insect parasitoids that develop at the expense of primary parasitoids, thereby representing a highly-evolved fourth trophic level. They are called secondary parasitoids if they attack primary parasitoids and tertiary parasitoids if they attack secondary parasitoids. Secondary parasitism is usually harmful, as it reduces the efficacy of primary parasitoids (Sullivan and Volkl, 1999).

The structure of hyperparasitoid communities has been studied for a spectrum of phytophagous insects (Hawkins, 1994). There are two multitrophic systems with a well-studied parasitism ecology, *i.e.*, the heteronomous hyperparasitoids of the genus *Encarsia* attacking scale insects and whiteflies (Godfray, 1994) and the aphid parasitoid hyperparasitoid community (Volkl, 1997). In heteronomous hyperparasitoids or adelphoparasitoids, only males develop as hyperparasitoids, attacking females of their own or another parasitoid species (Williams, 1991). Aphid hyperparasitoids generally develop as obligate parasitoids (Sullivan, 1987). It is of utmost importance to develop a conservation policy in biocontrol program in order to exclude exotic facultative hyperparasitoids so that they do not interfere drastically with biocontrol by exotic primary parasitoids.

4. BIOCONTROL STRATEGIES

Biocontrol of insect pests in agricultural fields involves three elementary tactics, namely, classical biocontrol, augmentative biocontrol, and conservative biocontrol. Classical biocontrol comprises foreign exploration of parasitoids, their import, augmentation, and release to a place where they are accidentally or intentionally introduced. Augmentative biocontrol involves multiplication of the parasitoid population and their field release. Conservative biocontrol concerns preserving existing populations of parasitoids by the manipulation of agroecosystems.

4.1 Classical Biocontrol

Classical biocontrol of insects, where exotic natural enemies are introduced to control foreign pests, has been applied for more than 120 years and the release of more than 2000 species of natural enemies has resulted in a permanent reduction of at least 165 pest species worldwide (Greathead, 1995; Gurr and Wratten, 2000). The intentional introduction of an exotic, usually coevolved biocontrol agent for stable establishment and long-term pest control is referred to as classical biocontrol. The objective of the approach is to introduce safe and effective parasitoids to suppress pest populations. It has been applied in a wide variety of natural, agricultural, and urban localities. The basic objective in this technique is to identify parasitoids that control a pest in its home location and introduce them to the pest's new location. This tactic is employed in cases of pests invading a new geographical location.

The first step involved in classical biocontrol is the correct identification of the pest. Improper identification can seriously hamper exploration effects and delay the success of the program. Once the pest is identified with its nativity and potential parasitoids in the native country, it should be explored and subsequently imported into the target country for use against exotic pest species, after being quarantined. Further trials in the laboratory and semi-field conditions are undertaken while methods for mass rearing are developed. Once the pest population and agroclimatic conditions are stable for continuous releases in the field, it is then mass-released. Finally, the released parasitoids should be evaluated for their performance by comparing the pest populations before and after the release, comparing the population on plots with and without parasitoids, and correlating population changes of the parasitoids and the hosts by analyzing the life and mortality data of the parasitoids. Despite the success of classical biocontrol, complete control of a pest is seldom achieved by this method (Greathead, 1986; Hoy, 1994).

The successful introduction of the parasitoid *Pediobius foveolatus* (Crawford) from India against the Mexican bean beetle, *Epilachna varivestis* Mulshant, in the United States Atlantic coastal plains is an example of successful classical biocontrol within an integrated pest management (IPM) program in soybeans (Kogan and Turnipseed, 1987).

Cassava, native to tropical Central and South America, is an exotic African crop introduced over 500 years ago. In the early 1970s the cassava mealybug, *Phenacoccus manihoti* Matile-Ferrero, was accidentally introduced from South America to Africa and spread throughout Africa causing tuber yield losses of up to 84% and foliage loss of nearly 100%. In the mid-1970s scientists began exploring South America for natural enemies of the cassava mealybug and, in 1981, the encyrtid wasp, *Epidinocarsis lopezi* (De Santis), was identified as its parasitoid. *Epidinocarsis lopezi* was first released in Nigeria in 1981 and in less than 1.5 years it spread to areas as far as 170 km from the original site. Currently the pest is virtually eliminated from African countries and no longer poses a serious threat to most cassava-growing regions.

The papaya mealybug, *Paracoccus marginatus* (Williams and Granara de Willink), invaded India. The first record on papaya plants was from Coimbatore (Tamil Nadu) in July 2008; however, it then spread to Kerala, Karnataka, Maharashtra, and Tripura presumably due to the movement of infested fruits. The National Bureau of Agricultural Insect Resources (NBAIR) along with other national institutes initiated pioneering studies to understand the biology and ecology of the pest as well as natural biocontrol (Shylesha et al., 2010). Three species of encyrtid parasitoids, *Acerophagus papayae* (Noyes and Schauff), *Pseudleptomastix mexicana* (Noyes and Schauff), and *Anagyrus loecki* (Noyes and Menezes), are known to suppress the papaya mealybug in its native range and have effectively controlled papaya mealybugs when introduced into Guam, Palau islands, and recently to Sri Lanka. They were imported, mass multiplied, and tested. They were found effective in controlling the mealybugs (Shylesha et al., 2010). NBAIR mass multiplied the parasitoids and found it to be successful in controlling the papaya mealybugs and thus providing a solution to the great menace in papaya.

4.2 Augmentative Biocontrol

Augmentative biocontrol is focused on enhancing the numbers and/or activity of natural enemies in agroecosystems. This strategy involves mass multiplication and periodic release of parasitoids so that they may multiply during the growing season. It requires the greatest amount of efforts and resources for its implementation. Augmentation requires mass propagation to produce parasitoids in large numbers and mass handling techniques for distribution and release of these parasitoids infields (DeBach and Hagen, 1964; King et al., 1985a). Augmentative release of parasitoids provides a biological solution to pest problems in ephemeral crops where naturally occurring beneficial organisms fail to respond quickly enough to control the populations of primary pests (King, 1993).

Augmentation of natural enemies to achieve suppression of pest populations has been reviewed by Beglyarov and Smetnik (1977), Huffakar (1977), Knippling (1977), Ridgeway and Vinson (1977), Shumakov (1997) and Starter and Ridgeway (1977). The process of augmentation or inoculation is done where parasitoids are absent or where they exist at levels that are ineffective for pest control. The periodic releases may either be inoculative or inundative. Inoculative releases may be made as infrequent as once a year to reestablish a species of parasitoid that is periodically killed in an area by unfavorable conditions during part of the year but operates very effectively the rest of the year. Inundative release involves mass multiplication and release of parasitoids to suppress the pest populations directly as in case of conventional insecticides.

Some of the requirements for promoting the cause of augmentation are careful selection of a parasitoid or its strain that has the ability to adapt to the agroclimatic conditions where it is to be used, development of an efficient and cost-effective mass rearing technique, production of good quality parasitoids, and mass multiplication and storage for use as per needs. Impediments to augumentive biocontrol can be lack of artificial rearing media and systems for efficient delivery, properly designed facilities, limited automation, effective quality control, and improved management systems. There are numerous efficacious species for which no artificial rearing media or procedures for storage are available.

Control of lepidopteran pests by mass rearing and release of *Trichogramma* spp. has been practiced for many decades. The pioneering research of Howard and Fiske (1911) and

Flanders (1930) stimulated research on *Trichogramma* spp. and a number of successes in the reduction in insect populations by augmentation of *Trichogramma* spp. have been reported. Several *Trichogramma* spp. have been put to commercial use in different countries of the world (Hassan, 1993; Smith, 1996). In Europe, the greatest use of parasitoids in augmentative release has been for pest control in greenhouses (Hoy, 1994). Massive releases have been attempted in several programs involving various parasitoids.

4.2.1 Rice

In Malaysia, augmentative release of *Trichogrammatoidea nana* Zehntner at 39,500–44,500/ha resulted in a reduction of rice stem borer damage by 51–82% (Khoo, 1990). For the biological suppression of borers (*Scirpophaga incertulas*, *Chilo* spp. and *Sturmiopsis inferens*), inunda-tive release of *Trichogramma japonicum* at 50,000/ha/week during the egg laying period is advocated. The release starts when one egg mass or adult/m^2 or first moth in the light trap is recorded. With the release of *Trichogramma* the overall egg parasitism increases considerably (Jalali et al., 2010).

For the suppression of rice gall midge, *Orseolia oryzae* (Wood Manson), conservation of the parasitoid, *Platygaster oryzae* (Cameron) is suggested (Muthuswami and Gunathilagaraj, 1989). Unfortunately, techniques for mass multiplication of *P. oryzae* is not available, other-wise augmentative release of the parasitoids during the early phase of the crop growth could have been attempted (Gunathilagaraj and Ganesh Kumar, 1997).

The braconids, *Aulosaphes* spp., *Bracon* spp., and *Cardiochiles philippinensis* (Metcalf) were reported to parasitize the larvae of the rice leaf folder, *Cnaphalocrocis medinalis* (Guenee). The braconids, *Cotesia flavipes* (Cameron), *Myosoma chinensis* (Szepligeti), and *Tropobracon* spp. were reported to parasitize the larvae of the rice stem borer, *Scirpophaga incertulus* (Walker) (Bhattacharya et al., 2006). The braconid, *Scutibracon hispae* (Quickee and Walker) parasitizes the larvae of rice hispa, *Dicladispa armigera* (Olivier) and the icheumonids, *Isotima javensis* (Rohwer) and *Ischnojoppa luteator* (Fabricius) attack the larvae of *S. incertulus* (Bhattacharya et al., 2006).

The hymenopteran parasitoids, *Telenomus dignus* (Gahan) (Scelionidae), *Tetrastichus schoenobii* (Ferriere) (Eulophidae), and *T. japonicum* (Ashmead) (Trichogrammatidae) were recorded as important egg parasitoids of *S. incertulus* with peak parasitization ranging from 75.29% to 97.56% in Andra Pradesh, India (Varma et al., 2013). Two parasitoids, *Trissolcus* sp. (Hymenoptera: Scelinoidae) and *Ooencyrtus utetheisa* (Hymenoptera: Encrytidae) of the rice gundhi bug, *Leptocorisa* sp. (Hemiptera: Alydidae) were recorded. The eggs masses of the gundhi bugs were found to be parasitized during the 40th to the 42nd week (the first to the third week of October) with a peak of 16% parasitism in 41st week under Indian conditions (Purohit et al., 2014).

4.2.2 Sugar cane

Sugar cane crops are infested by a number of borers and sucking pests from sowing till harvest. For the control of sugar cane pests, beneficial insects have been successfully manipu-lated with a variety of augmentation and conservation strategies (Ashraf et al., 1999; Mohyuddin, 1991; Nordlund, 1984) and *Trichogramma* species (Hymenoptera: Trichogrammitidae) are among the most extensively used biocontrol agents against sugar cane borers. The relatively stable sugar cane agroecosystem provides ideal conditions for the colonization of parasitoids.

Sugar mills have their own cooperative parasitoid production units and have contributed in a major way. Doomra et al. (1994) recorded least infestation of *Scirpophaga excerptalis* and highest yield in areas where *T. japonicum* was released at 50,000/ha/10 days during April–June with mechanical control of top borer-infested shoots in Punjab, India. Six releases of *Trichogramma chilonis* at 40,000/ha at fortnightly intervals resulted in a lower borer infestation in the treated area (Kalyanasundaram et al., 1993). Varma et al. (1991) released *T. chilonis* in an area of 500 ha and reported highly significant results in reducing the incidence of *Chilo auricilius*. Release of the parasitoid at 125,000/ha has been recommended against shoot borer in Andhra Pradesh. Weekly releases at 125,000 parasitic wasps per hectare from the 4th to 11th week of the crop provided effective control of the internode borer, *Chilo sacchariphagus indicus* in Tamil Nadu. Release of *T. chilonis* (Ishii) and *E. melanoleuca* (Fletcher) have been made for the control of sugar cane borers and pyrilla, respectively, in Pakistan (Mohyuddin, 1994). Sajid Nadeem and Muhammed Hamed (2011) reported that releases of *T. chilonis* (Hymenoptera: Trichogrammatidae) controlled the sugar cane borers *Scirpophaga nivella* F., *Chilo infuscatellus* Snellen, *Emmalocera depressella* Swin., and *Acigona steniella* (Hampson) in Pakistan.

In addition to egg parasitoids, field release of the larval parasitoid, *I. javensis* (Rohwer) exercised effective control of top borers in Tamil Nadu and Karnataka. Release of the dipteran larval parasitoid, *S. inferens* (Townsend) at 312 gravid females per hectare provided effective control of *C. infuscatellus* in coastal areas of Tamil Nadu. David (1980) reported a reduced population of *C. infuscatellus* in Tamil Nadu, Orissa, and Haryana after the release of *S. inferens*.

Among sucking pests, notable success has been achieved in the biosuppression of the hopper, *P. perpusilla*, by the colonization *E. melanoleuca*. It has been effectively controlled by redistribution and periodic release of 8000–10,000 cocoons or 800,000–1,000,000 eggs of *E. melanoleuca* per hectare. In 1999, there was an outbreak of *Pyrilla* in 6,00,000 ha in North India, perhaps due to indiscriminate burning of trash. However, the parasitoid overpowered the pest within 20 days and the spraying cost of about Rs 480 million crores was saved (Dhaliwal et al., 2013). Misra and Pawar (1983) released 8000–12,000 cocoons or 800,000–1,000,000 eggs/ha of parasitoid *E. melanoleuca* when the *Pyrilla* density was five to seven individuals/leaf and obtained significant control all over India. Elizangela et al. (2011) recorded the parasitoid, *Tetrastichus howardi* (Hymenoptera: Eulophidae) parasitizing *Diatraea* sp. (Lepidoptera: Crambidae) in sugar cane fields in Brazil.

4.2.3 Maize

Hassan (1982) reported a 65–93% reduction in larval infestation of European corn borer, *O. nubilalis*, following *Trichogramma* releases during the late 1970s in Germany. In China, a significant reduction in populations of *Ostrinia* spp. and *Helicoverpa* spp. reduced crops to a large extent (Li, 1984). In India, the release of *T. chilonis* at 100,000/ha resulted in parasitism of *Chilo partellus* eggs to the tune of 75.2% and 62.6% in first generation and 90.4% and 78.4% in second generation egg laying at 3- and 5-day interval parasitoid releases, respectively. Stem tunneling by larvae was significantly less compared with untreated control plots (Jalali and Singh, 2003).

4.2.4 Cotton

Cotton suffers from a number of major insect pests like, bollworms, sucking pests, and leaf-eating caterpillars. Therefore, management becomes tedious as pests are associated with the crop till harvest. The farmers spray insecticides, approximately 10 rounds in rain-fed cotton and 20 rounds of insecticide in irrigated cotton in India. A number of promising parasitoids have been found attacking cotton pests. However, large-scale use of insecticides has reduced the population of most of the parasitoids to significant level. Releases of *Trichogramma* spp. at 150,000 parasitized eggs per hectare at weekly intervals have proved promising for boll-worm control in Indian conditions. A new parasitoid, *Aenasius bambawalei* (Hayat), has been reported in India to cause 50–80% parasitization of the nymphs of the mealybug, *Phenacoccus solenopsis* (Tinsley) (Jalali et al., 2010). The parasitoid kills the mealybugs within a week, turning them into reddish brown mummies. The parasitoid population could be augmented in cotton through collection of mealybug mummies from various host plants and releasing them onto the cotton plants. Augmentative releases of *Trichogramma pretiosum* were tested in Arkansas at 30,600 emerged adults/ha/release and yielded significantly more cotton than check fields (King et al., 1985b). Simmons and Minkenberg (1994) evaluated the potential of the parasitoid, *Eretmocerus californicus* against silverleaf whitefly, *Bemisia argentifolii* Bellows and Pering, and reported a higher rate of parasitization against the pest with higher seed cotton yield. King et al. (1995) reported the feasibility of large-scale rearing of *Catolaccus grandis* on artificial diet-reared boll weevil third instars and augmentative releases in commercially-managed South Texas cotton fields to suppress boll weevils yielded up to 90% mortality.

4.2.5 Pulses

There has been considerable effort on the use of *Trichogramma* spp. for the control of the gram pod borer, *Helicoverpa armigera*. Parasitization of the eggs of *H. armigera* by *T. chilonis* was reported in up to 85% in cowpea (Bhatnagar and Davies, 1979). The parasitoid, *Campoletis chlorideae* Uchida, is an early larval parasitoid whereas the dipteran *Corcelia illiota* Curron is most common on late larval population of *H. armigera* in legumes. Parasitization by *C. chlorideae* on chickpea ranged from 7.5% to 10.1% in different varieties in the Chattisgarh region (Bharatwaj et al., 1987). Six genera of parasitoids were found parasitizing the pigeon pod fly, *Melanagromyza obtusa* Mallach, of which two genera, *Omytes* and *Euderus*, are potential candidates for the biosuppression of the pest in Andhra Pradesh (Sithanandam et al., 1987). The hymenopteran parasitoid, *Dinarmus basalis*, was found to be effective against pulse beetle, *Callosobruchus chinensis* (Southgate, 1979; Islam et al., 1985; Monge and Huignard, 1991; Isalm, 1998; Duraimurugan et al., 2014).

4.2.6 Fruit Crops

4.2.6.1 APPLE

The San Jose scale, *Quadrispidiotus perniciuos*, is one of the most major pests of the apple. Augmentative and inoculative releases of two exotic parasitoids, *Encarsia perniciosi* (Tower) and *Aphytis proclia* (Walker)/*Aphytis diaspidis* (Howard) at 2000 per infested tree has given promising results for the suppression of the San Jose scale on apples (Rao et al., 1971; Singh, 1989). In Kashmir, releases of *E. perniciosi* and *A. proclia* resulted in increase of parasitism from 8.9% to 64.3%. Studies on the biology of *E. perniciosi* revealed the multiplication rate of parasitoids over 10 times.

Apple woolly aphid, *Eriosoma lanigera*, is another pest of apple. In India, the first consignment of the exotic parasitoid *Aphelinus mali* (Haldeman) to control woolly aphid was imported in 1930 from England. It has established in most of the woolly aphid-infested orchards (Singh, 1989).

The apple codling moth, *C. pomonella* (L.), is another major pest of apple and fruit crops. Two exotic egg parasitoids, *Trichogramma embryophagum* (Hartig) and *Trichogramma cacoaciae pallidum* (Mayer), recommended (Jalali et al., 2010) to control the apple codling moth.

4.2.6.2 STRAWBERRY

The repeated release of the egg parasitoid, *Anaphes iole* (Girault), against *Lygus hesperus* Knight populations and fruit damage in commercial strawberry fields was evaluated. A release rate of 37,000 parasitoids per hectare caused 50% parasitization of *L. hesperus* eggs (Norton and Welter, 1996).

4.2.6.3 CITRUS

The citrus butterfly, *Papilio demoleus* (Linnaeus), a common pest in citrus, was parasitized *T. chilonis* (up to 75.86%) and *Telenomus* spp. (up to 78%) (Krishnamoorthy and Singh, 1988). *Distatrix papilionis* (Vierick) is the dominant parasitoid of the caterpillars and *T. chilonis*, *Trichogramma incommodus*, and *D. papilionis* caused a cumulative parasitism of 88% (Krishnamoorthy and Singh, 1988). Most of the parasitoids of *P. demoleus* also attack another species of butterfly, *Papilio polytes*. The parasitoid *Telenomus incommodatus* (Nixon) was found to be effective against citrus butterflies (Krishnamoorthy and Singh, 1988).

The eggs of the fruit sucking moth, *Othreis fullonia*, were successfully parasitized by *T. chilonis*, suggesting the possibility of utilization of *T. chilonis* for the control of this pest (Dodhia et al., 1986).

Similarly, releases of exotic parasitoid, *Leptomastix dactylopii* Howard, from the West Indies have proved effective for the control of the citrus mealybug, *Planococcus citri* Mulsant, in India. It is a specific parasitoid of *P. citri*, possessing excellent host-searching ability. Field release of *L. dactylopii* resulted in the establishment in mixed plantations of citrus and coffee, and in citrus orchards in several parts of Karnataka, resulting in the control of *P. citri* within 3–4 months. No insecticidal sprays were required for the control of *P. citri* in the following season (Manjunath, 1985; Krishnamoorthy and Singh, 1987; Nagarkatti and Jayanth, 1992). The oriental mealybug, *Planococcus lilacinus*, was controlled by the local parasitoid, *Tetranemoidea indica* (Ayyar). The parasitoid *Tetrastichus radiatus* (Waterstone) was found effective against the citrus psylla, *Diaphorina citri* (Kuwayama). Eight species of parasitoids were found to be effective against the citrus leaf miner, *Phyllocnistis citrella* (Staint). *Tetrastichus phyllocnistis* (Narayanan) and *Ageniapsis* spp. were reported to control the citrus leaf miner. The parasitoids, *Encarsia divergence* (Silv.) and *Encarsia merceti* (Silv.), were found effective against the citrus black fly, *Aleurocanthus woglumi* (Ashby) (Krishnamoorthy, 2010).

The California red scale, *Aonidiella aurantii* (Mask.), was found to be parasitized by *Aphytis* spp. and recommended at 10–15 per tree. Release of *Aphytis melinus* at 2000/tree was effective (Krishnamoorthy, 1993). The citrus green scale, *Coccus viridis* (Green), was effectively parasitized by *Aneristus ceroplastae* and *Encyrtus lecaniorum* (Mayr.) (Tandon, 1985).

4.2.6.4 GUAVA

The striped mealybug, *Ferrisia virgata*, is a serious pest of guava in South India. The parasitoids, *Aenasius advena* (Comp.) and *Blepharis insularis* (Cam.), were found to control the mealybug (Mani et al., 1989). The wax scale, *Drepanococcus chiton* (Green), was found to be controlled by *Anicetus ceylonensis* (How.) and *Cephaleta brunniventris* (Mani, 1995). Three parasitoids, *Coccophagus cowperii* (Grlt.), *Coccophagus bogoriensis* (Con.), and *Aneristis* spp.

were effective against the green shield scale, *Chloropulvinaria psidii* (Mask.). The encyrtids *Coccidoxenoides peregrine* (Timberlake) and *Allotropa* spp. regulate *P. citri* in nature (Mani, 1994).

4.2.6.5 GRAPES

Indundative releases of *Trichogramma* spp. within commercial orchards were used in Australia for effective control of the tortricid leafroller, *Epiphyas postvittana* (Walker), in wine grapes (Glenn and Hoffman, 1997). Approximately 75% of egg masses in the vineyard where parasitized by *Trichogramma carverae* at release rates of 70,000 *Trichogramma* per hectare. Inundative releases of *Trichogramma platneri* (Nagarkatti) resulted in significantly greater parasitism of the oblique banded leafroller, *Chloristoneura rosaceana* (Harris), than in control plots (Lawson et al., 1997). An encyrtid parasitoid, *Anagyrus dactylopii* (How.), was found to be effective against grapevine mealybugs under Indian conditions (Krishnamoorthy, 2010).

4.2.6.6 MANGO

Several parasitoids, *Brachymeria lasus* (Walker), *Tetrastichus* spp., *Pediobius bruchida* (Rondani), *Hormius* spp., *Cathartoides* spp., and *Tetrastichus* spp. were found to control mango leaf webber, *Orthaga eudrusalis* (Walker), effectively (Krishnamoorthy, 2010).

4.2.6.7 POMEGRANATE

The fruit borer, *Deudorix isocrates* (Fabricius), was found to be parasitized by the egg parasitoids, *Telenomus* spp. and *Oenocyrtus papilionis* (Ashmead). Another borer, *Deudorix epijarbas* (Moore), is parasitized by the parasitoids *Telenomuscyrus* (Nixon), *Anastatus kashmirensis* (Matur), and *T. chilonis* (Ishii) (Krishnamoorthy, 2010). The aphelinid *Encarsia azimi* (Hayat) was found to parasitize ash whitefly, *Siphoninus philyreae* (Haliday) and *Leptomastrix dactyloppi* was found to parasitize mealybugs, *P. lilacinus*, *P. citri*, *C. viridis*, and *Macronellicoccus hirsutus* (Mani, 1994). The moth, *Apomyelois ceratoniae* (Zeller) (Lepidoptera, Pyralidae), is the most important pomegranate pest in Iran. The larval parasitoids, *Apanteles myeloenta* (Wilkinson) (Braconidae) and *Bracon hebetor* (Say) (Braconidae), and the pupal parasitoid, *Brachymeria minuta* (Linnaeus) (Chalcididae) were reported to be natural parasitoids in orchards (Nobakht et al., 2015).

4.2.6.8 BANANA

Eight primary parasitoids were recorded against the banana skipper, *Erionota thrax*, namely, *Ooencyrtus erionotae* (Hymenoptera: Encyrtidae), *Pediobius erionotae* (Hymenoptera: Eulophidae), *Agiommatus sumatraensis* (Hymenoptera: Pteromalidae), and *Palexosita solensis* (Diptera: Tachinidae), *Charops* spp. (Hymenoptera: Ichneumonidae), *Cotesia* (*Apanteles*) *erionotae* (Hymenoptera: Braconidae), *Xanthopimpla gampsura* (Hymenoptera: Ichneumonidae), and *Brachymeria thracis* (Hymenoptera: Chalcididae) as major egg, larval, and pupal parasitoids in Indonesia (Ahmad et al., 2008). The parasitization of *P. erionotae* was the highest in all stages, *i.e.*, eggs, larvae, and pupae, followed by *Brachymeria* sp.

4.2.7 Vegetables

4.2.7.1 CABBAGE

Cotesia plutellae is most common larval parasitoid of *Plutella xylostella* on cabbage and cauliflower in Indian conditions. *Diadegma fenestrale* and *Diadegma semiclausum* are the dominant larval parasitoids at high altitudes under Indian conditions. *Cotesia glomerata* parasitizes up to 65% of larvae of *Pieris brassicae*. *Diaeretiella rapae* is an important bioagent against cabbage aphids.

The larval parasitoid *C. plutellae* and the egg parasitoid *Trichogramma bactrae* can be conserved in the plains and *D. semiclausum* at higher altitudes to avoid pesticidal sprays (Singh, 1993; Jalali et al., 2001). Larval parasitoids *C. glomerata* (L.) and *Hypersota ebeninius* (Grav.) that parasitize *P. brassicae* larvae. *T. chilonis* and *Telenomus remus* were effective against the leaf-eating caterpillar, *S. litura*. When *Chelonus helopae* (Gupta) was released four times at 75,000 adults/ha (Gopala Krishnan and Mohan, 1993), they effectively managed *S. litura*. *Trichogramma* spp. are utilized in the Netherlands for the control of lepidopteran pests in cruciferous plants. Inundative release of *Trichogramma* was feasible for the control of *Mamestra brassicae* (Linn.) on Brussel sprouts (van der Schaff et al., 1984) and reduced larval infestation of *P. xylostella* (Linn.) in cabbage crop (Glas et al., 1981).

4.2.7.2 TOMATO

Trichogramma spp. and *C. chlorideae* are important bioagents of *H. armigera*. Inundative release of *Trichogramma brasiliensis* at 4000 adults/ha for 6 weeks suppressed the attack of the fruit borer on tomato (Dhaliwal et al., 2013). *Helicoverpa armigera* is suppressed by inundative releases of *T. chilonis*/*T. brasilensis*/*T. pretiosum* at 50,000/ha/week (Singh, 1993). The parasitoids *Hemiptarsenus varicornis* (Girault), *Gonotoma* spp., *Chrysonotomia*, and *Cirrospilus variegates* (Masi) controlled the serpentine leaf miner, *Liriomyza trifolii* (Burgess) (Krishnamoorthy, 2010).

4.2.7.3 BRINJAL

The brinjal shoot and fruit borer, *Leucinodes orbonalis* (Guen.), is parasitized by *Pristomerus testaceus* (Mori.), *Diadegma apostata* (G.), *Trathala flavoorbitalis* (Cameron), *Eriborus sinicus* (Holmgren), and *Phanerotoma* spp. (Krishnamoorthy, 2010). Oviposition and development of the ichneumonid *T. flavoorbitalis* (Cam.), a parasitoid of *L. orbonalis* (Guen.) was reported by Sandanayake and Edirsinghe (1992).

4.2.7.4 CHILLI

The eulophid parasitoid, *Ceranisus* spp. parasitizes the chilli thrips, *Scirtothrips dorsalis*. The gall midge, *Asphondylia gennadi* was parasitized by *Eurytoma* spp. (Krishnamoorthy, 2010).

4.2.7.5 ONION

Two parasitoids, *Ceranisus* spp. and *Thripobius* spp., are effective biocontrol agents against onion thrips (Krishamoorthy et al., 2006).

4.2.8 Plantation Crops

4.2.8.1 COCONUT

Two larval parasitoids wasps, *Goniozus nephantidis* (Muesebeck) (Bethylidae) and *Bracon brevicornis* (Wesmael) (Braconidae), are widely employed, either singly or in combination, for the biocontrol of *Opisina arenosella* in India (Venkatesan et al., 2009). Two egg parasitoids, *Chrysochalcissa oviceps* (Boucek) and *Gryon* sp. of coreid bug, *Paradasynus rostratus* (Distant) of coconut was reported by Priya and Faizal (2004).

4.2.8.2 TEA

Telenomus spp. is an important parasitoid of the tea mosquito bug, *Helopeltis antonii*. The purple scale, *Chrysomphalus aonidum* is found to be parasitized by *Comperiella bifasciata* and *Aphytis chrysomphali*. The Parasitoids, *Aphidius, Lipolexis, Aphelinus* regulate the aphid populations in different seasons (Jalali et al., 2010).

4.2.8.3 COFFEE

The parasitoids *C. bogoriensis* and *C. cowperi* are effective against the green scale, *C. viridis*. Two ectoparasitoids, *Cephalonomia stephanoderis* (Betrem) and *Prorops nasuta* (Waterston) (Hymenoptera: Bethylidae) and an adult endoparasitoid, *Phymastichus coffea* (La Salle) (Hymenoptera: Eulophidae) are known to suppress the coffee berry borer, *Hypothenemus hampei* (Ferrari) in Columbia (Aristizabel et al., 2011).

4.2.8.4 PEPPER

The parasitoid *Encarsia lounsburyi* can be released for effective management of black pepper scale, *Aspidiotus destructor* (Singh, 1996).

4.2.8.5 CARDAMOM

Friona spp., *Eriborus trochantertaus*, *Agrypon* spp., and *Xanthopimpla australis* are important parasitoids against shoot and capsule borers of cardamom (Singh, 1996).

4.2.8.6 FLOWER CROPS

Inundative releases of *Encarsia formosa* (Gahan) have been successful in some instances for the control of whitefly *Trialeurodes vaporariorum* (Westwood) on poinsettia. However, control of this species has been reported with lower weekly release at two parasitoids per plant or when *T. vaporariorum* cooccurred in the crop (Hoddle et al., 1998).

4.3 Conservative Biocontrol

Conservation of biocontrol agents is achieved by the modification of agroecosystem or existing practices to protect and enhance natural enemies or other organisms to reduce the effects of pests. It means the activities that preserve or protect the parasitoids and involves manipulation of agroecosystems to enhance the survival, fecundity, longevity, and behavior of parasitoids to increase their effectiveness. Conservation practices can be further categorized as those that focus on supplementary resources, controlling secondary enemies, or manipulating host plant attributes to benefit natural enemies (Rabb et al., 1976; Van den Bosch and Telford, 1964). Pest outbreaks tend to be less common in polycultures than in monocultures (Andow, 1991). Habitat management can be considered as a subset of conservative biocontrol that alters habitats to improve availability of the resources required by parasitoids for optimal performance. Habitat management may occur at within-crop, within-farm, or landscape levels (Landis et al., 2000).

Many proximate factors that limit the effectiveness of natural enemies in agricultural systems include pesticides, lack of adult food, and lack of alternate hosts (Powell, 1986; Dutcher, 1993; Rabb et al., 1976). Hence, the most important strategies to conserve parasitoids within an agroecosystem are cultural control by habitat manipulation and rational use of pesticides (Morales-Ramos and Rojas, 2003; O'Neil et al., 2003). A focus of many past conservation efforts has been to seek more selective pesticides to time and use pesticides to minimize their negative impacts on parasitoids (Ruberson et al., 1998). Increasing attention has been paid to conservation practices that seek to alter the quality of parasitoids habitat (Landis et al., 1999).

Diversity in agroecosystem may favor reduced pest pressure and enhanced activity of natural enemies (Altieri, 1991; Andow, 1991; Stamps and Linit, 1998). Intercropping chickpea with coriander was found to increase the activity of *C. chlorideae* and to decrease the population of

H. armigea (Shanta Kumari, 2003). A strip row of sesame in cotton field is reported to reduce the incidence of *H. armigera* by increasing the population of *Campoletis sonorensis* (Ambrose, 2007). Planting Sudan grass, as trap crop around maize, enhances the activity of the parasitoid *Apanteles flavipes* and reduces the occurrence of *C. partellus* (Khan and Pickett, 2004).

While some parasitoids obtain needed resources from hosts (Jervis and Kidd, 1986), others require nonhost foods. Floral nectar taken by many species (Jervis et al., 1993) results in an increased rate of parasitism (Powell, 1986). Several plants, like sunflower, have extra floral nectaries, nectar-producing glands on stems or foliage that attract and feed natural enemies (Kumar et al., 2013). Extra floral nectar produced by various plants, such as faba beans (*Vicia faba* L.) and cotton (*Gossypium hirsutum* L.), serves as an important food source for adult parasitoids (Bugg et al., 1989). Pollen may be consumed directly (Irvin et al., 1999) or as a contaminant within nectar. The presence of honey dew-producing insects has been suggested as desirable for some parasitoids. Experiments with the parasitoid *Pimpla turionellae* L. showed rapid weight loss for insects caged on flowers with inactive nectaries, illustrating an important advantage of nectar over pollen in providing water as well as nutrition (Orr and Pleasants, 1996). Floral architecture has influenced the selection of nectar plants by *Edovum puttleri* (Grisell), *P. foveolatus* (Crawford), *Trichogramma brassicae* (Bezdenko), *Aphidius colemani* (Vierck), *Lysiphlebus teatceipes* (Cresson), parasitoids of the Colarado potato beetle, *Leptinotarsa decemlineata*. *Edovum puttleri* feeds effectively upon flowers with exposed nectaries, while *P. foveolatus* utilizes flowers with partially concealed nectaries (Patt et al., 1997). The weed plant, *Erygium* sp., found around apple orchards, was found to increase the population of *A. mali* against the apple woolly aphid, *E. lanigera* (Kumar et al., 2013). Zhu et al. (2015) assessed the effect of sesame as a nectar plant raised on rice bunds and found increased survival of *Apanteles ruficrus*, *Cotesia chilonis*, and *T. chilonis* increased due to the presence of *Sesamum indicum* flowers providing nectar for the parasitoids.

Natural enemies, especially parasitoids, require shelter from environmental hazards, and a lack of shelter during extreme climates, rain, or pesticide application may be highly detrimental to their survival. Excessive wind is thought to limit foraging by adult hoverflies (Beane and Bugg, 1998). Orr et al. (1997) interplanted rye grass (*Lolium multiflorum* Lambert) in seed maize (*Zea mays* L.) fields to reduce the temperature of soil surface, increasing survival of augmentatively released *Trichogramma brassicae* (Bezdenko). Ground covers or intercrops have influence parasitoid density in cabbage, *Brassica oleracea capitata* (Theunissen et al., 1995).

Providing sufficient alternate prey is another way of habitat conservation. Parasitoids are mostly host-specific and they do not feed on other hosts (Powell, 1986). However, the ability of some parasitoids to parasitize more than one host species has been explored where the aphid *Myzus persicae* was found to be parasitized by *Aphidius colemani* (Vierck) and *L. testaceipes* (Cresson) with *S. graminum* (Rond.) as alternate host, other than their primary hosts (Stary, 1993).

5. MASS REARING OF PARASITOIDS

The overall performance of a parasitoid depends largely on its quality which is mainly determined by the mass rearing program. Mass production of an entomophagous species is a skillful and highly refined process. Natural or factitious hosts and rearing media through insectary procedures result in economic bulk production of that species.

5.1 Egg Parasitoid, *Trichogramma* spp.

The egg parasitoid, *Trichogramma* spp., is commonly distributed worldwide, parasitizing over 200 species of insects belonging mainly to Lepidoptera, Coleoptera, Neuroptera, and Diptera. They are found in diverse habitats ranging from aquatic to arboreal. Due to its amenability to mass production, this group of parasitoids has the distinction of being the maximally-produced and released natural enemy in the world (Manjunath et al., 1988).

Large-scale production of *Trichogramma* is directly dependent on the successful mass production of its factitious host. Eggs of *Sitotroga cerealella* (Olivier) are largely used for the mass production of *Trichogramma* in countries like Russia, the USA, France, and Germany, where large-scale releases are practiced. In China, eri and oak silkworm eggs are used for the mass production of *Trichogramma* (Wang et al., 2014). In India, the rice moth, *Corcyra cephalonica* (Stainton), is used as the laboratory host of *Trichogramma*. Eggs of *C. cephalonica* were found to be superior to other hosts in the quality of mass-reared parasitoids.

The first step in the mass rearing of the parasitoids is to obtain healthy host eggs free from infestation of mites and other foreign pathogens. By following the improved mass production technology (Paul and Sree Kumar, 1998), clean eggs of *C. cephalonica* are obtained, which are exposed to ultraviolet (UV) rays (15 wt. UV tube) for 45 min in a closed chamber maintaining a distance of 12.5–15 cm between the eggs and the tube. Using this treatment, the embryo is killed whereas the egg quality is not affected. This prevents the destruction of parasitized eggs by *Corcyra* larvae if they are not treated with UV rays. The eggs could also be made inviable by exposing them to very low temperatures 0–2 °C in the freezer for 3–4 h. By freezing, the quality of the host eggs are found to shrink and the eggs form lumps due to moisture released by the frozen eggs. The host eggs treated with UV rays can be stored in the lower chamber of the refrigerator for up to 5 days and can be used for parasitization. The treated eggs are placed in plastic vials (9 × 5 cm) a lid fitted with wire mesh (40 mesh) to enable uniform spreading of the eggs on the cards.

The egg cards used for pasting the eggs are 15 × 7.5 cm, containing segments of 2.5 × 1.25 cm. Each segment can hold approximately 0.1 ml of *Coryra* eggs when pasted evenly on a single layer. Dilute acacia gum is used for pasting the eggs. While pasting the eggs, a small area is left free at the anterior end to write the details of the culture. Culture details are also written on the back to prevent mixing of cultures. In laboratories where more than one species of *Trichogramma* is mass produced, different color codes may be used to prevent possible mixing. As far as possible, different species may be maintained by different persons or on different dates to prevent contamination. Loose eggs attached to the egg cards may be removed by gently brushing the dried cards. The egg cards are then placed in suitable glass vials, bottles, or polythene bags in which freshly emerged parasitoids are present. To avoid escape of adult parasitoids during introduction of cards for parasitization, parasitized egg cards from which adult emergence is expected the subsequent day are kept in vials, bottles, or polythene bags kept at 27 ± 2 °C along with fresh egg cards. Host parasitoid ratio of 6:1 or more is to be followed to avoid superparasitism. Normally, the eggs are exposed for parasitization for 24–48 h under a source of cold light. The eggs can also be exposed for parasitization as loose eggs by moistening the inner walls of the vials or bottles and adding a measured quantity of eggs into the bottle to form a uniform layer on the inner sides by rotation. After successful parasitization, indicated by blackening, the host eggs can be brushed off and stored or transported.

Parasitized eggs start turning black on the third day after parasitization, and complete blacken on the fourth day. Normally 80–90% success is expected in healthy cultures. The life cycle of *Trichogramma* is completed in 7–8 days whereas in *Trichogrammatoidea* it is completed in 8–9 days. Parasitized egg cards can be stored in the refrigerator at 12–15 °C for 10–15 days after blackening is complete. The best stage for storage is the pupal stage. However, during storage, the quality of the emerging adults is affected. Prolonged storage beyond 50 days will impair emergence, longevity, and fecundity of the resultant progeny (Figure 1).

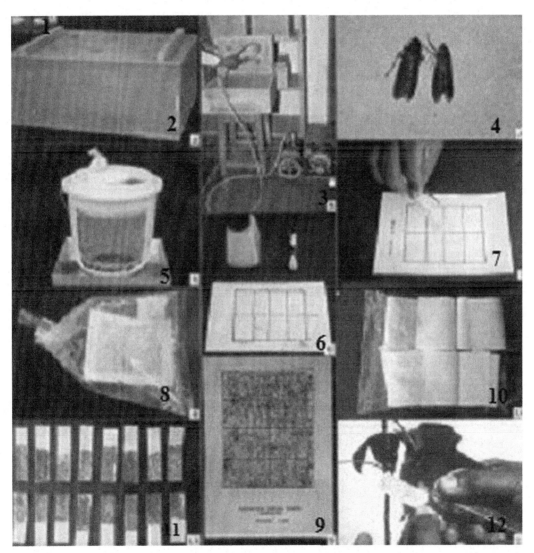

FIGURE 1 **Steps in the mass production of *Trichogramma* in the laboratory.** (1) and (2) *Corcyra* rearing box; (3) Adult collection with vacuum pump; (4) *Corcyra* male and female moths; (5) Oviposition cage; (6) Preparation of egg card; (7) Gluing of eggs over the egg card; (8) Parasitization of *Corcyra* eggs; (9) Fully-parasitized egg cards; (10) Packing and forwarding of parasitized cards; (11) Separation of small units of parasitized egg cards; (12) Release of parasitoids.

Parasitized egg cards can be easily transported in the pupal stage (3–4 days after parasitization) by folding the cards and kept in cardboard boxes or kept in plastic vials and sent by the fastest mode of transport with proper labels, such as "Live beneficial insects" (Ram, 2003). The parasitoids may be released as adults or the parasitized egg cards may be stapled to the underside of the leaves of the crop plants so that the parasitoids may emerge and attack the host eggs. Inundative releases are normally followed. In case of adult release, the parasitoid can be fed with 50% honey offered as fine streaks on the side of the container released by holding the container open upwards and move in all directions in the field to ensure effective dispersal (Hoy et al., 1991). Small plastic release boxes fitted with wire mesh at the bottom can be distributed in different parts and parasitized egg cards can be kept before emergence (Krishnamoorthy, 2012). Such release boxes may be either fixed on wooden poles or hung on the host plant. This method will reduce the loss of parasitized eggs due to predators or rain.

5.2 Egg Parasitoid, *Telenomus remus* Nixon (Scelionidae: Hymenoptera)

The egg parasitoid, *T. remus* is reared on eggs of *Spodoptera mauritia* (Boid) in Malaya as described by Nixon (1937). It parasitizes eggs of different species of *Spodoptera*, even on overlapping layer egg masses (Bueno et al., 2008). Each female of *T. remus* produces about 270 eggs during its reproductive life span (Morales et al., 2000). This parasitoid has been released in corn fields, as part of IPM programs in Venezuela, with a parasitism rate up to 90% (Ferrer, 2001), demonstrating the high potential of biocontrol for several species of Spodoptera. Its facticious hosts of the insect are *S. litura*, *Spodoptera frugiperda* (Smith), *Spodoptera exigua* (Hubner.), *Agrotis spinifera* (Hubner.), and *C. cephalonica*.

The eggs of the hosts (0–24h old) are pasted on egg cards and exposed to freshly emerged parasitoids kept in 10×2.5 cm size glass vials. A fine streak of honey is provided as food on the inner walls of the vials. The parasitized eggs will turn greyish black on the seventh day of parasitization and on the eighth day in the case of *C. cephalonica*. The total life cycle is completed in 11–13 days in *C. cephalonica* and in 10 days in other hosts. Optimum conditions for rearing are 27±2 °C and 75±5% relative humidity (RH). After 6–8 days, parasitized eggs can be safely stored in a refrigerator at 10±1 °C for 7 days without affecting the efficacy of parasitoid. Storage in refrigerator beyond 7 days may affect the quality of the parasitoids (Ram, 2003; Goulart et al., 2011).

5.3 Egg–Larval Parasitoid, *Chelonus blackburni* Cameron (Braconidae: Hymenoptera)

The egg–larval parasitoid *Chelonus blackburni* (Cameron) (Figure 2) attacks the egg stage and completes development in the larval stage of the host. It parasitizes the pink bollworm in Hawaii, North America, and Egypt and all three bollworms, *Earias* spp., *H. armigera*, and *Pectinophora gossypiella* (Saunders) in India. The factitious hosts of the parasitoid are *C. cephalonica* and *P. opercullella*.

Fresh eggs of *C. cephalonica* (0–24h old) are pasted on an egg card (15×7.5 cm) to accommodate 1 ml of eggs per card and exposed to newly-emerged parasitoids. The parasitized eggs are transferred to another plastic jar (20×15 cm) containing approximately 500 g of broken sorghum grains. The uniparental, egg–larval parasitoid develops inside a parasitized host

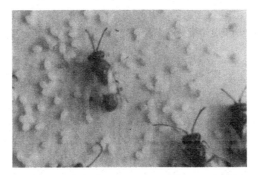

FIGURE 2 **Egg larval parasitoid,** *Chelonus blackburni.*

larva until it reaches last instar. The final instar larva of the parasitoid cuts out of the host and finishes devouring it externally. It then spins a cocoon usually adjacent to the carcass of the consumed host. The parasitized host is considerably smaller than an unparasitized host of the same age. After the cocoon is formed, the parasitoid becomes a prepupa and discharges its meconium prior to pupation. The adult emerges through a hole cut at the anterior part of the cocoon. Optimum conditions for rearing are $27 \pm 2\,°C$, $65 \pm 5\%$ RH and 14 h photoperiod (Chanda and Chakravorty, 2001; Ram, 2003).

A host parasitoid ratio of 100:1 should be maintained to avoid superparasitism. Host eggs more than 48 h old are less preferred for parasitization compared with fresh eggs. Target pests of this parasitoid are *P. gossypiella* and *Earias* spp.

5.4 Larval Parasitoid, *Cotesia flavipes* Cameron (Braconidae: Hymenoptera)

Cotesia flavipes is an endoparsitoid, occurring naturally on all important insects of sugar cane borers, except the root borer in India. In Mauritius, it is specific to *C. sacchariphagus* (Boger), and in Pakistan strains of this parasitoid are reported attacking *Chilo* spp. Within 2 h after sunrise both sexes emerge simultaneously. Light induces immediate copulation which lasts for 45–60 s. Both sexes show polygamous mating. The female oviposits immediately after emergence but the percent of oviposited eggs increase with time. As many as 86 eggs have been observed to be deposited in one insertion with an average of 25. The insemination ratio for wild females is 98.0%. In *D. saccharalis* (Fabricius) host finding is mediated by a water-soluble substance present in the fresh frass of larvae. The egg period is completed in 24–48 h. The larval stage is completed in 11 days and pupal stage in 5 days. Adult longevity is 5–7 days. About 21% of adults are males.

Cotesia flavipes are multiplied using rearing cage consisted of a glass jar (25×12 cm), blackened from outside and containing a column of water (about 1 cm high) at the bottom. A circular piece of fine wire gauze (about 12 cm) is placed horizontally inside the jar about 5 cm above the bottom. The mouth of the jar is covered with a piece of rough muslin cloth (or cotton netting) and anesthetized borer larvae are placed for parasitization. A glass sheet is placed over the mouth of the jar to keep the larvae in position on the underneath muslin cloth. Adult parasitoids are released inside the jar, where opened raisins are placed over the wire gauze portion. They parasitize the borer larvae through muslin cloth. Fresh lots of host larvae are exposed for parasitization at 24 h intervals and parasitized larvae are reared separately in cane setts till the parasitoid pupae emerge from them. Parasitoid cocoons when formed are transferred to rearing jars and placed on the wire gauze sheet till the emergence of wasps (Singh, 1994). With this technique, 6–8 *C. auricilius* (Ddgn.) or *C. partellus* larvae could be parasitized by a single female wasp and up to 40 adult parasitoids could be reared from single host larvae.

5.5 Papaya Mealybug Parasitoid, *Acerophagus papayae*

Two-month-old robin eyed healthy potatoes were procured, washed in water, and disinfected with a 5% sodium hypochlorite solution. Later, a 2 cm incision was given using a sharp blade and treated with 100 ppm of gibberellic acid solution for 30 min. The potatoes were air-dried and transferred to plastic trays of 18 inch diameter containing solarized sand and covered with black cloth. The potatoes were kept for germination. Sprouting was observed within a week and the potatoes were ready for infestation with mealybugs.

Paracoccus marginatus (Williams and Granara de Willink.) was collected from papaya (*Carica papaya* L.) fields and transferred using a camel hair brush over the potato sprouts (each potato with three to five ovisacs). The rootlets of the potatoes were clipped using sharp scissors to arrest the further growth of potato sprouts. Twenty five sprouted potatoes colonized with mealybugs were transferred to oviposition cages (45 × 45 × 45 cm). Twenty pairs of *A. papayae* (Figure 3) were allowed inside the cages for parasitization. Providing 10% honey streaked in a wax paper and water in glass bottles with sponge served as adult food. After 10 days, the sprouts were removed and transferred to plastic containers. The emerged parasitoids were collected using an aspirator and used for further mass production and field release (Figure 4) (Sankar et al., 2012).

5.6 Impact of Volatiles on Tritrophic Interaction of Parasitoids

The relationship between host plant insect pests and natural enemies is based on two hypotheses. The first states that vegetationally-diverse habitats support a greater diversity of prey and thus have more stable populations of enemies. The second hypothesis points to the relative

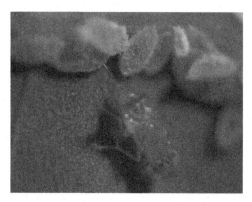

FIGURE 3   Insertion of ovipositor of *Acerophagus papayae* into the mealybug, *Paracoccus marginatus*.

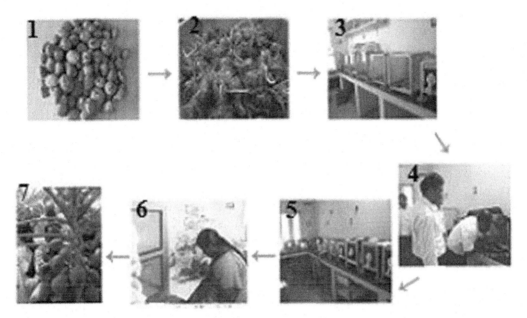

FIGURE 4 **Mass production of** *Acerophagus papayae* **in the laboratory.** (1) Seed healthy potatoes with Brownian eyes; (2) Sprouted potatoes; (3) Mass culturing of mealybugs over sprouted potatoes inside culture cages; (4) Parasitoid inoculation; (5) Mass culturing of parasitoids; (6) Collection of parasitoids using aspirator; (7) Field release.

attractiveness of a habitat to a particular arthropod, based on the concentration of host plants or prey species. Plant-produced chemicals play a key role in host/prey selection for parasitoids (Nordlund et al., 1998). The parasitization of *Helicoverpa zea* (Boddie) eggs by *T. pretiosum* Riley females was relatively high in plots of tomato, but almost nonexistent in the adjacent corn plots, revealing that the presence of volatile stimuli from the tomato plants in the corn fields resulted in an increased rate of parasitization by *T. pretiosum* (Whiteman and Nordlund, 1994).

Parasitoids not only respond to the volatiles or chemicals produced by the plants but also the chemicals produced or released by the herbivore themselves. Volatile compounds emanating from the scales of *H. armigera* and *C. cephalonica* were identified as hextriacontane, decosane, and monoocasane, which increased the activity of *T. chilonis*, an egg parasitoid of Lepidoptera. Similarly, kairamones from *H. armigera* eggs influence *T. chilonis* (Ananthakrishnan, 1993). These chemicals could be applied to the crop in order to attract the natural enemies, thereby increasing the likelihood of an encounter with pest hosts and consequently improving the rate of parasitism. The potential for manipulation of tritrophic interactions to enhance parasitism by applications of low doses of toxicants also needs to be explored (Wright and Verkerk, 1995).

5.7 Artificial Diets in Mass Culturing

The artificial rearing of parasitoids started long ago, with the main goal to obtain a means to multiply and produce parasitoids for biocontrol of pests and also to conduct studies on the biology, physiology, and behavior of parasitoids (Grenier et al., 2005).

Parasitoids are carnivorous species, and they need a protein-rich diet, sometimes with some special requirements in aromatic amino acids for cuticle tanning, concentrated energy from carbohydrates, and lipids. All needs have to be fulfilled, taking into account osmotic pressure and dietary balances. According to the presence or absence of insect components (host egg content, hemolymph or holotissue extract) artificial media can be divided into two main categories. Egg content of the host is generally used as artificial medium (Volkaff et al., 1992). Host egg homogenates increase yield of *T. pretiosum* (Xia et al., 1986). In artificial media, contamination by microorganisms can easily develop. Therefore, preservatives like antibiotics and antifungals are necessary for preparing artificial culture media for mass production. The antibiotics used are mainly penicillin, streptomycin, and gentamycin (Grenier and Liu, 1990).

Brachymeria intermedia is a solitary pupal endoparasitoid of a wide range of lepidopterans, including *Lymantria dispar* (L.). Artificial diets of this parasitoid contain host components, mainly extract of *G. mellonella* pupae (Bratti, 1989). In some media, the pupal extract is replaced by a larval extract obtained by the same method or with pupal or larval homogenate obtained by squeezing pupae in a syringe and then removing the large pieces of cuticle. In some of the media, the host content is partially or totally replaced with different amounts of chicken egg yolk, yeast, and wheat germ (Dindo et al., 1997).

Simmonds (1944) proposed that the development of artificial diets for entomophagous insects would greatly improve their augmentation and thereby the potential for biocontrol. However, even after six decades of research, a number of publications on artificial diets for mass rearing, about 135 bioagents involved in mass rearing of biocontrol agents in North America alone, and viable commercial mass rearing programs for entomophages based on artificial diets are still not available (Cohen, 2003).

The development of artificial diets for the mass rearing of parasitoids has numerous impediments. The artificial diet must be nutritionally adequate and meet phagostimulatory needs, maintain adequate moisture, be free from microbial contamination, and should be efficiently packed to maintain moisture (Cohen, 1992; Cohen et al., 1999). Artificial diets for parasitoids are often difficult to develop because of the constraints of chemical cues that encourage feeding as necessary ingredients for sustained growth and reproduction (Berlinger, 1992). The host-specific kairamones involved in bitrophic or tritrophic interactions of host plant–host insect/prey biocontrol agent should be incorporated in the artificial diet. Hence artificial diets should address the nutritional demands and phagostimulatory needs of parasitoids (Ambrose, 2010).

6. FUTURE AIMS

- Outstanding case-specific success in the biocontrol of crop pests by exotic parasitoids advocate importing parasitoids between countries.
- Identification and development of improved virulent strains of parasitoids for different pests in different cropping systems and different agroecosystems.
- Standardization of artificial diets and low-cost production technology for mass culture of potential parasitoids. Methods should be devised for scaling, mechanization, and improvement of efficacy of parasitoids and to increase their the shelf-life.

- Development of pesticide and temperature-resistant strains suitable for different agroecosystems.
- The use of pheromones to determine time of release of parasitoids and kairamones to increase efficacy of parasitoids.
- Awareness among farmers of the use of parasitoids in the suppression of agricultural and horticultural pests. Many more large-scale multiplication units should be started to ensure the supply of proven parasitoids to farmers.

7. CONCLUSIONS

Biocontrol of pest is an important component of IPM, where bioagents, exclusively parasitoids, play a key role and have received much more attention in recent years. India is a country with a rich biodiversity and rich wealth of parasitoids of crop pests. Avoidance of disruptive practices and suitable manipulation of ecosystem will enhance the scope of naturally-occurring biocontrol. To achieve the twin objectives of pest management and environmental protection, the focus is on biocontrol, the success of which will solely depend on the demonstration of parasitoid-based technology and assured availability of quality parasitoids to the farmers. Methods of conserving existing natural enemies should also receive proper attention.

Acknowledgment

The authors greatly acknowledge Tamil Nadu Agricultural University for providing necessary assistance to pay attention and to carry out the necessary actions in bringing out this chapter.

References

Ahmad, I., Maramis, R., Sastrodihardjo, S., Permana, A.D., 2008. Abundant parasitoids of *Erionota thrax* (Lepidoptera; Hesperidae) in four banana plantations around bandung areas. Presented at International Conference of Mathematics and Natural Sciences (ICMNS) Institute Teknologi Bandung, 28–30, October, 2008.

Altieri, M.A., Nicholls, C.I., 1998. Biological Control in Agro Ecosystems through Management of Entomophagous Insects. Commonwealth Publishers, New Delhi, pp. 67–86.

Altieri, M.A., 1991. Increasing bio-diversity to improve insect pest management in agroecosystems. In: Hawksworth, D.L. (Ed.), Bio-Diversity of Microorganisms and Invertebrates: Its Role in Sustainable Agriculture. CAB International, Wallingford, UK. 302 p.

Ambrose, D.P., 2007. The Insects: Beneficial and Harmful Aspects. Kalyani Publishers, p. 801.

Ambrose, D.P., 2010. Mass rearing of entomophagous insects for biological control: success, bottlenecks and strategies - a review. In: Ignatchi Muthu, S.J., David, B.V. (Eds.), Non Chemical Insect Pest Management. Elite Publishing House Private Ltd, New Delhi, India, pp. 156–169.

Ananthakrishnan, T.N., 1993. Changing dimensions in the chemical ecology of phytophagous insects: role of infochemicals on the behavioural diversity. In: Anantha Krishnan, T.N., Raman, A. (Eds.), Chemical Ecology of Phytophagous Insects. Oxford and IBH Publishing, New Delhi, pp. 1–20.

Andow, D., 1991. Vegetational diversity and arthropod population response. Annual Review of Entomology 36, 561–586.

Aristizabel, L.F., Jimenz, M., Bustilo, A.E., Arthurs, S.P., 2011. Introduction of parasitoids of *Hypothenemus hampei* (Coleoptera: Curculionidae: Scolytinae) on small coffee plantations in Colombia through farmer participatory methods development. Florida Entomologist 94 (3), 690–693.

Arnaud, P.H., 1978. A Host-Parasite Catalog of North American Tachinidae (Diptera). USDA Misc. Publ. No. 13199. 860 pp.

Ashraf, M., Fatima, B., Hussain, T., Ahmad, N., 1999. Biological control: an essential component of IPM programme for sugarcane borers. In: Symposium on Biological Control in the Tropics, MARDI Training Centre, Serdang, Selangor, Malaysia, 18–19 March, 1999, pp. 38–42.

Bajpai, N.K., 2010. Biological control of insect pests in crops. In: Bio Pest Management. Agrotech Publishing Academy, Udaipur, pp. 352–369.

Beane, K.A., Bugg, R.L., 1998. Natural and artificial shelter to enhance arthropod biological control agents. In: Pickett, C.H., Bugg, R.L. (Eds.), Enhancing Biological Control: Habitat Management to Promote Natural Enemies of Agricultural Pests. University of California, Berkeley, pp. 239–254.

Beglyarov, G.A., Smetnik, A.I., 1977. Seasonal colonization of the entomophagous of U.S.S.R. In: Ridgeway, R.L., Vinson, S.B. (Eds.), Biological Control by Augmentation of Natural Enemies. Plenum Press, New York, pp. 283–328.

Belshaw, R., 1993. Tachinid flies (Diptera: Tachinidae). London: Royal Entomological Society of London, 10 Part 4a (i), 170 pp.

Berlinger, M.J., 1992. Importance of host plant or diet on the rearing of insects and mites. In: Anderson, T., Eppla, N.L. (Eds.), Advances in Insect Rearing for Research and Pest Management. Westview Press, Boulder, Colorado, pp. 237–249.

Bharatwaj, D., Kaushikb, U.K., Pawar, A.D., 1987. Incidence of *Heliothis armigera* (Hubner) and parasitization by *Campoletis chloridae* (Uchida) in Chhattisgarh, Madhya Pradesh. Journal of Biological Control 1 (2), 79–82.

Bhatnagar, V.S., Davies, J.C., 1979. Pest management in intercrop subsistence farming. In: Proceedings of International Workshop in Intercropping. ICRISAT, Patancheru, AP, India, pp. 249–257.

Bhattacharya, B., Basit, A., Saikia, D.K., 2006. Predators and parasitoids of rice insect pests of Jorhat district of Assam. Journal of Biological Control 20 (1), 37–44.

Bratti, A., 1989. Techiche di allevamento in vitro per gli stadi larvali di insetti entomofagi parassitoidi. Bolletino dell' Istituto di Entomologica della Universita degli studi di Bologna 44, 169–220.

Bueno, R.C.O.F., Carneiro, T.R., Pratissolli, D., Bueno, A.R., Fernandes, O.A., 2008. Biology and thermal requirements of *Telenomus remus* reared on fall armyworm *Spodoptera frugiperda* eggs. Ciência Rural 38, 16.

Bugg, R.L., Ellis, R.T., Carlson, R.W., 1989. Ichneumonidae (Hymenoptera) using extra floral nectar of faba bean (*Vicia faba* L. Fabaceae) in Massachusetts. Biological Agriculture & Horticulture 6, 107–114.

Chanda, S., Chakravorty, S., 2001. Egg-larval parasitoid: *Chelonus blackburni* Cameron (Braconidae: Hymenoptera). Indian Journal of Experimental Biology 39, 143–147.

Charnov, E.L., 1982. The Theory of Sex Allocation. Princeton University Press, Princeton, NJ. 335 pp.

Cohen, A.C., Nordlund, D.A., Smith, R.A., 1999. Mass rearing of entomophagous insects and predaceous mites, are the bottle necks, biological engineering, economic or cultural? Biocontrol News and Information 20 (3), 85–90.

Cohen, A.C., 1992. Using a systematic approach to develop artificial diet for predators. In: Anderson, T.E., Neppla, N.C. (Eds.), Advances in Insect Rearing for Research and Pest Management. Westview Press, Boulder, CO, USA, pp. 77–91.

Cohen, A.C., 2003. The past and present of insect diets and rearing as precursors to a bright future. In: 10th Workshop of the IOBC Global Working Group on Arthropod Mass Rearing and Quality Control. Agropolis International, Montpellier, France 21–25 September, 2003, Global IOBC Bulletin, 2:12.

David, H., 1980. Tachinid parasites to control moth borers in India. In: Sithanandam, S., Solayappan, A.R. (Eds.), Biological Control of Sugarcane Pests in India. Tamil Nadu State Federation of Co-operative Sugar Factories Ltd, Chennai, pp. 61–66.

DeBach, P., Hagen, K.S., 1964. Manipulation of entomophagous species. In: DeBach, P. (Ed.), Biological Control of Insect Pests and Weeds. Chapman and Hall, London, pp. 429–458.

DeBach, P., 1974. Biological Control by Natural Enemies. Cambridge University Press, London, p. 323.

DeBach, P., Rosen, D., 1991. Biological Control by Natural Enemies. Cambridge University Press, Cambridge, UK.

DeBach, P., 1964. Successes, trends and future possibilities. In: DeBach, P. (Ed.), Biological Control of Insect Pests and Weeds. Chapman and Hall, London, pp. 429–455.

Dhaliwal, G.S., Singh, R., Jindal, V., 2013. A Text Book of Integrated Pest Management. Kalyani Publishers, New Delhi, pp. 140–169.

Dindo, M.L., Farneti, R., Gardenghi, G., 1997. Artificial culture of the pupal parasitoid *Brachymeria intermedia* (Nees) (Hymenoptera: Chalcididae) on oligidic diets. Boletin de la Asociation Espanola de Entomologia 21, 11–15.

Disney, R.H.L., 1994. Two remarkable new species of shuttle fly (Diptera: Phoridae) that parasitize termites (Isoptera) in Sulawesi. Systematic Entomology 11, 413–422.

Dodhia, J.F., Yadav, D.N., Patel, C., 1986. Management of fruit sucking moth, *Othreis fullonica* L. (Lepidoptera: Noctuidae) in citrus orchards at Anand (Gujarat). GAU Research Journal 11, 72–73.

Doomra, S., Maninder, S., Varma, G.C., 1994. Studies on the effect of *Trichogramma japonicum* Ashmead (Hymenoptera: Trichogrammatidae) on the control of *Scirpophaga excerptalis* (Walker) (Lepidoptera: Pyralidae). In: Biological Control of Insect Pests, pp. 143–144.

Doutt, R.L., 1964. The historical development of biological control. pp. 21–42. In: DeBach, P. (Ed.), Biological Control of Insect Pests and Weeds. Chapman and Hall Ltd, London. 844 pp.

Doutt, R.L., Annecke, D.P., Tremblay, E., 1976. Biology and host relationships of parasitoids. pp. 143–168. In: Huffaker, C.B., Messenger, P.S. (Eds.), Theory and Practice of Biological Control. Academic Press, New York. 788 pp.

Duraimurugan, P., Mishra, A., Pratap, A., Singh, S.K., 2014. Toxicity of spinosad to the pulse beetle, *Callosobruchus chinensis* (Coleoptera: Bruchidae) and its parasitoid, *Dinarmus basalis* (Hymenoptera: Pteromalidae). The Ecoscan 8 (1 & 2), 17–21.

Dutcher, G.D., 1993. Recent examples for conservation of arthropod natural enemies in agriculture. In: Lumsden, R.D., Vaughn, J.L. (Eds.), Pest Management: Biologically Based Technologies. American Chemical Society, Washington, DC, pp. 101–108.

Eggleton, P.B.R., 1992. Insect parasitoids: an evolutionary overview. Philosophical Transactions of the Royal Society of London Series 337, 1–20.

Elizangela, L.V., Pereira, F.R., Tavares, M.T., Pastori, P.L., 2011. Record of *Tetrastichus howardi* (Hymenoptera: Eulophidae) parasitizing *Diatraea* sp. (Lepidoptera: Crambidae) in sugarcane crop in Brazil. Entomotropica 26 (3), 143–146.

Ferrer, F., 2001. Biological control of agricultural insect pest in Venezuela; advances, achievements, and future perspectives. Biocontrol News and Information 22, 67–74.

Flanders, S.E., 1930. Mass production of egg parasitoids of the genus *Trichogramma*. Hilgardia 4, 464–501.

Flint, M.L., Dreistadt, S.H., 1998. Natural Enemies Handbook: The Illustrated Guide to Biological Pest Control. Univ. Calif. Div. Agric. Nat. Res. Publ. 3386, Oakland.

Gauld, D., 1986. Taxonomy, its limitations and its role in understanding parasitoid biology. In: Waage, J., Greathead, D. (Eds.), Insect Parasitoids. Academic Press Inc., Orlando, FL, pp. 1–19.

Glas, P.C., Smiths, P.H., Vlaming, V., van Lenteren, J.C., 1981. Biological control of lepidopteran pests in cabbage crops by means of inundative release of *Trichogramma* species, combination of field and laboratory experiments. Mededelingen Faculteit Landbouwwetenschappen Rijksuniversiteit Gent 46, 487–497.

Glenn, D.C., Hoffman, A.A., 1997. Developing a commercially viable system for biological control of light brown apple moth (Lepidoptera: Torticidae) in grapes under endemic Trichogramma (Hymenoptera: Trichogrammatidae). Journal of Economic Entomology 90, 372–383.

Godfray, H.C.J., 1994. Parasitoids: Behavioral and Evolutionary Ecology. Princeton University Press, Princeton, NJ.

Gopala Krishnan, C., Mohan, K.S., 1993. Epizootics of *Erynia neoaphidis* (Zygomycetes: Entomophthorales) in field population of *Brevicoryne brassicae* (Aphididae: Homoptera) on cabbage. Entomon 18, 21–26.

Goulart, M.M.P., Bueno, A.D.F., Bueno, R.C.B., Diniz, A.F., 2011. Host preference of the egg parasitoids *Telenomus remus* and *Trichogramma pretiosum* in laboratory. Revista Brasileira de Entomologia 55 (1), 129–133.

Greathead, D.J., 1995. Benefits and Risks of Classical Biological Control. Cambridge University Press, Cambridge, UK, pp. 55–63.

Greathead, D.J., 1986. Parasitoids in classical biological control. In: Waage, J., Greathead, D. (Eds.), Insect Parasitoids. Academic Press, Orlando, FL, pp. 289–318.

Grenier, S., Liu, W.H., 1990. Antifungals: mold control and safe levels in artificial media for *Trichogramma* (Trichogrammatidae, Hymenoptera). Entomophaga 35, 283–291.

Grenier, S., Gomes, S., Febvay, G., Bolland, P., Parra, J.R.P., 2005. Artificial Diet for Rearing *Trichogramma* Wasps on Biological Control of Arthropods. M.S. Hoddle Compiler, Davos, Switzerland, pp. 481–487.

Gunathilagaraj, K., Ganesh Kumar, M., 1997. Rice pest complex. An appraisal. The Madras Agricultural Journal 84, 249–262.

Gurr, G., Wratten, S., 2000. Measures of Success in Biological Control. Kluwer, Dordrecht. 48 pp.

Hagen, K.S., Franz, J.M., 1973. A history of biological control. pp. 433–476. In: Smith, R.F., Mittler, T.E., Smith, C.N. (Eds.), A History of Entomology. Annu. Rev. Inc., Palo Alto, California. 517 pp.

Hagen, K.W., van den Bosch, R., Dahlsten, D.L., 1971. The importance of naturally-occurring-biological control in western United States. In: Huffaher, C.B. (Ed.), Biological Control. Plenum Press, New York, pp. 253–293.

Hagler, J.R., 2000. Biological control of insects. In: Recheigl, J.E., Recheigl, N.A. (Eds.), Insect Pest Management: Techniques for Environmental Protection. Lewis Publishers, New York, pp. 207–241.

Hassan, S.A., 1982. Mass production and utilization of *Trichogramma*. Results of some research projects related to the practical use in the Federal republic of Germany. Les *Trichogramma*. Colloque INRA 9, 213–218.

Hassan, S.A., 1993. The mass rearing and utilization of *Trichogramma* to control lepidopterous pests: achievements and outlook. Pesticide Science 37, 387–391.

Hawkins, B., 1994. Pattern and Process in Host-Parasitoid Interactions. Cambridge University Press, Cambridge, UK.

Hawkins, B.A., Sheehan, W., 1994. Parasitoid Community Ecology. Oxford University Press, Oxford. 516 pp.

Hertig, B., 1960. Biologie der westpalaarktischen Raupenfliegen (Dipt. Tachinidae). Monographien zur angewandten Entomologie 16, 1–188.

Hoddle, M.S., Van Driesche, R.G., Sanderson, J.P., 1998. Biology and use of the whitefly parasitoid, *Encarsia formosa*. Annual Review of Entomology 43, 645–669.

Hoffmann, M.P., Frodsham, A.C., 1993. Natural Enemies of Vegetable Insect Pests. Cooperative Extension. Cornell University, Ithaca, NY. 63 p.

Hokkanen, H.M.T., 1993. New approaches in biological control. In: Pimental, D. (Ed.), Biotechnology in Integrated Pest Management. CBS Publishers and Distributors, New Delhi, pp. 185–198.

Howard, L.O., 1930. A History of Applied Entomology. Smithsonian Misc. Coll., vol. 84, pp. 1–564.

Howard, L.O., Fiske, W.F., 1911. The importation into the United states of the parasites of the gypsy moth and the brown tail moth. US Department of Agriculture, Bureau of Entomology Bulletin 9.

Hoy, M.A., Nowierski, R.M., Johnson, M.W., Flexner, H.K., 1991. Issues and ethics in commercial releases of arthropod natural enemies. American Entomology 37, 74–75.

Hoy, M.A., 1994. Parasitoids and predators in management of arthropod pests. In: Metcalf, R.L., Luckmann, W.H. (Eds.), Introduction to Insect Pest Management. Wiley, New York, pp. 129–198.

Hrcek, J., Miller, S.E., Whitfield, J.B., Shima, H., Novotny, V., 2013. Parasitism rate, parasitoid community composition and host specificity on exposed and semi-concealed caterpillars from a tropical rainforest. Oecologia 521–532.

Huffakar, C.B., 1977. Augmentation of natural enemies in People's Republic of China. In: Ridgeway, R.L., Vinson, S.B. (Eds.), Biological Control by Augmentation of Natural Enemies. Plenum Press, New York, pp. 3291–3440.

Huffaker, C.B., Messenger, P.S., 1976. Theory and Practice of Biological Control. Academic Press, New York.

Irvin, N., Wratten, S.D., Frampton, C.M., 1999. Understory management or the enhancement of leaf roller parasitoid *Dolichogenidea tasmanica* Cameron in Canterbury, New Zealand Orchards. In: The Hymenoptera: Evolution, Bio-Diversity and Biological Control. CSIRO, Melbourne, Australia.

Isalm, W., 1998. Rearing and release of the pulse weevil parasitoid, *Dinarmus basalts* (Rond.) (Hymenoptera: Pteromalidae). Tropical Agricultural Research and Extension 1 (2), 131–135.

Islam, W., Ahmed, K.N., Malek, M.A., 1985. Hymenopterous parasites on insect pests of two stored pulses. Bangladesh Journal of Zoology 13, 81–85.

Jalali, S.K., Singh, S.P., 2003. Bio-ecology of *Chilo partellus* (Swinhoe) (Pyralidae) and evaluation of its natural enemies. Agricultural Reviews 24, 79–100.

Jalali, S.K., Singh, S.P., Venkatesan, T., 2001. Choice of suitable Trichogrammatid species for suppression of diamond back moth, *Plutella xylostella* (Linnaeus) (Lepidoptera: Yponomeutidae) on cabbage. Shashpa 8, 161–166.

Jalali, S.K., Venkatesan, T., Wahab, S., 2010. Parasitoids and predators in pest management with special reference to Trichogrammatids. In: Bio Pest Management (Entomopathogenic Nematodes, Microbes and Bio-agents). Agrotech Publishing Academy, Udaipur, pp. 39–61.

Jervis, M.A., Kidd, N.A.C., 1986. Host feeding strategies in hymenopteran parasitoids. Biological Reviews 61, 395–434.

Jervis, M.A., Kidd, N.A.C., Fitten, M.G., Huddleston, T., Dawah, H.A., 1993. Flower visiting by hymenopteran parasitoids. Journal of Natural History 27, 67–105.

Kalyanasundaram, M., Justin, C.G.L., Swamiappan, M., Sundarababu, P.C., Jayaraj, S., 1993. Efficacy of *Trichogramma chilonis* against sugarcane internode borer *Chilo sacchariphagus* in Tamil Nadu. Indian Journal of Plant Protection 21, 31–33.

Khan, Z.R., Pickett, J.A., 2004. The 'push-pull' strategy for stemborer management: a case study in exploiting biodiversity and chemical ecology. In: Gurr, G.M., Wratten, S.D., Altieri, M.A. (Eds.), Ecological Engineering for Pest Management: Advances in Habitat Manipulations for Arthropods. CSIRO and CABI Publishing, pp. 155–164.

Khoo, S.G., 1990. Use of natural enemies in controlling agricultural pests in Malaysia. In: Bay-Peterson, J. (Ed.), The Use of Natural Enemies to Control Agricultural Pests. Food and Fertilizer Technology Centre for the Asian and Pacific Region, Taipei, Taiwan, pp. 30–39.

King, E.G., Hopper, K.R., Powell, J.E., 1985a. Analysis of systems for biological control of crop arthropod pests in the US by augmentation of predators and parasitoids. In: Hoy, M.A., Herzog, E.C. (Eds.), Biological Control in Agricultural IPM Systems. Academic Press, Orlando, FL, pp. 201–227.

King, E.G., Bull, D.L., Bouse, L.F., Phillips, J.R., November 8, 1985b. Biological control of bollworm and tobacco bud worm in cotton by augmentative releases of *Trichogramma*. Southwest Entomologist 8 (suppl.), 1–10.

King, E.G., Coleman, R.J., Morales-Ramos, J.A., Summy, K.R., 1995. Biological control. In: King, E.G., Phillips, J.R., Coleman, R.J. (Eds.), Cotton Insects and Mites: Characterization and Management. The Cotton Foundation Reference Book Series, No.3. The Cotton Foundation Publisher, Memphis, TN, pp. 511–538.

King, E.G., 1993. Augmentation of parasites and predators for suppression of arthropod pests. In: Lumbsten, R.D., Vaughen, J.L. (Eds.), Pest Management: Biologically Based Technologies Conference Proceeding Series. American Chemical Society, Washington, DC, pp. 90–100.

Knippling, E.F., 1977. The theoretical basis for augmentation of natural enemies. In: Ridgeway, R.L., Vinson, S.B. (Eds.), Biological Control by Augmentation of Natural Enemies. Plenum Press, New York, pp. 79–124.

Kogan, M., Turnipseed, S.G., 1987. Ecology and management of soybean arthropods. Annual Review of Entomology 32, 507–538.

Krishnamoorthy, A., Ganga Visalakshi, P.N., Mani, M., 2006. A new record of *Thripobius* spp. a parasitoid of onion thrips, *Thrips tobaci* Linderman in India. Paper Presented in Natl. Sym. Biol. Control of Sucking Pests in India, IAT, Bangalore, May 26–27, pp. 33–34.

Krishnamoorthy, A., Singh, S.P., 1987. Biological control of citrus mealy bug with an introduced parasite, *Leptomastix dactylopii* in India. Entomophaga 32, 143–148.

Krishnamoorthy, A., Singh, S.P., 1988. Observational studies on the occurrence of parasitoids of *Papilio* spp. in citrus. Journal of Plant Protection 16, 79–81.

Krishnamoorthy, A., 1993. Binomics and Development of Integrated Management Strategy for California Red Scale, *Aonidiella aurantii* (Maskell) (Ph.D. thesis). University of Agricultural Sciences, Bangalore, India.

Krishnamoorthy, A., 2010. Biological control of pests in horticultural ecosystems. In: Organic Pest Management: Potentials and Applications. Sathish Serial Publishing House, New Delhi, India.

Krishnamoorthy, A., 2012. Exploitation of egg parasitoids for control of potential pests in vegetable ecosystems in India. Comunicata Scientiae 3 (1), 1–15.

Kumar, L., Yogi, M.K., Jagdish, J., 2013. Habitat manipulation for biological control of insect pests: a review. Research Journal of Agriculture and Forestry Sciences 1 (10), 27–31.

Kumarasinghe, N.C., Wratten, S.D., 1996. The sugarcane lophopid planthopper *Pyrilla perpusilla* (Homoptera: Lophopidae): a review of its biology, pest status and control. Bulletin of Entomological Research 86, 485–498.

Landis, D.A., Menalled, F.D., Lee, J.C., Carmona, D.M., Perez Valdez, A., 1999. Habitat modification to enhance biological control in IPM. In: Kennedy, G.G., Sutton, T.B. (Eds.), Emerging Technologies for Integrated Pest Management: Concepts, Research and Implications. APS Press, St. Paul (in press).

Landis, D.A., Wratten, S.D., Gurr, G.M., 2000. Habitat management to conserve natural enemies of arthropod pests in agriculture. Annual Review of Entomology 45, 175–201.

LaSalle, J., 1993. Parasitic Hymenoptera, biological control and biodiversity. In: LaSalle, J., Gauld, I.D. (Eds.), Hymenoptera and Biodiversity. CAB International, Wallingford, UK.

Lawson, D.S., Nyrop, J.P., Reissing, W.H., 1997. Assay with commercially produced *Trichogramma* (Hymenoptera: Trichogrammatidae) to determine suitability for oblique banded leaf roller (Lepidoptera: Torticidae) control. Environmental Entomology 26, 684–693.

Li, L., 1984. Research and utilization of *Trichogramma* in China. In: Atkinson, P.L., Shijunn, Ma. (Eds.), Proceedings Chinese Academics of Science-US National Academy of Science Joint Symposium on Biological Control of Insects. Science Press, Beijing, China, pp. 204–223.

Mani, M., 1994. Effectiveness of the exotic parasitoid, *Leptomastix dactylopii* in the control of *Planococcus citri* in guava orchards. Journal of the Entomological Research 18, 351–355.

Mani, M., Krishnamoorthy, A., Singh, S.P., 1989. The impact of the predator, *Cryptolaemus montrouzieri* Muls. on the pesticide resistant populations of the striped mealy bug, *Ferrisia virgata* (Ckll.) on guava in India. Insect Science and Its Applications 11, 167–170.

Mani, M., 1995. Studies on the natural enemies of the wax scale *Drepanococcus chiton* (Green) (Coccidae: Homoptera) on ber and guava. Journal of Biological Control 12, 128–129.

Manjunath, T.M., 1985. *Leptomastix dactylopii* in India. In: King, E.G., Jackson, A.D. (Eds.), Biocontrol News and Information. FERRO, USDA, New Delhi, pp. 197–198.

Manjunath, T.M., Bhatnagar, V.S., Pawar, C.S., Sithanandam, S., 1988. Economic importance of *Heliothis* species in India and an assessment of their natural enemies and host plants. In: King, Jackson, R.D. (Eds.), Proceedings of the Workshop on Biological Control of *Heliothis*. Increasing the Effectiveness of Natural Enemies, 11–15 November, 1985, New Delhi, India, pp. 197–228.

Morales-Ramos, J.A., Rojas, M.G., 2003. Natural enemies and pest control: an integrated pest management concept. In: Koul, O., Dhaliwal, G.S. (Eds.), Predators and Parasitoids. Taylor and Francis, London, pp. 17–39.

McAlpine, J.F., 1981. Manual of Nearctic Diptera, vols (1–3).

Mellini, E., 1990. Sinossi di biologia dei Ditteri Larvevoridi. Bollettino dell'Istituto di Entomologia 45, 1–38.

Misra, M.P., Pawar, A.D., 1983. Bio control trials against sugarcane Pyrilla, *Pyrilla perpusilla* Walker in eastern Uttar Pradesh and elsewhere in India. In: Sithanandam, Solayappan, A.R. (Eds.), Biological Control of Sugarcane Pests in India. Tamilnadu State Federation of Co-operative Sugar Factories Ltd., Chennai, pp. 75–78.

Mohyuddin, A.I., 1991. Utilization of natural enemies for the control of insect pests of sugarcane. Insect Science and Its Application 12, 19–26.

Mohyuddin, A.I., 1994. Commercial sugarcane. IPM Working Development 2, 4–5.

Monge, J.P., Huignard, J., 1991. Population fluctuation of two bruchid species *Callosobruchus maculatus* F. and *Bruchidius atrolineatus* Pic (Coleoptera: Bruchidae) and their parasitoids, *Dinarmus basalis* Rondani and *Eupelmus milled* Crawford (Hymenoptera: Pteromalidae, Eupelmidae) in a storage condition in Niger. Journal of African Zoology 105, 197–206.

Morales, J., Gallardo, J.S., Vásquez, C., Rios, Y., 2000. Patrón de emergência, longevidad, parasitismo y proporción sexual de *Telenomus remus* (Hymenoptera, Scelionidae) com relación al cogollero del maíz. Instituto de Biotecnologia Aplicada à Agropecuária 12, 47–54.

Muthuswami, M., Gunathilagaraj, K., 1989. Effect of gall midge resistance in parasitic behavior of *Platygaster oryzae* Cameron. International Rice Research Newsletter 14 (4), 19.

Nagarkatti, S., Jayanth, K.P., 1992. Population dynamics of major insect pests of cabbage and their natural enemies in Bangalore district (India). In: Proceedings of International Conference of Plant Protection in Tropics. Malaysian Plant Protection Society, Kuala Lumpur, Malaysia, pp. 325–347.

Nobakht, Z., Karimzadeh, J., Shakaram, J., Jafari, S., 2015. Identification of parasitoids of *Apomyelois ceratoniae* (Zeller) (Lepidoptera, Pyralidae) on pomegranate in Isfahan province. Journal of Entomology and Zoology Studies 3 (1), 287–289.

Nordlund, D.A., 1984. Biological control with entomophagous insects. Journal of Entomology Society 19, 14–27.

Nordlund, D.A., Chalfont, R.B., Lewis, W.J., 1998. Arthropod populations, yield and damage in monocultures and polycultures of corn bears and tomatoes. Agriculture, Ecosystems & Environment 11, 353–367.

Norton, A.P., Welter, S.C., 1996. Augmentation of the egg parasitoid, *Anaphes iole* (Hymenoptera, Myrmaridae) for *Lygus hesperus* (Heteroptera: Miridae) management in strawberries. Environmental Entomology 25, 1406–1414.

Nadeem, S., Hamed, M., 2011. Biological control of sugarcane borers with inundative release of *Trichogramma chilonis* (Ishii) (Hymenoptera: Trichogrammatidae) in farmer fields. Pakistan Journal of Agricultural Science 48 (1), 71–74.

Nixon, G.E.J., 1937. Some Asiatic *Telenominae* (Hymenoptera, Proctotrupoidea). Annals and Magazines of Natural History 10 (20), 444–475.

O'Neil, R.J., Yaninek, J.S., Landis, D.A., Orr, D.B., 2003. Biological control and integrated pest management. In: Marediuo, K.M., Dako, D., Mota-Sancheza, D. (Eds.), Integrated Pest Management in Global Arena. CAB International, Wallingford, UK, pp. 19–30.

Orr, D.B., Suh, C.P.C., 1998. Parasitoids and predators. In: Rechcigl, J.E., Rechcigl, N.A. (Eds.), Biological and Biotechnological Control of Insect Pests. CRC Press LLC, Boca Raton, FL, pp. 3–34.

Orr, D.B., Pleasants, J.M., 1996. The potential of native prairie plant species to enhance the effectiveness of the *Ostrinia nubilalis* parasitoid, *Macrocentrus grandii*. Journal of the Kansas Entomological Society 69, 133–143.

Orr, D.B., Landis, D.A., Mutch, D.R., Manley, G.V., Stuby, S.A., King, R.L., 1997. Ground cover influence on the microclimate and *Trichogramma* (Hymenoptera: Trichogrammatidae) augmentation in seed corn production. Environmental Entomology 26, 433–438.

Pape, T., 1986. A phylogenic analysis of the wood louse flies (Diptera; Rhinophoridae). Tijdschrift voor Entomologie 129, 15–34.

Pape, T., 1987. The Sarcophagidae (Diptera) of Fennoscandia and Denmark. Brill/Scand. Sci., Leiden, The Netherlands.

Pape, T., 1992. Phylogeny of the Tachinidae family group (Diptera: Calyptratae). Tijdschrift voor Entomologie 135, 43–86.

Patt, J.M., Hamilton, G.C., Lashomb, J.H., 1997. Foraging success of foraging parasitoids on flowers: interplay of insect morphology, flora architecture and searching behavior. Entomologia Experimentalis et Applicata 83, 21–30.

Paul, A.V.N., Sree Kumar, K.M., 1998. Improved technology for mass rearing of Trichogrammatids and their facetious hosts *Corcyra cephalonica* St. In: Ananthakrishnan, T.N. (Ed.), Technology in Biological Control. Oxford & IBH Publishing Co. Pvt. Ltd, New Delhi, pp. 99–111.

Powell, W., 1986. Enhancing parasite activity within crops. pp. 314–340. In: Waage, J.K., Greathead, D. (Eds.), Insect Parasitoids. London Academic Press, London, 389 pp.

Priya, M., Faizal, M.H., 2004. Seasonal occurrence of two egg parasitoids, *Chrysochalcissa oviceps* Boucek and Gryon sp. of coreid bug (*Paradasynus rostratus* Distant) of coconut in Kerala. Entomon 29 (4), 383–388.

Purohit, M.S., Patel, H.V., Chavan, S.M., Shinde, C.U., Radadia, G.G., 2014. Egg parasitoids of rice gundhi bug, *Leptocorisa* sp. (Hemiptera: Alydidae) in South Gujarat. Insect Environment 20 (3), 74–75.

Quicke, D.L.G., 1997. Parasitic Wasps. Chapman and Hall, London. 470 pp.

Rabb, R.L., Stinner, R.E., van den Bosch, R., 1976. Conservation and augmentation of natural enemies. In: Huffaker, Messenger, P.S. (Eds.), Theory and Practice of Biological Control. Academic Press, New York, pp. 23–254.

Ram, P.S., 2003. Bio Pesticides and Bio Agents. In: Integrated Pest Management of Agricultural Crops. International Book Distributing Company Lucknow, Uttar Pradesh, pp. 571–590.

Rao, V.P., Ghani, M.A., Sankaran, T., Mathur, K.C., 1971. A Review of the Biological Control of Insects and Other Pests in South East Asia and the Pacific Region. CIBC Technical Communication, West Indies. No. 6, 149 pp.

Ridgeway, R.L., Vinson, S.B., 1977. Biological Control by Augmentation of Natural Enemies. Plenum Press, New York.

Rognes, K., 1991. Blowflies (Diptera, Calliphoridae) from Fennoscandia and Denmark. Fauna Entomologica Scandinavica 24, 1–23.

Ruberson, J.R., Nemoto, H., Hirose, Y., 1998. Pesticides and Conservation of Natural Enemies. Academic Press, San Diego, CA, pp. 207–220.

Sandanayake, W.R.M., Edirsinghe, J.P., 1992. *Trathala flavoorbitalis*: parasitization and development in relation to host-stage attacked. International Journal of Tropical Insect Science 3, 1393.

Sankar, C., Saravanan, C., Kathiravan, J., Marimuthu, R., Prabhu Kumar, S., 2012. Classical biological control of papaya mealybug, *Paracoccus marginatus* Williams and Granara de Willink (Hemiptera: Pseudococcidae) in tapioca by an encyrtide parasitoid, *Acerophagus papayae* in Perambalur district of Tamilnadu, India. Pest Management in Horticultural Ecosystems 18 (2), 213–216.

van der Schaff, P.A., Kaskens, J.W.M., Kole, M., Noldus, L.P.J., Pak, G.A., 1984. Experimental releases of two strains of *Trichogramma* spp. against lepidopteran pests in Brussel sprouts field crop in the Netherlands. Mededelingen Faculteit Landbouwwetenschappen Rijksuniversiteit Gent 46, 803–813.

Shantha Kumari, P., 2003. Biological Control of Crop Pests in India. Kalyani Publishers, New Delhi, p. 365.

Shumakov, E.M., 1997. Ecological principles associated with augmentation of natural enemies. In: Ridgeway, R.L., Vinson, S.B. (Eds.), Biological Control by Augmentation of Natural Enemies. Plenum Press, New York, pp. 39–78.

Shylesha, S.N., Joshi, S., Rabindra, R.J., Bhumannavar, B.S., 2010. Classical Biological Control of Papaya Mealy Bug. Technical Bulletin of National Bureau of Agriculturally Important Insects (NBAII), Bangalore, India.

Simmonds, F.J., 1944. The propagation of insect parasites on unusual hosts. Bulletin of Entomological Research 35, 219–226.

Simmons, G.S., Minkenberg, O.P.J.M., 1994. Field cage-release of augmentative biological control of *Bemisia argentifolli* (Homoptera: Aleyrodidae) I Southern California cotton with the parasitoid, *Eretmocerus ne californicus* (Hymenoptera: Aphelinidae). Environmental Entomology 23, 1552–1557.

Singh, S.P., 1989. Achievements of AICRP on Biological Control. Tech. Bull. No. 2. Biological Control Centre, NCIPM, Faridabad. 20 p.

Singh, S.P., 1993. Integrated pest management in horticultural crops. Indian Horticulture 38 (25–28) 37, 39–40.

Singh, S.P., 1994. Technology for Production of Natural Enemies. Project Directorate of Biological Control. Parishree Printers, Bangalore.

Singh, S.P., 1996. Biologica control in India. In: Upadhyay, R.K., Mukherjee, K.G., Rajak, R.L. (Eds.), IpM System in Agriculture. Biocontrol in Emerging Biotechnology, vol. 2. Adithya Books Pvt. Ltd, New Delhi, India, pp. 216–328. 620 pp.

Sithanandam, S., Rao, V.R., Reed, W., 1987. Parasites of the pigeon pea pod fly, *Melanagromyza obtusa* (Malloch). Indian Journal of Biological Control 1 (1), 10–16.

Smith, S.M., 1996. Biological control with *Trichogramma*: advances, successes in IPM. Annual Review of Entomology 47, 375–406.

Southgate, B.J., 1979. Biology of the Bruchidae. Annual Review of Entomology 24, 465–466.

Stamps, W.T., Linit, M.J., 1998. Plant diversity and arthropod communities: implications for temporary agroforestry. Agroforestry Systems 39, 73–89.

Starter, R.H., Ridgeway, R.L., 1977. Economic and social considerations for the utilization of augmentation of natural enemies. In: Ridgeway, R.L., Vinson, S.B. (Eds.), Biological Control by Augmentation of Natural Enemies. Plenum Press, New York, pp. 431–450.

Stary, P., 1993. Alternate host and parasitoid in first method in aphid pest management in glass houses. Journal of Applied Entomology 116, 187–191.

Sullivan, D.J., Volkl, 1999. Hyperparasitism multitrophic ecology and behavior. Annual Review of Entomology 44, 291–315.

Sullivan, D.J., 1987. Insect hyper parasitism. Annual Review of Entomology 32, 49–70.

Tandon, P.L., 1985. Spatial Distribution Sampling Technique and Population Dynamics of Citrus Green Scale *Coccus viridis* (Green) on Mandarin (Ph.D. thesis). UAS, Bangalore.

Theunissen, J., Booji, C.J.H., Lotz, L.A.P., 1995. Effects of intercropping white cabbage with clovers on pest infestation and yield. Entomologia Experimentalis et Applicata 74, 7–16.

Van Driesche, R.G., Bellows, T.S., 1996a. Biological Control. Chapman and Hall, London, UK.

Van Driesche, R.G., Bellows, T.S., 1996b. Biological Control. Kluwer Academic Publishers, Norwell, MA. 112.

Van den Bosch, R., Telford, A.D., 1964. Environmental modification and biological control. In: DeBach, P. (Ed.), Biological Control of Pests and Weeds. Reinhold, New York, pp. 459–488.

Varma, G.C., Rataul, H.S., Shenmar, M., Singh, S.P., Jalali, S.K., 1991. Role of inundative releases of *Trichogramma chilonis* Ishii In the control of *Chilo auricilius* Dudgeon on sugarcane. Journal of Insect Science 4, 165–166.

Varma, R.G.N., Jagadeeshwar, R., Shanker, C., 2013. Relative composition of egg parasitoids of rice yellow stem borer, *Scirpophaga incertulas* (Walker). Journal of Rice Research 6 (2), 53–58.

Venkatesan, T., Jalali, S.K., Srinivasamurthy, K., 2009. Competitive interactions between *Goniozus nephantidis* and *Bracon brevicornis*, parasitoids of the coconut pest, *Opisina arenosella*. International Journal of Pest Management 55 (3), 257–263.

Vinson, S.B., 1976. Host suitability for insect parasitoids. Annual Review of Entomology 25, 397–419.

Volkaff, N., Vinson, S.B., Wu, Z.X., Nettles, W.C., 1992. Invitro rearing of *Trissolcus basalis* (Scelionidae: Hymenoptera), an egg parasitoid of *Nezara viridula* (Pentatomidae, Hemiptera). Entomophaga 37, 141–148.

Volkl, W., 1997. Interactions between ants and aphid parasitoids: patterns and consequences for research utilization. Ecological Studies 130, 225–230.

Waage, J.K., Greathead, D.J., 1986. Insect Parasitoids. Academic, London. 389 pp.

Waage, J.K., 1989. The population ecology of pest-pesticide-natural enemy interactions. In: Jepson, P.C. (Ed.), Pesticides and Non-Target Invertebrates. Intercept, Wimborne, Dorset, UK, pp. 81–93.

Wang, Z., Lai-He, K., Zhang, F., Lu, X., Babendreier, D., 2014. Mass rearing and release of *Trichogramma* for biological control of insect pests of corn in China. Biological Control 68, 136–144.

Whiteman, D.W., Nordlund, D.A., 1994. Plant chemicals and the location of phytophagous arthropods by their natural enemies. In: Anantha Krishnan, T.N. (Ed.), Functional Dynamics of Phytophagous Insects. Oxford and IBH Publishing, New Delhi, pp. 133–159.

Williams, T., 1991. Host selection and sex ratio in a heteronomous hyper parasitoid. Ecological Entomology 16, 377–386.

Wright, D.J., Verkerk, R.H.J., 1995. Integration of chemical and biological control systems for arthropods, evaluation in a multitrophic context. Pesticide Science 44, 207–218.

Xia, Z.N., Nettles, W.C., Morrison, R.K., Irie, K., Inson, S.B., 1986. Three methods for the in vitro culture of *Trichogramma pretiosum* Riley. Journal of Entomological Science 21, 133–138.

Yeates, D.K., 1994. The cladistics and classification of the Bombyliidae (Diptera: Asiloidea). Bulletin of the American Museum of Natural History 219, 1–191.

Zhu, P., Wang, G., Zheng, X., Tian, U., Lu, Z., Heong, K.L., Xu, K., Chen, G., Yang, Y., Gurr, G.M., 2015. Selective enhancement of parasitoids of rice Lepidoptera pests by sesame (*Sesamum indicum*) flowers. BioControl 60, 157–167.

Trichogrammatids

S.K. Jalali[1], Prashanth Mohanraj[2], B.L. Lakshmi[3]

[1]Division of Molecular Entomology, National Bureau of Agricultural Insect Resources, Bengaluru, India; [2]Division of Insect Systematics, National Bureau of Agricultural Insect Resources, Bengaluru, India; [3]Priority Setting, Monitoring and Evaluation Cell, National Bureau of Agricultural Insect Resources, Bengaluru, India

1. INTRODUCTION

Trichogrammatidae are egg parasitoids parasitizing the eggs of several insects belonging to more than eight orders in aquatic and terrestrial habitats. They are among the smallest insects ranging in size from 0.2 to 1.5 mm, resulting in extreme difficulty in collecting and studying them. They have, therefore, not received adequate attention from taxonomists (Noyes, 2003).

Currently Trichogrammatidae comprise about 800 species belonging to 90 genera. While most genera in the family are small are and restricted in their known distributions, the larger genera are cosmopolitan. The only exception to this distributional pattern is *Trichogramma-toidea* which is poorly distributed in the Palearctic and the Nearctic (Pinto and Stouthamer, 1994). With about 230 species described worldwide *Trichogramma* is the largest genus in the family constituting over a quarter of the known genera in this family. This predominance is not necessarily a reflection of the dominance of the genus in the family but a consequence of its importance in pest management.

It was the discovery by Flanders (1929) establishing the feasibility of rearing *Trichogramma* on the eggs of *Sitotroga cerealella* that enabled the large-scale multiplication of this parasitoid which could then be used in inundative and augmentative releases in biocontrol programs. They have since been used extensively in biocontrol programs for the management of lepidopteran pests around the world and are increasingly being advocated for use in a number of agroecosystems. Various methods involving the use of chemicals and breeding strategies have helped enhance the efficacy of these parasitoids as bioagents in the ecofriendly management of many crops. Hence, the focus of this chapter is on *Trichogramma*, the most useful of the trichogrammatids in agricultural situations.

2. TAXONOMY

A large number of species and subspecies of *Trichogramma* (over 230) are distributed throughout the world parasitizing the eggs of over 200 insect species belonging to 70 families and eight orders in diverse habitats from aquatic to arboreal. In 1968, the epoch-making paper on *Trichogramma* taxonomy by Nagarkatti and Nagaraja (1968) from the Indian station of the Commonwealth Institute of Biological Control at Bangalore put the taxonomy of this group on an objective and empirical footing for the first time. The highest number of species has been recorded in the USA (58), followed by India, Brazil, China, and Russia. Some important species recorded from several insect eggs include *Trichogramma minutum* (374), *Trichogramma evanescens* (256), *Trichogramma chilonis* (152), *Trichogramma pretiosum* (145), and *Trichogramma dendrolimi* (111). Other important species for host insects records include *Trichogramma embryophagum* (78), *Trichogramma japonicum* (68), *Trichogramma semblidis* (59), *Trichogramma cacaeciae* (58), *Trichogramma exiguum* (55), *Trichogramma deion* (46), *Trichogramma semifumatum* (37), *Trichogramma chilotraeae* (33), *Trichogramma brassicae* (32), *Trichogramma euproctidis* (29), *Trichogramma platneri* (29), *Trichogramma fasciatum* (28), *Trichogramma perkinsi* (26), *Trichogramma ostriniae* (23), *Trichogramma australicum* (19), *Trichogramma closterae* (17), *Trichogramma achaeae* (16), *Trichogramma fuentesi* (16), *Trichogramma principium* (16), *Trichogramma galloi* (15), *Trichogramma bruni* (13), *Trichogramma cordubense* (13), *Trichogramma interius* (13), *Trichogramma papilionis* (13), *Trichogramma funestum* (12), *Trichogramma rojasi* (12), *Trichogramma retorridum* (10), *Trichogramma stampae* (10), and *Trichogramma thalense* (10).

The host plant record is highest for *T. chilonis* and *T. pretiosum* (70 each), *T. minutum* (53), and *T. deion* (51). Other species for which most of host plants are recorded include *T. dendrolimi* (40), *T. evanescens* (39), *T. exiguum* (30), *T. embryophagum* (23), *T. japonicum* (19), *T. platneri* (16), *T. semblidis* (15), and *T. interius* (15) (Table 1).

3. MOLECULAR CHARACTERIZATION

Morphological identification of most species is based on subtle differences in male genitalia (Nagarkatti and Nagaraja, 1971, 1977). Many important species share similar genital structures, forcing workers to rely on less dependable characters that are often intraspecifically variable (Pinto et al., 1989). Given the economic importance of trichogrammatids in biocontrol programs, species identification must be quick, simple, and widely applicable. Novel approaches that used DNA sequences of the internal transcribed spacer (ITS) 2 and mitochondrial cytochrome c oxidase I (COX1) helped to solve this difficulty (Stouthamer et al., 1999; Hebert et al., 2003). Sequence and restriction analysis of the ITS2 region have been used to distinguish *Trichogramma* species collected from tomato fields in Portugal (Silva et al., 1999).

Archana et al. (2005) characterized four trichogrammatids: *T. japonicum*, *T. chilonis* (Indian strain), *T. chilonis* (German strain), and *T. pretiosum* (thelytokous) based on polymerase chain reaction (PCR) amplification of the polymorphism and sequencing of the ITS region of the ribosomal DNA. The ITS1 region was amplified in the range of 500 and 700 bp, the ITS2 region ranging between 600 and 700 bp and the entire ITS region ranging between 1100 and 1300 bp. *Trichogramma japonicum* and *T. pretiosum* (thelytokous) showed maximum variation from the two strains of *T. chilonis*, i.e., *T. chilonis* (Indian strain) and *T. chilonis* (German strain), and these two strains indicated a close relationship by occupying the same branch in

TABLE 1 Worldwide Distribution and Host Range of *Trichogramma*[a]

Sl. no.	*Trichogramma* species	Distribution	Primary hosts (insects)	Secondary hosts (plants)
1	*Trichogramma acacioi* Brun, Gomez de Moraes & Soares	Brazil	4	4
2	*Trichogramma acantholydae* Pintureau & Kenis	Italy	2	1
3	*Trichogramma achaeae* Nagaraja & Nagarkatti	Argentina, Barbados, Cape Verde Islands, Chile, France, Hawaii, India, New Caledonia, China, Russia, Trinidad and Tobago, USA	16	14
4	*Trichogramma acuminatum* Querino & Zucchi	Brazil	–	–
5	*Trichogramma acutovirilia* Pinto	Canada, USA	–	–
6	*Trichogramma adashkevitshi* Sorokina	Uzbekistan	–	–
7	*Trichogramma agriae* Nagaraja	India, Trinidad and Tobago	2	2
8	*Trichogramma agrotidis* Voegelé & Pintureau	Argentina, Bulgaria, France, Russia, Switzerland	5	3
9	*Trichogramma aldanense* Sorokina	Russia	1	1
10	*Trichogramma alloeovirilia* Querino & Zucchi	Brazil	–	–
11	*Trichogramma alpha* Pinto	Canada, USA	3	3
12	*Trichogramma aomoriense* Honda	Japan	1	1
13	*Trichogramma arcanum* Pinto	Canada, USA, Sweden	–	–
14	*Trichogramma artonae* Chen & Pang	China	1	–
15	*Trichogramma atopovirilia* Oatman & Platner	Brazil, Colombia, El Salvador, Guatemala, Honduras, Mexico, USA, Venezuela	14	8
16	*Trichogramma atropos* Pinto	Brazil, Venezuela	–	–
17	*Trichogramma aurosum* Sugonjaev & Sorokina	Bulgaria, Canada, Russia, USA,	6	7
18	*Trichogramma australicum* Girault	Australia, Bahamas, Barbados, Bolivia, Brazil, Colombia, Grenada, India (?), Indonesia, Japan, Madagascar, Malaysia, Mauritius, Pakistan, China, Peru, Philippines, Reunion, Sri Lanka, Taiwan, Thailand, Trinidad and Tobago, Venezuela	19	10

(Continued)

TABLE 1   Worldwide Distribution and Host Range of *Trichogramma*[a]—cont'd

Sl. no.	*Trichogramma* species	Distribution	Primary hosts (insects)	Secondary hosts (plants)
19	*Trichogramma bactrianum* Sugonjaev & Sorokina	Tadzhikistan	1	–
20	*Trichogramma ballmeri* Pinto	Canada, USA	5	4
21	*Trichogramma bellaunionense* Basso & Pintura	Uruguay	2	1
22	*Trichogramma bennetti* Nagaraja & Nagarkatti	Brazil, Colombia Guyana, Trinidad and Tobago, Venezuela	3	2
23	*Trichogramma bertii* Zucchi & Querin	Brazil	2	–
24	*Trichogramma bezdencovii* Bezdenko	Armenia, Belarus, Bulgaria, Chile, Russia, Ukraine	5	1
25	*Trichogramma bilingense* He & Pang	China	2	–
26	*Trichogramma bispinosum* Pinto	USA	–	–
27	*Trichogramma bistrae* (Kostadinov)	Bulgaria	–	–
28	*Trichogramma bourarachae* Pintureau & Babault	Morocco, Portugal, Tunisia	4	3
29	*Trichogramma bournieri* Pintureau & Babault	Coroners, Kenya, South Africa	4	2
30	*Trichogramma brassicae* Bezdenko	Argentina, Australia, many countries in Europe, Canada, Japan, China, USA	32	10
31	*Trichogramma brevicapillum* Pinto & Platner	Afghanistan, USA	17	8
32	*Trichogramma breviciliata* Yousuf & Hassan	India	1	–
33	*Trichogramma breviflagellata* Yousuf	India	1	–
34	*Trichogramma brevifringiata* Yousuf & Shafee	India	1	1
35	*Trichogramma browningi* Pinto & Oatman	USA	4	3
36	*Trichogramma bruni* Nagaraja	Argentina, Bolivia, Brazil, Chile, Costa Rica, Mexico, Trinidad and Tobago, Venezuela	13	5
37	*Trichogramma buesi* Voegelé	–	–	–

TABLE 1 Worldwide Distribution and Host Range of *Trichogramma*[a]—cont'd

Sl. no.	*Trichogramma* species	Distribution	Primary hosts (insects)	Secondary hosts (plants)
38	*Trichogramma buluti* Bulut & Kilincer	Turkey	1	–
39	*Trichogramma cacaeciae* Marchal	Argentina, Austria, Belarus, Bulgaria, Cuba, Czechoslovakia, Denmark, Estonia, France, Germany, Greece, Iran, Kazakhstan, Kirgizia, Latvia, Lithuania, Moldova, Morocco, the Netherlands, China, Peru, Switzerland, Syria, Tunisia, Ukraine, UK, USA, Uzbekistan,	58	13
40	*Trichogramma californicum* Nagaraja & Nagarkatti	India, Mexico, USA	9	7
41	*Trichogramma canadense* Pinto	Canada	–	–
42	*Trichogramma canariense* del Pino & Polaszek	Canary Islands	1	1
43	*Trichogramma carina* Walker	France	–	–
44	*Trichogramma carverae* Oatman & Pinto	Australia	7	11
45	*Trichogramma castrense* Velasquez de Rios & Teran	Venezuela	1	1
46	*Trichogramma cephalciae* Hochmut & Martinek	Czech Republic, Slovakia, Germany, Italy, Poland	5	2
47	*Trichogramma chilonis* Ishii	Antilles, Bahamas, Bangladesh, Cape Verde Islands, Comoros, Germany, Grenada, Hawaii, India, Indonesia, Japan, South Korea, Madagascar, Malaysia, Mauritius, Micronesia, Montserrat, Nepal, New Caledonia, Pakistan, Papua New Guinea, China, Philippines, Rèunion, Romania, Solomon Islands, South Africa, Taiwan, Thailand, USA, Vietnam	152	70
48	*Trichogramma chilotraeae* Nagaraja & Nagarkatti	Antilles, Bahamas, Brazil, Colombia, India, Indonesia, Malaysia, Peru, Philippines, Thailand, Trinidad and Tobago, Russia, Venezuela	33	11
49	*Trichogramma choui* Chan & Chou	Taiwan	1	4
50	*Trichogramma chusniddini* Sorokina & Atamirzaeva	Taiwan	1	4

(Continued)

TABLE 1　Worldwide Distribution and Host Range of *Trichogramma*[a]—cont'd

Sl. no.	*Trichogramma* species	Distribution	Primary hosts (insects)	Secondary hosts (plants)
51	*Trichogramma closterae* Pang & Chen	China	17	2
52	*Trichogramma clotho* Pinto	Costa Rica	–	–
53	*Trichogramma colombiense* Velasquez de Rios & Teran	Colombia, Venezuela	2	2
54	*Trichogramma cordubense* Vargas & Cabello	Azores, Iran, Madeira, Portugal, Spain	13	5
55	*Trichogramma cultellus* Jose, Hirose & Honda	Japan	1	–
56	*Trichogramma cuttackense* Nagaraja	India	1	1
57	*Trichogramma danaidiphagum* Nagaraja & Prasanth	India	4	4
58	*Trichogramma danausicida* Nagaraja	India	2	2
59	*Trichogramma danubiense* Birova & Kazimirova	Slovakia, UK	3	1
60	*Trichogramma daumalae* Dugast & Voegelé	France, Bulgaria, UK	2	–
61	*Trichogramma deion* Pinto & Oatman	Canada, Cuba, Hawaii, USA	46	51
62	*Trichogramma demoraesi* Nagaraja	Brazil	7	2
63	*Trichogramma dendrolimi* Matsumura	Austria, Belarus, Bulgaria, Chile, Germany, Greece, Hungary, India, Iran, Italy, Japan, Kazakhstan, South Korea, Latvia, Lithuania, Moldova, the Netherlands, Pakistan, China, Poland, Romania, Russia, Taiwan, Turkey, Ukraine, Russia, Vietnam	111	41
64	*Trichogramma dianae* Pinto	USA	–	–
65	*Trichogramma diazi* Velásquez de Rios & Terán	Venezuela	1	1
66	*Trichogramma dissimile* Zucchi	Brazil	1	1
67	*Trichogramma distinctum* Zucchi	Brazil, Uruguay	3	1
68	*Trichogramma drepanophorum* Pinto & Oatman	USA	2	1

TABLE 1 Worldwide Distribution and Host Range of *Trichogramma*[a]—cont'd

Sl. no.	*Trichogramma* species	Distribution	Primary hosts (insects)	Secondary hosts (plants)
69	*Trichogramma elegantum* Sorokina	Turkmenistan	1	–
70	*Trichogramma embryophagum* (Hartig)	Albania, Algeria, Armenia, Austria, Belarus, Bulgaria, Chile, Czech Republic, France. Georgia, Germany, Greece, Hungary, India, Iran, Israel, Italy, Israel, Italy, Kazakhstan, Kirgizia, Latvia, Moldova, Morocco, the Netherlands, Norway, China, Poland, Portugal, Romania, Russia, Slovakia, Spain, Taiwan, Turkey, Turkmenistan, Ukraine, USA, Russia, Vietnam	78	23
71	*Trichogramma erebus* Pinto	Bahamas, Colombia, Costa Rica, Guatemala, Mexico, USA	1	1
72	*Trichogramma erosicorne* Westwood	Sri Lanka	5	–
73	*Trichogramma esalqueanum* Querino & Zucchi	Brazil	2	1
74	*Trichogramma ethiopicum* (Risbec)	Cameroon, West Africa	7	1
75	*Trichogramma euproctidis* (Girault)	Argentina, Armenia, Belarus, Bulgaria, Caucasus, Chile, France, Germany, Greece, Italy, Japan, Moldavia, Morocco, Neotropical, China, Peru, Russia, Tadzhikistan, Turkmenistan, Ukraine, USA, Russia, Uzbekistan, Vietnam	29	6
76	*Trichogramma evanescens* Westwood	Armenia, Austria, Azerbaijan, Belgium, Bulgaria, Canada, Chile, Comoros, Cuba, Czech Republic, Denmark, Egypt, France, Georgia, Germany, Hungary, India, Iran, Italy, Kazakhstan, Libya, Lithuania, Madagascar, Madeira, Mauritius, Moldova, Morocco, the Netherlands, China, Philippines, Poland, Portugal, Romania, Russia, Slovakia, Spain, Sri Lanka, Sweden, Switzerland, Tselinograd Obl., Tunisia, Turkey, Turkmenistan, Ukraine, UK, USA, Russia, Uzbekistan, Vietnam, Serbia	256	39

(Continued)

TABLE 1   Worldwide Distribution and Host Range of *Trichogramma*[a]—cont'd

Sl. no.	*Trichogramma* species	Distribution	Primary hosts (insects)	Secondary hosts (plants)
77	*Trichogramma exiguum* Pinto & Platner	Antilles, Argentina, Barbados, Bermuda, Brazil, Canada, Central America, Chile, Colombia, Cuba, El Salvador, Grenada, Guatemala, Guyana, India, Jamaica, Mexico, Neotropical, Peru, St. Christopher (Kitts) & Nevis, St. Vincent & Grenadines, Trinidad & Tobago, USA, Uruguay, Venezuela	55	30
78	*Trichogramma falx* Pinto & Oatman	New Zealand	–	–
79	*Trichogramma fasciatum* (Perkins)	Argentina, Bahamas, Barbados, Bolivia, Brazil, Canada, Chile, Colombia, Costa Rica, Cuba, Czech Republic, Ecuador, India, Indonesia, Jamaica, Japan, Madagascar, Mexico, Montserrat, Neotropical, Peru, Russia, Trinidad and Tobago, Turkmenistan, USA, Venezuela	28	6
80	*Trichogramma flandersi* Nagaraja & Nagarkatti	India	4	2
81	*Trichogramma flavum* Ashmead	Nearctic, China, USA	1	1
82	*Trichogramma forcipiforme* Zhang & Wang	China	1	–
83	*Trichogramma fuentesi* Torre	Antilles, Argentina, Barbados, Colombia, Cuba, Dominican Republic, Mexico, Peru, USA, Venezuela	16	7
84	*Trichogramma funestum* Pinto & Oatman	USA	12	11
85	*Trichogramma funiculatum* Carver	Australia, New Zealand	3	2
86	*Trichogramma fuzhouense* Lin	China	–	–
87	*Trichogramma gabrielino* Pinto	USA	–	–
88	*Trichogramma galloi* Zucchi	Bolivia, Brazil, Colombia, Paraguay, Peru, Uruguay	15	6
89	*Trichogramma gicai* Pintureau & Stefanescu	Madeira, Spain	2	4
90	*Trichogramma gordhi* Pinto	Costa Rica, Mexico	–	–
91	*Trichogramma guariquense* Velasquez de Rios & Teran	Venezuela	1	1

TABLE 1   Worldwide Distribution and Host Range of *Trichogramma*[a]—cont'd

Sl. no.	*Trichogramma* species	Distribution	Primary hosts (insects)	Secondary hosts (plants)
92	*Trichogramma hebbalensis* Nagaraja	India	1	1
93	*Trichogramma hesperidis* Nagaraja	India	2	1
94	*Trichogramma higai* Oatman & Platner	Hawaii	1	1
95	*Trichogramma huberi* Pinto	Guatemala, Mexico	–	–
96	*Trichogramma infelix* Pinto	Costa Rica	–	–
97	*Trichogramma ingricum* Sorokina	Russia	1	1
98	*Trichogramma interius* Pinto	Mexico, USA	13	15
99	*Trichogramma inyoense* Pinto & Oatman	Canada, USA	3	4
100	*Trichogramma iracildae* Querino & Zucchi	Brazil	1	1
101	*Trichogramma itsybitsi* Pinto & Stouthamer	USA	4	4
102	*Trichogramma ivelae* Pang & Chen	Australia, China	7	–
103	*Trichogramma jalmirezi* Zucchi	Brazil	2	1
104	*Trichogramma japonicum* Ashmead	Antilles, Australia, Bahamas, Bangladesh, Barbados, Brazil, Colombia, Grenada, Hawaii, India, Indonesia, Japan, South Korea, Malaysia, Montserrat, Myanmar, China, Peru, Philippines, Taiwan, Thailand, Trinidad and Tobago, Venezuela, Vietnam	68	19
105	*Trichogramma jaxarticum* Sorokina	Turkmenistan	1	1
106	*Trichogramma jezoense* Ishii	Japan	3	–
107	*Trichogramma julianoi* Platner & Oatman	Canada, USA	2	1
108	*Trichogramma kalkae* Schulten & Feijen	Malawi, Togo	3	1
109	*Trichogramma kankerense* Yousuf & Hassan	India	1	–
110	*Trichogramma kashmiricum* Nagaraja, Ahmad & Gupta	India	1	1

(Continued)

TABLE 1 Worldwide Distribution and Host Range of *Trichogramma*[a]—cont'd

Sl. no.	*Trichogramma* species	Distribution	Primary hosts (insects)	Secondary hosts (plants)
111	*Trichogramma kaykai* Pinto & Stouthamer	Japan, USA	6	2
112	*Trichogramma kilinceri* Bulut & Kilincer	Turkey	1	1
113	*Trichogramma koehleri* Blanchard	Argentina	5	–
114	*Trichogramma kurosuae* Taylor, Yashiro, Hirose & Honda	Japan	1	–
115	*Trichogramma lachesis* Pinto	Costa Rica, Mexico	–	–
116	*Trichogramma lacustre* Sorokina	Bulgaria, France, Russia, UK	1	1
117	*Trichogramma lasallei* Pinto	Bolivia, Brazil, British Virgin Islands, Costa Rica, Cuba, Mexico, USA, Uruguay, US Virgin Islands, Venezuela,	8	5
118	*Trichogramma lenae* Sorokina	Russia	1	–
119	*Trichogramma leptoparameron* Dyurich	Moldova, Russia	1	1
120	*Trichogramma leucaniae* Pang & Chen	China	2	–
121	*Trichogramma leviculum* Pinto	Costa Rica, Guatemala, Mexico	–	–
122	*Trichogramma lingulatum* Pang & Chen	Japan, China	2	1
123	*Trichogramma longxishanense* Lin	China	–	–
124	*Trichogramma lopezandinense* Sarmiento	Colombia	5	2
125	*Trichogramma maltbyi* Nagaraja & Nagarkatti	USA	9	5
126	*Trichogramma mandelai* Pintureau & Babault	Chad	1	1
127	*Trichogramma manicobai* Brun, Gomez de Moraes & Soare	Brazil	1	–
128	*Trichogramma manii* Nagaraja & Gupta	India	1	1
129	*Trichogramma maori* Pinto & Oatman	New Zealand	2	1
130	*Trichogramma marandobai* Brun, Gomez de Moraes & Soares	Brazil	1	–

TABLE 1 Worldwide Distribution and Host Range of *Trichogramma*[a]—cont'd

Sl. no.	*Trichogramma* species	Distribution	Primary hosts (insects)	Secondary hosts (plants)
131	*Trichogramma margianum* Sorokina	Turkmenistan	1	1
132	*Trichogramma marthae* Goodpasture	USA	4	2
133	*Trichogramma marylandense* Thorpe	Canada, USA	1	–
134	*Trichogramma maxacalii* Voegelé & Pointel	Brazil	6	4
135	*Trichogramma meteorum* Vincent	USA	1	1
136	*Trichogramma minutum* Riley	Antigua, Antilles, Argentina, Australia, Barbados, Bolivia, Brazil, Canada, Chile, Colombia, Comoros, Costa Rica, Cuba, Czech Republic, Dominican Republic, Egypt, France, Germany, Greece, Grenada, Guadeloupe, Guam, Guatemala, Guyana, Haiti, Hawaii, Indonesia, Italy, Jamaica, Japan, South Korea, Madagascar, Malaysia, Martinique, Mexico, Nearctic, New Zealand, China, Peru, Puerto Rico, Reunion, Russia, Saint Lucia, Sri Lanka, St Christopher (Kitts) and Nevis, St Vincent and Grenadines, Sudan, Tadzhikistan, Taiwan, Trinidad and Tobago, UK, USA, Uruguay, Uzbekistan, Venezuela, Virgin Islands	374	53
137	*Trichogramma mirabile* Dyurich	Moldova	1	–
138	*Trichogramma mirum* Girault	Australia	–	–
139	*Trichogramma misiae* Kostadinov	Bulgaria	1	–
140	*Trichogramma mullensi* Pinto	USA	1	1
141	*Trichogramma mwanzai* Schulten & Feijen	Malawi	1	1
142	*Trichogramma nemesis* Pinto	Canada, USA	2	–
143	*Trichogramma nerudai* Pintureau & Gerding	Argentina, Chile	8	10
144	*Trichogramma nestoris* (Kostadinov)	Bulgaria	–	–

(Continued)

TABLE 1   Worldwide Distribution and Host Range of *Trichogramma*[a]—cont'd

Sl. no.	*Trichogramma* species	Distribution	Primary hosts (insects)	Secondary hosts (plants)
145	*Trichogramma neuropterae* Chan & Chou	Taiwan	2	2
146	*Trichogramma niveiscapus* (Morley)	UK	1	1
147	*Trichogramma nomlaki* Pinto & Oatman	Canada, Chile, USA	1	1
148	*Trichogramma nubilale* Ertle & Davis	China, USA	14	6
149	*Trichogramma oatmani* Torre	Cuba	4	1
150	*Trichogramma obscurum* Pinto	Costa Rica, Guatemala, Mexico, Venezuela	1	1
151	*Trichogramma offella* Pinto & Oatman	USA	1	1
152	*Trichogramma okinawae* Honda	Japan	1	1
153	*Trichogramma oleae* Voegelé & Pointel	Argentina, France, Greece, Italy	7	1
154	*Trichogramma ostriniae* Pang & Chen	Japan, South Korea, China, South Africa, Taiwan, USA	23	9
155	*Trichogramma pallidiventris* Nagaraja	India	3	1
156	*Trichogramma panamense* Pinto	Panama	–	–
157	*Trichogramma pangi* Lin	China	–	–
158	*Trichogramma papilionidis* Viggiani	Angola	1	1
159	*Trichogramma papilionis* Nagarkatti	Hawaii, Japan, Malaysia	13	5
160	*Trichogramma parkeri* Nagarkatti	Canada, USA	8	5
161	*Trichogramma parnarae* Huo	China	1	1
162	*Trichogramma parrai* Querino & Zucchi	Brazil	–	–
163	*Trichogramma parvum* Pinto	USA	–	–
164	*Trichogramma pelovi* Kostadinov	Bulgaria	1	1
165	*Trichogramma perkinsi* Girault	Argentina, Bahamas, Bolivia, Central America, Chile, Colombia, Cuba, Guatemala, Hawaii, India, Mexico, Neotropical, Peru, Venezuela	26	4

TABLE 1 Worldwide Distribution and Host Range of *Trichogramma*[a]—cont'd

Sl. no.	*Trichogramma* species	Distribution	Primary hosts (insects)	Secondary hosts (plants)
166	*Trichogramma piceum* Dyurich	Italy, Moldova	3	1
167	*Trichogramma pinneyi* Schulten & Feijen	Malawi	1	1
168	*Trichogramma pintoi* Voegelé	–	–	–
169	*Trichogramma pintureaui* Rodríguez & Galán	Cuba	2	1
170	*Trichogramma plasseyense* Nagaraja	India, Papua New Guinea	5	5
171	*Trichogramma platneri* Nagarkatti	Canada, Hawaii, Israel, Nearctic, USA	29	16
172	*Trichogramma pluto* Pinto	Costa Rica, Guatemala, Mexico	–	–
173	*Trichogramma poliae* Nagaraja	India	7	3
174	*Trichogramma polychrosis* Chen & Pang	China	1	–
175	*Trichogramma pratissolii* Querino & Zucchi	Brazil	2	1
176	*Trichogramma pratti* Pinto	USA	1	1
177	*Trichogramma pretiosum* Riley	Antilles, Argentina, Australia, Bolivia, Brazil, British Virgin Islands, Canada, Caribbean, Chile, China, Colombia, Costa Rica, Cuba, Dominican Republic, Ecuador, El Salvador, Greece, Guatemala, Hawaii, Honduras, India, Mexico, Nearctic, Neotropical, Nicaragua, Paraguay, Peru, South Africa, Spain, Sudan, Taiwan, USA, Uruguay, Venezuela	145	70
178	*Trichogramma primaevum* Pinto	Australia	–	–
179	*Trichogramma principium* Sugonjaev & Sorokina	Bulgaria, France, Iran, Kazakhstan, Syria, Taiwan, Tselinograd Ob., Turkmenistan, Russia, Uzbekistan	16	4
180	*Trichogramma psocopterae* Chan & Chou	Taiwan	–	1
181	*Trichogramma pusillum* Querino & Zucchi	Brazil	–	–
182	*Trichogramma raoi* Nagaraja	India, China	5	5
183	*Trichogramma retorridum* (Girault)	Canada, Mexico, Nearctic, USA	10	7

(Continued)

TABLE 1   Worldwide Distribution and Host Range of *Trichogramma*[a]—cont'd

Sl. no.	*Trichogramma* species	Distribution	Primary hosts (insects)	Secondary hosts (plants)
184	*Trichogramma rojasi* Nagaraja & Nagarkatti	Argentina, Brazil, Chile, Cuba, Peru	12	6
185	*Trichogramma rossicum* Sorokina	Russia	1	1
186	*Trichogramma sankarani* Nagaraja	India	2	3
187	*Trichogramma santarosae* Pinto	Costa Rica	–	–
188	*Trichogramma sathon* Pinto	Mexico, USA	5	2
189	*Trichogramma savalense* Sorokina	Russia, Uzbekistan	3	3
190	*Trichogramma sembeli* Oatman & Platner	Hawaii	3	2
191	*Trichogramma semblidis* (Aurivillius)	Bulgaria, Canada, France, Germany, Hungary, India, Iran, Italy, Kazakhstan, Nearctic, the Netherlands, Norway, Poland, Russia, Siberia, Spain, Sweden, Switzerland, Syria, Tselinograd Obl., Ukraine, UK, USA	59	15
192	*Trichogramma semifumatum* (Perkins)	Bolivia, Brazil, Chile, Colombia, Costa Rica, Cuba, France, Hawaii, India, Mexico, Neotropical, Peru, Russia, Trinidad and Tobago, USA, Venezuela	37	8
193	*Trichogramma sericini* Pang & Chen	China, Russia	2	–
194	*Trichogramma shaanxiense* Huo	China	1	–
195	*Trichogramma shchepetilnikovae* Sorokina	Tadzhikistan	1	1
196	*Trichogramma sibiricum* Sorokina	Canada, Russia, USA	6	9
197	*Trichogramma siddiqi* Nasir & Schöller	Pakistan	1	1
198	*Trichogramma silvestre* Sorokina	Russia	2	–
199	*Trichogramma singularis* Girault	Australia	–	–
200	*Trichogramma sinuosum* Pinto	USA	1	–
201	*Trichogramma sogdianum* Sorokina	Uzbekistan	1	1
202	*Trichogramma sorokinae* Kostadinov	Bulgaria	–	–

TABLE 1 Worldwide Distribution and Host Range of *Trichogramma*[a]—cont'd

Sl. no.	*Trichogramma* species	Distribution	Primary hosts (insects)	Secondary hosts (plants)
203	*Trichogramma stampae* Vincent	Colombia, Ghana, Iran, Malaysia, Mexico, USA	10	8
204	*Trichogramma sugonjaevi* Sorokina	Uzbekistan	1	–
205	*Trichogramma suorangelica* Pinto	Mexico	–	–
206	*Trichogramma taiwanense* Chan & Chou	Taiwan	1	5
207	*Trichogramma tajimaense* Yashiro, Hirose & Honda	Japan	2	4
208	*Trichogramma talitzkii* Dyurich	Moldova, UK	1	1
209	*Trichogramma tenebrosum* Oatman & Pinto	Australia	–	3
210	*Trichogramma terani* Velásquez de Rios & Terán	Venezuela	1	1
211	*Trichogramma thalense* Pinto & Oatman	USA	10	7
212	*Trichogramma tielingense* Zhang & Wang	China	1	–
213	*Trichogramma trjapitzini* Sorokina	Russia	2	–
214	*Trichogramma tshumakovae* Sorokina	Iran, Kirgizia	2	–
215	*Trichogramma tupiense* Querino & Zucchi	Brazil	–	–
216	*Trichogramma turkeiense* Bulut & Kilincer	Turkey	5	1
217	*Trichogramma turkestanicum* Meyer	Denmark, Kazakhstan, Portugal, Tselinograd Obl., Turkey	3	2
218	*Trichogramma umerus* Jose, Hirose & Honda	Japan	1	–
219	*Trichogramma urquijoi* Cabello Garcia	Spain	1	1
220	*Trichogramma ussuricum* Sorokina	Russia	1	–
221	*Trichogramma valentinei* Pinto & Oatman	New Zealand	–	–
222	*Trichogramma vargasi* Oatman & Platner	Hawaii	1	–

(Continued)

TABLE 1   Worldwide Distribution and Host Range of *Trichogramma*[a]—cont'd

Sl. no.	*Trichogramma* species	Distribution	Primary hosts (insects)	Secondary hosts (plants)
223	*Trichogramma viggianii* Pinto	Mexico	–	–
224	*Trichogramma yabui* Honda & Taylor	Japan	2	1
225	*Trichogramma yawarae* Hirai & Fursov	Japan	6	2
226	*Trichogramma zahiri* Polaszek	Bangladesh	1	–
227	*Trichogramma zeirapherae* Walter	Germany	2	–
228	*Trichogramma zeta* Pinto	Costa Rica, Dominican Republic, Mexico, USA	–	–
229	*Trichogramma zucchii* Querino	Brazil	–	–

[a] *After Noyes J.S. (2014).*

the phylogenetic tree. The results showed that these species of *Trichogramma* can be clearly differentiated from one another by sequencing the ITS region.

Sharma et al. (2008) studied the comparative random amplified polymorphic DNA (RAPD)–PCR profiles of seven species of *Trichogramma* with different RAPD primers, resulting in identification of 25 polymorphic DNA bands that were unique to individual species. Aligning of nucleotide sequences of single genomic region (ITS2) of these species showed a greater degree of divergence between species compared with RAPD–PCR data. The nucleotide sequence of the aligned ITS2 regions from different *Trichogramma* species ranged between 351 (*T. japonicum*) and 419 (*T. pretiosum* (thelytokous)) bases, out of which only 75 were unchanged thus showing a polymorphism of more than 80%. The results showed that RAPD–PCR and conserved ITS2 regions of ribosomal DNA can be sources of molecular markers for identification and phylogeny of *Trichogramma* species.

Molecular characterization of 31 species of *Trichogramma/Trichogrammatoidea*, including both indigenous and exotic species, was carried out to enable quick and reliable species identification. Variation in base pair sizes was used as a basis to categorize the species. The species analyzed fell into three groups: 800–900bp, 570–600bp, and 500–550bp. The restriction enzymes *Eco*R1, SacI, *Mse*I, and *Mva*I gave reproducible profiles and proved useful in distinguishing the various species. Based on the restriction patterns, a dichotomous key was constructed to enable easy identification (Ashok Kumar et al., 2009; Jalali, 2010).

In Punjab, (Singh and Shenhmar, 2008a,b) used RAPD markers to investigate genetic variation among different temperature-tolerant strains of *T. chilonis* and reported that nine primers showed genetic differentiation between the strains.

4. GENETIC IMPROVEMENT OF TRICHOGRAMMATIDS

Trichogrammatids are the most widely used species in biocontrol programs in India and other parts of the world. Genetic improvement of *Trichogramma* for tolerance to various abiotic stresses is considered very useful for its survival and ability to manage insect pests in high temperature conditions and in crop fields sprayed with insecticides. Generally, this parasitoid does not exert good control when exposed to these stresses.

Jalali and Singh (1993) were the first to attempt the selection of a superior strain of *T. chilonis* from parasitoid populations collected from all over India. The mean fecundity after rearing them for 40 generations was higher in BioC1 (Gujarat) and BioC2 (Punjab) strains. Results indicated that BioC1 and BioC2 are superior to other ecotypes.

In Maharashtra, after selection for 15 generations for developing an endosulfan-tolerant strain, LC_{50} (50% lethal dose) values increased to 0.05% from 0.044% in base colonies and studies on mode of inheritance indicated that the gene for endosulfan tolerance is completely dominant (Ingle et al., 2007).

In Punjab, Singh and Shenhmar (2008a,b) and Singh et al. (2008c) conducted a study during 2003–2005 to evaluate augmentative releases of two genetically improved strains of *T. chilonis*, i.e., Ludhiana (Punjab) and genetically improved high temperature-tolerant for the management of stalk borer (*Chilo auricilius*). Borer incidence was reduced by 39.05 and 55.09%, respectively, compared with the control thus revealing that genetic improvement of biocontrol agents may help in increasing their efficiency in the field. However, more field studies are required.

An effort was made to develop a low temperature-adapted form of *T. chilonis*, adapted to 18–24 °C, with a view to utilizing them in the cooler months. In cage studies for parasitoid-searching ability, 58% of *Corcyra cephalonica* eggs were parasitized by the low temperature-adapted strain compared with 17.2% by the nonadapted strain. The host searching test indicated that adaptation at low temperatures led to better host searching and that such a strain could be successfully utilized under a low temperature regime (Jalali et al., 2006a).

The imidacloprid-tolerant strain of *T. chilonis* was likewise developed by subjecting the parasitoids to selection pressure. At F62 generation, tolerance was developed to one-quarter of the field dose of imidacloprid, with 27.9% mortality at 6 h of exposure and 76.7% parasitism and 78.9% adult emergence from the eggs of *C. cephalonica* (Maurya et al., 2008). An attempt was made to develop a strain of *T. chilonis* tolerant to cartap hydrochloride in Uttarakhand. The parasitoid population took 66 generations to develop tolerance against one-quarter of the field recommended dose, with 80.3% parasitism and 79.7% emergence from the eggs of *C. cephalonica* (Maurya and Khan, 2007a). Maurya and Khan (2007b) made another attempt to develop a chlorpyrifos-tolerant strain of *T. chilonis* and reported that after 70 generations of selection pressure, a strain was developed exhibiting tolerance to one-quarter of the field recommended dose, with parasitism and emergence by the tolerant strain up to 83.3 and 79.0%, respectively, whereas in the susceptible strain 15.6 and 19.0% parasitism and adult emergence, respectively, was recorded.

Ashok Kumar et al. (2008) developed a strain of *T. chilonis* having combined tolerance to three major groups of insecticides: endosulfan (organochlorine), monocrotophos (organophosphate), and fenvalerate (synthetic pyrethroid) and to high temperature (32–38 °C) through selection. The studies suggested that the improved strain will provide effective control of the pest even under harsh climatic conditions and under high insecticide pressure in different economically important crops.

The study was conducted to examine if a strain of *T. chilonis* developed for its tolerance to newer insecticides has cross-tolerance to other commonly used insecticides. When both survival and parasitism were taken into consideration, the new insecticide-tolerant strain showed a moderate degree of tolerance to other insecticides, such as endosulfan, monocrotophos, fenvalerate, dichlorvos, and decamethrin for which tolerance was not induced. Thus release of tolerant strains will be more useful against different pests on crops with high insecticide use (Devi et al., 2007).

An effort was made to develop a strain for its tolerance to endosulfan. It was subjected to insecticidal stress across 325 generations and a strain tolerant to the recommended field doses of endosulfan was developed. This strain parasitized 56% of eggs in the presence of endosulfan as compared to a mere 3% by the susceptible strain (Jalali et al., 2006a,b,c). Ballal et al. (2009) evaluated endosulfan-tolerant strain under net house conditions and reported it to be superior to susceptible strain in parasitizing *Helicoverpa armigera* eggs. Later, a multiple insecticides-tolerant strain was developed and it was evaluated all over India in cotton ecosystems during 2004–2005. The results indicated that this strain was superior to the susceptible strain in fields sprayed with insecticides (Jalali et al., 2006b). In another study in Karnataka, thermal tolerance studies with *T. chilonis* were carried out at a constant temperature of 32, 34, 36, 38, and 40°C and at variable temperatures (32–40°C and a relative humidity of 30–50%) to work out temperature tolerance. *T. chilonis* Uttar Pradesh strain 1 (TcUP1), *T. chilonis* Uttar Pradesh strain 2 (TcUP2), *T. chilonis* Tamil Nadu strain 1 (TcTN1), and *T. chilonis* Tamil Nadu strain 2 (TcTN2) were identified as temperature-tolerant with high fecundity. These strains provided parasitism up to 65%, 50% females, and longevity of 3 days compared with 15% parasitism, 20–30% females, and <1 day longevity in the susceptible strain. The insecticide-tolerant strains of *T. chilonis*, TcT1, TcT3, TcT4, TcT5, and TcT6, recorded good parasitism of 60, 40, 90, and 85%, respectively, compared with 5–15% by the susceptible laboratory-reared population. Resistance factors for TcT1, TcT3, TcT4, TcT5, and TcT6 were 9.55, 2.05, 5.07, 3.69, and 150.02, respectively, to five groups of insecticides, namely, organochlorine, organophosphate, synthetic pyrethroid, oxadiazine, and spinosyn (Jalali, 2014).

5. EVALUATION OF TEMPERATURE-TOLERANT STRAINS OF TRICHOGRAMMATIDS

In an experiment conducted in Ludhiana, Punjab, from April to October 2005, the dispersal/host searching range of genetically improved (through selection) heat-tolerant strain and local Ludhiana strain of *T. chilonis* was up to 10 and 8 m from the release point, respectively. Parasitism efficiency of both strains was found to be negatively correlated with temperature and positively correlated with relative humidity. The searching range of the heat-tolerant strain was superior to that of the Ludhiana strain, irrespective of the change in temperature from April to October. The mean parasitism rate in each month was also higher for the former (Singh and Shenhmar, 2008a,b). In another study in Punjab, the efficacy of genetically improved high temperature-tolerant (PDBC strain) and Ludhiana (Punjab) strain of *T. chilonis* against early shoot borer, *Chilo infuscatellus*, was evaluated in farmers' field during 2004 and 2005. Nine releases of the PDBC and Ludhiana strains were as effective as application of cartap hydrochloride (Padan 4G) in reducing the incidence by 52.7 and 50.5%, respectively, over untreated control. Based on reduced pest incidence, higher parasitism, and yield, releases of *T. chilonis* (PDBC strain/Ludhiana) at 50,000 parasitoids/ha at 10-day intervals from April to June can be successfully used for the suppression of the early shoot borer (Singh et al., 2007).

6. EFFECT OF PLANT EXTRACTS ON TRICHOGRAMMATIDS

The effect of extracts of *Oryza sativa*, *Triticum aestivum*, *Zea mays*, *Saccharum officinarum*, *Vigna sinensis*, *Cajanus cajan*, *Helianthus annuus*, and *Ricinus communis* on the parasitism efficiency of *Trichogramma* spp. was studied. The highest percentage parasitism by *T. chilonis*, *T. japonicum*, and *Trichogramma poliae* was observed in the presence of extracts of *S. officinarum* (61.3%), *C. cajan* (54.0%), and *C. cajan* (43.3%). The lowest parasitism (24.0%) by *T. japonicum* was recorded with *V. sinensis* and *R. communis* extracts (Shankarganesh and Khan, 2006).

Rao and Raguraman (2005) tested five neem products and methyl demeton for their toxicity against *T. chilonis*, and *Chrysoperla zastrowi sillemi*. All neem products proved to be relatively safe to both the natural enemies. Methyl demeton (control) was toxic to the predatory grubs and adult parasitoids. Neem oil (3.0%) and NSP (neem + sweet flag (*Acorus calamus*) + pungam (*Pongamia glabra* (*Pongemia pinnata*)) in 1:1:1, v/v) (0.42%) were relatively toxic compared with other forms of neem. In a field trial with bhendi cv. Arka Anamika in Madurai, Tamil Nadu, during summer 2001, a slight (2.1–20.2%) and high (22–49%) reduction in the population of predators was observed in the botanical and chemical (methyl demeton)-treated plots, respectively. Recolonization of predators was observed in the plots sprayed with botanicals, while this was not the case in those sprayed with chemicals. The use of neem can be a substantial contribution toward preservation of biodiversity in agroecosystems.

7. ROLE OF SYNOMONES IN EFFICACY OF TRICHOGRAMMATIDS

Trichogramma species use a variety of host plant-produced chemical stimuli for host location and host acceptance. Thus, it is evident that chemical stimuli from plants (synomones) also play a role in host habitat location behavior by these parasitoids. The findings of Nordlund et al. (1984) led to a series of studies on the role of plant semiochemicals in host–habitat location behavior by *Trichogramma* species.

In a study on chickpea (*Cicer arietinum*), 10 different cultivars were assayed for synomonal response in *T. chilonis*. Heneicosane in the vegetative phase and heneicosane and tricosane in the flowering phase seemed to be critical for the synomonal activity. The cultivar Pusa 256 RL: 1985 at the vegetative phase and Pusa 1003 at the flowering phase elicited the highest response from the parasitoids, indicating that they could be used effectively in integrated pest management (IPM) programs (Srivastava et al., 2004). However, Shanmugam et al. (2005) reported that pigeon pea (*C. cajan*), chickpea (*C. arietinum*), and silk Leaf (*Lagascea mollis*) leaf extracts did not show any increase in parasitism, but cotton flower extract at various concentrations increased the parasitism between 78.9% and 83.0% over the control, which is valuable to synchronize the *T. chilonis* release with the oviposition by bollworms at peak flowering stage to get maximum benefit. The study was carried out on the allomonic effect of pigeon pea plant extracts on the parasitization potential. Results revealed that leaf and twig extracts exhibited strong antiparasitism effects on *T. chilonis*, causing more than 50% reduction in parasitism over untreated control. This could be one of the reasons why this egg parasitoid could not be exploited in pigeon pea (Boomathi et al., 2005).

In an experiment on tritrophic interaction between *T. chilonis*, *H. armigera*, and pigeon pea genotypes, the extent of egg parasitism on pods in different cultivars/genotypes varied from 1.2% to 8.3% and on leaves parasitism varied from 5.0% to 29.0% (Tandon and Bakthavatsalam, 2003).

In a study on the volatile compounds from the leaves and flowers of *Tagetes erecta* and the leaves of *Solanum viarum*, olfactometeric responses of *H. armigera* and its egg parasitoid, *T. chilonis* showed that the parasitoid showed maximum net response to hexane extract of *T. erecta* flower buds (47.5%), followed by floral and leaf volatiles. Studies revealed that the trap crop *T. erecta* is *Trichogramma*-friendly (Tandon and Bakthavatsalam, 2005).

Paramasivan and Paul (2005) worked on development and field evaluation and reported that semiochemical dust formulation prepared using equal proportions of flowering phase leaf extract of Mahekanchan, sunflower hybrid (TCSH-1), and egg of *Chilo partellus* registered the highest parasitism in the field, followed by the dust formulation prepared using the flowering phase leaf extract of TCSH-1. It is suggested that these semiochemical formulations could be used for increasing the efficiency of *T. chilonis* in the field to enhance their parasitism.

The tritrophic studies were carried out under net cage conditions for different tomato cultivars, L-15, PKM-1, ArkaVikas, Arka Sourabh, Arka Ashish, using *T. chilonis* and *T. pretiosum* on *H. armigera*. Irrespective of genotypes, *T. pretiosum* recorded higher parasitism than *T. chilonis*. Furthermore, it was observed that trichome density played an important role in efficacy of egg parasitoids (Karabhantanal and Kulkarni, 2002). In their study on tomatoes, Tandon and Bakthavatsalam (2007) evaluated 15 tomato genotypes (varieties/hybrids) for their influence on the parasitizing efficiency of *T. chilonis* on *H. armigera* eggs under screen house conditions and recorded significant differences for different genotypes. The least parasitism was recorded on Arka Abha (20%) and highest on Arka Ahuti (50%), followed by Anand-1 (46.66%). Paul et al. (2008) advocated that tomato varieties with favorable semiochemicals could be exploited in an IPM program to enhance the effectiveness of the egg parasitoid *T. chilonis* against the fruit borer *H. armigera*.

In a study on brinjal, onion, and cluster bean were identified as cost-effective intercrops. The highest parasitism of 82.7 and 74.3% on *Leucinodes orbonalis* eggs was recorded in the acetone extract of leaves of onion and flowers of cluster bean by *T. chilonis* on the seventh day after introduction of parasitoids compared with control with acetone alone (50.3%).

Rani et al. (2008) studied the leaf surface chemicals of *R. communis* damaged by *Achaea janata* (castor semilooper) feeding, as well as a nonhost, the Serpentine leaf miner, *Liriomyza trifolii*, and reported maximum egg parasitism in *A. janata*-infested castor leaf extracts compared with the leaf miner-infested or normal healthy castor leaf extracts. The results are interesting in the context of tritrophic interactions between the pest, parasitoid, and host plant and are useful in the biocontrol of insect pests.

Olfactometer studies on the effect of hexane washings of host plants such as sugar cane, tomato, soybean, pearl millet, finger millet, cotton, sorghum, and maize in attracting *T. chilonis* indicated that all the plant species attracted more parasitoids in "no-choice" compared with "choice" tests. Parasitism recorded was 50.7–69.8% and volatiles from maize leaves were the most attractive followed by other monocots including sugar cane.

Studies carried out at Dharwad (Karnataka) on the relationships among chili cultivars, the fruit borer, *H. armigera*, and different *Trichogramma* spp. revealed that among the chili cultivars, Byadagikaddi was the preferred host for oviposition by *H. armigera* as well as egg parasitoids as higher parasitism was recorded by all the four species of *Trichogramma*. Among the various species, significantly higher parasitism was recorded by *T. chilonis* followed by *T. achaeae*, *T. pretiosum*, and *T. japonicum*.

Bioassays of the hexane extracts of nine rice varieties and a cultivar revealed that the variety Pusa Sugandh-2 elicited maximum parasitism in vegetative phase for both *T. japonicum* and *T. chilonis*, whereas in the flowering phase variety Pusa Basmati-1 elicited maximum mean parasitism in yellow stem borer (Singh et al., 2009).

Ganesh et al. (2002) reported the response of different *Trichogramma* spp., *T. chilonis*, *T. japonicum*, and *T. poliae*, on its parasitic behavior on sprayed acetone leaf extracts of cruciferous crops. *T. chilonis* parasitized significantly more eggs (70.0%) on Indian mustard treated cards, and the lowest (23.7%) on ornamental Rye (*Secale cereale* L.); *T. poliae* parasitized 42.3% eggs on cauliflower extract but only 16.7% on ornamental Rye, Indian mustard and turnip; while *T. japonicum* parasitized 40% eggs on Chinese cabbage but 11.7% on turnip extracts.

8. EFFECT OF TEMPERATURE ON TRICHOGRAMMATIDS

One of the key factors for field performance of trichogrammatids is their ability to tolerate and survive in high temperature conditions, which influences insect physiology and behavior. Several studies have been conducted to document the effect of high and low temperature on performance of *Trichogramma* species. Jalali et al. (2006a,b,c) worked on adaptive performance of *T. chilonis* at low temperature (18–24°C) and reported that a strain adapted to low temperature parasitized 58% of the *C. cephalonica* eggs compared with 17.2% by the nonadapted strain in an insect cage study. Tests indicated that adaptation to low temperatures led to better host searching and that such a strain could be successfully utilized under a low temperature regime.

Singh and Ram (2006) observed a drastic reduction in various biological parameters when *T. chilonis* was subjected to temperature shock at 30–45°C for 6h at temperatures beyond 35°C. These results may help in developing augmentative release strategies of this parasitoid against lepidopterous pests in cotton by devising suitable release techniques.

In a study on performance of various ecotypes of *T. chilonis*, ecotypes collected from the Udumalpet and Palladam areas recorded significantly higher parasitism of *Plutella xylostella* eggs than those collected from other areas in Tamil Nadu, India. It was suggested that such ecotypes could be exploited for large-scale field releases against *P. xylostella* in the field in narrow range of temperature (24–28°C) on cole crops (Justin and Thangaselvabai, 2006).

In a study on the temperature-dependent biology of *C. partellus* and its natural enemies, it was observed that the optimum range lies between 25 and 32°C and it was suggested that natural enemies like *T. chilonis* can perform effectively in areas where the temperature ranges between 18 and 32°C rather than in other areas (Jalali and Singh, 2006).

9. EFFECT OF INSECTICIDES ON TRICHOGRAMMATIDS

The unsystematic and repeated use of pesticides has a tendency to destroy the ecological balance by eliminating the beneficial fauna, which keeps the pest population below threshold level, resulting in pest resurgence, secondary pest outbreak, and higher yield losses (Singh and Jalali, 1998). Prior to the release of trichogrammatids in an IPM system, it is essential to know their compatibility with chemical pesticides. As the majority of the farmers use pesticides extensively, it is essential to know the effects of chemicals on trichogrammatids. Such information will assist in the timing of parasitoid releases regarding the application of chemical pesticides.

9.1 Laboratory Tests

In Karnataka, Giraddi and Gundannavar (2006) studied the toxicity of emamectin benzoate on three species, *T. chilonis*, *T. pretiosum*, and *T. japonicum*, and reported it to be moderately toxic to all three species, with emergence ranging from 58% to 88.7% at various concentrations. In a study to test the safety of spiromesifen 240 SC on *T. chilonis*, it was observed that adult emergence and parasitism was less affected (Kavitha et al., 2006). In a laboratory experiment in Tamil Nadu, the effect of emamectin 5 SG on the egg parasitoid *T. chilonis* and on *C. zastrowi sillemi* was investigated and it was observed that application at high concentration (10.0 g a.i./ha) resulted in 79% adult emergence and 76% parasitism for *T. chilonis*, thus proving to be slightly toxic (Kannan and Chandrasekaran, 2006).

In Tamil Nadu, a laboratory experiment was carried out to assess the safety of abamectin 1.9 EC along with spinosad, cypermethrin, and endosulfan to the egg parasitoid *T. chilonis* and other parasitoids. The results revealed that abamectin 1.9 EC at all the doses tested had a lesser adverse effect on the emergence of the *Trichogramma* adults compared with cypermethrin and endosulfan and also on its parasitizing potential (Jasmine et al., 2007).

Shanmugam et al. (2006a,b) studied the safety of some newer insecticides against *T. chilonis* under laboratory conditions and observed that neem products were nontoxic, spinosad was safer than the newer insecticides, imidacloprid, indoxacarb, and thiamethoxam, which were moderately toxic with 47 and 37% parasitism. Ramesh and Manickavasagam (2006) reported that the biopesticides Achook, Nimbicidine, and Dipel were found to be safer to the parasitoids, monocrotophos and phosphamidon were moderately toxic, while cypermethrin was found to be highly toxic to *T. pretiosum* (thelytokous), *T. pretiosum*, *T. chilonis*, and *T. japonicum*. Rao (2005) tested various neem products and insecticides and reported that neem formulations are relatively safe against *T. chilonis* and suggested that synthetic insecticides should not be used in bio control programs involving the parasitoid.

The toxicity of biopesticides and synthetic insecticides to natural enemies was investigated under laboratory conditions at Hyderabad. Most of the botanical pesticides were safe to *T. chilonis* and among the insecticides, carbosulfan was highly toxic to *T. chilonis* followed by acetamiprid, whereas imidacloprid was relatively less toxic, and the microbial biopesticide, *Bacillus thuringiensis* was safe.

In a laboratory study conducted in Uttarakhand on the effect of different concentrations of endosulfan on parasitization ability of *T. chilonis*, *T. japonicum*, and *T. poliae*, it was determined that there was a reciprocal relationship with percentage parasitism and concentrations. Among the three species, the lowest percent parasitism was recorded for *T. japonicum* (Tiwari and Khan, 2004). In the laboratory, the safety of imidacloprid as foliar treatment against *T. chilonis* and *C. zastrowi sillemi* was evaluated. The results revealed that there was no significant adverse effect on adult emergence and percentage parasitism by *T. chilonis* (Kumar and Santharam, 1999).

Dhanasekaran et al. (2007) tested four combinations of insecticides: 5% cypermethrin + 25% monocrotophos, 5% cypermethrin + 20% dimethoate, 3% cypermethrin + 57% acephate, and 25% monocrotophos + 11% dichlorvos, against some beneficial insects. The order of toxicity for *T. chilonis* was 3% cypermethrin + 57% acephate Dry flowables (DF) > 25% monocrotophos + Soluble (liquid) concentrate (SL) dichlorvos 11% = 5% cypermethrin + 25% monocrotophos > 35% endosulfan emulsifiable concentrate (EC).

Side effects of the five most commonly recommended insecticides for sugar cane in Punjab were evaluated for their impact on *T. japonicum*. Endosulfan, imidacloprid, and triazophos were observed to be moderately toxic, while malathion and chlorpyrifos were highly toxic (Singh and Shenhmar, 2008a,b).

Toxicity of several insecticides was carried out on *T. chilonis* in order to determine their safety for rice ecosystems for an IPM trial. Based on risk quotient, which is the ratio between the field recommended doses and the LC_{50} of the beneficial, only chlorantraniliprole was found to be harmless to *T. chilonis*, while thiamethoxam, imidacloprid, VirtakoReg. Etofenprox, and BPMC were found to be highly toxic to the parasitoid. Because *T. chilonis* is an important egg parasitoid of leaf folders, reported to reduce the pest population considerably and is often released augmentatively in rice IPM programs, these dangerous chemicals should be avoided in the rice ecosystem.

Laboratory experiments were conducted to evaluate the effect of different insect growth regulators on the parasitoid *T. chilonis*. After 72h, 1.5 and 1ml lufenuron/liter showed the lowest mortality (16.76% and 19.62%, respectively) while 1.5g and 1ml diafenthiuron/liter recorded 70.29% mortality (Alexander et al., 2006).

9.2 Field Tests

Field experiments conducted in Punjab during 2002 to evaluate the effect of *T. chilonis* in combination with insecticides for the management of *H. armigera* in cotton indicated that mean parasitism did not differ among the treatments, but it was highest in the control (6.78%). The lowest pest incidence and high yields were recorded in insecticide-treated plots compared with the parasitoid or its combination (Virk and Brar, 2005). In another field study in Punjab on sunflower, the effect of carbaryl, chlorpyrifos, and endosulfan was studied in a farmer's field at Pattar Kalan village in Jalandhar district, Punjab, during winter 2002 against *H. armigera* and sucking pests and their natural enemies. It was observed that 3 days after spraying, endosulfan was comparatively safer to *T. chilonis*; however, the toxic effect decreased 7 days after spraying (Kaur and Brar, 2003).

In a study in West Bengal, Samanta et al. (2006) reported that, based on residual toxicity, both *T. chilonis* and *T. japonicum* could be released on the brinjal crop for the management of *L. orbonalis* 3–5 days after α-cypermethrin had been sprayed, 4–6 days after methomyl, and 6–7 days after quinalphos. Nair et al. (2008) conducted a field trial on pigeon pea in Nadia, West Bengal, India, during two consecutive seasons, August 2003–April 2004 (first season) and August 2004–April 2005 (second season), to evaluate the efficacy of pyridalyl 10 EC, along with three other insecticides against *H. armigera* on red gram (pigeon pea) and its effect on natural enemies including *Trichogramma* sp. Pyridalyl was highly effective in controlling the pest of red gram and significantly superior to the other insecticides. It was found to be relatively safe to the natural enemies of the pest. Persistent and residual toxicity of eight insecticides was carried out in West Bengal in 2003 on *Trichogramma* species and a larval parasitoid to identify relatively safer insecticides to formulate guidelines regarding introduction of natural enemies in a pesticide-treated guava orchard. Tolerance of *T. chilonis* was much more than *T. japonicum* irrespective of the pesticide. Considering the retention period of toxic residues, both the *Trichogramma* species could be released in the crop ecosystem 3–5 days after spraying α-cypermethrin and fluvalinate, 4–6 days after endosulfan, monocrotophos, and deltamethrin sprays, and 7–8 days after methomyl, quinalphos, and cypermethrin sprays. Dey et al. (2012) conducted field studies for two seasons during 2009 and 2010 in West Bengal to determine the efficacy of flubendiamide 480 SC against pod borer complex of blackgram (*Vigna mungo*) cv. "Kalindi (B-54)" in West Bengal. The results revealed that the chemical had low toxicity to the common natural enemies including *Trichogramma* sp.

In a study, the relative efficacy of *T. chilonis* and *T. pretiosum* (thelytokous) alone and combination with endosulfan on chickpea and pigeon pea for the control of *H. armigera* was studied in Uttarakhand. The study revealed no significant differences in parasitism with maximum parasitism of 10.6% in both chickpea and pigeon pea, which reduced significantly in endosulfan-treated plots (Kumar et al., 2009).

In Tamil Nadu, laboratory/field experiments were conducted to study the effects of spinosad on cotton bollworms and their natural enemies. The concentrations (0.005–0.02%) were safe to natural enemies including *T. chilonis* in the laboratory through contact and feeding experiments. This study suggests that spinosad can be a beneficial component in biointensive pest management programs for cotton.

10. UTILIZATION OF TRICHOGRAMMATIDS

10.1 Sugar cane

Large-scale field demonstrations in Punjab by Shenhmar et al. (2003) using *T. chilonis* showed that 11–12 releases of *T. chilonis* at 50,000/ha from July to October reduced stalk borer incidence by 55–60%.

Bhat et al. (2004) found that inundate releases of *T. chilonis* in Andhra Pradesh remarkably reduced the incidence of sugar cane early shoot borer (*C. infuscatellus*) to 82%, brought down the use of pesticides and fungicides, making farmers realize better yield and develop a positive attitude toward adoption of *T. chilonis* as a component of IPM.

Manisegaran (2004) found that six inundative releases of *T. chilonis* at 75,000/ha in Tamil Nadu reduced the incidence and intensity of internode borer, *Chilo sacchariphagus indicus*, recording higher cane weight (1.72 kg), yield (121.5 t/ha), 35.6% increase in yield over the farmers' practice, highest net return (Rupees 62,919/ha) and cost-to-benefit ratio (CBR) (2.07).

A study conducted by Narayan (2004) in Uttar Pradesh revealed that the egg parasitoid *T. chilonis* brought about a reduction in the infestation index of the sugar cane stalk borer, *C. auricilius* and yields of the treated plots were significantly higher by 19.672 and 16.367 tonnes/acre during 1999–2000 and 2000–2001, respectively.

The investigations carried out in Karnataka by Rachappa and Naik (2004) on the impact of various intercrops and *T. chilonis* on *C. infuscatellus* indicated that marked control of *C. infuscatellus* and highest activity of *T. chilonis* was found in the combination of sugar cane + coriander

(*Coriandrum sativum*), resulting in highest number of millable canes (113, 130/ha), cane yield (92.2 t/ha), cane equivalent yield (100.5 t/ha), and maximum net return (Rupees 63,498).

In a study conducted in Tamil Nadu by Thirumurugan and Koodalingam (2005) sugar cane crops sprayed with 5% tomato extract + *T. chilonis* at fortnightly intervals recorded the lowest cumulative percentage (10.34%) of internode borer incidence and the highest sugar yield of 12.10 t/ha from cane yield of 93.25 t/ha with commercial cane sugar (CCS) of 12.98%.

One of the components of the IPM practiced by Rajendran (2006) in Tamil Nadu included the release of the egg parasitoid *T. chilonis* at 2.5 cc/ha six times at fortnightly after 5 months of planting. The IPM practice increased the millable canes and yield of the crop by 36.45% with reduced the infestation by shoot borers.

Rao et al. (2006) studied the impact of *T. chilonis* on early shoot borer, *C. infuscatellus*, in Andhra Pradesh and the results showed that the *T. chilonis*-released plots recorded less incidence (0.16%) of early shoot borer, a cane yield of 120.3 t/ha and 19.0% juice sucrose.

Singh (2006) evaluated parasitoids against borer pests of sugar cane (stalk borer, *C. auricilius*, internode borer, *C. sacchariphagus indicus*, and top borer, *Scirpophaga excerptalis*) in Uttar Pradesh and found that nine alternative releases of *T. chilonis* at weekly intervals recorded the highest CBR (1.61).

A field experiment conducted by Thirumurugan et al. (2006) in Tamil Nadu to study the effect of intercrops and plant extracts on the activities of *T. chilonis* against sugar cane shoot borer showed that the cane intercropped with soybean (5:1) + *T. chilonis* at 2.5 cc/ha at fortnightly intervals from days 30–75 after planting scored the highest CBR of 1:2.77.

A study on the effect of partitioned growth stage protection on the incidence of sugar cane borers by Pandya and Patel (2007) in Gujarat showed that lowest incidence of top borer (*S. excerptalis*), root borer (*Polyocha depressella*), internode borer (*C. sacchariphagus indicus*), and stalk borer (*C. auricilius*) were recorded in maximum protection treatment, which comprised release of *T. chilonis* seven times at 40,000 parasites/ha at 15-day intervals beginning from 135 days after planting (DAP) as an important component.

Thirumalai and Selvanarayanan (2009) found that the mean reduction in the internode borer population over the unreleased control in individual release of *T. chilonis* was 22.92 and 24.92%, respectively, during the first and second years in Tamil Nadu.

Multilocation trials conducted by Yalawar et.al. (2010) in Karnataka to evaluate the performance of *T. chilonis* in managing *C. sacchariphagus indicus* showed that irrespective of the locations, the performance of *T. chilonis* was superior to untreated controls during different months.

10.2 Rice

Field experiments conducted in Tamil Nadu and Andhra Pradesh by Katti et al. (2001) revealed that natural biocontrol with no insecticidal application throughout the crop season maintained the highest levels of parasitism by yellow stem borer, gall midge, and leaf folder as well as populations of some predatory spiders.

Beevi et al. (2003) found that the yield obtained during spring 2001 in Kerala, upon treatment with both *T. japonicum* and *T. chilonis* at 100,000 parasitoids/ha was highest (4512 kg/ha).

Field experiments conducted during the 1999 dry and wet seasons in Assam revealed that the release of *T. chilonis* during the dry season played a significant role in the reduction of stem borers on different cultivars while it had no significant effect on the leaf folder (*Cnaphalocrocis medinalis*) infestation (Das et al., 2004).

Kumar and Khan (2005) evaluated the bioefficacy of *T. japonicum* and *T. chilonis* against yellow stem borer (*Scirpophaga incertulas*) and leaf folder (*C. medinalis*) of rice in Uttaranchal and found that releases at the rate of 100,000/ha were superior to the lower dose.

During a study on the efficacy of *T. japonicum* Ashmead at different rates of release on suppression of yellow stem borer, *S. incertulas*, in Nagaland, Riba and Sarma (2006) found that a maximum increase of 12.62% in yield was achieved from the treatment with *T. japonicum* at 100,000 adults/ha applied in two equally split doses.

Karthikeyan et al. (2007) found that the release of *T. japonicum* at 100,000/ha followed by application of 1% azadirachtin against yellow stem borer reduced both dead hearts and white ears, while the release of *T. chilonis* reduced leaf folder damage in Kerala. Release of egg parasitoids in rice resulted in an increase of yield by 25.79–45.13%, with a mean CBR of 1:2.6 in parasitoid-released plots.

During a study on the impact of biointensive pest management strategies in Uttaranchal, Kumar et al. (2007) found that highest net return was obtained in bio control plot where *Trichogramma* spp. were released at 100,000/ha with Rupees 42,388.50 on a per hectare basis.

Organic and integrated practices (seven releases of *T. chilonis* and *T. japonicum* at 100,000 each at weekly intervals starting 30 days after transplanting (DAT)) proved to be effective in Punjab in the management of rice leaf folder and stem borer (Kaur et al., 2008).

A study undertaken by Mishra and Kumar (2009) in Uttar Pradesh to evaluate the bioefficacy of *T. japonicum* and *T. chilonis* revealed that all the doses (50,000/ha, 75,000/ha, and 100,000/ha) used in the inundative releases were effective but releases at 100,000/ha were found to be superior to the lower dose.

Field experiments conducted in farmers' rice fields in Kerala by Karthikeyan et al. (2010) with three rice IPM modules showed that the module comprising release of egg parasitoids recorded highest grain yield (4489 kg/ha) and CBR (1:1.30). Stem borer incidence was reduced by 61.6%. Leaf folder damage also showed a reduction of 64.8%.

Studies conducted by Pandey and Choubey (2012) in Uttar Pradesh showed that the release of *T. japonicum* at 50,000 eggs/ha was effective in reducing the incidence of *S. incertulas*.

10.2.1 Basmati Rice

Two to four releases of *T. japonicum* provided effective management of both yellow stem borer (*S. incertulas*) and leaf folder (*C. medinalis*) in Uttar Pradesh (Garg et al., 2002).

Pandey et al. (2003) showed that release of *Trichogramma* against insect eggs as one of the components of pest management in Uttar Pradesh resulted in higher grain yields (51.4 q/ha) and lower pest populations.

The integration of one application of 1.0 kg a.i./ha cartap hydrochloride 4G at 30 days after transplanting with weekly tagging of trichocards (*T. chilonis* and *T. japonicum* at 20,000/0.4 ha each) gave higher grain yield and showed maximum CBR (1:5.08) in Punjab (Mahal et al., 2006).

In a demonstration on biointensive pest management of leaf folder and stem borer on basmati rice in Punjab by Kaur et al. (2007) including one application of cartap hydrochloride and 7 weekly releases of *T. chilonis* and *T. japonicum* at 100,000/ha, each starting from 30 days after transplantation, resulted in a CBR of 1:4.01.

Kaur and Brar (2008) evaluated different doses of *Trichogramma* species against leaf folder and stem borer on Basmati rice in farmers' field in Punjab and found that seven releases of *T. chilonis* Ishii and *T. japonicum* Ashmead each at 100,000/ha, can be used both *Trichogramma* species the control both the pests.

Singh et al. (2008b) found that the release of *T. japonicum* at 100,000 eggs/ha with botanicals reduced the populations of both *S. incertulas* and *C. medinalis* and conserved the spider population in Uttar Pradesh.

10.3 Cotton

Two IPM modules, i.e., IPM module-I (release of *T. chilonis* at 150,000/ha/week from July to October) and IPM module-II (release of *T. chilonis* at 150,000/ha/week from July to August) were evaluated for the control of bollworm complex by Brar et al. (2001) along with other components of IPM in Punjab. Module-I reduced the incidence of bollworms, increased egg parasitization, and gave a higher yield of seed cotton, while module-II could prove very useful in curtailing the excessive and indiscriminate use of insecticides and ensuring sustainable cotton yield.

A study on the biocontrol-based management of cotton bollworms in Punjab by Brar et al. (2002) revealed that integration of *T. chilonis* (Bathinda strain) with insecticides reduced damage by bollworm by 70.3% and increased yields by 44.5% over insecticides alone.

A biointensive insect pest management module comprising eight releases of 150,000 *T. chilonis*/ha/week, synchronized with bollworm egg appearance studied by Rahman et al. (2003) in Andhra Pradesh recorded higher gross returns, net returns, and CBR.

Yadav and Jha (2003) observed that interspersing *Cassia occidentalis* with cotton Hybrid-8 at a ratio of 1:3 resulted in very high populations of *T. chilonis* and *T. achaeae* and caused an appreciable reduction in the population of the cotton bollworm, *Earias vittella*, in Gujarat.

Virk and Brar (2003) recommended that cotton is sown between April 10–30 at moderate spacing (67.5 × 45.0 cm) and *T. chilonis* released at 50,000/ha from July to September at weekly intervals for better yields in Punjab.

In Tamil Nadu, Balakrishnan et al. (2004) found that two releases of *T. chilonis* with two sprays of *B. thuringiensis* var. *kurstaki* was the optimum against *H. armigera* (Hubner) in a rainfed cotton ecosystem and recorded higher yield (782 kg/ha).

Bhat et al. (2004) found that inundative releases of *T. chilonis* remarkably reduced the incidence of cotton bollworm (*H. armigera*) to 43% in Andhra Pradesh. The use of pesticides and fungicides also came down and farmers realized better yield and developed a positive attitude toward adoption of *T. chilonis* as an IPM strategy.

The field experiment carried out in Maharashtra to study the efficacy of *T. chilonis* against bollworms revealed that combined releases of *T. chilonis* at 150,000 parasitized eggs/ha with different rates of *C. zastrowi sillemi* contributed to a higher yield of seed cotton (Panchbhai et al., 2004).

In a study by Virk et al. (2004) in Punjab on the role of trap crops on the parasitization efficiency of *T. chilonis* on *H. armigera*, it was found that when sorghum was used as a trap crop, the highest mean parasitization (10.48%), lowest mean incidence of *H. armigera* among green bolls (2.74%), and significantly highest yield (14.8 q/ha) were observed.

In an evaluation on the efficacy of two natural enemies of *H. armigera*, namely *T. chilonis* and *C. zastrowi sillemi*, in Maharashtra by Panchbhai et al. (2005), it was found that a high yield of 14.96 q/ha was obtained in combined releases of *T. chilonis* (150,000 parasitized eggs/ha) + *C. zastrowi sillemi* (four eggs/plant).

The validation and popularization of IPM technology in cotton through farmers' participatory approach in tribal areas of Southern Rajasthan was studied by Ameta et al. (2006). The study, comprising two releases of *T. chilonis* at 150,000/ha at days 75 and 85 after germination as one of the IPM components, showed that the module reduced the population of *Earias insulana* and *H. armigera*. The mean percentage damage to square, flower, green boll, open boll, and locule was less and the authors recorded a higher mean seed cotton yield of 1918 and 1804 kg/ha during 2003 and 2004, respectively.

Mohapatra and Patnaik (2006) validated a biointensive IPM module in Orissa, a major component of which was two releases of *T. chilonis* at 150,000/ha 55 days after germination. The module registered significantly lower bollworm damage with increasing activity of natural enemies and gave higher monetary return over farmers' practices.

An on-farm validation of a biointensive IPM module in rain-fed cotton in Maharashtra by Puri et al. (2006), consisting of release of the egg parasitoid *T. chilonis* at 150,000/ha as one of the components, significantly lowered the pest population and resulted in higher net returns and yields.

A study conducted in Tamil Nadu by Rajaram et al. (2006) to evaluate an IPM strategy, including the release of *T. chilonis* at 5 cc/release/week as one of the components, recorded a lower leafhopper (*Amrasca* sp.) incidence (1.4/plant), lower damage (11–14%) by bollworm (*Helicoverpa* sp.), lower stem weevil (*Pempherulus affinis*) infestation (15.56%), and a high yield of 145 kg/ha.

An IPM module developed for rain-fed cotton by Bhosle (2007) in Maharashtra, including the release of *T. chilonis* as one of the components, effectively lowered the bollworm damage and conserved natural enemies.

Karabhantanal et al. (2007) found that a biointensive IPM module consisting of two releases of *Trichogramma* sp. recorded significantly lower bollworm damage, minimum bad opened bolls and maximum good opened bolls, minimum locule damage, and higher yield in Karnataka.

Large-scale farmers' participatory field validation trials conducted by Tanwar et al. (2007) in Haryana to evaluate the impact of IPM consisting of release of *T. chilonis* as a component substantially reduced the number of chemical insecticidal sprays to two to three sprays against six to seven single or mixed sprays in farmers practice (FP) and recorded better parasitization of *H. armigera* ova by *T. chilonis* (22.4%).

Field studies conducted in Punjab by Khosa et al. (2008) to evaluate sorghum (*Sorghum bicolor*) as a trap crop in cotton to increase the parasitizing efficiency of *T. chilonis* against bollworms revealed that bollworm incidence among intact fruiting bodies and green bolls was the lowest (1.37%) and mean parasitization (18.74%) was highest where sorghum was planted after five rows of cotton.

In an evaluation of the field efficacy of biorational IPM modules against bollworm complexes on rain-fed cotton, including release of *T. chilonis* 150,000/ha as an important component of both the modules, Lande et al. (2008a,b,c) found that both the models recorded high seed cotton yield of 11.20 and 10.93 q/ha, with minimum bad seed cotton.

The results of two years' experiments conducted by Lande et al. (2008a,b,c) in cotton ecosystems in Maharashtra revealed that a biorational IPM module consisting of release of *T. chilonis* at 150,000/ha as one of the components recorded a higher number of *Chrysopa* eggs, *Chrysopa* larvae, and ladybird beetles, and field parasitization of *H. armigera* larvae was equally effective with the untreated control.

The pooled data over the two seasons collected by Lande et al. (2008a,b,c) in Maharashtra revealed that an IPM module comprising the release of *T. chilonis* at 150,000/ha as one of its components recorded the highest seed cotton yield (11.95 q/ha).

Singh and Saini (2008) studied the effect of different management practices on bollworm incidence in Haryana, in which weekly releases of *T. chilonis* were also included. The results revealed that the combination of practices resulted in ~50% reduction in the number of insecticidal sprays and proved successful with respect to bollworm (including *Earias* sp., *H. armigera* and *Pectinophora gossypiella*) incidence.

An overview of the findings of Sivanarayana et al. (2008) on the awareness and adoption of IPM practices by farmers in Andhra Pradesh indicated that farmers had medium to low levels of awareness and adoption. Therefore, it stressed the need for the extension agency to gear up its various programs to improve the adoption level of cotton IPM practices.

A field experiment conducted in Gujarat by Godhani et al. (2009) revealed that the highest seed yield (2111 kg/ha) was obtained from plots with a single release of *T. chilonis* and *C. zastrowi sillemi*, which also registered the highest net return (8685/ha).

Comparative evaluation of different pest management strategies undertaken by Swamy and Prasad (2010) revealed that the CBR was highest in the IPM strategy (4.42) and the IPM program successfully resulted in lower production costs and ultimately helped in conserving natural enemy species and sustaining biodiversity.

10.3.1 Bacillus thuringiensis (Bt) *Cotton*

Shanmugam et al. (2006a,b) found that a biointensive pest management (BIPM) module comprising release of *T. chilonis* as one of the components, recorded a high seed cotton yield of 1920 kg/ha and reduced insecticide usage in the Mahyco cotton hybrid (MECH 162 *Bt* cotton) in Tamil Nadu.

Field studies conducted by Nagesh (2009) in Andhra Pradesh to evaluate the IPM module comprising release of *T. chilonis*, recorded low incidence of American bollworm, *H. armigera*, in terms of percentage square damage and green boll damage, reduced the number of insecticide sprayings, and resulted in a higher net income and higher incremental CBR of 1:4.94 compared with 1:0.8 in the farmers' method.

Habitat manipulation in *Bt* cotton (cv. Vikram-5) was studied in Gujarat by Godhani et al. (2010) consisting of release of *T. chilonis* at 150,000/ha as one of the components of the treatment resulted in the highest seed cotton yield (2458 kg/ha).

10.4 Maize

Jalali and Singh (2003) studied the efficacy of *T. chilonis* and *Cotesia flavipes* released at different time intervals in controlling *C. partellus* infesting fodder maize in Karnataka and found that the crop yield was higher when *T. chilonis* was released at 3-day intervals compared with 5-day intervals, and *T. chilonis* was more efficient in controlling the pest compared to *C. flavipes*.

Studies conducted in Punjab by Kanta et al. (2008) for the control of maize stem borer, *C. partellus*, with *T. chilonis* revealed that single release of *T. chilonis* in the form of parasitized eggs of *C. cephalonica* at 100,000/ha on 13-day-old crops proved very effective in the control of maize stem borer (65% egg parasitization).

10.5 Chickpea

Evaluation of some biocontrol agents against *H. armigera* (Hubner) in chickpea by Mandal and Mishra (2004) indicated that *T. chilonis* was not effective in controlling *H. armigera* in chickpea. The reduction in mean larval population and pod damage was minimum, ranging from 9.08% to 12.10%, and the increase in mean yield was the lowest (13.45%).

10.6 Sesame (*Sesamum indicum* L.)

Trichogramma chilonis Ishii was evaluated in Karnataka at various dosages for management of *Acherontia styx* (Kanaburgi et al., 2012). *Trichogramma* at 225,000/ha was found to be superior with 81.90% parasitism compared with releases of *Trichogramma* at 150,000 and 75,000/ha.

10.7 Tomato (*Solanum lycopersicum* L.)

In Punjab, a study was undertaken to determine the efficacy of *T. pretiosum* (five releases weekly at 50,000/ha) as a component of IPM for the management of the tomato fruit borer, *H. armigera*. Egg parasitism was very high (36.32–61.00%) in plots where *T. pretiosum* was released. It was concluded that the treatment combination *T. pretiosum* + *H. armigera* nucleo-polyhedrosis virus (HaNPV) + endosulfan was the most effective for *H. armigera* control (Brar et al., 2003).

Kumar et al. (2003) conducted experiments in Ranchi, Bihar, to determine the effect of IPM modules on tomato pests and diseases. Two releases of *T. pretiosum* (thelytokous—having only females) at 100,000 parasitized eggs/ha at 15-day intervals was found to be effective in controlling *H. armigera*.

In Tamil Nadu, Praveen and Dhandapani (2003) studied the economic feasibility of biocontrol based pest management comprising the release of *Trichogramma chilonis*, *C. zastrowi sillemi*, *B. thuringiensis*, and HaNPV. This module effectively controlled *H. armigera* with higher fruit yield (23,292 kg/ha) than untreated control (13, 689 kg/ha) with higher CBR (1.00:3.21).

Singh et al. (2003) evaluated *T. pretiosum* (thelytokous) (five releases at 50,000/ha weekly) for *H. armigera* control. The mean parasitism was highest (43.20%) in plots where *T. pretiosum* (thelytokous) was released five times at 50,000/ha/week. *Trichogramma pretiosum* (thelytokous) + HaNPV + endosulfan treatment recorded the highest yield (226.67 q/ha). It was concluded that integrating *T. pretiosum* (thelytokous), HaNPV, and endosulfan was the most effective treatment for the control of *H. armigera* on tomato.

In Punjab, a field experiment to determine the efficacy of *T. chilonis*, *T. pretiosum*, and *T. pretiosum* (thelytokous) was carried out for the management of *H. armigera* on tomato. The highest yield (261.07 q/ha) was obtained in plots where *T. chilonis* was released at a rate of 100,000/ha, followed by *T. chilonis* at 75,000/ha (248.27 q/ha) (Kumar et al., 2004).

During an investigation carried out in Karnataka, an IPM module consisting of *T. pretiosum* (45,000/ha) as one of the components was found to be significantly superior to the rest of the modules, such as farmers' practice, insecticides tested in restricting the larval population of

H. armigera, lowering fruit damage (11.87%), increasing marketable fruit yield (224.56q/ha), and yielding additional net profit (Rupees 22,915/ha) (Karabhantanal et al., 2005).

In Tamil Nadu, Amutha and Manisegaran (2006) evaluated *T. chilonis* released six times at weekly interval and recorded minimal *H. armigera* damage (10.0%), the highest yield (29.90t/ha), with highest CBR (1:2.99) compared with insecticides alone.

In Uttar Pradesh, safer management tools against major insect pests of tomato and garden pea were evaluated and it was found that four releases of *T. chilonis* at 50,000/ha at 10-day intervals from the flower initiation stage was most promising (Sushil et al., 2006).

Field experiments carried out in Tamil Nadu, to evaluate BIPM comprising five releases of *T. pretiosum* at 50,000 adults/ha from flower initiation period was found to be effective in reducing the larval population and in increasing yield (Vijayalakshmi, 2007).

Sharma and Bhardwaj (2008) studied the efficacy of *T. chilonis* against *H. armigera* and found that the lowest net profit (Rupees 18,175/ha) and lowest CBR (3.11:1) was recorded for *T. chilonis* in Rajasthan.

Evaluation of *T. pretiosum* (thelytokous), HaNPV, and endosulfan was carried out alone and in combination for the control of tomato fruit borer. Based on the mean fruit damage and marketable yield, it was found that the bioagents and endosulfan alone were less effective but the combination of *T. pretiosum* (thelytokous) (five releases at weekly intervals at 50,000/ha), HaNPV (three sprays at 10-day intervals at 1.5×10^{12} polyhedral occlusion bodies (POBs)/ha) and endosulfan (three sprays at 15-day intervals at 700g a.i./ha) proved most effective for the management of *H. armigera* (Siddique et al., 2010).

Efficacy and economics of biopesticide combinations and insecticides against tomato fruit borer was studied by Tyagi et al. (2010). Four sprays of *B. thuringiensis* at 1kg/ha with release of *T. pretiosum* at 50,000 parasitoids at 10-day intervals proved to be the most effective treatment in terms of reduction in fruit damage, net return, and yield but four sprays of NPV at 250 larval equivalent (LE)/ha along with release of *T. pretiosum* at 50,000 parasitoids at 10-day intervals proved to be the most cost-effective treatment for management of tomato fruit borer.

When different doses of *T. chilonis* were tested on tomato at six locations in Kashmir against *H. armigera*, it was found that *T. chilonis* released at 130,000/ha resulted in the least fruit damage (Khan, 2011).

10.8 Brinjal

In Uttar Pradesh, Satpathy et al. (2005) released *T. chilonis* at 250,000/ha in alternation with endosulfan at 350g a.i./ha, 4% neem seed kernel extract (NSKE), and weekly shoot clippings for controlling shoot and fruit borer (*L. orbonalis*) starting from day 35 after transplanting and found that infestation of fruits was higher in control plots (84.32%), compared with *T. chilonis*+endosulfan (58%) and *T. chilonis*+shoot clippings+NSKE (60%).

Alternative measures evaluated against brinjal shoot and fruit borer (*L. orbonalis*) under the coastal agroclimatic conditions of Bhubaneswar, Orissa showed that treatment comprising six releases of *T. chilonis* at 50,000 at 15-day intervals, starting from day 30 after transplanting, failed to show a significant impact on pest damage (Singh et al., 2005).

Yadav and Sharma (2005) assessed the efficacy of bioagents in Rajasthan in relation to malathion against *L. orbonalis* and found that the bioagents were not superior to malathion 50EC (0.05%); however, *T. chilonis* was statistically superior to control in suppressing the infestation.

In Tamil Nadu, Suresh et al. (2007) assessed the effect of release of the egg parasitoid *T. chilonis* at 50,000/ha on 30 and 60 DAT and the results indicated that the percentage shoot and fruit borer damage was significantly less and fruit yield was maximum in the treatment with farm yard manure (FYM) (12.5 t/ha) + biofertilizers (2 kg/ha) + neem cake (800 kg/ha) followed by two releases of *T. chilonis*.

Elanchezhyan and Baskaran (2008) showed that an IPM-based module consisting of six releases of *T. chilonis* at 50,000/acre 15, 22, 29, 36, 43, and 50 days after transplanting was the best in managing *L. orbonalis* in Tamil Nadu.

Wider area validation and economic analysis of adaptable IPM technology in eggplant was carried out by Sardana et al. (2008a,b) in Uttar Pradesh and revealed that five releases of *T. pretiosum* (thelytokous) as one of the components of the technology was very effective in reducing the incidence of pests, minimizing yield losses, reducing the number of sprays, producing higher fruit yields, and increasing income.

In a field evaluation of IPM tools against brinjal shoot and fruit borer, *L. orbonalis*, in Orissa Singh et al. (2008a) found that *T. chilonis* alone or in combination with *Bracon hebetor* did not have significant effects on control of the pest.

In Uttar Pradesh, Sardana et al. (2009) carried out large-scale IPM farm trials on brinjal, comprising releases of *Trichogramma* spp. at 150,000/ha as one of the components and found that although IPM technology resulted in lower yields and a consequent lower CBR, it yielded a clean and better quality produce. Therefore, IPM technology was not only directly environmentally friendly but also more sustainable as it resulted in increased biodiversity (natural enemies, soil flora and fauna).

10.9 Cauliflower

Efficacy of bioagents was assessed in Maharashtra against *P. xylostella* (L.) (Lepidoptera: Plutellidae) on cauliflower by Lad et al. (2009) and the results revealed that *Trichogrammatoidea bactrae* at 150,000 eggs/ha and *T. chilonis* at 150,000 eggs/ha were superior to quinalphos and effected 53.94 and 45.81% larval reduction, respectively.

10.10 Cabbage

In a study on the seasonal activity of *T. chilonis* Ishii on diamondback moth, *P. xylostella* (Linn.), infesting cabbage, it was found that the activity of *T. chilonis* peaked during standard meteorological weeks 5 and 9, with 50.2 and 33.8% egg parasitism, respectively, and the parasitoid showed a positive correlation with host eggs (Shukla and Kumar, 2004).

10.11 Okra

Sardana et al. (2008a,b) studied the validation and economic viability of the IPM technology for okra crops comprising five releases of *T. chilonis* based on pheromone monitoring in Uttar Pradesh and found that the IPM technology was very effective in reducing the incidence of pests and minimizing yield losses. It resulted in reducing the number of sprays to 4 or 5 from 11 or 12, higher fruit yields, higher CBR, and an increase in net income.

In combinations of treatments in Uttar Pradesh including 50,000 *Trichogramma*/ha, Yadav et al. (2008a,b) found that the modules with *Trichogramma* gave the highest economic return and the application of endosulfan–*Trichogramma* followed by application of neemarin–*Trichogramma* recorded the highest CBR.

Yadav et al. (2008a,b) evaluated the efficacy of biocontrol treatments against pests of okra in Uttar Pradesh and found that the application of *Bt*–neem (*Azadirachta indica*) formulation with azadirachtin–endosulfan–*Trichogramma* applied at 15-day intervals reduced fruit and shoot borer (*E. vittella*) infestation to 1.93% and gave the highest yield (79.70 q/ha).

Investigations carried out to assess the potential of the parasitoids against okra fruit borers revealed that the combination of *T. chilonis* + *Chelonus blackburni* + *Bracon brevicornis* was more effective in lowering larval populations of *E. vittella* and *H. armigera* after three releases/spray and recorded higher yield (Thanavendan and Jeyarani, 2009).

10.12 Potato

Rajneesh and Nandihalli (2009) evaluated the efficacy of BIPM systems in Karnataka against potato shoot borer, *L. orbonalis* L., consisting of four releases of *Trichogramma* at 50,000/ha at weekly intervals. The biointensive system was superior on week 5 after planting. A higher net profit was also obtained with the biointensive module.

10.13 Chili

Among the four IPM modules evaluated by Gundannavar et al. (2007) against chili pests in Karnataka, the module including the release of *T. chilonis* at 100,000/ha (9 weeks after transplanting) was effective against *H. armigera* and appeared to be a promising strategy as it did not include any chemical intervention.

10.14 Castor

Basappa and Lingappa (2002) developed and evaluated IPM modules in Karnataka involving the release of *T. chilonis* at 200,000/ha at 30 days after sowing followed by fenvalerate (0.01%) spray at 45 days after spraying (DAS) and second release of *T. chilonis* at 300,000/ha at 65 DAS offered good protection to castor from *A. janata* and proved to be cost-effective with the highest CBR (1:4.99).

Basappa and Lingappa (2004) studied the effects of four IPM modules for the management of castor semilooper, *A. janata* Linn., in Karnataka. The module, comprising release of *T. chilonis* (200,000/ha) at 60 DAS as one of the components, resulted in the lowest yield and CBR but conserved the larval parasitoid, *Microplitis maculipennis* Zepligate.

Singh et al. (2006) conducted field experiments in Andhra Pradesh to determine the comparative performance of different pest management treatments for castor + pigeon pea intercropping (4:1) system under rain-fed conditions. The treatment comprising release of the egg parasitoid *T. chilonis* at 125,000/ha, proved effective in obtaining higher castor and pigeon pea seed yields, cost: benefit ratio, higher larval parasitism of insect pests and reducing the incidence of insect pests and diseases on the castor crop.

Field trials conducted in Andhra Pradesh for the validation of an IPM package for castor and pigeon pea intercropping (4:1) system in a rain-fed agroecosystem, comprising the release of egg the parasitoid *T. chilonis* at 125,000/ha as one of the components, proved effective in reducing the incidence of insect pests and diseases on the castor crop. Higher castor seed yield (9.1–12.5 q/ha) and CBR (1.94–3.13) were also obtained (Singh et al., 2008d).

In an evaluation of biointensive IPM modules utilizing promising natural enemies for the management of castor semilooper, *A. janata* Linnaeus (Lepidoptera: Noctuidae), *T. achaeae* at 150,000/ha at 30 DAS + fenvalerate 20 EC at 45 DAS + *T. achaeae* at 60 DAS showed superiority in suppression of the pest, safety to natural enemies, higher yield (15.6 q/ha), and CBR (2.08) (Naik et al., 2010).

10.15 Sugar Beet

Shivankar et al. (2008) studied the field bioefficacy of *T. chilonis* against *Spodoptera litura* Fab. and found that release of *T. chilonis* at 50,000/ha and spraying of azadirachtin at 3000 ppm (5 ml/l) gave 89.7 and 89.3% reduction, respectively, of larval population of *Spodoptera* and was found to be an effective and safe treatment to combat sugar beet pests. Azadiractin and *Trichogramma* release also resulted in more root yield.

10.16 Mulberry

The efficacy of *T. chilonis* in controlling *Diaphania pulverulentalis* H. (Lepidoptera: Pyralidae) in three southern sericultural states, Karnataka, Andhra Pradesh, and Tamil Nadu, was evaluated in mulberry gardens by Prasad et al. (2006). The parasitoid suppressed the occurrence of the leaf roller, *D. pulverulentalis* H., in the field by 23.26%.

10.17 Teak

Field trials conducted in Madhya Pradesh to develop an ecofriendly, nontoxic, and practically feasible method of minimizing the population of the most serious teak leaf skeletonizer, *Eutectona machaeralis*, revealed that the optimum dose of the egg parasitoid *T. pretiosum* (thelytokous) is 125,000 wasps/ha to minimize the attack of the pest (Joshi et al., 2007a,b).

Field evaluation of the wasps of *T. pretiosum* (thelytokous) by Joshi et al. (2007a,b) in different quantities for minimizing the attack of teak skeletonizer, *E. machaeralis* (Walk.), in Madhya Pradesh proved that introduction of the parasitoid at 125,000–300,000 wasps/ha was highly effective in minimizing the attack of the pest.

10.18 Lac Insect, *Kerria lacca* (Hemiptera: Kerridae)

A study was carried out by Bhattacharya et al. (2003) to evaluate the performance of three egg parasitoids, *T. pretiosum* (thelytokous), *T. chilonis*, and *T. pretiosum* (at 100,000, 150,000, and 200,000/ha), against two lepidopteran predators (*Eublemma amabilis* and *Pseudohypatopa pulverea*) of *Kerria lacca* raised on the bushy lac host plant bhalia (*Flemingia macrophylla*) in Bihar. The highest suppression was recorded in *T. pretiosum* (thelytokous) at 150,000 and 200,000/ha

(69.90% and 69.81%, respectively). This reduction in predator population positively affected the yield of lac sticks and increase of yield ratio.

Efficacy of three species of egg parasitoids (*T. achaeae*, *T. exiguum*, and *T. ostrinae*) were evaluated in the field in Jharkhand for the management of *E. amabilis* Moore (Lepidoptera: Noctuidae), a major insect predator of Indian lac insect *K. lacca* on two host plants *Butea monosperma* and *Albizia lucida*. The increase in lac crop yield was significantly higher when all three egg parasitoids were released at 100 parasitoids/plant for the kusmi biotype. However, for rangeeni, only *T. exiguum* and *T. ostrinae* could provide significant difference in yield (Bhattacharya et al., 2006).

Bhattacharya et al. (2007) carried out evaluation of three species of trichogrammatid egg parasitoids, namely, *T. achaeae* Nagaraja and Nagarkatti, *T. exiguum* Pinto and Platner, and *T. ostriniae* Pang et Chen for the suppression of the lepidopteran insect predator, *E. amabilis* Moore, in lac culture on *F. macrophylla* O. Kytze. All the three parasitoids were found equally effective in suppressing the population of *E. amabilis*. The reduction in the population of *E. amabilis* was 77–86% in the rangeeni crop and 52–72% in the kusmi crop with a dose of 20 egg parasitoids per bush. In general, trichogrammatids were found more effective for rangeeni lac than kusmi.

Bhattacharya et al. (2008) conducted field experiments and found that *T. achaeae*, *T. exiguum*, *T. pretiosum* (thelytokous), *T. chilonis*, *T. poliae*, *T. ostrinae*, and *T. pretiosum* were highly effective against the eggs of *E. amabilis* and *P. pulverea*. The population of insect predators was reduced by more than 75% using the egg parasitoids.

11. CONCLUSIONS

There are several groups of natural enemies commonly employed for biological control globally, among them the egg parasitoids constitute an important component of IPM practices due to the advantages they offer in the effective management of pests at its restive phase. A large number of species and subspecies of trichogrammatids (over 230) are distributed throughout the world parasitizing the eggs of over 200 insect species belonging to 70 families and 8 orders in diverse habitats, from aquatic to arboreal. Among the egg parasitoids, *Trichogramma* retain their importance as biological control agents for the management of lepidopterous pests across the globe due to its easy mass production technique available and usefulness on many crops and insect pests. The main factors underlying the increasing use of *Trichogramma* is their adaptability across a wide range of crop ecosystems, high searching ability, and their amenability for laboratory mass production with a high intrinsic rate of natural increase. Given the economic importance of trichogrammatids in biocontrol programs, species identification must be quick, simple, and widely applicable. Novel approaches that have used DNA sequences of the ITS2 and mitochondrial COX1 have helped to solve this difficulty, as well as use of restriction analysis of ITS2 region to distinguish *Trichogramma*. Genetic improvement of *Trichogramma* for tolerance to various abiotic stresses has resulted in development of strains tolerant to insecticides and high temperature. The developed strains were found to exert good control under field stress conditions. The effect of plant extracts and synomones have been documented. Studies on insecticide tolerance have been conducted and have resulted in devising IMP in different crop ecosystems. An account of the utilization of *Trichogramma* spp. on many crops have indicated their usefulness for the management of crop pests.

References

Alexander, B., Rajavel, D.S., Suresh, K., 2006. Effect of new insect growth regulators (IGRS) on *Trichogramma chilonis* Ishii. Hexapoda 13 (1/2), 66–69.

Ameta, O.P., Sharma, K.C., Rana, B.S., Bambawale, O.M., 2006. Validation and popularization of IPM technology in cotton through farmers' participatory approach in tribal area of Southern Rajasthan. Annals of Agricultural Research 27 (2), 162–166.

Amutha, M., Manisegaran, S., 2006. Evaluation of IPM modules against *Helicoverpa armigera* (Hubner). Annals of Plant Protection Sciences 14 (1), 22–26.

Archana, P.A.V.N., Malathi, V.G., Rani, K.U., Singh, A.K., 2005. Utilization of internal transcribed spacer region sequences of ribosomal DNA for molecular differentiation of four trichogrammatids. Indian Journal of Entomology 67 (4), 321–327.

Ashok Kumar, G., Jalali, S.K., Venkatesan, T., Stouthamer, R., Niranjana, P., Lalitha, Y., 2009. Internal transcribed spacer-2 restriction fragment length polymorphism (ITS-2-RFLP) tool to differentiate some exotic and indigenous trichogrammatid egg parasitoids in India. Biological Control 49 (3), 207–213.

Ashok Kumar, G., Jalali, S.K., Venkatesan, T., Nagesh, M., Lalitha, Y., 2008. Genetic improvement of egg parasitoid *Trichogramma chilonis* Ishii for combined tolerance to multiple insecticides and high temperature. Journal of Biological Control 22 (2), 347–356.

Balakrishnan, N., Baskaran, R.K.M., Mahadevan, N.R., 2004. Efficacy of *Trichogramma chilonis* Ishii in combination with biopesticides against *Helicoverpa armigera* (Hubner) in rainfed cotton ecosystem. Journal of Biological Control 18 (2), 121–127.

Ballal, C.R., Srinivasan, R., Jalali, S.K., 2009. Evaluation of an endosulfan tolerant strain of *Trichogramma chilonis* on cotton. BioControl 54 (6), 723–732.

Basappa, H., Lingappa, S., 2002. Management strategies for castor semilooper, *Achaea janata* Linn. (Lepidoptera: Noctuidae) in castor. Indian Journal of Plant Protection 30 (1), 51–54.

Basappa, H., Lingappa, S., 2004. IPM modules for the management of castor semilooper, *Achaea janata* Linn. and their impact on larval parasitoid, *Microplitis maculipennis* Zepligate. Indian Journal of Plant Protection 32 (2), 131–132.

Beevi, S.P., Lyla, K.R., Karthikeyan, K., 2003. Impact of the inundative release of egg parasitoids, *Trichogramma* spp. in rice pest management. In: Tandon, P.L., et al. (Eds.), Biological Control of Lepidopteran Pests-Proceedings of the Symposium of Biological Control of Lepidopteran Pests. Society for Biocontrol Advancement, Bangalore, pp. 329–332.

Bhat, B.N., Ramprasad, S., Mathivanan, N., Srinivasan, K., 2004. Management of soil borne diseases and insect pests with bioagents – a case study. Progressive Agriculture 4 (1), 38–40.

Bhattacharya, A., Jaiswal, A.K., Singh, J.P., 2008. Management of lac insect predators through IPM based biorational approaches. Emerging trends of researches in insect pest management and environmental safety I, 221–226.

Bhattacharya, A., Kumar, S., Jaiswal, A.K., 2007. Evaluation of *Trichogramma* species for the suppression of lepidopteran insect predator, *Eublemma amabilis* Moore, in lac culture on *Flemingia macrophylla*. Journal of Biological Control 21 (2), 267–270.

Bhattacharya, A., Kumar, S., Jaiswal, A.K., Kumar, K.K., 2006. Efficacy of the egg parasitoids, *Trichogramma* spp. for the management of *Eublemma amabilis* Moore (Lepidoptera: Noctuidae) – a predator of Indian lac insect. Entomon 31 (2), 121–124.

Bhattacharya, A., Mishra, Y.D., Sushil, S.N., Jaiswal, A.K., Kumar, K.K., 2003. Relative efficacy of some *Trichogramma* spp. for management of lepidopteran predators of lac insect, *Kerria lacca* (Kerr) under field conditions. In: Tandon, P.L., et al. (Eds.), Biological Control of Lepidopteran Pests-Proceedings of the Symposium of Biological Control of Lepidopteran Pests. Society for Biocontrol Advancement, Bangalore, pp. 301–303.

Bhosle, B.B., More, D.G., Bambawale, O.M., Sharma, O.P., Patange, N.R., 2007. Effectiveness of cotton IPM module in rainfed Marathwada Region. Annals of Plant Protection Sciences 15 (1), 21–25.

Boomathi, N., Raguraman, S., Sivasubramanian, P., Ganapathy, N., 2005. Allomonic effect of pigeon pea (*Cajanus cajan* (L.) Millsp.) plant extracts on egg parasitoid, *Trichogramma chilonis* Ishii parasitization. Entomon 30 (1), 101–103.

Brar, K.S., Sekhon, B.S., Singh, J., Shenhmar, M., Bakhetia, D.R.C., 2001. Evaluation of biocontrol based IPM modules for the control of bollworm complex on cotton. Journal of Insect Science, Ludhiana 14 (1/2), 19–22.

Brar, K.S., Sekhon, B.S., Singh, J., Shenhmar, M., Singh, J., 2002. Biocontrol based management of cotton bollworms in the Punjab. Journal of Biological Control 16 (2), 121–124.

Brar, K.S., Singh, J., Shenhmar, M., Kaur, S., Kaur, S., Joshi, N., Singh, I., 2003. Integrated management of *Helicoverpa armigera* (Hubner) on tomato. In: Tandon, P.L., et al. (Eds.), Biological Control of Lepidopteran Pests-Proceedings of the Symposium of Biological Control of Lepidopteran Pests. Society for Biocontrol Advancement, Bangalore, pp. 271–274.

Das, D.J., Basit, A., Bhattacharya, B., Saikia, K., Barooah, D., 2004. Establishment and recovery of *Trichogramma chilonis* Ishii on certain rice varieties. Shashpa 11 (1), 45–50.

Devi, P.S., Jalali, S.K., Venkatesan, T., 2007. Evidence of cross-tolerance in newer insecticides tolerant strain of *Trichogramma chilonis* Ishii to other insecticides. Indian Journal of Entomology 69 (2), 101–104.

Dey, P.K., Chakraborty, G., Somchoudhury, A.K., 2012. Bioefficacy of flubendiamide 480 SC against pod borers of black gram (*Vigna mungo* (L.) Hepper). Journal of Interacademicia 16 (3), 609–613.

Dhanasekaran, S., Joseph, S.A., Elumalai, K., 2007. Comparative toxicity of combination insecticides to some beneficial insects. Journal of Applied Zoological Researches 18 (2), 166–169.

Elanchezhyan, K., Baskaran, R.K.M., 2008. Evaluation of intercropping system based modules for the management of major insect pests of brinjal. Pest Management in Horticultural Ecosystems 14 (1), 67–73.

Flanders, S.E., 1929. The mass production of *Trichogramma minutum* Riley and observations on the natural and artificial parasitism of the codling moth egg. In: Jordan, K., Horn, W. (Eds.), Proceedings of the 4th International Congress of Entomology. Gottfr., Patz, Naumburg, pp. 110–130.

Ganesh, K.S., Khan, M.A., Tiwari, S., 2002. Effect of cruciferous plant extracts on parasitic behaviour of ooparasitoids, *Trichogramma* spp. (Hym.: Trichogrammatidae). Cruciferae Newsletter 24, 83 + 3 pages.

Garg, D.K., Kumar, P., Singh, R.N., Pathak, M., 2002. Role of parasitoid *Trichogramma japonicum* and other natural enemies in the management of yellow stem borer and leaf folder in basmati rice. Indian Journal of Entomology 64 (2), 117–123.

Giraddi, R.S., Gundannavar, K.P., 2006. Safety of emamectin benzoate, an avermectin derivative to the egg parasitoids, *Trichogramma* spp. Karnataka Journal of Agricultural Sciences 19 (2), 417–418.

Godhani, P.H., Patel, R.M., Jani, J.J., Yadav, D.N., Korat, D.M., Patel, B.H., 2009. Impact of habitat manipulation on insect pests and their natural enemies in hybrid cotton. Karnataka Journal of Agricultural Sciences 22 (1), 104–107.

Godhani, P.H., Patel, R.M., Patel, B.H., Korat, D.M., 2010. Impact of habitat manipulation on insect pests and their arthropod natural enemies in *Bt* cotton. Karnataka Journal of Agricultural Sciences 23 (2), 366–368.

Gundannavar, K.P., Giraddi, R.S., Kulkarni, K.A., Awaknavar, J.S., 2007. Development of integrated pest management modules for chilli pests. Karnataka Journal of Agricultural Sciences 20 (4), 757–760.

Hebert, P.D., Cywinska, A., Ball, S.L., de Waard, J.R., 2003. Biological identifications through DNA barcodes. Proceedings of the Royal Society B: Biological Sciences 270 (1512), 313–321.

Ingle, M.B., Ghorpade, S.A., Baheti, H.S., Kamdi, S.R., 2007. Genetic improvement in an egg parasitoid *T. chilonis* for tolerance to pesticides. Asian Journal of Bioscience 2 (1–2), 196–197.

Jalali, S.K., Singh, S.P., 1993. Superior strain selection of the egg parasitoid *Trichogramma chilonis* Ishii – biological parameters. Journal of Biological Control 7 (1), 57–60.

Jalali, S.K., Singh, S.P., 2003. Biological control of *Chilo partellus* (Swinhoe) in fodder maize by inundative releases of parasitoids. Indian Journal of Plant Protection 31 (2), 93–95.

Jalali, S.K., Singh, S.P., 2006. Temperature dependent developmental biology of maize stalk borer *Chilo partellus* (Swinhoe) (Lepidoptera: Pyralidae) and its natural enemies. Indian Journal of Entomology 68 (1), 54–61.

Jalali, S.K., 2010. Biological and Molecular Characterization of Inter and Intra Specific Variation in Trichogrammatids Final Project Technical Report. National Bureau of Agriculturally Important Insects, Bangalore. 40 pp.

Jalali, S.K., 2014. Effect of abiotic stresses on natural enemies of crop pests and mechanism of tolerance to these stresses. Indian Farming 64 (2), 102–105.

Jalali, S.K., Murthy, K.S., Venkatesan, T., Lalitha, Y., Devi, P.S., 2006a. Adaptive performance of *Trichogramma chilonis* Ishii at low temperature. Annals of Plant Protection Sciences 14 (1), 5–7.

Jalali, S.K., Singh, S.P., Venkatesan, T., Murthy, K.S., Lalitha, Y., 2006b. Development of endosulfan tolerant strain of an egg parasitoid *Trichogramma chilonis* Ishii (Hymenoptera: Trichogrammatidae). Indian Journal of Experimental Biology 44 (7), 584–590.

Jalali, S.K., Venkatesan, T., Murthy, K.S., Rabindra, R.J., Lalitha, Y., Udikeri, S.S., Bheemanna, M., Sreenivas, A.G., Balagurunathan, R., Yadav, D.N., 2006c. Field efficacy of multiple insecticides tolerant strain of *Trichogramma chilonis* Ishii against American bollworm, *Helicoverpa armigera* (Hübner) on cotton. Indian Journal of Plant Protection 34 (2), 173–180.

Jasmine, R.S., Kuttalam, S., Stanley, J., 2007. Relative toxicity of abamectin 1.9 EC to egg parasitoid, *Trichogramma chilonis* Ishii and egg larval parasitoid, *Chelonus blackburni* (Cam.). Asian Journal of Bioscience 2 (1/2), 92–95.

Joshi, K.C., Sambath, S., Chander, S., Roychoudhury, N., Yousuf, M., Kulkarni, N., 2007a. Evaluation of *Trichogramma* spp. to minimise the attack of teak skeletonizer, *Eutectona machaeralis* (Walk.). Indian Journal of Forestry 30 (3), 267–271.

Joshi, K.C., Sambath, S., Yousuf, M., Chander, S., Roychoudhury, N., Kulkarni, N., 2007b. Evaluation of *Trichogramma* spp. to minimise the attack of teak leaf skeletonizer. Indian Forester 133 (4), 527–533.

Justin, C.G.L., Thangaselvabai, T., 2006. Influence of temperature and preference of *Trichogramma chilonis* Ishii on eggs of the diamondback moth, *Plutella xylostella* (L.), and rice moth, *Corcyra cephalonica* Stainton. Pest Management and Economic Zoology 14 (1/2), 25–28.

Kanaburgi, K., Patil, R.R., Chandaragi, M., 2012. Management of sesame sphingid, *Acherontia styx* Westwood using *Trichogramma chilonis* Ishii egg parasitoid. Journal of Experimental Zoology, India 15 (2), 499–501.

Kannan, S.S., Chandrasekaran, S., 2006. Safety of emamectin 5 SG on *Trichogramma chilonis* and *Chrysoperla carnea* under laboratory conditions. Journal of Ecotoxicology and Environmental Monitoring 16 (6), 509–514.

Kanta, U., Kaur, S., Singh, D.P., Brar, K.S., 2008. Bio-suppression of *Chilo partellus* with *Trichogramma chilonis* on kharif maize. Journal of Insect Science, Ludhiana 21 (1), 87–89.

Karabhantanal, S.S., Kulkarni, K.A., 2002. Implication of tritrophic interactions in the management of the tomato fruit borer, *Helicoverpa armigera* (Hubner). Pest Management and Economic Zoology 10 (2), 183–186.

Karabhantanal, S.S., Awaknavar, J.S., Patil, R.K., Patil, B.V., 2005. Integrated management of the tomato fruit borer, *Helicoverpa armigera* Hubner. Karnataka Journal of Agricultural Sciences 18 (4), 977–981.

Karabhantanal, S.S., Bheemanna, M., Patil, B.V., 2007. Management of sucking pests and bollworms in cotton. Journal of Cotton Research and Development 21 (2), 253–256.

Karthikeyan, K., Jacob, S., Purushothaman, S.M., 2007. Field evaluation of egg parasitoids, *Trichogramma japonicum* Ashmead and *Trichogramma chilonis* Ishii, against rice yellow stem borer and leaf folder. Journal of Biological Control 21 (2), 261–265.

Karthikeyan, K., Jacob, S., Beevi, P., Purushothaman, S.M., 2010. Evaluation of different integrated pest management modules for the management of major pests of rice (*Oryza sativa*). Indian Journal of Agricultural Sciences 80 (1), 59–62.

Katti, G., Pasalu, I.C., Varma, N.R.G., Dhandapani, N., 2001. Quantification of natural biological control in rice ecosystem for possible exploitation in rice IPM. Indian Journal of Entomology 63 (4), 439–448.

Kaur, R., Brar, K.S., 2008. Evaluation of different doses of *Trichogramma* species for the management of leaf folder and stem borer on basmati rice. Journal of Biological Control 22 (1), 131–135.

Kaur, R., Brar, K.S., Kaur, R., 2008. Management of leaf folder and stem borer on coarse and basmati rice with organic and inorganic practices. Journal of Biological Control 22 (1), 137–141.

Kaur, R., Brar, K.S., Singh, J., Shenhmar, M., 2007. Large-scale evaluation of bio-intensive management for leaf folder and stem borer on basmati rice. Journal of Biological Control 21 (2), 255–259.

Kaur, S., Brar, K.S., 2003. Effect of insecticides on *Helicoverpa armigera* (Hubner) and its natural enemies on sunflower. Pest Management and Economic Zoology 11 (2), 153–157.

Kavitha, J., Kuttalam, S., Chandrasekaran, S., Ramaraju, K., 2006. Effect of spiromesifen 240 SC on beneficial insects. Annals of Plant Protection Sciences 14 (2), 343–345.

Khan, A.A., 2011. Exploitation of *Trichogramma chilonis* Ishii for suppression of *Helicoverpa armigera* (Hubner) in tomato. Journal of Insect Science, Ludhiana 24 (3), 254–258.

Khosa, J., Virk, J.S., Brar, K.S., 2008. Role of sorghum as trap crop for increasing parasitizing efficiency of *Trichogramma chilonis* Ishii against cotton bollworms. Journal of Insect Science, Ludhiana 21 (1), 79–83.

Kumar, A., Kumar, S., Khan, M.A., 2009. Relative efficacy of *Trichogramma chilonis* Ishii and *Trichogramma brasiliensis* (Ashmead) alone and combination with endosulfan on chickpea and pigeon pea for control of *Helicoverpa armigera* Hubner. Journal of Entomological Research 33 (1), 41–43.

Kumar, D., Krishna, G., Sengupta, A., Sahay, R.N., 2003. Effectiveness of IPM module for tomato in the farmers' field of Ranchi district. Journal of Research, Birsa Agricultural University 15 (1), 125–128.

Kumar, K., Santharam, G., 1999. Laboratory evaluation of imidacloprid against *Trichogramma chilonis* Ishii and *Chrysoperla carnea* (Stephens). Journal of Biological Control 13 (1/2), 73–78.

Kumar, P., Shenhmar, M., Brar, K.S., 2004. Field evaluation of trichogrammatids for the control of *Helicoverpa armigera* (Hubner) on tomato. Journal of Biological Control 18 (1), 45–50.

Kumar, S., Khan, M.A., 2005. Bio-efficacy of *Trichogramma* spp. against yellow stem borer and leaf folder in rice ecosystem. Annals of Plant Protection Sciences 13 (1), 97–99.

Kumar, S., Maurya, R.P., Khan, M.A., 2007. Impact of biointensive pest management strategies on yellow stem borer and leaf folder in rice and their effect on the economics of production. Journal of Entomological Research 31 (1), 11–13.

Lad, S.K., Peshkar, L.N., Baviskar, S.S., Jadhav, R.S., 2009. Efficacy of microbials and bioagents for the management of *Plutella xylostella* (L.) on cauliflower. Journal of Soils and Crops 19 (1), 129–134.

Lande, G.K., Bhalkare, S.K., Nehare, S.K., 2008a. Field efficacy of biorational IPM modules against bollworm complex on rainfed cotton. Journal of Entomological Research 32 (3), 201–205.

Lande, G.K., Bhalkare, S.K., Rathod, P.K., 2008b. Effect of different IPM modules on the bioagents in cotton ecosystem. Journal of Applied Zoological Researches 19 (1), 37–41.

Lande, G.K., Katole, S.R., Bhalkare, S.K., Bisane, K.D., 2008c. Evaluation of biorational IPM modules against bollworm complex on cotton. Journal of Cotton Research and Development 22 (1), 91–96.

Mahal, M.S., Kajal, V.K., Kaur, R., Singh, R., 2006. Integration of chemical and biocontrol approaches for the management of leaffolder, *Cnaphalocrocis medinalis* Guenee and stem borer, *Scirpophaga incertulas* Walker on basmati rice. Journal of Biological Control 20 (1), 1–6.

Mandal, S.M.A., Mishra, B.K., 2004. Evaluation of some biocontrol agents against *Helicoverpa armigera* (Hubner) in chickpea. Journal of Plant Protection and Environment 1 (1/2), 27–29.

Manisegaran, S., 2004. Re-validation of *Trichogramma chilonis* Ishii for the control of internode borer *Chilo sacchariphagus indicus* Kapur in sugarcane. Indian Journal of Entomology 66 (1), 24–26.

Maurya, R.P., Khan, M.A., 2007a. Development of cartap hydrochloride tolerant strain of egg parasitoid, *Trichogramma chilonis* Ishii through artificial selection. Pantnagar Journal of Research 5 (2), 44–51.

Maurya, R.P., Khan, M.A., 2007b. Development of chlorpyriphos tolerant strain of *Trichogramma chilonis* Ishii and its efficacy against *Helicoverpa armigera*. Annals of Plant Protection Sciences 15 (2), 345–353.

Maurya, R.P., Khan, M.A., Kumar, S., Thakur, S.S., 2008. Development of imidacloprid tolerant strain of *Trichogramma chilonis* Ishii. Journal of Entomological Research 32 (3), 217–223.

Mishra, D.N., Kumar, K., 2009. Field efficacy of bio-agent *Trichogramma* spp. against stem borer and leaf folder in rice crop under mid-western plain zone of UP. Environment and Ecology 27 (4A), 1885–1887.

Mohapatra, L.N., Patnaik, R.K., 2006. Validation of IPM technology for rainfed cotton in western Orissa. Journal of Cotton Research and Development 20 (1), 102–108.

Nagarkatti, S., Nagaraja, H., 1968. Biosystematic studies on *Trichogramma* species. I. Experimental hybridization between *Trichogramma australicum* Girault, *T. evanescens* Westwood and *T. minutum* Riley. Technical Bulletin Commonwealth Institute of Biological Control 10, 81–96.

Nagarkatti, S., Nagaraja, H., 1971. Redescription of some known species of *Trichogramma* showing the importance of male genitalia as a diagnostic character. Bulletin of Entomological Research 61 (1), 13–31.

Nagarkatti, S., Nagaraja, H., 1977. Biosystematics of *Trichogramma* and *Trichogrammatoidea* species. Annual Review of Entomology 22, 157–176.

Nagesh, M., 2009. Evaluation of integrated pest management modules against American bollworm, *Helicoverpa armigera* Hub. in non-*Bt* cotton. Journal of Cotton Research and Development 23 (2), 286–288.

Naik, M.I., Kumar, M.A.A., Manjunatha, M., Shivanna, B.K., 2010. Evaluation of bio-intensive IPM modules utilizing promising natural enemies for the management of semilooper *Achaea janata* Linnaeus (Lepidoptera: Noctuidae) in castor. Environment and Ecology 28 (1B), 552–557.

Nair, N., Sekh, K., Dhar, P.P., Somchoudhury, A.K., 2008. Field efficacy of pyridalyl-10 EC against *Helicoverpa armigera* Hub. and its effect on its natural enemies. Journal of Entomological Research 32 (3), 237–240.

Narayan, D., 2004. Eco-friendly approach for minimising population of sugarcane stalk borer *Chilo auricilius* Linn. in Terai belt of Uttar Pradesh. Indian Journal of Entomology 66 (1), 14–16.

Nordlund, D.A., Chalfant, R.B., Lewis, W.J., 1984. Arthropod populations, yields and damage in monocultures and polycultures of corn, beans and tomatoes. Agriculture Ecosystem and Environment 11 (4), 353–367.

Noyes, J.S., 2003. Universal Chalcidoidea Database. At: http://www.nhm.ac.uk/entomology/chalcidoids.

Noyes, J.S., 2014. Universal Chalcidoidea Database. World Wide Web electronic publication. http://www.nhm.ac.uk/chalcidoids.

Panchbhai, P.R., Sharnagat, B.K., Dhawad, C.S., Wadaskar, R.M., Bagade, I.B., 2005. Effect of augmentation of *Trichogramma chilonis* Ishii and *Chrysoperla carnea* (Stephens); two natural enemies of *Helicoverpa armigera* (Hubner) on larval infestation and yield in cotton. Pest Management and Economic Zoology 13 (1), 77–81.

Panchbhai, P.R., Sharnagat, B.K., Nemade, P.W., Bagade, I.B., Nandanwar, N.R., 2004. Efficacy of independent and combined releases of *Trichogramma chilonis* and *Chrysoperla carnea* against cotton bollworms. Journal of Soils and Crops 14 (2), 371–375.

Pandey, R.K., Vats, A.S., Singh, G.R., 2003. Effect of alternate pest management practices on rice pests in eastern Uttar Pradesh. Crop Research (Hisar) 25 (1), 127–131.

Pandey, S., Choubey, M.N., 2012. Management of yellow stem borer, *Scirpophaga incertulas* in rice. Agricultural Science Digest 32 (1), 7–12.

Pandya, H.V., Patel, M.B., 2007. Effect of partitioned growth stage protection on incidence of sugarcane borers in Gujarat. Cooperative Sugar 38 (9), 41–44.

Paramasivan, A., Paul, A.V.N., 2005. Use of semiochemical formulations for management of the egg parasitoid *Trichogramma chilonis* Ishii (Hymenoptera: Trichogrammatidae). Shashpa 12 (1), 31–34.

Paul, A.V.N., Srivastava, M., Dureja, P., Singh, A.K., 2008. Semiochemicals produced by tomato varieties and their role in parasitism of *Corcyra cephalonica* (Lepidoptera: Pyralidae) by the egg parasitoid *Trichogramma chilonis* (Hymenoptera: Trichogrammatidae). International Journal of Tropical Insect Science 28 (2), 108–116.

Pinto, J.D., Stouthamer, R., 1994. Systematics of the Trichogrammatidae with emphasis on *Trichogramma*. In: Wajnberg, E., Hassan, S.A. (Eds.), Biological Control with Egg Parasitoids. CABI, UK, pp. 1–36.

Pinto, J.D., Velten, R.K., Platner, G.R., Oatman, E.R., 1989. Phenotypic plasticity and taxonomic characters in *Trichogramma*. Annals of the Entomological Society of America 85 (4), 413–422.

Prasad, K.S., Rajadurai, S., Shekhar, M.A., Kariappa, B.K., 2006. Field evaluation of *Trichogramma chilonis* Ish. (Hymenoptera: Trichogrammatidae) – an egg parasitoid of leaf roller, *Diaphania pulverulentalis* H. (Lepidoptera: Pyralidae). Plant Archives 6 (2), 597–598.

Praveen, P.M., Dhandapani, N., 2003. Development of biocontrol based pest management in tomato, *Lycopersicon esculentum* (Mill.). In: Tandon, P.L., et al. (Ed.), Biological Control of Lepidopteran Pests-Proceedings of the Symposium of Biological Control of Lepidopteran Pests. Society for Biocontrol Advancement, Bangalore, pp. 267–270.

Puri, S.N., Sharma, O.P., Lavekar, R.C., Murthy, K.S., Dhandapani, A., 2006. On-farm validation of bio-intensive IPM module in rainfed cotton in southern Maharashtra. Indian Journal of Plant Protection 34 (2), 248–249.

Rachappa, V., Naik, L.K., 2004. Integrated management of early shoot borer, *Chilo infuscatellus* (Snellen) in sugarcane. Annals of Plant Protection Sciences 12 (2), 248–253.

Rahman, S.J., Rao, A.G., Reddy, P.S., 2003. Potential and economics of biointensive insect pest management (BIPM) module in cotton for sustainable production. In: Tandon, P.L., et al. (Eds.), Biological Control of Lepidopteran Pests-Proceedings of the Symposium of Biological Control of Lepidopteran Pests. Society for Biocontrol Advancement, Bangalore, pp. 279–283.

Rajaram, V., Mathirajan, V.G., Krishnasamy, S., 2006. IPM in cotton under dry farming condition. International Journal of Agricultural Sciences 2 (2), 557–558.

Rajendran, B., 2006. A benefit analysis of evaluation of integrated pest management practices for sugarcane. Indian Sugar 56 (1), 19–24.

Rajneesh, H., Nandihalli, B.S., 2009. Evaluation of integrated pest management modules in the management of *Leucinodes orbonalis* L. in potato. Karnataka Journal of Agricultural Sciences 22 (3), 716–717.

Ramesh, B., Manickavasagam, S., 2006. Non-killing effects of certain insecticides on the development and parasitic features of *Trichogramma* spp. Indian Journal of Plant Protection 34 (1), 40–45.

Rani, P.U., Jyothsna, Y., Lakshminarayana, M., 2008. Host and non-host plant volatiles on oviposition and orientation behaviour of *Trichogramma chilonis* Ishii. Journal of Biopesticides 1 (1), 17–22.

Rao, C.V.N., Rao, N.V., Bhavani, B., 2006. Efficacy of *Trichogramma chilonis* Ishii against sugarcane early shoot borer, *Chilo infuscatellus* Snellen under sugar factory operational areas of coastal Andhra Pradesh. Journal of Biological Control 20 (2), 225–228.

Rao, N.B.V.C., 2005. Preliminary evaluation of insecticides and neem derivatives on egg parasitoid *Trichogramma chilonis* Ishii. Insect Environment 11 (1), 25–27.

Rao, N.S., Raguraman, S., 2005. Influence of neem based insecticides on egg parasitoid, *Trichogramma chilonis* and green lace-wing predator, *Chrysoperla carnea*. Journal of Ecobiology 17 (5), 437–443.

Riba, T., Sarma, A.K., 2006. Efficacy of *Trichogramma japonicum* Ashmead against yellow stem borer, *Scirpophaga incertulas* walk on rice in Nagaland. Journal of Applied Zoological Researches 17 (2), 196–200.

Samanta, A., Chowdhury, A., Somchoudhury, A.K., 2006. Residues of different insecticides in/on brinjal and their effect on *Trichogramma* spp. Pesticide Research Journal 18 (1), 35–39.

Sardana, H.R., Bambawale, O.M., Batra, P., 2008a. Wider area validation and economic analysis of adaptable IPM technology in eggplant *Solanum melongena* L. in a farmers' participatory approach. Pesticide Research Journal 20 (2), 204–209.

Sardana, H.R., Bambawale, O.M., Jalali, S.K., 2009. Validation of non-chemical IPM and INM technologies on brinjal through farmers' participatory approach. Indian Journal of Plant Protection 37 (1/2), 20–23.

Sardana, H.R., Bambawale, O.M., Singh, D.K., Kadu, L.N., 2008b. Large area validation of adaptable integrated pest management technology in okra (*Hibiscus esculentus*) through farmers' participation. Indian Journal of Agricultural Sciences 78 (12), 1063–1066.

Satpathy, S., Shivalingaswamy, T.M., Kumar, A., Rai, A.B., Rai, M., 2005. Biointensive management of eggplant shoot and fruit borer (*Leucinodes orbonalis* Guen.). Vegetable Science 32 (1), 103–104.

Shankarganesh, K., Khan, M.A., 2006. Bio-efficacy of plant extracts on parasitisation of *Trichogramma chilonis* Ishii, *T. japonicum* Ashmead and *T. poliae* Nagaraja. Annals of Plant Protection Sciences 14 (2), 280–282.

Shanmugam, P.S., Balagurunathan, R., Sathiah, N., 2006a. Biointensive integrated pest management for *Bt* cotton. International Journal of Zoological Research 2 (2), 116–122.

Shanmugam, P.S., Balagurunathan, R., Sathiah, N., 2006b. Safety of some newer insecticides against *Trichogramma chilonis* Ishii. Journal of Plant Protection and Environment 3 (1), 58–63.

Shanmugam, P.S., Kumar, K.T., Satpute, U.S., 2005. Synomonic effects of plant extracts on parasitisation of *Corcyra* eggs by *Trichogramma chilonis* Ishii. Journal of Applied Zoological Researches 16 (1), 13–14.

Sharma, K.C., Bhardwaj, S.C., 2008. Management of *Helicoverpa armigera* (Hubner) in tomato. Annals of Plant Protection Sciences 16 (1), 33–39.

Sharma, R.K., Gupta, V.K., Singh, S., Dilawari, V.K., 2008. Analysis of the RAPD profiles and ITS-2 sequence of *Trichogramma* species for molecular identification and genetic diversity studies. Pest Management and Economic Zoology 16 (1), 1–8.

Shenhmar, M., Singh, J., Singh, S.P., Brar, K.S., Singh, D., 2003. Effectiveness of *Trichogramma chilonis* Ishii for the management of *Chilo auricilius* Dudgeon on sugarcane in different sugar mill areas of the Punjab. In: Tandon, P.L., et al. (Eds.), Biological Control of Lepidopteran Pests-Proceedings of the Symposium of Biological Control of Lepidopteran Pests. Society for Biocontrol Advancement, Bangalore, pp. 333–335.

Shivankar, S.B., Magar, S.B., Shinde, V.D., Yadav, R.G., Patil, A.S., 2008. Field bio-efficacy of chemical, botanical and bio-pesticides against *Spodoptera litura* Fab. in sugarbeet. Annals of Plant Protection Sciences 16 (2), 312–315.

Shukla, A., Kumar, A., 2004. Seasonal activity of egg parasitoid, *Trichogramma chilonis* Ishii on diamondback moth, *Plutella xylostella* (Linn.) infesting cabbage (Lepidoptera: Plutellidae). Uttar Pradesh Journal of Zoology 24 (3), 277–280.

Siddique, S.S., Babu, R., Arif, M., 2010. Efficacy of *Trichogramma brasiliense*, nuclear polyhedrosis virus and endosulfan for the management of *Helicoverpa armigera* on tomato. Journal of Experimental Zoology, India 13 (1), 177–180.

Silva, I.M.M.S., Honda, J., van Kan, F., Hu, S.J., Neto, L., Pintureau, B., Stouthamer, R., 1999. Molecular differentiation of five *Trichogramma* species occurring in Portugal. Biological Control 16 (2), 177–184.

Singh, A., Paul, A.V.N., Jain, A., 2009. Synomonal effect of nine varieties and one cultivar of rice on *Trichogramma japonicum* Ashmead and *Trichogramma chilonis* (Ishii) (Hymenoptera: Trichogrammatidae). Acta Entomologica Sinica 52 (6), 656–664.

Singh, B., Saini, R.K., 2008. Effect of different management practices on bollworms incidence and yield of seed cotton. Journal of Insect Science, Ludhiana 21 (1), 17–23.

Singh, H.S., Krishnakumar, N.K., Pandey, V., Naik, G., 2008a. Field evaluation of IPM tools against brinjal shoot and fruit borer, *Leucinodes orbonalis*. Indian Journal of Plant Protection 36 (2), 266–271.

Singh, H.S., Sridhar, V., Naik, G., 2005. Evaluation of some alternative measures against brinjal shoot and fruit borer, *Leucinodes orbonalis* Guen. under Bhubaneswar climatic conditions. Journal of Applied Zoological Researches 16 (2), 123–125.

Singh, J., Brar, K.S., Shenhmar, M., Kaur, S., Kaur, S., Joshi, N., Singh, I., 2003. Evaluations of *Trichogramma brasiliense* Ashmead, nuclear polyhedrosis virus and endosulfan for the management of *Helicoverpa armigera* (Hubner) on tomato. In: Tandon, P.L., et al. (Ed.), Biological Control of Lepidopteran Pests-Proceedings of the Symposium of Biological Control of Lepidopteran Pests. Society for Biocontrol Advancement, Bangalore, pp. 337–340.

Singh, M.R., 2006. Field evaluation of sequential releases of *Trichogramma chilonis* Ishii and *Cotesia flavipes* Cameron against borer pests of sugarcane. Indian Journal of Entomology 68 (4), 408–410.

Singh, R., Ram, P., 2006. Effect of high temperature shocks on performance of *Trichogramma chilonis* Ishii (Hymenoptera: Trichogrammatidae). Journal of Biological Control 20 (2), 127–133.

Singh, S., Shenhmar, M., 2008a. Host searching ability of genetically improved high temperature tolerant strain of *Trichogramma chilonis* Ishii in sugarcane. Annals of Plant Protection Sciences 16 (1), 107–110.

Singh, S., Shenhmar, M., 2008b. Impact of some insecticides recommended for sugarcane insect pests on emergence and parasitism of *Trichogramma japonicum* (Ashmead). Pesticide Research Journal 20 (1), 87–88.

Singh, S., Prasad, C.S., Nath, L., Tiwari, G.N., 2008b. Eco-friendly management of *Scirpophaga incertulas* (Walk.) and *Cnaphalocrocis medinalis* (Guen.) in basmati rice. Annals of Plant Protection Sciences 16 (1), 11–16.

Singh, S., Sharma, R.K., Kaur, P., Shenhmar, M., 2008c. Evaluation of genetically improved strain of *Trichogramma chilonis* for the management of sugarcane stalk borer (*Chilo auricilius*). Indian Journal of Agricultural Sciences 78 (10), 868–872.

Singh, S., Shenhmar, M., Brar, K.S., Jalali, S.K., 2007. Evaluation of different strains of *Trichogramma chilonis* Ishii for the suppression of sugarcane early shoot borer, *Chilo infuscatellus* Snellen. Journal of Biological Control 21 (2), 247–253.

Singh, S., Singh, S.K., Sudhakar, R., 2006. Comparative performance of different pest management treatments for castor + pigeonpea intercropping system under rainfed conditions. Research on Crops 7 (2), 548–551.

Singh, S., Singh, S.K., Sudhakar, R., 2008d. Validation of integrated pest management module for castor and pigeonpea intercropping system for rainfed agroecosystem. Pesticide Research Journal 20 (2), 217–220.

Singh, S.P., Jalali, S.K., 1998. Impact of pesticides on natural enemies of agricultural pests. In: Dhaliwal, G.S., Randhawa, N.S., Arora, R., Dhawan, A.K. (Eds.), Ecological Agriculture and Sustainable Development, vol. 2. Indian Ecological Society, Punjab Agricultural University/Centre for Research in Rural & Industrial Development, Ludhiana, Chandigarh, pp. 162–175.

Sivanarayana, G., Ramadevi, M., Ramaiah, P.V., 2008. Awareness and adoption of cotton (*Gossypium hirsutum* L.) integrated pest management practices by the farmers of Warangal district in Andhra Pradesh. Journal of Research ANGRAU 36 (4), 33–40.

Srivastava, M., Paul, A.V.N., Singh, A.K., Dureja, P., 2004. Synomonal effect of chickpea varieties on the egg parasitoid, *Trichogramma chilonis* Ishii (Trichogrammatidae: Hymenoptera). Indian Journal of Entomology 66 (4), 332–338.

Stouthamer, R., Hu, J., van Kan, F.J.P.M., Platner, G.R., Pinto, J.D., 1999. The utility of internal transcribed spacer 2 DNA sequences of the nuclear ribosomal gene for distinguishing sibling species of *Trichogramma*. BioControl 43 (4), 421–440.

Suresh, K., Rani, B.U., Rajendran, R., Baskaran, R.K.M., 2007. Management of brinjal shoot and fruit borer through nutritional manipulation. Journal of Entomological Research 31 (3), 191–196.

Sushil, S.N., Mohan, M., Hooda, K.S., Bhatt, J.C., Gupta, H.S., 2006. Efficacy of safer management tools against major insect pests of tomato and garden pea in northwest Himalayas. Journal of Biological Control 20 (2), 113–118.

Swamy, S.V.S.G., Prasad, N.V.V.S.D., 2010. Comparative evaluation of different pest management strategies on cotton. Annals of Plant Protection Sciences 18 (1), 131–135.

Tandon, P.L., Bakthavatsalam, N., 2003. Parasitization efficiency of *Trichogramma chilonis* Ishii on *Helicoverpa armigera* (Hubner) eggs influence of pigeon pea genotypes. In: Tandon, P.L., et al. (Ed.), Biological Control of Lepidopteran Pests-Proceedings of the Symposium of Biological Control of Lepidopteran Pests. Society for Biocontrol Advancement, Bangalore, pp. 75–78.

Tandon, P.L., Bakthavatsalam, N., 2005. Electro-physiological and olfactometric responses of *Helicoverpa armigera* (Hubner) (Lepidoptera: Noctuidae) and *Trichogramma chilonis* Ishii (Hymenoptera: Trichogrammatidae) to volatiles of trap crops – *Tagetes erecta* Linnaeus and *Solanum viarum* Dunal. Journal of Biological Control 19 (1), 9–15.

Tandon, P.L., Bakthavatsalam, N., 2007. Plant volatile diversity in different tomato genotypes and its influence on parasitization efficiency of *Trichogramma chilonis* Ishii on *Helicoverpa armigera* (Hubner). Journal of Biological Control 21 (2), 271–281.

Tanwar, R.K., Bambawale, O.M., Jeyakumar, P., Dhandapani, A., Kanwar, V., Sharma, O.P., Monga, D., 2007. Impact of IPM on natural enemies in irrigated cotton of North India. Entomon 32 (1), 25–32.

Thanavendan, G., Jeyarani, S., 2009. Biointensive management of okra fruit borers using braconid parasitoids (Braconidae: Hymenoptera). Tropical Agricultural Research 21 (1), 39–50.

Thirumalai, M., Selvanarayanan, V., 2009. Suppression of sugarcane internode borer (*Chilo sacchariphagus indicus* Kapur) by sequential release of parasitoids. Hexapoda 16 (2), 114–117.

Thirumurugan, A., Koodalingam, K., 2005. Enhancing efficacy of *Trichogramma chilonis* against internode borer, *Chilo sacchariphagus indicus* in sugarcane. Cooperative Sugar 37 (1), 67–71.

Thirumurugan, A., Joseph, M., Sudhagar, R., Ganesan, N.M., 2006. Improving efficacy of *Trichogramma chilonis* against shoot borer, *Chilo infuscatellus* (Snellen) in sugarcane ecosystem of tropical India. Sugar Tech 8 (2/3), 155–159.

Tiwari, S., Khan, M.A., 2004. Effect of endosulfan on percent parasitisation by three species of *Trichogramma*. Indian Journal of Entomology 66 (2), 135–137.

Tyagi, A., Gaurav, S.S., Prasad, C.S., Mehraj-ud-din, M., 2010. Efficacy and economics of biopesticides combinations and insecticide against tomato fruit borer. Annals of Plant Protection Sciences 18 (1), 148–152.

Vijayalakshmi, D., 2007. Evaluation of BIPM Module on tomato fruit borer (*Helicoverpa armigera*) larval population. Madras Agricultural Journal 94 (1/6), 130–133.

Virk, J.S., Brar, K.S., 2003. Effect of cultural practices on the efficiency of *Trichogramma chilonis* Ishii for the management of cotton bollworms. In: Tandon, P.L., et al. (Ed.), Biological Control of Lepidopteran Pests-Proceedings of the Symposium of Biological Control of Lepidopteran Pests. Society for Biocontrol Advancement, Bangalore, pp. 341–346.

Virk, J.S., Brar, K.S., 2005. Evaluation of *Trichogramma chilonis* Ishii in combination with insecticides for the management of cotton bollworms. Indian Journal of Entomology 67 (1), 75–78.

Virk, J.S., Brar, K.S., Sohi, A.S., 2004. Role of trap crops in increasing parasitisation efficiency of *Trichogramma chilonis* Ishii in cotton. Journal of Biological Control 18 (1), 61–64.

Yadav, D.N., Jha, A., 2003. Encouraging *Trichogramma* spp. (Hymenoptera: Trichogrammatidae) by providing alternate host and its impact on population of *Earias vittella* (Fabr.) in cotton. Pest Management and Economic Zoology 11 (2), 193–197.

Yadav, D.S., Sharma, M.M., 2005. Comparative efficacy of bioagents, neem products and malathion against brinjal shoot and fruit borer, *Leucinodes orbonalis* Guenee. Pesticide Research Journal 17 (2), 46–48.

Yadav, J.B., Singh, R.S., Tripathi, R.A., 2008a. Economics of IPM modules against okra pests. Annals of Plant Protection Sciences 16 (1), 204–206.

Yadav, J.B., Singh, R.S., Tripathi, R.A., 2008b. Evaluation of bio-pesticides against pest complex of okra. Annals of Plant Protection Sciences 16 (1), 58–61.

Yalawar, S., Pradeep, S., Kumar, M.A.A., Hosamani, V., Rampure, S., 2010. Performance of egg parasitoid, *Trichogramma chilonis* against sugarcane internode borer, *Chilo sacchariphagus indicus* (Kapur). Karnataka Journal of Agricultural Sciences 23 (1), 142–143.

6

Anthocorid Predators

Chandish R. Ballal[1], Kazutaka Yamada[2]

[1]ICAR-National Bureau of Agricultural Insect Resources, Bangalore, India; [2]Tokushima Prefectural Museum, Tokushima, Japan

1. INTRODUCTION

Anthocorid predators are recognized as potential biocontrol agents and are represented in all the zoogeographical regions of the world. They feed on small lepidopteran larvae, small grubs, psocids, mites, thrips, aphids, and storage pests and are commonly known as minute flower bugs or minute pirate bugs (Barber, 1936; Oku and Kobayashi, 1966; Muraleedharan and Ananthakrishnan, 1978). The majority of anthocorids are predaceous at nymphal and adult stages and few are zoophytophagous, while one species *Paratriphleps laeviusculus* Champion is known to be strictly phytophagous (Bacheler and Baranowski, 1975). There has been some debate on the "zoophytophagous" nature of *Orius insidiosus* (Say). Armer et al. (1998) reported that facultative phytophagy by *O. insidiosus* provides the insect with water and nutrients that help its survival when prey are not available, while Zeng and Cohen (2000) detected the presence of amylase in this anthocorid and inferred that they could be more committed to plant feeding than other species of predators that lack this enzyme. Natural populations of anthocorid predators have been successful in maintaining pest populations at low levels. In countries, like France, the United Kingdom, the Netherlands, Germany, etc., several species of anthocorid predators are available commercially and are released in greenhouses and fields for the management of insect pests, especially sucking pests such as thrips and mites. In India, research has generally focused on identifying indigenous anthocorids on different pests infesting different crops. Until the late 1990s, only around 4% of the total world literature on basic studies on the production and evaluation of anthocorid predators was from India, and very few attempts were made to rear and evaluate these potential predators. However, a considerable amount of information has been generated at the National Bureau of Agricultural Insect Resources (ICAR-NBAIR), Bangalore, on the diversity of indigenous anthocorid predators; protocols for mass rearing them have been devised and some potential ones have been field evaluated.

Ecofriendly Pest Management for Food Security
http://dx.doi.org/10.1016/B978-0-12-803265-7.00006-3

2. TAXONOMY OF INDIAN ANTHOCORIDAE

Anthocoridae (*sensu stricto*) is a large and diverse family within Cimicoidea. There is a great need for additional taxonomic and phylogenetic studies to be conducted on this family. Detailed faunal investigations have to be conducted in tropical and subtropical areas. Accurate identification of anthocorids is essential if we intend to use them for biocontrol. The phytophagous *P. laeviusculus* (perhaps misidentified to be *Paratriphleps pallida* (Reuter)) was introduced from Peru into Texas for control of bollworms. If it had established in Texas, the misidentification would have proved to be costly (Lattin, 1999). Initial faunal studies on the anthocorids of the Oriental Region were done by Distant (1904, 1906, 1910) and Poppius (1909, 1913) and covered the taxa known from India, Ceylon, and Burma (Myanmar), followed by publications by Ghauri (1964, 1972), Rajasekhara (1973), Muraleedharan (1975, 1977a,b,c); Muraleedharan and Ananthakrishnan (1974a,b,c, 1978), Muraleedharan (1977a,b). Muraleedharan and Ananthakrishnan (1974a,b) described some species of *Orius, Buchananiella*, and *Scoloposcelis* from India. Muraleedharan and Ananthakrishnan (1978) stated that 50 species of anthocorids have been recorded from the Oriental region and since then no comprehensive work has been done for about 40 years. Considerable progress has been made on the taxonomic studies of Indian Anthocoridae and some species were added to the list of Indian Anthocoridae (Yamada et al., 2008, 2010a,b,c, 2011; Ballal and Yamada, 2014). The following are some of the new species described from India:

Buchananiella pseudococci pseudococci (Wagner, 1951): *Buchananiella* Reuter, 1884 is a small genus of the family Anthocoridae, currently known by 10 described species. The members of this genus are primarily distributed in the tropics and subtropics, except for a single species, *Buchananiella continua* (White), which appears to have been introduced to many zoogeographical regions. *Buchananiella pseudococci pseudococci* was recorded from India as *Cardiastethus pseudococci pseudococci* by Yamada et al. (2008). Subsequently, Ghahari et al. (2009) transferred it to the genus *Buchananiella*. *Buchananiella pseudococci pseudococci* was associated with *O. arenosella* Walker (Lepidoptera: Xylorictidae) that infested coconut leaves in the northern parts of Kerala State, India. In India, four more species were recorded under the genus *Buchananiella*: *Buchananiella crassicornis* Carayon, *Buchananiella carayoni* Muraleedharan and Ananthakrishnan, *Buchananiella sodalis* Buchanan-White, and *Buchananiella indica* Muraleedharan.

Anthocoris muraleedharani Yamada, 2010: *Anthocoris* Fallén, 1814 is the second largest genus in the family Anthocoridae, comprising more than 70 species worldwide. The majority of species occur in the Holarctic Region, but the genus is most speciose in Asia, as about 40 species have been reported from China. In India, Poppius (1909) described *Anthocoris annulipes* Poppius and *Apelaunothrips indicus* Poppius, both from Sikkim, northeastern territory of the country. Subsequently, Muraleedharan (1977b) described *Anthocoris nilgiriensis* Muraleedharan from Tamil Nadu, India. *Anthocoris muraleedharani* was described from Bangalore, Karnataka by Yamada et al. (2010a).

Rajburicoris keralanus Yamada, 2010: The genus *Rajburicoris* Carpintero and Dellapé, 2008 was first recorded from India based on *R. keralanus* collected from Kerala. *Rajburicoris keralanus* was found within the leaf-curl galls induced by the thrips, *Liothrips karnyi* (Bagnall, 1914) (Thysanoptera: Phlaeothripidae) on the leaves of black pepper,

Piper nigrum Linnaeus (Piperaceae). *Rajburicoris keralanus* was observed preying on the nymphs and adults of *L. karnyi* in the laboratory. This anthocorid was found in very low numbers within the leaf-curl galls compared with *Montandoniola* sp. The eggs of *R. keralanus* were found in the inner portion of the leaf-curl gall rather than being inserted into the leaf tissue as seen for the species of *Montandoniola*. The authors also recorded *Rajburicoris stysi* Carpentero and Dellape from India.

Montandoniola indica Yamada, 2011: *Montandoniola* Poppius, 1909 is a well-known genus containing efficient predators of economically important thrips. The genus is now represented by 10 species in the world. In India, *Montandoniola moraguesi* (Puton) has been known to occur in agroecosystems and the bionomics of this species has been studied (Muraleedharan and Ananthakrishnan, 1971). However, Indian *M. moraguesi* probably refers to *M. indica*. *Montandoniola indica* is now recorded as an efficient predator of gall-forming thrips, *L. karnyi* infesting black pepper leaves, and *Gynaikothrips uzeli* Zimmerman infesting *Ficus* (Yamada et al., 2011; Ballal et al., 2012d).

Sexual dimorphism is well marked in most of the anthocorids, such as *Orius, Cardiastethus, Blaptostethus, Anthocoris,* etc. For instance, in *Cardiastethus exiguus* Poppius the posterior margin of the last four abdominal sternites is skewed toward lateral and the anterior margin of the last sternite is laterally excised from the penultimate sternite in the male, and the posterior margin of the seventh sternite has a v-shaped median projection bisecting the eighth sternite in the female. The antennae are also sexually dimorphic; segment II is distinctly stout and sausage-shaped in the male, but thinner and gradually enlarged toward the apex only in the distal half in the female. Segments III and IV are also thick and spindle-shaped in the male, but relatively more slender in the female (Ballal et al., 2003a). In addition to traditional taxonomy, molecular techniques have also been used for identifying anthocorids. Using molecular tools, Gomez-Polo et al. (2013) were able to discriminate between seven *Orius* species such as *Orius majusculus* Reuter, *Orius laevigatus* (Fieber), *Orius minutus* (Linnaeus), *Orius laticollis* (Reuter), *Orius horvathi* (Reuter), *Orius albidipennis* (Reuter), and *Orius niger* (Wolff) commonly present in Mediterranean vegetable crops. Male genitalia characters are generally utilized to identify anthocorids. Southwood (1956) and Sands (1957) have attempted to use egg structure (especially egg operculum) and characters of immature stages for taxonomic studies of anthocorids. Based on structural differences in egg opercula, Schuldiner-Harpaz and Coll (2012) identified four species of *Orius*, such as *O. albidipennis, O. niger, O. laevigatus,* and *O. horvathi*.

3. ANTHOCORIDS AS POTENTIAL BIOAGENTS

Orius species are reported as the most promising among anthocorids. Around 70 species of *Orius* are known all over the world. *Orius minutus* controlled aphids infesting apple orchards in China (Qin, 1985) and *Orius sauteri* (Poppius) was recorded as an important predator of thrips, spider mites, aphids and eggs of lepidopteran insects in fields and orchards in Japan, China, Korea, and Russian Far East (Yano and van Lenteren, 1996). In Switzerland, *O. majusculus* suppressed the populations of thrips on sweet pepper and cucumber (Fischer et al., 1992). Gomez-Polo et al. (2012) reported that *O. majusculus* effectively prey upon the two main pests in lettuce crops, western flower thrips *Frankliniella occidentalis* (Pergande) and the aphid *Nasonovia ribisnigri* (Mosely). Through diagnostic molecular gut-content analysis studies, they suggested that when the populations of the above two pests were not abundant, springtails of the genus *Entomobrya* sp. were alternate prey for *O. majusculus*. *O. insidiosus* was observed to prey heavily on *Helicoverpa virescens* (Fabricius), *Ostrinia nubilalis* (Hübner), and *Helicoverpa zea* (Boddie) in the USA (Lingren, 1977; Lingren et al., 1978; McDaniel et al., 1981; Reid, 1991). Gao (1987) observed *Dufouriella ater* (Dufour) to be a potential predator of Japanese pine bast scale *Matsucoccus matsumarae* (Nuwana) in China. *Blaptostethus pallescens* Poppius (originally described as *Blaptostethus piceus* Fieber var. *pallescens* Poppius from Cele-bes) was assessed as a potential predator of pests in the maize ecosystem in Egypt (Tawfik and El-Sherif, 1969; Tawfik and El-Husseini, 1971; Tawfik et al., 1974). Al-Jboory et al. (2012) recorded *O. albidipennis* and *Orius* sp. as predators of *Tuta absoluta* (Meyrick) in Jordan. *Orius pumilio* (Champion) and *O. insidiosus* are major predators of *Frankliniella bispinosa* (Morgan) infesting Queen Anne's lace (*Daucus carota* L.) and False Queen Anne's lace (*Ammi majus* L.) in Florida (Shapiro et al., 2009). The potential role played by *O. insidiosus* in suppressing natural populations of *F. occidentalis* infesting field peppers was reported by Funderburk et al. (2000).

In central Italy, the continued and abundant presence of *Lyctocoris campestris* Fabricius in stores of *Triticum spelta* L., its adaptability to mass rearing conditions, and its broad range of prey species indicate its potential as a biocontrol agent of stored product pests (Trematerra and Dioli, 1993). *Blaptostethus pallescens* was recorded from Madagascar (Muraleedharan, 1977b) and from grain warehouses in Egypt, where mites were common (Tawfik and El-Husseini, 1971). *Orius vicinus* was recorded as a potential predator of European red mite and two-spotted spider mite infesting apple orchards in New Zealand (Wearing et al., 2014).

Jerinić-Prodanović and Protić (2013) recorded *Anthocoris amplicollis* (Horváth), *Anthocoris confusus* Reuter, *Anthocoris nemoralis* (Fabricius), *Anthocoris nemorum* (Linnaeus), *O. majusculus*, *O. minutus*, and *O. niger* as predators of psyllids in Serbia. *Orius tristicolor* (White) was recorded as one of the key natural enemies of *Bactericera cockerelli* (Sulc) (Hemiptera: Triozidae), a major pest of potato, tomato, and peppers in California (Butler and Trumble, 2012). *Orius majusculus* was recorded as a potential predator of *Thrips tabaci* Lindeman, a major pest of leek, *Allium porrum* L. (Alliaceae), in Piedmont, northwest Italy, which is thus worth conserving (Bosco and Tavella, 2010).

Several anthocorids have been recorded on different crop ecosystems in India (Table 1). However, systematic work is lacking on the seasonal occurrence of different anthocorid pred-ators in India. Information is also lacking on the extent of control of pests exerted by the natu-ral populations of anthocorid predators in the different agroecosystems. This information is necessary to identify the potential anthocorids that need to be conserved.

Indian work on anthocorid predators has been reviewed by Muraleedharan and Ananthakrishnan (1978), Nasser and Abdurahiman (2004), Ballal and Gupta (2011) and Ballal and Yamada (2014). Anthocorids have been recorded as potential bioagents of vari-ous pests like *T. tabaci* (Lindeman), *Thrips flavus* Schrank, *Thrips hawaiiensis* (Morgan), *Thrips palmi* Karny, *Frankliniella schultzei* (Tryb.), *Anaphothrips sudanensis* Tryb., *Caliothrips gramini-cola* (Bagn. and Cam.), *Scirtothrips dorsalis* Hood, *Haplothrips ganglbaueri* Schmutz., *Baliothrips biformis* (Bagn.), *Caliothrips indicus* (Bagn.), *Microcephalothrips abdominalis* (D. L. Crawford), *Megalurothrips distalis* (Karny), *Gynaikothrips flaviantennatus* Moult, *Schedothrips orientalis* (Ananthakrishnan), *Gynaikothrips bengalensis* Ananthakrishnan, *Tribolium castaneum* (Hbst.), *O. arenosella*, *Pineus* sp., *Lipaphis erysimi* Kaltenbach (Singh et al., 1991; Saxena, 1975; Veer, 1984;

TABLE 1 Some Potential Anthocorid Predators Recorded in India

Anthocorid	Place	Host plant	Host insect	Source
ON THRIPS				
Orius sp.	New Delhi	Groundnut	Thrips	Singh et al. (1991)
Orius albidipennis (Reut.) (could be _Orius maxidentex_)	India	Onion	_Thrips tabaci_ Lind.	Saxena (1975)
Orius bifilarus Ghauri	Uttar Pradesh	Polyphagous	_Thrips flavus_ Schrank and _Thrips hawaiiensis_ (Morgan)	Veer (1984)
	Himachal Pradesh	Jasmine	Thrips	Nisha Devi and Gupta (2010)
Orius niger (Wolff)	Himachal Pradesh	Jasmine	Thrips	Nisha Devi and Gupta (2010)
Orius indicus	India		_Taeniothrps nigricornis_	Rajasekhara and Chatterji (1970)
O. maxidentex Ghauri	Tamil Nadu	Sesame	_Thrips palmi_ Karny _Frankliniella schultzei_ (Tryb.)	Kumar and Ananthakrishnan (1984)
		Croton sparsiflorus Morong	_T. palmi_	Ananthakrishnan and Thangavelu (1976)
		Panicum maximum Jacq.	_Anaphothrips sudanensis_ Tryb. _Caliothrips graminicola_ (Bagn. and Cam.)	
		Prosopis spicigera L.	_Scirtothrips dorsalis_ Hood	
		Rice	_Haplothrips ganglbaueri_ Schmutz.	
O. maxidentex Ghauri and _Orius tantillus_ Motsch.	India	Onion and garlic	_T. tabaci_ Lind.	Muraleedharan and Ananthakrishnan (1978)
		Chili, tea, and other plants	_S. dorsalis_ Hood	
		Rice	_Baliothrips biformis_ (Bagn.)	
		Groundnut, sesbania, and bajra	_Caliothrips indicus_ (Bagn.)	
		Chrysanthemum, dahlia, and marigold	_Microcephalothrips abdominalis_ (D. L. Crawford)	
		Various plants	_H. ganglbaueri_	
Orius minutus (L.)	Tamil Nadu	_Gliricidia sepium_ (_maculata_) (Jacq.)	_Megalurothrips distalis_ (Karny) _F. schultzei_ (Tryb.) _H. ganglbaueri_	Viswanathan and Ananthakrishnan (1974)

(Continued)

TABLE 1　Some Potential Anthocorid Predators Recorded in India—cont'd

Anthocorid	Place	Host plant	Host insect	Source
Carayonocoris indicus Muraleedharan	India	*Cassia marginata* Roxb. and mango	*Haplothrips ganglbaueri* F. *schultzei* (Tryb.)	Muraleedharan and Ananthakrishnan (1978) and Kumar and Ananthakrishnan (1984)
Montadoniola moraguesi (Put.) (could be *Montandoniola indica*)	India	Different plants	Gall makers, *G. flaviantennatus* Moult *Schedothrips orientalis* (Ananthakrishnan)	Ananthakrishnan and Varadarasan (1977)
			Gynaikothrips bengalensis Ananthakrishnan	Muraleedharan and Ananthakrishnan (1978)
		Black pepper	*Liothrips karnyi*	Devasahayam and Koya (1994) Devasahayam (2000)
M. indica (Yamada)	Karnataka	Ficus	*Gynaikothrips uzeli*	Ballal et al. (2012a)
	Kerala	Pepper	*L. karnyi*	Yamada et al. (2011)
Rajburicoris keralanus Yamada	Kerala	Black pepper	*L. karnyi*	Yamada et al. (2010b)
Scoloposcelis parallelus (Motsch.)	Tamil Nadu	Decaying bark of *Erythrina indica* Lam.	Fungus thrips, *Ecacanthothrips sanguinens* Bagn.	Vishwanathan and Ananthakrishnan (1973) and Muraleedharan and Ananthakrishnan (1978)
Xylocoris clarus (Dist.)	India	Leaf litter	Litter insects and mites including *Dexiothrips madrasensis* (Ananthakrishnan) and *Apelaunothrips indicus* (Ananthakrishnan)	Muraleedharan and Ananthakrishnan (1978)
ON MIDGES				
O. maxidentex Ghauri	Karnataka	Sorghum	*Contarinia sorghicola* Coquillett	Thontadarya and Rao (1987)
O. albidipennis (Reut.) (could be *O. maxidentex*)	Delhi	Sorghum	*C. sorghicola*	Rao (1976)
Orius sp.	Gujarat	Sorghum	*C. sorghicola*	Patel et al. (1975)

TABLE 1 Some Potential Anthocorid Predators Recorded in India—cont'd

Anthocorid	Place	Host plant	Host insect	Source
ON STORED GRAIN PESTS				
Orius sp.	Punjab	Wheat in grain storage structures	*Sitophilus oryzae* (L.) *Rhyzopertha dominica* (F.) *Tribolium castaneum* (Hbst.) *Trogoderma garanarium* Everts *Sitotroga cerealella* (Ol.) *Ephestia cautella* (Wlk.) (*Cadra cautella*) *Cryptolestes pusillus* (Schonh.) (*Laemophloeus minutus* (Ol.)) *Liposcelis* sp.	Dhaliwal (1976, 1977) and Battu et al. (1975)
Xylocoris flavipes (Reut.)	India	Stored cereals	*T. castaneum* (Hbst.)	Mukherjee et al. (1971)
ON LEPIDOPTERANS				
Cardiastethus sp.	Kerala	Coconut	*Opisina arenosella* Walker	Abdurahiman et al. (1982)
Cardiastethus exiguus Poppius, *Cardiastethus affinis* Poppius, *Buchananiella sodalis* Buchanan-White				Nasser and Abdurahiman (1990)
C. exiguus pauliana, *Cardiastethus pygmaeus pauliniae* Lansbury				Pillai and Nair (1993)
Buchananiella pseudococci pseudococci Wagner (as *C. pseudococci pseudococci*)				Yamada et al. (2008)
C. affinis Poppius	Kerala	Mango	*Orthaga exvinacea* (Hampson)	Yamada et al. (2008)
ON APHIDS				
Tetraphleps raoi Ghauri	Shillong	Benguet pine	*Pineus* sp.	Chacko (1973)

(Continued)

TABLE 1   Some Potential Anthocorid Predators Recorded in India—cont'd

Anthocorid	Place	Host plant	Host insect	Source
Bilia castanea (Carvalho) *O. albidipennis* (Reut.) (could be *O. maxidentex*)	West Bengal	*Brassica nigra* (L.)	*Lipaphis erysimi* Kaltenbach	Ghosh et al. (1981)
Blaptostethus pallescens Poppius	Karnataka	Maize	*Rhopalosiphum maidis*	Ballal et al. (2009)
ON MEALYBUGS				
C. exiguus Poppius	Karnataka	Papaya	Papaya mealybug	Chandish R. Ballal (unpublished)
Anthocoris muraleedharani Yamada	Karnataka	*Bauhinia purpurea*	*Ferrisia virgata*	Yamada et al. (2010a) and Ballal et al. (2015)
ON SPIDER MITES				
B. pallescens Poppius	Karnataka	Castor	*Tetranychus urticae*	Ballal et al. (2009)
B. castanea Carvalho	Karnataka	Grape *Nyctanthes arbor-tristis*	*T. urticae*	Chandish R. Ballal (unpublished)

Kumar and Ananthakrishnan, 1984; Ananthakrishnan and Thangavelu, 1976; Muraleedharan and Ananthakrishnan, 1978; Viswanathan and Ananthakrishnan, 1973, 1974; Ananthakrishnan and Varadarasan, 1977; Mukherjee et al., 1971; Abdurahiman et al., 1982; Nasser and Abdurahiman, 1990; Pillai and Nair, 1993; Chacko, 1973; Ghosh et al., 1981).

Orius spp. are the most common anthocorids that have been recorded from different crop ecosystems in India; *Orius maxidentex* Ghauri and *Orius tantillus* Motschulsky being the most common ones. *Orius maxidentex* was observed to be a potential predator of *Helicoverpa armigera* (Hübner) in the sunflower ecosystem (Ballal and Singh, 2001). Duffield (1995) has also reported a seasonal abundance and within plant distribution pattern of *Orius* spp. monitoring that of *H. armigera* eggs. *Orius tantillus* has been observed to be an active predator of *H. armigera* eggs and first instar larvae on the reproductive parts of pigeon pea and sorghum plants (Sigsgaard and Esbjerg, 1994). The records on *O. albidipennis* in India are doubtful and need to be confirmed. This species is known to be restricted to southern Europe, Africa, and the Middle East. It is likely that the species that have previously been reported from India as *O. albidipennis* could in fact be *O. maxidentex*. Surveys by the authors led to further documentation of *Orius* spp., such as *O. niger*, *Orius dravidiensis* Muraleedharan, *Orius shyamavarna* Muraleedharan and Ananthakrishnan, *O. niger aegypitiacus* Wagner, *O. minutus*, and *Orius amnesius* Ghauri from different host plants; the last species being a new distribution record for India.

In the coconut ecosystem, anthocorids *C. exiguus*, *Cardiastethus affinis* Poppius, and *B. sodalis* Buchanan-White were recorded to be potential predators of *O. arenosella*, a serious pest of coconut plantations in south India (Nasser and Abdurahiman, 1990; Abdurahiman et al., 1982). Though *C. exiguus* is recorded as a potential predator of *O. arenosella*, the authors could record its association with a wide variety of pests, such as thrips, mites, and mealybugs on cashew, papaya, rose, mango, jamun, *Tecoma stans*, *Butea monosperma*, *Thespesia*, *Cassia javanica*, *Caesalpinia pulcherrima*, *Aegle marmelos*, *Areca*, *Ligustrum*, and the dry fruits of *Adenanthera pavonina* and *Delonix regia*. *Cardiastethus affinis* Poppius (1909) known as a predator of *O. arenosella* was observed by the authors to be associated with *Orthesia* (on Lantana) and *Hemiberlesia lataniae* (on Agave) and could also be recorded from fallen leaves and flowers of *Spathodea campanulata*. Yamada et al. (2008) recorded *C. affinis* preying on *Orthaga exvinacea* infesting mango in Kerala state.

The information available on the genus *Blaptostethus* in India is scanty. The only records available are that of *Blaptostethus kumbi* Rajasekara collected from sugarcane fields in Mysore (Rajasekhara, 1973) and *B. pallescens* from Tamil Nadu and Bombay (Muraleedharan, 1977b). In Karnataka, Jalali and Singh (2002) and Ballal et al. (2003b) recorded *B. pallescens* as a potential predator of maize stem borer, *Chilo partellus* (Swinhoe) and two spotted spider mite *Tetranychus urticae* Koch, respectively. *Anthocoris muraleedharani* was recorded preying on the mealybug, *Ferrisia virgata* (Cockerell) infesting purple orchid tree *Bauhinia purpurea* L. (Yamada et al., 2010a), while *M. indica* was recorded as a predator of *G. uzeli* infesting *Ficus retusa* Linnaeus in Karnataka (Ballal et al., 2012d).

Species of *Xylocoris* are known to occur beneath tree bark, in leaf litter, under plants, and also amid stored grains. The genus *Xylocoris* was represented by two species in India: *Xylocoris (Arrostelus) flavipes* (Reuter, 1875) and *Xylocoris (Proxylocoris) clarus* (Distant). Muraleedharan and Ananthakrishnan (1978) have studied the biology of *X. clarus*. *Xylocoris flavipes*, also known as the Warehouse Pirate bug, is recorded as a potential bioagent of moth pests in storage and utilized in different countries in warehouses. We could record *Xylocoris (Proxylocoris) afer* (Reuter, 1884) (from the dry fruits of *Ficus* and Lagerstromia), *Xylocoris (Arrostelus) ampoli* Yamada 2013 and *Xylocoris (Proxylocoris) confusus* Carayon 1972 (from maize plants), which were first recorded from India. Yamada et al. (2013) recorded *Xylocoris (Proxylocoris) cerealis* (Yamada and Yasunaga) (which was generally found in rice mill factories in Thailand) for the first time in natural conditions. From this they inferred that species like *X. flavipes* may also occur in natural conditions. *Xylocoris flavipes* (Reuter) is recorded as a common predator of stored grain pests in India (Ballal et al., 2013). We could also record *X. flavipes* from natural field situations in India, thus supporting the observations made by Yamada et al. (2013). Other *Xylocoris* spp. like *X. afer, X. ampoli*, and *X. confusus* could also have functional roles in different crop ecosystems/storage systems for pest management.

Anthocorids have been recorded in the hilly elevated areas of our country. *Tetraphleps raoi* Ghauri was found on Benguet pine trees that were infested by *Pineus* sp., growing at about 5000 ft in Shillong (Chacko, 1973). Bhagat (2015) recorded *Anthocoris flavipes* Reuter and *Temnostethus (Ectemnus)? paradoxus* Hutchinson from the Ladakh region and *Anthocoris* sp., *O. minutus*, and *Orius lindbergi* Wagner from the Kashmir region. *Anthocoris minki* Dohrn and *A. confusus* Reuter were recorded in Himachal Pradesh (Gupta P.R., Personal communication).

4. BASIC STUDIES

The biological information on anthocorids from India is restricted to a few species, such as *Orius indicus* (Reuter) (Rajasekhara and Chatterji, 1970), *M. moraguesi* (Muraleedharan and Ananthakrishnan, 1971), *C. exiguus* (Ballal et al., 2003a), *O. tantillus* (Gupta and Ballal, 2006), *B. pallescens* (Ballal et al., 2003b). The only work on population dynamics of anthocorids was by Viswanathan and Ananthakrishnan (1974). Muraleedharan and Ananthakrishnan (1978) studied the bioecology of four species of anthocorids, such as *Carayanocoris indicus* Muraleedharan, *M. moraguesi* (Puton), *X. clarus* (Distant), and *Scoloposcelis parallelus* (Mots.).

A number of basic studies have been conducted on *C. exiguus* (Abdurahiman et al., 1982; Nasser and Abdurahiman, 1993; Ballal et al., 2003b). The reproductive biology of *C. exiguus* was studied by Nasser and Abdurahiman (1990) and the field efficacy of the natural populations of anthocorids was also studied (Nasser and Abdurahiman, 1998). Abdurahiman et al. (1982) studied the biology of *Cardiastethus* sp. (probably *C. exiguus*). The biological parameters of this species when reared on *Corcyra cephalonica* Stainton eggs was studied by Ballal et al. (2003a, 2012a), based on which a rearing protocol was developed. Sexual dimorphism is clear in this species; abdominal and antennal characters of adults could be used for separating the sexes (Ballal et al., 2003a). *Cardiastethus exiguus* exhibits cannibalism when food is scarce indicating that it is very important to provide adequate food while mass rearing this anthocorid. Figure 1 depicts the different stages of *C. exiguus*.

Investigations were conducted on the biology and feeding potential of *B. pallescens* Poppius (Ballal et al., 2003b). This anthocorid was originally collected by the author from maize fields and a laboratory culture initiated at the NBAIR. Figure 2 depicts the developmental stages of *B. pallescens*. The nymphal period was 15–16 days and adult longevity was 42.4 and 58.2 days in the case of male and female, respectively. The mean feeding potential was recorded as 100 eggs during the nymphal stage and 630 eggs during the adult stage and mean fecundity was 150 eggs. The high longevity and fecundity of *B. pallescens*, when reared on ultraviolet (UV)-irradiated *C. cephalonica* eggs, indicates that this anthocorid is amenable to continuous and large-scale rearing. Gupta et al. (2011) studied the preferential feeding of *B. pallescens* on the different stages of cotton mealybug *Phenacoccus solenopsis* Tinsley.

Ballal et al. (2013) studied the biology and feeding potential of *X. flavipes*. Figure 3 depicts the developmental stages of *X. flavipes*. The nymphal stage lasted for 19 days. The feeding potential of the nymphal stage was 3.4 eggs per day, while that of the adult was 10.3 per day. The male longevity was 10.8 days and female longevity 28 days. The fecundity was recorded as 61.96. Approximate rate of increase (r_c), precise intrinsic rate of increase (r_m), and finite rate of increase (λ) were calculated as 0.058, 0.045, and 1.14, respectively.

Laboratory investigations indicated that *A. muraleedharani*, originally collected from colonies of striped mealybug *F. virgate*, could feed on the cotton mealybug *P. solenopsis* while it showed no preference for papaya mealybug, *Paracoccus marginatus* Williams and Granara de Willink (Ballal et al., 2012b). The biology and feeding potential of *M. indica*, *C. affinis*, *B. indica*, and *Amphiareus constrictus* were also studied by the authors and led to the development of production protocols.

5. REARING

Successful rearing of anthocorids depends on various factors, like the type of host (Calixto et al., 2013), oviposition substrate (Zhou et al., 1991), and founder population size (Castane et al., 2014).

Lyctocoris campestris, a predator of stored product pests, could be reared on *Plodia interpunctella* (Hübner). The first two nymphal instars survived well on eggs or small host larvae, whereas the later instars fed on active larvae of any size (Parajulee and Phillips, 1992)

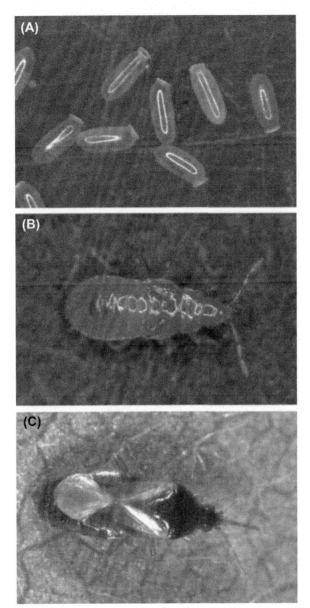

FIGURE 1   Life stages of *Cardiastethus exiguus*. (A) Eggs, (B) Nymph, (C) Adult.

and moisture was important for egg laying and hatching. In central Italy, the continued and abundant presence of *L. campestris* in stores of *T. spelta* L., its adaptability to mass rearing conditions, and its broad range of prey species indicates its potential as a biocontrol agent of stored product pests (Trematerra and Dioli, 1993). The majority of the rearing systems developed for *Orius* spp. utilize UV-irradiated or frozen eggs of storage pests (which are more

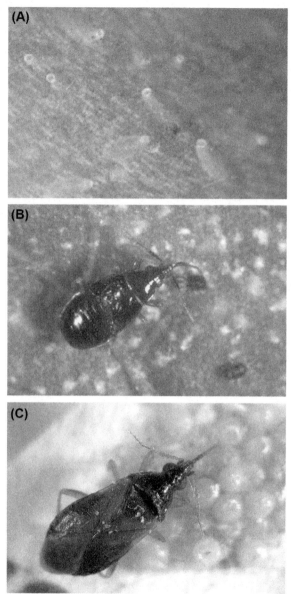

FIGURE 2   Life stages of *Blaptostethus pallescens*. (A) Eggs inserted into plant tissue with operculum visible, (B) Nymph predating on red spider mite, (C) Adult predating on eggs of tobacco caterpillar.

amenable to rearing) as alternate hosts (Table 2). However, there are some exceptions. The eggs of *H. virescens* were recorded as the ideal host for rearing *O. insidiosus* (Bush et al., 1993). Wang et al. (1996) observed that the survival rate of *O. sauteri* was 90% when reared on sorghum aphid, *Melanaphis sacchari* Zehntner.

In India, extensive work has to go into standardization of mass rearing and release techniques for all the identified potential anthocorid predators. Technologies are now available with the

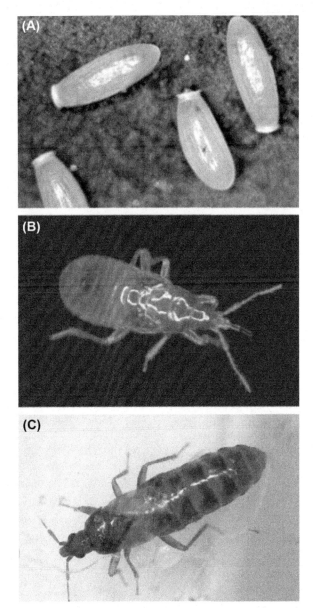

FIGURE 3   Life stages of *Xylocoris flavipes*. (A) Eggs, (B) Nymph, (C) Adult.

NBAIR for rearing some of the potential anthocorid predators, such as *O. tantillus*, *B. pallescens*, *C. exiguus*, *C. affinis*, *X. flavipes*, *M. indica*, and *A. muraleedharani* (Ballal et al., 2015).

Cardiastethus exiguus can be reared on the UV-irradiated eggs of *C. cephalonica* Stainton. Utilizing 1cc of *C. cephalonica* eggs, 450 *C. exiguus* adults could be produced (Ballal et al., 2003a). Rejuvenation of the culture with wild population may be required after rearing in the laboratory for more than a year. *Cardiastethus affinis* could also be reared successfully on

TABLE 2   Rearing of Anthocorid Predators

Anthocorid species	Host insects used for rearing	References
Anthocoris gallarum ulmi (DeG)	*Myzus persicae* Sulzer *Acyrthosiphun pisum* (Harris) *Aphis fabae* Scopoli	Ruth and Dwumfour (1989)
Xylocoris flavipes (Reuter)	Larvae of *T. confusum* Duv. or *Tribolium castaneum* (Hbst.)	Awadallah and Tawfik (1972)
	Eggs and adult remains of *Corcyra cephalonica* Stainton	Zhou et al. (1986)
	UV-irradiated eggs of *C. cephalonica*	Ballal et al. (2013)
Xylocoris sordidus (Reuter)	*Tribolium confusum*	Awadallah et al. (1986)
Xylocoris galactinus (Fieber)	*Musca domestica* Linnaeus	Afifi and Ibrahim (1991)
Cardiastethus nazarenus (Wied.)	Eggs of *Icerya purchasi* Maskell Gamma-irradiated or frozen eggs of *P. interpunctella* (Hübner)	Awadallah et al. (1976)
Anthocoris nemorum (L.)	Eggs of *Sitotroga cerealella* (Ol.)	Niemczyk (1970, 1978)
	Eggs of *E. kuehniella* Zeller	Fauvel et al. (1984)
Orius minutus (L.)	Eggs of *S. cerealella*	Niemczyk (1970, 1978)
Orius albidipennis (Reut.)	Paralyzed larvae of *E. kuehniella*, *Phthorimaea operculella* Zeller and *T. confusum*	Zaki (1989)
Orius laevigatus (Fieber)	Paralyzed larvae of *E. kuehniella*, *P. operculella* Zeller and *T. confusum* Eggs of *E. kuehniella* Eggs of *M. domestica*	Zaki (1989) Ito (2007), Schenikova and Stepanycheva (2005), and Bonte and de Clercq (2011)
Orius sauter (Poppius)	*Melanaphis sacchari* (Zehntner) Eggs of *E. kuehniella* Acarid mites Pollen and *A. gossypii* Glover	Wang et al. (1996), Yano and van Lenteren (1996), and Funao and Yoshiyasu (1995)
Orius majusculus Reuter	Eggs of *E. kuehniella*	Alauzet et al. (1992)
	Acarid mite *Tyrophagus putrescentiae* (Shrank)	Husseini et al. (1993)
Orius insidiosus (Say)	Eggs of *Dacus dorsalis* Hend.	Takara and Nishida (1978)
	Eggs of *H. virescens* (Fabricius)	Bush et al. (1993)
	Eggs of *E. kuehniella*	Bueno et al. (2006) and Calixto et al. (2013)
Orius strigicollis (Poppius)	Frozen eggs of *E. kuehniella*	Ito (2007)
Orius pumilio (Champion)	Eggs of *E. kuehniella*	Thomas et al. (2012)
Orius tantillus Motschulsky	Eggs of *E. kueniella*	Nagai et al. (1998)
	Eggs of *P. interpunctella*	Kawamoto et al. (1999)
	Eggs of *C. cephalonica* and *S. cerealella*	Gupta and Ballal (2009)

TABLE 2 Rearing of Anthocorid Predators—cont'd

Anthocorid species	Host insects used for rearing	References
Dufouriella ater (Dufour)	Eggs of *C. cephalonia* and *Galleria mellonella* Linnaeus	Gao (1987)
Lyctocoris campestris Fabricius	Eggs of *P. interpunctella*	Parajulee and Phillips (1992)
Cardiastethus exiguus Poppius	Eggs of *C. cephalonica*	Ballal et al. (2003a)
Blaptostethus pallescens Poppius	Eggs of *C. cephalonica*	Ballal et al. (2003b)
Anthocoris muraleedharani Yamada	Eggs of *C. cephalonica* and crawlers of mealybug *F. virgata* (Cockerell)	Yamada et al. (2010a)
Montandoniola indica Yamada	Eggs of *C. cephalonica*	Ballal et al. (2012d)

C. cephalonica eggs. *Cardiastethus exiguus* and *C. affinis* do not require a plant substrate for oviposition as in the case of *Orius* and *Anthocoris* spp.

The authors have developed simple production protocols for some other anthocorid predators that also do not require a plant substrate for oviposition, such as *X. flavipes*, *B. indica* Muraleedharan, and *A. constrictus* Stal. The last two species were collected from dry leaves of sugarcane and dry *Crossandra* flowers, respectively. However, it is worth investigating whether these two predators could be used as potential bioagents for the management of some of the sucking or lepidopteran pests. Manjunath et al. (1978) recorded *A. constrictus* as a predator of brown plant hopper *Nilaparvata lugens* (Stål) and green leafhopper *Nephotettix nigropictus* (Stål) infesting rice in Karnataka.

Tawfik and El-Husseini (1971) attempted to rear *B. pallescens* on lepidopteran larvae, aphids, and mites and reported that the nymphs were unable to molt when the food was restricted to aphids or plant sap. Ballal et al. (2003b) developed a protocol for rearing *B. pallescens* on UV-irradiated *C. cephalonica* eggs and French bean pods were used as oviposition substrates. Problems of cannibalism in the rearing containers and excessive handling could be avoided through the adoption of this protocol and 1000 *B. pallescens* could be produced utilizing 9 cc of *C. cephalonica* eggs. *Blaptostethus pallescens* is an ideal candidate that can be taken up for commercialization.

Rearing of *Orius* spp. on natural hosts, like thrips, is difficult and expensive as it involves rearing of thrips on host plants necessitating the continuous maintenance of host plants. In India, though *O. tantillus* and *O. maxidentex* are the most common anthocorids, we observed that the former is more amenable to laboratory multiplication and hence detailed studies on biology and feeding potential of *O. tantillus* and the effect of continuous rearing on alternate laboratory hosts were conducted by Gupta and Ballal (2006) and Ballal et al. (2012c). Although this anthocorid was amenable to rearing on the eggs of *C. cephalonica* and *Sitotroga cerealella*, the latter was the preferred host for rearing. Saini et al. (2003) reported on the usefulness of *S. cerealella* eggs as effective laboratory hosts for the multiplication of *O. insidiosus*. The eggs of *E. kueniella* and *P. interpunctella* were found to be suitable for rearing *O. tantillus* by Nagai et al. (1998) and Kawamoto et al. (1999), respectively. The cost of rearing *O. tantillus* on *S. cerealella* eggs was worked out as 4.67 Indian Rupees for 100 adults (Gupta and Ballal, 2009). However, there is a need to improve the rearing efficiency of *O. tantillus* as problems were encountered with respect to continuous rearing of quality culture and scaling up the production. Nisha Devi and Gupta (2010) successfully reared *Orius bifilarus* and *O. niger* on the eggs of *C. cephalonica*.

The other anthocorids that are amenable to rearing and that require plant substrates for oviposition are *M. indica* (Ballal et al., 2012d), *A. muraleedharani* (Yamada et al., 2010a; Ballal et al., 2015), *X. afer*, and *Blaptostethoides pacificus* Herring (Chandish R. Ballal, unpublished). *Anthocoris muraleedharani* was reared on the eggs of *C. cephalonica* and crawlers of *F. virgata* and bean pods and potato sprouts were used as oviposition substrates. *Montandoniola indica* can be efficiently reared on *C. cephalonica* eggs, with bean pods used as oviposition substrates. Earlier studies indicated that releases of predators from the temperate region was not always effective; hence, *M. indica* which is a tropical species would be an ideal candidate for evaluation trials against Cuban Laurel thrips, a serious pest on ornamental *Ficus* in several countries.

Pollen is used in many of the rearing systems for anthocorids. In China, *O. sauteri* was reared on a wide range of arthropods, pollen, or an artificial diet (Zhou and Wang, 1989). The survival rate was 72–83% when reared on the artificial diet and higher survival of 90.9% on pollen. However, Funao and Yoshiyasu (1995) reported that maize pollen alone was not sufficient for rearing *O. sauteri*, whereas a combination of pollen and *Aphis gossypii* Glover was a suitable diet. Calixto et al. (2013) also reported that pollen did not improve the performance parameters of *O. insidiosus* when offered exclusively or as a complementary food.

In anthocorid rearing, selection of an ideal oviposition substrate is one of the most important steps. The plant species is known to affect the oviposition behavior of *O. insidiosus* in that it prefers plants where newly hatched nymphs perform best. Several of the anthocorids can lay eggs on French bean pods. Seagraves and Lundgren (2010) reported that *O. insidiosus* laid significantly more eggs on prey-free common bean, *Phaseolus vulgaris* L. than prey-free soybean *Glycine max* (L.). Oviposition rate was higher on *P. vulgaris* with prey compared with that without prey. However, they observed that plant suitability overrides the importance of prey availability for this zoophytophagous insect. Soybean seedlings were observed to be ideal oviposition substrates for *Orius strigicollis* and *O. laevigatus* (Ito, 2007) and *O. sauteri* (Zhou et al., 1991). The egg-laying preferences of *O. niger* to different substrates was studied by Ahmadi (2010) and among crop plant species bean was preferred, among greenhouse plant species cucumber was preferred, and among orchard species apple was preferred. In Switzerland, *O. majusculus* could be successfully multiplied on a diet based on the eggs of *Ephestia kuehniella* and the leaves of geranium, *Pelargonium peltatum* (L.) L'Hér, were used as oviposition substrate (Alauzet et al., 1992). For the mass rearing of *O. insidiosus*, bean stems were used as oviposition substrates by Richards and Schmidt (1996), while Bueno et al. (2006) used inflorescence of Farmer's friend *Bidens pilosa* Linnaeus as oviposition substrate.

Researchers have attempted to rear anthocorids on artificial media or using artificial oviposition substrates, with limited success, either because of the high cost involved or due to the low quality individuals produced when reared on artificial diets. *Orius sauteri* was reared on an artificial diet (Zhou and Wang, 1989). Castañé and Zalom (1994) found that an artificial oviposition medium made of carrageenan salt of potassium chloride and covered with paraffin wax was suitable for the rearing of *O. insidiosus*, while Martin et al. (1978) reared some developmental stages of *O. insidiosus* on the diet used for rearing the chrysopid predator *Chrysoperla carnea* Stephens. *Anthocoris nemorum* L. and *A. nemoralis* were reared on an artificial diet, comprising one part water and two parts sweetened condensed milk (Bronnimann, 1964). Mukherjee et al. (1971) tested various diets for the rearing of *X. flavipes*. An aqueous mixture of dextrose, soluble starch, sodium caseinate, yeast powder, fish extract, cholesterol, and a salt mixture (derived from bonemeal, limestone, iodized salt, manganese sulfate, and zinc sulfate) was found to perform well for the rearing of *X. flavipes*. Lee and Lee (2007) reared *O. strigicollis* on an artificial diet.

If anthocorids have to be effectively utilized in biocontrol or integrated pest management (IPM) systems, packaging methodology for shipping and cost of production are important aspects to be considered. In Ontario, the cost of production of *O. insidiosus* could be reduced from 0.73 to 0.03 dollars per bug and up to 100,000 bugs could be reared per week. A method has been standardized for rearing 200–600 bugs in one ziplock plastic bag (Schmidt, 1994). The cost of rearing *O. sauteri* on *Trichogramma* pupae was 0.24 dollar per insect (JianYing et al., 2002) and for *O. insidiosus* when reared on *E. kuehniella* eggs was US$0.069 (Mendes et al., 2005). Shapiro and Ferkovich (2006) devised an oviposition substrate for *O. insidiosus* by making water-filled domes from Parafilm M. To improve the cost effectiveness of *O. laevigatus* rearing, a plantless rearing system was developed by Puysseleyr et al. (2014), wherein wax paper was used as a walking substrate, water encapsulated in Parafilm domes as water source, a substrate made of Parafilm, moist cotton wool as oviposition substrate, and *E. kuehniella* eggs as food source. Though body weight of *O. laevigatus* reared under this system was lower by 11% and the preoviposition period was prolonged by 29%, other biological parameters and predation rates were not negatively affected.

It is important to maintain the quality of mass-reared anthocorid predators. Carvalho et al. (2011) evaluated the performance of the wild and laboratory reared populations of two anthocorids, *O. laevigatus* and *O. insidiosus*, based on their searching behavior and orientation toward prey using an olfactometer. Though the searching behavior of the laboratory reared *O. laevigatus* seemed to be inferior to the wild population; such a difference was not observed in *O. insidiosus*. This indicates that some of the species could be more amenable to continuous rearing and maintain their quality attributes, like searching and predation efficiency, through generations. Such species are potential candidates for the biocontrol of pests.

6. DIAPAUSE

Overwintering during adult stages occurs in anthocorid bugs from the temperate region. Embryonic diapause also occurs in some cases (Saulich and Musolin, 2008). Some species of *Orius* are known to enter into reproductive diapause when the preoviposition period exceeds 14 days (Ruberson et al., 1991). Continuous development during all seasons occurs in *L. campestris* and some species of *Xylocoris*. Multivoltine species are known to enter into diapause during short day conditions, while summer diapause is rarely recorded in anthocorids. Termination of diapause has not been studied in anthocorids. The seasonal development of tropical and subtropical species is also not well studied. The species of anthocorids that develop without winter diapause are considered as potential bioagents.

7. COLD STORAGE

The storability of different species of anthocorid predators is an aspect to be investigated as this would enable planning for large scale production and timely availability for field/polyhouse releases. Kim (2009) reported that *O. laevigatus* adults could be stored at 10 °C for up to 36 days, with 70% survival. *Orius sauteri* eggs laid in soybean sprouts could be stored at 5 °C for up to 19 days with 75% hatching and the adults were reared at 15 °C for 4 days and then stored at 5 °C. The male and female adults lived for 98 and 29 days, respectively, with no reduction in fecundity compared with the adults maintained at room temperature (Guo et al., 2002). The eggs of *C. exiguus* could be safely stored for up to 5 days at 10 °C and 10 days at 15 °C and adult females for up to 15 days at 15 °C (Ballal et al., 2012a).

Long-term storage of parasitoids and predators through diapause induction is an extremely useful option in rearing, for which it is important to understand the mechanism to induce diapause. Limited information is available on this aspect with reference to anthocorid predators. Naturally overwintered adults of *O. sauteri* could be successfully stored for 2 months with a survival rate of 55% by Wang et al. (1994). Parker (1981) collected *A. nemorum* as adults from fields in early autumn and allowed them to overwinter artificially. When they recovered in early spring, they were reared to reach the desired stage and then cold stored till the time of requirement.

8. PRACTICAL UTILITY

Table 3 gives a detailed account of the different anthocorid species that have been successfully used against various pests in different countries. The anthocorid species that have been commonly and effectively used in as augmentative biocontrol are: *A. nemoralis*, *X. flavipes*, and *Orius* spp. (Bonte and de Clercq, 2008; Funderburk, 2009; Bonte and de Clercq, 2010).

The commercial availability of anthocorids in countries, like the United Kingdom, Canada, the Netherlands, the USA, Israel, Italy, Germany, France, Poland, etc., has led to their extensive utility against polyhouse and field crop pests. *Orius sauteri* was effective in suppressing *T. palmi* Karny in eggplants in Japan, China, and Korea (Yano and van Lenteren, 1996). In Poland, *O. insidiosus*, *O. laevigatus*, *O. majusculus*, and *O. strigicollis* are released against vegetable crop pests in greenhouses (Fiedler and Sosnowska, 2009). In the province of Almeira, Spain, known as the most densely concentrated greenhouse area in the world, biocontrol is implemented in about 50% of the important greenhouse crops and the key beneficial species in this system is *O. laevigatus* (van der Blomm et al., 2009). In pear orchards in the Province of Venice, Italy, *Psylla pyri* L. was well controlled by *A. nemoralis* (in the absence of broad spectrum insecticides) and was released at 200–300 anthocorids/ha two or three times per year. This anthocorid could also prey on aphids (Mori and Sancassani, 1984).

Anthocorids have also been used in classical biocontrol attempts. *Montandoniola moraguesi* was introduced from the Philippines into Hawaii and Bermuda for classical biological control of leaf gall thrips of *Ficus*, *Gynaikothrps ficorum* (Marchal). Pluot-Sigwalt et al. (2009) opined that *M. moraguesi* may be restricted to the Mediterranean region and Africa, while the species from Guadeloupe, West Indies, Hawaii, Bermuda, Australia, and Florida appeared to be *Montandoniola confusa* Streito and Matocq and inferred that the species originally introduced from Philippines might have been *M. confusa* and not *M. moraguesi*. *Tetraphleps raoi* was sent from India to Kenya for release against *Pineus pini* Gmelin on several *Pinus* spp. It established and resulted in the decline of the *P. pini* population (Karanja and Aloo, 1990).

Orius spp. have been widely and efficiently used for the suppression of the western flower thrips *F. occidentalis* infesting various polyhouse crops. In the United Kingdom, releases of one to two *O. laevigatus* per plant controlled *F. occidentalis* on pepper (Chambers et al., 1993),

TABLE 3 Successful Field Releases of Anthocorids

Anthocorid	Host plant/ material	Host insect	Place	Source
Anthocoris nemoralis (Fabricius)	Pear	*Cacopsylla* (*Psylla*) *pyri* Linnaeus	Venice, Italy Mount Etna	Mori and Sancassani (1984) Beninato and Morella (2000)
A. nemoralis and *A. nemorum* L.	Pear	*C. pyri*		Sigsgaard et al. (2006)
Lyctocoris sp.	Pine	*Orthotomicus erosus* (Wollaston)	South Africa	Kfir (1986)
Xylocoris flavipes (Reuter)	Groundnut	*Ephestia cautella* (Wlk.), *P. interpunctella* (Hübner)	Georgia, USA	Brower and Mullen (1990)
	Rolled oats in warehouse	*E. cautella* (Wlk.), *Oryzaephilus surinamensis* (L.) and *Tribolium castaneum* (Hbst.)		LeCato et al. (1977)
	Whole wheat flour	*Tribolium confusum* Jacquelin du Val	Japan	Murata et al. (2007)
	Groundnut in warehouses	*E. cautella* and *P. interpunctella* (Hubner)	Georgia, USA	Keever et al. (1986)
	Simulated trials in stored rice and wheat	*Corcyra cephalonica* Stainton and *Sitotroga cerealella* (Ol.)	India	Ballal et al. (2013), Bhumannavar et al. (2010), Bhumannavar and Rabindra (2011), Sreerama Kumar et al. (2012), and Kaur and Virk (2011)
Montandoniola moraguesi (Put.)	*Ficus retusa*	*Gynaikothrips ficorum* (Marchal)	Bermuda	Groves (1974)
Montandoniola confusa Streito and Matocq	*Ficus* spp.	*Gynaikothrips uzeli* Zimmerman	Hawaii	Arthurs et al. (2011)
Orius sp.	Cucumber	*Frankliniella occidentalis* Pergande	Germany	Buhl and Bassler (1992)
Orius insidiosus Say	Maize	*Ostrinia nubilalis* Hubner and *H. zea* Boddie	USA	Reid (1991)
	Chrysanthemum	*F. occidentalis*	The Netherlands	Fransen and Tolsma (1992)
	Ornamentals	*F. occidentalis*	Sweden	Sorensson and Nedstam (1993)
	Chrysanthemum	*F. occidentalis*	Brazil	Silveira et al. (2004)

(Continued)

TABLE 3   Successful Field Releases of Anthocorids—cont'd

Anthocorid	Host plant/ material	Host insect	Place	Source
Orius laevigatus (Fieber)	Strawberry	*F. occidentalis*	France	Villevieille and Millot (1991)
	Sweet pepper	*F. occidentalis*	UK Spain	Chambers et al. (1993) Van der Blomm et al. (2009)
	Sweet pepper	*F. occidentalis, Thrips tabaci* (Lindeman)	Italy	Tommassini and Maini (2002)
		F. occidentalis	Tunisia	Elimem and Chermiti (2012a,b)
		F. occidentalis	Russia	Schenikova and Stepanycheva (2005)
Orius majusculus (Reuter)	Cucumber	*F. occidentalis*	France	Trottin Caudal et al. (1991)
	Chrysanthemum	*F. occidentalis*	UK Spain	Buxton and Finlay (1993)
Orius tantillus Motschulsky	Different crops Onion	*F. occidentalis* *T. tabaci*	Australia	http://www.biologicalservices.com.au/products/orius-29.html
Orius strigicollis (Poppius)	Egg plant and adzuki bean	Thrips	Taiwan	Wang et al. (2001)
Orius minutus Linnaeus	Potato	*Tetranychus urticae* Koch.	Iran	Fathi (2014)
Blaptostehus pallescens Poppius	Okra, chili, rose, brinjal	*Tetranychus urticae*	India	Ballal et al. (2009), Bhumannavar et al. (2010), Sreerama Kumar et al. (2012), NBAII (2013), and NBAII (2014)
	Capsicum	*Scirtothrips dorsalis* Hood		Bhumannavar and Rabindra (2011)
	Stored rice	*C. cephalonica* Stainton *S. cerealella* (Olivier)		Ballal et al. (2009), Kaur and Virk (2011), Bhumannavar et al. (2010), Bhumannavar and Rabindra (2011), and NBAII (2014)
Cardiastethus exiguus Poppius	Coconut	*Opisina arenooosella* Walker	India	Lyla et al. (2006) Venkatesan et al. (2008)

while in Germany good results were achieved with two releases of *Orius* at 2-week intervals (0.5–1 insect/m^2) in cucumbers (Buhl and Bassler, 1992). In France, *O. majusculus* could efficiently suppress *F. occidentalis* on greenhouse cucumbers in greenhouses (Trottin Caudal et al., 1991). *Orius majusculus* was also successful in the control of *F. occidentalis* infesting chrysanthemums in a greenhouse in the United Kingdom, while in the Netherlands, *O. insidiosus* was released on chrysanthemums, roses, and *Saintpaulia* in greenhouses. Pest numbers were reduced on chrysanthemum and *Saintpaulia*, but not on roses (Fransen et al., 1993). In untreated chrysanthemum plots, 90% damage by *F. occidentalis* to Pink Pompon and 30–50%

damage to D-Mark was recorded. However, when *O. insidiosus* was released, damage to Pink Pompon (susceptible chrysanthemum variety) did not exceed 20% in the last 8 weeks, while damage to D-Mark (resistant variety) did not exceed 5% throughout the growing period (Fransen and Tolsma, 1992). *Orius laevigatus* was recorded as a potential predator for the control of *F. occidentalis* on strawberries in greenhouses in France (Villevieille and Millot, 1991).

The success of biocontrol trials using anthocorid predators can depend on several factors such as insecticide sprays, pest species present, level of infestation, etc. Research during the 1970s and 1980s demonstrated that releases of *X. flavipes* resulted in a 79–100% suppression of moth populations in small storages of peanuts, up to 99% reduction of sawtoothed grain beetle populations in 35 quarts of corn, and a 90–98% suppression of red flour beetles in a simulated peanut warehouse. In Georgia, USA, releases of the predaceous bug, *X. flavipes*, prevented the increase of populations of *Ephestia cautella* (Wlk.) (*Cadra cautella*), *Oryzaephilus surinamensis* (L.) and *T. castaneum* (Hbst.) in small quantities of rolled oats in metal drums in the laboratory or scattered over the floor of a room in a warehouse (LeCato et al., 1977). Also populations of *E. cautella* and *P. interpunctella* were 54–83% less in the warehouse containing groundnuts with only traces of malathion residues that received periodic releases of *X. flavipes* and the larval parasitoid, *Bracon hebetor* Say, compared with that receivng no natural enemies and in which groundnuts were conventionally treated with malathion (Keever et al., 1986). Alternative prey present on the target crop could also affect the performance of anthocorid predators. Fitzgerald and Jay (2013) observed that *O. laevigatus* had maximum preference for strawberry aphid (*Chaetosiphon fragaefolii* (Cockerell), followed by European tarnished plant bug (*Lygus rugulipennis* Poppius) and *F. occidentalis*, when all three pests were present on strawberry.

In India, small scale trials were conducted utilizing the potential anthocorid predators. *Cardiastethus exiguus* was evaluated against the coconut black-headed caterpillar infesting coconut trees in the states of Karnataka and Kerala. The trials indicated the effectiveness of *C. exiguus* against eggs and neonate larvae of *O. arenosella* (Lyla et al., 2006; Venkatesan et al., 2008).

Releases of *B. pallescens* against red spider mites infesting okra plants in nethouses led to a significant reduction in the population of the mites and percent damage (Ballal et al., 2009). Trials conducted in Orissa, Punjab, Tamil Nadu, and Himachal Pradesh indicated that this predator is a potential bioagent of *T. urticae* infesting okra, brinjal, rose, and chili (Bhumannavar et al., 2010; Sreerama Kumar et al., 2012; NBAII, 2013; NBAII, 2014). Though this predator could cause significant mortality of onion thrips, *T. tabaci* infesting onion and garlic, it could not establish well in the onion ecosystem. The rearing and evaluation trials using *X. flavipes* targeting storage pests has been done in India (Kaur and Virk, 2011; Sreerama Kumar et al., 2012; Ballal et al., 2013). *Xylocoris flavipes* could be effectively utilized for the suppression of rice moth infesting rice and wheat in the states of Kerala, Assam, Tamil Nadu, Andhra Pradesh, and Himachal Pradesh (Sreerama Kumar et al., 2012; NBAII, 2013; NBAII, 2014). Releases of *X. flavipes* against *S. cerealella* infesting paddy resulted in an 89% reduction in pest emergence in the treatment batch compared with control. However, research needs to be done on the dosages and methods of effective deployment of *X. flavipes* and its efficacy against different pests infesting various commodities and stored seeds. The use of natural enemies against stored product pests is legally permissible in developed countries like the USA. In 1992, the Environmental Protection Agency exempted biocontrol agents from having a requirement for tolerance levels in commodities, and essentially allowed their use in any stored product situation, where they are not expected to become a component of food (http://www.entomology.wisc.edu/mbcn/fea210.html).

We have also observed that *X. afer*, an anthocorid that was collected from dry *Ficus* fruits, is highly amenable to rearing. Vieira et al. (2014) have recorded the predatory efficiency of *X. afer* on the diamondback moth, *Plutella xylostella* (Linnaeus). Hence, it would be worth evaluating this anthocorid against some of the lepidopteran pests. It would also be useful to evaluate the mass-reared *M. indica* against *L. karnyi* infesting black pepper.

9. METHODS TO IMPROVE THE PERFORMANCE OF ANTHOCORID PREDATORS

Banker plants are used to breed predators or parasites within a crop, especially useful in commercial greenhouses. Bankers allow beneficials to establish before the target pests arrive and also where the crop is not favorable for beneficials to establish. In crops such as cucumbers, gerbera, and roses, the flowers do not supply suitable pollen and in such crops, *Orius* spp. are not able to survive in the absence of an adequate pollen source. In these crops, *Orius* needs to be established with the use of banker plants. *Orius* spp. can establish on these plants and can then move from the banker plants to the crop in search of thrips. It is important to choose potential banker plants for specific anthocorids. Black Pearl pepper and Purple Flash pepper have been suggested as potential banker plants for *O. insidiosus* (Wong and Frank, 2013; Waite et al., 2014), while Pumarino and Alomar (2014) reported that *Lobularia maritime* (Alyssum) can be selected as a banker plant for *O. majusculus*. Nagai and Hikawa (2012) reported Black-Eyed Susan, *Rudbeckia hirta* L. as an insectary plant as large populations of *O. sauteri* could migrate from this plant to tomatoes. Intercropping of bell pepper plants with flowering plants like dill, *Anethum graveolens* L.; coriander, *Coriandrum sativum* L.; and buckwheat, *Fagopyrum esculentum* Moench could improve the seasonal predation of European corn borer (*O. nubilalis* (Hübner)) by *O. insidiosus* (Bickerton and Hamilton, 2012). In Turkey, *O. niger* was found on 53 weed species in the winter–spring and summer–fall periods; thus, these weeds could be considered as potential banker plants (Atakan and Tunc, 2010). *Orius niger* was observed to be the most frequent predatory insect on alfalfa and stinging nettle and hence these two plants are potential reservoirs of this anthocorid predator (Ban et al., 2010).

Use of supplemental food such as dry cysts of *Artemia* sp. and dry bee pollen was observed to have a positive effect on longevity and reproduction of *O. majusculus* and *O. laevigatus* (Oveja et al., 2012).

For biocontrol in greenhouses during winter, it is important to use supplementary lighting to prevent short day induced reproductive diapauses. Bahşi and Tunç (2012) recorded that a strain of *O. majusculus* from Antalya had a shorter critical day-length compared with northern strains and thus diapauses could be averted. This also indicates that besides identifying important species of anthocorids to target specific pests, it would also be important to select potential strains of each species for specific pests/crops/situations.

The dosage and mode of release are important factors for the success of biocontrol attempts using anthocorid predators. Commercial insectaries use buckweed or vermiculite as a medium for releasing anthocorids. Garzia et al. (2012) used a mechanical device for releases of the predatory mite *Phytoseiulus persimilis* and *O. laevigatus* on sweet pepper and chrysanthemum to control the two-spotted spider mite and *F. occidentalis*. The two bioagents were well distributed and pest control was achieved sooner than in the case of manual releases.

Studies on chemical ecology of anthocorid predators indicate that some volatiles from plants or arthropods attract some species of anthocorids such as *Anthocoris* (Lattin, 1999). This information can be utilized effectively in IPM programmes utilizing anthocorids.

10. COMPATIBILITY OF ANTHOCORIDS WITH OTHER BIOAGENTS AND/OR OTHER METHODS OF PEST MANAGEMENT

1. Insecticides

Some synthetic insecticides were found to be relatively safe to *O. insidiosus* like cyantraniliprole, flonicamid, spirotetramat, and terpenes (Srivastava et al., 2014). Some selected fungicides, such as azoxystrobin, were reported to be compatible with *Aphidius colemani* Viereck and *O. laevigatus* for IPM of cotton aphid and western flower thrips infesting cucumber in polyhouse cultivation in Korea (Choi et al., 2013). Abamectin, λ-cyhalothrin, acrinahtrin, and Azadirachtin were not compatible with *Orius* spp. (Bosco et al., 2012). However, Bosco et al. (2012) and Pascual et al. (2014) reported that Spinosad had clear deleterious effect on *Orius* spp. on crops like olive, pepper, and strawberries. Some products like Etofenprox gave contradictory results in the field and laboratory experiments, corroborating the need of multiple testing methods in evaluating the effects of pesticides on beneficial insects.

2. Bt Crops

Transgenic crops (cotton, maize, brinjal) are of prime importance as integral to IPM systems. There was a speculation whether transgenic maize (or other transgenic crops) would have a detrimental effect on the indigenous anthocorids. No nontarget effects of Bt maize on these indigenous natural enemies, including anthocorid predators, were reported in southern Bohemia (Habustova et al., 2014). Lumbierres et al. (2012) reported that ingestion of Bt protein by *O. majusculus* via plant leaves or pollen or via the food web has no negative effects on its biological parameters, while a positive effect with respect to increased fecundity and reduced developmental time was observed.

3. Other Bioagents

Studies were conducted to investigate the compatibility of anthocorids with trichogrammatids (Gupta and Ballal, 2007). Given a choice of unparasitized and parasitized (by *Trichogramma chilonis* Ishii.) eggs of both *C. cephalonica* and *H. armigera*, *B. pallescens*, and *O. tantillus* preferred to feed on unparasitized eggs. This indicates that it may be possible to integrate releases of anthocorids with trichogrammatids for biocontrol of thrips/lepidopteran pests infesting different crop ecosystems.

The interaction between *O. majusculus* and entomopathogenic fungi *Metarhizium brunneum* (Petch) and *Neozygites floridana* (Weiser & Muma) (bioagents of *T. urticae* infesting strawberry) was studied by Jacobsen et al. (2015). It was observed that searching time by *O. majusculus* was lower on leaf discs with presence of *M. brunneum* spores compared with no fungal spores and higher on leaf discs with presence of *N. floridana* spores. Ladurner et al. (2012) reported that *O. laevigatus* is compatible with *Beauveria bassiana* (strain ATCC 74040) and Gao et al. (2012) also reported on the compatibility of *O. sauteri* with *Beauveria bassiana* (Bals-Criv) Vuill. (strain Bb-RSB). In a study by Pourian et al. (2011), *O .albidipennis* was observed to be able to detect and avoid areas treated with entomofungal bioagent *Metarhizium anisopliae* (Metchnikoff).

Combined releases of *O. laevigatus* and *Macrolophus pygmaeus* Rambur appeared to be an excellent option for controlling thrips and aphids in greenhouse-grown sweet pepper (Messelink and Janssen, 2014). The combined use of *O. majusculus*, predatory midges, predatory thrips, and parasitoids clearly enhanced the suppression of aphids and thrips infesting sweet pepper, thus supporting the view that intraguild predation, which is potentially negative for biocontrol, could be compensated by positive effects of generalist predators for control of multiple pests (Messelink et al., 2011, 2013). In Tunisia, combined releases of *O. laevigatus* and predatory mite *Amblyseius swirskii* Athias-Henriot were effective in the management of *F. occidentalis* infesting greenhouse pepper (Elimem and Chermiti, 2012a). In protected strawberry crops, the combined use of *Neoseiulus* spp. with *Orius* sp. successfully controlled *F. occidentalis* in France and Spain (Sampson et al., 2011). In Sweden, the predators *Amblyseius* (=*Neoseiulus*) *cucumeris* (Oudemans) and *O. insidiosus* were very effective when they were released together on four ornamental plants (*Saintpaulia, Impatiens, Gerbera,* and *Brachyscome multifida*) infested with *F. occidentalis* in greenhouses (Sorensson and Nedstam, 1993).

Husseini et al. (2010) evaluated the effect of plant quality on intraguild predation between *Aphidoletes aphidimyza* (Rondani) (Diptera: Cecidomyiidae) and *O. laevigatus* with *A. gossypii* as shared prey on greenhouse cucumber *Cucumis sativus* L. (with different N fertilization levels). Regardless of N fertilization levels, *O. laevigatus* alone was more effective in aphid suppression than *A. aphidimyza* alone. However, they concluded that there was weak asymmetric intraguild predation among the two predators that did not affect their predatory efficiency.

The warehouse pirate bug, *X. flavipes* and reduviid bug *Amphibolus venator* (Klug) were evaluated against *Tribolium confusum* infesting wheat flour by Murata et al. (2007). Release of *X. flavipes* alone recorded 96.9% suppression of *T. confusum, A. venator* alone 76.2% suppression, and both bugs together 95.6% suppression. *Amphibolus venator* attacked *X. flavipes* adults but not *X. flavipes* nymphs, thus indicating that combined releases could be planned for managing different storage pests.

11. CONCLUSIONS

It has now been realized that anthocorids are effective predators and play an important role in the natural control of many insect pests in the different parts of the world. Techniques have evolved in other countries to mass rear anthocorid predators and release them in the field or greenhouses for management of several pests, especially thrips and mites. Thrips and mites are a major problem in India too on ornamental and polyhouse crops; therefore, there is a need to evaluate the available anthocorid predators systematically against these pests in different crop ecosystems. The problem of stored grain pests is very serious in India and since chemical treatments are not providing a solution to the problem, anthocorid predators need to be evaluated against them.

The following are the research areas that need attention for the better use of anthocorids in biocontrol and integrated management of different insect pests.

- Search for potential anthocorid predators for release against major crop pests and storage pests.
- Taxonomy on anthocorid predators to be strengthened.
- Studies on the interaction between pests and anthocorids in different crop ecosystems during different seasons.
- Mass rearing technologies to be standardized for anthocorid predators and available technologies to be disseminated to commercial insectaries.
- Compatibility studies between anthocorid predators and other potential bioagents (such as insect pathogens and parasitoids).
- Extensive evaluation studies on available anthocorid predators in culture and the novel studies that are amenable to mass rearing against field and polyhouse crop and storage pests.

References

Abdurahiman, U.C., Mohammmed, U.V.K., Remadevi, O.K., 1982. Studies on the biology of a predator, *Cardiastethus* sp. (Hemiptera: Anthocoridae) found in the galleries of *Nephantis serinopa* Meyr. (Lepidoptera: Xylorictidae). Current Science 51, 574–576.

Afifi, A.I., Ibrahim, A.M.A., 1991. Effect of prey on various stages of the predator *Xylocoris galactinus* (Fieber) (Hemiptera: Anthocoridae). Bulletin of Faculty of Agriculture, University of Cairo 42, 139–150.

Ahmadi, K., 2010. Egg-laying preferences of the predatory flower bug *Orius niger* (Wolff) to different substrates of oviposition. Julius-Kühn-Archiv (428), 444.

Alauzet, C., Dargagnon, D., Hatte, M., 1992. Production of the heteropteran predator: *Orius majusculus* (Het.: Anthocoridae). Entomophaga 37, 249–252.

Al-Jboory, I.J., Katbeh-Bader, A., Shakir, A., 2012. First observation and identification of some natural enemies collected from heavily infested tomato by *Tuta absoluta* (Meyrick) (Lepidoptera: Gelechiidae) in Jordan. Middle-East Journal of Scientific Research 11, 787–790.

Ananthakrishnan, T.N., Thangavelu, K., 1976. The cereal thrips *Haplothrips ganglbaueri* Schmutz with particular reference to the trends of infestation on *Oryzasativa* and the weed *Echinochloa crusgalli*. Proceedings of the Indian Academy of Sciences - B 83, 196–201.

Ananthakrishnan, T.N., Varadarasan, S., 1977. *Androthrips flavipes* Schmutz (Insecta: Thysanoptera), a predatory inquiline in thrips galls. Entomon 2, 105–107.

Armer, C.A., Wiedenmann, R.N., Bush, D.R., 1998. Plant feeding site selection on soybean by the facultatively phytophagous predator *Orius insidiosus*. Entomologia Experimentalis et Applicata 86, 109–118.

Arthurs, S., Chen, J., Dogramaci, M., Ali, A., Mannion, C., 2011. Evaluation of *Montandoniola confusa* Streito and Matocq sp. nov. and *Orius insidiosus* Say (Heteroptera: Anthocoridae) for control of *Gynaikothrips uzeli* Zimmerman (Thysanoptera: Phlaeothripidae) on *Ficus benjamina*. Biological Control 57, 202–207.

Atakan, E., Tunç, İ., 2010. Seasonal abundance of hemipteran predators in relation to western flower thrips *Frankliniella occidentalis* (Thysanoptera: Thripidae) on weeds in the eastern Mediterranean region of Turkey. Biocontrol Science and Technology 20, 821–839.

Awadallah, K.T., Tawfik, M.F.S., 1972. The biology of *Xylocoris* (=*Peizostethus*) *flavipes* (Reut.) (Hemiptera: Anthocoridae). Bulletin de la Societe Entomologique d'Egypte 56, 177–189.

Awadallah, K.T., Tawfik, M.F.S., Swailen, S.M., El Meghraby, M.M.A., 1976. The effect of feeding on various preys on the nymphal stage and oviposition of *Cardiastethus nazarenus* Reuter. Bulletin of Entomological Society of Egypt 60, 251–255.

Awadallah, K.T., Tawfik, M.F.S., El Husseini, M.M., 1986. Bio-cycle of the anthocorid predator *Xylocoris flavipes* (Reuter) in association with rearing on major pests of stored drug materials. Bulletin de la Societe Entomologique d'Egypte 66, 27–33.

Bacheler, J.S., Baranowski, R.M., 1975. *Paratriphleps laeviusculus*, a phytophagous anthocorid new to the United States (Hemiptera: Anthocoridae). Florida Entomologist 58, 157–163.

Bahşi, Ş.Ü., Tunç, İ, 2012. The response of a southern strain of *Orius majusculus* (Reuter) (Hemiptera: Anthocoridae) to photoperiod and light intensity: biological effects and diapause induction. Biological Control 63 (2), 157–163.

Ballal, C.R., Gupta, T., 2011. Anthocorids as potential predators. pp. 92–111. In: Bhardwaj, S.C., Singh, S., Bhatnagar, A. (Eds.), Plant Protection through Eco-friendly Techniques. Pointer Publishers, Jaipur, 272 pp.

Ballal, C.R., Singh, S.P., 2001. Effect of insecticide applications on the populations of *Helicoverpa armigera* (Hübner) (Lepidoptera: Noctuidae) and natural enemies in sunflower ecosystem. Entomon (Spl. Issue) 26, 37–42.

Ballal, C.R., Yamada, K., 2014. Diversity of Indian Anthocoridae. p. 89. In: World Biodiversity Congress – 2014. November 24th to 27th 2014, Colombo, Sri Lanka, 212 pp.

Ballal, C.R., Singh, S.P., Poorani, J., Gupta, T., 2003a. Feasibility of mass multiplication and utilization of *Cardiastethus exiguus* Poppius, a potential anthocorid predator of *Opisina arenosella* Walker (Lepidoptera: Oecophoridae). pp. 29–33. In: Tandon, P.L., Ballal, C.R., Jalali, S.K. (Eds.), Biological Control of Lepidopteran Pests, 354 pp.

Ballal, C.R., Singh, S.P., Poorani, J., Gupta, T., 2003b. Biology and rearing requirements of an anthocorid predator, *Blaptostethus pallescens* Poppius (Heteroptera: Anthocoridae). Journal of Biological Control 17, 29–33.

Ballal, C.R., Gupta, T., Joshi, S., Chandrashekhar, K., 2009. Evaluation of an anthocorid predator *Blaptostethus pallescens* against two-spotted spider mite *Tetranychus urticae*. In: Castane, C., Perdikis, D. (Eds.), Integrated Control in Protected Crops. IOBC/WPRS Bulletin, vol. 49, pp. 127–132.

Ballal, C.R., Gupta, T., Joshi, S., 2012a. Production protocols for and storage efficacy of an anthocorid predator *Cardiastethus exiguus* Poppius. Journal of Environmental Entomology 34 (1), 50–56.

Ballal, C.R., Gupta, T., Joshi, S., 2012b. Predatory potential of two indigenous anthocorid predators on *Phenacoccus solenopsis* Tinsley and *Paracoccus marginatus* Williams and Granara de Willink. Journal of Biological Control 26, 18–22.

Ballal, C.R., Gupta, T., Joshi, S., 2012c. Effect of different laboratory hosts on the fertility table parameters and continuous rearing of an anthocorid predator *Orius tantillus* (Motsch.). Pest Management in Horticultural Ecosystems 18, 24–28.

Ballal, C.R., Gupta, T., Joshi, S., 2012d. Morphometry and biology of a new anthocorid *Montandoniola indica*, a potential predator of *Gynaikothrips uzeli*. In: Integrated Control in Protected Crops, Mediterranean Climate, IOBC - WPRS Bulletin, vol. 80, pp. 79–84.

Ballal, C.R., Gupta, T., Joshi, S., 2013. Production and evaluation of the warehouse pirate bug *Xylocoris flavipes* (Reuter) (Hemiptera: Anthocoridae). In: Proceedings of the International Congress on Insect Science 2013, 14th to 17th February, 2013 at GKVK, Bangalore.

Ballal, C.R., Yamada, K., Verghese, A., 2015. Potential Indian Anthocorid Predators at the ICAR-NBAIR Live Insect Repository. *Technical Folder*. ICAR-National Bureau of Agricultural Insect Resources, Bangalore, India.

Bán, G., Fetykó, K., Tóth, F., 2010. Predatory arthropod assemblages of alfalfa and stinging nettle as potential biological control agents of greenhouse pests. Acta Phytopathologica et Entomologica Hungarica 45 (1), 159–172.

Barber, G.W., 1936. *Orius insidiosus* (Say), an important natural enemy of the cornear worm. United States Department of Agriculture Technical Bulletin 504, 1–24.

Battu, G.S., Bains, S.S., Atwal, A.S., 1975. Natural enemies of *Trogoderma granarium* Everts infesting wheat in the rural stores in the Punjab. Bulletin of Grain Technology 13, 52–55.

Beninato, S., la Morella, S., 2000. Control of *Cacopsylla pyri* with massive releases of *Anthocoris nemoralis* in pear orchards. GF 2000. Atti, Giornate Fitopatologiche, Perugia, 16–20 aprile, 2000, volume primo 2000, pp. 367–372.

Bhagat, R.C., 2015. Diversity and updated systematic checklist of cimicomorpha bugs (Heteroptera: Hemiptera) of Jammu, Kashmir and Ladakh Himalaya (India). Indian Journal of Fundamental. Indian Journal of Fundamental and Applied Life Sciences 5 (1), 275–279.

Bhumannavar, B.S., Rabindra, R.J. (Eds.), 2011. Annual Report 2010–11. National Bureau of Agriculturally Important Insects, Bangalore, India, 136 pp.

Bhumannavar, B.S., Murthy, K.S., Rabindra, R.J. (Eds.), 2010. Annual Report 2009–10. National Bureau of Agriculturally Important Insects, Bangalore, India, 124 pp.

Bickerton, M.W., Hamilton, G.C., 2012. Effects of intercropping with flowering plants on predation of *Ostrinia nubilalis* (Lepidoptera: Crambidae) eggs by generalist predators in bell peppers. Environmental Entomology 41, 612–620.

van der Blomm, J., Robledo, A., Torres, S., Sáncez, J.A., 2009. Consequences of the wide scale implementation of biological control in greenhouse horticulture in Almeira, Spain. In: Integrated Control of Protected Crops, Mediterranean Climate, IOBC/WPRS Bulletin, vol. 49, pp. 9–13.

Bonte, M., De Clercq, P., 2008. Developmental and reproductive fitness of *Orius laevigatus* (Hemiptera: Anthocoridae) reared on factitious and artificial diets. Journal of Economic Entomology 101, 1127–1133.

Bonte, M., De Clercq, P., 2010. Influence of diet on the predation rate of *Orius laevigatus* on *Frankliniella occidentalis*. Biocontrol 55, 625–629.

Bonte, M., de Clercq, P., 2011. Influence of predator density, diet and living substrate on developmental fitness of *Orius laevigatus*. Journal of Applied Entomology 135, 343–350.

Bosco, L., Tavella, L., 2010. Population dynamics and integrated pest management of *Thrips tabaci* on leek under field conditions in northwest Italy. Entomologia Experimentalis et Applicata 135, 276–287.

Bosco, L., Bodino, N., Baudino, M., Tavella, L., 2012. Insecticides and beneficial predators: side effects on *Orius* spp. on IPM pepper and strawberries. IOBC/WPRS Bulletin 80, 187–192.

Bronnimann, H., 1964. Rearing anthocorids on an artificial medium. CIBC Technical Bulletin 4, 147–150.

Brower, J.H., Mullen, M.A., 1990. Effects of *Xylocoris flavipes* (Hemiptera: Anthocoridae) releases on moth populations in experimental peanut storage. Journal of Entomological Science 25, 268–276.

Bueno, V.H.P., Mendes, S.M., Carvalho, L.M., 2006. Evaluation of a rearing method for the predator *Orius insidiosus*. Bulletin of Insectology 59, 1–6.

Buhl, R., Bassler, R., 1992. Biological control of thrips, aphids and leafminers. Gemuse Munchen 28, 155–158.

Bush, L., Kring, T.J., Ruberson, J.R., 1993. Suitability of greenbugs, cotton aphids and *Heliothis virescens* eggs for development and reproduction of *Orius insidiosus*. Entomologia Experimentalis et Applicata 67, 217–222.

Butler, C.D., Trumble, J.T., 2012. Identification and impact of natural enemies of *Bactericera cockerelli* (Hemiptera: Triozidae) in Southern California. Journal of Economic Entomology 105 (5), 1509–1519.

Buxton, J.H., Finlay, R., 1993. Integrated pest management in AYR chrysanthemums. Bulletin OILB SROP 16, 33–41.

Calixto, A.M., Bueno, V.H.P., Montes, F.C., Silva, A.C., van Lenteren, J.C., 2013. Effect of different diets on reproduction, longevity and predation capacity of *Orius insidiosus* (Say) (Hemiptera: Anthocoridae). Biocontrol Science and Technology 23, 1245–1255.

Carvalho, L.M., Bueno, V.H.P., Castane, C., 2011. Olfactory response towards its prey *Frankliniella occidentalis* of wild and laboratory-reared *Orius insidiosus* and *Orius laevigatus*. Journal of Applied Entomology 135, 177–183.

Castañé, C., Zalom, F.G., 1994. Artificial oviposition substrate for rearing *Orius insidiosus* (Hemiptera: Anthocoridae). Biological Control 4, 88–91.

Castane, C., Bueno, V.H.P., Carvalho, L.M., van Lenteren, J.C., 2014. Effects of founder population size on the performance of *Orius laevigatus* (Hemiptera: Anthocoridae) colonies. Biological Control 69, 107–112.

Chacko, M.J., 1973. Observations on some natural enemies of *Pineus* sp. (Hem.: Adelgidae) at Shillong (Meghalaya), India, with special reference to *Tetraphleps raoi* Ghauri (Hem.: Anthocoridae). Technical Bulletin of the Commonwealth Institute of Biological Control 16, 41–46.

Chambers, R.J., Long, S., Helyer, N.L., 1993. Effectiveness of *Orius laevigatus* (Hem.: Anthocoridae for the control of *Frankliniella occidentalis* on cucumber and pepper in the UK. Biocontrol Science and Technology 3, 295–307.

Choi, Y.S., Whang, I.S., Han, I.S., Kim, Y.C., Choe, G.R., 2013. A case study for intergrated pest management of *Frankliniella occidentalis* and *Aphis gossypii* by simultaneously using *Orius laevigatus* and *Aphidius colemani* with azoxystrobin in cucumber plants. Korean Journal of Applied Entomology 52, 379–386.

Devasahayam, S., Koya, A., 1994. Natural enemies of major insect pests of black pepper (*Piper nigrum* L.) in India. Journal of Spices and Aromatic Crops 3, 50–55.

Devasahayam, S., 2000. Bioecology of Leaf Gall Thrips, *Liothrips karnyi* Bagnall Infesting Black Pepper (Ph.D. thesis). Department of Zoology, University of Calicut, Kerala, India.

Devi, N., Gupta, P.R., 2010. Anthocorid bugs encountered on cultivated crops and ornamentals, and an attempt to rear *Orius niger* Wolff under laboratory conditions. Pest Management and Economic Zoology 18, 313–320.

Dhaliwal, G.S., 1976. Intensity of insect infestation under rural storage conditions in the Punjab. Entomologists' Newsletter 6, 8–9.

Dhaliwal, G.S., 1977. Incidence of storage insect pests in rural areas. Indian Journal of Entomology 39, 114–116.

Distant, W.L., 1904. Rhynchotal notes. XXV. Heteroptera. Fam. Anthocoridae. The Annals and Magazine of Natural History 13, 673–688.

Distant, W.L., 1906. Order Rhynchota. Suborder Heteroptera. Family Anthocoridae. In: Bingham, C.T. (Ed.), The Fauna of British India including Ceylon and Burma, vol. 3. Tayler and Francis, London, pp. 1–10.

Distant, W.L., 1910. Order Rhynchota. Suborder Heteroptera. Family Anthocoridae. In: Bingham, C.T. (Ed.), The Fauna of British India including Ceylon and Burma, vol. 5. Tayler and Francis, London, pp. 295–309.

Duffield, S.J., 1995. Crop-specific differences in the seasonal abundance of four major predatory groups on sorghum and short duration pigeonpea. International Chickpea Pigeonpea Newsletter 2, 74–76.

Elimem, M., Chermiti, B., 2012a. Effect of *Orius laevigatus* and *Amblyseius swirskii* releases on *Frankliniella occidentalis* populations in pepper crop greenhouses in the Bekalta region of Tunisia. IOBC/WPRS Bulletin 80, 147.

Elimem, M., Chermiti, B., 2012b. Use of the predators *Orius laevigatus* and *Aeolothrips* spp. to control *Frankliniella occidentalis* populations in greenhouse peppers in the region of Monastir, Tunisia. IOBC/WPRS Bulletin 80, 141–146.

Fathi, A.A.A., 2014. Efficiency of *Orius minutus* for control of *Tetranychus urticae* on selected potato cultivars. Biocontrol Science and Technology 24 (8), 936–949.

Fauvel, G., Thiry, M., Cotton, D., 1984. Progress towards the artificial rearing of *Anthocoris nemoralis* F. Bulletin SROP 7, 176–183.

Fiedler, Z., Sosnowska, D., 2009. Biological control of important pests on vegetable crops in Polish greenhouses. In: Integrated Control of Protected Crops, Mediterranean Climate, IOBC/WPRS Bulletin, vol. 49, pp. 21–24.

Fischer, S., Linder, C., Freuler, J., 1992. Biology and utilization of *Orius majusculus* Reuter (Heteroptera, Anthocoridae) for the control of the thrips *Frankliniella occidentalis* Perg. and *Thrips tabaci* Lind., in greenhouses. Revue Suisse de Viticulture d'Arboriculture et d'Horticulture 24, 119–127.

Fitzgerald, J., Jay, C., 2013. Implications of alternative prey on biocontrol of pests by arthropod predators in strawberry. Biocontrol Science and Technology 23, 448–464.

Fransen, J., Tolsma, J., 1992. Releases of the minute pirate bug, *Orius insidiosus* (Say) (Hemiptera: Anthocoridae), against western flower thrips, *Frankliniella occidentalis* (Pergande), on chrysanthemum. Medelingen van de Faculteit Landbouwwetenschappen, Universiteit Gent 57, 479–484.

Fransen, J.J., Boogaard, M., Tolsma, J., 1993. Minute pirate bug, *Orius insidiosus* (Say) (Hemiptera: Anthocoridae), as a predator of western flower thrips, *Frankliniella occidentalis* (Pergande), in chrysanthemum, rose and Saintpaulia. Bulletin, OILB-SROP 16 (8), 73–77.

Funao, T., Yoshiyasu, Y., 1995. Development and fecundity of *Orius sauteri* (Poppius) (Hemiptera: Anthocoridae) reared on *Aphis gossypii* Glover and corn pollen. Japanese Journal of Applied Entomology and Zoology 39, 84–88.

Funderburk, J., Stavisky, J., Olson, S., 2000. Predation of *Frankliniella occidentalis* (Thysanoptera: Thripidae) in field peppers by *Orius insidiosus* (Hemiptera: Anthocoridae). Environmental Entomology 29, 376–382.

Funderburk, J.E., 2009. Management of the Western flower thrips (Thysanoptera: Thripidae) in fruiting vegetables. Florida Entomologist 92, 1–6.

Gao, Y.L., Reitz, S.R., Wang, J., Tamez-Guerra, P., Wang, E.D., Xu, X.N., Lei, Z.R., 2012. Potential use of the fungus *Beauveria bassiana* against the western flower thrips *Frankliniella occidentalis* without reducing the effectiveness of its natural predator *Orius sauteri* (Hemiptera: Anthocoridae). Biocontrol Science and Technology 22, 803–812.

Gao, W.C., 1987. The anthocorid *Dufouriella ater* (Dufour) - a new natural enemy of Japanese pine bast scale *Matsucoccus matsumarae* (Nuwana). Acta Entomologica Sinica 30, 271–276.

Garzia, G.T., Failla, S., Manetto, G., Siscaro, G., Zappalà, L., 2012. Mechanical release of *Phytoseiulus persimilis* and *Orius laevigatus* on protected crops. IOBC/WPRS Bulletin 80, 253–259.

Ghahari, H., Carpentiro, D.L., Ostovan, H., 2009. An annotated catalogue of the Iranian Anthocoridae (Hemiptera: Heteroptera: Cimicomorpha). Acta Entomologica Musei Nationalis Pragae 49, 43–58.

Ghauri, M.S.G., 1964. Notes on the Hemiptera from Pakistan and adjoining areas. Annals and Magazine of Natural History 6, 409–421.

Ghauri, M.S.G., 1972. The identity of *Orius tantillus* (Motschusky) and notes on other Oriental Anthocoridae (Hemiptera, Heteroptera). Journal of Natural History 6, 409–421.

Ghosh, D., Poddar, S., Raychaudhuri, D.N., 1981. Natural enemy complex of *Aphis craccivora* Koch. and *Lipaphis erysimi* (Kalt.) in and around Calcutta, West Bengal. Science and Culture 47, 58–60.

Gomez-Polo, P., Alomar, O., Castañé, C., Agustí, N., 2012. Analysing predation of *Orius majusculus* (Heteroptera: Anthocoridae) in lettuce crops by PCR. IOBC/WPRS Bulletin 80, 103–108.

Gomez-Polo, P., Alomar, O., Castane, C., Riudavets, J., Agustí, N., 2013. Identification of *Orius* spp. (Hemiptera: Anthocoridae) in vegetable crops using molecular techniques. Biological Control 67, 440–445.

Groves, G.R., 1974. Report for the Year - 1974 of the Bermuda Department of Agriculture and Fisheries. 29 pp.

Guo, J.Y., Wu, M., Wan, F.H., 2002. Effects of cold storage on the adults and eggs of *Orius sauteri*. Chinese Journal of Biological Control 18, 10–12.

Gupta, T., Ballal, C.R., 2006. Biology and feeding potential of an anthocorid predator, *Orius tantillus* (Heteroptera: Anthocoridae) on *Sitotroga cerealella*. Indian Journal of Plant Protection 34, 168–172.

Gupta, T., Ballal, C.R., 2007. Feeding preference of anthocorid predators for parasitised and un-parasitised eggs. Journal of Biological Control 21, 73–78.

Gupta, T., Ballal, C.R., 2009. Protocols for commercial production of *Orius tantillus* (Motschulsky) (Hemiptera: Anthocoridae). Journal of Biological Control 23 (3), 385–391.

Gupta, T., Ballal, C.R., Joshi, S., 2011. Preferential feeding of an anthocorid predator *Blaptostethus pallescens* on different stages of cotton mealybug. Journal of Environmental Entomology 33 (4), 423–428.

Habuštová, O., Doležal, P., Spitzer, L., Svobodová, Z., Hussein, H., Sehnal, F., 2014. Impact of Cry1Ab toxin expression on the non-target insects dwelling on maize plants. Journal of Applied Entomology 138, 164–172.

Husseini, M., Ashouri, A., Enkegaard, A., Weisser, W.W., Goldansaz, S.H., Mahalati, M.N., Moayeri, H.R.S., 2010. Plant quality effects on intraguild predation between *Orius laevigatus* and *Aphidoletes aphidimyza*. Entomologia Experimentalis et Applicata 135, 208–216.

Husseini, M., Schumann, K., Sermann, H., 1993. Rearing immature feeding stage of *Orius majesculus* Reut. (Het., Anthocoridae) on the acarid mite *Tyrophagus putriscentiae* Schr. as new alternative prey. Journal of Applied Entomology 116, 113–117.

Ito, K., 2007. A simple mass rearing method for predaceous *Orius* bugs in the laboratory. Applied Entomology and Zoology 42, 573–577.

Jacobsen, S., Eilenberg, J., Klingen, I., Sigsgaard, L., 2015. Different behavioral responses in specialist and generalist natural enemy interactions (predators and fungi) in a strawberry-mite pest system. In: Proceedings of the IOBC/WPRS Working Group "Integrated Control in Oilseed Crops", Vigalzano di Pergine Valsugana, Italy, 26–28 May 2014. IOBC/WPRS Bulletin, vol. 109, pp. 89–91.

Jalali, S.K., Singh, S.P., 2002. Seasonal activity of stem borers and their natural enemies on fodder maize. Entomon 27, 137–146.

Jerinić-Prodanović, D., Protić, L., 2013. True bugs (Hemiptera, Heteroptera) as psyllid predators (Hemiptera, Psylloidea). ZooKeys 319, 169–189.

JianYing, G., FangHao, W., Min, W., 2002. Comparison of successive rearing of *Orius sauteri* with aphids and *Trichogramma* pupae reared in artificial host eggs. Chinese Journal of Biological Control 18, 58–61.

Karanja, M.K., Aloo, T.C., 1990. The Introduction and Establishment of *Tetraphleps raoi* Ghauri as a Control Agent of Woolly Aphid in Kenya. *Technical Note - Kenya Forestry Research Institute* No. 12: 12 pp.

Kaur, R., Virk, J.S., 2011. Role of *Blaptostethus pallescens* Poppius and *Xylocoris flavipes* (Reuter) in the suppression of *Corcyra cephalonica* Stainton in stored rice grain. Journal of Biological Control 25, 329–332.

Kawamoto, K., Sasaki, M., Nakaima, H., Onuma, M., Uchida, N., 1999. Development and reproductive ability of *Orius tantillus* (Motschulky) (Hemiptera: Anthocoridae) reared on egg of *Plodia interpunctella* (Hubner). Research Bulletin of the Plant Protection Service, Japan 35, 87–91.

Keever, D.W., Mullen, M.M.A., Press, J.W., Arbogast, R.T., 1986. Augmentation of natural enemies for suppressing two major insect pests in stored farmers stock peanuts. Environmental Entomology 15, 767–770.

Kfir, R., 1986. Releases of natural enemies against the pine bark beetle *Orthotomicus erosus* (Wollaston) in South Africa. Journal of the Entomological Society of Southern Africa 49, 391–392.

Kim, J.H., 2009. Cold storage effect on the biological characteristics of *Orius laevigatus* (Fieber) (Hemiptera: Anthocoridae) and *Phytoseiulus persimilis* Athias-Henriot (Acari: Phytoseiidae). Korean Journal of Applied Entomology 48, 361–368.

Kumar, N.S., Ananthakrishnan, T.N., 1984. Predator-thrips interactions with reference to *Orius maxidentex* Ghauri and *Carayanocoris indicus* Muraleedharan (Anthocoridae: Heteroptera). Proceedings of the Indian National Science Academy, B 50, 139–145.

Ladurner, E., Benuzzi, M., Franceschini, S., Sterk, G., 2012. Efficacy of *Beauveria bassiana* strain ATCC 74040 against whiteflies on protected tomato and compatibility with *Nesidiocoris tenuis* and *Orius laevigatus*. IOBC/WPRS Bulletin 80, 277–282.

Lattin, J.D., 1999. Bionomics of Anthocoridae. Annual Review of Entomology 44, 207–231.

LeCato, G.L., Collins, J.M., Arbogast, R.T., 1977. Reduction of residual populations of stored product insects by *Xylocoris flavipes* (Hemiptera: Anthocoridae). Journal of Kansas Entomological Society 50, 84–88.

Lee, K.S., Lee, J.H., 2007. Rearing of *Orius strigicollis* (Heteroptera: Anthocoridae) on artificial diet. Entomological Research 34, 299–303.

Lingren, P.D., Lukefahr, M.J., Diaz Jr., M., Harstack Jr., A., 1978. Tobacco budworm control in caged cotton with a resistant variety, augmentative releases of *Campoletis sonorensis*, and natural control by other beneficial species. Journal of Economic Entomology 71, 739–745.

Lingren, P.D., 1977. *Campoletis sonorensis*: maintenance of a population on tobacco budworms in a field cage. Environmental Entomology 6, 72–76.

Lumbierres, B., Albajes, R., Pons, X., 2012. Positive effect of Cry1Ab-expressing Bt maize on the development and reproduction of the predator *Orius majusculus* under laboratory conditions. Biological Control 63, 150–156.

Lyla, K.R., Beevi, P., Ballal, C.R., 2006. Field evaluation of anthocorid predator, *Cardiastethus exiguus* Poppius against *Opisina arenosella* Walker (Lepidoptera: Oecophoridae) in Kerala. Journal of Biological Control 20, 229–232.

Manjunath, T.M., Rai, P.S., Gowda, G., 1978. Natural enemies of brown planthopper and green leafhopper in India. International Rice Research Newsletter 3, 11.

Martin, P.B., Ridgeway, R.L., Schuetze, C.E., 1978. Physical and biological evaluations of an encapsulated diet for rearing *Chrysopa carnea*. Florida Entomologist 61, 145–152.

McDaniel, S.G., Sterling, W.L., Dean, D.A., 1981. Predators of tobacco budworm larvae in Texas cotton. Southwestern Entomologist 6, 102–108.

Mendes, S.M., Bueno, V.H.P., Carvalho, L.M., Reis, S.P., 2005. Production cost of *Orius insidiosus* as biological control agent. Pesquisa Agropecuaria Brasileira 40, 441–446.

Messelink, G.J., Janssen, A., 2014. Increased control of thrips and aphids in greenhouses with two species of generalist predatory bugs involved in intraguild predation. Biological Control 79, 1–7.

Messelink, G.J., Bloemhard, C.M.J., Vellekoop, R., 2011. Biological control of aphids by the predatory midge *Aphidoletes aphidimyza* in the presence of intraguild predatory bugs and thrips. Acta Horticulturae 915, 171–177.

Messelink, G.J., Bloemhard, C.M.J., Sabelis, M.W., Janssen, A., 2013. Biological control of aphids in the presence of thrips and their Enemies. BioControl 58, 45–55.

Mori, P., Sancassani, G.P., 1984. Study on the integrated control of the pear psyllid (*Psylla pyri*) in Venetia. Bulletin SROP 7, 354–357.

Mukherjee, A.B., Chaudhury, A.K.S., Bronniman, H., Chu, Y., 1971. Artificial diets for rearing of *Xylocoris flavipes* (Reuter), a predator of some pests of stored cereals. Indian Journal of Entomology 33, 356–358.

Muraleedharan, N., Ananthakrishnan, T.N., 1971. Bionomics of *Montandoniola moraguesi* (Puton) (Heteroptera, Anthocoridae) a predator on gall thrips. Bulletin of Entomology 12, 4–10.

Muraleedharan, N., Ananthakrishnan, T.N., 1974a. New and little known species of *Orius* Wolff from South India. Oriental Insects 8 (1), 37–41.

Muraleedharan, N., Ananthakrishnan, T.N., 1974b. The genus *Buchananiella* Reuter from India with description of a new species (Heteroptera: Anthocoridae). Oriental Insects 8, 33–35.

Muraleedharan, N., Ananthakrishnan, T.N., 1974c. A new species of the genus *Scoloposcelis* Fieber (Hem. Anthocoridae) from South India. Journal of Natural History 8, 511–512.

Muraleedharan, N., Ananthakrishnan, T.N., 1978. Bioecology of four species of Anthocoridae (Hemiptera: Insecta) predaceous on thrips, with key to genera of anthocorids from India. Occasional Paper, Records of the Zoological Survey of India (11), 1–32.

Muraleedharan, N., 1975. New record of *Cardiastethus pygmaeus pauliani* Lansbury) from India. Newsletter, Zoological Survey of India (1), 67.

Muraleedharan, N., 1977a. Notes on a collection of Lyctocorinae (Hemiptera: Anthocoridae) from Garo Hills (Meghalaya). Records of the Zoological Survey of India 73, 239–246.

Muraleedharan, N., 1977b. Some genera of Anthocorinae (Heteroptera: Anthocoridae) from south India. Entomon 2, 231–236.

Muraleedharan, N., 1977c. A new genus of Anthocoridae (Heteroptera) from South India. Oriental Insects 11, 463–466.

Murata, M., Imamura, T., Miyanoshita, A., 2007. Suppression of the stored-product insect *Tribolium confusum* by *Xylocoris flavipes* and *Amphibolus venator*. Journal of Applied Entomology 131, 559–563.

Nagai, K., Hikawa, M., 2012. Evaluation of black-eyed Susan *Rudbeckia hirta* L. (Asterales: Asteraceae) as an insectary plant for a predacious natural enemy *Orius sauteri* (Poppius) (Heteroptera: Anthocoridae). Japanese Journal of Applied Entomology and Zoology 56, 57–64.

Nagai, K., Takagi, M., Nakashima, Y., Hiramatsu, T., 1998. Selection of alternative prey for rearing *Orius tantillus* (Motschulsky). Japanese Journal of Applied Entomology and Zoology 42, 85–87.

Nasser, M., Abdurahiman, U.C., 1993. Cannibalism in *Cardiastethus exiguus* Poppius (Hemiptera: Anthocoridae), a predator of coconut caterpillar, *Opisina arenosella* Walker (Lepidoptera: Xylorictidae). Journal of Advanced Zoology 14, 1–6.

Nasser, M., Abdurahiman, U.C., 1998. Efficacy of *Cardiastethus exiguus* Poppius (Hemiptera: Anthocoridae), as a predator of the coconut caterpillar, *Opisina arenosella* Walker (Lepidoptera: Xylorictidae). Journal of Entomological Research 22, 361–368.

Nasser, M., Abdurahiman, U.C., 2004. Anthocorid predators and their biocontrol potential. pp. 91–114. In: Sahayaraj, K. (Ed.), Indian Insect Predators in Biological Control. Daya Publishing House, Delhi, 336 pp.

Nasser, M., Abdurahiman, U.C., 1990. Reproductive biology and predatory behaviour of the anthocorid bugs (Anthocoridae: Hemiptera) associated with the coconut caterpillar, *Opisina arenosella* (Walker). Entomon 15, 3–4.

NBAII, 2013. Annual Report 2012–13. National Bureau of Agriculturally Important Insects, Bangalore, India, 130 pp.

NBAII, 2014. Annual Report 2013–14. National Bureau of Agriculturally Important Insects, Bangalore, India, p iv + 120 pp.

Niemczyk, E., 1970. The development and fecundity of the bark bug - *Anthocoris nemorum* (L.) (Heteroptera: Anthocoridae) reared on the eggs of the Angoumois grain moth - *Sitotroga cerealella* Oliv. (Lep., Gelechiidae). Polskie Pismo Enotomologiczne 40, 857–865.

Niemczyk, E., 1978. *Orius minutus* (L.) (Heteroptera: Anthocoridae): the occurrence in apple orchards, biology and effect of different food on the development. Polskie Pismo Enotomologiczne 48, 203–209.

Oku, T., Kobayashi, T., 1966. Influence of predation of *Orius* sp. on aphid population in a soybean field. An example of interrelation between a polyphagous predator and its principal prey. Japanese Journal of Applied Entomology and Zoology 10, 89–94.

Oveja, M.F., Arnó, J., Gabarra, R., 2012. Effect of supplemental food on the fitness of four omnivorous predator species. IOBC/WPRS Bulletin 80, 97–101.

Parajulee, M.N., Phillips, T.W., 1992. Laboratory rearing and field observations of *Lyctocoris campestris* (Heteroptera: Anthocoridae), a predator of stored product insects. Annals of the Entomological Society of America 85, 736–743.

Parker, N.J.B., 1981. A method for mass rearing the aphid predator *Anthocoris nemorum*. Annals of Applied Biology 99 (3), 217–223.

Pascual, S., Cobos, G., Seris, E., Sánchez-Ramos, I., González-Núñez, M., 2014. Spinosad bait sprays against the olive fruit fly (*Bactrocera oleae* (Rossi)): effect on the canopy non-target arthropod fauna. International Journal of Pest Management 60, 258–268.

Patel, H.K., Patel, J.R., Patel, S.N., 1975. Records of predators and their parasites from Gujarat. Entomologists' Newsletter 5, 8–9.

Pillai, G.B., Nair, K.R., 1993. A checklist of parasitoids and predators of *Opisina arenosella* Wlk. on coconut. Indian Coconut Journal, Cochin 23, 2–9.

Pluot-Sigwalt, D., Streito, J.C., Matocq, A., 2009. Is *Montandoniola moraguesi* (Puton, 1896) a mixture of different species? (Hemiptera, Heteroptera, Anthocoridae). Zootaxa 2208, 25–43.

Poppius, B., 1909. Beiträge zur Kenntnis der Anthocoriden. Acta Societatis Scientiarum Fennicae 37, 1–43.

Poppius, B., 1913. Zur Kenntnis des Miriden, isometopiden, Anthocoriden, Nabiden und Schizopteriden Ceylons. Entomologisk tidskrift 34, 239–260.

Pourian, H.R., Talaei-Hassanloui, R., Kosari, A.A., Ashouri, A., 2011. Effects of *Metarhizium anisopliae* on searching, feeding and predation by *Orius albidipennis* (Hem., Anthocoridae) on *Thrips tabaci* (Thy., Thripidae) larvae. Biocontrol Science and Technology 21, 15–21.

Pumariño, L., Alomar, O., 2014. Assessing the use of *Lobularia maritima* as an insectary plant for the conservation of *Orius majusculus* and biological control of *Frankliniella occidentalis*. In: Proceedings of the IOBC/WPRS Working Group "Landscape Management for Functional Biodiversity", IOBC/WPRS Bulletin, vol. 100, pp. 113–116.

Puysseleyr, V.D., Höfte, M., De Clercq, P., 2014. Continuous rearing of the predatory anthocorid *Orius laevigatus* without plant materials. Journal of Applied Entomology 138, 45–51.

Qin, S.H., 1985. Mass rearing of predatory insects in the field for control of injurious mites in apple orchards. Natural Enemies of Insects Kunchong Tiandi 7, 117–124.

Rajasekhara, K., Chatterji, S., 1970. Biology of *Orius indicus* a predator of *Taeniothrips nigricornis*. Annals of Entomological Society of America 63, 364–367.

Rajasekhara, K., 1973. A new species of *Blaptostethus* (Hemiptera: Anthocoridae) from Mysore, India. Annals of Entomological Society of America 66, 86–87.

Rao, S.V.R., 1976. Studies on the biology, bionomics and chemical control of the sorghum midge, *Contarinia sorghicola* (Coquillett). Entomologists' Newsletter 6, 14–15.

Reid, C.D., 1991. Ability of *Orius insidiosus* (Hemiptera: Anthocoridae) to search for, find and attack European corn borer and corn earworm eggs on corn. Journal of Economic Entomology 84, 83–86.

Richards, P.C., Schmidt, J.M., 1996. The suitability of some natural and artificial substrates as oviposition sites for the insidious flower bug, *Orius insidiosus*. Entomologia Experimentalis et Applicata 80, 325–333.

Ruberson, J.R., Bush, L., Kring, T.J., 1991. Photoperiodic effect on diapause induction and development in the predator *Orius insidiosus* (Heteroptera: Anthocoridae). Environmental Entomology 20, 786–789.

Ruth, J., Dwumfour, E.F., 1989. Laboratory studies on the suitability of some aphid species as prey for the predatory flower bug *Anthocoris gallarum-ulmi* (DeG.) (Het., Anthocoridae). Journal of Applied Entomology 108, 321–327.

Saini, E.D., Cervantes, V., Alvarado, L., 2003. Effect of diet, temperature and crowding on fecundity, fertility and longevity of *Orius insidiosus* (Say) (Heteroptera: Anthocoridae). RIA Revista de Investigaciones Agropecuarias 32, 21–32.

Sampson, C., Boullenger, A., Puerto Garcia, F., Hernandez Parra, R., 2011. Implementing Integrated Pest Management programmes in protected strawberry crops across Europe. IOBC/WPRS Bulletin 70, 171–180.

Sands, W.A., 1957. The immature stages of some British Anthocoridae (Hemiptera). Transactions of Royal Entomological Society, London 109, 295–310.

Saulich, A.Kh., Musolin, D.R., 2008. Seasonal development and ecology of anthocorids (Heteroptera: Anthocoridae). Entomological Review 89, 501–528.

Saxena, R.C., 1975. Integrated approach for the control of *Thrips tabaci* Lind. Indian Journal of Agricultural Sciences 45, 434–436.

Schenikova, A., Stepanycheva, E.N., 2005. Rearing of predator bug *Orius laevigatus* (Fieb.) (Heteroptera: Anthocoridae) with alternative food and its application against *Frankliniella occidentalis* (Pergande). Bulletin OILB/SROP 28 (1), 233–236.

Schmidt, J., 1994. Pirate bugs plunder greenhouse pests. Agricultural Food Research in Ontario 17, 12–15.

Schuldiner-Harpaz, T., Coll, M., 2012. Identification of *Orius* (Heteroptera: Anthocoridae) females based on egg operculum structure. Journal of Economic Entomology 105, 1520–1523.

Seagraves, M.P., Lundgren, J.G., 2010. Oviposition response by *Orius insidiosus* (Hemiptera: Anthocoridae) to plant quality and prey availability. Biological Control 55, 174–177.

Shapiro, J.P., Ferkovich, S.M., 2006. Oviposition and isolation of viable eggs from *Orius insidiosus* in a parafilm and water substrate: comparison with green beans and use in enzyme linked immunosorbent assay. Annals of Entomological Society of America 99, 586–591.

Shapiro, J.P., Shirk, P.D., Reitz, S., Koenig, R., 2009. Sympatry of *Orius insidiosus* and *O. pumilio* (Hemiptera: Anthocoridae) in North Central Florida. Florida Entomologist 92, 362–366.

Sigsgaard, L., Esbjerg, P., 1994. *Orius tantillus*, a predator of *Helicoverpa armigera* in south India. In: Abstracts, Third International Conference on Tropical Entomology Nairobi, Kenya, Abstract C36.

Sigsgaard, L., Esbjerg, L., Philipsen, H., 2006. Experimental releases of *Anthocoris nemoralis* F. and *Anthocoris nemorum* (L.) (Heteroptera: Anthocoridae) against the pear psyllid *Cacopsylla pyri* L. (Homoptera: Psyllidae) in pear. Biological Control 39 (1), 87–95.

Silveira, L.C.P., Bueno, V.H.P., van Lenteren, J.C., 2004. *Orius insidiosus* as biological control agent of Thrips in greenhouse chrysanthemums in the tropics. Bulletin of Insectology 57, 103–109.

Singh, T.V.K., Singh, K.M., Singh, R.N., 1991. Influence of inter cropping: III. Natural enemy complex in groundnut. Indian Journal of Entomology 53, 333–368.

Sorensson, A., Nedstam, B., 1993. Effect of *Amblysieus cucumeris* and *Orius insidiosus* on *Frankliniella occidentalis* in ornamentals. Bulletin IOBC/WPRS 16, 129–132.

Southwood, T.R.E., 1956. The structure of the eggs of Terrestrial Heteroptera and its relationship to the classification of the group. Transactions of the Royal Entomological Society, London 108 (6), 163–221.

Sreerama Kumar, P., Bhumannavar, B.S., Veenakumari, K., Mohanraj, P., Ballal, C.R., Venkatesan, T., Sriram, S., Murthy, D.S., 2012. Annual Report 2011–12. National Bureau of Agriculturally Important Insects, Bangalore, India, 103 pp.

Srivastava, M., Funderburk, J., Olson, S., Demirozer, O., Reitz, S., 2014. Impacts on natural enemies and competitor thrips of insecticides against Western flower thrips (Thysanoptera: Thripidae) in fruiting vegetables. Florida Entomologist 97, 337–348.

Takara, J., Nishida, T., 1978. Eggs of the oriental fruit fly for rearing the predaceous anthocorid *Orius insidiosus* (Say). Proceedings of the Hawaiian Entomological Society 23, 441–445.

Tawfik, M.F.S., El-Husseini, M.M., 1971. The life history of the anthocorid predator, *Blaptostethus piceus* Fieber var. *pallescens* Poppius (Hemiptera: Anthocoridae). Bulletin de la Societe Entomologique d' Egypte 55, 239–252.

Tawfik, M.F.S., El-Sherif, S.I., 1969. The biology of *Pyroderces simplex* Wlsm., an important pest on maize plants in UAR (Lepidoptera: Lavernidae). Bulletin de la Socite Entomologique d' Egypte 53, 615–628.

Tawfik, M.F.S., Kira, M.T., Metwally, S.M.I., 1974. On the abundance of major pests and their associated predators in corn plantations. Bulletin de la Societe Entomologique d' Egypte 58, 168–177.

Thomas, J.M.G., Shirk, P.D., Shapiro, J.P., 2012. Mass rearing of tropical minute pirate bug *Orius pumilio* (Hemiptera: Anthocoridae). Florida Entomologist 95, 202–204.

Thontadarya, T.S., Rao, K.J., 1987. Biology of *Orius maxidentex* Ghauri (Hemiptera: Anthocoridae), a predator of the sorghum earhead midge, *Contarinia sorghicola* (Coquillet). Mysore Journal of Agricultural Sciences 21, 27–31.

Tommasini, M.G., Maini, S., 2002. Thrips control on protected sweet pepper crops: enhancement by means of *Orius laevigatus* releases. In: Thrips and Tospovirus: Proceedings of the International Symposium on Thysanoptera, pp. 249–256.

Trematerra, T., Dioli, P., 1993. *Lyctocoris campestris* (F.) (Heteroptera, Anthocoridae) in stores of *Triticum spelta* L. in Central Italy. Bolletino di Zoologia Agraria e di Bachicoltura 25, 251–257.

Trottin Caudal, Y., Grasselly, D., Trapateau, M., Dobelin, H., Millot, P., 1991. Biological control of *Frankliniella occidentalis* with *Orius majusculus* on cucumber. Bulletin SROP 14, 50–56.

Veer, V., 1984. On some new records of predators of *Thrips flavus* Schrank and *Thrips hawaiiensis* Morgan (Thysanoptera: Thripidae) from Dehradun (India). Indian Journal of Forestry 7, 245–246.

Venkatesan, T., Ballal, C.R., Rabindra, R.J., 2008. Biological Control of Coconut Black-Headed Caterpillar *Opisina arenosella* Using *Goniozus nephantidis* and *Cardiastethus exiguus*. Technical Bulletin No. 39. Project Directorate of Biological Control, Bangalore, 14 pp.

Vieira, N.F., Truzi, C., Veiga, A.C.P., Vacari, A.M., De Bortoli, S., 2014. Biological aspects of *Xylocoris afer* (Reuter, 1884) preying on eggs of *Plutella xylostella* (L., 1758). In: Conference XXV Brazilian Congress of Entomology, Goiana.

Villevieille, M., Millot, P., 1991. Biological control of *Frankliniella occidentalis* with *Orius laevigatus* on strawberry. Bulletin SROP 14, 57–64.

Viswanathan, T.R., Ananthakrishnan, T.N., 1973. Observations on the biology, ecology and behaviour of *Ecacanthothrips sanguineus* Bagnall (Tubulifera: Thysanoptera). Current Science 42, 727–728.

Viswanathan, T.R., Ananthakrishnan, T.N., 1974. Population fluctuations of 3 species of anthophilous Thysanoptera in relation to the numerical response of their predator, *Orius minutus* L. (Anthocoridae: Hemiptera). Current Science 43, 19–20.

Waite, M.O., Scott-Dupree, C.D., Brownbridge, M., Buitenhuis, R., Murphy, G., 2014. Evaluation of seven plant species/cultivars for their suitability as banker plants for *Orius insidiosus* (Say). BioControl 59, 79–87.

Wang, F.H., Zhou, W.R., Wang, R., 1994. Studies on the overwintering survival rate of *Orius sauteri* Hem.: Anthocoridae in Beijing. Chinese Journal of Biological Control 10, 100–102.

Wang, F.H., Zhou, W.R., Wang, R., 1996. Studies on the method of rearing *Orius sauteri*. Chinese Journal of Biological Control 12, 49–51.

Wang, C.L., Lee, P.C., Wu, Y.J., 2001. Field Augmentation of *Orius strigicollis* (Heteroptera: Anthocoridae) for the Control of Thrips in Taiwan. Extension Bulletin 500. Food and Fertiliser Technology Center, 9 pp.

Wearing, C.H., Marshall, R.R., Colhoun, C., Attfield, B.A., 2014. Phytophagous mites and their predators during the establishment of apple orchards under biological and integrated fruit production in Central Otago, New Zealand. New Zealand Journal of Crop and Horticultural Science 42, 127–144.

Wong, S.K., Frank, S.D., 2013. Pollen increases fitness and abundance of *Orius insidiosus* Say (Heteroptera: Anthocoridae) on banker plants. Biological Control 64, 45–50.

Yamada, K., Bindu, K., Nasser, M., 2008. Taxonomic and biological notes on *Cardiastethus affinis* and *C. pseudococci pseudococci* (Hemiptera: Heteroptera: Anthocoridae) in India. Zootaxa 1910, 59–68.

Yamada, K., Ballal, C.R., Gupta, T., Poorani, J., 2010a. Description of a new species of *Anthocoris* (Hemiptera: Heteroptera: Anthocoridae) from southern India, associated with striped mealybug on purple orchid tree. Acta Entomologica Musei Nationalis Pragae 50, 415–424.

Yamada, K., Bindu, K., Nasser, M., 2010b. The second species of the genus *Rajburicoris* Carpintero and Dellapé (Hemiptera: Heteroptera: Anthocoridae) from southern India, with reference to autapomorphies and systematic position of the genus. Proceedings of the Entomological Society, Washington 112, 464–472.

Yamada, K., Bindu, K., Nasreem, A., Nasser, M., Ballal, C.R., Poorani, J., 2010c. The flower bugs found in agroecosystems of Southern India (Heteroptera Anthocoridae). Presented at the Fourth Quinquennial Meeting of the International Heteropterists Society, July 12–12, 2010 in Tianjin, China.

Yamada, K., Bindu, K., Nasreem, A., Nasser, M., 2011. A new flower bug of the genus *Montandoniola* (Hemiptera: Heteroptera: Anthocoridae), a predator of gall-forming thrips on black pepper in southern India. Acta Entomologica Musei Nationalis Pragae 51, 1–10.

Yamada, K., Yasunaga, T., Artchawakom, T., 2013. The genus *Xylocoris* found from plant debris in Thailand, with description of a new species of the subgenus *Arrostelus* (Hemiptera: Heteroptera: Anthocoridae). Acta Entomologica Musei Nationalis Pragae 53, 493–504.

Yano, E., van Lenteren, J.C., 1996. Biology of *Orius sauteri* (Poppius) and its potential as a biocontrol agent for *Thrips palmi* Karny. Bulletin OILB SROP 19, 203–206.

Zaki, F.N., 1989. Rearing of two predators, *Orius albidipennis* (Reut.) and *Orius laevigatus* (Fieber) (Hem.: Anthocoridae) on some insect larvae. Journal of Applied Entomology 107, 107–109.

Zeng, F., Cohen, A.C., 2000. Demonstration of amylase from the zoophytophagous anthocorid *Orius insidiosus*. Archives of Insect Biochemistry 44, 136–139.

Zhou, W., Wang, R., 1989. Rearing of *Orius sauteri* (Hem.: Anthocoridae) with natural and artificial diets. Chinese Journal of Biological Control 5, 9–12.

Zhou, W.R., Zhang, S.F., Zhang, Z.F., Long, S.D., 1986. Comparative studies on mass rearing *Xylocoris flavipes* on different media. Chinese Journal of Biological Control 2, 63–66.

Zhou, W., Wang, R., Qiu, S., 1991. Use of soybean sprouts as the oviposition material in mass rearing of *Orius sauteri* (Het.: Anthocoridae). Chinese Journal of Biological Control 7, 7–9.

7

Reduviid Predators

Dunston P. Ambrose, A. Ganesh Kumar

Entomology Research Unit, St. Xavier's College (Autonomous), Palayamkottai, Tamil Nadu, India

1. INTRODUCTION

The Reduviidae is the largest family of predaceous land Heteroptera containing about 7000 species and subspecies in 913 genera and 25 subfamilies (Froeschner and Kormilev, 1989; Cassis and Gross, 1995; also see Ambrose, 2000a). The reduviids are abundant and occur worldwide. They are voracious predators (hence the name, "assassin bugs") and most of them are generalist predators. Ambrose (2006) listed 14 subfamilies with 144 genera and 464 species of Indian assassin bugs. Later, Ambrose et al. (2007a) published a separate checklist of subfamily Peiratinae and added two more species. Being larger than many other predaceous land bugs and encompassing in their development a greater range of size, they consume not only more prey but also a wider array of prey (Schaefer, 1988). There are more than 182 reduviids belonging to 69 genera and 10 subfamilies, preying upon about 261 known domestic, agricultural, and forestry insect pests. Because they are polyphagous, they may not be useful as predators on specific insect pests but are valuable predators in situations where a variety of insect pests occur. They possess following biocontrol characteristics: (1) limited prey range, (2) positive functional and numerical responses, (3) good host searching efficiency, (4) multiply faster with a high fecundity and short life cycle, (5) female-biased population, (6) possess good pest suppression efficacy, (7) amenable for mass culturing in the laboratory, (8) adaptability to new environmental conditions, and (9) free from parasites or parasitoids and predators. Thus, assassin bugs are important mortality factors and should be conserved and augmented for their utilization in biocontrol programs (Ambrose, 2002, 2003; also see Ambrose, 2000a).

In addition, assassin bugs exhibit a wide range of ecological diversity in terms of their microhabitats and habitats such as ecotypes; morphological diversity as morphs; functional diversity in terms of their predatory, reproductive, offensive, and defensive; and biological diversity (Ambrose, 1999, 2000a; Hwang and Weirauch, 2012; also see Ambrose, 2003).

The family Reduviidae's composition and relationships remain unsettled and there is an absolute need for a comprehensive reassessment of the family's higher level classification and phylogenetic relationships (Ambrose, 2004). In addition, a multidisciplinary biosystematics covering morphology, ecology, behavior, biology, cytology, biochemistry, and

217

molecular aspects of family Reduviidae has been undertaken (Ambrose, 1999, 2000a, 2004, 2015; Khokar, 2000; Ambrose and Ambrose, 2003, 2004, 2009a; George et al., 2005; Poggio et al., 2007; Weirauch, 2008; Baskar et al., 2011, 2012a,b, 2013, 2014; Singh et al., 2011; Kaur et al., 2012; Kaur and Kaur, 2013; Ambrose et al., 2014).

Despite the abundance of the world's reduviid fauna with its rich taxonomic, geographical, ecological, trophic, morphological, biological, and behavioral diversity and its rich and diverse prey records and biocontrol potential against numerous insect pests, they have not been fully utilized in integrated pest management (IPM) (Ambrose, 2002, 2003; also see Ambrose, 2000a). Their conservation and augmentation for IPM can be achieved only if their biosystematics and ecology are studied thoroughly. One must know what the insect is, who are its relatives, and what are its phylogenetic relationships? Such knowledge broadens and deepens the biological information and makes it more useful (Schaefer and Ahmad, 1987; Schaefer, 1988; also see Ambrose, 2000a). Because the success of utilizing biocontrol agents mainly depends upon the knowledge we have on its phylogeny, biology, and ecology covering distribution and diversity, an understanding enables to evolve strategies to effectively use the biocontrol agent (Ambrose, 1999; Raja et al., 2011).

2. BIOECOLOGY

2.1 Ecology

Knowledge on microhabitats of 238 Indian assassin bugs reveals that 85 species of reduviids (35%) live exclusively under boulders, 31 species (13%) on shrubs, 15 under the bark (6%), and 11 in litter (5%). Many species dwell in more than one microhabitat (Hwang and Weirauch, 2012; also see Ambrose, 1999).

Of 303 Indian reduviids, 128 reduviids (36%) exclusively dwell in tropical rain forests, 26 species in the semiarid zones (9%), and 24 species in scrub jungles (8%). Many species were found in all the three major ecosystems (19 species, 6%) and in adjacent agroecosystems. Two triatomines are exclusively present in human dwellings. Twelve species were found attracted to light and two of them were also found in agroecosystems (see Ambrose, 1999).

Tropical rain forest bugs are soft cuticled, alate, devoid of warning coloration, and their instars are armored with straight and club-shaped hairs and often feign death. They lack tibial pads or fossula spongiosae. They pin and lift their prey by the straight or slightly curved rostrum. The eggs are laid in clusters and are exposed. Fecundity and hatchability are higher. Eclosion and ecdysis occur usually in daytime. Precopulatory riding is common.

The scrub jungle and semiarid zone bugs are hard cuticled, pterygopolymorphic, possess warning coloration, and their instars are armored with setose hairs, spines, and tubercles. They possess both fore and midtibial pads. They chase and pounce upon their prey and pierce with their curved or acutely curved rostrum. They withstand prolonged starvation. The eggs are laid singly, unexposed, in crevices, or deep in the soil. Fecundity and hatchability are lower. Eclosion and ecdysis occur usually at dusk and dawn. Nymphal camouflaging is common. Such information is not available for extra Indian reduviids (see Ambrose, 1999, 2003).

Adaptive ecotypic as well as polymorphic diversities in relation to color, size, fecundity, hatchability, sex ratio, longevity, efficacy in predation, resistance to insecticides, and genetic constitution have been reported in reduviids. Knowledge on ecotypes and polymorphs are very useful to evolve strategies to use them in biocontrol of pests. For instance, the niger morph of *Rhynocoris marginatus* (Fabricius) is more efficient predator, more resistant to insecticides and possesses better biological characteristics than the sanguineous and the nigrosanguineous morphs (George, 1999a, 2000a,b; Singh, 2012; Lenin, 2014; Manimuthu, 2014; also see Ambrose, 1999).

Population dynamics studies on entomophagous reduviids indicate that their populations are directly regulated by their prey populations, which in turn are indirectly governed by abiotic factors (Thanasingh, 2002; Rajan, 2011; Raja et al., 2011; also see Ambrose, 1999).

2.2 Biology and Behavior

The biology of apiomerines, ectrichodiines, harpactorines, peiratines, reduviines, salyavatines, and stenopodaines, covering information on preoviposition period, fecundity, incubation period, hatchability, stadial period, eclosion and ecdysis periodicities, adult longevity, sex ratio, nymphal morphology, nymphal mortality, nymphal food requirement, postembryonic development, life table, growth patterns, and parental care have been studied.

The preoviposition period is usually longer in Reduviinae, shorter in Ectrichodinae and Salyavatinae, and intermediate in Harpactorinae, Peiratinae, and Triatominae. The harpactorines lay the highest number of eggs followed by triatomines and reduviines. Stenopodines lay the least number of eggs. The ectrichodines, peiratines and salyavatines lay moderate number of eggs. The incubation period is the shortest in Harpactorinae and the longest in Peiratinae. The hatchability is the highest in Triatominae, followed by Harpactorinae and the least in Peiratinae. The fifth instar is the longest stadium and the highest nymphal mortality is recorded during I nymphal instars and II stadium is shortest followed by the III stadium. The nymphal mortality was highly prevalent among Ectrichodiinae followed by Salyavatinae, Reduviinae and Harpactorinae. They are univoltine (e.g., Peiratinae), bivoltine (e.g., Reduviinae) (Figure 1), and multivoltine (e.g., Harpactorinae)

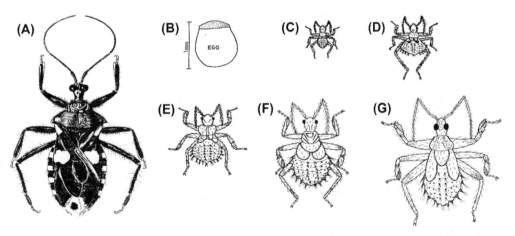

FIGURE 1 Life stages of *Acanthaspis siva*: (A) adult, (B) egg, (C–G) first to fifth nymphal instars. *From Ambrose and Livingstone (1987a).*

(George et al., 1998; George, 1999a,b, 2000a,b,c,d,e,f; Sathiamma et al., 1999; Venkatesan et al., 1999; Ambrose et al., 2003, 2006a, 2007b, 2009; Claver et al., 2004; Ambrose and Ravichandran, 2007; Das and Ambrose, 2008a,b; Das et al., 2008a,b; Nagarajan et al., 2010a; Rajan et al., 2011; Kalidas and Sahayaraj, 2012; Srikumar et al., 2014; also see Ambrose, 1999, 2000a, 2003).

The impacts of ecophysiological conditions, such as crowding, space, camouflaging of reduviine instars, prey species, prey density, prey deprivation, flooding, soil moisture, sensory inputs, female parental age, and insecticides on oviposition pattern, hatchability, and postembryonic development of reduviids were studied to develop strategies to mass rear these predators.

Crowding decreased body weight, size of eggs, nymphal instars and adults, extended preoviposition period, enhanced the hatchability, delayed eclosion, and extended stadia and adult longevity of reduviids. An increase in rearing space increased fecundity, hatchability, and prey capturing time and decreased stadial period and adult longevity (see Ambrose, 1999, 2003). Though camouflaging primarily gives protection from predators and coinstars, it also promotes predatory efficiency and survival rate. The decamouflaging decreased the total stadial period and increased the nymphal mortality in *Acanthaspis pedestris* (see Ambrose, 1999). Prey species and amount of prey consumed by reduviids affect their postembryonic development (George, 2000c; Sahayaraj, 2001; see Ambrose, 1999, 2000a). Prey deprivation increased stadial period, nymphal mortality, predatory efficiency, and prey consumption (see Ambrose, 1999). Flooding of eggs reduced the hatchability and prolonged the incubation (Ambrose et al., 1994; Ambrose and George, 1996a). The soil moisture affected the incubation period, hatchability, and egg mortality (Vennison and Ambrose, 1990). The female parental aging prolonged postembryonic development with increased egg and nymphal mortalities, reduced adult longevity, and increased male-biased sex ratio (Ambrose et al., 1988; Das and Ambrose, 2008c; also see Ambrose, 1999, 2000a). The females that mated with several males of different ages had shorter preoviposition period, larger oviposition days, higher fecundity and hatchability, and shorter incubation period (see Ambrose, 1999, 2003). Insecticide-treated reduviids exhibited either delayed or abnormal postembryonic development (Ambrose, 2001; George and Ambrose, 2004a; Ambrose et al., 2006b, 2007c, 2010a).

Though the biology of Oriental reduviids has been studied, many non-Oriental species have to be explored (Sahayaraj, 2012; Muthupandi et al., 2014). The conservation and augmentation of reduviids can be achieved with a comprehensive and elaborate understanding of their biology (also see Ambrose, 1999, 2000a, 2003).

2.2.1 Courtship Behavior

Though courtship behavior of reduviids is much like that of other insects; it has some diagnostic features. The sensory-mediated sequential acts of courtship behavior are arousal, approach, riding over (Harpactorinae), nuptial clasp, extension of genitalia and connection, copulation, and postcopulatory acts of many Oriental reduviids was studied. Vision, antennal sensillae, and tibial pad/comb sensillae influence the mating. The role of pheromones on mating was also established (Ambrose et al., 2006a, 2007b, 2009; Das and Ambrose, 2008a,b; Das et al., 2008a,b; Ambrose and Ambrose, 2009b; also see Ambrose, 1999). In addition, diversity of spermatophore capsules and postcopulatory cannibalism of certain reduviids as a prerequisite for normal fecundity, oviposition, and hatchability were reported (see Ambrose, 1999, 2000a).

2.2.2 *Oviposition Behavior*

Oviposition behavior of reduviids along with description and diversity of eggs were reported for a number of species along with typical oviposition pattern of certain reduviids (Harpactorinae). There are five major types of oviposition in reduviids: (1) eggs are laid in clusters and cemented to each other and the substratum (Harpactorinae), (2) eggs are laid in clusters but not glued to each other longitudinally rather glued to the substratum (Ectrichodiinae), (3) each egg is separately cemented to the substratum (Holoptilinae), (4) eggs are buried deep inside the soil (Peiratinae and Stenopoainae), and (5) eggs are loosely strewn around erratically without any pattern (Reduviinae, Salyavainae, and Triatominae) (Ambrose and Ambrose, 2009b; also see Ambrose, 1999).

3. PEST PREY RECORD AND ASSOCIATED BEHAVIOR

3.1 Pest Prey Records

One hundred and eighty three reduviid predators belonging to 70 genera and 10 subfamilies have been found to prey upon a wide array of insect pests (Table 1).

3.2 Predation

Predatory behavior, prey records, and food requirements of reduviids have been reported worldwide by several authors (see Ambrose, 1999). Predatory behavior is mediated by sensory responses and the sequences are categorized as arousal, approach, capture, rostral probing, injection of toxic saliva and paralyzing, sucking, and postpredatory behavior (Ambrose et al., 2006a, 2007b, 2009; Das and Ambrose, 2008a,b; Das et al., 2008a,b, 2009; Kumar et al., 2009, 2011; Evangelin et al., 2014; also see Ambrose, 1999).

Reduviidae have adapted to a wide range of terrestrial habitats and diversified in their prey choices while developing a wide repertoire of innovative prey capture strategies (Forero et al., 2011; Soley et al., 2011; Wignall et al., 2011; Zhang and Weirauch, 2011; also see Ambrose, 1999, 2000a). The diversity of prey capture strategies that has evolved in assassin bugs is remarkable. It ranges from luring ants and injecting saliva to paralyze and kill the prey in feather-legged bugs (Holoptilinae) (Jacobson, 1911) to sticky trap the prey by extending resin-coated forelegs in front of prey in Apiomerinae and Ectrichodiinae (Wolf and Reid, 2001; Ambrose et al., 2006a, 2007b, 2009; Weirauch, 2006, 2008; Das and Ambrose, 2008a,b; Das et al., 2008a,b, 2009; Kumar et al., 2009, 2011; also see Ambrose, 1999). Emesinae, or thread-legged bugs, wait and capture the prey with their long raptorial forelegs armed with spines and tubercles. Some emesinae cut through webs to reach their spider prey (Soley et al., 2011) or lure spiders using aggressive mimicry (Wignall et al., 2011). The reduviines, salyavatines, and stenopodaines wait and grab their prey. The harpactorines pin and jab the prey with their long rostrum. The peiratines, reduviines, and some ectrichodines chase, pounce, and grab their prey with their well-developed tibial pads (Claver and Ambrose, 2001a; Forero et al., 2011; also see Ambrose, 1999).

The impact of sensory inputs, such as vision and chemosense, and structural adaptations, such as tibial pads and tibial combs, sex, nymphal camouflaging, prey species,

TABLE 1 Records of Reduviid Predators and Their Insect Pest Prey Species

S.No	Reduviid species	Pest prey species	References
I. APIOMERINAE			
1.	*Apiomerus crassipes* (F.)	*Aphis gossypii* Glover, *Chauliognathus marginatus* (F.), *Conderus vespertinus* (F.), *Diabrotica undecimpunctata howardi* Barber, *Epilachna borealis* (F.), *Gaurotes cyannipennis* (Say), *Nabis alternates* Oarshky, *Photinus pyralis* (L.), *Podisus maculiventris* Say, and *Poelabrus rugulosus* Leconte	Blatchley (1926), Swadener and Yonke (1973a), Taylor (1949), Thompson and Simmonds, (1964), and Morrill (1975)
2.	*Apiomerus flavipes* Stål	*Eutectona machaeralis* Walker and *Trigona fulviventris* (Guerin)	Misra (1975) and Weaver (1978)
3.	*Apiomerus lanipes* (Fabricius)	*Ceralitis capitula* (Wiedemann) and *Drosophila* sp.	Amaral et al. (1994)
4.	*Apiomerus mutabilis* Costa Lima, Seabra & Hathaway	*Trigona* sp.	Gil-Santana and Forero (2010)
5.	*Apiomerus nigricollis* Stål	*Trigona* sp.	Grossi et al. (2012)
6.	*Apiomerus pilipes* (Fabricius)	*Melipona compressipes manausensis* Schwarz and *Melipona seminigra merrillae* Cockerell	Silva and Gil-Santana (2004)
7.	*Apiomerus spissiper* (Say)	*Anthonomous grandis* Boheman and *Pogonomyrmex barbatus* Buckley	Morgan (1907) and Cade et al. (1978)
8.	*Apiomerus* sp.	Aphids, caterpillars, eucalyptus defoliator, and locusts	Zanunico et al. (1994)
II. ECTRICHODINAE			
9.	*Ectrychotes dispar* Reuter	*Helicoverpa armigera* Hübner	Bhatnagar et al. (1983) and Pawar et al. (1986)
10.	*Ectrychotes crudelis* (F.)	*A. gossypii, Amrasca devastans* (Ishida), and rice pests	Ambrose (2007) and Lam et al. (2008)
11.	*Haematoloecha delibuta* (Distant)	*Musci* spp.	Takeno (1998)
12.	*Neohaematorrhophus therasii* Ambrose and Livingstone	*Camponotus compressus* F., *Corcyra cephalonica* (Stainton), *Dysdercus cingulatus* (Fabricius), *H. armigera, Odontotermes assumthi* Holmgren, *Odontotermes obesus* Rambur, and *Oxycarnis hyalinipennis* Costa	Ambrose (1980, 1987), Ambrose and Livingstone (1991), Sahayaraj and Ambrose (1994, 1996)
13.	*Scadra annulipes* Reuter	*Agrotis ipsilon* (Hufnagel)	Singh and Gangrade (1975)

TABLE 1 Records of Reduviid Predators and Their Insect Pest Prey Species—cont'd

S.No	Reduviid species	Pest prey species	References
III. EMESINAE			
14.	*Schidium marcidum* (Uhler)	*Filices* spp.	Takeno (1998)
IV. HARPACTORINAE			
15.	*Acholla multispinosa* (DeGeer)	Walnut pest *Leptinotarsa decemlineata* Say	Readio (1927)
16.	*Alcmena spinifex* (Thunberg)	*C. compressus, Colemania* sp., *Ergolis merinone* (Cramer), *Erosomyia* sp., *Euproctis fraterna* Moore, *Euproctis scintillans* Walker, *O. obesus, Oecophylla* sp., *Poekilocerus* sp., and scales	Ambrose (1999) and Das et al. (2008a)
17.	*Alcmena* sp.	*D. undecimpunctata howardi, Diabrotica puncta* Oliv., and *E. machaeralis*	Chittenden (1905) and Misra (1975)
18.	*Agriosphodrus dohrni* (Stål)	*Lymantria dispar* (Linnaeus), *Clostera anachoreta* (Dennis & Schiffermüller), *Apocheima cinerarius* Erschoff, *Pristiphora beijingensis* Zhou and Zhang, *Stilpnotia salicis* (L.), and *Aphis robiniae* Macchiati	Yao et al. (1995)
19.	*Arilus carinatus* (Forster)	*Costalimaita ferruginea* (Fabricius), *Hedypathes betulinus* (Klug), aphids, eucalyptus defoliators, and locusts	Zanunico et al. (1994), Soares and Iede (1997), Barbosa (2002)
20.	*Arilus cristatus* (L.)	*Acrosternum hilare* (Say), *Cyllene robiniae* (Forst.), *Hyphantria cunea* (Drury), *Papilio cresphontes* Cramer, *Pontia rapae* Linn., *Chrysodeixis includens* (Walker), *Spodoptera exigua* (Walker), tortricids, and caterpillars	Chittenden (1916), Garman (1916), Swadener and Yonke (1973b), Richman et al. (1980), and Ambrose (1999)
21.	*Amphibolus venator* (Klug)	*Alphitobius diaperinus* (Panzer), *C. cephalonica, Latheticus oryzae* Waterhouse, *Oryzaephilus surinamensis* (L.), *Tenebroides mauritanicus* (L.), *Tribolium castaneum* (Herbst), *Tribolium confusum* Duv., *Tribolium freemani* Hinton, and *Trogoderma granarium* Everts	Pingale (1954), Hussain and Aslam (1970), Haines (1991), Imamura et al. (2006), and Youssef and Abd-Elgayed (2015)
22.	*Atrachelus cinereus* (F.)	*S. exigua, Thaumastocoris peregrinus* Carpintero & Dellapé	Ambrose (1999) and Santadino et al. (2013)

(Continued)

TABLE 1 Records of Reduviid Predators and Their Insect Pest Prey Species—cont'd

S.No	Reduviid species	Pest prey species	References
23.	*Brassivola hystrix* Distant	*Achea janata* L., *Dysdercus koenigii* (F.), *Liaphis erysimi* Kalt., *Macrotermes estherae* (Desneux), and *Sylepta derogata* (F.)	Lakkundi (1989)
24.	*Brassivola* sp.	*C. cephalonica* and *H. armigera*	Ambrose (1999)
25.	*Brassivola* sp.	*Dasychira mendosa* Hb., *Eumeta cramerii* (Westwood), and *Euproctis lunata* (Walker)	Babu et al. (1995) and Kuppusamy and Kannan (1995)
26.	*Coranus aegypticus* (F.)	*Amrasca biguttula biguttula* (Ishida)	Singh et al. (1987)
27.	*Coranus nodulosus* Ambrose and Sahayaraj	*C. cephalonica, D. cingulatus, Oxycarenus hyalinipenns*, and *Pectinophora gossypiella* Saunders	Sahayaraj (1991) and Ambrose (2007)
28.	*Coranus pallidus* Reuter	*Aphanus apicalis* Dall, *Aphanus sordidus* F., and *Dieuches patruelis* Stål	Risbec (1941)
29.	*Coranus papillosus* Thunberg	*Cydia pomonella* L. and *H. armigera*	Pettey (1919), Nei (1942), and Ambrose (2007)
30.	*Coranus siva* Kirkaldy	*C. cephalonica* and *O. obesus*	Kumar (1993) and Ambrose (2007)
31.	*Coranus soosaii* Ambrose & Vennison	*C. cephalonica, Earias insulana* Biosdual, *H. armigera, O. hyalinipennis*, and *P. gossypiella*	Kumaraswami (1991)
32.	*Coranus spiniscutis* Reuter	*Anomis flava* (F.), *Chilo zonellus* Swinhoe, *D. cingulatus, Eysarcoris ventralis* (Westwood), *H. armigera, Leptocorisa varicornis* Fabricius, *Margaronia pyloalis* Walk., *Myllocerus blandis* Faust, *Myllocerus laetivirens* Marshall, *Myllocerus maculosus* Desbr, *O. obesus, Papilio polytes* Linnaeus, *Plecoptera reflexa* Guenee, *Spodoptera litura* (F.), *Spodoptera* spp., and *Utetheisa pulchella* L.	Bose (1949), Ren (1984), Ambrose and Claver (2001a), Ambrose (2003), and Claver and Reegan (2010)
33.	*Coranus subapterus* (DeGeer)	*Diacrisia (Spilosoma) obliqua* (Walker), *Forficula auricularia* L., and Aphids	Butler (1918) and Wallace (1953)
34.	*Coranus vitellinus* Distant	*O. assumthi* and *O. obesus*	Ambrose (1987)
35.	*Coranus* sp.	Stored grain pests	Ambrose (2003)

TABLE 1 Records of Reduviid Predators and Their Insect Pest Prey Species—cont'd

S.No	Reduviid species	Pest prey species	References
36.	*Coranus* sp.	*S. derogata*	Joseph (1959)
37.	*Coranus* sp.	*E. insulana, Earias vitella* (F.), *H. armigera* and *Naranga* sp.	Krishnaswamy et al. (1984)
38.	*Coranus* sp.	*A. biguttula biguttula* and *P. gossypiella*	Singh et al. (1987)
39.	*Coranus* sp.	*D. cingulatus, E. insulana, E. vitella, H. armigera, M. maculosus, S. derogate,* and *S. litura*	Lakkundi (1989)
40.	*Coranus* sp.	*O. obesus*	Kumar (1993)
41.	*Coranus* sp.	*A. janata, C. cephalonica, H. armigera,* and *O. obesus*	Ambrose (1999)
42.	*Coranus* sp.	Pigeon pea pests	Ambrose and Claver (2001a)
43.	*Cosmolestus picticeps* (Stål)	*Helopeltis bradyi* Waterhouse, *Helopeltis cinchonae* Mann, *Mahasena corbetti* Tams	Corbett and Pagden (1941)
44.	*Cydnocoris gilvus* (Burmeister)	*E. machaeralis, Galleria mellonella* L. *Helopeltis antonii* Sign., *O. obesus, Oxya nitidula* (Walker), *Pyrausta machaeralis* Walker, and *S. litura*	Misra (1975), Patil and Thontadarya (1983), Venkatesan et al. (1999), and Srikumar et al. (2014)
45.	*Endochus cameronicus* Miller	*H. bradyi* and *H. cinchonae*	Corbett and Pagden (1941)
46.	*Endochus inornatus* Stål	*A. janata, C. cephalonica, Dysdercus koenigii* F., *H. antonii, H. armigera, M. pusutulata, O. obessus, Poecilocerus* sp., *S. litura,* and *Stephanitis typical* (Distant)	Singh et al. (1982), Sundararaju (1984), Lakkundi (1989), Sathiamma et al. (1999), and Chandral and Sinazer (2011)
47.	*Endochus parvispinus* Distant	*A. janata, H. armigera, Oxycarenus laetus* Kirby, and *S. litura*	Lakkundi (1989)
48.	*Endochus umbrinus* (Distant)	*A. janata, C. cephalonica, D. cingulatus, H. armigera,* and *S. litura*	Sahayaraj (1991), Sahayaraj and Ambrose (1992), and Ambrose (2007)
49.	*Endochus* sp.	*Hyblaea puera* Cramer and *Poekilocerus pictus* F.	Mohanadas (1996) and Ambrose (2007)
50.	*Euagoras plagiatus* (Burmeister)	*C. cephalonica, Colemania sphenarioides* (Bol.), *H. armigera, H. puera, Mylabris pustulata* (Thunberg), *O. obesus, Oecophylla smaragdina* F., *Poecilocerus* sp., and *Stephanitis typical*	Vennison (1988), Mohanadas (1996), Sathiamma et al. (1999), and Ambrose et al. (2008a)

(Continued)

TABLE 1   Records of Reduviid Predators and Their Insect Pest Prey Species—cont'd

S.No	Reduviid species	Pest prey species	References
51.	*Euagoras sordidatus* Stål	*A. janata, C. cephalonica, H. armigera, H. puera,*and *Leptocorisa acuta* Thunberg	Taylor (1932) and Nayar et al. (1976)
52.	*Hediocoris tibialis* (Stål)	*Ascotis reciprocoria* Walker	Abasa (1981)
53.	*Irantha armipes* (Stål)	*A. biguttula biguttula, A. janata, C. cephalonica, H. antonii, H. armigera, O. obesus, S. litura, Teleonemia scrupulosa* Stål, *Tephrina pulinda* (Walker), and pigeon pea pests	Sundararaju (1984), Livingstone and Yacoob (1986), Lakkundi (1989), Babu et al. (1995), Ambrose and Claver (2001a), and Ambrose (2003)
54.	*Irantha consobrina* Distant	*C. cephalonica*	Kumaraswami (1991)
55.	*Irantha* sp.	*L. acuta* and *L. varicornis*	Austin (1922) and Hutson (1922)
56.	*Isyndus heros heros* (F.)	*D. koengii, E. insulana, H. armigera, H. bradyi, H. cinchonae, Idocerus* sp., *Idocerus* sp., *M. corbetti, Setothosea asigna* (Moore), and *S. litura*	Corbett and Pagden (1941), Lakkundi (1989), Singh (1992), and Ambrose (2007)
57.	*Isyndus obscurus* Dallas	*H. cinchonae, M. corbetti, Malacosoma neustria* (L.), *Oxycetonia jucunda* (Falderman), *Parasa lepida* (Cramer), and *S. asigna*	Corbett and Pagden (1941), Yasumatsu and Watanabe (1964), Yano (1990), and Singh (1992)
58.	*Isyndus reticulatus* Stål	*Gossampinus malabaria* (Dc) Men., *Halyabbas unicolor* Distant, *Pyrus calleryana* DeCne, bamboo slug caterpillar, pear slug caterpillar, stink bug, and tree cotton tufted caterpillar	Hoffmann (1935)
59.	*Isyndus* sp.	*H. bradyi* and *H. cinchonae*	Corbett and Pagden (1941)
60.	*Lanca* sp.	*E. insulana, E. vitella, H. antonii, H. armigera, P. gossypiella*, and *S. litura*	Kumaraswami (1991)
61.	*Lophocephala guerini* Laporte	*H. puera*	Loganathan and David (1999)
62.	*Montina confusa* (Stål)	Aphids, eucalyptus defoliators, and locusts	Zanunico et al. (1994)
63.	*Occamus typicus* Distant	*H. antonii* and *S. typical*	Sundararaju (1984) and Sathiamma et al. (1999)
64.	*Panthous bimaculatus* Distant	*Atteva fabriciella* (Swederus), *C. cephalonica*, and *S. litura*	Varma (1989) and Muthupandi et al. (2014)

TABLE 1   Records of Reduviid Predators and Their Insect Pest Prey Species—cont'd

S.No	Reduviid species	Pest prey species	References
65.	*Phonoctonus fasciatus* (de Beauvois)	*Dysdercus superstitiosus* (F.), *Dysdercus fasciatus* Signoret, *Odontopus* sp.	Evans (1962) and Parker (1972)
66.	*Phonoctonus subimpictus* Stål	*Dysdercus melanoderes* Karsch	Stride (1956a,b)
67.	*Phonoctonus* sp.	*Odontopus sexpunctatus* Castalnau	Stride (1956a,b)
68.	*Phonoctonus* spp.	*Dysdercus* spp.	Evans (1962) and Fadare (1978)
69.	*Pislius tibuliformis* F.	*Dysdercus* sp.	Parker (1965)
70.	*Platerus* sp.	*Selepa celtis* (Moore), *T. pulinda*	Balu et al. (1995)
71.	*Polididus armatissimus* (Stål)	*H. armigera*	Ambrose and Vennison (1989)
72.	*Pristhesancus papuensis* (Stål)	*Biprorulus bibax* Breddin, *Helicoverpa virescens* (F.), *Helicoverpa zea* (Boddie), *P. includensi*, and *S. exigua*	Shepard et al. (1982), Martin and Brown (1984), and James (1992, 1994a,b)
73.	*Pristhesancus plagipennis* Walker	*B. bibax*, *Creontiades dilutus* (Stål), *Helicoverpa* spp., *Nezera viridula* (L.), *Phaenacantha australica* (Kirkaldy), and *Spodoptera* spp.	Illingworth (1921), Summerville (1931), James (1992), James et al. (1994), Pyke and Brown (1996), Smith et al. (1997), and Grundy et al. (2000a)
74.	*Pselliopus cinctus* (F.)	*Blissus leucopterus leucopterus* (Say), *L. decemlineata*, *S. celtis*, and Chinch bug	Chittenden (1907), Webster (1907), and Ambrose (2007)
75.	*Rhaphidosoma atkinsoni* Bergroth	*S. celtis* and *O. obesus*	Babu et al. (1995) and Ambrose (2003)
76.	*Rhaphidosoma* sp.	*O. obesus*	Ambrose (2003)
77.	*Rhynocoris albopunctatus* (Stål)	*Antestiopsis lineaticollis intricata* (Ghesv. and Carayon), *Dysdercus* sp., *E. insulana*, *Earias biplaga* (Walker), *H. armigera*, and *P. gossypiella*	Taylor (1932), Carayon et al. (1956), and Nyirra (1970)
78.	*Rhynocoris annulatus* L.	*Neodiprian sertifier* Bird	DeBach and Hagen (1964)
79.	*Rhynocoris bicolor* (F.)	*Clavigralla shadabi* Dolling and *D. superstitiosus*	Parker (1969) and Ambrose (2007)
80.	*Rhynocoris carmelita* (Stål)	*Ephestia kuehniella* Zeller	Edwards (1961)

(Continued)

TABLE 1   Records of Reduviid Predators and Their Insect Pest Prey Species—cont'd

S.No	Reduviid species	Pest prey species	References
81.	*Rhynocoris costalis* (Stål)	*D. cingulatus* and *S. litura*	Maxwell-Lefroy and Howlett (1909), Imms (1965), Sitaramaiah et al. (1975), Nayar et al. (1976), David and Kumaraswami (1988), and Sitaramaiah and Satyanarayana (1976)
82.	*Rhynocoris fuscipes* (F.)	*A. janata, Aulacophora foveicollis* (Lucas), *C. cephalonica, C. compressus, Clavigralla gibbosa* Spinola, *Calocoris angustatus* Leth., *Clavigralla horrens* D., *Coptosilla pyranthe* L., *D. cingulatus, Dicladispa armigera* (Oliver), *Dolycoris indicus* Stål, *E. fraterna, Eurema hecabe* L., *E. insulana, E. merinone, E. scintillans, E. vitella, Epilachna dodecastigma* Mulsant, *Epilachna vigintioctopunctata* F., *Exelastis atomosa* Walshingham, *H. armigera, H. puera, Tephrina pulinda* (Walker), *S. exigua, Lygaeus hospes* F., *M. pustulata, M. indica* Thunberg, *Myllocerus curvicoronis* F., *Myzus persicae* (Sulz.), *N. viridula* (L.) var. *smaragdula* (F.), *Patanga succincta* (L.), *Pelopidas mathias* (F.), *Perigrinus maidis* Ashm., *Plutella xylostella* L., *Rhaphidopalpa foveicollis* Lucas, *Ripotortus clavatus* Thunberg, *S. litura, S. typical, Semiothisa pervolgata* Walker, *D.* (*Spilosoma*) *obliqua* (Walker), and pigeon pea pests	Cherian (1937), Cherian and Kylasam (1939), Cherian and Brahmachari (1941), Rao (1974), Singh and Gangrade (1975), Ponnamma et al. (1979), Ambrose (1980, 1987, 1995, 2007), Nagarkatti (1982), Hiremath and Thontadarya (1983), Singh (1985), Singh and Singh (1987), David and Natarajan (1989), Babu et al. (1995), Mohanadas (1996), Sathiamma et al. (1999), Ambrose and Claver (2001a), and Nagarajan (2010)
83.	*Rhynocoris iracundus* (Poda)	*Carpocoris pudicus* Poda, *Gonocerus acuteangulatus* Geoze, and *Palomena prasina* L.	Romeo (1927) and Ambrose (2007)
84.	*Rhynocoris kumarii* Ambrose and Livingstone	*Ariadne merione* (Cramer), *C. cephalonica, C. compressus, C. gibbosa, D. cingulatus, D. indicus, E. atomosa, E. insulana, Euproctis mollifera* (Thunberg), *E. scintillans, E. vitella, H. armigera, M. indica, M. pustulata, O. hyalinipennis, P. gossypiella, P. succincta, Papilio demoleus* L., *R. clavatus, S. litura, T. confusum,* and pigeon pea pests	Ambrose (1980, 1995), Ambrose and Livingstone (1987b), Kumaraswami (1991), and Ambrose and Claver (1995, 2001a)

229

TABLE 1   Records of Reduviid Predators and Their Insect Pest Prey Species—cont'd

S.No	Reduviid species	Pest prey species	References
85.	*Rhynocoris lapidicola* Samuel and Joseph	*Bagrada cruciferarum* Kirkaldy, *C. cephalonica, H. armigera, S. exigua, M. persicae,* and *Myzus inconspicuous* Distant	Joseph (1959), Mukerjee (1971), Ambrose (1987), and David and Kumaraswami (1988)
86.	*Rhynocoris longifrons* (Stål)	*A. janata, C. compressus, D. cingulatus, E. atomosa, H. armigera, M. indica, M. pustulata, N. viridula, O. obesus, Riptortus pedestris* Thunberg, *S. litura,* and pigeon pea pests	Ambrose (1980), Kumar (1993), Ambrose and Claver (2001a), Ambrose (2003), and Kumar (2011)
87.	*Rhynocoris marginatus* (F.)	*A. janata, C. angustatus, C. cephalonica, C. compressus, C. gibbosa, C. horrens, D. cingulatus, D. indicus, E. atomosa, E. fraterna, E. insulana, E. scintillans, E. vitella, H. armigera, M. indica, M. pustulata, O. hyalinipennis, P. demoleus, P. gossypiella, R. clavatus, S. litura,* and pumpkin beetle	Imms (1965), Nayar et al. (1976), Ambrose (1980, 1987, 1999), Bhatnagar et al. (1983), Pawar et al. (1986), Kumaraswami (1991), and Ambrose (2003, 2007)
88.	*Rhynocoris marginellus* (F.)	*H. bradyi* and *H. cinchonae*	Corbett and Pagden (1941)
89.	*Rhynocoris nysiiphagus* Samuel and Joseph	*Antigastra catalaunalis* (Duponchel), *C. cephalonica, C. zonellus, Dichocrocis punctiferalis* (Guen.), *E. insulana, E. vitella, M. inconspicuous, M. persicae,* and *S. derogata*	Joseph (1959)
90.	*Rhynocoris pusvisculatus* (Distant)	*Gonipterus scutellatus* Gyll.	Mossop (1927)
91.	*Rhynocoris segmantarius* (Germar)	*C. shadabi, D. fasciatus, Dysdercus nigrofasciatus* Stål, *D. superstitiosus , H. armigera,* and *Phenacocons janidotimatile* Ferrero	Ullyett (1930), Taylor (1932), Neuenschwander et al. (1975), and Ambrose (2007)
92.	*Rhynocoris squalis* (Distant)	*S. litura*	Rao et al. (1981)
93.	*Rhynocoris tropicus* (Herrich-Schaefer)	*D. superstitiosus*	Parker (1969)
94.	*Rhynocoris* sp.	*O. sexpunctatus*	Fuseini and Kumar (1975)
95.	*Scipinia areanacea* Distant	*Heteropsylla cubana* Crawford and *Leucaena leucocephala* (Lam.)	Uthamasamy (1995) and Ambrose (2007)
96.	*Scipinia horrida* (Stål)	*A. merione, C. compressus, E. fraterna, E. scintillans, Erosomyia indica* Grover and Prasad., *H. virescens, L. decemilineata, L. leucocephala,* and *O. obesus*	Barrion et al. (1987), Uthamasamy (1995), and Ambrose (2007)

(Continued)

TABLE 1　Records of Reduviid Predators and Their Insect Pest Prey Species—cont'd

S.No	Reduviid species	Pest prey species	References
97.	*Sinea confusa* Caudell	*A. gossypii, Conenurgia erecheta* (Cram.), *H. virescens, H. zea, L. decemilineata, P. gossypiella, S. exigua*, caterpillars, *Anthonomous grandis grandis* Boheman, pear slug, and *Anthonomus eugenii* cano	Chittenden (1907), Morgan (1907), Webster (1912), Henneberry and Clayton (1985), Cohen (1990), and Uthamasamy (1995)
98.	*Sinea diadema* F.	*S. exigua*, caterpillars on clover, *Leptinotarsa decemlineata* (Say), *A. grandis grandis, A. eugenii* cano, *Caliroa cerasi* Li, and *Acalyma vittatum* (F.)	Readio (1927) and Ambrose (1999)
99.	*Sinea spinipes* (Herrich-Schaefer)	*Chorocora sayi* and *S. exigua*	Caffrey and Barber (1919) and Ambrose (1999)
100.	*Sphedanolestes aurescens* Distant	*Nephantis serinopa* Mayr.	Mohamed et al. (1982)
101.	*Sphedanolestes himalayensis* (Distant)	*C. cephalonica* and *O. obesus*	Ambrose (1999)
102.	*Sphedanolestes impressicollis* (Stål)	*Altica cirsicolu* Ohno, *Basileptu fulvipes* (Motschulsky), *Chrysolinu urtichalcea* (Mannerheim), *Coccinella septempunctuta* L., *Duucus curotu* L. var. *sutivu* DC., *Hemipyxis plugioderoides* (Motschulsky), and *Linaeideu aenea* (Linnaeus)	Takeno (1998)
103.	*Sphedanolestes minusculus* (Bergroth)	*A. janata, C. cephalonica, C. sphenarioides, Erosomyia indica* Grover and Prasad, *H. armigera, M. pustulata, O. obesus, O. smaragdina, Poekilocerus* sp., and *S. litura*	Kumar (1993) and Chandral et al. (2005)
104.	*Sphedanolestes pubinotum* (Reuter)	*A. janata, Colemania sphenariodes* Bolivar, *C. cephalonica, C. compressus, E. insulana, H. armigera, H. puera, O. obesus, O. smaragdina*, and *S. litura*	Ambrose (1980, 1999), Kumaraswami (1991), and Mohanadas (1996)
105.	*Sphedanolestes signatus* (Distant)	*Earias vitella* and *H. antonii*	Sundararaju (1984)
106.	*Sphedanolestes* sp.	*S. litura*	Rao et al. (1981)
107.	*Sphedanolestes* sp.	*C. cephalonica, O. obesus*, and *S. litura*	Ambrose (2007)
108.	*Sphedanolestes* sp.	*C. cephalonica* and *O. obesus*	Ambrose (1999)

TABLE 1   Records of Reduviid Predators and Their Insect Pest Prey Species—cont'd

S.No	Reduviid species	Pest prey species	References
109.	*Sycanus affinis* Reuter	*Acherontia styx* Westwood., *Amsacta albistriga* W., *Anomis sabulifera* (Guenée), *C. cephalonica*, *Cirphis albistigma* M., *N. serinopa, Opisina arenosella* Walker, *P. demoleus, S. litura,* and *Sesamia inferens* Walker	Satpathy et al. (1975) and Haines (1991)
110.	*Sycanus collaris* (F.)	*Clostera* (= *Pygaera*) *fulgurita* (Walker), *Clostera* spp., *E. machaeralis, H. antonii, H. bradyi, H. cinchonae, H. puera, P. machaeralis, R. clavatus, S. litura,* and *Sinohala helleri* Ohs	Beeson (1941), Corbett and Pagden (1941), Sundararaju (1984), Singh et al. (1987), Pillai (1988), George et al. (1998), and Ambrose (1999, 2007)
111.	*Sycanus dichotomus*	*Darna trima* (Moore), *Matisa plasia, Pteroma psendula* Hmps., and *S. asigna*	Singh (1992)
112.	*Sycanus indagator* Stål	*A. janata, C. cephalonica, E. machaeralis, H. armigera, H. puera, P. machaeralis, P. xylostella,* and *S. litura*	Beeson (1941), Rao (1974), Misra (1975), Nagarkatti (1982), and Patil and Thontadarya (1983)
113.	*Sycanus macracanthus* Stål	*Setothosea asigna*	Singh (1992)
114.	*Sycanus pyrrhomelas* (Walker)	*E. insulana, E. vitella, P. succincta,* and pigeon pea pests	Ambrose and Paniadima (1988) and Ambrose and Claver (2001a)
115.	*Sycanus reclinatus* (Dohrn)	*C. cephalonica, E. insulana, E. vitelli,* and *H. armigera*	Vennison and Ambrose (1992)
116.	*Sycanus versicolor* (Dohrn)	*H. armigera, P. gossypiella,* and *S. litura*	Kumaraswami (1991)
117.	*Sycanus* sp.	*E. machaeralis*	Misra (1975)
118.	*Sycanus* sp.	*E. machaeralis*	Patil and Thontadarya (1983)
119.	*Sycanus* sp.	*M. corbetti*	Singh (1992)
120.	*Velinus nodipes* (Uhler)	*L. aenea*	Takeno (1998)
121.	*Vesbius sanguinosus* (Stål)	*C. cephalonica* and *O. obesus*	Ambrose (1999)
122.	*Vestula lineaticeps* (Signoret)	*Dysdercus* sp.	Parker (1971)
123.	*Villanovanusi* sp.	*T. pulinda* and *S. celtis*	Balu et al. (1995)
124.	*Zelus ceruivalis* (Stål)	*Bucculatrix thurberiella* Busck, *H. zea, S. exigua,* and *Trichopulisa ni* Hübner	Whitcomb and Bell (1964), Ables (1978), and Ambrose (1999)

(Continued)

TABLE 1　Records of Reduviid Predators and Their Insect Pest Prey Species—cont'd

S.No	Reduviid species	Pest prey species	References
125.	*Zelus exanguis* (Stål)	*B. thurberiella, H. zea, Portheiria dispar*, and *S. exigua*	Kirkland (1896), Whitcomb and Bell (1964), Ables (1978), and Ambrose (1999)
126.	*Zelus longipes* (L.)	*Anastrepha suspensa* (Loew), *Diaphorina citri* Kuwayama, *Drosophila* spp., *Saccharosynde saccharivora*, and *Tamarixia radiata* (Waterson)	Gifford (1964), Michaud (2002), Hall et al. (2008), and Navarrete et al. (2014)
127.	*Zelus luridus* Stål	*Drosophila melanogaster* Meigen	Weirauch (2006)
128.	*Zelus renardii* Kolenati	*Chlorochroa ligata* (Say), *Conchuela, Euschistus impictiventris* Stål, *Euschistus, H. ivrescens, L. leucocephala, S. exigua, Sirtothrips citri* Moulten, bollworm eggs, defoliating caterpillar, nymphs of brown stink bugs, and nymphs of stink bug	Morrill (1910), Horton (1918a,b), Readio (1927), Clancy (1946), Ables (1978), Cohen (1990), Zanunico et al. (1994), and Uthamasamy (1995)
129.	*Zelus socius* Uhler	*Circulifer tenellus* (Baker), beet leafhopper, and bollworms	Zanunico et al. (1994)
130.	*Zelus tetracanthus* Stål	*B. thurberiella, Eutettix tenella* Barker, and *H. zea*	Whitcomb and Bell (1964) and Severein (1924)
131.	*Zelus* sp.	*E. tenella*	Severein (1924)
132.	*Zelus* sp.	*Alabama argillacea* Hübner	Gravana and Sterling (1983)
133.	*Zelus* sp.	*A. biguttula*	Singh et al. (1987)
134.	*Zelus* sp.	*Anticarsia gemmatalis* Hübner	Godfrey et al. (1989)
135.	*Zelus* sp.	*S. exigua*	Ruberson et al. (1994)
V. HOLOPTILINAE			
136.	*Ptilocerus ochraceus* Montandon	*Dolichoderus* (*Hypoclinea*) *bituberculatus* (Mayr)	Cade et al. (1978)
137.	*Ptilocerus venosus* (Walker)	*Hypoclinea bituberenlata* (Meyer)	Weaver (1978)
VI. PEIRATINAE			
138.	*Catamiarus brevipennis* (Serville)	*A. janata, D. cingulatus, E. insulana, H. armigera, M. pustulata, P. succincta*, and pigeon pea pests	Bhatnagar et al. (1983), Pawar et al. (1986), Ambrose (1987), Sahayaraj (1991), and Ambrose and Claver (2001a)
139.	*Ectomocoris atrox* Stål	Pests of rice	Ambrose (2007)
140.	*Ectomocoris biguttulus* Stål	Pests of rice	Ambrose (2007)

TABLE 1 Records of Reduviid Predators and Their Insect Pest Prey Species—cont'd

S.No	Reduviid species	Pest prey species	References
141.	*Ectomocoris cordatus* Wolff	*Dieuches* sp.	Thangavelu and Ananthasubramanian (1981)
142.	*Ectomocoris cordiger* Stål	*E. machaeralis, H. armigera*, and *P. machaeralis*	Misra (1975)
143.	*Ectomocoris ochropterus* Stål	*H. armigera*	Ambrose (1999)
144.	*Ectomocoris quadriguttatus* F.	*H. armigera*	Ambrose (1999)
145.	*Ectomocoris tibialis* (Distant)	*A. janata, C. cephalonica, C. compressus, D. cingulatus, E. vitella, H. armigera, M. estherae, M. pustulata, P. gossypiella*, and *S. litura*	Ambrose (1980, 1987), Ambrose and Sahayaraj (1990), and Sahayaraj (1991)
146.	*Ectomocoris xavieri* Vennison and Ambrose	*H. armigera*	Vennison (1988)
147.	*Eumenes maxillosus* Deg	*H. armigera*	Taylor (1932)
148.	*Lestomerus affinis* (Serville)	*E. insulana, H. armigera, Omphora atrata* Klug, and *Omphora pilosa* Klug	Haridass and Ananthakrishnan (1980), Ambrose (1987), and Evangelin et al. (2014)
149.	*Melanolestes picipes* (Herrich-Schaefer)	May beetles and their grubs, and June bugs	Readio (1927)
150.	*Peirates* sp.	*C. pomonella*	Pettey (1919)
151.	*Rasahus hamatus* (F.)	*Scapteriscus borelli* Giglio-Tos	Flower et al. (1987)
152.	*Sirthenea carinata* (F.)	*Scapteriscus* spp.	Flower et al. (1987)
153.	*Sirthenea flavipes* (Stål)	*H. armigera* and pests of rice	Ambrose (2007)
154.	*Sirthenea* sp.	*S. borelli*	Flower et al. (1987)
155.	*Sirthenea stria* (F.)	*Scapteriscus* spp.	Ambrose (2007)
156.	*Spilodermus quadrinotatus* (F.)	*H. armigera*	Ambrose (2007)

VII. REDUVIINAE

S.No	Reduviid species	Pest prey species	References
157.	*Acanthaspis pedestris* (Stål)	*A. janata, C. cephalonica, Chilo paretellus, E. insulana, E. vitella, H. armigera, M. estherae, M. indica, M. pustulata, O. assumthi, O. hyalinipennis, Odontotermes wallonensis* Weisman, *P. gossypiella, P. succinct*, and *S. litura*	Butani (1958), Rajagopal (1984), Ambrose (1987, 1988), Sahayaraj (1991), Ambrose (2003), and Ravichandran (2004)

(Continued)

TABLE 1 Records of Reduviid Predators and Their Insect Pest Prey Species—cont'd

S.No	Reduviid species	Pest prey species	References
158.	*Acanthaspis quinquespinosa* (F.)	*C. cephalonica, C. compressus, D. koengii, Dysdercus laetus* Kirby, *H. armigera, S. exigua, M. pustulata, O. obesus, O. wallonensis, P. gossypiella,* and *S. litura*	Butani (1958), Ambrose (1980, 1987, 1988), Rajagopal and Veeresh (1981), Lakkundi (1989), and Sahayaraj (1991)
159.	*Acanthaspis siva* (Distant)	*C. cephalonica* and *C. compressus*	Ambrose, 1980 and Lakkundi, 1989
160.	*Acanthaspis subrufa* (Distant)	*C. partellus* and *O. wallonensis*	Rajagopal (1984)
161.	*Acanthaspis* sp.	*C. cephalonica, O. obesus,* and *S. litura*	Sahayaraj (1991)
162.	*Alloeocranum guadrisignatum* (Reuter)	*D. cingulatus, O. hyalinipennis,* and *O. obesus*	Sahayaraj (1991) and Ambrose (2007)
163.	*Edocla slateri* (Distant)	*C. compressus, E. insulana,* and *H. armigera*	Ambrose (1980, 1987)
164.	*Empyrocoris annulata* (Distant)	*C. compressus, E. insulana, O. assumthi, P. gossypiella,* and *S. litura*	Ambrose (1980) and Kumaraswami (1991)
165.	*Opithacidius pertinax* Breddin	*Periplaneta americana* L.	Pinto (1927a,b)
166.	*Pasira perpusilla* (Walker)	*Elasmolomus lineosus* (Distant)	Thangavelu and Ananthasubramanian (1981)
167.	*Peregrinator biannulipes* (Mont. and Sign.)	*Anagasta kuehniella* (Zeller), *C. cephalonica,* and *T. confusum*	Awadallah et al. (1984)
168.	*Platymeris rhadamanthus* Gerstaecker	*P. americana*	Edwards (1961)
169.	*Platymeris laevicollis* Distant	*Oryctes boas* (F.), *Oryctes monocerous* (Oliv.), *Oryctes rhinocerous* L., and *P. americana*	Vanderplank (1958), Nayar et al. (1976), and Antony et al. (1979)
170.	*Reduviolus* sp.	*Calocoris angustatus* Letheirry	David and Ramamurthry (2011)
171.	*Reduvius personatus* (Linn.)	*Paralipsa gularis* Zeller and domestic pests	Readio (1927) and Trematterra and Tozzia (1985)
172.	*Reduvius* sp.	*C. angustatus*	Nayar et al. (1976) and David and Kumaraswami (1988)
173.	*Reduvius* sp.	*C. cephalonica*	Mukerjee (1971)
174.	*Velitra sinensis* (Walker)	*P. succincta*	Ambrose (1987)

TABLE 1   Records of Reduviid Predators and Their Insect Pest Prey Species—cont'd

S.No	Reduviid species	Pest prey species	References
175.	*Zelurus angularis* (Stål)	*Leptinopterus femoratus* (Olivier) and some species of Triatominae	Grossi et al. (2012)
VIII. SALYAVATINAE			
176.	*Paralisarda malabarica* Miller	*O. obesus*	Ambrose (2003)
IX. STENOPODAINAE			
177.	*Oncocephalus annulipes* (Stål)	*E. insulana, H. armigera, O. obesus,* and *S. litura*	Ambrose (1987) and Vennison (1988)
178.	*Oncocephalus impudicus* Reuter	*E. machaeralis, H. armigera,* and *P. machaeralis*	Misra (1975), Patil and Thontadarya (1983), and Ambrose (1999)
179.	*Oncocphalus modestus* (Reuter)	*H. armigera*	Ambrose (1999)
180.	*Oncocephalus notatus* (Klug)	*H. armigera*	Ambrose (1999)
181.	*Staccia diluta* (Stål)	*H. armigera* and pests of rice	Ambrose (1999, 2007)
X. TRIBELOCEPHALINAE			
182.	*Opistoplatys* sp.	*Cydia leucostoma* Myrick	Muraleedharan et al. (1988)
183.	*Tribelocephala indica* (Walker)	*H. armigera*	Ambrose (1999)

prey deprivation, competition, and space on the predation have been investigated (see Ambrose, 1999). Experimentally antennectomized, eye-blinded, and tibial pad or tibial comb-coated reduviids exhibited delayed arousal, approach, capturing, sucking responses, and exhibited shorter postpredatory activities (Claver and Ambrose, 2001a). The female reduviids, which are comparatively larger than males with higher food requirements, exhibited quicker predatory acts. The camouflaged nymphal instars prey more efficiently than the experimentally decamouflaged nymphal instars by gaining deception toward their prey. Prey species influenced not only the predatory behavior but also the prey consumption (Ravichandran, 2004; Nagarajan, 2010; Kumar, 2011; also see Ambrose, 1999). Prey deprivation up to a certain period, intraspecific competition up to an optimum level (both varies from species to species) (Das et al., 2009, 2010; Kumar et al., 2009), and reduction in the predation arena also promote predation (see Ambrose, 1999, 2000a, 2003).

3.3 Defense and Offense

Assassin bugs have an array of defensive and offensive behaviors, accompanied by morphological adaptations. These adaptive protective behaviors threaten and help to escape from their enemies and larger prey and gain protection from cannibalism.

Camouflaging of nymphs has been found in Catherinae, Reduviinae, and Salyavatinae. The nymphs use both their hind tarsomeres as shovels and actively throw and accumulate sand and debris. They cover the entire body except antennal flagellum, rostrum, the ventral surface of head, prosternum, and tibial pads with sand and debris. The fine felt-like dorsal hairs and serrated cuticle hold the camouflaging materials. The hypodermal cells secrete a sticky material, which hardens onto fine silvery filaments that intertwine with the camouflage material. Camouflaging efficiency (camouflage material carrying capacity) decreases as nymphs grow older and disappears in adults. Camouflaging also aids to prey capture by deceiving the prey and gaining protection from enemies and cannibalism (Brandt and Mahsberg, 2002; also see Ambrose, 1999, 2000a).

The harpactorine instars often feign death. The harpactorine and ectrichodiine instars, by withdrawing their cephalic and thoracic appendages, also roll their body into a ball and lie motionless. Nodding of the head and rubbing of the rostrum against the transversely striated prosternal groove to produce a characteristic sound is better developed in reduviids with acutely curved rostrums (Peiratinae) than in reduviids with straight rostrum, which produce an obscure sound. Extension of the rostrum, spitting of watery saliva, and stinging also occur. Emission of a volatile secretion with a pungent odor from the dorsolateral abdominal scent gland or Brindley's gland is well pronounced in Harpactorinae and Reduviinae (see Ambrose, 1999, 2000a).

A combination of Batesian and Mullerian mimicry between lygaeids and similarly appearing reduviids provides protection from vertebrate predators. *Rhaphidosoma* species (Harpactorinae) and stenopodaines have elongate slender bodies that resemble closely with the elongate slender grass and sledge stems and leaves among which they live (Quicke et al., 1992; Wappler et al., 2013; also see Ambrose, 1999, 2000a).

3.4 Prey Preference

Although reduviids feed on a wide variety of arthropods (generalist predators), they exhibit prey and stage preferences. The ectrichodiines feed exclusively on millipedes; the peiratines prefer beetles and grasshoppers; the harpactorines usually prey upon soft prey, such as caterpillars, grubs, and termites; the reduviines prefer ants, bees, and termites; the holoptilines, stenopodaines, salyavatines, saicines, and tribelocephalines prefer ants, termites, and blattids and the emesines prefer flies (Ambrose, 2003; Hwang and Weirauch, 2012; also see Ambrose, 1999, 2000a).

Thus, the prey preference of reduviids is determined by the size, shape, texture, and chemical nature and defensive functions of the prey and the size and offensive functions of the predators (George, 1999b, 2000c; Grundy and Maelzer, 2000a; Ambrose and Claver, 2001b; Ravichandran, 2004; Nagarajan, 2010; Balakrishnan et al., 2011; Kumar, 2011; Muniyandi et al., 2011; also see Ambrose, 1999, 2000a).

The size of the predator and prey determine the stage preference of reduviids. Stage preference studies by capture success and choice experiments reveal that the stage preference is greater in younger instars, which gradually fades as they grow (Sahayaraj, 1999a; Grundy and Maelzer, 2000a; Cogni et al., 2002; Ravichandran, 2004; Nagarajan, 2010; Balakrishnan et al., 2011; Kumar, 2011; Muniyandi et al., 2011; Sahayaraj et al., 2012; also see Ambrose, 1999, 2000a).

4. BIOCONTROL

4.1 Functional and Numerical Responses

Reduviids positively respond to the increasing prey density by effectively suppressing them with decreased searching time and increased discovery, attack, and predation rate. The functional response of reduviids confirms type II model of Holling (1959). The age, sex, and size of the predators and the size of the prey also determine functional response. For instance, the female predators are more vigorous in responding to increasing prey density. Handling time of the predator decreases with decreasing prey size (Sahayaraj, 2000; Ravichandran et al., 2003; Ravichandran, 2004; Ravichandran and Ambrose, 2006; Ambrose et al., 2008b,c, 2009, 2013; Ambrose and Nagarajan, 2010; Nagarajan, 2010; Nagarajan et al., 2010b; Balakrishnan et al., 2011; Kumar, 2011; Manimuthu et al., 2011; Muniyandi et al., 2011; Sahayaraj et al., 2012; also see Ambrose, 1999, 2000a, 2003).

Reduviid predators respond to prey increase densities by increasing their own population. This response of predator is termed as numerical response. Reduviids exhibit positive numerical response in two ways. They kill a greater number of prey in terms of available prey population per predator at a given time and they increase their population size in response to higher prey density (Holling, 1966; Claver, 1998; Ravichandran, 2004; Nagarajan, 2010; Kumar, 2011; Claver and Ambrose, 2012; also see Ambrose, 1999, 2000a, 2003).

4.2 Important Species

The biocontrol potential of four species of Harpactorinae, namely *Rhynocoris fuscipes* (Fabricius), *Rhynocoris kumarii* Ambrose and Livingstone, *Rhynocoris longifrons* (Stål) and *R. marginatus* Fabricius; a peiratine reduviid, *Ectomocoris tibialis* (Distant); and a reduviine species, *A. pedestris* (Stål) have been investigated against a wider array of insect pests of cotton, vegetables, castor, groundnut, and cereals in India (Ravichandran, 2004; Nagarajan, 2010; Kumar, 2011; also see Ambrose, 1999, 2000a, 2003).

Despite many reduviids were identified preying upon a wider array of insect pests, insufficient consistent investigations on their distribution and their biology have been carried out. One such important group of reduviid predators neglected is *Phonoctonus* spp. preying upon species of *Dysdercus* Guerin Meneville and *Odontopus* Silbermann. Other promising groups of reduviid predators are the species of *Sinea* and *Zelus*, which regulate insect pest complex of cotton (see Ambrose, 1999, 2000a, 2003).

4.3 Conservation and Augmentation

Reduviids have to be conserved and augmented to effectively use them as biocontrol agents of in IPM various pest species on various crops. Hence, mass rearing of reduviid predators for augmentative release is imperative. However, it possesses numerous challenges, such as the cost of maintaining insect prey cultures and the risk of prey shortages due to disease or facility mismanagement (Cohen, 1985; DeClercq and Degheele, 1992; Grundy et al., 2000a; Ambrose, 2010). The artificial diets have not been successfully substituted for prey insects of reduviid predators. The artificial diets for predatory reduviids are often difficult to develop because of the constraints of including chemical cues that encourage feeding and the necessary ingredients for sustained growth (Berlinger, 1992). Cannibalism is another important impediment to mass rearing of generalist predators, like reduviids (DeClercq and Degheele, 1992; George and Ambrose, 2000a; Grundy et al., 2000a; Ambrose, 2010). Mortality due to cannibalism is further aggravated by crowded rearing conditions (Ambrose, 2010; George and Ambrose, 2000a; Grundy et al., 2000a).

Reduction in fecundity, longevity, and prey searching ability are other obstacles in mass rearing due to inbreeding (Ambrose, 2010). George et al. (2007) attempted the ecotype selection of *A. pedestris* for mass multiplication using unweighted pair group method with arithmetic mean analysis. The selective breeding of distantly-related ecotypes of reduviid species reduces the loss of genetic variability due to inbreeding during mass rearing. The mass reared insects are often more susceptible to pathogens due to stress of poor diet, constant temperature, unsuitable relative humidity, and crowding (Soares, 1992). Thus incidence of disease occurrence is another major impediment (Helms and Raun, 1971). Another major constraint on mass rearing of reduviids is cannibalism that causes high mortality. Hence, the nymphs have to be reared individually, which makes the mass rearing costly as well as impractical (Ambrose, 2010). The reduviid predators strike at moving prey and therefore they are provided with live prey to facilitate predation (Ambrose, 2010; Grundy et al., 2000a). The factors that influence the biology of reduviids should be taken into consideration while mass rearing them.

In India, five reduviid predators, namely, *R. marginatus*, *R. fuscipes*, *R. kumarii*, *E. tibialis*, and *A. pedestris*, have been successfully mass reared in the authors' laboratory on head-crushed larvae of rice meal moth, *Corcyra cephalonica*, in order to avoid the entangling of reduviids in the webbing of the larvae in large plastic containers (7×15 cm). The containers were provided with small stones and leaves simulating their concealed microhabitats and covered with muslin cloth. Mass rearing of reduviids in containers simulating environmental conditions using materials, such as plant shoots, leaves, and sand with stones as substrata increased their predatory value, conversion rate, and body weight and reduced the postembryonic development period; quickened oviposition; increased fecundity and promoted female-biased sex ratio (Ambrose and Claver, 1999; Ambrose, 2000b, 2002, 2010; Claver and Ambrose, 2001b; Ravichandran, 2004; Kumar et al., 2009; Nagarajan, 2010; Kumar, 2011).

In Australia, assassin bug, *Pristhesancus plagipennis* (Walker), was successfully mass reared on hot water killed larvae of yellow mealworm, *Tenebrio molitor* (L.), and American bollworm, *Helicoverpa armigera* (Hübner) (Grundy et al., 2000a). Their rearing method minimizes cannibalism, avoids the need for live insect prey, and is space and labor efficient. Between the two prey provided the larva of *T. molitor* was the most suitable prey for minimizing postembryonic developmental period and mortality and producing adults with higher body weight. Using this method, 20–27 nymphs can be mass reared in a 5l container. This optimum rearing density reduces the nymphal mortality to 16–22%. The optimum predator density of oviposition is 16 adults per 5l container. They mass reared up to 100,000 predators per year (Grundy et al., 2000a).

4.4 Release, Synchronization, and Stabilization

The augmentative release of *R. kumarii*, *R. fuscipes*, *R. longifrons*, and *A. pedestris* in cotton agroecosystems in India significantly reduced the population of cotton stainer, *Dysdercus cingulatus* (Fabricius), *H. armigera*, and *Spodoptera litura*. Their augmentative release uniformly reduced the plant damage in cotton agroecosystems, enhanced the yield of quality of cotton and seed cotton and did not alter the population of other arthropod predators (Claver and Ambrose, 2001c,d; Ravichandran, 2004; Nagarajan, 2010; Kumar, 2011).

In Australia, Grundy and Maelzer (2000b) released third instar nymphs of *P. plagipennis* in cotton and soybean agroecosystems and recorded its potential to control *Helicoverpa* spp. larvae in cotton and green mirid, *Creontiades dilutes* (Stål), and looper caterpillars, *Chrysodeixis* spp., in soybean fields. The release of *P. plagipennis* increased the potential yield of cotton and soybean. Field release and recovery studies also indicated their ability in different agroecosystems to synchronize (Ravichandran, 2004; Nagarajan, 2010; Kumar, 2011).

Provision of suitable harboring materials or refugia for predators and intercropping for the sustenance of insect pests enhanced the synchronization and stabilization of released reduviid predators (Claver, 1998; Claver and Ambrose, 2001d; Sahayaraj, 2001; Claver et al., 2003a). For instance, the provision of different mulches, such as sorghum trash, caryota leaves, palmyra leaves, banana leaf, coconut spathe, and stone in cotton agroecosystems increased the predator abundance, decreased the pest population and pod damage, and enhanced the yield of cotton (Claver and Ambrose, 2003; Claver et al., 2003a; Ravichandran, 2004; Nagarajan, 2010; Kumar, 2011). Although, augmentation and conservation of reduviids have been attempted in a small scale, further studies have to be encouraged to achieve the economical large-scale mass rearing and augmentative release with special reference to the role of substrata on mass rearing (Sahayaraj, 1999b, 2002; Grundy and Maelzer, 2000b; Claver and Ambrose, 2001c; Sahayaraj and Martin, 2003; Grundy, 2004; Ambrose, 2010).

4.5 Chemical Ecology

Knowledge on behavioral dynamics of predators to chemical cues of the prey is imperative to promote the utilization of assassin bugs in IPM. Hence, their pheromonal and kairomonal ecology have to be studied to understand the intricate mechanisms governing the prey preference of the predators and the predators' prey suppression efficacy (Maran, 1999; Sahayaraj and Paulraj, 2001; Lenin et al., 2011; Nagarajan and Ambrose, 2013; Kumar and Ambrose, 2014). Similarly, understanding pheromonal ecology of reduviids will enable to effectively mass-trap the potential predators from the agroecosystem and to utilize them in IPM (Aldrich, 1998, 1999).

Preliminary studies on allelochemical interaction of reduviids, such as *A. pedestris*, *R. fuscipes*, *R. longifrons*, *R. kumarii*, and *R. marginatus* and their prey by monitoring the prey location of these predators to the hexane extracts of prey species have been attempted. The studies uniformly reveal the preferences for soft cuticle prey, such as caterpillar, over hard-cuticled hemipteran and coleopteran prey species. The, reduviids preferred prey species without defensive secretions and offensive characteristics (Ravichandran, 2004; Sahayaraj, 2008; Nagarajan, 2010; Kumar, 2011; Lenin et al., 2011; Sujatha et al., 2012; Nagarajan and Ambrose, 2013; Kumar and Ambrose, 2014).

4.6 Biochemistry of Saliva

Biochemistry of saliva of reduviids, especially toxic potential, enables to evaluate their pest suppression efficacy (Ambrose, 1999, 2002; Ambrose and Maran, 1999, 2000a,b;

Maran, 1999; Maran and Ambrose, 2000; Morey et al., 2006; Sahayaraj et al., 2007, 2013; Maran et al., 2011). Studies on biochemistry and composition of their venomous saliva are meager and confined to preliminary investigations in the species *Peirates affinis* Serville (Edwards, 1960), *Platymeris rhadamanthus* Gerstaecker (Edwards, 1961), *Holotrichius innesi* Horrvath (Zerachia et al., 1973), *A. pedestris* (Morrison, 1989), *Zelus renardii* Kolenati (Cohen, 1993), *Haematorrhophus nigroviolaceous* Reuter (Ambrose, 1999), *Peirates turpis* Walker, *Agriosphordus dohrni* Stål, *Isyndus obscurus* Dallas (Gerardo et al., 2001), *Catamiarus brevipennis* Serville, and *R. marginatus* (Sahayaraj et al., 2006). The detailed studies on saliva in predatory reduviids may be helpful to utilize them as effective biocontrol agents and to explore and formulate novel salivary toxins as insecticides. However, the major problem is that reduviids have a meager quantity of saliva owing to their smaller size of salivary glands and the difficulty in getting the required amount of saliva for analysis (Maran et al., 2011; Sahayaraj et al., 2013).

4.7 Biocontrol Potential Evaluation

The biocontrol potential of several reduviids has been investigated (Grundy et al., 2000b; Sahayaraj, 2014; also see Ambrose, 1999, 2003). Pest suppression by reduviids was evaluated by comparing the mortality of pests in field cages with and without predators and in terms of reduction in leaf, flower, and boll damage. Certain species, like *R. fuscipes*, *R. kumarii*, *R. longifrons*, *R. marginatus*, and *A. pedestris*, suppressed various pests species, such as *D. cingulatus*, *H. armigera*, *S. litura*, and *Mylabris indica* Thunberg in cotton agroecosystems even at low prey and predator densities suggesting that they could be employed in pest management (Claver, 1998; Ravichandran, 2004; Nagarajan, 2010; Kumar, 2011). The evaluation of *R. marginatus* on groundnut pests and aphids (Sahayaraj, 2000, 2003) and of *R. kumarii* on *H. armigera*, and *Anomis flava* (Fabricius) further supported their biocontrol potential (Ambrose, 2000c). Pest suppression efficacy varies from predator to predator and from pest to pest. The reduviids kill more number of immatures than adults and their differential biocontrol potential is correlated to their searching behavior, which is governed by their dispersal ability (Ambrose, 1999, 2000a; Claver et al., 2002).

5. INSECTICIDAL IMPACT

Many insecticides used in agroecosystems not only affect the target pests but also the nontarget beneficial organisms, i.e., natural enemies including assassin bugs. Almost all the tested insecticides prolonged the postembryonic development, altered the sex ratio, and reduced the size and the weight of life stages. They also altered the fecundity and longevity, prey consumption, haphazard movement, loss of orientation, and restlessness, and caused altered hemogram; transformation of hemocytes, accelerated mitosis and reduced carbohydrate and protein. The lipid and water contents were increased, alimentary canal, testes and ovaries were disintegrated, and sperm count was reduced with distorted sperm (Ambrose, 2003).

The studies on the impact of insecticides on reduviids reveal that they did affect the morphology, behavior, predatory efficiency, body weight, longevity, fecundity, life–table parameters, physiology, biocontrol potential, and biochemistry (Claver et al., 2003b; George and Ambrose, 2004b; Ambrose et al., 2006b, 2008d, 2010a,b). Hence, it is recommended that "soft and ecofriendly" insecticides should be used with reduviid predators in IPM (George and Ambrose, 1998, 1999, 2000b, 2001, 2004c; Ambrose, 2001; Ghelani et al., 2000; Grundy et al., 2000b; also see Ambrose, 1999, 2000a). Polymorphic adaptation of reduviids to insecticides was also reported (George and Ambrose, 2001; see Ambrose, 1999). Though botanicals are relatively safe to reduviids (Sahayaraj, 2001; Sujatha et al., 2012), certain botanicals also affect them (Sahayaraj and Paulraj, 1999a, b; Essakiammal, 2011; Kumari, 2011; Maniyammal, 2011). Therefore, screening of insecticides and botanicals is imperative. However, the field dosage of insecticides, such as dimethoate, quinalphos, and endosulfan, were found safe to *R. fuscipes* and *A. pedestris*, which also promoted the postembryonic development, i.e., hormoligosis (Ambrose and George, 1995, 1996b; Ambrose, 2001)

6. REDUVIIDS AS GENERAL PREDATORS

Because reduviids are generalist predators, they may attack beneficial arthropods and few reduviids appear to specialize on beneficial insects. For instance, stingless bees, *Trigona* (Jurine), are useful pollinators. Several apiomerines wait for these bees at flowers or at sites where the bees visit to collect the resin for their nest construction and prey upon them. They may smear the resin from the site on their forelegs to seize the prey. The kairomones of apiomerines also attract the bees (see in Pionar, 1992). Fossil "resin bugs" and stingless bees have been found together in Dominican amber from at least 25 million years ago indicating their old relationship (see Ambrose, 2000a).

In India, *Acanthaspis siva* (Distant), preys upon the Indian honey bee, *Apis indica* F., and *P. plagipennis* on the other honey bees in their hives (McKeown, 1942; Ambrose, 2000a). Another bug, *Sycanus collaris* (Fabricius) preys upon the tasar silkworm, *Antheraea mylitta* D. (David and Ramamurthy, 2011).

The role of "generalist" and "specialist" reduviid predators against insect pests needs to be studied. Some generalists may attack certain specialist predators, with potentially negative effects on pest control. Hence, a proper assessment of role of reduviid predators in the regulation of insect pests in diverse crop systems and management of environment and habitat to increase reduviid predators need attention. The reduviids are known to prey upon ladybird beetles in cotton fields and the generalist, *Z. renardii* (both nymphs and adults) preying upon lacewing, *Chrysoperla carnea* (Stephens), in cotton fields in the San Joaquin Valley (see Ambrose, 2000a). *Zelus renardii* imposes heavy mortality (up to 10%) on *C. carnea* and thereby renders it ineffective as biocontrol agent. The decreased lacewing survival has been attributed to predation by reduviids than competition for aphid prey (Rosenheim and Wilhoit, 1993; Rosenheim et al., 1993; Cisenros and Rosenheim, 1997). *Zelus renardii* not only influences the prevalence of intraguild predation but also the intensity of disruption of aphid biocontrol; while, none of their immature stages was found effective biocontrol agent of cotton aphid. Though such situation has not yet been recorded in India; a proper assessment of the role of reduviids predators in the regulation of insect pests in diverse crop systems and in the management of environment and habitat is necessary and important (Ambrose, 2002; Ravichandran, 2004; Nagarajan, 2010; Kumar, 2011; also see Ambrose, 1999, 2000a).

7. CONCLUSION

The worldwide distribution, abundance, diversity, pest prey records, amenability to mass rearing, synchronization, and absence of parasites/parasitoids and predators are the merits of reduviids as biocontrol agents. A holistic approach to their biosystematics and ecology with subsequent efforts on economical mass rearing, augmentative release, and biocontrol potential evaluation should be undertaken to realize their biocontrol potential.

Acknowledgments

The authors are grateful to the authorities of St. Xavier's College (Autonomous), Palayamkottai, Tamil Nadu, India for facilities and to the Council of Scientific and Industrial Research (CSIR), Government of India, New Delhi, for financial assistance (Ref. No. 21(0865)/11/EMR-II, 2012–2013 dated 28.12.2011).

References

Abasa, R.O., 1981. *Harpactor tibialis* Stål (Hemiptera: Reduviidae) insect pests and their arthropod parasites and predators. In: Heinrichs, E.A. (Ed.), Biology and Management of Rice Insects. Wiley Eastern Ltd., New Delhi, India, pp. 13–359.

Ables, J.R., 1978. Feeding behaviour of an assassin bug, *Zelus renardii*. Annals of the Entomological Society of America 71, 476–478.

Aldrich, J.R., 1998. Status of semiochemical research on predatory Heteroptera. In: Coll, M., Ruberson, J.R. (Eds.), Predatory Heteroptera: Their Ecology and Use in Biological Control. Entomological Society of America, Lanham, Maryland.

Aldrich, J.R., 1999. Predators. In: Hardie, J., Minks, A.K. (Eds.), Pheromons of Non-lepidopteran Insects Associated with Agricultural Plants. CAB International Publication.

Amaral, B.F., Gióia, I., Waib, C.M., Mendeleck, E., Cônsoli, F.L., 1994. Observations on the biology of *Apiomerus lanipes* (Fabricius) (Hemiptera, Reduviidae). Revista Brasileira de Zoologia 11 (2), 283–288.

Ambrose, A.D., Ambrose, D.P., 2003. Linear regression coefficient (r) of post-embryonic developmental morphometry as a tool in the biosystematics of Reduviidae (Insecta: Hemiptera). Shashpa 10 (1), 57–66.

Ambrose, A.D., Ambrose, D.P., 2004. Post-embryonic developmental growth index as a tool in the biosystematics of Reduviidae (Insecta: Hemiptera). Shashpa 11 (2), 105–115.

Ambrose, A.D., Ambrose, D.P., 2009a. Morphometric indices in the biosystematics of Reduviidae (Insecta: Hemiptera). Indian Journal of Entomology 71 (1), 18–28.

Ambrose, D.P., Ambrose, A.D., 2009b. Predation, copulation, oviposition and functional morphology of tibia, rostrum and eggs as tools in the biosystematics of Reduviidae (Hemiptera). Indian Journal of Entomology 71 (1), 1–17.

Ambrose, D.P., Claver, M.A., 1995. Food requirement of *Rhynocoris kumarii* Ambrose and Livingstone (Heteroptera, Reduviidae). Journal of Biological Control 9 (1), 47–50.

Ambrose, D.P., Claver, M.A., 1999. Substrata impact on mass rearing of the reduviid predator *Rhynocoris marginatus* (Fabricius) (Insecta: Heteroptera: Reduviidae). Pakistan Journal of Biological Sciences 2 (4), 1088–1091.

Ambrose, D.P., Claver, M.A., 2001a. Survey of reduviid predators in seven pigeonpea agroecosystems in Tirunelveli, Tamil Nadu, India. International Chickpea and Pigeonpea Newsletter 8, 44–45.

Ambrose, D.P., Claver, M.A., 2001b. Prey preference of the predator *Rhynocoris kumarii* (Heteroptera: Reduviidae) to seven cotton insect pests. Journal of Applied Zoological Research 12 (2&3), 129–132.

Ambrose, D.P., George, P.J.E., 1995. Relative toxicity of three insecticides to Acanthaspis pedestris Stål, a potential predator of insect pests (Insecta: Heteroptera: Reduviidae). Journal of Advanced Zoology 16 (2), 88–91.

Ambrose, D.P., George, P.J.E., 1996a. Impact of flooding on hatching in three *Rhynocoris* reduviid predators (Insecta: Heteroptera: Reduviidae). Fresenius Environment Bulletin 5, 185–189.

Ambrose, D.P., George, P.J.E., 1996b. Effect of monocrotophos, dimethoate and methylparathion on the differential and the total haemocyte counts of *Acanthaspis pedestris* Stål (Insecta: Heteroptera: Reduviidae). Fresenius Environment Bulletin 5, 190–195.

Ambrose, D.P., Livingstone, D., 1987a. Biology of *Acanthaspis siva* Distant, a polymorphic assassin bug (Insecta, Heteroptera, Reduviidae). Journal of Mitter and Zoological Museum, Berlin 63 (2), 321–330.

Ambrose, D.P., Livingstone, D., 1987b. Biology of a new harpactorine assassin bug *Rhinocoris kumarii* (Hemiptera – Reduviidae) in South India. Journal of Soil Biology and Ecology 7 (1), 48–58.

Ambrose, D.P., Livingstone, D., 1991. Biology of *Neohaematorrhophus therasii* (Hemiptera: Reduviidae). Colemania Insect Biosystematics and Ecology 1 (1), 5–9.

Ambrose, D.P., Maran, S.P.M., 1999. Quantification, protein content and paralytic potential of saliva of fed and prey deprived reduviid *Acanthaspis pedestris* Stål (Heteroptera: Reduviidae: Reduviinae). Indian Journal of Environmental Sciences 3 (1), 11–16.

Ambrose, D.P., Maran, S.P.M., 2000a. Polymorphic diversity in salivary and haemolymph proteins and digestive physiology of assassin bug *Rhynocoris marginatus* (Fab.) (Het., Reduviidae). Journal of Applied Entomology 124, 315–317.

Ambrose, D.P., Maran, S.P.M., 2000b. Haemogram and haemolymph protein of male and female (mated and oviposited) *Rhynocoris fuscipes* (Fabricius) (Heteroptera: Reduviidae: Harpactorinae). Advances in Biosciences 19 (II), 39–46.

Ambrose, D.P., Nagarajan, K., 2010. Functional response of *Rhynocoris fuscipes* (Fabricius) (Hemiptera: Reduviidae) to teak skeletonizer *Eutectona machaeralis* Walker (Lepidoptera: Pyralidae). Journal of Biological Control 24 (2), 175–178.

Ambrose, D.P., Paniadima, A., 1988. Biology and behaviour of a harpactorine assassin bug *Sycanus pyrrhomelas* Walker (Hemiptera: Reduviidae) from South India. Journal of Soil Biology and Ecology 8 (1), 37–58.

Ambrose, D.P., Ravichandran, B., 2007. Redescription and postembryonic development of *Paralisarda malabarica* Miller (Reduviidae: Salyavatinae). Entomologia Croatica 11 (1–2), 53–61.

Ambrose, D.P., Sahayaraj, K., 1990. Effect of space on the postembryonic development and predatory behaviour of *Ectomocoris tibialis* Distant (Insecta: Heteroptera: Reduviidae). Uttar Pradesh Journal of Zoology 10 (2), 163–170.

Ambrose, D.P., Vennison, S.J., 1989. Biology of a migratory assassin bug, *Polididus armatissimus* Stål (Insecta, Heteroptera, Reduviidae). Arquivos Do Museu Bocage 1 (17), 271–280.

Ambrose, D.P., Gunaseeli, I.M., Vennison, S.J., 1988. Impact of female parental age on the development and size of offspring of assassin bug *Rhinocoris kumarii*. Environment and Ecology 6 (4), 948–955.

Ambrose, D.P., George, P.J.E., Das, S.S.M., 1994. Impact of flooding on the hatchability of *Acanthaspis pedestris* Stal eggs (Insecta: Heteroptera: Reduviidae). Journal of Soil Biology and Ecology 14 (2), 136–139.

Ambrose, D.P., Kumar, S.P., Subbu, G.R., Claver, M.A., 2003. Biology and prey influence on the postembryonic development of *Rhynocoris longifrons* (Stål) (Hemiptera: Reduviidae), a potential biological control agent. Journal of Biological Control 17 (2), 113–119.

Ambrose, D.P., Kumar, S.P., Nagarajan, K., Das, S.S.M., Ravichandran, B., 2006a. Redescription, biology, life table, behaviour and ecotypism of *Sphedanolestes minusculus* Bergroth (Hemiptera: Reduviidae). Entomologia Croatica 10 (1–2), 47–66.

Ambrose, D.P., Rajan, S.P.N., Ravichandran, B., 2006b. Impact of deltamethrin on the biology and life-table parameters of a non-target biological control agent *Rhynocoris marginatus* (Fabricius) (Hemiptera: Reduviidae). Indian Journal of Environment and Toxicology 16 (2), 48–54.

Ambrose, D.P., Krishnan, S.S., Jebasingh, V., 2007a. An annotated checklist of Indian Peiratinae (Hemiiptera: Reduviidae) with ecological and morphological characteristics. Biosystematica 1 (1), 45–57.

Ambrose, D.P., Gunaseelan, S., Krishnan, S.S., Jebasingh, V., Ravichandran, B., Nagarajan, K., 2007b. Redescription, biology and behaviour of a harpactorine assassin bug *Endochus migratorius* Distant. Hexapoda 14 (2), 89–98.

Ambrose, D.P., Nambirajan, S.P., Ravichandran, B., 2007c. Impact of cypermethrin on the biology and lifetable parameters of a nontarget biological control agent *Rhynocoris marginatus* (Fabricius) (Hemiptera: Reduviidae). Hexapoda 14 (1), 38–45.

Ambrose, D.P., Rajan, S.J., Singh, V.J., Krishnan, S.S., 2008a. Development, intrinsic rate of natural increase and ecotypism of *Euagoras plagiatus* (Burmeister) (Insecta: Hemiptera: Reduviidae). Hexapoda 15 (1), 32–37.

Ambrose, D.P., Raja, J.M., Rajan, S.J., 2008b. Functional response of *Acanthaspis quinquespinosa* (Fabricius) (Hemiptera: Reduviidae) on *Coptotermes heimi* (Wasmann). Journal of Biological Control 22 (1), 163–165.

Ambrose, D.P., Rajan, S.J., Raja, J.M., 2008c. Functional response of *Rhynocoris kumarii* Ambrose and Livingstone and normal and Synergy-505 exposed *Rhynocoris marginatus* (Fab.) to larva of *Euproctis fraternal* (Moore). Indian Journal of Entomology 70 (3), 206–216.

Ambrose, D.P., Ambrose, A.D., Jeyanthi, N., Maran, S.P.M., 2008d. Relative toxicity of two organophosphorous insecticides and polymorphic resistance in *Rhynocoris marginatus* (Fabricius) (Hemiptera: Reduviidae). Entomon 33 (4), 227–232.

Ambrose, D.P., Sebastirajan, J., Nagarajan, K., Singh, V.J., Krishnan, S.S., 2009. Biology, behaviour and functional response of *Sphedanolestes variabilis* Distant (Insecta: Hemiptera: Reduviidae: Harpactorinae), a potential predator of lepidopteroan pests. Entomologia Croatica 13 (2), 33–44.

Ambrose, D.P., Rajan, S.J., Kumar, A.G., Raja, J.M., 2010a. Impact of Synergy-505 (chlorpyriphos 50% + cypermethrin 5%) on the biology and life table parameters of the assassin bug, *Rhynocoris marginatus* (Fabricius) (Hemiptera: Reduviidae). Entomon 35 (3), 209–212.

Ambrose, D.P., Rajan, S.J., Raja, J.M., 2010b. Impacts of Synergy-505 on the functional response and behavior of the reduviid bug, *Rhynocoris marginatus*. Journal of Insect Science 10, 1–10.

Ambrose, D.P., Nagarajan, K., Kumar, A.G., 2013. Interaction of reduviid predator, *Rhynocoris marginatus* (Fabricius) (Hemiptera: Reduviidae) with its prey teak skeletonizer, *Eutectona machaeralis* Walker (Lepidoptera: Pyralidae) as revealed through functional response. Journal of Entomological Research 37 (1), 55–59.

Ambrose, D.P., Lenin, E.A., Kiruba, D.A., 2014. Intrageneric phylogenetics based on mitochondrial DNA variation among fifteen harpactorine assassin bugs with four ecotypes and three morphs (Hemiptera: Reduviidae: Harpactorinae). Zootaxa 3779 (5), 540–550.

Ambrose, D.P., 1980. Bioecology, Ecophysiology and Ethology of Reduviids (Heteroptera) of the Scrub Jungles of Tamil Nadu, India (Ph.D. thesis). University of Madras, Madras, India.

Ambrose, D.P., 1987. Assassin bugs of Tamil Nadu and their role in biological control (Insecta-Heteroptera-Reduviidae). In: Joseph, K.J., Abdurahiman, U.C. (Eds.), Advances in Biological Control Research in India. M/S Printex Ltd, Calicut, India, pp. 16–28.

Ambrose, D.P., 1988. Biological control of insect pests by augmenting assassin bugs (Insecta – Heteroptera – Reduviidae). In: Ananthasubramanian, K.S., Venkatesan, P., Sivaraman, S. (Eds.), Bicovas Proceedings, vol. 2. Loyola College, Madras, India, pp. 25–40.

Ambrose, D.P., 1995. Reduviids as predators: their role inbiologial control. In: Ananthakrishnan, T.N. (Ed.), Biological Control of Social Forests and Plantation Crop Insects. Oxford & IBH Pub. Co. Pvt. Ltd., New Delhi, India, pp. 153–170.

Ambrose, D.P., 1999. Assassin Bugs. Oxford & IBH Publ. Co. Pvt. Ltd., New Delhi, India and Science Publishers, Inc., New Hampshire, USA, p. 337.

Ambrose, D.P., 2000a. Assassin bugs (Reduviidae excluding Triatominae). In: Schaefer, C.W., Panizzi, A.R. (Eds.), Heteroptera of Economic Importance. CRC Press, Boca Raton, Washington DC, pp. 695–712.

Ambrose, D.P., 2000b. Substrata impact on mass rearing of the reduviid predator, *Rhynocoris kumarii* Ambrose and Livingstone (Heteroptera: Reduviidae). Journal of Entomological Research 24 (4), 337–342.

Ambrose, D.P., 2000c. Suppression of *Helicoverpa armigera* (Hubner) and *Anomis flava* (Fabricius) infesting okra by the reduviid predator *Rhynocoris kumarii* Ambrose and Livingstone in field cages. Pest Management in Horticultural Ecosystems 6 (1), 32–35.

Ambrose, D.P., 2001. Friendly insecticides to conserve beneficial insects. Zoos' Print Journal 16 (8), 561–572.

Ambrose, D.P., 2002. Assassin bugs (Heteroptera: Reduviidae) in integrated pest management programme: success and strategies. In: Ignacimuthu, S., Sen, A. (Eds.), Strategies in Integrated Pest Management. Phoenix Pub. House Pvt. Ltd., New Delhi, pp. 73–85.

Ambrose, D.P., 2003. Biocontrol potential of assassin bugs (Hemiptera: Reduviidae). Indian Journal of Experimental Zoology 6 (1), 1–44.

Ambrose, D.P., 2004. The status of biosystematics of Indian Reduviidae (Hemiptera: Heteroptera). In: Rajmohana, K., Sudheer, K., Kumar, P.G., Santhosh, S. (Eds.), Perspectives on Biosystematics and Biodiversity. T.C. Narendran Comemorative Volume, University of Calicut, Kerala, India, pp. 441–459.

Ambrose, D.P., 2006. A checklist of Indian assassin bugs (Insecta: Hemiptera: Reduviidae) with taxonomic status, distribution and diagnostic morphological characteristics. Zoos' Print Journal 21 (9), 2388–2406.

Ambrose, D.P., 2007. The Insects: Beneficial and Harmful Aspects. Kalyani Publishers, New Delhi. p. 801.

Ambrose, D.P., 2010. In: Ignacimuthu, S., David, B.V. (Eds.), Mass Rearing of Entomophages Insects for Biological Control: Success, Bottlenecks and Strategies. Non-chemical Pest Management. Elite Publishing House Pvt Ltd, New Delhi, India, pp. 156–163.

Ambrose, D.P., 2015. The Insects: Structure, Function and Biodiversity, second ed. Kalyani Publishers, New Delhi. p. 626.

Antony, M., Daniel, J., Kurian, C., Pillai, G.B., 1979. Attempts in introduction and colonization of the exotic reduviid predator *Platymeris laevicollis* Distant for the biological suppression of the coconut rhinocerous beetle, *Oryctes rhinocerous*. Proceedings of Plant Crops Symposium II, 445–454.

Austin, G.D., 1922. A preliminary report on paddy fly investigation. Ceylon Department of Agriculture Bulletin 59, 1–22.

Awadallah, K.T., Tawfik, M.F.S., Abdellah, M.M.H., 1984. Suppression effect of the reduviid predator, *Alloeocranum biannulipes* (Montr. Et Sign.) on population of some stored product insect pests. Zeitschrift Fur Angewandte Entomologie 97, 249–253.

Babu, A., Seenivasagam, R., Karuppasamy, C., 1995. Biological control resources in social forest stands. pp. 7–24. In: Ananthakrishnan, T.N. (Ed.), Biological Control of Social Forest and Plantation Crop Insects. Oxford & IBH Publishing Co. Pvt. Ltd., New Delhi, India, p. 225.

Balakrishnan, P., Nagarajan, K., Kumar, A.G., Ambrose, D.P., 2011. Host preference, stage preference and functional response of *Acanthaspis pedestris* Stål (Hemiptera: Reduviidae) to its most preferred prey the acridid grasshopper, *Orthacris maindroni* Boliver. In: Ambrose, D.P. (Ed.), Insect Pest Management, a Current Scenario. Entomology Research Unit, St. Xavier's College, Palayamkottai, India, pp. 210–217.

Balu, A., Pillai, S.R.M., Sasidharan, K.R., Deeparaj, B., Sunitha, B., 1995. Natural enemies of Babul (*Acacia nilotica* L.) (Wild. ex. Del. ssp. Indica (Bth) defoliators *Selepa celtis* and *Tephrnia pulinda* (Insecta: Lepidoptera). In: Ananthakrishnan, T.N. (Ed.), Biological Control of Social Forest and Plantation Crop Insects. Oxford & IBH Publishing Co. Pvt. Ltd., New Delhi, India, pp. 43–53.

Barbosa, F.R., 2002. Pragas secundárias da mangueira: besouro amarelo. Embrapa, Brasília. Disponível em www.agencia.cnptia.embrapa.br.

Barrion, A.T., Anguda, R.M., Litsinger, J.A., 1987. The natural enemies and chemical control of the Leucaena psyllid *Heteropsyila cubana* Crawford in the Phillipines. Leucaena Research Report 7, 45–49.

Baskar, A., Fleming, A.T., Tirumurugaan, K.G., Ambrose, D.P., 2011. Genetic deiversity in the assassin bug, *Rhynocoris longifrons* (Stål) (Heteroptera: Reduviidae) based on RAPD analysis. In: Ambrose, D.P. (Ed.), Insect Pest Management, a Current Scenario. Entomology Research Unit, St. Xavier's College, Palayamkottai, India, pp. 28–37.

Baskar, A., Ambrose, D.P., Thirumurugaan, K.G., Fleming, A.T., 2012a. Ecotypic deiversity in the assassin bug *Rhynocoris fuscipes* (Fabricius) (Heteroptera: Reduviidae). Journal of Advanced Zoology 33 (2), 113–121.

Baskar, A., Ambrose, D.P., Thirumurugaan, K.G., Fleming, A.T., 2012b. Ecotypic deiversity in the assassin bug *Rhynocoris marginatus* Fabricius (Heteroptera: Reduviidae). Hexapoda 19 (1), 38–46.

Baskar, A., Ambrose, D.P., Tirumurugaan, K.G., Fleming, A.T., 2013. Genetic deiversity in the assassin bug *Rhynocoris marginatus* Fabricius (Heteroptera: Reduviidae), based on RAPD analysis. Indian Journal of Entomology 75 (2), 146–153.

Baskar, A., Mathew, L.P., Ambrose, D.P., Tirumurugaan, K.G., Vignesh, A.R., Dhanasekaran, S., 2014. Genetic variation in genus *Rhynocoris* (Heteroptera: Reduviidae) based on mitochondrial cytochrome c oxidase subunit I gene. Journal of Insect Science 27 (1), 44–51.

Beeson, C.F.G., 1941. Ecology and Control of Forest Insect Pests of India and Neighbouring Countries. Vasane Press, Dehradun, India. p. 767.

Berlinger, M.J., 1992. Importance of host plant or diet on the rearing of insects and mites. In: Anderson, T., Leppla, N. (Eds.), Advances in Insect Rearing for Research and Pest Management. Westview Press, Boulder, Colorado, pp. 237–249.

Bhatnagar, V.S., Sithanantham, S.S., Pawar, C.S., Jadhav, D., Rao, V.R., Reed, W., 1983. Conservation and augmentation of natural enemies with reference to integrated pest managements in chickpea (*Cicer arietinum* L.) and pigeonpea (*Cajanus cajan* (L.). Millsp.). In: Proceedings of the International Workshop on Integrated Pest Control for Grain Legumes, Goiani, Oia's Brazil, pp. 157–180.

Blatchley, W.S., 1926. Heteroptera or True Bugs of Eastern North America. Nature Publishing Co., Indianoplis, USA. p. 116.

Bose, M., 1949. On the biology of *Coranus spiniscutis* Reuter, an assassin bug (Heteroptera: Reduviidae). Indian Journal of Entomology 11, 203–208.

Brandt, M., Mahsberg, D., 2002. Bugs with a backpack: the function of nymphal camouflage in the West African assassin bugs *Paredocla* and *Acanthaspis* spp. Animal Behaviour 63, 277–284.

Butani, D.K., 1958. Parasites and predators recorded on sugarcane pests in India. Indian Journal of Entomology 20, 270–282.

Butler, E.A., 1918. Note on the *Coranus subapterus* DeGeer. Entomologist Monthly Magazine, London 644, 16–17.

Cade, W.H., Simpson, P.H., Breland, O.P., 1978. *Apiomerus spissipes* (Hemiptera: Reduviidae): a predator of harvester ants in Texas. The Southwestern Entomologist 3 (3), 195–197.

Caffrey, D.T., Barber, G.W., 1919. The grain bug *Zelus renardii*. USDA Bulletin of Entomology 799, 35.

Carayon, J., Usinger, R.L., Wygodzinsky, P., 1956. Notes on the higher classification of the Reduviidae with the description of a new tribe of the Phymatinae. Revue de Zoologie et de Botanique Africaines 57, 256–281.

Cassis, G., Gross, G.F., 1995. Hemiptera: Heteroptera (Coleorrhyncha to Cimicomorpha). In: Houston, W.W.K., Maynard, G.V. (Eds.), Zoological Catalogue of Australia, Houston, vol. 27. CSIRO, Melbourne, Australia, pp. 1–506.

Chandral, S., Sinazer, R.L., 2011. Influence of prey on the development and reproduction of *Endochus inornatus* Stål. Journal of Biopesticides 4 (2), 112–117.

Chandral, S., Latha, S.R., Kumar, S.P., 2005. Influence of prey on the development, reproduction and size of the assassin bug, *Sphedanolestes minusculus* Bergroth (Heteroptera: Reduviidae), a potential biological control agent. Journal of Entomological Research 29 (2), 93–98.

Cherian, M.C., Brahmachari, K., 1941. Notes on three predatory hemipteran from South India. Indian Journal of Entomology 3, 115–118.

Cherian, M.C., Kylasam, M.S., 1939. On the biology and feeding habits of *Rhynocoris fuscipes* (Fabr.) (Heteroptera: Reduviidae). Journal of Bombay Natural History Society 61, 256–259.

Cherian, M.C., 1937. Administration Report of Government Entomologist, Coimbatore for 1936–37 Report. Department of Agriculture, Madras. pp. 126–133.

Chittenden, F.H., 1905. The corn root worm. USDA Circular 59, 5.

Chittenden, F.H., 1907. The Colarado potato beetle. USDA Circular 87, 11.

Chittenden, F.H., 1916. The common cabbage worm. USDA Farmer's Bulletin 766, 9.

Cisenros, J.J., Rosenheim, A.J., 1997. Ontogenetic change of prey preference in the generalist predator *Zelus renardii* and its influence on predator-predator interactions. Ecological Entomology 22 (4), 339–407.

Clancy, D.W., 1946. Natural enemies of some Arizona cotton insects. Journal of Economic Entomology 39, 326–328.

Claver, M.A., Ambrose, D.P., 2001a. Suitability of substrata for the mass rearing of *Rhynocoris fuscipes* (Heteroptera: Reduviidae), a key predator of pod borer *Helicoverpa armigera* (Hubn.). Entomon 26 (2), 141–146.

Claver, M.A., Ambrose, D.P., 2001b. Impact of antennectomy, eye blinding and tibial comb coating on the predatory behaviour of *Rhynocoris kumarii* Ambrose and Livingstone (Het., Reduviidae) on *Spodoptera litura* Fabr. (Lep., Noctuidae). Journal of Applied Entomology 125, 519–525.

Claver, M.A., Ambrose, D.P., 2001c. Evaluation of *Rhynocoris kumarii* Ambrose & Livingstone (Hemiptera: Reduviidae) as a potential predator of some lepidopteran pests of cotton. Journal of Biological Control 15 (1), 15–20.

Claver, M.A., Ambrose, D.P., 2001d. Impact of augmentative release of *Rhynocoris kumarii* Ambrose & Livingstone (Heteroptera: Reduviidae) on *Dysdercus cingulatus* (Fabricius) (Hemiptera: Pyrrhocoridae) population and damage on cotton. Journal of Biological Control 15 (2), 119–125.

Claver, M.A., Ambrose, D.P., 2003. Influence of mulching and intercropping on the abundance of the reduviid predator, *Rhynocoris fuscipes* (F.). In: Ignacimuthu, S., Jayaraj, S. (Eds.), Biological Control of Insect Pests. Phoenix Pub. House Pvt. Ltd., New Delhi, India, pp. 183–191.

Claver, M.A., Ambrose, D.P., 2012. Numerical response of *Rhynocoris kumarii* (Insecta: Heteroptera: Reduviidae). Journal of Soil Biology and Ecology 32 (1 & 2), 93–100.

Claver, M.A., Reegan, A.D., 2010. Biology and mating behaviour of *Coranus spiniscutis* Reuter (Hemiptera: Reduviidae), a key predator of rice gandhi bug *Leptocorisa varicornis* Fabricius. Journal of Biopesticides 3 (2), 437–440.

Claver, M.A., Ramasubbu, G., Ravichandran, B., Ambrose, D.P., 2002. Searching behaviour and functional response of *Rhynocoris longifrons* (Stål) (Heteroptera: Reduviidae), a key predator of pod sucking bug, *Clavigralla gibbosa* Spinola. Entomon 27, 339–346.

Claver, M.A., Kalyanasundaram, M., David, P.M.M., Ambrose, D.P., 2003a. Abundance of boll worm, flower beetle, predators and field colonization by *Rhynocoris kumarii* (Het., Reduviidae) following mulching and shelter provisioning in cotton. Journal of Applied Entomology 127, 383–388.

Claver, M.A., Ravichandran, B., Khan, M.M., Ambrose, D.P., 2003b. Impact of cypermethrin on the functional response, predatory and mating behaviour of a nontarget potential biological control agent *Acanthaspis pedestris* (Stål) (Het., Reduviidae). Journal of Applied Entomology 127, 18–22.

Claver, M.A., Muthu, M.S.A., Ravichandran, B., Ambrose, D.P., 2004. Behaviour, prey preference and functional response of *Coranus spiniscutis* (Reuter), a potential predator of tomato insect pests. Pest Management in Horticultural Ecosystems 10 (1), 19–27.

Claver, M.A., 1998. Biocontrol Potential Evaluation of the Predator *Rhynocoris Kumarii* Ambrose and Livingstone (Heteroptera: Reduviidae) against Three Cotton Insect Pests (Ph.D. thesis). Manonmaniam Sundaranar University, Tirunelveli, Tamil Nadu, India.

Cogni, R., Frietas, A.V.L., Filho, B.F.A., 2002. Influence of prey size on predation success by *Zelus longipes* L. (Het., Reduviidae). Journal of Applied Entomology 126, 74–78.

Cohen, A.C., 1985. Simple method for rearing the insect predator *Geocoris punctipes* (Heteroptera: Lygaeidae) on a meat diet. Journal of Economic Entomology 78, 1173–1175.

Cohen, A.C., 1990. Feeding adaptations of some predaceous heteropterans. Annals of the Entomological Society of America 83, 1215–1223.

Cohen, A.C., 1993. Organization of digestion and preliminary characterization of salivary trypsin like enzymes in a predaceous heteropteran, *Zelus renardii*. Journal of Insect Physiology 39, 823–829.

Corbett, G.H., Pagden, H.T., 1941. A review of some recent entomological investigations and observations. Malaysian Agricultural Journal 29, 347–375.

Das, S.S.M., Ambrose, D.P., 2008a. Redescription, biology and behaviour of a harpactorine assassin bug *Vesbius sanguinosus* Stål (Insecta, Hemiptera, Reduviidae). Polish Journal of Entomology 77, 11–29.

Das, S.S.M., Ambrose, D.P., 2008b. Redescription, biology and behaviour of the harpactorine assassin bug *Irantha armipes* (Stål) (Hemiptera: Reduviidae). Acta Entomologica Slovenica 16 (1), 37–56.

Das, S.S.M., Ambrose, D.P., 2008c. Impact of female parental age on the postembryonic development and reproductive potential of *Sphedanolestes pubinotum* Reuter (Hemiptera: Reduviidae). Hexapoda 15 (2), 93–96.

Das, S.S.M., Krishnan, S.S., Singh, V.J., Ambrose, D.P., 2008a. Redescription, biology and behaviour of a harpactorine assassin bug *Alcmena spinifex* Thunberg (Insecta: Hemiptera: Reduviidae). Hexapoda 15 (1), 17–24.

Das, S.S.M., Krishnan, S.S., Singh, V.J., Ambrose, D.P., 2008b. Redescription, postembryonic development and behaviour of a harpactorine assassin bug *Sphedanolestes himalayensis* Distant (Hemiptera: Reduviidae). Entomologia Croatica 12 (1), 37–54.

Das, S.S.M., Kumar, A.G., Ambrose, D.P., 2009. Impact of intraspecific competition on the predation of *Irantha armipes* (Stål) (Hemiptera: Reduviidae) on cotton bollworm, *Helicoverpa armigera* (Hübner). Journal of Biological Control 23 (4), 381–384.

Das, S.S.M., Kumar, A.G., Ambrose, D.P., 2010. Impact of intraspecific competition in the predation of the cotton bollworm, *Helicoverpa armigera* (Hübner) by *Scipinia horrida* (Stål) (Hemiptera: Reduviidae). Entomon 35 (3), 203–208.

David, B.V., Kumaraswami, T., 1988. Elements of Economic Entomology. Popular Book Dept, Madras, India. p. 536.

David, P.M.M., Natarajan, S., June 21, 1989. Bug that Destroys Chilli Fruits. The Hindu, (Indian National Daily). p. 24.

David, B.V., Ramamurthy, V.V., 2011. Elements of Economic Entomology. Revised and Enlarged, sixth ed. Namrutha Publications, Porur, Chennai. p. 386.

DeBach, P., Hagen, K.S., 1964. Manipulation of entomophagous species. In: DeBach, P. (Ed.), Biological Control of Insect Pests and Weeds. Chapman and Hall Ltd., London, pp. 429–458.

DeClercq, P., DeGheele, D., 1992. A meat-based diet for rearing the predatory stink bugs *Podisus maculiventris* and *Podisus sagitta* (Heteroptera: Pentatomidae). Entomophaga 37, 149–157.

Edwards, J.S., 1960. Spitting as a defensive mechanism in a predatory reduviid. In: Proceeding of International Congress of Entomology, Vienna, pp. 259–263.

Edwards, J.S., 1961. The action and composition of saliva of an assassin bug *Platymeris rhadamanthus* Gaerst. (Hemiptera: Reduviidae). Journal of Experimental Biology 38, 61–77.

Essakiammal, E., 2011. Impact of Biopesticide Vijay Neem on the Haemocytes and Haemogram of *Rhynocoris Fuscipes* (Fabricius) (Hemiptera: Reduviidae) (M.Sc. thesis). Manonmaniam Sundaranar University, Tirunelveli, India. p. 31.

Evangelin, G., Bertrand, H., Muthupandi, M., William, J.S., 2014. Feeding behaviour of the predatory reduviid, *Rhynocoris kumarii* (Hemiptera: Reduviidae). International Journal of Life Sciences 3 (2), 64–69.

Evans, D.E., 1962. The food requirement of *Phonoctonus nigrofasciatus* stål (Hemiptera: Reduviidae). Entomologia Experimentalis et Applicata 5, 33–39.

Fadare, T.A., 1978. Efficiency of *Phonotonus* spp. (Hemiptera: Reduviidae) and regulators of populations of *Dysdercus* spp. (Hemiptera: Pyrrchocoridae). Nigerian Journal of Entomology 1, 45–48.

Flower, H.G., Crestana, L., Camargode, M.T.V., Junior, J.J., Costa, M.M.L., Saes, N.B., Camargo, D., Pinto, J.C.A., 1987. Functional response of the most prevalent potential predators to variation in the density of mole crickets (Orthoptera, Gryllotalpidae: *Scapteriscus borlli*). Naturalia 11/12, 47–52.

Forero, D., Choe, D.H., Weirauch, C., 2011. Resin gathering in neotropical resin bugs (Insecta: Hemiptera: Reduviidae): functional and comparative morphology. Journal of Morphology 272, 204–229.

Froeschner, R.C., Kormilev, N.A., 1989. Phymatinae or ambush bugs of the world: a synonymic list with keys to species, except *Lophoscutus* and *Phymata* (Hemiptera). Entomography 6, 1–76.

Fuseini, B.A., Kumar, R., 1975. Biology and immature stage of cotton stainers (Heteroptera: Pyrrhocoridae) found in Ghana. Biological Journal of Linnean Society 7, 83–111.

Garman, H., 1916. The locust borer. Kentucky Bulletin 200, 121.

George, P.J.E., Ambrose, D.P., 1998. Relative toxicity of five insticides to *Rhynocoris fuscipes* (Fabricius), a potential predator of insect pests (Insecta: Heteroptera: Reduviidae). Shashpa 5 (2), 197–202.

George, P.J.E., Ambrose, D.P., 1999. Biochemical modulations by insecticides in a non-target harpactorine reduviid *Rhynocoris kumarii* Ambrose and Livingstone (Heteroptera: Reduviidae). Entomon 24, 61–66.

George, P.J.E., Ambrose, D.P., 2000a. Nymphal cannibalism in reduviids-a constraint in mass rearing. In: Ignacimuth, S., Sen, A., Janarthanan, S. (Eds.), Biotechnological Applications for IPM. Oxford & IBH Pub. Co. Pvt. Ltd., New Delhi, India, pp. 119–123.

George, P.J.E., Ambrose, D.P., 2000b. Impact of five insecticides on the differential and the total haemocyte counts of *Rhynocoris marginatus* (Fabricius) (Insecta: Heteroptera: Reduviidae). Indian Journal of Environmental Sciences 4 (2), 169–173.

George, P.J.E., Ambrose, D.P., 2001. Polymorphic adaptive insecticidal resistance in *Rhynocoris marginatus* (Fabr.) (Het., Reduviidae) a non-target biocontrol agent. Journal of Applied Entomology 125 (4), 207–209.

George, P.J.E., Ambrose, D.P., 2004a. Insecticidal impact on the life table parameters of a harpactorine reduviid predator *Rhynocoris marginatus* (Fabricius), Heteroptera. Entomologia Croatica 8 (1–2), 13–23.

George, P.J.E., Ambrose, D.P., 2004b. Toxic effects of insecticides in the histomorphology of alimentary canal, testis and ovary in a reduviid *Rhynocoris kumarii* Ambrose and Livingstone (Hemiptera: Reduviidae). Journal of Advanced Zoology 25 (1 & 2), 46–50.

George, P.J.E., Ambrose, D.P., 2004c. Impact of insecticides on the haemogram of *Rhynocoris kumarii* Ambrose and Livingstone (Hem., Reduviidae). Journal of Applied Entomology 128 (9–10), 600–604.

George, P.J.E., Seenivasagan, R., Karuppasamy, R., 1998. Life table and intrinsic rate of natural increase of *Sycanus collaris* Fabricius (Reduviidae: Heteroptera) a predator of *Spodoptera litura* Fabricius (Lepidoptera: Noctuidae). Journal of Biological Control 12, 113–118.

George, P.J.E., Celin, J.A., Antonysamy, A., Ambrose, D.P., 2005. UPGMA cluster analysis as a tool in the biosystematics of Reduviidae (Insecta: Hemiptera). Shashpa 12 (2), 73–78.

George, P.J.E., Celin, J.A., Antonysamy, A., Ambrose, D.P., 2007. Ecotype selection of *Acanthaspis pedestris* Stål (Heteroptera: Reduviidae) for mass multiplication by UPGMA cluster analysis. Indian Journal of Science and Technology 1 (1), 1–5.

George, P.J.E., 1999a. Development, life table and intrinsic rate of natural increase of three morphs of *Rhynocoris marginatus* (Fabricius) (Heteroptera: Reduviidae) on cotton leaf worm *Spodoptera litura* (Fabricius) (Lepidoptera: Noctuidae). Entomon 24, 339–343.

George, P.J.E., 1999b. Biology and life table studies of the reduviid *Rhynocoris marginatus* (Fabrcius) (Heteroptera: Reduviidae) on three lepidopteran insect pests. Journal of Biological Control 13, 33–38.

George, P.J.E., 2000a. Life tables and intrinsic rate of natural increase of three morphs of *Rhynocoris marginatus* (Fabricius) (Heteroptera: Reduviidae) on *Corcyra cephalonica* Stainton. Journal of Experimental Zoology, India 3, 59–63.

George, P.J.E., 2000b. Polymorphic adaptation in reproductive strategies of *Rhynocoris marginatus* (Fabricius) on *Earias vitella* (Fabricius). Journal of Biological Control 14, 35–39.

George, P.J.E., 2000c. Reproductive performance of a harpactorine reduviid *Rhynocoris fuscipes* (Fabricius) (Heteroptera: Reduviidae) on three lepidopteran insect pests. Insect Science Application 20, 269–273.

George, P.J.E., 2000d. Nymphal cannibalistic behaviour of *Rhynocoris kumarii* Ambrose and Livingstone (Heteroptera: Reduviidae) to varied prey deprivation and mass rearing conditions. Journal of Biological Control 14, 1–3.

George, P.J.E., 2000e. Impact of antennectomy on the predatory efficiency of *Rhynocoris marginatus* Fabricius (Insecta: Heteroptera: Reduviidae). Advances in Biosciences 19 (11), 133–140.

George, P.J.E., 2000f. The intrinsic rate of natural increase of a harpactorine reduviid *Rhynocoris kumarii* Ambrose and Livingstone on three lepidopteran insect pests. Entomon 25, 281–286.

Gerardo, C., Salumi, A., Akahane, T.W., Yoshisa, K., Tomni, W., 2001. Novel peptides from assassin bugs (Hemiptera: Reduviidae): isolation, chemical and biological characterization. FEBS Letter 499, 256–2621.

Ghelani, Y.H., Viyas, H.N., Jhala, R.C., 2000. Toxicological impact of certain insecticides to *Rhynocoris fuscipes* Fabricius (Heteroptera: Reduviidae) – a potential predator. Journal of Applied Zoological Researches 11 (1), 54–57.

Gifford, T.R., 1964. A brief review of sugar cane insects research in Florida 1960–1964. Proceedings of Soil Crop Science Society of Florida 24, 449–453.

Gil-Santana, H.R., Forero, D., 2010. Taxonomical and biological notes on neotropical Apiomerini (Hemiptera: Heteroptera: Reduviidae: Harpactorinae). Zootaxa 2331, 57–68.

Godfrey, K.E., Whitcomb, W.H., Stimac, J.L., 1989. Arthropod predators of velvetbean caterpillar, *Anticarsia gemmatalis* Hübner (Lepidoptera: Noctuidae) eggs and larvae. Environmental Entomology 18, 118–123.

Gravana, S., Sterling, W.L., 1983. Natural predation of the cotton leaf worm (Lepidoptera: Noctuidae). Journal of Economic Entomology 76, 779–784.

Grossi, P.C., Koike, R.M., Gil-Santana, H.R., 2012. Predation on species of *Leptinopterus* Hope (Coleoptera, Lucanidae) by three species of Reduviidae (Hemiptera, Heteroptera) in the Atlantic Forest, Brazil. EntomoBrasilis 5 (2), 88–92.

Grundy, P.R., Maelzer, D.A., 2000a. Predation by the assassin bug *Pristhesancus plagipennis* (Walker) (Hemiptera: Reduviidae) of *Helicoverpa armigera* (Hübner) (Lepidoptera: Noctuidae) and *Nezara viridula* (L.) (Hemiptera: Pentatomidae) in the laboratory. Australian Journal of Entomology 39, 280–282.

Grundy, P.R., Maelzer, D.A., 2000b. Assessment of *Pristhesancus plagipennis* (Walker) (Hemiptera: Reduviidae) as an augmentated biological control in cotton and soybean crops. Australian Journal of Entomology 39, 305–309.

Grundy, P.R., Maelzer, D.A., Bruce, A., Hassan, E., 2000a. A mass-rearing method for the assassin bug *Pristhesancus plagipennis*. Biological Control 18, 243–250.

Grundy, P.R., Maelzer, D.A., Collins, P.J., Hassan, E., 2000b. Potential for integrating eleven agricultural insecticides with the predatory bug *Pristhesancus plagipennis* (Hemiptera: Reduviidae). Journal of Economic Entomology 93, 584–589.

Grundy, P., 2004. Impact of low release rates of the assassin bug *Pristhesancus plagipennis* (Walker) (Hemiptera: Reduviidae) on *Helicoverpa* spp. (Lepidoptera: Noctuidae) and *Creontiades* spp. (Hemiptera: Miridae) in cotton. Australian Journal of Entomology 43, 77–82.

Haines, C.P., 1991. Insects and Arachnids of Tropical Stored Products: Their Biology and Identification (A Training Manual), second ed. Natural Resource Institute, Chatham, UK. p. 246.

Hall, D.G., Hentz, M.G., Adair Jr., R.C., 2008. Population ecology and phenology of *Diaphorina citri* (Hemiptera: Psyllidae) in two Florida citrus groves. Environmental Entomology 37, 914–924.

Haridass, E.T., Ananthakrishnan, T.N., 1980. Models for the feeding behaviour of some reduviids from South India (Insecta: Heteroptera: Reduviidae). Proceedings of Indian Academy of Science (Animal Sciences) 89, 387–402.

Helms, T.J., Raun, E.S., 1971. Perennial culture of disease-free insects. In: Burgess, H.D., Hussey, N.W. (Eds.), Microbial Control of Insects and Mites. Academic Press, New York, pp. 639–654.

Henneberry, T.J., Clayton, T.E., 1985. Consumption of pink bollworm (Lepidoptera: Gelechiidae) and tobacco budworm (Lepidoptera: Noctuidae) eggs by some predators commonly found in cotton fields. Environmental Entomology 14, 416–419.

Hiremath, I.G., Thontadarya, T.S., 1983. Natural enemies of sorghum earhead bug *Calocoris angustatus* Lethierry (Hemiptera: Miridae). Current Research 2, 10–11.

Hoffmann, W.E., 1935. The bionomics and morphology of *Isyndus reticulates* Stål (Hemiptera: Reduviidae). Lingnam Science Journal 14, 145–153.

Holling, C.S., 1959. Some characteristics of simple type of predation and parasitism. Canadian Entomologist 91, 385–395.

Holling, C.S., 1966. The functional response of invertebrate predators to prey density. Memoirs of the Entomological Society of Canada 48, 1–87.

Horton, J.R., 1918a. The citrus thrips. USDA Bulletin 616, 25.

Horton, J.R., 1918b. Argentine ant in relation to citrus groves. USDA Bulletin 647, 1–73.

Hussain, S., Aslam, N.A., 1970. Some observations on a beneficial reduviid bug: *Amphibolus venator* Klug (Fam. Reduviidae: Hemiptera). Agriculture Pakistan 21 (1), 37–42.

Hutson, J.C., 1922. Report of the entomologist. Ceylon Department of Agriculture Bulletin 1921, 23–26.

Hwang, W.S., Weirauch, C., 2012. Evolutionary history of assassin bugs (Insecta: Hemiptera: Reduviidae): Insights from divergence dating and ancestral state reconstruction. PLoS One 7 (9), e45523. http://dx.doi.org/10.1371/journal.pone.0045523.

Illingworth, J.F., 1921. The linear bug *Phadnacantha australia* Kirkaldy: a new pest of sugar cane in Queensland. Queensland Bureau of Sugar Experimental Station, Division of Entomology Brisbane Bulletin 14.

Imamura, T., Nishi, A., Takahashi, K., Visarathanonth, P., Miyanoshita, A., 2006. Life history parameters of *Amphibolus venator* (Klug) (Hemiptera: Reduviidae), a predator of stored-product insects. Representatives of National Food Research Institute 70, 19–22.

Imms, A.D., 1965. A General Textbook of Entomology. The English Language Book Society and Mathuen Co. Ltd., London. pp. 459–460.

Jacobson, E., 1911. Biological notes on the hemipteran *Ptilocerus ochraceus*. Tijdschrift voor Entomologie 54, 175–179.

James, D.G., Moore, C.J., Aldrich, J.R., 1994. Identification, synthesis and bioactivity of a male produced aggregation pheromone in assassin bug, *Pristhesancus plagipennis* (Hemiptera: Reduviidae). Journal of Chemical Ecology 20, 3281–3295.

James, D.G., 1992. Effect of temperature on development and survival of *Pristhesancus plagipennis* (Hemiptera: Reduviidae). Entomophaga 37, 259–264.

James, D.G., 1994a. The development of suppression tactics for *Biprorulus bibax* (Hemiptera: Pentatomidae) as part of an integrated pest management programme in inland citrus of Southeastern Australia. Bulletin of Entomological Research 84, 31–38.

James, D.G., 1994b. Prey consumption by *Pristhesancus plagipennis* (Hemiptera: Reduviidae) during development. Austral Entomology 21, 43–47.

Joseph, M.T., 1959. Biology binomics and economic importance of some reduviids collected from Delhi. Indian Journal of Entomology 21, 46–58.

Kalidas, S., Sahayaraj, K., 2012. Comparative biology and life table traits of *Rhynocoris longifrons* Stål (Hemiptera: Reduviidae) on factitious host and four natural cotton pests. In: Proceedings of International Conference on Science and Technology for Clean and Green Environment, pp. 142–149.

Kaur, R., Kaur, H., 2013. Chromosomes and their meiotic behavior in twelve species of the subfamily Harpactorinae (Hemiptera: Heteroptera: Reduviidae) from north India. Zootaxa 3694 (4), 358–366.

Kaur, H., Kaur, R., Patial, N., 2012. Studies on internal male reproductive organs and course of meiosis in a predator species *Repipta taurus* (Fabricius, 1803) (Heteroptera: Reduviidae: Harpactorinae). Journal of Entomological Research 36 (3), 247–250.

Khokar, S., 2000. Cladistic analysis of the Reduviidae (Hemiptera: Heteroptera). In: Vth National Conference of Applied Zoologists Research Association (AZRA), India. Abs. No.80.

Kirkland, A.H., 1896. Predaceous Hemiptera-Heteroptera. In: Forbush, E.H., Fernald, C.H. (Eds.), The Gypsy Moth, *Porthetria Dispar* (Linn.). Wright and Potter Printing Co., State Printers, Boston, Mass, pp. 392–403.

Krishnaswamy, N., Chowhan, O.P., Das, R.K., 1984. Some common predators of rice pests in Assam, India. IRRN 9, 15–16.

Kumar, A.G., Ambrose, D.P., 2014. Olfactory response of an assassin bug, *Rhynocoris longifrons* (Insecta: Hemiptera: Reduviidae) to the hexane extracts of different agricultural insect pests. Insects Review 1 (1), 12–19.

Kumar, S.P., Kumar, A.G., Ambrose, D.P., 2009. Impact of intraspecific competition in the predation of *Rhynocoris longifrons* (Stål) (Hemiptera: Reduviidae) on camponotine ant *Camponotus compressus* Fabricius. Hexapoda 16 (1), 1–4.

Kumar, A.G., Rajan, K., Rajan, S.J., Ambrose, D.P., 2011. Predatory behaviour of an assassin bug, *Coranus spiniscutis* (Reuter) on rice meal moth, *Corcyra cephalonica* (Stainton) and leaf armyworm, *Spodoptera litura* (Fabricius). In: Ambrose, D.P. (Ed.), Insect Pest Management, a Current Scenario. Entomology Research Unit, St. Xavier's College, Palayamkottai, India, pp. 404–408.

Kumar, S.P., 1993. Biology and Behavior of Chosen Assassin Bugs, (Insecta: Heteroptera: Reduviidae) (Ph.D. thesis). Madurai Kamaraj University, Madurai, India.

Kumar, A.G., 2011. Mass Multiplication, Large Scale Release and Biocontrol Potential Evaluation of a Reduviid Predator *Rhynocoris Longifrons* (Stål) (Insecta: Heteroptera: Reduviidae) against Chosen Agricultural Insect Pests (Ph.D. thesis). Manonmaniam Sundaranar University, Triunelveli, Tamil Nadu, India.

Kumaraswami, N.S., 1991. Bioecology and Ethology of Chosen Predatory Bugs and Their Potential in Biological Control (Ph.D. thesis). Madurai Kamaraj University, Madurai, India.

Kumari, R.M.V., 2011. Impact of the Biopesticide Azadirachtin on the Haemocytes and Haemogram of *Rhynocoris Kumarii* Ambrose and Livingstone (Hemiptera: Reduviidae) (M.Sc. thesis). Manonmaniam Sundaranar University, Tirunelveli, India. p. 31.

Kuppusamy, A., Kannan, S., 1995. Biological control of bagworm (Lepidoptera: Psychidae) in social forest stands. In: Ananthakrishnan, T.N. (Ed.), Biological Control of Social Forests and Plantation Crop Insects. Oxford & IBH Publishing Co. Pvt. Ltd., New Delhi, India, pp. 25–41.

Lakkundi, N.H., 1989. Assessment of Reduviids for Their Predation and Possibilities of Their Utilization in Biological Control (Ph.D. thesis). IARI, Division of Entomology, New Delhi.

Lam, P.V., Lan, L.P., Hilbeck, A., Tuat, N.V., Lang, A., 2008. Invertebrate predators in *Bt* cotton in Vietnam: techniques for prioritizing species and developing risk hypothesis for risk assessment. In: Andow, D.A., Hilbeck, A., Tuat, N.V. (Eds.), Environmental Risk Assessment of Genetically Modified Organisms: Challenges and Opportunities with *Bt* cotton in Vietnam, vol. 4. CABI, pp. 176–208.

Lenin, E.A., Anbarasi, R., Nagarajan, K., Rajan, K., Kumar, A.G., Ambrose, D.P., 2011. Kairomonal ecology of assassin bug, *Rhynocoris kumarii* Ambrose and Livingstone (Hemiptera: Reduviidae) and chosen prey species. In: Ambrose, D.P. (Ed.), Insect Pest Management, a Current Scenario. Entomology Research Unit, St. Xavier's College, Palayamkottai, India, pp. 326–332.

Lenin, E.A., 2014. Biosystematics of Chosen Reduviine Assassin Bugs (Insecta: Hemiptera: Reduviidae) (Ph.D. thesis). Manonmaniam Sundaranar University, Tirunelveli, Tamil Nadu, India.

Livingstone, D., Yacoob, M.H.S., 1986. Natural enemies and biologies of the egg parasitoids of Tingidae of southern India. Uttar Pradesh Journal of Zoology 3, 1–12.

Loganathan, J., David, P.M.M., 1999. Predator complex of the teak defoliator, *Hyblaea puera* Crumer (Lepidoptera: Hyblaeidae) in an intensively managed teak plantation at Veeravanallur, Tamil Nadu. Entomon 24, 259–263.

Manimuthu, M., Rajan, S.J., Raja, J.M., Kumar, A.G., Ambrose, D.P., 2011. Functional response of assassin bug, *Acanthaspis pedestris* Stål to rice meal moth larva, *Corcyra cephalonica* (Stainton) and termite, *Odontotermes obesus* Rambur. In: Ambrose, D.P. (Ed.), Insect Pest Management, a Current Scenario. Entomology Research Unit, St. Xavier's College, Palayamkottai, India, pp. 302–308.

Manimuthu, M., 2014. Biosystematics of Chosen Peiratine Assassin Bugs (Insecta: Hemiptera: Reduviidae) (Ph.D. thesis). Manonmaniam Sundaranar University, Tirunelveli, Tamil Nadu, India.

Maniyammal, S., 2011. Effect of a Chosen Neem Pesticide on the Haemocytes and Haemogram of *Rhynocoris Marginatus* (Fabricius) (Hemiptera: Reduviidae) (M.Sc. thesis). Manonmaniam Sundaranar University, Tirunelveli, India, p. 31.

Maran, S.P.M., Ambrose, D.P., 2000. Paralytic Potential of *Catamiarus brevipennis* (Serville), a potential biological control agent (Insecta: Heteroptera: Reduviidae). In: Ignacimuth, S., Sen, A., Janarthanan, S. (Eds.), Biotechnological Applications for IPM. Oxford & IBH Pub. Co. Pvt. Ltd., New Delhi, India, pp. 125–131.

Maran, S.P.M., Kumaraswami, N.S., Rajan, K., Kiruba, D.A., Ambrose, D.P., 2011. The salivary protein profile and paralytic potential of three species of *Rhynocoris* (Hemiptera: Reduviidae) to three insect pests. In: Ambrose, D.P. (Ed.), Insect Pest Management, a Current Scenario. Entomology Research Unit, St. Xavier's College, Palayamkottai, India, pp. 346–361.

Maran, S.P.M., 1999. Chosen Reduviid Predators–Prey Interaction: Nutritional and Pheromonal Chemical Ecology (Insecta: Heteroptera: Reduviidae) (Ph.D. thesis). Manonmaniam Sundaranar, University, Tirunelveli, India.

Martin, W.R.J.R., Brown, J.M., 1984. The action of acephate in *Pseudoplusia includes* (Lepidoptera: Noctuidae) and *Pristhesancus papuensis* (Hemiptera: Reduviidae). Entomologia Experimentalis et Applicata 35, 3–9.

Maxwell-Lefroy, H., Howlett, F.M., 1909. Indian insect life. Thacker, Spink & Co., London.

McKeown, K.C., 1942. Australian insects. In: Proceedings of the Royal Zoological Society of New South Wales, Sydney.

Michaud, J., 2002. Biological control of Asian citrus psyllid, *Diaphorina citri* (Hemiptera: Psyllidae) in Florida: a preliminary report. Entomological News 113, 216–222.

Misra, R.M., 1975. Notes on *Anthia sexguttata* Fabr. (Carabidae: Coleoptera) a new predator of *Pyrausta machaeralis* Walker and *Hyblea puera* Crammer. Indian Forester 101, 604.

Mohamed, U.V.K., Abdurahiman, U.C., Remadevi, O.K., 1982. Coconut caterpillar and its natural enemies. In: A Study of the Parasites and Predators of *Nephantis serinopa* Myrick. University of Calicut, India, p. 23. Zoological Monograph, No. 2.

Mohanadas, 1996. New records of some natural enemies of the teak defoliator *Hyblaea puera* Crammer (Lepidoptera: Hybiaeidae) from Kerala, India. Entomon 21, 251–254.

Morey, S.S., Kiran, K.M., Gadag, J.R., 2006. Purification and properties of hyaluronidase from *Palamneus gravimanus* (Indian black scorpion) venom. Toxicon 47, 188–195.

Morgan, A.C., 1907. A predatory bug *Apiomerus spisspies* Say reported as an enemy of the cotton bollweevils. USDA Bulletin of Entomology 63, 1–54.

Morrill, A.W., 1910. Plant bugs injurious to cotton bolls. USDA Bulletin of Entomology 86, 110.

Morrill, W.L., 1975. An unusual predator of the Florida harvester ant. Journal of the Georgia Entomological Society 10, 50–51.

Morrison, N.M., 1989. Gel electrophoretic studies with reference to functional morphology of the salivary glands of *Acanthaspis pedestris* Stål (Insecta: Heteroptera: Reduviidae). Proceedings of the Indian Academy of Sciences (Animal Sciences) 98, 167–173.

Mossop, M.C., 1927. Insect enemies of the eucalyptus snout beetle. Farming South Africa 11, 430–431.

Mukerjee, A.B., 1971. Observations on the feeding habit of *Reduvius* sp. Predators of *Corcyra cephalonica* (Stainton). Indian Journal of Entomology 33, 230–231.

Muniyandi, J., Kumar, A.G., Nagarajan, K., Ambrose, D.P., 2011. Host preference, stage preference and functional response of assassin bug, *Rhynocoris kumarii* Ambrose and Livingstone (Hemiptera: Reduviidae) to its most preferred prey tobacco cutworm, *Spodoptera litura* (F.). In: Ambrose, D.P. (Ed.), Insect Pest Management, a Current Scenario. Entomology Research Unit, St. Xavier's College, Palayamkottai, India, pp. 240–248.

Muraleedharan, N., Selvasundaram, R., Radhakrishnan, B., 1988. Natural enemies of certain tea pests occurring in southern India. Insect Science Application 9, 647–654.

Muthupandi, M., Evangelin, G., Horne, B., William, S.J., 2014. Biology of the harpactorine assassin bug, *Panthous bimaculatus* Distant (Hemiptera: Reduviidae) on three different diets. Halteres 5, 11–16.

Nagarajan, K., Ambrose, D.P., 2013. Chemically mediated prey-approaching behaviour of the reduviid predator *Rhynocoris fuscipes* (Fabricius) (Insecta: Heteroptera: Reduviidae) by Y-arm olfactometer. Pakistan Journal of Biological Sciences 16, 1363–1367.

Nagarajan, K., Baskar, A., Kumar, S.P., Ambrose, D.P., 2010a. Development, reproductive performance and ecotypic diversity of *Coranus siva* Kirkaldy (Hemiptera: Reduviidae). Hexapoda 17 (1), 27–34.

Nagarajan, K., Rajan, K., Ambrose, D.P., 2010b. Functional response of assassin bug, *Rhynocoris fuscipes* (Fabricius) (Hemiptera: Reduviidae) to cucumber leaf folder, *Diaphania indicus* Saunders (Lepidoptera: Pyraustidae). Entomon 35 (1), 1–7.

Nagarajan, K., 2010. Mass Multiplication, Large Scale Release and Biocontrol Potential Evaluation of a Reduviid Predator *Rhynocoris Fuscipes* (Fabricius) (Insecta: Heteroptera: Reduviidae) against Chosen Insect Pests (Ph.D. thesis). Manonmaniam Sundaranar University, Tirunelveli, Tamil Nadu, India.

Nagarkatti, S., 1982. The utilization of biological control of *Heliothis* management in India. In: Reed, W., Kumble, V. (Eds.), Proceedings of International Workshop on *Heliothis* Management. ICRISAT, Patencheru, India, pp. 159–167.

Navarrete, B., Carrillo, D., Reyes-Martinez, A.Y., Sanchez-Peña, S., Lopez-Arroyo, J., Mcauslane, H., Peña, J.E., 2014. Effect of *Zelus longipes* (Hemiptera: Reduviidae) on *Diaphorina citri* (Hemiptera: Liviidae) and its parasitoid *Tamarixia radiata* (Hymenoptera: Eulophidae) under controlled conditions. Florida Entomologist 97 (4), 1537–1543.

Nayar, K.K., Ananthakrishnan, T.N., David, B.V., 1976. General and Applied Entomology. Tata McGraw-Hill Publ. Co. Ltd., New Delhi, India. p. 589.

Nei, R.K., 1942. Biological control of the codling moth in South Africa. Journal of Entomological Society of South Africa 5, 118–137.

Neuenschwander, P., Gagen, K.S., Smith, R.F., 1975. Predation on aphid in California's alfalfa fields. Hilgarida 43, 53–78.

Nyirra, Z.M., 1970. The biology and behaviour of *Rhynocoris albopunctatus* (Hemiptera, Reduviidae). Annals of the Entomological Society of America 63, 1224–1227.

Parker, A.H., 1965. The predatory behaviour and life history of *Pisilus tipuliformis* Fabr. (Hemiptera: Reduviidae). Entomologia Experimentalis et Applicata 8, 1–12.

Parker, A.H., 1969. The predatory and reproduction behaviour of *Rhinocoris bicolor* and *R. tropicus* (Hemiptera: Reduviidae). Entomologia Experimentalis et Applicata 12, 107–117.

Parker, A.H., 1971. The predatory and reproductive behaviour of *Vestula lineaticeps* (Sign.) (Hemiptera: Reduviidae). Bulletin of Entomological Research 61, 119–124.

Parker, A.H., 1972. The predatory and sexual behaviour of *Phonoctomus fasciatus* and *P. subimpictus* Stål (Hemiptera: Reduviidae). Bulletin of Entomological Research 62, 139–150.

Patil, B.V., Thontadarya, T.S., 1983. Natural enemy complex of the teak skeletonizer *Pyrausta mechaeralis* Walker (Lepidoptera: Pyralididae) in Karnataka. Entomon 8, 249–255.

Pawar, C.S., Bhatnagar, V.S., Jadhav, D.R., 1986. *Heliothis* species and their natural enemies, with their potential in biological control. Proceedings of Indian Academy of Science (Animal Sciences) 95, 695–703.

Pettey, F.W., 1919. Insect enemies of the codling moth in South Africa and their relation to its control. South African Journal of Science 16, 239–257.

Pillai, G.B., 1988. Biological control of pests of plantation crops. In: Summer Institute on Techniques of Mass Production of Biological Agents for Management of Pest and Diseases. Centre for plant protection studies, TNAU, Coimbatore, pp. 102–111.

Pingale, S.V., 1954. Biological control of some stored grain pests by the use of a bug predator, Amphibolus venator Klug. Indian Journal of Entomology 16, 300–302.

Pinto, C.F., 1927a. *Crithidia spinigeri* n. sp. Parasite do apparelho digestive de *Spiniger domesticus* (Hemiptera: Reduviidae). Bol. Biol. 7, 86–87.

Pinto, C.F., 1927b. *Spiniger domesticus* n. sp. Hemiptera succrd' insects (families des Reduviidae Sousfamille des Reduviinae). Comptes Rendus Societe de Biologie (Paris) 97, 833–834.

Pionar Jr., G.O., 1992. Fossil evidence of resin utilization by insects. Biotropica 24, 466–468.

Poggio, M.G., Bressa, M.J., Papeschi, A.G., 2007. Karyotype evolution in Reduviidae (Insecta: Heteroptera) with special reference to Stenopodainae and Harpactorinae. Comparative Cytogenetics 1 (2), 159–168.

Ponnamma, K.N., Kurian, C., Koya, K.M.A., 1979. Record of *Rhinocoris fuscipes* (Fabr.) (Heteroptera: Reduviidae) as a predator of *Myllocerus curicornis* (F.) (Coleoptera: Curculionidae) the ash weevil, pest of the coconut palm. Agriculture Research Journal of Kerala 17, 91–92.

Pyke, B.A., Brown, E.H., 1996. The Cotton Pest and Beneficial Guide. CRDC, Woolloongabba, Australia. pp. 41–43.

Quicke, D.L.J., Ingram, S.N., Proctor, J., Huddleston, T., 1992. Batesian and Müllerian mimicry between species with connected life histories, with a new example involving braconid wasp parasites of *Phoracantha* beetles. Journal of Natural History 26 (5), 1013–1034.

Raja, J.M., Rajan, S.J., Rajan, K., Ambrose, D.P., 2011. Reduviids (Insecta: Hemiptera: Reduviidae) in Courtallam and Pilavaikal tropical rainforests and Marunthuvazmalai and Muppanthal scrub jungles, Tamil Nadu, South India. In: Ambrose, D.P. (Ed.), Insect Pest Management, a Current Scenario. Entomology Research Unit, St. Xavier's College, Palayamkottai, India, pp. 152–161.

Rajagopal, D., Veeresh, G.K., 1981. Termitophiles and termitariophiles of *Odontermes wallonensis* (Isoptera: Termitidae) in Karnataka, India. Colemania 1, 129–130.

Rajagopal, D., 1984. Observation on the natural enemies of *Odontotermes wallonensis* (Wassmann) (Isoptera: Termitidae) in South India. Journal of Soil Biology and Ecology 4, 102–107.

Rajan, S.J., Raja, J.M., Krishnan, S.S., Rajan, K., Ambrose, D.P., 2011. Biology of assassin bug, *Acanthaspis pedestris* stål (Hemiptera: reduviidae) on two prey species, *Corcyra cephalonica* (Stainton) (Lepidoptera: Pyralidae) and *Odontotermes obesus* Rambur (Isoptera: Termitidae). In: Ambrose, D.P. (Ed.), Insect Pest Management, a Current Scenario. Entomology Research Unit, St. Xavier's College, Palayamkottai, India, pp. 130–133.

Rajan, K., 2011. Bioecology and Ethology of Chosen Haematophagous Assassin Bugs and Their Medical Importance (Insecta: Heteroptera: Reduviidae) (Ph.D. thesis). Manonmaniam Sundaranar University, Triunelveli, Tamil Nadu, India.

Rao, R.S.N., Satyanarayana, S.V.V., Soundarajan, V., 1981. Notes on new addition to the natural enemies of *Spodoptera litura* F. and *Myzus persicae* Sulz. on fine cored tobacco in Andhra Pradesh. Science Culture 47, 98–99.

Rao, V.P., 1974. Biology and Breeding Techniques for Parasites and Predators of *Ostrina* Sp. And Heliothis Sp. CIBC Final Technical Report PL 480 project, Bangalore, India, p. 86.

Ravichandran, B., Ambrose, D.P., 2006. Functional response of a reduviid predator *Acanthaspis pedestris* Stål (Hemiptera: Reduviidae) to three lepidopteran insect pests. Entomon 31 (3), 149–157.

Ravichandran, B., Claver, M.A., Ambrsoe, D.P., 2003. Functional response of the assassin bug *Rhynocoris longifrons* (Stål) (Heteroptera: Reduviidae) to cotton boll worm *Helicoverpa armigera* (Hübner). In: Ignacimuthu, S., Jayaraj, S. (Eds.), Biological Control of Insect Pests. Phoenix Publishing House Pvt. Ltd., New Delhi, India, pp. 155–159.

Ravichandran, B., 2004. Biocontrol Potential Evaluation of a Reduviid Predator *Acanthaspis Pedestris* Stål (Insecta: Heteroptera: Reduviidae) on Three Chosen Lepidopteran Insect Pests (Ph.D. thesis). Manonmaniam Sundaranar University, Tirunelveli, Tamil Nadu, India. p. 102.

Readio, P.A., 1927. Studies on the biology of the Reduviidae of America north of Mexico. University of Kansas Science Bulletin 17, 5–291.

Ren, S.Z., 1984. Studies on the genus *Coranus* Curtis from China (Heteroptera: Reduviidae). Entomotaxonomia 6, 279–284.

Richman, D.P., Hemanway, R.C., Whitcomb, W.H., 1980. Field cage evaluation of predators of the soybean looper *Pseudoplusia includes* (Lepidoptera: Noctuidae). Environmental Entomology 9, 315–317.

Risbec, J., 1941. Les insects de I' arachide. Trav. Lab. Entomol. Sect. Sudon. Rub. Agron. 22.

Romeo, A., 1927. Ossseruazional suakuni pentatomidie loreidi neidintorni di Rando zoo (Catania). Agraria di Portici, Annual Report, I Supplement (3) ii, 261–268.

Rosenheim, J.A., Wilhoit, L.R., 1993. Predators that eat other predators disrupt cotton aphid control. California Agriculture 47, 7–9.

Rosenheim, J.A., Wilhoit, L.R., Armer, C.A., 1993. Influence of intraguild predation among generalist predators on the suppression of an herbivore population. Oecologia 96, 439–449.

Ruberson, J.R., Herzog, G.A., Lambert, W.R., Lewis, W.J., 1994. Management of the beet armyworm: integration of control approaches. In: Proceedings, Beltwide Cotton Prod. Conference, pp. 857–859.

Sahayaraj, K., Ambrose, D.P., 1992. Biology and predatory potential of *Endochus umbrinus* Reuter (Heteroptera: Harpactorinae: Reduviidae) from South India. Bulletin of Entomology 33 (1–2), 42–55.

Sahayaraj, K., Ambrose, D.P., 1994. Prey influence on laboratory mass rearing of *Neohaematorrhophus therasii* Ambrose and Livingstone a potential biocontrol agent Insecta: Heteroptera: Reduviidae. Bioscience Research Bulletin 10 (1), 35–40.

Sahayaraj, K., Ambrose, D.P., 1996. Biocontrol potential of the reduviid predator *Neohaematorrhophus therasii* Ambrose and Livingstone (Heteroptera, Reduviidae). Journal of Advanced Zoology 17 (1), 49–53.

Sahayaraj, K., Martin, P., 2003. Assessment of *Rhynocoris marginatus* (Fab.) (Hemiptera: Reduviidae) as augmented control in groundnut pests. Journal of Central European Agriculture 4 (2), 103–110.

Sahayaraj, K., Paulraj, M.G., 1999a. Toxicity of some plant extracts against life stages of a reduviid predator. Indian Journal of Entomology 61, 3242–3344.

Sahayaraj, K., Paulraj, M.G., 1999b. Effect of plant products on the eggs of *Rhynocoris marginatus* Fab. (Hemiptera: Reduviidae). Insect Environment 5, 23–24.

Sahayaraj, K., Paulraj, M.G., 2001. Behaviour of *Rhynocoris marginatus* (Fabricius) (Heteroptera: Reduviidae) to chemical cues from three lepidopteran pests. Journal of Biological Control 15 (1), 1–4.

Sahayaraj, K., Borgio, J.F., Muthukumar, S., Anndh, G.P., 2006. Antibacterial activity of *Rhynocoris marginatus* (Fab.) and *Catamiarus brevipennis* (Serville) (Hemiptera: Reduviidae) venoms against human pathogens. Journal of Venomous Animal Toxins Including Tropical Diseases 12 (suppl. 3), 487–496.

Sahayaraj, K., Sankaralinkam, S.K., Balasubramanian, R., 2007. Prey influence on the salivary gland and gut enzymes qualitative profile of *Rhynocoris marginatus* (Fab.) and *Catamiarus brevipennis* (Serville) (Heteroptera: Reduviidae). Journal of Insect Science 4, 331–336.

Sahayaraj, K., Kalidas, S., Tomson, M., 2012. Stage preference and functional response of *Rhynocoris longifrons* (Stål) (Hemiptera: Reduviidae) on three hemipteran cotton pests. Brazilian Archives of Biology and Technology 55 (5), 733–740.

Sahayaraj, K., Muthukumar, S., Rivers, D., 2013. Biochemical and electrophoretic analyses of saliva from the predatory reduviid species *Rhynocoris marginatus* (Fab.). Acta Biochemica Polonica 60 (1), 91–97.

Sahayaraj, K., 1991. Bioecology, Ecophysiology and Ethology of Chosen Predatory Hemipterans and Their Potential in Biological Control (Insecta: Heteroptera: Reduviidae) (Ph.D. thesis). Madurai Kamaraj University, Madurai, India.

Sahayaraj, K., 1999a. Effect of prey and their ages on the feeding preferences of *Rhynocoris marginatus* (Fab.). International Arachis Newsletter 19, 39–40.

Sahayaraj, K., 1999b. Field evaluation of the predator, *Rhynocoris marginatus* (Fab.) on two groundnut defoliators. International Arachis Newsletter 19, 41–42.

Sahayaraj, K., 2000. Evaluation of biological control potential of *Rhynocoris marginatus* on four groundnut pests under laboratory condition. International Arachis Newsletter 20, 72–73.

Sahayaraj, K., 2001. A qualitative study of food consumption, growth and fecundity of a reduviid predator in relation to prey density. Entomologia Croatica 5, 19–30.

Sahayaraj, K., 2002. Field bio-efficacy of a reduviid predator *Rhynocoris marginatus* (Fab.) and plant products against *Aproaerema modicella* Dev. and *Spodoptera litura* (Fab.) of groundnut. Indian Journal of Entomology 64 (3), 292–300.

Sahayaraj, K., 2003. Biological control potential of aphidophagous reduviid predator *Rhynocoris marginatus*. International Arachis Newsletter 23, 29–30.

Sahayaraj, K., 2008. Approaching and rostrum protrusion behaviours of *Rhynocoris marginatus* on three prey chemical cues. Bulletin of Insectology 61 (2), 233–237.

Sahayaraj, K., 2012. Artificial rearing on the nymphal developmental time and survival of three reduviid predators of Western Ghats, Tamil Nadu. Journal of Biopesticide 5 (2), 218–221.

Sahayaraj, K., 2014. Reduviids and their merits in biological control. In: Sahayaraj, K. (Ed.), Basic and Applied Aspects of Biopesticides. Springer, pp. 195–214.

Santadino, M.V., Virgala, M.B.R., Coviella, C.E., 2013. First record of native predators on the invasive species *Thaumastocoris peregrinus* (Hemiptera: Thaumastocoridae) in eucalyptus in Argentina. Revista de la Sociedad Entomológica Argentina 72 (3–4), 219–222.

Sathiamma, B., Nair, K.R.C., Soniya, V.P., 1999. Additional record of insect predators of coconut lace bug *Stephanitis typical* (Distant) and studies on biology and feeding potential of *Euagorus plagiatus* Burmeister (Hemiptera: Reduviidae). Entomon 24, 165–171.

Satpathy, J.M., Patnaik, N.C., Samalo, A.P., 1975. Observations on the biology and habits of *Sycanus affinis* Reuter (Hemiptera: Reduviidae) and its status as a predator. Journal of Bombay Natural History Society 72, 589–595.

Schaefer, C.W., Ahmad, I., 1987. Parasites and predators of Pyrrhocoroidea (Hemiptera) and possible control of cotton strainers *Phonoctonus* spp. (Hemiptera: Reduviidae). Entomophaga 32, 269–275.

Schaefer, C.W., 1988. Reduviidae (Hemiptera: Heteroptera) as agents of biological control. In: Ananthasubramanian, K.S. (Ed.), Bicovas I. Loyola College, Madras, pp. 27–33.

Severein, H.H.P., 1924. Natural enemies of beet leaf hopper (*Eutettix tenella* Baker). Journal of Economic Entomology 17, 369–377.

Shepard, M., McWhorter, R.E., King, E.W., 1982. Life history and illustrations of *Pristhesancus papuensis* (Hemiptera: Reduviidae). Canadian Entomologist 114, 1089–1092.

Silva, A.C., Gil-Santana, H.R., 2004. Predation of *Apiomerus pilipes* (Fabricius) (Hemiptera, Reduviidae, Harpactorinae, Apiomerini) over Meliponinae bees (Hymenoptera, Apidae), in the state of Amazonas, Brazil. Revista Brasileira de Zoologia 21 (4), 769–774.

Singh, O.P., Gangrade, G.A., 1975. Parasites, predator and diseases of larvae of *Diacrisia obliqua* Walker (Lepidoptera: Arctiidae) on soybean. Current Science 44, 481–482.

Singh, O.P., Singh, K.J., 1987. Record of *Rhynocoris fuscipes* Fabricius as a predator of Green stink bug, *Nezara viridula* Linn., infesting soybean in India. Journal of Biological Control 1, 143–146.

Singh, V., Naik, B.C., Sundararaju, D., 1982. CPCRI Annual Report for 1979. Central Plantation Crops Research Institute, Kasargod, India. pp. 132–134.

Singh, J., Arora, R., Singh, A.S., 1987. First record of predators of cotton pests in the Punjab. Journal of Bombay Natural History Society 84, 456.

Singh, V.J., Rajan, S.J., Ambrose, D.P., 2011. Linear regression coefficient values (r) of postembryonic developmental morphometry as a tool in the biosystematics of four *Rhynocoris* species (Hemiptera: Reduviidae: Harpactorinae). In: Ambrose, D.P. (Ed.), Insect Pest Management, a Current Scenario. Entomology Research Unit, St. Xavier's College, Palayamkottai, India, pp. 123–129.

Singh, O.P., 1985. New record of *Rhynocoris fuscipes* Fabr. as a predator of *Dicladispa armigera* (Oliver). Agricultural Science Digest 5, 179–180.

Singh, G., 1992. Management of oil palm pests and diseases in Malaysia in 2000. In: Aziz, A., Kadir, S.A., Barlon, H.S. (Eds.), Pest Management and the Environment in 2000. CAB International, Wallingford Oxon, UK, pp. 195–212.

Singh, V.J., 2012. Biosystematics of Chosen Reduviids (Insecta: Hemiptera: Reduviidae: Harpactorinae) (Ph.D. thesis). Manonmaniam Sundranar Univesity, Tirunelveli, Tamil Nadu, India.

Sitaramaiah, S., Satyanarayana, S.V.V., 1976. Biology of *Harpactor costalis* Stål (Heteroptera: Reduviidae) on tobacco caterpillar *Spodoptera litura* F. Tobacco Research 2, 134–136.

Sitaramaiah, S., Joshi, B.G., Prasad, G.R., Satyanarayana, S.V.V., 1975. *Harpactor costalis* Stål (Reduviidae: Heteroptera) new predator of the tobacco caterpillar (*Spodoptera litura* Fabr.). Science Culture 41, 545–546.

Smith, D., Beattie, G.A.C., Broadley, R., 1997. Citrus Pests and Their Natural Enemies. Department of Primary Industries and Horticultural Research and Development, Queensland. p. 272.

Soares, C.M.S., Iede, E.T., 1997. Prospects for the control of mate beetle *Hedypathes betulinus* Klug, 1825 Col.: Cerambycidae. Documents National Center of Research of Forests 33, 391–400.

Soares, G.G., 1992. Problems with entomopathogens in insect rearing. In: Anderson, T., Leppla, N. (Eds.), Advances in Insect Rearing for Research and Pest Management. Westview Press, Boulder, Colorado, pp. 289–314.

Soley, F.G., Jackson, R.R., Taylor, P.W., 2011. Biology of *Stenolemus giraffa* (Hemiptera: Reduviidae), a web invading, araneophagic assassin bug from Australia. New Zealand Journal of Zoology 38, 297–316.

Srikumar, K.K., Bhat, P.S., Raviprasad, T.N., Vanitha, K., Saroj, P.L., Ambrose, D.P., 2014. Biology and behavior of six species of reduviids (Hemiptera: Reduviidae: Harpactorinae) in a cashew ecosystem. Journal of Agricultural and Urban Entomology 30, 65–81.

Stride, G.O., 1956a. On the biology of certain West African species of *Phonoctonus* (Hemiptera: Reduviidae) mimetic predators of the Pyrrhocoridae. Journal of Entomological Society of South Africa 19, 52–69.

Stride, G.O., 1956b. On the mimetic association between certain species of *Phonoctonus* (Hemiptera: Reduviidae) and the Pyrrhocoridae. Journal of Entomological Society of South Africa 19, 12–27.

Sujatha, S., Vidya, L.S., Sumi, G., 2012. Prey–predator interaction and info-chemical behavior of *Rhynocoris fuscipes* (Fab.) on three agricultural pests (Heteroptera: Reduviidae). Journal of Entomology 9 (2), 130–136.

Summerville, W.A.T., 1931. The larger horned citrus bug. Queensland Department of Agriculture Stock Division of Entomology. Plant Pathology Bulletin 8.

Sundararaju, D., 1984. Cashew pests and their natural enemies in Goa. Journal of Plant Crops 12, 38–46.

Swadener, S.O., Yonke, T.R., 1973a. Immature stages and biology of *Apiomerus crassipes*(Hemiptera: Reduviidae). Annals of the Entomological Society of America 66, 188–196.

Swadener, S.O., Yonke, T.R., 1973b. Immature stages and biology of *Sinea Complexa* with notes on four additional reduviids (Hemiptera: Reduviidae). Journal of Kansas Entomological Society 46, 124–136.

Takeno, K., 1998. Enumeration of the Heteroptera in Mt. Hikosan, western Japan with their hosts and preys. Esakia 38, 29–53.

Taylor, J.S., 1932. Report on cotton insect and disease investigations Part II. Notes on the American bolt worm (*Heliothis obsoleta* Fabr.) on cotton, and on its parasite, *Microbracoss brevicornis* Western. Science Bulletin Department of Agriculture South Africa 113, 18.

Taylor, E.J., 1949. Biology of the damsel bug, Nabis alternates. In: The Utah Academy of Science Arts and Letters, Proceedings, vol. 26, pp. 132–133.

Thanasingh, P.D., 2002. Biodiversity of Entomofauna (In Part) in an Agroecosystem a Semiarid Zone and a Scrub Jungle in Tamil Nadu South India (Ph.D. thesis). Manonmaniam Sundaranar University, Tirunelveli, Tamil Nadu, India.

Thangavelu, K., Ananthasubramanaian, K.S., 1981. Natural enemies of *Rhypara chrominae* (Lygaeidae: Heteroptera) from southern India. Proceedings of Indian Academy of Science (Animal Sciences) 47, 632–636.

Thompson, W.R., Simmonds, F.J., 1964. A Catalogue Parasites and Predators of Insect Pests. Sect: 3 Predator Host Catalogue Common Wealth Agri. Bureau Books, England. 2048.

Trematterra, P., Tozzia, G.C., 1985. Arthropodi in ricoveri larval di *Paralipsa gularis* (Zeller) (Lepidoptera: Galleridae). In: Atti. XIV Congresso Naziohale Italians di Entemologia soffo gli auspicu due Academia Nazionale Italinna di Entomologia, pp. 535–541.

Ullyett, G.C., 1930. The life history, bionomics and control of cotton stainer (*Dysdercus* spp.) in South Africa. Science Bulletin Department of Agriculture South Africa 94, 3–9.

Uthamasamy, S., 1995. Biological control of the introduced psyllid, *Hetropsylla cubana* Crawford (Homoptera: Psyllidae), infesting subabul *Leucaena leueocephala* (Lam.) de wit-progress and prospects. In: Ananthakrishnan, T.N. (Ed.), Biological Control of Social Forest and Plantation Crops Insects. Oxford & IBH Publ. Co. Pvt. Ltd., New Delhi, India, pp. 65–75.

Vanderplank, F.L., 1958. The assassin bug *Platymeris rhadamanthus* Gerst. (Hemiptera: Reduviidae) a useful predator of the rhinoceros beetles *Oryctes boas* F. and *Oryctes monoceros* (Oliv.). Review of Entomological Society of South Africa 1, 309–314.

Varma, R.V., 1989. New record of *Panthous bimaculatus* (Hemiptera: Reduviidae) as predator of pests of *Alantha triphysa*. Entomon 14, 357–358.

Venkatesan, S., Seenivasagan, R., Karuppasamy, G., 1999. Nutritional influence of prey on the predatory potential and reproduction of *Sycanus collaris* Fabricius (Reduviidae: Heteroptera). Entomon 24, 241–252.

Vennison, S.J., Ambrose, D.P., 1990. Egg development in relation to soil moisture in two species of reduviids (Insecta: Heteroptera). Journal of Soil Biology and Ecology 10 (2), 116–118.

Vennison, S.J., Ambrose, D.P., 1992. Biology, behaviour and biocontrol effeciency of a reduviid predator, *Sycanus reclinatus* Dohrn (Heteroptera: Reduviidae) from southern India. Journal of Mitterand Zoological Museum, Berlin 68 (1), 143–156.

Vennison, S.J., 1988. Bioecology and Ethology of Assassin Bugs (Insecta: Heteroptera: Reduviidae) (Ph.D. thesis). Madurai Kamaraj University, Madurai, India.

Wallace, H.R., 1953. Notes on the biology of *Coranus subapterus* DeGeer (Hemiptera: Reduviidae). Proceedings of the Royal Entomological Society London (A) 28, 100–110.

Wappler, T., Garrouste, R., Engel, M.S., Nel, A., 2013. Wasp mimicry among Palaeocene reduviid bugs from Svalbard. Acta Palaeontologica Polonica 58 (4), 883–887.

Weaver, N., 1978. Chemical control of behavior-interspecific. In: Rockstein, M. (Ed.), Biochemistry of Insects. Academic Press, New York, pp. 392–418.

Webster, F.M., 1907. The chinch bug. USDA Bulletin of Entomology 69, 95.

Webster, R.L., 1912. The pearl slug. Sinea diadema. Lowa Bulletin 130, 190.

Weirauch, C., 2006. Observations on the sticky trap predator *Zelus luridus* Stål (Heteroptera, Reduviidae, Harpactorinae), with the description of a novel gland associated with the female genitalia. Denisia 19, 1169–1180.

Weirauch, C., 2008. Cladistic analysis of Reduviidae (Heteroptera: Cimicomorpha) based on morphological characters. Systematic Entomology 33, 229–274.

Whitcomb, W.H., Bell, K., 1964. Predaceous insects, spiders and mites and Arkansas cotton fields. Bulletin of Arkansas Agriculture Experimental Station 690, 1–84.

Wignall, A.E., Jackson, R.R., Wilcox, R.S., Taylor, P.W., 2011. Exploitation of environmental noise by an araneophagic assassin bug. Animal Behaviour 82 (5), 1037–1042.

Wolf, K.W., Reid, W., 2001. Surface morphology of legs in the assassin bug *Zelus longipes* (Hemiptera: Reduviidae): a scanning electron microscopy study with an emphasis on hairs and pores. Annals of the Entomological Society of America 94, 457–461.

Yano, 1990. A host record of *Isyndus obscures* (Dallas) (Hemiptera: Reduviidae). Japanese Journal of Entomology 58, 204.

Yao, D., Yan, J., Li, G., Liu, H., 1995. A study on the morphology and bionomic of predatory bug *Agriosphodrus dohrni* (Signoret). Forest Research 8, 442–446.

Yasumatsu, K., Watanabe, C., 1964. Reduviidae in a Tentative Catalogue of Insect Natural Enemies of Injurious Insects in Japan, Part I Parasite, Predators Host Catalogue. Ent. Lab. Fac. Agri. Kyushu University, Fukoka. pp. 4–6.

Youssef, N.A., Abd-Elgayed, A.A., 2015. Biological parameters of the predator, *Amphibolus venator* Klug (Hemiptera: Reduviidae) preying on larvae of *Tribolium confusum* Duv. (Coleoptera: Tenebrionidae). Annals of Agricultural Sciences 60 (1), 41–46.

Zanunico, J.C., Alves, J.B., Zanunico, T.V., Garcia, J.F., 1994. Hemipteran predators of eucalyptus defoliator caterpillars. Forest Ecology and Management 65, 65–73.

Zerachia, T., Bergmann, F., Shulov, A., 1973. Pharmacological activities of the venom of the predaceous bug *Holotrichius innessi* (Heteroptera: Reduviidae). In: Kaiser, E. (Ed.), Animal and Plant Toxins, pp. 143–146.

Zhang, G., Weirauch, C., 2011. Sticky predators: a comparative study of sticky glands in harpactorine assassin bugs (Insecta: Hemiptera: Reduviidae). Acta Zoologica 94 (1), 1–10.

Syrphid Flies (The Hovering Agents)

Omkar, Geetanjali Mishra

Centre of Excellence in Biocontrol of Insect Pests, Ladybird Research Laboratory,
Department of Zoology, University of Lucknow, Lucknow, India

1. INTRODUCTION

The need to produce more and more crops to feed the exponentially increasing world population has led to massive intensification of the agricultural process, which in turn has led to a concurrent increase in pest attacks. As a consequence, this has led to heavy unrestrained use of pesticides, a process that is neither economical nor ecologically sustainable. Despite the world spending around $40 billion on pesticides (Popp, 2011; Popp et al., 2013), and current worldwide pesticide use having increased to more than 3 million tons per annum (Pan-UK, 2003; Pimentel, 2005), more than 35–40% of crop yields are lost to pests, pathogens, and weeds (Oerke, 2005; Pimentel, 2005). Amazingly, this persistent loss in crop yield can be attributed largely to the pesticides themselves, whose excessive use combined with their nonselective action and nonbiodegradable nature has led to resistance in target pests, resurgence of new pest species, secondary pest outbreaks (Kenmore, 1980; Settle et al., 1996), toxic effects on nontarget beneficial species, environmental pollution, and hazards to human health. Excessive and repeated use of pesticides has also led to the development of resistance in pests to such an extent that some of them are near impossible to control (Georghiou, 1986; Naylor and Ehrlich, 1997; Palumbi, 2001).

Scientists and agriculturists have consensually agreed to use and propagate pest management solutions that involve natural enemies that are capable of coevolving with the pest and can thus respond effectively to pest population dynamics, along with being environmentally friendly and cost-effective, in other words, the biological control of pest species. Numerous natural enemies in the form of predators, parasitoids, parasites, and pathogens are well recognized as providers of ecosystem services to the agriculture sector by suppressing or reducing pest populations. These ecosystem services are worth $417 billion annually the world over (Costanza et al., 1997), with the potential of being further enhanced.

Of the many natural enemies being investigated across the globe as potential biological control agents, syrphid flies (Diptera: Syrphidae) are one of the important ones. Other than being potential biocontrol agents, syrphids are well-recognized as a major component of the pollinator complex and indicators of agricultural pollution, habitat disturbance, and quality (Sommaggio, 1999; Burgio and Sommaggio, 2007; Bugg et al., 2008; Jauker et al., 2009). Their potential as biological control agents is specifically of interest in management of aphid populations because larval stages of most predatory syrphids feed quite voraciously on these tiny sap sucking insects.

For agriculturists and horticulturists the world over, aphids form one of the most notorious groups of pest, with their highly polyphenic nature, the ability to reproduce both sexually and asexually, intricate life cycles, telescopic generations, and close associations with their host plants (Dixon, 1998; van Emden and Harrington, 2007). They launch a three-pronged attack on the host plant, the first being the extracting of sap using piercing and sucking mouthparts. This leads to the stunting of plant growth, deformation and discoloration of leaves and fruits, or formation of galls on leaves, stems, and roots (Hamman, 1985). Secondly, they cause fungal growth on the plants due to honeydew secreted by them, and finally, in the last attack, they are capable of transmitting many plant viruses. Their potential for causing losses can be gauged by the 68–96% loss in oil yield from *Brassica campestris* Linnaeus due to heavy infestation by the aphid *Lipaphis erysimi* Kaltenbach (Suri et al., 1988). More than 4000 aphid species have been recognized throughout the world (Dixon, 1998). Owing to the immense damage caused by them, efforts to manage aphid populations have attracted a lot of attention, with special emphasis being placed on predators primarily, ladybirds, Chrysopidae, and syrphids. Compared with the former two, syrphids have been highly underrated, despite being an integral part of predator complexes of aphids as well as the presence of strong supportive evidence (Tenhumberg and Poehling, 1995; Winder et al., 1994; Brewer and Elliott, 2004; Freier et al., 2007; Haenke et al., 2009).

Thus, in this chapter, we place emphasis on these relatively underplayed bioagents to shed light on their reproductive, developmental, and predation biology, position in guild and the methods by which they can be utilized more effectively in the biological control of pests.

2. WHAT ARE SYRPHIDS?

Syrphids are commonly known as hover flies worldwide, with the exception of the USA, where they are also known as flower flies. The adults are highly active with usually brightly banded abdomens in black and yellow, ranging in lengths from 0.25–0.75 inch, large eyes, an aptitude for hovering, as well as the ability to fly sideways. They are active pollinators. Their larvae are usually predatory. These flies are well-known and well investigated as Batesian mimics of bees (Golding et al., 2001; Penney et al., 2012; Edmunds and Reader, 2014). The family Syrphidae is one of the most species-rich families of the order Diptera, with about 6000 described species belonging to 180 genera from three subfamilies (Syrphinae, Milesiinae, and Microdontinae) (Vockeroth and Thompson, 1987; Sommaggio, 1999; Katzourakis et al., 2001). These small bee-like flies can be found almost everywhere with reports of their presence on all continents except Antarctica.

3. GENERAL BIOLOGY

Syrphids in general mate 1–13 days after emergence (Tawfik et al., 1974a), and then proceed to lay numerous long solitary whitish to gray oblong eggs (~1 mm in length), lying on their sides, usually near aphid patches (Figure 1). These eggs hatch within 2–3 days, releasing small transparent to greenish colored legless maggots which are about 1–13 mm long depending on the stage and the species (Gilkeson and Klein, 1981; Figure 1). Syrphid larvae are easily identified

FIGURE 1 Life stages of *Episyrphus balteatus*: (A) adult female, (B) ovipositing female, (C) egg, (D) third larval instar, (E) early pupa, and (F) late pupa.

by the pulsating internal organs beneath partially transparent skin. These neonate larvae have very limited stored resources and must find food quickly to avoid starvation. However, owing to poor defenses, many of the carnivorous neonates are killed by their intended first meal itself (Schmutterer, 1972). The larvae are mobile and move around by sticking their rear end to a substrate. They undergo three molts prior to turning into pupa after 2–3 weeks and having consumed up to 400 aphids. The pupa is enclosed in a hardened, teardrop-shaped case, which is the hardened skin of the last larval stage (Figure 1). During the favorable season, adult hover flies emerge from the pupal case in 1–2 weeks, otherwise they overwinter as pupae. Based on various abiotic factors, syrphids can have five to seven generations each season.

Syrphids are life-history omnivores (Polis and Strong, 1996). Adult syrphids feed mostly on pollen and nectar from flowers but can also feed on honeydew produced by the aphids. Pollen is essential for the maturation of the female reproductive system (Schneider, 1948) which is synovigenic (i.e., immature at emergence). Nectar in itself cannot lead to egg production but can only provide energy. In the presence of a suitable and sustaining food source, females are capable of laying eggs from the end of pre-oviposition period, which lasts about a week after emergence (Stürken, 1964; Geusen-Pfister, 1987) until death (Geusen-Pfister, 1987; Gilbert, 1993; Branquart and Hemptinne, 2000a). Male hover flies, once mature, do not require many proteins and amino acids, while the females require these persistently for batches of maturing eggs (Gilbert, 1993).

Unlike adults, larvae possess quite varied feeding habits, which can be broadly categorized into four classes: (1) phytophagous, that feed on plant tissue and plant products, (2) saprophagous or coprophagous that scavenge on or filter decaying matter, (3) aquatic filter feeders that live in the nests of social insects (bees, wasps, and ants), and (4) carnivorous, primarily aphidophagous. Most of these aphidophagous species are found in two tribes of the subfamily Syrphinae: Syrphini and Melanostomini, both of worldwide distribution. While most of the aphidophagous syrphids prey on free-living aphids, the pipizine syrphids show a preference for wax-secreting aphids (families Thexalidae, Pemphigidae, and Adelgidae) and feed on aphids that live in galls or roots (Stubbs and Falk, 1993).

4. REPRODUCTIVE BIOLOGY

The reproductive behavior of syrphids is simple and straightforward with no reported dramatic displays of courtship. Males of many syrphid species are known to congregate consistently at specific times and in specific locations such as under trees or in clear air spaces (Gilbert, 1984; Waldbauer, 1990). These places have been called leks (Heinrich and Pantle, 1975), although no dominance hierarchy or territoriality is exhibited. Swarming behavior is in itself quite rare in syrphids and where found does not assist greatly in facilitating matings (Downes, 1969). Male syrphids show two types of mate search: near feeding places and near oviposition plots. Some males (Syrphinae, Pipizinae, Sphegini, and Cheilosiini) hover for long periods in the air at both feeding and oviposition plots, while others (Milesiini, Xylotini, and Eristalinae) either search for females on flowering plants or near sites of emergence (Mutin, 1996).

In-flight mating behavior shows both male grasping and female carrying behavior in *Pseudodoros clavatus* (F.) (van Rijn et al., 2006) and *Merodon equestris* (F.) (Conn, 1979). Attempts to compete for females and dislodging attempts at copulating males have also been reported (Gilbert, 1984; Waldbauer, 1990; van Rijn et al., 2006). In *Eristalis* sp. males spin in the air, moving together higher and faster and with increasing diameter of turns. Then one male gives up and flies away and the other returns to the place of the scuffle. Sometimes the males are joined together and fall with the dominant male taking the copulation (Mutin, 1996). Such observations, along with the repeated matings and long periods spent in copula, indicate behaviors consistent with sperm competition and mate-guarding, respectively (Belliure and Michaud, 2001).

The average number of matings in female syrphids ranges from 5.4 to 9.2 times, with up to 20 matings in a lifetime (Tawfik et al., 1974b). However, to begin oviposition, the ovaries of the females need to be mature, a process that is dependent on the presence and consumption of pollen; with nectar being sustaining in nature (Schneider, 1948; Chambers, 1988; Gilbert, 1993). However, the presence of pollen is not an everyday need, since the syrphids have the capability to store food reserves for several days (van Rijn et al., 2006). Thus, female syrphids are able to switch from an initial foraging site search to an oviposition site search later. Once the food reserves run low, syrphids can switch between foraging and ovipositing grounds. In the absence of oviposition sites, females resorb their eggs instead of ovipositing at unsuitable sites (Branquart and Hemptinne, 2000b).

The number of eggs laid by syrphids is also known to be dependent on aphid density, with high aphid densities leading to longer stays (Sanders, 1979) and higher number of eggs (Ito and Iwao, 1977; Chambers, 1991; Tenhumberg and Poehling, 1991). The potential fecundity varies isometrically with the size of females and is expressed by ovariole number, reproductive biomass, and abdomen volume. However, female size has no influence on egg size (Branquart and Hemptinne, 2000a). Under laboratory conditions, females of *Episyrphus balteatus* (De Geer) lay around 2000–4500 eggs in a lifetime (Branquart and Hemptinne, 2000a). The variation in fecundity is also dependent on the prey source, with females of *Xanthogramma aegyptium* Wied. laying an average of 67.9 and 105.5 eggs on aphids, *Aphis gossypii* Glover and *Myzus persicae* Sulzer, respectively (Tawfik et al., 1974a). While high fecundity such as 3225, 2774, 553, and 1119 eggs on oak leaves, mud soils, fermented rice straw, and sawdust, respectively, by *Eristalis cerealis* F. has been reported (Kim et al., 1994), *Paragus aegypticus* Macq. has a low average fecundity of 28.2 eggs (Tawfik et al., 1974a).

Not only the prey patch but also the host plant of the prey patch has sizable effects on the fitness of syrphids (reviewed by Almohamad et al., 2009). Laubertie et al. (2012) demonstrated that variation in flowering plants had significant effect on the longevity and several reproductive parameters of *E. balteatus*. Coriander was most efficient in terms of proportion of females laying eggs, while buckwheat yielded the highest mean longevity. However, the flowering plant *phacelia*, *Phacelia tanacetifolia* Bentham, increased oviposition rate and lifetime fecundity the most leading to the maximum reproductive potential. Despite clear responses of syrphids to different flowering plants, no correlation between pollen and nectar consumption or between the quantity of pollen ingested and the resulting female performance was observed (Laubertie et al., 2012).

Unlike ladybird beetles, where increased number of eggs laid led to reduction in longevity, the same has not been proven true in syrphids. In fact females of *Sphaerophoria flavicauda* Zetterstedt that mated and oviposited lived longer than those that did not. Unexpectedly, mated males have been found to have a shorter life span than unmated males (Tawfik et al., 1974c; Makhmoor and Verma, 1987).

While the effect of various biotic and abiotic factors on syrphid reproduction has received adequate attention, studies on oviposition site selection have also revealed interesting insights. Female syrphids are known to select their oviposition sites by a four step process (Table 1) involving: (1) assessment of long range optical cues, including the size, density, and color of vegetation (Sanders, 1981a,b, 1983a), (2) short range optical cues, which involve aphid colony size recognition (Dixon, 1959; Kan and Sasakawa, 1986; Kan, 1988a,b) in terms of aphid density (Tamaki et al., 1967) and quality (Kan, 1988a,b), (3) processing of olfactory stimuli either aphid-produced (Aphidozetic) or plant-produced (Phytozetic) (Chandler, 1968a; Shonouda et al., 1998; Togashi, 1987), and (4) utilization of gustatory

TABLE 1   Role of Semiochemical Cues from the Host Plants and Aphids in Searching and Utilization of Oviposition Sites in Syrphids (Almohamad et al., 2009)

Sense involved	Influences	References
Visual cues	1. Size of plant patch	Chandler (1968a) and Sanders (1983a,b)
	2. Density of plant patch	Chandler (1966, 1968c)
	3. Color of plants	Sanders (1982), Sutherland et al. (1999), and Laubertie et al. (2006)
	4. Form of plants	Chandler (1968a) and Sutherland et al. (1999)
	5. Size and position of aphid colony	Chandler (1968b), Ito and Iwao (1977), Bargen et al. (1998), Scholz and Poehling (2000), Sutherland et al. (2001), and Almohamad et al. (2006)
	6. Shape of aphids	Chandler (1968b)
	7. Movement of aphids	Chandler (1968b) and Ito and Iwao (1977)
Olfactory cues	8. Smell of plants	Sutherland et al. (1999) and Verheggen et al. (2008)
	9. Smell of plants	Volk (1964), Almohamad et al. (2008), and Verheggen et al. (2008)
	10. Smell of aphid associated with plants	Volk (1964), Harmel et al. (2007), and Verheggen et al. (2008)
Gustatory cues	11. Honey dew	Dixon (1959), Kan and Sasakawa (1986), Budenberg and Powell (1992), Bargen et al. (1998), Scholz and Poehling (2000), and Sutherland et al. (2001)
Touch	12. Acute site for eggs	Dixon (1959) and Schneider (1969)

Response by females	Influences involved
Habitat selection	1, 2, 3
Plant selection	2, 3, 8
Aphid colony selection	5, 6, 7, 9, 11
Egg site selection	5, 6, 7, 12

stimuli, with the female using her labellum to assess honeydew (Dixon, 1959; Kan and Sasakawa, 1986), which is an important oviposition stimulus for syrphids (Bombosch and Volk, 1966; Budenberg and Powell, 1992).

Females of *E. balteatus* and *Syrphus ribesii* (L.) prefer to oviposit in the presence of specific aphid species (Sadeghi and Gilbert, 2000a), and female age and host deprivation is known to change the magnitude of preference but not its order (Sadeghi and Gilbert, 2000b). The hierarchy threshold model (Sadeghi and Gilbert, 2000a,b) suggests that a gravid female searching among five potential preys (A–E) will follow an intrinsically possessed preference for each food, leading to the evolution of an order that will not change in the entire lifetime (Figure 2). Acceptance of food will always be in the rank order, however, settling for prey lower in order will depend on whether the stimulus of the food exceeds the current motivational threshold (variable with age or egg load) (Almohamad et al., 2009). The model is interesting because it fuses two processes and explanations of specialization: (1) slow, based on trade-offs and coevolution and (2) individual behavioral flexibility in response to prevailing ecological conditions (Almohamad et al., 2009).

The preference for an aphid patch as an oviposition site may also be influenced by semiochemicals emitted by the prey species and the host plants (Samuel et al., 2013; Table 1). Females of *P. clavatus* and *E. balteatus* do not oviposit on clean plant tissues but accept plants with residues of honeydew from which aphids have been removed (Bhat and Ahmad, 1988). However, plants treated with artificial honeydew did not elicit a response and aphid presence was essential for *Sphaerophoria cylindrica* (Say), *Syrphus* sp. (Ben Saad and Bishop, 1976), and *Platycheirus fulviventris* (Macq.) (Rotheray, 1987) to oviposit. Females of *Metasyrphus corollae* (F.), a syrphid predator of *Aphis fabae* Scopoli, respond to aqueous extract containing kairomones of the prey applied on to clean *Vicia faba* leaves. The aqueous extract of aphids caused females of *M. corollae* to lay eggs at all aphid densities, the number of eggs being directly proportional to aphid kairomone concentration (Shonouda, 1996).

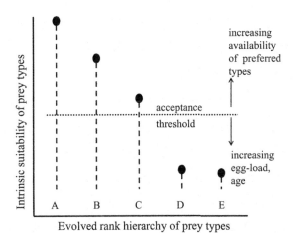

FIGURE 2   Hierarchy threshold model of host choice (Courtney et al., 1989) on application to a gravid syrphid females searching among a set of five possible preys (A–E; Almohamad et al., 2009).

There is sufficient evidence that oviposition in syrphids is generally positively prey density-dependent (Geusen-Pfister, 1987; Tenhumberg and Poehling, 1991; Bargen et al., 1998; Scholz and Poehling, 2000; Sutherland et al., 2001). However, field studies indicate that this density dependence is not absolute and large or aging aphid colonies are usually avoided by female syrphids (Kan and Sasakawa, 1986), a move that is known as future oviposition tactics. This tactic is a means of the syrphid to ensure that its larvae hatch and develop to adulthood in the presence of a prey source that will not crash in the future. This is essential because when the larvae of *E. balteatus* are reared on a below-optimal food resource, larvae have prolonged larval period and the adults thus produced have reduced fecundity and longevity (Ruzicka and Gonzales Cairo, 1976; Cornelius and Barlow, 1980; Samuel et al., 2005). The assessment of the likelihood of the aphid colony crashing may be as signals in the form of oviposition deterring pheromones.

The presence of conspecific eggs or larvae was not inhibitory for oviposition by *P. clavatus* and *E. balteatus* (Bargen et al., 1998). However, studies show evidence contrary to *E. balteatus*, where oviposition is deterred in the physical as well as the chemical presence of eggs (Scholz and Poehling, 2000).

5. FORAGING BEHAVIOR

Location of prey sources by syrphids is influenced by visual as well as olfactory cues. *E. balteatus* shows enhanced behavioral responses to yellow-colored smaller flowers with greater nectar concentration rather than pollen, whose quantity had no effect (Sutherland et al., 1999).

On locating the prey source, syrphids begin to forage, the efficiency of which declines with an increase in aphid abundance and plant size (Scott and Barlow, 1986) while it is not influenced by temperature (Brodsky and Barlow, 1986). Syrphids are also suggested to follow the optimal forging theory (Hemptinne et al., 1993), which suggests that there is an optimal window of oviposition, i.e., syrphids oviposit in regulated numbers in optimal density of an aphid patch so as to successfully ensure their offspring survival. Early (in a young aphid colony) or late (in an old aphid colony) oviposition is likely to lead to higher mortality in early and late instars, respectively, leading to suboptimal offspring production. However, evidence suggests that syrphid species vary with respect to the size of aphid colonies they prefer for oviposition, some species preferring the smallest (Kan and Sasakawa, 1986) and others the largest sized colony (Sanders, 1979; Rotheray, 1987). Some species specifically avoid ovipositing in colonies having maturing alate forms (Kan, 1988a).

The first instars of *E. balteatus* exhibit aphid olfactory cue and not honeydew-guided short distance prey search, while no response has been observed in later instars to aphid–plant complex odors. On the other hand, these cues act as feeding stimulants for larvae (Bargen et al., 1998).

The prey consumption efficacy of syrphids differs with species, with larvae of *E. balteatus* largely recognized as the most common and efficient aphidophagous predator (Sharma and Bhalla, 1988; Tenhumberg and Poehling, 1994; Kumar et al., 1996; Sobhani et al., 2013) consuming up to 100 species of aphids (Sadeghi and Gilbert, 2000c). The predation rate, however, changes with the change in instar stage (Patro and Behera, 1992; Zeki and Kilincer, 1994; Lakhanpal and Raj, 1998; She et al., 1998; Joshi et al., 1999; Dayal and Gupta, 2001) and prey species (Hamid, 1986). Hopper et al. (2011) indicated that there are species-specific differences in predation capacity relative to body size for syrphids feeding on the same aphid species. This was against their assumption that there would be a strong correlation between larval size and predation capacity regardless of species. The results thus suggest that syrphid species cannot be grouped by body size alone to extrapolate laboratory-based predation potential estimates to the field.

Laboratory observations have shown that prey consumption of syrphids increases curvilinearly with increasing prey density, exhibiting a type II functional response to varying aphid densities (Ekukole and Misari, 1993). However, some studies indicate that while the first instar of some syrphids show a type II functional response, later instars commonly exhibit type III response is more common in later instars (Singh and Mishra, 1988; Tenhumberg and Poehling, 1994). These functional responses also change in response to the physical traits of host plant surface (Sobhani et al., 2013). The prey-handling time and amount of prey consumed are determined by the size and hunger of the syrphid larvae and the degree of depletion of prey contents (Barlow and Whittingham, 1986).

6. GROWTH AND DEVELOPMENT

In general, the developmental period of most syrphids usually ranges from 20.10 to 25.80 days at $26 \pm 2\,°C$ (Joshi et al., 1999). The relatively few investigations that have been conducted on development of syrphids have focused largely on the effect of various prey and host plants. The results of these studies are not path-breakingly different from that on other predators. The developmental durations, survival, and weight of immature stages vary with the aphid species consumed (Du and Chen, 1993; Lakhanpal and Raj, 1998; Alhmedi et al., 2008) and also the host plant that is infested (Hindayana, 2001; Vanhaelen et al., 2002). When provided with more aphids, syrphid instars consumed more and resulted in heavier pupae, which did not have an effect on the reproductive output of the emerging females that strangely depends on the adult longevity (Scott and Barlow, 1984). But on the other hand, Amorós-Jiménez et al. (2014) found that the floral preference of the adults is linked to developmental performance of their offspring. The flower most frequently visited by adults was sweet alyssum, which lends to enhanced adult body size and thus egg to adult survival.

The development of eggs, larvae, and pupae of *Melangyna viridiceps* and *Symosyrphus grandicornis* on *Macrosiphum rosae* was studied at constant temperatures of 4, 8, 10, 15, 20, 25, or 30 °C. Although eggs hatched at 4 °C, first instar larvae died at this temperature. The developmental rates were directly proportional to the temperatures (Soleyman-Nezhadiyan and Laughlin, 1998). Syrphids per se have a lower development threshold (LDT; 4–5 °C) than coccinellids, which enables them to remain active at lower temperatures and to develop faster between 10 °C and 27 °C than coccinellids, whose LDT is 10 °C. As a consequence, early in the year, when temperatures are low but increasing, syrphids appear before and complete their development more quickly than coccinellids, and in the latter half of the year, when temperatures are generally lower and decreasing, only syrphids are likely to be able to complete their development before the aphids disappear (Dixon et al., 2005; Figure 3).

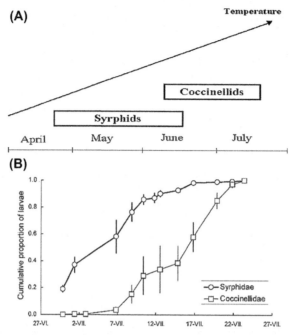

FIGURE 3   (A) Occurrence of immature stages of coccinellids and syrphids with seasons. (B) Increase in cumulative numbers of syrphids and coccinellids in a crop of winter wheat in 2004 in Prague, Czech Republic (Dixon et al., 2005).

7. INTRAGUILD PREDATION

In a study by Agarwala and Yasuda (2001) assessing the effectiveness of wax covers on larvae of coccinellid *Scymnus* against syrphid predators, it was observed that an increase in wax cover increased protection against intraguild predation by syrphids. In another study, eggs and first and second instar larvae of *E. balteatus* were highly susceptible to predati1on by the other predators, while only the larvae of *Chrysoperla carnea* preyed upon the pupae of *E. balteatus*. Eggs and first and second instar larvae of *E. balteatus* sustained intraguild predation, irrespective of arena size and extraguild prey, while predation on third instar larvae of *E. balteatus* was significantly reduced (Hindayana et al., 2001). In an interesting result when individuals of *Adalia bipunctata* and *E. balteatus* were released together in field experiments, for the management of *Dysaphis plantaginea*, additive results were obtained. The two species did not significantly interact; an additive model best explains their joint effect (Wyss et al., 1999). Organisms involved in intraguild predation (IGP) do not always follow classical one to one rules, with the same individual having the potential to either be a predator or a prey depending on prevailing circumstances (Polis et al., 1989). The interactions may show different results depending on relative size, mobility, and density, and by the densities of extraguild prey (Sengonça and Fringe, 1985; Polis et al., 1989; Lucas et al., 1998; Hindayana et al., 2001).

In a plantless as well as planted arena, intraguild predation between mirid *Macro-lophus caliginosus* Wagner (Hemiptera: Miridae), and syrphids, *Sphaerophoria rueppellii*

(Wiedemann), *Sphaerophoria scripta* (L.), and *E. balteatus*, were investigated. Mutual intra-guild predation was observed with syrphid eggs and mirid nymphs and adults being susceptible to predation by the other species. All syrphid species showed the same propensity to predation. *Sphaerophoria scripta* eggs were less susceptible to predation than those of the two other syrphid species. However, plant presence changed the results with predation by syrphid larvae on mirids never exceeding 10%, while 100% of syrphid eggs were preyed upon after 48 h. Presence or absence of aphids on the plants did not influence the predation (Fréchette et al., 2007).

8. IMPROVING BIOCONTROL

Mathematical models (Winder et al., 1994) have predicted the capability of syrphids as major biocontrol agents in cereal crops, a premise that has found extensive support from field based studies (e.g., Chambers and Sunderland, 1983; Chambers et al., 1986; Chambers and Adams, 1986; Entwistle and Dixon, 1989; Tenhumberg and Poehling, 1995). Syrphids are some of the most persistent and consistent predators recovered from aphid infested crops and are therefore considered primarily responsible for the regulation of aphid populations to below economic injury levels (Smith et al., 2008).

The urge to further enhance the efficacy of these syrphid predators has prompted investigations into numerous methods. Chief among these measures is habitat management, a concept that has taken the field of conservative biological control (CBC) by storm. The goal of conservative biological control per se is to maximize ecosystem services through the creation or improvement of habitat for native or at least naturally-occurring natural enemies. In practice, CBC is influenced by a more targeted and systematically careful use of pesticides or by habitat manipulation to enhance natural enemy fitness and effectiveness (Barbosa, 1998; Eilenberg et al., 2001; Gurr et al., 2004). Habitat manipulation or management involves the removal or addition of plants in and around agricultural and horticultural fields. This may be achieved through plantation of hedgerows or weed strips along field edges, creation of strips of unmanaged habitat within the field, leaving adjacent fields fallow, interplanting crops to provide additional floral resources, and/or undertaking larger-scale efforts such as restoration or conservation of perennial habitat surrounding the farm (Fiedler et al., 2008). The purpose of these changes to the habitat is primarily to make available increased amounts of food, alternative prey, and hiding and mating places to allow them to thrive (van Emden, 1990; White et al., 1995; Hickman and Wratten, 1996; Landis et al., 2000; Gurr et al., 2004; Nicholls and Altieri, 2004; Heimpel and Jervis, 2005). Identification of plant refuges in different cropping systems allows the biocontrol practitioners to reduce or eliminate source of pests and enhance populations of natural enemies, either through attraction of more enemies or their better reproduction.

Landscape complexity, often measured by landscape composition or diversity, has been widely shown to benefit populations of natural enemies of agricultural pests (reviewed in Bianchi et al., 2006; Kremen and Chaplin-Kramer, 2007). It has been found that the predators and parasitoids increase in both abundance (Schmidt et al., 2008; Gardiner et al., 2009a,b) and diversity (Purtauf et al., 2005; Schmidt and Tscharntke, 2005; Drapela et al., 2008; Schmidt et al., 2008; Werling and Gratton, 2008) when the surrounding landscapes are noncrop habitats. The surrounding noncrop habitats have been observed to cause a shift in enemy community assemblages (Elliott et al., 2002; Gardiner et al., 2009b), which may allow synergistic interaction among the different new enemy complexes produced (Cardinale et al., 2003). While most studies have supportive evidence for the effects of non-crops on natural enemies, some studies are contrary (Menalled et al., 1999; Prasifka et al., 2004; Schmidt et al., 2008; Oberg et al., 2008).

A number of studies on habitat management and its efficacy on biocontrol by syrphids have been conducted and have proved beneficial (Molthan and Ruppert, 1988; Sengonça and Fringe, 1988; Lövei et al., 1992; MacLeod, 1992; White et al., 1995; Morris and Li, 2000; Tenhumberg and Poehling, 1995; Schmidt et al., 2003; Brewer and Elliott, 2004). It has been observed that addition of flowering species in and adjacent to cropped lands increases syrphid activity (Chambers and Adams, 1986; Tenhumberg, 1995; Tenhumberg and Poehling, 1995) as well as that of both ground-dwelling and flying predators and parasitoids (Lys and Nentwig, 1992; Salveter and Nentwig, 1993; Frank, 1999; Thies and Tscharntke, 1999). Hickman and Wratten (1996) found that the addition of *phacelia* along the boundary of the cropped fields led to significantly higher numbered of syrphid eggs and fewer aphids. This is presumed to have an impact as adults of all syrphid groups feeding on pollen and nectar of flowering plants are thus expected to harvest benefits (Chambers and Aikman, 1988; Cowgill et al., 1992; Hickman and Wratten, 1996).

Chaplin-Kramer (2010) studied the effects of landscape composition on syrphid larval abundance, aphid population growth, and subsequently the state of infestation of broccoli. They found that while natural enemies were enhanced, pest population growth was slowed, and pest densities were weakly but significantly reduced. In broccoli, where only severe infestation involving the head leads to loss in revenues (aphid densities >1000/plant), only five instances were noted over a 3 year period, all of which happened in habitats that had <50% natural habitat.

An in-depth study by Haenke et al. (2009) has further enhanced the scope of study of habitat management on syrphid fly populations by including the influences and interactions between local and landscape scale diversity. The authors compared syrphid abundance and species richness in four ecotones: broad sown flower strips, narrow sown flower strips, naturally developed grassy strips, and the boundary of adjoining wheat fields lacking such strips as a control adjacent to winter wheat fields across a landscape whose complexity increased with gradient from 30% to 100% arable land. Contrary to expectations that species richness would increase in flower strips and that this effect would be stronger in simple rather than complex landscapes (i.e., seminatural habitats in complex landscapes), the results revealed that syrphids showed higher densities with increasing proportion of arable land. In complex landscapes, the effects of sown flower strips were hardly visible, whereas in simple landscapes, they were most effective. These results provide a general guideline that promoting landscape heterogeneity might be economically more efficient in simple landscapes (Roschewitz et al., 2005; Tscharntke et al., 2005; Holzschuh et al., 2007). This is despite the evidence that syrphids have excellent vision abilities (Lunau and Wacht, 1994) and are capable of detecting remote resource patches in monotonous and nonnutritious environments. But, on the contrary, Harwood et al. (1994) indicated that the probability of syrphids crossing areas with breaks in vegetation ground cover is quite low.

Haenke et al. (2014) observed the potential of hedgerows (forest edges, forest connected hedges, and isolated hedges) in enhancing the syrphid populations and in managing their spillover to adjacent crop fields, especially when connected to forests. Predatory syrphids were more abundant not only in hedgerows that were connected to forests than forest edges but also in crop fields adjacent to hedgerows than adjacent to forest edges, indicating spillover from seminatural habitats. This spillover was also tempered by the coverage of mass-flowering crops in the surrounding landscapes. It was thus suggested that increasing the total amount of hedgerows under moderate landscape-scale proportions of mass-flowering crops may serve best for the conservation of biodiversity and augmentation of important ecosystem services (Haenke et al., 2014).

The addition of floral resources as a tool for enhancing local syrphid populations are quite extensive (MacLeod, 1992; White et al., 1995; Morris and Li, 2000), with studies determining which flower is more preferred by the syrphids (Cowgill et al., 1992; Hickman et al., 1995; Colley and Luna, 2000). However, while flowers are capable of increasing the visit frequencies of syrphids, it does not necessarily translate into increased number of eggs (Chandler, 1968a; Sengonça and Fringe, 1988), though contrary evidence is also on record (White et al., 1995; Hickman and Wratten, 1996).

9. CONCLUSION

Syrphids form one of the most common predators present in aphid patches. They are the first and last predators to exploit prey patches. This is owing to the extremely close oviposition to prey patches by the adult females, allowing their relatively immobile larvae to find food easily without risking starvation. It is also known that syrphid females continue to oviposit almost continuously on aphid patches. Despite only their larval stages being predatory in nature, the almost continuous oviposition by adults along with larvae that show type III functional responses make syrphids potentially formidable biocontrol agents. Their lower developmental threshold allows the syrphids to be the first and last ones to exploit prey patches. However, the syrphid stages are much prone to predation by other members of the prey guild, a drawback for their use in biological control. The use of habitat management by providing different landscapes has further improved the efficacy of syrphids. Creative use of flower strips and other vegetation has shown not only increased the presence of and oviposition by adult females, leading to potentially more effective biocontrol, but also may lead to better pollination prospects for the crop under consideration. Thus, these hovering biocontrol agents may prove to be effective managers of pest populations, particularly but probably not exclusively in isolation with the rest of the predators. They are likely to be more effective as tools of conservation biological control that increases their populations in tune with the growth and decline of prey patch as well as that of other predators, rather than in classical biological control.

Acknowledgments

The authors are thankful to the Department of Higher Education, Govt. of Uttar Pradesh, Lucknow for generous funding under Centre of Excellence Programme on Biocontrol of Insect Pests.

References

Agarwala, B.K., Yasuda, H., 2001. Larval interactions in aphidophagous predators: effectiveness of wax cover as defence shield of *Scymnus* larvae against predation from syrphids. Entomologia Experimentalis et Applicata 100 (1), 101–107.

Alhmedi, A., Haubruge, E., Francis, F., 2008. Role of prey–host plant associations on *Harmonia axyridis* and *Episyrphus balteatus* reproduction and predatory efficiency. Entomologia Experimentalis et Applicata 128, 49–56.

Almohamad, R., Verheggen, F.J., Francis, F., Haubruge, É., 2006. Evaluation of hoverfly *Episyrphus balteatus* DeGeer (Diptera: Syrphidae) oviposition behavior toward aphid-infested plants using a leaf disc system. Communications in Agricultural and Applied Biological Sciences 71 (2 Pt B), 403–412.

Almohamad, R., Verheggen, F.J., Francis, F., Haubruge, É., 2008. Impact of aphid colony size and associated induced plant volatiles on searching and oviposition behaviour of a predatory hoverfly. Belgian Journal of Entomology 10, 17–26.

Almohamad, R., Verheggen, F.J., Haubruge, É., 2009. Searching and oviposition behavior of aphidophagous hoverflies (Diptera: Syrphidae): a review. Biotechnologie, Agronomie, Société et Environnement 13 (3), 467–481.

Amorós-Jiménez, R., Pineda, A., Fereres, A., Marcos-García, M.Á., 2014. Feeding preferences of the aphidophagous hoverfly *Sphaerophoria rueppellii* affect the performance of its offspring. BioControl 59 (4), 427–435.

Barbosa, P., 1998. Conservation Biological Control. Academic Press, San Diego, California, USA, p. 396.

Bargen, H., Saudhof, K., Poehling, H.M., 1998. Prey finding by larvae and adult females of *Episyrphus balteatus*. Entomologia Experimentalis et Applicata 87, 245–254.

Barlow, C.A., Whittingham, J.A., 1986. Feeding economy of larvae of a flower fly, *Metasyrphus corollae* (Dip.: Syrphidae): partial consumption of prey. Entomophaga 31 (1), 49–57.

Belliure, B., Michaud, J.P., 2001. Biology and behavior of *Pseudodorus clavatus* (Diptera: Syrphidae), an important predator of citrus aphids. Annals of the Entomological Society of America 94 (1), 91–96.

Ben Saad, A., Bishop, G.W., 1976. Effect of artificial honeydews on insect communities in potato fields. Environmental Entomology 5, 453–457.

Bhat, M.R., Ahmad, D., 1988. Observations on the biology of *Episyrphus balteatus* (Degeer) (Diptera: Syrphidae) in Kashmir. Bulletin of Entomology, New Delhi 29, 216–217.

Bianchi, F.J., Booij, C.J., Tscharntke, T., 2006. Sustainable pest regulation in agricultural landscapes: a review on landscape composition, biodiversity and natural pest control. Proceedings of the Royal Society B-Biological Sciences 273, 1715–1727.

Bombosch, S., Volk, S., 1966. Selection of oviposition site by *Syrphus corollae* F. In: Hodek, I. (Ed.), Ecology of Aphidophagous Insects. Academia, Prague, pp. 117–119.

Branquart, E., Hemptinne, J.L., 2000a. Development of ovaries, allometry of reproductive traits and fecundity of *Episyrphus balteatus* (Diptera: Syrphidae). European Journal of Entomology 97 (2), 165–170.

Branquart, E., Hemptinne, J.L., 2000b. Selectivity in the exploitation of floral resources by hoverflies (Diptera: Syrphinae). Ecography 23, 732–742.

Brewer, M.J., Elliott, N.C., 2004. Biological control of cereal aphids in North America and mediating effects of host plant and habitat manipulations. Annual Review of Entomology 49, 219–242.

Brodsky, L.M., Barlow, C.A., 1986. Escape responses of the pea aphid, *Acyrthosiphon pisum* (Harris) (Homoptera: Aphididae): influence of predator type and temperature. Canadian Journal of Zoology 64 (4), 937–939.

Budenberg, W.J., Powell, W., 1992. The role of honeydew as an ovipositional stimulant for two species of syrphids. Entomologia Experimentalis et Applicata 64, 57–61.

Bugg, R.L., Colfer, R.G., Chaney, W.E., Smith, E.A., Cannon, J., 2008. Flower Flies (Syrphidae) and Other Biological Control Agents for Aphids in Vegetable Crops. UCANR Publication 8285.

Burgio, G., Sommaggio, D., 2007. Syrphids as landscape bioindicators in Italian agroecosystems. Agriculture Ecosystem & Environment 120, 416–422.

Cardinale, B.J., Harvey, C.T., Gross, K., Ives, A.R., 2003. Biodiversity and biocontrol: emergent impacts of a multienemy assemblage on pest suppression and crop yield in an agroecosystem. Ecology Letters 6, 857–865.

Chambers, R.J., Adams, T.H.L., 1986. Quantification of the impact of hoverflies (Diptera: Syrphidae) on cereal aphids in winter wheat: an analysis of field populations. Journal of Applied Ecology 23, 895–904.

Chambers, R.J., Aikman, D.P., 1988. Quantifying the effects of predators on aphid populations. Entomologia Experimentalis et Applicata 46, 257–265.

Chambers, R.J., Sunderland, K.D., 1983. The abundance and effectiveness of natural enemies of cereal aphids on two farms in Southern England. In: Cavalloro, R. (Ed.), Aphid Antagonists. A.A. Balkema, Rotterdam, pp. 83–87.

Chambers, R.J., Sunderland, K.D., Stacey, D.L., Wyatt, I.J., 1986. Control of cereal aphids in winter wheat by natural enemies: aphid-specific predators, parasitoids and pathogenic fungi. Annals of Applied Biology 108, 219–231.

Chambers, R.J., 1988. Syrphidae. In: Minks, A.K., Harrewijn, P. (Eds.), World Crop Pests: Aphids, Their Biology, Natural Enemies and Control. Elsevier, Amsterdam, Netherlands, pp. 259–270.

Chambers, R.J., 1991. Oviposition by aphidophagous hoverflies (Diptera: Syrphidae) in relation to aphid density and distribution in winter wheat. In: Polgar, L., Chambers, R.J., Dixon, A.F.G., Hodek, I. (Eds.), Behaviour and Impact of Aphidophaga. SPB Academic, The Hague, pp. 115–121.

Chandler, A.E.F., 1966. Some aspects of host plant selection in aphidophagous Syrphidae. In: Hodek, I. (Ed.), Ecology of Aphidophagous Insects. Junk: The Hague; Academia, Prague, pp. 113–115.

Chandler, A.E.F., 1968a. Some host–plant factors affecting oviposition by aphidophagous Syrphidae (Diptera). Annals of Applied Biology 61 (3), 415–423.

Chandler, A.E.F., 1968b. The relation between aphid infestations and oviposition by aphidophagous Syrphidae (Diptera). Annals of Applied Biology 61 (3), 425–434.

Chandler, A.E.F., 1968c. Some factors influencing the occurrence and site of oviposition by aphidophagous Syrphidae (Diptera). Annals of Applied Biology 61 (3), 435–446.

Chaplin-Kramer, R.E., 2010. The landscape ecology of pest control services: cabbage aphid-syrphid trophic dynamics on California's Central Coast Ph.D. thesis. University of California, Berkeley, USA.

Colley, M.R., Luna, J.M., 2000. Relative attractiveness of potential beneficial insectary plants to aphidophagous hoverflies (Diptera: Syrphidae). Environmental Entomology 29 (5), 1054–1059.

Conn, D.L.T., 1979. Morphological and behavioural differences in populations of *Merodon equestris* (F.) (Dipt., Syrphidae). Entomologist's Monthly Magazine 114, 1364–1367.

Cornelius, M., Barlow, C.A., 1980. Effect of aphid consumption by larvae on development and reproductive efficiency of a hover fly, *Syrphus corollae* (Diptera: Syrphidae). Canadian Entomologist 112, 989–992.

Costanza, R., d'Arge, R., de Groot, R., Farber, S., Grasso, M., Hannon, B., Limburg, K., Naeem, S., O'Neill, R.V., Paruelo, J., Raskin, R.G., Sutton, P., van den Belt, M., 1997. The value of the world's ecosystem services and natural capital. Nature 387, 253–260.

Courtney, S.P., Chen, G.K., Gardner, A., 1989. A general model for individual host selection. Oikos 55, 55–65.

Cowgill, S.E., Sotherton, N.W., Wratten, S.D., 1992. The selective use of floral resources by the hoverfly *Episyrphus balteatus* (Diptera: Syrphidae) on farmland. Annals of Applied Biology 122, 223–231.

Dayal, K., Gupta, P.R., 2001. Predatory potential of the larvae of the hoverfly, *Betasyrphus serarius* (Wiedemann) on *Aphis fabae solanella* (Theobald). Insect Environment 6 (4), 150–151.

Dixon, A.F., Jarosik, V., Honek, A., 2005. Thermal requirements for development and resource partitioning in aphidophagous guilds. European Journal of Entomology 102 (3), 407.

Dixon, T.J., 1959. Studies on the oviposition behaviour of Syrphidae (Diptera). Transactions of the Royal Entomological Society, London 111, 57–80.

Dixon, A.F.G., 1998. Aphid Ecology. Chapman & Hall, p. 300.

Downes, J.A., 1969. The swarming and mating flight of Diptera. Annual Review of Entomology 14 (1), 271–298.

Drapela, T., Moser, D., Zaller, J., Frank, T., 2008. Spider assemblages in winter oilseed rape affected by landscape and site factors. Ecography 31, 254–262.

Du, Y.Z., Chen, X.Z., 1993. Influence of different aphid prey on the development of *Metasyrphus corollae* (Dip.: Syrphidae). Chinese Journal of Biological Control 9 (3), 111–113.

Edmunds, M., Reader, T., 2014. Evidence for batesian mimicry in a polymorphic hoverfly. Evolution 68 (3), 827–839.

Eilenberg, J., Hajek, A., Lomer, C., 2001. Suggestions for unifying the terminology in biological control. BioControl 46, 387–400.

Ekukole, G., Misari, S.M., 1993. Effect of prey density on the predatory behaviour of *Ischiodon aegyptius* Wied. and *Scymnus* sp. on the cotton aphid, *Aphis gossypii* Glover in Cameroon. Samaru Journal of Agricultural Research 10, 55–62.

Elliott, N., Kieckhefer, R., Beck, D., 2002. Effect of aphids and the surrounding landscape on the abundance of Coccinellidae in cornfields. Biological Control 24, 214–220.

van Emden, H.F., Harrington, R., 2007. Aphids as Crop Pests. Cabi Publishing, London, UK.

van Emden, H.F., 1990. Plant diversity and natural enemy efficiency in agroecosystems. In: Mackauer, M., Ehler, L.E., Roland, J. (Eds.), Critical Issues in Biological Control. Intercept, Andover, UK, pp. 63–80.

Entwistle, J.C., Dixon, A.F.G., 1989. The effect of augmenting grain aphid numbers in a field of winter wheat in spring on the aphid's abundance in summer and its relevance to the forecasting of outbreaks. Annals of Applied Biology 114, 397–408.

Fiedler, A.K., Landis, D.A., Wratten, S.D., 2008. Maximizing ecosystem services from conservation biological control: the role of habitat management. Biological Control 45, 254–271.

Frank, T., 1999. Density of adult hoverflies (Dipt., Syrphidae) in sown weed strips and adjacent fields. Journal of Applied Entomology 123, 351–355.

Fréchette, B., Rojo, S., Alomar, O., Lucas, É., 2007. Intraguild predation between syrphids and mirids: who is the prey? Who is the predator? BioControl 52 (2), 175–191.

Freier, B., Triltsch, H., Mowes, M., Moll, E., 2007. The potential of predators in natural control of aphids in wheat: results of a ten-year field study in two German landscapes. BioControl 52, 775–788.

Gardiner, M.M., Landis, D.A., Gratton, C., DiFonzo, C.D., O'Neal, M., Chacon, J.M., Wayo, M.T., Schmidt, N.P., Mueller, E.E., Heimpel, G.E., 2009a. Landscape diversity enhances biological control of an introduced crop pest in the north-central USA. Ecological Applications 19, 143–154.

Gardiner, M.M., Landis, D.A., Gratton, C., Schmidt, N., O'Neal, M., Mueller, E., Chacon, J., Heimpel, G.E., DiFonzo, C.D., 2009b. Landscape composition influences patterns of native and exotic lady beetle abundance. Diversity and Distributions 15, 554–564.

Georghiou, G.P., 1986. The magnitude of the resistance problem. In: Pesticide Resistance: Strategies and Tactics for Management. National Academy Press, Washington DC, USA, pp. 14–44.

Geusen-Pfister, H., 1987. Studies on the biology and reproductive capacity of *Episyrphus balteatus* (Diptera: Syrphidae) under greenhouse conditions. Journal of Applied Entomology 104, 261–270.

Gilbert, F.S., 1984. Thermoregulation and the structure of swarms in *Syrphus ribesii* (Syrphidae). Oikos 42, 249–255.

Gilbert, F.S., 1993. Hoverflies, second ed. Naturalists' Handbooks, vol. 5. Richmond Press, Slough, UK.

Gilkeison, L., Klein, M., 1981. A Guide to the Biological Control of Greenhouse Aphids. http://eap.mcgill.ca/publications/EAP53.htm#Syrphid Files (Hover Flies or Flower Flies).

Golding, Y.C., Ennos, A.R., Edmunds, M., 2001. Similarity in flight behaviour between the honeybee *Apis mellifera* (Hymenoptera: Apidae) and its presumed mimic, the dronefly *Eristalis tenax* (Diptera: Syrphidae). Journal of Experimental Biology 204 (1), 139–145.

Gurr, G.M., Scarrat, S.L., Wratten, S.D., Berndt, L., Irvin, N., 2004. Ecological engineering, habitat manipulation and pest management. In: Gurr, G.M., Wratten, S.D., Altieri, M.A. (Eds.), Ecological Engineering for Pest Management. CSIRO Publishing, Collingwood, Australia, pp. 1–12.

Haenke, S., Kovács–Hostyánszki, A., Fründ, J., Batáry, P., Jauker, B., Tscharntke, T., Holzschuh, A., 2014. Landscape configuration of crops and hedgerows drives local syrphid fly abundance. Journal of Applied Ecology 51 (2), 505–513.

Haenke, S., Scheid, B., Schaefer, M., Tscharntke, T., Thies, C., 2009. Increasing syrphid fly diversity and density in sown flower strips within simple vs. complex landscapes. Journal of Applied Ecology 46 (5), 1106–1114.

Hamid, S., 1986. Aphid control potential of *Episyrphus balteatus* (DeGeer). In: Proceedings of the 5th Pakistan Congress of Zoology, University of Karachi, Karachi, January 8–11, 1987, pp. 153–158.

Hamman, P.J., 1985. Aphids on Trees and Shrubs. L-1227. Texas Agricultural Extension Service House and Landscape Pests, College Station, Texas, p. 2.

Harmel, N., Almohamad, R., Fauconnier, M.-L., Du Jardin, P., Verheggen, F., Marlier, M., Haubruge, E., Francis, F., 2007. Role of terpenes from aphid-infested potato on searching and oviposition behavior of *Episyrphus balteatus*. Insect Science 14 (1), 57–63.

Harwood, R.J.W., Hickman, J.M., Macleod, A., Sherratt, T., Wratten, S.D., 1994. Managing field margin for hover flies. In: Boatman, N. (Ed.), Margins: Integrating Agriculture and Conservation. BritishCrop Protection Monograph No. 58. BCPC, ThorntonHeath, Surrey, pp. 147–152.

Heimpel, G.E., Jervis, M.A., 2005. An evaluation of the hypothesis that floral nectar improves biological control by parasitoids. In: Wäckers, F.L., van Rijn, P.C.J., Bruin, J. (Eds.), Plant-Provided Food for Carnivorous Insects: A Protective Mutualism and Its Applications. Cambridge University Press, Cambridge, UK.

Heinrich, B., Pantle, C., 1975. Thermoregulation in small flies (*Syrphus* sp.): basking and shivering. Journal of Experimental Biology 62 (3), 599–610.

Hemptinne, J.L., Dixon, F.G., Doucet, J.L., Petersen, J.E., 1993. Optimal foraging by hoverflies (Diptera: Syrphidae) and lady birds (Coleoptera: Coccinellidae): mechanisms. European Journal of Entomology 90 (4), 451–455.

Hickman, J.M., Wratten, S.D., 1996. Use of *Phacelia tanacetifolia* strips to enhance biological control of aphids by hoverflies larvae in cereal fields. Journal of Economical Entomology 89, 832–840.

Hickman, J.M., Lovei, G.L., Wratten, S.D., 1995. Pollen feeding by adults of the hoverflies *Melanostoma fasciatum* (Diptera: Syrphidae). New Zealand Journal of Zoology 22, 387–392.

Hindayana, D., Meyhofer, R., Scholz, D., Poehling, H.M., 2001. Intraguild predation among the hoverfly *Episyrphus balteatus* de Geer (Diptera: Syrphidae) and other aphidophagous predators. Biological Control 20 (3), 236–246.

Hindayana, D., 2001. Resource Exploitation by *Episyrphus balteatus* (De Geer) (Diptera: Syrphidae) and Intraguild Predation (Unpublished Ph.D. thesis). University of Hannover, Germany.

Holzschuh, A., Steffan-Dewenter, I., Kleijn, D., Tscharntke, T., 2007. Diversity of flower-visiting bees in cereal fields: effects of farming system, landscape composition and regional context. Journal of Applied Ecology 44, 41–49.

Hopper, J.V., Nelson, E.H., Daane, K.M., Mills, N.J., 2011. Growth, development and consumption by four syrphid species associated with the invasive lettuce aphid, *Nasonovia ribisnigri*, in California. BioControl 58, 271–276.

Ito, K., Iwao, S., 1977. Oviposition behaviour of a syrphid, *Episyrphus balteatus*, in relation to aphid density on the plant. Japanese Journal of Applied Entomology and Zoology 21, 130–134.

Jauker, F., Diekotter, T., Schwarzbach, F., Wolters, V., 2009. Pollinator dispersal in an agricultural matrix: opposing responses of wild bees and hoverflies to landscape structure and distance from main habitat. Landscape Ecology 24, 547–555.

Joshi, S., Venkatesan, T., Ballal, C.R., Rao, N.S., 1999. Comparative biology and predatory efficiency of six syrphids on *Aphis craccivora* Koch. Pest Management in Horticultural Ecosystems 5 (1), 1–6.

Kan, E., Sasakawa, M., 1986. Assessment of the maple aphid colony by the hover fly, *Episyrphus balteatus* (de Geer) (Diptera: Syrphidae) I. Journal of Ethology 4, 121–127.

Kan, E., 1988a. Assessment of aphid colonies by hoverflies. I. Maple aphids and *Episyrphus balteatus* (de Geer) (Diptera: Syrphidae). Journal of Ethology 6, 39–48.

Kan, E., 1988b. Assessment of aphid colonies by hoverflies. II. Pea aphids and three syrphid species; *Betasyrphus serarius* (Wiedemann), *Metasyrphus frequens* Matsumura, and *Syrphus vitripennis* (Meigen) (Diptera: Syrphidae). Journal of Ethology 6, 135–142.

Katzourakis, A., Purvis, A., Azmeh, S., Rotheray, G., Gilbert, F., 2001. Macroevolution of hoverflies (Diptera: Syrphidae): the effect of using higher-level taxa in studies of biodiversity, and correlates of species richness. Journal of Evolutionary Biology 14, 219–227.

Kenmore, P.E., 1980. Ecology and Outbreaks of a Tropical Insect Pest of the Green Revolution, the Rice Brown Planthopper, *Nilaparvata lugens* Stal. University of California, Berkeley.

Kim, I.S., Uhm, K.B., Goh, H.G., Choi, K.M., 1994. Selection of artificial diet for mass rearing of the pollinator, *Eristalis cerealis* Fabricius. RDA Journal of Agricultural Sciences and Crop Protection 36 (2), 369–372.

Kremen, C., Chaplin-Kramer, R., 2007. Insects as providers of ecosystem services: crop pollination and pest control. In: Stewart, A.J., New, T.R., Lewis, O.T. (Eds.), Insect Conservation Biology: Proceedings of the Royal Entomological Society's 23rd Symposium. CABI Publishing, Wallingford, UK, pp. 349–382.

Kumar, A., Kapoor, V.C., Mahal, M.S., 1996. Feeding behaviour and efficacy of three aphidophagous syrphids. Journal of Insect Science 9 (1), 15–18.

Lakhanpal, G.C., Raj, D., 1998. Predation potential of coccinellid and syrphid on important aphid species infesting rapeseed in Himachal Pradesh. Journal of Entomology Research 22 (2), 181–190.

Landis, D.A., Wratten, S.D., Gurr, G.M., 2000. Habitat management to conserve natural enemies of arthropod pests in agriculture. Annual Review of Entomology 45, 175–201.

Laubertie, E.A., Wratten, S.D., Sedcole, J.R., 2006. The role of odour and visual cues in the pan-trap catching of hoverflies (Diptera: Syrphidae). Annals of Applied Biology 148 (2), 173–178.

Laubertie, E.A., Wratten, S.D., Hemptinne, J.L., 2012. The contribution of potential beneficial insectary plant species to adult hoverfly (Diptera: Syrphidae) fitness. Biological Control 61 (1), 1–6.

Lövei, G.L., McDougall, D., Bramley, G., Hodgson, D.J., Wratten, S.D., 1992. Floral resources for natural enemies: the effects of *Phacelia tanacetifolia* (Hydrophyllaceae) on within-field distribution of hoverflies (Diptera: Syrphidae). In: Proceeding of the 45th New Zealand Plant Protection Conference: Vegetables and Ornamentals, New Zealand, pp. 60–61.

Lucas, E., Coderre, D., Brodeur, J., 1998. Intraguild predation among aphid predators: characterization and influence of extraguild prey density. Ecology 79, 1084–1092.

Lunau, K., Wacht, S., 1994. Optical releasers of innate proboscis extension in the hoverfly *Eristalis tenax* L. (Diptera: Syrphidae). Journal of Comparative Physiology 174, 575–579.

Lys, J.A., Nentwig, W., 1992. Augmentation of beneficial arthropods by strip-management. Oecologia 92, 373–382.

MacLeod, A., 1992. Alternative crops as floral resources for beneficial hoverflies (Diptera: Syrphidae). In: Proceeding of the Brighton Crop Protection Conference: Pests and Diseases, Brighton, UK, pp. 997–1002.

Makhmoor, H.D., Verma, A.K., 1987. Bionomics of major aphidophagous syrphids occurring in mid-hill regions of Himachal Pradesh. Journal of Biological Control 1 (1), 23–31.

Menalled, F.D., Marino, P.C., Gage, S.H., Landis, D.A., 1999. Does agricultural landscape structure affect parasitism and parasitoid diversity? Ecological Applications 9, 634–641.

Molthan, J., Ruppert, V., 1988. Significance of flowering wild herbs in boundary strips and field for flower-visiting beneficial insects. Mitteilungen aus der Biologischen Bundesanstalt fuer Land und Forstwirtschaft Berlin-Dahlem 247, 85–99.

Morris, M.C., Li, F.Y., 2000. Coriander (*Coriandrum sativum*) "companion plants" can attract hoverflies, and may reduce pest infestation in cabbages. New Zealand Journal of Crop and Horticultural Science 28, 213–217.

Mutin, V.A., 1996. The mate seeking behaviour of male syrphids. In: AI Kurentsovs Annual Memorial Meetings, vol. 7, pp. 117–124.

Naylor, R., Ehrlich, P., 1997. Natural pest control services and agriculture. In: Daily, G. (Ed.), Nature's Services. Island Press, Washington, pp. 151–174.

Nicholls, C.I., Altieri, M.A., 2004. Agroecological bases of ecological engineering for pest management. In: Gurr, G.M., Wratten, S.D., Altieri, M.A. (Eds.), Ecological Engineering for Pest Management. CSIRO Publishing, Collingwood, Australia, pp. 33–54.

Oberg, S., Mayr, S., Dauber, J., 2008. Landscape effects on recolonisation patterns of spiders in arable fields. Agriculture, Ecosystems & Environment 123, 211–218.

Oerke, E.C., 2005. Crop losses to pests. Journal of Agricultural Science 144, 31–43.

Palumbi, S.R., 2001. Humans as the world's great evolutionary force. Science 293, 1786–1790.

Pan-UK, January 18, 2003. Current Pesticide Spectrum, Global Use and Major Concerns. http://www.pan-uk.org/briefing/SIDA_Fil/Chap1.htm.

Patro, B., Behera, M.K., 1992. Bionomics of *Paragus* (*Paragus*) *serratus* (Fabricius) (Diptera: Syrphidae), a predator of the bean aphid, *Aphis craccivora* Koch. Tropical Science 33, 131–135.

Penney, H.D., Hassall, C., Skevington, J.H., Abbott, K.R., Sherratt, T.N., 2012. A comparative analysis of the evolution of imperfect mimicry. Nature 483 (7390), 461–464.

Pimentel, D., 2005. Environmental and economic costs of the application of pesticides primarily in the United States. Environment, Development and Sustainability 7 (2), 229–252.

Polis, G.A., Strong, D.R., 1996. Food web complexity and community dynamics. The American Naturalist 147, 813–846.

Polis, G.A., Myers, C.A., Holt, R.D., 1989. The ecology and evolution of intraguild predation: potential competitors that eat each other. Annual Review of Ecology and Systematics 20, 297–330.

Popp, J., Pető, K., Nagy, J., 2013. Pesticide productivity and food security. A review. Agronomy for Sustainable Development 33 (1), 243–255.

Popp, J., 2011. Cost-benefit analysis of crop protection measures. Journal of Consumer Protection and Food Safety 6 (Suppl. 1), 105–112.

Prasifka, J.R., Heinz, K.M., Minzenmayer, R.R., 2004. Relationships of landscape, prey and agronomic variables to the abundance of generalist predators in cotton (*Gossypium hirsutum*) fields. Landscape Ecology 19, 709–717.

Purtauf, T., Roschewitz, I., Dauber, J., Thies, C., Tscharntke, T., Wolters, V., 2005. Landscape context of organic and conventional farms: influences on carabid beetle diversity. Agriculture, Ecosystems & Environment 108, 165–174.

van Rijn, P.C., Kooijman, J., Wackers, F.L., 2006. The impact of floral resources on syrphid performance and cabbage aphid biological control. IOBC WPRS Bulletin 29 (6), 149.

Roschewitz, I., Gabriel, D., Tscharntke, T., Thies, C., 2005. The effects of landscape complexity on arable weed species diversity in organic and conventional farming. Journal of Applied Ecology 42, 873–882.

Rotheray, G.E., 1987. Aphidophagy and the larval and pupal stages of the syrphid *Platycheirus fulviventris* (Macquart). Entomologist's Gazette 38, 245–251.

Ruzicka, Z., Gonzales Cairo, V., 1976. The effect of larval starvation on the development of *Metasyrphus corollae* (Diptera). Vestnik Ceskoslovenske Spolecnosti Zoologicke 40, 206–213.

Sadeghi, H., Gilbert, F., 2000a. Oviposition preferences in aphidophagous hoverflies. Ecological Entomology 25, 91–100.

Sadeghi, H., Gilbert, F., 2000b. The effect of egg load and host deprivation on oviposition behaviour in aphidopha-gous hoverflies. Ecological Entomology 25, 101–108.

Sadeghi, H., Gilbert, F., 2000c. Aphid suitability and its relationship to oviposition preference in predatory hoverflies. Journal of Animal Ecology 69 (5), 771–784.

Salveter, R., Nentwig, W., 1993. Schwebfliegen (Diptera, Syrphidae) inder Agrarlandschaft: Phänologie, Abundanz und Markierungsversuche. Mitteilungen der Naturforschenden Gesellschaft in Bern NF 50, 147–191.

Samuel, R.N., Dass, I.J., Singh, R., 2005. Feeding potential and its effect on development of an aphid predator, Epi-syrphus balteatus (De Geer) (Diptera: Syrphidae) vis-à-vis variable prey density. Journal of Aphidology 19, 93–100.

Samuel, R.N., Dass, I.J., Singh, R., 2013. Studies on oviposition behaviour and egg hatching pattern of an aphid predator, Episyrphus balteatus (De Geer) (Diptera: Syrphidae): a promising biocontrol agent. Journal of Aphidology 27, 45–52.

Sanders, W., 1979. Das Eiablageverhalten der Schweb Riege Syrphus corallae Fabr. in Abhangigkeit von der Grosse der Blattlauskolonie. Zeitschrift fur Angewandte Entomologie 66, 217–232.

Sanders, W., 1981a. The oviposition behaviour of the hover-fly Syrphus corollae Fabr. in relation to light and shadow on the aphid colony. Zeitschrift fur Angewandte Zoologie 68, 1–12.

Sanders, W., 1981b. The influence of white and black surfaces on the behaviour of the hoverfly Syrphus corollae Fabr. Zeitschrift fur Angewandte Zoologie 68, 307–314.

Sanders, W., 1982. Der Einfluß von Farbe und Beleuchtung des Umfeldes auf die Eiablagehandlung der Schwebfliege Syrphus corollae Fabr. Zeitschrift fur Angewandte Zoologie 69, 283–297.

Sanders, W., 1983a. The searching behaviour of gravid Syrphus corollae Fabr. (Dipt. Syrphidae) and its dependence on optical cues. Zeitschrift fur Angewandte Zoologie 70, 235–247.

Sanders, W., 1983b. The searching behaviour of gravid Syrphus corollae Fabr. (Dipt. Syrphidae) in relation to variously designed plant models. Zeitschrift fur Angewandte Zoologie 70, 449–462.

Schmidt, M.H., Tscharntke, T., 2005. Landscape context of sheetweb spider (Araneae: Linyphiidae) abundance in cereal fields. Journal of Biogeography 32, 467–473.

Schmidt, M.H., Lauer, A., Purtauf, T., Thies, C., Schaefer, M., Tscharntke, T., 2003. Relative importance of predators and parasitoids for cereal aphid control. Proceedings of the Royal Society B-Biological Sciences 270, 1905–1909.

Schmidt, M.H., Thies, C., Nentwig, W., Tscharntke, T., 2008. Contrasting responses of arable spiders to the landscape matrix at different spatial scales. Journal of Biogeography 35, 157–166.

Schmutterer, H., 1972. Zur Beutespezifität polyphager, räuberischer Syrphiden Ostafrikas. [On the food specificity of polyphagous predatory syrphids of East Africa.] Zeitschrift fur Angewandte Entomologie 71, 278–286.

Schneider, F., 1948. Beitrag zur Kenntnis der Generatiansverhältnisse und Diapause rauberischer Schweb-Hiegen. Mitteilungen der Schweizerischen Entomologischen Gesellschaft 21, 249–285.

Schneider, F., 1969. Bionomics and physiology of aphidophagous syrphidae. Annual Review of Entomology 14, 103–124.

Scholz, D., Poehling, H.M., 2000. Oviposition site selection of Episyrphus balteatus. Entomologia Experimentalis et Applicata 94 (2), 149–158.

Scott, S.M., Barlow, C.A., 1984. Effect of prey availability during development on the reproductive output of Metasyrphus corollae (Diptera: Syrphidae). Environmental Entomology 13 (3), 669–674.

Scott, S.M., Barlow, C.A., 1986. Effect of prey availability on foraging and production efficiencies of larval Metasyr-phus corollae (Dipt.: Syrphidae). Entomophaga 31 (3), 243–250.

Sengonça, C., Fringe, B., 1985. Interference and competitive behaviour of the aphid predators, Chrysoperla carnea and Coccinella septempunctata in the laboratory. Entomophaga 30, 245–251.

Sengonça, C., Fringe, B., 1988. Einfluss von Phacelia tanacetifolia auf Schadlings und Nutzlings-populationen in Zuck-errubenfeldern. Pedobiologia 32, 311–316.

Settle, W.H., Ariawan, H., Astuti, E.T., Cahyana, W., Hakim, A.L., Hindayana, D., Lestari, A.S., Pajarningsih, S., 1996. Managing tropical rice pests through conservation of generalist natural enemies and alternative prey. Ecology 77, 1975–1988.

Sharma, K.C., Bhalla, O.P., 1988. Biology of six syrphid predators of cabbage aphid (Brevicoryne brassicae) on seed crop of cauliflower (Brassica oleracea var. botrytis). Indian Journal of Agricultural Science 58 (8), 652–654.

She, C.R., Pan, R.Y., Yang, S.H., LiLing, L., Chen, G., 1998. Studies on the biology of Sphaerophoria macrogaster (Diptera: Syrphidae). Wuyi Science Journal 14, 73–77.

Shonouda, M.L., Bombosch, S., Shalaby, A.M., Osman, S.I., 1998. Biological and chemical characterization of a kairo-mone excreted by the bean aphid, Aphis fabae Scop. (Hom., Aphididae), and its effect on the predator Metasyrphus corollae Fabr. II. Behavioural response of the predator M. corollae to the aphid kairomone. Journal of Applied Entomology 122, 25–28.

Shonouda, M.L., 1996. Crude aqueous-extract (kairomone) from *Aphis fabae* Scop. (Hom., Aphidae) and its effect on the behaviour of the predator *Metasyrphus corollae* Fabr. (Dipt., Syrphidae) female. Journal of Applied Entomology 120 (8), 489–492.

Singh, R., Mishra, S., 1988. Development of syrphid fly, *Ischiodon scutellaris* (Fabricius) on *Rhopalosiphum maidis* (Fitch). Journal of Aphidology 2 (1–2), 28–34.

Smith, H.A., Chaney, W.E., Bensen, T.A., 2008. Role of syrphid larvae and other predators in suppressing aphid infestations in organic lettuce on California's Central Coast. Journal of Economic Entomology 101 (5), 1526–1532.

Sobhani, M., Madadi, H., Gharali, B., 2013. Host plant effect on functional response and consumption rate of *Episyrphus balteatus* (Diptera: Syrphidae) feeding on different densities of *Aphis gossypii* (Hemiptera: Aphididae). Journal of Crop Protection 2 (3), 375–385.

Soleyman-Nezhadiyan, E., Laughlin, R., 1998. Voracity of larvae, rate of development in eggs, larvae and pupae, and flight seasons of adults of the hoverflies *Melangyna viridiceps* Macquart and *Symosyrphus grandicornis* Macquart (Diptera: Syrphidae). Australian Journal of Entomology 37 (3), 243–248.

Sommaggio, D., 1999. Syrphidae: can they be used as environmental bioindicators? Agriculture, Ecosystems & Environment 74, 343–356.

Stubbs, A.E., Falk, S.J., 1993. British hoverflies. An illustrated identification guide. British Entomological and Natural History Society 253.

Stürken, K., 1964. Die Bedeutung der imaginalernährung für das reproduktion svermögen der syrphiden. Zeitschrift fur Angewandte Zoologie 25, 385–417.

Suri, S.M., Singh, D., Brar, K.S., 1988. Estimation of losses in yield of brown sarson due to aphids in Kangra Valley. I, effect of crop growth stage and aphid feeding exposure. Journal of Insect Science 1, 162–167.

Sutherland, J.P., Sullivan, M.S., Poppy, G.M., 1999. The influence of floral character on the foraging behaviour of the hoverfly, *Episyrphus balteatus*. Entomologia Experimentalis et Applicata 93 (2), 157–164.

Sutherland, J.P., Sullivan, M.S., Poppy, G.M., 2001. Distribution and abundance of aphidophagous hoverflies (Diptera: Syrphidae) in wildflower patches and field margin habitats. Agricultural and Forest Entomology 3, 57–64.

Tamaki, G., Landis, B.J., Weeks, R., 1967. Autumn populations of green peach aphid on peach trees and the role of syrphid flies in their control. Journal of Economic Entomology 60, 433–436.

Tawfik, M.F.S., Azab, A.K., Awadallah, K.T., 1974a. Studies on the life-history and description of the immature forms of the Egyptian aphidophagous Syrphids. II - *Paragus aegyptius* Macq. (Diptera: Syrphidae). Bulletin de la Societe Entomologique d'Egypte 58, 35–44.

Tawfik, M.F.S., Azab, A.K., Awadallah, K.T., 1974b. Studies on the life-history and descriptions of the immature forms of the Egyptian aphidophagous Syrphids. III - *Xanthogramma aegyptium* Wied. (Diptera: Syrphidae). Bulletin de la Societe Entomologique d'Egypte 58, 73–83.

Tawfik, M.F.S., Azab, A.K., Awadallah, K.T., 1974c. Studies on the life-history and description of the immature form of the Egyptian aphidophagous Syrphids. IV - *Sphaerophoria flavicauda* Zett. (Diptera: Syrphidae). Bulletin de la Societe Entomologique d'Egypte 58, 103–115.

Tenhumberg, B., Poehling, H.M., 1991. Studies on the efficiency of syrphid larvae, as predators of aphids on winter wheat. In: Polgár, L., Chambers, R.J., Dixon, A.F.G., Hodek, I. (Eds.), Behaviour and Impact of Aphidophaga. SPB Academic Publishing, The Hague, The Netherlands, pp. 281–288.

Tenhumberg, B., Poehling, H.M., 1994. Quantification of the predation efficacy of *E. balteatus* (Diptera: Syrphidae) with the help of traditional models. Bulletin OILB/SROP 17 (4), 112–126.

Tenhumberg, B., Poehling, H.M., 1995. Syrphids as natural enemies of cereal aphids in Germany: aspects of their biology and efficacy in different years and regions. Agriculture, Ecosystems & Environment 52, 39–43.

Tenhumberg, B., 1995. Estimating predatory efficiency of *Episyrphus balteatus* (Diptera: Syrphidae) in cereal fields. Environmental Entomology 24 (3), 687–691.

Thies, C., Tscharntke, T., 1999. Landscape structure and biological control in agroecosystems. Science 285, 893.

Togashi, I., 1987. Insects associated with aphid, *Toxoptera odinae*, and honeydew secreted by *T. odinae*. Transactions of the Shikoku Entomological Society 18, 315–326.

Tscharntke, T., Klein, A.M., Kruess, A., Steffan-Dewenter, I., Thies, C., 2005. Landscape perspectives on agricultural intensification and biodiversity- ecosystem service management. Ecology Letters 8, 857–874.

Vanhaelen, N., Gaspar, C., Francis, F., 2002. Influence of prey host plant on a generalist aphidophagous predator, *Episyrphus balteatus* (Diptera: Syrphidae). European Journal of Entomology 99, 561–564.

Verheggen, F.J., Arnaud, L., Bartram, S., Gohy, M., Haubruge, E., 2008. Aphid and plant volatiles induce oviposition in an aphidophagous hoverfly. Journal of Chemical Ecology 34 (3), 301–307.

Vockeroth, J.R., Thompson, F.C., 1987. Syrphidae. In: McAlpine, J.F. (Ed.), Manual of Nearctic Diptera, Volume 2, Chapter 52. Research Branch. Agriculture Canada Hull, Quebec, pp. 713–743.

Volk, S., 1964. Untersuchungen Zur Eiablage von *Syrphus corollae* Fabr. (Diptera: Syrphidae). Zeitschrift fun Angewandte Entomologie 54, 365–386.

Waldbauer, G.P., 1990. Hilltopping by males of *Eupeodes volucris* (Diptera: Syrphidae). Great Lakes Entomology 23, 175–176.

Werling, B.P., Gratton, C., 2008. Influence of field margins and landscape context on ground beetle diversity in Wisconsin (USA) potato fields. Agriculture, Ecosystems & Environment 128, 104–108.

White, M.H., Wratten, S.D., Berry, N.A., Weigmann, U., 1995. Habitat manipulation to enhance biological control of Brassica pests by hoverflies (Diptera: Syrphidae). Journal of Economic Entomology 88 (5), 1171–1176.

Winder, L., Hirst, D.J., Carter, N., Wratten, S.D., Sopp, P.I., 1994. Estimating predation of the grain aphid *Sitobion avenae* by polyphagous predators. Journal of Applied Ecology 31, 1–12.

Wyss, E., Villiger, M., Müller-Schärer, H., 1999. The potential of three native insect predators to control the rosy apple aphid, *Dysaphis plantaginea*. BioControl 44 (2), 171–182.

Zeki, C., Kilincer, N., 1994. Studies on the starvation period and prey selection of larvae of *Metasyrphus corollae* (F.) and *Episyrphus balteatus* (De Geer) (Diptera: Syrphidae). In: Turkiye III. Biyolojik Mucadele Kongresi Bildirileeri, 25–28 Ocak 1994. Ege Universitesi Ziraat Fakultesi, Bitki Koruma Bolumu, Izmir, pp. 59–66.

Ladybird Beetles

Omkar[1], Ahmad Pervez[2]

[1]Centre of Excellence in Biocontrol of Insect Pests, Ladybird Research Laboratory, Department of Zoology, University of Lucknow, Lucknow, India; [2]Department of Zoology, Radhey Hari Government Post Graduate College, Kashipur, India

1. INTRODUCTION

The family Coccinellidae of the insect order Coleoptera comprises six subfamilies, namely, Chilocorinae, Coccidulinae, Scymninae, Coccinellinae, Sticholotidinae, and Epilachninae (Nedved and Kovář, 2012). They are holometabolous insects and possess five stages in their life cycle i.e., egg, larva, prepupa, pupa, and adult. There are three molting and four larval instars (Figure 1). Members of this family are commonly known as ladybirds, ladybugs, or lady beetles. Approximately 6000 known species of Coccinellidae have been recorded. At one time, this group caught the attention of people, especially children, primarily because of their colors but for more than a century it has been established that most of the species are natural enemies of many insect and acarine pests. However, the subfamily Epilachninae is mostly phytophagous. In addition to the vedalia beetle, *Rodolia cardinalis* Mulsant, used in the management of cottony cushion scale, *Icerya purchasi* Maskell, in the Californian orchards, we have listed more than 100 ladybird species that have immense potential as biocontrol agents in pest management programs. The Indian predator–prey catalog of ladybirds comprises 261 predaceous species (Omkar and Pervez, 2004a).

Since its first celebrated success story, the ladybird's journey has been a like a rollercoaster ride. History narrates some of the success stories of various ladybirds (DeBach, 1964). However, their failed attempts are also sizeable; not all ladybirds have a good history of success. Those that are specialists and coccidophagous have greater potential. Exhaustive documentation on the ecology (Hodek and Honek, 1996; Hodek et al., 2012), prey–predator dynamics (Dixon, 2000; Hirose, 2006), distribution, and native ladybirds (Majerus, 1994) have been published since the year 2000. Certain ladybirds, such as *Harmonia axyridis* (Pallas) (Koch, 2003; Pervez and Omkar, 2006; Roy et al., 2006; Osawa, 2011), *Chilocorus nigritus* (Fabr.) (Omkar and Pervez, 2003), *Coccinella septempunctata* L. (Omkar and Pervez, 2002a), *Adalia bipunctata* (L.) (Omkar and Pervez, 2005a), *Cryptolaemus montrouzieri* (Mulsant) (Kairo et al., 2013), and the

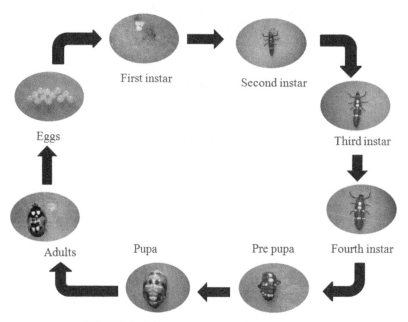

FIGURE 1   Life cycle of *Propylea dissecta* (Mulsant).

genus *Propylea* (Pervez and Omkar, 2011) have largely been the focus of interest as these are eurytopic and could assure biocontrol at some stage of pest infestation (i.e., low, medium, or high infestations). Ladybirds are uniquely referred to as a component of *"biological services"* (Landis et al., 2008). Ladybirds are the best models to better address the questions related to ecology and evolution. Hence, there is a plethora of information on their ecological and evolutionary aspects. We have discussed only those aspects directly related to their biocontrol efficacy. Thus, in this chapter we review the biocontrol aspects of ladybirds, foods and feeding habits, mating activity, status in predatory guild, semiochemical-dependent coccinellid behavior, and the effects of abiotic factors on ladybirds and their biocontrol potential.

2. FOOD AND FEEDING HABITS IN LADYBIRDS

Food or diet is an important criterion for ladybirds that tag them with their prey types. Certain phytophagous species, e.g., *Epilachna* spp. are crop pests. It also forms the basis of the creditability of its biocontrol potential of predatory ladybirds. The diet of predaceous ladybirds includes aphids, coccids, psyllids, diaspids, pentatomids, aleyrodids, and other insects and acarines. Nonpredaceous ladybird species feed on fungi, pollens, honey dew, etc. Diet breadth in predaceous ladybirds is an outcome of seasonal abundance and synchrony of their potential prey, prey attributes in terms of morphology, prey chemistry and behavior, the efforts involved in reaching to prey, host plant architecture, and level of threats or challenges imposed by intraguild predators and other natural enemies. Some ladybirds are *stenophagic* (narrow prey range) while others are *euryphagic* (broader prey range). Ladybirds are also categorized into *specialists* (which feed on monospecific or few prey species) and *generalists* (polyphagous) according to dietary habits.

Food for ladybirds is directly related to the reproduction. Among the *accepted foods* of ladybirds, only certain foods support both development and reproduction (*essential foods*) and the rest are meant only for their survival (***alternative foods***) (Hodek et al., 2012). The classical approach of prey classification for the ladybirds has been challenged (Rana et al., 2002). The concept that certain diets can only be the alternative foods, suboptimal or ***toxic foods*** (harmful to survival) (Hodek and Honek, 1996), is somehow elusive if understood in light of prey specialization.

2.1 Prey Specialization

Ladybirds have a tendency for prey specialization, which could be both diet and habitat-related. The concept prey specialization elucidates that ladybirds reared on suboptimal diets for a few generations specialize and condition themselves for the suboptimal diet. These conditioned ladybirds perform better on a suboptimal diet than that on optimal diets. Prey switching after a few generations of rearing on either optimal or suboptimal diets cause deterioration in their performance and fitness (reproduction and development). This has been explained as cost of prey specialization (Rana et al., 2002).

Prey specialization could also be habitat-related, where the habitats may be a microcosm, ovipositional microhabitat, patch choice, plant type, or some other feature of the biotic or abiotic environment (Sloggett, 2008a). This habitat selection may be considered as independent of the presence or absence of natural enemies, as ladybirds with differently restricted lateral distributions on the vegetation share the common natural enemies (Sloggett and Majerus, 2000). However, certain ladybirds are safe in their selected habitats, e.g., a specialist myrmecophilous ladybird, *Coccinella magnifica* Redt., inhabiting ant colonies had low rates of parasitization by a parasitoid, *Dinocampus coccinellae* Schrank, unlike the generalist *C. septempunctata* (Sloggett and Majerus, 2003; Sloggett et al., 1998, 2004).

Prey specialization has been argued as a function of ladybird size, prey size, and prey density (Sloggett, 2008b). Ladybird body size provides an important trade-off determining dietary breadth and prey specialization. Specialists more closely match the size of their limited prey species, have higher overall capture efficiencies, and can easily reproduce at lower prey densities for longer. However, generalists adopt a one-size-fits-all strategy, which results in lower capture efficiencies making them difficult to sustain at lower prey density. Specialists rarely move to other habitats because the prey in other habitats would be difficult to exploit. Unlike generalists, whose adults frequently move between patchy aphid habitats, specialists stay in the patches and lead a more sedentary life due to their greater tolerance of lower aphid densities (Sloggett et al., 2008). Specialists detaining a longer patch residence time is a consequence of its diet specialization and also a reason for their better biocontrol potential. These specialists exploit lower prey densities better than generalists and even utilize the patch prey habitats to reproduce, while generalists usually refrain from such activity in low prey density habitats (Sloggett and Majerus, 2000; Sloggett, 2008b). Early arrival and detentions of ladybirds in a patchy prey resource habitat is the key to their success in the biocontrol. Generalists are expected to arrive early but specialists are frequently more committed in action. Specialists even start to reproduce earlier than generalists and also reproduce in the later stage of declining prey patch. Their sedentary, stubborn, and nondispersing behavior makes them better biocontrol agents.

2.2 Prey Suitability

Many authors credited prey biochemical contents and palatability for the prey suitability (see Hodek et al., 2012). Ladybirds show a marked preference for certain aphid foods, which is also responsible for its performance in terms of voracity and fitness (Omkar and Mishra, 2005c). Prey suitability is largely dependent on aphid biochemicals, aphid morphology, nymph size, aphid activity, and the host plant. Generalists resorting to various kinds of food could be considered as an emergency arrived due to the shortage of the food they best processed. Mixed foods improve the performance of generalists owing to dietary self-selection (Soares et al., 2004). Furthermore, mixing various diets together has a positive impact on the reproductive performance along with various other bioattributes of ladybirds (Omkar, 2006). This may lead to food differentiation, mentioned above, in terms of essential, alternative, and toxic foods. The poorly preferred foods that are not rejected may contain some plant-sequestered growth-inhibiting substances and toxins that adversely affect the growth rates (Noriyuki et al., 2012b). The saps of host plants may contain some chemical feeding deterrents and toxins as plant secondary compounds. There are some allelochemicals, like glucosinolates in the aphids sequestered from plants which cause toxic effects to the aphidophagous ladybirds (Pratt et al., 2008).

2.3 Food Economics

Ladybirds undergo a voracious larval stage (comprising four instars) and the food consumed during this stage could be considered as capital, as capital resources acquired during larval feeding influence reproductive success via effects on adult body size and other traits. The food consumed after eclosion and onwards could be referred to as income. Poor capital investment results in diminished reproductive outputs (Santos-Cividanes et al., 2011). Poor capital decreases the body size and to some extent the initial egg size laid by the resultant capital females; however, later on, the egg size of both poor and rich capital females become equal (Vargas et al., 2012a). The capital benefits to rich females provide them with the luxuries of early and quantitative egg production and oviposition; however, they become exhausted earlier than females with poor capitals, who compensated that loss by having prolonged reproductive period and age (Vargas et al., 2012b).

Mostly, ladybirds are income breeders because egg maturation depends on resources available during reproduction. Daily income fluctuations largely affect the oviposition. However, certain ladybirds could be unaffected by income fluctuation but appear constrained by capital (Vargas et al., 2013). Generally, large females compensate for income fluctuations better than small females, likely because of their greater capital and possibly an ability to consume larger meals when food was available. Specialists have a greater sensitivity to income which reflects a higher degree of prey specificity compared with the generalists.

2.4 Prey Density and Predator Performance

2.4.1 Functional Response

Analysis of how a ladybird responds to varying pest populations and how it affects pest management can be understood in terms of functional response. It is the predator's feeding response by increased prey consumption with prey density that may be analytically explained by Holling's type I (linear), type II (curvilinear), and type III (sigmoidal) responses (Holling, 1959, 1965; Figure 2). Ladybirds usually respond in a type II manner where there is an initial increase in the rate of prey consumption at lower prey densities but this rate decreases with a further increase in prey density (Omkar and Pervez, 2004b; Pervez and Omkar, 2005). This happens due to satiation, as there is a threshold of prey consumption and the prey density-dependent curve reaches an asymptote (Pervez and Omkar, 2003). Natural enemies exhibiting type III response are considered better biocontrol agents as prey mortality is exponential, which is usually followed by the parasitoids.

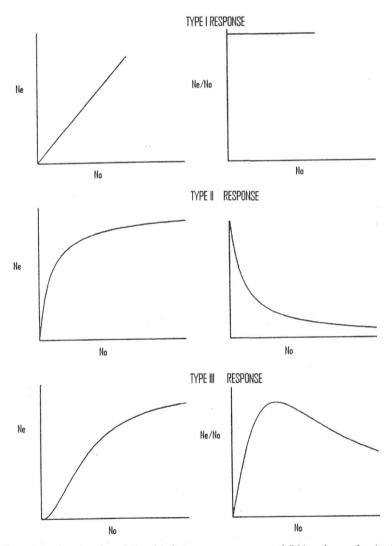

FIGURE 2 The graphs showing the relationship between prey consumed (N_e) and prey density (N_o) in terms of type I, II, and III responses. The figures adjacent to every type response shows the relationship between proportion of prey eaten (N_e/N_o) and N_o. *After Pervez and Omkar (2010).*

Interestingly, a plethora of literature supports the type II response in ladybirds (Pervez and Omkar, 2005; Cabral et al., 2009; Omkar and Kumar 2013) with a few reports on the type III response (Messina and Hanks, 1998; Barbosa et al., 2014). Functional response outcomes are dependent on factors, like generation time ratio of predator and prey, size disparity between them, walking speed, predatory stage, gender, etc. A type II response indicates that the predator may be useful as a biocontrol agent at lower densities. This also reveals that the constants of this system would be ephemeral or a young aphid colony and the ladybird has to be introduced earlier. However, biocontrol is needed when aphid colonies have already reached a peak. To combat this, subtle modifications were done in functional response approaches by exploiting ecological niches and introducing more than one predator, both interspecific and intraspecific in prey microcosm (Snyder and Ives, 2003; Pervez and Omkar, 2011; Kumar et al., 2014a). The responses somehow deviate from the established monotonic prey–predator outcomes.

Snyder and Ives (2003) combined the ladybird with a parasitoid and found an additive effect, though the parasitoid pupae were victimized by the ladybirds. Parasitoids initially induced little reduction in aphid population growth but later on caused a large decline showing a type I response. Had the response been type II, there would have been a synergistic effect on prey mortality as also predicted by stage structured model (Snyder and Ives, 2003). Omkar and Pervez (2011) synergistically exploited the ecological niches of two ladybird species and inferred that biocontrol would be benefitted if the tandemly released predators are noninterfering. This system could further be enhanced if predatory species differ spatiotemporally. Heterospecific ladybird combination also had a type II response but the magnitude of the two functional response parameters, i.e., coefficient of attack rate (a) and prey handling time (T_h, i.e., time taken to attack, eat, and digest a prey) skewed in favor of biocontrol of aphid at low and moderate densities. Kumar et al. (2014a) made subtle modifications and tested cogeneric ladybird species with similar ecological niches and altered prey number with prey biomass. They found a modified type II response (i.e., in between type II and III). Predators showed an initial type II at the low prey biomass but functional response curve showed an exponential elevation at the higher prey biomass level. The cogeneric duo combination, however, had an antagonistic effect due to high interferences. The authors therefore claim that generalists and aphidophagous ladybirds (which are usually not considered as better biocontrol agents as compared to specialists and coccidophagous ladybirds) are highly useful if released in appropriate combinations.

Host plants also affect the functional response outcomes and may alter the foraging pattern of ladybirds from type II to type III, as substratum ease the foraging success of the predator (Messina and Hanks, 1998). This is not usually the case with other ladybirds but it is likely that the magnitude of the response will probably be affected by the substrata and by the same prey obtained from different host plants (Kumar et al., 1999). Prey activity can also influence this response as ladybird preying on stable or immobilized prey would have a better impact on the prey mortality than the mobile and unstable prey (Provost et al., 2006).

A review on functional responses of other natural enemies, like parasitoids, showed that only about a quarter of studies showed type III functional responses, providing further evidence that the functional responses are one of the basic elements in selection of efficient biocontrol agents (Fernandez-Arhex and Corley, 2003). Questions on ladybird's biocontrol potential are neither answered by type III response success nor posed by type II response.

Other aspects, like prey preference, switching, intrinsic rate of increase of prey and predator, mean generation time, consumption rate, prey patchiness, predator patch allocation time, host plant, abiotic factors, and intra and interspecific predator competition also have major impact on the ladybird's biocontrol efficacy.

2.4.2 Numerical Response

Numerical response is the response of predator in terms of increasing its number in response of the increasing prey density, which could be both aggregative and reproductive. In response to increasing prey density, predaecous ladybirds show aggregative numerical response by increasing the cumulative prey consumption; however, the rate of prey consumption decreases curvilinearly (Omkar and Srivastava, 2003; Omkar and Pervez, 2004b) due to mutual interference. The reproductive numerical response is a consequence of the functional response in predaceous ladybirds. Females lay a high number of eggs at high prey densities which is a direct implication of functional response. Increase in food consumption facilitates the development of more numbers of ovarioles (Evans, 2000). However, this also has a threshold limit and reaches a saturation point at very high prey density similar to the asymptote of functional response curve. Both functional and reproductive numerical responses are interlinked having similar shapes in ladybirds exhibiting a type II response (Omkar and Pervez, 2004b). Chaudhary et al. (2015) explained the size-dependent response of ovipositing females against size-dependent prey density. They found that large females may capture and consume both small and large aphid instars, but later aphid instars are better in fulfilling the high energy requirements and oviposition costs of large female ladybirds.

2.4.3 Efficiency of Conversion of Ingested Materials

Efficiency of conversion of ingested materials (ECI) also measures the impact of predator in terms of percentage prey biomass converted into predator biomass. It is dependent on prey quality and it is high on suitable foods rather than factitious foods (Jalali et al., 2009). ECI is also dependent on the age, stage, and gender of predator, prey type, and the host plant (Shannag and Obeidat, 2008). It reaches the high value at the instar levels, as there exists of severity pressure of ecdysis into the next stage. This increases voracity of the growing instars and they resort to increased predation in a bid to attain a critical weight needed for the ecdysis or pupation. Earlier instars have higher ECI values than the later instars because of the intense survival pressure on the former. Lower values of ECI in the later instars also indicate the metabolic costs needed for the preparation of pupation (Isikber and Copland, 2001). Younger instars could have low ECI values after encountering bigger-sized aphids, which could not be easily attacked (Jalali et al., 2009). Here, the handling time increases and younger instars spent more energy in prey capturing.

Adults have a comparatively reduced survival pressure and they can resort to prolonged starvation, overwintering, and switching to nonessential foods in prey scarcity. It was hypothesized that coccidophagous ladybirds have better conversion efficiency than aphidophagous ladybirds (Mills, 1982). However, experimental evidence did not support this (Dixon, 2000). ECI also depends on metabolism and allocation of food energy into growth, development, and reproduction. A moderate part of prey biomass is transformed for growth in ladybirds, as marked similarity exist in biochemical profiling, e.g., proteins and lipids, of both ladybird and its prey (Specty et al., 2003). Adults have an age-dependent pattern in the prey consumption rate, i.e., it initially increased, reached a threshold value, and thereafter declined. This pattern, however, has no effect on ECI which possibly may be due to the enhanced digestive capabilities of adult ladybirds in a bid to compensate for their increased nutritional needs (Mishra et al., 2012a).

ECI might not be a direct indicator of biocontrol potential of a ladybird. However, skewness towards the higher ECI suggests that predator is capable of enhancing its fitness in terms of increased body mass that may ease and elevate its prey capturing and assimilation tendencies. This may cumulatively affect its biocontrol potential. ECI is sometimes also referred to as conversion of prey biomass into progeny biomass, as an outcome of reproductive numerical response. It is a relationship in terms of percent egg biomass produced against the prey biomass consumed. In aphidophagous ladybirds, ECI in terms of reproduction has a negative relationship, i.e., it decreases with the increase in prey density (Omkar and Pervez, 2004b; Omkar and Kumar 2013). Interestingly, egg biomass increases with consumption of extra prey biomass but rate of egg biomass production decreases with prey density. This explains the failure of aphidophagous ladybirds in establishing itself in growing aphid colonies, as they do not keep pace with exponential increase in prey density.

3. MATING AND REPRODUCTION

Morphologically, ladybirds could probably be sexed owing to subtle patterns on the body (Omkar and Pervez, 2000; Omkar and Pathak, 2007). However, this dimorphism is not common in ladybirds. Nevertheless, they have a size-dependent sexual dimorphism with females being larger than males. The males are consistently smaller in body size than females despite of the similar developmental period (Dixon, 2000). Male body size is a trade-off to its gonadal requirements needed for the sperm production and sexual activity (Yasuda and Dixon, 2002). Females have a larger body size for a quantitative egg production and accommodation, as larger females have more ovarioles in their gonads than smaller females (Dixon and Guo, 1993). Smaller males could be better mates in the limited food scenario and will spend less time eating and more time mating.

3.1 Mating

Mostly, ladybirds are protandrous, i.e., males mature earlier than females and males exhibit subtle courtship (approach, watch, examine, embrace, and mount) before entering into mating (Omkar and Pervez, 2005b). Certain coccidophagous female ladybirds are choosy and prefer previously-mated males than virgins (Jiaqin et al., 2014). Choosy females possibly identify previously-mated males by marking them with their cuticular hydrocarbons during copulations. Many polymorphic aphidophagous female ladybirds preferentially select melanic and darker morphs as potential mates, especially during winter (Srivastava and Omkar, 2005a; Mishra and Omkar, 2014; Figure 1). The choice for a particular morph in certain ladybirds is also dependent on the seasonal variations (Wang et al., 2009). This choice has a direct significant effect on their reproductive performance. Many ladybird species are promiscuous and females mate multiple times (Srivastava and Omkar, 2005b). This occurs even when sperm is not limited. Females benefit from sperm provided by multiple mates.

In laboratory conditions, polyandry benefits females for both quantitative progeny and paternity (Haddrill et al., 2007). The apparent reproductive benefits of polyandry with mate choice in ladybirds are more likely a result of additive fecundity stimulation as a function of multiple matings, rather than a result of females choosing better males (Pervez and Maurice, 2011; Bayoumy and Michaud, 2014). Interspecific matings are also conspicuous in ladybirds (Noriyuki et al., 2012a).

Copulation is prolonged if heavier adults mate compared with lighter adults with increased paternity (Pervez and Singh, 2013). This duration-dependent higher paternity for males can possibly be ascribed to the transfer of larger numbers of sperm in multiple spermatophores (Haddrill et al., 2008). Multiple matings benefit ladybirds (Omkar and Srivastava, 2002; Omkar and James, 2005; Omkar and Pervez, 2005b; Omkar et al., 2010a; Bista and Omkar, 2012) especially polyandry and promiscuity (Omkar and Mishra, 2005a). The sperm transfer mechanism is both direct (Omkar et al., 2006) and indirect via a spermatophore, which could also be proposed as a "nuptial gift" as most females devour it after copulation (Perry and Rowe, 2010). However, mating and spermatophore transfer is costly in some ladybirds as there is a subtle trade-off between mating and the longevity (Mishra and Omkar, 2006a). The spermatophore costs involve a trade-off between current and future reproductive investment, as the efficiency of sperm transfer decreases or halted in successive matings (Perry and Tse, 2013). Males ingest more food following spermatophore transfer indicating that spermatophore transfer depletes male nutritional reserves and has postmating implications. Nevertheless, in directly inseminating ladybirds without a spermatophore, food depletion is compensated by producing a higher relative sperm content in the ejaculate (Perry and Rowe, 2010).

3.2 Reproduction

Adult body size has a prominent effect on reproduction in ladybirds. As mentioned above, heavier males mate for a longer duration, which is positively correlated with oviposition period, fecundity, and fertility in predaceous ladybirds (Pervez and Singh, 2013). As ladybirds have a direct sperm transfer mechanism, they can produce fertile eggs even after 1 min of copulation, indicating that the sperm transfer mechanism is initiated very early in the copulation act (Omkar et al., 2006). However, these ladybirds mate for a few hours on a single mating event, which suggests that sperm transfer mechanism is early and continuous. Ladybirds have multiple ejaculations during a single complete mating which accounts for their increased fecundity and fertility. While an increase in the number of matings leads to increased egg production, it has a reproductive cost in terms of longevity (Omkar and Mishra, 2005b).

3.3 Age, Aging, and Rhythmicity

Both paternal and maternal age influence mating performance and reproductive outcome. It is likely that maternal age affects fecundity, oviposition rate, and the reproductive and nonreproductive periods and paternal age affects percent egg viability. Nonetheless, parental age per se has an indirect implication on these aspects probably due to age-influenced variation in seminal proteins that affects both oviposition and fecundity (Omkar et al., 2010b). However, paternal effects diminish as a function of mating history, suggesting that transgenerational signals of male origin are also subject to depletion (Michaud et al., 2013). Ladybirds have sex-dependent aging trajectories with reproductive performance declining later in females than in males (Mishra and Omkar, 2006a). These trajectories are unsynchronized with sex; female virility diminishes later than that of males. Optimization of age difference between mates maximizes the reproductive output; females that are five days older than the males during early life produce better progeny in quantitative terms (Omkar and Mishra, 2009). This was the first study on insects.

4. LADYBIRDS IN A PREDATORY GUILD

Heterospecific predators, including ladybirds, combat for a common food resource habitat (guild). As the common food (extraguild prey) diminishes they start attacking each other. In this struggle, one becomes dominant (intraguild predator) and overpowers the other (intraguild prey). Intraguild interactions are not the simplest and may give rise to series of complex interactions, which could both be advantageous and disadvantageous to the fitness of ladybirds and to the fate of biocontrol.

4.1 Isolation of the Guild Members

Occasionally, intraguild members share common food and space resources but they are isolated and rarely bother each other. Few ladybirds differ spatially in the guild by sharing different lateral or vertical spaces in prey habitat (Flowers et al., 2007). Plant morphology and other characteristics may also act as a barrier between the intraguild members (Janssen et al., 2007). This spatial isolation of intraguild predators might also have a mutual effect on biocontrol, if a predator facilitates its other spatially isolated guild member in fetching itself to the food resource (Losey and Denno, 1998). Ladybirds also have a difference in the temporal distribution and are isolated in different time zones (Toda and Sakuratani, 2006).

4.2 Ladybirds in the Guild

Ladybirds involved in intraguild combat are more conspicuous if the extraguild prey is a limiting factor. They exploit the food resource by attacking the extraguild prey and reducing their abundance for their competitors. This *exploitative competition* of food resource is more advantageous to the smaller-sized ladybird as they have lesser food demands (Obrycki et al., 1998a,b). As the density of extraguild prey decreases, frequency of intraguild predation (IGP) increases (Gagnon, 2010). On the contrary, an increase in extraguild prey density creates a dilution effect which manages the IGP and subsequently lowers its possibilities (Lucas and Brodeur, 2001). IGP is responsible for habitat dislocation (including both emigration and immigration) of intraguild members and affects the biocontrol events. An increase in the density of intraguild prey also increases the frequency of IGP, as the probability of encounters with the competitors will also increase (Noia et al., 2008). IGP is directed towards the main intraguild predator as the status of intraguild predator and prey will never change regardless of differences in the extraguild prey and its habitats (Yasuda and Ohnuma, 1999).

The relative size and stage, mobility of species, aggressive strategies, mandibular structure, degree of feeding and habitat specificity, defense strategies, and density of extraguild prey determine the outcome of IGP (Yasuda et al., 2001; Lucas, 2005; Roy et al., 2006). Disparity in body size also increases the likelihood of IGP and it skews towards larger and voracious species. Even their lower instars attack higher instars of inferior and less voracious species. Frequency of encounters of same-stage instars is usually lower. However, the older instars frequently encounter each other more than the younger ones (Yasuda et al., 2004). IGP also provides nutritional benefits to the developing stages if the extraguild prey lack nutritional values (Ware et al., 2009).

4.3 Interaction within Ladybirds

The majority of ladybirds attack, prey upon, and displace the other members of their family in a limited food resource struggle. Invasion and establishment of aggressive species following displacement of indigenous ones may be an outcome of IGP. One such invader is the Harlequin ladybird, *H. axyridis*. This invader has a competitive advantage over indigenous species, such as *C. septempunctata* (Yasuda et al., 2001), *Coleomegilla maculata* DeGeer (Labrie et al., 2006), *Hippodamia variegata* Goeze, and *A. bipunctata* (Lanzoni et al., 2004) due to its vast prey range, higher predation, and foraging potential (Alhmedi et al., 2010a). A review of 24 studies on impact of this invader suggested that 15 of them had described an intense negative impact on indigenous fauna, particularly due to exploitative competition or IGP (Lucas et al., 2007). It is acclaimed to be the most invasive ladybird on Earth (Roy et al., 2006). This, along with other successful invaders, may affect the dynamics and composition of indigenous species guilds.

This invasive species indulges in ***interference competition*** resulting from direct interaction of competitors or from the modification of a resource by one competitor such that it becomes unusable by another. *Harmonia axyridis* is acclaimed as top predator because of magnitude, direction, and symmetry of IGP (Felix and Soares, 2004). There is, however, no reported IGP between adults of two species; still the adults of *H. axyridis* may interact negatively by interfering with each other's foraging and oviposition capabilities. They even interfere with other competitors when food resource is not undergoing shortage (Soares et al., 2008). The dominant species, especially *H. axyridis*, also indulge in asymmetric reproductive interference that may compel the inferior indigenous species to become specialist predators of less preferred prey in nature (Noriyuki et al., 2012a).

The invasive species also gets benefitted by ***apparent competition***, results from the indirect interaction of two co-occurring species sharing a common resource. Adult *H. axyridis* have the intrinsic advantage of being the least vulnerable to parasitism by braconid parasitoid, *D. coccinellae*, and fungus, *Beauveria bassiana* (Balsamo) Vuillemin (Roy et al., 2008) compared with other co-occurring species. The vulnerability of other ladybird species toward parasitism negatively affects their fitness, decreasing their competitive ability. Apart from being a victor in competitions, *H. axyridis* is the biggest winner in almost all IGP interactions within the family. Its eggs are more chemically defended thereby skewing IGP of eggs in its favor (Cottrell, 2007). Morphological traits, like larger body size, the presence of spines on the back of the third and fourth instars provides extra protection (Ware et al., 2008). Additionally, size, strength of the integument, and distastefulness of its pupae reduce its vulnerability

(Felix and Soares, 2004). Larvae of this invasive species are larger and more aggressive than those of other ladybirds (Yasuda et al., 2004). When paired with a larger larva of a different species, *H. axyridis* can either win (Michaud, 2002) or lose (Ware and Majerus, 2008). Other ladybirds that fall victim to *H. axyridis* may be superior and top intraguild predators in their respective guilds (Rondoni et al., 2014).

4.4 Interaction with Other Predators

Ladybirds also struggle with other group of insects outside the family for food resource. They usually co-occur with chrysopid (Neuroptera: Chrysopidae) larvae and share limited resources. However, IGP interactions between them are asymmetrical which could have positive, negative, or neutral impacts on aphid biocontrol (Gardiner and Landis, 2007). *Harmonia axyridis* mostly acts as an intraguild predator and declines local populations of chrysopids and predatory gall midge along with that of aphids (Gardiner and Landis, 2007). A few top ladybird species fall victim to voracious chrysopid larvae, especially to *Chrysoperla carnea* (Stephens) (Noppe et al., 2012) despite being equipped with strong defense. Ladybird larvae wriggle and bite and may use armor of their dorsal and lateral spines (Michaud, 2003) to attack the enemy. Some ladybird pupae flip when perturbed which may sniff off the antennae of the attackers (Lucas et al., 2000). Both chrysopids and ladybirds use chemical defenses. Coccinellid larvae (Lucas, 2005), pupae (Lucas et al., 2000), and adults all use chemical substances containing alkaloids to deter predators. Ladybirds could be victimized by other predators, e.g., big instars of syrphids, pentatomids, and spiders in the guild.

5. SEMIOCHEMICALS AND LADYBIRDS

Semiochemicals play key roles in the conspecific, heterospecific, and prey interactions of ladybirds. Their main role is related to attack, food, sex, and defense (Hemptinne and Dixon, 2000). Pattanayak et al. (2014, 2015) employed headspace solid-phase microextraction to determine sex-specific volatile hydrocarbon profile of five ladybirds from India and found methyl-branched saturated and unsaturated hydrocarbons. They listed 26 alkane-based compounds, which varied in composition in ladybirds, including *Menochilus sexmaculatus* (Fabr.), *C. septempunctata*, *Coccinella transversalis* Fab., and *Propylea dissecta* (Mulsant). In *Anegleis cardoni* (Weise) the two sexes differ in major chemicals, but both the sexes have similar hydrocarbon (HC) profiles and **9- and 7-methyl tricosane** predominate the cuticular HC profile in of *A. bipunctata* (Hemptinne et al., 1998). The HCs are key volatiles and are found in the footprints (Magro et al., 2010) and cuticular lipids (Durieux et al., 2012).

5.1 Kairomones Driven Foraging

Volatile infochemicals generated by prey, their host plants, or from the interaction between host plant and herbivore may be used by predators in prey location. Ladybirds are more attracted toward the plants where the aphids are disturbed than those where aphids are undisturbed or toward healthy plants (Ninkovic et al., 2001). Plants and aphids release volatiles on disturbance which are perceived by the ladybirds. Plants induce two classes of volatile compounds (IVCs) (Dudareva et al., 2006). Damaged leaves release class I IVCs mainly consisting of C6 green volatile compounds, while class II IVCs (mainly terpenes) are synthesized and released after a few hours of plant injury (Holopainen, 2004). Ladybirds are attracted toward chemicals, like limonene and β-caryophyllene, for prey location and oviposition (Francis et al., 2005; Alhmedi et al., 2010b). Limonene is a potential kairomone for *H. axyridis* and females are found more in limonene zones rather than in control zones. This explains why they hover more in citrus orchards that are a rich source of limonene (Michaud, 2002).

When attacked, aphids release a small droplet from their abdominal cornicles containing an alarm pheromone, identified as the sesquiterpene or **(E)-β-farnesene** (Francis et al., 2005). Conspecific aphids trigger their escape behaviors after perceiving it. This kairomone attracts adult ladybirds, who display electrophysiological responses toward a group of semiochemicals including both host and herbivore odors. This additively mitigates the attraction of ladybirds toward aphid colonies (Joachim and Weisser, 2013). Nonetheless, feeding by ladybirds is largely driven by this kairomone.

5.2 Oviposition Deterring Pheromones

Oviposition-deterring pheromones (ODP) are chemicals that are released from the invaders that deter the mother ladybird to oviposit in such sites (Ruzicka, 2002, 2003, 2006; Michaud and Jyoti, 2007; Ruzicka and Zemek, 2008). It is a part of maternal strategy in a bid to protect her offspring from the enemies. Females search for the most suitable oviposition sites that may favor their progeny (Kindlmann and Dixon, 1993). This site selection depends on abiotic conditions, nutritional requirements of offspring and resource availability (Dixon, 2000), absence of predators, and plant architecture (Seagraves, 2009).

ODPs of both conspecifics and heterospecifics invaders are indicators of age and future of prey colony. These are persistent in both larval (Mishra et al., 2012a) and adult footprints (Mishra and Omkar, 2006b; Mishra et al., 2012b) and are released from anal disk on the tenth abdominal segment of larvae (Laubertie et al., 2006). These are a mixture of HCs dominated by **n-pentacosane** producing a persistent signal (Hemptinne et al., 2001). Additionally, **n-tricosane** and **n-heptacosane** are also important footprint semiochemicals in ladybirds (Nakashima et al., 2006). **n-Z-12-pentacosene** is a principal compound in *M. sexmaculatus* (Pattanayak et al., 2014) and is considered as an important ODP (Klewer et al., 2007). These footprints also act as foraging-deterrent pheromones and reduce feeding activity and delay development in ladybirds (Kumar et al., 2014b).

The effects of ODP are largely density and stage-dependent (Doumbia et al., 1998; Mishra and Omkar, 2006b). However, the possibility of the sex-dependent sensitivity toward these chemicals is not totally ignored (Hodek and Michaud, 2008). Aged and experienced mothers may modify the response to ODP (Frechette et al., 2004). Female ladybirds usually deter from ovipositing on plants contaminated with conspecific larval tracks but not with heterospecific tracks (Yasuda et al., 2000).

5.3 Sex Pheromones

Ladybirds have chemically differentiated sexual dimorphism and both sexes differ in cuticular chemistry on the body surface (Fassotte et al., 2014). Volatile hydrocarbons differ between the sexes of ladybirds (Pattanayak et al., 2014). Female ladybirds release sex pheromones in the form of mixture of volatile compounds. Sex pheromones of *H. axyridis* had **β-caryophyllene** as major constituent along with β-elemene, methyl-eugenol, α-humulene, and α-bulnesene (Brown et al., 2006; Fassotte et al., 2014). Not much is known about the organ responsible for sex pheromone production in ladybirds, however, it is likely that specific abdominal glands are involved in pheromone production.

Volatile supplement analogues could be used artificially for retaining ladybirds in the aphid vicinity. Biocontrol could be compromised if the adult ladybirds disperse elsewhere and leave their task in between. The retention of ladybirds in aphid vicinity could be improvised by providing nutritional resources (e.g., pollen sources, nectar-producing plants, or artificial food supplements (Lundgren, 2009)). Genetically modified plants using higher levels of herbivore-induced plant volatiles can attract natural enemies of phytophagous pests at potentially increasing densities, thereby improving the biocontrol (Schnee et al., 2006).

Information pertaining to semiochemicals revealed that most of the ecological and behavioral events occurring in ladybirds are chemically driven, especially, foraging, premating behavior, mate search, oviposition, and oviposition deterrence. Even parental care and kin recognition are also largely affected by the surface chemistry, especially surface alkanes of the ladybird stages (Omkar et al., 2005; Pervez et al., 2005). Furthermore, some of the chemicals have many roles in ladybird ecology. Much attention is being paid to identify and characterize the semiochemicals that are related to the series of biological events occurring in ladybirds.

6. EFFECTS OF ABIOTIC FACTORS

6.1 Temperature

Temperature is the most crucial abiotic factor affecting ecological, functional, and behavioral attributes of predaceous ladybirds. It sets the limits of biological activity in form of low and high temperature thresholds. The developmental rate (DR) is zero at lower development threshold (LDT), which increases with temperature, reaches a peak value, and then decreases rapidly as the high thermal threshold is achieved. Not much LDT variation occurs in ladybirds with similar dietary habits and this is also a reason for its successful establishment in different habitats and countries (Nedved and Honek, 2012). Both ladybirds and their prey have different thermal limits and an effective temperature range for survival, development, reproduction, and mobility. These limits, like LDT, DR, and sum of effective temperature (SET) in day degrees are considered as thermal constants (Jarošík et al., 2011). Being thermal constants, LDT and SET may predict the intrinsic rate of increase and generation time of the ladybirds as the development is fastest if the LDT is high and SET is low. Thus, there is a negative relationship between these two thermal constants and the negative slope may be a result of biological variation. However, Jarošík et al. (2014) questioned whether thermal constants are constant, as they found these thermal constants may also be considered as diet-dependent variables in aphidophagous ladybirds.

Temperature also has a significant impact on reproduction and demography. The energy partitioning model elucidated the impact of temperature along with age in shaping the fecundity function in insects, especially ladybirds (Kindlmann et al., 2001). Temperature-driven age-specific fecundity function of ladybirds is triangular in shape with the initial increase in the daily oviposition rate, reaching a peak and thereafter declining with further aging (Omkar and Pervez, 2002b; Pervez and Omkar, 2004). A gradual increase in temperature expedited this peak of oviposition rate toward the younger side of the aging ladybirds. At the optimum temperature, around 27–30 °C in most aphidophagous ladybirds, the peak is attained at the youngest age of the ladybird demography (Pervez and Omkar, 2004; Maia et al., 2014). Thereafter, with further rises in temperature this peak is delayed and shortened. It is a similar case with the scale predators (Senal, 2006; Ponsonby, 2009). However, in acarophagous ladybirds the optimum temperature is 30–35 °C, resulting in high values for the demographic parameters, like net reproductive rate (R_o) and intrinsic rate of increase (r_m) (Roy et al., 2003; Taghizadeh et al., 2008). Optimization of temperature can be the booster for the quantitative augmentative rearing of potentially important ladybirds. Thus, biocontrol programs will benefit if the ladybirds are reared at the optimum temperatures.

6.2 Light

Light immensely affects ladybird bioattributes, particularly variables like intensity, quality (wavelength), and duration of exposure (photoperiod). Ladybirds have a wide range of tolerance limits to these variables. They are primarily diurnal insects and depend on visual cues and presence/absence of light to undergo various essential activities, like mating, molting, and pupation (Mishra and Omkar, 2005). Most ladybirds are highly sensitive to light, particularly its photoperiod (Wang et al., 2013; Mishra and Omkar, 2005; Omkar and Pathak, 2006) and wavelength (Omkar et al., 2006; Nalepa, 2013). Shorter day lengths with an intensity of 1500 lux are beneficial for reproductive activities as it may shorten the pursuit times followed by increase in the duration of copula (Wang et al., 2014). The likelihood of female ladybirds accepting the males increases in the dark because females are unable to evaluate visual criteria to select male and thereby mating rejection displays are also minimized. Long day photoperiods support better intrinsic rates of increase and net reproductive rates in acarophagous ladybirds (Aksit et al., 2007). However, prolonged light days could have a negative effect on ladybird physiology (Mishra and Omkar, 2005).

There exists a photoperiod-dependent bimodal or two-peak pattern in the development of certain ladybirds, where the first peak represents fast developing and the second shows slow-developing individuals. The slow-developing individuals were more in numbers in short day lengths; however, long day lengths promoted fast developing adults having heavier body masses and more capable of producing quantitative progeny (Singh et al., 2014).

White light is more suitable for essential activities compared with its red or blue components (Omkar and Pathak, 2006) of the visible spectrum and these may influence their mating activity and performance (Omkar et al., 2005). No gender-based difference existed in the morphology of compound eyes (Yan et al., 2006) and it could be inferred that both genders have same visual perception (Wang et al., 2014). Information on ecology of ladybirds in relation to light also benefits their mass rearing along with enhancing their fitness and quality.

7. BIOCONTROL PROSPECTS

The impact of ladybirds in terms of successful pest biocontrol is largely dependent on their voracity, prey specificity, intrinsic rate of increase (Dixon, 2000), and the mean generation time ratio between prey and predator (Dixon, 2005). Interestingly, size of the ladybirds attacking similar kind of prey does affect the biocontrol with the bigger-sized ladybirds being the better biocontrol agents. However, if the prey type is different, the smaller-sized ladybird with specialization on that particular prey type is a more promising biocontrol agent than the bigger but less specialized ladybird. There are many examples of smaller sized ladybirds for the biocontrol of mites, whiteflies, diaspids, and coccids, while bigger-sized ladybirds have less preference for such prey. This may be explained by the concept of resource partitioning of prey and predator, where small predatory species can better exploit small-sized prey species and vice versa (Dixon, 2007). These species may easily displace other species from a particular portion of the resource space as they are better adapted to exploit that particular species of prey in that resource space. Prey resource partitioning, both within and between the species, occurs in ladybirds, for instance, small-sized ladybirds show increased oviposition after consuming small aphid instars at higher prey density (Chaudhary et al., 2015). Additionally, specialists are better biocontrol agents than the generalists because of their selective feeding and persistence in the target prey habitats (Snyder and Ives, 2003). However, the invasion of generalists in their resource space is an issue of serious concern as they become intraguild prey or they are forced to emigrate (Rondoni et al., 2012).

The response of ladybirds to various prey types for successful biocontrol is mentioned below.

7.1 Coccids and Diaspids

Coccidophagous ladybirds are successful biocontrol agents of both coccids and diaspids. Ever since the first mega success story of vedalia beetle, *R. cardinalis* against *I. purchasi* in California, numerous ladybird species were successfully used in the biocontrol of scale insects and mealybugs. *Chilocorus nigritus* is another highly successful and effective generalist predator of numerous species of Diaspididae with equal effects on some species of Coccidae and Asterolecaniidae. Reviews on ecology (Omkar and Pervez, 2003) and factors affecting the utility of *C. nigritus* (Ponsonby, 2009) gave lists of successful releases along with the list of both essential and supplement prey types. Congregation of *C. nigritus* at the prey site results in mass destruction of scales, with minimal mutual interference. Varying predator densities do not affect feeding rates. The minimal mutual interference behavior is similar to those of parasitoids, as they do not disperse with increased crowding and there is no temporary cessation in the foraging pattern after encountering conspecifics. The Indian ladybird, *C. montrouzieri*, is also a generalist predator of various scales and mealybugs and has been commercially exploited in both classical and augmentative biocontrol programs (Table 1). Michaud (2012) stated that the introduction of coccidophagous ladybirds has been successful in both tropical and subtropical habitats. A review of classical programs in Africa listed nine coccinellid species successfully established (Greathead, 2003). Coccidophagous ladybirds have been considered as the most successful because of their great successful rate in declining the prey

TABLE 1 Certain Examples of the Successful Biocontrol Programs of Various Pests Using Predaceous Ladybirds

Ladybird	Pest	Host plant	Country	References
APHIDS				
Coccinella undecimpunctata undecimpunctata	*Aphis fabae*	Soybean	Egypt	Zaki et al. (1999)
	Aphis gossypii	Okra	Egypt	Zaki et al. (1999)
Coccinella septempunctata and *H. variegata*	*Aphis gossypii*	Cucumber	Morocco	El Habi et al. (1999, 2000)
Harmonia axyridis	*Macrosiphum euphorbiae* (L.)	Cut roses	USA	Snyder et al. (2004)
	Macrosiphum rosae	Amenity roses	France	Ferran et al. (1996)
	Phorodon humuli	Hops	France	Trouve et al. (1997)
	Aphis gossypii	Cucumber	Japan	Kuroda and Miura (2003)
	Aphis craccivora Koch	Faba bean	Egypt	Ferran et al. (2000)
Hippodamia convergens	*Monellia caryella* (Fitch) *Monelliopsis pecanis* Bissell	Pecan orchard	USA	LaRock and Ellington (1996)
C. septempunctata	*Aphis gossypii*	Cotton	China	Zhang (1992)
	Lipaphis pseudobrassicae Davis	Mustard	India	Shenhmar and Brar (1995)
Synonycha grandis Thunberg and *Coleophora biplagiata* Crotch	*Ceratovacuna lanigera* Zehntner	Sugar cane	China	Deng et al. (1987)
C. septempunctata brucki Mulsant	*Acyrthosiphon kondoi* Shinji	Lucerne (alfalfa)	Japan	Takahashi (1997)
SCALE INSECTS				
Chilocorus nigritus	Coccid, *Coccus viridis* (Green)	Acid lime	India	Mani and Krishnamoorthy (1998)
	Aonidiella aurantii Maskell	Citrus	India and South Africa	Ponsonby (2009)
	Aspidiotus destructor Signoret	*Cocos nucifera* L.	Sri Lanka	Ponsonby (2009)
Chilocorus stigma (Say)	*Chionaspis pinifoliae* (Fitch)	Christmas tree	Michigan	Fondren and McCullough (2005)
C. septempunctata and *H. convergens*	*Toumeyella pini* (King) and *Toumeyella parvicornis* (Cockerell)	*Pinus* sp.	USA	Cooper and Cranshaw (2004)

(Continued)

TABLE 1   Certain Examples of the Successful Biocontrol Programs of Various Pests Using Predaceous Ladybirds—cont'd

Ladybird	Pest	Host plant	Country	References
MEALYBUGS				
Cryptolaemus montrouzieri	*Planococcus citri* (Risso) *Nipaecoccus viridis* (Newstead)	Pummel	India	Mani and Krishnamoorthy (2008)
	Maconellicoccus hirsutus (Green) *P. citri*	Grapevine	India	Srinivasan and Sundera Babu (1989)
	Chloropulvinaria aurantii Cockerell	Citrus	Azerbaijan	Prokopenko (1982)
	Chloropulvinaria floccifera Westwood	Tea	Russia	Mzhavandze (1984)
	Ferrisia virgata (Cockerell)	Tobacco	India	Gautam et al. (1988)
WHITEFLIES				
Delphastus pussilus (LeConte) *Nephaspis bicolor* Gordon *Nephaspis oculatus* (Blatchley)	*Aleurodicus disperses* Russell	Almond	Hawaii	Nechols (1982) and Kumar et al. (1987)
Serangium parcesetosum Sicard	*Bemisia tabaci* (Gennadius)	Citrus	USA	Legaspi et al. (2001)
	Bemisia argentifolii Bellows	Poinsettia	USA	Ellis et al. (2001)
	Dialeurodes citri Ashmead	Citrus	Turkey	Yigit et al. (2003)
THRIPS				
Hippodamia variegata	*Thrips tabaci* Lindeman	Tobacco	Bulgaria	Dimitrov (1976)

populations and their exotic establishments. This was mainly because rate of natural increase ratio between the predator and prey was skewed in favor of the introduced ladybird (Hodek and Honek, 2009).

7.2 Aphids

The association between ladybirds and aphids was recognized centuries ago. Kirby and Spence (1846) discussed the concept of augmentative releases of ladybirds in greenhouses and appreciated the growers who conserved aphidophagous ladybird from bird predation. In 1874, *Coccinella undecimpunctata* L. was introduced in New Zealand for aphid biocontrol, where it established successfully (Dixon, 2000). A need for aphid biocontrol led to the introduction of 179 ladybird species in North America since 1900, but only 18 have successfully established (Gordon, 1985; Obrycki and Kring, 1998). A few ladybirds took many years to establish. However, many established after accidental introduction, including *C. septempunctata*, *H. axyridis*, and *Propylea quatuordecimpunctata* (L.). The introduction and invasion of certain ladybird species has been implicated in the decline of some native species in the USA and elsewhere (Elliott et al., 1996; Obrycki and Kring, 1998).

Despite several unsuccessful attempts, a few successful examples gave hope for aphid biocontrol practitioners. Adult and larval releases of the ladybirds *C. septempunctata* and *H. variegata* provide satisfactory control against *Aphis gossypii* (Glover) in glasshouse cucumbers in Morocco. However, this was dependent on the timing of releases and on a high predator-to-aphid ratio (El Habi et al., 1999, 2000). Similarly, adults of *C. undecimpunctata undecimpunctata* were needed at rates of up to 200 per 140 plants for the optimal biocontrol of *Aphis fabae* Scopoli on soybeans (Zaki et al., 1999). Duo releases of *Hippodamia convergens* Guérin-Méneville with a parasitoid *Aphidius colemani* Viereck yielded favorable results against *A. gossypii* (Sterk and Meesters, 1997). Similarly, dual releases of *P. dissecta* and *C. transversalis* may give better results in *A. gossypii* biocontrol (Omkar and Pervez, 2011).

In orchards in Switzerland, *A. bipunctata* was more potent predator of aphid, *Dysaphis plantaginea* (Passerini) on apple trees compared with coexisting gall midge, *Aphidoletes aphidimyza* Rondani and syrphid, *Episyrphus balteatus* (DeGeer) (Wyss et al., 1999b). *Adalia bipunctata* and *E. balteatus* imposed additive effect on aphid mortality. Larvae of *A. bipunctata* were largely accountable for negative impact on the spring population of *D. plantaginea* (Wyss et al., 1999a). Some success was also noticed in duo releases of adult *H. convergens* and lacewing larvae against aphids in pecan orchards in New Mexico (LaRock and Ellington, 1996), while those of *H. axyridis* and *Cycloneda sanguinea* (L.) were effective against citrus aphids (Michaud, 2000). Trials of *H. axyridis* were highly effective against aphid, *Macrosiphum euphorbiae* (Thomas) on glasshouse-grown roses (Snyder et al., 2004).

Some subtle modifications were done to use the ladybirds that were thought to be inactive for aphid biocontrol, like overwintered or overly-flying adult ladybirds. Overwintered adult sites were preconditioned by exposing them to fly-in screen cages and providing them with water before release onto aphid-infested potted plants and were found useful in nurseries and glasshouses (Flint et al., 1995). A flightless form of *H. axyridis* was produced using a chemical mutagen followed by selective breeding (Tourniaire et al., 2000). The inundative releases of egg masses and such flightless adults against *A. gossypii* on glasshouse cucumbers were highly successful (Kuroda and Miura, 2003). However, in huge agriculture fields they may not be successful because of their impaired foraging on account of being flightless. However, these adults had lower reproductive fitness and had fewer offspring despite ovipositing for a longer period (Ferran et al., 1998). Fifty larvae of *H. axyridis* released per plant-infesting aphid, *Phorodon humuli* (Schrank), gave encouraging results; however, after becoming adults they migrated elsewhere (Trouvé et al., 1997). These larvae were also successful in rose aphid biocontrol in Paris (Ferran et al., 1996). Larvae of *C. undecimpunctata undecimpunctata* were successful against *A. gossypii* in China (Zaki et al., 1999), while the larvae of *C. septempunctata* and *M. sexmaculatus* effectively managed mustard aphid, *Lipaphis pseudobrassicae* Davis (Shenhmar and Brar, 1995). Ladybirds in cereal crops were the reasons for the suppression of aphids, *Schizaphis graminum* (Rondani) and *Diuraphis noxia* (Kurdj.) (Brewer and Elliot, 2004).

Aphid biocontrol could benefit if the prey is targeted early in the season, i.e., prey suppression initiation could be done when the aphid colony is young. This could further be modified by increasing their fitness on the intermediate or nonpest patchy aphid hosts. For instance, initial populations of nonpest cereal aphids enable ladybird fitness and elevated densities, which then suppressed aphid densities in grain sorghum and wheat (Michels and Matis, 2008).

7.3 Other Pests

Certain specialist ladybirds belonging to genus *Stethorus* are potential biocontrol of tetranychid mites, especially at high density of mites (Biddinger et al., 2009). The *Stethorus vegans* (Blackburn) population negatively affected that of the mite *Tetranychus urticae* Koch (Ullah, 2000). Michaud (2012) pointed out that ladybirds specializing on one kind of prey may not be necessarily potential biocontrol agents for that prey. For instance, *Clitostethus oculatus* (Blatchley) is credited for biocontrol of whitefly, *Aleurodicus dispersus* (Russell) in Hawaii (Waterhouse and Norris, 1989) and India (Ramani et al., 2002); however, *Nephaspis* is a specialized predator of said prey. The latter needs mutual assistance of other natural enemies including parasitoids for the biocontrol of *A. dispersus* (Legg et al., 2003).

Aphidophagous ladybirds acclaim criticism for not being better biocontrol agents largely due to significant differences in their intrinsic rates of increase and mean generation time ratios, although the relative development rates of aphidophagous ladybirds are lower than those of aphids.

8. CONCLUSIONS

The predaceous ladybirds have promising future in the biocontrol of various insect pests of crops, cereal, vegetable and oilseed crops, and various fruit trees. Some are specialists and numerous are generalists. Even invasive ladybirds, like *H. axyridis*, could be exploited in the biocontrol for various prey types. Evidence of prey specialization reveals how prey preference and prey suitability have evolved. There are lots of arguments on prey specialization in predaceous ladybirds. Resource partitioning and consistent exposure of a single prey type could have evolved prey specialization in predaceous ladybirds. Additionally, prey specialization has been argued as a function of ladybird body size, prey size, and prey density. Analytical studies pertaining to functional response of certain ladybirds indicate a few predators although had a negative density-dependent type II response but the magnitude of response, especially at the initial stages assure the biocontrol of pest at the lower density, i.e., they could be successful for the suppression of young aphid colonies. Certain combinations of ladybirds showing modified type II, i.e., an intermediate response between type II and III, is a promising breakthrough supporting promising biocontrol future of ladybird predators. These had an exponential increased rate of prey mortality during the middle prey densities indicating that such combinations could have negative effects on the aphid colonies that are growing mature but not old. Female ladybirds are larger in body size than males. This helps them in quantitative egg production and with a better reproductive numerical response in biocontrol program.

Ladybirds could be used as efficient models for reproductive, aging, ecological, and evolutionary studies as they have short life cycle with some species having prominent sexual dimorphism. Mating and reproductive studies provide knowledge on the optimal conditions pertaining to number and quality of mates to produce better progeny both in terms of quantity and quality. Similarly, the information pertaining to age, aging trajectories and age differences between mates not only increases the level of knowledge on ladybird physiology in general but would also help in ladybird multiplication by allowing mating with optimal age individuals.

Intraguild interactions upset the coexistence of ladybirds and displace many indigenous ladybird fauna. Biological invasions of certain dominant species, especially, *H. axyridis* and *C. septempunctata*, resulted in complete disappearance of numerous native ladybird species. These invasions had a tremendous negative impact on the specialists, especially at low prey density. Only the smarter ladybirds, which may better exploit the resources, can coexist with these invaders. The complexity of intraguild interaction of ladybirds has always troubled those using mathematical modeling to predict the intensity of such interactions on the outcome of biocontrol program. Even functional response models need to be reconsidered under the light of intraguild interactions. Nevertheless, for theoretical ecologists, intraguild combat is operational only when the prey density is low and biological control is off the table.

Ladybirds' life events are mostly chemical-dependent and in this chapter we have highlighted some of the essential chemically driven functions of ladybirds. Still, the knowledge pertaining to semiochemicals and ladybird chemistry is meager and further research is needed to explore the role of chemicals on certain unaddressed issues, like kin recognition, mate recognition, selective cannibalism, prey synchrony, etc. Among the abiotic factors, temperature has a major impact on the ladybird's life events. Similar temperature considerations should be done prior to the ladybird's importations, augmentative rearing, and biocontrol releases. Optimization of abiotic conditions, like temperature and light is a prerequisite for the augmentative rearing of ladybirds. This chapter also highlights the point that aphidophagous ladybirds are underestimated in biocontrol success stories. There is an urgent need to upgrade our understanding of ladybirds in pest management through comprehensive ecological and ethological studies supported with laboratory experimentation, and glasshouse and field studies.

Acknowledgments

Omkar is thankful to the Department of Higher Education Govt. of U.P. for financial assistance under Center of Excellence program. Ahmad Pervez is thankful to University Grants Commission, New Delhi, India for financial assistance under Major Research Project (No. F. 41-18/2012 [SR]).

References

Aksit, T., Cakmak, I., Ozer, G., 2007. Effect of temperature and photoperiod on development and fecundity of an acarophagous ladybird beetle, *Stethorus gilvifrons*. Phytoparasitica 35 (4), 357–366.

Alhmedi, A., Haubruge, E., Francis, F., 2010a. Intraguild interactions implicating invasive species: *Harmonia axyridis* as a model species. Biotechnological and Agronomical Society of Environment 14 (1), 187–201.

Alhmedi, A., Haubruge, E., Francis, F., 2010b. Identification of limonene as a potential kairomone of the harlequin ladybird *Harmonia axyridis* (Coleoptera: Coccinellidae). European Journal of Entomology 107, 541–548.

Barbosa, P.R.R., Oliveira, M.D., Giorgi, J.A., Silva-Torres, C.S.A., Torres, J.B., 2014. Predatory behavior and life history of *Tenuisvalvae notata* (Coleoptera: Coccinellidae) under variable prey availability conditions. Florida Entomologist 97 (3), 1026–1034.

Bayoumy, M.H., Michaud, J.P., 2014. Female fertility in *Hippodamia convergens* (Coleoptera: Coccinellidae) is maximized by polyandry, but reduced by continued male presence. European Journal of Entomology 111 (4), 513–520.

Biddinger, D.J., Weber, D.C., Hull, L.A., 2009. Coccinellidae as predators of mites: Stethorini in biological control. Biological Control 51, 268–283.

Bista, M., Omkar, 2012. Impact of multiple mating on behavioural patterns and reproductive attributes of seven spotted ladybird beetle, *Coccinella septempunctata* (L.). Journal of Applied Bioscience 38, 159–164.

Brewer, M.J., Elliott, N.C., 2004. Biological control of cereal aphids in North America and mediating effects of host plant and habitat manipulations. Annual Review of Entomology 49, 219–242.

Brown, A.E., Riddick, E.W., Aldrich, J.R., Holmes, W.E., 2006. Identification of – (−) β-caryophyllene as a gender-specific terpene produced by the multi-coloured Asian lady beetles. Journal of Chemical Ecology 32, 2481–2499.

Cabral, S., Soares, A.O., Garcia, P., 2009. Predation by Coccinella undecimpunctata L. (Coleoptera: Coccinellidae) on Myzus persicae Sulzer (Homoptera: Aphididae): effect of prey density. Biological Control 50, 25–29.

Chaudhary, D.D., Kumar, B., Mishra, G., Omkar, 2015. Resource partitioning in a ladybird, Menochilus sexmaculatus: function of body size and prey density. Bulletin of Entomological Research 105, 121–128.

Cooper, D.D., Cranshaw, W.S., 2004. Seasonal biology and associated natural enemies of two Toumeyella spp. in Colorado. Southwestern Entomologist 29, 39–45.

Cottrell, T.E., 2007. Predation by adult and larval lady beetles (Coleoptera: Coccinellidae) on initial contact with lady beetles eggs. Environmental Entomology 36, 390–401.

DeBach, P., 1964. Biological Control of Insect Pest and Weeds. Chapman and Hall, London. pp. 844.

Deng, G.R., Yang, H.H., Jin, M.X., 1987. Augmentation of coccinellid beetles for controlling sugarcane woolly aphid. Chinese Journal of Biological Control 3, 166–168.

Dimitrov, A., 1976. Control of tobacco thrips. Rastitelna Zashchita 24, 34–37.

Dixon, A.F.G., Guo, Y., 1993. Egg and cluster size in ladybird beetles (Coleoptera: Coccinellidae): the direct and indirect effects of aphid abundance. European Journal of Entomology 90, 457–463.

Dixon, A.F.G., 2000. Insect Predator–Prey Dynamics: Ladybird Beetles and Biological Control. Cambridge University Press, Cambridge.

Dixon, A.F.G., 2005. Insect Herbivore–Host Dynamics, Tree – Dwelling Aphids. Cambridge University Press, Cambridge, UK, pp. 199.

Dixon, A.F.G., 2007. Body size and resource partitioning in ladybirds. Population Ecology 49, 45–50.

Doumbia, M., Hemptinne, J.L., Dixon, A.F.G., 1998. Assessment of patch quality by ladybirds: role of larval tracks. Oecologia 113, 197–202.

Dudareva, N., Negre, F., Nagegowda, D.A., Orlova, I., 2006. Plant volatiles: recent advances and future perspectives. Critical Reviews in Plant Sciences 25, 417–440.

Durieux, D., Fischer, C., Brostaux, Y., Sloggett, J.J., Deneubourg, J.L., Vandereycken, A., Joie, E., Wathelet, J.P., Lognay, G., Haubruge, E., Verheggen, F.J., 2012. Role of long-chain hydrocarbons in the aggregation behaviour of Harmonia axyridis (Pallas) (Coleoptera: Coccinellidae). Journal of Insect Physiology 58, 801–807.

El Habi, M., El Jadd, L., Sekkat, A., Boumezzough, A., 1999. Lutte contre Aphis gossypii Glover (Homoptera: Aphididae) sur concombre sous serre par Coccinella septempunctata Linnaeus (Coleoptera: Coccinellidae). Insect Science and Its Application 19, 57–63.

El Habi, M., Sekkat, A., El Jadd, L., Boumezzough, A., 2000. Biology of Hippodamia variegata Goeze (Col., Coccinellidae) and its suitability against Aphis gossypii Glov. (Hom., Aphididae) on cucumber under greenhouse conditions. Journal of Applied Entomology 124, 365–374.

Elliott, N.C., Kieckhefer, R., Kauffman, W., 1996. Effects of an invading coccinellid on native coccinellids in an agricultural landscape. Oecologia 105, 537–544.

Ellis, D., McAvoy, R., AbuAyyash, L., Flanagan, M., Ciomperlik, M., 2001. Evaluation of Serangium parcesetosum (Coleoptera: Coccinellidae) for biological control of silverleaf whitefly, Bemisia argentifolii (Homoptera: Aleyrodidae), on poinsettia. Florida Entomologist 84, 215–221.

Evans, E.W., 2000. Egg production in response to combined alternative food by the predator Coccinella transversalis. Entomologia Experimentalis et Applicata 94 (2), 141–147.

Fassotte, B., Fischer, C., Durieux, D., Lognay, G., Haubruge, E., Francois, F., Verheggen, J., 2014. First evidence of a volatile sex pheromone in lady beetles. PLoS One 9 (12), e115011. http://dx.doi.org/10.1371/journal.pone.0115011.

Felix, S., Soares, A.O., 2004. Intraguild predation between the aphidophagous ladybird beetles Harmonia axyridis and Coccinella undecimpunctata (Coleoptera: Coccinellidae): the role of body weight. European Journal of Entomology 101, 237–242.

Fernandes-Arhex, V., Corley, J., 2003. The functional response of parasitoids and its implications for biological control. Biocontrol Science and Technology 13, 403–413.

Ferran, A., Niknam, H., Kabiri, F., Picart, J.L., DeHerce, C., Brun, J., Iperti, G., Lapchin, L., 1996. The use of Harmonia axyridis larvae (Coleoptera: Coccinellidae) against Macrosiphum rosae (Hemiptera: Sternorrhyncha: Aphididae) on rose bushes. European Journal of Entomology 93, 59–67.

Ferran, A., Giuge, L., Tourniaire, R., Gambier, J., Fournier, D., 1998. An artificial non-flying mutation to improve the efficiency of the ladybird *Harmonia axyridis* in biological control of aphids. BioControl 43, 53–64.

Ferran, A., El-Arnaouty, S.A., Beyssat-Arnaouty, V., Galal, H., 2000. Introduction and release of the coccinellid *Harmonia axyridis* Pallas for controlling *Aphis craccivora* Koch on faba beans in Egypt. Egypt Journal of Biological Pest Control 10, 129–136.

Flint, M.L., Dreistadt, S.H., Rentner, J., Parrella, M.P., 1995. Lady beetle release controls aphids on potted plants. California Agriculture 49, 5–8.

Flowers, R.W., Salomb, S.M., Kokc, L.T., Mullins, D.E., 2007. Behavior and daily activity patterns of specialist and generalist predators of the hemlock woolly adelgid, *Adelges tsugae*. Journal of Insect Science 7 (44), 1–20.

Fondren, K.M., McCullough, D.G., 2005. Phenology, natural enemies and horticultural oil for control of pine needle scale (*Chionaspis heterophylla*) (Fitch) (Homoptera: Diaspididae) on Christmas tree plantations. Journal of Economic Entomology 98, 1603–1613.

Francis, F., Vandermoten, S., Verheggen, F., Lognay, G., Haubruge, E., 2005. Is the (E)-beta-farnesene only volatile terpenoid in aphids? Journal of Applied Entomology 129, 6–11.

Frechette, B., Dixon, A.F.G., Alauzet, C., Hemptinne, J.L., 2004. Age and experience influence patch assessment for oviposition by an insect predator. Ecological Entomology 29, 578–583.

Gagnon, A.E., 2010. Predation intraguilde chez les Coccinellidae: impact sur la lutte biologique au puceron du soya (Unpubl. Ph.D. dissertation). Universite Laval, Quebec, Canada.

Gardiner, M.M., Landis, D.A., 2007. Impact of intraguild predation by adult *Harmonia axyridis* (Coleoptera: Coccinellidae) on *Aphis glycines* (Hemiptera: Aphididae) biological control in cage studies. Biological Control 40, 386–395.

Gautam, R.D., Navarajan Paul, A.V., Srivastava, K.P., 1988. Preliminary studies on *Cryptolaemus montrouzieri* Muls. against the white-tailed mealybug *Ferrisia virgata* (Cockerell) infesting tobacco plants. Journal of Biological Control 2, 12–13.

Gordon, R.D., 1985. The Coccinellidae (Coleoptera) of America north of Mexico. Journal of New York Entomological Society 93, 1–912.

Greathead, D.J., 2003. A historical overview of biological control in Africa. In: Neuenschwander, P., Borgemeister, C., Langewald, J. (Eds.), Biological Control in IPM Systems in Africa. CABI, Wallingford, UK, pp. 414.

Haddrill, P.R., Shuker, D.M., Mayes, S., Majerus, M.E.N., 2007. Temporal effects of multiple mating on components of fitness in the two-spot ladybird, *Adalia bipunctata* (Coleoptera: Coccinellidae). European Journal of Entomology 104, 393–398.

Haddrill, P.R., Shuker, D.M., Amos, W., Majerus, M.E.N., Mayes, S., 2008. Female multiple mating in wild and laboratory populations of the two-spot ladybird, *Adalia bipunctata*. Molecular Ecology 17, 3189–3197.

Hemptinne, J.L., Dixon, A.F.G., 2000. Defence, oviposition and sex: semiochemical parsimony in two species of ladybird beetles (Coleoptera: Coccinellidae)? A short review. European Journal of Entomology 97, 443–447.

Hemptinne, J.L., Lognay, G., Dixon, A.F.G., 1998. Mate recognition in the two-spot ladybird beetle, *Adalia bipunctata*: role of chemical and behavioural cues. Journal of Insect Physiology 44, 1163–1171.

Hemptinne, J.L., Lognay, G., Doumbia, M., Dixon, A.F.G., 2001. Chemical nature and persistence of the oviposition deterring pheromone in the tracks of the two spot ladybird, *Adalia bipunctata* (Coleoptera: Coccinellidae). Chemoecology 11, 43–47.

Hirose, Y., 2006. Biological control of aphids and coccids: a comparative analysis. Population Ecology 48, 307–315.

Hodek, I., Honek, A. (Eds.), 1996. Ecology of Coccinellidae. Kluwer Academic Publishers, Dordrecht, Boston, London. pp. 464.

Hodek, I., Honek, A., 2009. Scale insects, mealybugs, whiteflies and psyllids (Hemiptera, Sternorrhyncha) as prey of ladybirds. Biological Control 51, 232–243.

Hodek, I., Michaud, J.P., 2008. Why is *Coccinella septempunctata* so successful? (A point-of-view). European Journal of Entomology 105, 1–12.

Hodek, I., van Emden, H.F., Honek, I., 2012. Ecology and Behavior of the Ladybird Beetles (Coccinellidae). Wiley-Blackwell, Oxford, UK.

Holling, C.S., 1959. Some characteristics of simple types of predation and parasitism. The Canadian Entomologist 91, 385–398.

Holling, C.S., 1965. Functional response of predators to prey density and its role in mimicry and population regulation. Memoirs of Entomological Society of Canada 45, 3–60.

Holopainen, J., 2004. Multiple functions of inducible plant volatiles. Trends in Plant Science 9, 529–533.

Isikber, A.A., Copland, M.J.W., 2001. Food consumption and utilization by larvae of two coccinellid predators, *Scymnus levaillanti* and *Cycloneda sanguinea*, on cotton aphid, *Aphis gossypii*. BioControl 46, 455–467.

Jalali, M.A., Tirry, L., De Clercq, P., 2009. Food consumption and immature growth of *Adalia bipunctata* (Coleoptera: Coccinellidae) on a natural prey and a factitious food. European Journal of Entomology 106, 193–198.

Janssen, A., Sabelis, M.W., Magalhaes, S., Montserrat, M., van der Hammen, T., 2007. Habitat structure affects intraguild predation. Ecology 88 (11), 2713–2719.

Jarošík, V., Honěk, A., Magarey, R.D., Skuhrovec, J., 2011. Developmental database for phenology models: related insect and mite species have similar thermal requirements. Journal of Economic Entomology 104, 1870–1876.

Jarošík, V., Kumar, G., Omkar, Dixon, A.F.G., 2014. Are thermal constants constant? A test using two species of ladybird. Journal of Thermal Biology 40, 1–8.

Jiaqin, X., Yuhong, Z., Hongsheng, W., Ping, L., Congshuang, D., Hong, P., 2014. Effects of mating patterns on reproductive performance and offspring fitness in *Cryptolaemus montrouzieri*. Entomologia Experimentalis et Applicata 153 (1), 20–23.

Joachim, C., Weisser, W.W., 2013. Real-time monitoring of (E)-β-farnesene emission in colonies of the pea aphid, *Acyrthosiphon pisum*, under lacewing and ladybird predation. Journal of Chemical Ecology 39, 1254–1262.

Kairo, M.T.K., Paraiso, O., Gautam, R.D., Peterkin, D.D., 2013. *Cryptolaemus montrouzieri* (Mulsant) (Coccinellidae: Scymninae): a review of biology, ecology, and use in biological control with particular reference to potential impact on non-target organisms. CAB Reviews 8 (5), 1–20.

Kindlmann, P., Dixon, A.F.G., 1993. Optimal foraging in ladybird beetles (Coleoptera: Coccinellidae) and its consequences for their use in biological control. European Journal of Entomology 90, 443–450.

Kindlmann, P., Dixon, A.F.G., Dostalkova, I., 2001. Role of aging and temperature in shaping the reaction norms and fecundity function in insects. Journal of Evolutionary Biology 14, 835–840.

Kirby, W., Spence, W., 1846. An Introduction to Entomology, sixth ed. Lea and Blanchard, Philadelphia, Pennsylvania.

Klewer, N., Ruzicka, Z., Schulz, S., 2007. (Z)-Pentacos-12-ene, an oviposition-deterring pheromone of *Cheilomenes sexmaculata*. Journal of Chemical Ecology 33, 2167–2170.

Koch, R.L., 2003. The multicolored Asian lady beetle, *Harmonia axyridis*: a review of its biology, uses in biological control, and nontarget impacts. Journal of Insect Science 3, 1–16.

Kumar, K., Singh, R., Lal, H.K., 1987. Control of Spiraling Whitefly. Report Presented on 1 to 3 December 1987 at Annual Research Meeting of the Ministry of Primary Industries, Fiji.

Kumar, A., Kumar, N., Siddiqui, A., Tripathi, C.P.M., 1999. Prey–predator relationship between *Lipaphis erysimi* Kalt. (Hom., Aphididae) and *Coccinella septempunctata* L (Col., Coccinellidae). II. Effect of host plants on the functional response of the predator. Journal of Applied Entomology 123, 591–596.

Kumar, B., Mishra, G., Omkar, 2014a. Functional response and predatory interactions in conspecific and heterospecific combinations of two congeneric species (Coleoptera: Coccinellidae). European Journal of Entomology 111 (2), 257–265.

Kumar, B., Mishra, G., Omkar, 2014b. Larval and female footprints as feeding deterrent cues for immature stages of two congeneric ladybird predators (Coleoptera: Coccinellidae). Bulletin of Entomological Research 104, 652–660.

Kuroda, T., Miura, K., 2003. Comparison of the effectiveness of two methods for releasing *Harmonia axyridis* (Pallas) (Coleoptera: Coccinellidae) against *Aphis gossypii* Glover (Homoptera: Aphididae) on cucumbers in a greenhouse. Applied Entomology and Zoology 38, 271–274.

Labrie, G., Lucas, E., Coderre, D., 2006. Can developmental and behavioral characteristics of the multicolored Asian lady beetle *Harmonia axyridis* explain its invasive success? Biological Invasions 8, 743–754.

Landis, D.A., Gardiner, M.M., van der Werf, W., Swinton, M., 2008. Increasing corn for biofuel production reduces biocontrol services in agricultural landscapes. Proceedings of the National Academy of Sciences 105, 20552–20557.

Lanzoni, A., Accinelli, G., Bazzocchi, G.G., Burgio, G., 2004. Biological traits and life table of the exotic *Harmonia axyridis* compared with *Hippodamia variegata*, and *Adalia bipunctata* (Col., Coccinellidae). Journal of Applied Entomology 128, 298–306.

LaRock, D.R., Ellington, J.J., 1996. An integrated pest management approach, emphasizing biological control, for pecan aphids. Southwestern Entomologist 21, 153–166.

Laubertie, E., Martini, X., Cadena, C., Treilhou, M., Dixon, A.F.G., Hemptinne, J.L., 2006. The immediate source of the oviposition-deterring pheromone produced by larvae of *Adalia bipunctata* (L.) (Coleoptera: Coccinellidae). Journal of Insect Behavior 19, 231–240.

Legaspi, J.C., Ciomperlik, M.A., Legaspi, B.C., 2001. Field cage evaluation of *Serangium parcesetosum* (Col., Coccinellidae) as a predator of citrus blackfly eggs (Hom., Aleyrodidae). Southwest Entomological Science Notes 26, 171–172.

Legg, J., Gerling, D., Neuenschwander, P., 2003. Biological control of whiteflies in Sub – Saharan Africa. In: Neuenschwander, P., Borgemeister, C., Langewald, J. (Eds.), Biological Control in IPM Systems in Africa. CABI, Wallingford, UK, pp. 87–100.

Losey, J.E., Denno, R.F., 1998. Positive predator–predator interactions: enhanced predation rates and synergistic suppression of aphid populations. Ecology 79, 2143–2152.

Lucas, E., Brodeur, J., 2001. A fox in a sheep-clothing: dilution effect for a furtive predator living inside prey aggregation. Ecology 82, 3246–3250.

Lucas, E., Coderre, D., Brodeur, J., 2000. Selection of molting and pupation sites by *Coleomegilla maculata* (Coleoptera: Coccinellidae): avoidance of intraguild predation. Environmental Entomology 29, 454–459.

Lucas, E., Labrie, G., Vincent, C., Kovach, J., 2007. The multicolored Asian ladybeetle, *Harmonia axyridis*, beneficial or nuisance organism? In: Vincent, C., Goettel, M., Lazarovitz, G. (Eds.), Biological Control: A Global Perspective. CABI Publishing, Wallingford, UK.

Lucas, E., 2005. Intraguild predation among aphidophagous predators. European Journal of Entomology 102, 351–364.

Lundgren, J.G., 2009. Relationships of Natural Enemies and Non-prey Foods. Springer, Dordrecht, The Netherlands.

Magro, A., Ducamp, C., Ramon-Portugal, F., Lemopte, E., Crouau-Roy, B., Frederick, A., Dixon, A.F.G., Hemptinne, J.L., 2010. Oviposition deterring infochemicals in ladybirds: the role of phylogeny. Evolutionary Ecology 24, 251–271.

Maia, A.H.N., Pazianotto, R.A.A., Luiz, A.J.B., Prado, J.S.M., Pervez, A., 2014. Inference on arthropod demographic parameters: computational advances using R. Journal of Economic Entomology 107 (1), 432–439.

Majerus, M.E.N., 1994. Ladybirds. Harper Collins, London, pp. 367.

Mani, M., Krishnamoorthy, A., 1998. Suppression of the soft green scale *Coccus viridis* (Green) on acid lime in India. Advances in IPM for horticultural crops. In: Proceedings of the 1st National Symposium on Pest Management in Horticultural Crops: Environmental Implications and Thrusts, 15–17 October 1997, Bangalore, India. Indian Institute of Horticultural Research, Bangalore, pp. 210–212.

Mani, M., Krishnamoorthy, A., 2008. Biological suppression of the mealybugs *Planococcus citri* (Risso), *Ferrisia virgata* (Cockerell) and *Nipaecoccus viridis* (Newstead) on pummelo with *Cryptolaemus montrouzieri* Mulsant in India. Journal of Biological Control 22, 169–172.

Messina, F.J., Hanks, J.B., 1998. Host plant alters the shape of the functional response of an aphid predator (Coleoptera: Coccinelidae). Environmental Entomology 27, 1196–1202.

Michaud, J.P., Jyoti, J.L., 2007. Repellency of conspecific and heterospecific larval residues to *Hippodamia convergens* (Coleoptera: Coccinellidae) ovipositing on sorghum plants. European Journal of Entomology 104, 399–405.

Michaud, J.P., Bista, M., Mishra, G., Omkar, 2013. Sexual activity diminishes male virility in two *Coccinella* species: consequences for female fertility and progeny development. Bulletin of Entomological Research 103, 570–577.

Michaud, J.P., 2000. Development and reproduction of ladybeetles (Coleoptera, Coccinellidae) on the citrus aphids *Aphis spiraecola* Patch and *Toxoptera citricida* (Kirkaldy) (Homoptera, Aphididae). Biological Control 18, 287–297.

Michaud, J.P., 2002. Invasion of the Florida citrus ecosystem by *Harmonia axyridis* (Coleoptera: Coccinellidae) and asymmetric competition with a native species, *Cycloneda sanguinea*. Environmental Entomology 31, 827–835.

Michaud, J.P., 2003. A comparative study of larval cannibalism in three species of ladybirds. Ecological Entomology 28, 92–101.

Michaud, J.P., 2012. Coccinellids in biological control. In: Hodek, I., van Emden, H.F., Honek, A. (Eds.), Ecology and Behaviour of the Ladybird Beetles (Coccinellidae), first ed. Wiley-Blackwell, Oxford, UK, pp. 488–519.

Michels Jr., G.J., Matis, J.H., 2008. Corn leaf aphid, *Rhapalosiphum maidis* (Hemiptera: Aphididae), is a key to greenbug, *Schizaphis graminum* (Hemiptera: Aphididae), biological control in grain sorghum, *Sorghum bicolor*. European Journal of Entomology 105, 513–520.

Mills, N.J., 1982. Voracity, cannibalism, and coccinellid predation. Annals of Applied Biology 101, 144–148.

Mishra, G., Omkar, 2005. Influence of components of light on the life attributes of an aphidophagous ladybird, *Propylea dissecta* (Coleoptera: Coccinellidae). International Journal of Tropical Insect Science 25, 32–38.

Mishra, G., Omkar, 2006a. Ageing trajectory and longevity trade-off in an aphidophagous ladybird, *Propylea dissecta* (Coleoptera: Coccinellidae). European Journal of Entomology 103, 33–40.

Mishra, G., Omkar, 2006b. Conspecific interference by adults in an aphidophagous ladybird *Propylea dissecta* (Coleoptera: Coccinellidae): effect on reproduction. Bulletin of Entomological Research 96, 407–412.

Mishra, G., Omkar, 2014. Phenotype-dependent mate choice in *Propylea dissecta* and its fitness consequences. Journal of Ethology 32, 165–172.

Mishra, G., Omkar, Kumar, B., Pandey, G., 2012a. Stage and age-specific predation in four aphidophagous ladybird beetles. Biocontrol Science and Technology 22 (4), 463–476.

Mishra, G., Singh, N., Shahid, M., Omkar, 2012b. Effect of presence and semiochemicals of conspecific stages on oviposition by ladybirds (Coleoptera: Coccinellidae). European Journal of Entomology 109, 363–371.

Mzhavanadze, V.I., 1984. *Cryptolaemus* against the camellia scale. Zashchita Rastenii 7, 26.

Nakashima, Y., Birkett, M.A., Pye, B.J., Powell, W., 2006. Chemically mediated intraguild predator avoidance by aphid parasitoids: interspecific variability in sensitivity to semiochemical trails of ladybird predators. Journal of Chemical Ecology 32 (9), 1989–1998.

Nalepa, C.A., 2013. Coccinellidae captured in blacklight traps: seasonal and diel pattern of the dominant species *Harmonia axyridis* (Coleoptera: Coccinellidae). European Journal of Entomology 110, 593–597.

Nechols, J.R., 1982. Entomology: Biological Control. pp. 33–49, Annual Report 1982, Guam Agricultural Experiment Station.

Nedved, O., Honek, A., 2012. Life history and development. In: Hodek, I., van Emden, H.F., Honek, A. (Eds.), Ecology and Behaviour of the Ladybird Beetles (Coccinellidae), first ed. Wiley-Blackwell, Oxford, UK, pp. 54–109.

Nedved, O., Kovář, I., 2012. Phylogeny and classification. In: Hodek, I., van Emden, H.F., Honek, A. (Eds.), Ecology and Behaviour of Ladybird Beetles (Coccinellidae). Wiley-Blackwell, pp. 1–12.

Ninkovic, V., Al Abassi, S., Petersson, J., 2001. The influence of aphid-induced plant volatiles on ladybirds beetle searching behaviour. Biological Control 21, 191–195.

Noia, M., Borges, I., Soares, A.O., 2008. Intraguild predation between the aphidophagous ladybird beetles *Harmonia axyridis* and *Coccinella undecimpunctata* (Coleoptera: Coccinellidae): the role of intra and extraguild prey densities. Biological Control 46, 140–146.

Noppe, C., Michaud, J.P., de Clercq, P., 2012. Intraguild predation between lady beetles and lacewings: outcomes and consequences vary with focal prey and arena of interaction. Annals of the Entomological Society of America 105, 562–571.

Noriyuki, S., Osawa, N., Nishida, T., 2012a. Asymmetric reproductive interference between specialist and generalist predatory ladybirds. Journal of Animal Ecology. http://dx.doi.org/10.1111/j.1365-2656.2012.01984.x.

Noriyuki, S., Osawa, N., Nishida, T., 2012b. Intrinsic prey suitability in specialist and generalist *Harmonia* ladybirds: a test of the trade-off hypothesis for food specialization. Entomologia Experimentalis et Applicata 144, 279–285.

Obrycki, J.J., Kring, T.T., 1998. Predaceous Coccinellidae in biological control. Annual Reviews of Entomology 58, 839–845.

Obrycki, J.J., Giles, K.L., Ormord, A.M., 1998a. Experimental assessment of interactions between larval *Coleomegilla maculata* and *Coccinella septempunctata* (Coleoptera: Coccinellidae) in field cages. Environmental Entomology 27, 1280–1288.

Obrycki, J.J., Giles, K.L., Ormord, A.M., 1998b. Interactions between an introduced and indigenous coccinellid species at different prey densities. Oecologia 117, 279–285.

Omkar, James, B.E., 2005. Reproductive behaviour of an aphidophagous ladybeetle *Coccinella transversalis* (Coleoptera: Coccinellidae). International Journal of Tropical Insect Science 25, 96–102.

Omkar, Kumar, G., 2013. Responses of an aphidophagous ladybird beetle, *Anegleis cardoni*, to varying densities of *Aphis gossypii*. Journal of Insect Science 13, 24 pp. 1–12.

Omkar, Mishra, G., 2005a. Evolutionary significance of promiscuity in an aphidophagous ladybird, *Propylea dissecta* (Coleoptera: Coccinellidae). Bulletin of Entomological Research 95, 527–533.

Omkar, Mishra, G., 2005b. Mating in aphidophagous ladybirds: costs and benefits. Journal of Applied Entomology 129, 432–436.

Omkar, Mishra, G., 2005c. Preference–performance of a generalist predatory ladybird: a laboratory study. Biological Control 34, 187–195.

Omkar, Mishra, G., 2009. Optimization of age difference between mates maximizes reproductive output. BioControl 54, 637–650.

Omkar, Pathak, S., 2006. Effects of different photoperiods and wavelengths of light on the life-history traits of an aphidophagous ladybird, *Coelophora saucia* (Mulsant). Journal of Applied Entomology 130, 45–50.

Omkar, Pathak, S., 2007. Sexual dimorphism in an aphidophagous ladybird beetle, *Coelophora saucia*. Journal of Applied Bioscience 33 (2), 180–181.

Omkar, Pervez, A., 2000. Sexual dimorphism in *Propylea dissecta* (Mulsant), (Coccinellidae: Coleoptera). Journal of Aphidology 14 (1&2), 139–140.

Omkar, Pervez, A., 2002a. Ecology of aphidophagous ladybird beetle, *Coccinella septempunctata* Linn. (Coleoptera: Coccinellidae): a review. Journal of Aphidology 16 (1&2), 175–201.

Omkar, Pervez, A., 2002b. Influence of temperature on age specific fecundity of a ladybeetle, *Micraspis discolor* (Fabricius). Insect Science and Its Application 22 (1), 61–65.

Omkar, Pervez, A., 2003. Ecology and biocontrol potential of a scale-predator, *Chilocorus nigritus*. Biocontrol Science and Technology, UK 13 (4), 379–390.

Omkar, Pervez, A., 2004a. Predaceous coccinellids in India: predator-prey catalogue. Oriental Insects 38, 27–61.

Omkar, Pervez, A., 2004b. Functional and numerical responses of *Propylea dissecta* (Mulsant) (Col., Coccinellidae). Journal of Applied Entomology 128 (2), 140–146.

Omkar, Pervez, A., 2005a. Ecology of two-spotted aphidophagous ladybird, *Adalia bipunctata*: a review. Journal of Applied Entomology 129 (9&10), 465–474.

Omkar, Pervez, A., 2005b. Mating behavior of an aphidophagous ladybird beetle, *Propylea dissecta* (Mulsant). Journal of Insect Science 12, 37–44.

Omkar, Pervez, A., 2011. Functional response of two aphidophagous ladybirds searching in tandem. Biocontrol Science and Technology 21 (1), 101–111.

Omkar, Srivastava, S., 2002. The reproductive behaviour of an aphidophagous ladybeetle, *Coccinella septempunctata* (Coleoptera: Coccinellidae). European Journal of Entomology 99, 465–470.

Omkar, Srivastava, S., 2003. Predation and searching efficiency of a ladybird beetle, *Coccinella septempunctata* Linnaeus in laboratory environment. Indian Journal of Experimental Biology 41, 82–84.

Omkar, Mishra, G., Singh, K., 2005. Effects of different wavelengths of light on the life attributes of two aphidophagous ladybirds (Coleoptera: Coccinellidae). European Journal of Entomology 102, 33–37.

Omkar, Singh, S.K., Pervez, A., 2006. Influence of mating duration on fecundity and fertility in two aphidophagous ladybirds. Journal of Applied Entomology 130, 103–107.

Omkar, Singh, S.K., Mishra, G., 2010a. Multiple matings affect the reproductive performance of the aphidophagous ladybird beetle, *Coelophora saucia* (Coleoptera: Coccinellidae). European Journal of Entomology 107, 177–182.

Omkar, Singh, S.K., Mishra, G., 2010b. Parental age at mating affects reproductive attributes of the aphidophagous ladybird beetle, *Coelophora saucia* (Coleoptera: Coccinellidae). European Journal of Entomology 107, 341–347.

Omkar, 2006. Suitability of different foods for a generalist ladybird, *Micraspis discolor* (Coleoptera: Coccinellidae). International Journal of Tropical Insect Science 26, 35–40.

Osawa, N., 2011. Ecology of *Harmonia axyridis* in natural habitats within its native range. BioControl 56, 613–621.

Pattanayak, R., Mishra, G., Omkar, Chanotiya, C.S., Rout, P.K., Mohanty, C.S., 2014. Does the volatile hydrocarbon profile differ between the sexes: a case study on five aphidophagous ladybirds. Archives of Insect Biochemistry and Physiology 87 (3), 105–125.

Pattanayak, R., Mishra, G., Chanotiya, C.S., Rout, P.K., Mohanty, C.S., Omkar, 2015. Semiochemical Profile of Four Aphidophagous Indian Ladybird Beetles. The Canadian Entomologist, Cambridge Universty Press, UK. http://dx.doi.org/10.4039/tce.2015.45.

Perry, J.C., Rowe, L., 2010. Condition-dependent ejaculate size and composition in a ladybird beetle. Proceedings of Royal Society London B 277, 3639–3647.

Perry, J.C., Tse, C.T., 2013. Extreme costs of mating for male two-spot ladybird beetles. PLoS One 8 (12), e81934. http://dx.doi.org/10.1371/journal.pone.0081934.

Pervez, A., Omkar, 2003. Predation potential and handling time estimates of a generalist aphidophagous ladybird, *Propylea dissecta*. Biological Memoirs 29 (2), 91–97.

Pervez, A., Omkar, 2004. Temperature dependent life attributes of an aphidophagous ladybird, *Propylea dissecta* (Mulsant). Biocontrol Science and Technology 14 (6), 587–594.

Pervez, A., Gupta, A.K., Omkar, 2005. Kin recognition and avoidance of kin cannibalism in aphidophagous ladybirds: a laboratory study. European Journal of Entomology 102 (3), 513–518.

Pervez, A., Omkar, 2005. Functional response of coccinellid predators: an illustration of a logistic approach. Journal of Insect Science 5 (05), 1–6.

Pervez, A., Omkar, 2006. Ecology and biological control application of multicoloured Asian ladybird, *Harmonia axyridis*: a review. Biocontrol Science and Technology 16 (02), 112–128.

Pervez, A., Omkar, 2010. Innovations in the aphid biocontrol programme using predaceous ladybirds. Proceedings of the National Symposium on Modern Approaches to Insect Pest Management 95–105.

Pervez, A., Maurice, N., 2011. Polyandry affects the reproduction and progeny of a ladybird beetle, *Hippodamia variegata* (Goeze). European Journal of Environmental Sciences 1 (1), 19–23.

Pervez, A., Omkar, 2011. Ecology of an aphidophagous ladybird *Propylea*: a review. Journal of Asia Pacific Entomology 14 (3), 357–365.

Pervez, A., Singh, S., 2013. Body size dependent mating patterns of an aphidophagous ladybird, *Hippodamia variegata*. European Journal of Environmental Sciences 3 (2), 109–112.

Ponsonby, D.J., 2009. Factors affecting utility of *Chilocorus nigritus* (F.) (Coleoptera: Coccinellidae) as a biocontrol agent. CAB Reviews: Perspectives in Agriculture, Veterinary Science, Nutrition and Natural Resources 4 (46), 1–20.

Pratt, C., Pope, T.W., Powell, G., Rossiter, J.T., 2008. Accumulation of glucosinolates by the cabbage aphid *Brevicoryne brassicae* as a defense against two coccinellid species. Journal of Chemical Ecology 34, 323–329.

Prokopenko, A.I., 1982. *Cryptolaemus* suppresses *Chloropulvinaria*. Zashchita Rastenii 3, 25.

Provost, C., Lucas, E., Coderre, D., Chouinard, G., 2006. Prey selection by the lady beetle *Harmonia axyridis*: the influence of prey mobility and prey species. Journal of Insect Behavior 19, 265–277.

Ramani, S., Poorani, J., Bhummanvar, B.S., 2002. Spiralling whitefly, *Aleurodicus dispersus*, in India. Biocontrol News and Information 23, 55–62.

Rana, J.S., Dixon, A.F.G., Jarošík, V., 2002. Costs and benefits of prey specialization in a generalist insect predator. Journal of Animal Ecology 71, 15–22.

Rondoni, G., Onofri, A., Ricci, C., 2012. Differential susceptibility in a specialised aphidophagous ladybird, *Palynapis luteorubra* (Coleptera: Coccinellidae), facing intraguild predation by exotic and native generalist predators. Biocontrol Science and Technology 22 (11), 1334–1350.

Rondoni, G., Ielo, F., Ricci, C., Conti, E., 2014. Intraguild predation responses in two aphidophagous coccinellids identify differences among juvenile stages and aphid densities. Insects 5, 974–983.

Roy, M., Brodeur, J., Cloutier, C., 2003. Effect of temperature on intrinsic rates of natural increase (r_m) of a coccinellid and its spider mite prey. BioControl 48, 57–72.

Roy, H.E., Brown, P., Majerus, M.E.N., 2006. *Harmonia axyridis*: a successful biocontrol agent or an invasive threat? In: Eilenberg, J., Hokkanen, H. (Eds.), An Ecological and Societal Approach to Biological Control. Kluwer Academic Publishers, Dordrecht, The Netherlands.

Roy, H.E., Brown, P.M.J., Rothery, P., Ware, R.L., Majerus, M.E.N., 2008. Interactions between the fungal pathogen *Beauveria bassiana* and three species of coccinellid: *Harmonia axyridis*, *Coccinella septempunctata* and *Adalia bipunctata*. BioControl 5, 265–276.

Ruzicka, Z., Zemek, R., 2008. Deterrent effects of larval tracks on conspecific larvae in *Cycloneda limbifer*. Biocontrol 53, 763–771.

Ruzicka, Z., 2002. Persistence of deterrent larval tracks in *Coccinella septempunctata*, *Cycloneda limbifer* and *Semiadalia undecimnotata* (Coleoptera: Coccinellidae). European Journal of Entomology 99, 471–475.

Ruzicka, Z., 2003. Perception of oviposition-deterring larval tracks in aphidophagous coccinellids *Cycloneda limbifer* and *Ceratomegilla undecimnotata* (Coleoptera: Coccinellidae). European Journal of Entomology 100, 345–350.

Ruzicka, Z., 2006. Oviposition-deterring effects of conspecific and heterospecific larval tracks on *Cheilomenes sexmaculata* (Coleoptera: Coccinellidae). European Journal of Entomology 103, 757–763.

Santos-Cividanes, T.M., dos Anjos, A.C.R., Cividanesm, F.J., Dias, P.C., 2011. Effects of food deprivation on the development of *Coleomegilla maculata* (De Geer) (Coleoptera: Coccinellidae). Neotropical Entomology 40, 112–116.

Schnee, C., Kollner, T.G., Held, M., Turlings, T.C.J., Gershenzon, J., Degenhardt, J., 2006. The products of a single maize sesquiterpene synthase form a volatile defense signal that attracts natural enemies of maize herbivores. Proceedings of the National Academy of Sciences 103, 1129–1134.

Seagraves, M.P., 2009. Lady beetle oviposition behavior in response to the trophic environment. Biological Control 51, 313–322.

Senal, D., 2006. Avci bocek *Chilocorus Nigritus* (Fabricius) (Coleoptera: Coccinellidae)'un bazi biyolojik ve ekolojik ozellikleri ile dogaya adaptasyonu üzerinde aras, tirmalar (Ph.D. thesis). Department of Plant Protection, University of Cukurova, Turkey.

Shannag, H.K., Obeidat, W.M., 2008. Interaction between plant resistance and predation of *Aphis fabae* (Homoptera: Aphididae) by *Coccinella septempunctata* (Coleoptera: Coccinellidae). Annals of Applied Biology 152, 331–337.

Shenhmar, M., Brar, K.S., 1995. Biological control of mustard aphid, *Lipaphis erysimi* (Kaltenbach) in the Punjab. Journal of Biological Control 9, 9–12.

Singh, N., Mishra, G., Omkar, 2014. Effect of photoperiod on slow and fast developing individuals in aphidophagous ladybirds, *Menochilus sexmaculatus* and *Propylea dissecta* (Coleoptera: Coccinellidae). Insect Science 1–18. http://dx.doi.org/10.1111/1744-7917.12182.

Sloggett, J.J., Majerus, M.E.N., 2000. Habitat preferences and diet in the predatory Coccinellidae (Coleoptera): an evolutionary perspective. Biological Journal of Linnaeus Society 70, 63–88.

Sloggett, J.J., Majerus, M.E.N., 2003. Adaptations of *Coccinella magnifica*, a myrmecophilous coccinellid to aggression by wood ants (*Formica rufa* group). II. Larval behaviour, and ladybird oviposition location. European Journal of Entomology 100, 337–344.

Sloggett, J.J., Wood, R.A., Majerus, M.E.N., 1998. Adaptations of *Coccinella magnifica* Redtenbacher, a myrmecophilous coccinellid, to aggression by wood ants (*Formica rufa* group). I. Adult behavioral adaptation, its ecological context and evolution. Journal of Insect Behavior 11, 889–904.

Sloggett, J.J., Webberley, K.M., Majerus, M.E.N., 2004. Low parasitoid success on a myrmecophilous host is maintained in the absence of ants. Ecological Entomology 29, 123–127.

Sloggett, J.J., 2008a. Habitat and dietary specificity in aphidophagous ladybirds (Coleoptera: Coccinellidae): explaining specialization. Proceedings of Netherland Entomological Society Meeting 19, 95–113.

Sloggett, J.J., 2008b. Weighty matters: body size, diet and specialization in aphidophagous ladybird beetles (Coleoptera: Coccinellidae). European Journal of Entomology 105, 381–389.

Sloggett, J.J., Zeilstra, I., Obrycki, J.J., 2008. Patch residence by aphidophagous ladybird beetles: do specialists stay longer? Biological Control 47, 199–206.

Snyder, W.E., Ives, A.R., 2003. Interactions between specialist and generalist natural enemies: parasitoids, predators, and pea aphid biocontrol. Ecology 84 (1), 91–107.

Snyder, W.E., Clevenger, G.M., Eigenbrode, S.D., 2004. Intraguild predation and successful invasion by introduced ladybird beetles. Oecologia 140, 559–565.

Soares, A.O., Coderre, D., Schanderl, H., 2004. Dietary self-selection behaviour by the adults of the aphidophagous ladybeetle *Harmonia axyridis* (Coleoptera: Coccinellidae). Journal of Animal Ecology 73, 478–486.

Soares, A.O., Borges, I., Borges, P.A.V., Labrie, G., Lucas, E., 2008. *Harmonia axyridis*: what will stop the invader? BioControl 53, 127–145.

Specty, O., Febvay, G., Grenier, S., Delobel, B., Piotte, C., Pageaux, J.F., Ferran, A., Guillaud, J., 2003. Nutritional plasticity of the predatory ladybeetle *Harmonia axyridis* (Coleoptera: Coccinellidae): comparison between natural and substitution prey. Archives of Insect Biochemistry and Physiology 52, 81–91.

Srinivasan, T.R., Sundara Babu, P.C., 1989. Field evaluation of *Cryptolaemus montrouzieri* Mulsant, the coccinellid predator against grapevine mealybug, *Maconellicoccus hirsutus* (Green). South Indian Horticulture 37, 50–51.

Srivastava, S., Omkar, 2005a. Mate choice and reproductive success of two morphs of the seven spotted ladybird, *Coccinella septempunctata* (Coleoptera: Coccinellidae). European Journal of Entomology 102, 189–194.

Srivastava, S., Omkar, 2005b. Short- and long-term benefits of promiscuity in the seven-spotted ladybird *Coccinella septempunctata* (Coleoptera: Coccinellidae). International Journal of Tropical Insect Science 25, 176–181.

Sterk, G., Meesters, P., 1997. IPM on strawberries in glasshouses and plastic tunnels in Belgium, new possibilities. Acta Horticulturae 439, 905–911.

Taghizadeh, R., Fathipour, Y., Kamali, K., 2008. Influence of temperature on life-table parameters of *Stethorus gilvifrons* (Mulsant) (Coleoptera: Coccinellidae) fed on *Tetranychus urticae* Koch. Journal of Applied Entomology 132, 638–645.

Takahashi, K., 1997. Use of *Coccinella septempunctata brucki* Mulsant as a biological agent for controlling alfalfa aphids. Japan Agricultural Research Quarterly 31, 101–108.

Toda, Y., Sakuratani, Y., 2006. Expansion of geographical distribution of an exotic ladybird beetle, *Adalia bipunctata* (Coleoptera: Coccinellidae), and its interspecific relationships with native ladybird beetles in Japan. Ecological Research 21, 292–300.

Tourniaire, R., Ferran, A., Guige, L., Piotte, C., Gambier, J., 2000. A natural flightless mutation in the ladybird, *Harmonia axyridis*. Entomologia Experimentalis et Applicata 96, 33–38.

Trouvé, C., Ledee, S., Ferran, A., Brun, J., 1997. Biological control of the damson–hop aphid, *Phorodon humuli* (Hom.: Aphididae), using the ladybeetle *Harmonia axyridis* (Col.: Coccinellidae). Entomophaga 42, 57–62.

Ullah, I., 2000. Aspects of the Biology of the Ladybird Beetle *Stethorus vagans* (Blackburn) (Coleoptera, Coccinellidae) (Ph.D. thesis). University of Western Sydney, Richmond, Australia.

Vargas, G., Michaud, J.P., Nechols, J.R., 2012a. Maternal effects shape dynamic trajectories of reproductive allocation in the ladybird *Coleomegilla maculata*. Bulletin of Entomological Research 102, 558–565.

Vargas, G., Michaud, J.P., Nechols, J.R., 2012b. Larval food supply constrains female reproductive schedules in *Hippodamia convergens* (Coleoptera: Coccinellidae). Annals of the Entomological Society of America 105, 832–839.

Vargas, G., Michaud, J.P., Nechols, J.R., 2013. Trajectories of reproductive effort in *Coleomegilla maculata* and *Hippodamia convergens* (Coleoptera: Coccinellidae) respond to variation in both income and capital. Environmental Entomology 42 (2), 341–353.

Wang, S., Michaud, J.P., Zhang, R.Z., Zhang, F., Liu, S., 2009. Seasonal cycles of assortative mating and reproductive behaviour in polymorphic populations of *Harmonia axyridis* in China. Ecological Entomology 34, 483–494.

Wang, S., Tan, X., Guo, X., Zhang, F., 2013. Effect of temperature and photoperiod on the development, reproduction, and predation of the predatory ladybird *Cheilomenes sexmaculata* (Coleoptera: Coccinellidae). Journal of Economical Entomology 106, 2621–2629.

Wang, S., Wang, K., Michaud, J.P., Zhang, F., Xiao-Ling, T., 2014. Reproductive performance of *Propylea japonica* (Coleoptera: Coccinellidae) under various light intensities, wavelengths and photoperiods. European Journal of Entomology 111 (3), 341–347.

Ware, R.L., Majerus, M.E.N., 2008. Intraguild predation of immature stages of British and Japanese coccinellids by the invasive ladybird *Harmonia axyridis*. BioControl 53, 169–188.

Ware, R.L., Ramon-Portugal, F., Magro, A., Ducamp, C., Hemptinne, J.L., Majerus, M.E.N., 2008. Chemical protection of *Calvia quatuordecimguttata* eggs against intraguild predation by the invasive ladybird *Harmonia axyridis*. BioControl 53, 189–200.

Ware, R.L., Yguel, B., Majerus, M., 2009. Effects of competition, cannibalism and intraguild predation on larval development of the European coccinellid *Adalia bipunctata* and the invasive species *Harmonia axyridis*. Ecological Entomology 34 (1), 12–19.

Waterhouse, D.F., Norris, K.R., 1989. Aleurodicus Dispersus. Biological Control, Pacific Prospects – Supplement 1. ACIAR monograph no.12. Australian Center for International Agricultural Research, Canberra. pp. 13–22.

Wyss, E., Villiger, M., Muller-Scharer, H., 1999a. The potential of three native insect predators to control the rosy apple aphid, *Dysaphis plantaginea*. BioControl 44, 171–182.

Wyss, E., Villiger, M., Hemptinne, J.L., Muller-Scharer, H., 1999b. Effects of augmentative releases of eggs and larvae of the ladybird beetle, *Adalia bipunctata*, on the abundance of the rosy apple aphid, *Dysaphis plantaginea*, in organic apple orchards. Entomologia Experimentalis et Applicata 90, 167–173.

Yan, H.Y., Wei, G.S., Yan, H.X., Feng, L., 2006. The morphology and fine structure of the compound eye of *Propylea japonica*. Chinese Bulletin of Entomology 43, 344–348.

Yasuda, H., Dixon, A.F.G., 2002. Sexual size dimorphism in the two spot ladybird beetle *Adalia bipunctata*: developmental mechanism and its consequences for mating. Ecological Entomology 27, 493–498.

Yasuda, H., Ohnuma, N., 1999. Effect of cannibalism and predation on the larval performance of two ladybird beetles. Entomologia Experimentalis et Applicata 93, 63–67.

Yasuda, H., Takagi, T., Kogi, K., 2000. Effects of conspecific and heterospecific larval tracks on the oviposition behaviour of the predatory ladybird, *Harmonia axyridis* (Coleoptera: Coccinellidae). European Journal of Entomology 97, 551–553.

Yasuda, H., Kikuchi, T., Kindlmann, P., Sato, S., 2001. Relationships between attacks and escape rates, cannibalism, and intraguild predation in larvae of two predatory ladybirds. Journal of Insect Behavior 14, 373–384.

Yasuda, H., Evans, E.W., Kajita, Y., Urakawa, K., Takizawa, T., 2004. Asymmetric larval interactions between introduced and indigenous ladybirds in North America. Oecologia 141, 722–731.

Yigit, A., Canhilal, R., Eckmekci, U., 2003. Seasonal population fluctuations of *Serangium parcesetosum* (Col., Coccinellidae), a predator of citrus whitefly, *Dialeurodes citri* (Hom., Aleyrodidae) in Turkey's eastern Mediterranean citrus groves. Environmental Entomology 32, 1105–1114.

Zaki, F.N., El-Shaarawy, M.F., Farag, N.A., 1999. Release of two predators and two parasitoids to control aphids and whiteflies. Journal of Pest Science 72, 19–20.

Zhang, Z.Q., 1992. The natural enemies of *Aphis gossypii* Glover (Hom., Aphididae) in China. Journal of Applied Entomology 114, 251–262.

10

Chrysopids

N. Dhandapani[1], Pallavi Sarkar[1], Geetanjali Mishra[2]

[1]Department of Entomology, Tamil Nadu Agricultural University, Coimbatore, Tamil Nadu, India;
[2]Centre of Excellence in Biocontrol of Insect Pests, Ladybird Research Laboratory, Department of Zoology, University of Lucknow, Lucknow, India

1. INTRODUCTION

Chrysopids (Chrysopidae: Neuroptera), commonly known as lacewings, occur in numerous agricultural and horticultural zones of the northern hemisphere (Greve, 1984; New, 1984; Canard et al., 1984; Duelli, 2001; Szentkiralyi, 2001a,b). As most of them are morphologically indistinguishable from each other, they were originally considered to be a single species with a holarctic distribution. However, these have now been identified as a complex of many cryptic, sibling species (Henry et al., 2015). While morphologically indistinct, lacewings possess highly species-specific songs (Henry et al., 2015) that they use to communicate with each other, especially during courtship, which helps in maintaining effective reproductive isolation.

Chrysopids are regarded as important aphid predators of a number of agricultural and horticultural crops, such as cotton crops in Russia and Egypt, sugar beet in Germany, and vineyards in Europe. It is the chrysopid larvae that are of primary importance in pest management, particularly aphids. They are active predators of a wide variety of pests, including aphids, chinch bugs, mealybugs, scales, whiteflies, leafhoppers, lepidopterous eggs and larvae, and mites (Hydorn, 1971; Syed et al., 2005; Satpathy et al., 2012; Mansoor et al., 2013). Owing to their predatory nature, lacewing larvae are also known as aphid lions. Adults, on the other hand, largely feed on nectar, pollen, and aphid honeydew.

Lacewings are considered as potential biocontrol agents and as possible ecofriendly alternatives to hazardous pesticides used in pest management (Mansoor et al., 2013). While the known hazardous effects of pesticides to human and environmental well-being may be mitigated by use of natural enemies, further benefits, such as improvement of crop yield and quality, may also be achieved. While the basic characteristics of chrysopid biology have been well documented (Canard et al., 1984; McEwen et al., 2007), it is our aim here to outline specifically how chrysopids can be harnessed in the biocontrol of pests and the means of effectively achieving it.

Ecofriendly Pest Management for Food Security
http://dx.doi.org/10.1016/B978-0-12-803265-7.00010-5

2. GENERAL BIOLOGY

Adult chrysopids are cosmopolitan in distribution. They are medium-sized (7–15 mm), yellowish-green to gray insects with red, yellow, or brown markings. They have characteristic large copper-colored compound eyes, long filiform antennae, and large lacy wings that are held roof-like over the body (Figure 1). Except for a few adults that are predatory in nature, most depend upon pollen, nectar, and insect honeydew for sustenance. Sexually mature adults contact each other through their songs. Post mating, they deposit eggs on long, slender stalks, either singly or in groups, with each group usually having 20–30 eggs (Figure 1). The size of the greenish-colored eggs usually ranges between 0.7 and 2.3 mm and are borne on stalks that range from 2 to 26 mm in length. The eggs usually turn pale whitish and then black just before hatching, which usually takes 3–4 days. The larvae molt twice undergoing three larval stages within a duration of 8–10 days. The larvae of lacewings are alligator-shaped with mottled, grayish-brown skin (Figure 1). They usually reach a size of 3/8 inch before they pupate. Larvae are active predators and seize prey in their pincers and suck the juices out of them (Figure 2). The third instar larva spins a cocoon in which it pupates for 5–7 days (Figure 1), after which the adult emerges. Usually pupation takes place in silken cocoons, in concealed places, and in the soil. Adult chrysopids usually begin to oviposit after 3–7 days. Oviposition reaches its peak somewhere between 9 and 23 days after emergence with around 600–800 eggs/female. Virgin females also lay eggs albeit infertile with most lacking pedicels (Ribeiro and Carvalho, 1991). Males survive for 30–35 days, while females can live up to 60 days. Biotic and abiotic factors strongly influence the life history traits of these insects.

Adult

Pupa

Larva

Eggs

FIGURE 1   Life cycle of a chrysopid.

FIGURE 2   A chrysopid larva preying on aphid.

3. INFLUENCE OF BIOTIC AND ABIOTIC FACTORS

It is widely reported that unsuitable food can extend the preimaginal development of chrysopids and decrease the survival, fecundity, and longevity of the adults (Principi and Canard, 1984; Obrycki et al., 1989; Zheng et al., 1993). The biology of *Chrysoperla zastrowi sillemi* (Esben-Petersen) was evaluated against five prey aphids (*Lipaphis erysimi* Kaltenbach; *Myzus persicae* (Sulzer); *Brevicoryne brassicae* L.; *Aphis craccivora* Koch; and *Aphis spiraecola* Patch) and eggs of *Corcyra cephalonica* (Stainton). Lacewing larvae had the highest growth index, larval survival, larval weight, pupal weight, multiplication rate, and fecundity when fed on *M. persicae*, but it was on *C. cephalonica* eggs that lacewings developed and reproduced best, significantly better than the aphid prey (Halder and Rai, 2014). Similarly, duration of each preimaginal stage development and total development time of *Chrysoperla carnea* (Stephens), one of the most common and effective chrysopid predators, was significantly affected by species of prey tested with fastest development and maximum reproductive output on *M. persicae*, followed by *Aphis gossypii* Glover (Takalloozadeh, 2015).

In another study, the developmental and reproductive biology of *C. carnea* was assessed in relation to the various life stages of its prey, cotton jassid, *Amrasca devastans* (Dist.) (Homoptera: Cicadellidae). Of all the five nymphal stages and adults tested as prey for *C. carnea*, second instar nymphs of *A. devastans* were consumed the most. While all stages were suitable for development, consumption of third nymphal instar resulted in fastest development and maximum reproductive output. Adults were least suitable as food (Saeed and Razzaq, 2015). A similar lack of preference and suitability of adults has been reported in other chrysopids too (Huang and Enkegaard, 2010).

Chrysopids develop well not only on their actual prey but are also amenable to artificial diets, with some being more suitable than others. In a study, the effect of six different diets (a mixture of 30% concentrations of glucose, fructose, and sucrose (1:1:1); glucose, fructose, sucrose plus extract of *Sitotroga cerealella* (Olivier) eggs (1:1); glucose, fructose, sucrose plus extract of *Anagasta kuehniella* (Zeller) eggs (1:1); a mixture of honey, yeast, and distilled water

(1:1:1); honey, yeast plus extract of *S. cerealella* eggs (1:1:1); and honey, yeast plus extract of *A. kuehniella* eggs (1:1:1)) was evaluated on multiple developmental and reproductive parameters of *C. carnea*. The mixture of honey, yeast, and extract of *A. kuehniella* eggs (1:1:1) was the most suitable diet for the biological attributes of *C. carnea* (Sarailoo and Lakzaei, 2014).

Different aphid species infesting the same plant were also found to be different in their suitability to *C. carnea*, though no significant difference was observed in their preference. Overall, predator development was delayed and surviving adults weighed less when provided with *L. erysimi* or *B. brassicae*, which sequestered high levels of indole glucosinolates from winter canola. Qualitative differences in nutritional suitability mediated by levels of sequestered plant compounds were found between the *Brassica*-specialist aphids and generalist *M. persicae* (Jessie et al., 2015).

Not only prey quality but also quantity affects the developmental and reproductive attributes of chrysopids. Increased prey consumption with the shortest developmental time, maximum fecundity, and longest adult longevity were observed in *C. carnea* with increased prey density (Batool et al., 2014).

Adult density also influences the reproductive attributes of lacewings. Ahmed et al. (2014) observed that the maximum reproductive output was attained when five pairs were kept in glass chimneys of 1750 cm³, while a density of 20 adult pairs was most detrimental.

Studies on the effect of temperature revealed the best overall rearing of *C. carnea* at $28 \pm 1\,°C$. The observed developmental, reproductive characteristics, and life table parameters were found to be better at this temperature, which was the best among the other tested rearing temperatures ($20 \pm 1, 24 \pm 1, 28 \pm 1$, and $32 \pm 1\,°C$). However, fastest development was observed at $32 \pm 1\,°C$, which can be useful if quick development is desired in laboratories (Saljoqi et al., 2015). Similar trends of varying temperatures were seen even when fed on different prey, such as *A. gossypii* and *M. persicae* (Mannan et al., 1997).

Similarly, when the effects of five constant relative humidity (RH) regimes (12, 33, 55, 75, and 94%) were studied, the results revealed differential effects on the preimaginal development, adult longevity, and reproduction of chrysopid, *Dichochrysa prasina* Burmeister. Preimaginal developmental time from egg to adult for females and males was significantly shorter at 75% and 94% RH than at other RHs. When the females had periodic access to a water source a significant increase in both longevity and egg production occurred and the estimated intrinsic rate of increase was high, irrespective of RH (Pappas et al., 2008). Such results on the differential effects of various abiotic and biotic factors have the potential to be useful for the mass-rearing of chrysopids and for better understanding their population dynamics under field conditions.

4. PREDATION EFFICIENCY

The daily predation rates of *C. carnea* increase slowly during the first two instars and reached at the peak in the third instar (Batool et al., 2014). The third instars are known to consume a major portion (60–85%) of the total number of prey (Balasubramani and Swamiappan, 1994; Klingen et al., 1996). This seems to be a universal trend, as also witnessed in *Chrysoperla externa* (Hagen), where first, second, and third instars consume up to 4.57, 16.56, and 78.87% of total prey consumption, respectively (Guerra et al., 1999). The total larval consumption by *C. externa* has been reported to be 846.85 nymphs of *A. gossypii* and 253.20 individuals of whitefly, *Bemisia argentifolii* (Bellows & Perring) (Costa et al., 1999). Latham and Mills (2009) have found that field observation is the much better way to determine daily per capita consumption than laboratory arena method for chrysopids. This has been attributed to the fact that chrysopids in general kill more biomass than they consume, leading to underestimation in laboratory conditions.

Not only does the actual consumption of prey differ with instars, so does the searching efficiency. Second instars have been reported to have the lowest searching efficiency, with handling time in general showing a progressive decrease with larval development (Fonseca et al., 2000).

Preferential prey consumption depends on the developmental stage of both the prey and the predator. Under no choice conditions, first instars of brown lacewing, *Sympherobius fallax* (Navas), were found to consume second-stage mealybugs more than any other stages while actively discriminating against adult stage by not consuming them at all. On the other hand, second and third instars of *S. fallax* preferred third-stage mealybugs. However, in choice experiments, the first instar larval predators preferred second-stage mealybugs significantly more than the other two stages, while the second and third instar predators preferred third-stage mealybugs significantly more than the second and the fourth stages (Gillani et al., 2009).

Prey consumption by the lacewings follows a predicted path in the face of increasing prey densities. The lacewings usually respond to increasing prey densities in the manner predicted by Holling (1959). Holling (1959) proposed three functional responses, the initial response of which differed in all the three, though a saturation threshold was later obtained. The consumption is directly proportional to the prey density in the type I response resulting in a prominent linear graph. The type II response is curvilinear in nature with an initial rapid increase in consumption followed by a subsequent slowing down of this increase. The type III response has a sigmoidal graph, which is obtained as a result of the increasing prey consumption over a limited range of prey densities. The mortality in prey was density-independent in type I, negatively density-dependent in type II, and positively density-dependent in type III responses. New studies further add two more modifications of functional responses, i.e., type IV (dome-shaped; Bressendorff and Toft, 2011) or type V (negative exponential; Sabelis, 1992) shapes.

The type II functional response is most commonly observed in chrysopids (Mirabzadeh et al., 1998; Yüksel and Gocmen, 1992; Zhou et al., 1991; Fonseca et al., 2000). Both chrysopids, *Chrysoperla nipponensis* (Okamoto) and *C. carnea*, have shown a type II functional response based on logistic regression analysis. *Chrysoperla carnea* was more predaceous than *C. nipponensis*, with handling time higher for *C. nipponensis* than for *C. carnea*. However, the attack coefficient of *C. nipponensis* was slightly higher than *C. carnea* (Montoya-Alvarez et al., 2010). First and second larval instars of *C. carnea* exhibit type II functional responses, while third instar larvae showed type III functional response on the two-spotted spider mite, *Tetranychus urticae* Koch (Hassanpour et al., 2009). The functional response of *C. carnea*, preying upon eggs and first instar larvae of the cotton bollworm, *Helicoverpa armigera* Hubner, has also been evaluated. The first and second instar larvae of *C. carnea* exhibited type II functional responses against both prey stages. However, the third instar larvae of *C. carnea* showed a type II functional response to the first instar larvae of *H. armigera*, but a type III functional response to the eggs (Hassanpour et al., 2011).

The effects of prey types on consumption capacity was studied in the common lacewing, *C. carnea*, using four aphid species i.e., *A. craccivora*, *A. gossypii*, *M. persicae*, and *L. erysimi*, as prey at five constant temperature regimes (i.e., 10, 15, 20, 25, and 30 °C), under laboratory conditions. The data revealed that the consumption capacity of *C. carnea* differs significantly, when its larvae were reared separately on the four aphid species at different temperatures. Consumption was maximum at 15 °C when *A. craccivora* was given as prey to the larvae of this predator and minimum at 30 °C when reared on *L. erysimi* (Yadav and Pathak, 2010).

5. EXPERIMENTAL RELEASES

Chrysopids seem to be efficient predators of a number of insect pests. To check their effectiveness as bioagents, test releases have been made to assess the size of initial releases, kind of release (eggs, larvae, or adults), and release rates. Results of a few such releases depict the effectiveness of chrysopids in pest management.

In greenhouse and field experiments in Utah, USA that assessed the ability of *C. carnea* as a generalist predator, aphid colonies were reduced to zero or near zero on slender leaved grasses, *Oryzopsis hymenoides* (Roem. and Schult.) (Indian ricegrass) and *Pseudoroegneria spicata* (Pursh) (blue bench wheatgrass), but were maintained at moderate densities on intermediate wheatgrass, *Elymus lanceolatus* (Scribn. & J.G. Sm.) (Messina et al., 1995). This indicates the modification of predation efficacy on plants with differential architecture. The area of residence of the prey is influenced by the plant structure, thus modifying the degree of accessibility to the predator.

Studies of predation by larvae of *Chrysoperla rufilabris* (Burmeister) on eggs and young larvae of Colorado potato beetle, *Leptinotarsa decemlineata* (Say), revealed that when 5–10 chrysopid larvae were released per cage, there was a 97.7% reduction in prey population. Ten *C. rufilabris* and fifty *C. rufilabris* caused 79.9% and 97.5% reduction in prey population, respectively. The augmentative release of 80,940 *C. rufilabris* larvae/ha caused 84% reduction in prey population. These results indicate that these larvae might be useful in periodic release programs against *L. decemlineata* (Nordlund et al., 1991).

First and second instars of *C. rufilabris* were also evaluated as biocontrol agents of *Bemisia tabaci* (Gennadius) on Chinese rose, *Hibiscus rosa-sinensis* (L.), in a greenhouse. Inundative releases of 25 or 50 *C. rufilabris* larvae/plant at an interval of 2 weeks led to production of marketable plants. One hundred larvae released at the center of 12 plants also maintained the quality of the plants. Even slight treatment of five larvae/plant was more effective in increasing marketability of plants compared with the untreated plants that were not suitable for market by the end of the experimental duration (Breene et al., 1992).

In studies carried out in Braunschweig, Germany during 1992, the efficiency of *C. carnea* in controlling aphids, *Aulacorthum solani* (Kaltenbach) (75%), *Macrosiphum euphorbiae* (Thomas) (12%), *Nasonovia ribisnigri* (Mosley) (2%), and *M. persicae* (11%) on lettuces in greenhouses was tested. Application of *C. carnea* (eggs) alone led to significant control of aphid populations. Addition of eggs to the young lettuce plants prior to their transplantation followed by three sprays of *C. carnea* eggs at weekly intervals (25–30 eggs/m²) was the most efficient means of managing aphid populations (Quentin et al., 1995).

The effective stage and rate of releases of *C. carnea* against safflower aphid, *Dactynotus carthami* (Hille Ris Lambers), were evaluated. The results revealed the highest aphid population reduction in the plots treated with dimethoate 0.03%. Larval releases of *C. carnea* at two or three larvae at the population level of 10 or 20 aphids per shoot reduced the population between 40% and 50%. Egg releases of *C. carnea* did not help in satisfactory reduction of aphid population. On the basis of yield, promising treatments were dimethoate 0.03% (15.8 q/ha), three larvae of *C. carnea* released at 10 aphids (13.43 q/ha), and 20 aphids (13.20 q/ha) population levels (Badgujar et al., 2000).

Green lacewings are highly predaceous on cardamom aphid, *Pentalonia nigronervosa* Coquerel. Second instars and adults of *C. carnea* were released at ratios of 1:5, 1:10, 1:15, 1:120, 1:25, and 1:50. A predator–prey ratio of 1:5 was superior with a 78.33% population reduction after 3 days (Mathew et al., 1999). Interactions between *C. carnea* larvae and spider mite, *Tetranychus neocaledonicus* (Andre), were also observed on okra. The ratios were 1:25, 1:50, 1:100, 1:200, and 1:500 in the laboratory while in the field 25,000, 50,000, and 100,000 *C. carnea* larvae were released in Dharwad (India). At 1:25 and 1:50, the prey were eliminated completely. At 1:100, 1:200, and 1:500 the populations were maintained even 13 days after release. Release of 100,000 larvae/ha were the most effective (Sharanabasava and Manjunatha, 2001).

Trial release of larvae of *C. carnea* on strawberry aphids, *Chaetosiphon fragaefolii* (Cockerell), on potted strawberry plants in gauze houses led to significant reduction in their numbers. This was also investigated in open field experiments with release rates of eight larvae per plant. However, a release experiment on larger strawberry plants grown under protection was not successful even at a release rate of 25 larvae per plant (Easterbrook et al., 2006).

The test releases support the high efficacy of the chrysopids as biocontrol agents. There is need to properly mass-produce the chrysopids and then to release them in appropriate quantities at the appropriate time by using the most effective application techniques so as to obtain best results.

6. REARING TECHNIQUES

After testing the effectiveness of the predators through test releases, attempts for their mass-production are done prior to field releases. The mass-production and commercialization of biocontrol agents depends upon the ability of insectaries to produce and market highly reliable and relatively inexpensive supply of natural enemies. An efficient and standardized mass-rearing procedure requires: (1) the use of inexpensive and nutritious diets; (2) mechanized and space-efficient production systems; (3) reliable storage methods; and (4) periodic evaluation of natural enemy quality (Tauber et al., 2000). Research has made practical and economically beneficial advances in mass-rearing of *Chrysoperla* sp.

Chrysoperla sp. are ranked as some of the most commonly used and commercially available natural enemies. For many years, *C. carnea* and *C. rufilabris* have been mass-reared and marketed commercially in North America and Europe. Two additional species, *C. externa* and *C. nipponensis*, are used in Latin America and Asia. In response to a questionnaire in 1992, members of American Association of Applied Ecologists ranked *Chrysoperla* sp. as unrivaled on the list of commonly applied and commercially available predators (Tauber et al., 2000).

Rearing of larvae is most costly in mass-production of *Chrysoperla* largely because all three instars are predaceous. Most of the insectaries depend on mass-produced insect prey as food, which is relatively expensive compared with artificial diets (Tauber et al., 2000).

Mass-rearing of insects, especially the cannibals, is both space-demanding and labor-intensive. Space-efficient, automated mass-rearing systems for *Chrysoperla* are the need of the hour. Compact holding units for adults, with mechanical devices for feeding adults and larvae and harvesting eggs, as well as automated systems for packaging are much required and would make production of these bioagents very cost-effective while greatly enhancing production. This requires the getting together of engineers as well as biologists.

In the traditional and conventional methods of rearing, 200 pairs of adults are kept in oviposition units measuring 75×30 cm, with sides lined by nylon wire mesh. The sliding top cover is slid over a comb and is fitted with a black cloth as a substrate for oviposition. The sliding top cover is replaced everyday starting from the fourth day. In the meantime, adults are provided daily with cotton swabs dipped in water, 50% honey, castor pollen, and equal quantities of protinex, fructose, honey, and powdered yeast dissolved and mixed together in little water. One day after laying, the eggs are removed by gently working with sponge and then stored for future use at low temperature (Anonymous, 1995).

In larval rearing, 120 three-day-old eggs of chrysopids are mixed with 0.75 ml of *Corcyra* eggs, to provide instant nutrition to the neonates. Three-day-old larvae are then transferred to 2.5 cm cubical cells of plastic louvers, with each louver capable of holding 192 larvae. Total quantity of *Corcyra* eggs required for rearing 100 chrysopid larvae is 4.25 ml. The louvers are secured on one side by organdy or brown paper to facilitate pupation. These louvers are stacked in racks. A single $2 \times 1 \times 0.45$ m angle iron rack can hold 100 louvers containing 19,200 larvae (TNAU Agritech Portal, 2013).

Newer and less space-consuming units are being developed worldwide. Two cylindrical and cubical adult feeding and oviposition units for *Chrysoperla* have been developed that contain 5.4 and 9.3 times internal surface area (with higher internal surface: volume, ratios) of the traditional 11.30 l cardboard units. These new units can hold 2500 and 4500 adults, respectively, compared with 500 adults in traditional units. Harvesting of eggs from both new units involves the use of a sodium hypochlorite based egg harvest procedure which takes only 1 min/unit and yields stalk-free eggs (Nordlund and Correa, 1995a).

A cylindrical adult feeding and oviposition unit (AFOU) unit has been developed to ease the rearing of chrysopids. This can be unrolled to facilitate egg harvesting with a new harvest system, the hot wire system. The hot wire egg harvesting system consists of two devices with a pneumatic cylinder to pull the cylinder portion of the AFOU past the hot wire, which is used to harvest eggs from the top of the AFOU. This improved system requires less manual labor. Egg recovery was found to be higher with a hot wire egg harvesting system than with the traditional loose ball of nylon netting. The hot wire method did not affect the hatching percent of the recovered eggs (Nordlund and Correa, 1995b).

An even more compact method of mass-production is the Biosec Bioassay tray. Each tray (44×21 cm) comprises eight blocks, each block containing 16 wells; thus each tray has a total of 128 wells. The dimension of each well is 2×1.5 cm. The eggs of *C. carnea* are transferred along with 0.045 cc *Corcyra* eggs. The Pull-N-Peel tab is pulled firmly over the wells and the trays stacked one over the other. It takes 12–16 days for the adults to emerge. The recovery is approximately 95%. These trays give better adult recovery, prevent the larvae from escaping, occupy less space, and are also relatively cheaper. This method is being tried out at the

National Bureau of Agricultural Insect Resources, Bangalore, India. Glass cages have been found to be better for rearing than Perspex and wooden cages (Sattar and Abro, 2011).

The filler material used in the units and the degree of filling also influence the development of chrysopids. The percentage pupation is significantly higher in most of the filler materials when used at 75% of the container's capacity. Maximum pupation (88%) is obtained in larvae reared in wood shavings and was at par with those reared in injection vials (93%) and plastic louvers (91%). Pupal weight was highest when larvae were reared in dried plant leaves; however, it was on a par with those reared in injection vials (9.53 mg/pupa) and plastic louvers (9.37 mg/pupa). Thermocol pieces and plastic chips when used as filler material resulted in low pupal weight (Mahabaleshwar and Kulkarni, 2000).

7. STORAGE TECHNIQUES

Medium and long-term storage of natural enemies is often a much-ignored aspect while planning their cost-effective production and distribution (Tauber and Helgesen, 1978; Tauber et al., 2000). Well-researched and innovative storage allows the insectaries to stock natural enemies for use during periods of high demand. Other than facilitating this, it also allows the development of alternative methods of distribution and permits long-term, low-cost maintenance of stock for use in mass-rearing or research.

Studies on *C. externa*, *C. nipponensis*, and *C. carnea* indicate that simple and economical long-term storage of adults can be easily achieved without loss of quality (Tauber et al., 1993; Wang and Nordlund, 1994; Chang et al., 1995; Tauber et al., 1997). Such post storage adults reproduce effectively in a quick predictable and synchronous manner, thus allowing producers with two alternate distribution systems: (1) supply of eggs or larvae from post storage adults and (2) supply of cold-stored adults that can be brought into production as required.

Post mass-rearing, the biocontrol stage obtained for release in fields needs to be stored till the time of application. It is easiest to store chrysopid eggs, but what is of utmost importance in the storage technique is that the viability of the egg must not be reduced. Studies show that it is possible to store eggs in solution or at low temperatures.

Eggs of *C. carnea* can be stored for 1 day at 4 °C in a 0.125% agar solution in water without any reduction in egg viability or larval survival. However, storage in the same solution for two or more days negatively affects egg viability without affecting larval survival. None of the eggs were found viable when stored by this technique for 8 days, thereby making this method an effective for short-term storage but not for long-term (McEwen, 1996).

Studies have demonstrated that *C. carnea* eggs remain fully viable for up to 3 weeks when stored at 8 °C (Osman and Selman, 1993) and up to 14 days at 10 °C and 75% RH with minor egg loss (Anonymous, 2002b). Newly laid and 1, 2, and 3-day-old eggs of *Ceraeochrysa cubana* (Hagen), *C. externa*, and *Ceraeochrysa smithi* (Navas) stored at 15.6 °C for up to 21 days had high rates of hatching (>70%). *Chrysoperla externa* eggs can be stored at 12.8 °C for 21 days with hatching percent greater than 95% (López-Arroyo et al., 2000). Storage did not influence post storage reproductive success (López-Arroyo et al., 2000). The maximum egg storage time is an average of 2 weeks (Anonymous, 2002b). Storage usually lengthens the pupal period and reduces the rate of post storage cocoon spinning by 30–40%, thereby causing significant reductions in the rate of larval development.

Pupae stored for up to 15 days at 8 °C had 74.8% adult emergence (Osman and Selman, 1993). At 6 °C storage of pupae can be done up to several months (López-Arroyo et al., 2000). Such relatively long-term storage of eggs and pupae with high hatchability and adult emergence, respectively, allows more flexibility in storage methods (Osman and Selman, 1993). Larval storage is not recommended for more than 16 h at low temperature because of the delicate constitution and the probability of incidence of cannibalism (López-Arroyo et al., 2000).

Adults of *C. carnea* can be stored for less than 3 weeks at 8 °C and for 2 weeks at 10 °C and 75% RH if put into storage when 1-day-old (Anonymous, 2002a). The length of storage of diapausing adults can be changed based on manipulation of photoperiod, temperature, and diet. Short-term storage (of up to 10 weeks) is effective under short day lengths (Light: Day 10:14–8:16) at temperatures of up to 21 °C (Chang et al., 1995).

Storage from 10–18 weeks is accomplished at relatively low temperature (5–10 °C). Storage from 18–31 weeks or even longer is most effective when pupae or adults are exposed to a decrease in day lengths and diapausing adults are maintained under short day lengths (LD 8:16) at low temperatures (5 °C) (Chang et al., 1995). Short day lengths (LD 10:14) at 5 °C allow effective and low maintenance storage of adults for up to 31 weeks (Tauber et al., 1997).

Rates of survival are high (~97%) with both males and females retaining their high reproductive potential, which has been found equal to that of unstored adults. Shortly after transfer to warm and long day conditions more than 90% of the pairs produce fertile eggs. When the adults were acclimatized to reduced temperatures and then stored at 10 °C, there was no significant reduction in survival after 120 days. A significant increase in preoviposition period and a significant decrease in ovipositional rate after storage for 120 days were reported at 10 °C, although egg fertility in the first 10 days of oviposition was unaffected. None of the above variables were unduly affected by 60 days storage (Tauber et al., 1997). A high carbohydrate diet before storage and a rich carbohydrate–protein diet during and after storage are key to achieving effective storage (Chang et al., 1995).

Such effective storage techniques aid biocontrol efforts by allowing flexibility and thereby increased efficiency in mass-production to meet peak seasonal demands for eggs or young larvae. It provides an alternative method of shipping to distributors and also makes standardized stock available for use in long-term ecological, physiological, or genetic research (Tauber et al., 1993).

8. RELEASE TECHNIQUES

Proper release of reared chrysopids is essential as not doing so in a well-thought-out manner can largely influence the efficacy of biocontrol achieved. It is important to decide the stage, quantity, and time of releases. Eggs and larvae are usually the recommended stages for biocontrol application since adults are liable to fly away. Eggs, however, delay the start of biocontrol by a week because that time is required to reach the voracious second and third instar stages. The application of larvae avoids this delay. Also more eggs are needed than larvae to compensate for the delay in onset of biocontrol mechanisms.

Chrysopid eggs have been tested extensively for field application since released adults tend to disperse and leave the target crop prior to laying eggs. However, it is believed that release of pre-fed adults may mitigate this problem since they would be ready to oviposit when released.

Traditionally, chrysopid eggs were dispensed manually, mixed with a solid medium such as rice hulls or vermiculite to enable uniform distribution (Tauber et al., 1997). Application of eggs through these solid carriers causes eggs to fall off the leaves whereas liquid carriers help attach eggs to the targeted plants. For example, distributing *C. carnea* eggs in an agar solution rather than sucrose-based carriers has the advantage of lowered attractiveness to ants and other predators. In "prototype" applicators, *C. rufilabris* eggs were immersed in a commercial liquid carrier (BioCarrier™), pneumatically agitated to create uniform egg suspension and discharged into the targeted crop without damage to the eggs and with good retention on the leaves (Anonymous, 2002c).

Special spray techniques for applying mass-reared eggs of *C. carnea* in fields need to be developed. Water is a suitable liquid medium in which the eggs could be suspended for up to 12 h. A pressure of up to 4.5 bar liquid pressure was found to be suitable for application of eggs, as the hatching percentage was not affected. A flat jet-producing Flood Jet TK 1.5 nozzle is suitable for the application of eggs at a pressure of about 3 bar (Löchte and Sengonca, 1995).

A mechanical system has been developed to meter and discharge liquid suspensions of insect eggs for release of bioagents. The system consists of a reservoir pressurized by compressed air, which distends the bottom surface of the reservoir, agitates the suspension and forces it through a liquid supply tube, and terminates in a tapered orifice. A pulse-width modulated valve provides intermittent flow, which produces large droplets (around 2 mm diameter) of suspension. The duration of and intervals between liquid pulses, which determines application rate of and spacing between discharged eggs, respectively, is controlled by a microprocessor. Liquid droplets can be propelled outward to 3 m and at flow rates ranging from 25–250 ml/min per orifice. In operation, discharge distance of the eggs is controlled by the reservoir pressure, ranging from 10–50 kPa and the flow rate by the duty cycle, or relative open time of the metering valve (10–100%). The system is capable of providing adequate agitation to discharge uniform suspensions of eggs in liquid carriers for 50 min with no reduction in viability of chrysopid eggs (Giles and Wunderlich, 1998).

A new solid application method is directly related to the means of transportation. Chrysopids are shipped as eggs on cards, eggs on tapes, or as prefed larvae in hexcel units. There are 5000 or 10,000 lacewings eggs per card with each 1-inch square holding an average of approximately 170 and 330 eggs, respectively. Chrysopid tapes are made up in units of 20 tapes, each 6 inches long and 3/4 inches wide, with waxed paper on each end so that the ends can be conveniently attached to form a bracelet around an infested limb or vine. Each tape holds an average of approximately 250 chrysopid eggs with equal volume of moth egg diet and corn grit to cover the exposed sticky areas on the tapes. These tapes and cards can be directly placed firmly on the foliage as soon as larvae start crawling (Anonymous, 2002a). The release method for hexcel units simply involves peeling off the organdy cloth cover and lifting or brushing each larva onto the plant with a fine paintbrush. The eggs go from green to gray over a 5 day period and then white on hatch. Tiny larvae can be seen crawling on sides of container. These can be applied by hand or hand-held blower, ground rig/hopper system, helicopter, or aircraft (fixed wing). The release of newly hatched larvae in vines, shrubs, and trees is made easier when chrysopid eggs are shipped glued on cards or masking tape or as 500 prefed larvae in hexcel units (Anonymous, 2002a).

It was observed that the vibrational mass-mechanical release of second instars of *C. rufilabris* mixed into vermiculite and rice hull carriers is more effective than rotational release as the latter results in slightly greater larval damage (Morisawa and Giles, 1995).

Weekly releases of larvae are recommended to create overlapping generations in the field. Releases should start when pest levels are low, and then should be repeated every 1–4 weeks as needed. Release rates depend on the degree of infestation, the presence of other bioagents, and other factors, but the guideline is 5000 to 50,000 larvae per acre per season or 1000 larvae per 2500 square feet of garden area. The high propagation efficiency of larvae is well demonstrated by the fact that a single lacewing can produce 200 offspring in 30 days and 40,000 in 60 days. Lacewing larvae require high humidity when young, so plants need to be well watered after the releases. Some researchers suggest that release during cool temperatures, particularly in the evening, and misting foliage enhances survival of larvae.

After the release of bioagents there is need to increase the attractiveness of the fields by spraying food substances and attractants to augment the local populations of chrysopids. Augmentation of *C. externa* in soybeans and corn did not affect the resulting number of chrysopid larvae but applying wheat and sugar at the time of the release gave a two to sixfold increase in the densities of adults and eggs of *C. externa* in cornfields (Anonymous, 2002c). Food sprays that imitate honeydew can attract or retain adults and stimulate egg laying. Ice plant pollen is among the best for increasing *C. carnea* egg laying, while corn pollen is used to prepare for autumn migration. Commercial yeast hydrolyzates, like wheat; 4:7:10 brewer's or baker's yeast hydrolyzate (or whey substitute): sugar:water in a weekly spray onto alfalfa attracts more lacewing adults and eggs. A solution of 4.5 kg of fermented whey plus 4.5 kg sugar in 25–40 l of water can be sprayed every 1–2 weeks to attract and induce oviposition in *C. carnea* (Anonymous, 2002a).

However, the effectiveness of food sprays in manipulating field populations varies and studies emphasize the need of further basic and applied research in this area. The best commercially available food sprays contain both enzymatic protein hydrolyzates and sugar or honey. In most trials, protein sprays without sugar fail to increase the number of chrysopids and sugar sprays without protein attract lacewing adults but do not stimulate egg laying. By using food sprays to attract and induce chrysopids to lay eggs before natural honeydew becomes abundant, it is also possible to suppress honeydew-producing or other pests before their numbers become large (Anonymous, 2002c). Therefore, two areas of research could be of great value: (1) the seasonal variation in the reproductive responses of lacewings to food and (2) the impact of food sprays on nontarget organisms that could reduce lacewing effectiveness.

Proper and timely releases are essential for the success of a biocontrol program. A lot of basic research is needed before the mass-rearing, storage, and release techniques become standardized enough to make each biocontrol release a success.

9. INTEGRATED PEST MANAGEMENT PROGRAMS USING CHRYSOPIDS

Owing to their efficacy, chrysopids have great potential to be part of pest management programs. In fact numerous integrated pest management (IPM) programs for various important crops have been identified with chrysopids as important and sometimes even central tools. The paragraphs below enlist many of these effective IPM programs.

A biointensive IPM module has been developed in Tamil Nadu, India for the cotton crop, with recommended application of *Chrysopa scelestes* at 50,000/ha with adjuvants of 2.5 kg crude sugar and 0.3 kg cotton seed kernel powder/ha. This resulted in a cost benefit ratio of 1:4.3. Release of *C. carnea* along with 2% neem oil spray on cotton was more effective than most insecticides. Release of the predators *C. carnea* has been found to suppress the population of larvae of American bollworm. Also, encouraging the activities of predators *C. carnea* along with bacterial formulations of *Bacillus thuringiensis kurstatki* (1 kg/ha) has been found to effective against pink bollworm (Dhandapani et al., 1992).

Similarly on okra release of the predator, *C. carnea* (25,000 larvae/ha/release) and Econeem 0.3% (0.5 l/ha) for three times at 15 days interval starting from 45 days after sowing was found to be effective in reducing the population of sucking pests as well as the fruit borers, such as leafhopper, *Amrasca biguttula biguttula*, sweet potato whitefly, *B. tabaci*, and cotton aphid, *A. gossypii*, as well as fruit borers, *H. armigera*, and *Earias vitella* in Coimbatore, Tamil Nadu, India. The percentage fruit damage by *H. armigera* (8.61%) and *E. vitella* (9.21%) was also reduced compared with untreated crop (22.56 and 22.6%, respectively). The fruit yield (10,326 kg/ha) and cost benefit ratio (1:2.60) were also higher when *C. carnea* and Econeem 0.3% were combined, compared to either *C. carnea* (9643 kg fruit/ha and 1:2.39) or Econeem 0.3% (9533 kg fruit/ha and 1:2.44) alone (Praveen and Dhandapani, 2001).

Studies carried out at the International Crop Research Institute for Semi-Arid and Tropics (ICRISAT), Hyderabad, India indicate as much as 85% parasitization of *H. armigera* eggs on cow pea. Field experimentation with *C. carnea* (1.0 lakh/ha) with *Bacillus thuringiensis (Bt)* and *Helicoverpa armigera* nucleopolyhedrosis virus (*Ha*NPV) in pigeon pea as part of integrated management of the pod borer, increased the pod yield significantly than when released alone (Kalyanasundaram et al., 1994).

An IPM-based module consisting of brinjal (PKM 1) + cluster bean (4:1) + six releases of *Trichogramma chilonis* (2.5 cc/acre) on 15, 22, 29, 36, 43, and 50 days after transplantation (DAT) + two releases of *Chrysoperla* eggs (20,000 eggs/acre) on 60 and 70 DAT + yellow sticky trap (25/acre) + *Leucinodes orbonalis* pheromone trap (5/acre) (Intercropping system based module I) was found to be the best in managing major pests of brinjal (Elanchezhyan and Baskaran, 2008).

Field experiments conducted in Tamil Nadu revealed that release of *T. chilonis*, 1,00,000/ha and *C. carnea*, 50,000/ha, on 40 and 55 days after sowing on groundnut (*Arachis hypogaea* L.) (Kalyanasundaram et al., 1994) and 90, 105, and 120 days after sowing on cotton (Dhandapani et al., 1992) could effectively check the population of *H. armigera* and *B. tabaci*.

10. CONCLUSIONS

Based on the above information, it is safe to conclude that chrysopids are quite effective biocontrol agents for number of insect pests. Also the advancement of techniques for their mass-production, storage, and release make them easily available for use in effective pest management. These tiny insects are especially strong predators of aphids and can be used much more effectively against them. Lots of pest management successes have been achieved with chrysopids but still there are major lacunae in landscape management studies as a means of further improving the efficacy. This will be especially important because the major problem in chrysopid use is to keep the adults in the desired area. More work on landscape management will make biocontrol of pests easier and the chrysopids more effective.

References

Ahmed, Q., Muhammad, R., Ursani, T.J., Ahmad, N., Rashdi, S.M.S., Depar, N., 2014. Effect of population density on fecundity of Chrysoperla Carnea (Neuroptera: Chrysopidae) under laboratory Conditions. Sindh University Research Journal (Science Series) 46 (2).

Anonymous, 1995. Technology for production of natural enemies. In: Singh, S.P. (Ed.), Technical Bulletin No. 4. ICAR.

Anonymous, 2002a. Lacewing Little Helpers. www.rain.org/~sals/lace.html.

Anonymous, 2002b. Advances in the Commercialization of Green Lacewings. Part I: Introduction, Systematics and Mass Production. www.entomology.wicu/mbcn/fea702.html.

Anonymous, 2002c. Advances in the Commercialization of Green Lacewings. Part II: Field Application. www.entomology.wicu/mbcn/fea703.html.

Badgujar, M.P., Deotale, V.Y., Sharnagat, B.K., Nandanwar, V.N., 2000. Performance of Chrysoperla carnea against safflower aphid, Dactynotus carthami (HRL). Journal of Soils and Crops 10 (1), 125–127.

Balasubramani, V., Swamiappan, M., 1994. Development and feeding potential of the green lacewing Chrysoperla carnea Steph. (Neur., Chrysopidae) on different insect pests of cotton. Anzeiger für Schädlingskunde, Pflanzenschutz, Umweltschutz 67 (8), 165–167.

Batool, A., Abdullah, K., Mamoon-ur-Rashid, M., Khattak, M.K., Abbas, S.S., 2014. Effect of prey density on biology and functional response of Chrysoperla carnea (Stephens) (Neuroptera: Chrysopidae). Pakistan Journal of Zoology 46 (1), 129–137.

Breene, R.G., Meagher Jr., R.L., Nordlund, D.A., Wang, Y.T., 1992. Biocontrol of Bemisia tabaci (Homoptera: Aleyrodidae) in a greenhouse using Chrysoperla rufilabris (Neuroptera: Chrysopidae). BioControl 2 (1), 9–14.

Bressendorff, B.B., Toft, S., 2011. Dome-shaped functional response induced by nutrient imbalance of the prey. Biology Letters 7, 517–520. http://dx.doi.org/10.1098/rsbl.2011.0103.

Canard, M., Séméria, Y., New, T.R., 1984. Biology of Chrysopidae. Dr. W. Junk Publishers.

Chang, Y.F., Tauber, M.J., Tauber, C.A., 1995. Storage of the mass-produced predator Chrysoperla carnea (Neuroptera: Chrysopidae): influence of photoperiod, temperature, and diet. Environmental Entomology 24 (5), 1365–1374.

Costa, R.I.F., da, Macedo, L.P.M., de, Almeida, S.A., de, Soares, J.J., 1999. Oviposition potential and longevity of Chrysoperla externa (Hagen, 1861) (Neuroptera: Chrysopidae) in the laboratory. Anais II Congresso Brasileiro de Algodão: O algodão no século XX, perspectivar para o século XXI, Ribeirão Preto, SP, Brasil, 5–10 Setembro 1999. pp. 253–255.

Dhandapani, N., Kalyanasundaram, M., Swamiappan, M., Sundara Babu, P.C., Jayaraj, S., 1992. Experiments on management of major pests of cotton with biocontrol agents in India. Journal of Applied Entomology 114 (1), 52–56.

Duelli, P., 2001. Lacewings in field crops. In: McEwen, P., New, T.R., Whittington, A.E. (Eds.), Lacewings in the Crop Environment. Cambridge University Press, United Kingdom, pp. 158–171.

Easterbrook, M.A., Fitzgerald, J.D., Solomon, M.G., 2006. Suppression of aphids on strawberry by augmentative releases of larvae of the lacewing Chrysoperla carnea (Stephens). Biocontrol Science and Technology 16 (9), 893–900.

Elanchezhyan, K., Baskaran, R.K., 2008. Evaluation of intercropping system based modules for the management of major insect pests of brinjal. Pest Management in Horticultural Ecosystems 14 (1), 67–73.

Fonseca, A.R., Carvalho, C.F., Souza, B., 2000. Functional response of Chrysoperla externa (Hagen) (Neuroptera: Chrysopidae) fed on Schizaphis graminum (Rondani) (Hemiptera: Aphididae). Anais da Sociedade Entomológica do Brasil 29 (2), 309–317.

Giles, D.K., Wunderlich, L.R., 1998. Electronically controlled delivery system for beneficial insect eggs in liquid suspensions. Transactions of the ASAE 41 (3), 839–847.

Gillani, W.A., Copland, M., Raja, S., 2009. Studies on the feeding preference of brown lacewing (Sympherobius fallax Navas) larvae for different stages of long-tailed mealy bug (Pseudococcus longispinus) (Targioni and Tozzetti). Pakistan Entomologist 31 (1), 1–4.

Greve, L., 1984. Chrysopid distribution in northern latitudes. In: Canard, M., Seamearia, Y., New, T.R. (Eds.), Biology of Chrysopidae. Junk Publishers, The Hague, pp. 180–186.

Guerra, C.L., da Costa, R.I.F., de Almeida, S.A., Soares, J.J., 1999. Food consumption of Chrysoperla externa (Hagen, 1861) (Neuroptera: Chrysopidae) on nymphs of cotton aphid Aphis gossypii (Glover, 1877) (Homoptera: Aphididae). In: Anais II Congresso Brasileiro de Algodão: O algodão no século XX, perspectivar para o século XXI, Ribeirão Preto, SP, Brasil, 5–10 Setembro 1999, pp. 247–249.

Halder, J., Rai, A.B., 2014. Suitability of different prey aphids on the growth, development and reproduction of Chrysoperla zastrowi sillemi (Esben-Petersen)(Chrysopidae: Neuroptera). In: Proceedings of the Zoological Society. Springer, India, pp. 1–7.

Hassanpour, M., Nouri-Ganbalani, G., Mohaghegh, J., Enkegaard, A., 2009. Functional response of different larval instars of the green lacewing, *Chrysoperla carnea* (Neuroptera: Chrysopidae), to the two-spotted spider mite, *Tetranychus urticae* (Acari: Tetranychidae). Journal of Food Agriculture & Environment 7 (2), 424–428.

Hassanpour, M., Mohaghegh, J., Iranipour, S., Nouri–Ganbalani, G., Enkegaard, A., 2011. Functional response of *Chrysoperla carnea* (Neuroptera: Chrysopidae) to *Helicoverpa armigera* (Lepidoptera: Noctuidae): effect of prey and predator stages. Insect Science 18 (2), 217–224.

Henry, C.S., Brooks, S.J., Johnson, J.B., Haruyama, N., Duelli, P., Mochizuki, A., 2015. A new East–Asian species in the *Chrysoperla carnea*-group of cryptic lacewing species (Neuroptera: Chrysopidae) based on distinct larval morphology and a unique courtship song. Zootaxa 3918 (2), 194–208.

Holling, C.S., 1959. Some characteristics of simple types of predation and parasitism. Canadian Entomologist 91, 385–398.

Huang, N., Enkegaard, A., 2010. Predation capacity and prey preference of *Chrysoperla carnea* on *Pieris brassicae*. BioControl 55 (3), 379–385.

Hydorn, S.B., 1971. Food Preferences of *Chrysopa rufilabris* (Burmeister) in North Central Florida (M.S. thesis). University of Florida, Tallahassee, Florida. 79pp.

Jessie, W.P., Giles, K.L., Rebek, E.J., Payton, M.E., Jessie, C.N., Mccornack, B.P., 2015. Preference and performance of *Hippodamia convergens* (Coleoptera: Coccinellidae) and *Chrysoperla carnea* (Neuroptera: Chrysopidae) on *Brevicoryne brassicae*, *Lipaphis erysimi*, and *Myzus persicae* (Hemiptera: Aphididae) from winter-adapted canola. Environmental Entomology. http://dx.doi.org/10.1093/ee/nvv068.

Kalyanasundaram, M., Dhandapani, N., Swamiappan, M., Sundara Babu, P.C., Jayaraj, S., 1994. A study on the management of some pests of groundnut (*Arachis hypogaea* L.) with biocontrol agents. Journal of Biocontrol 8 (1), 1–4.

Klingen, I., Johansen, N.S., Hofsvang, T., 1996. The predation of *Chrysoperla carnea* (Neurop., Chrysopidae) on eggs and larvae of *Mamestra brassicae* (Lep., Noctuidae). Journal of Applied Entomology 120 (6), 363–367.

Latham, D.R., Mills, N.J., 2009. Quantifying insect predation: a comparison of three methods for estimating daily per capita consumption of two aphidophagous predators. Environmental Entomology 38 (4), 1117–1125.

Löchte, C., Sengonca, C., 1995. Applying the eggs of *Chrysoperla carnea* (Stephens) using a spray technique for biocontrol of aphids in the field. Mitteilungen der Deutschen Gesellschaft für Allgemeine und Angewandte 10 (1–6), 243–246.

López-Arroyo, J.I., Tauber, C.A., Tauber, M.J., 2000. Storage of lacewing eggs: post-storage hatching and quality of subsequent larvae and adults. BioControl 18 (2), 165–171.

Mahabaleshwar, H., Kulkarni, K.A., 2000. Mass rearing of *Chrysoperla carnea* (Stephens) by using wood shavings as filler material. Karnataka Journal of Agricultural Sciences 13 (1), 184–186.

Mannan, V.D., Varma, G.C., Brar, K.S., 1997. Biology of *Chrysoperla carnea* (Stephens) on *Aphis gossypii* Glover and *Myzus persicae* (Sulzer). Journal of Insect Science 10 (2), 143–145.

Mansoor, M.M., Abbas, N., Shad, S.A., Pathan, A.K., Razaq, M., 2013. Increased fitness and realized heritability in emamectin benzoate-resistant *Chrysoperla carnea* (Neuroptera: Chrysopidae). Ecotoxicology 22 (8), 1232–1240.

Mathew, M.J., Venugopal, M.N., Saju, K.A., 1999. Predatory potential of green lacewing on cardamom aphid. Insect Environment 4 (4), 152–153.

McEwen, P.K., New, T.R., Whittington, A.E., 2007. Lacewings in the Crop Environment. Cambridge University Press.

McEwen, P.K., 1996. Viability of green lacewing *Chrysoperla carnea* Steph. (Neuropt., Chrysopidae) eggs stored in potential spray media, and subsequent effects on survival of first instar larvae. Journal of Applied Entomology 120 (3), 171–173.

Messina, F.J., Jones, T.A., Nielson, D.C., 1995. Host plant affects the interaction between the Russian wheat aphid and a generalist predator, *Chrysoperla carnea*. Journal of the Kansas Entomological Society 68 (3), 313–319.

Mirabzadeh, A., Sahragard, A., Azma, M., 1998. Functional response of *Chrysoperla carnea* Steph. larvae to *Pterochloroides persicae*, *Uroleucon (Urol.) cichorii* and *Aphis craccivora*. Seed and Plant 14 (3), 29–34.

Montoya-Alvarez, A.F., Ito, K., Nakahira, K., Arakawa, R., 2010. Functional response of *Chrysoperla nipponensis* and *C. carnea* (Neuroptera: Chrysopidae) to the cotton aphid *Aphis gossypii* Glover (Homoptera: Aphididae) under laboratory conditions. Applied Entomology and Zoology 45 (1), 201–206.

Morisawa, T., Giles, D.K., 1995. Effects of mechanical handling on green lacewing larvae (*Chrysoperla rufilabris*). Applied Engineering in Agriculture 11 (4), 605–607.

New, T.R., 1984. Chrysopidae: ecology on field crops. In: Canard, M., Semeria, Y., New, T.R. (Eds.), Biology of Chrysopidae. Dr. W. Junk Publishers, The Hague, The Netherlands, pp. 160–166.

Nordlund, D.A., Correa, J.A., 1995a. Description of green lacewing adult feeding and oviposition units and a sodium hypochlorite-based egg harvesting system. Southwestern Entomologist 20 (3), 293–301.

Nordlund, D.A., Correa, J.A., 1995b. Improvements in the production system for green lacewings: an adult feeding and oviposition unit and hot wire egg harvesting system. BioControl 5 (2), 179–188.

Nordlund, D.A., Vacek, D.C., Ferro, D.N., 1991. Predation of Colorado potato beetle (Coleoptera: Chrysomelidae) eggs and larvae by *Chrysoperla rufilabris* (Neuroptera: Chrysopidae) larvae in the laboratory and field cages. Journal of Entomological Science 26 (4), 443–449.

Obrycki, J.J., Hamid, M.N., Sajap, A.S., Lewis, L.C., 1989. Suitability of corn insect pests for development and survival of *Chrysoperla carnea* and *Chrysopa oculata* (Neuroptera: Chrysopidae). Environmental Entomology 18 (6), 1126–1130.

Osman, M.Z., Selman, B.J., 1993. Suitability of different aphid species to the predator *Chrysoperla carnea* Stephens (Neuroptera: Chrysopidae). University Journal of Zoology, Rajshahi University 12, 101–105.

Pappas, M.L., Broufas, G.D., Koveos, D.S., 2008. Effect of temperature on survival, development and reproduction of the predatory lacewing *Dichochrysa prasina* (Neuroptera: Chrysopidae) reared on *Ephestia kuehniella* eggs (Lepidoptera: Pyralidae). BioControl 45 (3), 396–403.

Praveen, P.M., Dhandapani, N., 2001. Eco-friendly management of major pests of okra (*Abelmoschus esculentus* (L.) Moench). Journal of Vegetable Crop Production 7 (2), 3–12.

Principi, M.M., Canard, M., 1984. Feeding habits. In: Canard, M., Semeria, Y., New, T.R. (Eds.), Biology of Chrysopidae. Dr. W. Junk Publishers, The Hague, pp. 76–92.

Quentin, U., Hommes, M., Basedow, T., 1995. Studies on the biocontrol of aphids (Hom., Aphididae) on lettuce in greenhouses. Journal of Applied Entomology 119 (3), 227–232.

Ribeiro, M.J., Carvalho, C.F., 1991. Aspectos biológicos de Chrysoperla externa (Hagen, 1861) (Neuroptera: Chrysopidae) em diferentes condições de acasalamento. Revista Brasileira de Entomologia 35 (2), 423–427.

Sabelis, M.W., 1992. Predatory arthropods. In: Crawley, M.J. (Ed.), Natural Enemies: The Population Biology of Predators, Parasites and Diseases. Blackwell. http://dx.doi.org/10.1002/9781444314076.ch10.

Saeed, R., Razaq, M., 2015. Effect of prey resource on the fitness of the predator, Chrysoperla carnea (Neuroptera: Chrysopidae). Pakistan Journal of Zoology 47 (1), 103–109.

Saljoqi, Ur-R., Khan, N.A., Ehsan-ul-Haq, J., Rehman, S., Saeed, Z.H., Nadeem, H.G.M., Zada, H., 2015. The impact of temperature on biological and life table parameters of *Chrysoperla carnea* Stephens (Neuroptera: Chrysopidae) fed on cabbage aphid, *Brevicoryne brassicae* (Linneaus). Journal of Entomology and Zoology Studies 3 (2), 238–242.

Sarailoo, M.H., Lakzaei, M., 2014. Effect of different diets on some biological parameters of *Chrysoperla carnea* (Neuroptera: Chrysopidae). Journal of Crop Protection 3 (4), 479–486.

Satpathy, S., Kumar, A., Shivalingaswamy, T.M., Rai, A.B., 2012. Effect of prey on predation, growth and biology of green lacewing (*Chrysoperla zastrowi sillemi*). Indian Journal of Agricultural Sciences 82 (1), 55.

Sattar, M., Abro, G.H., 2011. Mass rearing of Chrysoperla carnea (Stephens) (Neuroptera: Chrysopidae) adults for integrated pest management programmes. Pakistan Journal of Zoology 43 (3), 483–487.

Sharanabasava, H., Manjunatha, M., 2001. In: Halliday, R.B., Walter, D.E., Proctor, H.C., Norton, R.A., Colloff, M.J. (Eds.), Predator Prey Interactions between *Chrysoperla Carnea* Stephens (Neuroptera: Chrysopidae) and *Tetranychus neocaledonicus* (André) (Acari: Tetranychidae) on Okra, pp. 423–426.

Syed, A.N., Ashfaq, M., Khan, S., 2005. Comparison of development and predation of *Chrysoperla carnea* (Neuroptera: Chrysopidae) on different densities of two hosts (*Bemisia tabaci* and *Amrasca devastans*). Pakistan Entomologist 27 (1), 231–234.

Szentkiralyi, F., 2001a. Lacewings in fruit and nut crops. In: McEwen, P.K., New, T.R., Whittington, A.E. (Eds.), Lacewings in the Crop Environment. Cambridge University Press, Cambridge, pp. 172–238.

Szentkiralyi, F., 2001b. Lacewings in vegetables, forests, and other crops. In: McEwen, P.K., New, T.R., Whittington, A.E. (Eds.), Lacewings in the Crop Environment. Cambridge University Press, Cambridge, pp. 239–290.

Takalloozadeh, H.M., 2015. Effect of different prey species on the biological parameters of *Chrysoperla carnea* Stephens (Neuroptera: Chrysopidae) in laboratory conditions. Journal of Crop Protection 4 (1), 11–18.

Tauber, M.J., Helgesen, R.G., 1978. Implementing biological control systems in commercial greenhouse crops. Bulletin of the Entomological Society of America 24 (4), 424–426.

Tauber, M.J., Tauber, C.A., Gardescu, S., 1993. Prolonged storage of *Chrysoperla carnea* (Neuroptera: Chrysopidae). Environmental Entomology 22 (4), 843–848.

Tauber, M.J., Albuquerque, G.S., Tauber, C.A., 1997. Storage of nondiapausing *Chrysoperla externa* adults: influence on survival and reproduction. BioControl 10 (1), 69–72.

Tauber, M.J., Tauber, C.A., Daane, K.M., Hagen, K.S., 2000. Commercialization of predators: recent lessons from green lacewings (Neuroptera: Chrysopidae: Chrysoperla). American Entomologist 46, 26–37.

TNAU Agritech Portal, 2013. TNAU Agritech Portal. http://agritech.tnau.ac.in.

Wang, R., Nordlund, D.A., 1994. Use of Chrysoperla spp. (Neuroptera: Chrysopidae) in augmentative release programmes for control of arthropod pests. Biocontrol News and Information 15, 51–57.

Yadav, R., Pathak, P.H., 2010. Effect of temperature on the consumption capacity of *Chrysoperla carnea* (Stephens) (Neuroptera: Chrysopidae) reared on four aphid species. The Bioscan 5 (2), 271–274.

Yüksel, S., Gocmen, H., 1992. The effectiveness of *Chrysoperla carnea* (Stephens) (Neuroptera, Chrysopidae) as a predator on cotton aphid *Aphis gossypii* Glov. (Homoptera, Aphididae). In: Proceedings of the Second Turkish National Congress of Entomology, pp. 209–216.

Zheng, Y., Hagen, K.S., Daane, K.M., Mittler, T.E., 1993. Influence of larval dietary supply on the food consumption, food utilization efficiency, growth and development of the lacewing *Chrysoperla carnea*. Entomologia Experimentalis et Applicata 67 (1), 1–7.

Zhou, C.A., Zou, J.J., Peng, J.C., Ouyang, Z.Y., Hu, L.C., Yang, Z.L., Wang, X.B., 1991. Predation of major natural enemies on *Panonychus citri* and its comprehensive evaluation in citrus orchards in Hunan, China. Acta Phytophylacica Sinica 18 (3), 225–229.

Mite Predators

Yaghoub Fathipour, Bahador Maleknia

Department of Entomology, Faculty of Agriculture, Tarbiat Modares University, Tehran, Iran

1. INTRODUCTION

Currently, most crops in many countries are grown with the aid of various synthetic chemicals including pesticides and fertilizers. Conventional agricultural practices contribute to pollution affecting not only human health but also the environment. Only 1% of pesticides applied to crops actually reach the target pests, and 99% of these chemicals enter the environment, killing nontarget natural enemies more than mite pests. Some mites are serious pests in indoor and outdoor agricultural ecosystems. They also attack stored products in warehouses and make them unusable. Organic agriculture is a more ecofriendly option and reduces the negative effects of conventional agriculture. It protects both humans and the environment from these unknown risks by using sustainable methods such as crop rotation, natural soil enrichment, and pest natural enemies. In cases where the production of organic crops is impossible, ecofriendly pest management programs would be appropriate strategies for food safety.

Thousands of mite species can survive on land and in water. Although most species do not cause economic damage, some of these are pests. Plant mites feed by piercing plant cells, making leaves necrotic. The effects are more severe in greenhouses because the enclosure inadvertently separates them from natural enemies. Greenhouses create favorable environmental conditions for pests as well as plants. Several kinds of pest mites create profound damage in cropping systems. Spider mites (Tetranychidae) are the most serious pests in many greenhouses and filed plants. False spider mites (Tenuipalpidae) may become serious pests on indoor and outdoor crops. Several species of tarsonemid and eriophyid mites injure both vegetable crops and ornamental plants. Acarid mites damage the bulbs of flowers and stored roots of many crops as well as stored grains. Other pest mites have minor economic importance such as Siteroptidae and Penthaleidae (Zhang, 2003).

Considerable knowledge about the biology, ecology, and behavior of these pests is needed to develop mite management programs. We need detailed life histories of key mite pests under natural conditions, including the effects of climate and host plant type. A key gap in our knowledge is the availability of comprehensive management programs that consolidate all the available necessary techniques in a unified program to manage pest mite populations in such a manner that economic damage is avoided and adverse side effects are minimized. Integrated pest management (IPM) is a systemic approach in which interacting components (mainly control measures) act together to maximize the advantages (mainly enhancing the crop yield) and minimize the disadvantages (mainly causing risk to human and environment) of pest management programs (Fathipour and Sedaratian, 2013). Among mite biocontrol measures, predators receive priority. Several kinds of mite predator occur in nature, belonging to several families of insects and mites. The fundamentals of a biocontrol program are to precisely recognize a pest and to find efficient predators. In this chapter, we review important pest mites and mite predators, and then we suggested relatively comprehensive methods to evaluate the efficacy of mite predators to choose them for management programs. The future of biocontrol of mites and marketing of biocontrol agents is also discussed.

2. IMPORTANT PEST MITE FAMILIES

Several mite families cause economic damage in agricultural ecosystems. The most important pest mites are spider mites (family Tetranychidae), false spider mites (family Tenuipalpidae), thread footed mites (family Tarsonemidae), and gall mites (family Eriophyidae).

2.1 Tetranychidae

Mites of the family Tetranychidae (commonly known as spider mites) are important pests in agricultural and forestry ecosystems, and can be found on many field crops, fruit trees, vegetables, and ornamental plants. Many spider mites naturally inhabit ephemeral and patchily distributed resources such as weeds. A few species (e.g., *Tetranychus urticae* Koch, *Tetranychus turkestani* (Ugarov & Nikolskii), *Tetranychus cinnabarinus* (Boisduval), *Tetranychus pacificus* McGregor, *Tetranychus kanzawai* Kishida, *Tetranychus ludeni* Zacher, *Panonychus ulmi* (Koch), *Panonychus citri* (McGregor), *Oligonychus punicae* (Hirst), *Oligonychus coffeae* (Nietner), *Oligonychus afrasiaticus* (McGregor), *Eutetranychus orientalis* Klein, and *Eotetranychus lewisi* (McGregor)) are major pests in multiple locations and on multiple crops. The most notorious tetranychid mite is the globally-distributed two-spotted spider mite (*T. urticae*). It passes through egg, larva, protochrysalis, protonymph, deutochrysalis, deutonymph, and teliochrysalis stages before becoming an adult. It is a major pest of ornamental and fruit and vegetable crops grown worldwide, and is found on approximately 1200 plant species in 70 genera (Sedaratian et al., 2011). The European red mite, *P. ulmi*, a major fruit pest attacking apples, stone fruits, and pears, is an important apple pest throughout the world. Mites that feed on leaves cause injury to the tree by removing leaf tissue. The most serious injury occurs in early summer when trees produce fruit buds for the following season. Moderate to heavily-infested trees produce fewer and less vigorous fruit buds. The mites feeding on leaves also reduce the ability of leaves to manufacture enough food for desirable sizing of fruit. The citrus red mite, *P. citri*, is a key pest of citrus worldwide causing characteristic stippling mostly on upper leaf surfaces. When damage is serious, leaves may bleach or burn at the tips and leaf drop may occur. On fruit, citrus red mites cause a stippling and later silvering on the rind of mature oranges and lemons.

This stippling of the peel does not hurt the quality of the fruit inside. *Tetranychus cinnabarinus* or carmine spider mite is also known as the red spider mite, cotton spider mite, and carnation mite. It has widely spread in subtropical areas of the world and infested more than 100 host plants, including cotton crop (Sarwar et al., 2012). *Tetranychus ludeni* Zacher commonly known as dark-red spider mite, red-legged spider mite, or bean mite is distributed in the tropics and has been recorded in over 300 plant species. It is a serious pest of bean, eggplant, hibiscus, pumpkin, and other cucurbitaceous plants in warm areas. It is also quite common on greenhouse plants in temperate areas (Adango et al., 2006). *Tetranychus kanzawai* Kishida, or Kanzawa spider mite, is also known as tea red spider mite because its main host is tea and is a notorious pest mite in Japan (Shah et al., 2011).

Short life span, high fecundity, and the ability to resist many acaricides have made chemical control of mites substantially difficult. Harmful effects of residual toxicity of synthetic pesticides on humans and environment increase the need for safe, ecofriendly techniques (Sedaratian et al., 2009). Control of spider mites is obviously difficult and acaricides provide primary control of these mites. However, there are problems associated with the use of pesticides such as the development of resistance, their impact on the environment, high costs, and human safety. Therefore, there is growing interest in alternative control measures especially biocontrol, use of resistant cultivars, or an integration of these methods (Fathipour and Sedaratian, 2013). Host plant resistance can be a valuable component of an IPM system that is compatible with other control measures such as chemical control, and it makes beneficial natural enemies more effective. In many cases, even partially resistant cultivars are useful to enhance the efficacy of biocontrol agents (Khanamani et al., 2014).

2.2 Tenuipalpidae

The false spider mite, Tenuipalpidae, belongs to the superfamily Tetranychoidea of the order Prostigmata, as does the spider mite family Tetranychidae. They are not true spider mites because they do not produce silk webbings on plants. They are also known as flat mites because most species are dorsoventrally flattened. About 30 genera and over 600 species of Tenuipalpidae have been described in the most economically important genera: *Brevipalpus* and *Tenuipalpus* (Carrillo et al., 2012). Tenuipalpids feed on stems, fruits, flowers, and leaves (often on the lower surface), and some species form galls. They cause serious damage to many crops such as citrus, tea, grapes, fruit trees, ornamentals, orchids, grasses, and pineapple. Additionally, several tenuipalpids are confirmed vectors of virus or virus-like diseases of plants. *Brevipalpus phoenicis* (Geijskes) vectors coffee ringspot and passion fruit green spot viruses in Brazil and Costa Rica and *Brevipalpus californicus* (Banks) vectors orchid fleck virus in many parts of the world; tenuipalpids also are associated with citrus necrosis. Most false spider mites attack outdoor plants; *Brevipalpus* has approximately 300 described species, but only a few are agricultural pests, including *B. californicus*, *Brevipalpus obovatus* Donnadieu, *Brevipalpus lewisi* McGregor, and *B. phoenicis* (Geijskes) (Childers and Rodrigues, 2011). Another tenuipalpid, the red palm mite (*Raoiella indica* Hirst), is in the process of colonizing tropical and subtropical areas of the Western Hemisphere and has caused severe economic injury to palms (many types), bananas, and plantains (Hoy, 2011). *Cenopalpus irani* Dosse is also a false spider mite distributed in apple orchards located in Western Iran (Darbemamieh et al., 2009). Different control methods for tetranychoid mites are resistant varieties, chemical, and biocontrol. Current control for these pests involves use of acaricides on calendar-based programs, resulting in problems such as pest resistance and residue on the harvested and consumed products (Peña et al., 2012).

2.3 Tarsonemidae

The family Tarsonemidae contains about 40 genera and more than 500 described species. Important plant parasites are of genera *Polyphagotarsonemus*, *Hemitarsonemus*, *Phytonemus*, and *Steneotarsonemus*. Damage by species of *Steneotarsonemus* is typically associated with fungal infections. In addition to the plant-parasitic tarsonemids, others are pests of bees, forest trees, and mushroom culture. The life cycle of tarsonemid mites consists of egg, larva, and adult stages, but there is a quiescent nymph inside the larval cuticle. Eggs are often laid singly, but some species lay eggs in small clusters. In this family the cyclamen mite, *Phytonemus pallidus* (Banks) or strawberry mite cause the greatest damage to agriculture, closely followed by *Polyphagotarsonemus latus* (Banks). Both species have large host–plant ranges including many commercially important crops (Abou-Awad et al., 2014). The cyclamen mite is important as a pest of strawberries, cyclamen, gerbera, begonia, African violets, and ivy (Hoy, 2011). A few other pest species that occur occasionally in greenhouses are: *Hemitarsonemus tepidariorum* (Warburton), *Steneotarsonemus laticeps* (Halbert), *Xenotarsonemus belemnitoides* Weis-Fogh, *Tarsonemus confuses* Ewing, and *Tarsonemus bilobatus* Suski (Zhang, 2003). These mites can be controlled using a combination of cultural practices, releases of predators, and modification of spray practices to conserve endemic predators.

2.4 Eriophyidae

The mites of family Eriophyidae are elongate, annulate and phytophagous, which are referred to as blister mites, rust mites, bud mites, or gall mites, depending on the type of damage they cause (Amrine, 2013). Most are tiny vermiform animals and bear only two pairs of legs in all active stages (larva, nymph, and adult). Some species have two adult female forms, a normal feeding form (protogyne) and an overwintering or otherwise estivating form (deutogyne) (Michalska et al., 2010). Of the several 100 described eriophyids, relatively few are serious pests. Eriophyids are vectors of plant viruses.

Invasion of bud tissue by mites may cause partial or complete arrest of bud development or in abnormal increase in bud size. *Phytoptus avellanae* Nalepa, *Phyllocoptes gracillis* (Nalepa), and *Cecidophyopsis vermiformis* (Nalepa) are examples of eriophyid bud mites. Gall-forming eriophyids cause a type of tissue "pocketing" on affected leaves in which the pocket is virtually closed, and in which abnormal growth of leaf hairs provides a protective mat for the eriophyid inhabitants. *Eriophyes tilliae* (Pgst.), *Eriophyes emarginatae* Keifer, and *Vasates quadripedes* Shimer are the examples of the eriophyid gall mites. Unlike the gall mites which pocket leaf tissue to form a protected feeding site, blister mites actually invade the leaf mesophyll, causing an internal deformity of the leaf that is expressed externally as a discolored blister. The pear leaf blister mite, *Eriophyes pyri* (Pgst.), is an important member of this group. This species causes russeting and deformation of fruit. A similar blister mite occurs on apples. Feeding injury by rust mites is expressed as a bronzing, browning, or silvering of the affected leaf surface. Secondary effects include leaf edge rolling and folding. Rust mites may feed on either leaf surface but commonly are encountered on the undersides of leaves. The peach silver mite, *Aculus fockeui* (Nalepa and Trouessart), apple rust mite, *Aculus schlechtendali* (Nalepa), and pear mite, *E. pyri* (Nalepa) are rust mites (Amrine, 2013).

Mites of the family Rhyncaphytoptidae seem to have affinities to Eriophyidae, but they are aberrant in rostral structure and in habit. They evoke no injurious symptoms in their hosts and are suspected to be polyphagous.

3. CONTROL OF MITE PESTS

Spider mites, whiteflies, and thrips were acknowledged serious pests of fruits, field, and greenhouse crops, and ornamental plants after the 1950s, with the extensive use of organophosphate and synthetic organic pesticides and chemical fertilizers (Huffaker et al., 1970). It is common for a new pesticide to lose its efficacy within a few years after use against target pests therefore, in the absence of powerful and selective pesticides, crop protection experts may rely on the knowledge about pest biology and cultural practices to design multitactical control strategies.

3.1 Integrated Mite Management

Several methods have been applied to control mite pests in order to improve the quality and quantity of crop production in agricultural systems. Of these, synthetic acaricides are the main method for global mite control. This wide use of acaricides is of environmental concern and has developed acaricide resistance in mites. Furthermore, the deleterious effects of acaricides on nontarget beneficial organisms including mite predators are among the major causes of pest outbreaks. It is therefore necessary to reduce the usage of hazardous synthetic chemicals and follow IPM, which is a component of sustainable agriculture with the most robust ecological foundation (Kogan, 1998). It also serves as a model for the practical application of ecological theory and provides a paradigm for the development of other agricultural system components. It is becoming a practicable and acceptable global approach among acarologists and focuses on the history, concepts, and integration of available control measures into IPM programs. However, it advocates an integration of all possible or at least some of the known natural means of control with or without insecticides and acaricides to obtain the best pest management in terms of economics and maintenance of pest population below economic injury level. IPM is a decision support system for the selection and use of pest management tactics, singly or harmoniously coordinated into a management strategy, based on benefit/cost analyses that take into account the interests of and impacts on producers, society, and the environment (Fathipour and Sedaratian, 2013). It reflects a strategy of relying on multiple tactics of cultural, biological, genetic, and chemical controls (Figure 1). Mite predators clearly play an important role in IPM of mites, particularly in complex cropping systems where they may remove the need for any chemical intervention. Further information on IPM definitions and history can be found (e.g., Kogan, 1998; Ehler, 2006; Fathipour and Sedaratian, 2013).

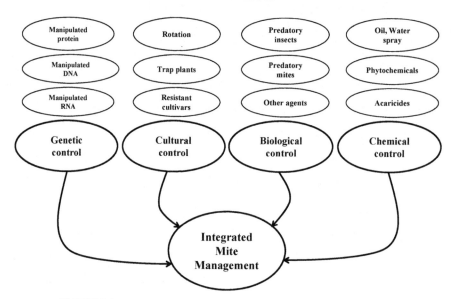

FIGURE 1   Different control strategies in integrated mite management.

3.2 Biological Control

Biological control, or biocontrol, is the use of an organism to reduce the population density of another organism and it is the core component of IPM that is growing in popularity, especially among organic growers. Two types of biocontrol, natural biocontrol and applied biocontrol, are often distinguished. Natural biocontrol is the reduction of native pest organisms by their indigenous natural enemies. In contrast, applied biocontrol is achieved through human efforts or intervention, and consists of three main approaches: conservation, inoculative (classical), and augmentative biocontrol.

Conservation biocontrol is manipulation of natural enemies and involves changing agricultural production practices to enhance the effectiveness of natural enemies in the cropping systems. Conservation is most often achieved by management of soil water and crop residue, modification of cropping patterns, manipulation of noncrop vegetation, and direct provision of resources to natural enemies. In general, these approaches are aimed at enhancing the density of resident natural enemies to increase their effectiveness in pest suppression (Dhawan and Peshin, 2009). Using ground cover as a conservation tactic was studied for the control of tetranychid mites in citrus orchards (Aguilar-Fenollosa et al., 2011). Using this tactic caused a reduction in pesticide use and an increase in populations of biocontrol agents.

In inoculative biological control, exotic natural enemies are imported and released in a new area for permanent establishment where the pest was accidentally introduced. It has been used frequently against introduced pests, which presumably arrived in a new area without their natural enemies. The aim is that the offspring of the released natural enemies build up populations that are large enough for suppression of pest populations over subsequent years. As it was the first type of biocontrol practiced widely, it is also called "classical" biological control. Economic assessment of the use of introduced natural enemies is not common, but has been made for several arthropod pests. Compared with insect pests, relatively little work has been done with mites as targets of classical biocontrol, with the exception of phytoseiid predators in orchard systems (McMurtry, 1983).

Augmentative biocontrol concerns the periodical release of natural enemies. In commercial augmentative biocontrol, natural enemies are mass-reared in biofactories for release in large numbers to obtain an immediate control of pests (van Lenteren, 2012). Augmentative releases may not always be successful because shipping, handling, and release methods (including rates and timing of releases) may not be appropriate. Therefore, it is necessary to carry out comprehensive experiments regarding the augmentative releases of natural enemies in new crops and different environmental circumstances to achieve reliable pest suppression.

4. MITE PREDATORS IN NATURAL SYSTEMS

Biocontrol uses parasitoids, predators, and pathogens to manage insect and mite pest populations. For mite pests, it started primarily by predators than other bioagents. Two types of predators, specialists, and generalists are often distinguished. A specialist feeds on a specific prey, while a generalist has a wider prey range.

Mite pests are often controlled by two main arthropod groups: predatory insects and predatory mites. Both predatory insects and mites are potential bioagents of mite pests with latter being more potent because of their specific diet. Predatory insects are usually generalists and most of them show switching behavior that reduces their biocontrol efficacy. Nevertheless, their larger body size and high food requirement tends to decrease pest populations rapidly. Mite pests are the main diet of the predatory mites and it seems that these predators would be a core component of the biocontrol programs in integrated mite management.

4.1 Predatory Insects

Predatory insects of the orders Coleoptera, Thysanoptera, Hemiptera, Diptera, and Neuroptera are potential biocontrol agents of mite pests. These predators are often less effective than phytoseiid predators because most of them are generalists and they do not remain in a crop field unless adequate food is present and will disperse to find food.

4.1.1 Coleoptera

The order Coleoptera contains more species than any other animal species, and makes up 40% of all insect species, which are diverse with keen hunting instincts. For example, beetles in the family Coccinellidae consume aphids, scale insects, thrips, mites, and other arthropods that damage crops. While most predatory beetles are generalists, a few species have more specific prey requirements.

4.1.1.1 COCCINELLIDAE

Coccinellids, ladybirds, ladybugs, and lady beetles are common names used for insects of this family. They are mostly potential predators of aphids, scale insects, mealybugs, and mite pests, although a few are phytophagous. Many coccinellids feed on mites, but these prey are not sufficiently nourishing to allow coccinellids complete their life span and reproduction process (Kianpour et al., 2011). Some generalist coccinellids feeding on mites include *Coleomegilla maculata* (DeGeer), *Hippodamia convergens* Guérin-Méneville, and *Harmonia axyridis* (Pallas), and the beetles of the genus *Stethorus*, such as *Stethorus punctum* (LeConte), *Stethorus gilvifrons* Mulsant, and *Stethorus punctillum* Weise are specialist spider mite predators. All known species of the genus *Stethorus* are predators of spider mites, and several species have been reported as biocontrol agents of tetranychid pests. *Stethorus* spp. are found in many geographic areas and climates, including temperate and tropical, and they are especially important in orchards. Adults actively fly and aggregate near mite colonies. An abundance of *Stethorus* on a crop usually indicates that spider mite populations are high. If *Stethorus* eggs, immature stages, and adults are on many leaves they should be able to clean up the pest population in a few days (Hoy, 2011).

Pesticides are highly detrimental to lady beetles and have both direct and indirect (through consumption of treated prey) effects. International Organization of Biological Control has developed standardized test methods for evaluating the effects of pesticides on natural enemies, and the modification of spray practices is expected to result in enhanced biocontrol of mite pests by *Stethorus* species (Hoy, 2011). Other methods for conserving lady beetles include providing refuges from toxic sprays, providing suitable sites for overwintering of diapausing beetles, using crop cultivars (containing waxes, trichomes, secondary plant compounds) that are suitable for lady beetle searching behavior, and providing food sprays containing sugars and proteins to crops (Obrycki and Kring, 1998).

4.1.1.2 STAPHYLINIDAE

Rove beetles are primarily distinguished by their short elytra that leave more than half of their abdomens exposed. Because many species have a pair of short projections at the end of the abdomen, people often mistake them for earwigs. The life cycle includes an egg, three larval stages, prepupa, and pupa, with pupation taking place in the soil. This pupation site may be an important consideration in modifying cultural practices to enhance the ability of these beetles to persist in crops (Frank et al., 1992).

Biocontrol practitioners have observed predation by various nonspecialist Staphylinidae on fly larvae and other invertebrates. *Oligota* (Holobus) feeds on all life stages of spider mites, although some instars of the beetles perform better on different stages of mites. *Oligota pygmaea* (Solier) is an important predator of spider mites, such as *O. coffeae*, *Oligonychus yothersi* (McGregor), *Bryobia arborea*, *T. urticae*, and *P. ulmi*. The rove beetles *Oligota oviformis* Casey and *Oligota flavicornis* (Boisduval and Lacordaire) are other predators of spider mites.

4.1.2 Thysanoptera

Thrips species feed on a large variety of plants and arthropods by puncturing them and sucking up the contents. A large number of thrips species are considered pests because they feed on economically important plants. Some species of thrips feed on other insects or mites and are considered beneficial, while some feed on fungal spores or pollen.

Three families (Phlaeothripidae, Aeolothripidae, and Thripidae) contain species that prey on spider mites or other small arthropods such as thrips, coccids, and whiteflies. Phlaeothripidae is the largest family of Thysanoptera in which at least 50% of the species are associated with dead plant tissues, feeding on fungi or the products of fungal decay; a large proportion of the remainder are leaf-feeding, with a smaller number breeding in flowers, and a few species predatory on other small arthropods including mites. Adults and larvae of many species in family Aeolothripidae appear to be facultative predators of other small arthropods, in that they feed on both floral tissues as well as on thrips and mites that live in flowers. Many thrips of the family Thripidae feed on both flowers and leaves, and some of them are the most common pest species and virus vectors on crops. Curiously, a few of these pest species, including *Thrips tabaci* Lindeman and *Frankliniella occidentalis* (Pergande), may at times act as beneficial thrips, in that they will also feed on spider mites that can be so damaging to plants. A few Thripidae, such as those of the genus *Scolothrips*, are obligate predators of mites. One of the most common predatory thrips are six-spotted thrips, *Scolothrips sexmaculatus* (Pergande), that prey on *Eotetranyhcus sexmaculatus* (Riley), *O. punicae*, *P. citri*, *P. ulmi*, and *T. urticae*. This species, commercially available from beneficial producers, is consistently the earliest, most abundant, and most significant natural control agent of spider mites on strawberries and peaches. Other *Scolothrips* species are *Scolothrips takahashii* Priesner, a common and important predator in fields of bean, and orchard of citrus, pear, and tea in some parts of the world, and *Scolothrips longicornis* Priesner, a native beneficial thrips in Mediterranean and Middle East areas. The latter species is common in crops suffering from infestations with spider mites such as bean, cucumber, and eggplant (Pakyari et al., 2011) and it can be used with combination of phytoseiid mites in management programs (Farazmand et al., 2015a).

4.1.3 Hemiptera

Insects in this order are extremely diverse in their size, shape, and color. This order is divided into three suborders; true bugs (Heteroptera); hoppers (Auchenorrhyncha); and aphids, scale insects, psyliid insects and mealybugs (Sternorrhyncha). Members of the suborder Heteroptera are known as "true bugs" that have become adapted to a broad range of habitats, terrestrial, aquatic, and semiaquatic. Plant-feeding, blood-feeding (bed bugs), and predatory bugs are found in this suborder. Predatory species of Heteroptera are generally regarded as beneficial insects, and they may be more common in row crops (such as corn, soybean, alfalfa, wheat, peanut, sorghum, and cotton) than in orchards. Heteropteran predators, with a relatively long developmental time, are limited as predators of rapidly reproducing spider mites, which may develop from egg to adult in less than a week (Hoy, 2011). The species of the genera *Orius* (Anthocoridae), and *Geocoris* (Lygaeidae) and some species of the family Miridae are predators of spider mites, but mostly they are phytophagous.

4.1.3.1 ANTHOCORIDAE

The genera of *Anthocoris*, *Orius*, and *Tetraphleps* are predators of small arthropods, such as mites, although some feed on plants. *Orius* species are generalist predators feeding on a variety of small arthropods, including spider mites, thrips, aphids, and insect eggs and larvae. *Orius insidiosis* (Say), *Orius albidipennis* (Reuter), and *Orius tristicolor* (White) are used in biocontrol programs as the predators of *T. urticae*, *P. ulmi*, and *P. citri* (Madadi et al., 2009; Rosenheim, 2005; Weintraub et al., 2011). *Orius vicinus* (Ribaut) is recorded as a predator of the mites *A. schlechtendali*, *P. ulmi*, and *T. urticae* (Wearing and Colhoun, 1999). *Orius* species are commercially produced, and efforts are being made to rear on artificial diet (Ferkovich and Shapiro, 2004).

4.1.3.2 MIRIDAE

Most mirids are important agricultural pests and some of them are predators. Mirids are known to be predators of aphids, whiteflies, and mites (Chouinard et al., 2006). The mirid, *Macrolophus caliginosus* (Wagner), is found in plants infested with the two-spotted spider mite or green peach aphid, *Myzus persicae*, and is a useful predator to use in greenhouses and orchards (Moayeri et al., 2006). It is used in European greenhouses to control multiple pests, including aphids, whiteflies, and mites (Put et al., 2012).

4.1.3.3 LYGAEIDAE

The bugs of the genus *Geocoris* in this family Lygaeidae are capable of feeding on a wide variety of prey such as insect eggs, aphids, mealybugs, spider mites, leafhoppers, bugs, and small caterpillars or beetle larvae (Rosenheim, 2005). Each nymph of the Lygaeidae family may consume up to 1600 spider mites during its immature stages, and 80 mites a day as an adult.

4.1.4 Diptera

A famous family in the order Diptera is Cecidomyiidae, known as gall midges or gall gnats. As the name implies, the larvae of most gall midges feed within plant tissue, creating abnormal plant growths called galls but a large number of species are natural enemies of other crop pests. Cecidomyiidae includes phytophagous, saprophagous, or zoophagous species; their larvae are predaceous, and some are even reported as parasitoids. The most common prey for this family are aphids and spider mites, followed by scale insects, whiteflies, and thrips (Gagne, 1995). *Feltiella acarisuga* (Vallot) is the most famous species of this family used in biocontrol programs. The eggs are deposited on plant tissue infested with tetranychid mites. Upon hatching, the larvae crawl to spider mites and feed on all life stages. *Feltiella acarisuga* is now commercially available as a biocontrol agent for use against tetranychid mites (Zhang, 2003).

4.1.5 Neuroptera

The three families Chrysopidae, Coniopterygidae, and Hemerobiidae are considered important natural enemies of a variety of pests. All three families contain generalist species that can feed on spider mites.

4.1.5.1 CHRYSOPIDAE

The members of the family Chrysopidae, green lacewings, are mainly predators of aphids, although they also feed on pollen, honey, and mites. *Chrysoperla* species are important worldwide as commercially available natural enemies that are released often in home gardens, row crops, orchards, and greenhouses. Although *Chrysoperla* species are important natural enemies of small arthropods, none is known to specialize on spider mites, so augmentative releases are rarely made for this purpose (Hoy, 2011). *Chrysoperla* is widely recommended for biocontrol programs. The larvae feed on aphids, scales, early caterpillars, spider mites, etc. that infest a variety of plants. Adults generally feed on nectar, pollen, or honeydew but a few of them are predatory (Corrales and Campos, 2004). The green lacewings, *Chrysoperla carnea* (Stephens), *Mallada boninensis* Okamoto, and *Mallada basalis* (Walker) are potential predators of some mite pests, such as *T. urticae*, *T. ludeni*, *T. kanzawai*, and *P. citri* (Vasanthakumar et al., 2012).

4.1.5.2 CONIOPTERYGIDAE

In the family Coniopterygidae, dusty-wings, adults and larvae feed on aphids, mealy-bugs, spider mites, scale insects, whiteflies, and other soft-bodied arthropods. Adults have been reported to feed on other substances, e.g., honeydew, scale secretions, and honey water. Coniopterygids in the genus *Conwentzia* may prey on spider mites, but dense webbing can be detrimental to survival of larvae. *Conwentzia psociformis* (Curtis) is most commonly found preying on spider mites in some agricultural systems (García-Marí and González-Zamora, 1999).

4.1.5.3 HEMEROBIIDAE

Hemerobiidae are known as "brown lacewings" and are distributed worldwide. *Hemerobius* species prey upon spider mites, although they are mainly aphidophagous (Solomon et al., 2000).

4.2 Predatory Mites

Predatory mites are increasingly used in biocontrol of pest mites, thrips, and nematodes. They are found in different families such as Aceosejidae, Ameroseiidae, Antennoseidae, Anystidae, Ascidae, Bdellidae, Cheyletidae, Cunaxidae, Eviphidae, Laelapidae, Macrochelidae, Parasitidae, Phytoseiidae, Stigmaeidae, Tydeidae, and Veigaiaidae.

4.2.1 Phytoseiidae

Phytoseiid mites are well-known natural enemies of phytophagous arthropods on cultivated and noncultivated plants. They have mostly been used to control pest spider mites, but some can control thrips in greenhouses and fields. Certain phytoseiids consume large numbers of prey and maintain plant-feeding mites at low densities. They have a high reproductive rate, a rapid developmental rate comparable to their prey, a female-biased sex ratio equivalent to their prey allowing them to respond numerically to increased prey density, and can easily be mass-reared (Hoy, 2011). Their life cycle comprises egg, larva, protonymph, deutonymph, and adult stages, and feeding by larvae in some species is facultative.

Phytoseiid mites feed on a variety of food and have developed different feeding habits. Some may feed on pollen, fungal spores, and plant exudates, and some can feed, develop, and reproduce on these plant materials. Others are obligatory predators that can survive on these alternative foods but fail to deposit eggs or fail to develop on such food. McMurtry et al. (2013) proposed the new classification of lifestyles of phytoseiid mites as:

Type I lifestyle—**Specialized mite predators**
Subtype I-a—Specialized predators of *Tetranychus* (Tetranychidae)
Subtype I-b—Specialized predators of web nest-producing mites (Tetranychidae)
Subtype I-c—Specialized predators of tydeoids (Tydeoidea)
Type II lifestyle—**Selective predators of tetranychid mites**
Type III lifestyle—**Generalist predators**
Subtype III-a—Generalist predators living on pubescent leaves
Subtype III-b—Generalist predators living on glabrous leaves
Subtype III-c—Generalist predators living in confined space on dicotyledonous plants

Subtype III-d—Generalist predators living in confined spaces on monocotyledonous plants
Subtype III-e—Generalist predators from soil/litter habitats
Type IV lifestyle—**Pollen feeding generalist predators**

4.2.1.1 *PHYTOSEIULUS PERSIMILIS* ATHIAS-HENRIOT

This is a type I lifestyle (subtype I-a) phytoseiid mite that feeds exclusively on *Tetranychus* species specifically *T. urticae* and whose survival depends on the presence and quality of its prey. *Phytoseiulus persimilis* has been used as a biocontrol agent in greenhouses since 1968. It is highly voracious, with adults able to eat 20–30 eggs or 3–5 females and 10–20 juveniles of *T. urticae* per day (Escudero and Ferragut, 2005).

4.2.1.2 *NEOSEIULUS CALIFORNICUS* (MCGREGOR)

This predatory mite has characteristics of both type II specialist predatory mites and type III generalist predatory mites. *Neoseiulus californicus* prefers tetranychid mites as food, but consumes other mites, thrips, and even pollen in absence of primary prey. It is often used to control *T. urticae* and other phytophagous mites on various crops in temperate and subtropical regions around the world.

4.2.1.3 *NEOSEIULUS CUCUMERIS* (OUDEMANS)

This is a type III lifestyle (subtype III-e) phytoseiid mite, which is an efficient biocontrol agent of small arthropod pests worldwide. It is also easy to rear and to establish on crops, where it commonly reaches high population densities. It is reared commercially and is successfully used for biocontrol of phytophagous mites and various thrips species. Its primary targets are thrips including western flower thrips, onion thrips, and plague thrips. *Neoseiulus cucumeris* is a generalist that also feeds on pollen and mites, particularly broad mite.

4.2.1.4 *AMBLYSEIUS SWIRSKII* (ATHIAS-HENRIOT)

This is a type III lifestyle (subtype III-b) phytoseiid mite that has been commercialized since 2005 and is presently being globally used as an augmentative bioagent against thrips and whiteflies in a range of greenhouse crops. It can develop and reproduce on a variety of other food sources including spider mites, eriophyid mites, broad mites, and pollen (Nguyen et al., 2014).

4.2.1.5 *NEOSEIULUS BARKERI* HUGHES

This is a type III lifestyle (subtype III-e) phytoseiid mite that feeds on storage mites, spider mites, thrips, broad mites, whitefly eggs, and also various pollens (Jafari et al., 2012). It has been reported in many parts of the world including Asia, Africa, Europe, America, and Australia. Furthermore, it has been used in augmentative biocontrol of onion thrips (*T. tabaci*) and *T. urticae* (Jafari et al., 2010).

4.2.1.6 *IPHISEIUS DEGENERANS* (BERLESE)

This is a type III lifestyle (subtype III-c) phytoseiid mite that is considered to be a generalist feeding both on live prey and pollen (Vantornhout et al., 2005). In 1994, it was first commercialized in Belgium to control thrips in greenhouse crops.

4.2.2 Laelapidae

The family Laelapidae is a morphologically and ecologically diverse group of mites, which currently comprises nine subfamilies and more than 1300 described species of parasites of vertebrates and invertebrates, as well as free-living forms and arthropod symbionts. Few species of this family are biocontrol agents (Kazemi, 2014).

Species of *Hypoaspis* are predators common in soil, mostly generalists, and have been employed in the biocontrol of soil-inhabiting mites and insects. Only two species, *Hypoaspis aculeifer* (Canestrini) and *Hypoaspis miles* (Berlese), have been used in greenhouses to control a number of soil-inhabiting pest insects and mites. Presently, these are commercially available for the control of mushroom flies and western flower thrips (Premachandra et al., 2003). Greenhouse experiments provide convincing evidence that the predatory mite, *H. aculeifer*, can suppress populations of bulb mites on intact bulbs in the soil. It is commercially used to control the greenhouse sciarids. It also feeds on thrips pupae, shore fly larvae, and acarid mites (*Rhizoglyphus* and *Tyrophagus*) and can be released to reduce densities of these pests in the soil or growing media (Jess and Schweizer, 2009). This evidence showed that there is great potential for soil-dwelling predatory mites in the control of arthropod pests in the soil and in compost.

4.2.3 Parasitidae

The cosmopolitan mites of family Parasitidae are free-living predators feeding on eggs and immature stages of other soil-inhabiting microarthropods and nematodes (Kazemi et al., 2013). *Parasitus fimetorum* (Berlese) is commonly found in association with bulb mites in soil and can suppress *Rhizoglyphus robini* Claparede on lily bulb propagation when peat is used as the growing medium. Several species of *Pergamasus* feed on *Tyrophagus* mites in the soil.

4.2.4 Ascidae

Ascid mites live in soil, leaf litter, subcortical situations, and are often associated with other animals. The mites are mostly predators, but can also feed on fungal mycelium; some are parasitic. Several species of Ascidae are promising bioagents of soil-inhabiting pests in greenhouses and field. *Lasioseius fimetorum* Karg feeds on larvae and pupae of *Tyrophagus putrescentiae* and *F. occidentalis* (Enkegaard and Brødsgaard, 2000). These mites are voracious predators of nematodes and other small invertebrates like mites in the soil and are potential biocontrol agents.

4.2.5 Stigmaeidae

Stigmaeids live on plants and in the soil feeding on tetranychids, tenuipalpids, and eriophyids. After phytoseiid mites, stigmaeids, especially the genera *Agistemus* and *Zetzellia* are important spider mite predators (Kheradmand et al., 2007). The predatory mite *Zetzellia mali* (Ewing) feeds on various phytophagous mites in apple orchards. The predatory mite, *Agistemus exsertus* Gonzalez is the most common stigmaeid on some fruit trees, vegetables, ornamentals, field crops, and wild plants and is a generalist predator feeding on phytophagous mites (tetranychids, tenuipalpids, and eriophyids), some scale insects, whiteflies, stored product moths, and pollen grains (Saber and Rasmy, 2010). Some studies showed that a combination of stigmaeids and phytoseiids have greater efficacy than either predator alone over a wide range of prey densities (Croft and MacRae, 1993).

4.2.6 Anystidae

Members of this family are large, reddish, and soft-bodied mites that are fast runners and appear to depend on chance encounters for finding their prey. *Anystis agilis* (Banks) and *Anystis baccarum* (L.) are abundant generalists of this family that feed on a wide variety of arthropods including mites.

4.2.7 Other Predators

Several other mite families are predators of mite pests such as Erythraeidae, Cunaxidae, Hypoaspidae, and Hemisarcoptidae. Erythraeidae is a cosmopolitan family in which their larvae are parasites of arthropods but deutonymphs and adults are free-living predators of small arthropods (Zhang, 2003). *Balaustium hernandezi* Von Heyden is a generalist predator of several arthropod pests, including the two-spotted spider mite. Cunaxids are generalist predators of small arthropods and nematodes. *Cunaxa setirostris* (Hermann) feeds on active stages of *T. ludeni* and has good potential as biocontrol agent of tetranychid pests (Arbabi and Singh, 2000).

5. FINDING EFFICIENT PREDATORS FOR INTEGRATED MITE MANAGEMENT

The use of high quality biocontrol agents for release is a fundamental step in the successful implementation of biocontrol programs. Before using a biocontrol agent in an IPM program, it is essential to know about the efficacy of the biocontrol agent (Fathipour et al., 2006). The main methods that are used in the evaluation of predators' effectiveness are life table parameters, foraging behavior, and multitrophic interactions (Figure 2). Studying the characteristics of predators helps to understand their influence on the population dynamics of prey and their influence on the structure of the mite communities in which they exist (Jervis and Kidd, 1996). It is thus a necessary prerequisite for the selection of predators for biocontrol programs and for the evaluation of their performance after release. Among life table parameters, the intrinsic rate of increase (r) is a key parameter in the prediction of population growth potential and has been widely used to evaluate efficiency of predators (Rahmani et al., 2009; Kianpour et al., 2011). In addition to the life table parameters, foraging behaviors including functional response, numerical response, mutual interference, preference, and switching are useful tools for evaluation of mite predators. Moreover, studies on multiple interactions like intraguild predation (predator–predator) (Farazmand et al., 2015b; Maleknia et al., 2015), olfactory response (plant–predator) (Maleknia et al., 2013), tritrophic interaction (plant–prey–predator) (Khanamani et al., 2014), cannibalism, and competition (predator–predator) (Maleknia et al., 2012; Farazmand et al., 2014) could be helpful in evaluation of mite predators (Figure 2).

After the selection of the efficient predators and period to their release in agricultural systems, it is necessary the pest density and release ratio of the predator(s) to be determined. Monitoring process involves taking multiple samples over time and evaluating predator and prey densities, which should be compared with those in the no-release plots over time. Estimating the crop damage level and crop yield in treated and nontreated (control) plots at the end of the biocontrol program would have a key role in selection of efficient predators for integrated mite management programs.

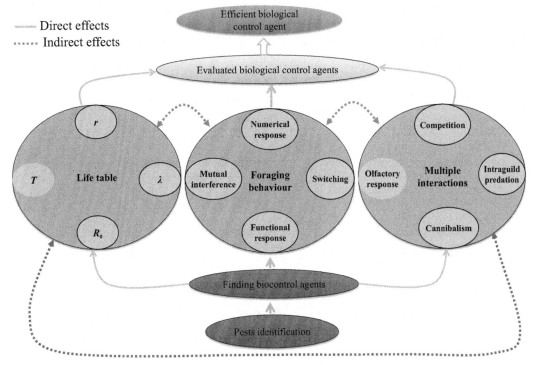

FIGURE 2 Important factors for evaluation of biological control agents.

5.1 Life Table Parameters

Life table parameters influence population growth rates of a mite in the current and next generations. In the female life table, the number of female progeny, survival rate of immature and female adult stages, daily fecundity, and sex ratio were used for the estimation of different life table parameters (Chi, 1988). The estimated parameters were the age-specific survival rate (l_x), age-specific fecundity (m_x), intrinsic rate of natural increase (r), net reproductive rate (R_0), mean generation time (T), finite rate of increase (λ), and gross reproductive rate (GRR). The equations and life table construction were adopted from Birch (1948) and Carey (1993). In the construction of a female age-specific life table, it is necessary to calculate age-specific survival rate (l_x) and age-specific fecundity (m_x) based on female individuals, where l_x shows the probability that a newborn individual will survive to age x, and m_x is the mean number of female eggs laid per female adult at age x (Table 1).

However, most economically important species of pest arthropod are bisexual and both sexes may cause economical loss. Moreover, there are many differences between male and female in longevity, survival rate, predation rate, and pesticide susceptibility. Also, neglecting the variable developmental rate and male population may cause errors in calculating demographic parameters (Chi, 1988). The susceptibility of an individual to both chemical and biological control agents may vary widely with sex and developmental stage; therefore,

TABLE 1 Equations Used in Calculation of Female and Two-Sex Life Table Parameters

Female life table	Two-sex life table
x (day) Age	x (day) Age
N_x Number of surviving individuals (only females in adult stage) entering the age x	N_x Number of surviving individuals (females and males in adult stage) entering the age x
M_x Daily mean number of eggs produced per female of age x	f_{xj} Age/stage-specific fecundity (daily number of eggs produced per female of age x)
$l_x = \dfrac{N_x}{N_0}$ The age-specific survivorship; N_0=number of individuals at the age $x=0$	s_{xj} ; $l_x = \sum_{j=1}^{k} s_{xj}$ Age/stage-specific survival rate (x=age and j=stage); k=number of stages
m_x Daily mean number of female eggs produced per female of age x	$m_x = \dfrac{\sum_{j=1}^{k} s_{xj} f_{xj}}{\sum_{j=1}^{k} s_{xj}}$ Age-specific fecundity (daily number of eggs produced per individual i.e., this number is divided by all individuals (males and females) of age x); k=number of stages
$R_0 = \sum_{x=\alpha}^{\beta} l_x m_x$ The net reproductive rate (female eggs per female)	$R_0 = \sum_{x=0}^{\omega} \sum_{j=1}^{k} s_{xj} f_{xj}$ The net reproductive rate (eggs per individual)
$\sum_{x=0}^{\omega} e^{-r(x+1)} l_x m_x = 1$ Intrinsic rate of increase (r) (number of females added to the population per female per day, i.e., the intrinsic birth rate (b) minus the intrinsic death rate (d)) (day^{-1})	$\sum_{x=0}^{\omega} e^{-r(x+1)} l_x m_x = 1$ Intrinsic rate of increase (r) (number of individuals added to the population per individual per day, i.e., the intrinsic birth rate (b) minus the intrinsic death rate (d)) (day^{-1})
$\lambda = e^r$ Finite rate of increase (the rate at which the population (only females) increases from one day to the next day) (day^{-1})	$\lambda = e^r$ Finite rate of increase (the rate at which the population (females and males) increases from one day to the next day) (day^{-1})
$GRR = \sum_{x=\alpha}^{\beta} m_x$ The gross reproductive rate (female eggs per female)	$GRR = \sum_{x=\alpha}^{\beta} m_x$ The gross reproductive rate (eggs per individual)
$T = \dfrac{\ln R_0}{r}$ Mean generation time (day)	$T = \dfrac{\ln R_0}{r}$ Mean generation time (day)

constructing an age/stage, two-sex life table to take the stage differentiation and the male population in to consideration would be useful (Chi, 1988). In a two-sex life table, the life history raw data of all individuals (males, females, and those dying before the adult stage) are taken into consideration. The age/stage-specific survival rate (s_{xj}) (where x = age in days and j = stage); the age/stage-specific fecundity (f_{xj}) (daily number of eggs produced per female of age x); the age-specific survival rate (l_x); the age-specific fecundity (m_x) (daily number of eggs produced per individual, i.e., this number is divided by all individuals (males and females) of age x); and the population growth parameters (the intrinsic rate of increase (r), the finite rate of increase (λ), the gross reproductive rate (GRR), the net reproductive rate (R_0), and the mean generation time (T) were calculated accordingly (Table 1). More details on two-sex life table parameters are found in the relevant references (e.g., Chi, 1988; Khanamani et al., 2013; Safuraie-Parizi et al., 2014; Goodarzi et al., 2015; Nikooei et al., 2015).

All life table parameters indicate the reproduction and population growth potential of mite predators, r is more important because other parameters are closely related to this parameter and it indicates the appropriate potential of the mite predator to increase its population; therefore, a higher intrinsic rate of increase shows higher potential of the predator for increase in its population. The intrinsic rate of increase can be affected by different factors, such as sublethal effects of acaricides (Hamedi et al., 2010, 2011), prey species (Escudero and Ferragut, 2005), host plant type (Khanamani et al., 2013, 2014), and temperature (Taghizadeh et al., 2008; Ganjisaffar et al., 2011) (Table 2). It seems that an efficient predator should have a higher intrinsic rate of increase than its prey, although other traits, like predation capacity, are determinant in assessing the effectiveness of predators.

5.2 Foraging Behavior

A basic understanding of the **foraging behavior of predators** and their prey aids predictions about predator function. Foraging behavior of predators, like functional response, numerical response, mutual interference, and switching are usually affected by a number of factors, such as temperature, host plant, prey stage, experimental condition, and pesticides (Pakyari et al., 2009; Xiao and Fadamiro, 2010; Sakaki and Sahragard, 2011; Jafari et al., 2012; Seiedy et al., 2012).

5.2.1 Functional Response

Functional response is the number of prey successfully attacked per predator as a function of prey density (Solomon, 1949). It describes the way a predator responds to the changing density of its prey. Holling (1959) considered three types of functional response. In type I there is a linear relation between prey density and the maximum number of prey killed, while in type II the proportion of prey consumed declines monotonically with prey density. Type III is described by a sigmoid relation in which the proportion of the prey consumed is positively density-dependent over some regions of prey density. A functional response models help to evaluate two vital parameters: handling time (i.e., the time needed to attack, consume, and digest the prey), and attack rate or searching efficiency (i.e., the rate at which a predator searches for finding its prey). Many predators that have been released as biocontrol agents have shown to exhibit a type II response on their prey (Xiao and Fadamiro, 2010).

TABLE 2   The Life Table Parameters of Some Important Mite Predators Derived from Selected Literature

Species	Prey	Temp (°C)	r (day^{-1})	R_0 (offspring)	T (day)	References
Agistemus cyprius	Panonychus citri	25	0.056	6.360	32.50	Goldarazena et al. (2004)
Agistemus industani	P. citri	25	0.103	18.70	37.80	Goldarazena et al. (2004)
Agistemus floridanus	P. citri	25	0.100	7.901	29.90	Goldarazena et al. (2004)
Agistemus olivi	Aceria mangiferae	30	0.263	92.77	17.18	Momen (2012)
A. olivi	Aculops fockeui	30	0.249	83.37	17.72	Momen (2012)
A. olivi	Aculops lycopersici	30	0.178	35.36	20.01	Momen (2012)
Amblyseius swirskii	Tetranychus urticae	25	0.147	7.520	13.59	Alipour et al. (2014)
A. swirskii	A. lycopersici	25	0.201	24.77	15.56	Park et al. (2011)
A. swirskii	Corn pollen	25	0.185	17.93	15.99	Park et al. (2011)
A. swirskii	Typha pollen	18	0.015	1.830	38.52	Lee and Gillespie (2011)
A. swirskii	Typha pollen	25	0.135	11.14	17.81	Lee and Gillespie (2011)
A. swirskii	Typha pollen	32	0.159	6.630	11.81	Lee and Gillespie (2011)
Euseius scutalis	P. citri	25	0.234	26.00	17.5	Kasap and Sekeroglu (2004)
Neoseiulus barkeri	T. urticae	15	0.036	5.510	47.18	Jafari et al. (2010)
N. barkeri	T. urticae	25	0.221	22.02	13.59	Jafari et al. (2010)
N. barkeri	T. urticae	35	0.247	14.26	10.74	Jafari et al. (2010)
Neoseiulus californicus	T. urticae	25	0.274	28.60	15.31	Gotoh et al. (2004)
N californicus	Thrips tabaci	25	0.041	1.950	18.62	Rahmani et al. (2009)
N. californicus	T. urticae	25	0.283	49.20	17.51	Escudero and Ferragut (2005)
N. californicus	Tetranychus turkestani	25	0.267	42.90	17.91	Escudero and Ferragut (2005)
N. californicus	Tetranychus evansi	25	0.840	3.301	14.51	Escudero and Ferragut (2005)
Phytoseiulus persimilis	T. turkestani	25	0.367	43.01	12.81	Escudero and Ferragut (2005)
P. persimilis	Tetranychus evansi	25	0.106	4.401	14.21	Escudero and Ferragut (2005)
P. persimilis	T. urticae	25	0.217	20.45	13.83	Alipour et al. (2014)
P. persimilis	T. urticae	25	0.238	33.30	14.72	Bahari et al. (2014)

TABLE 2   The Life Table Parameters of Some Important Mite Predators Derived from Selected Literature—cont'd

Species	Prey	Temp (°C)	r (day^{-1})	R_0 (offspring)	T (day)	References
Phytoseius plumifer	*T. urticae*	25	0.110	13.67	14.41	Rasmy et al. (2011)
P. plumifer	*T. urticae*	15	0.056	8.701	48.75	Kouhjani-Gorji et al. (2012)
P. plumifer	*T. urticae*	25	0.187	24.25	17.06	Kouhjani-Gorji et al. (2012)
P. plumifer	*T. urticae*	35	0.244	19.98	11.02	Kouhjani-Gorji et al. (2012)
P. plumifer	Corn pollen	27	0.112	4.40	12.99	Khodayari et al. (2013)
Scolothrips longicornis	*T. urticae*	15	0.056	18.43	52.19	Pakyari et al. (2011)
S. longicornis	*T. urticae*	26	0.201	31.09	17.04	Pakyari et al. (2011)
S. longicornis	*T. urticae*	35	0.310	17.65	9.271	Pakyari et al. (2011)
S. longicornis	*T. turkestani*	26	0.261	37.63	12.43	Gheibi and Hesami (2011)
S. longicornis	*Schizotetranychus smirnovi*	25	0.168	24.62	19.07	Heidarian et al. (2010)
Stethorus japonicus	*T. urticae*	20	0.093	328.8	62.25	Mori et al. (2005)
S. japonicus	*T. urticae*	25	0.156	270.4	51.10	Mori et al. (2005)
S. japonicus	*T. urticae*	30	0.241	405.7	39.87	Mori et al. (2005)
S. gilvifrons	*T. urticae*	25	0.066	72.22	64.85	Perumalsamy et al. (2010)
S. gilvifrons	*T. urticae*	15	0.035	10.08	64.33	Taghizadeh et al. (2008)
S. gilvifrons	*T. urticae*	25	0.133	31.33	25.99	Taghizadeh et al. (2008)
S. gilvifrons	*T. urticae*	35	0.240	59.27	17.01	Taghizadeh et al. (2008)
S. gilvifrons	*Tetranychus cinnabarinus*	25	0.159	59.36	25.63	Aksit et al. (2007)
Typhlodromus aripo	*Mononychellus tanajoa*	25	0.153	6.201	11.91	Gnanvossou et al. (2003)
Typhlodromus bagdasarjani	*T. urticae*	20	0.065	7.101	28.80	Ganjisaffar et al. (2011)
T. bagdasarjani	*T. urticae*	25	0.129	13.60	19.41	Ganjisaffar et al. (2011)
T. bagdasarjani	*T. urticae*	30	0.156	13.04	16.60	Ganjisaffar et al. (2011)
T. bagdasarjani	*T. urticae*	25	0.143	12.91	17.81	Moghadasi et al. (2014)
T. bagdasarjani	*T. urticae*	25	0.153	7.550	12.97	Khanamani et al. (2015)
Typhlodromus monihoti	*Oligonychus gossypii*	25	0.154	3.401	11.71	Gnanvossou et al. (2003)
Zetzellia mali	*T. urticae*	25	0.146	7.250	12.88	Khodayari et al. (2008)

In functional response experiments, biocontrol agents are limited in the experimental patches applying one prey species while in nature they freely attack varying prey range. In some functional response studies, predators are allowed to leave patches forever or for a short time period, but they return after a while and continue searching for prey.

The data on functional response are analyzed in two steps, at first to discriminate the type of the functional response, the positive or negative sign of the linear coefficient is determined by logistic regression of the proportion of prey consumed (N_a/N_t) as a function of prey density (N_t):

$$\frac{N_a}{N_t} = \frac{\exp\left(P_0 + P_1 N_t + P_2 N_t^2 + P_3 N_t^3\right)}{1 + \exp\left(P_0 + P_1 N_t + P_2 N_t^2 + P_3 N_t^3\right)}$$

where N_t is the initial prey density, N_a is the number of prey eaten, and N_a/N_t is the probability of being eaten. P_0, P_1, P_2, and P_3 are the intercept, linear, quadratic, and cubic coefficients, respectively, estimated using the method of maximum likelihood. If $P_1 < 0$, it describes a type II functional response because the proportion of prey consumed declines gradually as increasing the initial prey offered. If $P_1 > 0$ and $P_2 < 0$, the proportion of prey consumed is positively density-dependent and it shows a type III functional response (Juliano, 2001). After determining the type of functional response, the random predator equation is used to estimate the searching efficiency (a) and handling time (T_h, the amount of time that predator handles each prey individual) (Royama, 1971; Rogers, 1972) as the following model:

$$N_a = N_t \{1 - \exp\left[-aTP_t / (1 + aT_h N_t)\right]\}$$

where N_t is the initial number prey density, N_a is the number of prey eaten, T is the total time available for predators during the experiment (usually 24 h), a is the searching efficiency, T_h is the amount of time that predator handles each prey individual (handling time), and P_t is the number of predators (Rogers, 1972). More details on the equations and design of the functional response experiments are found in the relevant references (e.g., Juliano, 2001; Fathipour et al., 2006; Pakayari et al., 2009; Heidarian et al., 2012; Jafari et al., 2012; Farazmand et al., 2012, 2013).

Predators with higher searching efficiency (a) and lower handling time (T_h) are better agents. Predators exhibiting the type III response are efficient biocontrol agents; nevertheless, many of the predators that have been successfully released as biocontrol agents have shown the type II functional response on their prey (Xiao and Fadamiro, 2010). Functional response parameters of some mite predators are given in Table 3.

5.2.2 Numerical Response

The numerical response of a predator is a progressive change in the number of its progeny in relation to prey density (Solomon, 1949). It may be considered as a strategy of female predators to augment their offspring at different prey densities (Cédola et al., 2001). In numerical response experiments, adult female predators are exposed to different densities of the prey, and after the experimental time (usually 24 h) the number of prey consumed and the number

TABLE 3 The Functional Response Parameters of Some Important Mite Predators Derived from Selected Literature

Species	Prey	Prey stages	T (°C)	a	T_h	References
Amblyseius swirskii	*Tetranychus urticae*	Egg	23	$0.054\,h^{-1}$	$0.70\,h$	Maleknia et al. (2015)
A. swirskii	*T. urticae*	Egg	25	$0.11\,h^{-1}$	$0.83\,h$	Karimi et al. (2014)
A. swirskii	*T. urticae*	Larva	23	$0.064\,h^{-1}$	$0.99\,h$	Maleknia et al. (2015)
Chileseius camposi	*Panonychus ulmi*	Egg	20	$0.24\,day^{-1}$	$0.16\,day$	Sepúlveda and Carillo (2008)
Euseius finlandicus	*T. urticae*	Egg	24	$0.0375\,h^{-1}$	$0.69\,h$	Shirdel (2003)
E. finlandicus	*T. urticae*	Larva	24	$0.0456\,h^{-1}$	$0.45\,h$	Shirdel (2003)
Euseius hibisci	*T. urticae*	Larva	25	$0.460\,day^{-1}$	$0.133\,day$	Badii et al. (2004)
E. hibisci	*T. urticae*	Protonymph	25	$0.351\,day^{-1}$	$0.197\,day$	Badii et al. (2004)
Iphiseius degenerans	*Eutetranychus orientalis*	Nymph	25	$0.041\,h^{-1}$	$1.366\,h$	Fantinou et al. (2012)
I. degenerans	*T. urticae*	Nymph	25	$0.017\,h^{-1}$	$3.441\,h$	Fantinou et al. (2012)
Kampimodromus aberrans	*T. urticae*	Egg	25	$1.452\,day^{-1}$	$0.061\,day$	Kasap and Atlihan (2011)
K. aberrans	*T. urticae*	Larva	25	$1.415\,day^{-1}$	$0.038\,day$	Kasap and Atlihan (2011)
K. aberrans	*T. urticae*	Protonymph	25	$1.564\,day^{-1}$	$0.094\,day$	Kasap and Atlihan (2011)
K. aberrans	*T. urticae*	Deotonymph	25	$0.769\,day^{-1}$	$0.138\,day$	Kasap and Atlihan (2011)
Neoseiulus barkeri	*T. urticae*	Egg	23	$0.058\,h^{-1}$	$0.580\,h$	Maleknia et al. (2015)
N. barkeri	*T. urticae*	Larva	23	$0.048\,h^{-1}$	$0.870\,h$	Maleknia et al. (2015)
N. barkeri	*T. urticae*	Nymph	20	$0.036\,day^{-1}$	$0.921\,day$	Jafari et al. (2012)
N. barkeri	*T. urticae*	Nymph	25	$0.064\,day^{-1}$	$0.824\,day$	Jafari et al. (2012)
N. barkeri	*T. urticae*	Nymph	30	$0.073\,day^{-1}$	$0.597\,day$	Jafari et al. (2012)
Neoseiulus californicus	*T. urticae*	Egg	25	$1.30\,day^{-1}$	$0.0290\,day$	Gotoh et al. (2004)
N. californicus	*T. urticae*	Egg	25	$0.0936\,h^{-1}$	$1.64\,h$	Farazmand et al. (2012)
N. californicus	*T. urticae*	Larva	25	$0.0693\,h^{-1}$	$1.73\,h$	Farazmand et al. (2012)
Phytoseiulus persimilis	*T. urticae*	Egg	25	$0.130\,h^{-1}$	$0.49\,h$	Karimi et al. (2014)
P. persimilis	*T. urticae*	Egg	23	$0.120\,h^{-1}$	$2.51\,h$	Maleknia et al. (2015)
P. persimilis	*T. urticae*	Larva	23	$0.990\,h^{-1}$	$3.07\,h$	Maleknia et al. (2015)
P. persimilis	*T. urticae*	Larva	25	$0.114\,h^{-1}$	$3.15\,h$	Seiedy et al. (2012)

(Continued)

TABLE 3 The Functional Response Parameters of Some Important Mite Predators Derived from Selected Literature—cont'd

Species	Prey	Prey stages	T (°C)	a	T_h	References
Phytoseius plumifer	*T. urticae*	Nymph	15	$0.027\,h^{-1}$	$0.492\,h$	Kouhjani-Gorji et al. (2009)
P. plumifer	*T. urticae*	Nymph	20	$0.037\,h^{-1}$	$0.506\,h$	Kouhjani-Gorji et al. (2009)
P. plumifer	*T. urticae*	Nymph	25	$0.059\,h^{-1}$	$0.651\,h$	Kouhjani-Gorji et al. (2009)
Scolothrips longicornis	*T. urticae*	Egg	20	$0.032\,day^{-1}$	$2.42\,day$	Pakyari et al. (2009)
S. longicornis	*T. urticae*	Egg	26	$0.041\,day^{-1}$	$1.50\,day$	Pakyari et al. (2009)
S. longicornis	*T. urticae*	Egg	30	$0.048\,day^{-1}$	$1.41\,day$	Pakyari et al. (2009)
S. longicornis	*T. urticae*	Egg	25	$0.11\,h^{-1}$	$2.40\,h$	Farazmand et al. (2013)
S. longicornis	*T. urticae*	Nymph	25	$0.10\,h^{-1}$	$2.46\,h$	Farazmand et al. (2013)
Scolothrips takahashii	*Tetranychus viennensis*	Egg	20	$0.066\,h^{-1}$	$0.69\,h$	Ding-Xu et al. (2007)
S. takahashii	*T. viennensis*	Egg	25	$0.114\,h^{-1}$	$0.91\,h$	Ding-Xu et al. (2007)
S. takahashii	*T. viennensis*	Egg	30	$0.168\,h^{-1}$	$0.89\,h$	Ding-Xu et al. (2007)
Stethorus gilvifrons	*E. orientalis*	Egg	30	$0.046\,h^{-1}$	$0.082\,h$	Imani and Shishehbor (2011)
S. gilvifrons	*T. urticae*	Nymph	25	$0.046\,h^{-1}$	$0.113\,h$	Mehrkhou et al. (2006)
S. gilvifrons	*Tetranychus turkestani*	Larva	25	$0.0012\,day^{-1}$	$0.415\,day$	Sohrabi and Shishehbor (2007)
S. gilvifrons	*T. turkestani*	Nymph	25	$0.322\,h^{-1}$	$0.103\,h$	Karami and Shishehbor (2012)
Typhlodromus bagdasarjani	*T. urticae*	Egg	24	$0.0453\,h^{-1}$	$0.38\,h$	Shirdel (2003)
T. bagdasarjani	*T. urticae*	Egg	25	$0.0893\,h^{-1}$	$1.80\,h$	Farazmand et al. (2012)
T. bagdasarjani	*T. urticae*	Nymph	25	$0.0473\,h^{-1}$	$2.39\,h$	Farazmand et al. (2012)
T. bagdasarjani	*T. urticae*	Larva	24	$0.0466\,h^{-1}$	$0.30\,h$	Shirdel (2003)

of eggs laid are recorded. The efficiency of conversion of ingested food (ECI) (in number) into egg biomass (in number) at different prey densities could be determined is following Omkar and Pervez (2004):

$$\text{ECI} = \frac{\text{Number of eggs laid}}{\text{Number of prey consumed}} \times 100$$

The data on oviposition and ECI at different prey densities are fitted using regression analysis to determine the relationship between (1) oviposition and prey density and (2) ECI of female predator and prey density. The ECI reveals the relationship between conversion of prey biomass and prey density in which the ECI is more at low prey density and subsequently decrease at higher prey densities. This probably indicates that female predators at low prey density probably invest most of their energy in egg production and, in the process, invest less in maintenance and metabolic activities. The decreased ECI at higher prey densities possibly suggests that well-fed females laid large numbers of eggs, besides investing much in maintenance and metabolic costs (Omkar and Pervez, 2004). More details on functional response are found in relevant references (e.g., Jervis and Kidd, 1996; Omkar and Pervez, 2004; Sabaghi et al., 2011; Carrillo and Peña, 2012).

5.2.3 *Mutual Interference*

Aggregation by predators to prey patches is an integral component of models of prey–predator population dynamics. In discrete-time models, aggregation in space to prey patches stabilizes the prey–predator interaction by causing searching efficiency to decrease with increasing parasitoid density. Inverse density dependence in searching efficiency is known as mutual interference. Hassell and Varley (1969) noted an inverse relationship between the predator density and searching efficiency. At increased predator density, individual predators will waste an increasing proportion of their searching time to encounter other conspecifics rather than handling prey. Reduced predation, running away, and hiding are some outcomes of interrupting a predator during search or capture of prey. Mutual interference occurs commonly in the laboratory (Pakyari and Fathipour, 2009; Farazmand et al., 2012, 2013), but it has rarely been reported in field studies. Understanding this mutual interference is necessary to predict the success of biocontrol programs, as it assists with mass-rearing efforts and can facilitate the explanation of observed outcomes in the field.

In the mutual interference experiments, a constant density of the prey is offered to different densities of the predator, and after a given time period (usually 24h), the number of prey killed in each predator density is counted. The per capita searching efficiency of each predator is calculated by the following equation (Nicholson, 1933):

$$a = \left(\frac{1}{PT}\right) \ln \left(\frac{N_t}{(N_t - N_a)}\right)$$

Where N_t is the total number of available prey, N_a is the total number of prey killed, P is the number of predators and T is the duration of the experiment (set to 1.0 for 1 day).

Searching efficiency is plotted against predator density, both on a logarithmic scale. The points are fitted to a linear regression by the least square method, using the following inductive model given by Hassell and Varley (1969):

$$a = QP^{-m} \quad \text{or} \quad \log a = \log Q - m \log P$$

where a is the per capita searching efficiency of the predators, Q is the quest constant, m is the mutual interference constant, and P is the predator density.

The per capita predation rate is usually decreased with increasing predator density and this is the result of mutual interference through intraspecific competition. More details on mutual interference are found in the relevant references (e.g., Nicholson, 1933; Hassell and Varley, 1969; Jervis and Kidd, 1996; Fathipour et al., 2006; Pakyari and Fathipour, 2009; Farazmand et al., 2012).

5.2.4 Switching

In predator–prey systems, switching plays an important role to increase the persistence of predator–prey systems in the long term (van Baalen et al., 2001). Switching occurs when predators change to alternative prey and the density of preferred prey starts decreasing (Aggelis et al., 2005). Several factors can lead to a type III functional response; one of the most important of these factors is the presence of alternative prey which can lead to a type III response through switching behavior (Buckel and Stoner, 2000; Heidarian et al., 2012).

Switching can be tested by offering predators mixtures of different prey species (or prey stages) in single-patch experiments. The combined density of the two prey species should be kept constant, but the relative abundance of the two species should vary among treatments (e.g., 10:40, 15:35, 20:30, 25:25, 30:20, 35:15, and 10:40 $prey_1$:$prey_2$). After a given time, the number of prey consumed is counted. Murdoch (1969) described the null or no switch model for a system with two prey species as:

$$P_1 = cF_1 / (1 - F_1 + [cF_1])$$

where F_1 is the proportion of $prey_1$ in the environment, P_1 is the proportion of $prey_1$ among all prey consumed and c is a parameter described in the following equation:

$$N_1/N_2 = c (E_1/E_2)$$

where E_1/E_2 is the ratio of the number of two prey species eaten, N_1/N_2 is the ratio of the number of two prey species available in the environment, and c is constant; c measures preference and can be defined as the ratio of $prey_1$ to $prey_2$ eaten when the two prey species are equally abundant. When $c = 1$, there is no preference; when $c > 1$, there is a preference for $prey_1$; when $c < 1$, there is a preference for $prey_2$ (Murdoch, 1969). Finally, to test the hypothesis of switching, the observed ratio should be compared with the expected ratio based on the ratios given. When switching occurs, the observed ratio E_1/E_2 is higher than the expected ratio at high values of N_1/N_2. Murdoch (1969) reported that no switch behavior usually occurs except under particular circumstances. First, when the predator has a weak preference for one of two prey, it has a chance to switch toward abundant prey. Second, when the total prey density is constant and mortality in each prey is density-dependent. When switching occurred, the relationship between the numbers of prey eaten versus the numbers given will be sigmoid, resembling a type III functional response. However, there is a difference between sigmoid preference and a sigmoid functional response. Sigmoid preferences translate into a functional response with an accelerating part, but they do not necessarily lead to a satiating functional response (van Baalen et al., 2001). Switching behavior of predators has been obtained under laboratory conditions; however, switching in nature is rare and predators in general do not stabilize prey populations by this mechanism (Murdoch, 1969). Moreover, the effects of alternative food may help to design effective strategies for biocontrol involving a supply of alternative food or release predator in infested plants with several pests or just with a pest. More details on switching are found in the relevant references (e.g., Murdoch, 1969; Murdoch and Marks, 1973; van Baalen et al., 2001; Heidarian et al., 2012).

5.2.5 Predation Capacity

The use of standard parameters for describing and comparing potential of predators is important in biocontrol programs. The predation potential of predators cannot be properly described using intrinsic rate of increase (r) alone (Chi et al., 2011). To accurately evaluate the effect of predation in a predator-prey system, we need not only to assess the growth potential of the predator, but also its predation potential. The finite predation rate can be the standard parameter by linking the finite rate of increase (λ), stable age/stage distribution (a_{xj}), and age/stage-specific predation rate (c_{xj}) (Chi et al., 2011). The finite predation rate takes both the intrinsic rate of increase of predator and age/stage-specific predation rate into consideration and then can be used to describe and compare the predation potentials of natural enemies used in biocontrol programs. The finite predation rate can be used for comparison of predation capacity among different predators under the same condition or predation of a predator under different conditions (Yu et al., 2013).

The daily consumption of all individuals, including males, females, and those dying before the adult stage, was used to calculate the age/stage-specific consumption rate (c_{xj}). This is the mean number of prey consumed by individual predator age x and stage j. The age-specific predation rate (k_x) is the mean number of prey consumed by predator at age x (Chi and Yang, 2003) and is calculated as follows:

$$k_x = \frac{\sum_{j=1}^{\beta} s_{xj} c_{xj}}{\sum_{j=1}^{\beta} s_{xj}}$$

where s_{xj} is the age/stage-specific survival rate (x = age in days and j = stage) and β is the number of life stages. In addition, the age-specific net predation rate (q_x) is calculated as follows:

$$q_x = l_x k_x$$

The net predation rate (C_0) gives the mean number of prey consumed by an average individual predator during its entire life span, and is calculated as:

$$C_0 = \sum_{x=0}^{\infty} \sum_{j=1}^{\beta} s_{xj} c_{xj}$$

According to these, the total number of prey consumed by a cohort of size N is calculated as NC_0.

The transformation rate from prey population to predator offspring (Q_p) is the mean number of prey that a predator needs to consume to produce an offspring (Chi et al., 2011), and is calculated as:

$$Q_p = \frac{C_0}{R_0}$$

The stable predation rate (Ψ) is the total predation capacity of a stable population which total size is one and is calculated as follows:

$$\Psi = \sum_{x=0}^{\infty} \sum_{j=1}^{\beta} a_{xj} c_{xj}$$

where a_{xj} is the proportion of individuals belonging to age x and stage j in a stable age/stage distribution.

Because the predator population itself will increase at the finite rate λ, the total number of prey consumed will increase as $^\lambda\psi$. The finite predation rate ($\lambda\psi = \omega$) describes the predation potential of a predator population by combining its growth rate (λ), age/stage consumption rate (c_{xj}), and stable age/stage structure (a_{xj}), and this is calculated as follows:

$$\omega = \lambda\psi = \lambda\sum_{x=0}\sum_{j=1}^{\beta} a_{xj}c_{xj}$$

Considering this, the intrinsic predation rate is calculated as $\ln(\omega)$. In other words, the predation capacity will increase exponentially, $\left(\omega = e^{(\text{intrinsic predation rate})}\right)$ (Khanamani et al., 2015). More details on predation capacity are found in relevant references (e.g., Chi and Yang, 2003; Chi et al., 2011; Khanamani et al., 2015).

5.3 Multiple Interactions

Trophic interactions among primary producers (plants) and consumers (herbivorous and carnivorous mites) in the food web are the main fitness indicators of energy and nutrient cycle patterns through ecosystems. Plant quality can affect the fitness of herbivorous mites directly, as their food source, and indirectly, as foraging cues for their predators. The bottom-up effect of host plants can be extended to the third (first carnivores) and even fourth (second carnivores) trophic levels. Knowledge on bottom-up force of plants and top-down forces of predators is important for planning an IPM program, especially through decision making. Community modules, referring to pair-wise interactions inside a complex food web, mainly include plant↔herbivore, herbivore↔predator, plant↔predator, predator↔predator, and herbivore↔herbivore modules. Many other complicated relationships are predictable such as plant↔(predator↔competitor)↔herbivore↔competitor relationship.

In complex food web systems (Figure 3), predators can potentially interact negatively through competition, interference, and intraguild predation. All these interactions can affect directly or indirectly the success of biocontrol programs.

Intraguild predation (IGP) is a widespread phenomenon among arthropod food webs where more than one species feed on the same prey and therefore competitors feed on each other (Rosenheim et al., 1995). IGP may be unidirectional or bidirectional. In unidirectional, one of the two predator species preys on the other, and in bidirectional, both predators prey on each other. IGP occurs among wide varieties of natural enemies, such as phytoseiids and phytoseiids, and phytoseiids and thripids and can be affected by several factors, such as environmental conditions, host plant characteristics, mobility of prey, vulnerability of prey, feeding specificity, and presence of extraguild (EG) prey (Farazmand et al., 2015b). IGP is usually negligible when the EG prey density is high. More information on IGP can be found in the relevant references (e.g., Schausberger and Croft, 2000; Farazmand et al., 2015b).

Cannibalism, the consumption of conspecific individuals, is a common phenomenon that occurs in insects. In the case of prey scarcity, feeding on conspecifics can be a choice for some species as alternative food source in order to survive and eventually reproduce. This phenomenon is important in the biology and ecology of species and can affect the

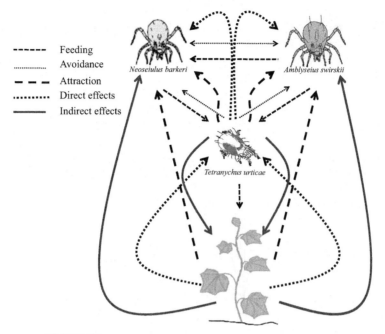

FIGURE 3   Tritrophic interactions in cucumber greenhouses.

distribution and structure of populations (Farazmand et al., 2014). The most obvious benefits of cannibalism may be (1) to obtain nutrients for maintaining metabolism during periods of low availability of alternative food sources and/or gaining energy for juvenile development and/or egg production, (2) to decrease potential intraspecific competition for food, (3) to decrease potential intraspecific competition for physical resources, such as egg laying sites and/or shelters, (4) to reach superiority in reproductive competition by reducing the fitness of other individuals of the same sex, and (5) to eliminate a potential mutual predator (Schausberger, 2003). Cannibalism could be affected by kin recognition, i.e., the ability to discriminate conspecific individuals on the basis of genetic relatedness (Schausberger, 2007).

Several invertebrates are passive dispersers, hence they do not have control over where they land. To minimize the risk of dispersing in an environment with low availability of profitable patches, dispersers may use volatile cues to infer the quality of patches in the surrounding environment. In the case of predators that feed on herbivores, they can use volatiles emitted by plants triggered by the feeding of prey (herbivore-induced plant volatiles, HIPVs), to locate plants with prey (Dicke et al., 1998). These HIPVs are beneficial to the predators as they increase the probability that predators find a patch with prey, but also to plants, recruiting bodyguards to defend them from herbivores. Indeed, plants with prey are expected to benefit more from the presence of predators than clean plants or plants with competitors of such predators. Hence, if signal production is costly enough, only plants with prey are expected to attract predators, and the interests of predators and plants are aligned. However, if the signal cost is low, all plants are expected to produce it, and predators will be equally attracted to plants with prey as well as to plants without prey and/or with competitors. Assuming that other resources are equally distributed among plants with or without prey or competitors, it is clearly disadvantageous for predators not to discriminate among these types of plants. Hence, the ability to discriminate among different types of plants is expected to evolve in predators, provided that signals are available to perform this discrimination. Therefore, whether predators are able to discriminate between clean plants and plants with prey and/or competitors depends on the cost of the plant signal and on the stage of the coevolutionary arms race between predators and plants. The ability to discriminate between signals may also depend on the individual experience of predators (Maleknia et al., 2013, 2014). Knowledge about olfactory response of predators could be helpful for landing, dispersal, finding prey, and survival of predator on the plants specially when predatory mite reared on a plant and released on a different plants systems.

6. THE FUTURE OF BIOLOGICAL CONTROL OF MITE PESTS

Agricultural crops produced in a specific area are often exported to different parts of the world; therefore, growers are obliged to consider the approved international standards regarding the production of safe and residue-free crops. Demand for organic and pesticide-free crops is rising worldwide in both developed and developing countries because of increasing the public awareness on adverse effects of pesticides, food-borne diseases, and food safety and security. To achieve this goal and to approach the safety in crop production, biocontrol practices should be taken into consideration worldwide. Based on the high demand for safe food, a future high demand for biocontrol agents, especially mite predators, is predicted (Figure 4). Therefore, more research and activity on different aspects of biocontrol as an ecofriendly practice is inevitable.

Commercial biocontrol is an industry that efficient natural enemies are mass-reared in biofactories for release in large numbers. This industry includes mass production, shipment, and release of natural enemies. Using natural enemies is often a commercial activity because of the need for mass production and large-scale releases of natural enemies to control noxious pests. Biocontrol of mite pests is an ecofriendly and economic strategy that decreases pest populations not only in the indoor and small-scale agricultural systems, such as greenhouses, but also in outdoor and large-scale systems, such as orchards, crop fields, and forests. Over recent decades the focus on crop production has moved from yields to food quality, safety, and, sustainability. To achieve this goal, the practical implementation of IPM strategies based on bioagents is indispensible. Measuring the level of development and implementation of biocontrol programs is not a simple matter because of: (1) a lack of relevant databases on production, usage, and outcomes of biocontrol agents, (2) a lack of responsible organizations for scouting the implementation of biocontrol programs around the world, (3) a nonexisting direct relationship between the number of natural enemies produced and total area treated by biocontrol agents, (4) indistinct borders between natural biocontrol (conservation) and applied biocontrol (release) worldwide, and (5) the number

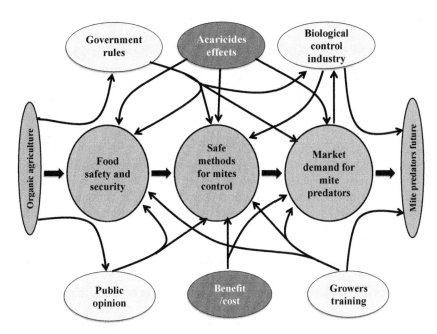

FIGURE 4   Factors affecting the future of mite predators.

of natural enemies produced not giving a good indication of implementation of biocontrol programs.

Most natural enemies are produced in low or medium numbers per week, but some of them are produced in noticeable numbers. A large number of predatory mites is required for efficient biocontrol. In the past decade, biocontrol has been used on a small scale, but demand from consumers and retailers for safe food has changed pest management systems with increased demand and supply of biocontrol agents. However, this high demand for biocontrol agents, especially for use in greenhouses, may cause a shock in market pull and this shock would lead to develop the biocontrol industry by introducing and using new and efficient agents. Pilkington et al. (2010) expressed that in greenhouse systems, *Encarsia formosa* (Gahan) accounts for 25% use in greenhouse, followed by the predatory phytoseiids *P. persimilis* and *N. cucumeris* (12%). This division of agents available to the greenhouse industry will, however, have changed significantly in recent years with the success of another phytoseiid, *A. swirskii* in Europe and North America.

Indisputably, increasing public awareness on safe food and organic agriculture in the world will increase the use of biocontrol agents, like mite predators. There are two important challenges in expanding the commercial biocontrol strategies: (1) problems with marketing include benefit to cost ratio and producers not being able to offer production and (2) problems with market demands that are affected by issues like public opinion and consumers demand. Governments and international organizations have to play an important role to enhance the marketing of biocontrol agents by framing stringent rules to improve the program, thus producing the safe field crops and fruits.

The benefit to cost ratio is an essential tool for appropriate control strategies. This ratio for chemical control will be higher than for biocontrol if we consider the indirect costs of chemical control, like environmental pollution, human health problems, harmful side effects, and risk of pest resistance. Biocontrol is safe, sustains food security, improves food quality, reduces pesticide use, is safe to human health (especially for farmers and farm workers), controls invasive alien species, protects biodiversity, and maintains ecosystem services (Cock et al., 2010). Although biocontrol is an ecofriendly strategy that leads to food safety and security, it is still an expensive and time-consuming method initially but in the long term it is more effective, self-sustaining, and self-propagating. In high-value crops such as strawberries, the costs of releasing mite predators can be justified, while in other crops it will only be justifiable if the pest mites become resistant to all of the registered acaricides. Therefore, this may be merely a control option to keep the mite population below the economic injury level. Furthermore, the indirect costs of pesticides can alter the benefit to cost ratio, even in low-value crops (Hoy, 2011).

Exaggeration of the results of biocontrol programs in order to improve the marketing of biocontrol agents may lead to an increase in public tendency towards these agents in a short time period but the inverse impact of such exaggeration would be indispensible in the long run, negatively affecting the marketing of produced natural enemies. In some cases, natural enemies are sold to growers not to control pests but to provide an incentive. Such releases are termed placebo and are costly because they impose unnecessary costs to growers and spoil their trust in biocontrol. It is more ethical to train growers to enhance their knowledge and use safe control measures besides biocontrol as it is the main concept of IPM (Hoy et al., 1991).

Commercial biocontrol agents are less acceptable, available, and applicable than chemical pesticides for controlling pests worldwide. Biocontrol is less acceptable because the costs of using synthetic pesticides are lower than biocontrol agents and people, especially in developing countries, do not consider the indirect costs of pesticides. Taking these costs into account, pesticides should be at least three times more expensive and, as a result, realistic pricing of pesticides would more often lead choosing biocontrol (Pimentel, 2005). Several countries (e.g., Denmark, Norway, and Sweden) are levying pesticide taxes, which are partly used to develop IPM programs and incentives for growers to encourage "low-pesticide" farming. These pesticide taxes also resulted in an immediate decrease in use of pesticides (Cannell, 2007). Pilkington et al. (2010) expressed that formulating documents such as the Sixth Environmental Action Programme, by the European Commission reinforces and validates consumer sentiment and assists pesticide manufacturers and suppliers in this regulation by suggesting actions aimed at substituting dangerous active ingredients with safer or nonchemical substances. The document also encourages the uptake of low input or pesticide-free agriculture for the industry as a whole and encourages the use of IPM strategies, limiting use of dangerous pesticides by creating monetary drivers such as taxes and duties to drive up the cost of these chemicals.

Enactment of strict laws on producing safe crops may increase the acceptability of biocontrol. Biocontrol is less available and the pesticide industry is not interested in this measure because (1) natural enemies cannot be patented, (2) they cannot be stored for long periods, (3) the act very specifically, (4) they can often not be combined with chemical control, and (5) they require extra training of sales personnel and growers (van Lenteren, 2012).

Furthermore, the limited number of retailers to pass bioagents from producers to growers is another reason for poor biocontrol of mite pests. Making the necessary rules to support and encourage the producers of natural enemies may solve the problems associated with less availability of biocontrol. Biocontrol is less applicable because selecting appropriate agents, using them in crop-producing systems, and evaluating the efficiency of the control measure are quite professional and complicated compared with pesticides. Holding the necessary workshops, technical sessions, and training courses for growers as well as preparing appropriate brochures and agent-specific guidelines may enhance the applicability of biocontrol. A guideline on the release ratio of predatory mites (Hoy, 2011) points out the need of approximately 1 predator female (healthy) to 5 or 10 spider mites (all stages) for a rapid and effective biocontrol. If our crop consists of 1000 plants with 30 leaves each, and each leaf averages 10 spider mites, we have approximately 300,000 spider mites and we need to release approximately 30,000 healthy predators with a 1:10 release ratio. In contrast, if we release only 0.1 spider mite per leaf, only 3000 predators would be needed, thereby reducing costs and damage to the crop.

Sometimes the appearance of agricultural crops, especially fruits, is more important than their quality and safety and this manner of crop marketing forces the growers to use pesticides as a rapid action measure. Although biocontrol is a safe and potentially effective method, it is not sufficient alone for protecting the crops from pest damage (Gerson and Weintraub, 2007). To solve this problem and facilitate the use of natural enemies, public opinion on agricultural crops should change to accept superficial damage by pests.

7. CONCLUSIONS

In both natural and agricultural ecosystems, many natural enemies, especially predators, could control herbivore populations. Prior to practicing biocontrol programs, a preliminary survey should be made on the life table parameters and foraging behavior of target pests and their natural enemies. This survey is a prerequisite for the selection of effective natural enemies for biocontrol programs as well as the evaluation of their performance after release in agricultural systems. Today, the commercial biocontrol industry is well organized and improved mass production, shipment, and release methods of biocontrol agents as well as adequate guidance for users. Although biocontrol guarantees production of safe food, it receives less attention because of its lower benefit to cost ratio, and its lower availability, acceptability, and applicability compared with chemical control. Although pesticides are generally considered profitable in agriculture, their use is not always beneficial to growers and consumers because of direct costs and indirect adverse effects. The costs of the indirect effects of pesticide application including threat to food safety, environmental pollution, human health problems, harmful side effects, and risk of pest resistance are borne by the public. The costs of indirect effects of pesticides should be estimated by responsible organizations and pesticide manufacturers should be forced to pay these costs. In order to gain the appropriate results, and considering the ethical and moral values, making strict rules in accordance with the relevant national and international organizations is compulsory. Enforcement of rules will minimize the use of pesticides in agricultural systems and encourage and expand the biocontrol strategies in IPM programs. There is no doubt that the biological control industry and IPM-based strategies in agricultural production are stronger than the past because of the food safety concerns of the public.

References

Abou-Awad, B.A., Hafez, S.M., Farahat, B.M., 2014. Bionomics and control of the broad mite *Polyphagotarsonemus latus* (Banks) (Acari: Tarsonemidae). Archives of Phytopathology and Plant Protection 47, 631–641.

Adango, E., Onzo, A., Hanna, R., Atachi, P., James, B., 2006. Comparative demography of the spider mite, *Tetranychus ludeni*, on two host plants in West Africa. Journal of Insect Science 6, 32–53.

Aggelis, G., Vayenas, D.V., Tsagou, V., Pavlou, S., 2005. Prey–predator dynamics with predator switching regulated by a catabolic repression control mode. Ecological Modelling 183, 451–462.

Aguilar-Fenollosa, E., Pascual-Ruiz, S., Hurtado, M.A., Jacas, J.A., 2011. Efficacy and economics of ground cover management as a conservation biological control strategy against (*Tetranychus urticae*) in clementine mandarin orchards. Crop Protection 30, 1328–1333.

Aksit, T., Cakmak, I., Ozer, G., 2007. Effect of temperature and photoperiod on development and fecundity of an acarophagous ladybird beetle, *Stethorus gilvifrons*. Phytoparasitica 35, 357–366.

Alipour, Z., Fathipour, Y., Farazmand, A., 2014. Demographic Parameters of *Phytoseiulus persimilis* and *Amblyseius swirskii* (Phytoseiidae) on Two-Spotted Spider Mite Fed on Resistant and Susceptible Rose Cultivars (M.Sc. thesis). Tarbiat Modares University, Tehran, p. 96.

Amrine, J.W., 2013. Eriophyid Mites. Avaiable from: http://puyallup.wsu.edu/plantclinic/resources/pdf/pls89eriophyidmites.pdf (accessed 10.03.15.).

Arbabi, M., Singh, J., 2000. Studies on biological aspects of predaceous mite *Cunaxa setirostris* on *Tetranychus ludeni* at laboratory condition in Varanasi, India. Journal of Agriculture and Rural Development 2, 13–23.

van Baalen, M., Krivan, V., van Rijn, P.C.J., Sabelis, M.W., 2001. Alternative food, switching predators, and the persistence of predator–prey systems. The American Naturalist 157, 512–524.

Badii, M., Hernández-Ortiz, E., Flores, A., Landeros, J., 2004. Prey stage preference and functional response of *Euseius hibisci* to *Tetranychus urticae* (Acari: Phytoseiidae, Tetranychidae). Experimental and Applied Acarology 34, 263–273.

Bahari, F., Fathipour, Y., Talebi, A.A., 2014. The Effect of Long Term Exposure to Antibiotic Resistance of Greenhouse Cucumber on Demographic Parameters of *Tetranychus urticae* and its Predator *Phytoseiulus persimilis* (M.Sc. thesis). Tarbiat Modares University, Tehran, p. 104.

Birch, L.C., 1948. The intrinsic rate of natural increase of an insect population. Journal of Animal Ecology 17, 15–26.

Buckel, J.A., Stoner, A.W., 2000. Functional response and switching behavior of young-of-the-year piscivorous bluefish. Journal of Experimental Marine Biology and Ecology 245, 25–41.

Cannell, E., 2007. European farmers plough ahead: pesticide use reduction. Pesticides News 78.

Carey, J.R., 1993. Applied Demography for Biologists with Special Emphasis on Insects. Oxford University Press, New York, p. 206.

Carrillo, D., Peña, J., 2012. Prey-stage preferences and functional and numerical responses of *Amblyseius largoensis* (Acari: Phytoseiidae) to *Raoiella indica* (Acari: Tenuipalpidae). Experimental and Applied Acarology 57, 361–372.

Carrillo, D., Frank, J.H., Rodrigues, J.C.V., Peña, J.E., 2012. A review of the natural enemies of the red palm mite, *Raoiella indica* (Acari: Tenuipalpidae). Experimental and Applied Acarology 57, 347–360.

Cédola, C., Sánchez, N., Liljesthröm, G., 2001. Effect of tomato leaf hairiness on functional and numerical response of *Neoseiulus californicus* (Acari: Phytoseiidae). Experimental and Applied Acarology 25, 819–831.

Chi, H., Yang, T.-C., 2003. Two-sex life table and predation rate of *Propylaea japonica* Thunberg (Coleoptera: Coccinellidae) fed on *Myzus persicae* (Sulzer) (Homoptera: Aphidae). Environmental Entomology 32, 327–333.

Chi, H., Huang, Y.-B., Allahyari, H., Yu, J.-Z., Mou, D.-F., Yang, T.-C., Farhadi, R., Gholizadeh, M., 2011. Finite predation rate: a novel parameter for the quantitative measurement of predation potential of predator at population level. Nature Proceedings hdl:10101/npre.2011.6651.1.

Chi, H., 1988. Life-table analysis incorporating both sexes and variable development rates among individuals. Environmental Entomology 17, 26–34.

Childers, C.C., Rodrigues, J.C., 2011. An overview of *Brevipalpus* mites (Acari: Tenuipalpidae) and the plant viruses they transmit. Zoosymposia 6, 168–180.

Chouinard, G., Bellerose, S., Brodeur, C., Morin, Y., 2006. Effectiveness of (*Hyaliodes vitripennis*) (Say) (Heteroptera: Miridae) predation in apple orchards. Crop Protection 25, 705–711.

Cock, M.J., van Lenteren, J.C., Brodeur, J., Barratt, B.I., Bigler, F., Bolckmans, K., et al., 2010. Do new Access and Benefit Sharing procedures under the Convention on Biological Diversity threaten the future of biological control? BioControl 55, 199–218.

Corrales, N., Campos, M., 2004. Populations, longevity, mortality and fecundity of *Chrysoperla carnea* (Neuroptera, Chrysopidae) from olive-orchards with different agricultural management systems. Chemosphere 57, 1613–1619.

Croft, B., MacRae, I., 1993. Biological control of apple mites: impact of *Zetzellia mali* (Acari: Stigmaeidae) on *Typhlodromus pyri* and *Metaseiulus occidentalis* (Acari: Phytoseiidae). Environmental Entomology 22, 865–873.

Darbemamieh, M., Kamali, K., Fathipour, Y., 2009. Bionomics of *Cenopalpus irani*, *Bryobia rubrioculus* and their egg predator *Zetzellia mali* (Acari: Tenuipalpidae, Tetranychidae, Stigmaeidae) in natural conditions. Munis Entomology & Zoology 4, 378–391.

Dhawan, A.K., Peshin, R., 2009. Integrated pest management: concept, opportunities and challenges. Experimental and Applied Acarology 38, 51–81.

Dicke, M., Takabayashi, J., Posthumus, M.A., Schutte, C., Krips, O.E., 1998. Plant-phytoseiid interactions mediated by herbivore-induced plant volatiles: variation in production of cues and in responses of predatory mites. Experimental and Applied Acarology 22, 311–333.

Ding-Xu, L., Juan, T., Zuo-Rui, S., 2007. Functional response of the predator *Scolothrips takahashii* to hawthorn spider mite, *Tetranychus viennensis*: effect of age and temperature. BioControl 52, 41–61.

Ehler, L.E., 2006. Integrated pest management (IPM): definition, historical development and implementation, and the other IPM. Pest Management Science 62, 787–789.

Enkegaard, A., Brødsgaard, H., 2000. *Lasioseius fimetorum*: a soil-dwelling predator of glasshouse pests? BioControl 45, 285–293.

Escudero, L.A., Ferragut, F., 2005. Life-history of predatory mites *Neoseiulus californicus* and *Phytoseiulus persimilis* (Acari: Phytoseiidae) on four spider mite species as prey, with special reference to *Tetranychus evansi* (Acari: Tetranychidae). Biological Control 32, 378–384.

Fantinou, A.A., Baxevani, A., Drizou, F., Labropoulos, P., Perdikis, D., Papadoulis, G., 2012. Consumption rate, functional response and preference of the predaceous mite *Iphiseius degenerans* to *Tetranychus urticae* and *Eutetranychus orientalis*. Experimental and Applied Acarology 58, 133–144.

Farazmand, A., Fathipour, Y., Kamali, K., 2012. Functional response and mutual interference of *Neoseiulus californicus* and *Typhlodromus bagdasarjani* (Acari: Phytoseiidae) on *Tetranychus urticae* (Acari: Tetranychidae). International Journal of Acarology 38, 369–378.

Farazmand, A., Fathipour, Y., Kamali, K., 2013. Functional response and mutual interference of *Scolothrips longicornis* (Thysanoptera: Thripidae) on two spotted spider mite. IOBC/WPRS Bulletin 93, 30–44.

Farazmand, A., Fathipour, Y., Kamali, K., 2014. Cannibalism in *Scolothrips longicornis* (Thysanoptera: Thripidae), *Neoseiulus californicus* and *Typhlodromus bagdasarjani* (Acari: Phytoseiidae). Systematic and Applied Acarology 19, 471–480.

Farazmand, A., Fathipour, Y., Kamali, K., 2015a. Control of the spider mite *Tetranychus urticae* using phytoseiid and thrips predators under microcosm conditions: single-predator versus combined-predators release. Systematic and Applied Acarology 20, 162–170.

Farazmand, A., Fathipour, Y., Kamali, K., 2015b. Intraguild predation among *Scolothrips longicornis* (Thysanoptera: Thripidae), *Neoseiulus californicus* and *Typhlodromus bagdasarjani* (Acari: Phytoseiidae) under laboratory conditions. Insect Science 22, 263–272.

Fathipour, Y., Sedaratian, A., 2013. Integrated management of *Helicoverpa armigera* in soybean cropping systems. In: El-Shemy, H. (Ed.), Soybean-Pest Rresistance. InTech, pp. 231–280.

Fathipour, Y., Hosseini, A., Talebi, A.A., Moharramipour, S., 2006. Functional response and mutual interference of *Diaeretiella rapae* (Hymenoptera: Aphidiidae) on *Brevicoryne brassicae* (Homoptera: Aphididae). Entomologica Fennica 17, 90–97.

Ferkovich, S., Shapiro, J., 2004. Comparison of prey-derived and non-insect supplements on egg-laying of (*Orius insidiosus*) maintained on artificial diet as adults. Biological Control 31, 57–64.

Frank, J., Bennett, F., Cromroy, H., 1992. Distribution and prey records for *Oligota minuta* (Coleoptera: Staphylinidae), a predator of mites. Florida Entomologist 75, 376–380.

Gagne, R.J., 1995. Revision of tetranychid (Acarina) mite predators of the genus *Feltiella* (Diptera: Cecidomyiidae). Annals of the Entomological Society of America 88, 16–30.

Ganjisaffar, F., Fathipour, Y., Kamali, K., 2011. Temperature-dependent development and life table parameters of *Typhlodromus bagdasarjani* (Phytoseiidae) fed on two-spotted spider mite. Experimental and Applied Acarology 55, 259–272.

García-Marí, F., González-Zamora, J.E., 1999. Biological control of *Tetranychus urticae* (Acari: Tetranychidae) with naturally occurring predators in strawberry plantings in Valencia, Spain. Experimental and Applied Acarology 23, 487–495.

Gerson, U., Weintraub, P.G., 2007. Mites for the control of pests in protected cultivation. Pest Management Science 63, 658–676.

Gheibi, M., Hesami, S., 2011. Life table and reproductive table parameters of *Scolothrips longicornis* (Thysanoptera: Thripidae) as a predator of two-spotted spider mite, *Tetranychus turkestani* (Acari: Tetranychidae). World Academy of Science, Engineering and Technology 60, 262–264.

Gnanvossou, D., Yaninek, J.S., Hanna, R., Dicke, M., 2003. Effects of prey mite species on life history of the phytoseiid predators *Typhlodromalus manihoti* and *Typhlodromalus aripo*. Experimental and Applied Acarology 30, 265–278.

Goldarazena, A., Aguilar, H., Kutuk, H., Childers, C.C., 2004. Biology of three species of *Agistemus* (Acari: Stigmaeidae): life table parameters using eggs of *Panonychus citri* or pollen of *Malephora crocea* as food. Experimental and Applied Acarology 32, 281–291.

Goodarzi, M., Fathipour, Y., Talebi, A., 2015. Antibiotic resistance of canola cultivars affecting demography of *Spodoptera exigua* (Lepidoptera: Noctuidae). Journal of Agricultural Science and Technology 17, 23–33.

Gotoh, T., Yamaguchi, K., Mori, K., 2004. Effect of temperature on life history of the predatory mite *Amblyseius (Neoseiulus) californicus* (Acari: Phytoseiidae). Experimental and Applied Acarology 32, 15–30.

Hamedi, N., Fathipour, Y., Saber, M., 2010. Sublethal effects of fenpyroximate on life table parameters of the predatory mite *Phytoseius plumifer*. BioControl 55, 271–278.

Hamedi, N., Fathipour, Y., Saber, M., 2011. Sublethal effects of abamectin on the biological performance of the predatory mite, *Phytoseius plumifer* (Acari: Phytoseiidae). Experimental and Applied Acarology 53, 29–40.

Hassell, M.P., Varley, G.C., 1969. New inductive population model for insect parasites and its bearing on biological control. Nature 223, 1133–1137.

Heidarian, M., Fathipour, Y., Kamali, K., 2010. Biology and Efficiency of *Scolothrips longicornis* (Thysanoptera: Thripidae) Preying on Almond Spider Mite Collected from Shahrekord Almond Orchards (M.Sc. thesis). Islamic Azad University, p. 65.

Heidarian, M., Fathipour, Y., Kamali, K., 2012. Functional response, switching, and prey-stage preference of *Scolothrips longicornis* (Thysanoptera: Thripidae) on *Schizotetranychus smirnovi* (Acari: Tetranychidae). Journal of Asia-Pacific Entomology 15, 89–93.

Holling, C.S., 1959. Some characteristics of simple types of predation and parasitism. The Canadian Entomologist 91, 385–398.

Hoy, M., Nowierski, R., Johnson, M., Flexner, J., 1991. Issues and ethics in commercial releases of arthropod natural enemies. American Entomologist 37, 74–75.

Hoy, M.A., 2011. Agricultural Acarology: Introduction to Integrated Mite Management. CRC Press, p. 407.

Huffaker, C.B., van de Vrie, M., McMurtry, J., 1970. Ecology of tetranychid mites and their natural enemies: a review. II. Tetranychid populations and their possible control by predators: an evaluation. Hilgardia 40, 391–458.

Imani, Z., Shishehbor, P., 2011. Functional response of *Stethorus gilvifrons* (Col.: Coccinellidae) to different densities of *Eutetranychus orientalis* (Acari: Tetranychidae) in laboratory. Journal of Entomological Society of Iran 31, 29–40.

Jafari, S., Fathipour, Y., Faraji, F., Bagheri, M., 2010. Demographic response to constant temperatures in *Neoseiulus barkeri* (Phytoseiidae) fed on *Tetranychus urticae* (Tetranychidae). Systematic and Applied Acarology 15, 83–99.

Jafari, S., Fathipour, Y., Faraji, F., 2012. The influence of temperature on the functional response and prey consumption of *Neoseiulus barkeri* (Phytoseiidae) on two-spotted spider mite. Journal of Entomological Society of Iran 31, 39–52.

Jervis, M., Kidd, N., 1996. Insect Natural Enemies. Practical Approaches to Their Study and Evaluation. Chapman & Hall, p. 491.

Jess, S., Schweizer, H., 2009. Biological control of *Lycoriella ingenua* (Diptera: Sciaridae) in commercial mushroom (*Agaricus bisporus*) cultivation: a comparison between *Hypoaspis miles* and *Steinernema feltiae*. Pest Management Science 65, 1195–1200.

Juliano, S., 2001. Nonlinear curve fitting: predation and functional response curves. In: Scheiner, S.M., Gurevitch, J. (Eds.), Design and Analysis of Ecological Experiments. Oxford University Press, NewYork, pp. 178–196.

Karami-Jamour, T., Shishehbor, P., 2012. Host plant effects on the functional response of *Stethorus gilvifrons* to strawberry spider mites. Biocontrol Science and Technology 22, 101–110.

Karimi, M., Fathipour, Y., Talebi, A.A., 2014. Life Time Functional Response and Predation Rate of *Phytoseiulus persimilis* and *Amblyseius swirskii* on Two-Spotted Spider Mite (M.Sc. thesis). Tarbiat Modares University, Tehran, p. 104.

Kasap, I., Atlihan, R., 2011. Consumption rate and functional response of the predaceous mite *Kampimodromus aberrans* to two-spotted spider mite *Tetranychus urticae* in the laboratory. Experimental and Applied Acarology 53, 253–261.

Kasap, I., Sekeroglu, E., 2004. Life history of *Euseius scutalis* feeding on citrus red mite *Panonychus citri* at various temperatures. BioControl 49, 645–654.

Kazemi, S., Arjomandi, E., Ahangaran, Y., 2013. A review of Iranian Parasitidae (Acari: Mesostigmata). Persian Journal of Acarology 2, 169–180.

Kazemi, S., 2014. A new mite species of *Pseudoparasitus* Oudemans (Acari: Mesostigmata: Laelapidae), and a key to known Iranian species of the genus. Persian Journal of Acarology 3, 41–50.

Khanamani, M., Fathipour, Y., Hajiqanbar, H., 2013. Population growth response of *Tetranychus urticae* to eggplant quality: application of female age-specific and age-stage, two-sex life tables. International Journal of Acarology 39, 638–648.

Khanamani, M., Fathipour, Y., Hajiqanbar, H., Sedaratian, A., 2014. Two-spotted spider mite reared on resistant eggplant affects consumption rate and life table parameters of its predator, *Typhlodromus bagdasarjani* (Acari: Phytoseiidae). Experimental and Applied Acarology 63, 241–252.

Khanamani, M., Fathipour, Y., Hajiqanbar, H., 2015. Assessing compatibility of the predatory mite *Typhlodromus bagdasarjani* (Acari: Phytoseiidae) and resistant eggplant cultivar in a tritrophic system. Annals of the Entomological Society of America 108, 501–512.

Kheradmand, K., Ueckermann, E.A., Fathipour, Y., 2007. Mite of the genera *Zetzellia* and *Eustigmaeus* from Iran (Acari: Stigmaeidae). Acarina 15, 143–147.

Khodayari, S., Kamali, K., Fathipour, Y., 2008. Biology, life table, and predation of *Zetzellia mali* (Acari: Stigmaeidae) on *Tetranychus urticae* (Acari: Tetranychidae). Acarina 16, 191–196.

Khodayari, S., Fathipour, Y., Kamali, K., 2013. Life history parameters of *Phytoseius plumifer* (Acari: Phytoseiidae) fed on corn pollen. Acarologia 53, 185–189.

Kianpour, R., Fathipour, Y., Kamali, K., Omkar, 2011. Effects of mixed prey on the development and demographic attributes of a generalist predator, *Coccinella septempunctata* (Coleoptera: Coccinellidae). Biocontrol Science and Technology 21, 435–447.

Kogan, M., 1998. Integrated pest management: historical perspectives and contemporary developments. Annual Review of Entomology 43, 243–270.

Kouhjani-Gorji, M., Fathipour, Y., Kamali, K., 2009. The effect of temperature on the functional response and prey consumption of *Phytoseius plumifer* (Acari: Phytoseiidae) on the two-spotted spider mite. Acarina 17, 231–237.

Kouhjani-Gorji, M., Fathipour, Y., Kamali, K., 2012. Life table parameters of *Phytoseius plumifer* (Phytoseiidae) fed on two-spotted spider mite at different constant temperatures. International Journal of Acarology 38, 377–385.

Lee, H.-S., Gillespie, D.R., 2011. Life tables and development of *Amblyseius swirskii* (Acari: Phytoseiidae) at different temperatures. Experimental and Applied Acarology 53, 17–27.

van Lenteren, J.C., 2012. The state of commercial augmentative biological control: plenty of natural enemies, but a frustrating lack of uptake. BioControl 57, 1–20.

Madadi, H., Enkegaard, A., Brødsgaard, H.F., Kharrazi-Pakdel, A., Ashouri, A., Mohaghegh-Neishabouri, J., 2009. Interactions between *Orius albidipennis* (Heteroptera: Anthocoridae) and *Neoseiulus cucumeris* (Acari: Phytoseiidae): effects of host plants under microcosm condition. Biological Control 50, 137–142.

Maleknia, B., Golpayegani, A.Z., Farhoudi, F., Mirkhalilzadeh, S.R., Allahyari, H., 2012. Effect of a heterospecific predator on the oviposition behavior of *Phytoseiulus persimilis*. Persian Journal of Acarology 1, 17–24.

Maleknia, B., Zahedi Golpayegani, A., Saboori, A., Magalhaes, S., 2013. Olfactory responses of *Phytoseiulus persimilis* to rose plants with or without prey or competitors. Acarologia 53, 273–284.

Maleknia, B., Golpayegani, A.Z., Saboori, A., Mohammadi, H., 2014. *Phytoseiulus persimilis* olfactory responses to rose leaves: starvation and previous host plant experience. Persian Journal of Acarology 3, 77–90.

Maleknia, B., Fathipour, Y., Soufbaf, M., 2015. Tritrophic Interactions in a System of Cucumber, Two-Spotted Spider Mite and Phytoseiid Mites (Ph.D. thesis). Tarbiat Modares University, p. 165.

McMurtry, J.A., Moraes, G.J.D., Sourassou, N.F., 2013. Revision of the lifestyles of phytoseiid mites (Acari: Phytoseiidae) and implications for biological control strategies. Systematic and Applied Acarology 18, 297–320.

McMurtry, J.A., 1983. Phytoseiid predators in orchard systems: a classical biological control success story. In: Hoy, M.A., et al (Eds.), Biological Control of Pests by Mites. University of California, Berkeley, pp. 21–26.

Mehrkhou, F., Fathipour, Y., Talebi, A.A., 2006. Foraging Behavior of *Stethorus gilvifrons* (Col.: Coccinellidae) on *Tetranychus uryicae* (Acari: Tetranychidae) (M.Sc. thesis). Tarbiat Modares University, Tehran, p. 81.

Michalska, K., Skoracka, A., Navia, D., Amrine, J.W., 2010. Behavioural studies on eriophyoid mites: an overview. Experimental and Applied Acarology 51, 31–59.

Moayeri, H., Ashouri, A., Brødsgaard, H., Enkegaard, A., 2006. Odour-mediated responses of a predatory mirid bug and its prey, the two-spotted spider mite. Experimental and Applied Acarology 40, 27–36.

Moghadasi, M., Saboori, A., Allahyari, H., Zahedi Golpayegani, A., 2014. Life table and predation capacity of *Typhlodromus bagdasarjani* (Acari: Phytoseiidae) feeding on *Tetranychus urticae* (Acari: Tetranychidae) on rose. International Journal of Acarology 40, 501–508.

Momen, F.M., 2012. Influence of life diet on the biology and demographic parameters of *Agistemus olivi* Romeih, a specific predator of eriophyid pest mites (Acari: Stigmaeidae and Eriophyidae). Tropical Life Sciences Research 23, 25–34.

Mori, K., Nozawa, M., Arai, K., Gotoh, T., 2005. Life-history traits of the acarophagous lady beetle, *Stethorus japonicus* at three constant temperatures. BioControl 50, 35–51.

Murdoch, W.W., Marks, J.R., 1973. Predation by coccinellid beetles: experiments on switching. Ecology 54, 160–167.

Murdoch, W.W., 1969. Switching in general predators: experiments on predator specificity and stability of prey populations. Ecological Monographs 39, 335–354.

Nguyen, D.T., Bouguet, V., Spranghers, T., Vangansbeke, D., De Clercq, P., 2015. Beneficial effect of supplementing an artificial diet for *Amblyseius swirskii* with *Hermetia illucens* haemolymph. Journal of Applied Entomology 139, 342–351.

Nicholson, A.J., 1933. The balance of animal populations. Journal of Animal Ecology 2, 132–178.

Nikooei, M., Fathipour, Y., Javaran, M.J., Soufbaf, M., 2015. How different genetically manipulated *Brassica* genotypes affect life table parameters of *Plutella xylostella* (Lepidoptera: Plutellidae). Journal of Economic Entomology 108, 515–524.

Obrycki, J.J., Kring, T.J., 1998. Predaceous Coccinellidae in biological control. Annual Review of Entomology 43, 295–321.

Omkar, Pervez, A., 2004. Functional and numerical responses of *Propylea dissecta* (Col., Coccinellidae). Journal of Applied Entomology 128, 140–146.

Pakyari, H., Fathipour, Y., 2009. Mutual interference of *Scolothrips longicornis* Priesner (Thysanoptera: Thripidae) on *Tetranychus urticae* Koch (Acari: Tetranychidae). IOBC/wprs Bulletin 50, 65–68.

Pakyari, H., Fathipour, Y., Rezapanah, M., Kamali, K., 2009. Temperature-dependent functional response of *Scolothrips longicornis* (Thysanoptera: Thripidae) preying on *Tetranychus urticae*. Journal of Asia-Pacific Entomology 12, 23–26.

Pakyari, H., Fathipour, Y., Enkegaard, A., 2011. Effect of temperature on life table parameters of predatory thrips *Scolothrips longicornis* (Thysanoptera: Thripidae) fed on twospotted spider mites (Acari: Tetranychidae). Journal of Economic Entomology 104, 799–805.

Park, H.-H., Shipp, L., Buitenhuis, R., Ahn, J.J., 2011. Life history parameters of a commercially available *Amblyseius swirskii* (Acari: Phytoseiidae) fed on cattail (*Typha latifolia*) pollen and tomato russet mite (*Aculops lycopersici*). Journal of Asia-Pacific Entomology 14, 497–501.

Peña, J.E., Bruin, J., Sabelis, M.W., 2012. Biology and control of the red palm mite, *Raoiella indica*: an introduction. Experimental and Applied Acarology 57, 211–213.

Perumalsamy, K., Selvasundaram, R., Roobakkumar, A., Rahman, V.J., Muraleedharan, N., 2010. Life table and predatory efficiency of *Stethorus gilvifrons* (Coleoptera: Coccinellidae), an important predator of the red spider mite, *Oligonychus coffeae* (Acari: Tetranychidae), infesting tea. Experimental and Applied Acarology 50, 141–150.

Pilkington, L.J., Messelink, G., van Lenteren, J.C., Le Mottee, K., 2010. Protected biological control-biological pest management in the greenhouse industry. Biological Control 52, 216–220.

Pimentel, D., 2005. Environmental and economic costs of the application of pesticides primarily in the United States. Environment, Development and Sustainability 7, 229–252.

Premachandra, W.T.S.D., Borgemeister, C., Berndt, O., Ehlers, R.U., Poehling, H.M., 2003. Combined releases of entomopathogenic nematodes and the predatory mite *Hypoaspis aculeifer* to control soil-dwelling stages of western flower thrips *Frankliniella occidentalis*. BioControl 48, 529–541.

Put, K., Bollens, T., Wäckers, F.L., Pekas, A., 2012. Type and spatial distribution of food supplements impact population development and dispersal of the omnivore predator *Macrolophus pygmaeus* (Rambur) (Hemiptera: Miridae). Biological Control 63, 172–180.

Rahmani, H., Fathipour, Y., Kamali, K., 2009. Life history and population growth parameters of (Acari: Phytoseiidae) fed on *Thrips tabaci* (Thysanoptera: Thripidae) in laboratory conditions. Systematic and Applied Acarology 14, 91–100.

Rasmy, A.H., Osman, M., Abou-Elella, G., 2011. Temperature influence on biology, thermal requirement and life table of the predatory mites *Agistemus exsertus* Gonzalez and *Phytoseius plumifer* (Can. & Fanz.) reared on *Tetranychus urticae* Koch. Archives of Phytopathology and Plant Protection 44, 85–96.

Rogers, D., 1972. Random search and insect population models. Journal of Animal Ecology 41, 369–383.

Rosenheim, J.A., 2005. Intraguild predation of *Orius tristicolor* by *Geocoris* spp. and the paradox of irruptive spider mite dynamics in California cotton. Biological Control 32, 172–179.

Rosenheim, J.A., Kaya, H.K., Ehler, L.E., Marois, J.J., Jaffee, B.A., 1995. Intraguild predation among biological control agents-theory and evidence. Biological Control 5, 303–335.

Royama, T., 1971. A comparative study of models for predation and parasitism. Researches on Population Ecology 13, 1–91.

Sabaghi, R., Sahragard, A., Hosseini, R., 2011. Functional and numerical responses of *Scymnus syriacus* Marseul (Coleoptera: Coccinellidae) to the black bean aphid, *Aphis fabae* Scopoli (Hemiptera: Aphididae) under laboratory conditions. Journal of Plant Protection Research 51, 423–428.

Saber, S.A., Rasmy, A.H., 2010. Influence of plant leaf surface on the development, reproduction and life table parameters of the predacious mite, *Agistemus exsertus* Gonzalez (Acari: Stigmaeidae). Crop Protection 29, 789–792.

Safuraie-Parizi, S., Fathipour, Y., Talebi, A.A., 2014. Evaluation of tomato cultivars to *Helicoverpa armigera* using two-sex life table parameters in laboratory. Journal of Asia-Pacific Entomology 17, 837–844.

Sakaki, S., Sahragard, A., 2011. A new method to study the functional response of *Scymnus syriacus* (Coleoptera: Coccinellidae) to different densities of *Aphis gossypii*. Journal of Asia-Pacific Entomology 14, 459–462.

Sarwar, M., Xu, X., Wu, K., 2012. Management of spider mite *Tetranychus cinnabarinus* (Boisduval) (Tetranychidae) infestation in cotton by releasing the predatory mite *Neoseiulus pseudolongispinosus* (Xin, Liang and Ke) (Phytoseiidae). Biological Control 65 (1), 37–42.

Schausberger, P., Croft, B., 2000. Cannibalism and intraguild predation among phytoseiid mites: are aggressiveness and prey preference related to diet specialization? Experimental and Applied Acarology 24, 709–725.

Schausberger, P., 2003. Cannibalism among phytoseiid mites: a review. Experimental and Applied Acarology 29, 173–191.

Schausberger, P., 2007. Kin recognition by juvenile predatory mites: prior association or phenotype matching? Behavioral Ecology and Sociobiology 62, 119–125.

Sedaratian, A., Fathipour, Y., Moharramipour, S., 2009. Evaluation of resistance in 14 soybean genotypes to *Tetranychus urticae* (Acari: Tetranychidae). Journal of Pest Science 82, 163–170.

Sedaratian, A., Fathipour, Y., Moharramipour, S., 2011. Comparative life table analysis of *Tetranychus urticae* (Acari: Tetranychidae) on 14 soybean genotypes. Insect Science 18, 541–553.

Seiedy, M., Saboori, A., Allahyari, H., Talaei-Hassanloui, R., Tork, M., 2012. Functional response of *Phytoseiulus persimilis* (Acari: Phytoseiidae) on untreated and *Beauveria bassiana*-treated adults of *Tetranychus urticae* (Acari: Tetranychidae). Journal of Insect Behavior 25, 543–553.

Sepúlveda, F., Carrillo, R., 2008. Functional response of the predatory mite *Chileseius camposi* (Acarina: Phytoseiidae) on densities of it prey, *Panonychus ulmi* (Acarina: Tetranychidae). Revista de Biología Tropical 56, 1255–1260.

Shah, M., Suzuki, T., Ghazy, N.A., Amano, H., Ohyama, K., 2011. Night-interrupting light inhibits diapause induction in the Kanzawa spider mite, *Tetranychus kanzawai* Kishida (Acari: Tetranychidae). Journal of Insect Physiology 57, 1185–1189.

Shirdel, D., 2003. Species Diversity of Phytoseiidae (Acari: Mesostigmata) in East Azarbaijan, Iran and Comparison of Preying Efficiencies of Two Species on *Tetranychus urticae* Koch (Ph.D. thesis). Islamic Azad University, Tehran, p. 192.

Sohrabi, F., Shishehbor, P., 2007. Functional and numerical responses of *Stethorus gilvifrons* Mulsant feeding on strawberry spider mite, *Tetranychus turkestani* Ugarov and Nikolski. Pakistan Journal of Biological Sciences 10, 4563–4566.

Solomon, M., Cross, J., Fitzgerald, J., Campbell, C., Jolly, R., Olszak, R., Niemczyk, E., Vogt, H., 2000. Biocontrol of pests of apples and pears in northern and central Europe-3. Predators. Biocontrol Science and Technology 10, 91–128.

Solomon, M.E., 1949. The natural control of animal populations. Journal of Animal Ecology 18, 1–35.

Taghizadeh, R., Fathipour, Y., Kamali, K., 2008. Influence of temperature on life-table parameters of *Stethorus gilvifrons* (Mulsant) (Coleoptera: Coccinellidae) fed on *Tetranychus urticae* Koch. Journal of Applied Entomology 132, 638–645.

Vantornhout, I., Minnaert, H.L., Tirry, L., De Clercq, P., 2005. Influence of diet on life table parameters of *Iphiseius degenerans* (Acari: Phytoseiidae). Experimental and Applied Acarology 35, 183–195.

Vasanthakumar, D., Roobakkumar, A., Rahman, J.V., Kumar, P., Sundaravadivelan, C., Babu, A., 2012. Enhancement of the reproductive potential of *Mallada boninensis* Okamoto (Neuroptera: Chrysopidae), a predator of red spider mite infesting tea: an evaluation of artificial diets. Archives of Biological Sciences 64, 281–285.

Wearing, C.H., Colhoun, K., 1999. Development of *Orius vicinus* (Ribaut) (Heteroptera: Anthocoridae) on different prey. Biocontrol Science and Technology 9, 327–334.

Weintraub, P.G., Pivonia, S., Steinberg, S., 2011. How many *Orius laevigatus* are needed for effective western flower thrips, *Frankliniella occidentalis*, management in sweet pepper? Crop Protection 30, 1443–1448.

Xiao, Y., Fadamiro, H.Y., 2010. Functional responses and prey-stage preferences of three species of predacious mites (Acari: Phytoseiidae) on citrus red mite, *Panonychus citri* (Acari: Tetranychidae). Biological Control 53, 345–352.

Yu, J.-Z., Chi, H., Chen, B.-H., 2013. Comparison of the life tables and predation rates of *Harmonia dimidiata* (F.) (Coleoptera: Coccinellidae) fed on *Aphis gossypii* Glover (Hemiptera: Aphididae) at different temperatures. Biological Control 64, 1–9.

Zhang, Z.-Q., 2003. Mites of Greenhouses: Identification, Biology and Control. CABI Publishing, UK, p. 244.

12

Entomopathogenic Nematodes

S. Subramanian, M. Muthulakshmi

Department of Nematology, Tamil Nadu Agricultural University, Coimbatore, Tamil Nadu, India

1. INTRODUCTION

The economic importance of entomopathogenic nematodes (EPNs) belonging to the genera *Steinernema* Travassos (1927), *Heterorhabditis* Poinar (1976), and *Neosteinernema* Nguyen and Smart (1994) (Rhabditida: Nematoda) is increasing because of their potential use in biocontrol of a wide range of insect pests. These nematodes search for or ambush a suitable insect host, enter through natural orifices and cuticles, and release their symbiotic bacterium *Xenorhabdus* or *Photorhabdus* into the insect hemolymph. Proliferation of these bacteria leads to death of the insect host due to septicemia.

About 250 species of insects belonging to 75 families of 10 orders have been reported to be susceptible to EPNs (Poinar, 1979). EPNs have certain advantages over chemicals as biocontrol agents. The nematodes are nonpolluting and thus environmentally safe and acceptable compared with organic insecticides. They are more virulent and cause mortality of the host within 48–72h after infection. Infective juvenile (IJs) can be applied with conventional equipment (Georgis, 1990) and they are compatible with most pesticides (Rovesti and Deseo, 1991). They find their hosts either actively or passively in soil and cryptic habitats—sometimes, in soil, they have proven superior to chemicals in controlling the target insects (Gaugler, 1981). Nematode production can easily be accomplished in vivo in insect hosts or in vitro on solid or liquid media (Dutky et al., 1964; House et al., 1965; Bedding, 1981, 1984; Friedman, 1990). A greater number of juveniles can be produced per unit area in fermentation tanks, up to 150,000l, which makes this method especially suited for large-scale commercial production of steinernematids (Friedman et al., 1991). Hundreds of researchers in than 40 countries are working to develop the use of nematodes as bioinsecticides. Nematodes are produced in the USA, Europe, Australia, Japan, and China for the control of insect pests in high-value horticulture, agriculture, and home and garden niche markets. Yet the share of the pest control market captured by EPNs is probably less than 1% (Smart, 1995). The share will increase due to more efficient production methods and the demands of the public for safer, virulent, and environmentally acceptable products. The future of nematode-based products for insect control is excellent. More efficient methods of production and formulation will lower the cost

Ecofriendly Pest Management for Food Security
http://dx.doi.org/10.1016/B978-0-12-803265-7.00012-9

of nematode products and make them more competitive economically. The search for indigenous strains is warranted because collection of such strains may provide:

- Isolates more suitable for inundative release against local insect pests because of adaptation to local climate;
- Information on indigenous fauna prior to possible introduction of insect pests (Bedding, 1990); and
- Reduced risk of significant impact on nontarget organisms as they have ecological compatibility with pest species.

In India, the occurrence of *Heterorhabditis bacteriophora* (Poinar, 1976) in Tamil Nadu was reported by Sivakumar et al. (1988), followed by the discovery of a new species, *Heterorhabditis indica*, from Coimbatore, Tamil Nadu, by Poinar et al. (1992). Ganguly and Singh (2000) described a new steinernematid, *Steinernema thermophilum*, from the Indian Agricultural Research Institute, New Delhi.

EPNs are used as biocontrol agents against insects of the orders Lepidoptera, Coleoptera, Hemiptera, Dictyoptera, and Orthoptera. They are called EPNs because they are associated with symbiont bacteria (*Xenorhabdus* spp. in the case of *Steinernema* and *Photorhabdus* spp. in the case of *Heterorhabditis*) which are entomopathogenic and the bacteria kill the host. To date, 61 species have been described for *Steinernema* and *Neosteinernema longicurvicauda*, and 16 species of *Heterorhabditis* have been described (http://entnemdept.ufl.edu/nguyen/morph/steinsp1.htm).

EPNs in the genera *Heterorhabditis* and *Steinernema* have a global distribution (Hominick et al., 1996; Hominick, 2002). They have been reported in every continent except Antarctica (Griffin et al., 1990). Mostly, they are collected through *Galleria* baiting techniques (Bedding and Akhurst, 1975); however, they have also been collected from insects in their natural environment. These nematodes have a broad host range and can infect a variety of hosts in the laboratory where host contact is assured. For instance, *Steinernema carpocapsae* has infected 250 insect species in laboratory tests (Peters, 1996). In the field, however, they have a narrower range of hosts than in the laboratory which adds to their safety as biocontrol agents (Akhurst, 1990; Georgis et al., 1991; Bathon, 1996).

2. TAXONOMY

EPNs belong to the families Steinematidae (*Steinernema* spp. and *Neosteinernema* spp.) (Figure 1) and Heterorhabditidae (*Heterorhabditis* spp.) (Figure 2).

2.1 Taxonomy of Steinernematidae

Steiner (1923) described the first entomopathogenic nematode, *Aplectana kraussei*. The generic name was later changed to *Steinernema* by Travassos (1927). Two years later Steiner described *Neoaplectana glaseri* which resembled *Steinernema kraussei*. Filipjev (1934) placed *Steinernema* and *Neoaplectana* in the new subfamily Steinernematinae. Chitwood and Chitwood (1937) raised the subfamily Steinernematinae to family Steinernematidae. Wouts et al. (1982) concluded that both *S. kraussei* and *N. glaseri* were identical and that *Neoplectana* was a junior synonym of *Steinernema*. In 1994, Nguyen and Smart described a new genus

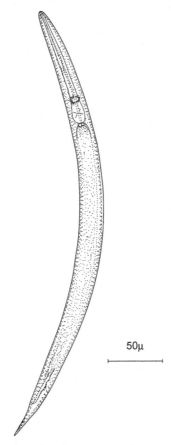

50μ

FIGURE 1 Third stage infective juvenile of *Steinernema siamkayai*.

Neosteinernema and added it to the family. Currently, the family contains 61 species under *Steinernema* and one under *Neosteinernema*, namely, *N. longicurvicauda*.

3. LIFE CYCLE OF EPNs

Steinernema undergo two sexual or amphimictic generations in insect hosts. *Heterorhabditis* has one asexual or parthenogenetic generation followed by a sexual or amphimictic generation. EPNs cause mortality to insects in a three-step process. Firstly, the nematode has to migrate, searching for the target host. Secondly, upon contact with the host, it has to penetrate through natural body openings, such as the anus, mouth, spiracles, and/or through the cuticle. Finally, the nematode–bacteria complex should be able to overcome the insect's immune system and multiply, producing new generations of IJs.

EPNs have a life cycle that includes the egg stage, four juvenile stages, and the adult stage. Both *Steinernema* and *Heterorhabditis* species have a free-living third stage juvenile (J3), termed

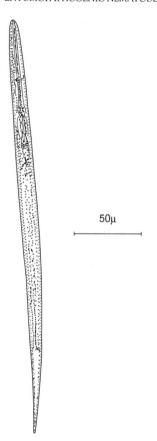

50μ

FIGURE 2   Third stage infective juvenile of *Heterorhabditis indica*.

a dauer juvenile, that is, the infective stage and is therefore also called an IJ. These IJs harbor a symbiotic bacterium (Endo and Nickle, 1994) that plays an essential role in subsequent stages of the life cycle. Bacterial symbionts are found in a modified ventricular part of the intestine in *Steinernema*, as well as throughout the intestinal lumen in *Heterorhabditis* (Hominick, 1990). The IJs seek out a suitable host and instigate the infection process.

The IJs of *Steinernema* enter the insect body through natural openings, such as the mouth, anus, and spiracles. IJs of *Heterorhabditis*, however, enter through natural openings and the body wall. These nematodes possess a dorsal tooth in the anterior region of head, which helps them to gain entry into the hemocoel by breaking the thin cuticle of interseg-mental membrane (Bedding and Molyneux, 1982; Cui et al., 1993). Once in the hemocoel of the insect, the IJ releases the cells of symbiotic bacteria from its intestine (Akhurst, 1982). These bacterial cells are released by regurgitation (Ciche and Ensign, 2003) through the mouth during infection of the insect larva. The IJs recover or exit from the devel-opmentally arrested third, nonfeeding stage, triggered by either bacterial or insect food signals (Strauch and Ehlers, 1998). The insect hemolymph provides a rich medium for the bacterial cells and these begin to grow, release toxins and exoenzymes, and kill the insect within two days (Han and Ehlers, 2000). These bacteria also produce antibiotics and other noxious substances that protect the host cadaver from other microbes in the soil (Webster et al., 2002). Nematodes resume development, start feeding on bacteria, and molt to the fourth stage reaching adulthood within 2–3 days. Nematode development continues over two to three generations until the nutrient status of the cadaver deteriorates whereupon adult development is suppressed and IJs accumulate. These nonfeeding infective stages emerge into the soil where they may survive for several months in the absence of a suitable host (Hominick, 1990).

Steinernematids exhibit amphimictic reproduction (Poinar, 1990). The IJ of *Steinernema* matures to become either a male or a female and sex determination appears to be of the XX/XO type, typical of nematodes (Dix et al., 1994). After mating, females lay eggs that hatch as first stage juveniles that molt successively to second, third, and fourth stage juveniles and to males and females of the second generation. Depending upon the food available, up to three generations can develop inside the dead host (Wang and Bedding, 1996). Upon depletion of the food resources or overcrowding inside the host cadaver, the formation of developmentally-arrested third stage IJs occurs which leaves the cadaver in search of a new host (San-Blas et al., 2008).

By contrast, in *Heterorhabditis*, the IJs mature to give first generation hermaphrodite females, but these females give rise to a second generation of amphimictic males and females and to self-fertile hermaphroditic females and IJs (Dix et al., 1992; Strauch et al., 1994). The development into amphimictic adults is induced by favorable nutritional conditions, whereas the development of hermaphrodites is induced by low concentrations of nutrients (Strauch et al., 1994). In both genera, *Steinernema* and *Heterorhabditis*, as long as abundant nutrients are available, additional adult generations develop. When the nutrients are consumed, the late second stage juveniles cease feeding, incorporate a pellet of symbiotic bacteria in their bacterial chamber (Popiel et al., 1989), and molt to the third stage juvenile. They retain the cuticle of the second stage as a sheath, and leave the cadaver in search of new hosts. They may survive for several months in the absence of a suitable host. The cycle from entry of IJs into a host to emergence of new IJs is temperature dependent and varies for different species and strains. It generally takes about 6–18 days at temperatures ranging from 18–28 °C in *Galleria mellonella* (Poinar, 1990; Nguyen and Smart, 1992).

Bacteria multiply rapidly in hemolymph and produce toxins and other secondary metabolites that contribute to the weakening of the host's defense mechanism. Hosts attacked by EPNs usually die due to poisoning or failure of certain organs in 24–72 h after the infection (Forst and Clarke, 2002). Two developmental cycles occur in the host that of nematodes and that of bacteria. The first generation nematodes pass into the second generation. After larvae cast off the fourth sheath, and in the adult period, nematodes pass into the third generation, which thrives in the host as long as there is food. The host is by then already dead, killed by the toxins secreted by bacteria. The third generation nematodes are thus already saprophagic. Bacteria also produce such toxins (3,5 dihydroxy-4-isopropyl-stilben) that deter other microorganisms from settling in the carcass. When the developmental cycle is finished, nematodes leave the parts of the carcasses that have not decomposed and they return into the ground. Nematodes cannot develop without a host (an insect) (Kaya and Gaugler, 1993), and without a host they live in the ground for only a very brief period. The importance of EPNs and biological plant protection against harmful organisms was first established in the USA in the 1930s. In 1923, Glaser and Fox discovered a nematode that attacked and caused the death of the beetle *Popillia japonica* Newman (Glaser and Farell, 1935). Glaser introduced a method of growing EPNs in vitro. Using such nematodes carried out the first field experiment in New Jersey to suppress *P. japonica* (Kaya and Gaugler, 1993). When EPNs were first discovered, the hypothesis was proposed that nematodes alone cause death of the attacked insects. In 1937, Bovien first mentioned the possibility for the existence of symbiotic bacteria which live with EPNs in a mutualist relationship. Boemare et al. (1982) proved that nematodes from the genus *Steinernema* produce toxic substances that negatively influence the immune system of infected insects and can cause death of the host without the presence of symbiotic bacteria. It has not yet been established whether EPNs from the genus *Heterorhabditis* can alone produce toxic substances that could diminish the vitality of infected insects (Kaya and Gaugler, 1993). The use of EPNs in biological plant protection was employed years ago and is still traditionally connected with suppressing soil-inhabiting insect pests (Ishibashi and Choi, 1991). Research since the mid1990s indicates that EPNs also have the potential to suppress above-ground insect pests, but only in certain conditions (Trdan et al., 2008). The lower efficiency of EPNs in suppression of above-ground insect pests is primarily due to insufficient moisture, exposure to thermal extremes, and ultraviolet (UV) radiation. These factors are of crucial importance for the survival of nematodes (Kaya and Gaguler, 1993). For this reason, nematodes are less efficient against above-ground insect pests outdoors, although previous laboratory tests showed much higher efficiency (Laznik et al., 2010). To introduce nematodes to plants we can use equipment intended for spraying plant protection products, manuring, or irrigation. For this purpose backpack manual or tractor sprayers, sprinklers, and also planes are suitable. Infective larvae can be passed through spray tubes with diameter of at least 500 µm, capable of withstanding pressure up to 2000 kPa (Kaya and Gaugler, 1993). EPN IJs can tolerate short-term exposure (2–24 h) to many chemical and biological insecticides, fungicides, herbicides, fertilizers, and growth regulators, and can thus be tank-mixed and applied together (De Nardo and Grewal, 2003). Nematode–chemical combinations in tank mixes could offer a cost-effective alternative to foliar integrated pest management (IPM) systems. Due to the sensibility of nematodes to UV radiation, they have to be applied to plants in the evening, early in the morning, or in cloudy weather, when the radiation is not so intense (Kaya and Gaugler, 1993). Nematode survival and efficacy on foliage has also been shown to be enhanced to varying degrees by the addition of various adjuvants to the spray mixture, which have antidesiccant (e.g., glycerol, various polymers) or UV-protective (brighteners) actions (Grewal, 2002) although more needs to be done to enhance postapplication survival. The greatest potential for using EPNs against foliar pests is almost certain in IPM programs, in conjunction with other biocontrol agents or selective chemicals (Rovesti and Deseo, 1990). EPNs are considered exceptionally safe bioagents (Akhurst and Smith, 2002). Because their activity is specific, their environmental risk is considerably lower than that of chemical agents for plant protection. Since the first use of EPNs for suppressing *P. japonica* Newman in the USA (Glaser and Farrell, 1935), no cases of environmental damage due to these bioagents have been documented. The use of nematodes is safe for users and EPNs and their bacteria are not harmful to mammals and plants (Akhurst and Smith, 2002).

4. HOST SPECIFICITY

4.1 Infectivity

Information on the infectivity of progeny produced in vitro is meager. The infectivity of EPNs is generally reduced when the storage time increases. Thus infectivity of *Steinernema feltiae* assessed by mortality of *G. mellonella* showed a trend to decrease as storage time increases at 5°C and 15°C (Fan and Hominick, 1991). The persistence and efficacy of EPNs raised in liquid media was poor (Gaugler and Georgis, 1991) due to low lipid assimilation. Xu and Xie (1992) reported that reproduction of *S. anomali* was three to five times greater and its virulence was three times higher when cultured in the larvae of a pyralid compared with those cultured in chicken broth. Westerman (1992) reported that an increase storage time caused a decrease in quality of *Heterorhabditis* sp. with respect to persistence and efficacy. Factors such as oxygen supply, nematode concentration, pH, accumulation of waste products, and presence of contaminants like bacteria and fungi in the storage medium affect the nematodes during storage.

Yang et al. (1997) showed that 2.8% IJ from plant origin media penetrated into *G. mellonella* within 24 h whereas around 6% IJ from animal origin media penetrated. Vyas et al. (1999) showed that *Steinernema glaseri* produced in vitro caused 100 and 71.25% mortality of *Spodoptera litura* and *Holotrichia consanguinea*, respectively, after 48 h exposure using 32 IJ/larva. IJs treated with antidesiccants, applied at a concentration of 100,000/m^2, caused 83.8% mortality of *H. consanguinea*.

5. MODE OF INFECTION

Generally, several EPNs infect a single insect host. IJ nematodes penetrate the insect's body cavity either through natural body openings (such as the mouth, anus, genital pore, or breathing pore or by breaking the outer cuticle) of the insect; heterorhabditis do this using a dorsal "tooth" or hook. Once inside the body cavity of the host, the IJs release bacteria that live symbiotically within the entomopathogenic nematode's gut but do not harm the nematode. The nematode–bacterium relationship is highly specific: only *Xenorhabdus* spp. bacteria coexist with steinernematids, and only *Photorhabdus* bacteria coexist with heterorhabditis. Once released into the host, the bacteria multiply quickly and under optimal conditions cause the host to die within 24–48 h.

EPNs feed both on the bacteria they release and host insect tissue. After a few days inside the host, EPNs mature to the adult stage. These adult EPNs produce hundreds of thousands of new juveniles that may undergo several life cycles within a single host. When the host has been consumed, the IJs, armed with a fresh supply of bacteria, emerge from the empty shell of the host, move into the soil, and begin the search for a new host. A protective exterior cuticle surrounds the IJ, protecting it from the environment and predators. Under ideal conditions, steinernematids emerge 6–11 days after initial infection and heterorhabditis emerge 12–14 days after initial infection (Kaya and Koppenhöfer, 1999). The duration of IJ survival in soil is unknown because they can become prey to invertebrates and microorganisms.

Steinernematids and *Heterorhabditis* have similar life histories. The nonfeeding, developmentally arrested IJ seeks out insect hosts and initiates infections. When a host has been located, the nematodes penetrate into the insect body cavity, usually via natural body openings (mouth, anus, spiracles) or areas of thin cuticle. Once in the body cavity, a symbiotic bacterium (*Xenorhabdus* for steinernematids, *Photorhabdus* for heterorhabditis) is released from the nematode gut, multiplies rapidly, and causes rapid insect death. The nematodes feed upon the bacteria and liquefying host, and mature into adults. Steinernematid IJs may become males or females, whereas heterorhabditis develop into self-fertilizing hermaphrodites although subsequent generations within a host produce males and females.

The life cycle is completed in a few days, and hundreds of thousands of new IJs emerge in search of fresh hosts. Thus, EPNs are a nematode–bacterium complex. The nematode may appear as little more than a biological syringe for its bacterial partner, yet the relationship between these organisms is one of classic mutualism. Nematode growth and reproduction depend upon conditions established in the host cadaver by the bacterium. The bacterium further contributes antiimmune proteins to assist the nematode in overcoming host defenses, and antimicrobials that suppress colonization of the cadaver by competing secondary invaders. Conversely, the bacterium lacks invasive powers and is dependent upon the nematode to locate and penetrate suitable hosts.

6. HOST RESISTANCE

EPNs are highly virulent against a wide range of insect pests causing rapid mortality within 48–72h. Pathogenicity of *S. feltiae*, *S. glaseri*, and *Heterorhabditis heliothidis* was evaluated for *Alphitobius diaperinus*. All stages of the beetle were highly susceptible to *S. feltiae*, the median lethal dose (LD_{50}) ranging from 9–56IJ/host but only the adult beetles were susceptible to *S. glaseri*. Late instar larvae were highly susceptible to both *N. carpocapsae* and *H. heliothidis* in sandy loam and clay soils, the LD_{50} ranging from 1–14 nematodes/larva (Geden et al., 1985). Kaya (1985) studied the susceptibility of *Spodoptera exigua* and *Pseudaletia unipuncta* at concentrations of 0, 10, 25, 60, 100, or 200 nematodes per larva. Mortalities of neonate larvae of *S. exigua* exposed to 50 or more nematodes ranged from 68–74% while mortalities of 3 and 8-day-old larvae ranged from 91–100%. The results with *P. unipuncta* showed 34–44% mortality of neonate larvae exposed to 50 of more nematodes and 32–91% mortality for 7-day-old larvae.

Kondo and Ishibashi (1986) studied the infection effectiveness of *S. feltiae* (DD-136) to *S. litura* on soil in the laboratory and found that about 50% of the insects were infected after being exposed for 1h on soil inoculated with 1000IJs and 100% mortality was recorded for insects exposed for 6h or longer. Under laboratory conditions, *S. feltiae* killed third instar larvae of fall webworm *Hyphantria cunea* within 24h at 1000 and 5000IJs/insect larva, and within 48h at 100 and 500 J3/larva at 25°C (Yamanaka et al., 1986). Fuxa et al. (1998) determined median lethal concentrations (LC_{50}) for four nematode populations (two strains of *S. feltiae* hybrid and *Steinernema bibionis*) against fifth instar armyworm *Spodoptera frugiperda*. The LC_{50} ranged from 7.6–33.3 nematodes/larva. First instar armyworms suffered virtually 100% mortality from the *S. feltiae* Mexican strain at 1.0 nema/larva. LC_{50} values were 2.3 and 7.9 nematodes/larva in third and fifth instar larvae, respectively. Pupae had 7–20% mortality at doses ranging from 30–60 nematodes/pupa. Glazer and Wyoski (1990) determined the pathogenicity of Steinernematid and Heterorhabditid nematodes to geometrid *Boarmia selenaria*. Over 80% mortality was achieved in vitro with a minimum of 20,000IJ of *S. feltiae*. The LD_{50} was 4250IJ/larva.

Shanti and Sivakumar (1991) compared the virulence of *S. glaseri* (NC-34) and *S. carpocapsae* (DD-136) against *Holotrichia* grubs. LC_{50} value was considerably lower for *S. glaseri* (13.38) than for *S. carpocapsae* (33.8). Median lethal time (LT_{50}) values for *S. glaseri* were 85.0–152.0h and 119.1–223.8h for *S. carpocapsae*. The pathogenicity of *S. carpocapsae* against red imported fire ant, *Solenopsis invicta*, was studied by Drees et al. (1992). At 103–105 IJ/petri dish, the mortality of larvae, pupae, and alates ranged from 28 to 100% at higher doses after 96h at 23–25 °C. Sosa et al. (1993) stated that application of *S. carpocapsae*, *H. bacteriophora*, and *S. glaseri* at 5000 IJ/ml to the diet of sugar cane borer, *Diatraea saccharalis*, resulted in 100, 100, and 30% mortality, respectively. Exposure of *Helicoverpa zea* at 10, 20, 40, 80, and 1000 IJ/prepupa resulted in mortalities of 40, 55, 85, 90, and 100%, respectively. The LC_{50} of *Steinernema riobravis* for *H. zea* prepupae was 13 nematodes per prepupa (Cabanillas and Raulston, 1994). Henneberry et al. (1995) reported that *S. riobravis* and Kapow strain of *S. carpocapsae* were highly virulent on second, third and last instar larvae of *Pectinophora gossypiella*. The LC_{50} for last instar larvae treated with *S. riobravis* (1.84) was significantly lower than LC_{50} for larvae with *S. carpocapsae* (3.07).

7. FACTORS AFFECTING EFFICACY

Ambusher	Sit and wait strategy "nictation"
	Steinernema carpocapsae
Cruisers	Active foragers
	Steinernema glaseri and *H. bacteriophora*
Intermediary	*Steinernema feltiae*, *S. riobravis*

8. BEHAVIORAL DEFENSE STRATEGIES

1. Walling off IJs to avoid infection (termites).
2. High defecation rate to reduce infection via anus (scarabaeids).
3. Low CO_2 output or release of CO_2 in bursts to minimize chemical cues (lepidopterous pupae).
4. Boring into roots to escape from infection by IJ in soil (rootworms).
5. Preening to remove IJs (scrabaeids).

9. MORPHOLOGICAL DEFENSE STRATEGIES

1. Pupal cells of dense soil particles,
2. Silken cocoons,
3. Sieve plates over spiracles,
4. Tightly closed spiracles,
5. Hard body surface.

10. PHYSIOLOGICAL DEFENSE STRATEGIES

1. Encapsulation of invading nematodes by rootworms—hemocytes,
2. Humoral response against mutualistic bacterium (melanization)—granular cells.

EPNs use two search strategies: ambushers or cruisers (Grewal et al., 1994). An understanding of host-finding strategies will help to properly match EPN species to pest insects to ensure infection and control (Gaugler, 1999). Only EPNs in the IJ stage will survive in the soil and find and penetrate insect pests. IJ EPNs locate their hosts in soil.

10.1 Ambushing

EPNs that use the ambushing strategy tend to remain stationary at or near the soil surface and locate host insects by direct contact (Campbell and Gaugler 1993). An ambusher searches by standing on its tail so that most of its body is in the air, referred to as "nictation." The nictating nematode attaches to and attacks passing insect hosts. Ambusher EPNs most effectively control insect pests that are highly mobile at the soil surface, such as cutworms, armyworms, and mole crickets.

10.2 Cruising

EPNs that use the cruising strategy are highly mobile and able to move throughout the soil profile. Cruisers locate their host by sensing carbon dioxide or other volatiles released by the host. Cruiser EPNs are most effective against sedentary and slow-moving insect pests at various soil depths, such as white grubs and root weevils.

11. PESTICIDES

11.1 Effect on Nematodes

Prakasa Rao et al. (1975) reported that up to 10 ppm pyridimine phosphate and monocrotophos, up to 500 ppm chlorfenvinphos, phosaline, and fenitrothion, up to 250 ppm formothian and binapacryl, and up to 500 ppm vamidothion, phenthioate, and fenthion were well tolerated (with <10% mortality) by *Steinernema dutkyi* (DD-136) up to 24h. Fedorko et al. (1977) studied the effect of oxamyl on *S. carpocapsae*. The LD$_{50}$ was 0.5% and the IJs were more sensitive after 24h exposure to the chemical. Srivastava (1978) reported that concentrations of 0.01–0.04% diazion, monocrotophos, and dichlorvos did not affect the survival of *S. carpocapsae*. Exposure of IJs of *S. carpocapsae* to mevinphos and trichlorfon at 0.1 and 1.0 mg/ml for 120h caused 51.2, 83.3, and 69.5% mortality respectively. The percentage mortality of *S. carpocapsae* in methomyl at 1.0 mg/ml increased from 2.6% (72h) to 21.3% (168h). Phenamiphos at 0.1 and 0.5 mg/ml for 120 or 168h showed significant increase in mortality (Hara and Kaya, 1983). Das and Divaka (1987) observed that *S. feltiae* was not affected 72h after exposure to carbaryl, dimethoate, endosulfan, and malathion.

Lindane, parathion, etrimfos, propoxur, chlorphriphos, endosulfan, carbofuran, methomyl, and oxamyl were toxic to *Heterorhabditis* sp. at 0.5, 1.0, and 2.0% (a.i. concentrations)

(Heungens and Buysse, 1987). Rovesti et al. (1988) reported that the movement of IJ of *H. bacteriophora* was adversely affected by terbufos and fonofos between 2500 and 10,000 ppm for 72 h exposure. Phorate caused complete mortality of *H. bacteriophora* IJs at 10,000 ppm and inhibited movement at all other concentrations (312–5000 ppm). Aldicarb, carbofuran, and methomyl prevented the mobility of IJs. Exposure of IJs to metham sodium (Vapam) at 2500–1000 ppm for 3 days caused complete mortality. After 24 h exposure, 0.2% dimethoate and bromophos caused 12.3 and 14.7% mortality, respectively, of IJs of *S. feltiae*. Diazion and dimethoate at 0.2% cused 16.3 and 16.7% mortality of *H. bacteriophora* (Jaworska, 1990). Gaugler and Campbell (1991) reported that a 30 μg/ml solution of oxamyl reduced the proportion of nonmobile *S. carpocapsae* and *H. bacteriophora* by 35–66.67%, while stimulating a 7.5-fold increase in sinusoidal movement. Movement of IJs decreased at 10,000 μg/ml. Exposure at highest the concentration (1000 and 10,000 μg/ml) caused complete paralysis of IJs after 48 h.

Fujiie et al. (1993) observed that *Steinernema kushidai* was adversely affected the movement of IJ by 500–2000 ppm diazinon and fenthion for 48 h of exposure. Profenofos and pyroclofos caused 57.1 and 47.8% mortality of *S. carpocapsae*, respectively, at 100 μg/ml in 48 h while dichlorvos, diazinon, fenthion, malathion, propetamphos, trichlorfon, and prothiofos showed week toxicity (<20% mortality) (Zhang et al., 1994). IJs of *S. carpocapsae* and *S. feltiae* were equally sensitive to fenoxycarb as the LD_{50} values were 0.027 and 0.028 mg/ml, respectively. IJs of *S. carpocapsae* were more sensitive to carbofuran as LD_{50} values were 0.017 and 0.029 mg/ml for 72 and 120 h, respectively, but LD_{50} values of *S. feltiae* were 0.00003 and 0.00008 mg/ml for 72 and 120 h, respectively (Gordon et al., 1996). Malathion at concentrations of 200 and 400 ppm caused behavioral disruption of IJ of *H. bacteriophora* as indicated by decreased mortality (Baweja and Sehgal, 1997). The IJs of *H. bacteriophora* were not significantly affected by imidacloprid at 10, 40, or 160 mg (a.i)/l. The mortality after 24 h of exposure in imidacloprid ranged between 0.7 and 3% (Koppenhöfer and Kaya, 1998). Nishimatsu and Jackson (1998) observed an additive effect on corn rootworm (*Diabrotica* spp.) mortality with combinations of terbufos and fonofos and *S. carpocapsae*.

Endosulfan (0.05%), cypermethrin (0.003%), and malathion (0.01%) caused 42–49% mortality of *S. carpocapsae*. Phosphamidon (0.05%) and monocrotophos (0.05%) caused only 2–6% mortality after 48 h exposure. Phorate at 0.05% caused highest mortality (91%) of *S. carpocapsae* after 48 h (Gupta and Siddiqui, 1999). Bednarek et al. (2000) reported that carbosulfan and carbofuran at 3 mg/kg soil exhibited mortality of *S. glaseri*, *S. feltiae*, and *Heterorhabditis megidis* that did not differ significantly from control. Endosulfan and fenvalerate at field-recommended doses caused 2.4–3.2 and 1.6–33.6% mortality of *Steinernema* sp. and *H. indica* isolates of Project Directorate of biological control (PDBC), respectively, after 5 days of exposure. Quinalphos was found to be deleterious to some isolates of EPN as it caused 16–100% mortality (Hussaini et al., 2001).

EPNs are agents that can be used for the biocontrol of pests associated with insecticides in a tank mix. Compatibility studies need to be conducted to analyze which products are compatible with nematodes. The aim of this work was to evaluate the compatibility between EPNs and the insecticides that are most used on tomato crops, and to correlate the toxicological classification of the chemical products with two species of EPNs that have the potential to control tomato leaf miner, *Tuta absoluta*. Among the products tested, Certero® (triflumuron), Decis® (deltamethrin), Previcur® (dimethylamino-propyl), Ampligo (lambdacyhalothrin + chlorantraniliprole), Premio® (chlorantraniliprole), Engeo Pleno® (thiamethoxam + lambdacyhalothrin) were compatible (International organization for biological control (IOBC) class 1) with both nematode species (Paulo Henrique et al., 2014).

11.2 Infectivity

Fedorko et al. (1977) reported that doses of oxamyl recommended for control of plant parasitic nematodes in field were no threat to *S. carpocapsae* which were able to invade and kill their insect hosts after treatment. Rovesti et al. (1988) reported that aldicarb, carbofuran, and methomyl prevented infectivity of *H. bacteriophora* to *G. mellonella* at concentrations 5000–10,000 ppm after 3 days of exposure.

Kaya and Burlando (1989) studied the infectivity of *S. feltiae* to *G. mellonella* in phenamiphos treated sand at concentrations 4.5, 9.0, and 18 mg a.i./kg dry sand exposed for 4 days. They reported that the extracted nematodes caused 92.2, 75.0, and 76.7% mortality of *G. mellonella*, respectively. Rovesti and Deseo (1990) reported that parathion at 2500 rpm caused a reduction in parasitism of both *S. carpocapsae* and *S. feltiae* bioassays. But the effect of phosphamidon, diazinon, chlorpyrifos, and endosulfan in the infectivity test was negligible. When exposed to aldicarb and methomyl at 1000 and 5000 rpm, both the IJ failed to infect *G. mellonella* after 3 days. Gaugler and Campbell (1991) observed that 30 μg/ml oxamyl did not affect *S. carpocapsae* pathogenicity, but reduced the ability of *H. bacteriophora* to cause lethal infection in *G. mellonella* larvae. The reduction of *H. bacteriophora* virulence was apparent as early as 2 days posttreatment, when oxamyl-treated juveniles caused 13.5% host mortality compared with 56.2% in untreated nematodes. Vainio (1994) studied the effect of isofenphos, permethrin, fluazifop butyl, and iprodione on long-term survival of *S. feltiae* in the field. The mortality of bait larvae of *Tenebrio molitor* due to the nematode was greatest in soil samples collected from plots treated with both nematodes and pesticides.

Gordon et al. (1996) reported that using 0.001 mg/ml fenoxycarb, the LD_{50} values for insecticide-treated IJ of *S. carpocapsae* and *S. feltiae* was the same as in the corresponding control. The infectivity of both IJ exposed to 0.01 mg/ml carbofuran for 24, 48, 72, 120, and 168 h showed variation in mortality of *G. mellonella*. Baweja and Sehgal (1997) observed that *H. bacteriophora* parasitization rate of *S. litura* larvae decreased with increasing concentration of malathion with 80% mortality at 50–200 ppm, 40% at 400 ppm, and none at 800 ppm. The percentage mortality of *Corcyra cephalonica* due to *S. carpocapsae* exposed to 0.05% phorate for 48 h was not significantly different (82–97% mortality) from control (100% mortality) (Gupta and Siddiqui, 1999).

12. FUNGICIDES

The effect of direct IJ exposure to fungicides for 24 h was tested in vitro at 15, 20, and 25 °C. The results showed that the compatibility of *S. feltiae* with azoxystrobin was high, and similar findings were obtained for *S. carpocapsae* (strain C67) and all of the tested fungicides, except for tebuconazole + spiroxamine + triadimenol, maneb, dinocap, and copper (II) hydroxide + metalaxil-M. Nematode *H. downs* (strain 3173) suffered the highest mortality rate when IJs were mixed with tebuconazole + spiroxamine + triadimenol (Laznik et al., 2012).

13. MASS PRODUCTION

Mass multiplication of EPNs in vitro
1. **Animal origin**
    4 duck offal + 25 ml water
    4 chicken offal + 1 foam
    4 goat/beet offal
    Period, 2 months
2. Plant origin
    0.44 g nutrient broth
    0.16 g yeast extract
    7.2 g soy flour
    7.2 g Bengal gram
    5.2 g cotton/corn oil
    1 g foam
    29 ml water
    1000 IJ/flask/100 g medium and incubated at 25 °C
    Yield, 1.26×10^6 IJ/100 g
    Period, 1 month

In vivo-suited hosts include:
    Bombyx mori
    Galleria mellonella
    Agrotis ipsilon
    Helicoverpa armigera
    Spodoptera litura
    Inoculum level, 20 n/insect/7–15 days
    Yield, 12,000–62,000/insect larva

13.1 In vitro Mass Production of EPNs

Glaser (1931) recognized the value of developing culture techniques for EPNs and was able to produce the entire life cycle of *S. glaseri* on artificial media. Glaser et al. (1940a) established the first large-scale production process by using a variety of media preseeded with living yeast and supplemented with antimicrobials for additional protection against contamination. Glaser et al. (1942) successfully multiplied *S. glaseri* in media with homogenate animal tissues in standard potato dextrose agar. *Steinernema glaseri* developed well on piece of fresh, sterile rabbit kidney placed on agar and on semisolid gels of autoclaved ground beef kidney or liver in 0.5% sodium chloride agar (Glaser, 1940a). Passage through an insect (Glaser, 1931) or the addition of powdered Japanese beetle grubs, bovine ovaries (Glaser, 1940b), or raw liver extract prevented decline in nematode reproduction (Stoll, 1953). *Steinernema glaseri* and *S. carpocapsae* were multiplied by Jackson (1965) using rabbit kidney on agar slants.

Cheaper substrates were developed after the discovery of bacteria (Dutky, 1937) associated with EPNs. House et al. (1965) mixed powdered dog biscuit with equal quantity of distilled water by volume contained in 9 cm petri dish, autoclaved and inoculated with 1–2×10^3 IJ of DD-136. After 20 days of incubation at 24 °C, each petri dish yielded 3–7.1×10^6 IJs. Kondo and Ishibashi (1984) compared the growth and multiplication of *S. feltiae* on plain agar, nutrient agar, and dog food media. On plain agar, the IJ did not molt and died due to starvation. Normal development of IJs into adults was observed both in nutrient agar and dog food media. Very few progeny were produced on nutrient agar whereas in dog food medium, the nematode population rapidly increased from 5–15 days after inoculation and gradually decreased. Abe (1897) evaluated wheat bran as a medium constituent for the culture of *S. feltiae*. The productivity of IJs were in the order of 10^4/kg wet medium with wheat bran alone and 10^7 kg when the medium was amended with salad oil and or fermented with *Bacillus* sp. Ritter (1988) determined the minimum amount of cholesterol in the lipid-defined artificial medium to support the development of *S. carpocapsae* as 0.0025%. Hypolipidemic agents such as cholestyramine resin, clofibrate, and niacin affected the growth and development of the nematode.

The propagation of *S. kushidai* was vigorous in dog food agar medium supplemented with peptone (Ogura and Mamiya, 1989). Multiplication of *S. carpocapsae* in solid culture media was found to be dependent on the level of inoculum as reported by Wang et al. (1990). Increase in inoculum levels resulted in a reduction in adult size and rate of multiplication. The optimum level was found to be 100 IJ/80 g medium. Richou et al. (1997) studied the culture parameters for production of *S. carpocapsae* A24 and *H. bacteriophora* H06. They found that the yield of both the strains was positively correlated with nematode inoculum size and culture time. Ogura and Haraguchi (1993) successfully cultured *S. kushidai* on the following two artificial media:

Medium I
    Dog food peptone/agar
    8.8% powdered dog food
    1.2% peptone
    0.2% agar
    (mixed with Sorensen phosphate buffer, pH 6.5)
Medium II
    Pig intestine peptone/agar
    8.8% chopped, steamed pig intestine
    1.2% peptone
    0.3% agar
    (mixed with Sorensen phosphate buffer, pH 6.5)

Inoculation of 1000 IJ and incubation at 25 °C for 20 days yielded 344,300 and 244,000 IJ in dog food peptone/agar and pig intestine/agar, respectively. Josephrajkumar and Sivakumar (1998) observed that MacConkey's agar medium containing soy flour and dextrose supported 868 times more multiplication of a native *Steinernema* sp. in 15 days at 25 ± 1 °C but it was not suitable for *S. glaseri*. Yamanaka et al. (2002) studied the influence of temperature on the growth and propagation of *S. glaseri* in vivo (*G. mellonella*) and in vitro (dog food peptone, DFP). *Steinernema glaseri* neither developed nor survived on DFP or *G. mellonella*, or at temperatures above 35 °C on DFP and *G. mellonella*, and progeny production was greatest at 25 and 28 °C, respectively. In vitro mass production of native *Steinernema* sp. was attempted by Vyas et al. (2001) using 21 animal and plant protein-based media. Maximum production of nematodes was recorded in hen egg yolk medium which was economically better than universally used dog food biscuit agar. Of the nine artificial media evaluated in vitro for production of *Steinernema* spp. Wouts medium, modified egg yolk, sunflower + cholesterol, and modified dog biscuit yielded the highest number of IJs of two isolates of *S. carpocapsae* (PDBC EN 6.11 and 6.61), one isolate of *Steinernema bicornutum* (PDBC EN 2.1), and one isolate of

H. indica (PDBC EN 6.71). Modified egg yolk was highly suitable for *S. bicornutum* PDBC EN 2.1 (69.7×10^5 IJs/flash). For *S. carpocapsae* isolates PDBC EN 6.61 and 6.1, modified dog biscuit medium was highly suitable which yielded 64.66×10^5 and 40.28×10^5 IJ per flask, respectively. *Heterorhabditis indica* recorded a maximum yield on Wouts medium (Hussaini et al., 2002).

13.2 Solid Culture Media

Development of *S. glaseri* on pieces of fresh sterile rabbit kidney placed on agar and on semisolid gels of autoclaved ground beef kidney or liver in 0.5% sodium chloride, agar has been observed (McCoy and Glaser, 1936). Glaser et al. (1942) first successfully formulated a medium having homogenate animal tissue in standard potato dextrose agar for *S. glaseri*. Stoll (1953) found that successive subculturing on the same medium caused a decline in reproduction. This is prevented by supplementing the medium with raw liver extract. Rabbit kidney on agar slants were used to mass multiply for *S. glaseri* and *S. carpocapsae* on powdered dog biscuit mixed with equal quantity of distilled water by volume contained in 9cm petri dishes. Each plate yielded $3–7.1 \times 10^6$ IJ in about 20 days after inoculation. Axenic culturing of *S. carpocapsae* in a medium containing hemin supplemented with α-globulin was successfully reported by Buecher et al. (1970).

The medium contains the following composition:

Caenorhabditis briggsae maintenance medium (CbMM) ($2 \times$ concentration).

0.05M phosphate buffer, pH 7	50 mg/ml
Hemin chloride 0.1M KOH, pH 7.6–7.8	1 mg/ml
Protoporphyrin in 0.1M KOH, pH 7.6–7.8	1 mg/ml
Ferric chloride in water	1 mg/ml

The above medium was supplemented with protein and the final concentration of CbMM adjusted to $1 \times$ concentration. The medium was inoculated with 10 IJ of *S. carpocapsae*, incubated at 30°C, and harvested after 3 weeks. It recorded a yield of 25X. Bedding (1981) successfully mass-produced *H. bacteriophora* in a medium containing 20g nutrient broth, 10g yeast extract, 260g maize oil, 385g soy flour, and 1200g distilled water. Fifteen grams of the medium was coated on 0.8g of foam and inoculated with the symbiotic bacteria. The medium yielded about 1.64 million IJ/g. Rearing methods based on the animal medium is laborious and generated an offensive odor. Therefore, a medium based on Difco Bacto nutrient broth which is easier and more pleasant to work was developed by Wouts (1981). *Heterorhabditis heliothidis* was reared monoxenically on this medium containing the following composition.

Difco Bacto nutrient broth	0.44 g
Difco Bacto yeast extract	0.16 g
Soy flour	7.20 g
Corn oil	5.20 g
Water	27.0 g
Foam	1.0 g

The medium was coated in to polyether polyurethane sponge. The primary symbiotic bacterium was cultured on nutrient broth or yeast extract. The flask was left in the dark, and 2–3 days later the IJs were inoculated. It yielded about 10 million IJ/250 ml flask in a month. Bedding (1984) devised simple and economical animal tissue rearing medium for *Steinernema* sp. and *Heterorhabditis* sp. monoxenically. The medium was coated on crumbled polyether polyurethane sponge in a conical flask. The medium contained the following:

Composition (%)	*Steinernema* sp.	*Heterorhabditis* sp.
Homogenated pig kidney	70	60
Beef fat	10	20
Tap water	20	20

This medium yielded about half a million IJ per gram of medium in 4–5 weeks for *Heterorhabditis* sp. and 2–3 weeks for *Steinernema* sp. at 20–28 °C. Kondo and Ishibashi (1984) reported that the propagation of *S. feltiae* on nutrient agar and dog food media. They found that the medium, which favors bacterial growth, was most suitable for culturing nematodes. Development of the nematode population rapidly increased in dog food medium till day 15 and then gradually decreased. Lis (1984) defined a medium containing beef extract, peptone, and corn meal with foam. Abe (1897) evaluated wheat bran, rice bran, chicken feces, bagasse, and water in different proportion as a medium for culturing *S. feltiae*. The medium containing 25 g wheat bran, 10 g rice shell, 2–3 g salad oil, and 75 g water gave the best yielded. Nematode production increased from 10^4–10^7 when the medium was amended with salad oil or fermented with *Bacillus*. Yang and Xu (1989) evaluated culture conditions for mass production of EPNs. Best results were obtained with a 1 l vessel autoclaved for a period of 20 min 121 °C and addition of nematodes 2–4 days after adding the bacteria.

Gray and Webster (1989) carried out series of experiments to determine the optimum physical and chemical components of the medium for maximum production of *Neoplectana* (=*Sterinernema*) *carpocapsae* DD-136 and *H. heliothidis*. The optimum temperature for growth was 30 °C. The symbiotic bacteria grew best at pH 6.5. Tryptic soya broth (15 g/l) and yeast extract (5 g/l) enhanced the yields. D-glucose (14 mM) enhanced the yield of both nematodes. Corn syrup (4 g/l) is a good source of carbon. Vitamin supplement did not improve yields.

Heterorhabditis heliothidis production was enhanced by $MgCl_2$ (10 mM), KCl (10 mM), and KNO_3 (1 mM). Ogura and Mamiya (1989) propagated *S. kushidai* on 10 different combinations of dog food agar, peptone, distilled water, and buffer. The medium containing 8.5% dog food, 0.2% agar, 1.5% peptone, and buffer (Sorensen phosphate buffer, pH 6.47) to 100 ml yielded 1.8×10^5 IJ after 20 days after inoculation (DAI). The yield was higher when the diet was supplemented with peptone. They also found that bacteria and/or nematodes metabolized large amounts of protein to propagate. Wang et al. (1990) found that 100 IJ/80 g of medium was optimum for multiplication of *S. carpocapsae* in solid culture medium. An increased inoculum size resulted in reduction of adult size and rate of multiplication. Kondo and Ishibashi (1991) compared the growth and propagation of *S. carpocapsae*, *S. feltiae*, and *S. glaseri* on dog food agar and nutrient agar media. They found that dog food was superior. They showed multiplication of 1×10^4 for *S. carpocapsae* and *S. glaseri* and 1×10^3 for *S. feltiae*. Han et al. (1992) evolved successful cultivation of Steinermatids and Heterorhabditis in medium containing corn flour, fishmeal, and lipid. The growth of nematode population decreased with increase in nematode level. The medium yielded 0.69×10^6 IJ/g. Ogura and Haraguchi (1993) reported two artificial media for *S. kushidai* and the composition are as follows:

Dog Food Peptone Agar	
Powdered dog food	8.8%
Peptone	1.2%
Agar	0.2%
Sorensen phosphate buffer, pH 6.5	
Pig Intestine Peptone Agar	
Chopped steamed pig intestine	26.4%
Peptone	1.2%
Agar	0.3%
Sorensen phosphate buffer pH 6.5	

Pig intestine peptone agar medium yielded 344,300, and later 444,000 IJ in 20 days at 25 °C.

Han et al. (1993) studied the effect of inoculum size, temperature, and time on in vitro production of *S. carpocapsae* on medium containing yeast extract, soy flour, egg, and lard. The medium was inoculated with primary form of symbiotic bacterium. The highest yield of 38×10^6/50 g was obtained with inoculum of 3×10^5 in 20 days at 19 °C. Richou et al. (1992) studied the interaction effect of inoculum size, temperature, and time on in vitro production of *S. carpocapsae*, which was established in sponge culture medium containing 1% peptone, 15% soy flour, 20% egg, and 5% lard in a 500 ml flask. Nematodes were added to the medium 3 days after inoculation with primary form of bacterium. The highest yield of 38×10^6 IJ/50 g of medium was obtained in 29 days at 19 °C. Hong (1995) tested several culture media for the growth of *S. carpocapsae* and *S. glaseri*. The medium containing 2% chicken liver, 20% silkworm pupae, 5% lard, 9% soy bean powder, 18% flour, 1% yeast, and 24% water yielded 2×10^6 IJ/flask.

Yang Huaiwen et al. (1997) studied the quality of *S. carpocapsae* produced on plant protein, animal protein, plant and animal protein media, and in vivo culture. The medium containing 87.8% duck intestine and 12.2% silkworm pupae yielded about 5.8×10^5 IJ/gm. Comparisons on fatty acids content of *S. carpocapsae* cultured on two plant-based and two animal-based media showed that palmitic and oleic acids were found to be increased in nematodes cultured in animal media. High levels of linoleic and linolenic acids and low levels of palmitic acid were found in nematodes cultured in plant media. Joseph Rajkumar and Sivakumar (1998) standardized a medium for *S. glaseri*. MacConkey's agar containing soy flour and dextrose which supported 868 times more multiplication of a native *Steinernema* sp. in 15 days at 25 °C, was not suitable for *S. glaseri*.

In vitro mass production techniques for *S. glaseri* and *Steinernema* sp. was been developed by Vyas et al. (1999). The medium containing 10 ml of 10% chicken extract, 0.5 g yeast extract, 0.5 g beef extract, and 90 ml distilled water (pH 6.5) yielded about 50,000 nematodes from 1.5 cm foam incorporated with 5 ml medium in 15 days. The symbiotic bacterium *Xenorhabdus* grew and multiplied rapidly in the medium. Hussaini et al. (2000) mass-produced native *Steinernema* sp. in four different combinations of dog biscuit medium. The medium containing 1.25 g dog biscuit, 0.25 g peptone, 0.0625 g beef extract, and 25 ml water yielded about 24.5×10^5 IJ 30 DAI.

Vyas et al. (2001) mass-produced native *Steinernema* sp. using 21 animal and plant protein-based media. Maximum production of nematodes was recorded in chicken egg yolk medium, which was highly economical and better than universally-used dog food biscuit agar. Hatab and Gaugler (2001) compared the growth, lipid, and fatty acid content of *H. bacteriophora* produced in cultures supplemented with lipids, olive oil, canola oil, safflower oil, and cod liver oil, while beef and lard-supplemented diet yielded 40–45% less than insect lipid. The culture time was increased by 30–35% in media supplied with beef fat and lard. Hussaini et al. (2002) tried nine different plant and animal protein media for production of three indigenous isolates of *S. carpocapsae*, two indigenous isolates of *S. tricornutum* and *H. indica*. Of nine different media evaluated, modified egg yolk medium containing 7 g egg yolk, 20 g soy flour, 0.8 g NaCl, 5 g oil, and 60 ml distilled water yielded the best. The yield was 69.70×105 IJ/flask. *S. bicornutum* had not multiplied on any of the media 25 DAI.

13.3 Liquid Culture Media

McCoy and Glaser (1936) used fermented potato mash medium to culture *S. glaseri*. The medium contains two-thirds chopped potato and one-third sweet potato. Pure culture of baker's yeast was inoculated on potato gruel and allowed to ferment for 20–40 h. The medium produced $>4 \times 10^6$ IJ/200 mm agar plate. McCoy ad Girth (1938) cultured *S. glaseri* in a liquid medium containing grounded veal pulp. The veal was chopped in a food chopper and then infused for 48 h with twice sterile-distilled water (SDW). The infusion was filtered using a flannel cloth. This produced about 20 million IJ/tray. Glaser (1940a) cultured *S. glaseri* in a liquid medium in Erlenmeyer flask using pieces of sterile rabbit kidney of 3–4 mm, in 0.5% NaCl or kidney extract devoid of particulate matter. Biotechnology Australia Pvt. Ltd. (International No:86/1074) (1986) developed a culture medium for *S. feltiae*. The medium contained 100 g of ox kidney homogenate per liter and 10 g of yeast/liter was fermented by *Xenorhabdus nematophilus* at 28 °C at pH 6.8 and an oxygen concentration of 50% for 24 h. Two thousand IJs were inoculated at 23 °C and with 20% oxygen saturation. The medium yielded 90,000 IJ/ml in 10 days (Wouts, 1991). Mass production of *S. glaseri* was described by Stoll (1951). The medium contained 9 ml of heart infusion broth and 25 mg dextrose in 0.5% NaCl at pH 6–6.5 in 22×180 mm shaking tubes. The medium produced $25–75 \times 10^3$ nematodes in 3 weeks. Buecher and Popiel (1989) reared *S. feltiae* in a liquid culture along with its bacterial symbiont.

A technology for production of the biocontrol agent complex *H. megidis–Photorhabdus luminescens* in monoxenic liquid culture in laboratory-scale bioreactors was described by Ehlers et al. (1998). The dauer juvenile yields varied between 21 and 68 million IJ/liter. The effect of nutrition on fecundity and yield of *S. carpocapsae* in continued culture was evaluated by Ehlers et al. (2000) using a liquid culture medium developed by Ehlers et al. (1998). The medium contains the following composition:

Trypticase soy broth	10.0 g
Nutrient broth	10.0 g
Yeast extract	5.0 g
Casein	5.0 g
KCl	0.35 g
CaCl$_2$	0.21 g
NaCl	5.0 g
Thistle oil	10.0 g
Tap water	1.0 g

The medium yielded about 12,000 IJ/ml in 14 days.

Gil (2002) tested various carbon sources for liquid culture to improve the yield of *H. bacteriophora*. Canola oil was considered to be the suitable carbon source compared with carbohydrate when applied as a sole source of carbon. However, when applied together they significantly increased nematode yield. The results indicated that glucose is superior carbon source for bacteria and canola oil is optimum for nematodes. The application of fed batch culture provides significant enhancement of nematode yield in liquid culture. When 25 mg of glucose/ml was supplemented to 25 mg oil-based liquid culture, the highest yield of 3.52×10^5 IJ was obtained in 12 days. Nematode growth was suppressed by 75 mg glucose/ml. EPNs can be mass-produced in vivo and in vitro (Friedman, 1990; Shapiro-Ilan and Gaugler, 2002).

The most commonly used insect host for in vivo production is the greater wax moth, *G. mellonella*, larvae (Lindegren et al., 1993; Flanders et al., 1996; Kaya and Stock, 1997) because of its high susceptibility, commercial availability as a food for birds, lizards, and fish baits, and high yield (Woodring and Kaya, 1988). While in vitro mass production is achieved via a solid medium (Bedding, 1981, 1984) or in liquid culture (Friedman, 1990; Ehlers, 2001), in vitro mass production has advantage in terms of economy of scale over in vivo production. However, in vivo culturing is still useful for research and can also be used in developing countries as a cottage industry for commercial application (Gaugler and Han, 2002). The first attempt at in vitro mass production was not very successful due to a lack of knowledge about the symbiotic bacteria (Glaser, 1932).

However, much later, when the bacterial symbionts were discovered and their importance realized for the reproduction of *S. carpocapsae*, a foundation was laid down for the in vitro mass production of EPNs (Poinar and Thomas, 1966). EPNs were grown on artificial medium consisting of nutrient broth, yeast extract, and vegetable oil, cooked with flour and coated onto polyether polyurethane sponge and incubated for 3 days on 25°C in a flask. A yield of up to 10 million was obtained in one month (Wouts, 1981). However, Bedding developed a monoxenic solid culture, which was successfully commercialized for the first time

(Bedding, 1981, 1984). In his method, he used the same ingredients as that of Wouts (1981) but fortified the medium additionally with emulsified beef fat and pig's kidneys generating a higher yield of approximately 6×10^5–10×10^5 IJs/g of medium (Bedding, 1984). Liquid culture production was first attempted by Stoll (1953) by incubating raw liver extract in a shaker. He achieved a yield of 400 IJs/ml. In order to provide the nematodes with an adequate supply of oxygen, a sterile air was bubbled through the liquid media in bottles with shear stresses tolerable for the nematodes (Buecher and Hansen, 1971).

Their study paved the way for mass production in liquid media. The problems of aeration and shear forces were resolved by Pace et al. (1986) and increased the yield further up to a level of 40,000 IJs/ml after 10 days. Friedman (1990) reported the development of technique for large-scale production in liquid culture using 80,000 l fermenters and achieved a yield of 50×10^{12} IJs/month. His method decreased the cost of production as well. The first liquid culture production and commercialization of EPNs was initiated by the company Biosys in 1992 (Ehlers, 1996). Nowadays, most of the EPNs are produced by companies situated in Europe and the USA using liquid culture fermentation.

Recent improvements in liquid culture fermentation have not only increased the yield but also has commercialized more nematode species, such as *S. carpocapsae*, *S. feltiae*, *S. glaseri*, *Steinernema scapterisci*, *S. riobravis*, *S. kraussei*, *H. bacteriophora*, *H. indica*, *H. megidis*, and *Heterorhabditis marelatus* (Kaya et al., 2006). The next step following mass production is formulation. EPNs are available in different formulations (Grewal, 2002). While formulating nematodes, two things are deemed necessary; a long shelf-life of nematodes and their ease in application. The simplest method is impregnating a moist substrate (such as sponge, cedar shavings, peat, vermiculite, etc.) with nematodes.

The sponge needs to be squeezed in water before application, to release the nematodes, whereas nematodes from other carriers could be applied directly to the soil as a mulch. However, these formulations are labor-intensive, require continuous refrigeration and lack economy of scale and therefore can be used on small scale only. Some of the better formulations developed include activated charcoal (Yukawa and Pitt, 1985), calcium alginate (Kaya and Nelsen, 1985; Georgis, 1990), flowable gels (Georgis and Manweiler, 1994), wheat flour (Connick et al., 1993), and attapalgite clay chips (Bedding, 1988). The calcium alginate formulation increases the shelf-life of *S. carpocapsae* up to 3–4 months. However, the disposal of thin sheets of plastics and time-consuming extraction steps were some of the problems which rendered this formulation unsuitable for commercial application. One of the better formulations is water dispersible granules (WDG) which gives a maximum shelf-life of up to 6 months for the comparatively storage-stable nematode, *S. carpocapsae* (Silver et al., 1995). The improvement in shelf-life was attributed to the induction of partial anhydrobiosis by a mixture of ingredients in the formulation such as silica, clays, cellulose, lignin, and starches (Grewal, 2000b). Nematode-infected cadavers have also been used (Shapiro-Ilan et al., 2001, 2003; Ansari et al., 2009), but these formulations lack economy of scale as they involve in vivo production and there are difficulties in dose determination, storage, and application (Gaugler and Han, 2002).

EPNs are potential biocontrol agents of a variety of insect pests, they have proven efficacy, kill their host quickly, are safe to the environment, mammals and plants, can be mass-produced both in vivo and in vitro, are compatible with many other control measures, can be applied easily with conventional spray equipment, and are exempt from registration in many countries (Kaya and Gaugler, 1993). However, despite of all these advantages, the limited shelf-life places a major barrier to their long-term commercial application. The shelf-life of partially desiccated IJs in the WDG formulation at 25 °C was 5–6 months for *S. carpocapsae*, 2 months for *S. feltiae*, and 1 month for *Steinernema riobrave* (Grewal, 2000a). This formulation, however, is no longer available.

The limited shelf-life of EPNs has been addressed by many authors (Selvan et al., 1993; Grewal, 2000a; Ehlers et al., 2005). These studies have focused on increasing the longevity and storage of EPNs through improved formulations and the induction of partial desiccation and potentially anhydrobiosis. Anhydrobiosis slows down metabolic activity and enables the nematode to preserve their fat stores, allowing them to live for longer (Grewal, 2000a). However, a long-term storage method based on desiccation has not been met with complete success (Womersley, 1990). EPNs can be stored for long time in liquid nitrogen, but only in small quantities (Popiel and Vasquez, 1991; Curran et al., 1992). Nematode cryopreservation allows EPNs to be stored for experimental purposes but not in commercial quantities and at high cost. This necessitates the search for alternative methods for the long-term storage of EPNs.

14. COMMERCIAL FORMULATION

Although the IJs of EPNs can be stored for several months in water in refrigerated bubbled tank, high cost and difficulties of maintaining quality preclude the deployment of this method. Therefore, nematodes are usually formulated into solid or semiliquid substrates soon after they are produced.

14.1 Formulations with Actively Moving Nematodes

Nematode placement on or in inert carriers provide a convenient means to store and ship small quantities. The inert carrier formulations are easy and less expensive to make (Grewal, 2002).

14.1.1 Sponge

Georgis (1990) made the sponge formulation placing steinernematids and heterorhabditis onto clean, polyether–polyurethane sponges at rates of 50–100 nematodes per cm² surface area. An aqueous suspension was prepared by hand-squeezing the sponge in a small volume of water to extract nematodes form the sponge. Nematodes on sponges could be stored for 1–3 months at 5–10 °C. About $5–25 \times 10^6$ IJs were placed on a sheet of sponge that was then placed in a plastic bag. The bags were placed on ice packs for shipping. *H. bacteriophora* and *H. marelatus* in sponge formulations were found effective against *Otiorhynchus sulcatus*, *P. japonica*, and *Chrysoteuchia topiaria*. Owing to the time-consuming and physically demanding removal method, this formulation is not suitable (Grewal, 2002).

14.1.2 Vermiculite

Grewal (2002) reported that the vermiculite formulation was a significant improvement over sponge. An aqueous nematode suspension was mixed homogeneously with vermiculite and the mixture was placed in thin polyethylene bags. The vermiculite–nematode mixture was added to the spray tank directly, mixed in water and applied as spray or drench. The shelf-life of *S. carpocapsae* in vermiculite formulation was 0.1–0.2 months at 25 °C and 5.0–6.0 months at 2–10 °C (Grewal, 2002).

14.2 Formulations with Reduced Mobility Nematodes

Since the nematode activity was high in inert carriers and the stored energy reserves of IJs rapidly depleted, formulations have been developed in which the mobility of nematodes is minimized either through physical trapping or by using metabolic inhibitors (Grewal, 2002).

14.2.1 *Physical Trapping*

14.2.1.1 ALGINATE GEL

Georgis (1990) trapped nematodes in thin sheets of calcium alginate spread over plastic screens. The nematodes were released from alginate gel matrix by dissolving it in water with the aid of sodium citrate. Grewal (1998) reported that alginate-based *S. carpocapsae* products were the first to possess room temperature shelf-life (3–4 months) and led to increased acceptability of nematodes in high-value niche markets.

14.2.1.2 FLOWABLE GELS

A formulation in which nematodes were mixed in a viscous flowable gel or paste to reduce activity was reported by Georgis (1990). However, nematode shelf-life was shorter than in alginate gels. Chang and Gehert (1995) described a paste formulation in which nematodes were mixed in a hydrogenated oil and acrylamide. The nematodes could be easily entrapped in calcium alginate matrix by a fast, gentle, aqueous, and room temperature process. Kaya and Nelsen (1985) encapsulated *S. feltiae* and *H. heliothidis* in calcium alginate and fed to larvae of *S. exigua* in petri dishes. The nematodes were released when the insect larvae were fed on the gel beads. When moisture was present, larval mortality was nearly 100%. *Steinernema feltiae* in the capsules survived for eight months with no detectable loss in infectivity when stored at 4 °C. Capinera and Hibbard (1987) tested the alginate gel capsules as bait formulation against grass hoppers, *Melanoplus* spp. and the black cutworm, *A. ipsilon* (Capinera et al., 1988), with moderate success. Another potential use of capsules was to incorporate nematodes with seeds for prophylactic control of insects, such as the cabbage root maggot (*Delia radicum*), onion maggot (*D. antique*), or corn rootworms, *Diabrotica* spp. (Kaya et al., 1987). This approach was test-verified by placing a quantity of *S. carpocapsae* and a single tomato seed into each alginate gel bead. Nematodes were released from the beads during seed germination and were in position to protect seedlings from insect attack.

Navon et al. (1999) evaluated edible-to-insects calcium alginate capsules for *S. riobrave*. They reported a high nematode survival for 48 h in the gel at 61% relative humidity. Chang and Gehert (1992) developed macrogels containing encapsulated *S. carpocapsae* along with a water-holding acrylamide in gellan gum. Gellan gum is liquid at room temperature and can form gel with the addition of a divalent cation such as calcium.

14.2.2 *Metabolic Arrest*

Yukawa and Pitt (1985) described a formulation wherein nematodes were trapped in powdered activated charcoal that served as an adsorbent. The charcoal and nematode mixture was stored in sealed containers. Grewal (1998) reported that the addition of a proprietary metabolic inhibitor reduces oxygen demand and enables storage of concentrated *S. carpocapsae*, *S. feltiae*, and *S. riobrave* without bubbling air for extended periods at room temperature. Over 7×10^9 IJs of *S. carpocapsae* could be stored for 6 days at room temperature in a 10 l container without loss in viability. The liquid concentrate formulation is used for shipping *S. riobrave* for application on citrus against *Diaprepes* root weevil.

14.3 Formulations with Anhydrobiotic Nematodes

Many formulations of EPNs are prepared by desiccating IJs in order to reduce their rate of consumption of stored energy reserves. Womersely (1990) referred to EPN as quiescent anhydrobiotes as they are capable of only partial anhydrobiosis. Induction of partial anhydrobiotes had been achieved successfully by controlling water activity of the formulations (Bedding, 1988; Silver et al., 1995; Grewal, 2000a) (Table 1).

14.3.1 Bait

Georgis et al. (1989) tested bait formulations of EPNs on black cutworms, *A. ipsilon*, and tawny mole crickets, *Scapteriscus vicinus*, by adding hydroxyethyl cellulose and glycerin to nematodes to minimize their migration from the bait. However, the bait formulations did not outperform the aqueous nematode suspension when applied against black cutworms, although it produced a significant pest reduction compared with control (Capinera et al., 1988). When trap stations were used that ensured nematode contact with the target pest and protected nematodes from sunlight and desiccation, the baits outperformed the standard chemical insecticides against adult houseflies in pig units (Renn, 1998) and German cockroaches in apartments (Appel et al., 1993).

TABLE 1 Commercial Use of EPN *Steinernema* and *Heterorhabditis* as Bioinsecticides

EPN species	Major pest(s) targeted—as recommended by various commercial companies
Steinernema glaseri	White grubs (scarabs, especially Japanese beetle, *Popillia* sp.)
Steinernema kraussei	Black vine weevil, *Otiorhynchus sulcatus*
Steinernema carpocapsae	Turf grass pests—billbugs, cutworms, armyworms, sod webworms, chinch bugs. Orchard, ornamental, and vegetable pests—codling moth, cranberry girdler, dogwood borer, and other clearwing borer species, black vine weevil, peach tree borer, shore flies (*Scatella* spp.)
Steinernema feltiae	Fungus gnats (*Bradysia* spp.), shore flies, western flower thrips
Steinernema scapterisci	Mole crickets (*Scapteriscus* spp.)
Steinernema riobrave	Citrus root weevils (*Diaprepes* spp.)
Heterorhabditis bacteriophora	White grubs (scarabs), cutworms, black vine weevil, flea beetles, corn rootworm
Heterorhabditis megidis	Weevils
Heterorhabditis indica	Fungus gnats, root mealybug, grubs
Heterorhabditis marelatus	White grubs (scarabs), cutworms, black vine weevil

14.3.2 Gels

Steinernematids and heterorhabditis are compatible with many gel-forming polyacrylamides. Nematodes were placed on the gel and packed inside a mesh bag from which they are easily extracted in water (Georgis, 1990). Bedding and Butler (1994) developed a formulation in which nematode slurry was mixed in anhydrous polyacrylamide so that the resulting gel attained a water activity between 0.800 and 0.995. The nematodes were partially desiccated, but survival at room temperature was low.

14.3.3 Powders

Bedding (1988) described a formulation in which nematodes were mixed in clay to remove excess surface moisture and to produce partial desiccation. The formulation, termed a "sandwich," consisted of a layer of nematodes between two layers of clay. This formulation was commercialized by Biotechnology Australia Ltd., but was later discontinued due to inconsistent storage stability and clogging of spray nozzles. Grewal (1998) developed an improved wettable powder formulation that enabled storage of heterorhabditis and steinernematids at room temperature (2.0–3.5 months). In this formulation S. carpocapsae was effective against S. frugiperda, Sphenophorus parvulus, Otiorhynchus ovatus, Cyclocephala borealis, and Fumibotrys fumalis (Grewal, 2002).

14.3.4 Granules

Capinera and Hibbard (1987) described a granular formulation in which nematodes were partially encapsulated in Lucerne meal and wheat flour. Later Connick et al. (1993) described an extruded or formed granule in which nematodes were distributed throughout a wheat gluten matrix. This "Pesta" formulation included a filler and a humectants to enhance nematode survival. However, granules rapidly dry out during storage resulting in poor nematode survival. A WDG formulation was developed in which IJs were encased in 10–20 cm diameter granules consisting of mixtures of various types of silica, clay, cellulose, lignin, and starch (Georgis et al., 1995; Silver et al., 1995). The granular matrix allowed access of oxygen to nematodes during storage and shipping. At optimum temperature, the nematodes entered into a partial anhydrobiotic state owing to the slow removal of their body water by the formulation (Grewal, 2000a,b). This is the first commercial formulation that enabled storage of S. carpocapsae for over 6 months at 25 °C.

15. MASS MULTIPLICATION AND FORMULATION

EPNs are currently produced by different methods either in vivo or in vitro (solid and liquid culture) (Shapiro-Ilan and Gaugler, 2002). In vivo production is a simple process of culturing specific EPNs in live insect hosts and requires minimal technology and the use of a surrogate host (typically larvae of wax moth (G. mellonella)), trays, and shelves. In vivo production uses a White trap (White, 1927), which takes advantage of the juvenile stage's natural migration away from the host cadaver. However, this method is not cost-effective for scaled-up productions and may be only ideal for small markets (Shapiro-Ilan et al., 2002a). In vitro culturing of EPNs is based on introducing nematodes to a pure culture of their symbiont in a nutritive medium. Significant improvements to in vitro culture utilizing large fermenters are used by to produce large quantities of EPNs for commercial use. Nematodes can be stored and formulated in different ways including the use of polyurethane sponge, WDGs, vermiculite, alginate gels, and baits. Formulated EPNs can be stored for 2–5 months depending on the nematode species and storage media and conditions. Unlike other microbial control agents (fungi, bacteria, and viruses) EPNs do not have a fully dormant resting stage and they will use their limited energy during storage. The quality of the nematode product can be determined by nematode virulence and viability assays, age, and the ratio of viable to nonviable nematodes (Grewal et al., 2005).

EPNs, such as *S. glaseri*, *Steinernema siamkayai*, and *H. indica* were formulated in alginate gel capsules and tested in vitro at two temperatures (5 and 25 °C) for their storage and infectivity against rice moth larva, *C. cephalonica* Stainton. The nematode *S. glaseri* survived longer up to 24 weeks and 100% survival was observed up to 4 weeks at 5 °C. However, their infectivity was 100% up to 4 weeks at 5 °C. It was followed by *S. siamkayai* which survived up to 22 weeks at 5 °C and 14 weeks at 25 °C. *Heterorhabditis indica* revealed 100% infectivity up to 2 weeks at 5 and 25 °C (Umamaheswari et al., 2006). In many commercial EPN-based biopesticide companies, formulations ranging from the impregnation of EPNs on artificial sponge to highly advanced granular formulations have been developed. Major challenges have included the development of room temperature shelf-stability, ease of use, and contamination control. The aim of this work was to evaluate the suitability of two indigenous EPNs, *H. bacteriophora* (BA1) and *S. carpocapsae* (BA2), for formulation and storage on hydrogel, kaolinite, and calcium alginate at room temperature. The survival and virulence of both nematodes in the three formulations were discussed. Comparing the two nematodes, it was found that the storage potential of BA2 juveniles was only superior to that of BA1 in case of formulation with calcium alginate. The pathogenicity of the two EPNs in the three formulations was tested against the wax moth, *G. mellonella*. It was concluded that BA2 was more virulent than BA1 to the larvae of the wax moth, *G. mellonella*, in all tested formulas (Hussein and Abdel-Aty, 2012).

16. STORAGE

EPNs should be stored away from sunlight. Exposure to direct sunlight for 30–60 min reduced the pathogenicity (Gaugler and Boush, 1978). Bedding (1984) stored 250 million IJ of *S. glaseri* in 100 g of dry sponge loaded in 12 cm diameter polyethylene tubes aerated with aquarium pump at 1–2 °C and the mortality was less than 10%. *Steinernema feltiae* and *Heterorhabditis* sp. had an inverse relationship between survival time and temperature whereas *S. glaseri* survived for many months (Molyneux, 1985). The nematode can also be stored in activated charcoal as reported by Yukawa and Pitt (1985). Woodring and Kaya (1988) reported the storage of *Steinernema* and *Heterorhabditis* at 4–10 °C for 6–12 months and 2–4 months, respectively. Clumps are formed while storing *Heterorhabditis*, which can be prevented by adding sodium bicarbonate without adversely affecting the nematode (Woodring and Kaya, 1988). The survival of IJ of *S. carpocapsae* and *S. glaseri* gradually decreased when the pH of water decreased (Kung et al., 1990). Smith et al. (1990) tried cryopreservation of *S. feltiae* found best for maintenance of IJ. Westerman (1992) studied the influence of time of storage on efficacy, persistence, and migration rate of *Heterorhabditis* sp. He found that increase in time of storage caused decrease in quality with respect to persistence and efficiency against black vine weevil. Curran et al. (1992) cryopreserved IJ of *Steinernema* and *Heterorhabditis* using two stages incubated in glycerol and 70% methanol before storage in cryotubes in liquid nitrogen. Optimum conditions for acceptable survival of nematode was obtained by using 15% glycerol and incubated for 48h. The maximum survival recorded was 97% for *S. feltiae* in liquid nitrogen for 12 months. Lindegren et al. (1993) devised a modified method for storage of *S. carpocapsae*. Nematodes were concentrated by applying negative pressure with a medium grade aquarium pump. Nematode wool concentrate was produced by substitution of nylon mesh (10 μm opening) instead of fiberglass filter. Nematodes were stimulated to intervene forming a mat or "nematode wool" and stored at 7 °C. Ogura and Nakashima (1997) reported the poor survival of *S. kushidai* after recovery from artificial culture. They suggested preconditioning of IJ at 10 °C for 8 days before storage and survival of the IJ was above 50% for 100 days after storage at 5 °C.

Boff et al. (2000) studied the effect of storage temperatures and time on infectivity, reproduction, and development of *H. medigis* in *G. mellonella*. The IJ were stored at 5, 10, 15, and 20 °C for a period of 70 days. As the storage period increased, the number of active nematodes decreased. Independent of the dose and storage period, the highest infectivity and optimal development was observed when IJ were stored at 10 and 15 °C. Lee et al. (2000) studied the cryopreservation of IJ of *S. carpocapsae* at −190 °C in liquid nitrogen and its efficacy was analyzed on survival and pathogenicity with glycerol pretreatment and storage periods. They used 22% glycerol for 6, 12, and 24h followed by 70% methanol at 0 °C for 10 min. These treatments did not show any change in pathogenicity during cryopreservation. The viability of *S. bicornutum* and *H. indica* under storage in distilled water and 0.1% formalin was evaluated by Hussaini et al. (2000). The storage temperature at 8 and 30 °C at a population density of 500 IJ/ml was tested for 6 weeks. At both the temperatures the suspended isolates were viable in distilled water than in 0.1% formalin. After storage under 8 °C the virulence of IJ stored in distilled water was tested against *G. mellonella* which showed reduction in virulence by 13–20%. The infectivity of *H. indica* was not reduced by storage. Karunakar et al. (2001) studied the influence of storage temperature population density and duration of storage on survival of *S. feltiae*, *S. glaseri*, and *H. indica*. The results showed that both the steinernematids survived best at 7.5 °C and *Heterorhabditis* at 10 °C. The percentage of survival was higher when stored at 250 IJ/ml compared with 500 and 1000 IJ/ml. Negative correlation was observed between the duration and survival. *S. feltiae*, *S. glaseri*, and *H. indica* survived for 120, 90, and 90 days, respectively.

Tetra Pak containers were evaluated as an alternative to tissue culture flasks for nematode storage. Tetra Pak containers were an excellent alternative to tissue culture flasks for storage of *H. bacteriophora* and will more than likely be useful for other EPN species (Gulcu and Hazir, 2011). EPNs are mass produced for use as biopesticides using in vivo or in vitro methods (Shapiro-Ilan and Gaugler, 2002). In vivo production (culture in live insect hosts) requires a low level of technology, has low startup costs, and the resulting nematode quality is generally high, yet cost efficiency is low. The approach can be considered ideal for small markets. In vivo production may be improved through innovations in mechanization and streamlining. A novel alternative approach to in vivo methodology is production and application of nematodes in infected host cadavers; the cadavers (with nematodes developing inside) are distributed directly to the target site and pest suppression is subsequently achieved by the

IJs that emerge. In vitro solid culture, i.e., growing the nematodes on crumbled polyurethane foam, offers an intermediate level of technology and costs. In vitro liquid culture is the most cost-efficient production method but requires the largest startup capital. Liquid culture may be improved through progress in media development, nematode recovery, and bioreactor design. A variety of formulations have been developed to facilitate nematode storage and application including activated charcoal, alginate and polyacrylamide gels, baits, clay, paste, peat, polyurethane sponge, vermiculite, and WDGs. Depending on the formulation and nematode species, successful storage under refrigeration ranges from one to seven months. Optimum storage temperature for formulated nematodes varies according to species; generally, steinernematids tend to store best at 4–8 °C whereas heterorhabditis persist better at 10–15 °C. Lipids represent the main source of energy in EPNs. In the IJ phase, the level of such reserves can be influenced by storage, and this may affect the infectivity. The aim of this study was to evaluate the percentage of lipids and the associated infectivity in IJs of *S. carpocapsae*, *S. riobrave*, *Heterorhabditis* sp. JPM4, *Heterorhabditis* sp. CCA, and *Heterorhabditis* sp. PI that had been stored at different temperatures (8–28 °C) for various times (0–180 days). The amount of lipids present in IJs was evaluated histologically using a colorimetric method, while infectivity was assayed against *G. mellonella* larvae. Lipid levels diminished with increasing storage time for all nematodes, but the rates of decrease varied according to storage temperature and species. Lipid reserves were conserved for longer storage periods at 8, 16, and 20 °C, while at 24 and 28 °C the percentage of lipids decreased rapidly. The infectivity of IJs of *Heterorhabditis* spp. was less tolerant than those of *Steinernema* spp. to temperatures of 8, 24, and 28 °C. Thus, while storage at 8 °C was optimal for conserving lipid reserves, infectivity was best preserved at temperatures of 16 and 20 °C giving rise to the least reduction in infectivity after 180 days of storage. In this way, lipids and infectivity are influenced by different storage temperatures for the species tested (Andalo et al., 2011).

17. APPLICATION TECHNOLOGY

Application technology plays a major role in determining the efficacy of EPNs.

17.1 Foliar Application

Spray distribution and droplet size are important considerations when applying pesticides in the foliar environment. Lello et al. (1996) reported that the higher output hydraulic nozzles (standard fan and full cone) deposited greater numbers of nematodes onto leaves and gave up to 98% mortality of *Plutella xylostella* on Chinese cabbage. Mason et al. (1998a) reported that over 90% of the drops never had nematodes in case of spinning disc nozzles. Nematode deposition on foliage was generally increased by the addition of adjuvants (Mason et al., 1998b). However, the major reason for the lack of success of foliar application of nematodes is the intolerance of IJs to extremes of desiccation (Lello et al., 1996), temperature (Grewal et al., 1994), and UV radiation (Gaugler and Boush, 1978; Gaugler et al., 1992). Effective protection of *S. carpocapsae* all strain from sunlight was achieved using fluorescent brightener, Tinopal LPW, and viral enhancer Blankophor BBH, as UV protectants (Nickle and Shapiro, 1992, 1994). They retained 95 and 85%, respectively, of original nematode infectivity to *G. mellonella* after 4h exposure to direct sunlight. Addition of antidesiccants Biosys 727 (20% w/w), natural wax (18% w/w), or Folicote (6% w/w) with *S. carpocapsae* at 125 IJ/ml reduced the damage to cotton foliage by *S. litura* by 46% (Glazer et al., 1992). Sezhian et al. (1996) demonstrated that addition of an insect phagostimulant (Entice) to nematode suspension at 0.032% in addition to 0.1% antidesiccant glycerine and the surfactant 0.05% Triton enhanced the effectiveness of *S. carpocapsae* (DD-136) against the fourth instar larvae of *S. litura* feeding on sunflower heads in laboratory and field tests. Schroeder and Sieburth (1997) studied the impact of various surfactants, such as Agri Sc, Silwet L-77 and Tegopren 5840 on control of root weevil *Diaprepes abbreviates* larvae with *S. riobrave*. All surfactants increased the larval mortality of four days for the filter paper bioassay and one week for the potted plant bioassay using citrus seedlings.

Lello et al. (1996) stated that the effects of sunlight can be minimized by applying the nematodes at dusk, but maintaining high humidity (>80% relative humidity) and free water on leaf surfaces was more difficult to achieve. However, shading from direct sunshine when spraying the nematodes in vegetable fields was found to improve mortality rate of *P. xylostella* by 33.9% compared with the use of a humidifier (Ping et al., 1999). Seob et al. (2000) evaluated the efficacy of several commercial antidesiccants, namely, alkyl glucoside, carboxyl methyl cellulose, glycerol, Keltrol-F, Kunipia-G, and Laponite. Leaf spraying of nematodes at 5000 IJs/ml of distilled water containing 0.1% Keltrol-F resulted in 87.7% control efficacy of *S. exigua* on groundnut.

Hanounik et al. (2000) stated that the use of *Heterorhabditis* sp. isolate HSP-17 with the antidesiccant, leaf-shield, induced 87.5% mortality of adult red palm weevil, *Rhynchophorus ferrugineus*, compared with 65% mortality when HAS-17 was used with the antidesiccant Liqua-gel. In vitro technology is used when large-scale production is needed at reasonable quality and cost. IJs of EPNs are usually applied using various spray equipment and standard irrigation systems. Enhanced efficacy in EPN applications can be facilitated through improved delivery mechanisms (e.g., cadaver application) or optimization of spray equipment. Substantial progress has been made in recent years in developing EPN formulations, particularly for above-ground applications, e.g., mixing EPNs with surfactants or polymers or with sprayable gels. Bait formulations and insect host cadavers can enhance EPN persistence and reduce the quantity of nematodes required per unit area. This chapter provides a summary and analysis of factors that affect production and application of EPNs and offers insights for their future in biological insect suppression (Shapiro-Ilan et al., 2012).

18. FIELD EFFICACY

1. Application through irrigation water
2. Soil drenches
3. Spraying—soil, foliar—high volume and low volume sprayer, knapsack/rocker sprayers
4. Tractor mounted with nematode application filtered in tank dose 2–4 billion IJ/ha soil insects, 4–8 billion IJ/ha foliar insects.

The ability of the EPNs to kill an intended insect host can be tested through several indices like LC_{50} (Georgis, 1991), penetration rate (Caroli et al., 1996), invasion efficiency (Epsky and Capinera, 1993), establishment efficiency (Hominick, 1990; Epsky and Capinera, 1993), and pathogenicity and it may be related to the persistence of the soil-applied nematodes (Forschler and Gardner, 1991). EPNs are tested against a wide variety of insects with rather inconsistent results (Klien, 1990). While there are success stories of efficacy of the EPN, certain factors also add to their failure (Georgis et al., 2006). The efficacy of the EPNs against insects varies with insect and nematode species (Shapiro-Ilan et al., 2002b; Ansari et al., 2006), the developmental stage of the insect host (Ansari et al., 2008), and is influenced by various biotic and abiotic factors. The selection of the correct species of nematode is of primary importance. For example, the use of *S. riobrave* against the key insect pest *Diabrotica abbreviatus* (Shapiro-Ilan et al., 2002a), and *S. glaseri* against *Hoplia philanthus* (Ansari et al., 2008). Abiotic environmental factors including soil type (Kung et al., 1990; Shapiro et al., 2000; Koppenhöfer and Fuzy, 2006), soil moisture (Molyneux and Bedding, 1984; Koppenhöfer et al., 1995; Grant and Villani, 2003; Shapiro-Ilan et al., 2006; Koppenhöfer and Fuzy, 2007), soil texture (Georgis and Poinar, 1983; Molyneux and Bedding, 1984; Barbercheck and Kaya, 1991), temperature (Grewal et al., 1994; Shapiro et al., 1999; Chen et al., 2003; Wetchayunt et al., 2009), light (Gaugler et al., 1992; Nickle and Shapiro, 1994; Lello et al., 1996; Fujiie and Yokoyama, 1998), and bulk density (Ames, 1990; Portillo-Aguilar et al., 1999) can influence the efficacy of nematodes.

In addition to these abiotic factors, culture medium and propagation temperature (Buecher and Popiel, 1989; Hazir et al., 2001), targeted insect pest (Koppenhöfer and Fuzy, 2003; Grewal et al., 2004; Koppenhöfer et al., 2004), phenology of the targeted pest (Smith et al., 1993; Jackson and Brooks, 1995), nematode species and strains (Fuxa et al., 1998; Ebssa et al., 2004; Ansari et al., 2006), and host plant (Barbercheck and Wang, 1996) can also have an influence on the efficacy. Dose of nematode application (Selvan et al., 1993; Wang and Bedding, 1996), production technique (Gaugler and Georgis, 1991; Grewal et al., 1999), and synergism with other control agents like pesticides (Koppenhöfer and Kaya, 1998; Koppenhöfer et al., 2000), entomopathogenic fungi (Pereault et al., 2009), and bacteria like *Bacillus thuringiensis* (Koppenhöfer et al., 1999; Shapiro-Ilan et al., 2004) can also affect the efficacy of EPNs. Among biotic factors, predation from natural enemies (Gilmore and Potter, 1993), competition with other species (Everard et al., 2009), and antagonism of microorganisms pose a great threat to the efficacy of EPNs (Kaya and Koppenhöfer, 1996). Furthermore, application method can greatly influence the efficacy of EPN (Perez et al., 2003). Some other factors related to the nematode, host, and/or environment (Kaya and Gaugler, 1993) are also suggested to influence the efficacy (Smith, 1999). Several contributing factors influencing the efficacy of the EPNs are recognized by different authors in different experimental situations for different insect pests. However, some promising field efficacies have been described against many insects such as black cutworm, sugar beet weevil, sweet potato weevil, and house fly adult in animal-rearing farms (Georgis et al., 2006).

18.1 Long-Term Control

Stanislav et al. (2005) revealed that the efficacy of four species of EPNs (*S. feltiae, S. carpocapsae, H. bacteriophora, H. megidis*) for the control of granary weevil adults (*Sitophilus granarius*) was tested under laboratory condition. Suspensions of nematodes were applied in three concentrations (5000, 10,000, and 20,000 IJs per ml) at three temperatures (15, 20, and 25 °C). After 1 week, mortality of the beetles was assessed. No significant differences in the percentage of mortality were determined between treatments with *S. feltiae*, *S. carpocapsae*, and *H. bacteriophora*, while *H. megidis* was the least efficient. Mortality of the beetles was statistically significantly higher at 20 and 25 °C.

Commercially produced EPNs are currently in use for control of scarab larvae in lawns and turf, fungus gnats in mushroom production, invasive mole crickets in lawn and turf, black vine weevil in nursery plants, and diaprepes root weevil in citrus, in addition to other pest insects. However, demonstrated successful control of several other insects, often has not led to capture of a significant share of the pesticide market for these pests (Lawrence and Ramon, 2012). Adult and nymphs of German cockroaches *Blattella germanica* were infested with two EPNs, *S. carpocapsae* and *H. bacteriophora*, in vitro at 15, 20, and 25 °C. Results showed that *S. carpocapsae* was more virulent than *H. bacteriophora* at all temperatures over the exposure times of 24, 48, and 72 h. Infection of nymphs by *S. carpocapsae* at 20 °C caused 100% mortality after 72 h and 100% mortality of adults at 25 °C after 72 h. Reproduction of *S. carpocapsae* significantly increased at 25 °C after 72 h compared with *H. bacteriophora* (Baker et al., 2012).

EPNs were screened for efficacy against the cottonwood borer, *Plectrodera scalator* (Fabricius). *Steinernema feltiae* SN and *S. carpocapsae* all killed 58% and 50% of larvae, respectively, in Whatman filter paper bioassays but less than 10% in diet cup bioassays. *S. glaseri* NJ, *S. riobrave* TX, and *H. indica* MG-13 killed less than 10% of larvae in both assays. *Heterorhabditis marelata* IN was infective in the diet cup bioassay and killed 12.9% of larvae in a Whatman No.1 filter paper bioassay. The nematode isolates tested are not suitable for use as biological control agents against *P. scalator* (Fallon et al., 2006). A survey was conducted for EPNs in various agricultural fields in the Eastern Black Sea region of Turkey. A total of 77 soil samples were collected from 15 distinct geographic areas during 2010–2011. Seven entomopathogenic nematode isolates (ZET02, ZET04, ZET09, ZET28, ZET31, ZET35, and ZET76) were detected from the soil samples (9.1% positive) using the *Galleria* baiting technique. Morphological and molecular characterizations of the isolates were performed for species identification. Five isolates were identified as *H. bacteriophora* (ZET02, ZET04, ZET09, ZET28, and ZET35) and two isolates were identified as *S. feltiae* (ZET31 and ZET76). The efficacy of all isolates was tested on *Melolontha melolontha* larvae in plastic boxes and pot experiments. Different concentrations of nematodes at 0, 500, 1000, or 2000 IJ/ml and two different temperature regimes (15 and 25 °C) were used. One hundred percent mortality was obtained from the ZET09 and ZET35 isolates at a concentration of 2000 IJs/ml at 25 °C. The same isolates also provided 100% protection with 100 IJs/cm in strawberry planted pot experiments (Erbas et al., 2014) (Table 2).

18.2 Short-Term Control

The infectivity of three Egyptian Heterorhabditid nematode species *H. bacteriophora* (Poinar), *H. indica* (Poinar), and *Heterorhabditis baujardi* (Poinar) (Rhabditida: Heterorhabditilidae) and one imported Steinernematid species, *Steinernema abbasi* (Elawad) (Rhabditida: Steinernematidae) were evaluated against preadults of the citrus mealybug, *Planococcus citri* (Risso) (Hemiptera: Pseudococcidae) and adults of Egyptian mealybug, *Icerya aegyptiaca* (Douglas) (Hemiptera: Monophlebidae) under laboratory conditions. Also, laboratory and semifield evaluation of the efficiency of EPN alone or mixed with wax remover oleyl polypeptide, against adults and nymphs of the mealybug *Ferrisia virgata* (Cockerell) (Hemiptera: Pseudococcidae) were carried out. Nematode reproduction in their cadavers

TABLE 2 Use of Nematodes as Biological Insecticides (Shapiro-Ilan and Gaugler, 2010)

Pest common name	Pest scientific name	Key crop(s) targeted	Efficacious nematodes[a]
Artichoke plume moth	*Platyptilia carduidactyla*	Artichoke	Sc
Armyworms	Lepidoptera: Noctuidae	Vegetables	Sc, Sf, Sr
Banana moth	*Opogona sachari*	Ornamentals	Hb, Sc
Banana root borer	*Cosmopolites sordidus*	Banana	Sc, Sf, Sg
Billbug	*Sphenophorus* spp. (Coleoptera: Curculionidae)	Turf	Hb, Sc
Black cutworm	*Agrotis ipsilon*	Turf, vegetables	Sc
Black vine weevil	*Otiorhynchus sulcatus*	Berries, ornamentals	Hb, Hd, Hm, Hmeg, Sc, Sg
Borers	*Synanthedon* spp. and other sesiids	Fruit trees and ornamentals	Hb, Sc, Sf
Cat flea	*Ctenocephalides felis*	Home yard, turf	Sc
Citrus root weevil	*Pachnaeus* spp. (Coleoptera: Curculionidae	Citrus, ornamentals	Sr, Hb
Codling moth	*Cydia pomonella*	Pome fruit	Sc, Sf
Corn earworm	*Helicoverpa zea*	Vegetables	Sc, Sf, Sr
Corn rootworm	*Diabrotica* spp.	Vegetables	Hb, Sc
Cranberry girdler	*Chrysoteuchia topiaria*	Cranberries	Sc
Crane fly	Diptera: Tipulidae	Turf	Sc
Diaprepes root weevil	*Diaprepes abbreviates*	Citrus, ornamentals	Hb, Sr
Fungus gnats	Diptera: Sciaridae	Mushrooms, greenhouse	Sf, Hb
Grape root borer	*Vitacea polistiformis*	Grapes	Hz, Hb
Iris borer	*Macronoctua onusta*	Iris	Hb, Sc
Large pine weevil	*Hylobius albietis*	Forest plantings	Hd, Sc
Leaf miners	*Liriomyza* spp. (Diptera: Agromyzidae)	Vegetables, ornamentals	Sc, Sf
Mole crickets	*Scapteriscus* spp.	Turf	Sc, Sr, Scap
Navel orange worm	*Amyelois transitella*	Nut and fruit trees	Sc
Plum curculio	*Conotrachelus nenuphar*	Fruit trees	Sr
Scarab grubs[b]	Coleoptera: Scarabaeidae	Turf, ornamentals	Hb, Sc, Sg, Ss, Hz
Shore flies	*Scatella* spp.	Ornamentals	Sc, Sf
Strawberry root weevil	*Otiorhynchus ovatus*	Berries	Hm
Small hive beetle	*Aethina tumida*	Bee hives	Yes (Hi, Sr)
Sweet potato weevil	*Cylas formicarius*	Sweet potato	Hb, Sc, Sf

[a] At least one scientific study reported 75% suppression of these pests using the nematodes indicated in field or greenhouse experiments. Subsequent/other studies may reveal other nematodes that are virulent to these pests. Nematode species used are abbreviated as follows: Hb = Heterorhabditis bacteriophora, Hd = Heterorhabditis downesi, Hi = Heterorhabditis indica, Hm = Heterorhabditis marelata, Hmeg = Heterorhabditis megidis, Hz = Heterorhabditis zealandica, Sc = Steinernema carpocapsae, Sf = Steinernema feltiae, Sg = Steinernema glaseri, Sk = Steinernema kushidai, Sr = Steinernema riobrave, Sscap = Steinernema scapterisci, Ss = Steinernema scarabaei.
[b] Efficacy of various pest species within this group varies among nematode species.

was detected. Moreover, the effect of *S. carpocapsae* alone or with medical additive (Docusate Sodium) comparing with Super Misrona oil were studied on the grape mealybug, *Planococcus ficus* (Signoret). Three methods of formulation of Egyptian were tested for their effectiveness on *H. bacteriophora* and *S. carpocapsae*. These formulations were WDG, calcium alginate gel, and infected cadavers. The infectivity of the formulated nematodes was tested against *I. aegyptiaca* adults (Abd El Rahman et al., 2012).

This study reinvestigated the use of EPNs for their effect against *Eldana saccharina* and other pests of sugar cane, including thrips (*Fulmekiola serrata*) and white grubs *H. consanguinea* (Coleoptera: Scarabaeidae). EPNs were isolated from soils sugar cane cultivation from 10 locations in KwaZulu-Natal, South Africa. These were subsequently cultured in vivo on *E. saccharina* larvae and then identified as being either *Heterorhabditis* or *Steinernema* using polymerase chain reaction-based molecular methods. Results showed that, of the 10 isolates, 5 were *Heterorhabditis* and 5 were *Steinernema* species. Two *Steinernema* isolates (EST3D and GING13G) were used to conduct pathogenicity tests and thereafter in pot and field trials. Results of the pathogenicity tests showed that 100% mortality was achieved with both isolates within 48 h. The isolates were then used at a high rate (2000 IJs/m²) and a low rate (1000 IJs/m²) in a field trial to determine their efficacy against thrips. Results from the first sampling, three weeks after application, showed that isolate EST3D at the high rate resulted in significantly fewer thrips on sugar cane leaves. A pot trial was also conducted using white grubs identified as *Hypopholis* sp. Susceptibility of the white grub to the IJs was seen when application involved the addition of an insecticide containing the active ingredient, imidacloprid (Pillay et al., 2009).

Laboratory bioassays were conducted to establish the potential of EPNs as biocontrol agents of *P. ficus* (Signoret). Six indigenous and two commercially available nematode species were screened for their efficacy in killing adult female *P. ficus*. The two indigenous species with the most promising results were *Heterorhabditis zealandica* and *Steinernema yirgalemense*, which were responsible for 96 and 65% mortality, respectively. Tests were conducted to compare the efficacy of *H. bacteriophora* and *S. feltiae* produced in vivo and in vitro. *Heterorhabditis bacteriophora* showed no significant difference in efficacy between the two production methods, but in vivo-cultured *S. feltiae* produced a significantly higher mean mortality of 40%, in contrast to a 19% mean mortality with in vitro produced IJs. The capability of both *H. zealandica* and *S. yirgalemense* to complete their life cycles in the host and to produce a new cohort of IJs was demonstrated. Bioassays indicated a concentration-dependent susceptibility of *P. ficus* to *H. zealandica*, *S. yirgalemense* and commercially produced *H. bacteriophora*, with LC_{50} and LC_{90} values of 19 and 82, 13 and 80, and 36 and 555, respectively. Both *H. zealandica* and *S. yirgalemense* were able to move 15 cm vertically downward and infect *P. ficus* with a respective mortality of 82 and 95%. This study showed *P. ficus* to be a suitable host for *H. zealandica* and *S. yirgalemense*, with both nematode species showing considerable potential for future use in the field control of *P. ficus* (Le Vieux and Malan, 2013).

Two species of were studied in terms of their survivability, detectability by the subterranean termite *Heterotermes aureus*, and their ability to induce mortality in *H. aureus*. *Heterorhabditis bacteriophora* (Poinar) and *S. carpocapsae* (Weiser) are nematodes sold commercially as a means of biocontrol for termites. Laboratory tests showed a difference in the survivability between nematode species and also ability to kill termites. It was shown that *H. aureus* had no ability to detect either nematode species when given a choice between arenas infested with nematodes and not. Though nematodes might have some limited capacity for termite control, those considering using nematodes to control *H. aureus* may consider the species of nematode before making a purchase.

The pathogenic effect of indigenous EPNs, *H. indica*, and commercial biopesticides of three fungal pathogens (*Metarhizium anisopliae*, *Beauveria bassiana*, and *Trichoderma viride*), one antagonistic bacteria (*Pseudomonas fluorescens*), and two neem-based biopesticides (Neem and Nimor) were tested on the greater wax moth, *G. mellonella*, larva under laboratory conditions. The efficacy of biopesticides was tested individually or in combination with *H. indica*. Pathogenic interaction on *G. mellonella* larva by *H. indica* and biopesticides was assessed at 12 h intervals after storage. Significant differences in the percentage of larval mortality were determined among the biopesticide treatments. When tested in isolation, *B. bassiana* imposed greater mortality on host larva (40%) compared with other biopesticides; while *P. fluorescens* and *H. indica* combination proved to be the most efficient causing 100% mortality of *G. mellonella* larvae after 24 h of storage. Progeny produced by *H. indica* on single *G. mellonella* was found to be more (140,108 IJs/larva) in the combination treatment with *T. viride*. Pathogenicity influence of *H. indica* when exposed with other biopesticides on host larva, have proved to be more virulent and compatible. The results on pathogenicity of EPNs *H. indica* on *G. mellonella* larvae are a novelty in the field of biological control (Sankar et al., 2009).

19. CONCLUSION

The soil environment is the ideal location for EPNs and insects. Soil is the natural reservoir for these nematodes and their potential for use as biopesticides is great. Many commercial firms are selling EPNs for control of black vine weevil larvae in nurseries throughout the world, in cranberry bogs in the USA, and for citrus weevil control in Florida. In many places, poor insect control by conventional chemical insecticides has spurred the commercialization of nematode production. Storage capabilities of the most effective nematode species and strains need to be developed so that consistent field results are achieved. A better understanding of the complex interactions in the soil ecosystem will enable us to maximize use of EPNs for biocontrol of insect pests.

References

Abd El Rahman, R.M., Abd El Razzik, M.I., Osman, E.A., Mangoud, A.A.H., 2012. Efficacy of the entomopathogenic nematodes and their based-product on some species of mealy bugs (Hemiptera: Pseudococcidae) in Egypt. Egyptian Academic Journal of Biological Science 5 (3), 193–196.

Abe, Y., 1897. Culture of *Steinernema feltiae* (DD 136) on bran media. Japanese Journal of Nematology 17, 131–134.

Akhurst, R.J., 1982. Antibiotic activity of *Xenohabdus* spp., bacteria symbiotically associated with insect pathogenic nematodes of the families Heterorhabditidae and Steinernematidae. Journal of General Microbiology 128, 3061–3065.

Akhurst, R.J., 1990. Safety to nontarget invertebrates of nematodes of economically important pests. In: Laird, L.A., Lacey, E.W., Davidson, E.W. (Eds.), Safety of Microbial Insecticides. CRC Press, Boca Raton, FL, pp. 233–240.

Akhurst, R., Smith, K., 2002. Regulation and safety. In: Gaugler, R. (Ed.), Entomopathogenic Nematology. CAB International, Wallingford, pp. 311–332.

Ames, L., 1990. The Role of Some Abiotic Factors in the Survival and Motility of *Steinernema scapterisci*. University of Florida, Gainesville.

Andalo, V., Monino, A., Maximiniano, C., Campos, V.P., Mendonca, L.A., 2011. Influence of temperature and duration of storage on the lipid reserves of entomopathogenic nematodes. Revista Colombiana de Entomologia 37 (2), 203–209.

Ansari, M.A., Shah, F.A., Moens, T.M., 2006. Field trials against Hoplia philanthus (Coleoptera: Scarabaeidae) with a combination of an entomopathogenic nematode and the fungus Metarhizium anisopliae CLO 53. Biological Control 39, 453–459.

Ansari, M.A., Shah, F.A., Butt, T.M., 2008. Combined use of entomopathogenic nematodes and Metarhizium anisopliae as a new approach for black vine weevil, Otiorhynchus sulcatus, control. Entomologia Experimentalis et Applicata, 129, 340–347.

Ansari, M.A., Hussain, M.A., Moens, M., 2009. Formulation and application of entomopathogenic nemaotode-infected cadavers for control of Hoplia philanthus in turfgrass. Pest Management Science 65, 367–374.

Appel, A.G., Benson, E.P., Ellenberger, J.M., Manweiler, S.A., 1993. Laboratory and field evaluations of an entomogenous nematode (Nematoda: Steinernematidae) for German cockroach (Dictyoptera: Blatellidae) control. Journal of Economic Entomology 86, 777–784.

Baker, N.R., Ali, H.B., Gowen, S., 2012. Reproduction of entomopathogenic nematodes Steinernema carpocapsae and Heterorhabditis bacteriophora on the German cockroach Blattella germanica at different temperatures. Iraqi Journal of Science 53 (3), 505–512.

Barbercheck, M.E., Kaya, H.K., 1991. Effect of host condition and soil texture on host finding by the entomogenous nematodes, Heterorhabditis bacteriophora (Rhabditida: Heterorhabditidae) and Steinernema carpocapsae (Rhabditida: Steinernematidae). Environmental Entomology 20, 582–589.

Barbercheck, M.E., Wang, J., 1996. Effects of cucurbitacin D on in vitro growth of Xenorhabdus and Photorhabdus spp., symbiotic bacteria of entomopathogenic nematodes. Journal of Insect Pathology 68, 141–145.

Bathon, H., 1996. Impact of entomopathogenic nematodes on non-target parasitism hosts. Biocontrol Science and Technology 6, 421–434.

Baweja, V., Sehgal, S.S., 1997. Potential of Heterorhabditis bacteriophora Poinar (Nematode, Heterorhabditidae) in parasitizing Spodoptera litura Fabricius in response to malathion treatment. Acta Parasitologica 42, 168–172.

Bedding, R.A., Butler, K.L., 1994. Method for Storage of Insecticidal Nematodes. World Patent No. WO 94/05150.

Bedding, R.A., 1981. Low cost in vitro mass production of Neoaplectana and Heterorhabditis species (Nematoda) for field control of insect pests. Nematologica 27, 109–114.

Bedding, R.A., 1984. Large scale production, storage and transport of the insect parasitic nematodes Neoaplectana spp. and Heterorhabditis spp. Annals of Applied Biology 104, 117–120.

Bedding, R.A., 1988. Storage of insecticidal nematodes. World Patent No. WO 88/08668.

Bedding, R.A., 1990. Logistics and strategies for introducing entomopathogenic nematode technology into developing countries. In: Gaugler, R., Kaya, H.K. (Eds.), Entomopathogenic Nematodes in Biological Control. CRC Press, Boca Raton, FL, pp. 233–245.

Bedding, R.A., Akhurst, R.J., 1975. A simple technique for the detection of insect parasitic nematodes in soil. Nematologica 21, 109–110.

Bedding, R.A., Molyneux, A.S., 1982. Penetration of insect cuticle by infective juveniles of Heterorhabditis spp. (Heterorhabditidae: Nematoda). Nematologica 28, 354–359.

Bednarek, A., Popowska-Nowak, E., Pezowicz, E., Kamionek, M., 2000. Influence of insecticides, carbosulfan and carbofuran on the mortality and pathogenicity of entomopathogenic nematodes and fungus. Bulletin OILB/SROP 23, 115–117.

Boemare, N.E., Laumond, C., Luciani, J., 1982. Mise en kvidence de la toxicogkn&se provoquke par le nkmatode entomophage Neoaplectana carpmapsue Weiser chez l'insecte Galleria mellonella. C. R. Academy of Science Series III 295, 543–546.

Boff, M., Wieger, G.L., Smuts, P.H., 2000. Effects of storage temperature and temperature on infectivity reproduction and development of H. megidis in G. mellonella. Nematology 2, 635–644.

Buecher, E.J., Popiel, I., 1989. Liquid culture of the entomopathogenic nematodes S. feltiae with its bacterial symbiont. Journal of nematology 21, 500–504.

Buecher, E.J., Hansen, E.L., 1971. Mass culture of axenic nematodes using continuous aeration. Journal of Nematology 3, 199–200.

Buecher, E.J., Hansen, E.L., Yarwood, E.A., 1970. Growth of nematode in defined medium containing hemin and supplemented with commercially available proteins. Nematologica 16, 403–409.

Cabanillas, H.E., Raulston, J.R., 1994. Pathogenicity of Steinernema riobravis against corn earworm Helicoverpa zea (Boddie). Fundamental and Applied Nematology 17, 219–223.

Campbell, J.F., Gaugler, R., 1993. Nictation behavior and its ecological implications in the host search strategies of entomopathogenic nematodes (Heterorhabditidae and teinernematidae). Behaviour 126, 155–169.

Capinera, J., Hibbard, B.E., 1987. Bait formulations of chemical and microbial insecticides for suppression of crop-feeding grasshoppers. Journal of Agricultural Entomology 4, 337.

Capinera, J., Pelissier, D., Menout, G.S., Epsky, N.D., 1988. Control of black cutworm, *Agrotis ipsilon* (Lepidoptera: Noctuidae), with entomogenous nematodes (Nematoda: Steinernematidae, Heterorhabditidae). Journal of Invertebrate Pathology 52, 427.

Caroli, L., Glazer, I., Gaugler, R., 1996. Entomopathogenic nematode infectivity assay: comparison of penetration rate into different hosts. Biocontrol Science and Technology 6, 227–234.

Chang, F.N., Gehert, M.J., 1992. Insecticide Delivery System and Attractant. US Patent No. 5: 141–744.

Chang, F.N., Gehert, M.J., 1995. Stabilized Insect Nematode Compositions. US Patent No. 5: 401–506.

Chen, S., Li, J., Han, X., Moens, M., 2003. Effect of temperature on the pathogenicity of entomopathogenic nematodes (*Steinernema and Heterorhabditis* spp.) *to Delia radicum*. BioControl 48, 713–724.

Ciche, T.A., Ensign, J.C., 2003. For the insect pathogen. Photorhabdus luminescens, which end of a nematode is out. Applied and Environmental Microbiology 69, 1890–1897.

Chitwood, B.G., Chitwood, M.B., 1937. An Introduction to Nematology. Monumental Printing Co., Baltimore, MD, p. 53.

Connick, W.J.J., Nickle, W.R., Vinyard, B.J., 1993. 'Pesta': new granular formulations for *Steinernema carpocapsae*. Journal of Nematology 25, 198–203.

Cui, L., Gaugler, R., Wang, Y., 1993. Penetration of steinernematid nematodes (Nematoda: Steinernematidae) into Japanese beetle larvae, *Popillia japonica* (Coleoptera: Scarabaeidae). Journal of Invertebrate Pathology 62, 73–78.

Curran, J., Gilbert, C., Butler, K., 1992. Routine cryopreservation of isolates of *Steinernema* and *Heterorhabditis* spp. Journal of Nematology 24, 269–270.

Das, J.N., Divaka, B.J., 1987. Compatibiligy of cetain pesticides with DD-136 nematode. Plant Protection Bulletin 39, 20–22.

De Nardo, E.A.B., Grewal, P.S., 2003. Compatibility of *Steinernema feltiae* (Nematoda: Steinernematidae) with pesticides and plant growth regulators used in glasshouse plant production. Biocontrol Science and Technology 13, 441–448.

Dix, I., Burnell, A.M., Griffin, C.T., Joyee, S.A., Nugent, M.J., Downes, M.J., 1992. The identification of biological species in the genus *Heterhabditis* (Nematoda: Heterorhabditidae) by cross-breeding second generation amphimictic adults. Parasitology 104, 509–518.

Dix, D.R., Bridgham, J.T., Broderius, M.A., Byersdorfer, C.A., Eide, D.J., 1994. The FET4 gene encodes the low affinity Fe (II) transport protein of Saccharomyces cerevisiae. Journal of Biological Chemistry 269, 26092–26099.

Drees, B.M., Miller, R.W., Vinson, S.B., Georgis, R., 1992. Susceptibility and behavioural response of red imported fire ant (Hymenoptera: Formicidae) to selected entomogenous nematodes (Rhabditida: Steinernematidae and Heterorhabditidae). Journal of Economic Entomology 85, 365–370.

Dutky, S.R., 1937. Investigation of the Disease of the Immature Stages of the Japanese Beetle. (thesis). Rutgers University, New Brunswick, NJ.

Dutky, S.R., Thompson, J.V., cantwell, G.E., 1964. A technique for the mass propagation of the DD-136 nematode. Journal of Insect Pathology 6, 417–422.

Ehlers, R.V., Lunau, S., Krasomio, O.K., Osterfeld, K.H., 1998. Liquid culture of the entomopathogenic bacterium complex *H. megidis/Photorhabdus luminesencs*. BioControl 43, 77–86.

Ehlers, R.U., 2001. Mass production of entomopathogenic nematodes for plant protection. Applied Microbiology and Biotechnology 56, 623–633.

Ehlers, R.U., Niemann, I., Hollmer, S., Strauch, O., Jende, M.D., Shanmugasundaram, M., Mehta, U.K., Easwaramoorthy, S.K., Burnell, A., 2000. Mass production potential of the bacto-helminthic biocontrol complex, *Heterorhabditis indica – Photorhabdus luminescens*. Biocontrol Science and Technology 10, 607–616.

Ehlers, R.U., 1996. Current and future use of nematodes in biocontrol: practice and commercial aspects with regard to regulatory policy issues. Biocontrol Science and Technology 6, 303–316.

Ehlers, R.U., Oestergaard, J., Hollmer, S., Wingen, M., Strauch, O., 2005. Genetic selection for heat tolerance and low temperature activity of the entomopathogenic nematode-bacterium complex *Heterorhabditis bacteriophora – Photorhabdus luminescens*. Biocontrol 50, 699–716.

Endo, B.Y., Nickle, W.R., 1994. Ultrastructure of the buccal cavity region and oesophagus of the insect parasitic nematode, *Heterorhabditis bacteriophora*. Nematologica 40, 379–398.

Epsky, N.D., Capinera, J.L., 1993. QuantiWcation of invasion of two strains of *Steinernema carpocapsae* (Weiser) into three lepidopteran larvae. Journal of Nematology 25, 173–180.

Erbas, Z., Gokce, C., Hazir, S., Demirbag, Z., Demir, I., 2014. Isolation and identification of entomopathogenic nematodes (Nematoda: Rhabditida) from the Eastern Black Sea region and their biocontrol potential against *Melolontha melolontha* (Coleoptera: Scarabaeidae) larvae. Turkish Journal of Agriculture and Forestry 38, 187–197.

Everard, A., Griffin, C.T., Dillon, A.B., 2009. Competition and intraguild predation between the braconid parasitoid *Bracon hylobii* and the entomopathogenic nematode *Heterorhabditis downesi*, natural enemies of the large pine weevil, *Hylobius abietis*. Bulletin of Entomological Research 99, 151–161.

Fallon, D.J., Solter, L.F., Bauer, L.S., Miller, D.L., Cate, J.R., McManus, M.L., 2006. Effect of entomopathogenic nematodes on *Plectrodera scalator* (Fabricius) (Coleoptera: Cerambycidae). Journal of Invertebrate Pathology 92, 55–57.

Fan, X., Hominick, W.M., 1991. Effects of low storage temperature survival and infectivity of two *Steinernema* sp. Revue de Nematology 14, 407–412.

Fedorko, A., Kamionek, M., Kozlowska, J., Mianowska, E., 1977. The effect of some carbamide herbicides on nematodes from different ecological groups. Polish Ecological Studies 3, 23–28.

Filipjev, I.N., 1934. The classification of the free-living nematodes and their relations to parasitic nematode. Smithsonian Miscellaneous Collections 89, 1–63.

Flanders, K.L., Miller, J.M., Shields, E.J., 1996. In vivo production of Heterorhabditis bacteriophora Oswego (Rhabditida: Heterorhabditidae), a potential biological control agent for soil inhabiting insects in temperate regions. Journal of Economic Entomology 89, 373–380.

Forschler, B.T., Gardner, W.A., 1991. Field efficacy and persistence of entomogenous nematodes in the management of white grubs (Coleoptera: Scarabaeidae) in turf and pasture. Journal of Economic Entomology 84, 1454–1459.

Forst, S., Clarke, D., 2002. Bacteria-nematode symbioses. In: Gaugler, R. (Ed.), Entomopathogenic Nematology. CAB International, Wallingford, pp. 57–77.

Friedman, M.J., 1990. Commercial production and development. In: Gangler, R., Kaya, H.K. (Eds.), Entomopathogenic Nematodes in Biological Control. CRC Press, Boca Raton, pp. 153–172.

Friedman, M.J., Langston, S.E., Pollitt, S., 1991. Mass Production in Liquid Culture of Insect Killing Nematodes. US Patent No. 5, 023, 183.

Fujiie, A., Yokoyama, T., 1998. Effects of ultraviolet light on the entomopathogenic nematode, *Steinernema kushidai* and its symbiotic bacterium, *Xenorhabdus japonicus*. Applied Entomology and Zoology 33, 263–269.

Fujiie, A., Yokoyama, T., Fujikata, M., Sawada, M., Hasegawa, M., 1993. Pathogenicity of an entomogenous nematode, *Steinernema kushidai* on *Anomala cuprea* (Coleoptera: Scarabaeidae). Japanese Journal of Applied Entomology and Zoology 37, 53–60.

Fuxa, J.R., Richter, A.R., Agudelo-Silva, F., 1998. Effect of host age and nematode strain on susceptibility of *Spodoptera frugiperda* to *Steinernema feltiae*. Journal of Nematology 20, 91–95.

Ganguly, S., Singh, L.K., 2000. *Steinernema thermophilum* sp. n. (Rhabditida: Steinernematidae) from India. International Journal of Nematology 10, 183–191.

Gaugler, R., 1981. Biological control potential of neoaplectanid nematodes. Journal of Nematology 13, 241–249.

Gaugler, R., 1999. Matching nematode and insect to achieve optimal field performance. In: Polavarapu, S. (Ed.), Optimal use of insecticidal nematodes in pest management. Rutgers University Press, New Brunswick, NJ, pp. 9–14.

Gaugler, R., Boush, G.M., 1978. Effects of ultraviolet radiation and sunlight on the entomogenous nematode. *Neoaplectana carpocapsae*. Journal of Invertebrate Pathology 59, 155–160.

Gaugler, R., Campbell, J.F., 1991. Behavioral response of the entomopathogenic nematodes *Steinernema carpocapsae* and *Heterorhabditis bacteriophora* to oxamyl. Annual of Applied Biology 119, 131–138.

Gaugler, R., Georgis, R., 1991. Culture method and efficacy of entomopathogenic nematode. Biological Control 1, 269–274.

Gaugler, R., Bednarek, A., Campbell, J.F., 1992. Ultraviolet inactivation of Heterorhabditid and Steinernematid nematodes. Journal of Invertebrate Pathology 59, 155–160.

Gaugler, R., Han, R., 2002. Production technology. In: Gaugler, R. (Ed.), Entomopathogenic Nematology. CABI Publishing, Wallingford, UK, pp. 289–310.

Geden, C.J., Axtell, R.C., Brooks, W.M., 1985. Susceptibility of the lesser mealworm, *Alphitobius diaperinus* (Coleoptera: Tenebrionidae) to the entomogeneous nematodes *Steinernema feltiae*, *S. glaseri* (Steinernematidae) and *Heterorhabditis heliothidis* (Heterorhabditidae). Journal of Entomological Science 20, 331–339.

Georgis, R., 1990. Formulation and application technology. In: Gaugler, R., Kaya, H.K. (Eds.), Entomopathogenic Nematodes in Biological Control. CRC Press, Boca Raton, pp. 173–191.

Georgis, R., Kaya, H., Gaugler, R., 1991. Effect of Steinernematid and Heterorhabditid nematodes on non target arthropods. Environmental Entomology 20, 815–822.

Georgis, R., Dunlop, D.B., Grewal, P.S., 1995. Formulation of entomopathogenic nematodes. In: Hall, F.R., Barry, J.W. (Eds.), Biorational Pest Control Agents: Formulation and Delivery. American Chemical Society, Bethesda, MD, pp. 197–205.

Georgis, R., Manweiler, S.A., 1994. Entomopathogenic nematodes: a developing biological control technology. Agricultural Zoology Reviews 6, 63–94.

Georgis, R., Poinar Jr., G.O., 1983. Effect of soil texture on the distribution and infectivity of Neoaplectana carpocapsae (Nematoda: Steinernematidae). Journal of Nematology 15, 308–311.

Georgis, R., Wojcik, W.F., Shetler, D.J., 1989. Use of *Steinernema feltiae* in a bait for the control of black cutworms (*Agrotis ipsilon*) and tawny mole crickets (*Scapteriscus vicinus*). Florida Entomologist 72, 203–204.

Georgis, R., Koppenhöfer, A.M., Lacey, L.A., Belair, G., Duncan, L.W., Grewal, P.S., Samish, M., Tan, L., Torr, P., Van Tol, R.W.H.M., 2006. Successes and failures in the use of parasitic nematodes for pest control. Biological Control 38, 103–123.

Gil, A.C., 2002. Como elaborar projetos de pesquisa, 4 ed. Atlas, São Paulo.

Gilmore, S.K., Potter, D.A., 1993. Potential role of Collembola as biotic mortality agents for entomopathogenic nematodes. Pedobiologia 37, 30–38.

Glaser, R.W., 1931. The cultivation of a nematode parasite of an insect. Science 73, 614.

Glaser, R.W., 1932. A poathogenic nematode of the Japanese beetle. Journal of Parasitology 18, 199–201.

Glaser, R.W., 1940a. The bacteria-free culture of a nematode parasite. Proceedings of Society in Experimental Biology and Medicine 43, 512–514.

Glaser, R.W., 1940b. Continued culture of a nematode parasite in a Japanese beetle. Journal of Experimental Zoology 84, 1–12.

Glaser, R.W., Farell, C.C., 1935. Field experiments with the Japanese beetle and its nematode parasite. Journal of the New York Entomological Society 43, 345–371.

Glaser, R.W., McCoy, E.E., Girth, H.B., 1940a. The biology and culture of *Neoaplectana chresima*, a new nematode parasitic in insects. Journal of Parasitology 28, 123–126.

Glaser, R.W., McCoy, E.E., Girth, H.B., 1940b. The biology and economic importance of a nematode parasitic in insects. Journal of Parasitology 26, 479.

Glaser, R.W., McCoy, E.E., Girth, H.B., 1942. The biology and culture of Neoaplectana chresima, a new nematode parasitic in insects. Journal of Parasitology 28, 123–126.

Glazer, I., Wyoski, M., 1990. Steinernematid and heterorhabditid nematodes for biological control of the giant looper, *Boarmia selenaria*. Phytoparasitica 18, 9–16.

Glazer, I., Nakache, Y., Klein, M., 1992. Use of entomopathogenic nematodes against foliage pests. Hassadeh 72, 626–630.

Gordon, R., Chippett, J., Tilley, J., 1996. Effects of two carbamates on infective juveniles of *Steinernema carpocapsae* all strain and *Steinernema feltiae* Umea strain. Journal of Nematology 28, 310–317.

Grant, J.A., Villani, M.G., 2003. Soil moisture effects on entomopathogenic nematodes. Environmenal Entomology 32 (1), 80–87.

Gray, B.D., Webster, J.M., 1989. The monoxenic culture of *N. carpocapsae* (DD 136) and *H. heliothidis*. Revue Nematology 12, 113–123.

Grewal, P.S., 1998. Formulations of entomopathogenic nematodes for storage and application. Japanese Journal of Nematology 28, 68–74.

Grewal, P.S., 2000b. Anhydrobiotic potential and long-term storage stability of entomopathogenic nematodes (Rhabditida: Steinernematidae). International Journal of Parasitology 30, 995–1000.

Grewal, P.S., Selvan, S., Gaugler, R., 1994. Thermal adaptation of entomopathogenic nematodes: niche breadth for infections, establishment, and reproduction. Journal of Thermal Biology 19, 245–253.

Grewal, P.S., 2002. Formulation and application technology. In: Gaugler, R. (Ed.), Entomopathogenic Nematology. CABI Publishing, Wallingford, UK, pp. 266–287.

Grewal, P.S., 2000a. Enhanced ambient storage stability of an entomopathogenic nematode through anhydrobiosis. Pest Management Science 56, 401–406.

Grewal, P.S., Converse, V., Georgis, R., 1999. Influence of production and bioassay methods on infectivity of two ambush foragers (Nematoda: Steinernematidae). Journal of Invertebrate Pathology 73, 40–44.

Grewal, P.S., Power, K.T., Grewal, S.K., Suggars, A., Haupricht, S., 2004. Enhanced consistency in biological control of white grubs (Coleoptera : Scarabaeidae) with new strains of Entomopathogenic nematodes. Biological Control 30, 73–82.

Grewal, P.S., Ehlers, R.-U., Shapiro-Ilan, D.I., 2005. Nematodesvas Biological Control Agents. CABI Publishing, Wallingford, UK.

Griffin, C.T., Downes, M.J., Block, W., 1990. Tests of Antarctic soils for insect parasitic nematodes. Antarctic Science 2, 221–222.

Gulcu, B., Hazir, S., 2011. An alternative storage method for entomopathogenic nematodes. Turkish Journal of Zoology 36 (4), 562–565.

Gupta, P., Siddiqui, M.R., 1999. Compatibility studies on *Steinernema carpocapsae* with some pesticidal chemicals. Indian Journal of Entomology 61, 220–225.

Han, R., Cao, L., Liu, X., 1993. Effects of inoculum size, temperature and time on in vitro production of *Steinernema carocapsae* Agriotos. Nematologica 39, 366–375.

Han, R., Cao, L., Liu, X., 1992. Relationship between medium composition, inoculums size, temperature and culture time in the yields of *Steinernema* and *Heterorhabditis* nematodes. Fundamental and Applied Nematology 15, 223–229.

Han, R., Ehlers, R.U., 2000. Pathogenicity, development and reproduction of *Heterorhabditis bacteriophora* and *Steinernema carpocapsae* under *axenic in vivo* conditions. Journal of Invertebrate Pathology 75, 55–58.

Hanounik, S.B., Saleh, M.M.E., Abuzuhairah, R.A., Alheiji, M., Aldhahir, H., Aljarash, Z., 2000. Efficacy of entomopathogenic nematodes with antidesiccants in controlling the red palm weevil *Rhynchophorus ferrugineus* on date palm trees. International Journal of Nematology 10, 131–134.

Hara, A.H., Kaya, H.K., 1983. Toxicity of selected organophosphate and carbamate pesticides to infective juveniles of the entomogenous nematode *Neoaplectana carpocapsae* (Rhabditida: Steinernematidae). Environmental Entomology 12, 496–501.

Hatab, A.M., Gaugler, R., 2001. Diet composition and lipid of in vitro produced Heterorhabditis bacteriophora. Biological Control 20, 1–7.

Hazir, S., Stock, S.P., Kaya, H.K., Koppenhöfer, A.M., Keskin, N., 2001. Developmental temperature eVects on Wve geographic isolates of the entomopathogenic nematode Steinernema feltiae (Nematoda: Steinernematidae). Journal of Invertebrate Pathology 77, 243–250.

Henneberry, T.J., Lindegren, J.E., Jech, L.F., Burke, R.A., 1995. Pink bollworm (Lepidoptera: Gelechiidae): effect of Steinernematid nematodes on larval mortality. South Western Entomologist 20, 25–32.

Heungens, A., Buysse, G., 1987. Toxicity of several pesticides in water solution on Heterorhabditis nematodes. Mededelingen Van de Faculteit Landbouwweterns-chappen Rijksuniversiteit Gent 52, 631–638.

Hominick, W.M., 1990. Entomopathogenic rhabditid nematodes and pest control. Parasitology Today 6, 148–152.

Hominick, W.M., Reid, A.P., Bohan, D.A., Briscoe, B.R., 1996. Entomopathonematodes: Biodiversity, geographical distribution and de conon Biological Diversity. Biocontrol Science and Technology 6, 317–331.

Hominick, W.H., 2002. Biogeography. In: Gaugler, R. (Ed.), Entomopathogenic Nematology. CABI Publishing, New York, pp. 115–143.

Hong, P.Y., 1995. Improved of culture medium for entomopathogenic nematodes. Journal of Jilin Agricultural University 17, 16–19.

House, H.L., Welch, H.E., Cleugh, T.R., 1965. Food medium of prepared dog biscuit for the mass production of the nematode DD 136 (Nematode: Steinernematidae). Nature 206, 847.

Hussaini, S.S., Kavitha, J.S., Hussain, M.A., 2001. Tolerance of some indigenous entomopathogenic nematodes in different UV protectants. Indian Journal of Plant Protection 31, 12–18.

Hussaini, S.S., Kavitha, J.S., Hussain, M.A., 2000. Mass production of a native *Steinernema* sp. (SSL2) PDBCEN 13.21 (Nematoda: Steinernematidae) on different artificial media. Indian Journal of Plant Protection 28, 94–96.

Hussaini, S.S., Singh, S.P., Parthasarathy, R., Shakeela, V., 2002. *In vitro* production of entomopathogenic nematodes in different artificial media. Indian Journal of Nematology 32, 44–46.

Hussein, M.A., Abdel-Aty, M.A., 2012. Formulation of two native entomopathogenic nematodes at room temperature. Journal of Biopesticides 5 (Suppl.), 23–27.

Ishibashi, N., Choi, D.R., 1991. Biological control of soil pests by mixed application of entomopathogenic and fungivorous nematodes. Journal of Nematology 23, 175–181.

Jackson, G.J., 1965. Differentiation of three species of *Neoaplectana* (Nematoda: Rhabditida) grown axenically. Parasitology 55, 571–578.

Jackson, J.J., Brooks, M.A., 1995. Parasitism of western corn root worm larvae and pupae by *Steinernema carpocapsae*. Journal of Nematology 27, 15–20.

Jaworska, M., 1990. Effect of some insecticides on entomophilic nematodes – wplyw niekto'rych insektycydo'w na entomopatogennenicienic. Zeszyty Problemowe Problemowe Postepo'w Nauk Rolniczych 391, 73–79.

Josephrajkumar, A., Sivakumar, C.V., 1998. Suitability of a standardized medium for entomopathogenic nematode *Steinernema glaseri*. Journal of Applied Zoological Research 9, 47.

Karunakar, G., Easwaramoorthy, S., David, H., 2001. Influence of storage temperature, population density and duration of storage on the survival on the survival of three species of entomopathogenic nematodes. Journal of Biological Control 15, 183–188.

Kaya, H.K., 1985. Susceptibility of early larva stages of *Pseudaletia unipuncta* and *Spodoptera exigua* (Lepidoptera: Noctuidae) to the entomogenous nematode *Steinernema feltiae* (Rhabditida: Steinernematidae). Journal of Invertebrate Pathology 46, 58–62.

Kaya, H.K., Burlando, T.M., 1989. Infectivity of *Steinernema feltiae* in Fenamiphos-treated sand. Journal of Nematology 21, 434–436.

Kaya, H.K., Gaugler, R., 1993. Entomopathogenic nematodes. Annual Review of Entomology 38, 181–206.

Kaya, H.K., Koppenhöfer, A.M., 1996. Effects of microbial and other antagonistic organism and competition on entomopathogenic nematodes. Biocontrol Science and Technology 6, 357–371.

Kaya, H.K., Koppenhöfer, A.M., 1999. Biology and ecology of insecticidal nematodes. In: Polavarapu, S. (Ed.), Proceeding of Workshop Optimal use of Insecticidal Nematodes in Pest Management. August 28030, 1999. New Brunswick, NJ, pp. 1–8.

Kaya, H.K., Stock, S.P., 1997. Techniques in insect nematology. In: Lacey, L.A. (Ed.), Manual of Techniques in Insect Pathology. Academic Press, Wapato, pp. 281–323.

Kaya, H.K., Nelsen, C.E., 1985. Encapsulation of Steinernematid and Heterorhabditid nematodes with calcium alginate: a new approach for insect control and other applications. Environmental Entomology 14, 572–574.

Kaya, H.K., Mannion, C.M., Burlando, T.M., Nelson, C.E., 1987. Escape of *Steinernema feltiae* from alginate capsules containing tomato seeds. Journal of Nematology 19, 287–291.

Kaya, H.K., Aguillera, M.M., Alumai, A., Choo, H.Y., Torre, M.L., Fodor, A., Ganguly, S., Hazar, S., Lakatos, T., Pye, A., Wilson, M., Yamanaka, S., Yang, H., Ehlers, R.U., 2006. Status of entomopathogenic nematodes and their symbiotic bacteria from selected countries or regions of the world. Biological Control 38, 134–155.

Klien, M.G., 1990. Efficacy against soil inhabiting insect pests. In: Gaugler, R., Kaya, H.K. (Eds.), Entomopathogenic Nematodes in Biological Control. CRC Press, Boca Raton, FL, pp. 195–214.

Kondo, E., Ishibashi, N., 1984. Growth and propagation of *Steinernema feltiae* (DD-136) on plain agar, nutrient agar and dog food media. Japanese Journal of Nematology 14, 40–48.

Kondo, E., Ishibashi, N., 1991. Dependency of three steinernematid nematodes on their symbiotic bacteria for growth and propagation. Japanese Journal of Nematology 21, 11–17.

Kondo, E., Ishibashi, N., 1986. Infection efficiency of *Steinernema feltiae* (DD-136) to the common cutworm *Spodoptera litura* (Lepidoptera: Noctuidae) on the soil. Applied Entomology and Zoology 21, 561–571.

Koppenhöfer, A.M., Fusy, E.M., 2003. *Steinernema scarabaei* for the control of white grubs. Biological Control 28, 47–59.

Koppenhöfer, A.M., Fuzy, E.M., 2007. Soil moisture effects on infectivity and persistence of the entomopathogenic nematodes, *Steinernema scarabaei*, *Steinernema glaseri*, *Heterorhabditis zealandica* and *Heterorhabditis bacteriophora*. Applied Soil Ecology 35, 128–129.

Koppenhöfer, A.M., Kaya, H.K., 1998. Synergism of imidacloprid and an entomopathogenic nematode: a novel approach to White grub (Coleoptera: Scarabaeidae) control in turfgrass. Journal of Economic Entomology 91, 618–623.

Koppenhöfer, A.M., Kaya, H.K., Taormino, S.P., 1995. Infectivity of entomopathogenic nematodes (Rhabditida: Steinernematidae) at different soil depths and moistures. Journal of Invertebrate Pathology 65, 193–199.

Koppenhöfer, A.M., Choo, H.Y., Kaya, H.K., Lee, D.W., Gelernter, W.D., 1999. Increased field and green house efficacy against scarab grubs with a combination of an entomopathogenic nematode and *Bacillus thuringiensis*. Biological Control 14, 37–44.

Koppenhöfer, A.M., Ganguly, S., Kaya, H.K., 2000. Ecological characterization of *Steinernema monticolum*, a cold adapted entomopathogenic nematode from Korea. Nematology 2, 407–416.

Koppenhöfer, A.M., Eugen, M., Fuzy, R., 2004. Pathogenicity of *Heterorhabditis bacteriophora*, *Steinernema glaseri*, and *S. scarabaei* (Rhabditida: Heterorhabditidae, Steinernematidae) against 12 white grub species (Coleoptera: Scarabaeidae). Biocontrol Science and Technology 14, 87–92.

Koppenhöfer, A.M., Fuzy, E.M., 2006. Effect of soil type on infectivity and persistence of the entomopathogenic nematodes Steinernemascarabaei, Steinernema glaseri, Heterorhabditis zealandica, and Heterorhabditis bacteriophora. Journal of Invertebrate Pathology 92, 11–22.

Kung, S.P., Gaugler, R., Kaya, H.K., 1990. Influence of soil pH and oxygen on persistence of *Steinernema* spp. Journal of Nematology 39, 149–152.

Lawrence, A.L., Ramon, G., 2012. Entomopathogenic nematodes for control of insects pests above and below ground with comments on commercial production. Journal of Nematology 44 (2), 218–225.

Laznik, Z., Toth, T., Lakatos, T., Vidrih, M., Trdan, S., 2010. Control of the Colorado potato beetle (*Leptinotarsa decemlineata* [Say]) on potato under field conditions: a comparison of the efficacy of foliar application of two strains of *Steinernema feltiae* (Filipjev) and spraying with thiametoxam. Journal of Plant Diseases and Protection 117, 129–135.

Laznik, Z., Vidrih, M., Trdan, S., 2012. The effects of different fungicides on the viability of entomopathogenic nematodes *Steinernema feltiae* (Filipjev), *S. carpocapsae* Weiser, and *Heterorhabditis downesi* Stock, Griffin & Burnell (Nematoda: Rhabditida) under laboratory conditions. Chilean Journal of Agricultural Research 72 (1), 62–67.

Le Vieux, P.D., Malan, A.P., 2013. The potential use of entomopathogenic nematodes to control *Planococcus ficus* (Signoret) (Hemiptera: pseudococcidae). South African Journal of Enology and Viticulture 34 (1), 296–306.

Lee, S., Kim, Y., Han, S., 2000. Cryopreservation of entomopathogenic nematode *S. carpocapsae*. Korean Journal of applied Entomology 39, 149–152.

Lello, E.R., Patel, M.N., Mathews, G.A., Wright, D.J., 1996. Application technology for entomopathogenic nematodes against foliar pests. Crop Protection 15, 567–574.

Lindegren, J.E., Meyer, K.F., Henneberry, T.J., Vail, P.V., Jech, L.J.F., Valero, K.A., 1993. Susceptibility of pink boll worm (Lepidoptera: Gelichiidae) soil associated stages to the entomopathogenic nematode, *Steinernema carpocapsae* (Rhabditidae: Steinernematidae). Southwestern Entomologist 18, 113–120.

Lis, S.C., 1984. A new method for mass rearing of parasitic nematodes *Neoaplectana* sp. Plant Protection Beijing China 10, 36–37.

Mason, J.M., Mathews, G.A., Wright, D.J., 1998a. Appraisal of spinning disc technology for the application of entomopathogenic nematodes. Crop Protection 17, 453–461.

Mason, J.M., Mathews, G.A., Wright, D.J., 1998b. Screening and selection of adjuvants for the spray application of entomopathogenic nematode against a foliar pest. Crop Protection 17, 461–470.

McCoy, E.E., Glaser, R.W., 1936. Nematode culture for Japanese beetle control. NJ Department of Agriculture Circular 265, 3–9.

McCoy, E.E., Girth, H.B., 1938. The culture of N. glaseri on veal pulp. NJ Department of Agriculture Circular 256, 3.

Molyneux, A.S., 1985. Survival of IJ of *Heterorhabditis* spp. and *Steinernema* spp. at various temperatures and their subsequent infectivity for insects. Revu de Nematologie 8, 165–170.

Molyneux, A.S., Bedding, R.A., 1984. Influence of soil texture and moisture on the infectivity of *Heterorhabditis* sp. D1 and *Steinernema glaseri* for larvae of the sheep blowfly, *Lucilia cuprina*. Nematologica 30, 358–365.

Navon, A., Keren, S., Salame, L., Glazer, I., 1999. An edible – to insects calcium alginate gel as a carrier for entomopathogenic nematode. Biocontrol Science and Technology 8, 429–437.

Nguyen, K.B., Smart, G.C., 1992. *Steinernema neocurtillis* n. sp. (Rhabditida – Steinernematidae and a key to species of the genus *Steinernema*. Journal of Nematology 24, 463–477.

Nguyen, K.B., Smart Jr., G.C., 1994. *Neosteinernema longicurvicauda* n. gen., n. sp. (Rhabditida: Steinernematidae), a parasite of the termite, *Reticulitermes flavipes* (Koller). Journal of Nematology 26, 162–174.

Nickle, W.R., Shapiro, M., 1992. Use of stilbene brightener, Tinopal LPW, as solar radiation protectants for *Steinernema carpocapsae*. Journal of Nematology 24, 371–373.

Nickle, W.R., Shapiro, M., 1994. Effects of eight brighteners as solar radiation protectants for *Steinernema carpocapsae*. All strain. Supplement to the Journal of Nematology 26, 784–786.

Nishimatsu, T., Jackson, J.J., 1998. Interaction of insecticides, entomopathogenic nematodes, and larvae of the western corn root worm (Coleoptera: Chrysomelidae). Journal of Economic Entomology 91, 410–418.

Ogura, Haraguchi, N., 1993. Xenic culture of *Steinernema kushidai* (Nematoda: Steinernematidae) on artificial media. Japanese Journal of Nematology 39, 266–273.

Ogura, N., Mamiya, Y., 1989. Artificial culture of an entomogenous nematode, *Steinernema kushidai* (Nematoda: Steinernematidae). Applied Entomology and Zoology 24, 112–116.

Ogura, N., Nakashima, T., 1997. Cold tolerance and acclimation of infective juveniles of *Steinernema kushidai* (Nematoda: Steinernematidae). Nematologica 43, 107–115.

Pace, W.G., Grote, W., Pitt, D.E., Pitt, J.M., 1986. Liquid culture of nematodes. International Patent WO 86/01074.

Paulo Henrique, S.S., Fernando, S.S., Elsa, J.G., Alcides, M., Cramer, F., 2014. Compatibility of entomopathogenic nematodes (Nematoda: Rhabditida) with insecticides used in the tomato crop. Nematoda 1, e03014.

Pereault, R.J., Whalon, M.E., Alston, D.G., 2009. Field efficacy of entomopathogenic fungi and nematodes targeting caged last-instar plum curculio (Coleoptera: Curculionidae) in Michigan cherry and apple orchards. Environmental Entomology 38, 1126–1134.

Perez, E.E., Lewis, E.E., Shapiro-Ilan, D.I., 2003. Impact of host cadaver on survival and infectivity of entomopathogenic nematodes (Rhabditida: Steinernematidae and eterorhabditidae) under desiccating conditions. Journal of Invertebrate Pathology 82, 111–118.

Peters, A., 1996. The natural host range of *Steinernema* and *Heterorhabditis* spp. and their impact on insect populations. Biocontrol Science and Technology 6, 389–402.

Pillay, U., Martin, L.A., Rutherford, R.S., Berry, S.D., 2009. Entomopathogenic nematodes in sugarcane in South Africa. Proceedings of the South African Sugar Technologists Association 82, 538–541.

Ping, Y., Xin, L.N., Ying, L.J., 1999. Study on application of entomopathogenic nematodes to control the diamondback moth (*Plutella xylostella*). Natural Enemies of Insects 21, 107–112.

Poinar, G.O., Karunakar, G.K., David, H., 1992. *Heterorhabditis indicus* n. sp. (Rhabditida: Nematoda) from India: separation of *Heterorhabditis* spp. by infective juveniles. Fundamental and Applied Nematology 54, 53–59.

Poinar Jr., G.O., 1976. Description and biology of a new insect parasitic rhabditoid. In: Gaugler, R., Kaya, H.K. (Eds.), Entomopathogenic Nematodes in Biological Control. CRC Press, Boca Raton, FL, pp. 23–60.

Poinar Jr., G.O., 1979. Nematodes for Biological Control of Insect. CRC Press, Boca Raton, FL. 277 p.

Poinar, G.O., 1990. Taxonomy and biology of Steinernematidae and Heterorhabditidae. In: Gaugler, R., Kaya, H.K. (Eds.), Entomopathogenic Nematodes in Biological control. CRC Press, Boca Raton, FL, pp. 23–60.

Poinar, G.O., Thomas, G.M., 1966. Significance of *Achromobacter nematophilus* Poinar & Thomas (*Achromobacteriaceae: Eubacteriales*) in the development of the nematode, DD136 (*Neoaplectana sp. Steinemematidae*). Parasitology 56, 385–390.

Popiel, I., Grove, D.L., Friedman, M.J., 1989. Infective juvenile formation in the insect-parasitic nematode *Steinernema fetiae*. Parasitol 99, 77–81.

Popiel, I., Vasquez, E.M., 1991. Cryopreservation of Steinernema carpocapsae and Heterorhabditis bacteriophora. Journal of Nematology 23, 432–437.

Portillo-Aguilar, C., Villani, M.G., Tauber, M.J., Tauber, C.A., Nyrop, J.P., 1999. Entomopathogenic nematode (Rhabditida: Heterorhabditidae and Steinernematidae) response to soil texture and bulk density. Environmental Entomology 28, 1021–1035.

Prakasa Rao, P.S., Das, P.K., Padhi, G., 1975. Note on compatibility of DD-136 (*Neoaplectana dutkyi*), an insect parasitic nematode with some insecticides and fertilizers. Indian Journal of Agricultural Science 45, 275–277.

Renn, N., 1998. The efficacy of entomopathogenic nematodes for controlling housefly infestations of intensive pig unit. Medical and Veterinary Entomolgy 12, 46–51.

Richou, H., Ying, L., Fei, P.X., 1997. Modelling of the culture parameters for production of *Steinernema carpocapsae* and *Heterorhabditis bacteriophora* in solid cultures. Natural Enemies of Insects 19, 75–83.

Richou, H., Li, C., Xiuling, L., 1992. Effects of inoculums size, temperature and time on in vitro production of *Steinernema carpocapsae* (Agriotis). Nematologica 39, 366–375.

Ritter, K.S., 1988. *Steinernema feltiae* (*Neoaplectana carpocapsae*): effect of sterols and hypolipidemic agents on development. Experimental Parasitology 67, 257–267.

Rovesti, L., Deseo, K.V., 1990. Compatibility of chemical pesticides with the entomopathogenic nematodes, *Steinernema carpocapsae* Weiser and *S. feltiae* Filipjev (Nematoda. Steinernematidae). Nematologica 36, 237–245.

Rovesti, L., Deseo, K.V., 1991. Compatibility of pesticides with the entomopathogenic nemaotdes, *Heterorhabditis heliothidis*. Nematologica 37, 113–116.

Rovesti, L., Heinzpeter, E.W., Tagliente, F., Deseo, K.V., 1988. Compatibility of pesticides with the entomopathogenic nematode *Heterorhabditis bacteriophora* Poinar (Nematoda: Heterorhabditidae). Nematologica 34, 462–476.

San-Blas, E., Gowen, S.R., Pembroke, B., 2008. *Steinernema feltiae*: Ammonia triggers the emergence of their infective juveniles. Experimental Parasitology 119, 180–185.

Sankar, M., Sethuraman, V., Palaniyandi, M., Prasad, J.S., 2009. Entomopathogenic nematode – *Heterorhabditis indica* and its compatibility with other biopesticides on the Greater wax moth – *Galleria mellonella* (L.). Indian Journal of Science and Technology 2 (1), 57–62.

Schroeder, W.J., Sieburth, P.J., 1997. Impact of surfactants on control of the root weevil *Diaprepes abbreviates* larvae with *Steinernema riobravis*. Journal of Nematology 29, 216–219.

Seob, L.S., Gyun, K.Y., Chan, H.S., 2000. Leaf spray control efficacy of the entomopathogenic nematode, *Steinernema carpocapsae* Weiser, supplemented with the selected antidesiccant, Keltrol-F, on the beet armyworm *Spodoptera exigua* (Hubner). Korean Journal of Applied Entomology 39, 199–205.

Selvan, S., Gaugler, R., Grewal, P.S., 1993. Water content and fatty acid composition of infective juvenile entomo-pathogenic nematodes during storage. Journal of Parasitology 79, 510–516.

Shapiro-Ilan, D.I., Lewis, E.E., Behle, R.W., McGuire, H.R., 2001. Formulation of entomopathogenic nematode infected Cadavers. Journal of Invertebrate Pathology 78, 17–23.

Shapiro-Ilan, D.I., Gaugler, R., Tedders, W.L., Brown, I., Lewis, E.E., 2002a. Optimization of inoculation for in vivo production of entomopathogenic nematodes. Journal of Nematology 34, 343–350.

Shapiro-Ilan, D.I., Gouge, D.H., Koppenhöfer, A.M., 2002b. Factors affecting commercial success: Case studies in cotton, turf and citrus. In: Gaugler, R. (Ed.), Entomopathogenic Nematology. CABI Publishing, Wallingford, UK, pp. 333–356.

Shapiro-Ilan, D.I., Lewis, E.E., Tedders, W.L., Son, Y., 2003. Superior efficacy observed in entomopathogenic nema-todes applied in infected-host cadavers compared with application in aqueous suspension. Journal of Inverte-brate Pathology 83, 270–272.

Shapiro-Ilan, D.I., Jackson, M., Reilly, C.C., Hotchkiss, M.W., 2004. Effects of combining an entomopathogenic fungi or bacterium with entomopathogenic nematodes on mortality of Curculio caryae (Coleoptera: Curculionidae). Biological Control 30, 119–126.

Shapiro-Ilan, D.I., Gouge, D.H., Piggott, S.J., Patterson Fife, J., 2006. Application technology and environmental con-siderations for use of entomopathogenic nematodes in biological control. Biological Control 38, 124–133.

Shapiro-Ilan, D.I., Gaugler, R., 2002. Production technology for entomopathogenic nematodes and their bacterial symbionts. Journal of Industrial Microbiology and Biotechnology 28, 137–146.

Shapiro-Ilan, D.I., Gaugler, R., 2010. Nematodes: Rhabditida: Steinernematidae & Heterorhabditidae. In: Biological Control: A Guide to Natural Enemies in North America. Shelton, A. (Ed.), Cornell University. http://www.biocontrol.entomology.cornell.edu/pathogens/nematodes.html.

Shapiro-Ilan, D., Han, R., Dolinksi, C., 2012. Entomopathogenic nematode production and application technology. Journal of Nematology 44 (2), 206–217.

Shapiro, D.L., Lewis, L.C., Obrycki, J.J., Abbas, M., 1999. Effects of fertilizers on suppression of Black cutworm (Agrotis ipsilon) Damage with Steinernema carpocapsae. Supplement to the Journal of Nematology 31, 690–693.

Shapiro, D.I., Mccoy, C.W., Fares, A., Obreza, T., Dou, H., 2000. Effects of soil type on virulence and persistence of entomopathogenic nematodes in relation to control of Diaprepes abbreviatus (Coleoptera: Curculionidae). Envi-ronmental Entomology 29, 1083–1087.

Sezhian, N., Sivakumar, C.V., Venugopal, M.S., 1996. Alteration of effectiveness of Steinernema carpocapsae Weiser (Steinernematidae: Rhabditida) against Spodoptera litura (F.) (Noctuidae: Lepidoptera) larvae on sunflower by addition of an insect phagostimulant. Indian Journal of Nematology 26, 77–81.

Shanti, A.N., Sivakumar, C.V., 1991. Comparative virulence of Steinernema glaseri Steiner and Steinernema carpocap-sae (Weiser) to the chafer Holotrichia consanguinea (Coleoptera: Scarabaeidae). Indian Journal of Nematology 21, 149–152.

Silver, S.C., Dunlop, D.B., Grove, D.I., 1995. Granular Formulations of Biological Entities with Improved Storage Stability. World Patent No. WO 95/0577.

Sivakumar, C.V., Jeyaraj, S., Subramanian, S., 1988. Observations on an Indian population of the entomophilic nema-tode Heterorhabditis bacteriophora. Journal of Biological Control 2, 112–113.

Smart Jr., G.C., 1995. Entomopathogenic nematodes for the biological control of insects. Journal of Nematology (Suppl.) 27, 529–534.

Smith, K., 1999. Factors affecting efficacy. In: Polavarapu, S., (Ed.), Proceedings of Workshop – Optimal Use of Insec-ticidal Nematodes in Pest Management, August 28–30, 1999.

Smith, B.S., Hodgson Smith, A., Popiel, I., Minter, D.M., James, E.R., 1990. Cryopreservation of entomogenous nema-tode parasite Steinernema feltiae. Cryobiology 27, 319–327.

Smith, M.T., Georgis, R., Nyczepir, A.P., Miller, R.W., 1993. Biological control of the Pecan weevil, Curculio caryae (Coleoptera : Curcueionidae) with entomopathoenic nematodes. Journal of Nematology 25, 78–82.

Sosa Jr., O., Hall, D.G., Schroeder, W.J., 1993. Mortality of sugarcane borer (Lepidoptera: Pyralidae) treated with entomopathogenic nematodes in field and laboratory trials. Journal of American Society of Sugarcane Technolo-gists 13, 18–21.

Srivastava, R.P., 1978. Studies on the feasibility of the use of microbial pesticides for the integrated control of Papilio demodocus Esper. In: Abstracts, XXth International Horticultural Congress, Sydney, Australia. International Soci-ety for Horticultural Science, Sydney, Australia, pp. 15–23. Abs. No. 1624 (En) Div. of Entomology, I.A.R.I. New Delhi, 110012, India.

Stanislav, T., Nevenka, V., Gregor, U., Lea, M., 2005. Concentration of suspension and temperature as factors of pathogenicity of entomopathogenic nematodes for the control of granary weevil, *Sitophilus granaries* (L.) (Coleoptera: Curculionidae). Acta Agriculturae Slovenica 85 (1), 117–124.

Steiner, G., 1923. *Aplectana kraussei* n. sp., eine in der Blattwespe *Lyda* sp. parasitierende Nematodenform, nebst Bomerkungen uber das Scitenorgan der parasitischen Nematoden. Zentralblatt fur Bakteriologie Parasitenkunde Infektionskrankheiten and Hygiene Abteiliong I Originale 59, 14–18.

Stoll, N.R., 1951. Axenic *N. glaseri* in fluid cultures. Journal of Parasitology 37 (Suppl.), 18.

Stoll, N.R., 1953. Infectivity of Japanese beetle grubs retained by *Neoaplectana glaseri* after seven years axenic culture. Journal of Parasitology 39 (Suppl.), 33.

Strauch, O., Stoessel, S., Anterson, J.M., 1994. UV-B damage and protection at the molecular level in plants. Photosynthesis Research 39, 475–489.

Strauch, O., Ehlers, R.U., 1998. Food signal production of *Photorhabdus luminescens* inducing the recovery of entomopathogenic nematodes *Heterorhabditis* spp. in liquid culture. Applied Microbiology and Biotechnology 50, 369–374.

Travassos, L., 1927. Sobre O genera *Oxystomatium*. Boletim Biologico (Sau Paulo) 5, 20–21.

Trdan, S., Vidrih, M., Vaalie, N., Laznik, Z., 2008. Impact of entomopathogenic nematodes in adults of *Phyllotreta* spp. under laboratory conditions. Acta Agriculturae Scandinavica B S P 58, 169–175.

Umamaheswari, R., Sivakumar, M., Subramanian, S., 2006. Survival and infectivity of entomopathogenic nematodes in alginate gel formulations against rice meal moth larva, *Corcyra cephalonica* Stainton. Natural Product Radiance 5 (2), 95–98.

Vainio, A., 1994. Effect of pesticides on long-term survival of *Steinernema feltiae* in the field. Bulletin OILB/SROP 17, 70–79.

Vyas, R.V., Yadav, P., Gheelani, Y.H., Chaudhary, R.K., Patel, N.B., Patel, D.J., 2001. In vitro mass production of native *Steinernema* sp. Annals of Plant Protection Sciences 9, 77–80.

Vyas, V.R., Patel, N.S., Patel, D.J., 1999. Mass production technology for entomopathogenic nematodes *Steinernema* spp. Indian journal of Nematology 29, 178–181.

Wang, J.X., Bedding, R.A., 1996. Population development of *Heterorhabditis bacteriophora* and *Steinernema carpocapsae* in the larvae of *Galleria mellonella*. Fundamental and Applied Nematology 19, 363–367.

Wang, J.X., Qiu, L.X., Bedding, R.A., 1990. Effects of varying inoculums size of the entomopathogenic nematode, *Steinernema carpocapsae* in solid culture. In: Proceedings of Fifth International Colloquium of Invertebrate Pathology and Microbial Control, Adelaide, Austrailia.

Webster, J.M., Chen, G., Hu, K., Li, J., 2002. Bacterial metabolites. In: Gaugler, R. (Ed.), Entomopathogenic Nematology. CABI Publishing, Wallingford, pp. 99–114.

Westerman, P.R., 1992. The influence of time of storage on performance of the insect parasitic nematode *Heterorhabditis* sp. Fundamental and Applied Nematology 15, 407–412.

Wetchayunt, W., Rattanapan, A., Phairiron, S., 2009. Temperature effect on novel entomopathogenic nematode, *Steinernema siamkayai* Stock, Somsook and Reid (n. sp.) and its efficacy against *Spodoptera* litura Fabricius (Lepidoptera: Noctuidae). Communications in Agriculture and Applied Biological Sciences 74, 587–592.

White, G.F., 1927. A method for obtaining infective nematode larvae from cultures. Science 66, 302–303.

Womersely, C.Z., 1990. Dehydration survival and anhydrobiotic potential. In: Gaugler, R., Kaya, K.H. (Eds.), Entomopathogenic Nematodes in Biological Control. CRC Press, Boca Raton, FL, pp. 117–137.

Woodring, J.L., Kaya, H.K., 1988. Steinernematid and Heterorhabditid Nematodes: A Handbook of Biology and Techniques. Arkansas Agricultural Experiment Station, Arkansas. 29 pp.

Wouts, W., Mracek, M.Z., Gerdin, S., Bedding, R.A., 1982. *Neoaplectana* Steiner, 1929. A junior synonym of *Steinernema* Travassos, 1927 (Nematoda: Rhabditida). Systematic Parasitology 4, 147–154.

Wouts, W.M., 1981. Mass production of the entomogenous nematode, *Heterorhabditis heliothidis* (Nematoda: Heterorhabditidae) on artificial media. Journal of Nematology 13, 467–469.

Wouts, W.M., 1991. Steinernema (Neoaplectana) and Heterorhabditis species. In: Nickle, W.R. (Ed.), Manual of Agricultural Nematology. Marcel Dekker. Inc., New York, pp. 555–597.

Xu, J.L., Xie, R.C., 1992. Biological comparison of two species of entomopathogenic nematodes in the wax moth and an artificial media. Natural Enemies of Insects 14, 76–82.

Yamanaka, S., Seta, K., Yasuda, M., 1986. Evaluation of the use of entomogenous nematode, *Steinernema feltiae* (str. Mexican) for the biological control of the fall webworm, *Hyphantria cunea* (Lepidoptera: Arctiidae). Japanese Journal of Nematology 16, 26–31.

Yamanaka, S., Tanabe, H., Takeuchi, K., 2002. Influence of temperature on growth and propagation of *Steinernema glaseri* (Nematoda: Steinernematidae). Japanese Journal of Nematology 30, 47–50.

Yang, H., Jian, H., Zhang, S., Zhang, G., 1997. Quality of the entomopathogenic nematode, *Steinernema carpocapsae* produced on different media. Biological Control 10, 193–198.

Yang, P., Xu, J.L., 1989. Discussion of some culture conditions affecting the output of entomogenous nematodes in mass production. Natural Enemies of Insects 11, 187–192.

Yukawa, T., Pitt, J.M., 1985. Nematode Storage and Transport. World Patent No. WO 85/03412.

Zhang, L., Shono, T., Yamanaka, S., Tanabe, H., 1994. Effects of insecticides on the entomopathogenic nematode *Steinernema carpocapsae* Weiser. Applied Entomology and Zoology 29, 539–547.

Further Reading

Dutky, S.R., 1974. Nematode parasites. In: Maxwell, F.G., Harris, F.A. (Eds.), Proceedings of the Summer Institute on Biological Control of Plant Insects and Diseases. University Press of Mississippi, pp. 576–590.

Forschler, B.T., Ali, J.N., Gardner, W.A., 1990. *Steinernema feltiae* activity and infectivity in response to herbicide exposure in aqueous and soil environments. Journal of Invertebrate Pathology 55, 375–379.

Sosa Jr., O., Beavers, J.B., 1985. Entomogenous nematodes as biological control organisms for *Ligyrus subtropicus* (Coleoptera: Scarabaeidae) in sugarcane. Environmental Entomology 14, 80–82.

Insect Viruses

Vivek Prasad, Shalini Srivastava

Molecular Plant Virology Laboratory, Department of Botany, University of Lucknow, Lucknow, India

1. INTRODUCTION

Insect pests, pathogens, and weeds are major causes of loss in food production. In an effort to mitigate these losses, control of pest populations has traditionally relied heavily on the use of synthetic broad-spectrum insecticides (organophosphates, carbamates, pyrethroids, neonicotinoids); however, prolonged use of insecticides is detrimental to the environment, a concern that has long been expressed (Carson, 1962; Epstein, 2014). Overuse of insecticides has led to contamination of the environment, killing of the pests' natural enemies, a decrease in biodiversity, and development of insecticide resistance in the pests under intense pressure. Another cause for concern is the possible effect of pesticides on neurodegenerative disorders in humans like Parkinson's disease (Kamel, 2013), and the linking of prenatal exposure to the insecticide, chlorpyrifos, with structural changes in the developing human fetal brain (Rauh et al., 2012). These correlations are not surprising at all as the majority of the pesticides exert their insecticidal effects as neurotoxins (Casida and Durkin, 2013).

A restricted use of insecticides, along with their revaluation for food security, has been strongly advocated (Verger and Boobis, 2013), and biocontrol as a means of insect control has thus come into focus. It is a safe and selective method of control and its environmentally friendly nature has made it an essential component of the integrated pest management (IPM) strategy. With biocontrol as its underlying pillar, IPM programs largely seek to curb indiscriminate use of chemical pesticides and focus on multiple tools for pest control, including breeding crop plants for insect resistance, transgenic strategies, and good agricultural practices. Although practiced for years, the potential of biocontrol is yet to emerge out of the shadow of chemical control.

A review highlights the economic value of biocontrol in IPM, in terms of investment and output (Naranjo et al., 2015). No doubt, regulated use of pesticides that are safe and show shorter environmental persistence will continue to play a major role as a primary and immediate management strategy in any effective and sustainable IPM program, but smarter methods of pest control are needed, which rely on good agricultural practices, thus reducing pesticide use. Reduced pesticide use always comes with benefits. In addition to resistance to pesticides that developed in the brown planthopper, *Nilaparvata lugens* (that devastated rice crops in regions of Asia), excess use of the recommended pesticide dose led to natural predators of this planthopper being killed as well. Without the ill-effects of overdose, a judicious use of pesticides helped keep the population of these planthoppers in check (Anonymous, 1984). Hence, biocontrol utilizing natural enemies of the insect pests is one such smart practice destined to move from an optional to a binding method of pest control, with a major role being played by entomopathogenic viruses.

2. STRATEGIES FOR BIOLOGICAL PEST CONTROL

Biocontrol has been practiced for decades and control of several important insect pests has been possible by using entomopathogenic fungi, nematodes, protozoa, bacteria (both spore-forming and nonspore-forming) and viruses, collectively termed as microbial control agents. In its broadest sense, biocontrol refers to regulating the pest population through their natural enemies that include predators, pathogens, and parasitoids, or products derived from them. As an agricultural practice, biocontrol comes in many forms.

In its most classical sense, biocontrol involves the release of an exotic (nonnative) natural enemy species to reduce and maintain pest populations at acceptable levels indefinitely (Williams et al., 2013). Thus, a classical biocontrol agent is one that is imported, augmented, and released in an area and given time to establish itself to exert a long-term controlling influence on the pest population. Augmentation may be required if the natural enemy is present in the indigenous pest population, but its numbers are too few or too slow in exerting their influence. Inoculative augmentation involves application and subsequent buildup of the biocontrol agent through repeated cycles, whereas inundative augmentation, the form most frequently utilized, involves mass killing through large-scale application of the agent for a short-term control (Lomer et al., 2001; Shah and Pell, 2003). Another strategy, called conservation, involves identification of an indigenous natural enemy of an insect pest and modification of farming practices to conserve and promote the former in the field (Fuxa, 1998; Landis et al., 2000). Sterile insect technique (SIT) programs involve the use of sterile male insects that are mass-reared and released into an infested area to outnumber the fertile (wild) males, and thus help reduce the pest population.

In this review, we first enumerate a few successful attempts at biocontrol with agents other than viruses, and then analyze in detail the use of insect viruses, particularly the Baculoviruses, both wild types and recombinants, in biocontrol of insect pest.

3. ENTOMOPATHOGENS IN BIOCONTROL

The importance of insect pathogens in regulating pest populations has been recognized and exploited for centuries, with *Bacillus thuringiensis* opening a modern chapter in biocontrol of insects (Hajek et al., 2007). Classical biocontrol was attempted in the late nineteenth century in the United States with the introduction of *Rodolia cardinalis*, the vedalia ladybeetle, a natural enemy of cottony cushion scale (*Icerya purchasi*), an invasive pest of citrus. A similar effort was put together in Mexico to combat cottony cushion scale, citrus mealybug and sugarcane stem borers (Williams et al., 2013).

Successful attempts at biocontrol of other pests soon followed, and only a few of the landmark studies are mentioned in this paragraph. A spore-forming protist called *Nosema locustae* (Microsporida: Nosematidae) was the first microbial agent to be developed as a biopesticide for locust and grasshopper (Orthoptera: Acrididae) control (Henry, 1981), and was extensively field-tested in Argentina (Lange and De Wysiecki, 1996). The spores of *N. locustae* were mass produced in vivo in the differential grasshopper, *Melanoplus differentialis*, and spore covered wheat bran was used as a bait to lure the grasshopper pests (Henry, 1985). *Scelio* (Hymenoptera: Scelionidae), which parasitizes the grasshopper egg pods, also proved to be important in biocontrol of locusts and grasshoppers (Lomer et al., 2001). Locust plagues were efficiently controlled in Sudan, with a massive reduction in population density, through oil formulations of spores of a mitosporic fungus, *Metarhizium anisopliae* (*Flavoviride*) var. *acridum* (Kooyman and Abdalla, 1998). The fungi are applied inundatively to achieve a rapid mass killing of the pest akin to a mycopesticide. The fungal conidia penetrate the insect cuticle, producing hyphal bodies that multiply in the hemocoel, and eventually prolific asexual reproduction follows with the death of the insect. In the field trials conducted in Niger, *M. anisopliae* var. *acridum* had no effect on the nontarget arthropod species, as compared to about 75% of nontarget insects belonging to Carabidae, Tenebrionidae, Formicidae and Ephydridae families getting reduced if the organophosphate pesticide, fenitrothion was used (Peveling et al., 1999). *M. anisopliae*, strains MT and E9 caused high mortality of the sugarcane borer *Diatraea saccharalis*, mortality being related to several morphological changes in the pupae of the borer (Schneider et al., 2013). *M. anisopliae* has also been effectively used as a biopesticide against several species of rhinoceros beetles (Coleoptera: Scarabaeidae: Dynastinae), which attack date, oil, and coconut palms, often leading to death of these hosts or severe reduction in productivity. Their population could be reduced at vermicomposting sites, with no effect on the earthworms (Gopal et al., 2006). The modus operandi of *Beauveria bassiana*, another entomopathogenic fungus used to control foliar pests, is the same. It also kills pests by producing toxins and proliferating within the insect body, finally covering the entire body of the dead insect, causing a disease called white muscardine in several agriculturally important pests, like whiteflies, aphids, grasshoppers, termites and beetles, to name a few. Various products incorporating *B. bassiana*, which can work as substitutes for insecticides, are available such as Mycotrol, BotaniGard, and Mycotrol O (Wraight et al., 2000). Conidia of *Beauveria brongniartii* have been employed in field trials in Switzerland against larvae of the cockchafer beetle, a serious pest of almost all plant species and trees in forests and grasslands, feeding on their roots (Keller et al., 1997).

Entomopathogens have also been tried in various combinations, and a combined formulation of *Metarhizium brunneum* and *B. bassiana* was found suitable for control of the sweetpotato weevil, *Cylas formicarius* (Reddy et al., 2014). An isolate of *Zoophthora radicans* from Israel, an entomophthoralean species, was released in Australia to control the population of the spotted alfalfa aphid, *Therioaphis trifolii* (Milner et al., 1982). Hyphomycetous fungi, *Verticillium lecanii*, *B. bassiana*, and *M. anisopliae* were also used to control aphids, an important vector for transmission of plant viruses. Commercial preparations of these fungi are available as mycopesticides for use in the glass house (Milner, 1997; Yeo et al., 2003). An invasive species of soybean aphid, *Aphis glycines*, affecting soybean production in North America, also serves as a vector for transmission of several plant viruses, including potato virus Y (Davis and Radcliffe, 2008; Ragsdale et al., 2011). Its natural enemies, the coccinellids, *Coccinella septempunctata* and *Harmonia axyridis* (Mignault et al., 2006; Xue et al., 2009), and the insidious flower bug, *Orius insidious* (Butler and O'Neil, 2008) have emerged as particularly important biocontrol agents. Asian citrus psyllid, *Diaphorina citri* (Hemiptera: Psyllidae), is an important vector for transmission of the Citrus greening organism, "*Candidatus liberibacter asiaticus*" (Bove, 2006). With no resistant varieties being available, entomopathogenic fungi, *B. bassiana* and *Hirsutella citriformis*, have been recruited to check the psyllid population (Casique-Valdes et al., 2011; Grafton-Cardwell et al., 2013). In general, entomopathogenic fungi are efficient at bringing down the pest population below a threshold that is unable to cause much harm.

Gamma proteobacteria, *Xenorhabdus* and *Photorhabdus*, which live in a mutualistic association with families of entomopathogenic nematodes (EPNs) Steinernematidae and Heterorhabditidae, respectively, are involved in the killing of a few important insect pests (Frost et al., 1997; Waterfield et al., 2009). They have been efficiently used for control of plum curculio larvae (Shapiro-Ilan et al., 2004) and codling moth in apple and pear orchards (Lacey and Shapiro-Ilan, 2008). Parasitoid wasps, *Trichogramma* (Trichogrammatidae), have also been evaluated as biopesticides (Smith, 1996) and in China, the Asian corn borer, *Ostrinia fumacalis*, and several other lepidopteran pest populations have been kept in check with its release (Wang et al., 2014).

Whitefly parasitoids, *Encarsia* and *Eretmocerus* (Hymenoptera), have been recruited in pest management programs of tomato and pepper (Stansly et al., 2005; Liu et al., 2015) to control the whitefly, *Bemisia tabaci*, an important agricultural pest that serves as a vector for transmission of several plant viruses. A highly beneficial attempt at biocontrol was witnessed when the sterile insect technique (SIT) was implemented to eradicate New World screwworm (*Cochliomyia hominivorax*) (Diptera: Calliphoridae) from North and Central America. Efficacy of SIT was also evaluated in India for a destructive internal borer of coconut palm, the red palm weevil, *Rhynchophorus ferrugineus* (Krishnakumar and Maheshwari, 2007).

Maximum success, however, has been achieved with *B. thuringiensis* that continues to be the most widely used biocontrol agent. A number of formulations of *B. thuringiensis*, chiefly consisting of spores and toxins, are commercially available and they show insecticidal activity against species of Lepidoptera, Coleoptera, and Diptera. The crystal protein toxins (Cry toxins or δ-endotoxins) are contained in the parasporal inclusions and are produced during sporulation. Following solubilization of the inclusion in the midgut of the insects, the toxins are released, activated by host proteases, and interact with midgut epithelial cells, disrupting the membrane integrity, leading to death of the insect (Gill et al., 1992; Schnepf et al., 1998). Cry toxins are completely biodegradable and are highly specific to their target insect.

Bacillus thuringiensis has been used to control leafrollers, budmoths, and fruitworms (Westigard et al., 1986; Blommers, 1994). Transgenic corn and cotton plants expressing insecticidal protein genes from *B. thuringiensis* (known as Bt crops) have been successfully grown in various countries, the economic impact of which has been calculated for India where Bt cotton has benefitted the farmers with an increased yield of 24% per acre and a profit of 50% among smallholders (Shelton et al., 2002; Kathage and Qaim, 2012).

Insect-resistant genetically modified rice has been developed in China for commercial production. This is an important breakthrough as there are over 200 insect pests of rice in China, with stem borers and planthoppers being the major pests (Huang et al., 2005; Chen et al.,

2011). Despite the sporadic development of field-resistance of insect pests to the Bt toxin (Gassman et al., 2009; Tabashnik et al., 2013), Bt crops have managed to usher economic as well as social development. Resistance management strategies involving release of sterile insects to mate with the resistant insects, the use of two-toxin Bt crops targeting the same insect, or the use of refuge host plants that do not make the toxin and are sporadically planted along with Bt crops, have come into play to delay evolution of pest resistance (Tabashnik et al., 2010; Carriere et al., 2015; Jin et al., 2015).

4. INSECT VIRUSES

Insect viruses are ubiquitous, and epizootics in insect populations rather common, but never before have they generated so much interest, because an effective natural control of pests by viruses has a promising potential in any pest management program. Because they are cheaper and locally accessible, a fair measure of success has been achieved through the use of insect viruses as biopesticides in crop protection. Major groups of insect viruses include RNA viruses such as cypovirus, dicistrovirus, nodavirus, and tetravirus, and DNA viruses including densovirus, entomopoxvirus, ascovirus, iridovirus, nudivirus, hytrosavirus, iflaviruses, and baculovirus. A large volume of information is available and continues to be generated on their phylogeny, molecular characteristics, and cytopathology (Asgari and Johnson, 2010; Vega and Kaya, 2012; Rohrmann, 2013; ICTV, 2014). A few groups of promising insect pathogenic viruses are described below, with emphasis on baculoviruses, which have proven their potential as biocontrol agents and have found their way into commercially available biopesticide formulations for control of insect pests in forests, greenhouses, and agricultural fields.

4.1 Dicistroviruses

Dicistroviruses (family Dicistroviridae) are so named because their genome includes two open reading frames or cistrons. These are small, nonenveloped isometric viruses with a monopartite, single-stranded, positive sense RNA genome (Hulo et al., 2011; Carrillo-Tripp et al., 2014). Species of this genus infect arthropods, primarily insects (Bonning and Miller, 2010), several being pathogenic to pests of agricultural importance, including major vectors of plant viruses. The potential use of the dicistroviruses as biopesticides is being studied against insects, such as the glassy-winged sharp-shooter (GWSS) (*Homalodisca coagulate*: Hemiptera: Cicadellidae), that is a polyphagous pest, and feeds on nearly 100 plant species. It also transmits several plant pathogens, including *Xylella fastidiosa*, the cause of Pierce's disease of grapes. For a long time GWSS was largely managed through application of pyrethroid and neonicotinoid insecticides; however, the dicistroviruses offered the possibility to switch to benign pest control methods. *Homalodisca coagulate* virus 1, a dicistrovirus, accumulates in the midgut of GWSS (Hunnicutt et al., 2008) and its infection has potential in the management of GWSS as the ensuing pathological effects include enhanced mortality rate, reduced fecundity, and growth of the pest. The absence of an appropriate method for large-scale production of the virus has proved to be a limitation in its use as a biopesticide, but the problem was tackled with the use of baculovirus expression system for in vitro production of dicistrovirus pesticides. The Dicistrovirus, *Rhopalosiphum padi virus* (RhPV), infects an aphid vector of important cereal viruses, and accumulates in the nuclei of baculovirus-infected Sf21 cells expressing the recombinant RhPV clone (Gildow and D'Arcy, 1990; Pal et al., 2007).

4.2 Densoviruses

Densoviruses (DNVs) (family: Parvoviridae, subfamily: Densovirinae) are icosashedral, nonenveloped viruses with a single-stranded linear DNA genome, ending in two hairpin structures. Included within densoviruses are *Myzus persicae* densovirus infecting the green peach aphid, *M. persicae* (van Munster et al., 2003) and *Acheta domesticus* densovirus infecting the European house cricket, *A. domesticus* (Szelei et al., 2011). Most DNVs cause serious diseases in their hosts and have been considered for the biocontrol of major insect pests owing to their high virulence and ease of transmission. Biocontrol was attempted with *Periplaneta fuliginosa* densovirus that infects the smokybrown cockroach (*P. fuliginosa*) (Jiang et al., 2008), while *Junonia coenia* densovirus was shown to be lethal for *Spodoptera frugiperda*, the fall armyworm (Mutuel et al., 2010).

4.3 Iridoviruses

Iridoviruses (family: Iridoviridae) are icosahedral, double-stranded DNA viruses that infect a range of vertebrates and invertebrates, including insects (Williams, 1996). They replicate in the cytoplasm and generate iridescence in the heavily infected tissues. There are very few reports on the potential of iridoviruses in biocontrol. Cell cultures of the *Diaprepes* root weevil, a major pest of citrus and ornamentals, were established and subsequently infected with the Invertebrate iridescent virus 6 (IIV-6), in order to evaluate its potential in biocontrol (Hunter and Lapointe, 2003). IIV6 infection in *Phyllophaga vandinei*, scarab beetle, led to altered feeding and mating behavior in adults, but with a low mortality of the pest (Jenkins et al., 2011).

4.4 Polydnaviruses

Polydnaviruses (family: Polydnaviridae) include two genera, *Bracovirus* (32 species) and *Ichnovirus* (21 species), which exhibit mutualism with two diverse families of parasitoid wasps, Braconidae and Ichneumonidae, respectively. Polydnaviruses are large double-stranded DNA viruses that are regarded as endogenous virus elements as their genomes get stably integrated into the genomes of parasitoid wasps, the primary hymenopteran hosts. In the nuclei of calyx cells of female wasps, these elements replicate and assemble into virions containing DNA encoding virulence genes. Virions are injected into the host, generally Lepidopterans, during wasp oviposition, and wasp larvae emerge from the host and grow into adults, independent of the host. The wasps use the bracovirus to ensure the survival of their offspring in the host, through modifications in host physiology, suppression of the immune system, and the eventual death of the host (Strand and Burke, 2014; Drezen et al., 2014). A bracovirus symbiotic endoparasitoid, *Cotesia plutellae*, and *B. thuringiensis* were used in combination to control the insecticide-resistant populations of the diamondback moth, *Plutella xylostella*. Due to immunosuppression induced by the wasp, the parasitized larvae of the moth showed enhanced susceptibility to *B. thuringiensis* (Park and Kim, 2012).

4.5 Nudiviruses

Nudiviruses (family: Nudiviridae) are rod-shaped enveloped double-stranded DNA viruses that infect arthropods. Nudiviruses were previously classified as baculoviruses with which they share a close phylogenetic relationship, but differ from them in being nonoccluded (Wang and Jehle, 2009; Theze et al., 2011). *Oryctes rhinoceros Nudivirus* (OrNV) is a natural enemy of the rhinoceros beetle, *O. rhinoceros*, which attacks coconut and oil palms (Hunger, 2005; Bedford, 2013). Biocontrol with OrNV has played a prominent role in the management of the rhinoceros beetle, especially because the virus is highly virulent on the insect larvae and also because the pest population could not be contained by chemicals alone.

4.6 Ascoviruses

Ascoviruses (family: Ascoviridae) are enveloped double-stranded DNA viruses, bacilliform to allantoid in shape. Transmitted by parasitoid wasps, their natural hosts are insects, mainly lepidopteran noctuids (owlet moths). Ascoviruses have a unique cytopathology, with virus-induced apoptosis leading to fragmentation of the cells and generation of virions containing vesicles that accumulate in the hemolymph of the infected host and are acquired by the parasitic wasps (Bideshi et al., 2005). The ascovirus is transmitted by female wasps during oviposition. The species of identified viruses include *S. frugiperda* ascovirus 1a, *Heliothis virescens* ascovirus 3h (HvAV-3h), *Trichoplusia ni* ascovirus-2a, and *Diadromus pulchellus* ascovirus-4a (Cheng et al., 2007). To our knowledge, there are no reports of their application in biocontrol of insects. The fact that HvAV-3h readily infects three important insect pests belonging to different subfamilies (*Helicoverpa armigera*, *Spodoptera exigua*, and *Spodoptera litura*) makes it a possible biocontrol agent (Huang et al., 2012).

4.7 Hytrosaviruses

Hytrosaviruses (family: Hytrosaviridae) are viruses identified from different dipteran species and the family includes two genera, Muscavirus and Glossinavirus (ICTV, 2014). These viruses are enveloped, rod shaped, nonoccluded, and possess a large circular double-stranded DNA genome. In adult dipterans they cause salivary gland hypertrophy (SGH) along with testicular and ovarian malformation leading to partial or complete sterility. Hytrosaviruses include MdSGHV and GpSGHV infecting the house fly, *Musca domestica*, and the tsetse fly, *Glossina pallidipes*, respectively. These viruses are phylogenetically related to the baculoviruses (Abd-Alla et al., 2009; Jehle et al., 2013) and have been renamed Musca hytrovirus and Glossina hytrovirus, respectively (ICTV, 2014). MdSGHV has potential use as an agent to control the domestic fly density (Geden et al., 2011), though modulation of the fecundity in tsetse fly population due to GpSGHV infection is seen as a major impediment in the implementation of area wide control program in Africa through the sterile insect technique (Abd-Alla et al., 2013a,b; Kariithi et al., 2013).

4.8 Entomopoxviruses

Entomopoxviruses (EPVs) (family: Poxviridae, subfamily: Entomopoxvirinae) include about 31 species distributed over three genera (ICTV, 2014). EPVs infect insects including members of Coleoptera, Diptera, Lepidoptera, and Orthoptera. The virions are occluded and contain a large linear double-stranded DNA genome rich in AT residues. The virus replicates in the cytoplasm and forms spindle bodies (Lai-fook and Dall, 2000). The genomes of several EPVs infecting Lepidoptera have been sequenced (Afonso et al., 1999; Bawden et al., 2000; Theze et al., 2013). Transgenic rice plants expressing an entomopoxvirus gene that encoded a virus enhancing factor resulted in enhanced susceptibility of armyworm larvae to nucleopolyhedrovirus (NPV) infection, thus providing a strategy for more effective pest management (Hukuhara et al., 1999). An entomopoxvirus which causes chronic infection of the German cockroach, *Blattella germanica*, and another that infects European spruce bark beetle, *Ips typographus*, may have possible uses in biocontrol (Radek and Fabel, 2000).

4.9 Cypovirus

Cypovirus (family Reoviridae, subfamily Spinareovirinae) are cytoplasmic polyhedrosis viruses that are commonly isolated from insects, particularly from Lepidoptera. Sixteen species have been reported so far (ICTV, 2014). The viruses replicate in the cytoplasm and have a linear double-stranded RNA genome composed of 10 segments. The virions are nonenveloped icosahedrons, embedded in a protein matrix to form occlusion bodies. All 10 genome segments have been sequenced for some cypoviruses including *H. armigera* cypovirus 5, *Bombyx mori* cypovirus, *Lymantria dispar* cypovirus, and *Dendrolimus punctatus* cypovirus (Li et al., 2006; Tan et al., 2008; Cao et al., 2012; Zhou et al., 2014). NoarCPV, a cytoplasmic polyhedrosis virus, was used to reduce the population of *Norape argyrrhorea*, an important oil palm defoliator pest in Peru (Zeddam et al., 2003).

4.10 Baculovirus

4.10.1 Structure and Replication

The family Baculoviridae includes a diverse group of viruses infecting arthropods and is traditionally represented by two genera of baculoviruses, nuclear polyhedrovirus (NPV) and granulovirus (GV), which infect the larval stage of insects primarily from four insect orders, Lepidoptera, Hymenoptera, Diptera, and Coleoptera. Those infecting the Lepidoptera (moths and butterflies) are particularly important, as major insect pests of agricultural importance belong to this order. Baculoviruses show evolutionary relatedness to the insect hosts (Jehle et al., 2006). Baculoviridae comprises four genera: α and β-baculoviruses, represented by NPV and GV, respectively, that infect the Lepidopterans, and γ and δ-baculoviruses include NPVs only and infect Hymenoptera and Diptera, respectively. Presently, α, β, δ, and γ-baculoviruses are represented by 32, 14, 1 and 2 species, respectively (ICTV, 2014). Complete genome sequences of 71 baculoviruses (strains included) are available with the rapidly expanding National Center for Biotechnology Information database at the time of writing this chapter (Table 1), and a comparison of the genome sequences of the baculoviruses has provided an insight into their biology and evolution (Herniou et al., 2003; Zhang et al., 2014).

Baculoviruses possess a double-stranded supercoiled circular DNA genome (80–180 kb). Viruses infecting insects belonging to Hymenoptera, possess smaller genomes (Lauzon et al., 2006). The genome is packaged in rod-shaped nucleocapsids (230–385 nm in length and 40–60 nm in diameter) (Minion et al., 1979). Nucleocapsids possess a pointed cap-like

TABLE 1   List of Baculoviruses with Known Complete Genome Sequences

Baculovirus	Accession ID (NCBI)	Insect host
α-*BACULOVIRUS (INFECTING MEMBERS OF LEPIDOPTERA)*		
Adoxophyes orana nucleopolyhedrovirus (AdorNPV)	NC_011423.1	Summer fruit tortrix
Agrotis ipsilon multiple nucleopolyhedrovirus (AgipMNPV)	NC_011345.1	Black cutworm
Agrotis segetum nucleopolyhedrovirus (AgseNPV)	NC_007921.1	Turnip moth
Agrotis segetum nucleopolyhedrovirus B isolate English (AgseNPV-B)	NC_025960.1	Turnip moth
Antheraea pernyi nucleopolyhedrovirus (AnpeNPV)	NC_008035.3	Chinese tussah silk moth
Anticarsia gemmatalis nucleopolyhedrovirus (AgMNPV)	NC_008520.1	Velvetbean caterpillar
Apocheima cinerarium nucleopolyhedrovirus (ApciNPV)	NC_018504.1	Geometer moth
Autographa californica nucleopolyhedrovirus (AcNPV)	NC_001623.1	Alfalfa looper
Bombyx mandarina nucleopolyhedrovirus (BomaNPV)	NC_012672.1	Wild silk moth
Bombyx mori NPV (BmNPV)	NC_001962.1	Silkworm
Buzura suppressaria nucleopolyhedrovirus isolate Hubei (BusuNPV)	NC_023442.1	Tea looper caterpillar
Choristoneura fumiferana DEF MNPV (CfDEFMNPV)	NC_005137.2	Eastern spruce budworm
Choristoneura fumiferana MNPV (CfMNPV)	NC_004778.3	Eastern spruce budworm
Choristoneura murinana alphabaculovirus strain Darmstadt (ChmuNPV)	NC_023177.1	European fir budworm
Choristoneura occidentalis alphabaculovirus (ChocNPV)	NC_021925.1	Western spruce budworm
Choristoneura rosaceana alphabaculovirus (ChroNPV)	NC_021924.1	Oblique banded leaf roller
Chrysodeixis chalcites nucleopolyhedrovirus (ChchSNPV)	NC_007151.1	Tomato looper
Clanis bilineata nucleopolyhedrosis virus (CbNPV)	NC_008293.1	Two-lined velvet hawkmoth
Condylorrhiza vestigialis MNPV (CoveMNPV)	NC_026430.1	Brazilian poplar moth
Ecotropis oblique NPV (EoNPV)	NC_008586.1	Tea looper
Epiphyas postvittana NPV (EppoMNPV)	NC_003083.1	Light brown apple moth
Euproctis pseudoconspersa nucleopolyhedrovirus (EupsNPV)	NC_012639.1	Tea tussock moth
Helicoverpa armigera multiple nucleopolyhedrovirus (HearMNPV)	NC_011615.1	Cotton bollworm
Helicoverpa armigera NPV (HearNPV)	NC_003094.2	Cotton bollworm
Helicoverpa armigera NPV NNg1 (HearNPV-NNg1)	NC_011354.1	Cotton bollworm
Helicoverpa armigera nucleopolyhedrovirus G4 (HearNPV-G4)	NC_002654.2	Cotton bollworm
Helicoverpa SNPV AC53 (HaSNPV-AC53)	NC_024688.1	Cotton bollworm
Helicoverpa zea SNPV (HzSNPV)	NC_003349.1	Corn earworm

(Continued)

TABLE 1 List of Baculoviruses with Known Complete Genome Sequences—cont'd

Baculovirus	Accession ID (NCBI)	Insect host
Hemileuca sp. nucleopolyhedrovirus (hesp)	NC_021923.1	Buck moth
Hyphantria cunea nucleopolyhedrovirus (HycuNPV)	NC_007767.1	Fall webworm
Lambdina fiscellaria nucleopolyhedrovirus isolate GR15 (LafiNPV)	NC_026922.1	Oak looper
Leucania separata nuclear polyhedrosis virus (LsNPV)	NC_008348.1	Armyworm
Lymantria dispar MNPV (LdMNPV)	NC_001973.1	Gypsy moth
Lymantria xylina MNPV (LyxyMNPV)	NC_013953.1	Casuarina moth
Mamestra brassicae MNPV strain K1 (MabrNPV-K1)	NC_023681.1	Cabbage moth
Mamestra configurata NPV-A (MacoNPV-A)	NC_003529.1	Bertha armyworm
Mamestra configurata NPV-B (MacoNPV-B)	NC_004117.1	Bertha armyworm
Maruca vitrata MNPV (MaviMNPV)	NC_008725.1	Pod borer of legumes
Orgyia leucostigma NPV (OrleSNPV)	NC_010276.1	Tussock moth
Orgyia pseudotsugata MNPV (OpMNPV)	NC_001875.2	Douglas-fir tussock moth
Peridroma alphabaculovirus isolate GR-167 (PespNPV)	NC_024625.1	Pearly underwing, cutworm
Plutella xylostella multiple nucleopolyhedrovirus (PlxyMNPV)	NC_008349.1	Diamondback moth, cabbage moth
Pseudoplusia includens SNPV IE (PsinSNPV-IE)	NC_026268.1	Soybean looper
Rachiplusia ou MNPV (RoMNPV)	NC_004323.1	Gray looper moth
Spodoptera exigua MNPV (SeMNPV)	NC_002169.1	Beet armyworm
Spodoptera frugiperda MNPV (SfMNPV)	NC_009011.2	Fall armyworm
Spodoptera litura NPV (SpltNPV)	NC_003102.1	Oriental leafworm moth
Spodoptera litura nucleopolyhedrovirus II (SpltNPV-II)	NC_011616.1	Cotton leafworm, tobacco cutworm
Thysanoplusia orichalcea NPV isolate p2 (ThorMNPV-p2)	NC_019945.1	Soybean looper, golden plusia
Trichoplusia ni SNPV (TniSNPV)	NC_007383.1	Cabbage looper
β-*BACULOVIRUS (INFECTING MEMBERS OF LEPIDOPTERA)*		
Adoxophyes orana granulovirus (AdorGV)	NC_005038.1	Summer fruit tortrix
Agrotis segetum granulovirus (AgseGV)	NC_005839.2	Turnip moth
Choristoneura occidentalis granulovirus (ChocGV)	NC_008168.1	Western spruce budworm
Clostera anachoreta granulovirus (ClanGV)	NC_015398.1	Scarce chocolate tip
Clostera anastomosis granulovirus isolate CaLGV-Henan	NC_022646.1	Black-back prominent moth

TABLE 1   List of Baculoviruses with Known Complete Genome Sequences—cont'd

Baculovirus	Accession ID (NCBI)	Insect host
Cryptophlebia leucotreta granulovirus (CrleGV)	NC_005068.1	False codling moth
Cydia pomonella granulovirus (CpGV)	NC_002816.1	Codling moth
Epinotia aporema granulovirus (EpapGV)	NC_018875.1	Bean shoot borer
Erinnyis ello granulovirus (ErelGV)	NC_025257.1	Cassava hornworm
Helicoverpa armigera granulovirus (HearGV)	NC_010240.1	Cotton bollworm
Phthorimaea operculella granulovirus (PhopGV)	NC_004062.1	Potato tuber moth
Pieris rapae granulovirus (PrGV)	NC_013797.1	Small white, small cabbage white
Plutella xylostella granulovirus (PlxyGV)	NC_002593.1	Diamondback moth
Pseudaletia unipuncta granulovirus (PsunGV)	NC_013772.1	White speck, common armyworm
Spodoptera frugiperda betabaculovirus isolate VG008 (SfGV-VG008)	NC_026511.1	Fall armyworm
Spodoptera litura granulovirus (SpliGV)	NC_009503.1	Tobacco cutworm
Xestia c-nigrum granulovirus (XcGV)	NC_002331.1	Setaceous hebrew character
γ-BACULOVIRUS (INFECTING MEMBERS OF HYMENOPTERA)		
Neodiprion abietis NPV (NeabNPV)	NC_008252.1	Balsam fir sawfly
Neodiprion lecontei NPV (NeleNPV)	NC_005906.1	Red-headed pine sawfly
Neodiprion sertifer NPV (NeseNPV)	NC_005905.1	European pine sawfly
δ-BACULOVIRUS (INFECTING MEMBERS OF DIPTERA)		
Culex nigripalpus NPV (CuniNPV)	NC_003084.1	Mosquito

structure at one end and are blunt at the other end, and capsids consist of subunit rings that are arranged perpendicular to its longitudinal axis (Fraser, 1986). Nucleocapsids are embedded in a crystalline protein matrix of occlusion bodies (OBs) called polyhedra, composed of polyhedrin protein in NPV, and granulin in GVs (Hooft van Iddekinge et al., 1983; Ackermann and Smirnoff, 1983). The traditional grouping of baculoviruses into NPV and GV was based on the morphology of the OBs formed in the infected cells.

The occluded form protects the virions from environmental degradation and hostile conditions of the insect gut (Ji et al., 2010). OBs of both NPV and GV, found within the nucleus of infected cells, are essentially dormant structures that can survive adverse conditions upon their release from dead insects. OBs formed by NPVs are polyhedral and 0.15–15 μm in size, while GVs generally form ovoid occlusion bodies that are 300–500 nm in length and 120–300 nm in width (Crook and Jarrett, 1991). NPVs are characterized as either SNPV (single nucleocapsids per envelope) or MNPV (5–15 nucleocapsids per envelope) (Ji et al., 2010). GVs also exhibit both phenotypes, but the multiple forms are rare. Nomenclature of these viruses follows a binomial standard, with the isolate of NPV or GV being named after the host from which it was identified. Thus the same virus may be isolated from more than one host which are related to one another or different viruses from a single host (Jakubowska et al., 2005).

To investigate variability among NPV isolates, viral DNA restriction patterns are generally compared. Most of what is known of these viruses comes from studies on *Autographa californica* multiple nucleopolyhedrovirus, AcMNPV, infecting the alfalfa looper moth, *A. californica* (Vail et al., 1971). Its genome was the first to be fully sequenced among the baculoviruses (Ayres et al., 1994). The virus exists as two distinct phenotypes, differing in the composition of the envelope and its role in the virus replication cycle. The virions occur as either budded virions (BVs) which are produced early in the infection cycle when the virus buds through the plasma membrane and multiplies and spreads within the host, or as multiple enveloped virions embedded in OBs that are formed later in the infection cycle.

The viral life cycle (Figure 1) begins with the ingestion of the occlusion bodies by the insect larvae, the primary route of infection being oral (*per os*). In the midgut, the highly alkaline pH, together with the digestive enzymes and the OB-associated alkaline proteases, causes the dissolution of polyhedron and the envelope which surrounds the OBs, thus releasing the occlusion derived virions (ODVs). The ODVs then cross the peritrophic

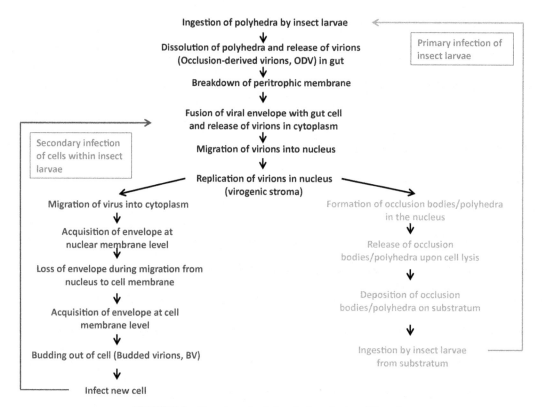

FIGURE 1 Flowchart depicting the baculovirus life cycle.

matrix, an acellular mesh-like structure, composed of chitin, proteoglycans, and mucin that lines and protects the gut (Tellam et al., 1999). The intact nucleocapsids interact with ODV-specific receptors on the microvilli of epithelial cells and gain access to the cell cytosol, finally reaching the nucleus through the nuclear pore complex and releasing their genome (Slack and Arif, 2007; Au and Pante, 2012).

After initial replication in the nucleus of midgut epithelial cells, the assembled nucleocapsids bud out of the nucleus, thus carrying the nuclear membrane as an envelope, the envelope being lost during their transit through the cytoplasm. The host cell membrane that is now acquired by the BVs is a modified membrane and includes GP64, a virion specific envelope fusion protein found in group I α-baculoviruses and is also involved in infection of other cells (Oomens and Blissard, 1999). The nucleocapsid buds off from the basal membrane and moves toward other susceptible tissues and spreads throughout the organism. There may even be a direct transformation of ODV to BV phenotype in AcMNPV, thus establishing secondary infections quickly in nonmidgut tissues (Zhang et al., 2004). Finally, late in the infection cycle, there is a shift from production of BVs, with virions moving to a peripheral area to become occluded, forming occlusion bodies in the nucleus following expression of genes coding for polyhedrin and p10. The polyhedrin crystallizes into a lattice that surrounds the virions and p10 fibrils align with the surface of the polyhedra and assemble into the smooth envelope of the occlusion body. Polyhedron envelope protein along with p10 appear to be involved in sealing the polyhedra in *Orgyia pseudotsugata* baculovirus, thus protecting and stabilizing the polyhedron from damage (Gross et al., 1994). A study has shown the requirement of *ac132* and *ac11* genes for production of BVs and for production of multiple enveloped ODVs, while *ac11* also appears to be essential for the envelopment of ODV (Yang et al., 2014; Tao et al., 2015).

Following infection with NPVs and GVs, the symptoms of pathogenesis in insect hosts are more or less similar, with a few exceptions. Diseased larvae appear swollen, glossy, and moribund, stop feeding and begin to die within 7–10 days. Baculoviruses encode chitinase and cathepsin protease and hence, owing to tissue liquefaction or "melting" and cuticle rupture, infected Lepidopteran insects end, literally, in a watery grave. The released liquid contains the OBs for further dissemination of the virus (Vega and Kaya, 2012).

Baculoviruses do not infect man or plants and are highly selective, generally being reported from Lepidopteran species and showing a narrow host range, infecting one or a few species of a particular genus or family of the host insect from which originally isolated (Doyle et al., 1990). Nevertheless, a few baculoviruses, including AcMNPV, *Anagrapha falcifera* multiple nucleopolyhedrovirus (AfMNPV) and *Mamestra brassicae* multiple nucleopolyhedrovirus, are known to infect a relatively large number of Lepidopteran species (Hostetter and Puttler, 1991). Baculoviruses are important not only as biopesticides (Fuxa, 1991; Moscardi, 1999) but also as vectors for protein expression, being utilized in basic research and medicine (Hu, 2006; van Oers, 2006). Several products that have been derived from the Baculovirus expression vector system have been approved for human as well as veterinary use (Felberbaum, 2015). A voluminous literature is available on the molecular biology and phylogeny of baculoviruses, their application as biopesticides in agriculture and forests, and their modification through biotechnology for insect control (Moscardi, 1999; Herniou et al., 2004; Inceoglu et al., 2006; Jehle et al., 2006; Miele et al., 2011; Theze et al., 2011), important examples of which are summarized here.

4.10.2 Baculoviruses in Biocontrol of Pests

Baculoviruses have contributed toward cyclical fluctuations and decline in population densities of myriad pests in various geographical locations, being associated with natural epizootics (Moreau, 2006). Figure 2 broadly depicts the baculovirus infection cycle and its relevance in protecting crops from insects. In 1892, control of the nun moth, *Lymantria monacha*, was attempted with NPV in the pine forests in Germany, and this appears to be, historically, the first attempt at biocontrol on a large scale. During late 1940s, successful control of alfalfa caterpillar, *Colias philodice eurytheme*, was carried out in the United States through artificial epizootics of polyhedrosis virus, with the virus suspension being applied as an aerial spray (Steinhaus and Thompson, 1949; Thompson and Steinhaus, 1950; Capinera, 2008). Similarly, a virus known to curb the population of European spruce sawfly, *Diprion hercyniae*, was artificially disseminated to control the sawfly density in Ontario (Bird and Burk, 1961).

After the initial momentum garnered from these early reports, the study of baculoviruses for insect control took a more systematic turn when an extensive control program was developed in Brazil to curb velvetbean caterpillar, *Anticarsia gemmatalis*, a pest of soybeans. This program was initiated in the 1980s, with an area of over 2 million hectares being treated with the occlusion bodies of *Anticarsia gemmatalis* nucleopolyhedrovirus, AgMNPV, a widely used viral pesticide, its genome being sequenced in 2006 (Oliviera et al., 2006). The larvae were fed on an artificial diet, and infected on a regular basis to produce enough virus. The infected larvae were used as a wet formulation on the soybean plants with positive results (Oliveira et al., 2006). Cotton bollworm (*H. armigera*), a major pest of cotton that has developed resistance to chemical pesticides in various parts of the world, was also controlled successfully by baculoviruses. The insects were grown on an artificial diet, largely composed of corn and wheat, and the *Helicoverpa armigera* nucleopolyhedrovirus (HearNPV)-infected larvae were processed to give purified OB suspension in distilled water for use on several hectares of cotton

FIGURE 2 A generalized representation of the baculovirus infection cycle. (A) The baculovirus polyhedra are ingested by larvae. Upon ingestion the gastric juices of the larvae digest the walls of the polyhedra and individual virions are released, replicated, and fresh polyhedra formed. (B) The release of new baculovirus polyhedra follows larval death (larva with no fill). (C) The polyhedra may remain on the foliage or (D) fall on the ground.

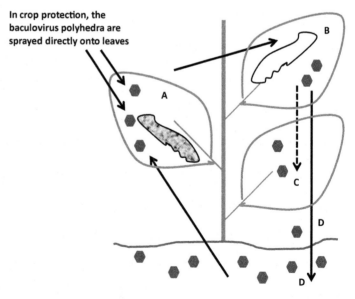

In crop protection, the baculovirus polyhedra are sprayed directly onto leaves

(Granados and Federici, 1986; Huber, 1986; Moscardi, 1999; Szewczyk et al., 2006; Luo et al., 2014). Several isolates of HearSNPV were characterized and compared with HearSNPV-G4, an isolate that was used on cotton in China. Following a comparative evaluation, the Iberian isolate HearSNPV-SP1 was identified as the fastest killing isolate and recommended for use as a biopesticide in Spain against the cotton bollworm (Arrizubieta et al., 2014).

Eastern tent caterpillar nuclear polyhedrosis virus, ETNPV, was used in managing eastern tent caterpillar, *Malacosoma americanum*, a defoliator of ornamental and fruit trees (Progar et al., 2010). In another landmark study, NPVs were used against populations of *S. exigua* (beet armyworm), a major polyphagous pest of a number of crops that have developed resistance to chemical pesticides due to their extensive use (Ahmad and Arif, 2010). Isolates of *Spodoptera exigua* multicapsid nucleopolyhedrovirus (SeMNPV) have been characterized and developed as the active ingredient of insecticidal products to control beet armyworm on sweet pepper crops under greenhouse conditions. The strain used in the formulation was isolated from a group of larvae during a natural epizootic in Almeria, Southern Spain (Caballero et al., 1992; Lasa et al., 2007). Efficiency of the product was also evaluated under field conditions (Kolodny-Hirsch et al., 1997). SeMNPV is very specific for its host with a very narrow host range. Unlike AcMNPV or *Helicoverpa zea* single nucleopolyhedrovirus, HzSNPV, which infect a relatively wider range of Lepidopteran species. The choice of a suitable virus–host combination that allows an optimal production of SeMNPV is important because host populations vary in their susceptibility to geographically distinct isolates of viruses (Erlandson et al., 2007). Insecticidal activity of two strains of SeMNPV, which are active ingredients of Vir-ex (Spain) and Spexit (Switzerland), were measured on the host colonies from which they were isolated as well as on colonies that differed in their geographical origin (Elvira et al., 2013).

Interestingly, viruses showed the highest infectivity toward the local host population and hence Vir-ex is produced in Almerian insects for use in Almerian greenhouse crops. Similar results were obtained while examining the efficacy of various NPVs on *Chrysodeixis chalcites* (golden twin spot moth or tomato looper), a pest on banana, which results in up to 30% losses in banana grown under greenhouse conditions in the Canary Islands. Its populations are vulnerable to infection by isolates of *Chrysodeixis chalcites* single nucleopolyhedrovirus (ChchSNPV), from various geographical locations, as well as by AcMNPV, a virus with a relatively wide range. The most prevalent strain of ChchSNPV was a local strain from Canary Islands (ChchSNPV-TF1) which was effective on tomato as well, with a larval mortality rate higher than either indoxacarb or *B. thuringiensis* var. *Kurstaki*. ChchSNPV-TF1 carries promise of being an effective biocontrol agent as it is as effective as other commercialized baculoviruses in terms of high pathogenicity and speed of kill (Bernal et al., 2013; Simon et al., 2015). *Spodoptera exempta* (African armyworm) is a major threat to food production in Africa, and like other pests it has learned to coordinate its own cycle with the life cycle of the host. High density outbreaks of the pest are known to graze down crops, such as rice, wheat, maize, barley, and sorghum. After completing its larval phase on the host, it pupates on the ground and emerges as adult moths in swarms which relocate with the rain-laden winds. *Spodoptera exempta* nucleopolyhedrovirus (SeNPV) is a natural pathogen of the African armyworm, infecting a wide range of the host tissues and its application effectively suppressed armyworm outbreaks (Cherry et al., 1997; Holt et al., 2006; Grzywaecz et al., 2008, 2014; Graham et al., 2012). Like all other potential agents, intensive research went into determining the host range of SeNPV, its replication cycle and interaction with the host. Microbial insecticides were evaluated for managing turfgrass pests under harsh golf course settings, with *Agrotis ipsilon* multicapsid nucleopolyhedrovirus used as a spray to control black cutworms on turf grasses (Bixby and Potter, 2010; Prater et al., 2006; Held and Potter, 2012).

Codling moth, *Cydia pomonella*, a very serious pest on pear and apple, is controlled by the commonly used broad-spectrum insecticide, azinphosmethyl, with a toll on the beneficial insects along with other drawbacks associated with the use of chemical pesticides. Shortly after hatching, the larvae penetrate the fruit, and eventually leave it once they are fully grown to pupate. Control of the Codling moth with *Cydia pomonella* granulovirus, CpGV, has been hugely successful in Europe, America, and other countries (Huber and Dickler, 1977; Glen and Payne, 1984; Blommers et al., 1987; Cross et al., 1999). CpGV preparations are highly virulent (median lethal dose (LD_{50}) as low as 1.2–5 granules per larva) (Sheppard and Stairs, 1977), and commercially available. One major concern, however, involves the sensitivity of CpGV to solar degradation (Arthurs et al., 2006). AoGV, the granulovirus of the summer fruit moth *Adoxophyes orana*, has been commercially produced and extensively tested in Europe and Japan (Sekita et al., 1984; Cross et al., 1999).

Culex nigripalpus nucleopolyhedrovirus (CuniNPV), a deltabaculovirus, deserves a special mention not only because it naturally infects the *Culex* species (Diptera), but especially because of development of insecticide resistance in these vectors for malarial diseases (Liu, 2015). CuniNPV produces globular OBs in the nucleus of infected cells and is highly pathogenic for *C. nigripalpus* and *C. quinquefasciatus*, both of which are vectors for the encephalitis virus. CuniNPV has been characterized at the molecular level and can be developed as a biocontrol agent specifically targeting *Culex* population (Perera et al., 2006).

4.10.3 *Baculoviruses Production and Bioassay for Pathogenicity Determination*

The advantage of baculoviruses over chemical insecticides is their narrow host range (Cory et al., 2000). Thus, the monospecific nature of *Euproctis chrysorrhoea* NPV infecting the brown-tail moth, *E. chrysorrhoea*, becomes a matter of concern at the monetory level as the farmer would need varied commercial preparations to target different pests, thus escalating the cost of crop protection. Hence, baculovirus preparations that incorporate AcMNPV or other broad host range NPVs as the active ingredient, would be more cost-effective, where more than one pest can be controlled with the same preparation. Presently, several baculovirus-based pesticides have been registered throughout the world and are available to control lepidopteran pest populations in particular, with "Elcar" being the first to be developed, registered by Sandoz Inc. in the United States in 1975, for use against *Helicoverpa zea* (Syn. *Heliothis zea*). HzSNPV, the active ingredient in Elcar, provided broad-spectrum protection against pests on several crops. It is being presently resold as "Gemstar" by Certis USA. All registered baculovirus-based products have been rigorously tested for human toxicity and infectivity on various animals and in doses that are several times over and above those that may be encountered during field applications (Lapointe et al., 2012), with OBs being inert and nonallergenic. Baculovirus-based pesticides are presently classified as "low-risk" biological control agents with no adverse effects on nontarget insects.

In all commercial preparations available (Table 2), the virus is mass produced in the larvae of the insect host. The infected cadavers are collected, a difficult task as the larvae tend to liquefy, but a task made easy in situations where the virus does not encode the chitinase or cathepsin genes and the dead larvae largely remain intact (Slack et al., 2004). This has

TABLE 2 Baculovirus-Based Insecticidal Products Available Commercially

Product	Supplier	Active ingredient	Target pest	Crop protected
Helicovex	Andermatt Biocontrol AG, Switzerland	*Helicoverpa armigera* nucleopolyhedrovirus (HearNPV)	African cotton bollworm (*H. armigera*) and other *Helicoverpa* species	Vegetables, fruits, etc.
Spexit	Andermatt Biocontrol AG, Switzerland	*Spodoptera exigua* multiple nucleopolyhedrovirus (SeMNPV)	Beet armyworm (*S. exigua*)	Vegetables, fruits, etc.
Littovir	Andermatt Biocontrol AG, Switzerland	*Spodoptera littoralis* nucleopolyhedrovirus (SpliNPV)	Egyptian cotton leafworm (*S. littoralis*)	Vegetables, fruits, etc.
Loopex	Andermatt Biocontrol AG, Switzerland	*Autographa californica* nucleopolyhedrovirus (AcNPV)	Cabbage looper (*T. ni*)	Brassicas, etc.
Madex	Andermatt Biocontrol AG, Switzerland	*Cydia pomonella* granulovirus (CpGV)	Codling moth (*C. pomonella*)	Apple, pear, walnut, etc.
Madex Max, Madex Plus Madex Top	Andermatt Biocontrol AG, Switzerland	Three different strains of *Cydia pomonella granulovirus* (CpGV)	Codling moth (*C. pomonella*) also CpGV-resistant populations	Apple, pear, walnut, etc.
Madex Twin	Andermatt Biocontrol AG, Switzerland	*Cydia pomonella* granulovirus (CpGV)	Oriental fruit moth (*Grapholita molesta*) and codling moth (*C. pomonella*)	Apple, pear, etc.
Capex	Andermatt Biocontrol AG, Switzerland	*Adoxophyes orana* granulovirus (AoGV)	Summer fruit tortrix (*A. orana*)	Apple, pear, rose, plum, etc.
Cryptex	Andermatt Biocontrol AG, Switzerland	*Cryptophlebia leucotreta* granulovirus (CrleGV)	False codling moth (*C. leucotreta*)	Citrus, avocado, tea, cotton, etc.
Abietiv	Andermatt Biocontrol AG, Switzerland	*Neodiprion abietis* nucleopolyhedrovirus (NeabNPV)	Balsam fir sawfly (*N. abietis*)	Balsam fir
Lymantria dispar MNPV	Andermatt Biocontrol AG, Switzerland	*Lymantria dispar* multiple nucleopolyhedrovirus (LdMNPV)	Gypsy moth (*L. dispar*)	Deciduous hardwood trees
Heliokill	Ajai Biotech, India	Nucleopolyhedrovirus	Tobacco budworm (*Heliothes* sp.)	Cotton, vegetables, pulses, grapes
Spodopterin	Ajai Biotech, India	Nucleopolyhedrovirus	Armyworm (*Spodoptera* sp.)	Cotton, vegetables, ornamentals, grapes
Spodo-Cide	Pest Control India	Nucleopolyhedrovirus	Tobacco caterpillar (*Spodoptera litura*)	Vegetables, ornamentals, etc.
Heli-Cide	Pest Control India	Nucleopolyhedrovirus	American bollworm (*H. armigera*)	Vegetables, ornamentals, etc.

(Continued)

TABLE 2 Baculovirus-Based Insecticidal Products Available Commercially—cont'd

Product	Supplier	Active ingredient	Target pest	Crop protected
Biovirus-H	Biotech International Ltd., India	*Helicoverpa armigera* nucleopolyhedrovirus (HearNPV)	American bollworm (*H. armigera*)	Vegetables, ornamentals, etc.
Biovirus-S	Biotech International Ltd., India	*Spodoptera litura* nucleopolyhedrosis virus (NPV)	Tobacco caterpillar (*S. litura*)	Vegetables, ornamentals, etc.
Spod-X	Certis USA	*Spodoptera exigua* multiple nucleopolyhedrovirus (SeMNPV)	Beet armyworm (*S. exigua*)	Vegetables, greenhouse flowers
Cyd-X HP, Cyd-X	Certis USA	*Cydia pomonella* granulovirus (CpGV)	Codling moth (*C. pomonella*)	Apple, pear, walnut, etc.
Madex HP	Certis USA	*Cydia pomonella* granulovirus (CpGV)	Codling moth (*C. pomonella*)	Apple, pear, walnut, etc.
Gemstar	Certis USA	*Helicoverpa zea* single nucleopolyhedrovirus (HzSNPV)	Corn earworm (*H. zea*), bollworm (*H. armigera*), tobacco budworm (*Heliothes* sp.)	Corn, cotton, tobacco, vegetables, etc.
Mamestrin	Natural Plant Protection, France	Baculovirus	Cabbage moth (*Mamestra brassicae*), American bollworm (*H. armigera*), diamondback moth (*P. xylostella*) potato tuber moth (*Phthorimaea operculella*), grape berry moth (*Lobesia botrana*)	Cabbage, cotton, potato, grapes
TM Biocontrol-1	USDA Forest Service	*Orgyia pseudotsugata* nuclear polyhedrovirus (OpNPV)	Douglas-fir tussock moth (*O. pseudotsugata*)	Forest trees
Gypchek	USDA Forest Service	*Lymantria dispar* multiple nucleopolyhedrovirus (LdMNPV)	Gypsy moth (*L. dispar*)	Forest use
Vir-ex	Biocolor SA, Spain	*Spodoptera exigua* multiple nucleopolyhedrosis virus (SeMNPV)	Beet armyworm (*S. exigua*)	Various crops
VPN Ultra 1,6 WP	Agricola El Sol, Guatemala	*Autographa californica* nucleopolyhedrovirus (AcNPV)	Several pests, including cotton bollworm (*H. armigera*), tobacco hornworm (*Manduca sexta*), cabbage looper (*T. ni*), beet armyworm (*S. exigua*)	Cotton, vegetables, fruits, ornamentals

Product	Company	Virus	Target pest	Use
Virosoft[CP4]	BioTEPP Inc., Canada	*Cydia pomonella* granulovirus (CpGV)	Codling moth (*C. pomonella*)	Apple, pear, walnut, etc.
Granupom	Neudorff, Germany	*Cydia pomonella* granulovirus (CpGV)	Codling moth (*C. pomonella*)	Apple and pear
Carpovirusine	Fargro Ltd, UK	*Cydia pomonella* granulovirus (CpGV-M)	Codling moth (*C. pomonella*)	Apple and pear
Dispavirus	Canada	*Lymantria dispar* multiple nucleopolyhedrovirus (LdMNPV)	Gypsy moth (*L. dispar*)	Forest use
Lecontvirus WP	Canada	*Neodiprion lecontei* nucleopolyhedrosis virus (NeleNPV)	Red-headed pine sawfly (*N. lecontei*)	Forest, ornamentals, etc.
500 Billion Pib/G *H. armigera* Nuclear Polyhedrosis Virus/Hanpv	Henan Jiyuan Baiyun Industry Co. Ltd., China	*Helicoverpa armigera* nucleopolyhedrovirus (HearNPV)	Cotton bollworm (*H. armigera*), oriental tobacco budworm (*H. assulta*), corn earworm (*H. zea*)	Cotton, corn, tobacco
200 Billion Pib/G Senpv *S. exigua* Nuclear Polyhedrosis Virus	Henan Jiyuan Baiyun Industry Co. Ltd., China	*Spodoptera exigua* multiple nucleopolyhedrovirus (SeMNPV)	Beet armyworm (*S. exigua*)	Fruits and vegetables
150 Billion Pib/G *S. litura* Nuclear Polyhedrosis Virus/Spltnpv	Henan Jiyuan Baiyun Industry Co. Ltd., China	*Spodoptera litura* nucleopolyhedrovirus (SpltNPV)	Tobacco caterpillar (*S. litura*)	Fruits and vegetables

facilitated the mass production of biopesticides as intact cadavers are then used to infect larvae reared on artificial diet made low-cost and ideal for lepidopteran larvae (Elvira et al., 2010). A relatively less expensive method involves in vitro production of virus in insect cell cultures (Goodman et al., 2001), but establishing the insect cell lines is itself no small task. Moreover, for γ and δ-baculoviruses there are no insect cell lines available at present. An added problem is that each cell line–virus combination requires a unique composition of the medium (Vega and Kaya, 2012). Efficient mass production of baculoviruses involves a balance between total OBs produced per infected cell, pathogenicity of the OBs produced (which may depend on their size and number of virions included), and the pathological effects of the infection on the larvae (Elvira et al., 2013). The virus strain and insect host population utilized will determine the efficiency of virus production within the host (Ribeiro et al., 1997; Cory et al., 2005).

To determine the insecticidal activity of the OBs, LD_{50} and mean time to death are usually estimated in second instars by using the droplet-feeding technique (Hughes et al., 1986). The late first instars are starved at $25 \pm 2\,°C$ and allowed to molt to second instar over a period of 10 h. The groups of second instars are then fed on sucrose plus droplets of fluorella blue (food dye) containing various concentrations of OBs per ml (usually between 2.8×10^3 and 2.3×10^5) which result in mortality of 10–90%. Control larvae are fed on a diet of sucrose and food dye alone. Twenty-five larvae that ingested the OB suspension within 10 min are then placed individually in a 25-well tissue culture plate, each containing a cube of diet and incubated at $25 \pm 2\,°C$, and monitored for mortality every 12 h for 7 days. The assays were repeated thrice for different batches of insects. Pathogenicity was expressed as 50% lethal dose (Lasa et al., 2008).

4.10.4 *Factors Affecting Pathogenicity and Virulence of Baculoviruses*

Biopesticides, in general, display an erratic performance under field conditions; hence represent a very small percentage of the global market in crop protection. The performance of the microbial insecticides, including baculoviruses, is compromised due to their sensitivity to desiccation and ultraviolet (UV) degradation; hence successful implementation of any pest control program involving baculoviruses would depend upon several factors, beginning with a detailed understanding of host–parasite biology and disease dynamics (Thomas, 1999). A major factor in biocontrol is the selection of the most virulent isolate of the virus for an insect pest in a given geographical location. Isolates differ in their insecticidal phenotypes and a study has identified genes in SeMNPV that are likely involved in the pathogenicity and virulence of OBs (Serrano et al., 2015). The next step is to estimate the correct dose, the latter necessitating intensive research into the pest infestation pattern on the plant (Moscardi, 1999). A low or inadequate dosage would lead to a failure of protection leading to abandoning of the program by the farmer in favor of other methods which fetch immediate rewards, such as chemical pesticides. Higher than the required dosage of the baculovirus formulation are not cost-effective. Virulence on the host is affected by the timing of the spray, as with age, the larvae lose their susceptibility to the formulation (Bianchi et al., 2002; Bernal et al., 2014).

Furthermore, factors mediated by the plants are also important, and foliage mediated changes to the peritrophic matrix of the insects can lead to reduced efficacy of baculovirus infection (Plymale et al., 2008). Baculoviruses also appear to be most active against larvae fed on the host plant from which the virus was isolated. Thus the host plants can affect the response of the larvae to the entomopathogens (Shikano et al., 2010).

As mentioned earlier, solar radiations affect the field persistence of the viruses, possibly through generation of reactive oxygen species which adversely affect the OBs. To overcome this problem, several protectants, adjuvants, optical brighteners, and dyes have been incorporated into the formulations (Vega and kaya, 2012; Bernal et al., 2014). For example, mixtures of HearNPV-SP1 and Tinopal UN PA-GX or Leucophor showed enhanced pathogenecity on *H. armigera* compared with OBs alone (Ibargutxi et al., 2008). Spray-dried lignin-encapsulated formulations of AfMNPV and CpGV provided solar stability under field conditions (Behle et al., 2003; Arthurs et al., 2006).

Several inexpensive, natural plant-based products that are rich in antioxidants, such as Moringa, rice bran, green tea, and cocoa, have also been tried as UV protective additives for *Spodoptera littoralis* nucleopolyhedrovirus (El-Helaly, 2013). Overexpression of *Bombyx mori* nucleopolyhedrovirus (BmNPV) Bm65 gene, with predicted endonuclease activity perhaps being involved in DNA repair, correlated with improved survival rate of BmNPV BVs after UV radiation (Tang et al., 2015). Thus efforts are on to reduce susceptibility of the baculoviruses to UV light.

Another problem stems from the large-scale application of virus formulations over several years. This may lead to development of resistance in the target pest, therefore, resistance management strategies for baculoviruses need to be developed to prevent its spread. Resistance to commercially applied CpGV has been described in Europe for certain populations of codling moth which had received regular applications of the biopesticides for several years (Eberle and Jehle, 2006; Asser-Kaiser et al., 2007).

The development of recombinant baculoviruses expressing toxins and hormones, discussed in the following section, could delay the onset of resistance in pests, but a reluctance to accept genetically modified organisms may prevent this approach from being adopted in several countries. Looking for isolates against which the local population of pests do not show resistance should not be difficult because baculoviruses show a lot of genetic variability in populations from different geographic locations (Hara et al., 1995; Redman et al., 2010; Rowley et al., 2011; Elvira et al., 2013). In Europe, all commercial formulations against codling moth were derived from the Mexican isolate CpGV-M. Hence, efficiency of other isolates was tested against the resistant codling moth strain CpRR1. Several naturally occurring CpGV isolates from different geographic regions were infectious to resistant CpRR1. Restored mutation in the *pe38* gene of CpGV using an isolate from a commercial product Virosoft could overcome the resistance of CpRR1 (Berling et al., 2009; Gebhardt et al., 2014).

4.10.5 Recombinant Baculovirus

Despite the widespread interest in baculoviruses, their actual use in biocontrol has been limited by their extreme specificity to insect populations, the complex procedures involved in their production, the adverse effect of the environment on the viruses, particularly their low tolerance to UV radiation, and most importantly, their slow speed of kill. Losses to a crop cannot be checked with as much speed as with chemical pesticides. Depending on the strain of virus and the insect species, it may take several days to weeks before the infected larvae die. Thus considerable defoliation may occur before the biopesticide takes effect. The production of baculovirus-based insecticides is labor-intensive, and this, coupled with poor economics, has limited their use on crop pests that have developed resistance to chemical pesticides.

Drawbacks associated with the use of wild type baculoviruses as biopesticides, necessitated the construction of recombinant baculoviruses expressing foreign genes under powerful promoters in the insect pests (Bonning and Hammock, 1996). Additional pesticidal genes have been included in the genome of *Autographa californica* multiple nucleopolyhedrovirus (AcMNPV), the baculovirus workhorse. Recombinants produced were often highly virulent and worked with greater efficiency. Other recombinants have been constructed expressing insect-specific toxin genes from scorpion, spider, parasitic wasp, and *B. thuringiensis*. Similarly, hormone genes such as juvenile hormone esterase, diuretic hormone, or enzymes have been expressed in baculoviruses to achieve this end.

In one of the earliest attempts, field trials were conducted with a genetically modified AcMNPV infecting the alfalfa looper, *A. californica*. Modified AcMNPV expressed *Androctonus australis* insect toxin (AaIT), a neurotoxin polypeptide derived from the venom of the scorpion. The modified baculovirus killed larvae faster and hence led to significantly improved crop protection. Larvae infected with the recombinant virus fell onto the soil and did not lyse. Thus reducing the likelihood of transmission which perhaps led to reduced secondary cycle of infection compared to the wild type virus (Cory et al., 1994). HearNPV expressing AaIT was also evaluated in China, with significant improvement in the yield from cotton plants over plants treated with the wild type virus (Sun and Peng, 2007). Apart from the AaIT toxin, the expression of other insect-specific toxins such as mite toxin Txp-1 (Burden et al., 2000) and scorpion toxins LqhIT1 and LqhIT2 (Gershburg et al., 1998) have led to enhanced pathogenicity. Recombinant AcMNPV coexpressing the latter toxins, rather than singly, demonstrated increased insecticidal activity on *H. virescens*, *H. armigera*, and *S. litura* larvae (Regev et al., 2003). ApPolh5-3006AvTox2, another recombinant AcMNPV, coexpressing insecticidal toxins from the spider *Araneus ventricosus* (Av-Tox2 toxin gene) and *B. thuringiensis* (Cry 1–5 crystal protein gene) possessed improved insecticidal capability (Jung et al., 2012). Similarly, recombinant *S. exigua* multicapsid nucleopolyhedrovirus (SeXD1) killed insects faster than wild type SeMNPV, and thus may provide better protection (Cai et al., 2010).

Normal hormonal balance of an insect can be disrupted by overexpression of insect hormones in the host larvae using recombinant baculoviruses. Recombinant BmNPV, named BmDH5, expressing the diuretic hormone killed larvae about 20% faster than wild type BmNPV (Maeda, 1989). A recombinant AcMNPV expressing the juvenile hormone esterase (JHE) gene from *H. virescens*, which regulates insect development, showed improvement in speed of kill, but was also cleared from the hemolymph with rapidity, necessitating studies into improving the in vivo half-life of JHE (Hammock et al., 1990). A recombinant AcMNPV, named ColorBtrus, has been generated that produces OBs that incorporate the *B. thuringiensis* insecticidal toxin, Cry1Ac, along with the green fluorescent protein, with enhanced speed of action and pathogenicity (Chang et al., 2003). Another recombinant AcMNPV, NeuroBactrus, was created by fusing polyhedron gene with the *B. thuringiensis* crystal protein gene, Cry1-5 and AaIT toxin from *A. australis*. NeuroBactrus showed enhanced insecticidal activity against *P. xylostella* and *S. exigua* as compared with the wild type AcMNPV (Shim et al., 2013).

The biological activity of the AcBt7 recombinant baculovirus expressing Cry 1Ab toxin protein was evaluated against *S. litura* larvae. The results from the bioassay revealed that the recombinants were five times more effective than the wild type virus (El-Menofy et al., 2014). Pathogenicity of *B. thuringiensis* subsp. *Kurstaki* was significantly enhanced when it was mixed with a recombinant AcMNPV expressing a polydnavirus genes, *Cotesia plutel-lae* bracovirus inhibitor-kB, with suppression of antimicrobial peptide synthesis by *P. xylostella* (Shrestha et al., 2009). Gene deletion studies were also carried out for increasing insecticidal activity of the viruses, few with a positive outcome. Deletion mutants of AcMNPV were generated for *pp34*, coding polyhedral membrane, *egt*, coding for ecdysteroid and *p10*, coding for fibrillar structures. None, however, showed adequate promise (Bianchi et al., 2000).

Recombinant baculoviruses do not come with a cure-all label. Often the increase in speed of killing is accompanied by a large reduction in the yield of progeny virus, thus in field situations the recombinants would be outcompeted by the wild type virus, bringing the story back to square one (Burden et al., 2000). Today, complete genome sequences of several baculoviruses are available, and this would allow the scientists to play with other unexplored genes in order to generate viruses which multiply rapidly and kill insect pests in a very short time.

5. CONCLUSIONS

Pest management strategies do not rely on any one single method, as is the case with chemicals. The biggest advantage that IPM programs have is their flexibility, which recognizes that the host, pathogen, and the environment interact and these interactions constantly change. Thus, each situation warrants a different strategy. Therefore, we need to play with several tools, along with chemical control as one of the components of any IPM program. Reduced usage of chemical pesticides will soon become the norm as commercialization and acceptability of food products which contain pesticide residues will become increasingly difficult. Abandoning the pesticides altogether will never happen, though reducing their usage and replacing the existing harmful ones with safer ones is more realistic. Indeed, the Codex Alimentarius Commission of the United Nations, with its mission to provide safe food, has established a maximum pesticide residue limit in a specific crop that is expected not to cause any harm to human health from consumption. As with any pesticide, it is important to understand the diverse effects (risks and benefits) of deployment of pest management with baculoviruses or their recombinants (Lapionte et al., 2012) with regulatory agencies stepping in to ensure safety of human beings and the environment. However, as things stand, baculoviruses have generated a space for themselves in an IPM program, along with other tools.

References

Abd-Alla, A.M.M., Vlak, J.M., Bergoin, M., Maruniak, J.E., Parker, A., Burand, J.P., Jehle, J.A., Boucias, D.G., 2009. Hytrosaviridae: a proposal for classification and nomenclature of a new insect virus family. Archives of Virology 154, 909–918.

Abd-Alla, A.M., Bergoin, M., Parker, A.G., Maniania, N.K., Vlak, J.M., Bourtzis, K., Boucias, D.G., Aksoy, S., 2013a. Improving sterile insect technique (SIT) for tsetse flies through research on their symbionts and pathogens. Journal of Invertebrate Pathology 112, S2–S10.

Abd-Alla, A.M., Kariithi, H.M., Mohamed, A.H., Lapiz, E., Parker, A.G., Vreysen, M.J., 2013b. Managing hytrosavirus infections in *Glossina pallidipes* colonies: feeding regime affects the prevalence of salivary gland hypertrophy syndrome. PLoS One 8, e61875.

Ackermann, H.W., Smirnoff, W.A., 1983. A morphological investigation of 23 baculoviruses. Journal of Invertebrate Pathology 41, 269–280.

Afonso, C.L., Tulman, E.R., Lu, Z., Kutish, G.F., Rock, D.L., 1999. The genome of *Melanoplus sanguinipes* entomopoxvirus. Journal of Virology 73, 533–552.

Ahmad, M., Arif, M.I., 2010. Resistance of beet armyworm *Spodoptera exigua* (Lepidoptera: Noctuidae) to endosulfan, organophosphorous and pyrethroid insecticides in Pakistan. Crop Protection 29, 1428–1433.

Anonymous, 1984. Proceedings of the FAO/IRRI Workshop on Judicious and Efficient Use of Pesticides on Rice. IRRI, Los Banos, Philippines (21–23 February, 1983).

Arrizubieta, M., Williams, T., Caballero, P., Simon, O., 2014. Selection of a nucleopolyhedrovirus isolate from *Helicoverpa armigera* as the basis for biological insecticide. Pest Management Science 70, 967–976.

Arthurs, S.P., Lacey, L.A., Behle, R.W., 2006. Evaluation of spray-dried lignin based formulations and adjuvants as UV light protectants for the granulovirus of the codling moth, *Cydia pomonella* (L.). Journal of Invertebrate Pathology 93, 88–95.

Asgari, S., Johnson, K.N. (Eds.), 2010. Insect Virology. Caister Academic Press, p. 436.

Asser-Kaiser, S., Fritsch, E., Undorf-Spahn, K., Kienzle, J., Eberle, K.E., Gund, N.A., Reineke, A., Zebitz, C.P.W., Heckel, D.G., Huber, J., Jehle, J.A., 2007. Rapid emergence of baculovirus resistance in codling moth due to dominant, sex-linked inheritance. Science 317, 1916–1918.

Au, S., Pante, N., 2012. Nuclear transport of baculovirus: revealing the nuclear pore complex passage. Journal of Structural Biology 177, 90–98.

Ayres, M.D., Howard, S.C., Kuzlo, J., Lovez-Ferber, M., Possee, R.D., 1994. The complete DNA sequence of *Autographa californica* nuclear polyhedron virus. Virology 202, 586–605.

Bawden, A.L., Glassberg, K.J., Diggans, J., Shaw, R., Farmerie, W., Moyer, R.W., 2000. Complete genome sequence of the *Amsacta moorei* entomopoxvirus: analysis and comparison with other poxviruses. Virology 274, 120–139.

Bedford, G.O., 2013. Biology and management of palm dynastid beetles: recent advances. Annual Review of Entomology 58, 353–372.

Behle, R.W., Tamez-Guerra, P., McGuire, M.R., 2003. Field activity and storage stability of *Anagrapha falcifera* nucleopolyhedrovirus (AfMNPV) in spray-dried lignin-based formulations. Journal of Economic Entomology 96, 1066–1075.

Berling, M., Rey, J.B., Ondet, S.J., Tallot, Y., Soubabere, O., Bonhomme, A., Sauphanor, B., Lopez-Ferber, M., 2009. Field trials of CpGV virus isolates overcoming resistance to CpGV-M. Virologica Sinica 24, 470–477.

Bernal, A., Williams, T., Hernandez-Suarez, E., Carnero, A., Caballero, P., Simon, O., 2013. A native variant of *Chrysodeixis chalcites* nucleopolyhedrovirus: the basis for a promising bioinsecticide for control of *C. chalcites* on Canary Islands' banana crops. Biological Control 67, 101–110.

Bernal, A., Simon, O., Williams, T., Caballero, P., 2014. Stage-specific insecticidal characteristics of a nucleopolyhedrovirus isolate from *Chrysodeixis chalcites* enhanced by optical brighteners. Pest Management Science 70, 798–804.

Bianchi, F.J.J.A., Snoeijing, I., van der Werf, W., Mans, R.M.W., Smits, P.H., Vlak, J.M., 2000. Biological activity of SeMNPV, AcMNPV, and three AcMNPV deletion mutants against *Spodoptera exigua* larvae (Lepidoptera: Noctuidae). Journal of Invertebrate Pathology 75, 28–35.

Bianchi, F.J.J.A., Vlak, J.M., van der Werf, W., 2002. Evaluation of the control of beet armyworm, *Spodoptera exigua*, with baculoviruses in greenhouses using a process-based simulation model. Biological Control 24, 277–284.

Bideshi, D.K., Tan, Y., Bigot, Y.A., Federici, B.A., 2005. Viral caspase contributes to modified apoptosis for virus transmission. Genes & Development 19, 1416–1421.

Bird, F.T., Burk, J.M., 1961. Artificially disseminated virus as a factor controlling the European spruce sawfly, *Diprion hercyniae* (Htg.) in the absence of introduced parasites. The Canadian Entomologist 93, 228–238.

Bixby-Brosi, A.J., Potter, D.A., 2010. Evaluating a naturally occurring baculovirus for extended biological control of the black cutworm (Lepidoptera: Noctuidae) in golf course habitats. Journal of Economic Entomology 103, 1555–1563.

Blommers, L., Vaal, F., Freriks, J., Helsen, H., 1987. Three years of specific control of summer fruit tortrix and codling moth on apple in the Netherlands. Journal of Applied Entomology 104, 353–371.

Blommers, L.H.M., 1994. Integrated pest management in European apple orchards. Annual Review of Entomology 39, 213–241.

Bonning, B.C., Hammock, B.D., 1996. Development of recombinant baculoviruses for insect control. Annual Review of Entomology 41, 191–210.

Bonning, B.C., Miller, W.A., 2010. Dicistroviruses. Annual Review of Entomology 55, 129–150.

Bove, J.M., 2006. Huanglongbing: a destructive, newly-emerging, century-old disease of citrus. Journal of Plant Pathology 88, 7–37.

Burden, J.P., Hails, R.S., Windass, J.D., Suner, M.M., Cory, J.S., 2000. Infectivity, speed of kill, and productivity of a baculovirus expressing the itch mite toxin txp-1 in second and fourth instar larvae of *Trichoplusia ni*. Journal of Invertebrate Pathology 75, 226–236.

Butler, C.D., O'Neil, R.J., 2008. Voracity and prey preference of insidious flower bug (Hemiptera: Anthocoridae) for immature stages of soybean aphid (Hemiptera: Aphididae) and soybean thrips (Thysanoptera: Thripidae). Environmental Entomology 37, 964–972.

Caballero, P., Zuidema, D., Santiago-Alvarez, C., Vlak, J.M., 1992. Biochemical and biological characterization of four isolates of *Spodoptera exigua* nuclear polyhedrosis virus. Biocontrol Science and Technology 2, 145–157.

Cai, Y., Cheng, Z., Li, C., Wang, F., Li, G., Pang, Y., 2010. Biological activity of recombinant *Spodoptera exigua* multicapsid nucleopolyhedrovirus against *Spodoptera exigua* larvae. Biological Control 55, 178–185.

Cao, G., Meng, X., Xue, R., Zhu, Y., Zhang, X., Pan, Z., Zheng, X., Gong, C., 2012. Characterization of the complete genome segments from BmCPV-SZ, a novel *Bombyx mori* cypovirus 1 isolate. Canadian Journal of Microbiology 58, 872–883.

Capinera, J.L. (Ed.), 2008. Encyclopedia of Entomology. Springer Science & Business Media, p. 4346.

Carriere, Y., Crickmore, N., Tabashnik, B.E., 2015. Optimizing pyramided transgenic Bt crops for sustainable pest management. Nature Biotechnology 33, 161–168.

Carrillo-Tripp, J., Bonning, B.C., Miller, W.A., 2014. Challenges associated with research on RNA viruses of insects. Current Opinion in Insect Science 6, 1–7.

Carson, R., 1962. Silent Spring. Houghton Mifflin, Boston.

Casida, J.E., Durkin, K.A., 2013. Neuroactive insecticides: targets, selectivity, resistance, and secondary effects. Annual Review of Entomology 58, 99–117.

Casique-Valdes, R., Reyes-Martinez, A.Y., Sanchez-Pena, S.R., Bidochka, M.J., Lopez-Arroyo, J.L., 2011. Pathogenicity of *Hirsutella citriformis* (Ascomycota: Cordycipitaceae) to *Diaphorina citri* (Hemiptera: Psyllidae) and *Bactericera cockerelli* (Hemiptera: Triozidae). Florida Entomologist 94, 703–705.

Chang, J.H., Choi, J.Y., Jin, B.R., Olszewski, J.A., Seo, S.J., O'Reilly, D.R., Je, Y.H., 2003. An improved baculovirus insecticide producing occlusion bodies that contain *Bacillus thuringiensis* insect toxin. Journal of Invertebrate Pathology 84, 30–37.

Chen, M., Shelton, A., Ye, G.-Y., 2011. Insect-resistant genetically modified rice in China: from research to commercialization. Annual Review of Entomology 56, 81–101.

Cheng, X.-W., Wan, X.-F., Xue, J., Moore, R.C., 2007. Ascovirus and its evolution. Virologica Sinica 22, 137–147.

Cherry, A.J., Parnell, M., Grzywaez, D., Brown, M., Jones, K.A., 1997. The optimization of *in vivo* nuclear polyhedrosis virus production of *Spodoptera exempta* (Walker) and *Spodoptera exigua* (Hubner). Journal of Invertebrate Pathology 70, 50–58.

Cory, J.S., Hirst, M.L., Williams, T., Halls, R.S., Goulson, D., Green, B.M., Carty, T.M., Possee, R.D., Cayley, P.J., Bishop, D.H.L., 1994. Field trial of a genetically improved baculovirus insecticide. Nature 370, 138–140.

Cory, J.S., Hirst, M.L., Sterling, P.H., Speight, M.R., 2000. Narrow host range nucleopolyhedrovirus for control of the browntail moth (Lepidoptera: Lymantriidae). Environmental Entomology 29, 661–667.

Cory, J.S., Green, B.M., Paul, R.K., Hunter-Fujita, F., 2005. Genotypic and phenotypic diversity of a baculovirus population within an individual insect host. Journal of Invertebrate Pathology 89, 101–111.

Crook, N.E., Jarrett, P., 1991. Viral and bacterial pathogens of insects. Society for Applied Bacteriology Symposium Series 20, 91S–96S.

Cross, J.V., Solomon, M.G., Chandler, D., Jarrett, P., Richardson, P.N., Winstanley, D., Bathon, H., Huber, J., Keller, B., Langenbruch, A., Zimmermann, G., 1999. Biocontrol of pests of apples and pears in northern and central Europe. 1. Microbial agents and nematodes. Biocontrol Science and Technology 9, 125–149.

Davis, J.A., Radcliffe, E.B., 2008. The importance of an invasive aphid species in vectoring a persistently transmitted potato virus: *Aphis glycines* is a vector of potato leaf roll virus. Plant Disease 92, 1515–1523.

Doyle, C.J., Hirst, M.L., Cory, J.S., Entwistle, P.F., 1990. Risk assessment studies: detailed host range testing of wild-type cabbage moth, *Mamestra brassicae* (Lepidoptera: Noctuidae) nuclear polyhedrosis virus. Applied and Environmental Microbiology 56, 2704–2710.

Drezen, J.-M., Chevignon, G., Louis, F., Huguet, E., 2014. Origin and evolution of symbiotic viruses associated with parasitoid wasps. Current Opinion in Insect Science 6, 35–43.

Eberle, K.E., Jehle, J.A., 2006. Field resistance of codling moth against *Cydia pomonella* granulovirus (CpGV) is autosomal and incompletely dominant inherited. Journal of Invertebrate Pathology 93, 201–206.

El-Helaly, A., 2013. Additives for a baculovirus against ultraviolet effect. Applied Science Reports 4, 187–191.

El-Menofy, W., Osman, G., Assaeedi, A., Salama, M., 2014. A novel recombinant baculovirus overexpressing a *Bacillus thuringiensis* Cry1Ab toxin enhances insecticidal activity. Biological Procedures Online 16, 7.

Elvira, S., Gorria, N., Munoz, D., Williams, T., Caballero, P., 2010. A simplified low-cost diet for rearing *Spodoptera exigua* (Lepidoptera: Noctuidae) and its effect on *S. exigua* nucleopolyhedrovirus production. Journal of Economic Entomology 103, 17–24.

Elvira, S., Ibargutxi, M.A., Gorria, N., Munoz, D., Caballero, P., Williams, T., 2013. Insecticidal characteristics of two commercial *Spodoptera exigua* nucleopolyhedrovirus strains produced on different host colonies. Biological and Microbial Control 106, 50–56.

Epstein, L., 2014. Fifty years since 'Silent spring'. Annual Review of Phytopathology 52, 377–402.

Erlandson, M., Newhouse, S., Moore, K., Janmaat, A., Myers, J., Theilmann, D., 2007. Characterization of baculovirus isolates from *Trichoplusia ni* populations from vegetable greenhouses. Biological Control 41, 256–263.

Felberbaum, R.S., 2015. The baculovirus expression vector system: a commercial manufacturing platform for viral vaccines and gene therapy vectors. Biotechnology Journal. http://dx.doi.org/10.1002/biot.201400438.

Fraser, M.J., 1986. Ultrastructural observations of virion maturation in *Autographa californica* nuclear polyhedrosis virus infected *Spodoptera frugiperda* cell cultures. Journal of Ultrastructure and Molecular Structure Research 95, 189–195.

Frost, S., Dowds, B., Boemare, N., Stackebrandt, E., 1997. *Xenorhabdus* and *Photorhabdus* spp.: bugs that kill bugs. Annual Review of Microbiology 51, 47–72.

Fuxa, J.R., 1991. Insect control with baculoviruses. Biotechnology Advances 9, 425–442.

Fuxa, J.R., 1998. Environmental manipulation for microbial control of insects. In: Barbosa, P. (Ed.), Conservation Biological Control. Academic Press, San Diego, pp. 255–289.

Gassman, A.J., Carriere, Y., Tabashnik, B.E., 2009. Fitness costs of insect resistance to *Bacillus thuringiensis*. Annual Review of Entomology 54, 147–163.

Gebhardt, M.M., Eberle, K.E., Radtke, P., Jehle, J.A., 2014. Baculovirus resistance in codling moth is virus isolate-dependent and the consequence of a mutation in viral gene *pe*38. Proceedings of the National Academy of Sciences of the United States of America 111, 15711–15716.

Geden, C.J., Steenberg, T., Lietze, V.U., Boucias, D.G., 2011. Salivary gland hypertrophy virus of house flies in Denmark: prevalence, host range, and comparison with a Florida isolate. Journal of Vector Ecology 36, 231–238.

Gershburg, E., Stockholm, D., Froy, O., Rashi, S., Gurevitz, M., Chejanovsky, N., 1998. Baculovirus-mediated expression of a scorpion depressant toxin improves the insecticidal efficacy achieved with excitatory toxins. FEBS Letters 422, 132–136.

Gildow, F.E., D'Arcy, C.J.D., 1990. Cytopathology and experimental host range of *Rhopalosiphum padi* virus, a small isometric RNA virus infecting cereal grain aphids. Journal of Invertebrate Pathology 55, 245–257.

Gill, S.S., Cowles, E.A., Pietrantonio, P.V., 1992. The mode of action of *Bacillus thuringiensis* endotoxins. Annual Review of Entomology 37, 615–634.

Glen, D.M., Payne, C.C., 1984. Production and field evaluation of codling moth granulosis virus for control of *Cydia pomonella* in the United Kingdom. Annals of Applied Biology 104, 87–98.

Goodman, C.L., Mcintosh, A.H., El Sayed, G.N., Grasela, J.J., Stiles, B., 2001. Production of selected baculoviruses in newly established lepidopteran cell lines. In Vitro Cellular & Developmental Biology-Animal 37, 374–379.

Gopal, M., Gupta, A., Thomas, G.V., 2006. Prospects of using *Metarhizium anisopliae* to check the breeding of insect pest *Oryctes rhinoceros* in coconut leaf vermicomposting sites. Bioresource Technology 97, 1801–1806.

Grafton-Cardwell, E.E., Stelinski, L.L., Stansly, P.A., 2013. Biology and management of Asian citrus psyllid, vector of the huanglongbing pathogens. Annual Review of Entomology 58, 413–432.

Graham, R.I., Grzywaez, D., Mushobozi, W.L., Wilson, K., 2012. *Wolbachia* in a major African crop pest increases susceptibility to viral diseases rather than protects. Ecology Letters 15, 993–1000.

Granados, R.R., Federici, B.A. (Eds.), 1986. The Biology of Baculoviruses. CRC Press, Boca Raton, Florida.

Gross, C.H., Russel, R.L.Q., Rohrmann, G.F., 1994. *Orgyia pseudotsugata* baculovirus p10 and polyhedron envelope protein genes: analysis of their relative expression levels and role in polyhedron structure. Journal of General Virology 75, 1115–1123.

Grzywaez, D., Mushobozi, W.L., Parnell, M., Jolliffe, F., Wilson, K., 2008. The evaluation of *Spodoptera exempta* nucleopolyhedrovirus (SpexNPV) for the field control of African armyworm (*Spodoptera exempta*) in Tanzania. Crop Protection 27, 17–24.

Grzywaez, D., Stevenson, P.C., Mushobozi, W.L., Belmain, S., Wilson, K., 2014. The use of indigenous ecological resources for pest control in Africa. Food Security 6, 71–86.

Hajek, A.E., McManus, M.L., Delalibera, I., 2007. A review of introductions of pathogens and nematodes for classical biological control of insects and mites. Biological Control 41, 1–13.

Hammock, B.D., Bonning, B.C., Possee, R.D., Hanzlik, T.N., Maeda, S., 1990. Expression and effects of the juvenile hormone esterase in a baculovirus vector. Nature 344, 458–461.

Hara, K., Funakoshi, M., Kawarabata, T., 1995. *In vivo* and *in vitro* characterization of several isolates of *Spodoptera exigua* nuclear polyhedrosis virus. Acta Virologica 39, 215–222.

Held, D.W., Potter, D.A., 2012. Prospects for managing turfgrass pests with reduced chemical inputs. Annual Review of Entomology 57, 329–354.

Henry, J.E., 1981. Natural and applied control of insects by protozoa. Annual Review of Entomology 26, 49–73.

Henry, J.E., 1985. Effect of grasshopper species, cage density, light intensity, and method of inoculation on mass production of *Nosema locustae* (Microsporidia: Nosematidae). Journal of Economic Entomology 78, 1245–1250.

Herniou, E.A., Olszewski, J.A., Cory, J.S., O'Reilly, 2003. The genome sequence and evolution of baculoviruses. Annual Review of Entomology 48, 211–234.

Herniou, E.A., Olszewski, J.A., O'Reilly, D.R., Cory, J.S., 2004. Ancient coevolution of baculoviruses and their insect hosts. Journal of Virology 78, 3244–3251.

Holt, J., Mushobozi, W., Day, R.K., Knight, J.D., Kimani, M., N'juki, J., Musebe, R., 2006. A simple Bayesian network to interpret the accuracy of armyworm outbreak forecasts. Annals of Applied Biology 148, 141–146.

Hooft van Iddekinge, B.J.L., Smith, G.E., Summers, M.D., 1983. Nucleotide sequence of the polyhedron gene of *Autographa californica* nuclear polyhedrosis virus. Virology 131, 561–565.

Hostetter, D.L., Puttler, B., 1991. A new broad host spectrum nuclear polyhedrosis virus isolated from a celery looper, *Anagrapha falcifera* (Kirby), (Lepidoptera: Noctuidae). Environmental Entomology 20, 1480–1488.

Hu, Y.-C., 2006. Baculovirus vectors for gene therapy. Advances in Virus Research 68, 287–320.

Huang, J.K., Rozelle, R.F., Pray, C., 2005. Insect-resistant GM rice in farmers fields: assessing productivity and health effects in China. Science 308, 688–690.

Huang, G.-H., Garretson, T.A., Cheng, X.-H., Holztrager, M.S., Li, S.-J., Cheng, X.-W., 2012. Phylogenetic position and replication kinetics of *Heliothis virescens* ascovirus 3h (HvAV3h) isolated from *Spodoptera exigua*. PLoS One 7, e40225.

Huber, J., Dickler, E., 1977. Codling moth granulosis virus: its efficiency in the field in comparison with organophosphorus insecticides. Journal of Economic Entomology 70, 557–561.

Huber, J., 1986. Use of baculoviruses in pest management programs. In: Granados, R.R., Federici, B.A. (Eds.), The Biology of Baculoviruses. Practical Applications for Insect Control, vol. II. CRC, Boca Raton, FL, pp. 181–202.

Hughes, P., van Beek, N.A.M., Wood, H.A., 1986. A modified droplet feeding method for rapid assay of *Bacillus thuringiensis* and baculoviruses in noctuid larvae. Journal of Invertebrate Pathology 48, 187–192.

Hukuhara, T., Hayakawa, T., Wijonarko, A., 1999. Increased baculovirus susceptibility of armyworm larvae feeding on transgenic rice plants expressing an entomopoxvirus gene. Nature Biotechnology 17, 1122–1124.

Hulo, C., de Castro, E., Masson, P., Bougueleret, L., Bairoch, A., Xenarios, I., Le Mercier, P., 2011. Viral zone: a knowledge resource to understand viral diversity. Nucleic Acids Research 39, D576–D582.

Hunger, A., 2005. The *Oryctes* virus: its detection, identification, and implementation in biological control of the coconut palm rhinoceros beetle, *Orycetes rhinoceros* (Coleoptera: Scarabacidae). Journal of Invertebrate Pathology 89, 78–84.

Hunnicutt, L.E., Mozoruk, J., Hunter, W.B., Crosslin, J.M., Cave, R.D., Powell, C.A., 2008. Prevalence and natural host range of *Homalodisca coagulate* virus-1 (HoCV-1). Archives of Virology 153, 61–67.

Hunter, W.B., Lapointe, S.L., 2003. Iridovirus infection of cell cultures from *Diaprepes* root weevil, *Diaprepes abbreviates*. Journal of Insect Science 3, 37.

Ibargutxi, M.A., Munoz, D., Bernal, A., de Escudero, I.R., Caballero, P., 2008. Effects of stilbene optical brighteners on the insecticidal activity of *Bacillus thuringiensis* and a single nucleopolyhedrovirus on *Helicoverpa armigera*. Biological Control 47, 322–327.

International Committee on Taxonomy of Viruses (ICTV), 2014. Virus Taxonomy: 2014 Release. Montreal, Canada http://ictvonline.org/virustaxonomy.asp.

Inceoglu, A.B., Kamita, S.G., Hammock, B.D., 2006. Genetically modified baculoviruses: a historical overview and future outlook. Advances in Virus Research 68, 323–360.

Jakubowska, A., van Oers, M.M., Ziemnicka, J., Lipa, J.J., Vlak, J.M., 2005. Molecular characterization of *Agrotis segetum* nucleopolyhedrovirus from Poland. Journal of Invertebrate Pathology 90, 64–68.

Jehle, J.A., Blissard, G.W., Bonning, B.C., Cory, J.S., Herniou, E.A., Rohrmann, G.F., Theilmann, D.A., Thiem, S.M., Vlak, J.M., 2006. On the classification and nomenclature of baculoviruses: a proposal for revision. Archives of Virology 151, 1257–1266.

Jehle, J.A., Abd-Alla, A.M., Wang, Y., 2013. Phylogeny and evolution of Hytrosaviridae. Journal of Invertebrate Pathology 112, S62–S67.

Jenkins, D.A., Hunter, W.B., Goenaga, R., 2011. Effects of invertebrate iridescent virus 6 in *Phyllophaga vandinei* and its potential as a biocontrol delivery system. Journal of Insect Science 11, 44.

Ji, X., Sutton, G., Evans, G., Axford, D., Owen, R., Stuart, D., 2010. How baculovirus polyhedra fit square pegs into round holes to robustly package viruses. The EMBO Journal 29, 505–514.

Jiang, H., Zhou, L., Zhang, J.-M., Dong, H.-F., Hu, Y.-Y., Jiang, M.-S., 2008. Potential of *Periplaneta fuliginosa* densovirus as a biocontrol agent for smoky-brown cockroach, *P. fuliginosa*. Biological Control 46, 94–100.

Jin, L., Zhang, H., Lu, Y., Yang, Y., Wu, K., Tabashnik, B.E., Wu, Y., 2015. Large-scale test of the natural refuge strategy for delaying insect resistance to transgenic Bt crops. Nature Biotechnology 33, 169–174.

Jung, M.P., Choi, J.Y., Tao, X.Y., Jin, B.R., Je, Y.H., Park, H.H., 2012. Insecticidal activity of recombinant baculovirus expressing both spider toxin isolated from *Araneus ventricosus* and *Bacillus thuringiensis* crystal protein fused to a viral polyhedron. Entomological Research 42, 339–346.

Kamel, F., 2013. Paths from pesticides to Parkinson's. Science 341, 722–723.

Kariithi, H.M., van Oers, M.M., Vlak, J.M., Vreysen, M.J.B., Parker, A.G., Abd-Alla, A.M.M., 2013. Virology, epidemiology and pathology of *Glossina* hytrosavirus, and its control prospects in laboratory colonies of the tsetse fly, *Glossina pallidipes* (Diptera: Glossinidae). Insects 4, 287–319.

Kathage, J., Qaim, M., 2012. Economic impacts and impact dynamics of Bt (*Bacillus thuringiensis*) cotton in India. Proceedings of the National Academy of Sciences of the United States of America 109, 11652–11656.

Keller, S., Schweizer, C., Keller, E., Brenner, H., 1997. Control of white grubs (*Melolontha melolontha* L.) by treating adults with the fungus *Beauveria brongniartii*. Biocontrol Science and Technology 7, 105–116.

Kolodny-Hirsch, D.M., Sitchawat, T., Jansiri, T., Chenrchaivachirakul, A., Ketunuti, U., 1997. Field evaluation of a commercial formulation of the *Spodoptera exigua* (Lepidoptera: Noctuidae) nuclearpolyhedrosis virus for control of beet armyworm on vegetable crops in Thailand. Biocontrol Science and Technology 7, 475–488.

Kooyman, C., Abdalla, O.M., 1998. Application of *Metarhizium flavoviride* (Deuteromycotina: Hyphomycetes) spores against the tree locust *Anacridium melanorhodon* (Orthoptera: Acrididae) in Sudan. Biocontrol Science and Technology 8, 215–219.

Krishnakumar, R., Maheshwari, P., 2007. Assessment of the sterile insect technique to manage red palm weevil *Rhynchophorus ferrugineus* in coconut. In: Vreysen, M.J.B., Robinson, A.S., Hendrichs, J. (Eds.), Area-Wide Control of Insect Pests - From Research to Field Implementation. Springer, pp. 475–485.

Lacey, L.A., Shapiro-Ilan, D.I., 2008. Microbial control of insect pests in temperate orchard systems: potential for incorporation into IPM. Annual Review of Entomology 53, 121–144.

Lai-Fook, J., Dall, D.J., 2000. Spindle bodies of *Heliothis armigera* entomopoxvirus develop in structures associated with host cell endoplasmic reticulum. Journal of Invertebrate Pathology 75, 183–192.

Landis, D.A., Wratten, S.D., Gurr, G.M., 2000. Habitat management to conserve natural enemies of arthropod pests in agriculture. Annual Review of Entomology 45, 175–201.

Lange, C.E., De Wysiecki, M.L., 1996. The fate of *Nosema locustae* (Microsporidia: Nosematidae) in Argentine grasshoppers (Orthoptera: Acrididae). Biological Control 7, 24–29.

Lapointe, R., Thumbi, D., Lucarotti, C., 2012. Recent advances in our knowledge of baculovirus molecular biology and its relevance for the registration of baculovirus-based products for insect pest population control. In: Soloneski, S. (Ed.), Integrated Pest Management and Pest Control-Current and Future Tactics. InTech. ISBN:978-953-51-0050-8.

Lasa, R., Pagola, I., Ibanex, I., Belda, J.E., Williams, T., Caballero, P., 2007. Efficacy of *Spodoptera exigua* multiple nucleopolyhedrovirus as a biological insecticide for beet armyworm control in greenhouse of southern Spain. Biocontrol Science and Technology 17, 221–232.

Lasa, R., Williams, T., Caballero, P., 2008. Insecticidal properties and microbial contaminants in a *Spodoptera exigua* multiple nucleopolyhedrovirus (SeMNPV, Baculoviridae) formulation stored at different temperatures. Journal of Economic Entomology 101, 42–49.

Lauzon, H.A.M., Garcia-Maruniak, A., Zanotto, P.M.A., Clemente, J.C., Herniou, E.A., Lucarotti, C.J., Arif, B.M., Maruniak, J.E., 2006. Genomic comparison of *Neodiprion sertifer* and *Neodiprion lecontei* nucleopolyhedroviruses and identification of potential hymenopteran baculovirus-specific open reading frames. Journal of General Virology 87, 1477–1489.

Li, Y., Tan, L., Li, Y., Chen, W., Zhang, J., Hu, Y., 2006. Identification and genome characterization of *Heliothis armigera* cypovirus types 5 and 14 and *Heliothis assulta* cypovirus type 14. Journal of General Virology 87, 387–394.

Liu, T.-X., Stansly, P.A., Gerling, D., 2015. Whitefly parasitoids: distribution, life history, bionomics, and utilization. Annual Review of Entomology 60, 273–292.

Liu, N., 2015. Insecticide resistance in mosquitoes: impact, mechanisms, and research directions. Annual Review of Entomology 60, 537–559.

Lomer, C.J., Bateman, R.P., Johnson, D.L., Langewald, J., Thomas, M., 2001. Biological control of locusts and grasshoppers. Annual Review of Entomology 46, 667–702.

Luo, S., Naranjo, S.E., Wu, K., 2014. Biological control of cotton pests in China. Biological Control 68, 6–14.

Maeda, S., 1989. Increased insecticidal effect by a recombinant baculovirus carrying a synthetic diuretic hormone gene. Biochemical and Biophysical Research Communications 165, 1177–1183.

Miele, S.A.B., Garavaglia, M.J., Belaich, M.N., Ghiringhelli, P.D., 2011. Baculovirus: molecular insights on their diversity and conservation. International Journal of Evolutionary Biology 15. Article ID 379424.

Mignault, M.P., Roy, M., Brodeur, J., 2006. Soybean aphid predators in Quebec and the sustainability of *Aphis glycines* as prey for three Coccinellidae. BioControl 51, 89–106.

Milner, R.J., Soper, R.S., Lutton, G.G., 1982. Field release of an Israeli strain of the fungus *Zoophthora radicans* (Brefeld) Batko for biological control of *Therioaphis trifolii* (Monell) *f. maculata*. Journal of the Australian Entomological Society 21, 113–118.

Milner, R.J., 1997. Prospects for biopesticides for aphid control. Entomophaga 42, 227–239.

Minion, F.C., Coons, L.B., Broome, J.R., 1979. Characterization of the polyhedral envelope of the nuclear polyhedrosis virus of *Heliothis virescens*. Journal of Invertebrate Pathology 34, 303–307.

Moreau, G., 2006. Past and present outbreaks of the balsam fir sawfly in Western Newfoundland: an analytical overview. Forest Ecology and Management 221, 215–219.

Moscardi, F., 1999. Assessment of the application of baculoviruses for control of Lepidoptera. Annual Review of Entomology 44, 211–234.

van Munster, M., Dullemans, A.M., Verbeek, M., van den Heuvel, J.F.J.M., Reinbold, C., Brault, V., Clerivet, A., van der Wilk, F., 2003. Characterization of a new densovirus infecting the green peach aphid *Myzus persicae*. Journal of Invertebrate Pathology 84, 6–14.

Mutuel, D., Ravellac, M., Chabi, B., Multeau, C., Salmon, J.-M., Fournier, P., Ogliastro, M., 2010. Pathogenesis of *Junonia coenia* densovirus in *Spodoptera frugiperda*: a route of infection that leads to hypoxia. Virology 403, 137–144.

Naranjo, S.E., Ellsworth, P.C., Frisvold, G.B., 2015. Economic value of biological control in integrated pest management of managed plant systems. Annual Review of Entomology 60, 621–645.

van Oers, M.M., 2006. Vaccines for viral and parasitic diseases produced with baculovirus vectors. Advances in Virus Research 68, 193–253.

Oliveira, J.V., Wolff, J.L., Garcia-Maruniak, A., Ribeiro, B.M., de Castro, M.E., de Souza, M.L., Moscardi, F., Maruniak, J.E., Zanotto, P.M., 2006. Genome of the most widely used viral pesticide: *Anticarsia gemmatalis* multiple nucleopolyhedrovirus. Journal of General Virology 87, 3233–3250.

Oomens, A.G., Blissard, G.W., 1999. Requirement for GP64 to drive efficient budding of *Autographa californica* multicapsid nucleopolyhedrovirus. Journal of Virology 254, 297–314.

Pal, N., Boyapalle, S., Beckett, R., Miller, W.A., Bonning, B.C., 2007. A baculovirus-expressed dicistrovirus that is infectious to aphids. Journal of Virology 81, 9339–9345.

Park, B., Kim, Y., 2012. Immunosuppression induced by expression of a viral RNase enhances susceptibility of *Plutella xylostella* to microbial pesticides. Insect Science 19, 47–54.

Perera, O.P., Valles, S.M., Green, T.B., White, S., Strong, C.A., Becnel, J.J., 2006. Molecular analysis of an occlusion body protein from *Culex nigripalpus* nucleopolyhedrovirus (CuniNPV). Journal of Invertebrate Pathology 91, 35–42.

Peveling, R., Attignon, S., Langewald, J., Quambama, Z., 1999. An assessment of the impact of biological and chemical grasshopper control agents on ground-dwelling arthropods in Niger, based on presence/absence sampling. Crop Protection 18, 323–329.

Plymale, R., Grove, M.J., Cox-Foster, D., Ostiguy, N., Hoover, K., 2008. Plant-mediated alteration of the peritrophic matrix and baculovirus infection in lepidopteran larvae. Journal of Insect Physiology 54, 737–749.

Prater, C.A., Redmond, C.T., Barney, W., Bonning, B., Potter, D.A., 2006. Microbial control of the black cutworm (Lepidoptera: Noctuidae) in turfgrass using *Agrotis ipsilon* multiple nucleopolyhedrovirus. Journal of Economic Entomology 99, 1129–1137.

Progar, R.A., Rinella, M.J., Fekedulegn, D., Butler, L., 2010. Nuclear polyhedrosis virus as a biological control agent for *Malacosoma americanum* (Lepidoptera: Lasiocampidae). Journal of Applied Entomology 134, 641–646.

Radek, R., Fabel, P., 2000. A new entomopoxvirus from a cockroach: light and electron microscopy. Journal of Invertebrate Pathology 75, 19–27.

Ragsdale, D.W., Landis, D.A., Brodeur, J., Heimpel, G.E., Desneux, N., 2011. Ecology and management of the soybean aphid in North America. Annual Review of Entomology 56, 375–399.

Rauh, V., Perera, F.P., Horton, M.K., Whyatt, R.M., Bansal, R., Hao, X., Liu, J., Barr, D.B., Slotkin, T.A., Peterson, B.S., 2012. Brain anomalies in children exposed prenatally to a common organophosphate pesticide. Proceedings of the National Academy of Sciences of the United States of America 109, 7871–7876.

Reddy, G.V.P., Zhao, Z., Humber, R.A., 2014. Laboratory and field efficacy of entomopathogenic fungi for the management of the sweetpotato weevil, *Cylas formicarius* (Coleoptera: Brentidae). Journal of Invertebrate Pathology 122, 10–15.

Redman, E.M., Wilson, K., Grzywacz, D., Cory, J.S., 2010. High levels of genetic diversity in *Spodoptera exempta* NPV from Tanzania. Journal of Invertebrate Pathology 105, 190–193.

Regev, A., Rivkin, H., Inceoglu, B., Gershburg, E., Hammock, B.D., Gurevitz, M., Chejanovsky, N., 2003. Further enhancement of baculovirus insecticidal efficacy with scorpion toxins that interact cooperatively. FEBS Letters 537, 106–110.

Ribeiro, H.C., Pavan, O., Moutri, A., 1997. Comparative susceptibility of two different hosts to genotypic variants of the *Anticarsia gemmatalis* nuclear polyhedrosis virus. Entomologia Experimentalis et Applicata 83, 233–237.

Rohrmann, G.F., 2013. Baculovirus Molecular Biology. National Library of Medicine (US), National Center for Biotechnology Information, Bethesda, MD. http://www.ncbi.nim.nih.gov/books/NBK114593.

Rowley, D.L., Popham, H.J.R., Harrison, R.L., 2011. Genetic variation and virulence of nucleopolyhedroviruses isolated world wide from the heliothine pests *Helicoverpa armigera*, *Helicoverpa zea* and *Heliothis virescens*. Journal of Invertebrate Pathology 107, 112–126.

Schneider, L.C.L., Silva, C.V., Pamphile, J.A., Conte, H., 2013. Infection, colonization and extrusion of *Metarhizium anisopliae* (Metsch) Sorokin (Deuteromycotina: Hyphomycetes) in pupae of *Diatraea saccharalis* F. (Lepidoptera: Crambidae). Journal of Entomology and Nematology 5, 1–9.

Schnepf, E., Crickmore, N., Van Rie, J., Lereclus, D., Baum, J., Feitelson, J., Zeigler, D.R., Dean, D.H., 1998. *Bacillus thuringiensis* and its pesticidal crystal proteins. Microbiology and Molecular Biology Reviews 62, 775–806.

Sekita, N., Kawashima, K., Aizu, H., Shirisaki, S., Yamada, M., 1984. A short term control of *Adoxophyces orana fasciata* Walsingham (Lepidoptera: Tortricidae) by a granulosis virus in apple orchards. Applied Entomology and Zoology 19, 498–508.

Serrano, A., Pijlman, G.P., Vlak, J.M., Munoz, D., Williams, T., Caballero, P., 2015. Identification of *Spodoptera exigua* nucleopolyhedrovirus genes involved in pathogenicity and virulence. Journal of Invertebrate Pathology 126, 43–50.

Shah, P.A., Pell, J.K., 2003. Entomopathogenic fungi as biological control agents. Applied Microbial Biotechnology 61, 413–423.

Shapiro-Ilan, D.I., Mizell, R.F., Cottrell, T.E., Horton, D.L., 2004. Measuring field efficacy of *Steinernema feltiae* and *Steinernema riobrave* for suppression of plum curculio, *Conotrachelus nenuphar*, larvae. Biological Control 30, 496–503.

Shelton, A.M., Zhao, J.Z., Roush, R.T., 2002. Economic, ecological, food safety, and social consequences of the deployment of Bt transgenic plants. Annual Review of Entomology 47, 845–881.

Sheppard, R.F., Stairs, G.R., 1977. Dosage-mortality and time mortality studies of a granulosis virus in a laboratory strain of the codling moth, *Laspeyresia pomonella*. Journal of Invertebrate Pathology 29, 216–221.

Shikano, I., Ericsson, J.D., Cory, J.S., Myers, J.H., 2010. Indirect plant-mediated effects on insect immunity and disease resistance in a tritrophic system. Basic and Applied Ecology 11, 15–22.

Shim, H.J., Choi, J.Y., Wang, Y., Tao, X.Y., Liu, Q., Roh, J.Y., Kim, J.S., Kim, W.J., Woo, S.D., Jin, B.R., Je, Y.H., 2013. NeuroBactrus, a novel, highly effective, and environmentally friendly recombinant baculovirus insecticide. Applied and Environmental Microbiology 79, 141–149.

Shrestha, S., Kim, H.H., Kim, Y., 2009. An inhibitor of NF-kB encoded in *Cotesia plutella* bracovirus inhibits expression of antimicrobial peptides and enhances pathogenicity of *Bacillus thuringiensis*. Journal of Asia-Pacific Entomology 12, 277–283.

Simon, O., Bernal, A., Williams, T., Carnero, A., Hernandez-Suarez, E., Munoz, D., Caballero, P., 2015. Efficacy of an alphabaculovirus-based biological insecticide for control of *Chrysodeixis chalcites* (Lepidoptera: Noctuidae) on tomato and banana crops. Pest Management Science. http://dx.doi.org/10.1002/ps.3969.

Slack, J., Arif, B.M., 2007. The baculoviruses occlusion-derived virus: virion structure and function. Advances in Virus Research 69, 99–165.

Slack, J.M., Ribeiro, B.M., Souza, M.L., 2004. The gp64 locus of *Anticarsia gemmatalis* multicapsid nucleopolyhedrovirus contains a 3′ repair exonuclease homologue and lacks v-cath and chiA genes. Journal of General Virology 85, 211–219.

Smith, S.M., 1996. Biological control with *Trichogramma*: advances, successes, and potential of their use. Annual Review of Entomology 41, 375–406.

Stansly, P.A., Calvo, J., Urbaneja, A., 2005. Release rates for control of *Bemisia tabaci* (Homoptera: Aleyrodidae) biotype "Q" with *Eretmocerus mundus* (Hymenoptera: Aphelinidae) in greenhouse tomato and pepper. Biological Control 35, 124–133.

Steinhaus, E.A., Thompson, C.G., 1949. Granulosis disease in the buckeye caterpillar *Junonia coenia* Hubner. Science 110, 276–278.

Strand, M.R., Burke, G.R., 2014. Polydnaviruses: nature's genetic engineers. Annual Review of Virology 1, 333–354.

Sun, X., Peng, H., 2007. Recent advances in biological control of pest insects by using viruses in China. Virologica Sinica 22, 158–162.

Szelei, J., Woodring, J., Goettel, M.S., Duke, G., Jousset, F.X., Liu, K.Y., Li, Y., Styer, E., Boucias, D.G., Kleespies, R.G., Bergoin, M., Tijsen, P., 2011. Susceptibility of North-American and European crickets to *Acheta domesticus* densovirus (AdDNV) and associated epizootics. Journal of Invertebrate Pathology 106, 394–399.

Szewczyk, B., Hoyos-Carvajal, L., Paluszek, M., Skrzeca, I., Lobo de Souza, M., 2006. Baculoviruses: re-emerging biopesticides. Biotechnology Advances 24, 143–160.

Tabashnik, B.E., Sisterson, M.S., Ellsworth, P.C., Dennehy, T.J., Antilla, L., Liesner, L., Whitlow, M., Staten, R.T., Fabrick, J.A., Unnithan, G.C., Yelich, A., Ellers-Kirk, C., Harpold, V.S., Li, X., Carriere, Y., 2010. Suppressing resistance to Bt cotton with sterile insect release. Nature Biotechnology 28, 1304–1307.

Tabashnik, B.E., Brevault, T., Carriere, Y., 2013. Insect resistance to Bt crops: lessons from the first billion acres. Nature Biotechnology 31, 510–521.

Tan, L., Zhang, J., Li, Y., Jiang, H., Cao, X., Hu, Y., 2008. The complete nucleotide sequence of the type 5 *Helicoverpa armigera* cytoplasmic polyhedrosis virus genome. Virus Genes 36, 587–593.

Tang, Q., Hu, Z., Yang, Y., Wu, H., Qiu, L., Chen, K., Li, G., 2015. Overexpression of Bm65 correlates with reduced susceptibility to inactivation by UV light. Journal of Invertebrate Pathology 127, 87–92.

Tao, X.Y., Choi, J.Y., Kim, W.J., An, S.B., Liu, Q., Kim, S.E., Lee, S.H., Kim, J.H., Woo, S.D., Jin, B.R., Je, Y.H., 2015. *Autographa californica* multiple nucleopolyhedrovirus ORF11 is essential for budded-virus production and occlusion-derived-virus development. Journal of Virology 89, 373–383.

Tellam, R.L., Wijffels, G., Willadsen, P., 1999. Peritrophic matrix proteins. Insect Biochemistry and Molecular Biology 29, 87–101.

Theze, J., Bezier, A., Periquet, G., Drezen, J.-M., Herniou, E.A., 2011. Paleozoic origin of insect large dsDNA viruses. Proceedings of the National Academy of Sciences of the United States of America 108, 15931–15935.

Theze, J., Takatsuka, J., Li, Z., Gallais, J., Doucet, D., Arif, B., Nakai, M., Herniou, E.A., 2013. New insights into the evolution of Entomopoxvirinae from the complete genome sequences of four entomopoxviruses infecting *Adoxophyes honmai*, *Choristoneura biennis*, *Choristoneura rosaceana*, and *Mythimna separata*. Journal of Virology 87, 7992–8003.

Thomas, M.B., 1999. Ecological approaches and the development of 'truly integrated' pest management. Proceedings of the National Academy of Sciences of the United States of America 96, 5944–5951.

Thompson, C.G., Steinhaus, E.A., 1950. Further tests using a polyhedrosis virus to control the alfalfa caterpillar. Hilgardia 19, 411–445.

Vail, P.V., Sutter, G., Jay, D.L., Gough, D., 1971. Reciprocal infectivity of nuclear polyhedrosis viruses of the cabbage looper and alfalfa looper. Journal of Invertebrate Pathology 17, 383–388.

Vega, F.E., Kaya, H.K. (Eds.), 2012. Insect Pathology. Academic Press, p. 490.

Verger, P.J.P., Boobis, A.R., 2013. Reevaluate pesticides for food security and safety. Science 341, 717–718.

Wang, Y., Jehle, J.A., 2009. Nudiviruses and other large, double-stranded circular DNA viruses of invertebrates: new insights on an old topic. Journal of Invertebrate Pathology 101, 187–193.

Wang, Z.-Y., He, K.-L., Zhang, F., Lu, X., Babendreier, D., 2014. Mass rearing and release of *Trichogramma* for biological control of insect pests of corn in China. Biological Control 68, 136–144.

Waterfield, N.R., Ciche, T., Clarke, D., 2009. *Photorhabdus* and a host of hosts. Annual Review of Microbiology 63, 557–574.

Westigard, P.H., Gut, L.J., Liss, W.J., 1986. Selective control program for the pear pest complex in Southern Oregon. Journal of Economic Entomology 79, 250–257.

Williams, T., Arredondo-Bernal, H.C., Rodriquez-del-Bosque, L.A., 2013. Biological pest control in Mexico. Annual Review of Entomology 58, 119–140.

Williams, T., 1996. The iridoviruses. Advances in Virus Research 46, 345–412.

Wraight, S.P., Carruthers, R.I., Jaronski, S.T., Bradley, C.A., Garza, C.J., Galaini-Wraight, S., 2000. Evaluation of the entomopathogenic fungi *Paecilomyces* spp. and *Beauveria bassiana* against the silverleaf whitefly, *Bemisia argentifolii*. Journal of Invertebrate Pathology 71, 217–226.

Xue, Y., Bahlai, C.A., Frewin, A., Sears, M.K., Schaafsma, A.V., Hallett, R.H., 2009. Predation by *Coccinella septempunctata* and *Harmonia axyridis* (Coleoptera: Coccinellidae) on *Aphis glycines* (Homoptera: Aphididae). Environmental Entomology 38, 708–714.

Yang, M., Wang, S., Yue, X.-L., Li, L.L., 2014. *Autographa californica* multiple nucleopolyhedrovirus ORF132 encodes a nucleocapsid-associated protein required for budded-virus and multiply enveloped occlusion-derived-virus production. Journal of Virology 88, 12586–12598.

Yeo, H., Pell, J.K., Alderson, P.J., Clark, S.J., Pye, B.J., 2003. Laboratory evaluation of temperature effects on the germination and growth of entomopathogenic fungi and on their pathogenicity to two aphid species. Pest Management Science 59, 156–165.

Zeddam, J.-L., Arroyo Cruzado, J., Rodriguez, J.L., Ravallec, H., Candiotti Subilete, E., 2003. A cypovirus from the South American oil-palm pest *Norape argyrrhorea* and its potential as a microbial control agent. BioControl 48, 101–112.

Zhang, J.H., Washburn, J.O., Jarvis, D.L., Volkman, L.E., 2004. *Autographa californica* M nucleopolyhedrovirus early GP64 synthesis mitigates developmental resistance in orally infected noctuid hosts. Journal of General Virology 85, 833–842.

Zhang, H., Yang, Q., Qin, Q.L., Zhu, W., Zhang, Z.F., Li, Y.N., Zhang, N., Zhang, J.H., 2014. Genomic sequence analysis of *Helicoverpa armigera* nucleopolyhedrovirus isolated from Australia. Archives of Virology 159, 595–601.

Zhou, Y., Qin, T., Xiao, Y., Qin, F., Lei, C., Sun, X., 2014. Genomic and biological characterization of a new cypovirus isolated from *Dendrolimus punctatus*. PLoS One 9, e113201.

Bacillus thuringiensis

G. Keshavareddy, A.R.V. Kumar

Department of Entomology, University of Agriculture Sciences, GKVK, Bangalore, India

1. INTRODUCTION

The burgeoning human population continues to exert pressure on agricultural production systems. In order to increase food production, natural ecosystems are being rapidly converted for human use, destroying forests, soil, and native plants and animals. To produce sufficient food, commercial and subsistence agriculture farming systems must be highly productive, sustainable, and nonpolluting. The excessive use of chemical insecticides has led to several environmental problems such as high levels of insecticide residues in food, contamination of ground water, and other deleterious effects on nontarget organisms including humans. The production and use of synthetic pesticides has been considered as a serious environmental risk.

In this context, there is now a greater emphasis on moving toward nonpolluting methods of biological pest control. Biocontrol of insects is not a recent innovation as the use of entomopathogenic microorganisms for regulating the populations of insect pests was proposed at the end of the nineteenth century by several pioneering scientists including Louis Pasteur. Since the mid-1990s there has been a major resurgence of interest in this approach of plant protection. The reasons for the increased importance to biological methods of pest control are only too obvious, yet two important facts need to be emphasized: (1) the severe backlash of excessive use of pesticides led to the emergence of a new class of pests euphemistically termed "pests of the green revolution," and (2) two attempts to deal with such pests found a new ally, in "weapons" of biological origin, such as botanical and the microbial "pesticides." The use of neem-based pesticides—nuclear polyhedrosis viruses and *Bacillus thuringiensis* Berliner (*Bt*)—are prime examples of success stories that have underscored the importance of biologically-based pest management strategies.

The "microbial" alternatives to chemical insecticides include a variety of biological agents such as bacteria, viruses, and fungi. Of all the various microbial agents that have been evaluated, the most successful by far has been the bacterium, *Bt*. This soil-borne sporulating bacterium has been commercially marketed for insect control since the late 1950s. *Bt* is a Gram positive, rod shaped, and aerobic spore forming, soil bacterium occurring naturally in the soils around the world. *Bt* was initially thought to be a subspecies of *Bacillus cereus* as spores of both shared common antigens (Yoder and Nelson, 1960), there was an overlap of the flagellar antigens between some *Bt* and *B. cereus* isolates (Krieg, 1970), and, most importantly, they have DNA homologies of 80–100% (Kaneko et al., 1978).

These bacteria generally produce insecticidal crystal proteins (ICPs) ranging in size from 70–139 kDa and depend on the alkalinity of insect gut to be converted to an active 60 kDa fragment. These ICPs are highly specific to various members of Lepidoptera, Diptera, and Coleoptera. Some *Bt* strains with toxicity to mites, nematodes, flatworms, and protozoa are also reported (de Maagd et al., 1999). These toxins are effective against Lepidoptera (caterpillars), Diptera (flies), and Coleoptera (beetles). A few *Bt* strains are effective against the insect orders Hymenoptera, Homoptera, Orthoptera, and Mallophaga (Schnepf et al., 1998) and some *Bt* strains are effective against nematodes, mites, and protozoans (Feitelson et al., 1992).

The use of *Bt* in pest management is more successful compared with other microbes such as viruses. This probably stems from the fact that bacteria are much easier to culture in large quantities in artificial media than viruses, which require live hosts or cell line culture. The use of *Bt* in pest management took a quantum leap with the development of transgenic crops incorporating the genes coding for the production of the ICPs (Vaeck et al., 1987).

The development of transgenic varieties of crops incorporating *Bt* genes, their commercial cultivation, and their huge success in offering resistance against pests in few crops has intensified the search for different strains that could be used in further molecular breeding programs aimed at delaying the evolution of resistance through pyramiding of genes coding for different toxin proteins. At the same time, efforts are being made to isolate *Bt* strains that could work in identical ways against other pests. In fact, all the work since 1915 when Berliner first isolated *Bt* from *Ephestia kuehniella* (Zell.) and the use of *Bt* in pest management was primarily focused on Lepidoptera. However, the isolation of the strains *israelensis* (effective against Diptera) and *tenebrionis* (effective against Coleoptera) led to an explosion of work aimed at isolating strains effective against other insects. These and similar studies later established the existence of a high genetic diversity both between and within *Bt* serotypes showing differential pathogenicity to different taxa of insects (Beegle and Yamamoto, 1992).

Several alternative approaches to pest control to overcome the sole reliance on chemicals have been suggested and attempted with some success. The toxin from *Bt*, a spore-forming bacterium (de Maagd et al., 1999), has come to occupy a place of prominence among the naturally occurring toxins.

2. HISTORY

Japanese bacteriologist Ishiwata Shigetane (1901) isolated the bacterium *Bt* from a diseased silkworm, *Bombyx mori*, and named the organism as *Satta bacillus*. However, later, the German biologist Berliner (1915) isolated and formally described the same bacterium as *Bt* from a dead Mediterranean flour moth, *E. kuehniella* (Zell.), larva from a mill in the province of Thuringia in Germany.

The first practical application of *Bt* was reported by Husz (1929) who isolated a *Bt* strain from *Ephestia* and tested it on European corn borer, *Ostrinia nubilalis*, in south eastern Europe. This work eventually led to the first commercial product, "Sporeine", in France in

1938 (Luthy et al., 1982). The Swiss company Sandoz followed suit and undertook the first large-scale production and marketing of a "Thuricide" in 1940. The first well-documented industrial procedure for producing a *Bt*-based product dates from 1959, with the manufacture of "Bactospeine" under the first French patent for a biopesticide formulation (Marvier et al., 2007). The pioneering research by Steinhaus (1951) on *Bt*, and a growing realization that organic insecticides are deleterious to the environment and human health, spurred a renewed interest in *Bt* in the 1960s. This led to the introduction of viable *Bt* biopesticides, like Thuricide and Dipel.

These developments were followed by intense research to find new and more effective strains of *Bt*. The *Bt* isolate from the diseased *Pectinophora gossypiella* was coded HD-I. Several other Lepidoptera specific isolates were identified as a new subspecies based on flagellar serotyping by De Barjac and Lemille (1970), which they named as *Bt* subsp. *kurstaki*. These developments were followed by two major discoveries relating to *Bt* strains. Goldberg and Margalit (1977) reported isolation of a serotype that was specifically effective against mosquito larvae. Similarly, a Coleoptera-specific strain was reported by Hernstadt et al. (1986) and named as *Bt* subsp. san diego. However, after a careful study of strains Krieg et al. (1987) established that subsp. san diego was not different from *tenebrionis*.

At present it has been estimated that more than 60,000 isolates of *Bt* are maintained in culture collections worldwide (Schnepf, 1995; Federici, 1999). Known *Bt* toxins kill subsets of insects among Lepidoptera, Coleoptera, Diptera (Hofte and Whiteley, 1989), mites, and nematodes (Feitelson et al., 1992). The host range of *Bt* has expanded considerably in recent years due to extensive screening programs. By virtue of the lack of toxicity toward other species of animals, human beings, and plants, there is tremendous potential for exploiting *Bt* as a biocontrol agent (Federici, 1999).

3. TOXINS

The principal insecticidal activity of *Bt* is mainly due to the presence of crystalline Cry proteins or endotoxins, which are present inside polyhedral parasporal bodies (Davidson and Meyers, 1981). Other workers, however, recognized the presence of a number of extracellular compounds that might contribute to the virulence these include various proteases (Lovgren et al., 1990), heat-labile and heat-tolerant exotoxins (Beegle and Yamamoto, 1992), lectinases (Kumar and Sharma, 1994), secreted vegetative insecticidal proteins (VIPs) (Estruch et al., 1996), and chitinases.

Exotoxins are produced in the vegetative phase and were first discovered by McConnel and Richard (1959). Levison et al. (1990) confirmed the existence of a second exotoxin, type II exotoxin. Exotoxins have a broad host range including invertebrates, vertebrates, and even microorganisms. The symptoms caused by exotoxins on insects include deformation, teratological changes, and death during molting or pupation (Luthy, 1980). VIP 3A encodes an 88 kDa protein which is toxic to a wide variety of pests including *Agrotis ipsilon*, *Spodoptera frugiperda*, and *Helicoverpa zea* (Estruch et al., 1996).

The endotoxins (crystal proteins) are generally produced as an inactive protoxin with a molecular weight in the range of 70–130 kDa, which are proteolytically cleaved to an activated core fragment of 55–60 kDa in the larval midgut (Kumar and Sharma, 1994). Activation of the toxin is generally carried out by various exogenous and endogenous proteases that play a vital role in determining the host range of the organism (Rukmini et al., 2000).

3.1 Cry Proteins

The protein crystals produced by *Bt* are popularly referred to as δ-endotoxins, Cry proteins, or crystal toxins, and have insecticidal activity. Cry protein expression could be sporulation-dependent or independent, and 80% of the crystal proteins are synthesized simultaneously during the process of sporulation (Lecadet and Dedonder, 1971). The *cry* genes are expressed mainly during stationary phase (Whiteley and Schnepf, 1986; Schnepf et al., 1998).

When ingested by the insect, the crystal proteins (130–140 kDa) undergo activation in the insect gut under the influence of an alkaline pH and gut proteases and become active toxins (60–70 kDa). The active toxic protein binds specifically to the brush border membrane vesicles of the midgut epithelium cells. This leads to formation of pores in the midgut epithelium cells, followed by osmotic imbalance in the cells, loss of cellular function, and finally the insect dies of septicemia. These toxins are effective against Lepidoptera (caterpillars), Diptera (flies), and Coleoptera (beetles). A few *Bt* strains are effective against the insect orders Hymenoptera, Homoptera, Orthoptera, and Mallophaga (Schnepf et al., 1998) and some *Bt* strains are effective against nematodes, mites, and protozoans (Feitelson et al., 1992).

4. IDENTIFICATION OF *Bt* STRAINS

A large number of *Bt* isolates are available in laboratories worldwide and exhibit high levels of genetic and biochemical diversity. New strains are added every year. *Bt* strains can be characterized by a number of techniques including serotyping, crystal serology, crystal morphology, protein profiles, peptide mapping, DNA probes, amplified fragment length polymorphism, and insecticidal activity (Pattanayak et al., 2000). de Barjac first attempted to classify *Bt* isolates based on flagellar agglutination (de Barjac and Bonnefoi, 1962).

Based on the homology of toxin gene sequences and the spectrum of insecticidal activity Hofte and Whiteley (1989) classified 42 *Bt* genes into 14 distinct types and grouped them into four major classes. These classes are *cryI* (Lepidoptera-specific), *cryII* (Lepidoptera and Diptera-specific), *cryIII* (Coleoptera specific), and *cryIV* (Diptera-specific). Following the analysis of toxin domains of 29 distinct *Bt* toxin proteins, Feitelson et al. (1992) added two new major classes, *cryV* and cry*VI*. The nomenclature of Hofte and Whiteley (1989), based mainly on insecticidal activity, failed to accommodate genes that were highly homologous to known genes but with a different insecticidal spectrum. Crickmore et al. (1998) introduced a systematic nomenclature for classifying the *cry* genes and their protein products. It is based on amino acid sequence homology, where each protoxin acquired a name consisting of the mnemonic Cry (or cyt) and four hierarchical ranks consisting of numbers, capital letters, lower case letters, and numbers (e.g., Cry6Aa3), depending on its place in a phylogenetic tree.

Thus, proteins with less than 45% sequence identity differ in primary rank (Cry1, Cry2, etc.) and 78% and 95% identity constitute the borders for secondary and tertiary rank, respectively.

5. STRUCTURE OF Bt δ-ENDOTOXIN PROTEINS AND GENES

The protoxins designated Cryl, Cry4, Cry5, Cry7, or Cry9 etc., contain 1100–1200 amino acids and the toxin is processed from within the amino half. The Cry2, Cry3, Cry6, Cry18, etc., protoxins are smaller (Li et al., 1991; Grochulski et al., 1995). The first 285 amino acid residues are present as a bundle of seven amphipathic α-helices, wherein six are arranged in a circle, and helix five is in the center (domain I). Residues 286–500 are organized as three β-sheets (domain II) and contribute to the toxin specificity. The remaining amino acids are also present as β-sheets and arranged like a sandwich (domain III). All three domains have specific functional roles. The first domain is required for toxicity, and domain II is important for specificity. Although the function of domain III near the carboxyl end was not defined, it is speculated that it may have a role in the processing of protoxin and channel-forming function. Reports indicate that domains II and III are both involved in receptor recognition and binding. Additionally, domain III is involved in pore formation (de Maagd et al., 2001).

6. MODE OF ACTION OF δ-ENDOTOXINS

The δ-endotoxins exert their toxicity by forming lytic pores in the larval midgut epithelial membranes. Initially the protoxins are activated to toxins in the midgut by trypsin-like proteases. In general, 500 amino acids from the C-terminus of 130 kDa protoxins and 28 amino acids from the N-terminus are cleaved leaving a 65–70 kDa protease-resistant toxic active core comprising the N-terminal half of the protoxin (Hofte and Whiteley, 1989). The active toxins bind to specific receptors located on the apical brush border membrane of the columnar cells. Various studies revealed that there are many different toxin-binding protein receptors (Gill et al., 1992). For instance, CrylAc δ-endotoxin-binding aminopeptidase in the *Manduca sexta* (Linnaeus) midgut has a glycosyl phosphatidylinositol anchor (Garczynski and Adang, 1995). After binding to the specific receptor, the toxin inserts irreversibly into the plasma membrane of the cell leading to lesion formation.

The formation of toxin-induced pores in the columnar cells of apical membrane allows rapid fluxes of ions and the pores are K+ selective, permeable to cations, anions, or small solutes, like sucrose, irrespective of the charge. The disruption of gut integrity results in the death of the insect from starvation or septicemia.

There seems to be a different mechanism of action with respect to Cry2A toxins. English et al. (1994) compared the differences in solubility, binding to the brush border membrane, and ion channels formed by Cry2A and CrylAc toxins in *H. zea* (Boddie). The results showed unique attributes in the mode of action of Cry2A, which was less soluble than CrylAc and failed to bind to a saturable binding component on the midgut brush border membrane. In addition, voltage-dependent, nonselective channels were formed by this toxin in planar lipid bilayers. It was suggested that the unique mode of action of Cry2A may provide a useful tool in managing field resistance to *Bt* toxins.

7. OCCURRENCES AND DISTRIBUTION

Bt can be readily isolated from soils, insects, stored product dust, straw compounds, and deciduous coniferous leaves and also from aquatic environment. The worldwide abundance and distribution of *Bt* has been reported by de Lucca et al. (1981) who consider *Bt* as an indigenous bacterium in many environments. In fact *Bt* has been isolated along with *B. cereus* from soils of very widely separated geographical areas of the world (de Lucca et al., 1981; Meadows et al., 1992).

7.1　Soils

Studies on the distribution of *Bt* and closely related *B. cereus* in Egyptian soils was carried out by Salma et al. (1986) and they observed high *Bacillus* counts in most fertile clay soils and low counts in sandy soils. Ohba and Aizawa (1986) isolated the *Bt* subspecies *japanensis* from forest soils of Fukuoka and also they reported that they were highly toxic to larvae of *B. mori*. Martin and Travers (1989) isolated *Bt* in 785 out of 1115 samples collected from the USA and 29 other countries. They observed that over 60% of isolates were toxic to lepidopteran and dipteran larvae. Furthermore, they also reported that *Bt* was found in 94% of soils from Asia and central and south Africa, 84% of soils from Europe, and the lowest recoveries (60%) were from New Zealand (56%) and the USA.

In similar surveys to isolate *Bt*, Karamanlidonou et al. (1991) observed that out of 80 soil samples collected from Olive groves in Greece, 24 of them were found to contain *Bt*. They also observed that the mortality levels of these isolates to olive fruit fly ranged from 7–87%. Bernhard et al. (1997) reported one of the largest screening studies on *Bt* using 5303 isolates from all the continents of the world. In the Philippines, only 12 out of 54 soil samples harbored *Bt* (Padua et al., 1981). Theunis et al. (1998) found that 45% soil samples collected from Philippines were active against *Mamestra configurata* Walker. Bravo et al. (1998) isolated a total of 496 strains of *Bt* from 503 soil samples that they collected from five different regions of Mexico. Morris et al. (1998) found that out of 159 soil samples collected from Manitoba in Canada, 53 were positive for the presence of *Bt*. Furthermore, they stated that organic rich soils had the highest frequency of positive samples.

Unfortunately, such intensive surveys have only recently been taken up in India. Reddy (2000) recovered 37 isolates from soil samples collected from different agroclimatic zones of India and 15 isolates were used in bioassays against *Helicoverpa armigera*. In continuation of this study, Karthik John (2001) examined five isolates from India by staining and conducted a bioassayed against *Plutella xylostella* and found that, 50E, a local isolate, induced higher mortality compared with other isolates.

7.2　Stored Grain Dust

In addition to soil, *Bt* was found in other insect habitats, such as the stored grain. The search for *Bt* strains from grain dust has been almost as rewarding as has been the search from soil (Beegle and Yamomoto, 1992).

Krieg (1982) reported the natural occurrence of *Bt* in cereals and cereal products from Germany. Morris et al. (1998) isolated *Bt* from dust samples collected from 29 wheat, barley, and storage bins. They observed the presence of *Bt* in 19 of the bin samples. Theunis et al. (1998) found that 60% of the grain samples collected in Philippines were positive for the presence of *Bt* and they also reported that grain dust from rice mills was the richest source of *Bt* compared with soil and other habitats.

7.3 Insect Cadavers

The search for *Bt* is only hastened if the focus is on the host rather than the habitat. Because *Bt* causes death of the host insect in about 3–7 days, one of the most rewarding and easy sources of *Bt* is the use of insect cadavers. The search for *Bt* from cadavers is relatively easy because cadavers, unlike soil, carry a lower microbial diversity and screening the colonies becomes easier. Benhardt et al. (1997) reported that *Bt* could be more readily isolated from insect cadavers rather from soils. *Bacillus thuringiensis* subsp. *thuringiensis* was isolated from the larch fly, *Pristiphora erichsonii* (Hartig) in Canada (Morris, 1983). Kuzmanova (1975) isolated 18 strains of *Bt* from Lepidopteran larvae collected in different regions of Bulgaria. Padua et al. (1981) isolated *Bt* from soil and silkworm litter collected from various parts of the Philippines. They isolated 15 subspecies of *Bt* of which five belonged to *Bt kurstaki* and one isolate belonged to *Bt* subsp. *morrisoni*.

Ohba et al. (1981) reported a new subspecies of var. *tohokuensis* isolated from silkworm litter of sericulture farm in Tohoku Dist. Honshu Japan. They reported two new subspecies of *Bt*, namely, *kumamotensis* and *tochigiensis*. Morris (1983) reported *Bt* subsp. *galleria* and subsp. *canadensis* from various forest insects.

In one of the largest studies attempting to isolate *Bt* from cadavers, Huang and Huang (1986) found 172 strains of *Bt* from 4454 accessions of soil and insect cadaver samples from 44 localities in China. The dominant strain in these studies was *Bt* subsp. *galleria*. Kaelin et al. (1994) isolated *Bt*. from dead tobacco beetles, *Lasioderma serricorne* (F.), they collected from various locations worldwide.

8. *Bt* DELIVERY SYSTEMS

8.1 Spray Formulations

Bt was the most popular biocontrol agent with worldwide sales of around $90 million in 1995 (Lambert and Peferoen, 1992). There were nearly 200 registered *Bt* products having more than 450 uses and formulations (Schnepf et al., 1998). *Bt* is the major pesticide against gypsy moth in forests. *Bt israelensis* is extensively used to control mosquitoes and blackflies (Becker and Margalit, 1993). *Bt jegathesan* encodes another potent mosquitocidal toxin immunologically related to Cry2A (Delecluse et al., 1995). Thus, *Bt* plays an important role not only in agriculture and forestry but also in human and animal health (Kumar et al., 1996).

The earliest commercial production of *Bt* began in France in 1938 under the trade name Sporeine (Luthy et al., 1982). During the 1960s, several industrial formulations of *Bt* were manufactured in the United States, France, Germany, and the Soviet Union. The isolation of the highly potent *kurstaki* variety by Kurstak in 1962 and by Dulmage in 1967 (Dulmage, 1970) provided a much-needed boost to the commercialization of *Bt*. The discovery of new strains of *Bt* widened the toxicity spectrum of bioinsecticides. The use of conventional *Bt* insecticides, however, was found to have limitations like narrow specificity, short shelf-life, low potency, lack of systemic activity, and the presence of viable spores (Lambert and Peferoen, 1992). These problems are now overcome by various approaches that utilize molecular biology and genetic engineering tools as well as conventional microbiological methods (Kumar et al., 1996). The plasmid location of *Bt* toxin genes enabled the construction of novel *Bt* strains with microbial genetic approaches, such as plasmid curing and conjugal transfer. Conjugational transfer of native *Bt* plasmids between species of *Bacillus* is known to occur. Expression of transformed plasmid-coded genes was analyzed by genotyping crystal proteins and flagellar antigenicity. Following the conjugational approach, scientists at Ecogen Corporation (USA) produced several biopesticides with a broadened spectrum of toxicity (Baum et al., 1998). For instance, the product "Foil" is made from a strain that carries toxin genes active against European corn borer (Lepidoptera) and Colorado potato beetle (Coleoptera).

Using the conjugational approach, Bora et al. (1994) transferred the *crylAa* gene of *Bt* into *Bacillus megaterium*, which resides in the cotton phyllosphere. Leaf bioassays of cotton plants inoculated with a single spray of the transcipient showed that there was protection of the cotton plants from *H. armigera*. The conjugational approach to create novel *Bt* strains has certain limitations: (1) not all the *Bt* toxin genes are located on transferable plasmids, (2) the toxin protein with useful insecticidal activity may be synthesized at low amounts, and (3) plasmid incompatibility could also be a problem.

Another interesting approach to expand the insecticidal host range of *Bt* is to make use of the in vivo genetic recombination property (Baum et al., 1990). Lereclus et al. (1992) used insertion sequence IS232 to deliver the *cry3A* gene into an isolate producing CrylAc toxin. Expression of the introduced gene did not alter the composition of the polypeptides normally produced by the strain. Novel *Bt* mutants, defective in sporulation but overproducers of toxin, have been isolated (Lereclus et al., 1995). They can be used safely as a biopesticide in silkworm-rearing areas.

Development of novel cloning vectors for *Bt* has made possible the construction of improved *Bt* strains for use as microbial insecticides. The use of *Bt* as the host organism offers many advantages. Native *Bt* strains can stably maintain and efficiently express several homologous *Bt* toxin genes. The ability to maintain multiple *Bt* toxin genes in a single recipient broadens the insecticidal activity in an additive or synergistic manner. Multiple toxin genes with differing modes of action or receptor-binding properties may reduce the chances of insects developing resistance (Tabashnik, 1994). An essential element in the successful engineering of *Bt* strains is the availability of suitable cloning vectors. A number of convenient shuttle vectors, functional in *Escherichia* and *Bacillus* species, have been constructed using replication origins from resident *Bt* plasmids (Gawron-Burke and Baum, 1991). Considering the stability of resident *Bt* plasmids, shuttle vectors derived from resident plasmids might exhibit good segregational stability.

8.2 Transgenic Microorganisms

Cloned *Bt* toxin genes were introduced into a number of microbial hosts to create more stable and/or compatible agents for toxin delivery. Monsanto scientists were the first to report the expression of the *cry1Ab* gene in a root-colonizing *Pseudomonas* at levels sufficient to kill lepidopteran larvae. The gene was cloned into transposon Tn5 and transposed into the chromosome of six corn root-colonizing strains of *Pseudomonas fluorescens* and *Agrobacterium radiobacter* (Obukowicz et al., 1986). Following this, many groups developed *Psuedomonas* strains carrying *Bt* toxin genes. The recombinant *Psuedomonas* is killed by a proprietary chemical treatment that cross-links the bacterial cell wall to yield a nonviable encapsulated bacterium surrounding the crystal protein (CellCap product of Mycogen; Gaertner et al., 1993). Such a product is stable and safe the environment.

An interesting example of a toxin gene in a foreign bacterium is the introduction of the *cry3A* gene into *Rhizobium leguminosarum* and *Rhizobium meliloti* using a broad host range vector, pRK311 (Bezdicek et al., 1994). The recombinant rhizobia expressed the toxin in sufficient quantities within root nodules to significantly reduce feeding damage by the nodule-feeding insects, *Sitona lineatus* on *Pisum sativum* and *Sitona hispidulus* on *Medicago sativa*. The pRK311 plasmid remained stable in the rhizobia that were either free-living or within nodules of the legumes. The engineered strains of *R. leguminosarum* were equally competitive with the wild type strain. Udayasuriyan et al. (1995) transferred an insecticidal protein gene of *Bt* into plant-colonizing *Azospirillum* that may be used to control root-feeding insects. The *cry4Aa* (*cryIVA*) gene of *Bt israelensis* was introduced into various unicellular cyanobacteria with the intent of providing a more accessible source of the toxin for filter-feeding dipteran larvae (Soltes-Rak et al., 1993).

8.3 Transgenic Crop Plants

An elegant and the most effective delivery system for *Bt* toxins is the transgenic plant. The major benefits of this system are economic, environmental, and qualitative. In addition to the reduced input costs to the farmer, the transgenic plants provide season-long protection independent of weather conditions, effective control of burrowing insects difficult to reach with sprays, and control at all of the stages of insect development. An important feature of such a system is that only insects eating the crop are exposed to the toxin. Genetic transformation of almost all the major crop species is now feasible with the development of an array of techniques ranging from the *Agrobacterium*-mediated approach to electric discharge-mediated particle acceleration procedure (Pattanayak et al., 2000).

Genetically modified plants expressing δ-endotoxin genes from *Bt*, protease inhibitors, and plant lectins have been successfully developed, tested, and demonstrated highly viable for pest management in different cropping systems since the year 2000 (Gatehouse, 2008). Insect-resistant crops have been one of the major successes of applying plant genetic engineering technology to agriculture. Most of the plant-derived genes produce chronic rather than toxic effects and many insect pests are less or not sensitive to most of these factors. Therefore, the genes for δ-endotoxins are expected to provide better solutions.

Advances in biotechnology have provided several unique opportunities that include access to various plant transformation techniques, novel and effective molecules, the ability to change the levels of gene expression, the capability to change the expression pattern of genes, and development of transgenics with different insecticidal genes.

Genes from *Bt* have been the most successful group of genes identified for use in genetic transformation of crops for pest control on a commercial scale (Gill et al., 1992; Charles et al., 1996). However, it is suggested that transgenic plants rarely result in 100% control, but tend to retard insect growth and development (Estruch et al., 1997).

Genes coding for *Bt* δ-endotoxins have been deployed in a wide range of crop plants with considerable success (Sharma et al., 1999). The first *Bt* toxin gene was cloned in 1981 and the first transgenic plant was produced in the 1980s. Since then, several crop species have been genetically engineered to produce *Bt* toxins to control the target insect pests. Genes conferring resistance to insects have been inserted into crop plants, such as cotton, maize, potato, tobacco, rice, broccoli, and soybean (Bennett, 1994; Federici, 1998; Griffiths, 1998). The list of crops and *Bt* gene(s) engineered into them have been compiled (Table 1; Keshavareddy, 2009). The first transgenic cotton crop was grown in 1994 and large-scale cultivation was taken up in 1996 in the USA (McLaren, 1998).

Transgenic plants display considerable potential to benefit both developed and developing countries expressing insecticidal *Bt* proteins alone or in conjunction with proteins providing tolerance to herbicide are revolutionizing agriculture (Shelton et al., 2002). The use of such crops with input traits for pest management, primarily insects and herbicide resistance, has risen dramatically since their first introduction in the mid-1990s.

India, the largest cotton growing country in the world, had 50,000 ha of *Bt* cotton in 2002. The *Bt* cotton area in India has soared to 11.0 m ha in 2013 and has been adopted by even small and resource-poor farmers (James, 2013). The spectacular growth in *Bt* cotton is that it has consistently delivered unprecedented benefits to farmers and to the nation. *Bt* cotton has increased productivity by up to 50% while reducing the insecticide sprays by half, with increased income to cultivators. Success achieved in cotton has served as an excellent model to emulate in many other crops, such as rice, wheat, pulses, and oilseeds, which has the potential to make agriculture a viable profession for the peasants of India.

The first transgenic tobacco plants (Barton et al., 1987; Fischhoff et al., 1987; Vaeck et al., 1987) expressing full-length or truncated *Bt* toxin genes *(cry1A)* under the control of constitutive promoters were developed in 1987. The expression of the toxin protein was very poor in tobacco plants expressing the full-length gene and the mortality of *M. sexta* larvae was only 20%. Truncated *cry1A* genes coding for the toxic N-terminal fragment provided better protection to the tobacco and tomato plants. Compared with the plants transformed with full-length genes, the plants expressing truncated genes were more resistant to the larvae, and the highest reported level of toxin protein expression was about 0.02% of total leaf-soluble protein. Despite these low levels of expression, many of the plants were shown to be insecticidal to the larvae of *M. sexta*. However, many of the noctuid lepidopterans, which constitute a very serious group of insect pests, need higher amounts of *Bt* toxins for effective control. Gene truncation as well as the use of different promoters, enhancer sequences, and fusion proteins resulted in only limited improvement in *Bt* gene expression (Barton et al., 1987; Carozzi et al., 1992; Vaeck et al., 1987).

8.3.1 Transgenic Crops with a Single Bt Toxin

Transgenic plants containing *Bt* genes control pests more effectively than *Bt* formulations. So far, few species of *Bt* crops (cotton, maize, soybean, rice, and potato) have become commercially available worldwide and commercialization of *Bt* crops has significantly reduced the use of synthetic insecticides (Ferre and van Rie, 2002). The first transgenic tobacco plants using *cry* genes were developed in 1987 (Vaeck et al., 1987; Barton et al., 1987; Fischhoff et al., 1987; Carozzi et al., 1992; Ranjekar et al., 2003). A significant breakthrough was made in 1990 by researchers at Monsanto (USA) who modified the *cry* genes, *cry1Ab* and *cry1Ac*, for better expression in plants (Perlak et al., 1990). The tobacco plants engineered with truncated genes

TABLE 1 Transgenic Crop Plants Developed Using Various *cry* Genes Against Insect Pests

Crop group	Target crop	*cry* gene(s)	Target pest(s)	References
Commercial crops	Cotton	*cry1Ac*	*Helicoverpa zea* (Boddie) *Pectinophora gossypiella* (Saunders)	Bacheler and Mott (1997)
		cry1Ab	*Helicoverpa virescens* (Fabr.), *H. zea*	Perlak et al. (1990)
		cry1Ac + cry2Ab	*H. zea, P. gossypiella, Spodoptera exigua* (Hubner), *Spodoptera frugiperda* (J.E. Smith) *Pseudoplusia includens* (Walker)	Adamczyk et al. (2001) and Chitkowski et al. (2003)
		*cry1Ac + CpTI**	*H. zea*	Wu and Guo (2005)
	Tobacco	*cry1Aa*	*Manduca sexta* (L.)	Barton et al. (1987)
		cry1Ab	*M. sexta*	Vaeck et al. (1987)
		*cry1Ab + CpTI**	*M. sexta*	Perlak et al. (1991)
		cry1Ab	*M. sexta*	Williams et al. (1993)
		cry1Ac	*H. virescens, H. zea, Spodoptera littoralis* (Boisduval)	McBride et al. (1995)
		cry1C	*S. littoralis*	Strizhov et al. (1996)
		cry2A	*Helicoverpa armigera* (Hubner)	Selvapandiyan et al. (1998)
		cry2Aa	*H. virescens, H. zea*	Kota et al. (1999)
		cry1Ia5	*H. armigera*	Selvapandian et al. (1998)
Cereals	Corn	*cry1Ab*	*Ostrinia nubilalis* (Hubner)	Koziel et al. (1993)
		cry9C	*O. nubilalis*	Jansen et al. (1997)
	Rice	*cry1Ab*	*Chilo suppressalis* Walker	Fujimoto et al. (1993)
		cry1Ab and *cry1Ac*	*C. suppressalis*	Cheng et al. (1998)
		cry1B	*C. suppressalis, Cnaphalocrocis medinalis* Guenee	Marfa et al. (2002)
		cry1Ac, cry2A and GNA	*C. medinalis, Scirpophaga incertulas, Nilaparvata lugens* Stal	Maqbool et al. (2001)
		cry3A	*Dicladispa armigera* (Oliv.) *Sitophilus oryzae* (L.)	Jhonson et al. (1996)
		cry2A	*S. incertulas* and *C. suppressalis*	Chen et al. (2005)
		cry1Ab/cry1Ac	*S. incertulas, C. suppressalis, Sesamia inferens* Walker, *C. medinalis*	Ye et al. (2001)
Pulses	Soybean	*cry1Ac*	*H. virescens, H. zea, P. includens* Walker, *Anticarsia gemmatalis* (Hubner)	Walker et al. (2000)
	Pigeon pea	*cry1E-C*	*Spodoptera litura*	Surekha et al. (2005)
	Chickpea	*cry2Aa*	*H. armigera*	Sarmah and Deka (2004)
		cry1Ac	*H. armigera*	Sanyal et al. (2005)

(Continued)

TABLE 1   Transgenic Crop Plants Developed Using Various *cry* Genes Against Insect Pests—cont'd

Crop group	Target crop	*cry* gene(s)	Target pest(s)	References
Vegetables	Tomato	*cry1Ab*	*H. virescens*	Fischhoff et al. (1987)
		cry1Ac	*H. armigera*	Mandaokar et al. (2000)
	Eggplant	*cry1Ab*	*Leucinodes orbonalis* Guenee	Kumar et al. (1998)
		cry3A	*Leptinotarsa decemlineata* (Say)	Jelenkovic et al. (1998)
	Cabbage	*cry1Ab*	*Plutella xylostella* (L.)	Bhattacharya et al. (2002)
	Broccoli	*cry1C*	*P. xylostella*	Zhao et al. (2001)
Sugars and starches	Sugarcane	*cry1A(b)*	*Diatraea saccharalis* (F.)	Arencibia et al. (1997)
	Potato	*cry1Ab*	*Phthorimaea operculella* (Zeller)	Rico et al. (1998)
		cry3Aa	*L. decemlineata*	Adang et al. (1993)
		cry9Aa2	*P. operculella*	Gleave et al. (1998)
		cry1Ia1 (*cryV*)	*P. operculella* *Symmetrischema tangolias* (Gyen)	Lagnaoui et al. (2000)
		cry1Ba/cry1Ia	*L. decemlineata* *P. operculella*	Naimov et al. (2003)
Other crops	Canola	*cry1Ac*	*Trichoplusia ni* (Hubner) *S. exigua, H. virescens, H. zea*	Stewart et al. (1996)
	Alfalfa	*cry1C*	*S. littoralis* *S. exigua*	Strizhov et al. (1996)

CpTI*, Cowpea trypsin inhibitor; GNA, *Galanthus nivalis* agglutinin.

encoding Cry1Ac and Cry1Ab toxins were found to be resistant to the larvae of *M. sexta* (Kota et al., 1999). Cotton cultivar Coker 312 was first transformed by using partially modified *cry1Ac* gene. The transformed plants showed total protection against *Trichoplusia ni*, *S. exigua*, and *H. zea*. The maximum level of toxin protein in the plants was 0.1% of the total soluble protein (Sharma and Anjaiah, 2000). Selvapandian et al. (1998) transformed tobacco plants using *cry1Ia5* insecticidal toxin from an Indian *Bt* strain that provided complete protection against *H. armigera*. The transgenic tobacco plants with partially modified *cry1Ab* gene had a 10-fold higher level of insect control protein and plants with fully modified *cry1Ab* had a 100-fold higher level of *cry1Ab* protein compared with the wild type gene and exhibited 100% larval mortality of *M. sexta* (Perlak et al., 1990). Tobacco and tomato plants expressing *cry1Ab* and *cry1Ac* genes have been developed to control lepidopteran insects (Salm et al., 1994). A synthetic *cry1C* gene in alfalfa and tobacco plants results in the production of 0.01–0.2% of total soluble proteins as Cry1C toxin and provides 100% protection against the Egyptian cotton leafworm (*Spodoptera littoralis*) and the beet armyworm, *S. exigua* (Strizhov et al., 1996).

Expression of modified genes *cry1Ac* in cotton and *cry3Aa* in potato conferred considerable protection against lepidopteran and coleopteran pests, respectively (Ranjekar et al., 2003). Successful control of pink bollworm (*P. gossypiella*) has been achieved through transgenic cotton using a truncated *cry1Ab* gene in transgenic cotton plants (Wilson et al., 1992; Arencibia et al., 1997). Scientists at the Bose Institute (Kolkata) have introduced a modified *cry1Ac* gene in rice (IR-64) for resistance to yellow stem borer (Nayak et al., 1997). A synthetic *cry1Ac* gene was introduced into rice lines (Pusa Basmathi 1, Karnal Local, and IR-64) exhibiting total protection against neonate larvae of yellow stem borer. Rice cultivars (*indica* and *japonica* types) with truncated *cry1Ab* gene caused 100% mortality of the yellow stem borer, *Scirpophaga incertulas* (Walker) (Datta et al., 1998). Transgenic sugarcane plants with *cry1Ab* showed significant larvicidal activity against neonate larvae of sugarcane borer, *Diatraea saccharalis* (Fabricius).

"Jack," a transgenic line of soybean, *Glycine max* (L.), expressing a synthetic *cry1Ac* gene (Jack-*Bt*) showed three to five times less defoliation from corn earworm, *H. zea*, and eight to nine times less damage from velvetbean caterpillar, *Anticarsia gemmatalis* (Hubner) (Walker et al., 2000). Transgenic broccoli with Synthetic *cry1C* was resistant to the cabbage looper (*T. ni*) and cabbage butterfly (*Pieris rapae*; Selvapandian et al., 1998). Vegetable crops like brinjal and tomato were transformed by synthetic/modified *cry1Ab* and *cry1Ac* genes, respectively, to confer resistance to fruit borers (*Leucinodes orbonalis*) and *H. armigera*, respectively (Kumar et al., 1998). The "New Leaf" potatoes with *Bt* protein, Cry3A, are season-long resistant to Colorado potato beetle (Duncan et al., 2002).

8.3.2 *Two Toxin/Hybrid Toxin* **Bt** *Crops*

Although no insect species resistant to *Bt* crops have been reported under natural conditions, the potential of insects to evolve resistance against *Bt* toxins is an inevitable threat to this technology. To meet this challenge, several strategies have been proposed to manage insect resistance, such as the high-dose/refuge strategy, gene stacking, and temporal or tissue-specific expression of the toxin (Roush, 1998; Frutos et al., 1999; Shelton et al., 2002). Among the strategies, only the high-dose/refuge strategy has been used in developed countries, such as the USA and Australia.

At the same time, small farmers in Asian countries could hardly devote their land to a refuge and, moreover, a high dose of a foreign protein could cause a phenotypic trade-off resulting in a yield penalty (Datta et al., 2002). Hence, efforts are being made to develop two toxin *Bt* crops (otherwise known as the pyramiding approach) because two toxin cultivars require smaller refuges to achieve successful resistance management and are expected to provide equally sustained long-term protection as against the single gene transgenics (Cohen et al., 2000). The use of multiple toxin genes with different modes of action has been proposed so that cross-resistance is likely to be a less serious problem. As a result, two *cry* genes for toxins with different receptors or a *cry* gene in combination with an altogether different, unrelated toxin gene are considered the ideal options (de Maagd et al., 1996; Frutos et al., 1999).

Salm et al. (1994) developed transgenic tobacco and tomato plants expressing two *Bt* genes, *cry1Ab* and *cry1C*, specific toward lepidopteran insects. Both of the genes were partially modified to remove sequence motifs that affect messenger RNA (mRNA) stability in plant cells. The expression of a *cry1Ab–cry1C* fusion gene resulted in protection against *S. exigua*, expressing translational fusions not only to broaden the insect resistance of transgenic plants, but also to simultaneously employ different gene classes in resistance management strategies. Already, the second generation *Bt* crops carrying multiple *Bt* genes have entered the market.

Furthermore, hybrid toxins produced through inclusion of a domain from another toxin results in increased potency of a fused protein by the shift in receptor binding (Bosch et al., 1994).

Alternate receptor ligand interaction may also be exploited to further broaden the host range of *Bt* toxins (Sivasubramanian and Federici, 1994). The dual toxin Bollgard-II genotype (*cry1Ac + cry2Ab*) was found to be highly effective against lepidopterous pests, *H. zea*, *Pseudoplusia includens* (Walker), and *S. frugiperda* compared with Bollgard I (*cry1Ac*) and conventional cotton (Chitkowski et al., 2003). Rice plants expressing *cry1Ab* and *cry1Ac* genes were highly toxic to striped stem borer (*Chilo suppressalis*) and yellow stem borer (*S. incertulas*) with mortalities of 97–100% within 5 days after infestation (Nayak et al., 1997).

Transgenic IR72 lines of rice, TT9-3 and TT9-4, carrying a fused *Bt* gene (*cry1Ab* and *cry1Ac*) demonstrated that both the transgenic lines were highly resistant against natural and artificial infestations of four lepidopteran species, including *Cnaphalocrocis medinalis* (Guenee) and green semilooper, *Naranga aenescens* Moore (Ye et al., 2001). The elite Vietnamese rice cultivars transformed with translationally fused *cry* genes (*cry1Ab-1B*) exhibit 100% mortality of the neonate larvae of yellow stem borer within a week of infestation (Ho et al., 2006).

Transgenic potato plants developed with a hybrid *Bt* gene *SN19* (domains I and III from *cry1Ba*, and domain II from *cry1Ia*) were shown to be resistant against Colorado potato beetle, tuber moth, and European corn borer. These are the first transgenic plants resistant to pests belonging to two different insect orders. In addition, the target receptor recognition of this hybrid protein is expected to be different from Cry proteins currently in use for these pests that makes it a useful tool for resistance management (Naimov et al., 2003). Thus, introduction of *Bt*-transgenic plants for commercial cultivation has launched a new era in agriculture. The expression of very effective insecticidal proteins by plants delivers a remarkable level of insect control unsurpassed by any other method of insect pest management. Therefore, current and novel *Bt* δ-endotoxins are also fully expected to be part of the transgenic plant approach to combat pests in the future.

In 1990, researchers at Monsanto made a significant advancement in the expression of *Bt* genes in plants (Perlak et al., 1990). They noticed that *Bt* genes were excessively AT-rich compared with normal plant genes. This bias in nucleotide composition of the DNA could have a number of deleterious consequences to gene expression because AT-rich regions in plants are often found in introns or have a regulatory role in determining polyadenylation. There are also instances in other eukaryotic systems in which AT-rich regions can signal rapid degradation of specific mRNAs. In addition, plants have a tendency to use G or C in the third base of redundant codons—A or T being rarer. *Bt* genes have the opposite tendency and because codon preference is thought to be linked to the abundance of the corresponding transfer RNAs, the overuse of rare codons would decrease the rate of synthesis of a *Bt* protein in plant cells.

Perlak et al. (1991) followed two approaches to modify the *cry1Ab* and *cry1Ac* genes. One approach included selective removal of DNA sequences predicted to inhibit efficient expression of *Bt* gene at both translational and mRNA levels by site-directed mutagenesis. These genes were termed partially modified (PM) genes. The other approach was to generate a synthetic gene with a fully modified (FM) nucleotide sequence, taking into account factors, such as codon usage in higher plants, potential secondary structure of mRNA, and potential regulatory sequences. The PM-*cry1Ab* gene is approximately 96% homologous to the native gene with a GC content of 41%, with the number of potential plant polyadenylation signal sequences (PPSS) reduced from 18 to 7 and the number of ATTTA sequences reduced from 13 to 7. The FM-*cry1Ab* is approximately 79% homologous to the native gene, with a GC content of 49% and the number of PPSS reduced to 1 and all ATTTA sequences removed. The toxin protein levels in transgenic tobacco and tomato harboring these modified genes increased up to 100-fold over levels seen with the wild type *Bt* gene in plants. Perlak et al. (1990) made a gene construct in which the first 1359 nucleotides were derived from the FM-*cry1Ab* gene and the remaining sequence from PM-*cry1Ac* gene. The variant gene was placed under the control of the cauliflower mosaic virus (CaMV) 35S promoter containing a duplicated enhancer region. Cotton variety Coker 312 was transformed and the transgenic plants were shown to have total protection from *T. ni*, *S. exigua*, and *H. zea*. The maximum level of toxin protein was 0.1% of total soluble protein. Transgenic tobacco plants expressing this gene provided a 10 to 20-fold increase in *cry1Ac* mRNA and protein compared with gene constructs in which the CaMV 35S promoter with duplicated enhancer region was used to express the same gene. The toxin protein was localized in the chloroplast of tobacco plants that produce the *Bt* protein, nearly 1% of the total leaf protein, and had the highest levels of *Bt* toxin proteins yet reported. The enhancement of *Bt* toxin protein levels in tissues in which ribulose-1,5-bisphosphate expression is highest may lead to very effective control of certain insect pests that feed on leaves and other green tissues.

More than 30 plant species have been transformed using a range of modified *Bt* genes, since two decades (de Maagd et al., 1999). Some of the modified genes were designed to suit the codon usage of particular plant species (Koziel et al., 1993). A variety of plant promoters have been used in combination with *cry* genes, such as CaMV 35S promoter, wound-inducible promoter, chemically inducible promoters and tissue-specific promoters (Kumar et al., 1996).

9. BIOSAFETY

Glare and O'Callaghan (2000) exhaustively reviewed the literature on the biosafety of *Bt*, with a view to assess the environmental and health impacts possible from broadcast application of various strains in formulated products. *Bt* occurs in a wide range of habitats worldwide and the strains used as microbial control agents are present as constituents of soil, leaf, and insect microflora. In nature *Bt* population density is relatively low. However, pest control strategies result in the periodic release of high numbers of one particular *Bt* strain into the environment and there may be increased exposure of sensitive nontarget organisms to *Bt*. In addition, the use of various formulations may lead to different groups of nontarget species exposed to *Bt* than would occur naturally. The use of *Bt* toxins is also increasing through the use of transgenic crops, raising new concerns over development of resistance, nontarget effects, and potential gene transfer.

A problem in the interpretation of the effects on nontarget organisms is the likelihood of an organism coming into contact with *Bt*. In some cases, organisms that have been shown to be susceptible to a particular *Bt* may be unlikely to ever be in the same area as a pest that *Bt* is used to treat. The rapid increase in *Bt* toxin-expressing transgenic plants may further complicate the situation as the toxins are expressed continuously. Therefore, defining nontarget impacts from laboratory and even field studies can be difficult. These limitations need to be kept in mind when considering nontarget impacts.

Bt strains are not specific to genera, families, or even orders of insects. For example, *Bt kurstaki* is reported to be toxic to over 300 species of Lepidoptera and various other species. While this is an advantage in terms that a *Bt*-based pesticide can be used against more than one pest, it poses a risk in terms of potential nontarget impacts. Most specific ecosystem studies on nontarget impacts of *Bt* have been conducted with *Bt kurstaki* and *Bt israelensis*. While some nontarget toxicity has been reported (certain caterpillars in the case of *Bt kurstaki* and chironomids for *Bt israelensis)*, no nontarget species has been shown to be at risk of eradication through use of *Bt*. *Bt* still rates very safe compared with other pesticides in terms of impacts on nontarget invertebrates.

Review of a number of studies on the effect of *Bt* application on insect predators, parasitoids, and pathogens and soil microfauna indicate that *Bt* is rarely directly harmful to beneficial invertebrates. While competition for hosts and reduced host food quality can indirectly retard development or cause premature death of mature stages of predators and parasitoids, *Bt* has been shown to be useful in integrated pest management (IPM) systems in most situations. No adverse effects on bees and earthworms have been reported from nonexotoxin-producing strains.

Exceptionally, few problems associated with human health have been reported and proven cases of *Bt* causing clinical disease in humans remain extremely rare, despite large-scale use of *Bt* for over 80 years. Extensive safety tests against small mammals have rarely shown toxicity; the few reports of toxicity were when artificially high levels of inoculums were used. Mammalian safety issues arise partly from the close genetic relationship between *Bt* and *B. cereus*, a species that can cause gastroenteritis in humans. The production of *B. cereus*-type diarrheal enterotoxins by *Bt* isolates has raised concerns that *Bt* may cause gastroenteritis outbreaks in humans. Studies conducted to date show that strains of *Bt*, including those found in many commonly used commercial products, are capable of producing diarrheal enterotoxins. No valid evidence has been found to link usage of *Bt* with episodes of diarrhea following ingestion of food and enterotoxin titers produced by *Bt* strains were significantly lower than those for *B. cereus*. However, this is an area which requires more research.

Studies by Hernandez et al. (1998) have drawn public attention and raised concerns about mammalian toxicity. The authors suggested that mortality of mice following pulmonary exposure to *Bt* in the laboratory resulted from activity of a *B. cereus*-like hemolytic toxin. While the doses used in the study were artificially high, the fact that a toxin produced by a *Bt* strain could cause effects in mice after inhalation of spores but this requires more detailed investigation. Overall, it can be concluded that *Bt* poses little risk to mammals at dosages equivalent to field level exposure.

While *Bt* sales make up less than 1% of all pesticide sales worldwide, there has been a move toward use of pesticides, such as *Bt*, with low environmental risk. It is pertinent to consider the relative safety, environment, and mammal of *Bt* strains compared with other pesticides used against the same pests. Although there is a strong movement toward reduced or even nil pesticide use in some developed countries, it will still be imperative to control insect pests. The world population is increasing rapidly and feeding the large populations of developing countries is a major problem. Pest outbreaks can devastate crops. Is *Bt* a safer option than other commonly used pesticides? In the absence of total guarantees of safety judgments need to be made on comparative safety. In a number of studies, *Bt* was shown to be the least toxic to nontarget organisms among a range of insecticides. An International Organization for Biological and Integrated Control/West Palaearctic Regional Section working group conducted a series of bioassays of 40 pesticides, including *Bt* (Dipel), on beneficial arthropods (Hassan et al., 1983). Dipel rated "harmless" to eight of nine test insects, making it the least harmful of the 20 insecticides tested. *Bt* was listed for use on cereals, corn, forage, roots, vegetables, fruits, and vineyard and forest crops. Generally, only one or two insecticides were listed for each crop, from over 65 tested in total, indicating a low comparative risk from *Bt*. This result is common to a number of other studies which report *Bt* as a safer insecticide than many currently in use.

10. MANAGEMENT OF INSECT RESISTANCE TO *Bt*

Past experience has shown that insects have developed resistance to many organic insecticides and it can be assumed that resistance to *Bt* in transgenic crops will also develop eventually. Numerous *Bt*-resistant populations were developed in the laboratory (Tabashnik et al., 1998). However, only one insect species (diamondback moth, *P. xylostella*) developed resistance under natural conditions following sprays of *Bt* products (Shelton et al., 2000).

10.1 Mechanisms of Resistance Development

Many possible mechanisms can be envisaged to explain resistance development in insects (Frutos et al., 1999). The first is the activation of the protoxins by gut proteinases. The proteolytic processing of protoxins into toxins is necessary for the insecticidal activity of *Bt* formulations. It is also required by some transgenic plants expressing combinations of *Bt* toxins as fusion proteins or toxins extending beyond the protease activation site.

This mechanism of resistance was first described in the Indian meal moth, *Plodia interpunctella* (Hubner) (Oppert et al., 1994). The *Bt*-resistant strain of meal moth displayed a slower processing of protoxins than the susceptible strain, and activation of Cry1 protoxins with midgut enzymes resulted in proteins of intermediate size. The slower protoxin activation of this resistant strain led to a reduced quantity of toxin resulting in a survival advantage. Two other strains of *P. interpunctella* resistant to *Bt* endotoxins and showing a lower protoxin activation than a susceptible strain have also been reported (Oppert et al., 1997). These two resistant strains lacked a major trypsin-like gut enzyme. The absence of the gut proteinase and resistance to the toxin were genetically linked (Oppert et al., 1997). Similarly, a *Bt*-resistant strain of *Helicoverpa virescens* (Fabricius) showed a slower activation of the protoxin as well as a faster rate of degradation of the toxin by midgut extracts (Forcada et al., 1996).

The most frequently observed mechanism of resistance among insect pests is the modification of the receptor site (Tang et al., 1996). This mechanism was studied extensively on a wide range of insect species and toxins, and most of these studies pointed out a change in the level of affinity of the receptor for the toxin or to a decreased number of receptor sites (Frutos et al., 1999).

The first study of a mechanism of resistance in a field evolved *Bt*-resistant strain was made by Ferre et al. (1991) using a colony of *P. xylostella* (L.) from the Philippines. In the resistant strain, a loss of specific binding to *cry1Ab* (Ferre et al., 1991; Bravo et al., 1992) suggested that the resistance was due to a change in the *cry1Ab* binding site. Resistance was also studied in a strain of *P. xylostella* from Hawaii selected for resistance to Dipel® (Tabashnik et al., 1994). This colony was highly resistant to *cry1Ac* but still susceptible to *cry1C*. The resistance was correlated with a total loss of *cry1Ac* binding, whereas no change in *cry1C* binding was reported (Tabashnik et al., 1994). A rapid reversal of resistance was associated with the restoration of the binding properties of *cry1Ac* (Tabashnik et al., 1994). The resistance was most likely related to the alteration of the *cry1Ac* binding site.

A field-collected strain of *P. xylostella* from Florida selected for resistance to HD-1 was highly resistant to *cry1Aa*, *cry1Ab*, and *cry1Ac* but not to *cry1B*, *cry1C*, or *cry1D* (Tang et al., 1996). Analysis of the binding characteristics of biotinylated toxins on brush border membrane vesicles showed a loss of binding to *cry1Ab* in the resistant strain, whereas binding properties of *cry1B* and *cry1C* remained unchanged (Tang et al., 1996).

The third mechanism of resistance, not related to reduced binding or altered protein activation, but corresponding to an unknown or several known mechanisms was reported for the first time in *H. virescens* (Gould et al., 1992). Although this *H. virescens* population was selected for resistance to *cry1Ac*, it also showed resistance to *cry1Ab* and *cry2Aa*. The resistant strain grew faster than the susceptible strain in the presence of *cry1Aa*, *cry1B*, and *cry1C*. This is a broad resistance mechanism in an insect population selected for resistance to only one toxin, *cry1Ac*, that can become resistant to other toxins sharing few sequence homologies (e.g., *cry1B* and *cry2Aa*). The lack of knowledge on the mode of action after receptor binding makes understanding of this mode of resistance difficult and the determination of the number of different mechanisms that could occur at this postbinding level. One can only assume that at least part of these reported cases of resistance corresponds to altered postbinding events when resistance cannot be related to a reduced binding or to a lower level of toxin activation. Such a mechanism could explain broad spectrum resistance if several toxins share common postbinding steps in their mode of action. A detailed account of stability and genetics of resistance was given by Frutos et al. (1999).

10.2 Resistance Management Strategies

Many strategies to manage resistance to *Bt* toxins have been proposed and reviewed (Gould, 1998; Frutos et al., 1999). Field data and experimental evidence are now available to help designing resistance management strategies (Roush, 1997; Shelton et al., 2000). These strategies fall into several categories:

1. Tissue or time-specific expression of toxins.
2. Gene stacking/pyramiding/hybridization.
3. Mixtures, rotation, or mosaics of transgenic plants.
4. Combination of toxins with different modes of action.
5. Refuge strategy.
6. "Trap plants" strategy.
7. High-dose/refuge strategy.

10.2.1 *Tissue or Temporal Expression of Toxins*

Tissue and temporal expression (including expression controlled by wound-inducible or chemical-inducible promoters) of toxins in transgenic plants were proposed to limit production of toxin to the most economically sensitive or most vulnerable parts of plant or to specific time (Roush, 1997). This strategy does not require external refuge, the plant itself acts as such. However, several problems can be associated with this approach. Efficient tissue or time-specific expression is not yet available (Roush, 1997). This strategy is dependent on the feeding behavior of the pests and can be influenced by the feeding deterrent effect of *Bt*-transgenic plants (Gould, 1994). Also, an "in planta" refugia may be relevant for protection against a given member of a group of pests but fail to provide protection against a secondary pest attacking the plant at the same time but feeding on a different part.

Several potential economical and sociological problems are related to the use of chemical-inducible promoters (Roush, 1997). Chemicals to be used for induction might be costly and might be negatively perceived with respect to environmental protection because the issue will be the spray of a potentially polluting chemical. This might also be perceived as a negation of the main interest of the transgenic strategy, which is the reduction of chemicals in the environment.

10.2.2 Gene Stacking/Pyramiding/Hybridization

The gene stacking/pyramiding strategy relies on the simultaneous delivery of toxins recognizing different binding sites. It is based on the assumption that the frequency of individuals with two independent and rare resistance alleles is lower than that of individuals with one rare resistance allele (Gould, 1998). The gene pyramiding strategy works best if there is no cross-resistance and, whenever possible, selection of proper Cry proteins is done to minimize this risk.

The gene stacking strategy or "redundant killing" relies on the simultaneous delivery of toxins recognizing different binding sites. However, gene hybridization was developed as a practical strategy to broaden the range of insect species that were not adequately controlled by a single toxin as in the case of the single gene Bollgard-I *Bt* cotton variety. The strategy of *Bt* gene hybridization rests on three core assumptions: (1) that insects resistant to only one toxin can be effectively controlled by a second toxin produced in the same plant, (2) that insects homozygous for one resistance gene are rare and insects homozygous for multiple resistance genes are extremely rare, and (3) that a single gene will not confer resistance to two toxins that are immunologically distinct and that have different binding targets (Gahan et al., 2005). However, *cry* gene hybridization is done by two methods: (1) fusion of two *cry* genes encoding entire toxin protein, and (2) fusion of toxin domains from different *cry* genes.

10.2.2.1 FUSION OF TWO *CRY* GENES ENCODING ENTIRE TOXIN PROTEIN

Cotton is grown in large area in the world. In India also it is cultivated both under irrigated as well as under rain-fed conditions. Important lepidopteran pests of cotton include *H. armigera* and *P. gossypiella* that cause severe damage to cotton regularly. Other lepidopteran pests like *S. frugiperda* (beet armyworm), *S. exigua*, and soybean loopers, *P. includens*, occasionally cause severe damage to cotton by defoliating the leaves.

Bollgard-II was developed using two *cry* genes, *cry1Ac* and *cry2Ab*. The two *cry* genes were fused by creating the common restriction site. Later, the recombinant DNA was transferred to the plasmid to construct the binary vector. The vector was transferred to the cotton to develop Bollgard-II, which was found to be highly effective against all lepidopteran pests. Larval populations of *H. zea* were significantly lower in Bollgard-II than in Bollgard-I and conventional cotton and the proportion of bolls damaged by *H. zea* was also lower. Laboratory bioassays were also done to compare mortality of *S. frugiperda* and *S. exigua* feeding on Bollgard-II, Bollgard-I, and non-*Bt* cotton plants. Mortality of both species was significantly greater on Bollgard-II plant than on either Bollgard-I or conventional cotton (Chitkowski et al., 2003).

10.2.2.2 FUSION OF TOXIN DOMAINS FROM DIFFERENT *CRY* GENES

Cry1 proteins are generally active against lepidopterans but a somewhat higher activity against coleopteran pest *Leptinotarsa decemlineata* was reported for *cry1Ia* (Tailor et al., 1992). A *cry1Ba–cry1Ia* hybrid gene (*SN19*) encoding a protein consisting of domains I and III of *cry1Ba* and domain II of *cry1Ia*, with higher activity against Colorado Potato Beetle (CPB), was constructed and described by Naimov et al. (2001). Both the parental proteins, Cry1Ba and Cry1Ia, are highly toxic for European corn borer and potato tuber moth, suggesting that the hybrid protein may also have these properties and it was found that transgenic potato with single hybrid gene *cry1Ba–cry1Ia* (*SN19*) encoding protein, consisting of domains I and III of *cry1Ba* and domain II of *cry1Ia*, is highly effective against two different insect orders (Naimov et al., 2003).

10.2.3 *Mixtures, Rotation, or Mosaics of Transgenic Plants*

The status of these strategies has evolved depending on conditions and availability of material. Although equivalent in efficiency to combinations of toxins based on sprays (Roush, 1989), they were shown to be weak in delaying resistance when applied to plants (Caprio, 1998). Rotation is based on the assumption that the frequency of resistance alleles will decrease when the selection pressure is removed (Tabashnik, 1994). However, many reports showed that resistance could remain stable or decrease slowly after removal of selection pressure (Liu et al., 1996), making the use of rotation inefficient. Mosaics correspond to the simultaneous deployment in separate fields of varieties of the same crop containing different single toxins. The mosaic strategy was not considered appropriate to deploy two toxin genes (Roush, 1997).

10.2.4 *Combination of Toxins with Different Modes of Action*

Protease inhibitors were shown to act synergistically with *Bt* toxins (Macintosh et al., 1990), thus prompting the recommendation of such combinations for resistance management. Transgenic cotton with combination of *cry1Ac* + CpTI (cowpea trypsin inhibitor) genes. However, the report of contradictory results (Tabashnik et al., 1994) rapidly limited the interest.

10.2.5 *Refuge Strategy*

Refuges, or refugia, are areas of crops or host plants free of *Bt* toxins or insecticide treatment that allow part of the pest population to survive and to act as a reservoir of wild type susceptible alleles. By maintaining a refuge area close to the transgenic field, surviving individuals that have been exposed to *Bt* toxins will mate with unselected individuals coming from the refuge, thus diluting resistance alleles and reducing the intensity of selection pressure. The refuge strategy was proposed for delaying resistance to chemical pesticides (Curtis, 1985) and *Bt* toxins (Gould, 1998). Experimental data confirmed the effectiveness of the refuge strategy (Liu and Tabashnik, 1997). The refuge should be maintained free from any treatment with pesticides to ensure the presence of a sufficient number of susceptible individuals for subsequent mating with survivors (Gould, 1998; Roush, 1997). Many of these requirements are species related and refuge implementation will require a case by case approach, depending on conditions and expectations.

Spatial organization of refuges has been considered in different ways and the most efficient spatial organization for refuge seems to be the external refuge. However, the solution will most likely be a case by case decision based on local conditions and on pest biology and mobility. The size of the refuge is also a matter of debate. The larger the refuge, the greater the delay of resistance. There is yet no clear evidence of the minimal size of the refuge and estimates range from 10% to 50%, depending on crop, pest, and simulation program (Liu and Tabashnik, 1997). Currently, a 4% refuge is mandated for Bt cotton by the US Environmental Protection Agency (EPA) to manage resistance in *H. virescens* (Gould, 1998).

10.2.6 Trap Plants Strategy

A resistance management approach closely related to the refuge system is the "trap plant" strategy proposed by Alstad and Andow (1995). In this approach, the transgenic crop is not considered as the main source of production but as a trap. The Bt-transgenic plants are planted earlier than the nontransgenic plants and, owing to their advanced maturation, attract the pests that are killed after feeding, leaving the nontransgenic varieties relatively unharmed (Alstad and Andow, 1995). There is, however, no evidence that the basic key assumptions on movement and absence of ovipositing preference will be verified in the field for target insect species. A similar computer simulation model on slightly different assumptions about feeding preference and movements led to the conclusion that this strategy was failure-prone (Ives, 1996). Indeed, avoidance of transgenic plants and ovipositing preferences were demonstrated (Arpaia et al., 1998) and even if an insect species may be suitable for the "trap plant" strategy owing to its feeding preferences, the presence of Bt toxins may adversely alter its feeding.

10.2.7 High-Dose/Refuge Strategy

This strategy is considered the most efficient and promising way of managing resistance to Bt toxins. This approach is, in theory (Gould, 1998) and in practice (Liu and Tabashnik, 1997), very effective provided that a high level of insecticidal protein is produced throughout the life span of the plant and all over the field. This may not happen because the production of toxin is expected to decrease over time due to plant senescence (Onstad and Gould, 1998). If the level of production of Bt toxin decreases toward the end of the growing season, heterozygous individuals, which may often be slightly more resistant than susceptible homozygotes, might be able to survive and transmit resistance alleles to the offspring. Another situation leading to the delivery of a moderate dose is that by which larvae displaying a high frequency of movement may feed alternatively on toxic and nontoxic plants, thus diluting the dose of the toxin. The importance of this effect will depend on the behavior and intensity of movement of the pests and the spatial organization of the refuge. Use of the combinations of toxins along with the high-dose/refuge strategy will be more effective. The use of the combinations of toxins, both delivered at high dose, will allow through the "redundant killing" effect to more efficiently control heterozygous insects that might display a sufficient resistance to resist a single toxin, but more rarely to two highly active insecticidal proteins. However, even in such a situation a refuge is necessary for a combination of toxins to delay resistance (Roush, 1998; Caprio, 1998).

A major problem for the high-dose/refuge strategy using combinations of toxins will be cross-resistance. The control of heterozygotes is an important factor and theoretically the simultaneous use of a combination of high-dose toxins and a refuge will be the most efficient answer. Computer modeling indicates that in the absence of cross-resistance, a combination of high-dose toxins with a refuge is expected to last over 160 generations, whereas the sequential use of the same toxins will delay resistance for only 12 generations, which means that there is a 10-fold advantage of using a combination of toxins (Roush, 1998).

A difficult situation might be that of multiple crops and multiple pests prevalent in tropical and subtropical countries like India. A toxin may be highly active against a given pest but moderately active against another pest also present on the same crop or in the same area, for which the dose delivered by the transgenic plant does not correspond to a high dose. In that case, the second pest will be exposed to a moderate dose and may develop resistance. This leads to another key feature of insect resistance management, the use of *Bt* plants as a component of IPM approach (Hoy, 1998). The use of *Bt*-based formulations within an IPM program was shown to be effective for controlling pests (Meade and Hare, 1995). As there is no single answer or strategy to delay resistance, only a logical combination of various means of pest control will provide sustainability. The high-dose/refuge strategy, and especially the association of this approach with combinations of toxins is advocated by scientists, seed companies, and administrations as the best way of controlling evolution of resistance (Peferoen, 1997; Gould, 1998; Roush, 1998; Fox, 1998). This composite strategy allows flexibility in terms of spatial organization of refuge areas and choice of toxins that will help adaptation and implementation in various agronomic systems. This approach is now required by the US EPA. (Fox, 1998). Although currently limited to the USA, this legal requirement for active resistance management by the producers will have to be followed by other countries.

11. ECONOMICS OF *Bt* CROPS

Currently, a few *Bt* crops (cotton, corn, potato, paddy, etc.) are under cultivation in countries, such as the USA, India, China, South Africa, Australia, Argentina, Mexico, etc. An important finding of the studies conducted in China on *Bt* cotton was that the smallest farmers, those farming less than 1 ha, gained more than twice as much income per unit of land ($400 per hectare) from *Bt* cotton as the larger farmers ($185 per hectare). This finding is important from an equity/distribution viewpoint and is deserving of further investigation for *Bt* cotton that offers promise to small resource poor farmers. It also has important implications in relation to the claim often made by critics of transgenic crops that they are inappropriate for small farmers. Indeed, by far the largest benefits reported to date have been for small farmers who can least afford the loss in yield due to pests, and stand to gain the most from increases in income and suffer less health hazards resulting from fewer applications of conventional insecticide.

12. CONCLUSIONS

Insect pests have major effects on agricultural production and food supply. Although the application of insecticides has helped to minimize the impact of insect pests, chemical control entails economic, health, and environmental costs. Therefore, the development of new strategies or improving the existing strategies other than chemicals for insect pest control is critical for sustaining agricultural production and improving our environment and health. One such strategy is the use of microbes or their product(s) for insect pest control. The "microbial" alternatives to chemical insecticides include a variety of biological agents, such as bacteria, viruses, and fungi. Of all microbial agents that have been evaluated, the most successful by far has been *Bt*, which probably stems from the fact that bacteria are much easier to culture in large quantities in artificial media. Today more than 200 *Bt* products are registered worldwide under various trade names. *Bt* products have become so popular by virtue of the lack of toxicity toward human beings, other mammals, and plants. *Bt* plays an important role not only in agriculture and forestry but also in human and animal health. The discovery of new strains of *Bt* widened the toxicity spectrum of bioinsecticides.

The use of *Bt* in pest management took a quantum leap with the development of transgenic crops incorporating the gene coding for the production of the insecticidal crystal protein which is an elegant and the most effective delivery system for *Bt* toxins. Genetic transformation of almost all the major crop species is now feasible with the development of an array of techniques ranging from the *Agrobacterium*-mediated approach to electric discharge-mediated particle acceleration procedure. The major benefits of *Bt* toxins delivered through transgenics are economic, environmental, and qualitative. In addition to the reduced input costs to the farmer by reducing the synthetic insecticide sprays, the transgenic plants provide season-long protection independent of weather conditions, effective control of burrowing insects difficult to reach with sprays, and control at all of the stages of insect development. An important feature of such a system is that only insects eating the crop are exposed to the toxin. However, the rapid increase in *Bt* toxin-expressing transgenic plants may further complicate the situation, as the toxins are expressed continuously. Therefore, defining nontarget impacts from laboratory and even field studies can be difficult. These limitations need to be kept in mind when considering nontarget impacts of *Bt* toxins. While some nontarget toxicity has been reported (certain caterpillars in the case of *Bt kurstaki*), no nontarget species has been shown to be at risk of eradication through use of *Bt* and *Bt* is still very safe compared with other pesticides in terms of impacts on nontarget invertebrates.

Past and present experience has shown that insects have developed resistance to many organic insecticides and it can be assumed that *Bt*-based products (sprays or transgenic plants) have the potential to trigger evolution of resistance in insect pests and it is essential that the farmers, authorities, and industry will be important players in the successful implementation of resistance management strategies and therefore sustainability of *Bt*-transgenic products. There is also a need for training, information to farmers, and extension workers to play a major role in the successful implementation of resistance management strategies.

Finally, the enormous knowledge about *Bt* and its importance in insect pest management cannot be covered in a book chapter but it is hoped that this chapter will provide greater impetus for research in this field, particularly in India.

Acknowledgments

We immensely thank the editors of the book for providing us an opportunity to contribute for this work. We are also indebted to Dr K. Chandrashekara, University of Agricultural Sciences, Bangalore, for valuable suggestions and for reviewing the chapter.

References

Adamczyk Jr., J.J., Adams, L.C., Hardee, D.D., 2001. Field efficacy and seasonal expression profiles for terminal leaves of single and double *Bacillus thuringiensis* toxin cotton genotypes. Journal of Economic Entomology 94, 1589–1593.

Adang, M.J., Brody, M.S., Cardineau, G., Eagan, N., Roush, R.T., Shewmaker, C.K., Jones, A., Oakes, J.V., McBride, K.E., 1993. The reconstruction and expression of a *Bacillus thuringiensis cryIIIA* gene in protoplasts and potato plants. Plant Molecular Biology 21, 1131–1145.

Alstad, D.N., Andow, D.A., 1995. Managing the evolution of insect resistance to transgenic plants. Science 268, 1894–1896.

Arencibia, A., Vazquez, R.I., Prieto, D., Tellez, P., Carmona, E.R., Coego, A., Hernandez, L., Riva, G.A.D.L., Housein, G.S., 1997. Transgenic sugarcane plants resistant to stem borer attack. Molecular Breeding 3, 247–255.

Arpaia, S., Chiriatti, K., Gioro, G., 1998. Predicting the adaptation of Colorado potato beetle to transgenic eggplants expressing CryIII toxin: the role of gene dominance, migration and fitness cost. Journal of Economic Entomology 91, 21–29.

Bacheler, J.S., Mott, D.W., 1997. Efficacy of grower managed *Bacillus thuringiensis* cotton in North Carolina. In: Dugger, P., Richard, D. (Eds.), Proceedings, Beltwide Cotton Conference National Cotton Council, Memphis, pp. 931–934.

Barton, K., Whitely, H., Yang, N.S., 1987. *Bacillus thuringiensis* δ-endotoxin in transgenic *Nicotiana tabacum* provides resistance to lepidopteran insects. Plant Physiology 85, 1103–1109.

Baum, J.A., Coyle, D.M., Gilbert, M.P., Jany, C.S., Burke, C.G., 1990. Novel cloning vectors for *Bacillus thuringiensis*. Applied and Environmental Microbiology 56, 3420–3428.

Baum, J.A., Johnson, T.B., Carlton, B.C., 1998. Natural and recombinant bioinsecticide products. In: Hall, F.R., Menn, J.J. (Eds.), Biopesticides: Use and Delivery. Humana Press, Totowa, pp. 189–209.

Becker, N., Margalit, J., 1993. Use of *Bacillus thuringiensis israelensis* against mosquitoes and blackflies. In: Entwistle, P.F., Cory, P.F., Margalit, M.J., Higgs, S. (Eds.), *Bacillus thuringiensis*, an Environmental Biopesticide: Theory and Practice. John Wiley & Sons, New York, pp. 145–170.

Beegle, C.C., Yamamoto, T., 1992. History of *Bacillus thuringiensis* Berliner: research and development. Canadian Entomologist 124, 587–616.

Bennett, J., 1994. DNA-based techniques for control of rice insects and diseases: transformation, gene tagging and DNA fingerprinting. In: Teng, P.S., et al. (Eds.), Rice Pest Science and Management. International Rice Research Institute, Los Banos, Philippines, pp. 147–172.

Berliner, E., 1915. Uber die Schaffsucht der Mehlmottenraupe. Zeitschrift für Angewandte Entomologie 2, 29–56.

Bernhardt, K., Jarrett, P., Meadows, M., Butt, J., Ellis, D.J., Roberts, G.M., Pauli, S., Rodgers, P., Burges, H.D., 1997. Natural isolates of *Bacillus thuringiensis*: world wide distribution, characterization and activity against insect pests. Journal of Invertebrate Pathology 70 (1), 55–66.

Bezdicek, D.F., Quinn, M.A., Forse, L., Heron, D., Kahn, M.L., 1994. Insecticidal activity and competitiveness of *Rhizobium* spp. containing the *Bacillus thuringiensis* subsp. *tenebrionis* δ-endotoxin gene (*cry III*) in legume nodules. Soil Biology and Biochemistry 26, 1637–1646.

Bhattacharya, R.C., Viswakarma, N., Bhat, S.R., Kirti, P.B., Chopra, V.L., 2002. Development of insect-resistant transgenic cabbage plants expressing a synthetic *cryIA(b)* gene from *Bacillus thuringiensis*. Current Science 83 (2), 146–150.

Bora, R.S., Murthy, M.G., Shenbagarathai, R., Sekar, V., 1994. Introduction of a lepidopteran-specific insecticidal crystal protein gene of *Bacillus thuringiensis* subsp. *kurstaki* by conjugal transfer into a *Bacillus megaterium* strain that persists in the cotton phyllosphere. Applied and Environmental Microbiology 60, 214–222.

Bosch, D., Schipper, B., van der Kleij, H., de Maagd, R., Stiekema, W.J., 1994. Recombinant *Bacillus thuringiensis* crystal proteins with new properties: possibilities for resistance management. Bio/Technology 12, 915–918.

Bravo, A., Jansens, S., Peferoen, M., 1992. Immunocytochemical localization of *Bacillus thuringiensis* crystal proteins in toxicated insects. Journal of Invertebrate Pathology 60, 237–246.

Bravo, A., Sarabia, S., Lopez, L., Ontiveros, H., Abarca, C., Ortiz, A., Ortiz, M., Lina, L., Villalobos, F.J., Pena, G., Valdez, M.N., Soberon, M., Quintero, R., 1998. Characterization of cry genes in Mexican *Bacillus thuringiensis* strain collection. Applied and Environmental Microbiology 64 (12), 4965–4972.

Caprio, M.A., 1998. Evaluating resistance management strategies for multiple toxins in the presence of external refuges. Journal of Economic Entomology 91, 1021–1031.

Carozzi, N.B., Warren, G.W., Desai, N., Jayne, S.M., Lotstein, R., Rice, D.A., Evola, S., Koziel, M.G., 1992. Expression of a chimeric CaMV35S *Bacillus thuringiensis* insecticidal protein in transgenic tobacco. Plant Molecular Biology 20, 539–548.

Charles, J.F., Leroux, C.N., Delecluse, A., 1996. *Bacillus sphaericus* toxins: molecular biology and mode of action. Annual Review of Entomology 41, 451–472.

Chen, H., Tang, W., Xu, C., Li, X., Lin, Y., Zhang, Q., 2005. Transgenic indica rice plants harboring a synthetic *cry2A** gene of *Bacillus thuringiensis* exhibit enhanced resistance against lepidopteran rice pests. Theoretical and Applied Genetics 111, 1330–1337.

Cheng, X.Y., Sardana, R., Kaplan, H., Altosaar, I., 1998. *Agrobacterium* transformed rice plants expressing synthetic *cry1Ab* and *cry1Ac* genes are highly toxic to striped stem borer and yellow stem borer. Proceedings of the National Academy of Sciences of the United States of America 95, 2767–2772.

Chitkowski, R.L., Turnipseed, S.G., Sullivau, M.J., Bridges Jr., W.C., 2003. Field and laboratory evaluations of transgenic cottons expressing one or two *Bacillus thuringiensis* var. *kurstaki* Berliner proteins for management of noctuid (Lepidoptera) pests. Journal of Economic Entomology 96 (3), 755–762.

Cohen, M.B., Gould, G., Bentur, J.S., 2000. *Bt* rice: practical steps to sustainable use. International Rice Research Notes 25, 4–10.

Crickmore, N., Zeigler, D.R., Feitelson, J., Schnepf, E., Rie, J.V., Lereclus, D., Baum, J., Dean, D.H., 1998. Revision of the nomenclature for *Bacillus thuringiensis cry* genes. Microbiology and Molecular Biology Review 62 (3), 807–813.

Curtis, C.F., 1985. Theoretical models of the use of insecticide mixtures for the management of resistance. Bulletin of Entomological Research 75, 259–265.

Datta, K., Vasquez, A., Tu, J., Torrizo, L., Alam, M.F., Oliva, N., Abrigo, E., Khush, G.S., Datta, S.K., 1998. Constitutive and tissue-specific differential expression of the *cryIA(b)* gene in transgenic rice plants conferring resistance to rice insect pests. Theoretical and Applied Genetics 97, 20–30.

Datta, K., Baisakh, N., Thet, K.M., Tu, J., Datta, S.K., 2002. Pyramiding transgenes for multiple resistance in rice against bacterial blight, yellow stem borer and sheath blight. Theoretical and Applied Genetics 106, 1–8.

Davidson, E.W., Meyers, P., 1981. Parasporal inclusions of *Bacillus sphaericus*. FEMS Microbiology Letters 10, 261–265.

de Barjac, H., Bonnefoi, A., 1962. Essai de classification biochimique et serologique de 24 sources de bacillus du type *B. thuringiensis*. Entomophaga 7, 5–31.

de Brarjac, H., Lemille, F., 1970. Presence of flagellar antigenic subfactors in serotype-3 of *Bacillus thuringiensis*. Journal of Invertebrate Pathology 15, 139–140.

de Lucca, A.J., Simonson, J.G., Larson, A.D., 1981. *Bacillus thuringiensis* distribution in soils of the United States. Canadian Journal of Microbiology 27, 865–870.

de Maagd, R.A., van der Kleij, H., Bakker, P.L., Stiekema, W.J., Bosch, D., 1996. Different domains of *Bacillus thuringiensis* δ-endotoxins can bind to insect midgut membrane proteins on ligand blots. Applied and Environmental Microbiology 62, 2753–2757.

de Maagd, R.A., Bosch, D., Stiekema, W., 1999. *Bacillus thuringiensis* toxin mediated insect resistance in plants. Trends in Plant Science 4 (1), 09–13.

de Maagd, R.A., Bravo, A., Crickmore, N., 2001. How *Bacillus thuringiensis* has evolved specific toxins to colonize the insect world. Trends in Genetics 17, 193–199.

Delecluse, A., Rosso, M.L., Ragani, A., 1995. Cloning and expression of a novel toxin gene from *Bacillus thuringiensis* subsp. *jegathesan* encoding a highly mosquitocidal protein. Applied and Environmental Microbiology 61, 4230–4235.

Dulmage, H.T., 1970. Insecticidal activity of HD-1, a new isolate of *Bacillus thuringiensis* var alesti. Journal of Invertebrate Pathology 15, 232–239.

Duncan, D.R., Hammond, D., Zalewski, J., Cudnohufsky, J., Kaniewski, W., Thornton, M., Bookout, J.T., Lavrik, P., Rogan, G.J., Riebe, J.F., 2002. Field performance of transgenic potato with resistance to Colorado potato beetle and viruses. HortScience 37 (2), 275–276.

English, L., Robbins, H.L., Von Tersch, M., Kulesza, C.A., Ave, D., Coule, D., Jany, C.S., Slatin, S.L., 1994. Mode of action of CryIIA: a *Bacillus thuringiensis* delta endotoxin. Insect Biochemistry and Molecular Biology 24, 1025–1035.

Estruch, J.J., Warren, G.W., Mullins, M.A., Nye, G.J., Craig, J.A., Koziel, M.G., 1996. Vip3A, a novel *Bacillus thuringiensis* vegetative insecticidal crystal protein with a wide spectrum of activities against Lepidopteran insects. Proceedings of the National Academy of Sciences of the United States of America 93, 5389–5394.

Estruch, J.J., Carozzi, N.B., Desai, N., Duck, N.B., Warren, G.W., Koziel, M.G., 1997. Transgenic plants: an emerging approach to pest control. Nature Biotechnology 15, 137–141.

Federici, B.A., 1998. Broad-scale leaf pest-killing plants to be true test. California Agriculture 52, 14–20.

Federici, B.A., 1999. *Bacillus thuringiensis* in biological control. In: Handbook of Biological Control. Academic Press, New York, pp. 575–593.

Feitelson, J.S., Payne, J., Kim, L., 1992. *Bacillus thuringiensis*: insects and beyond. Bio/Technology 10, 271–275.

Ferre, J., van Rie, J., 2002. Biochemistry and genetics of insect resistance to *Bacillus thuringiensis*. Annual Review of Entomology 47, 501–533.

Ferre, J., Real, M.D., van Rie, J., Jansens, S., Peferoen, M., 1991. Resistance to the *Bacillus thuringiensis* bioinsecticide in a field population of *Plutella xylostella* is due to a change in midgut membrane receptor. Proceedings of the National Academy of Sciences of the United States of America 88, 5119–5123.

Fischhoff, D.A., Bowdisch, K.S., Perlak, F.J., Marrone, P.G., McCormick, S.H., Niedermeyer, J.G., Dean, D.A., Kusano-Kretzmer, K., Mayer, E.J., Rochester, D.E., Rogers, S.G., Fraley, R.T., 1987. Insect tolerant transgenic tomato plants. Bio/Technology 5, 807–813.

Forcada, C., Alacer, E., Garcera, M.D., Martinez, R., 1996. Differences in the midgut proteolytic activity of two *Heliothis virescens* strains, one susceptible and one resistant to *Bacillus thuringiensis* toxins. Archives of Insect Biochemistry and Physiology 31, 257–272.

Fox, J.L., 1998. Science panel urges EPA to mandate *Bt* resistance management. ASM News 64, 379–380.

Frutos, R., Rang, C., Royer, M., 1999. Managing insect resistance to plants producing *Bacillus thuringiensis* toxins. Critical Reviews in Biotechnology 19, 227–276.

Fujimoto, H., Itoh, K., Yamamoto, M., Kyozuka, J., Shimamoto, K., 1993. Insect resistant rice generated by introduction of a modified δ-endotoxin gene *Bacillus thuringiensis*. Bio/Technology 11, 1151–1155.

Gaertner, F.H., Quick, T.C., Thompson, M.A., 1993. CellCap: an encapsulation system for insecticidal biotoxin proteins. In: Kim, L. (Ed.), Advanced Engineered Pesticides. Marcel Dekker, New York, pp. 73–83.

Gahan, L.J., Ma, Y.T., Coble, M.L.M., Gould, F., Moar, W.J., Heckel, D.G., 2005. Genetic basis of resistance to Cry1Ac and Cry2Aa in *Heliothis virescens* (Lepidoptera: Noctuidae). Journal of Economic Entomology 98, 1357–1368.

Garczynski, S.F., Adang, M.J., 1995. Cry1Ac delta endotoxin-binding aminopeptidase N in the *Manduca sexta* midgut has glycosyl-phosphotidylinositol anchor. Insect Biochemistry and Molecular Biolology 25, 409–415.

Gatehouse, J.A., 2008. Biotechnological prospects for engineering insect-resistant plants. Plant Physiology 146, 881–887.

Gawron-Burke, C., Baum, J.A., 1991. Genetic manipulation of *Bacillus thuringiensis* insecticidal crystal protein genes in bacteria. In: Setlow, J.K. (Ed.), Genetic Engineering: Principles and Methods, vol. 13. Plenum Press, New York, pp. 237–263.

Gill, S.S., Cowles, E.A., Pietrantonio, F.V., 1992. The mode of action of *Bacillus thuringiensis* endotoxins. Annual Review of Entomology 37, 615–636.

Glare, T.R., O'Callaghan, M., 2000. *Bacillus thuringiensis*: Biology, Ecology and Safety. John Wiley and Sons, Chichester, p. 368.

Gleave, P.A., Mitra, S.D., Markwick, P.N., Morris, B.A.M., Beuning, L.L., 1998. Enhanced expression of the *Bacillus thuringiensis cry9Aa2* gene in transgenic plants by nucleotide sequence modification confers resistance to potato tuber moth. Molecular Breeding 4, 459–472.

Goldberg, L.J., Margalit, 1977. A bacterial spore demonstrating rapid larvicidal activity against *Anopheles sergentii, Uranotaenia unguiculata, Culex univeritattus, Aedes aegypti* and *Culex pipiens*. Mosquitoes News 37, 355–358.

Gould, F., Martinez-Ramirez, A., Anderson, A., Ferre, J., Silva, F.J., Moar, W.J., 1992. Broad spectrum resistance to *Bacillus thuringiensis* toxins in *Heliothis virescens*. Proceedings of the National Academy of Sciences of the United States of America 89, 7986–7990.

Gould, F., 1994. Potential and problems with high- dose strategies for pesticidal engineered crops. Biocontrol Science and Technology 4, 451–461.

Gould, F., 1998. Sustainability of transgenic insecticidal cultivars: integrating pest genetics and ecology. Annual Review of Entomology 43, 701–726.

Griffiths, W., 1998. Will genetically modified crops replace agrochemicals in modern agriculture? Pesticide Outlook 9, 6–8.

Grochulski, P., Masson, L., Borisova, S., Pusztai-Carey, M., Schwartz, J.L., Brousseau, R., Cygler, M., 1995. *Bacillus thuringiensis Cry1Aa* insecticidal toxin: crystal structure and channel formation. Journal of Molecular Biology 254, 447–464.

Hassan, S.A., Bigler, F., Bogenschutz, H., Brown, J.U., Firth, S.I., Huang, P., Ledieu, M.S., Naton, E., Oomen, P.A., Overmeer, W.P.J., Rieckmann, W., Peterson, W.S., Viggiani, G., van Zon, A.Q., 1983. Results of the joint pesticide testing programme by the IOBC/WPRS-Working group "Pesticides and Beneficial Arthropods". Zeitschrift für Angewandte Entomologie 95, 151–158.

Hernandez, E., Ramisse, F., Ducoureau, J.P., Cruel, T., Cavallo, J.D., 1998. Bacillus thuringiensis subsp. konkukian (serotype H34) superinfection: case report and experimental evidence of pathogenicity in immunosupressed mice. Journal of Clinical Microbiology 36, 2138–2139.

Hernstadt, C., Soares, G.C., Wilcox, E.R., Edwards, D.L., 1986. A new strain of *Bacillus thuringiensis* with activity against coleopteran insects. Bio/Technology 4, 305–308.

Ho, N.H., Baisakh, N., Oliva, N., Datta, K., Frutos, R., Datta, S.K., 2006. Translational fusion of hybrid *Bt* genes confer resistance against yellow stem borer in transgenic elite Vietnamese rice (*Oryza sativa* L.) cultivars. Crop Science 46, 781–789.

Hofte, H., Whiteley, H.R., 1989. Insecticidal crystal proteins of *Bacillus thuringiensis*. Microbiology Reviews 53, 242–255.

Hoy, M.A., 1998. Myths: models and mitigation of resistance to pesticides. Philosophical Transactions of Royal Society London 353, 1787–1795.

Huang, Y.X., Huang, R.R., 1986. Investigation of *Bt* resources in Guangixi. Chinese Journal of Biocontrol 2 (2), 81–83.

Husz, B., 1929. On the use of *Bacillus thuringiensis* in the fight against the corn borer. International Corn Borer Investigations Scientific Reports 2, 99–110.

Ishiwata, S., 1901. On a kind of severe flaccherie (sotto disease). Dainihon Sanshi Kaiho 114, 1–5.

Ives, A.R., 1996. Evolution of insect resistance to *Bacillus thuringiensis*-transformed plants. Science 273, 1412–1413.

James, C., 2013. Global Status of Commercialized Biotech/GM Crops: 2007. ISAAA Brief No. 46. ISAAA, Ithaca, New York.

Jansen, S., vanVliet, A., Dickburt, C., Buysse, L., Piens, C., Saey, B., deWulf, A., Gossele, V., Paez, A., Gobel, E., 1997. Transgenic corn expressing a *cry9C* insecticidal protein from *Bacillus thuringiensis* protected from European corn borer damage. Crop Science 37, 1616–1624.

Jelenkovic, G., Billings, S., Chen, Q., Lashomb, J., Hamilton, G., Ghidiu, G., 1998. Transformation of eggplant with synthetic cry3A gene produces a high level of resistance to the Colorado potato beetle. Journal of American Society for Horticultural Science 123, 19–25.

John, K., 2001. Evaluation and Molecular Characterization of *Bacillus thuringiensis* for Control of Diamondback Moth (*Plutella Xylostella*) (M.Sc. thesis). pp. 88.

Johnson, T., Rishi, A.S., Nayak, P., Sen, S.K., 1996. Cloning of a *cry3A* endotoxin gene of *Bacillus thuringiensis* var. *tenebrionis* and its transient expression in indica rice. Journal of Bioscience 21 (5), 673–685.

Kaelin, P., Morel, P., Gadani, F., 1994. Isolation of *Bacillus thuringiensis* from stored tobacco and *Lasioderma serricorne* (F.). Applied and Environmental Microbiology 60, 19–25.

Kaneko, T., Nozaki, R., Aizawa, K., 1978. Deoxyribonucleic acid relatedness between *Bacillus anthracis*, *Bacillus cereus* and *Bacillus thuringiensis*. Microbiology and Immunology 22, 639–641.

Karamanlindonou, G., Lambropoulos, A.F., Koliais, S.I., Manousis, T., Ellar, D., Kastritsis, C., 1991. Toxicity of *Bacillus thuringiensis* to laboratory populations of the olive fruit fly (*Dacus oleae*). Applied and Environmental Microbiology 57 (8), 2277–2282.

Keshavareddy, G., 2009. Development and Characterization of Transgenics in Groundnut (*Arachis hypogaea* L.) Overexpressing *Cry1AcF* Gene for *Spodoptera litura* (F.) (Lepidoptera: Noctuidae) Tolerance (Ph.D. thesis). pp. 163.

Kota, M., Daniell, H., Varma, S., Garczynski, S.F., Gould, F., Moar, W.J., 1999. Overexpression of *Bacillus thuringiensis* (*Bt*) Cry2Aa2 protein in chloroplasts confers resistance to plants against susceptible and *Bt*-resistant insects. Proceedings of the National Academy of Sciences of the United States of America 96, 1840–1845.

Koziel, M.G., Beland, G.L., Bowman, C., Carozzi, N.B., Crenshaw, R., Crossland, L., Dawson, J., Desai, N., Hill, M., Kadwell, S., Launis, K., Lewis, K., Maddox, D., McPherson, K., Meghji, M.R., Merlin, E., Rhodes, R., Warren, G.W., Wright, M., Evola, S.V., 1993. Field performance of elite transgenic maize plants expressing an insecticidal protein derived from *Bacillus thuringiensis*. Bio/Technology 11, 194–200.

Krieg, A., Huger, A.M., Schnetter, W., Herrnstadt, C., 1987. *Bacillus thuringiensis* var *sandiego* strain M-7 is identical to *Bt* subspecies *tenebrionis* strain Bi 250-82, which was previously isolated in Germany and is pathogenic to beetles. Journal of Applied Entomology 104 (4), 417–424.

Krieg, A., 1970. In vitro determination of *Bacillus thuringiensis*, *Bacillus cereus* and related Bacilli. Journal of Invertebrate Pathology 15, 313–320.

Krieg, A., 1982. The natural occurrence of *Bt* in cereals and cereal products with regard to the microbial control of flour moth in stored product protection. Nachrichenblatt-Des-Deutschen-Pflanzen-Schutzdienstes 34 (10), 153–157.

Kumar, P.A., Sharma, R.P., 1994. Genetic engineering of insect resistant crop plants with *Bacillus thuringiensis* crystal protein genes. Journal of Plant Biochemistry and Biotechnology 3, 3–8.

Kumar, P.A., Sharma, R.P., Malik, V.S., 1996. Insecticidal proteins of *Bacillus thuringiensis*. Advances in Applied Microbiology 42, 1–43.

Kumar, P.A., Mandaokar, A., Sreenivasu, K., Chakrabarti, S.K., Bisaria, S., Sharma, S.R., Kaur, S., Sharma, R.P., 1998. Insect-resistant transgenic brinjal plants. Molecular Breeding 4, 33–37.

Kuzmanova, I., 1975. Isolation of local strains of *Bacillus thuringiensis* and study of their biological properties. Gradinarska-I-Lozarska-Nauka 12 (6), 79–89.

Lagnaoui, A., Canedo, V., Douches, D.S., 2000. Evaluation of *Bt*-cry1Ia1 (CryV) Transgenic Potatoes on Two Species of Potato Tuber Moth *Phthorimaea operculella* and *Symmetrischema tangolias* (Lepidoptera: Gelechiidae) in Peru CIP Program Report 1999–2000. CIP, LIMA, Peru, pp. 117–121.

Lambert, B., Peferoen, M., 1992. Insecticidal promise of *Bacillus thuringiensis*. Facts and mysteries about a successful biopesticide. BioScience 42, 112–122.

Lecadet, M.M., Dedonder, R., 1971. Biogenesis of the crystalline inclusion of *Bacillus thuringiensis* during sporulation. European Journal of Biochemistry 23, 282–294.

Lereclus, D., Vallade, M., Chaufaux, J., Arantes, O., Rambaud, S., 1992. Expansion of insecticidal host range of *Bacillus thuringiensis* by *in vivo* genetic recombination. Bio/Technology 10, 418–421.

Lereclus, D., Agaisse, H., Gominet, M., Chaufaux, J., 1995. Overproduction of encapsulated insecticidal crystal proteins in a *Bacillus thuringiensis spo0A* mutant. Bio/Technology 13, 67–71.

Levison, B.L., Kasyan, K.J., Chiu, S.S., Currier, T.C., Gonzales Jr., J.M., 1990. Identification of β exotoxin production, plasmids encoding β exotoxin in *Bacillus thuringiensis* by using high performance liquid chromatography. Journal of Bacteriology 172, 3172–3179.

Li, J., Carroll, J., Ellar, D.J., 1991. Crystal structure of insecticidal δ-endotoxin from *Bacillus thuringiensis* at 2.5 Å resolution. Nature 353, 815–821.

Liu, Y.B., Tabashnik, B.E., 1997. Experimental evidence that refuges delay insect adaptation to *Bacillus thuringiensis*. Philosophical Transanctions of Royal Society London 264, 605–610.

Liu, Y.B., Tabashnik, B.E., Pusztai-Carey, M., 1996. Field-evolved resistance to *Bacillus thuringiensis* toxin Cry1C in diamondback moth (Lepidoptera: Plutellidae). Journal of Economic Entomology 89, 798–804.

Lovgren, A., Zhang, M.Y., Engsteon, A., Dalhammar, G., Landen, R., 1990. Molecular characterization of immune inhibiter a secreted virulence protease from *Bacillus thuringiensis*. Molecular Microbiology 4, 2137–2146.

Luthy, P., Cordier, J., Fischer, J., 1982. *Bacillus thuringiensis* as a bacterial insecticide. In: Kurstak, E. (Ed.), Microbial and Viral Pesticides. Marcel and Decker, New York, pp. 35–74.

Luthy, P., 1980. Insecticidal toxins of *Bacillus thuringiensis*. FEMS Microbiology Letters 8, 1–7.

Macintosh, S.C., Kishore, G.M., Perlak, F.J., Marron, P.G., Stone, T.B., Sims, S.R., Fuchs, R.L., 1990. Potentiation of *Bacillus thuringiensis* insecticidal activity by serine protease inhibitors. Journal of Agriculture and Food Chemistry 38, 1145–1152.

Mandaokar, A.D., Goyal, R.K., Shukla, A., Bisaria, S., Bhalla, R., Reddy, V.S., Chaurasia, A., Sharma, R.P., Altosaar, I., Kumar, P.A., 2000. Transgenic tomato plants resistant to fruit borer (*Helicoverpa armigera* Hubner). Crop Protection 19, 307–312.

Maqbool, S.B., Riazuddin, S., Loc, N.T., Gatehouse, A.M.R., Gatehouse, J.A., Christou, P., 2001. Expression of multiple insecticidal genes confers broad resistance against a range of different rice pests. Molecular Breeding 7, 85–93.

Marfa, V., Mele, E., Gabarra, R., Vassal, J.M., Guiderdoni, E., Messeguer, J., 2002. Influence of the developmental stage of transgenic rice plants (cv. Senia) expressing the *cry1B* gene on the level of protection against the striped stem borer (*Chilo suppressalis*). Plant Cell Reports 20, 1167–1172.

Martin, P.A.W., Travers, R.S., 1989. Worldwide abundance and distribution of *Bacillus thuringiensis* isolates. Applied and Environmental Microbiology 55, 2437–2442.

Marvier, M., McCreedy, C., Regetz, J., Kareiva, P., 2007. A meta-analysis of effects of *Bt* cotton and maize on non-target invertebrates. Science 316, 1475–1477.

McBride, K.E., Svab, Z., Schaaf, D.J., Hogan, P.S., Stalker, D.M., Maliga, P., 1995. Application of a chimeric *Bacillus* gene in chloroplasts leads to extraordinary level of an insecticidal protein in tobacco. Bio/Technology 13, 362–365.

McConnel, E., Richard, A.G., 1959. The production of *Bacillus thuringiensis* Berliner of a heat stable substance toxic for insects. Canadian Journal of Microbiology 5, 161–168.

McLaren, J.S., 1998. The success of transgenic crops in the USA. Pesticide Outlook 9, 36–41.

Meade, T., Hare, J.D., 1995. Integration of host plant resistance and *Bacillus thuringiensis* insecticides in the management of lepidopterous pests of celery. Journal of Economic Entomology 88, 1787–1794.

Meadows, M.P., Ellis, D.J., Butt, J., Jarret, P., Burges, H.D., 1992. Distribution, frequency and diversity of *Bt* in an animal feed mill. Applied and Environmental Microbiology 58, 1344–1350.

Morris, O.N., Converbe, V., Kahagaratnam, P., 1998. Isolation and characterization and culture of *Bacillus thuringiensis* from soil and dust from grains storage bins and their toxicity for *Mamestra contigurata* (Lepidoptera: Noctuidae). Canadian Entomology 130, 515–557.

Morris, O.N., 1983. Micro-organism isolated from the forest insects of British Columbia. Journal of Entomological Society of British Columbia 80, 29–36.

Naimov, S., Hendriks, M.W., Dukiandjiev, S., de Maagd, R.A., 2001. *Bacillus thuringiensis* delta-endotoxin Cry1 hybrid proteins with increased activity against the Colorado potato beetle. Applied and Environmental Microbiology 67 (11), 5328–5330.

Naimov, S., Dukiandjiev, S., de Maagd, R.A., 2003. A hybrid *Bacillus thuringiensis* delta- endotoxin gives resistance against a coleopteran and a lepidopteran pest in transgenic potato. Plant Biotechnology Journal 1, 51–57.

Nayak, P., Basu, D., Das, S., Basu, A., Ghosh, D., Ramakrishnan, N.A., Ghosh, M., Sen, S.K., 1997. Transgenic elite *indica* rice plants expressing CryIAc delta-endotoxin of *Bacillus thuringiensis* are resistant against yellow stem borer (*Scirpophaga incertulas*). Proceedings of National Academy of Sciences 94, 2111–2116.

Obukowicz, M.G., Perlak, F.J., Kusano-Kretzmer, K., Mayer, E.J., Bolten, S.L., Watrud, L.S., 1986. Tn5-mediated integration of the δ-endotoxin gene from *Bacillus thuringiensis* into the chromosome of root-colonising pseudomonads. Journal of Bacteriology 168, 982–989.

Ohba, M., Aizawa, K., 1986. Insect toxicity of *Bacillus thuringiensis* isolates from soils of Japan. Journal of Invertebrate Pathology 47 (1), 12–20.

Ohba, M., Ono, K., Aizawa, K., Iwanami, S., 1981. Two new subspecies of *Bacillus thuringiensis* isolated in Japan *Bacillus thuringiensis* subsps. *Kumamotensis* (serotype 19). Journal of Invertebrate Pathology 38 (2), 184–190.

Onstad, D.W., Gould, F., 1998. Do dynamics of crop maturation and herbivorous insect life cycle influence the risk of adaptation to toxins in transgenic host plants? Environmental Entomology 27, 517–522.

Oppert, B., Kramer, K.J., Johnson, D.E., MacInstosh, S.C., McGaughey, W.H., 1994. Altered protoxin activation by midgut enzymes from a *Bacillus thuringiensis* resistant strain of *Plodia interpunctella*. Biochemical and Biophysical Research Communications 198, 940–947.

Oppert, B., Kramer, K.J., Beeman, R.W., Johnson, D., Mc Gaughey, W.H., 1997. Proteinase- mediated insect resistance to *Bacillus thuringiensis* toxins. Journal Biological Chemistry 272, 23473–23476.

Padua, L.E., Gabriel, B.P., Aizawa, Ohba, M., 1981. *Bacillus thuringiensis* isolated from the Philippines. Philippine Entomologist 20 (6), 35–36.

Pattanayak, D., Srinivasan, K., Mandaokar, A., Shukla, A., Bhalla, R., Kumar, P.A., 2000. AFLP fingerprinting and genetic characterization of *Bacillus thuringiensis* subspecies. World Journal of Microbiology and Biotechnology 16, 667–672.

Peferoen, M., 1997. Progress and prospects for field use of *Bt* genes in crops. Trends in Biotechnology 15, 173–177.

Perlak, F.J., Deaton, R.W., Armstrong, T.O., Fuchs, R.L., Sims, S.R., Greenplate, J.T., Fischhoff, D.A., 1990. Insect resistant cotton plants. Bio/Technology 8, 939–943.

Perlak, F.J., Fuchs, R.L., Dean, D.A., McPherson, S.L., Fischhoff, D.A., 1991. Modification of the coding sequence enhances plant expression of insect control protein genes. Proceedings of National Academy of Sciences of the United States of America 88, 3324–3328.

Ranjekar, P.K., Patankasr, A., Gupta, V., Bhatnagar, R., Bentur, J., Kumar, P.A., 2003. Genetic engineering of crop plants for insect resistance. Current Science 84 (3), 321–329.

Reddy, D.C.L., 2000. Isolation and Molecular Characterization of *Bacillus thuringiensis* for Lepidopteran Insect Control (M. Sc. thesis). pp. 83.

Rico, E., Ballester, V., Mensua, J.L., 1998. Survival of two strains of *Phthorimaea operculella* (Lepidoptera: Gelechiidae) reared on transgenic potatoes expressing a *Bacillus thuringiensis* crystal protein. Agronomie 18, 151–155.

Roush, R.T., 1989. Designing resistance management programs: how can you choose? Pesticide Science 26, 423–441.

Roush, R.T., 1997. Managing risk of resistance in pests to insect-tolerant transgenic crops. In: Evans, P.M.G., Gibbs, M.J. (Eds.), Commercialization of Transgenic Crops: Risks, Benefits and Trade Considerations, Waterhouse. Cooperative Research Center for Plant science and Bureau of Statistics, Canberra, Australia, pp. 259–271.

Roush, R.T., 1998. Two-toxin strategies for management of insecticidal transgenic crops: can pyramiding succeed where pesticide mixtures have not? Philosophical Transanctions of Royal Society B: Biological Sciences 353, 1777–1786.

Rukmini, V., Reddy, C.Y., Venkateshwerlu, G., 2000. *Bacillus thuringiensis* crystal δ-endotoxin: role of proteases in the conversion of the protoxin to toxin. Biochimie 82, 109–116.

Salm, T.V., Bosch, D., Honee, G., Feng, I., Munsterman, E., Bakker, P., Stiekema, W.J., Visser, B., 1994. Insect resistance of transgenic plants that express modified *cry1A(b)* and *cry 1C* genes: a resistance management strategy. Plant Molecular Biology 26, 51–59.

Salma, H.S., Foda, M., Zaki, F., Ragaei, 1986. On the distribution of *Bt* and closely related *B. cereus* in Egyptian soils and their activity against cotton insects. Angewandte Zoologie 3, 257–267.

Sanyal, I., Singh, A.K., Kaushik, M., Amla, D.V., 2005. Agrobacterium-mediated transformation of chickpea (*Cicer arietinum* L.) with *Bacillus thuringiensis cry1Ac* gene for resistance against pod borer insect, *Helicoverpa armigera*. Plant Science 168, 1135–1146.

Sarmah, B.K., Deka, P.C., 2004. *Agrobacterium* mediated genetic transformation of chickpea (*Cicer arietinum*) for the development of resistance to pod borer and storage pests. Project report, ISCB- Indo-Swiss Collaboration. In: Biotechnology.

Schnepf, E., Crickmore, N., Van Rie, J., Lereclus, D., Baum, J., Feitelson, J., Zeigler, D.R., Dean, D.H., 1998. *Bacillus thuringiensis* and its pesticidal crystal proteins. Microbiology and Molecular Biology Reviews 62, 775–806.

Schnepf, H.E., 1995. *Bacillus thuringiensis* toxins: regulation, activities and structural diversity. Current Opinion in Biotechnology 6, 305–312.

Selvapandian, A., Reddy, V.S., Kumar, P.A., Tewari, K.K., Bhatnagar, R.K., 1998. Transformation of *Nicotiana tabacum* with a native *cry1Ia5* gene confers complete protection against *Heliothis armigera*. Molecular Breeding 4, 473–478.

Sharma, K.K., Anjaiah, V., 2000. An efficient method for the production of transgenic plants of peanut (*Arachis hypogaea* L.) through *Agrobacterium tumefaciens*-mediated genetic transformation. Plant Science 159, 7–19.

Sharma, H.C., Kumar, P.A., Seetharama, N., Hariprasad, K.V., Singh, B.U., 1999. Role of transgenic plants in pest management in sorghum. In: Symposium on Tissue Culture and Genetic Transformation of Sorghum, 23–28 Feb 1999. ICRISAT Center, Patancheru, Andhra Pradesh, India.

Shelton, A.M., Juliet, D., Tang, J.D., Roush, R.T., Metz, T.D., Earle, E.D., 2000. Field tests on managing resistance to *Bt*-engineered plants. Nature Biotechnology 18, 339–342.

Shelton, A.M., Zhao, J.Z., Roush, R.T., 2002. Economic, ecological, food safety and social consequences of the deployment of *Bt* transgenic plants. Annual Review of Entomology 47, 845–881.

Sivasubramanian, N., Federici, B.A., 1994. Method and Means of Extending the Host Range of Insecticidal Proteins. US patent 5143905.

Soltes-Rak, E., Kushner, D.J., Williams, D.D., Coleman, J.R., 1993. Effects of promoter modification on mosquitocidal *cryIVB* gene expression in *Synechococcus* sp. strain 7942. Applied and Environmental Microbiology 59, 2404–2410.

Steinhaus, E.A., 1951. Possible use of *Bacillus thuringiensis* as an aid in the biological control of the alfalfa caterpillar. Hilgardia 20, 359–381.

Stewart, C.N., Adang, M.J., All, J.N., Raymer, P.L., Ramachandran, S., Parrott, W.A., 1996. Insect control and dosage effects in transgenic canola containing a synthetic *Bacillus thuringiensis cry1Ac* gene. Plant Physiology 112, 115–120.

Strizhov, N., Keller, M., Mathur, J., Kalman, K.K., Bosch, D., Prudovsky, E., Schell, J., Sneh, B., Koncz, C., Zilberstein, A., 1996. A synthetic *cryIC* gene, encoding a *Bacillus thuringiensis* δ-endotoxin, confers *Spodoptera* resistance in alfalfa and tobacco. Proceedings of National Academy of Sciences of the United States of America 93, 15012–15017.

Surekha, C.H., Beena, M.R., Arundhati, A., Singh, P.K., Tuli, R., Gupta, A.D., Kirti, P.B., 2005. *Agrobacterium* mediated genetic transformation of pigeon pea (*Cajanus cajan* (L.) Millsp.) using embryonal segments and development of transgenic plants for resistance against *Spodoptera*. Plant Science 169, 1074–1080.

Tabashnik, B.E., Groeters, F.R., Finson, N., Johnson, M.W., 1994. Instability of resistance to *Bacillus thuringiensis*. Biocontrol Science and Technology 4, 419–426.

Tabashnik, B.E., Liu, Y.B., Maivar, T., Heckel, D.G., Masson, L., ferre, J., 1998. Insect resistance to *Bacillus thuringiensis* uniform or diverse? Philosophical Transactions of the Royal Society of London series B353, 1751–1756.

Tabashnik, B.E., 1994. Evolution of resistance to *Bacillus thuringiensis*. Annual Review of Entomology 39, 47–79.

Tailor, R., Tippet, J., Gibb, G., Pells, S., Pike, D., Jordan, L., 1992. Identification and charecterization of a novel *Bacillus thuringiensis* delta-endotoxin entomocidal to coleopteran and lepidopteran larvae. Molecular Microbiology 6, 1211–1217.

Tang, J.D., Shelton, A.M., van Rie, J., de Roeck, S., Moar, W.J., Roush, R.T., Peferoen, M., 1996. Toxicity of *Bacillus thuringiensis* spore and crystal protein to resistant diamondback moth (*Plutella xylostella*). Applied and Environmental Microbiology 62, 564–569.

Theunis, W., Aguda, R.M., Cruz, W.T., Decock, C., Pefeoen, M., Lambert, B., Bottrell, D.G., Gould, F.L., Litsinger, J.A., Cohen, M.B., 1998. *Bacillus thuringiensis* isolates from the Philippines: habitat, distribution, δ-endotoxin diversity and toxicity to rice stem borers (Lepidoptera: Pycalidae). Bulletin of Entomological Research 88, 335–342.

Udayasuriyan, V., Nakamura, A., Masaki, H., Uozumi, T., 1995. Transfer of an insecticidal protein gene of *Bacillus thuringiensis* into plant-colonising *Azospirillum*. World Journal of Microbiology and Biotechnology 11, 163–167.

Vaeck, M., Reynaerts, A., Hoftey, H., Jansens, S., DeBeuckleer, M., Dean, C., Zabeau, M., vanMontagu, M., Leemans, J., 1987. Transgenic plants protected from insect attack. Nature 327, 33–37.

Walker, D.R., All, J.N., McPherson, R.M., Boerma, H.R., Parrott, W.A., 2000. Field evaluation of soybean engineered with a synthetic *cry1Ac* transgene for resistance to corn earworm, soybean looper, velvetbean caterpillar (Lepidoptera: Noctuidae) and lesser cornstalk borer (Lepidoptera: Pyralidae). Journal of Economic Entomology 93 (3), 613–622.

Whiteley, H.R., Schnepf, H.E., 1986. The molecular biology of parasporal crystal body formation in *Bacillus thuringiensis*. Annual Review of Microbiology 40, 549–576.

Williams, S., Friedrich, L., Dincher, S., Carozzi, N., Kessmann, H., Ward, E., Ryals, J., 1993. Chemical regulation of *Bacillus thuringiensis* δ-endotoxin expression in transgenic plants. Bio/Technology 7, 194–200.

Wilson, W.D., Flint, H.M., Deaton, R.W., Fischhoff, D.A., Perlak, F.J., Armstrong, T.A., Fuchs, R.L., Berberich, S.A., Parks, N.J., Stapp, B.R., 1992. Resistance of cotton lines containing a *Bacillus thuringiensis* toxin to pink bollworm (Lepidoptera: Gelechiidae) and other insects. Journal of Economic Entomology 85, 1516–1521.

Wu, K.M., Guo, Y.Y., 2005. The evolution of cotton pest management practices in China. Annual Review of Entomology 50, 31–52.

Ye, G.Y., Tu, J., Hu, C., Datta, K., Datta, S.K., 2001. Transgenic IR72 with fused *Bt* gene *cry1Ab/cry1Ac* from *Bacillus thuringiensis* is resistance against four lepidopteran species under field conditions. Plant Biotechnology 18 (2), 125–133.

Yoder, P.E., Nelson, E.L., 1960. Bacteriophage for *Bacillus thuringiensis* Berliner and *Bacillus anthracis* Cohn. Journal of Insect Pathology 2, 198–199.

Zhao, J.Z., Li, Y.X., Collins, H.L., Cao, J., Earle, E.D., Shelton, A.M., 2001. Different cross resistance patterns in the diamondback moth (Lepidoptera: Plutellidae) resistant to *Bacillus thuringiensis* toxin Cry1C. Journal of Economic Entomology 94, 1547–1552.

Entomopathogenic Fungi

Kaushal K. Sinha[1], Ajoy Kr. Choudhary[2], Priyanka Kumari[2]

[1]Department of Botany, TM Bhagalpur University, Bhagalpur, India; [2]Department of Botany and Biotechnology, TNB College, Bhagalpur, India

1. INTRODUCTION

Chemical pesticides have been used by farmers for the control of insect pests for many decades; however, the effect on nontarget organisms, groundwater contamination, residues on food crops, and development of insect resistance to chemicals have forced scientists to focus on the development of alternative eco-friendly measures. Among them biocontrol is one of the most effective alternatives. Myco-biocontrol is the use of fungi in biological processes to lower the insect density with the aim to reduce disease-producing activity and consequently crop damage. Biological plant protection with entomopathogenic fungi has a key role in sustainable pest management program. Entomopathogens as biocontrol agents have several advantages when compared with conventional insecticides. These include low costs, high efficiency, safety for beneficial organisms, reduction of residues in the environment, and increased biodiversity in human-managed ecosystems (Lacey et al., 2001; Asi et al., 2013; Ortiz-Urquiza and Keyhani, 2013; Gul et al., 2014). Fungal biocontrol agents have a unique mode of infection. Unlike bacteria and viruses they do not need to be ingested; rather, they have contact action and they invade their hosts directly through the cuticle.

However, in modern agricultural practices pest problems are inevitable and occur largely because of simplified agroecosystems and also due to creation of less stable natural ecosystems. Present-day management of agricultural land has in fact disrupted the ecological forces that regulate potential pest species in the natural ecosystem. Those ecological forces include physicochemical conditions, food availability, predation, and competition. Cultivation of crops in monocultures provides ideal food resources, which in turn allows pest populations to achieve higher densities than they would in natural conditions. Also, use of broad-spectrum pesticides destroys the natural predators that keep pests under control. Although effective, chemical pesticides are expensive and provide only temporary relief, as the explosive reproductive and evolutionary capacities of insects allow them to develop mechanism resistance to these and other control strategies. Apart from dangers posed upon human and nontarget

Ecofriendly Pest Management for Food Security
http://dx.doi.org/10.1016/B978-0-12-803265-7.00015-4

organisms, they cause irreversible damage to the environment by disturbing the ecological balance. More than 500 arthropod species now show resistance to one or more types of chemicals. Other serious problems are caused by alien (nonnative) species that are accidently introduced to a country or continent and that escape their coevolved natural predators. Nonnative species of insects cause highly significant damage to crops, biodiversity, and landscapes, valued at billions of dollars per annum. There are also threats of emerging pests, such as new strains of plant pathogens that evolve to overcome varietal resistance in crops.

The entomopathogenic fungi fill an extremely important niche in microbial control of insect pests. Virtually all insect orders are susceptible to fungal diseases. Fungi are potential biocontrol agents mainly due to their high reproductive capabilities, target-specific activity, short generation time, and resting stage or saprobic phase–producing capabilities that can ensure their survival for a longer time when no host is present. However, a primary requirement for the use of an entomopathogenic fungus as a myco-biocontrol agent is the susceptibility of the insect and also virulence of the fungus. Virulence of the fungus depends on the selection of the stable strain with specific efficacy for the target host. Genetic improvement of fungi for myco-biocontrol is now possible through modern biotechnological tools. For biocontrol to become an integral part of modern agriculture, a few goals must be considered, which are the selection and development of a fermentation system for biomass production and development of a formulation and delivery system that is compatible with the microorganism's requirements as well as with common agricultural practices. Molecular biology techniques coupled with cloning of genes encoding putative pathogenesis determinants will create more potential strains to control pest populations. Further research is desirable to monitor the environmental fate of recombinant strains of fungus under variable environmental conditions.

Throughout the world there is immense pressure on the farmers on one hand to reduce the use of chemical pesticides without sacrificing yield or quality and also on the other hand the control of pests is becoming increasingly difficult due to development of pesticide-resistant pests. There is a need to develop alternative methods for integrated pest management (IPM). Entomopathogens contribute to the natural regulation of many populations of arthropods. Much of the research in this area concerns the causal agents of insect diseases and their exploitation for biological pest control.

There are approximately 750 species of fungi that cause infection in insects and mites. As a group they attack a wide range of insect and mite species, but individual species and strains of fungus are very specific. The fungi produce spores that infect their host by germinating on its surface and then growing on its body. Death occurs between 4 and 10 days, depending on the type of fungus and number of infecting spores. After death, the fungus produces thousands of new spores on the dead body, which disperse and continue their life cycle on new hosts.

2. ENTOMOPATHOGENIC FUNGI

Entomopathogenic fungi are among the first organisms to be used for biocontrol of pests. Biocontrol has a well-recognized success story and was initiated in 1762 with the introduction of the mynah bird (*Acridotheres tristis*) from India into Mauritius for control of the sugarcane red locust *Nomadacris septemfasciata*. Entomopathogenic fungi with its cosmopolitan existence and rich diversity present a sustainable solution toward IPM. These entomopathogens, due to their ecofriendliness and biopersistence, are preferred to kill insects at various stages of their life cycle. A very diverse array of fungal species is found from different classes that infect insects. These insect pathogenic species are found in a wide range of adaptations and infecting capacities including obligate and facultative pathogens. Entomopathogenic fungi are found in the divisions Zygomycota, Ascomycota, and Deuteromycota (Samson et al., 1988; Humber, 1998) (Table 1). Many of the promising genera of entomopathogenic fungi belong either to the class Entomophthorales in Zygomycota or to the class Hyphomycetes in Deuteromycota. Biology and ecology of entomopathogenic fungi have been well documented (Steinhaus, 1964; Samson et al., 1988; Balazy, 1993).

2.1 Entomopathogens of Different Phyla

2.1.1 Phylum Oomycota

These fungi have cellulose in their coenocytic hyphae (without chitin) and they have biflagellate zoospores. Sexual reproduction takes place either on the same hyphae or different hyphae between gametangia. They are parasites on plants and animals, however, some of the species are saprophytes. *Lagenidium giganteum* is parasitic on mosquito larvae. Some species of *Lagenidium* are pathogenic to crabs and other aquatic crustaceans (Hatai et al., 2000).

2.1.2 Phylum Chytridiomycota

In this fungal group the hyphae are coenocytic, the cell wall is made up of chitin, and zoospores are with a single flagellum. Based on rRNA phylogenetic comparisons, this group of fungus is considered to be basal. Insect-infecting Chytridiomycetes is of the order Blastocladiales genus *Coelomomyces*, in which approximately 70 insect pathogenic species have been described (Barr, 2001). They infect mostly Hemipterans and Dipterans. *Myriophagus* (Chytridiales) infects the pupae of Diptera and *Coelomycidium* (Blastocladiales) infects mosquitoes.

2.1.3 Phylum Zygomycota

Mycelium is multicellular, nonseptate, gametangia that after fusion form zygospores. Within the Zygomycota class Trichomycetes consist of species that are mostly related with insects. The order Entomophthorales contains more than 200 insect-infecting species.

2.1.4 Phylum Ascomycota and Dueteromycota

Mycelia are septate, haploid, and ascospores (sexual spores) are formed in the fruiting body, ascus. Normally eight ascospores are produced in each ascus. More than 300 entomopathogenic species are present in *Cordyceps*. Most common among them are *Aspergillus, Metarhizium, Hirsutella, Beauveria, Aschersonia, Culicinomyces, Lecanicillium, Paecilomyces, Tolypocladium*, etc.

2.1.5 Phylum Basidiomycota

There are reports of only few Basidiomycetes that are pathogenic to insects. *Uredinella* and *Septobasidium* are entomopathogenic (Samson et al., 1988).

TABLE 1   Some Important Entomopathogenic Fungi and Their Hosts

Fungi	Host	References
(A) PHYLUM OOMYCOTA		
Lagenidium giganteum	Mosquitoes	Chapman (1985)
Leptolegnia chapmanii	Mosquitoes	
(B) PHYLUM CHYTRIDIOMYCOTA		
Order Blascadiales	Diptrans	Chapman (1985)
Coelomomyces	Hemipterans	
Order Chytridiales	Pupae of Diptera	
Myriophagus		
Order Blastocladiales	Mosquitoes	
Coelomycidium	Black flies	
(C) PHYLUM ZYGOMYCOTA		
Order Entomophthorales		
Family Entomophthoraceae		
Botkoa apiculata	Aphids and other hemipterans	Humber (1989) and Balazy (1993)
Botkoa major	Diverse flies, Lepidopterans	
Entomophaga grylli	Grasshopper	Macleod et al. (1980)
Entomophaga calopteri	Grasshopper	
Entomophaga maimaiga	Gypsy moth larvae	Soper et al. (1988)
Pandora neoaphidis	Homoptera (Aphididae)	Balazy (1993)
Pandora delphacis	Homoptera (Delphacidae Cicadellidae)	
Pandora blunckii	Lepidoptera	
Pandora bullata	Diptera	
Zoophthora radicans	Many aphids	Latge and Papierok (1988)
	Leafhoppers	Papierok et al. (1984)
	Psyllids	
	Leaf rollers	
Zoophthora phytonomi	Clover Leaf weevil	USDA (1956)
Family Ancylistaceae		
Conidiobolus obscurus		
Conidiobolus thromboids	Aphids and other hemiptera	Latge and Papierok (1988)
Family Neozygitaceae		

TABLE 1 Some Important Entomopathogenic Fungi and Their Hosts—cont'd

Fungi	Host	References
Neozygites fresenii	Homoptera (Aphididae)	Keller (1997)
Neozygites parvispora	Thysanoptera	
PHYLUM DEUTEROMYCOTA		
Class Sordoriomycetes		
Order Hypocereals		
Family Cordicipitaceae		
Cordyceps militaris	Lepidoptera	Kobayasi (1941, 1982) and Kobayasi and Shimizu (1983)
Cordyceps sphecocephela	Hymenoptera (wasps)	
Family Clavicipitaceae		
Aschersonia aleyrodis	Whitefly	Rombach and Gillispie (1988)
Beauveria bassiana	Pine caterpillar	Ferron (1981)
	Potato Beetle	Ferron (1981)
	Corn Borer	Marcandier and Khachatourians (1987)
	Coding Moth	
	Grasshoppers	Billlings and Glenn (1911) and Camargo et al. (1985)
	Chinch bug	Bell and Hamalle (1970)
	Boll weevil	Gottwald and Tedders (1984)
	Cowpea curculio	Chapman (1985)
	Pecan weevil	Dunn and Mechalas (1963)
	Mosquitoes	
	Lygus bug	Rombach et al. (1986)
	Granary weevil	Lai et al. (1982)
	Brown plant hopper	Rombach and Gillispie (1988)
	Termites	Alves et al. (1988)
	Spider mite	Keller et al. (1986)
	Fire ants	Lecuona and Alves (1988)
Beauveria brongniartii	European cockchafer	Romback et al. (1986)
	Sugarcane borer	Rombach and Gillispie (1988)
Hirsutella citriformis	Brown plant hopper	
Hirsutella thompsonii	Citrus rust mite	Roberts and Wraight (1986)
Metarhizium anisopliae	Spider mite	Alves (1986) and Chapman (1985)

(Continued)

TABLE 1   Some Important Entomopathogenic Fungi and Their Hosts—cont'd

Fungi	Host	References
	Spittle bugs	Latch and Falloon (1976)
	Mosquitoes	Camargo et al. (1985) and Rombach and Gillispie (1988)
	Rhinoceros beetle	Bell and Hamalle (1970)
	Boll weevil	Gottwald and Tedders (1984)
	Black vine weevil	
	Cowpea curculio	Rombach et al. (1986)
	Pecan weevil	Rombach and Gillispie (1988)
Metarhizium flavoviride	Brown plant hopper	Ignoffo (1981)
	Black vine weevil	
Nomurarea rileyi	Velvetbean caterpillar	
	Army worms	
	Corn ear worm	
Phylum Basidiomycota	Scale insects	Samson et al. (1988)
Order Septobasidiales		
Septobasidium		

Some of the potential candidates for myco-biocontrol of insect pests are as follows:

2.2 Some Important Entomopathogens

2.2.1 *Genus*—Beauveria

Beauveria sp. is a filamentous fungus, belonging to the class Deuteromycetes. Some important species of this genus are *Beauveria bassiana, Beauveria brongniartii, Beauveria amorpha*, and *Beauveria caledonica. Beauveria bassiana* is a fungus that grows naturally in soil throughout the world and acts as a pathogen on various insect species, causing white muscardine disease; therefore, it belongs to the entomopathogenic fungi. Many of the isolates of *Beauveria* sp. are highly host specific. On the host, mycelium emerges through the host exoskeleton to form a dense white covering on the surface, occasionally forming synnemata. Microscopic examination showed that *B. bassiana* had conidiogenous cells with globose bases and extended and globose conidia, denticulate rachis, and globose conidia (<3.5 μm diameters). The spore balls represented dense clusters of large numbers of conidiogenous cells with unicellular, hyaline conidia. A host of agricultural and forest significance includes the Colorado potato beetle, the codling moth, and several genera of termites, bollworm, *Helicoverpa armigera* (Sandhu and Vikrant, 2004), *Hyblaeapara* and *Eutectona machaeralis*. As this fungus causes epizootic, it is being used worldwide as a biopesticide to control a number of pests such as termites, whitefly, and malaria-transmitting mosquitoes (Hamlen, 1979; Donald and McNeil, 2005).

As insecticides, the spores are sprayed on affected crops as an emulsified suspension or wettable powder. *Beauveria bassiana* in general is considered as a nonselective pesticide because it parasitizes a very high range of arthropod hosts. This entomopathogenic fungus is also applied against the European corn borer, pine caterpillar, and green leafhoppers.

2.2.2 *Genus*—**Verticillium**

Verticillium lecanii and Verticillium chlamydosporium are the important species of this genus. Fungus *V. lecanii* is widely distributed, which can cause large epizootic in tropical and subtropical regions, as well as in warm and humid environments (Nunez et al., 2008). This fungus is characterized by verticillate branching of conidiophores, i.e., the branches arise in whorls on the upper portions of conidiophore. It is an effective biocontrol agent against *Trialeurodes vaporariorum* in greenhouses (Kim et al., 2002). This fungus attacks nymphs and adults and sticks to the leaf underside by means of filamentous mycelium (Nunez et al., 2008). *Verticillium lecanii* is considered to control whitefly and several aphid species.

2.2.3 *Genus*—**Metarhizium**

The genus *Metarhizium* has three important species, *Metarhizium anisopliae*, *Metarhizium album*, and *Metarhizium flavoviride*. *Metarhizium anisopliae* is a potential entomopathogenic fungus. This genus is characterized by a branching pattern of the conidiophores, densely interwind, and alignment of conidiogenous cells with rounded to conical apisces, arranged in dense hymenium. Conidia are aseptate, cylindrical or ovoid, forming chains usually aggregated into prismatic or cylindrical columns or a solid mass of parallel chains, pale to bright green to yellow green in color. Pathogenicity of *M. anisopliae* has been tested on teak skeletonizer, *E. machaeralis*, and found to be a potential myco-biocontrol agent of teak pest (Sandhu et al., 2000).

2.2.4 *Genus*—**Nomuraea**

Few important species of this genus are *Nomuraea rileyi* and *Nomuraea atypicola*. *Nomuraea rileyi*, another potential insect-infecting fungus, is a dimorphic hyphomycete that can cause epizootic death in various insects. The mycelium is septate, white, with flocculent overgrowth, sparse in culture to dense on insects (often completely covering the hosts), usually becoming green, or purple-gray to purple as sporulation proceeds. Its conidiogenous cells are short, with blunt apices. The host specificity of *N. rileyi* and its eco-friendly nature encourage its use in insect pest management. Its mode of infection and development have been reported for several insects hosts such as *Trichoplusia* sp., *Heliothis zea*, *Plathypena scabra*, *Bombyx mori*, etc.

2.2.5 *Genus*—**Paecilomyces**

Paecilomyces is a genus of phylum Deuteromycota with two important species, *Paecilomyces fumosoroseus and Paecilomyces lilacinus*. Conidiophores are usually well developed synnematous in many species, septate bearing whorls of divergent branches and conidiogenous cells (phialides). The conidiogenous cell is flask to oval shaped or subglobose with distinct neck, borne singly or in whorls, conidia are one-celled, hyaline to light colored. *Paecilomyces* is a genus of nematophagous fungus that kills harmful nematodes by pathogenesis causing disease in nematodes.

Paecilomyces fumosoroseus (Wize) Brown and Smith (Seryczynska and Bajan, 1975) is one of the most important natural enemies of whiteflies worldwide and causes the sickness called "yellow muscardine." *Paecilomyces* sp. is strong epizootic against *Bemisia* and *Trialeurodes* spp.

both in greenhouse and field conditions. Tropical and subtropical agricultural soils have been reported to have greatest potential for its application as a biocontrol agent.

2.2.6 Genus—Hirsutella

Genus *Hirsutella* includes three important species, *Hirsutella thompsonii*, *Hirsutella gigantea*, and *Hirsutella citriformis*. This genus has been one of the most difficult members for identification among all major genera of fungal entomopathogens largely because of the huge number of species and high variability among these species. Synnemata, if present, is erect, cylindrical or slightly tapered, varying from short and verrucose to long and hair-like, unbranched to sparingly branched, or with many short side branches. Conidiogenous cells with swollen base are of subglobose shape. *Hirsutella thompsonii* is used for the control of citrus rust mite. This fungus is also pathogenic to Acarida, Lepidoptera, and Hemiptera groups of insects.

2.3 Transmission and Conidia Dispersal

Abundant spores of Hyphomycetes are transmitted by wind, rain, and also through invertebrates. Wind is a major source of transmission. Hyphae growing out of insect cadaver are also a major source of conidial dispersal. In some of the Hyphomycetes viz. *Metarhizium* and *Beauveria* spp. conidia are hydrophobic and are passively spread from infected cadavers. Entomophthoralean conidia are mostly actively discharged under hydrostatic pressure. After discharge, conidia are carried on wind or by co-occurring insects (Hemmati et al., 2001; Roy et al., 2001). If the primary conidia of entomophthoralean species fail to find suitable host substrates, then most of them form secondary conidia (capilliconidia), e.g., *Zoophthora* spp. High humidity and moisture necessary for germination and sporulation are depending sources for transmission of conidia. Another, fungal activity to infect multiple stages of life of an insect is helpful in disease spreading. Winged insects can spread spores and ultimately disease in insect populations (Dromph, 2001). Apart from those, some other natural enemies of insects and nontarget insects also become the source of fungal spore dispersal.

2.4 Life Cycle of Entomopathogens

Entomopathogens are found in the environment as spores or resting spores. Insects become infected by these fungi when they come into the contact with spores on the surface of plants, in the soil, in air as windborne particles, or on the bodies of dead insects. The life cycle of insect pathogenic fungi begins with the spore germination on the host cuticle. Adhesive processes involve both physical and chemical interactions. Spores infect the insects by penetrating through the insect cuticle, often joints or creases where the insect's protective covering is thinner. Once inside, the fungus grows in the insect hemocoel. Many fungi also produce toxins in the host that increases the speed of killing or checks competition from other microbes. After the insects die, the fungus grows out through the exoskeleton of the insect usually at thinner areas like joints or creases and begin to produce spores. The spores of hyphomycetes are spread passively by the action of the wind, rain, or contact with other hosts or animals in the environment. Spores of Entomophthorales create natural outbreaks that are ejected from the dead insects (Figures 1 and 2). Many species of fungi infect insects, which, however, enhance the spread of infection. Fungus-killed insects often have a "fuzzy" appearance because of their appearance

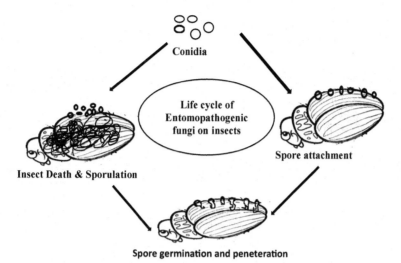

FIGURE 1   Sporulation of entomopathogens on insects.

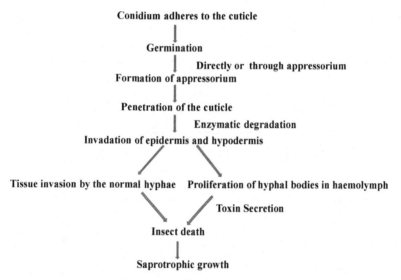

FIGURE 2   Diagrammatic representation of infection steps.

out of the exoskeleton. Spores that do not encounter hosts either die or persist in crop plants in the soil. Most spores are viable for a growing season. However, some species of fungus produce resting spores that become capable of infection under suitable environmental conditions.

2.5 Ecology

For developmental stages of entomopathogenic fungi high humidity is required, however, temperature is limiting within certain ranges (Meyling and Pell, 2006). The optimum temperature requirement for hyphomycetes is 20–30 °C, however, for Entomophthorales, it is 15–25 °C. In fact, optimum temperature for all stages of fungus development vary among the individuals of the same species. In addition, soil moisture, organic matter in soil, and pH are also important in determining infection level. Wind and sunlight are other abiotic factors influencing the infection process. Spore dispersal is assisted by wind. However, sunlight and ultraviolet (UV) rays can reduce the infection process.

The phylogenetic studies based on molecular systematics have revealed that fungal virulence toward insects has independently arisen several times (convergent evolution). They have evolved specialized mechanisms of biochemical degradation of insect integuments and also for outcoming insect defense compounds. Entomopathogenic fungi have diversity in nutritional mode varying from biotrophy (ability to derive nutrition only from living cells), to necrotrophy (utilizing dead insect tissues), and, thereby, they are homitrophy (initially biotrophy and then necrotrophy). They are also adapted to switch their nutritional mode from biotrophy to necrotrophy. These fungi exhibit a diverse array of adaptations to insect parasitism that include the general ability to overcome insect defenses. There is also substantial evidence that some entomopathogenic fungi, including *B. bassiana* (Ownley and Windham, 2007; Ownley et al., 2008a, 2008b) and species of *Lecanicillium* (Kim et al., 2007, 2008) are also antagonistic to plant pathogens. The mechanism of antagonism utilized by those fungi is antibiosis. Scale insects (Coccidae and Aleyrodidae) have the greatest diversity of fungal pathogens (Humber, 2008). These insects occur in dense and mainly immobile populations feeding on plants (Shahid et al., 2012). The sustained proximity between these insects, fungi, and other potential hosts may provide pathogenic fungi with the opportunity to move from plant to insect and beyond. In addition, various unexpected roles have been reported for fungal entomopathogens like their presence as endophytes, rhizosphere colonizers, plant growth-promoting fungi, mycoparasites, and plant disease antagonists.

Many fungi known as insect pathogens have been isolated as endophytes including *Acremonium*, *Beauveria*, *Cladosporium*, *Clonostachys*, and *Isaria* (Vega et al., 2008). There are reports that some fungal endophytes protect host cells against pathogens and herbivores (Arnold et al., 2003; Schulz and Boyle, 2005; Rudgers et al., 2007).

Soil is the reservoir of entomopathogenic fungi, Hypocreales belonging to the genera *Beauveria*, *Isaria* (Cordycipitaceae), and *Metarhizium* (Clavicipitaceae) (Meyling and Elenberg, 2007). In soil these fungi have to face both challenges and opportunities. Soil protects them from damage from solar radiation and acts as a buffer against extremes of temperature and water availability. Further, it is also the habitat of many potential insect hosts, some of which occur in high densities. Proximity to potential hosts is a factor in the evolution of fungal entomopathogenicity (Humber, 2008). Soil is in fact infused with antimicrobial metabolites secreted by microbes that can impair the ability of entomopathogenic fungi to infect their hosts. A dead or dying insect infected by an entomopathogenic fungus represents a potential source of energy for other microbes. Some of the hypocrealean entomopathogens produce secondary metabolites within their insect hosts that are supposed to help the fungus to outcompete other microbes during the saprophytic phase of insect utilization (Shahid et al., 2012).

There are reports that *Beauveria* and *Metarhizium* species that have infected and killed an insect in soil produce only limited somatic growth from fungus-infected cadavers suggesting that these fungi rely on insects rather than soil for carbon (Gottwald and Tedders, 1984; Inglis et al., 2001).

On the other hand there is also evidence that entomopathogenic fungi interact with the rhizosphere of the plant for growth or survival (St. Leger, 2008). However, it is still not clear whether this is a one-way interaction benefiting only microbial saprophytes or whether a mutualistic interaction has evolved in which the plant also benefits from the provision of mineral nutrients or protection from parasites and herbivores (Singh et al., 2004). For formulation and application of microbial biocontrol agents for management of insect pests, environmental variations within the agroecosystems need to be monitored properly. A more detailed investigation of pathogen insect ecology as well as other ecological interactions are required for developing consistent control strategy. Understanding of ecological factors will also be helpful in developing control measures through use of endophytic entomopathogenic fungi (Vidal, 2015).

2.6 Host Range and Specificity

The specificity of the insect-infecting fungus varies with genera, within genera, and also among species. Some of the entomopathogenic fungus like *M. anisopliae* and *B. bassiana* have wide host range (more than hundreds) including Coleoptera, Lepidoptera, Diptera, Homoptera, and Hymenoptera. However, some groups of insect-infecting fungi have limited host range. To formulate biopesticide isolate a strain is sometime more important than species. Specific strains have pronounced effects on the original host. Virulence of the same strain can also vary with the different stages of the insect's development.

3. FUNGAL INFECTION PROCESS

Fungi are a promising myco-biocontrolling agent for a number of crop pests. Several species belonging to orders Lepidoptera, Homoptera, Coleoptera, Hymenoptera, and Diptera are susceptible to various fungal infections. These fungi constitute a group with over 750 species that when dispersed in the environment cause fungal infection in insect populations. Fungi have a unique mode of infection; they reach the hemocoel through the cuticle or possibly through the mouthparts. Cuticle represents the first point of contact and barrier between the fungus and insect. The death of the fungus results from a combination of factors: mechanical damage resulting from tissue invasion, depletion of nutrient resources, mycoses, and production of toxins in the body (Figures 3 and 4). Entomopathogenic fungi have evolved mechanisms for adhesion and recognition of host surfaces. These adaptive responses include production of (1) hydrolytic, assimilatory, and detoxifying enzymes like lipase/esterase, catalyzes, cytochrome P_{450}, protease, and chitinase, (2) specialized infectious structures, e.g., appressoria or penetration tube, and (3) secondary and other metabolites that facilitate infection. On the other hand, besides immune responses insects have evolved a number of mechanisms to keep pathogens at bay. Those adaptions include (1) the production of epicuticular antimicrobial lipids, proteins, and metabolites, (2) shedding of cuticle during development, (3) biochemical-environmental adaptations such as induced fever, burrowing, and grooming, and (4) help of other competing microbes. All these adaptations of insects help them to stop the pathogen before it can breach the cuticle. Therefore, the selection pressure on the pathogen (for virulence) and target insects (for defense) have led to a coevolutionary arms race. Current evidence increasingly suggests that a major factor driving the coevolutionary arms

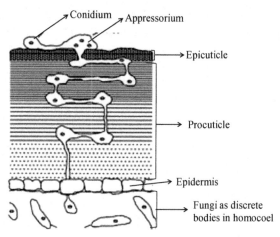

FIGURE 3   Infection process of *Beauveria bassiana*, according to Clarkson and Charnley (1996).

Entomopathogenic fungi- Mode of Action on Host insect.

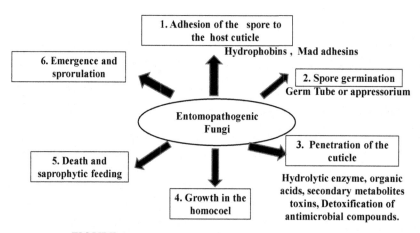

FIGURE 4   Steps of infection by entomopathogenic fungi.

race between the pathogen and the host occurred on the cuticular surface (Ortiz-Urquiza and Keyhani, 2013). Action on the surface represents defining interactions that lead either to mycoses by the pathogen or successful defense by the host.

3.1 Adhesion and Germination of Conidia

Attachment is a passive process with the aid of wind or water. Attachment of the spore (conidia or blastospores) to the cuticle surface of a susceptible host represents the initial event of establishment of mycosis. Penetration of cuticle by insect fungi can take place anywhere on the host cuticle, however, some preferential sites have been noted on various insects (Gabarty et al., 2014). Insect cuticle is a highly heterogenous structure that varies in composition even during the various stages of life. The outermost layer of cuticle is epicuticle followed by procuticle, which is divided into exo-, meso-, and endo-cuticular layer. The innermost layer is epidermis, which surrounds the internal structures.

It has been observed that dry spores of *B. bassiana* possess an outer layer composed of interwoven fascicles of hydrophobic rodlets, comprised of protein hydrophobins (Holder et al., 2007). This rodlet layer appears to be unique to the conidial stage and has not been detected on the vegetative cells. The adhesion of dry spores to the cuticle was suggested to be due to nonspecific hydrophobic forces exerted by the rodlets (Boucias et al., 1988). In *B. bassiana* two hydrophobins (Hyd1 and Hyd2) are responsible for rodlet layer assembly, which contribute to cell surface hydrophobicity, adhesion to hydrophobic surfaces, and virulence (Cho et al., 2007; Zhang et al., 2011). Two adhesion genes, *Mad1* and *Mad2* have been characterized in *M. anisopliae* (Wang and St. Leger, 2007). Loss of *Mad1* resulted in decreased adhesion to insect cuticle and also reduced germination, blastospore production, and virulence. It has been suggested that hydrophobins account for the relatively nonspecific (passive) adsorption stop and that *Mad2* adhesions perhaps are responsible for the more target-specific active stage (Ortiz-Urquiza and Keyhani, 2013).

Once a pathogen reaches and adheres to the host cuticle, it germinates, and further growth is profoundly influenced by the availability of nutrients, oxygen, water, pH, temperature, and also the effect of toxic host surface compound(s). St. Leger et al. (1989) have suggested that fungi with broad host range germinate in cultures in response to a wide range of nonspecific carbon and nitrogen sources. However, entomopathogenic fungi with restricted host range appear to have specific requirements for germination. Insect hydrocarbons, including aliphatic and methyl-branched alkanes have been shown to be growth substrates for *B. bassiana* (Napolitano and Juarej, 1997).

3.2 Infection Structure Formation

Insect-invading fungi infect the host by direct penetration of the cuticle. The outer cuticle (epicuticle) is very complex in structure that lacks chitin but contains phenol-stabilizing proteins and remains covered by a waxy layer containing fatty acids, lipids, and sterols (Hackmann, 1984). The inner cuticle (procuticle) forms a majority of the cuticle and contains chitin fibrils embedded into a protein matrix together with lipids and quinones (Neville, 1984). Protein accounts for a majority (70%) of the procuticle.

The penetration of fungi to the cuticle is accomplished by the germ tube or by the formation of an appressorium that attaches the cuticle and gives rise to a narrow penetration peg. Appressorium possibly represents an adaptation for concentrating physical and chemical energy over a very small area so that ingress can be achieved efficiently. Appressorium formation is influenced by host surface topography and also intracellular second messengers Ca_2^+ and cyclic AMP (cAMP).

3.3 Penetration of Host Cuticle

Entopathogenic fungi penetrate through the cuticle into the insect body to obtain nutrients for their growth and reproduction. Entry requires both mechanical pressure and enzymatic degradation. Penetration sites were often observed as dark, melanotic lesion in the epicuticle (Zacharuk, 1973). Proteins are major components of insect cuticle and a recyclable resource of the insects. Samuels and Paterson (1995) reported that both insects and entomopathogenic fungi produce a variety of cuticle-degrading proteases. Cuticle-degrading enzymes that are produced during penetration include proteases, lipases, and chitinase (Smith and Grula, 1982; Sanchez-Perez et al., 2014). A wide range of proteases have been identified including trypsin, chymotrypsin, esterase, collagenase, and chymoelastase. Among the first enzymes produced on the cuticle are endoproteases (PR1 and PR2) and aminopeptidase, which are, however, correlated with the formation of appressoria (Hepbrun, 1985). Cuticle active endoprotease acts as a key factor in the penetration process because 70% of the cuticle is comprised of protein. N-acetylglucosaminidase is produced at a slow rate as compared to proteolytic enzymes. Insect cuticle is a complex structure and thereby its penetration requires synergistic action of several different enzymes. For cuticular lipid degradation in *B. bassiana*, eight cytochrome P_{450} enzyme (CYP genes), four catalases, three lipases/esterases, long-chain alcohol and aldehyde dehydrogenases, and putative hydrocarbon carrier protein have been implicated to participate (Pedrini et al., 2010, 2013). In addition, environmental factors, e.g., temperature, humidity, and sunlight are likely to influence adhesion, germination, growth, and penetration on insect cuticle and their components. The fungus enters the hemocoel of the insect directly by penetrating the cuticle with the help of a range of extracellular enzymes (chitinases, lipases, esterases, and different classes of proteases) or through mouthparts. Penetration of cuticle requires synergistic action of several enzymes. After successful penetration, the fungus is then distributed into the hemolymph by formation of blastospores or as a discrete yeast-like structure and reaches to the insect's respiratory system in order to get maximum nutrition. Death of the insect results from a combination of factors, like mechanical damage resulting from tissue invasion, depletion of nutrients, and also due to production of toxins in the body of the insects (toxicosis).

3.4 Toxin Production

Deuteromycetes pathogens have been reported to produce different fungal toxins that have been implicated in insect health. These toxins have been shown to have diverse effects on various insect tissues (Table 2). The action of cytotoxins is suggested by cellular disruption prior to hyphae penetration. Charnley (1984) suggested that behavioral symptoms such as partial or general paralysis, sluggishness, and decreased irritability in fungal-infected insects are consistent with the action of neuromuscular toxins. Toxins such as beauvericin, beauverolides, bassianolide, and isarolides have been isolated from *B. bassiana* (Eyal et al., 1994). *Metarhizium anisopliae* has been reported to produce destruxins (DTXs) and cytochalasins in infected insects. DTX depolarizes lepidopteran muscle membrane by activating calcium channels and it also affects the function of insect hemocytes (Bradfish and Harmer, 1990).

Some of the important toxins produced by the entomopathogenic fungi are as follows.

3.4.1 Destruxins

Destruxins are cyclic depsipeptides consisting of five amino acids and a D-α hydroxyl acid (Suzuki et al., 1970; Pais et al., 1981) and were originally isolated from *M. anisopliae* (Kodaira, 1961, 1962). Currently, over 28 different destruxins have been described, mostly

TABLE 2 Toxins Produced by Different Entomopathogenic Fungi

Toxin	Fungi	Structure	Function
Efrapeptins	*Tolypocladium niveum* Syn. *Beauveria nivea*	Linear peptides rich in α-aminobutyric acid (Aib)	Inhibitors of mitochondrial oxidative phosphorylation and ATPase activity.
Destruxins	*Metarhizium anisopliae*	Cyclic depsipeptides consisting of five amino acids and D-α-hydroxy acid	Immunodepressant activity in insect and cytotoxic effect.
Beauvericin	*Beauveria bassiana*	Cyclic lactone trimer of amide of N-methyl L-phenylalanine and D-α-hydroxyisovaleric acid	Cytotoxic effect and insecticidal properties
Bassianolide	*B. bassiana Verticillium lecanii*	Cyclic polymer of D-α-hydroxy-isovaleryl L-N-methylleucinol	Acts as ionophore, toxic effect on insects
Leucinostatins	*Paecilomyces lilacinus Paecilomyces marquandii*	Linear peptide along with α-amino-isobutyric acid, L-leucine and β-alanine with three unusual amino acids.	Insecticidal activity by interfering with oxidative phosphorylation

from *Metarhizium* spp., with varying levels of activities against insects (Vey et al., 2001). The level of destruxin has been correlated with virulence (Al-Aidoors and Roberts, 1978) and host specificity (Amiri-Bisheli et al., 2000). Studies on activities of destruxins have also shown modulation of the host cellular immune system, including prevention of nodule formation (Vey et al., 2001) and inhibition of phagocytosis (Vilcinskas et al., 1977) among infected insects. Destruxins are produced after the mycelium grows inside the insect. Destruxins also have immune depressant activity in insects. Destruxin E was found to exhibit cytostatic and cytotoxic effect on mouse leukemia cells (Odier et al., 1987). Destruxins can also activate calcium channels in insect muscles (Samuels et al., 1988). The role of the destruxins may be to facilitate establishment of the pathogen in the host. These insect toxins apparently can be the cause of insect mortality after fungal infection (Samuels et al., 1988).

3.4.2 Beauvericin

Beauvericin has been isolated from *B. bassiana, P. fumosoroseus, Fusarium semitectum*, and *Fusarium moniliforme* var. *subglutinans* and plant pathogenic basidiomycetous fungus, *Polyporus sulphureus*. Chemically, beauvericin is cyclic lactone trimer of amide of N-methyl L-phenylalanine and D-α-hydroxyisovaleric acid. Beauvericin probably acts as an ionophore capable of making complexes with divalent cations. It is cytotoxic and has been reported to possess some insecticidal properties against mosquito larvae, blowflies, and Colorado potato beetle.

3.4.3 Bassianolide

Bassianolide is a toxic metabolite isolated from the strains of *B. bassiana* and *V. lecanii*, which were entomogenous on the cadavers of *B. mori* pupae. Chemically, bassianolide is a cyclic polymer of D-α-hydroxyisovaleryl L-N-methylleucinol. Bassianolide like other depsipeptides probably acts as ionophore. Bassianolide is toxic to insects (Eyal et al., 1994; Gupta et al., 1994). Injection of bassianolide, suspended in water, to *B. mori* larvae caused atonic symptoms at a dose of 2 μg/larvae. The toxin was lethal at a dose of >5 μg/larvae.

3.4.4 Leucinostatins

Leucinostatins have been isolated from submerged cultures of *P. lilacinus*, *Paecilomyces marquandii*, and *Paecilomyces farinosus*. Chemically, it is L-threo-b-hydroxy leucine, 2 amino-6-hydroxy-4 methyl-8-oxodecanoic acid (AHMOD) and *cis*-4-methyl-L-proline (Mori et al., 1982). Leucinostatins have antimicrobial properties against gram positive bacteria and a wide range of fungi (Fukushima et al., 1983a, 1983b) . This toxin has intraperitoneal and oral toxicity in mice. Roberts et al. (1991) observed that leucinostatins isolated as active principally from an extract of *P. farinsosus* showed insecticidal activity against Colorado potato beetle. This toxin also acts as an uncoupler of oxidative phosphorylation in mitochondria.

3.4.5 Efrapeptins

Efrapeptins are a complex mixture of peptide antibiotics produced by the fungi *Tolypocladium niveum* (Syn. *Tolypocladium inflatum*) and *Beauveria nivea*, a soil hyphomycetes. Efrapeptins showed toxicity against Colorado potato beetle, *Leptinotarsa decemlineata* (Coleoptera). All the peptides are strong inhibitors of mitochondrial oxidative phosphorylation and ATPase activity, when tested against preparations from entomopathogenic fungi (*M. anisopliae* and *T. niveum*). These peptides are probably catalytic-site competitive inhibitors that bind to the soluble F' part of the mitochondrial ATPase (Lardy et al., 1975).

3.5 Defense Systems in Insects

Insects have evolved a number of mechanisms to prevent the invasion of fungi. The defense arsenal of insects contains both passive structural barrier (cuticle) and a cascade of active responses to keep the pathogens at bay. The active responses include melanization, cellular reactions to recognize the nonself pathogen, production of protease inhibitors, and symbiotic and behavioral defenses against microbes (Figure 5).

3.5.1 Cuticular Lipid and Melanization

The hydrophobic nature of the epicuticle is generally considered a good substratum for adhesion of fungal spores. However, specific adaptations have been reported, such as the cuticular fatty amides of the booklouse, *Liposcelis bostrychophila* L., appear to prevent adhesion of conidia of entomopathogenic fungi to the insects (Lord and Howard, 2004). Cuticular lipids and aldehydes from the southern stinkbug, *Nezara viridula* L., have a fungistatic effect on *M. anisopliae* and cuticular extracts from *H. zea* Boddie display toxicity toward *B. bassiana* (Sosa-Gomez et al., 1997; Smith and Grula, 1982). The brassy willow leaf beetle, *Phratora vitellinae* (L.) releases volatile glandular secretions (salicylaldehyde), which exhibit toxicity against entomopathogenic fungi to help sanitize their microhabitats (Gross et al., 2008). Free fatty acids on the surface of various Lepidoptera species and fatty acids isolated from biting midge, *Forcipomyia nigra* Winnertz, were able to inhibit germination of different entomopathogenic fungi (Smith and Grula, 1982; Urbanek et al., 2012). It has been observed that cuticular pentane extracts derived from the European common cockchafer (*Melolontha melolontha* L.) inhibited spore germination and hyphal growth when tested against a nonpathogenic strain of *B. bassiana* but had no effect against a pathogenic strain (Lecuona et al., 1997).

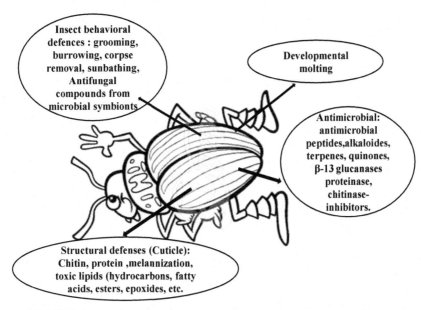

FIGURE 5 Possible defense mechanisms in insects against entomopathogens.

In response to fungal infection many insects have active responses where there is oxidation of phenolic compounds to dihydroxyphenylalanine, which in turn results in the production of melanic pigments. Melanine may partially shield cuticle from enzymatic attack or may be toxic to fungi. However, melanization is primarily an effective defense against weak or slow-growing pathogens but is ineffective against more virulent fungi.

3.5.2 Cellular Reactions

After penetrating through the cuticle and epidermis, the invading fungus is faced with the defense system of hemolymph. The responses to entomopathogenic fungi within the hemo-coel include phagocytosis, encapsulation, and nodulation. However, the effect of cellular reactions on fungus needs further investigation. The arbitrary infection method observed that hemocytes of the migratory grasshopper, *Melanoplus sanguinipes* encapsulated viable conidia of *B. bassiana*. However, it fails to suppress conidial germination within the nodule. Toxins and extracellular proteases of *B. bassiana* have been suggested to trigger evading encapsulation.

In addition, molting may provide a means for avoiding infection. Rapid ecdyses in aphids is supposed to be an important contributing factor to poor outcomes in applications of ento-mogenous fungi (Kim and Roberts, 2012). Some entomopathogenic fungi appear to inhibit molting of their hosts via oxidative inactivation of host.

3.5.3 Humoral Reactions

Humoral response represents a last ditch effort to thwart fungal pathogen. Vegetative growth of the entomopathogenic fungi in the insect hemocoel is common and is usually described by discrete yeast-like structures or hyphal bodies. This form of growth in contrast to the typical filamentous fungal mycelium allows the fungi to disperse rapidly and colonize the insect's circulatory system, increasing the surface area of nutrition. In response to fungal infection, insects elicit an acquired humoral immunity to subsequent infection (Bogus et al., 2007). Recognition of "nonself" is critical to initiation of the hemocytic defense reactions, and this selective response in insects depends on a specific chemical recognition by hemocytes (Boucias and Pendland, 1991). Serum and hemocyte cell membrane-bound lectins have been found in many insects. They could play a role in immune defense reactions since they agglutinate pathogens as well as fungi. Several species of Entomophthorales produce vegetative protoplasts (cells without cell walls) within the hemocoel, which in turn help the pathogen to escape detection by hosts' immune responses (Vilcinskas and Metha, 1997).

3.5.4 Protease Inhibitors and Other Defense Mechanisms

Protease inhibitors produced by the insects inhibit cuticle-degrading enzyme activities of pathogens. Such compounds have been isolated in the serum of *Anticarsia gemmatalis* larvae, which are resistant to infection by *N. rileyi* (Boucias et al., 1988). Surface antifungal defenses also include small-molecule toxins (including peptides) and proteins. Subterranean termites produce defensin-like antifungal peptides known as termicins. They have also been reported to secrete antifungal B-1,3-glucanase (also known as GNBPS, gram-negative bacteria binding protein) to the cuticle surface (Hamilton et al., 2011). There are reports that the inhibition of glucanase activity using D-δ-gluconolactone or transcriptional repression of the gene (through RNA interference) has resulted in increased susceptibility of termites to *M. anisopliae* (Bulmer et al., 2009; Hamilton and Bulmer, 2012).

Apart from cuticle and surface defensive compounds and proteins, the potential role of exogenous and endogenous (symbiotic) microbial communities in defending against fungal pathogens has been implicated (Zindel et al., 2011). Some insects bearing the endosymbiotic bacterium *Wolbachia* appear to display a generalized resistance to insect pathogens including *B. bassiana* (Panteleev et al., 2007). Some facultative symbionts manipulate host biology and behavior including feeding, reproduction, and homeostasis, and thus examining their potential protective effects against pathogenic fungi needs to be further investigated. *Blissus insularis* Barber have midgut crypts that contain high densities of *Burkholderia* bacterial species, which are known to produce potent antifungal compounds that help in enhancing resistance toward entomopathogenic fungi (Boucias et al., 2012).

In addition, interactions between entomopathogenic fungi and cuticle surface can also elicit specific behavioral responses in the insects meant to limit the ability of the pathogen to parasitize the host. Nest mate grooming is a well-recognized mechanism for minimizing potentially harmful microbes. Ants disinfect fungus-exposed brood through grooming and disinfection using formic acid and other chemical compounds. Temperature elevation either due to intrinsic biological mechanisms or by behavioral adaptations can also help to suppress infections (Heinrich, 1995). The phenomenon known as due to intrinsic biological mechanisms or by "behavioral fever" can be achieved by aggregation as seen in some ant and termite colonies or by heat seeking such as sunbathing as reported in different acridids (Gardener and Thomas, 2002).

Grooming, cuticular defense responses including encapsulation and termicin production, along with B-1,3-glucanase activity and elevated temperature, etc. are synergistic barriers to infection by entomopathogenic fungi (Chouvenc and Su, 2010). In response, entomopathogenic fungi must breach the cuticle, detoxify the host and/or endogenous microbial defenses, evade grooming and other behavioral responses, and potentially suppress other pathogens and parasites. Further research regarding genetic variability in cuticular degradative processes among virulent strains on the one hand, and cuticular defense responses in resistant insect species on the other hand, can be utilized for developing rational design strategies for improving the effectiveness of entomopathogenic fungi in field applications.

4. FORMULATION OF FUNGAL PESTICIDES

Virulence of the entomopathogenic fungal strains against target pests is an important characteristic and its ability for mass production at low cost is desirable. Over 750 different fungal species from at least 90 genera are known to be pathogenic to insects (Khachatourians and Sohail, 2008). However, a few fungal genera are well recognized as entomopathogens, including *Beauveria*, *Metarhizium*, *Isaria*, *Lecanicillium*, *Hirsutella*, and Entomophthorales. Various fungal-based products containing *B. bassiana*, *M. anisopliae*, *Isaria* spp., and *Lecanicillium* spp. have been developed for use against a wide variety of pests of field, forest, greenhouse, and household. At present commercial formulations of a wide range of fungi are available to farmers in most parts of the world. Extensive research is being conducted for improving fungal mass production and also to evaluate the effect of substrates, additives, and other factors on the virulence, viability, and thermotolerance of fungal spores (Kassa et al., 2008; Machado et al., 2010).

4.1 Mass Production

Diverse methods have been applied from time to time for practical application of entomopathogenic fungi to control insect pests. Cadavers of infected insects have been collected and introduced into fields either in the same season or after storage for further use in the next season. Insects have been artificially infected in the laboratory and released into pest populations while still active or as cadavers covered with sporulating fungus.

The most common methods, however, involve the production of the fungus on artificial media. For colonization, the placing of dishes of conidiating culture grown on artificial agar media in fields may be sufficient.

Most mass production schemes have utilized vegetable materials as the medium, e.g., rice or wheat bran, cracked barley, etc. Even the peat soil has been used with success in the People's Republic of China (PRC).

A millet-based fungal production system has been reported (Gouli et al., 2008). The millet provides nutrition to support fungal growth in the soil in the absence of an insect host. This type of formulation is comparatively simple. Millet grains are placed in a polyvinyl bag, which is soaked in water containing citric acid and boiled at 90 °C for 1 h, after which the mixture is autoclaved at 121 °C for 30 min. The bag is incubated at 25 °C with a 16:8 h (L/D) photoperiod for 3 weeks. The culture is then air dried until it reaches a moisture content of less than 5%. This process produces a granule with concentration of 1.1×10^8 *B. bassiana* conidia per gram of grain and a germination rate of 98.2% at 20 °C after 24 h.

Large-scale production of conidia in Russia is being accomplished by growing the fungi in a fermenter to produce large amounts of mycelia, which are then placed in shallow pans to a depth of approximately 1.0 cm where, after several days, conidia are produced.

Research has shown that the type of substrate on which the fungus is grown can affect the thermotolerance of the conidia produced. The amino acid composition of the medium is known to influence the storage and virulence of conidia.

Surface chemistry of entomopathogenic fungi is currently being investigated for proper understanding of their ecology. For example, *Isaria fumosorosea* blastospores were found to have a basic, monopolar, hydrophilic surface with an isoelectric point of 3.4 (Dunlap et al., 2005). The isoelectric point is the pH at which a surface or compound has a neutral charge. At a pH higher than 3.4, the surface charge of *I. fumosorosea* is negative and at more acidic pH the surface is positively charged. It is, therefore, the pH of the environment and of the formulation that can affect the charge of the spore surface and its ability to adhere to the insect cuticle or other surfaces.

Sometimes fungal-based baits are commonly produced on solid substrate that is attractive to the target insects. Insects with an active feeding stage in the soil are often targets for bait. However, some insects demonstrate an ability to detect the presence of entomopathogenic fungi and, thereby, avoid them. For example, the termites are generally susceptible to infection by *B. bassiana* and *M. anisopliae*, but it is difficult to achieve contact between the pest and sufficient concentrations of the fungus to obtain successful management. Some termite species, when they detect *M. anisopliae* in the colony, seal off the contaminated tunnel to prevent it from contacting them and others in the colony (Staples and Milner, 2000). However, when *M. anisopliae* was produced as a cellulose bait, repellency was overcome (Wang and Powell, 2004).

Another approach is to produce the mycelium in submerged culture. After culture the fungus is filtered from the medium to produce thin mats approximately 3–6 mm thick, which are treated with a sugar solution (10% maltose or sucrose) as desiccation protectant. The mat is air dried at room temperature and then milled to particles of 2 mm size. These particles when introduced into the field, in contact with the moisture, commence production of conidia. The particles of mycelium can produce new conidia for several consecutive days in the field, thereby, providing fresh inocula for a considerable period after introduction of the fungus. Mass-produced mycelium is filtered from the medium to produce mycelial mats, which are then dried at room temperature. Formulation of mycopesticide is one of the crucial issues as the half-life of the conidia of entomopathogens is less than 2 h in sunlight on plant leaves. Unlike chemical insecticides, use of organic solvents in formulation is not applicable to mycopesticides because they cause death of the organism. Microencapsulation of dried mycelium using alginate or pregelatinized starch is also in practice. However, most of the commercial formulations are in the form of wettable powders or emulsified oils.

4.2 Wettable Powders

Wettable powder formulations (WP) of entomopathogenic fungi are most commonly produced, which contain 50–80% technical powder, 15–45% filler, 1–10% dispersant, and 3–5% surfactant (Burges, 1998). These are mixed with water and applied to the foliage as a standard insecticidal spray with ultra-low volume or hydraulic applications. It can also be applied to the soil as a drench. Formulation is being developed using a wide array of compounds, each with unique properties that affect particular factors to enhance efficacy or spore survival. Additives are added that help to protect from UV light, enhance the ability of the spores to stick to the foliage (reduce spores from washing off), or increase humidity around the spore to promote germination under adverse environmental conditions. Photoactive dye, phloxine

B (0.005 g/m), has also been suggested (Kim et al., 2010b) to protect from phytotoxicity. Moisture has significant impact on conidial viability during storage and is also an important factor that affects shelf life. Potential application of moisture absorbent, like calcium chloride, silica gel, magnesium sulfate, white carbon, or sodium sulfate, has been suggested in 10% WP conidial powder formulations.

4.3 Oil Formulations

In most of the formulations, a variety of oils are added to improve the shelf life of fungal products and to increase their field efficacy in dry climate. The use of oil as carrier helps to wet the waxy hydrophobic or lipophilic surface of insects and most plant leaves. The simple addition of oil to spore powder increases the survival and viability of conidia (Moore et al., 1995). Oils also facilitate spore adhesion to the insect, stimulate germination, and assist in penetration by disrupting the waxy layer of the cuticle. Isoparaffinic hydrocarbon solvents, such as paraffin oil and mineral oil, are being used as carriers for oil-based formulations. Other oils such as vegetable oils, methyl oleate (wetting agent and emulsifier), corn, and cottonseed oil have also been suggested (Kim et al., 2011). The viability of conidia in corn oil was over 98% for up to 9 months of storage at 25 °C and 23% at 21 months of storage.

4.4 Methods of Application

Effectiveness of fungi as biocontrol agent depends upon the method of application. Several factors must be taken into consideration, including spore distribution and coverage to ensure that the pest comes in contact with the fungus and the climatic conditions that favor survival and germination of the fungus before, during, and after application. Four main methods of fungal applications are dipping the plant or roots, spraying the foliage, treating the soil, and indirect transmission by vectors.

Dipping of plant parts or roots in the spore suspension of entomopathogenic fungi is one of the possible applications. However, this method is not very common. Foliar spray of fungal spore to the site where the target insect occurs is commonly recommended to make fungal sprays in the late afternoon to maximize the higher relative humidity that occurs at that time, which facilitates fungal germination. In a formulation of liquid culture of *B. bassiana*, a thermotolerant enzyme chitinase was mixed in order to enhance the pathogenicity (Kim et al., 2010a and Kim et al., 2010c). In this formulation, polyoxyethylene 3 isotridecyl ether (TDE–3) is added as spreading agent.

Soil treatment of fungal spores is probably the most practicable application strategy. However, it is suitable only if the target insect has a susceptible soil phase. In soil, the moisture level is sufficient to promote conidial germination and mycelial growth allowing fungal inocula to sustain over time. Conidia in the soil are also protected from damaging UV light. Soil also helps to moderate large fluctuations in ambient temperatures, which often favors fungal growth and virulence. However, the antagonistic effect of soil pests (fungi, bacteria, arthropods) cannot be ruled out. There are reports that in nature entomopathogenic fungi are also readily transmitted by nonpest species to location where pests occur (Dromph, 2003).

Several commercial formulations of entomofungal pathogens have been developed for crop pest management. A list of commercially available fungi, target pests, and producer companies are presented in Table 3. Research on entomopathogens in India is carried out

TABLE 3   Commercial Formulations of Entomopathogenic Fungal Pesticides in Different Countries

Fungi	Target pest	Crop	Product and company	Formulation
1. *Aeschersorzia aleyrodis*	Whitefly	Cucumber, potato	Koppert/Holland	Wettable powder
2. *Beauveria bassiana*	Sucking insects	Cotton, glasshouse crops	Naturalis™, Tray Bioscience, USA	Liquid formulation
3. *B. bassiana*	Coffee berry borer	Coffee	Conidia, AgroEvo, Germany, Columbia	Suspendible granules
4. *B. bassiana*	Whiteflies/ Aphids/Thrips	Field crops	Mycontrol-WP/ Mycotech Corp, USA	Wettable powder
5. *B. bassiana*	Corn borer	Maize	Ostrinil/Natural plant protection/France	Microgranules of mycelium
6. *Beauveria brongniartii*	Scarab beetle larvae	Sugarcane	Betel/Natural plant protection/France	Microgranules of mycelium
7. *B. brongniartii*	Scarab beetle larvae	Pasture	Engerlingspilz/ Andermatt/Biocontrol	Barley kernels colonized with fungus
8. *Metarhizium anisopliae*	Termites	Houses	Bio-path™/ EcoScience/USA	Conidia on a mycelium placed in trap/chamber
9. *M. anisopliae*	Black vine weevil	Glasshouse ornamental crops, nursery stock houses	Biologie Bio 1020/ Bayer AG, Germany	Granules of mycelium
10. *M. anisopliae*	Locusts, Grasshoppers, Red-headed, cockchafer	Pasture/Turf	Biogreen/Biocare Technology Pvt. Ltd./ Australia	Conidia produced on grains.
11. *Pacilomyces fumosoroseus*	Whiteflies/ Thrips	Glasshouse crops	PFR-21™/WR Grace USA	Wettable powder
12. *P. fumosoroseus*	Mites	Wide range of crops	Priority/T. Stanes, India	Wettable powder
13. *Verticillium lecanii*	Aphids, Whiteflies and Thrips	Glasshouse crops	Mycotal/Koppert/ Netherlands	Wettable powder
14. *V. lecanii*	Aphids, whiteflies and Thrips	Glasshouse crops	Vertatec/Koppert/ Netherlands	Wettable powder

at institutes such as the National Bureau of Agricultural Insects Resources (NBAIR), Bangaluru, Directorate of Oilseed Research (DOR), Hyderabad, and Central Plantation Crop Research Institute (CPCRI), Kayangulam. At NBAIR, Bangaluru, there are ample collections of entomopathogenic fungi isolated from different insects from different geographic regions of the country. All those species were characterized with regard to morphology, internal transcribed spacer sequencing, cuticle-degrading enzyme-producing abilities, and virulence to insects. Commercially, the talc-based formulations of entomofungal pathogens are extensively marketed for pest management programs in India. These formulations have shelf life of 3–4 months. However, there is a need to develop formulations with longer shelf life of about 18 months. At NBAIR, Bangaluru, quality analysis of the products of microbial pesticides done periodically and monitored regularly reported that 50–70% of samples do not conform to the Central Insecticide Board Standards. Registration of microbial pesticides at Central Insecticide Board, India, is mandatory. Details of nontarget organisms, toxicological reports on laboratory animals, and ecotoxicological reports have to be provided along with other efficacy data at the time of registration. Entomopathogenic fungi should not be allergic, toxic, and pathogenic to nontarget organisms. However, the germplasm collections of entomopathogenic fungi have so far been made from the agroecosystem. A large wealth of microbial diversity from undisturbed natural ecosystem like forests, mangroves are yet to be exploited for biocontrol of insects pests.

Entomopathogenic fungi being a component of an integrated approach can provide significant and selective insect control. In the future there is need to monitor synergistic combinations of microbial control agents with other technologies (in combination with soft chemical pesticides, other natural enemies, resistant plants, chemigration, remote sensing, etc.) that will enhance the effectiveness and sustainability of integrated control strategies. Understanding how the insect pest or microbial pathogen interacts or survives in a given ecological condition is critical in developing appropriate formulations.

5. APPLICATION OF BIOTECHNOLOGY

Biotechnology provides an important tool for improving the virulence of the entomopathogenic fungi through gene manipulation. Genetically engineered strains of pathogens with wide host range, improved pathogenicity, reduced dose of inocula requirement to kill the pathogen, comparatively more adaptive to different environmental conditions, need to be developed. Genetic transformation systems for manipulation of virulent genes in vivo and in vitro have been established (Sandhu and Vikrant, 2006; Glare et al., 2008). Transformation techniques with selectable market have been used to isolate specific pathogenic genes to produce hyper-virulent *M. anisopliae* and *B. bassiana*. Considerable genetic diversity is found among various isolates of single species isolated from different hosts and localities. For this reason, it is desirable to collect, culture, and conserve germplasm from a wide variety of collection sites, hosts, and also from wide geographical regions. The entomopathogenic fungi currently play an important role in natural control of insect pests, and with the help of modern biotechnology, their level of use as microbial control agent should be sufficiently high to make them a significant component in insect control. In addition to naturally occurring fungal isolates, there is potential to modify genetic characteristics of entomopathogenic fungi through molecular biology techniques. One of the most common of these fungi, *M. anisopliae*, was transformed using such techniques (Goettel et al., 1990). A gene for benomyl resistance from another fungus, *Aspergillus nidulans*, was inserted by protoplast fusion into *Metarhizium*. The gene was expressed and caused *Metarhizium* to become resistant to fungicides. It is assumed that the transformation system can, therefore, be useful for other entomopathogenic fungi and other genes, thereby affording opportunity to produce improved strains of these fungi.

However, the availability of genomic data revealed that even in the closely related entomopathogenic fungi, like *B. bassiana* and *M. robertisii*, the molecular mechanisms of virulence vary although the overall processes of gene expression are similar (Xiao et al., 2012). This is also evident from phytogenetic studies suggesting that virulence toward insects has independently arisen several times (convergent evolution) (Ortiz-Urquiza and Keyhani, 2013). Overexpression of entomopathogen-driven hydrolytic enzymes (chitinase and protease), which are helpful in penetration of fungal hyphae through the cuticle, have yielded strains displaying increased virulence as compared to their wild-type parental isolates (Fang et al., 2009; Zhang et al., 2008). However, it has been found that in *M. anisopliae*, formulation of appressorium, hydrophobins and expression of cuticle-degrading proteases are triggered at low nutrient levels (St. Leger et al., 1992), demonstrating that the fungus senses environmental conditions or host cues at the initiation of infection.

Molecular biology studies have helped us to gain proper understanding of our knowledge about the phylogeny and species in entomopathogenic fungi. Polymerase Chain Reaction (PCR)-based Random Amplified Polymorphic DNA (RAPD) analysis has been used to separate sympatric isolates of *Beauveria*, and it was used to place different isolates of genetic groups (Meyling and Eilenberg, 2006). In an ecological investigation, PCR-based (RAPD) molecular systematics of *M. anisopliae* was studied in relation to source of isolations and geographical regions (Bidochka et al., 2001). A comparative study of protease zymography and RAPD analysis of 48 isolates of *B. bassiana* collected from Central India was made (Thakur et al., 2005). They observed that there was high genetic and biochemical diversity in the strains isolated from Lepidopteran and Coleopteran host insects. Various molecular techniques, like restriction fragment length polymorphism, amplified fragment length polymorphism, internal transcribed spacer, inter simple sequence repeats, simple sequence repeats, microsatellite, and rRNA gene sequencing have been used to evaluate fungal systematics and ecology (Obrnik et al., 2001; Inglis and Tigano, 2006; Arnold and Lewis, 2005; Bai et al., 2015). The ability to examine gene expression (protease and pathogenicity factor) during infection of various insect hosts was examined utilizing RT–PCR and ETS analysis (Cho et al., 2006). Through genetic engineering there is a need to develop super-virulent strains of entomopathogenic fungi, suited to different natural agroecosystem in order that they can withstand variable environmental conditions.

6. CONCLUSIONS

The application of entomopathogenic fungi in myco-biocontrol of insects is of immense significance because of environmental and food safety concerns. However, there are a number of constraints on the use of fungi as insecticides, such as: (1) the inoculum has short shelf life, (2) requires 2–3 weeks to kill the insects, (3) need for a fungicide-free period, (4) application needs to coincide with high relative humidity, and (5) availability of high-cost commercial formulations. In addition, the possibilities of contamination with mycotoxins (aflatoxins, trichothecenes, zearalenone, fumonisins, citrinin, etc.) produced by different saprophytic fungi as environmental pollutants cannot be ruled out. New techniques need to be developed that will help to manage the pests in a better way as the present pathogenesis mechanism of entomopathogens is slow and needs improvement. Modern techniques in molecular biology have the potential to manipulate desirable traits of these fungi to improve overall field activity. There is also the need for further understanding of the ecology of entomopathogens in order to develop rationally designed strategies for improving the effectiveness of these fungi in field applications.

Acknowledgments

AKC and P. Kumari are thankful to UGC (Sanction No. PSB—034/13-14) and Department of Environment & Forest, Government of Bihar (Sanction No. Yo. Wo. 29/215-849/14-17) for financial support.

References

Al-Aidoors, K., Roberts, D.W., 1978. Mutants of *Metarhizium anisopliae* with increased virulence toward mosquito larvae. Canadian Journal of Genetics and Cytology 11, 211–219.

Alves, S.B., 1986. Controle Microbiano de Insectos. Editora manole, Ltda., Sao Paulo, Brasil.

Alves, S.B., Stimac, J.L., Carmargo, M.T.V., 1988. Susceptibility of *Solenopsis invicta* Buren and *S. saevissima* Fr. Smith. to isolates of *Beauveria bassiana* (Bals.) Vuill. Anais da Soiedade Entomologica do Brasil 17, 379–388.

Amiri-Besheli, B., Khambay, B., Cameron, M., Deadman, M.L., Butt, T.M., 2000. Inter and intra-specific variations in destruxin production by the insect pathogenic *Metarhizium*, and its significance to pathogenesis. Mycological Research 104, 447–452.

Arnold, A.E., Lewis, L.C., 2005. Ecology and evolution of fungal edophytes and their roles against insects. In: Vega, F.E. (Ed.), Insect Fungal Associations: Ecology and Evolution. Oxford University Press, pp. 74–96.

Arnold, A.E., Mejia, L.C., Kyllo, D., Rojas, E.I., Maynard, Z., Robbins, N., Herre, E.A., 2003. Fungal endophytes limit pathogen damage in a tropical tree. Proceedings of the National Academy of Sciences 100, 15649–15654.

Asi, M.R., Bashir, M.H., Afzal, M., Zia, K., Akram, M., 2013. Potential of entomopathogenic fungi for biocontrol of *Spodoptera litura* Fabricius (Lepidoptera: Noctuidae). Journal of Animal and Plant Sciences 23 (3), 913–918.

Bai, N.S., Sasidharan, T.O., Remadevi, O.K., Dhramarajan, P., Pandian, S.K., Balaji, K., 2015. Morphology and RAPD analysis of certain potentially entomopathogenic isolates of *Metarhizium anisopliae* Metsch. (Deuteromycotina: Hypocreales). Journal of Microbiology and Biotechnology Research 5 (1), 34–40.

Balazy, S., 1993. Entomophthorales, Flora of Poland (Flora Polska), Fungi (Mycota), vol. 24. Polish Acad. Sci. W. Szafer Inst. Botany, Krakow, pp. 1–356.

Barr, D.J.S., 2001. Chytridiomycota. In: Systematics and Evolution. Springer Berlin, Heidelberg, pp. 93–112.

Bell, J.V., Hamalle, R.J., 1970. Three fungi tested for control of the cowpea curculio, *Chalcodermus aeneus*. Journal of Invertebrate Pathology 15, 447–450.

Bidochka, M.J., Kamp, A.M., Lavender, T.M., Dekoning, J., Decroos, J.N.A., 2001. Habitat association in two genetic groups of the insect – pathogenic fungus *Metarhizium aniosopliae*: uncovering cryptic species? Applied and Environmental Microbiology 67 (3), 1335–1342.

Billings, F.H., Glenn, P.A., 1911. Results of the Artificial Use of the White Fungus Disease in Kansas. U.S. Dept. Agric. Bull No. 107, 58 pp.

Bogus, M.I., Kedra, E., Bania, J., Szczepanik, M., Czygier, M., Jablonski, P., Pasztaleniec, A., Samborski, J., Mazgajska, J., Polanowski, A., 2007. Different defence strategies of *Dendrolimus Pini*, *Galleria mellonella*, and *Calliphora vicina* against fungal infection. Journal of Insect Physiology 53 (91), 909–922.

Boucias, D.G., Pendland, J.C., 1991. The fungal cell wall and its involvement in the pathogenic process in insect hosts. In: Latge, J.P., Boucias, D. (Eds.), Fungal Cell and Immune Response. Springer-Verlag Berlin, Heidelberg, pp. 303–316.

Boucias, D.G., Garcia-Maruniak, A., Cherry, R., Lu, H.T., Maruniak, J.E., Lietz, V.U., 2012. Detection and characterization of bacterial symbionts in the Heteropteran, *Blissus insularis*. FEMS Microbiology Ecology 82, 629–641.

Boucias, D.G., Pendland, J.C., Latge, J.P., 1988. Nonspecific factors involved in attachment of entomopathogenic deuteromycetes to host insect cuticle. Applied and Environmental Microbiology 54, 1795–1805.

Bradfish, G.A., Harmer, S.L., 1990. Omega conotoxinn GVIA and nifedipine inhibit the depolarizing action of the fungal metabolite, destruxin B on muscle from the tobacco budworm (*Heliothis virescens*). Toxicon 28 (11), 1249–1254.

Bulmer, M.S., Bachelet, I., Raman, R., Rosengaus, R.B., Sasisekharan, R., 2009. Targeting an antimicriobial effector function in insect immunity as a pest control strategy. Proceedings of the National Academy of Sciences of the United States of America 106, 12652–12657.

Burges, H.D., 1998. Formulation of Microbial Pesticides. Kluwer Academic Publishers, Dordrecht, The Netherlands.

Camargo, L.M.P.C.A., Filho, A.B., Cruz, B.P.B., 1985. Susceptibility of boll weevil (*Anthonomus grandis* Boheman) to fungi *Beauveria bassiana* (Bals.) *Vuillemin* and *Metarhizium anisopliae* (Metsch.). Sorokin Biologico 51, 205–208.

Chapman, H.C., 1985. Biological control of mosquitoes. Fresno: American Mosquito Control Association, Bulletin 6.

Charnley, A.K., 1984. Physiological aspects of destructive pathogenesis in insects by fungi: a speculative review. In: Andersonn, J.M., Rayner, A.D.M., Walton, D.W.H. (Eds.), Invertebrate–Microbial Interactions. British Mycological Society Symposium 6, Cambridge Univ. Press, London, pp. 229–271.

Cho, E.M., Kirkland, B.H., Holder, D.J., Keyhani, N.O., 2007. Phage display cDNA cloning and expression analysis of hydrophobins from the entompathogenic fungus *Beauveria* (*Cordyceps*) *bassiana*. Microbiology 153, 3438–3447.

Cho, E.M., Liu, L., Farmerie, W., Keyhani, N.O., 2006. EST analysis of cDNA libraries from the entomopathogenic fungus *Beauveria* (*Cordyceps*) *bassiana*. I. Evidence for stage-specific gene expression in aerial conidia, in vitro blastospores and submerged conidia. Microbiology 159 (9), 2843–2854.

Chouvenc, T., Su, N.Y., 2010. Apparent synergy among defense mechanisms in subterranean termites (Rhinotermitidae) against epizootic events: limits and potential for biological control. Journal of Economic Entomology 103, 1327–1337.

Clarkson, J.M., Charnley, A.K., 1996. New insights into mechanisms of fungal pathogenesis in insects. Trends in Microbiology 4 (5), 197–203.

Donald, G., McNeil, Jr., 2005. Fungus Fatal to Mosquito May Aid Global Was on Malaria. The New York Times.

Dromph, K.M., 2001. Dispersal of entomopathogenic fungi by collembolans. Soil Biology and Biochemistry 33 (15), 2047–2051.

Dromph, K.M., 2003. Cellembolans as vectors of entomopathogenic fungi. Pedobiologia 47, 245–256.

Dunlap, C.A., Biresaw, G., Jackson, M.A., 2005. Hydrophobic and electrostatic cell surface properties of blastospores of the entomopathogenic fungus *Paecilomyces fumosoroseus* colloids and surfaces. Biointerfaces 46, 261–266.

Dunn, P.H., Mechalas, B.T., 1963. The potential of *Beauveria bassiana* as a microbial insecticide. Journal of Insect Pathology 5, 451–459.

Eyal, J., Mabud, M.D.A., Fischbein, P.L., Walter, J.F., Osborne, L.S., Landa, Z., 1994. Assessment of *Beauveria bassiana* Nov. EO-1 stain, which produces a red pigment for microbial control. Applied Biochemistry and Biotechnology 44, 65–80.

Fang, W.G., Feng, J., Fan, Y.H., Zhang, Y.J., Bidochka, M.J., Leger, R.J.S., Pei, Y., 2009. Expression a fusion protein with protease and chitinases activities increase the virulence of the insect pathogen *Beauveria bassiana*. Journal of Invertebrate Pathology 102, 155–159.

Ferron, P., 1981. Pest control by the fungi *Beauveria* and *Metarhizium*. In: Burges, H.D. (Ed.), Microbial Control of Pests and Plant Diseases 1970–1980. Academic Press, London, pp. 465–482.

Fukushima, K., Arai, T., Mori, Y., Tsuboi, M., Suzuki, M., 1983a. Studies on peptide antibiotics, leminostalins I, separation, physio-chemical properties an biological activities of leucinostatins A and B. The Journal of Antibiotics 36 (12), 1606–1612.

Fukushima, K., Arai, T., Mori, Y., Tsuboi, M., Suzuki, M., 1983b. Studies on peptide antibiotics leucinostatins. II. The structures of leucinostatins A and B. The Journal of Antibiotics 36 (12), 1613–1630.

Gardener, S.N., Thomas, M.B., 2002. Costs and benefits of fighting infection in locusts. Ecological Research 4, 109–131.

Glare, T.R., Reay, S.D., Nelson, T.L., Moore, R., 2008. *Beauveria caledonica* is naturally occurring pathogen of forest beetles. Mycological Research 112 (3), 352–360.

Gabarty, A., Salem, H.M., Fouda, M.A., Abas, A.A., Ibrahim, A.A., 2014. Pathogenicity induced by the entomopathogenic fungi *Beauveria bassiana* and *Metarhizium anisopliae* in *Agrotis ipsilon* (Hufn.). Journal of Radiation Research and Applied Sciences 7, 95–100.

Goettel, M.S., Leger, R.T.S., Bhairi, S., 1990. Pathogenicity and growth of *Metarhizium anisopliae* stably transformed to benomyl resistance. Current Genetics 17 (2), 129–132.

Gottwald, T.R., Tedders, W.L., 1984. Colonization, transmission and longevity of *Beauveria bassiana* and *Metarhizium anisopliae* on pecan weevil larvae in the soil. Environmental Entomology 13, 557–560.

Gouli, V., Gouli, S., Skinner, M., Shternshis, M.V., 2008. Effect of the entomopathogenic fungi on mortality and injury level of western flower thrips, *Frankliniella occidentalis*. Archives of Phytopathology and Plant Protection 4 (2), 37–47.

Gross, J., Schumacher, K., Schmidtberg, H., Vilcinskas, A., 2008. Protected by fumigants: beetle perfumes in antimicrobial defense. Journal of Chemical Ecology 34, 179–188.

Gul, H.T., Saeed, S., Khan, F.Z.A., 2014. Entomopathogenic fungi as effective insect pest management tactic: a review. Applied Sciences and Business Economics 1, 10–18.

Gupta, S., Monitillot, C., Hwang, Y.S., 1994. Isolation of novel beavericin analogues from the fungus *Beauveria bassiana*. Journal of Natural Products 58, 733–738.

Hackman, R.H., 1984. Cuticle biochemistry. In: Bereiter, J. (Ed.), Biology of the Integument. Springer, Berlin, Germany, pp. 626–637.

Hamilton, C., Bulmer, M.S., 2012. Molecular antifungal defense in subterranean termites: RNA interference reveals in vivo roles of termicins and GNBPs against a naturally encountered pathogen. Developmental and Comparative Immunology 36, 372–377.

Hamilton, C., Lay, F., Bulmer, M.S., 2011. Subterranean termite prophylactic secretions and external antifungal defences. Journal of Insect Physiology 57, 1259–1266.

Hamlen, R.A., 1979. Biological control of insects and mites on European greenhouse crops: research and commercial implementation. Proceedings of the Florida State Horticultural Society 92, 367–368.

Hatai, K., Roza, D., Nakayama, T., 2000. Identification of lower fungi isolated from larvae of mangrove crab *Scylla serrata*, in Indonesia. Mycoscience 41 (6), 565–572.

Heinrich, B., 1995. Insect thermoregulation. Endeavour 19, 28–33.

Hemmati, F., Pell, J.K., McCartney, H.A., Deadman, M.L., 2001. Airborne concentrations of conidia of *Erynia neoaphidis* above cereal fields. Mycological Research 105, 485–489.

Hepbrun, H.R., 1985. Structure of the integument. In: Kerkut, G.A., Gilbert, L.I. (Eds.), Comprehensive Insect Physiology, Biochemistry and Pharmacology, vol. 3. Pergamon, Oxford, pp. 1–58.

Holder, D.J., Kirkland, B.H., Lewis, M.W., Keyhani, N.O., 2007. Surface characteristics of the entomopathogenic fungus *Beauveria* (*Cordyceps*) *bassiana*. Microbiology 153, 3448–3457.

Humber, R.A., 1989. Synopsis of revised classification for the entamophthorales (Zygomycotina). Mycotaxon 34, 441–460.

Humber, R.A., 1998. ARS collection of entomopathogenic fungal cultures. In: Lichtenfels, J.R., Kirkbride Jr., J.H., Chitwood, D.J. (Eds.), Systematic Collections of the Agricultural Research Service. USDA-ARS Misc. Publ. 1343, pp. 8–12.

Humber, R.A., 2008. Evolution of entomopathogenicity in fungi. Journal of Invertebrate Pathology 98, 262–266.

Ignoffo, C.M., 1981. The fungus *Nomuraea rileyi* as a microbial insecticide. In: Burges, H.D. (Ed.), Microbial Control of Pest and Plant Disease 1970–1980. Academic Press, London, pp. 513–538.

Inglis, G.D., Goettel, M.S., Butt, T.M., Strasser, H., 2001. Use of hypomycetous fungi for managing insect pests. In: Butt, T.M., Jackson, C.W., Magan, N. (Eds.), Fungi as Biocontrol Agents: Progress, Problems and Potential. CABI International/AAFC, Wallingford, United Kingdom, pp. 23–69.

Inglis, P.W., Tigano, M.S., 2006. Identification and taxonomy of some entomopathogenic *Paecilomyces* spp. (Ascomycota) isolates using rDNA-ITS sequences. Genetics and Molecular Biology 29 (1), 132–136.

Kassa, A., Brownbridge, M., Parker, B.L., Skinner, M., Gouli, V., Gouli, S., 2008. Why mass production of *Beauveria bassiana* and *Metarhizium anisopliae*. Mycological Research 112, 583–591.

Keller, S., 1997. The genus *Neozygites* (Zygomycetes, Entomophthorales) with special reference to species found in tropical regions. Sydowia 49, 118–146.

Keller, S., Keller, E., Auden, J.A.L., 1986. Ein Grossversuch zur Bekampfung des Maikafers (*Melolonotha melolonotha* L.) mit dem pilz *Beauveria brongniartii* (Sacc.) Petch. Bulletin de Pa Societe Entomologique Suisse 59, 47–56.

Khachatourians, G.G., Sohail, S.Q., 2008. Entomopathogenic fungi. In: Brakhage, A.A. (Ed.), Biochemistry and Molecular Biology, Human and Animal Relationships, The Mycota, vol. VI. second ed. Springer-Verlag, Berlin, Heidelberg.

Kim, J.J., Roberts, D.W., 2012. The relationship between conidial dose, moulting and insect development stage on the susceptibility of cotton aphid, *Aphis gossypii* to conidia of *Lecanicillium attenuatum*, an entomopathogenic fungus. Biocontrol Science and Technology 22, 319–331.

Kim, J.J., Goettel, M.S., Gillespie, D.R., 2007. Potential of *Lecanicillium* species for dual microbial control of aphids and the cucumber powdery mildew fungus, *Sphaerotheca fuliginea*. Biological Control 40, 327–332.

Kim, J.S., Je, Y.H., Woo, E.O., Park, J.S., 2011. Persistence of *Isaria fumosorosea* (Hypocreales: Cordycipitaceae) SFP-198 conidia in corn oil based suspension. Mycopathologia 171, 67–75.

Kim, J.S., Roh, J.Y., Choi, J.Y., Wang, Y., Shim, H.J., Je, Y.H., 2010a. Correlation of the aphicidal activity of *Beauveria bassiana* SFB-205 supernatant with enzymes. Fungal Biology 114 (1), 120–128.

Kim, J.S., Je, Y.H., Choi, J.Y., 2010b. Complementary effect of phloxine B on the insecticidal efficacy of *Isaria fumosorosea* SFP-198 wettable powder against green house white fly, *Trialeurodes vaporariorum* West. Pest Management Science 66, 1337–1343.

Kim, J.S., Skinner, M., Gouli, S., Parker, B.L., 2010c. Influence of tap-watering on the movement of *Beauveria bassiana*, GHA (Deuteromycota: Hyphomycetes) in potting medium. Crop Protection 29 (6), 631–634.

Kim, J.T., Geottel, M.S., Gillespie, D.R., 2008. Evaluation of *Lecanicillium longisporum*, vertalec for simultaneous suppression of cotton aphid, *Aphis gossypii* and cucumber powdery mildew, *Sphaerotheca fuliginea*, on potted cucumbers. Biological Control 45, 404–409.

Kim, J.T., Lee, M.H., Yoon, C.S., Kim, H.S., Yoo, T.K., Kim, K.C., 2002. Control of cotton aphid and greenhouse whitefly with a fungal pathogen. Journal of National Institute of Agricultural Science and Technology 7–14.

Kobayasi, Y., 1941. The genus *Cordyceps* and its allies. Science Reports of the Tokyo Bunrika Daigaku Section (B) 5, 53–260.

Kobayasi, Y., Shimizu, D. (Eds.), 1983. Iconography & Vegetable Wasps and Plant Worms. Hoikusha Publ. Co., Osaka.

Kobayasi, Y., 1982. Keys to the taxa of the genera *Cordyceps* and *Torrubiella*. Transactions of the Mycological Society of Japan 23, 329–364.

Kodaira, Y., 1961. Toxic substances to insects produced by Aspergillus ochraceus and Oopsora destructor. Agricultural and Biological Chemistry 25 (3), 261–262.

Kodaira, Y., 1962. Studies on the new substances to insects, destruxin A and B, produced by *Oospora destructor*. Part I. Isolation and purification of destruxin A and B. Agricultural and Biological Chemistry 26 (1), 36–42.

Lacey, L.A., Frutos, R., Kaya, H.K., Vail, P., 2001. Insect pathogens as biological control agents. Do they have a future? Biological Control 21, 230–248.

Lai, P.Y., Tamashiro, M., Fujii, J.K., 1982. Pathogenicity of six strains of entomogenous fungi *Coptotermes formosanus*. Journal of Invertebrate Pathology 39, 1–5.

Lardy, H., Reed, P., Lin, C.H.C., 1975. Antibiotic inhibitors of mitochondrial ATP synthesis. Biology Energy Transduction Federation Proceedings 34 (8), 1707–1710.

Latch, G.C.M., Fallon, R.E., 1976. Studies on the use of *Metarhizium anisopliae* to control *Oryctes rhinoceros*. Entomophaga 21, 39–48.

Latge, J.P., Papierok, B., 1988. Aphid pathogens. In: Minks, A.K., Harzewijn, P. (Eds.), Aphids: Their Biology, Natural Enemies and Control. Elesevier, pp. 323–335.

Lecuona, R., Clement, J.L., Riba, G., Joulie, C., Juarez, P., 1997. Spore germination and hyphal growth of *Beauveria* spp. on insect lipids. Journal of Economic Entomology 90, 119–123.

Lecuona, R.E., Alves, S.B., 1988. Efficiency of *Beauveria bassiana* (Bals) Vuill., *B.brongniartii* (Sacc.) Petch and granulose virus on *Diatraea saccharalis* (F. 1794) at different temperatures. Journal of Applied Entomology 105, 223–228.

Lord, J.C., Howard, R.W., 2004. A proposed role for the enticular fatty amides of *Liposcelis bostrychophila* (Psocoptera: Liposcelidae) in preventing adhesion of entomopathogenic fungi with dry conidia. Mycopathologia 158, 211–217.

Machado, A.C.R., Monteiro, A.C., Belasco de Almeida, A.M., Martins, M.I.E.G., 2010. Production technology for entomopathogenic fungus using a biphasic culture system. Pesquisa Agropecuaria Brasileira 45, 1157–1163.

Macleod, D.M., Tyrrell, B., Welton, M.A., 1980. Isolation and growth of the grasshopper pathogen, *Entomophthora grylli*. Journal of Invertebrate Pathology 36, 85–89.

Marcandier, S., Khachatourians, G.G., 1987. Susceptibility of the migratory grasshopper, *Melanoplus sanguinipes* to *Beauveria bassiana*: influence of relative humidity. The Canadian Entomologist 119, 901–907.

Meyling, N.V., Eilenberg, J., 2006. Isolation and characterisation of *Beauveria bassiana* isolates from phylloplanes of hedgerow vegetation. Mycological Research 110 (2), 188–195.

Meyling, N.V., Pell, J.K., 2006. Detection and avoidance of an entomopathogenic fungus by a generalist insect predator. Ecological Entomology 31 (2), 162–171.

Meyling, N., Eilenberg, J., 2007. Ecology of the entomopathogenic fungi *Beauveria bassiana* and *Metarhizium anisopliae* in temperate agroecosystems: Potential for conservation biological control. Biological Control 43, 145–155.

Moore, O., Bateman, R.P., Carey, M., Prior, C., 1995. Long-term storage of *Metarhizium flavoviride* conidia in oil formulation for the control of locusts and grasshoppers. Biocontrol Science and Technology 5, 193–199.

Mori, Y., Tsuboi, M., Suzuki, M., Fukushima, K., Aral, T., 1982. Structure of leucinostatin A, new peptide antibiotic from *Paecilomyces lilacinus*. Journal of the Chemical Society, D, Chemical Communications 94–96.

Napolitano, R., Juarej, M.P., 1997. Entomopathogenous fungi degrade epicuticular hydrocarbons of *Triatoma infestans*. Archives of Biochemistry and Biophysics 344, 208–214.

Neville, A.C., 1984. Cuticle: organization. In: Bereitor, J., et al. (Ed.), Biology of the Intgegument. Springer, Berlin, Germany, pp. 611–625.

Nunez, E., Iannacone, J., Gomez, H., 2008. Effect of two entomopathogenic fungi in controlling *Aleurodicus cocois* (Curtis, 1846) (Hemiptera: Aleyrodidae). Chilean Journal of Agricultural Research 68, 21–30.

Obrnik, M., Jirku, M., Dolezel, O., 2001. Phylogeny of mitosporic entomopathogenic fungi: is the genus *Paecilomyces* polyphyletic? Canadian Journal of Microbiology 47 (9), 813–819.

Odier, F., Vago, P., Quiot, T.M., Devauchelle, G., Bureau, J.P., 1987. Etude cytometrique des effects de la destruxine E surds cellules leucemiques de souris. Comptes Rendus del' Academic des Sciences 305, 575–578.

Ortiz-Urquiza, A., Keyhani, O., 2013. Action on the surface: entomopathogenic fungi versus the insect cuticle. Insects 4, 357–374.

Ownley, B.H., Windham, M., 2007. Biological control of plant pathogens. In: Trigiano, R.N. (Ed.), Plant Pathology: Concepts and Laboratory Exercises, second ed. CRC Press, Boca Raton, Florida, pp. 423–436.

Ownley, B.H., Dec, M.M., Gwinn, K.D., 2008a. Effect of conidial seed treatment rate of entomopathogenic *Beauveria bassiana* 11-98 on endophytic colonization of tomato seedlings and control of *Rhizoctonia disease*. Phytopathology 98, 5118.

Ownley, B.H., Griffin, M.R., Klingeman, W.E., Gwinn, K.D., Moulton, J.K., Pereira, R.M., 2008b. *Beauveria bassiana*: endophytic colonization and plant disease control. Journal of Invertebrate Pathology 98, 267–270.

Pais, M., Das, B.C., Ferron, P., 1981. Depsipeptides from *Metarhizium anisopliae*. Phytochemistry 20 (4), 715–723.

Panteleev, D.Y., Gloryacheva, I.I., Andrianov, B.V., Reznik, N.L., Lazebny, O.E., Kulikov, A.M., 2007. The endosymbiotic bacterium *Wolbachia* enhances the nonspecific resistance to insect pathogens and alters behaviour of *Drosophila melanogaster*. Russian Journal of Genetics 43, 1066–1069.

Papierok, B., Valadao, L., Torres, B., Arnault, M., 1984. Contribution a l'etude de la specificite parasitaire du champignon entomopathogene *Zoophthora radicans* (Zygomycetes, Entomophthorales). Entomophaga 29, 109–119.

Pedrini, N., Zhang, S., Juarez, M.P., Keyhani, N.O., 2010. Molecular characterization and expression analysis of a suite of cytochrome p450 enzymes implicated in insect hydrocarbon degradation in the entomopathogenic fungus *Beauveria bassiana*. Microbiology 156, 2549–2557.

Pedrini, N., Ortiz-Urquiza, A., Huarte-Bonnet, C., Zhang, S., Keyhani, 2013. Targeting of insect epicuticular lipids by the entomopathogenic fungus, *Beauveria bassiana*: hydrocarbon oxidation within the context of a host-pathogen interaction. Frontiers in Microbiology 4, 24.

Roberts, D.W., Wraight, S.P., 1986. Current status on the use of insect pathogens as biocontrol agents in agriculture fungi. In: Samson, R.A. (Ed.), Fundamental and Applied Aspects of Invertebrate PathologyWageningen: Invertebrate Pathology, pp. 510–513.

Roberts, D.W., Fuxa, J., Gaugler, J., Goettel, M., Jaques, R., Maddox, J., 1991. Use of pathogens in insect control. In: Pimentel, D. (Ed.), Handbook of Pest Management in Agriculture, vol. 2. second ed. CRC Press, Boca Raton, FL, pp. 243–278.

Rudgers, J.A., Holah, J., Orr, S.P., Clay, K., 2007. Forest succession suppressed by an introduced plant-fungal symbiosis. Ecology 88, 18–25.

Rombach, M.C., Gillespie, A.T., 1988. Entomogenous hyphomycetes for insect and mite control on greenhouse crops. Biocontrol News and Information 9, 7–18.

Rombach, M.C., Aguda, R.M., Shepard, B.M., Roberts, D.W., 1986. Infection of the rice brown planthopper, *Nilaparvata lugens* (Homoptera: Delphacidae) by field application of entomopathogenic hyphomycetes (Deuteromycotina). Environmental Entomology 15, 1070–1073.

Roy, H.E., Pell, J.K., Alderson, P.G., 2001. Targeted dispersal of the aphid pathogenic fungus *Erynia neoaphidis* by the aphid predator *Coccinella septempunctata*. Biocontrol Science and Technology 11, 99–110.

Samson, R.A., Evans, H.C., Latg, J.P., 1988. Atlas of Entomopathogenic Fungi. Springer, Berlin Heidelberg, New York.

Samuels, R.I., Reynolds, S.E., Charnley, A.K., 1988. Calcium channel activation of insect muscle by destruxins, insecticidal compounds produced by the entomopathogenic fungus. *Metarhizium anisopliae*. Comparative Biochemistry and Physiology 10 (2), 403–412.

Samuels, R.I., Paterson, I.C., 1995. Cuticle degrading proteinase from insect moulting fluid and culture filtrate of entomopathogenic fungi. Comparative Biochemistry and Physiology 110, 661–669.

Sanchez-Perez, L.C., Barranco-Floriab, J.E., Rodriguez-Navarro, S., Cervantes-Mayagoitia, J.F., Ramos-Lopez, M.A., 2014. Enzymes and entomopathogenic fungi: advances and insights. Advances in Enzyme Research 2, 65–76.

Sandhu, S.S., Vikrant, P., 2004. Myco-insecticide: control of insect pests. In: Gautam, S.P. (Ed.), Microbial Diversity: Opportunities & Challenges. India Publishers, New Delhi, India.

Sandhu, S.S., Vikrant, P., 2006. Evaluation of mosquito larvicidal toxins in the extra cellular metabolites of two fungal genera *Beauveria* and *Trichoderma*. In: Bagyanaryana, G. (Ed.), Emerging Trends in Mycology, Plant Pathology and Microbial Biotechnology Hyderabad, India.

Sandhu, S.S., Rajak, R.C., Hasija, S.K., 2000. Potential of entomopathogens for the biological management of medically important pest: progress and prospect. Glimpses in Plant Sciences 2000, 110–117.

Schulz, B., Boyle, C., 2005. The endophytic continuum. Mycological Research 104, 601–686.

Seryczynska, H., Bajan, C., 1975. Defensive reactions of L3, L4 larvae of the Colorado beetle to the insecticidal fungi *Paecilomyces farinosus* (Dicks) Brown et Smith, *Paecilomyces fumoso-roseus* (Wize), *Beauveria bassiana* (Bols/Vuill). (Fungi Imperfecti: Moniliales). Bulletin del' Academic Polonaise des Sciences Seriedes Sciences Biologiques 23, 267–271.

Shahid, A.A., Rao, A.Q., Bakhsh, A., Husnain, T., 2012. Entomopathogenic fungi as biological controllers. New insights into their virulence and pathogenicity. Archives of Biological Science Belgrade 64 (1), 21–42.

Singh, B.K., Millard, P., Whiteley, A.S., Murrell, J.C., 2004. Unravelling rhizosphere microbial intersections opportunities and limitations. Trends in Microbiology 12, 386–393.

Smith, R.J., Grula, E.A., 1982. Toxic components on the larval surface of the corn-earworm (*Heliothis zea*) and their effects on germination and growth of *Beauveria bassiana*. Journal of Invertebrate Pathology 39, 15–22.

Soper, R.S., Shimazu, M., Hunber, R.A., Ramos, M.E., Hajek, A.E., 1988. Isolation and characterization of *Entomophaga maimaiga* sp. nov., a fungal pathogen of gypsy moth, *Lymantria dispar*, from Japan. Journal of Invertebrate Pathology 51, 229–241.

Sosa-Gomez, D.R., Boucias, D.G., Nation, J.L., 1997. Attachment of *Metarhizium anisopliae* to the southern green stink bug *Nezara viridula* cuticle and fungistatic effect of cuticular lipids aldehydes. Journal of Invertebrate Pathology 69, 31–39.

St. Leger, R.J., 2008. Studies on adaptation of *Metarhizium anisopliae* to life in the soil. Journal of Invertebrate Pathology 98, 271–276.

St. Leger, R.J., Allee, L.L., May, B., Staples, R.C., Roberts, D.W., 1992. Worldwide distribution of genetic variation among isolates of *Beauveria* spp. Mycological Research 96, 1007–1015.

St. Leger, R.J., Butt, T.M., Staples, R.C., Roberts, D.W., 1989. Synthesis of proteins including a cuticle-degrading protease during differentiation of the entomopathogenic fungus *Metarhizium anisopliae*. Experimental Mycology 13, 253–262.

Staples, T.A., Milner, R.J., 2000. A laboratory evaluation of the repellency of *Metharizium anisopliae* conidia to *Coptotermes lacteus* (Isoptera: Rhinotermitidae). Sociobiology 36, 1–16.

Steinhaus, E.A., 1964. Microbial disease of insects. In: Debach, P. (Ed.), Biological Control of Insect Pest and Weeds. Chapman and Hall, London, pp. 515–547.

Suzuki, A., Taguchi, H., Tamara, S., 1970. Isolation and structure elucidation of three new insecticidal cyclodepsipeptides, destruxins C and D and desmethyldestruxin B, produced by *Metarhizium anisopliae*. Agricultural and Biological Chemistry 3 (5), 813–816.

Thakur, R., Rajak, R.C., Sandhu, S.S., 2005. Biochemical and molecular characteristics of indigenous strains of the entomopathogenic fungus *Beauveria bassiana* of Central India. Biocontrol Science and Technology 15 (7), 733–744.

U.S. Dept. Agriculture, 1956. The clover leaf weevil and its control. Farmer's Bulletin 1484.

Urbanek, A., Szadiewski, R., Stepnowski, P., Boros Majewska, J., Gabriel, I., Dawgul, M., Kamysz, W., Sosnowska, D., Golebiowski, M., 2012. Composition and antimicrobial activity of fatty acids detected in the hygroscopic secretion collected from the secretory setae of larvae of the biting midge *Forcipomyia nigra* (Diptera: Ceratopogonidae). Journal of Insect Physiology 58, 1265–1276.

Vega, F.E., Posada, F., Aime, M.C., Pava-Ripoll, M., Infante, F., Rehner, S.A., 2008. Entomopathogenic fungal endophytes. Biological Control 46, 72–82.

Vey, A., Hougland, R., Butt, T.M., 2001. Toxic metabolites of fungal biocontrol agent. In: Butt, T.M. (Ed.), Fungi as Biocontrol Agents. CAB-International, New York, pp. 311–345.

Vidal, S., 2015. Entomopathogenic fungi as endophytes: plant–endophyte–herbivore interactions and prospects for use in biological control. Current Science 108.

Vilcinskas, A., Metha, V., 1997. Effect of the entomopathogenic fungus *Beauveria bassiana* on the humoral immune response of *Galleria mellonella* larvae (Lepidoptera: Pyralidae). European Journal of Entomology 94, 461–472.

Vilcinskas, A., Matha, V., Gotz, P., 1977. Inhibition of phagocylic activity of plasmatocytes isolated from *Galleria mellonella* by entomogenous fungi and their secondary metabolites. Journal of Insect Physiology 43 (1), 33–46.

Wang, C.L., Powell, J.E., 2004. Cellulose bait improves the effectiveness of *Metarhizium anisopliae* as a microbial control of termites (Isoptera: Rhinotermitidae). Biological Control 30, 523–529.

Wang, C.S., St Leger, R.J., 2007. The MADI adhesin of *Metarhizium anisopliae* links adhesion with blastospore production and virulence to insects, and the MAD2 adhesin enables attachment to plants. Eukaryotic Cell 6, 808–816.

Xiao, G., Ying, S.H., Zheng, P., Wang, Z.L., Zhang, S., Xie, X.Q., Shang, Y., Stleger, R.J., Zhao, G.P., Wang, C., 2012. Genomic perspectives on the evolution of fungal entomopathogenicity in *Beauveria bassiana*. Scientific Reports 2, 483.

Zacharuk, R.Y., 1973. Electron–microscope studies of the histo-pathology of fungal infections by *Metarhizium anisopliae*. Miscellaneous Publication of the Entomological Society of America 9, 112.

Zhang, S.Z., Xia, Y.X., Kim, B., Keyhani, N.O., 2011. Two hydrophobins are involved in fungal spore coat rodlet layer assembly and each play distinct roles in surface interactions, development and pathogenesis in the entomopathogenic fungus, *Beauveria bassiana*. Molecular Microbiology 80, 811–826.

Zhang, Y.J., Feng, M.G., Fan, Y.H., Luo, Z.B., Yang, X.Y., Wu, D., Pei, Y., 2008. A cuticle–degrading protease (CDEP–1) of *Beauveria bassiana* enhances virulence. Biocontrol Science and Technology 18, 155–563.

Zindel, R., Gottlieb, Y., Aebi, A., 2011. Arthropod symbioses: a neglected parameter in pest and disease control programmes. Journal of Applied Ecology 48, 864–872.

Further Reading

Bidochka, M.J., Khachatourians, G.G., 1987. Hemocytic defence response to the entomopathogenic fungus *Beauveria bassiana* in the migratory grasshopper *Melanoplus sanguinipes*. Entomologia Experimentalis et Applicata 45, 151–156.

Boucias, D.G., Farmerie, W.G., Pendland, J.C., 1998. Cloning and sequencing the cDNA of the insecticidal toxin, hirsutellin A. Journal of Invertebrate Pathology 72, 258–261.

Fransen, J.J., Winkelman, K., Van Lenteren, J.C., 1987. The differential mortality at various life stages of the greenhouse whitefly, *Trialeurodes vaporariorum* (Homoptera: Aleyrodidae) by infection with the fungus, *Aschersonia aleyrodis*. Journal of Invertebrate Pathology 50, 158–165.

Kapongo, J.P., Shipp, L., Kevan, P., Broadbent, B., 2008. Optimal concentration of *Beauveria bassiana* vectored by bumble bees in relation to pest and bee mortality in greenhouse tomato and sweet pepper. Biocontrol 53, 799–812.

Li, H., Feng, M.G., 2005. Broomcorn millet grains cultures of the entomophthoralean fungus *Zoophthora radicans*: sporulation capacity and infectivity to *Plutella xytostella*. Mycological Research 109, 319–325.

Zheng, P., Xia, Y.L., Xiao, G.H., Xiong, C.H., Hu, X., Zhang, S.W., Cheng, H.J., Huang, Y., Zhou, Y., Wang, S.Y., 2011. Genome sequence of the insect pathogenic fungus *Cordyceps militaris* & valued traditional Chinese medicine. Genome Biology 12.

Plant Monoterpenoids

(Prospective Pesticides)

Arun K. Tripathi[1], Shikha Mishra[2]

[1]CSIR-Central Institute of Medicinal and Aromatic Plants, Lucknow, India; [2]CSIR-Central
Drug Research Institute, Lucknow, India

1. INTRODUCTION

Synthetic insecticides and fumigants are generally used as household pesticides both for stored product protection against insect damage and management of vectors of medical importance. Many active products that have been used as fumigants are being withdrawn from the insecticide market (Bell, 2000). Other pesticides being used are associated with numerous side effects such as residual problems, environment pollution, resistance, and mammalian toxicity. Under these circumstances, essential oil–based pesticides are the best alternatives to synthetic pesticides, in view of their low mammalian toxicity and high volatility (Shaaya et al., 1991). Essentials oils have multifold applications in the pharmaceutical, cosmetics, and food flavor industries, and approximately 300 essential oils have been commercialized by these industries (Bakkali et al., 2008) apart from being used as pesticides (Chang and Cheng, 2002).

Worldwide efforts have been made on developing prospective essential oils for insect-pest management (Tripathi et al., 1999; Tripathi, 2004; Isman et al., 2007; Bakkali et al., 2008), but approaches and prospects about commercial opportunities of bioactive lead monoterpenoids as household insecticides are meager. Details of such bioactive leads may facilitate the development of futuristic and commercially viable insect-pest control agents under household conditions. This chapter covers various issues ranging from essential oil chemistry to their bioactivity potential. The authors also examine the mode of action of monoterpenoids coupled with structure–activity relationships and commercial opportunities to develop them as household pesticides. For this purpose, this chapter is divided into the following contents.

2. CHEMISTRY OF ESSENTIAL OILS

Isoprenoids are mainly mono- and sesquiterpenes, which constitute the basic units of essential oils (Franzios et al., 1997). The monoterpenes are formed from the coupling of two isoprene units (C10). They are the most representative molecules, constituting 90% of the essential oils with a great structural diversity.

Monoterpenoids and related compounds are classified as: acyclic alcohols (e.g., linalool, geraniol, citronellol), cyclic alcohols (e.g., menthol, isopulegol, terpeniole), bicyclic alcohols (e.g., borneol, verbenol), phenols (e.g., thymol, eugenol), ketones (carvone, menthone, thujone), aldehydes (citronellal, cinnamaldehyde), acids (e.g., chrysanthemic acid), and oxides (cineole). However, the chemical profile of the essential oil products differs not only in the number of molecules but also in the stereochemical types of molecules extracted. The extraction product can vary in quality, quantity, and composition according to climate, soil composition, plant organ, age, and vegetative cycle stage (Masotti et al., 2003; Angioni et al., 2006).

Chirality and stereochemistry of components of the essential oils are of considerable importance because the spatial orientation of connective parts of a molecule can significantly influence the chemical behavior and pharmacological activity of the compound. In this regard, molecules with the same molecular formula and the same bonds between atoms, but different spatial arrangements of these atoms, are called stereoisomers (Sadgrove and Jones, 2015). In the context of essential oils, stereoisomers can be classified as diastereomers and enantiomers. Generally, a pair of stereoisomers is called a diastereomer. Some stereoisomers are exact mirror images of each other that cannot be superimposed and each of these is called as enantiomer of a chiral molecule. An example of a chiral compound is carvone (Figure 1) in which differences between enantiomers cannot be observed using routine gas chromatography (GC) or nuclear magnetic resonance (NMR). Thus, biological activity of monoterpenoids is related to functional groups and the bioactivities are dependent on the nature and position of the functional groups and molecular configuration.

2.1 Structure–Activity Relationships

The functional group of a monoterpenoid compound reflects its bioactivity potential and the structural modifications may lead to improved biological activity. The activities are dependent upon the nature and position of the functional groups and molecular configuration

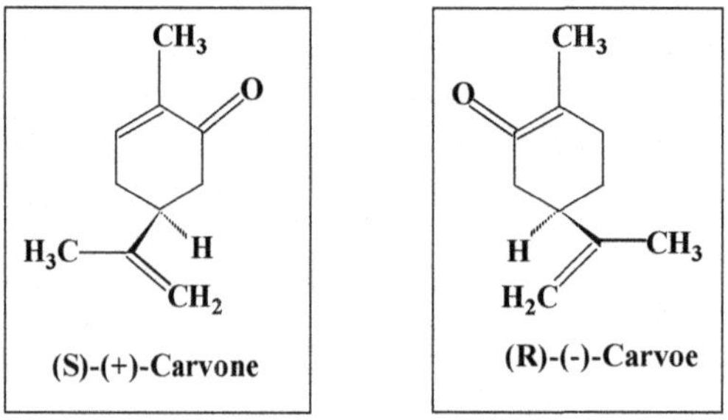

FIGURE 1   Enantiomers of a carvone.

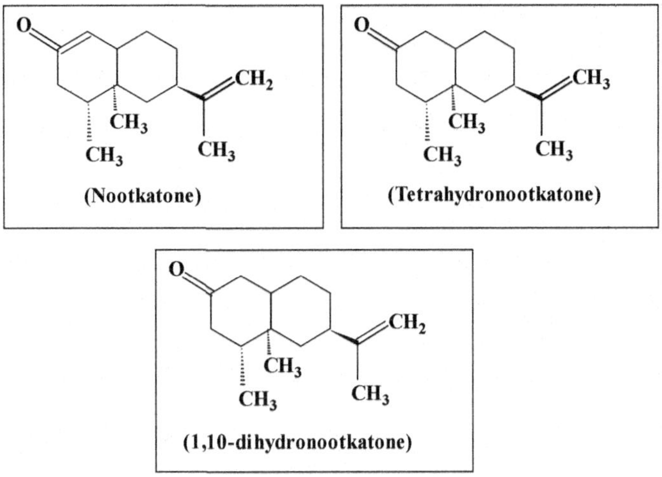

FIGURE 2 Structure of nootkatone and its derivatives.

(Kumbhar and Dewang, 2001). Therefore, knowledge of functional groups and optimal structure through biorational design is necessary to understand structure–activity relationships shown in the following example.

Nootkatone, a sesquiterpene ketone from the heartwood of yellow cedar (*Chamaecyparis nootkatensis*), is a repellent and a feeding deterrent to Formosan subterranean termites (Zhu et al., 2001). Results of the structure–activity relationships of nootkatone and its two derivatives, 1,10-dihydro- and tetrahydronootkatone (Figure 2), indicated that removal of the hydrogenated double bond present in nootkatone between carbon position 1 and 10 increased the repellency of 1,10-dihydronootkatone and tetrahydronootkatone, whereas nootkatone was weakly active as a repellent. Apart from this, the activity varies with the level of unsaturation of compounds as revealed in tetrahydronootkatone and 1,10-dihydronootkatone (Zhu et al., 2003).

2.2 Mode of Action

The mode of action of essential oils as insecticide and repellent are different at the biochemical and physiological levels. The insecticidal mechanism of essential oils is largely octopaminergic mediated in insects. Octopamine occurs in large amounts in the nervous systems of insects and has a broad spectrum of biological roles, e.g., as a neurotransmitter, neurohormone, and neuromodulator (Evans, 1980; Hollingworth et al., 1984) and acts as a putative central transmitter. However, octopamine is a peripheral transmitter in higher invertebrates (Sporhase-Eichmann et al., 1992).

The repellent mode of action of essential oils in insects is largely governed by olfactory receptors. The repellent molecules enter pores of the insect sensory receptors, e.g., female mosquito. Insects detect odors when the volatile odor (odorant) binds to odorant receptor (OR) proteins present on ciliated dendrites of specialized odor receptor neurons (ORNs) exposed to the external environment. The ORNs are usually present on the antennae and maxillary palps of the insect. Some ORNs, such as Or83b, which are highly conserved across insect species (Pitts et al., 2004), are important in olfaction and blocked by the repellents (Ditzen et al., 2008). In some of the arthropods, like ticks, repellents are detected by Haller's organ present on the tarsi of the first pair of legs.

3. MONOTERPENOIDS OF PESTICIDAL IMPORTANCE

Some monoterpenoids apparently seems to be of futuristic importance as elaborated below; they have not yet been commercialized as household pesticides.

3.1 D-Limonene

D-limonene (dextrorotatory-limonene) is a colorless liquid hydrocarbon classified as a cyclic terpene and possesses a strong citrus odor.

3.1.1 Chemistry

Synonyms of D-limonene are para methoxy phenylpropene, *p*-propenylanisole, and isoestragole.

CH₃

H₃C   CH₂
(d-Limonene)

3.1.2 Plant Sources

D-limonene is found in many plants like celery seed oil (*Apium graveolens*), *Citrus sinensis*, and *Anethum graveolens*.

3.1.3 Bioactivity

According to the US Environmental Protection Agency (EPA), D-limonene is a broad-based insecticide. The D-enantiomer is most active as an insecticide. It can be used for aphids, ants, mealybugs, gnats, ants, silverfish, roaches, and fleas. D-limonene has insecticidal effects toward stored grain insects, *Sitophilus oryzae*, *Tribolium castaneum*, *Oryzaephilus surinamensis* (Lee et al., 2003), and German cockroach (Karr and Coats, 1988) and termiticidal effects against the termite, *Heterotermes sulcatus* (Almeida et al., 2015). D-limonene destroys the wax coating of the insect's respiratory system.

D-limonene is an active ingredient of commercially available flea shampoo. Some of the commercially available insecticidal formulations are d'Bug—insecticide with 5.8% D-limonene (Athea® Laboratories, Inc., Cornell, USA) against crawling insects and Bug Killer (Zircon Industries, South Miles Road, Cleveland) useful against unwanted insects.

D-limonene is a potential chemopreventive agent (Crowell, 1999) with value as a dietary anticancer tool in humans (Tsuda et al., 2004). The International Agency for Research on

Cancer (IARC) classifies D-limonene as a Group 3 carcinogen—not classifiable as to its carcinogenicity to humans (IARC, 1999).

D-limonene has been designated as "generally recognized as safe" (GRAS) by the US Food and Drug Administration (FDA). The EPA has granted D-limonene an exemption from the requirement of a tolerance when it is an inert ingredient used as a solvent or fragrance in pesticide formulations. Toxicity profile of D-limonene as Oral: $LD_{50} > 5 \, g/kg$, rabbit, Dermal: $LD_{50} > 5 \, g/kg$, rabbit, Inhalation: $RD_{50} > 1000 \, ppm$, Skin: the skin irritancy of limonene in guinea pigs and rabbits is considered moderate and low, respectively. Sensitization: D-limonene is not a sensitizer. Improper storage and handling can lead to oxidation. The oxidized forms of D-limonene have been shown to be a skin sensitizer.

3.2 Anethole

Anethole is a derivative of the aromatic compound phenylpropene. It is a colorless, fragrant, and mildly volatile liquid.

3.2.1 Chemistry

Anethole is also known as para methoxy phenylpropene, *p*-propenylanisole, and isoestragole.

OCH_3

(Anethole)

3.2.2 Plant Sources

trans-Anethole is the principal constituent of star anise, *Illicium verum* (Ho et al., 1997) and *Foeniculum vulgare* (Rahimmalek et al., 2014). Closely related to anethole is its double-bond isomer estragole, abundant in tarragon (Asteraceae) and basil (Lamiaceae), which has a flavor reminiscent of anise.

3.2.3 Bioactivity

Several essential oils consisting of anethole have insecticidal action against larvae of the mosquito, *Ochlerotatus caspius* (Knio et al., 2008) and *Aedes aegypti* (Cheng et al., 2004; Morais et al., 2006); insect repellent action against mosquitoes (Padilha de Paula et al., 2003); fungicidal action against the fungus gnat, *Lycoriella ingenua* (Sciaridae) (Park et al., 2006); and acaricidal action against the mold mite *Tyrophagus putrescentiae* (Lee, 2005). Against the mite, anethole is a slightly more effective pesticide than DEET but anisaldehyde, a related natural compound that occurs with anethole in many essential oils, is 14 times more effective (Lee, 2005).

The insecticidal action of anethole is greater as a fumigant than as a contact agent. (E)-anethole is highly effective as a fumigant against the cockroach, *Blattella germanica* (Chang and Ahn, 2002) and against the adults of *S. oryzae, Callosobruchus chinensis,* and *Lasioderma serricorne* (Kim and Ahn, 2001).

In spite of such a diversified potential, anethole-based pesticidal products are lacking.

The Flavor and Extract Manufacturers Association approves anethole as GRAS (Newberne et al., 1999). According to a Joint FAO/WHO Expert Committee on Food Additives (JECFA), anethole has no safety concern at current levels of intake when used as a flavoring agent (IPCS, 2009).

3.3 Cinnamaldehyde

Cinnamaldehyde is a pale yellow, viscous liquid occurring naturally in the bark of cinnamon trees.

3.3.1 Chemistry

Synonyms of cinnamaldehyde are 3-phenylacrolein, cassia aldehyde, cinnamic aldehyde, and 3-phenyl-2-propenal.

The molecule cinnamaldehyde consists of a phenyl group attached to an unsaturated aldehyde.

Cinnamaldehyde

3.3.2 Plant Sources

Cinnamaldehyde is a major compound of *Cinnamomum zeylanicum* and *Cinnamomum cassia.* The essential oil of cinnamon bark has about 90% cinnamaldehyde.

3.3.3 Bioactivity

Cinnamaldehyde is an effective insecticide and repellent to animals, such as cats and dogs. It has been recognized as an effective mosquitocidal agent (Ma et al., 2014) and fumigant against adults of *S. oryzae* (Lee et al., 2008). Cinnamaldehyde is also used as a fungicide and typically applied to the root systems of plants. Its low toxicity and well-known properties make it ideal for agriculture. In spite of such diversified potential, cinnamaldehyde-based pesticidal products are not available.

Irritation data: 40 mg/48 h(s) skin-human severe, toxicity data: 2220 mg/kg oral-rat LD_{50}.

3.4 1,8-Cineole

1,8-cineole has a fresh camphor-like smell and a spicy, cooling taste. It is insoluble in water but miscible with ether, ethanol, and chloroform.

3.4.1 Chemistry

1,8-cineole is also known by a variety of synonyms: eucalyptol, cajeputol, 1,8-epoxy-p-menthane, 1,8-oxido-p-menthane, etc. It has unique compact chemical form, with the cyclic ether linkage spanning a structure based on cyclohexane.

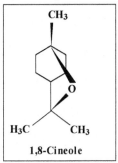

1,8-Cineole

3.4.2 Plant Sources

1,8-cineole comprises up to 90% of the essential oil of some species of the generic product Eucalyptus oil (Boland et al., 1991). 1,8-cineole is widely distributed in many plants, e.g., *Eucalyptus cinerea* (Silva et al., 2011), *Rosmarinus officinalis* and *Psidium* sp. (Goebel et al., 1995), *Artemisia annua* (Aggarwal et al., 2001), *Melaleuca linariifolia* (Padalia et al., 2015), etc.

3.4.3 Bioactivity

1,8-cineole is used as an insect repellent and insecticide to the American cockroach (Scriven and Meloan, 1984; Klocke et al., 1987; Sfara et al., 2009), as a mosquito larvicide (Corbet et al., 1995), as a mosquito ovipositional repellent (Klocke et al., 1987), as a fumigant toward adults of *Callosobruchus maculatus, Rhyzopertha dominica,* and *S. oryzae* (Shaaya et al., 1991; Aggarwal et al., 2001), and inhibitor of the enzyme acetylcholinesterase (Ryan and Bryan, 1988).

1,8-cineole-based pellets have been prepared by dry mixing method, which is effective against stored grain insects *T. castaneum* (Herbst), *C. maculatus* (F.), and *R. dominica* (F.) (Sharifian et al., 2011).

Cineole and other components of eucalyptus oil are readily biodegradable, unreactive, and relatively nontoxic (Webb and Pitt, 1993). Acute oral toxicity (LD_{50}) is 2480 mg/kg (rat) and toxic to humans. It is hazardous in case of skin contact (irritant), ingestion, and inhalation.

3.5 Linalool

Linalool is a colorless to yellow liquid with a smell similar to that of bergamot oil and lavender.

3.5.1 Chemistry

Other chemical names of linalool are 2,6-dimethylocta-2,7-dien-6-ol; β-linalool; linalyl alcohol; linaloyl oxide; p-linalool; allo-ocimenol; and 3,7-dimethyl-1,6-octadien-3-ol.

Linalool has a stereogenic center at C_3 and therefore there are two stereoisomers: (R)-(−)-linalool is also known as licareol and (S)-(+)-linalool is also known as coriandrol. Both enantiomeric forms are found in nature.

(Linalool)

3.5.2 Plant Sources

Linalool is found in a number of plants belonging to families Lamiaceae (mints, scented herbs), Lauraceae (laurels, cinnamon, rosewood), and Rutaceae (citrus fruits). Linalool is also found in essential oils of coriander (*Coriandrum sativum*) seed, palmarosa (*Cymbopogon martinii* var martini), sweet orange (*C. sinensis*) flowers, lavender (*Lavandula officinalis*), bay laurel (*Laurus nobilis*), and sweet basil (*Ocimum basilicum*) (Lis-Balchin and Hart, 1999).

3.5.3 Bioactivity

Linalool has been reported for insecticidal activity against stored product insects (Ryan and Byme, 1988; Weaver et al., 1991; Sanchez-Ramos and Castanera, 2001) and has potential against cat fleas (Hink et al., 1998). It can be used for the control of ticks and fleas (Hink and Duffey, 1990); mite, *T. putrescentiae* (Sanchez-Ramos and Castanera, 2001); fruit flies, *Ceratitis capitata* (Wiedemann) and *Bactrocera dorsalis* (Chang et al., 2009); and housefly, *Musca domestica* L. (Maganga et al., 1996). Linalool has antifungal (Duman et al., 2010; Ozek et al., 2010) and antimicrobial activities (Duman et al., 2010; Park et al., 2012). It has been demonstrated to reduce the presence of female mosquitoes by almost twice as much as citronella candles (Muller et al., 2008), and indoors, it repels mosquitoes by 93% (Muller et al., 2009). Linalool acts on the nervous system affecting ion transport and release of acetylcholinesterase (Lopez and Pascual-Villalobos, 2015).

This is an ingredient used in flea dips for dog and cats. Controlled-release formulation comprised of linalool has been reported as insecticidal (Lopez et al., 2012).

Linalool is considered GRAS for commercial purposes (Bickers et al., 2003). Additionally, linalool is an important intermediate in the manufacture of vitamin E (Ozek et al., 2010). However, linalool gradually breaks down when in contact with oxygen, forming an oxidized by-product that may cause allergic reactions such as eczema in susceptible individuals. The acute oral (rat) LD_{50} is quite high, i.e., 2.7 g/kg, suggesting substantial safety issues for mammals (Opdyke, 1979).

3.6 Eugenol

Eugenol is a phenylpropene, an allyl chain-substituted guaiacol. Eugenol is a member of the phenylpropanoids class of chemical compounds. It is a colorless to pale yellow oily liquid.

3.6.1 Chemistry

Synonyms of eugenol are 4-allyl-2-methoxyphenol, *p*-allylguaiacol; *p*-eugenol; caryophyllic acid; engenol; eugenic acid; 2-methoxy-1-hydroxy-4-allylbenzene.

Eugenol

3.6.2 Plant Sources

Clove, *Syzygium aromaticum*, is the richest source of eugenol. It is present in concentrations of 80–90% in clove bud oil and 82–88% in clove leaf oil (Barnes et al., 2007). Other plants having eugenol are *Artemisia absinthium* (Bullerman et al., 1977), *Cinnamomum tamala* (Dighe et al., 2005), *Myristica fragrans* (Bennett et al., 1988), *O. basilicum* (Johnson et al., 1999), and *Ocimum gratissimum* (Nakamura et al., 1999).

3.6.3 Bioactivity

Eugenol has been shown to exhibit insecticidal property toward *Sitophilus zeamais* (Huang et al., 2002), be toxic and repellent to the beetle *Dinoderus bifloveatus* (Ojimelukwe and Adler, 2000) and tick (*Ixodes ricinus*) (Bissinger and Roe, 2010), and a fumigant toward *C. maculatus* (Ajayi et al., 2014). Octopaminergic system-mediated insecticidal activity has been observed for eugenol (Enan, 2001). Eugenol-rich cinnamon oil makes the most effective natural mosquito repellent known to man.

Eugenol is used in insect attractant formulation developed for oriental fruit flies, *B. dorsalis*, and melon flies, *Bactrocera cucurbitae* (Gomez and Coen, 2013).

Eugenol is listed by the FDA as GRAS when consumed orally, in unburned form. It is nontoxic in food but toxic upon inhalation. High doses of eugenol may cause damage to the liver (Thompson et al., 1998).

3.7 Methyleugenol

Methyleugenol is produced by the methylation of eugenol (Burdock, 2005). It is the methyl ether of eugenol and belongs to a class of phenylpropanoids in which a benzene ring has an allyl group attached to it.

3.7.1 Chemistry

Synonyms of methyleugenol are 4-allyl-1,2-dimethoxybenzene, 4-allylveratrole, and *O*-methyleugenol.

Methyleugenol

3.7.2 Plant Sources

Methyleugenol is a constituent of a large number of essential oils obtained from, e.g., *Artemisia dracunculus, S. aromaticum, Daucus carota, M. fragrans*, and *R. officinalis*. Methyleugenol is a constituent of essential oil *Melaleuca bracteata* (90–95%) and *Cinnamomum oliveri* leaves (90–95%) (Burdock, 2005).

3.7.3 Bioactivity

Methyleugenol is most effective in terms of knockdown activity, as well as repelling and killing effects (Ngoh et al., 1998) apart from larvicidal activity against *Spodoptera litura* (Bhardwaj et al., 2010). It is a potent inhibitor of the enzyme acetylcholinesterase (Lee et al., 2001), responsible for the hydrolysis of the neurotransmitter acetylcholine, which can eventually lead to paralysis in insects.

Insect attractant formulation developed for Oriental fruit flies (*B. dorsalis*) and melon flies (*B. cucurbitae*) (Gomez and Coen, 2013).

Methyleugenol was registered as an active pesticide ingredient in the USA in 2006 (EPA, 2006).

4. COMMERCIAL ASPECTS

Essential oils are lipophilic in nature and interfere with basic metabolic, biochemical, physiological, and behavioral functions of insects (Brattsten, 1983), which facilitate their exploitation as insecticides or repellents. In spite of such properties, only limited pesticidal products based on plant essential oils have appeared in the market. The possible reasons behind this may be regulatory barriers to commercialization.

However, interest in essential oil–based pesticidal products has been considerable, particularly for control of domestic and veterinary pests. EcoSMART Technologies (Atlanta, US) has introduced insecticides containing eugenol and 2-phenethyl propionate controlling crawling and flying insects. An insecticide containing rosemary oil as the active ingredient has recently been introduced for use on horticultural crops under the name EcoTrol™. Another product based on rosemary oil is a fungicide sold under the name Sporan™, while a formulation of clove oil (major constituent: eugenol), sold as Matran™, is used for weed control. The primary active ingredients of these pesticidal products are classified as GRAS by the FDA. Menthol has been approved for use in North America for control of tracheal mites in beehives, and a product produced in Italy (Apilife VARTM) containing thymol and lesser amounts of cineole, menthol, and camphor is used to control Varroa mites in honeybees. The Israel-based company BotanoCap, Green Microencapsulation Solutions (Ganey Tikva, Israel) has developed a patented technology for the gradual release of essential oils and natural components.

Thus, essential oil–bearing plants hold much promise as pest control agents, especially repellents with a higher degree of safety than synthetics. Citronella is a popular aromatic ingredient in insect repellent formulations. Candles and incense containing oil of citronella are sold in the market as insect repellents. A number of natural products have been documented in the scientific literature for their effectiveness against common pests inhabiting houses. Neem oil from *Azadirachta indica*, when formulated as 2% in coconut oil, provided complete protection for 12 h from *Anopheles* mosquitoes (Sharma, 1993). Quwenling, a popular *Eucalyptus*-based repellent, contains a mixture of *p*-menthane-3, 8-diol; isopulegone; and citronellol. Quwenling has largely replaced dimethyl phthalate as the insect repellent of choice in China (Trigg and Hill, 1996).

Another product in the form of vanishing cream as mosquito repellent has been developed by the Indian Council of Medical Research center at Pondicherry; it contains N,N-diethyl phenyl acetamide (DEPA) synthesized in the laboratory. DEPA has been found to give promising repellency for 6–7 h at $1.0 \, mg/cm^2$ against mosquitoes, blackflies, and land leeches (Kalyanasundaram and Mathew, 2006). DEPA has been found safe in toxicological studies.

Many other commercial products like Buzz Away® (containing citronella, cedarwood, eucalyptus, and lemongrass), Green Ban® (containing citronella, cajuput, lavender, safrole-free sassafras, peppermint, and bergapten-free bergamot oil), and Skin-So-Soft® (containing various oils and stearates), although effective, they failed in scientific validation in an olfactometer assay against *A. aegypti* (Chou et al., 1997). Therefore, natural ingredients must be scientifically validated for their activity, efficacy, and safety before making any formulations.

Three major problems in the commercialization of new products of these types have been identified: (1) the scarcity of the natural resource; (2) the need for chemical standardization and quality control; and (3) difficulties in registration.

5. STABILITY OF ESSENTIAL OIL–BASED PESTICIDES

Major concerns about the essential oils to be used as pesticides are their high rate of volatility, chemical instability in the presence of air, light, moisture, and high temperatures. Such problems can be overcome by incorporating the essential oils in controlled-release formulations allowing the use of much less natural pesticides in an effective way. Microencapsulation, for example, is a method that is used to protect sensitive materials that can easily suffer degradation.

The delivery system should be designed in a way to protect the essential oil from the environmental degradation process and prevent removal of these natural pesticides from their target before they can take effect. Further, costs of the materials and processing formulation needed special attention.

6. ECONOMICS AND SUSTAINABILITY

Essential oils are presently regarded as a new class of ecological products for controlling household insects. The widespread range of activities of essential oils is being considered for both industrial and household uses. Since there is a large group of beneficiaries for the scientifically proven essential oil–based pesticides to be used, it is supposed that there will be continuous and bulk requirement of raw materials. To supply such large quantities of the herbs, large-scale cultivation would be required, which in turn will generate good business opportunities and human resource development.

7. HEALTH AND ENVIRONMENTAL IMPACTS

Demand for commercial herbal products for human and animal uses is growing rapidly in the market because of misconception that all natural things are safe. However, the herbal products appearing in the market, especially pesticidal, are not always subject to rigorous testing. Most of the essential oils used in the commercial products are listed as GRAS approved by the FDA and EPA in the USA for food and beverage consumption (EPA, 1993). However, no guarantee can be given for the products prepared from such essential oils about their efficacy as formulated or advertised by a specific producer. A number of nontrivial problems are associated with many of the herbal products for pest control, basically minor misinterpretations that provide misinformation to consumers, e.g., eucalyptus oil advertised for use against mosquitoes as a natural repellent, but it can also serve as an attractant for another blood-feeding pest, e.g., biting midges (Braverman et al., 1999).

Tea tree oil obtained from the plant *Melaleuca alternifolia* contains terpinen-4-ol; terpinene; 1,8-cineole; and terpinolene as active compounds and is used against insect bites (Budhiraja et al., 1999). But the maximum permissible limit for terpinen-4-ol and 1,8-cineole by the International Standard, ISO 4730 (ISO, 1996) is 30% and 15% concentration in the oil, respectively. These active compounds are also found in peppermint and rosemary essential oils (Veal, 1996). However, the use of fresh tea tree oil causes contact dermatitis in humans (Hausen et al., 1999) and further enhanced issues by formation of degradation products in the oil due to photodegradation (Hausen et al., 1999), which includes peroxides, epoxides, and endoperoxides such as ascaridole. Further, *Melaleuca* oil toxicosis in dogs and cats has been associated with depression, weakness, muscle tremors, and lack of coordination (Villar et al., 1994).

Another bioactive compound, D-limonene, is listed in the EPA's GRAS list (EPA, 1993). D-limonene is an active ingredient of flea dips for dogs and cats and as a pesticide used for indoor pest control. But D-limonene has been reported to cause dermatitis (Nilsson et al., 1999). Further, thujone, an active compound, *A. absinthium* essential oil, is a potent neurotoxin affecting the gamma-amino butyric acid system (Hold et al., 2000). Essential oil of *Mentha pulegium* is widely used as insect control agent. This oil contains pulegone as an active ingredient, which upon ingestion get oxidized by cytochrome P-450 system into toxic metabolites including menthofuran (Nelson et al., 1992). These metabolites bind to proteins (Thomassen et al., 1992) causing organ function failure, acute poisoning, and death (Burkhard et al., 1999).

Thus, simply assuming that all naturals are safe may be dangerous. Natural products are not always safer than synthetic pesticides. Just because a plant has been used for centuries does not mean it is safe or even desirable (Hinkle, 1995). Commercial success with these products lies with scientific validation of bioactivity data based on well-known chemistry, which may provide an impetus for the development and commercialization of future pesticides (Shaaya and Kostjukovsky, 1998).

8. FUTURE PERSPECTIVES

As discussed above, there is tremendous potential for growth in research and development on chemical transformation of monoterpenes, extraction technologies from plant material, and application technologies to improve efficacy. Systematic derivatization of natural monoterpenoids based on structure–activity relationships is the most promising potential for exploitation of these molecules as pest management agents (Kumbhar and Dewang, 2001). Further, factors such as production cost, resource availability, extraction, and formulation techniques needed to be considered along with innovative application technologies.

However, lack of availability of sufficient quantities of plant material, standardization and refinement of pesticide products, protection of technology (patents), and regulatory approval are the major constraints in the commercialization of the essential oil–based pesticides. They have greatest commercial application in urban pest control, public health, veterinary health, vector control, and in protection of stored commodities. These pesticides may also be useful for greenhouse crops, high-value row crops, and within organic food production systems where few alternative pesticides are available. Changing consumer preferences toward the use of natural over synthetic products and potential to extend the range of available products including new product development through biotechnology are additional opportunities in this field.

References

Aggarwal, K.K., Tripathi, A.K., Prajapati, V., Kumar, S., 2001. Toxicity of 1,8-cineole towards three species of stored product coleopterans. Insect Science and Its Application 21, 155–160.

Ajayi, O.E., Arthur, G.A., Henry, Y.F., 2014. Fumigation toxicity of essential oil monoterpenes to *Callosobruchus maculatus* (Coleoptera: Chrysomelidae: Bruchinae). Journal of Insects 1–7. http://dx.doi.org/10.1155/2014/917212.

Almeida, M.L.S., Oliveira, A.S., Rodrigues, A.A., Carvalho, G.S., Silva, L.B., Lago, J.H.G., Casarin, F.E., 2015. Antitermitic activity of plant essential oils and their major constituents against termite *Heterotermes sulcatus* (Isoptera: Rhinotermitidae). Journal of Medicinal Plants Research 9, 97–103.

Angioni, A., Barra, A., Coroneo, V., Dessi, S., Cabras, P., 2006. Chemical composition, seasonal variability, and antifungal activity of *Lavandula stoechas* L. ssp. *stoechas* essential oils from stem/leaves and flowers. Journal of Agricultural and Food Chemistry 54, 4364–4370.

Bakkali, F., Averbeck, S., Averbeck, D., Idaomar, M., 2008. Biological effects of essential oils – a review. Food and Chemical Toxicology 46, 446–475.

Barnes, J., Anderson, L.A., Phillipson, J.D., 2007. Herbal Medicines, third ed. Pharmaceutical Press, London, UK. ISBN:978-0-85369-623-0.

Bell, C.H., 2000. Fumigation in the 21st century Crop Protection. Crop Protection 19, 563–569.

Bennett, A., Stamford, I.F., Tavares, I.A., Jacobs, S., Capasso, F., Mascolo, N., Autore, G., Romano, V., Di Carlo, G., 1988. The biological activity of eugenol, a major constituent of nutmeg (*Myristica fragrans*): studies on prostaglandins, the intestine and other tissues. Phytotherapy Research 2, 124–130.

Bhardwaj, A., Tewary, D.K., Kumar, R., Kumar, V., Sinha, A.K., Shanker, A., 2010. Larvicidal and structure–activity studies of natural phenylpropanoids and their semisynthetic derivatives against the tobacco armyworm *Spodoptera litura* (Fab.) (Lepidoptera: Noctuidae). Chemistry and Biodiversity 7, 168–177.

Bickers, D., Calow, P., Greim, H., Hanifin, J.M., Rogers, A.E., Saurat, J.H., Sipes, I.G., Smith, R.L., Tagami, H., 2003. A toxicologic and dermatologic assessment of linalool and related esters when used as fragrance ingredients. Food Chemistry and Toxicology 41, 919–942.

Bissinger, B.W., Roe, R.M., 2010. Tick repellents: past, present, and future. Pesticide Biochemistry and Physiology 96, 63–79.

Boland, D.J., Brophy, J.J., House, A.P.N., 1991. Eucalyptus Leaf Oils: Use, Chemistry, Distillation and Marketing. Inkata Press, Melbourne, p. 6.

Brattsten, L.B., 1983. Cytochrome P-450 involvement in the interaction between plant terpenes and insect herbivores. In: Hedin, P.A. (Ed.), Plant Resistance to Insects. American Chemical Society, Washington, pp. 173–195.

Braverman, Y.A., Chizov-Ginsberg, A., Mullens, B.A., 1999. Mosquito repellent attracts *Culicoides imicola* (Diptera: Ceratopogonidae). Journal of Medical Entomology 36, 113–115.

Budhiraja, S.S., Cullum, M.E., Sioutis, S.S., Evangelista, L., Habanota, S.T., 1999. Biological activity of *Melaleuca alternifolia* (tea tree) oil component, terpinen-4-ol, in human myelocytic cell line HL-60. Journal of Manipulative and Physiological Therapeutics 22, 447–453.

Bullerman, L.B., Lieu, F.Y., Seier, S.A., 1977. Inhibition of growth and aflatoxin production by cinnamon and clove oils, cinnamic aldehyde and eugenol. Journal of Food Science 42, 1107–1109.

Burdock, G.A., 2005. Fenaroli's Handbook of Flavor Ingredients, fifth ed. CRC Press, Boca Raton, FL, pp. 672–673.

Burkhard, P.R., Burkhardt, K., Haenggeli, C.A., Landis, T., 1999. Plant induced seizures: reappearance of an old problem. Journal of Neurology 46, 667–670.

Chang, K.S., Ahn, Y.J., 2002. Fumigant activity of (E)-anethole identified in *Illicium verum* fruit against *Blattella germanica*. Pest Management Science 58, 161–166.

Chang, S.T., Cheng, S.S., 2002. Anti-termitic activity of leaf essential oils and components from *Cinnamomum osmophleum*. Journal of Agricultural and Food Chemistry 50, 1389–1392.

Chang, C.L., Cho, I.K., Li, Q.X., 2009. Insecticidal activity of basil oil, *trans*-anethole, estragole, and linalool to adult fruit flies of *Ceratitis capitata*, *Bactrocera dorsalis*, and *Bactrocera cucurbitae*. Journal of Economic Entomology 102, 203–209.

Cheng, S.S., Liu, J.Y., Tsai, K.H., Chen, W.J., Chang, S.T., 2004. Chemical composition and mosquito larvicidal activity of essential oils from leaves of different *Cinnamomum osmophloeum* provenances. Journal of Agricultural and Food Chemistry 52, 4395–4400.

Chou, J.T., Rossignol, P.A., Ayres, J.W., 1997. Evaluation of commercial insect repellents on human skin against *Aedes aegyptii* (Diptera: Culicidae). Journal of Medical Entomology 34, 624–630.

Corbet, S.A., Danahar, G.W., King, V., Chalmers, C.L., Tiley, C.F., 1995. Surfactant-enhanced essential oils as mosquito larvicides. Entomologia Experimentalis et Applicata 75, 229–236.

Crowell, P.L., 1999. Prevention and therapy of cancer by dietary monoterpenes. The Journal of Nutrition 129, 775S–778S.

Dighe, V.V., Gursale, A.A., Sane, R.T., Menon, S., Patel, P.H., 2005. Quantitative determination of eugenol from *Cinnamomum tamala* Nees and Eberm. Leaf powder and polyherbal formulation using reverse phase liquid chromatography. Chromatographia 61, 443–446.

Ditzen, M., Pellegrino, M., Vosshall, L.B., 2008. Insect odorant receptors are molecular targets of the insect repellent DEET. Science 319, 1838–1842.

Duman, A.D., Telci, I., Dayisoylu, K.S., Digrak, M., Demirtas, I., Alma, M.H., 2010. Evaluation of bioactivity of linalool-rich essential oils from *Ocimum basilucum* and *Coriandrum sativum* varieties. Natural Product Communications 5, 969–974.

EPA, 1993. Integrated Risk Information System, d-Limonene; CASRN 5989-27-5.

EPA, 2006. Methyl Eugenol. (ME) (203900) Fact Sheet. United States Environmental Protection Agency, Washington, DC. Available at: http://www.epa.gov/pesticides/biopesticides/ingredients/factsheets/factsheet-203900.htm.

Enan, E., 2001. Insecticidal activity of essential oils: octopaminergic sites of action. Comparative Biochemistry and Physiology-Part C 130, 325–337.

Evans, P.D., 1980. Biogenic amines in the insect nervous system. Advances in Insect Physiology 15, 317–473.

Franzios, G., Mirotson, M., Hatziapostolou, E., Kral, J., Scouras, Z.G., Mavragani, T.P., 1997. Insecticidal and genotoxic activities of mint essential oils. Journal of Agricultural and Food Chemistry 45, 2690–2694.

Goebel, H., Schmidt, G., Dworschak, M., Heiiss, D., 1995. Essential plant oils and headache mechanisms. Phytomedicine 2, 93–102.

Gomez, L.E., Coen, C.E., 2013. Insect Attractant Formulations and Insect Control. US Patent no. WO2013173300 A1.

Hausen, B.M., Reichling, J., Harkenthal, M., 1999. Degradation products of monoterpenes are the sensitizing agents in tea tree oil. American Journal of Contact Dermatitis 10, 68–77.

Hink, W.F., Liberati, T.A., Collart, M.G., 1998. Toxicity of linalool to life stages of the cat flea, *Ctenocephalides felis* (Siphonaptera: Pulicidae), and its efficacy in carpet and on animals. Journal of Medical Entomology 25, 1–4.

Hink, W.F., Duffey, T.E., (Inventors), 1990. Controlling Ticks and Fleas with Linalool. United States Patent No. 4933371.

Hinkle, N.C., 1995. Natural born killers. Pest Control Technology 54, 56–116.

Ho, S.H., Ma, Y., Huang, Y., 1997. Anethole, a potential insecticide from *Illicium verum* Hook F. against two stored product insects. International Pest Control 39, 50–51.

Hold, K.M., Sirisoma, M.S., Ikeda, T., Narahashi, T., Casida, E., 2000. Alpha-thujone (the active component of absinthe): gamma-aminobutyric acid type-A receptor modulation and metabolic detoxification. Proceedings of National Academy Sciences of the United States of America 97, 3826–3831.

Hollingworth, R.M., Johnstone, E.M., Wright, N., 1984. Aspects of the biochemistry and toxicology of octopamine in arthropods. In: Magee, P.S., Kohn, G.K., Menn, J.J. (Eds.), Pesticide Synthesis through Rational Approaches. ACS Symposium Series No. 255. American Chemical Society, Washington, DC, pp. 103–125.

Huang, Y., Shuit-Hung Ho, S.H., Lee, H.C., Yap, Y.L., 2002. Insecticidal properties of eugenol, isoeugenol and methyleugenol and their effects on nutrition of *Sitophilus zeamais* Motsch. (Coleoptera: Curculionidae) and *Tribolium castaneum* (Herbst) (Coleoptera: Tenebrionidae). Journal of Stored Products Research 38, 403–412.

IARC, 1999. Monographs on the Evaluation of Carcinogenic Risks to Humans, vol. 73. International Agency for Research on Cancer, WHO, Geneva, pp. 307–327.

IPCS, 2009. Summary of Evaluations Performed by the Joint FAO/WHO Expert Committee on Food Additives: Trans-Anethole. International Program on Chemical Safety (IPCS), WHO, Geneva. November 12, 2001.

Isman, M.B., Machial, C.M., Miresmailli, S., Bainard, L.D., 2007. Essential oil based pesticides: new insights from old chemistry. In: Ohkawa, H., Miyagawa, H., Lee, P.W. (Eds.), The Book: Pesticide Chemistry. Wiley-VCH Verlag GmbH & Co. KGaA. ISBN:9783527611249.

ISO, 1996. Essential Oils-Oil of *Melaleuca*, Terpen-4-ol Type (Tea Tree Oil). ISO (International Standards Organization), 4730, Oxford University Press, Oxford, UK.

Johnson, C.B., Kirby, J., Naxakis, G., Pearson, S., 1999. Substantial UV-B-mediated induction of essential oils in sweet basil (*Ocimum basilicum* L.). Phytochemistry 51, 507–510.

Kalyanasundaram, M., Mathew, N., 2006. N,N-diethyl phenylacetamide (DEPA): a safe and effective repellent for personal protection against hematophagous arthropods. Journal of Medical Entomology 43, 518–525.

Karr, L.L., Coats, J.R., 1988. Insecticidal properties of d-limonene. Journal of Pesticide Science 13, 287–290.

Kim, D.H., Ahn, Y.J., 2001. Contact and fumigant activities of constituents of *Foeniculum vulgare* fruit against three coleopteran stored product insects. Pest Management Science 57, 301–306.

Klocke, J.A., Darlington, M.V., Balandrin, M.F., 1987. 1,8-Cineole (Eucalyptol), a mosquito feeding and ovipositional repellent from volatile oil of *Hemizonia fitchii* (Asteraceae). Journal of Chemical Ecology 13, 21–31.

Knio, K.M., Usta, J., Daghar, S., Zournajian, H., Kreydiyyeh, S., 2008. Larvicidal activity of essential oils extracted from commonly used herbs in Lebanon against the seaside mosquito, *Ochlerotatus caspius*. Bioresource Technology 99, 763–768.

Kumbhar, P.P., Dewang, P.M., 2001. Monoterpenoids: the natural pest management agents. Fragrances and Flavours Association of India 3, 49–56.

Lee, H.S., 2005. Food protective effect of acaricidal components isolated from anise seeds against the stored food mite, *Tyrophagus putrescentiae* (Schrank). Journal of Food Protection 68, 1208–1210.

Lee, E.J., Kim, J.R., Choi, D.R., Ahn, Y.J., 2008. Toxicity of cassia and cinnamon oil compounds and cinnamaldehyde-related compounds to *Sitophilus oryzae* (Coleoptera: Curculionidae). Journal of Economic Entomology 101, 1960–1966.

Lee, S., Peterson, C.J., Coats, J.R., 2003. Fumigation toxicity of monoterpenoids to several stored product insects. Journal of Stored Products Research 39, 77–85.

Lee, S.E., Lee, B.H., Choi, W.S., Park, B.S., Kim, J.G., Campbell, B.C., 2001. Fumigant toxicity of volatile natural products from korean spices and medicinal plants towards the rice weevil, *Sitophilus oryzae* (L). Pest Management Science 57, 548–553.

Lis-Balchin, M., Hart, S., 1999. Studies on the mode of action of the essential oil of lavender (*Lavandula angustifolia* P. Miller). Phytotherapy Research 13, 540–542.

Lopez, M.D., Pascual-Villalobos, M.J., 2015. Are monoterpenoids and phenylpropanoids efficient inhibitors of acetylcholinesterase from stored product insect strains? Flavour and Fragrance Journal 30, 108–112.

Lopez, M.D., Maudhuit, A., Pascual-Villalobos, M.J., Poncelet, D., 2012. Development of formulations to improve the controlled release of linalool to be applied as an insecticide. Journal of Agricultural and Food Chemistry 60, 1187–1192.

Ma, W.B., Jun-Tao Feng, J.T., Jiang, Z.L., Zhang, X., 2014. Fumigant activity of 6 selected essential oil compounds and combined effect of methyl salicylate and trans-cinnamaldehyde against *Culex pipiens pallens*. Journal of the American Mosquito Control Association 30, 199–203.

Maganga, M.E., Gries, G., Gries, R., 1996. Repellency of various oils and pine oil constituents to house flies (Diptera: Muscidae). Environmental Entomology 25, 1182–1187.

Masotti, V., Juteau, F., Bessiere, J.M., Viano, J., 2003. Seasonal and phenological variations of the essential oil from the narrow endemic species *Artemisia molinieri* and its biological activities. Journal of Agricultural and Food Chemistry 51, 7115–7121.

Morais, S.M., Cavalcanti, E.S., Bertini, L.M., Oliveira, C.L., Rodrigues, J.R., Cardoso, J.H., 2006. Larvicidal activity of essential oils from Brazilian *Croton* species against *Aedes aegypti* L. Journal of the American Mosquito Control Association 22, 161–164.

Muller, G.C., Junnila, A., Butler, J., Kravchenko, V.D., Revay, E.E., Weiss, R.W., Schlein, Y., 2009. Efficacy of the botanical repellents geraniol, linalool, and citronella against mosquitoes. Journal of Vector Ecology 34, 2–8.

Muller, G.C., Junnila, A., Kravchenko, V.D., Revay, E.E., Butler, J., Orlova, O.B., Weiss, R.W., Schlein, Y., 2008. Ability of essential oil candles to repel biting insects in high and low biting pressure environments. Journal of the American Mosquito Control Association 24, 154–160.

Nakamura, C.V., Ueda-Nakamura, T., Bando, E., Melo, A.F.N., Cortez, D.A.G., Dias Filho, B.P., 1999. Antobacterial activity of *Ocimum gratissimum* L. essential oil. Memórias do Instituto Oswaldo Cruz 94, 675–678.

Nelson, S., McClanahan, R.H., Knebel, N., Thomassen, D., Gordon, W.P., Oishi, S., 1992. The metabolism of (*R*)-(+)-pulegone, a toxic monoterpene. Environmental Science Research 44, 287–296.

Newberne, P., Smith, R.L., Doull, J., Goodman, J.I., Munro, I.C., Portoghese, P.S., Wagner, B.M., Weil, C.S., Woods, L.A., Adams, T.B., Lucas, C.D., Ford, R.A., 1999. The FEMA GRAS assessment of trans-anethole used as a flavouring substance. Flavor and Extract Manufacturer's Association. Food Chemistry and Toxicology 37, 789–811.

Ngoh, S.P., Choo, L.E.W., Pang, F.Y., Huang, Y., Kini, M.R., Ho, S.H., 1998. Insecticidal and repellent properties of nine volatile constituents of essential oils against the American cockroach, *Periplaneta americana* (L.). Pesticide Science 54, 261–268.

Nilsson, U., Magnusson, K., Karlberg, O., Karlberg, A.T., 1999. Are contact allergens stable in patch test preparations? Investigation of the degradation of d-limonene hydroperoxides is in petrolatum. Contact Dermatitis 40, 127–132.

Opdyke, D.L., 1979. Monographs on Fragrance Raw Materials. Pergamon, New York.

Ozek, T., Tabanca, N., Demirci, F., Wedge, D.E., Başer, K.H.S., 2010. Enantiomeric distribution of some linalool containing essential oils and their biological activities. Records of Natural Products 4, 180–192.

Ojimelukwe, P.C., Adler, C., 2000. Toxicity and repellent effects of eugenol, thymol, linalool, menthol and other pure compounds on *Dinoderus bifloveatus* (Coleoptera: Bostrichidae). Journal of Sustainable Agriculture and Environment 2, 47–54.

Padalia, R.C., Verma, R.S., Chauhan, A., Goswami, P., Verma, S.K., Darokar, M.P., 2015. Chemical composition of *Melaleuca linarrifolia* Sm. from India: a potential source of 1,8-cineole. Industrial Crops and Products 63, 264–268.

Padilha de Paula, J., Gomes-Carneiro, M.R., Paumgartten, F.J., 2003. Chemical composition, toxicity and mosquito repellency of *Ocimum selloi* oil. Journal of Ethnopharmacology 88, 253–260.

Park, I.K., Choi, K.S., Kim, D.H., Choi, I.H., Kim, L.S., Bak, W.C., Choi, J.W., Shin, S.C., 2006. Fumigant activity of plant essential oils and components from horseradish (*Armoracia rusticana*), anise (*Pimpinella anisum*) and garlic (*Allium sativum*) oils against *Lycoriella ingenua* (Diptera: Sciaridae). Pest Management Science 62, 723–728.

Park, S.N., Lim, Y.K., Freire, M.O., Cho, E., Jin, D., Kook, J.K., 2012. Antimicrobial effect of linalool and α-terpineol against periodontopathic and cariogenic bacteria. Anaerobe 18, 369–372.

Pitts, R.J., Fox, A.N., Zwiebel, L.J., 2004. A highly conserved candidate chemoreceptor expressed in both olfactory and gustatory tissues in the malaria vector, *Anopheles gambiae*. Proceedings of National Science Academy of the United States of America 101, 5058–5063.

Rahimmalek, M., Maghsoudi, H., Sabzalian, M.R., Pirbalouti, A.G., 2014. Variability of essential oil content and composition of different Iranian fennel (*Foeniculum vulgare* Mill.) accessions in relation to some morphological and climatic factors. Journal of Agricultural Science and Technology 16, 1365–1374.

Ryan, M.F., Bryan, O., 1988. Plant-insect coevolution and inhibition of acetylcholinesterase. Journal of Chemical Ecology 14, 1965–1975.

Sadgrove, N., Jones, G., 2015. A contemporary introduction to essential oils: chemistry, bioactivity and prospects for Australian agriculture. Agriculture 5, 48–102.

Sanchez-Ramos, I.I., Castanera, P., 2001. Acaricidal activity of natural monoterpenes on *Tyrophagus putrescentiae* (Schrank), a mite of stored food. Journal of Stored Products Research 37, 93–101.

Scriven, R., Meloan, C.E., 1984. Determining the active component in 1,3,3-trimethyl-2-oxabicyclo [2,2,2] octane (Cineole) that repels the American cockroach, *Periplaneta americana*. Ohio Journal of Science 84, 85–88.

Sfara, V., Zerba, E.N., Alzogaray, R.A., 2009. Fumigant insecticidal activity and repellent effect of five essential oils and seven monoterpenes on first-instar nymphs of *Rhodnius prolixus*. Journal of Medical Entomology 46, 511–515.

Shaaya, E., Kostjukovsky, M., 1998. Efficacy of phyto-oils as contact insecticides and fumigants for the control of stored-product insects. In: Ishaaya, I., Degheele, D. (Eds.), Insecticides with Novel Modes of Action: Mechanisms and Application. Springer, Berlin, pp. 171–187.

Sharifian, I., Safaralizade, M.H., Najafi-Moghaddam, P., 2011. Investigation on the insecticidal efficacy of novel pellet formulation against stored products beetles. Munis Entomology & Zoology 6, 204–209.

Sharma, R.N., 1993. The utilization of essential oils and some common allelochemic constituent for non-insecticidal pest management strategies. In: Dhar, K.L., Thappa, R.K., Agarwal, S.G. (Eds.), Newer Trends Essential Oils Flavours. Tata McGraw Hill, New Delhi, India, pp. 341–351.

Silva, S.M., Simone, Y.A., Fabio, S.M., Gustavo, F., Fransisco, A.M., Nakashima, T., 2011. Essential oils from different plant parts of *Eucalyptus cinerea* F. Muell. ex Benth. (Myrtaceae) as a source of 1,8-cineole and their bioactivities. Pharmaceuticals 4, 1535–1550.

Shaaya, E., Ravid, U., Paster, N., Juven, B., Zisman, U., Pissarev, V., 1991. Fumigant toxicity of essential oils against four major stored-product insects. Journal of Chemical Ecology 17, 499–504.

Sporhase-Eichmann, U., Vullings, H.G.B., Bulis, R.M., Horner, M., Schurmann, F., 1992. Octopamine-immunoreactive neurons in the central nervous system of the cricket, *Gryllus bimaculatus*. Cell and Tissue Research 268, 287–304.

Thomassen, D., Knebel, N., Slattery, J.T., McClanahan, R.H., Nelson, S.D., 1992. Reactive intermediates in the oxidation of menthofuran by cytochromes P-450. Chemical Research in Toxicology 5, 123–130.

Thompson, D.C., Barhoumi, R., Burghardt, R.C., 1998. Comparative toxicity of eugenol and its quinone methide metabolite in cultured liver cells using kinetic fluorescence bioassays. Toxicology and Applied Pharmacology 149, 55–63.

Trigg, J.K., Hill, N., 1996. Laboratory evaluation of *Eucalyptus*-based repellent against four biting arthropods. Phytotherapy Research 10, 43–46.

Tripathi, A.K., 2004. Essential oils and herbal products in the management of household pests. In: Proceedings of the Training Programme on Pest Management in Buildings for Pest Management Professionals, November 16–18, 2004. CBRI, Roorkee, India.

Tripathi, A.K., Prajapati, V., Gupta, R., Kumar, S., 1999. Herbal material for the insect-pest management in stored grains under tropical conditions. Journal of Medicinal and Aromatic Plant Science 21, 408–430.

Tsuda, H., Ohshima, Y., Nomoto, H., Fujita, K.I., Matsuda, E., Iigo, M., Takasuka, N., Moore, M.A., 2004. Cancer prevention by natural compounds. Drug Metabolism and Pharmacokinetics 19, 245–263.

Veal, L., 1996. The potential effectiveness of essential oils as a treatment for headlice, *Pediculus humanus capitis*. Complementary Therapies in Nursing & Midwifery 2, 97–100.

Villar, D., Knight, M.J., Hansen, S.R., Buck, W.B., 1994. Toxicity of *Melaleuca* oil and related essential oils applied topically on dogs and cats. Veterinary and Human Toxicology 36, 139–142.

Weaver, D.K., Dunkel, F.V., Ntezurubanza, L., Jackson, L.L., Stock, D.T., 1991. The efficacy of linalool, a major component of freshly-milled *Ocimum canum* Sims (Lamiaceae), for protection against postharvest damage by certain stored product Coleoptera. Journal of Stored Products Research 27, 213–220.

Webb, N.J.A., Pitt, W.R., 1993. Eucalyptus oil poisoning in childhood: 41 cases in South-East Queensland. Journal of Paediatrics and Child Health 29, 368–371.

Zhu, B.C.R., Henderson, G., Chen, F., Maistrello, L., Laine, R.A., 2001. Nootkatone is a repellent for Formosan subterranean termite (*Coptotermes formosanus*). Journal of Chemical Ecology 27, 523–531.

Zhu, B.C.R., Henderson, G., Sauer, A.M., Yu, Y., Crowe, W., Laine, R.A., 2003. Structure-activity of valencenoid derivatives and their repellence to the formosan subterranean termite. Journal of Chemical Ecology 29, 2695–2701.

Further Reading

Butler, P., 2010. It's like magic; removing self-adhesive stamps from paper. American Philatelist 124, 910–913.

Carteau, D., Bassani, D., Pianet, I., 2008. The "Ouzo effect": following the spontaneous emulsification of trans-anethole in water by NMR. Comptes Rendus Chimie 11, 493–498.

Environment Canada, 2010. Screening Assessment for the Challenge Benzene, 1,2-dimethoxy-4-(2-propenyl)-(Methyl eugenol). Chemical Abstracts Service Registry Number 93-15-2 Environment Canada – Health Canada. Available at: http://www.ec.gc.ca/ese-ees/default.asp?lang=En&n=0129FD3C-1.

Fahlbusch, K.G., Hammerschmidt, F.J., Panten, J., Pickenhagen, W., Schatkowski, D., Bauer, K., Garbe, D., Surburg, H., 2002. Flavors and fragrances. In: Ullmann's Encyclopedia of Industrial Chemistry. Wiley-VCH, Weinheim. http://dx.doi.org/10.1002/14356007.a11-141.

Government of Canada, 2009. An Act to Amend the Tobacco Act. Statutes of Canada 2009. Chapter 27. Second Session, Fortieth Parliament, 57–58 Elizabeth II, 2009 Public Works and Government Services Canada, Ottawa, Ontario. Available at: http://laws.justice.gc.ca/PDF/Annual/2/2009_27.pdf].

Jack, L.F., 2001. Materials in Dentistry: Principles and Applications, second ed. Lippincott Williams & Wilkins. ISBN:0-7817-2733-2.

Jadhav, B.K., Khandelwal, K.R., Ketkar, A.R., Pisal, S.S., Khandelwal, K.P., 2004. Formulation and evaluation of mucoadhesive tablets containing eugenol for the treatment of periodontal diseases. Drug Development and Industrial Pharmacy 30, 195–203.

Park, H.M., 2011. Limonene, a natural cyclic terpene, is an agonistic ligand for adenosine A(2A) receptors. Biochemical and Biophysical Research Communications 404, 345–348.

Rao, P.V., Gan, S.H., 2014. Cinnamon: A Multifaceted Medicinal Plant. Evidence-Based Complementary and Alternative Medicine. Article ID 642942, 12 pages. http://dx.doi.org/10.1155/2014/642942.

Spernath, A., Aserin, A., 2006. Microemulsions as carriers for drugs and nutraceuticals. Advances in Colloid and Interface Science 128–130, 47–64.

Tarantino, A.S., De Benedictis, V.M., Marta Madaghiele1, M., Christian Demitri, C., Sannino, A., 2014. Innovative approach for active food packaging using cinnamaldehyde. Journal of Biotechnology 185S, S25.

Yamane, M.A., Williams, A.C., Barry, B.W., 1995. Effects of terpenes and oleic acid as skin penetration enhancers towards 5-fluorouracil as assessed with time; permeation, partitioning and differential scanning calorimetry. International Journal of Pharmaceutics 116, 237–251.

17

Antifeedant Phytochemicals in Insect Management

(so Close yet so Far)

Opender Koul

Insect Biopesticide Research Centre, Jalandhar, India

1. ANTIFEEDANT APPROACH

It is well known that phytochemicals are the compounds produced by plants that are not directly involved in growth and development and are traditionally referred to as secondary metabolites. These compounds have been evaluated in the search for new drugs, antibiotics, insecticides, herbicides, and behavior-modifying chemicals. Many of these compounds have been shown to have important adaptive significance in protection against herbivory (Croteau et al., 2000). Phytochemical diversity of insect defenses in tropical and temperate plant families has also been significantly established (Arnason et al., 2004), and this chapter will concentrate on the compounds that interfere with the feeding behavior of insects as well as update and expand the earlier comprehensive compilations (Koul, 2005, 2008; Isman, 2006). Such feeding behavior-modifying molecules belong mostly to the class of compounds called allomones (Nordlund, 1981), which are differentiated from the pheromones because they mediate interspecific, rather than intraspecific interactions. The tremendous diversity, coupled with the intensity, of allomone-mediated interspecific interactions makes allomonal chemicals potential agents for insect pest control (Koul, 2005).

Behavioral mechanisms provide a system of avoidance of nonhost chemicals by which insects select their food, though the molecular basis for action of chemical deterrents on both gustatory and olfactory sensory systems in insects is only poorly understood. Among plant antiherbivore chemistry, a strong link does not exist between feeding deterrence and internal toxicity in insects, suggesting that behavioral rejection is not an adaptation to ingested effects but more an outcome of deterrent receptors with wide chemical sensitivity (Mullin et al., 1991, 1994). Many of these substances are bitter, and acceptance of host plants by herbivores requires chemoreception of favorable levels of phagostimulants relative to plant antifeedants

Ecofriendly Pest Management for Food Security
http://dx.doi.org/10.1016/B978-0-12-803265-7.00017-8

(Dethier, 1980). This restricts the application of a very liberal definition for an antifeedant, namely, "any substance that reduces consumption by an insect" to a more precise definition, "a peripherally-mediated behavior modifying substance (i.e., acting directly on the chemosensilla in general and deterrent receptors in particular) resulting in feeding deterrence" (Isman, 1994). This definition, however, excludes chemicals that suppress feeding by acting on the central nervous system (following ingestion or absorption), or substances that have sublethal toxicity to the insect (Isman, 2002). Several definitions for the term *antifeedant* exist in the literature (Munakata, 1975; Norris, 1986; Frazier and Chyb, 1995; Glendinning, 1996; Messchendorp et al., 1998) suggesting that the definition of the term varies widely. A broader approach suggested by Mansson (2005) explains this where volatile and nonvolatile preingestive inhibitors have been regarded as antifeedant compounds. The reason not to include postingestive inhibitors in the antifeedant concept is that these inhibitors demand feeding during a longer period than preingestive inhibitors, which may already have caused significant and possibly mortal damage to the plant, when the insect finishes feeding. However, ingestive inhibitors could be borderline cases as chemosensory and accessory cells are involved in both preingestive and ingestive inhibition. Thus, one category could be of repellent and arrestant type of antifeedants where insects avoid feeding without coming in contact with plant material. Similarly, insects are suppressed from biting once contact has been made with plant material, leading to antifeedance (suppressants). The most accepted category is that of feeding deterrent phytochemicals, which deter insects from feeding after they have bitten the plant material, i.e., inhibition by gustatory responses. Feeding deterrents from plants with a wide diversity of structures are not known to directly interfere with insect taste cell responses to phagostimulants such as sugars (Lam and Frazier, 1991; Schoonhoven et al., 1992). Presently the mode of action of feeding modifying chemicals in insect gustatory systems is largely unknown (Frazier, 1992; Schoonhoven et al., 1992), though some molecular targets have been identified (Koul, 1997; De Bruyne and Warr, 2006). Taste receptor proteins are only now beginning to be biochemically purified and cloned. The determination of the molecular basis for action of feeding deterrents in the insect gustatory system is thus a primary goal among basic and applied entomologists interested in insect–plant interactions or in the control of herbivore pests. According to the theory of biochemical coevolution it should be possible to develop an evolutionary pattern of antifeedants on the basis of their distribution in different plant families and their biosynthetic pathways.

However, the pattern of distribution varies among families. One plant family may concentrate on one type of deterrent molecule like limonoids in the Rutales (Champagne et al., 1992), and within families individual members may have developed further barriers to feeding. For instance, it is clear that flavonoids in plants can modulate the feeding behavior of insects, though mechanisms associated with these behavioral responses are not clearly understood (Simmonds, 2001). Other families may diversify their deterrents; for example, nonprotein amino acids (e.g., L-canavanine), alkaloids, cyanogens, and isoflavones are found in the Fabaceae. Plants produce all these and many varied compounds in the first instance as protective devices against insect feeding. Thus, a majority of plant families rely on secondary plant metabolites for protection from phytophagous insects. One might surmise that within such a family the more advanced members are better protected than others. Berenbaum (1983) has pointed to good evidence in the Apiaceae where plant defense is based on hydroxycoumarins, linear furanocoumarins, and angular furanocoumarins, which are biosynthetically and toxicologically related.

2. SOURCES AND CHEMISTRY

Since less than 1% of all secondary plant substances, estimated to number 400,000 or more, have been tested against a limited number of insect species only, several effective compounds may remain to be discovered. Researchers, when testing candidate compounds, use only a few or even only one species for evaluation. Effective feeding deterrents to a particular insect will easily escape attention. For example, for the well-known antifeedant azadirachtin, tested against seven orthopterans, the interspecific differences span six orders of magnitude. Compounds known as insect antifeedants usually have a more oxidized or unsaturated structure. However, molecular size and shape as well as functional group stereochemistry also affect the antifeedant activity of a molecule. For example, sesquiterpene lactones are defensive compounds and can be divided into two groups based on the stereochemistry of their lactone ring junction, either *cis*-fused or *trans*-fused. Stereochemical variation in sesquiterpene lactone ring junctions can influence resistance to herbivorous insects as shown in controlled feeding trials with two pairs of diastereomeric sesquiterpene lactones deterring feeding by the polyphagous grasshopper *Schistocerca americana* (Drury). Sesquiterpene lactone stereochemistry and concentration significantly influenced feeding behavior with grasshoppers consuming less of the *trans*-fused compounds than the *cis*-fused compounds (Ahern and Whitney, 2014). Stereochemical trait polymorphism is widely distributed in nature and, therefore, could have substantial consequences for the ecology and evolution of large groups of plants, specifically the Asteraceae family. Flavonoids play an important role in the protection of plants against plant-feeding insects and herbivores. Many compounds of this class are known to deter feeding in insects (Mierziak et al., 2014). However, antifeedants can be found among all the major classes of secondary metabolites such as limonoids, quassinoids, diterpenes, sesquiterpenes, monoterpenes, coumarins, alkaloids, maytansinoids, ellagitannins, etc. However, the most potent antifeedants belong to the terpenoid group, which has the greatest number and diversity of known antifeedants. Many such compounds have been comprehensively dealt with earlier (Koul, 2008).

During the last few years several studies add to the kitty of such feeding-deterrent products from plants. Edible spices with strong smells or heavy tastes may be a promising resource of feeding deterrents. Feeding deterrence of the ethanol extracts of 21 common spices against the larvae of a generalist pest species, *Helicoverpa armigera*, using a multiple-choice leaf disc bioassay revealed that *Zanthoxylum bungeanum* extract (as a reference) always evoked significant feeding deterrence, while *Piper nigrum* (both black pepper and white pepper), *Piper longum*, and *Angelica dahurica* evoked the strongest and equivalent feeding deterrence. The potent feeding-deterrent activity of *Piper* species may be a common characteristic at genus level (Li et al., 2014).

Six different indigenous plants were screened for antifeedant and insecticidal activity against fourth instar larvae of *Epilachna* beetle, *Henosepilachna vigintioctopunctata*, which is a severe pest on brinjal. Among the plants screened, *Achyranthes aspera* showed higher activity against the selected pest. Ethyl acetate extracts of *A. aspera* showed higher antifeedant index and insecticidal activity against fourth instar larvae of *H. vigintioctopunctata*. Preliminary phytochemical analysis revealed that the presence of alkaloid and quinines in the ethyl acetate extracts indicate higher percentage of activity. Hence, it may suggest its use for controlling the vegetable insect pest *H. vigintioctopunctata* (Jeyasankar et al., 2014). Ethanol extracts obtained from aerial parts of 64 native plants from Central Argentina were tested for their insect antifeedant activity against *Epilachna paenulata* (Coleoptera: Coccinellidae) by choice test. Extracts derived from *Achyrocline satureioides* (Asteraceae), *Baccharis coridifolia* (Asteraceae), *Baccharis flabellata* (Asteraceae), *Ruprechtia apetala* (Polygonaceae) and *Vernonanthura nudiflora* (Asteraceae), showed more than 97% inhibition of the feeding of *E. paenulata* at $100 \mu g/cm^2$. These active extracts were further evaluated for their effectiveness against *Spodoptera frugiperda* (Lepidoptera: Noctuidae). All these extracts, except for that derived from *A. satureioides*, negatively influenced the feeding behavior of *S. frugiperda* at $100 \mu g/cm^2$ (Corral et al., 2014). Similarly, several other plants from Celastraceae, Rhamnaceae, Scrophulariaceae, Menispermaceae, Meliaceae, etc. have also been reported as strong antifeedant agents for lepidopteran larvae (Pavela, 2010a,b; Cespedes and Alarcon, 2011; Selvam and Ramakrishnan, 2014). Many essential oils have been reported as feeding deterrents against a variety of insects (Koul et al., 2008; Krishnappa et al., 2010).

Proteinase inhibitors from plants are also known to induce feeding deterrence in insects. Low-molecular-weight peptidyl proteinase inhibitors (PIs) including leupeptin, calpain inhibitor I, and calpeptin were potent antifeedants for adult western corn rootworm against the phagostimulation of cucurbitacin B or a corn pollen extract. Leupeptin was the strongest ($ED_{50} = 0.36$ and $0.55 nmol/disk$ for cucurbitacin B and corn pollen extract, respectively) among PIs tested with an antifeedant potency much stronger than the steroid progesterone ($ED_{50} = 2.29$ and $5.05 nmol/disk$ for cucurbitacin B and corn pollen extract, respectively), but slightly less than the reference alkaloid, strychnine ($ED_{50} = 0.17$ and $0.37 nmol/disk$ for Cucurbitacin B and CPE, respectively) (Kim and Mullin, 2003). All active PIs contain a di- or tripeptidyl aldehyde moiety, indicating that PIs exert their antifeedant effects by covalent interaction with putative sulfhydryl (SH) groups on taste receptors as do these PIs with cysteine proteinases. However, opposite inhibition potency against cucurbitacin B versus corn pollen extracts by two thiol group reducing agents, DTT and L-cysteine, and the results with other cysteine-modifying reagents obscure the net functional role of SH groups at western corn rootworm taste chemoreceptors. Surprisingly, the model phagostimulant for diabroticites, cucurbitacin B, was more easily counteracted by these feeding deterrents than the stimulants present in corn pollen extracts. Three-dimensional structure–antifeedant relationships for the PIs suggest that a novel taste chemoreception mechanism exists for these peptidyl aldehydes or that they fit partially into a strychnine binding pocket on protein chemoreceptors. Favorable economic benefit may be achieved if PIs are discovered to be useful in adult western corn rootworm control, since both pre- and postingestive sites would be targeted (Kim and Mullin, 2003).

The mode of action of these inhibitors is still under debate, and it remains unclear whether the deleterious effects of protease inhibitors stem from an antidigestive effect, through proteolytic inhibition, or from a toxic effect by inducing hyperproduction of protease, leading to a shortage in amino acids (Pandey and Jamal, 2014). However, the purified inhibitor of 21 kDa obtained from *Tamarindus indica* in this study clearly showed high antifeedant activity against

H. armigera suggesting its greater susceptibility as reported earlier for double-headed inhibitor from *Dolichos biflorus* seeds as well (Kuhar et al., 2013).

3. RECENTLY ISOLATED FEEDING DETERRENT MOLECULES

More than 1000 compounds have been isolated and evaluated against a variety of insect species as feeding deterrents and well documented (Koul, 2005, 2008). However, during the last 5 years some specific molecules have been isolated and their derivatives prepared to obtain potential compounds with enhanced activity. Terpenes from some tropical species of the Rutales were tested for insect antifeedant activity against rice weevil, *Sitophilus oryzae* (L.) using a flour disk bioassay that requires only small amounts of compounds (0, 0.05, 0.25, and 0.50% w/w). At 0.50% (w/w) five compounds isolated from *Lansium domesticum* (iso-onoceratriene, 3-keto-22-hydroxyonoceradiene, onoceradienedione, lansiolic acid, and lansiolic acid A) were shown to exhibit significant antifeedant activity. The most interesting results were obtained from the spirocaracolitones from *Ruptiliocarpon caracolito*, which produced total feeding inhibition at 0.50% and potent antifeedant activity at concentrations as low as 0.05%. In conclusion, the antifeedant bioassay provides a rapid and inexpensive method for screening novel compounds available in small quantities to assess their activity as insect antifeedants (Omar et al., 2007).

Onoceratriene

Onoceradienedione

3-Keto-22-hydroxyonoceradiene

3-Ketolansiolic acid

Lansiolic acid   **Lansiolid acid A**

Two novel limonoids, musidunin and musiduol, were isolated from a methanol extract of *Croton jatrophoides* by bioassay-guided fractionation. Their structures were established by extensive NMR experiments. Interestingly, A,B-seco limonoid (musidunin) contains a unique acetal annulation of A, A′, and B′ rings. Both limonoids exhibited antifeedant activities against two pests, *Pectinophora gossypiella* and *S. frugiperda* (Nihei et al., 2006).

Musidunin   **Musiduol**

Phytochemical investigation of the chloroform extract of *Tinospora cordifolia* yielded a new clerodane diterpenoid tincordin along with tinosporide, 8-hydroxytinosporide, columbin, 8-hydroxycolumbin, and 10-hydroxycolumbin. The structure of the new compound was elucidated comprehensively using 1D and 2D NMR methods. All major clerodane diterpenoids isolated were tested for their efficacy as insect antifeedants against *Earias vitella, Plutella xylostella*, and *Spodoptera litura* (Sivasubramanian et al., 2013). The activity of these compounds ranged between 65.0% and 80.0% at a concentration of $10.0\,\mu g/cm^2$ for all of the six compounds.

Tincordin   **Tinopsoride**   **8-Hydroxytinosporide**

| Columbin | 8-Hydroxycolumbin | 10-Hydroxycolumbin |

Three phragmalin-type limonoids, swietephragmin H, swietephragmin I and 11-hydroxyswietephragmin B, and a mexicanolide-type limonoid 2-hydroxy-6-deacetoxyswietenine together with known compounds, 6-O-acetyl-2-hydroxyswietenin, 2-hydroxyswietenine, swietemahonin G, methyl 6-hydroxyangolensate and 7-deacetoxy-7-oxogedunin were isolated from the leaves of *Swietenia mahogani* (Meliaceae). Their structures were established by extensive NMR experiments in conjunction with mass spectrometry. The antifeedant activity of the isolated compounds was evaluated against third instar larvae of *Spodoptera littoralis* and swietemahonin G was most active at 300 ppm while others were active in the range of 500–1000 ppm (Abdelgaleil et al., 2013).

Swietemahonin G

Pieris formosa is a poisonous plant to livestock and is used as an insecticide in rural areas of China. Two novel polyesterified 3,4-seco-grayanane diterpenoids, pierisoids A and B (1 and 2), were isolated from its flowers and were identified by spectroscopic analysis and X-ray diffraction. Both compounds showed obvious antifeedant activity against cotton bollworm, indicating their toxic properties, suggesting a defensive role of polyesterified 3,4-seco-grayanane diterpenoids for *P. formosa* against herbivores (Li et al., 2010).

Pierisoids A : R = COCH$_2$CH$_3$
Pierisoids B : R = COCH$_3$

Betulinic acid, a secondary metabolite isolated from *Ziziphus jujuba*, was evaluated for its effect on the feeding deterrence in relation to food utilization at concentrations of 50, 100, 150, and 200 ppm against the fourth instar larvae of *Papilio demoleus* following a nonchoice leaf disk method. Significant reduction in food consumption and digestion was observed that reduced the growth of larvae. The efficiency of the larvae to convert digested and ingested food into body tissues was observed. The antifeedant activity of Betulinic acid proved to be the most potent against all developmental stages of *P. demoleus* with antifeedant activity of 90–95% at 200 ppm in the first 24 h and 80–85% after 48 h exposure. Significant antifeedant activity found in 200 ppm concentration was 94.04%. This plant extract has the potential to serve as an alternate biopesticide in the management of *P. demoleus* larvae (Vattikonda et al., 2014).

Fraxinellone, a well-known and significant naturally occurring compound isolated from Meliaceae and Rutaceae spp., has been widely used as a drug for the treatment of tumors. On the other hand, fraxinellone exhibited a variety of insecticidal activities including feeding deterrent activity, inhibition of growth, and larvicidal activity. Antifeedant and larvicidal activities of fraxinellone against the larvae of Lepidoptera, including *Mythimna separata*, *Agrotis ipsilon*, *P. xylostella*, and one kind of sanitary pest, *Culex pipiens pallens*, are known. The ovicidal activities and the effects of fraxinellone on the larval development of *M. separata* were also observed. The LC_{50} values of fraxinellone against third instar larvae of *M. separata*, second instar larvae of *P. xylostella*, and fourth instar larvae of *C. pipiens pallens* were $15.95/6.43/3.60 \times 10^{-2}$ mg/ml, and its AFC_{50} values against fifth instar larvae of *M. separata*, second instar larvae of *P. xylostella*, and second instar larvae of *A. ipsilon* were 10.73/7.93/12.58 mg/ml, respectively. Compared with the control group, fraxinellone obviously inhibited the pupation rate and the growth of *M. separata*. Once *M. separata* was treated with fraxinellone at concentrations of 5.0, 10.0, and 20.0 mg/ml, respectively, the stages from the larvae to adulthood and the egg-hatching duration were prolonged to 1/2/3, and 4/3/4 days, respectively. Additionally, fraxinellone strongly inhibited the development rate and the egg hatch proportion of *M. separata* (Lu et al., 2013).

Fraxinellone

Studies on the feeding deterrent activity of some natural, cyclic terpenes (myrcene, (+)-3-carene, (+)-limonene, (Â±)-camphene), synthetic alkenes (2-methyl-1-pentene, 2,4,4-trimethyl-1-pentene, 1-methylcyclohexene, 1-tetradecene) and their derivatives Î±-methylenelactones in choice and no-choice tests against Colorado potato beetle with potato, *Solanum tuberosum* L. leaf discs as consuming food were carried out. Deterrent indexes determined in the tests show that the strongest antifeedant to larvae and adults of *Leptinotarsa decemlineata* was Î±-methylenelactone obtained from 2,4,4-trimethyl-1-pentene. The structure modification of (+)-3-carene and

(Â±)-camphene via introduction of Î±-methylenelactone moiety increased their deterrent activity, especially against larvae. Other starting alkenes and Î±-methylenelactones obtained from them were weak deterrents to both developmental stages of *L. decemlineata* (Kmiecik-Małecka et al., 2009).

Among three species of polygonaceous feeding leaf beetle, *Galerucella grisescens, Gallerucida bifasciata*, and *Gastrophysa atrocyanea*, only *G. bifasciata* did not use a polygonaceous plant, *Persicaria lapathifolia*, as a food. Second and third instar larvae and adults of *G. bifasciata* fed on the leaves, however, the feeding amount decreased day by day and all died. The first instar larvae hardly fed on the leaves and all larvae died. Biological tests and chemical analysis revealed that feeding deterrents were present in *P. lapathifolia* leaves. A feeding deterrent was isolated and identified as 3-hydroxy-5-methoxy-6,7-methylenedioxy flavanone. The compound deterred first instar larvae of *G. bifasciata* from feeding and significantly lessened the survival rate of the larvae at concentrations higher than 1.0 mg/ml. Larvae of *G. grisescens* were not deterred from feeding by the compound at the concentration of 5 mg/ml. The presence of other fractions increased the feeding deterrent activity of the compound to first instar larvae of *G. bifasciata*. These results revealed that *P. lapathifolia* leaves contain multiple feeding deterrents to *G. Bifasciata* (Abe et al., 2007).

3-Hydroxy-5-Methoxy-6,7-Methylenedioxy Flavanone

The Indian bhant tree, *Clerodendron infortunatum* L. (Lamialus: Lamiaceae), is a well-known medicinal plant, but little information about its bioefficacy against agricultural pests exists. Dried leaves of *C. infortunatum* were subjected to extraction with hexane and methanol and then partitioned using different solvents of varying polarity. Three pure compounds were isolated and identified as clerodin, 15-methoxy-14, 15-dihydroclerodin, and 15-hydroxy-14, 15-dihyroclerodin. The antifeedant activity of these compounds was studied using a choice as well as a no-choice test method with 24 and 48 h observation periods. In the choice test conditions, all three compounds and azadirachtin showed 100% antifeedant activity at the highest concentration. Antifeedant Index (AI_{50}) values were 6.0, 6.0, and 8.0 ppm in choice tests, and increased to 8.0, 9.0, and 11.0 ppm in the no-choice tests, respectively against *H. armigera* (Abbaszadeh et al., 2014).

Four prenylated flavonoids, isoglabratephrin, (+)-glabratephrin, tephroapollin-F and lanceolatin-A, were isolated from *Tephrosia apollinea* L. and tested against three stored grain insects. A nutritional bioassay, using a flour disc and test concentrations of 0.65, 1.3, and 2.6 mg/g, revealed a significant reduction in the relative growth rate, relative consumption rate, and efficiency of conversion of ingested food by all insects. The studies show that these compounds could be potential antifeedants and activity significantly varies in relation to structures among the tested flavonoids (Nenaah, 2014).

Isoglabratephrin   **(+)-Glabratephrin**

Tephroapollin-F   **Lanceolatin-A**

Antifeedant and larvicidal activities of rhein (1,8-dihydroxyanthraquinone-3-carboxylic acid) isolated from the ethyl acetate extract of *Cassia fistula* flower were studied against lepidopteran pests *S. litura* and *H. armigera*. Significant antifeedant activity was observed against *H. armigera* (76.13%) at 1000 ppm concentration. Rhein exhibited larvicidal activity also against *H. armigera* (67.5), *S. litura* (36.25%), and the LC_{50} values were 606.50 ppm for *H. armigera* and 1192.55 ppm for *S. litura*. The survived larvae produced malformed adults (Duraipandiyan et al., 2011).

Rhein

Some compounds when isolated from a plant source may be weak feeding deterrents and can be modified to a more potential active compound. β-damascone, for instance, is reported as a weak attractant against *Myzus persicae*, but modifications of its structure caused the avoidance of treated leaves by aphids during settling and reluctance to probe in simple choice and no-choice experiments. Here, the electrical penetration graph (EPG) technique, which allows monitoring of aphid probing within plant tissues, was applied to explore the biological background and localization in plant tissues of the deterrent activities of β-damascone and its analogues. Activity of β-damascone and β-damascone-derived compounds depended on their substituents, which was manifested in the variation in the potency of the behavioral effect and differences in aphid probing phases that were affected. β-damascone appeared as a behaviorally inactive compound. The moderately active β-damascone ester affected aphid activities only during the phloem phase. The highly active deterrents, dihydro-β-damascol, β-damascone acetate, δ-bromo-γ-lactone, and unsaturated γ-lactone, affected pre-phloem and phloem aphid probing activities. The most effective structural modification that evoked the strongest negative response from *M. persicae* was the transformation of β-damascone into δ-bromo-γ-lactone. The behavioral effect of this transformation was demonstrated in frequent interruption of probing in peripheral tissues, which caused repeated failures in finding sieve elements and reduction in the ingestion time during the phloem phase in favor of watery salivation. The inhibition of aphid probing at both the pre-phloem and phloem levels reveals the passage of the compounds studied through the plant surface and their distribution within plant tissues in a systemic way, which may reduce the risk of the transmission of nonpersistent and persistent viruses (Gabryś et al., 2015).

Similarly, the antifeedant properties of optically pure isomers of pulegone and isopulegol and some enantiomeric pairs of bicyclic terpenoid lactones with the *p*-menthane system, derived from these isomeric starting compounds, were studied in choice and no-choice tests with the lesser mealworm. The original monoterpenes and γ-spirolactones were weak feeding deterrents to larvae and adults of *Alphitobius diaperinus*. The δ-hydroxy-γ-spirolactones showed significant activity against adults, but substituting the hydroxy group with a ketone group considerably reduced the deterrent activity of the resulting δ-keto-γ-spirolactones. The bicyclic γ-hydroxy-δ-lactones with condensed rings and with the (1S,6R,8R)-(+) and (1R,6S,8S)-(−) configuration of the chiral centers, and δ-hydroxy-γ-lactone (1S,4S,6S)-(+) were very strong antifeedants to both larvae and adults. The hydroxy groups and the configuration of the chiral centers of the molecules were very important features for determining the antifeedant activity of the lactones tested. Generally, the compounds studied were better antifeedants to adults than to larvae (Szczepanik et al., 2008).

4. HABITUATION OF FEEDING DETERRENTS

There are several possible mechanisms for the decrease in efficacy, including sensory adaptation, motor fatigue, and habituation. Many investigators have used the term habituation liberally without actually proving that the decrement represents habituation and not sensory adaptation or motor fatigue. Habituation, perhaps the simplest form of learning, is defined as the waning of a response as a result of repeated or prolonged presentation of a stimulus, which is not due to sensory adaptation or motor fatigue (Carew and Sahley, 1986). It represents a loss of some particular responses rather than the acquisition of new ones (Bernays and Weiss, 1996). Habituation differs from sensory adaptation in its ability to be terminated or reversed immediately by a novel or noxious stimulus (Thompson and Spencer, 1966). Different mechanisms may be responsible for the waning of response. Szentesi and Bernays (1984) showed that decreased response to antifeedants following prolonged exposure might result from the effect of mouthpart chemosensory information on the central nervous system, during palpation and feeding, or involving persistent synaptic changes in specific neural pathways (Bernays and Chapman, 1994), or from effects that follow ingestion of the deterrent (e.g., induction of a detoxifying enzyme (Szentesi and Bernays, 1984; Bernays and Chapman, 2000; Bernays and Weiss, 1996). Decreased response to antifeedants following prolonged exposure occurs most readily when a single antifeedant provides a weak inhibitory stimulus (Jermy et al., 1982; Szentesi and Bernays, 1984), but not to mixtures of antifeedants (Jermy, 1987). Whether antifeedants (plant extracts or pure compounds) would decrease the feeding preference of insects on prolonged exposure to them; a study with *Trichoplusia ni*, a generalist herbivore, was conducted. (Akhtar et al., 2003). The selection of antifeedants was based upon their strong feeding deterrent and growth inhibiting properties on *T. ni* shown in initial screening bioassays (Akhtar and Isman, 2003). Other investigators have also reported similar findings for the chosen antifeedants (*Melia volkensii*, *Origanum vulgare*, digitoxin, cymarin, xanthotoxin, toosendanin, and thymol). The main objectives of the experiments have been the following: (1) to determine under what conditions (different concentrations and larval instars) feeding experience with antifeedants changed subsequent feeding preferences of the cabbage looper, *T. ni*, and (2) to determine if the decreased response to antifeedants following prolonged exposure was the result of habituation in *T. ni*. If so, it should then be possible to demonstrate dishabituation.

It is now known that taste receptor cells do discriminate between bitter stimuli (Caicedo and Roper, 2001). In a recent study, discriminating by habituating the caterpillars to salicin and then determining whether the habituation generalized to caffeine or aristolochic acid has enabled researchers to examine discrimination in caterpillars with a modified peripheral taste profile. It was found that the intact and lat-ablated caterpillars both generalized the salicin habituation to caffeine but not to aristolochic acid. It was also determined whether this pattern of stimulus generalization could be explained by salicin and aristolochic acid generating distinct ensemble, rate, temporal, or spatiotemporal codes. It was concluded that temporal codes from the periphery can mediate discriminative taste processing (Glendinning et al., 2001, 2006). Knowledge of these factors may have consequences for the use of antifeedants in pest management and might be helpful in understanding host–plant shifts in insects. Although decreases in response to other antifeedants in this study have been observed, it is difficult to ascertain that these are the result of habituation, as it would be necessary to demonstrate dishabituation as well. Further testing of each of the antifeedants using aversive stimuli is necessary before such conclusions can be drawn. A decrease in response to feeding deterrents might enable the insect to feed normally on plant species that belong to the potential host–plant spectrum but are in some degree deterrent to naïve inexperienced individuals (Szentesi and Jermy, 1989) and would permit broadening of diet if the need arises (Bernays and Weiss, 1996). This could be of adaptive value where there is not a strong correlation between deterrence and toxicity of plant phytochemicals (Jermy et al., 1982; Jermy, 1987; Bernays and Chapman, 1994). According to Akhatar et al. (2003), a decrease in response to antifeedants following prolonged exposure could have many disadvantages from the pest management point of view as their experiments have clearly indicated that continuous contact of a feeding insect with a deterrent-containing food source caused increased acceptance of that food over time, thereby decreasing efficacy of the deterrent, which points to habituation and could offer some solutions for pest management. Decreased deterrence resulting from habituation has different implications for pest management than does decreased deterrence resulting from increased tolerance to toxic substances. Compounds to which insects have become habituated can be made effective deterrents again through the process of dishabituation (Akhtar et al., 2003). However, this has substantial implication for integrated pest management and needs extensive experimentation and design of evaluation in order to make such an approach effective.

Feeding and oviposition deterrence of three secondary plant compounds and their 1:1 blends to adult female *Frankliniella occidentalis* Pergande (Thysanoptera: Thripidae) and the potential for habituation of the thrips to the pure compounds and the 1:1 blends at various concentrations have been studied. In choice assays, dose-dependent feeding and oviposition deterrence of the two fatty acid derivatives methyl jasmonate and *cis*-jasmone, the phenylpropanoid allylanisole, and their blends were investigated after they were directly applied to bean leaf discs. The concentration required to reduce the feeding damage by 50% relative to the control treatment (FDC_{50}) was lowest for *cis*-jasmone and highest for allylanisole. The feeding deterrent effect of both jasmonates was increased when blended with allylanisole. Feeding deterrence and oviposition deterrence were strongly correlated. In no-choice assays conducted over four consecutive days, it was discovered that dilutions at low concentrations (FDC_{15}) applied to bean leaves resulted in habituation to the deterrents, whereas no habituation occurred at higher concentrations (FDC_{50}). In fact, 1:1 blends reduced the probability that thrips habituate to the deterrent compounds. These studies could be useful in the development of integrated crop protection strategies with the implementation of allelochemicals as pest behavior-modifying agents (Egger et al., 2014).

5. COMMERCIAL CONCEPTS

Many feeding inhibitors from plant sources so far have yielded excellent results in laboratory conditions. In field situations only a few of them are satisfactory alternatives to traditional pest management. The chemical control is usually with broad-spectrum insecticides, and they have to be broad-spectrum by necessity. They have to sell in amounts large enough to accommodate financial development, research, and marketing. The class of antifeedants is tested against one or a small group of insects attacking a specific crop. As a compound, it inhibits the feeding of one species, but for another it may be ineffective or just an attractant. Thus, replacement of a traditional chemical with a specific allelochemical will make pest management more expensive. That is why as of today the only prospect among botanicals is neem. However, apart from neem products, there are few actual demonstrations of antifeedant efficacy in the field. Application of polygodial or methyl salicylate at the IARC Rothamsted have shown that aphid populations are reduced with concomitant increases in yields of winter wheat, in one case comparable to that achieved with the pyrethroid insecticide cypermethrin (Pickett et al., 1997). Similarly, toosendanin, an antifeedant limonoid from the bark of the trees *Melia toosendan* and *M. azedarach* (Meliaceae) has been subjected to considerable research as a botanical pesticide (Chiu, 1989; Chen et al., 1995; Koul et al., 2002). Vertebrate selectivity of this compound is very favorable (LD_{50} mice = 10 g/kg) (Isman, 1994). Production of a botanical insecticide based on toosendanin, using a refined bark extract containing approximately 3% toosendanin (racemic mixture) as the active ingredient. Toosendanin-based insecticides could become a potential commercial product worldwide as formulations based on the technical concentrate are under evaluation in Canada to assess its potential against pests of agriculture and forestry in North America.

The use of botanical insecticides in California between 2006 and 2011 grew by almost 50% (Miresmailli and Isman, 2014) due to the public perception that natural products are safer than synthetic chemicals, despite evidence to this that is contrary (Trumble, 2002). To put this in context, botanical insecticide use represents only 5.2% of biopesticides and only 0.04% of all pesticide use in California. Biopesticides represent approximately 2% of the US $60 billion global pesticide market (2012 estimate), but the segment is dominated by microbial insecticides led by products based on *Bacillus thuringiensis* (Koul, 2011). The biopesticide segment is currently growing at 16% per year, compared with conventional agrochemicals that are growing at a rate of 5.5% per year (Miresmailli and Isman, 2014).

Regulatory exemptions such as that provided by List 25(b) of the Federal Insecticide, Fungicide, and Rodenticide Act (FIFRA) of the US Environmental Protection Agency (EPA), which allows certain essential oils and inert materials to be used as pesticide active ingredients without regulatory review, has facilitated the commercialization of some essential oil–based pesticides in the US over the past decade. Many of the essential oils and their products are included in the "generally recognized as safe" (GRAS) list, which have been approved by the US Food and Drug Administration and the US Environmental Protection Agency (EPA) for food and beverage consumption. However, the question is how much we are benefiting from the full potential of these natural substances and what could be the strategies that need to be adopted in their natural context?

6. ANTIFEEDANT POTENTIAL: WHY SO CLOSE YET SO FAR?

Consumers view plant-based products as safe as compared to synthetic chemicals; thus one can assume that antifeedant compounds are potentially close to us in terms of their eco-friendly environmental and health impacts. This has also provided an impetus for the discovery and development of more environmentally benign and less hazardous insecticides. Examination of the scientific literature in the field of botanical insecticides covering the past 30 years demonstrates a strong and growing academic interest in this area. However, the question that cannot be summarily ignored is that have we achieved the desired goal of discovery and/or commercialization of effective plant-based insect control methods? Unfortunately, the published work so far does not suggest that, and apparently we are yet so far. The rationale for the relevance of chemical characterization or composition should be obvious—for a given plant species, plant defensive chemistry can be highly variable in time and space, and can also be affected by environmental conditions, such as soil type, history of predator and/or pathogen attack, and others (Trumble, 2002; Murtagh and Smith, 1996). The criteria for a potential plant product for insect control will depend on eco- and chemotypes and defensive chemistry. For example, chemical variation and chemotypes of the legume *Tephrosia vogelii*, a plant often used in sub-Saharan Africa for insect control, is particularly instructive in this regard (Belmain et al., 2012; Stevenson et al., 2012). This suggests that biological activity will depend on active ingredients and their standardization vis-à-vis the extraction procedures followed (Isman and Grieneisen, 2014). The initial biomass resource is generally outsourced and minimally monitored, in contrast to quality-control protocols that exist for synthetic pesticides. The extracts are usually specified based on the level of one or two marker compounds (putatively the active principles) even though the presence and level of other constituents in the mixture can significantly influence the overall toxicity and efficacy of the extract (Belmain et al., 2012). As a result of limited chemical standardization, the efficacy of botanical products may not be consistent (Bradford et al., 2013). These aspects do not influence synthetic pesticides due to their simpler structures and easy scalability compared with that of botanical insecticides. Formulations for botanicals is again complicated due to their lipophilic and highly volatile constituents and are known to be susceptible to conversion and degradation reactions, such as oxidative and polymerization processes, which can result in loss of quality and of certain properties (Bharathi, 2011). The stability of these substances is also affected by air, light, and elevated temperatures, and subsequently the residual effects of botanical insecticides can be limited and, in some cases, lacking entirely.

Another aspect is that compounds in botanical extracts originate from different biosynthetic pathways. Aromatic phenylpropanoids are formed via the shikimic acid pathway resulting in phenylalanine, whereas terpenoids are derived from the C5 building blocks isopentenyl diphosphate and its isomer dimethylallyl diphosphate. Once plant chemicals have been removed from their protective compartments as a result of destructive extraction methods, their constituents are prone to oxidative damage, chemical transformations, or polymerization reactions. All these aspects need a thorough investigation before any botanical insecticide can make an impact on a larger scale. For example, encapsulation techniques are being considered to prepare pesticide nanoemulsions that provide some level of controlled release of the botanical active ingredient (Sakulku et al., 2009). These microencapsulation techniques generally slow down the release or decay of the entire mixture that is obtained by the destructive extraction of plant tissues; however, no specific attention is paid to the behavior of individual constituents of the mixture. In order to bridge the gap from being so far to get closer to have a potential botanical, novel technologies that would consider the behavior and control level of individual constituents of botanical insecticides will pave the way for a new generation of feeding deterrents that could be applied in a manner that is closer to the natural defense methods used by plants against herbivores.

7. CONCLUSIONS

The practice of using feeding deterrents from plant sources allows us to develop and exploit naturally occurring plant defense mechanisms, thereby reducing the use of conventional pesticides. However, most of these new strategies need to be developed with four basic facts in mind: organize the natural sources, develop quality control, adopt standardization strategies, and modify regulatory constraints. In fact, substantial efforts are needed in all the four areas if plant-based products are to be successful and competitive. This will definitely give rise to a number of challenges and unexpected problems. For instance, limonene is known to be a bitter antifeedant, but at high levels of concentration it can cause irritation and allergic reactions when in contact with human skin. Therefore, deeper cooperation between industrial and academic research is required that could definitely accelerate the process and give us new environmentally safe methods in future plant protection via plant defense mechanisms of secondary metabolites. In fact, pesticides derived from plant essential oils do have several important benefits. Due to their volatile nature, there is a much lower level of risk to the environment than with current synthetic pesticides. Predator, parasitoid, and pollinator insect populations will be less impacted because of the minimal residual activity, making essential oil–based pesticides compatible with integrated pest management programs. It is also obvious that resistance will develop more slowly to essential oil–based pesticides owing to the complex mixtures of constituents that characterize many of these oils.

However, it is advisable to define the goal of research, whether it is the use of crude or semirefined plant extracts for resource-poor farmers in developing countries, or simply the first step in phytochemical prospecting for new lead chemistries for industrialized pesticide development. Many studies are basically first reports on the screening of (chemically uncharacterized) crude plant extracts from a single plant species to one insect pest in a laboratory bioassay, which are examples of preliminary research that may not have sufficient novelty or reproducibility to merit publication. However, even in the absence of chemical analysis, crude plant extracts can have valuable utility for resource-poor farmers, as has been demonstrated recently (Mugisha-Kamatenesi et al., 2008; Kareru et al., 2013). Scientists in developing countries need to be encouraged to make greater efforts to investigate the utility of plant extracts for crop protection in field trials, in collaboration with local farmers, because such studies should prove more valuable than laboratory-only studies. Miresmailli and Isman (2014) suggest three approaches to get potential products. The first, improved extraction methods based on physical, biological, and chemical techniques with specific attention to preserving the integrity of phytochemical mixtures are required. However, because of their complexity and cost, these methods have not yet been adopted for the mass production of plant extracts by most botanical product producers.

The second area suggested relates to novel formulation methods that mimic the chemical compartmentalization and storage capacity of plants that can be achieved by using nanotechnology. The main challenge for incorporating these techniques for industrial botanical insecticide production is, again, cost and complexity. Although the effectiveness and innovative aspects of a product are important components, cost and economic viability remain factors of concern if a product has to become successful globally and more so in developing countries. This implies that still substantial research is required in collaboration with industrial research centers to develop competitive products and formulations in order to make botanical insecticides a success story.

The third area is the development of advanced technologies and delivery methods that provide qualitative and quantitative release control at the level of individual constituents. In recent years, micro- and nanoencapsulation techniques have been investigated as means of providing controlled release of botanical insecticides (Cabral Marques, 2010; Fang and Bhandari, 2010). These technologies can extend the efficacy of botanical insecticides over longer periods of time. Miresmailli and Isman (2014) report that "a better understanding of the behavior and bioactivity of individual components of botanical insecticides coupled with more advanced methods of compartmentalization and formulation will allow greater degrees of control over the availability and activity of individual components of complex botanical mixtures and, consequently, should enhance the efficacy of botanical insecticides."

References

Abbaszadeh, G., Srivastava, C., Walia, S., 2014. Insecticidal and antifeedant activities of clerodane diterpenoids isolated from the Indian bhant tree, *Clerodendron infortunatum*, against the cotton bollworm, *Helicoverpa armigera*. Journal of Insect Science 14, 1–13.

Abdelgaleil, S.A.M., Doe, M., Nakatani, M., 2013. Ring B,D-seco limonoid antifeedants from *Swietenia mahagani*. Phytochemistry 69, 312–317.

Abe, M., Niizeki, M., Matsuda, K., 2007. A feeding deterrent from *Persicaria lapathifolia* (Polygonaceae) leaves to larvae of *Gallerucida bifasciata* (Coleoptera: Chrysomelidae). Applied Entomology and Zoology 42, 449–456.

Ahern, J.R., Whitney, K.D., 2014. Stereochemistry affects sesquiterpene lactone bioactivity against an herbivorous grasshopper. Chemoecology 24, 35–39.

Akhtar, Y., Isman, M.B., 2003. Larval exposure to oviposition deterrents after subsequent oviposition behavior in generalist, *Trichoplusia ni* and specialist, *Plutella xylostella* moths. Journal of Chemical Ecology 29, 1853–1870.

Akhtar, Y., Rankin, C.H., Isman, M.B., 2003. Decreased response to feeding deterrents following prolonged exposure in the larvae of generalist herbivore, *Trichoplusia ni* (Lepidoptera: Noctuidae). Journal of Insect Behavior 16, 811–831.

Arnason, J.T., Guillet, G., Durst, T., 2004. Phytochemical diversity of insect defenses in tropical and temperate plant families. In: Carde, R.T., Miller, G.J. (Eds.), Advances in Insect Chemical Ecology. Cambridge University Press, Cambridge, pp. 1–20.

Belmain, S.R., Amoah, B.A., Nyirenda, S.P., Kamanula, J.F., Stevenson, P.C., 2012. Highly variable insect control efficacy of *Tephrosia vogelii* chemotypes. Journal of Agricultural Food Chemistry 60, 10055–10063.

Berenbaum, M., 1983. Coumarin and caterpillars, a case for coevolution. Evolution 37, 163–179.

Bernays, E.A., Chapman, R.F., 1994. Host-Plant Selection by Phytophagous Insects. Chapman and Hall, New York.

Bernays, E.A., Chapman, R.F., 2000. Plant secondary compounds and grasshoppers: beyond plant defenses. Journal of Chemical Ecology 26, 1773–1794.

Bernays, E.A., Weiss, M.R., 1996. Induced food preferences in caterpillars: the need to identify mechanism. Entomologia Experimentalis et Applicata 78, 1–8.

Bharathi, D.G., 2011. Methodology for the evaluation of scientific journals: aggregated citations of cited articles. Scientometrics 86, 563–574.

Bradford, D.F., Stanley, K.A., Tallent, N.G., Sparling, D.W., Nash, M.S., Knapp, R.A., McConnell, L.L., Massey Simonich, S.L., 2013. Temporal and spatial variation of atmospherically deposited organic contaminants at high elevation in Yosemite National Park, California, USA. Environmental Toxicology and Chemistry 32, 517–525.

Cabral Marques, H.M., 2010. A review on cyclodextrin encapsulation of essential oils and volatiles. Flavour and Fragrance Journal 25, 313–326.

Caicedo, A., Roper, S.D., 2001. Taste receptor cells that discriminate between bitter stimuli. Science 291, 1557–1560.

Carew, T.J., Sahley, C.L., 1986. Invertebrate learning and memory: from behavior to molecules. Annual Review of Neuroscience 9, 435–487.

Cespedes, C.L., Alarcon, J., 2011. Biopesticides of botanical origin, phytochemicals and extracts from Celastraceae, Rhamnaceae and Scrophulariaceae. Plant Medicines 10, 175–181.

Champagne, D.E., Koul, O., Isman, M.B., Towers, G.H.N., Scudder, G.G.E., 1992. Biological activity of limonoids from the rutales. Phytochemistry 31, 377–394.

Chen, W., Isman, M.B., Chiu, S.F., 1995. Antifeedant and growth inhibitory effects of the limonoid toosendanin and *Melia toosendan* extracts on the variegated cutworm, *Peridroma saucia*. Journal of Applied Entomology 119, 367–370.

Chiu, S.F., 1989. Recent advances in research on botanical insecticides in China. In: Aranson, J.T., Philogene, B.J.R., Morand, P. (Eds.), Insecticides of Plant Origin. American Chemical Society Symposium Series 387, Washington, DC, pp. 69–77.

Corral, S.D., Diaz-Napal, G.N., Zaragoza, M., Carpinella, M.C., Ruiz, G., Palacios, S.M., 2014. Screening for extracts with insect antifeedant properties in native plants from central Argentina. Boletín of Latin Caribe Plantas Medicine and Aromáticas 13, 498–505.

Croteau, R., Kutchan, T.M., Lewis, N.G., 2000. Natural products (secondary metabolites). In: Buchanan, B., Gruissem, W., Jones, R. (Eds.), Biochemistry and Molecular Biology of Plants. American Society of Plant Physiologists, New York, pp. 1250–1318.

De Bruyne, M., Warr, C.G., 2006. Molecular and cellular organization of insect chemosensory neurons. BioEssays 28, 23–34.

Dethier, V.G., 1980. Evolution of receptor sensitivity to secondary plant substances with special reference to deterrents. American Naturalist 115, 45–66.

Duraipandiyan, V., Ignacimuthu, S., Paulraj, M.G., 2011. Antifeedant and larvicidal activities of rhein isolated from the flowers of *Cassia fistula* L. Saudi Journal of Biological Sciences 18, 129–133.

Egger, B., Spangl, B., Koschier, E.H., 2014. Habituation in *Frankliniella occidentalis* to deterrent plant compounds and their blends. Entomologia Experimentalis et Applicata 151, 231–238.

Fang, Z., Bhandari, B., 2010. Encapsulation of polyphenols – a review. Trends in Food, Science and Technology 21, 510–523.

Frazier, J.L., Chyb, S., 1995. Use of feeding inhibitors in insect control. In: Chapman, R.F., de Boer, G. (Eds.), Regulatory Mechanisms in Insect Feeding. Chapman & Hall, New York, pp. 364–381.

Frazier, J.L., 1992. How animals perceive secondary plant compounds. In: Rosenthal, G.A., Berenbaum, M.R. (Eds.), Herbivores: Their Interaction with Secondary Plant Metabolites, Evolutionary and Ecological Processes, vol. 2. Academic Press, San Diego, pp. 89–134.

Gabryś, B., Dancewicz, K., Gliszczyńska, A., Kordan, B., Wawrzeńczyk, C., 2015. Systemic deterrence of aphid probing and feeding by novel β-damascone analogues. Journal of Pest Science 88, 507–516.

Glendinning, J.I., Brown, H., Capoor, M., Davis, A., Gbedemah, A., Long, E., 2001. A peripheral mechanism for behavioral adaptation to specific bitter taste stimuli in an insect. Journal of Neuroscience 21, 3688–3696.

Glendinning, J.I., Davis, A., Rai, M., 2006. Temporal coding mediates discrimination of bitter taste stimuli by an insect. Journal of Neuroscience 26, 8900–8908.

Glendinning, J.I., 1996. Is chemosensory input essential for the rapid rejection of toxic foods? Journal of Experimental Biology 199, 1523–1534.

Isman, M.B., Grieneisen, M.L., 2014. Botanical insecticide research: many publications, limited useful data. Trends in Plant Science 19, 140–145.

Isman, M.B., 1994. Botanical insecticides and antifeedants: new sources and perspectives. Pesticide Research Journal 6, 11–19.

Isman, M.B., 2002. Insect antifeedants. Pesticide Outlook 13, 152–157.

Isman, M.B., 2006. Botanical insecticides, deterrents, and repellents in modern agriculture and an increasingly regulated world. Annual Review of Entomology 51, 45–66.

Jermy, T., Bernays, E.A., Szentesi, A., 1982. The effect of repeated exposure to feeding deterrents on their acceptability to phytophagous insects. In: Visser, J.H., Minks, A.K. (Eds.), Insect-Plant Relationships. PUDOC, Wageningen, pp. 25–32.

Jermy, T., 1987. The role of experience in the host selection of phytophagous insects. In: Chapman, R.F., Bernays, E.A., Stoffolano, J.G. (Eds.), Perspectives in Chemoreception and Behavior. Springer-Verlag, New York, pp. 143–157.

Jeyasankar, A., Premalatha, S., Elumalai, K., 2014. Antifeedant and insecticidal activities of selected plant extracts against *Epilachna* beetle, *Henosepilachna vigintioctopunctata* (Coleoptera: Coccinellidae). Advances in Entomology 2, 14–19.

Kareru, P., Rotich, Z.K., Maina, E.W., 2013. Use of Botanicals and Safer Insecticides, Designed in Controlling Insects: The African Case. Intech. http://dx.doi.org/10.5772/53924.

Kim, J.H., Mullin, C.A., 2003. Antifeedant effects of proteinase inhibitors on feeding behaviors of adult western corn rootworm (*Diabrotica virgifera virgifera*). Journal of Chemical Ecology 29, 795–810.

Kmiecik-Małecka, E., Małecki, A., Pawlas, N., Woźniakova, Y., Pawlas, K., 2009. The effect of Î±-methylenelactone group on the feeding deterrent activity of natural and synthetic alkenes against Colorado potato beetle, *Leptinotarsa decemlineata* Say. Polish Journal of Environmental Studies 18, 1107–1112.

Koul, O., Walia, S., Dhaliwal, G.S., 2008. Essential oils as green pesticides: potential and constraints. Biopesticides International 4, 63–84.

Koul, O., 1997. Molecular targets for feeding deterrents in phytophagous insects. In: Raman, A. (Ed.), Ecology and Evolution of Plant Feeding Insects in Nature and Man-Made Environments. International Scientific Publications, New Delhi and Backhuys Publishers, Leiden, The Netherlands, pp. 123–134.

Koul, O., 2005. Insect Antifeedants. CRC Press, Bota Racon, FL, pp. 1010.

Koul, O., 2008. Phytochemicals and insect control: an antifeedant approach. Critical Reviews in Plant Sciences 27, 1–24.

Koul, O., 2011. Microbial biopesticides: opportunities and challenges. CAB Reviews: Perspectives in Agriculture, Veterinary Science, Nutrition and Natural Resources 6 (056), 1–25.

Koul, O., Multani, J.S., Singh, G., Wahab, S., 2002. Bioefficacy of toosendanin from *Melia dubia* (syn. *M. azedarach*) against gram pod-borer, *Helicoverpa armigera* (Hubner). Current Science 83, 1387–1391.

Krishnappa, K., Anandan, A., Mathivanan, T., Elumalai, K., Govindarajan, M., 2010. Antifeedant activity of volatile oil of *Tagetes patula* against armyworm, *Spodoptera litura* (Fab.) (Lepidoptera: Noctuidae). International Journal of Current Research 4, 109–112.

Kuhar, K., Kansal, R., Subrahmanyam, B., Koundal, K.R., Miglani, K., Gupta, V.K., 2013. A Bowman-Birk protease inhibitor with antifeedant and antifungal activity from *Dolichos biflorus*. Acta Physiologia Plantarum 35, 1887–1903.

Lam, P.Y.-S., Frazier, J.L., 1991. Rational approach to glucose taste chemoreceptor inhibition as novel insect antifeedants. In: Baker, D.R., Fenyes, J.G., Moberg, W.K. (Eds.), Synthesis and Chemistry of Agrochemicals II. ACS Symposium Series 443, American Chemical Society, Washington, DC, pp. 400–412.

Li, C.-H., Niu, X.-M., Luo, Q., Xie, M.-J., Luo, S.-H., Zhou, Y.-Y., Li, S.-H., 2010. Novel polyesterified 3,4-secograyanane diterpenoids as antifeedants from *Pieris formosa*. Organic Letters 12, 2426–2429.

Li, W., Hu, J., Yang, J., Yuan, G., Guo, X., Luo, M., 2014. Feeding deterrence of common spices against *Helicoverpa armigera* larvae. Advances in Biosciences and Biotechnology 5, 1025–1031.

Lu, M., Wu, W., Liu, H., 2013. Insecticidal and feeding deterrent effects of fraxinellone from *Dictamnus dasycarpus* against four major pests. Molecules 18, 2754–2762.

Mansson, P.E., 2005. Host Selection and Antifeedants in *Hylobius abietis* Pine Weevils (Doctoral thesis). Swedish University of Agricultural Sciences, Alnarp.

Messchendorp, L., Smid, H.M., van-Loon, J.J.A., 1998. The role of an epipharyngeal sensillum in the perception of feeding deterrents by *Leptinotarsa decemlineata* larvae. Journal of Comparative Physiology A 183, 255–264.

Mierziak, J., Kostyn, K., Kulma, A., 2014. Flavonoids as important molecules of plant interactions with the environment. Molecules 19, 16240–16265.

Miresmailli, S., Isman, M.B., 2014. Botanical insecticides inspired by plant–herbivore chemical interactions. Trends in Plant Science 19, 29–35.

Mugisha-Kamatenesi, M., Deng, A.L., Ogendo, J.O., Omolo, E.O., Mihale, M.J., Otim, M., Buyungo, J.P., Bett, P.K., 2008. Indigenous knowledge of field insect pests and their management around Lake Victoria basin in Uganda. African Journal of Environmental Science and Technology 2, 342–348.

Mullin, C.A., Mason, C.H., Chou, J., Linderman, J.R., 1991. Phytochemical antagonism of γ-aminobutyric acid based resistance in *Diabrotica*. In: Mullin, C.A., Scott, J.G. (Eds.), Molecular Mechanisms of Insecticide Resistance: Diversity among Insects. ACS Symposium Series 505, American Chemical Society, Washington, DC, pp. 288–808.

Mullin, C.A., Chyb, S., Eichenseer, H., Hollister, B., Frazier, J.L., 1994. Neuroreceptor mechanisms in insect gustation: a pharmacological approach. Journal of Insect Physiology 40, 913–931.

Munakata, K., 1975. Insect antifeeding substances in plant leaves. Pure and Applied Chemistry 42, 57–66.

Murtagh, G.J., Smith, G.R., 1996. Month of harvest and yield components of tea tree. II. Oil concentration, composition and yield. Australian Journal of Agricultural Research 47, 817–827.

Nenaah, G.E., 2014. Toxic and antifeedant activities of prenylated flavonoids isolated from *Tephrosia apollinea* L. against three major coleopteran pests of stored grains with reference to their structure–activity relationship. Natural Product Research 28, 2245–2252.

Nihei, K., Asaka, Y., Mine, Y., Yamada, Y., Iigo, M., Yanagisawa, T., Kubo, I., 2006. Musidunin and musiduol, insect antifeedants from *Croton jatrophoides*. Journal of Natural Products 69, 975–977.

Nordlund, D.A., 1981. Semiochemicals: a review of terminology. In: Nordlund, D.A., Jones, R.L., Lewis, W.J. (Eds.), Semiochemicals: Their Role in Pest Control. John Wiley & Sons, New York, pp. 13–23.

Norris, D.M., 1986. Anti-feeding compounds. Chemical Plant Protection 1, 99–146.

Omar, S., Marcotte, M., Fieldsb, P., Sanchez, P.E., Poveda, L., Mata, R., Jimenez, A., Durst, Y., Zhang, J., MacKinnon, S., Leaman, D., Arnason, J.T., Philogéne, B.J.R., 2007. Antifeedant activities of terpenoids isolated from tropical Rutales. Journal of Stored Product Research 43, 92–96.

Pandey, P.K., Jamal, F., 2014. Bio-potency of a 21 kDa Kunitz-type trypsin inhibitor from *Tamarindus indica* seeds on the developmental physiology of *Helicoverpa armigera*. Pesticide Biochemistry and Physiology 116, 94–102.

Pavela, R., 2010a. Antifeedant activity of plant extracts on *Leptinotarsa decemlineata* Say and *Spodoptera littoralis* (Boisd.) larvae. Industrial Crops and Products 32, 246–253.

Pavela, R., 2010b. Antifeedant activity of plant extracts on *Leptinotarsa decemlineata* Say and *Spodoptera littoralis* (Boisd.) larvae. Industrial Crops and Products 32, 213–219.

Pickett, J.A., Wadhams, L.J., Woodcock, C.M., 1997. Developing sustainable pest control from chemical ecology. Agriculture Ecosystem and Environment 64, 149–156.

Sakulku, U., Nuchuchua, O., Uawongyart, N., Puttipipatkhachorn, S., Soottitantawat, A., Ruktanonchai, U., 2009. Characterization and mosquito repellent activity of citronella oil nanoemulsion. International Journal of Pharmacy 372, 105–111.

Schoonhoven, L.M., Blaney, W.M., Simmonds, M.S.J., 1992. Sensory coding of feeding deterrents in phytophagous insects. In: Bernays, E.A. (Ed.), Insect-Plant Interactions, vol. 4. CRC Press, Boca Raton, FL, pp. 59–99.

Selvam, K., Ramakrishnan, N., 2014. Antifeedant and ovicidal activity of *Tinospora cordifolia* willd (Menispermaceae) against *Spodoptera litura* (Fab.) and *Helicoverpa armigera* (Hubner) (Lepidoptera: Noctuidae). International Journal of Recent Scientific Research 5, 1955–1959.

Simmonds, M.S.J., 2001. Importance of flavonoids in insect-plant interactions: feeding and oviposition. Phytochemistry 56, 245–252.

Sivasubramanian, A., Narasimha, K.K.G., Rathnasamy, R., Oliveira Campos, A.M.F., 2013. A new antifeedant clerodane diterpenoid from *Tinospora cordifolia*. Natural Product Research 27, 1431–1436.

Stevenson, P.C., Kite, G.C., Lewis, G.P., Nyirenda, S.P., Forest, F., Belmain, S.R., Sileshi, G., Veitch, N.C., 2012. Distinct chemotypes of *Tephrosia vogelii* and implications for their use in pest control and soil enrichment. Phytochemistry 78, 135–146.

Szczepanik, M., Dams, I., Wawrzenczyk, C., 2008. Terpenoid lactones with the *p*-menthane system as feeding deterrents to the lesser mealworm, *Alphitobius diaperinus*. Entomologia Experimentalis et Applicata 128, 337–345.

Szentesi, A., Bernays, E.A., 1984. A study of behavioral habituation to a feeding deterrent in nymphs of *Schistocerca gregaria*. Physiological Entomology 9, 329–340.

Szentesi, A., Jermy, T., 1989. The role of experience in host plant choice by phytophagous insects. In: Bernays, E.A. (Ed.), Insect–Plant Interactions, vol. 11. CRC Press, Boca Raton, FL, pp. 39–74.

Thompson, R.F., Spencer, W.F., 1966. Habituation: a model phenomenon for the study of neuronal substrates of behavior. Psychology Reviews 73, 16–43.

Trumble, J.T., 2002. Caveat emptor: safety considerations for natural products used in arthropod control. American Entomologist 48, 7–13.

Vattikonda, S.R., Amanchi, N.R., Sangam, S.R., 2014. Effect of betulinic acid on feeding deterrence of *Papilio demoleus* L. (Lepidoptera: Papilionidae) larvae. Indian Journal of Fundamental and Applied Life Sciences 4, 43–48.

Further Reading

Andrew, R.L., Peakall, R., Wallis, I.R., Foley, W.J., 2007. Spatial distribution of defense chemicals and markers and the maintenance of chemical variation. Ecology 88, 716–728.

Ascher, K.R.S., Rones, G., 1964. Fungicide has residual effect on larval feeding. International Journal of Pest Control 6, 6–9.

Blum, M.S., Whitman, D.W., Severson, R.F., Arrendale, R.F., 1987. Herbivores and toxic plants: evolution of a menu of options for processing allelochemicals. Insect Science and Applications 8, 459–563.

Jermy, T., Metolcsy, G., 1967. Antifeedant effects of some systemic compounds on chewing phytophagous insects. Acta Phytopathologia Academy of Science Hungary 2, 219–224.

Jørgensen, K., Kvello, P., Jørgen Almaas, T., Mustaparta, H., 2006. Two closely located areas in the suboesophageal ganglion and the tritocerebrum receive projections of gustatory receptor neurons located on the antennae and the proboscis in the moth *Heliothis virescens*. Journal of Comparative Neurology 496, 121–134.

Koul, O., 1993. Plant allelochemicals and insect control: an antifeedant approach. In: Ananthakrishnan, T.N., Raman, A. (Eds.), Chemical Ecology of Phytophagous Insects. IBH & Oxford Publishers Pvt. Ltd, New Delhi, pp. 51–80.

Koul, O., Multani, J.S., Goomber, S., Daniewski, W.M., Berlozecki, S., 2004a. Activity of some non-azadirachtin limonoids from *Azadirachta indica* against lepidopteran larvae. Australian Journal of Entomology 43, 189–195.

Koul, O., Singh, G., Singh, R., Singh, J., Daniewski, W.M., Berlozecki, S., 2004b. Bioefficacy and mode-of-action of some limonoids of salannin group from *Azadirachta indica* A. Juss and their role in a multicomponent system against lepidopteran larvae. Journal of Biosciences 29, 409–416.

Schoonhoven, L.M., 1982. Biological aspects of antifeedants. Entomologia Experimentalis et Applicata 31, 57–69.

Van Drongelen, W., 1979. Contact chemoreception of host plant specific chemicals in larvae of various *Yponomenta* species (Lepidoptera). Journal of Comparative Physiology 134A, 265–279.

Wieczorek, H., 1976. The glycoside receptor of the larvae of *Mamestra brassicae* L. (Lepidoptera: Noctuidae). Journal of Comparative Physiology 106A, 153–176.

18

Neem Products

Rashmi Roychoudhury

Department of Botany, University of Lucknow, Lucknow, India

1. INTRODUCTION

The serious consequences of indiscriminate use of chemical pesticides resulting in environmental and health hazards were brought to the fore in 1962 by Rachel Carson's book *Silent Spring* and resulted in a popular movement for lesser and safer use of pesticides (www.en.Wikipedia.org/wiki/Rachel Carson). Some pesticides, such as DDT, were banned after a long struggle, and then a global search began for alternatives to toxic pesticides. Some botanical pesticides already in use were pyrethrum obtained from dried flowers of *Chrysanthemum cinerariifolium* (Family Asteraceae), rotenoids from the genera *Derris, Lonchocarpus, Tephrosia,* and *Mundulea* (Family Leguminosae), and nicotinoids from *Nicotiana* species (Family Solanaceae). These were not entirely safe and showed mammalian toxicity.

Plants biosynthesize a vast variety of chemicals of structural variety that exhibit equally diverse biological activities in the insects ranging from action on the nerve axons and synapses (e.g., pyrethrin, nicotine, picrotoxinine), muscles (e.g., ryanodine), respiration (e.g., rotenone, mammein), hormonal balance (e.g., juvenile and molting hormone analogues and antagonists), reproduction (e.g., β-asarone), and behavior (e.g., attractants, repellents, antifeedants) (Devkumar and Parmar, 1993). Thousands of plants and organisms have been screened for their pest and pathogen control potential leading to the vast accumulation of information (Ahmad et al., 1983; Devkumar and Parmar, 1993; Subramaniam, 1993), and several botanicals, horticultural mineral oils, detergents, and plant essential oils (Isman, 2006) are currently being used worldwide for pest control. Of all the botanical biopesticides, neem (Indian Lilac–*Azadirachta indica* A. Juss. Family Meliaceae, Order Rutales) has emerged as the best option. The pest control properties of this plant have been known in the Indian subcontinent for centuries for controlling storage pests and soil-borne pests. A chance discovery in 1959 by Heinrich Schmutterer of its locust-deterrent effect attracted worldwide attention and there was an upsurge of interest in neem and its unique pest control properties. The phagorepellent effect of neem on locusts had been noticed and reported in India (Pradhan et al., 1962), but investigations into the insecticidal components began only after the report by Schmutterer.

Ecofriendly Pest Management for Food Security
http://dx.doi.org/10.1016/B978-0-12-803265-7.00018-X

In the last few decades, neem has generated tremendous scientific activity and a great amount of work on its active constituents and their mode of action, leading to several international conferences and publications (Tewari, 1992; NRC, 1992; WNC, 1993; Schmutterer, 1995a; Mariappan, 1995; Singh et al., 1996a,b; Singh and Saxena, 1999; Randhawa and Parmar, 1996; Puri, 1999). The pesticidal constituents of neem are concentrated in the seed kernels and exert multitargeted behavioral and physiological effects on insects as compared to the chemicals with a single mode of action. They have low toxicity to mammals and to beneficial predators and parasitoids. These active constituents are present in all parts of the plant and have been isolated and characterized (Henderson et al., 1968; Butterworth et al., 1972; Champagne et al., 1992; Govindachari, 1992; Devkumar and Parmar, 1993; Nagasampagi, 1993; Kraus, 1995a,b; Devkumar and Sukhdev, 1996).

The bioactive constituents mainly concentrated in the seed kernels and oil cause behavioral effects (repellence, phagodeterrence/antifeedant, oviposition deterrence) and physiological effects (growth inhibition/regulation, molt retardation, reduced fecundity, reduced activity, and mortality). Collection and processing of seeds and oil extraction are thus crucial to the pesticide production from neem.

Neem is an integral part of integrated pest management (IPM), which is an ecofriendly technique of controlling pests using long-term preventive and environmental balancing strategies such as habitat management, use of resistant varieties, organic pesticides, crop rotation, and application of biorational products. Neem can be safely combined with chemical pesticides, botanical pesticides, microbial biopesticides, predators, and parasitoids of biocontrol.

2. THE TREE AND ITS INSECTICIDAL PARTS

Neem grows in arid, semiarid, tropical, and humid conditions in Pakistan, Bangladesh, Sri Lanka, Malaysia, Indonesia, Thailand, the Middle East, Sudan, and Niger. It has been introduced to the African countries, the United States (US), and Australia, and large-scale plantations of the tree have been developed in several south Asian and African countries (Tewari, 1992; Mohan Ram and Nair, 1996). In India the largest number of neem trees has been recorded from Uttar Pradesh in north India followed closely by the states of Tamil Nadu, Madhya Pradesh, Andhra Pradesh, and Karnataka (Jayaraj, 1993). The states of Maharashtra, Andhra Pradesh, Karnataka, Tamil Nadu, Gujarat, and Uttar Pradesh are foremost in the neem industry.

The availability and collection of fruits is vital for pesticide production and the extended flowering period of the tree in India ensures that neem fruits are available in different regions of the country at different times of the year. Flowering starts earliest in the southern region (in January in Kerala, in February and March in Karnataka, Tamil Nadu, and Andhra Pradesh), and extends by several months toward the northern parts (late March to April in the central part of India and up to May in the sub-Himalayan areas). A fully grown tree yields about 50 kg fruits and 150 kg leaves annually.

2.1 Seed Kernels and Oil

The fruits of neem are drupes, single seeded, green when immature, and bright yellow when ripe. The skin (epicarp) of the fruit is thin and easily removed exposing the light-yellow, sweet and edible pulp (mesocarp) often eaten by birds and bats. The seed shell (endocarp) is hard and encloses the brown kernel. The ripe fruits drop off and are collected, depulped, and dried in the villages. The seeds are sun dried and stored for 3–4 months for maturing. Improper depulping and drying might encourage fungal growth on the seeds, leading to aflatoxin production and loss of yield and poor quality of kernels and oil. The dry seeds are graded and sold to oil extraction units. The growing demand for neem seeds and oil by the pesticide industry has made the collection of leaves, fruits, seeds, and seed oil extraction commercially viable. Recently, the Government of India has made it mandatory that all domestic urea-producing units should make neem-coated urea to increase fertilizer efficiency and reduce cost of farm input (*Times of India*, May 15, 2015). This would be a boost to the neem industry. Tree felling has considerably reduced seed production, which is now 185,000 tonnes per annum, and large-scale neem plantations are needed to cater to the growing market (www.indiamart.com/impact//neempesticide).

The seeds are decorticated to get 55% shell and 45% kernels, which can be stored up to 8–12 months without deterioration. Kernel extracts are most bioactive as compared to extracts of seed shells and fallen leaves (Schmutterer, 1990). The maximum yield of oil from seeds is obtained after 3 months of storage. The seed kernels yield 40–50% oil, and the seeds yield 20–25% oil (Jattan et al., 1995). Oil is extracted from kernels or ripe dried seeds by cold pressing in expellers or by solvent extraction. Cold pressed oil is better suited for pesticides as solvent extraction damages the chemical nature of bioactive principles.

2.1.1 The Neem Cake

The neem cake obtained from the expeller contains 8–10% oil, which is recovered by solvent extraction. The residual oil and the limonoids in the cake cause insect repellence and solvents or water extracts make the product a good antifeedant and growth inhibitor. The octanortriterpenoids α- and β-nimolactones isolated from seed cake and fruit coat are moderately antifeedant. The seed cake is rich in plant nutrients (crude protein 13–18%; carbohydrate 24–50%; crude fiber 8–26%; fat 2–13%; ash 5–18%; and acid insoluble ash 1–17%, with nitrogen, phosphorous, calcium, and magnesium) and is used as manure for soil amendment and for urea coating (Puri, 1999).

2.1.2 The Seed Kernel

The seed kernel is the richest source of azadirachtin, which is 0.75 g/kg to 0.2 g/kg in Indian seeds and higher in seeds from Kenya (1 g/kg), Nigeria (2 g/kg), and Ghana (3.5 g/kg). On a small scale, azadirachtin up to 10 g/kg of seed kernel (Schmutterer, 1990) and even 20 g/kg seed can be extracted with solvents (Puri, 1999). The Flavour and Fragrance Development Centre (FFDC, Kannauj, India) has reported solvent extraction and cold extraction/leaching of seed kernel of 0.33% azadirachtin to yield fatty acids and 70% azadirachtin in bitters containing about 0.24% azadirachtin, which may be used for the preparation of pesticides (Anon., 1999). Azadirachtin content varies from 0.05% to 4.24% in seeds from different geographical regions in India. The content also varies with stages of maturity of seeds and with seasons, monsoon seeds yielding 1.53% azadirachtin mainly azadirachtin A, and in winter seeds 1.26% azadirachtin A and B in equal proportions and twice the amount of azadirachtin F was synthesized (Puri, 1999). The average yield of azadirachtin from kernels in India is 3.5% (www.indiamart.com/impact//neempesticide).

Seed kernels yield volatile organosulfur compounds, the limonoids and active constituents having strong antifeedant properties such as azadirachtin and its isomers, salannin, salannol, salanolacetate, 3-deacetylsalannin, 14-epoxyazaradione, azadiradione, gedunin, nimbinen, and deacetylnimbinen, nimbolinin ohchinolide, and 21-oxo-ohchinolide. Moderate antifeedants also isolated are salannolactame-21 and salannolactame-23. Compounds with insect growth regulating properties isolated from kernels are 7-deacetylgedunin, 22-23-dihydro-23-methoxy azadirachtin, 3-togloylazadirachtol, and 1-tigloyl-3-acetyl-11-methoxyazadirachtin (Schmutterer, 1990).

2.1.3 *Seed Oil*

Seed oil is greenish brown, odoriferous of garlic-like odor and contains organosulfur compounds, limonoids, and other active constituents. Strong antifeedants isolated from oil are azadirachtin, salannin, 3-deacetylsalannin, nimbin and its analogues 6-deacetylnimbin, 6-deacetylnimbinal, nimbolide, 28-deoxonimbolide, vilasinin derivatives 1,3-diacetylvilasinin, 1-tigloyl-3-acetylvilasinin, 1-senecioyl-3-acetylvilasinin 1-tigloyl-3-acetyl-12α-acetoxyvilasinin, meliantriol, azadiradione, 14-epoxyazaradione, 6-O-acetylnimbandiol (Schmutterer, 1990), azadirone, epoxyazaradione (nimbinin), nimbinene, and 6-deacetylnimbinene (Schmutterer, 1990; Tewari, 1992). The compounds deacetylazadirachtinol and 6-acetylnimbandiol are both antifeedants and insect growth inhibitors.

2.2 Leaves

These contain 0.13% of the essential oil containing organosulfur compounds making it odoriferous. Constituents isolated from leaves are azadirachtin A, nimbin and its derivatives, deacetylnimbinene, nimbandiol, a new tetranortriterpenoid nimocinol, nimocinolide, isonimocinolide, isonimbocinolide, isonimolicinolide nimbocinolide, nimbolide, dihydronimbic acid, and 28-deoxoimbolide (Tewari, 1992). The leaves are insect repellent and the solvent extracts antifeedant and growth inhibiting. Boiling the leaves destroys the active constituents; fresh ground leaf extracts are more effective.

3. ACTIVE CONSTITUENTS AND THEIR MODE OF ACTION

The most active constituents in neem are the tetranortriterpenoids (also called limonoids or melia) which are specific to plants of order Rutales. More than 300 limonoids are known today and one-third of these occur as meliacins in the two Meliaceae members, neem (*A. indica*) and China berry (*Melia azedarach*) (Devkumar and Sukhdev, 1996). The evolution of neem limonoids, particularly the highly specialized azadirachtin, with diverse actions against phytophagous insects, is rare and interesting (Rembold and Puhlman, 1995) as different insect species have shown diverse reactions to the active constituents of neem, some insects being more sensitive than the others.

The bioactivity of all the active constituents isolated from neem on the insect pests is species specific and dose dependent. The desert locust, *Schistocerca gregaria* (Caelifera) is the most sensitive to the phagodeterrent or antifeedant effect of neem, which does not even rest on host plants treated with the active constituent. Several Lepidopteran larvae and moths, the black armyworm *Spodoptera frugiperda*, tobacco budworm *Helicoverpa virescens*, cotton

TABLE 1 Bioactive Constituents in Neem (Schmutterer, 1990; Tewari, 1992)

Plant part	Bioactive constituent
Bark, root, stem	6 Diterpenoids, gedunin 12–16% tannins, 8–11% non-tannins, polysaccharides, 18 diterpenoids, gum polysaccharides, nimbin 0.04%, nimbinin 0.002%, nimbidin 0.4%
Leaves	Nimbin and its derivatives, 6-deacetylnimbinene, nimbandiol, nimocinol, nimbocinone, nimocinolide, nimbocinolide, nimbolide, quercetin
Seed kernel	25 active volatile organic sulfur compounds, Di-*n*-propyldisulfide (75.74%), azadirachtin, meliantriol, salanine, nimbin, nimbidin
Neem oil	Lipids, and several less polar triterpenoids bitter principles 7-acetyl-neotrichilenone, azadirachtin, its derivatives and analogues, meliantriol, derivatives of vilasinin, azadirones, gedunins, nimbins, salannins, and their derivatives
Seed cake/kernel (after oil extraction)	Rich in proteins, amino acids, phosphorous, calcium trypsin inhibitor (15 trypsin units inhibited/mg protein), lysine, triterpenoids, azadirachtin major tetranortriterpenoid salannin, nimbin, gedunin
Seed kernel/oil	<u>Antifeedants</u>–main: azadirachtin salannin, salannol, salannolacetate, 3-deacetylsalannin, azadiradione, 14-epoxyazaradion, gedunin, nimbinen, and deacetylnimbinen, vilasinin derivatives, meliantriol, azadiradione, and 14-epoxyazadiradione, 6-*O*-acetylnimbandiol, 3-deacetylsalannin <u>Growth inhibiting</u>: azadirachtin, 22-23-dihydro-23-methoxyazadirachtin, 3-tigloylazadirachtol, and 1-tigloyl-3-acetyl-11-methoxyazadirachtin, deacetylazadirachtinol similar in effect as azadirachtin

bollworm *Helicoverpa armigera*, etc. are deterred from feeding at low doses, while the Mexican bean beetle *Epilachna varivestis* (Coleoptera; Coccinellidae) and the termite *Reticulitermes speratus* are less sensitive and deterred from feeding at high doses. Table 1 lists the biologically active constituents in various parts of neem (Schmutterer, 1990; Tewari, 1992). The growth-inhibiting effect on the insects is also species specific and dose dependent. Table 2 lists the comparative biological activity of the various constituents isolated from neem (Krauss, 1995a; Devkumar and Sukhdev, 1996; Tewari, 1992; Champagne et al., 1992).

3.1 Systemicity of Active Constituents

Gill and Lewis (1971) first reported the systemic antifeedant action of azadirachtin, aqueous and alcoholic seed kernel extracts given as soil drench to plants against *S. gregaria*. Neem constituents are taken up systemically by the plants through the roots in soil drench treatments. The biological activity of the active components becomes diluted but remains unaltered and spreads uniformly to actively growing plant parts. Systemically treated plants exert the antifeedant and growth inhibiting effects on the insects. The extent of systemicity and the effect on insects varies with the test plant/insect species. The leaf-chewing lepidopterans are more sensitive to the systemic compounds than the sap-sucking phloem feeders, which could be due to the poor phloem mobility of the compounds as systemic antifeedant effect on aphids is seen with high azadirachtin content (>100 ppm). The systemic effect lasts for

TABLE 2   Biological Activity of Major Neem Isolates on Various Insects (Tewari, 1992; Krauss, 1995a; Devkumar and Sukhdev, 1996)

Active constituent	Plant part	Name of insect[a]/EC$_{50}$ = ppm	Biological activity
Protomeliacine Meliantriol	Fruit of *Melia azedarach* neem oil	*S. gregaria*/8 μg/cm^2 Whiteflies/2%	AF
Azadirone	Oil/fruit	*E. varivestis*/5500 ppm *P. japonica*/45% 5000 ppm	AF
Azadiradione	Oil/fruit	*E. varivestis*/320 ppm *P. japonica*/87% 5000 ppm *H. virescens*/560 ppm *R. speratus*/238 μg/disc	AF
Epoxyazadiradione (Nimbinin)	Oil	*E. varivestis*/1300 ppm *P. japonica*/23% 2500 ppm	AF
Nimocinolide Isonimocinolide	Leaves	*A. aegypti*/0.625 ppm *A. aegypti*/0.74 ppm	LM
Gedunin	Fruit, bark	*E. varivestis*/930 ppm *R. speratus*/49 μg/disc	AF
		S. frugiperda/47 ppm	GI
1,3-Diacetyvilasinin	Seed/oil	*E. varivestis*/13 ppm	AF
Nimbin	Seeds, leaves	*E. varivestis*/50 ppm	AF
Nimbolide	Leaves	*E. varivestis*/70 ppm *P. japonica*/1000 ppm	AF
Salannin	Seeds, Oil	*E. varivestis*/14 ppm *S. frugiperda*/13 μg/cm^2 *R. speratus*/19.5 μg/disc	AF
		H. virescens/170 ppm	GI
Azadirachtol Azadirachtin B	Seeds	*E. varivestis*/0.08 ppm	GI
		E. varivestis/85% at 2 ppm	GI
Azadirachtin	Seeds	*E. varivestis*/13 ppm *S. frugiperda*/100% 0.07 ppm	AF
		E. varivestis/69% at 1 ppm *H. virescens*/0.07 ppm *S. frugiperda*/0.04 ppm	GI

GI = growth inhibitor; AF = antifeedant; LM = larva mortality.

[a] Schistocerca gregaria, Epilachna varivestis, Popillia japonica, Helicoverpa virescens, Aedes aegypti, Spodoptera frugiperda, Reticulitermes speratus.

comparative longer duration than the foliar spray (Koul and Shankar, 1995), and neem as antifeedant can be effective when systemic. Soil drench with neem cake extracts also produce the systemic antifeedant effect. Systemicity is an important criterion for the viability of feeding deterrents in agricultural systems (Toscano et al., 1997) and neem products have the potential to be developed into effective systemic antifeedants.

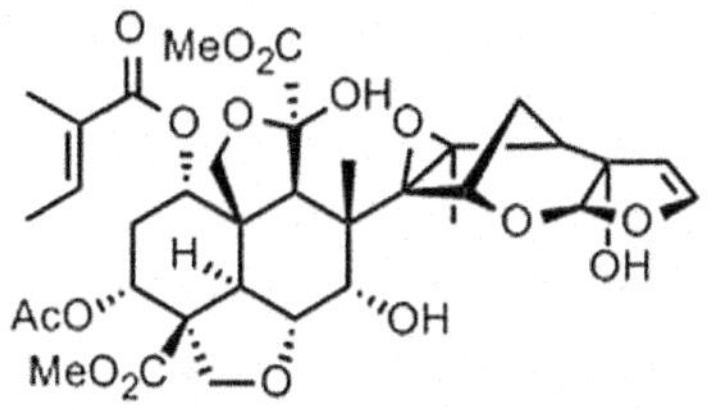

Azadirachtin

FIGURE 1 Molecular structure of Azadirachtin.

3.2 Azadirachtin

The major limonoid in seed kernel and seed oil is azadirachtin (aza) ($C_{36}H_{44}O_{16}$, m.p. 154–158 °C) (Figure 1) isolated by Butterworth and Morgan (1968) who reported its antifeedant properties. It has complex structure (Ley et al., 1991) and is both antifeedant and insect growth regulator. The maximum content of this important compound has been reported in the neem seed kernels from India, Myanmar, and Thailand (Ermel, 1995). Of the various isomers (A, B, D, E, F, H, I, K, L; Rembold and Puhlman, 1995), azadirachtin A is the main compound in terms of its quantity in the kernels, and aza E is regarded as the most effective insect growth regulator (Schmutterer, 1990). Azadirachtin is a complex compound and its exact mode of action is still not fully understood. It is insect antifeedant at high concentrations and growth inhibitor at low concentrations to all the test insects if applied topically or by feeding. The compound dihydroazadirachtin obtained by hydrogenation of azadirachtin A is considered more stable and as effective as azadirachtin and suitable for field applications (Mordue and Blackwell, 1993). Azadirachtol and its derivatives azadirachtin B, azadirachtin G, 2′, 3′-dihydrotigloylazadirachtol, 3-tigloyl-22-23-dihydroazadirachtol, 2′-3′-dihydrotigloyl-22-23-dihydroazadirachtol are strongly growth inhibiting and strong antifeedants to the *E. varivestis*. The derivatives azadirachtin D and azadirachtin I are also antifeedants. Their derivatives 3-deacetylazadirachtin, azadirachtin E, vepaol, and marrangin (azadirachtin L) are strong antifeedants, growth inhibitors, and larvicidal to Lepidopteran larvae.

The potential of neem as a broad-spectrum pesticide is mainly targeted toward control of agricultural pests of which the important phytophagous pests belong to the orders Orthoptera, Thysanoptera, Homoptera, Heteroptera, Coleoptera, and Lepidoptera (Schmutterer, 1990). Most insects show the behavioral and physiological effects of the various active constituents, and the specific response of the insects may be utilized for their control. Around 400 insect species of 15 orders are affected by neem constituents (Schmutterer and Singh, 1995).

3.3 Repellence and Oviposition Deterrence

Repellence is an olfactory effect of the insects' gustatory and olfactory sensilla (Blaney and Simmonds, 1995) to the repellent organosulfur compounds that causes an oriented movement of the insect at least 25 cm away from the stimulus (Saxena, 1995). Repellence also causes oviposition deterrence and the female insects do not land on the treated plant to lay eggs. Repellence is caused by the volatile organosulfur compounds in neem seed oil, kernels, and leaves.

These parts have yielded 25 such volatile compounds and their biogenetic precursors, the main constituent being Di-*n*-propyldisulfide. Neem volatiles are also present in the inner bark, timber, and heartwood of the tree giving them the characteristic odor and causing the insect repellent effect (Balandrin et al., 1988). Repellence is an important quality in pest control products. Neem seed "bitters" and azadirachtin lacking volatility do not generally repel homopterans (Saxena, 1995).

Neem seed oil, kernels, and leaves are traditionally used in several Asian countries to protect stored grain against pests. These are used as dry leaves or kernel extracts for soaking jute bags and dried or storage bins are treated with neem oil to keep away the pests *Sitophilus* spp. and *Tribolium* spp. for several months. Neem oil at 0.1–1% protects wheat, paddy, and legumes against *Callosobruchus chinensis*. Neem seed kernel powder (1–2 parts/100 parts) mixed with wheat, mung bean, gram, pea, cowpea, jowar, and sorghum effectively protects the grains against *Sitophilus oryzae* up to 9 months and suppresses *Trogoderma granarium*.

3.4 Phagodeterrence

This effect is seen when insects keep wandering over the host plant and may even settle, but they do not feed, probe, or suck sap. In some cases, the insects initially feed but later show aversion to food. Phagodeterrence is caused either by stimulation of specific deterrent neurons of the sensilla styloconica of the maxillae or by inhibition in firing of the phagostimulant neurons resulting in primary antifeedant effect that is mainly gustatory. Aversion to food or reduced feeding is nongustatory and could be caused by interference of the antifeedant with functioning of serotonergic cells of the stomatogastric nervous system, influenced by hormones involved in food metabolism (Blaney and Simmonds, 1995) and subsequent blocking of midgut peristalsis, as seen in *Locusta migratoria* (Dorn and Trumm, 1996). This secondary antifeedant effect is nongustatory. Antifeedant effect of neem constituents is seen against almost all species of insects tested and their larvae (Schoonoven, 1982; Simmonds and Blaney, 1985; Saxena and Khan, 1985; Blaney and Simmonds, 1995). The effect is influenced by the insect species, its developmental stage, host plant species, and plant age. Oligophagous species are more sensitive to antifeedants than the polyphagous ones (Simmonds and Blaney, 1985). The sensitive insects show feeding deterrence at low doses of the active constituents while the less sensitive insects are affected at comparatively high doses and often keep feeding on the treated plant.

3.5 Growth Inhibition and Fecundity Reduction

Insect growth regulation (IGR) effect is produced when a single low dose of azadirachtin, whether injected, fed, or topically applied, causes changes in sensitive insects due to disturbances in the endocrine mechanism and the juvenile hormones (Rembold, 1995; Koul, 1996) with interruption of the signal transfer from the hemolymph to the neurosecretory cells by binding irreversibly to a specific receptor, possibly on the neurolemma of the brain. The juvenile hormone production is disturbed, resulting in growth abnormalities. The effect is seen in the immature stages of larvae or nymphs; it is insect specific and does not affect mammals. The volatile organosulfur compounds also cause growth inhibition seen in larvae of *Helicoverpa zea* and *H. virescens* (Balandrin et al., 1988). The morphological effects of azadirachtin and its isomers on different insect species may vary, but the insect becomes progressively deformed or the stages in metamorphosis show various degrees of deformities, resulting in morbidity and mortality. In dose-dependent bioassays on *E. varivestis* (Rembold, 1995), the larvae fed on 1.25 ppm aza show interference with larval-pupal molt and those fed on 0.25 ppm aza show interference with pupal-adult molt. The surviving pupae and adults are severely malformed with shriveled wings and inability to enclose from pupal shell. Azadirachtins act as ecdysteroid agonists or antagonists affecting the neurotransmitters in the insects, their feeding behavior, growth, metamorphosis, and reproduction (Rembold, 1995; Schmutterer and Rembold, 1995; Schmutterer and Wilps, 1995).

Fecundity reduction is seen as reduced egg laying capacity in ovipositing insects due to inhibition by azadirachtin of ecdysteroid titer and vitellogenin synthesis (Schmutterer and Rembold, 1995). The effect of neem derivatives is not ovicidal, and either it does not or only slightly inhibits the emergence of nymphs/larvae from the laid eggs. The toxicity to eggs upon treatment is due to the choking effect and not to growth regulation. In the viviparously reproducing aphids, neem seed oil and azadirachtin reduced reproduction of several aphid species with fewer number of offspring and reduction in fertility (live offspring) and fecundity (live + dead offspring). Azadirachtin possibly inhibits the growth of the embryos at various stages of development. The effect varies with the aphid species and the host plant (Lowery and Isman, 1996a). Reduced fecundity is also reported in leafhoppers *Nilaparvata lugens*, *Spodoptera furcifera*, and *Nephotettix virescens* on exposure to neem oil and neem bitter-treated rice (Saxena and Khan, 1986; Saxena et al., 1987).

3.6 Toxicity

Neem derivatives do not cause contact toxicity on homopteran pests (Saxena, 1995) but soft-bodied insects and some insect larvae are controlled by ingredients from oil, kernels, or leaf extracts or by the organosulfur compounds. The toxic effect of neem oil on small, soft-bodied insects, thrips, etc. is due to the choking effect (Schmutterer, 1990) and cuticulogenesis and ecdysis (Lowery and Isman, 1996b); the effect being dose dependent. Aphids and whiteflies are controlled by sprays of aqueous or solvent extracts of neem seed kernels or seed cake at high concentrations and neem oil emulsions on infested plants causing mortality of the insects (Schauer, 1984; Lowery et al., 1993; Roychoudhury and Jain, 1996a; Facknath, 1999; Rizvi, 2003). The host plant species, age of the plants, and developmental stage of the aphids influence the persistence of toxic effect of neem (Lowery et al., 1993). Neem oil and seed kernel extracts are toxic to the preimaginal stages, pupae, and adult whiteflies (Roychoudhury and Jain, 1991, 1996b; Jain, 1992; Serra and Schmutterer, 1993; Nimbalkar et al., 1994). Neem oil is also ovicidal to stored grain pest *Callosobruchus maculatus* (Schmutterer, 1990) of legumes and cowpea seeds.

Neem cake and seed kernels are nematicidal and soil amendment with neem cake protect against nematodes for 1 year or several crop seasons. The mode of action is the slow solubility of the cake and release of nematicidal substances including ammonia, phenols, aldehydes, and fatty acids that possibly occupy the soil pore spaces inhabited by the nematodes (Parveen and Alam, 1996). Neem extracts inhibit the hatchability of eggs and reduce the number of juvenile nematodes and their penetration ability and gall formation (Mojumder, 1995).

Larvicidal properties of α- and β-nimolactones have been reported against nematodes. Neem seed kernel powder used for seed coating of cowpea, chickpea, and mung bean is nematicidal (Mojumdar and Raman, 1999). Soil amendment with neem cake has been used for protection against termites and nematodes in sugarcane and other crops (Tewari, 1992).

3.7 Vectoral Activity

Limited success has been achieved in control of virus vectors and virus diseases in agricultural crops with pesticides due to the complexity of the relations between insect vectors with their plant hosts (Perring et al., 1999), their host/virus specificity (Sulochna, 1977), and the behavioral biology and feeding biology (Ghosh and Raychaudhuri, 1980; Jain, 1992) of the vectors.

The repellent and antifeedant effect of neem has the potential to reduce vectoral activity, however, the strong antifeedants causing prolonged wanderings could increase aphid activity and migration to untreated fields and spread virus infection (Hunter and Ullman, 1992). Neem oil changed the feeding behavior of the leafhopper and planthopper vectors, *N. lugens* and *N. virescens*, by reduced feeding but increased probing, salivation, and xylem feeding with reduced vectoral activity and disease incidence. In membrane feeding tests, *N. virescens*, vector of tungro virus complex in rice, showed strong antifeedant activity against neem oil formulations containing azadirachtin or dihydroazadirachtin and inhibited virus acquisition with no infection on test plants (Biju et al., 2007). Neem cake, seed kernel extracts, and neem oil reduced the yeast-like symbionts in fat body of *N. lugens* (Saxena et al., 1987). The symbionts reportedly fulfill the lipid and sterol requirement and growth of the plant sap sucking homopteran insects.

Azadirachtin could have a strong effect on the host–parasite insect relationship (Rembold, 1995). Azadirachtin treatment causes homeostatic changes in the host affecting the recognition signal vital for the parasite to identify its host, so that the parasite is unable to settle and grow in the host. This significant finding may strongly influence the host-vector-pathogen interrelationship and play a vital role in vector-transmitted diseases. Endosymbiont bacteria (Douglas, 1998) also play an important role in virus transmission by aphids. Neem could inhibit virus transmission by aphids by either inhibiting the endosymbionts or the protein *Buchnera* GroEL necessary for virus integrity in the hemolymph or by affecting the binding sites for the virus on hemolymph–salivary glands interface (van den Heuvel et al., 1998). The exact mode of action of azadirachtin in host–parasite relationship is not yet discerned fully but it holds potential for controlling viral diseases.

3.8 Disease Control

Neem oil prevents germination and penetration of some fungal spores, and 1% neem oil is considered best for managing powdery mildew and black leaf spot caused by fungi on lilacs, hydrangeas, and phlox. Sprays of neem oil and seed kernel extract reduce fungal leaf spot of anthracnose, *Cercospora* leaf spot, and *Alternaria* leaf spot in black gram (Trivedi et al., 2014). Soil amendment with neem cake or foliar sprays with neem reduces fungal diseases of rice caused by *Helminthosporium oryzae*, *Drechslera oryzae*, and bacterial leaf blight on rice caused by *Xanthomonas campestris*, reduction being comparable with fungicides and bactericides

(Mariappan, 1995). Basal stem rot of coconut caused by *Ganoderma lucidum* is controlled by neem cake + *Trichoderma* basal treatment (Bhaskaran, 1995). Several neem constituents—mahmoodin, nimolide, nimbic acid, α- and β-nimolactones—have bactericidal effects. Nimbin and its derivatives are reportedly antimicrobial against bacteria and fungi (Schmutterer, 1995a).

Kernel extracts and oil formulations of neem are used to control virus disease incidence in treated plants (Eppler, 1995; Samuel and Mariappan, 1996; Bhatt, 2004; Trivedi et al., 2014) with delayed or mild symptoms of virus infection in some cases (Hunter and Ullman, 1992; Devi et al., 2008). The specific reasons for reduction in disease incidence or mild and delayed symptoms are not known.

Reduced vector incidence and inoculum density as reported with seed kernel extract, neem leaf extract, neem oil, and aza rich extracts on several aphids (Schauer, 1984; Roychoudhury and Jain, 1989; Lowery and Isman, 1996b; Roychoudhury and Jain, 1996a; Bhathal and Singh, 1996) is often correlated with reduced disease incidence. Neem products may reduce vector incidence due to contact toxicity or repellent/antifeedant effect (Eppler, 1995; Parveen and Alam, 1996) and/or the growth inhibiting and fecundity reducing effects.

3.9 Pest Control and Integrated Pest Management

The major agricultural pests are sensitive to neem and show the same antifeedant and growth-inhibiting effects on plants treated with the products or on plants in which the effect has become systemic (Koul and Shankar, 1995). The larvae of the Lepidopterans are particularly sensitive to neem, and at low concentrations the ingestion or contact with neem limonoid, particularly azadirachtin, produces strong growth inhibition and reduces pest population. Blocking the larvae from molting is neem's most important quality, and the larvicidal activity is used to suppress many pest species (Kumar, 1997). Neem products are therefore particularly suited for control of these pests. Crop treatment with neem cake extract, neem oil emulsion, or seed kernel extract effectively controls pests; the effect, however, varies with the azadirachtin content in the neem oil (Isman et al., 1990). Neem formulations from kernels and/or oil are being used to control more than 105 insect pests of 10 orders on crops and gardens. The target pests in the US, India, and other countries are aphids, whiteflies, jassids, thrips, beetles, caterpillars, lace bugs, leafhoppers, leaf miners, leaf folders, pod borers, fruit borers, diamond back moth, mealybugs, psyllids, bollworm, brown planthopper (BPH), and fungal and bacterial pathogens of sheath rot, sheath blight. In India, the crops on that neem products are used on are mainly cotton, rice, coconut, pigeon pea, chickpea, safflower, okra, cauliflower, cabbage, and tomato. In the US, neem products are approved for limited use on greenhouse crops, certain fruits, vegetables, and ornamental crops. In certain cases, it can be used on plants up to the day of harvest.

Protection of grains is as important for food security as pest control in the fields. Neem seed kernels and oil are used for protection of grains and legumes, and formulations have been made using defatted kernel powder or solvent-extracted neem cake (Singh et al., 1999) to control stored wheat grain, *T. granarium*, and the cigarette beetle *Lasioderma serricorne* that infest spices and stored tobacco.

Neem has become an integral part of IPM, the purpose of which is to reduce the use of toxic pesticides. IPM is recommended to farmers in US by the Environmental Protection Agency (EPA) and in several other countries including India to avoid damage. Neem pesticides are highly recommended as a biorational insecticide in IPM as it is efficacious against the target pests and less detrimental to natural enemies and thus not disruptive to biocontrol (Shuster and Stansly, 2015). Neem is safe to earthworms, soil bacteria, and young salmon (Schmutterer, 1995b). Several US universities, including the University of California, University of Connecticut, and University of Florida have fact sheets on neem oil pesticides. The impact and advisory on the use of neem oil as pesticide and its effect on environmental factors and on honeybees has been reported by the University of California IPM online as safe for the environment, water bodies and mammals, and some guidelines are stated for protection to honeybees (www.ipm.ucdavis.edu/Tools/PNA.pnaishow).

Neem products are compatible with IPM measures and can be used in combination with other biopesticides. *Trichoderma harzianum*, the beneficial soil fungus, which enhances growth of plants and inhibits soil fungal pathogens, is combined with neem seed cake to increase its efficacy (Bhaskaran, 1995). Neem products are safe with insect pathogenic virus, *Bacillus thuringiensis*, and entomopathogenic fungi. Treatment of crops with low concentration of seed kernel extract or neem oil does not harm the predatory spiders, mites, earwigs, crickets, ants, beetles, syrphid flies and lacewings. Neem products are also safe for the egg parasitoid *Trichogramma* spp. and endoparasitoid *Cotesia glomerata*. In some cases the nymphal-larval stages may be affected, however, the effect is comparable to that of chemical pesticides (Schmutterer, 1999). Intercropping of cabbage with coriander, tomato, and garlic combined with sprays with seed kernel extract and introduction of parasitoid *Cotesia plutellae* effectively controlled aphid populations, snails, and *Plutella xylostella* larvae on cabbage plants. The neem treatment had no adverse effect on parasitoids, and the treatment with neem improved the quality of marketable cabbage heads (Facknath, 1999). Neem product sprays on cotton under field conditions with Bt formulation and lower dose of broad-spectrum synthetic pyrethroid effectively controlled whitefly populations and *H. armigera* and also improved quality of cotton (Gupta et al., 1999). Treatment of cotton crop with neem seed kernel extracts or neem oil combined with nuclear polyhedrosis virus (NPV) controlled aphids, thrips, jassid, and bollworm and increased yield. Asian corn borer *Ostrinia furnacalis* was controlled on corn with combination of neem seed kernel powder, early release of endemic *Trichogramma* wasps and detasseling of corn (Breithaupt and Schmutterer, 1999). Combination of neem seed kernel and custard apple seed extracts effectively controls sorghum aphid, *Melanaphis sacchari*. It was nontoxic to the parasites and parasitoids of sorghum midge but reduced parasitism of *Mythimna separata* by *Apantales ruficrus*. Combination of seed kernel extract or neem oil with synthetic pesticide effectively controlled virus disease incidence and whitefly population on mung bean, and seed treatment of pigeon pea with neem cake significantly reduced fungal root rot and wilt (Mariappan and Jayaraj, 1995).

Neem pesticides are more effective in the presence of other chemical constituents rather than pure compounds. The activity of azadirachtin is increased in the presence of constituents such as nimbandiol, deacetylsalannin, deacetylnimbin, 6-acetylnimbandiol, salannin, and other components (Holmes and Hassan, 1999). The activity of azadirachtin may be increased with pesticides (Walter, 1999) or when combined with other botanicals for broad-spectrum activity and better efficacy. Combination of neem cake with groundnut cake and cake of Indian birch (*Pongamia pinnata*) in soil has been recommended in ancient texts (*Vrikshayurveda* or Ayurveda of Plants) for nematode control (Sridhar et al., 2001). Mixing pyrethroids in moderate amounts with neem gives effective pest protection (Gupta et al., 1999). Contact effect and antifeedant effect of neem is enhanced by combination with strong botanical such as *Pongamia glabra*, which is antifeedant and toxic and comparable in action with chemical pesticides. The formulation controls mites, mealybugs, and bollworms (Sharma et al., 1999). It is insecticidal and acaricidal and inhibits nitrification. Custard apple kernel extract, pyrethrum, meliantriol-rich china berry extract, rotenone, and wild castor (*Jatropha curcas*) are also toxic and insecticidal and used in moderate proportions with neem to control aphids, beetles, red scale, leaf roller, and other pests. China berry (*M. azedarach*) extract, a strong antifeedant, and neem oil formulations at 1.5–2% oil have shown maximum antifeedant and pest control activity (Chari et al., 1999). *Jatropha curcas* and custard apple (*Annona squamosa*) have growth-modifying effects on insects similar to neem, and these extracts can be used as additives with neem products (Vyas et al., 1999).

Citronella oil, used as mosquito repellent, is also a repellent and oviposition inhibitor to whiteflies (Roychoudhury and Jain, 1991) and can be formulated with neem extracts. Sesamum oil may be used to enhance activity and prevent early photo deactivation. It also has a sticker or synergistic effect and the effect of neem lasts for longer duration on treated leaves (Schauer, 1984). Azadirachtin is compatible with insecticidal soap and superior horticultural oil and is used in these combinations. Soil drench with neem seed extract may induce systemic antifeedant action and exert a mild toxic effect. The contact effect on leaves may be enhanced with the contact biopesticides.

4. NEEM INDUSTRY

Neem pesticides have gained popularity among farmers as safe botanical pesticides for pest control and IPM. There is growing demand for neem oil, seeds, kernels, and cake. Neem pesticides represent an excellent investment opportunity due to the ecofriendly nature of the product and its use in highly relevant and growth-oriented sectors such as pesticides, export prospects, and reasonable returns. A vast business opportunity exists for this industry in India according to the Export-Import Bank of India (EXIM bank) and a market exists for Rs 100 crore neem-based pesticides with 50% export potential. The market growth rate is projected at 7–9% per annum. The products for export are neem oil and azadirachtin, and 260 tonnes of oil is exported annually. A reservoir of 150 tonnes of aza per year is needed in India. Azadirachtin is the major limonoid for pest control and is exceptionally effective at low concentrations (<20 g a.i. per acre) for the growth inhibiting effects, but the expensive procedure of its extraction makes it expensive (Walter, 1999). Azadirachtin Technical of less purity is used for preparing ready-to-use formulations of pesticides. In azadirachtin-based formulations, the content of this limonoid varies from 0.09% azadirachtin to 4.5% in the US and 300 to 1500 ppm in India. Neem oil is used in approximately 100 pesticide products. Fact sheets on neem oil have been released by several agencies and universities in the US.

Several large- and small-scale units are engaged in commercial neem biopesticide production in India (Parmar, 1995) and also in other countries (Parmar and Ketkar, 1996). Neem biopesticide formulations are marketed throughout India and in other countries wherever the products can be registered and has regulatory approval. The first commercial product of neem Margosan-O (W. R. Grace & Co. Cambridge, MA, USA) was registered in the US in 1983. Early commercial preparations of neem products are Azatin (AgriDyne Technologies,

Salt Lake City, UT, USA), Bioneem and Neemesis (Ringer Corp. Minneapolis, MN, USA), Safer's ENI (Safer Ltd., Victoria, BC, Canada), RD-repelin (ITC Ltd, A.P., India), Neem Guard (Gharda Chemicals, Mumbai, India), Neemark (West Coast Herbochem, Mumbai, India), and Neemazal (Trifoliop M.GmBN, Germany). In India the major neem pesticide manufacturers are Pest Control (India) Pvt Ltd (PCI), Mumbai, Godrej Agrovet, Mumbai, SPIC Foundation, Coimbatore. A major exporter in India is EID Parry (I) Ltd Chennai, India. The major company for neem pesticides in the US is Thermal Trilogy, which has a joint venture with an Indian company. In the US, several trading companies are engaged in the business of neem.

Although India is in an enviable position for the neem industry because of its high quality of oil and seeds and abundant raw materials, the business has not grown as expected. Problems in exports exist due to regulatory restrictions in several countries and it is better to concentrate in the Indian market. Recurrent problems in exporting and marketing are polluted and contaminated products, with lead and aflatoxin in the oil. There is lack of quality control and an inability to fulfill Indian demand of 1.5 lakh tonnes/annum.

To keep up a steady supply of raw materials, large-scale plantations should be developed with intercropping on wastelands, roadsides, avenues, etc. Quality control of raw material should be done starting from fruit collection and drying of seeds to prevent fungal and soil contamination of the seeds. Strict compliance should be done with the regulatory specifications set by the Bureau of Indian Standards in (1) IS14299: 1993 Specification for Neem Extract Concentrate containing Azadirachtin. (2) ISI 14300: 1993 Specification for Neem Based EC Containing Azadirachtin. (3) IS 4765: 1975 Specification for Neem Kernel Oil and Depulped Neem seeds Oil (First Revision). (4) ISI 7787: 1975 Grading for Neem Kernels and Depulped Neem Seeds for Oil Milling. (5) IS 8558: 1997 Specification for Neem Cake for Manuring. Recommendations have also been made for quality assurance of the seed oil and the physical standards of the oil by the Panel of the Ministry of Industry, Government of India, the Solvent Extractor Association of India, and the Indian Standard Institute (Puri, 1999).

5. CONCLUSIONS

Health and environmental concerns have influenced public opinion in recent years for use of safe and nonhazardous pest control measures, and demand for organic products is rising. Neem has emerged as the best environmentally friendly botanical pesticide for its unique pest control properties, safety to mammals, and low toxicity to biocontrol agents, predators, and parasitoids. Azadirachtin is the major component and its growth regulatory effect of harmful pests ensures decrease in pest incidence. It is concentrated in neem oil and neem seed kernels, and these are basic to production of neem pesticide. India is in a position to dominate the neem pesticide industry due to its vast resources, and quality control of the products should be ensured to cater to the needs of exports and the Indian market.

References

Ahmad, S., Grainge, M., Hylin, J.W., Mitchell, W.C., Litsinger, J.A., 1983. Some promising plant species for use as pest control agents under traditional farming systems. In: GTZ (1984), pp. 565–580.

Anon., 1999. Neem An Information. Fragrance and Flavour Development Centre, (FFDC), Kannauj, India. p. 29 (In Hindi).

Balandrin, M.F., Lee, S.M., Klocke, I.F., 1988. Biologically active volatile organosulphur compounds from the seeds of the neem tree *Azadirachta indica* (Meliaceae). Journal of Agricultural and Food Chemistry 36, 1048–1054.

Bhaskaran, R., 1995. Use of neem products for the management of basal stem rot disease of coconut. In: Mariappan, V. (Ed.), Neem for the Management of Crop Diseases. Associated Publishing Co., New Delhi, pp. 43–47.

Bhathal, S.S., Singh, D., 1996. Toxic, developmental and reproductive effects of AZT-VR-K and some neem products against mustard aphid *Lipaphis erysimi* (Kaltenbach). In: Singh, et al. (Ed.), Neem and Environment, vol. 1. WNC 1993. Oxford & IBH Publishing Co. Pvt. Ltd, New Delhi, pp. 287–296.

Bhatt, K., 2004. Effect of Neem (*Azadirachta indica* A. Juss) and Bakayan (*Melia azedarach* Ach.) on Feeding Behaviour of Aphids and Whiteflies and on Their Vectoral Activities (Ph.D. thesis). Department of Botany, University of Lucknow, Lucknow, India, p. 162.

Biju, C.N., Niazi, F.R., Singh, J., 2007. Effect of azadirachtin and dihydro azadirachtin on *in vitro* acquisition of rice tungro viruses. Indian Phytopathology 60 (3), 362–365.

Blaney, W.H., Simmonds, M.S.J., 1995. Biological effects of neem and their modes of action. Feeding behaviour. In: Schmutterer, H. (Ed.), The Neem Tree. VCH, Weinheim, Federal Republic of Germany, pp. 171–176.

Breithaupt, J., Schmutterer, H., 1999. Aqueous neem seed kernel extracts for the Asian corn borer *Ostrinia furnacalis* in Papua New Guinea. In: Singh, R.P., Saxena, R.C. (Eds.), *Azadirachta indica* A. Juss. Oxford & IBH Co. Pvt. Ltd, New Delhi, pp. 192–198.

Butterworth, J.H., Morgan, E.D., 1968. Isolation of a substance that suppresses feeding in locusts. Journal of the Chemical Society, Chemical Communications 23–24.

Butterworth, J.H., Morgan, E.D., Percy, G.R., 1972. The structure of azadirachtin, the functional groups. Journal of the Chemical Society, Perkin Transactions 1 2445–2450.

Champagne, D.E., Koul, O., Isman, M.B., Scudder, G.G.E., Towers, G.H.N., 1992. Biological activity of limonoids from the Rutales. Phytochemistry 31 (2), 377–394.

Chari, M.S., Ramaprasad, G., Sitaramaiiah, S., Murthy, P.S.N., 1999. Laboratory and field evaluation of neem (*Azadirachta indica* A. Juss), pongamia (*Pongamia pinnata* L.), chinaberry (*Melia azedarach* L.) extracts and commercial neem formulations against tobacco caterpillar (*Spodoptera litura* F.). In: Singh, R.P., Saxena, R.C. (Eds.), *Azadirachta indica* A. Juss. Oxford & IBH publishing Co. Pvt. Ltd, New Delhi, pp. 111–129.

Devi, P.H.S., Devi, R.L., Singh, I.M., 2008. Evaluation of plant products against aphid (*Myzus persicae* Sultz.) transmission of mosaic disease of leaf mustard (*Brassica juncea* var. *rugosa*). Indian Phytopathology 61 (4), 514–517.

Devkumar, D., Parmar, B.S., 1993. Pesticides of higher plants and microbial origin. In: Parmar, B.S., Devkumar, D. (Eds.), Botanical and Biopesticides. SPS Publication No.4 Society of Pesticide Science, India and Westville Publishing House, New Delhi, pp. 1–73.

Devkumar, C., Sukhdev, 1996. Chemistry. In: Randhawa, N.S., Parmar, B.S. (Eds.), NEEM. Society of Pesticide Science, India. New Age International (P) Limited Publishers, New Delhi, pp. 77–110.

Dorn, A., Trumm, P., 1996. Effect of azadirachtin on neural regulation of midgut peristalsis in *Locusta migratoria*. In: Singh, et al. (Ed.), Neem and Environment. WNC 1993. Oxford & IBH Publishing Co. Pvt. Ltd, New Delhi, pp. 297–306.

Douglas, A.E., 1998. Nutritional interactions in insect-microbial symbiosis: aphids and their symbiotic bacteria. Annual Review of Entomology 43, 17–37.

Eppler, A., 1995. Effects on viruses and organisms. In: Schmutterer, H. (Ed.), The Neem Tree. VCH, Weinheim, Federal Republic of Germany, pp. 93–105.

Ermel, K., 1995. Biologically active ingredients. Azadirachtin content of neem seed kernels from different regions of the world. In: Schmutterer, H. (Ed.), The Neem Tree. VCH, Weinheim, Federal Republic of Germany, pp. 89–92.

Facknath, S., 1999. Application of neem extract and intercropping for the control of cabbage pests in Mauritius. In: Singh, R.P., Saxena, R.C. (Eds.), *Azadirachta indica* A. Juss. Oxford & IBH publishing Co. Pvt Ltd., New Delhi, pp. 165–175.

Ghosh, M.R., Raychaudhuri, D.N., 1980. Biological aspects of aphid vectors in the transmission of plant virus diseases. Proceedings of the Indian National Science Academy B46 (6), 822–826.

Gill, J.S., Lewis, C.T., 1971. Systemic action of an insect antifeedant. Nature 232, 402–403.

Govindachari, T.R., 1992. Chemical and biological investigations on *Azadirachta indica* (the neem tree). Current Science 63, 117–122.

Gupta, G.P., Katiyar, K.N., Sharma, K., 1999. Neem in the management strategies of insect pest of cotton. In: Singh, R.P., Saxena, R.C. (Eds.), *Azadirachta indica* A. Juss. Oxford & IBH Co. Pvt. Ltd, New Delhi, pp. 177–189.

Henderson, R., McCrindle, R., Melera, A., Overton, K.H., 1968. Tetranortriterpenoids IX. The constitution and stereochemistry of salannin. Tetrahedron 24, 1525–1528.

van den Heuvel, J.F.J.M., Hogenhout, S.A., Verbeek, M., van der Wilk, F., 1998. *Azadirachta indica* metabolites interfere with the host-endosymbiont relationship and inhibit the transmission of potato leafroll virus by *Myzus persicae*. Entomologia Experimentalis et Applicata 86, 253–260.

Holmes, M.S., Hassan, E., 1999. The contact and systemic action of neem seed extract against green peach aphid *Myzus persicae* Sulzer (Hemiptera: Aphididae). In: Singh, R.P., Saxena, R.C. (Eds.), *Azadirachta indica* A. Juss. Oxford & IBH Publishing Co. Pvt. Ltd, New Delhi, pp. 93–101.

Hunter, W.H., Ullman, D.E., 1992. Effect of neem product, RD-Repelin on settling behaviour and transmission of zucchini yellow mosaic virus by the pea aphid *Acyrthosiphon pisum* (Harris) (Homoptera: Aphididae). Annals of Applied Biology 120, 9–15.

Isman, M.B., 2006. Botanical insecticides, deterrents, and repellents in modern agriculture and an increasingly regulated world. Annual Review of Entomology 51, 45–66.

Isman, M.B., Koul, O., Luczynki, A., Kaminski, J., 1990. Insecticides and antifeedant bioactivities of neem oil and their relationships to azadirachtin content. Journal of Agricultural and Food Chemistry 38, 1406–1411.

Jain, R.K., 1992. Studies on the Epidemiology of Some Plant Virus Vectors and Virus Diseases and Their Control by Some Plant Products (Ph.D. thesis). Department of Botany, University of Lucknow, Lucknow, India, p. 311.

Jattan, S.S., Kumar, S., Pujar, G., Bisht, N.S., 1995. Perspectives in intensive management of neem plantations. The Indian Forester 12 (11), 981–988.

Jayaraj, S., 1993. Neem in pest control: progress and perspectives. In: World Neem Conference February 24–28, 1993, Bangalore, India. Indian Society of Tobacco Science, CTRI, Rajahmundry, A.P., India, pp. 37–43.

Koul, O., 1996. Mode of action of azadirachtin in insects. In: Randhawa, N.S., Parmar, B.S. (Eds.), NEEM. Society of Pesticide Science, India. New Age International (P) Limited Publishers, New Delhi, pp. 160–170.

Koul, O., Shankar, J.S., 1995. Systemic uptake of azadirachtin into *Ricinus communis* and its effects on *Spodoptera litura* larvae. Indian Journal of Experimental Biology 33, 865–867.

Kraus, W., 1995a. Biologically active ingredients. Azadirachtin and other triterpenoids. In: Schmutterer, H. (Ed.), The Neem Tree. VCH, Weinheim, Federal Republic of Germany, pp. 35–74.

Kraus, W., 1995b. Biologically active ingredients. Diterpenoid and nonterpenoidal compounds. In: Schmutterer, H. (Ed.), The Neem Tree. VCH, Weinheim, Federal Republic of Germany, pp. 74–88.

Kumar, K., 1997. Neem products and standardization—an overview. Neem Update II (3), 5–6 & 12.

Ley, S.V., Lovell, P.J., Smith, S.C., Wood, A., 1991. Chemistry of insect antifeedants from *Azadirachta indica* part 9. Oxidative reactions of azadirachtin derivatives leading to C8–C14 bond cleavage. Tetrahedron Letters 32, 6183–6186.

Lowery, D.T., Isman, M.B., 1996a. Inhibition of aphid (Homoptera: Aphididae) reproduction by neem seed oil and azadirachtin. Journal of Economic Entomology 89 (3), 602–607.

Lowery, D.T., Isman, M.B., 1996b. Effect of extracts from neem on aphids (Homoptera: Aphididae) and their natural enemies. In: Singh, et al. (Ed.), Neem and Environment, vol. 1. WNC 1993. Oxford & IBH Publishing Co. Pvt. Ltd, New Delhi, pp. 253–264.

Lowery, D.T., Isman, M.B., Brard, N.L., 1993. Laboratory and field evaluation of neem for the control of aphids (Homoptera: Aphididae). Journal of Economic Entomology 86, 864–870.

Mariappan, V. (Ed.), 1995. Neem for the Management of Crop Diseases. Associated Publishing Company, New Delhi, p. 220.

Mariappan, V., Jayaraj, S., 1995. Examples of integrated pest management programs including neem in Southern India. In: Schmutterer, H. (Ed.), The Neem Tree. VCH, Weinheim, Federal Republic of Germany, pp. 471–477.

Mohan Ram, H.Y., Nair, M.N.B., 1996. Botany. In: Randhawa, N.S., Parmar, B.S. (Eds.), NEEM. Society of Pesticide Science, India, New Age International (P) Limited Publishers, New Delhi, pp. 6–26.

Mojumder, V., 1995. Effects on viruses and organisms. Nematoda, nematodes. In: Schmutterer, H. (Ed.), The Neem Tree. VCH, Weinheim, Federal Republic of Germany, pp. 129–150.

Mojumdar, V., Raman, R., 1999. Nematicidal efficacy of Neema-SI, an experimental formulation for neem seed treatments against *Heterodera cajani* and *Meloidogyne incognita* in cowpea and chickpea respectively. In: Singh, R.P., Saxena, R.C. (Eds.), *Azadirachta indica* A. Juss. Oxford & IBH Co. Pvt. Ltd, New Delhi, pp. 217–222.

Mordue (Luntz) A.J., Blackwell, L.A., 1993. Review: Azadirachtin: an update. Insect Physiology Pergamon Press Ltd. Printed in Great Britain 39, 903–924.

Nagasampagi, B.A., 1993. Development of neem chemistry in India. In: World Neem Conference February 24–28, 1993, Bangalore, India. Indian Society of Tobacco Science, CTRI, Rajahmundry, A.P., India, pp. 59–68.

Nimbalkar, N., Yadav, D.B., Prabhune, A.G., 1994. Evaluation of bioefficacy, phototoxicity and compatibility of Neemax on okra cultivar Parbhani Kranti. Pestology 18, 10–19.

NRC, 1992. Neem - a Tree for Solving Global Problems. National Academy Press, Washington DC, USA, p. 141.

Parmar, B.S., 1995. Results with commercial neem formulations produced in India. In: Schmutterer, H. (Ed.), The Neem Tree. VCH, Weinheim, Federal Republic of Germany, pp. 453–471.

Parmar, B.S., Ketkar, C.M., 1996. Commercialization. In: Randhawa, N.S., Parmar, B.S. (Eds.), NEEM. Society of Pesticide Science, India. New Age International (P) Limited Publishers, New Delhi, pp. 318–332.

Parveen, G., Alam, M.M., 1996. Bioactivity against plant pathogens. In: Randhawa, N.S., Parmar, B.S. (Eds.), NEEM. Society of Pesticide Science, India. New Age International (P) Limited Publishers, New Delhi, pp. 192–201.

Perring, T.M., Gruenhagen, N.M., Farrar, C.A., 1999. Management of plant viral diseases through chemical control of insect vectors. Annual Review of Entomology 44, 457–481.

Pradhan, S., Jotwani, M.G., Rai, B.K., 1962. The neem seed deterrent to locusts. Indian Farming 12, 7–11.

Puri, H.S., 1999. NEEM: the Divine Tree *Azadirachta indica*. Harwood Academic Publishers, Australia, p. 182.

Randhawa, N.S., Parmar, B.S. (Eds.), 1996. Neem. Society of Pesticide Science, India, New Age International (P) Limited Publishers, New Delhi, pp. 332.

Rembold, H., 1995. Biological effects of neem and their modes of action. Growth and metamorphosis. In: Schmutterer, H. (Ed.), The Neem Tree, VCH, Weinheim, Federal Republic of Germany, pp. 177–194.

Rembold, H., Puhlman, I., 1995. Azadirachtins: Structure and activity relations in case of *Epilachna varivestis*. In: Schmutterer, H. (Ed.), The Neem Tree. VCH, Weinheim, Federal Republic of Germany, pp. 222–230.

Rizvi, S., 2003. Evaluation of Neem (*Azadirachta indica* A. Juss) for Biocontrol of Vector Borne Virus Diseases of Seasonal Crops of Lucknow Region (Ph.D. thesis). Department of Botany, University of Lucknow, Lucknow, India, p. 240.

Roychoudhury, R., Jain, R.K., 1989. Effect of neem and lemongrass products on some aphid vectors of plant viruses. Indian Phytopathology 42 (2), 280.

Roychoudhury, R., Jain, R.K., 1991. Effect of neem oil and citronella oil on the development of whitefly (*Bemisia tabaci*) and virus transmission. Recent Advances in Medicinal, Aromatic & Spice Crops 1, 159–161.

Roychoudhury, R., Jain, R.K., 1996a. Neem for the control of aphid and whitefly vectors and virus diseases of plants. In: Singh, et al. (Ed.), Neem and Environment. vol. 2. WNC 1993. Oxford & IBH Publishing Co. Pvt. Ltd, New Delhi, pp. 763–769.

Roychoudhury, R., Jain, R.K., 1996b. Evaluation of neem oil based formulation for the control of virus disease spread by the whitefly vector *Bemisia tabaci* Genn. In: Neem: A Tree for 21st Century. Int. Conf. on Plant and Environmental Pollution (ICPEP). National Botanical Research Institute, Lucknow, India. December 9–11 Abs, 17, p. 50.

Samuel, D.K.L., Mariappan, V., 1996. Effect of plant derivatives on the transmission of potato virus Y infecting chilli and its aphid vector. In: Singh, R.P., et al. (Ed.), Neem and Environment, vol. 2. WNC 1993. Oxford & IBH Publishing Co. Pvt. Ltd, New Delhi, pp. 777–782.

Saxena, R.C., 1995. Effects by order of insects. Homoptera: leaf and planthoppers, aphids, psyllids, whiteflies and scale insects. In: Schmutterer, H. (Ed.), The Neem Tree. VCH, Weinheim, Federal Republic of Germany, pp. 268–286.

Saxena, R.C., Khan, Z.R., 1986. Aberrations caused by neem oil odour in green leafhopper feeding on rice plants. Entomologia Experimentalis et Applicata 42, 279–284.

Saxena, R.C., Rueda, B.P., Justin Jr., H.D., Boncodin, M.E.M., Barrion, A.A., 1987. Evaluation and utilization of neem seed "bitters" for management of planthopper and leafhopper pests of rice. In: Proc. Pest Control Council of the Philippines, Davao City, Philippines, pp. 1–43.

Saxena, R.C., Khan, Z.R., 1985. Electronically recorded disturbances in feeding behaviour of Nephotettix virescense (Homoptera:Coccidellidae) on neem oil treated rice plants. Journal of Economic Entomology 78, 222–226.

Schauer, N., 1984. Effects of variously formulated neem seed extracts on *Acyrthosiphon pisum* and *Aphis fabae*. In: Schmutterer, H., Ascher, K.R.S. (Eds.), Natural Pesticides from the Neem Tree (*Azadirachta indica* A. Juss) and Other Tropical Plants, pp. 141–150.

Schmutterer, H., 1990. Properties and potential of natural pesticides from the neem tree: *Azadirachta indica*. Annual Review of Entomology 35, 271–297.

Schmutterer, H. (Ed.), 1995a. The Neem Tree. VCH, Weinheim, Federal Republic of Germany, pp. 696.

Schmutterer, H., 1995b. The tree and its characteristics. In: Schmutterer, H. (Ed.), The Neem Tree. VCH, Weinheim, Federal Republic of Germany, pp. 11–27.

Schmutterer, H., 1999. Side effects of neem products on insect pathogens and natural enemies of spider mites and insects. In: Singh, R.P., Saxena, R.C. (Eds.), *Azadirachta indica* A. Juss. Oxford & IBH Publishing Co. Pvt. Ltd, New Delhi, pp. 147–162.

Schmutterer, H., Rembold, H., 1995. Biological effects of neem and their modes of action. Reproduction. In: Schmutterer, H. (Ed.), The Neem Tree. VCH, Weinheim, Federal Republic of Germany, pp. 195–220.

Schmutterer, H., Singh, R.P., 1995. List of insect pests susceptible to neem products. In: Schmutterer, H. (Ed.), The Neem Tree. VCH, Weinheim, Federal Republic of Germany, pp. 326–351.

Schmutterer, H., Wilps, H., 1995. Biological effects of neem and their modes of action. Activity. In: Schmutterer, H. (Ed.), The Neem Tree. VCH, Weinheim, Federal Republic of Germany, pp. 204–210.

Schoonoven, L.M., 1982. Biological aspects of antifeedants. Entomologia Experimentalis et Applicata 31, 57–69.

Serra, C.A., Schmutterer, H., 1993. Control of the sweet potato whitefly *Bemisia tabaci* Gen. with neem extracts in tomato field in the Dominican Republic. Mitwandie Entomologist 8, 795–801.

Sharma, H.C., Sankaram, A.V.B., Nwanze, K.F., 1999. Utilization of natural pesticides derived from neem and custard apple in integrated pest management. In: Singh, R.P., Saxena, R.C. (Eds.), *Azadirachta indica* A. Juss. Oxford & IBH Co. Pvt. Ltd, New Delhi, pp. 199–216.

Shuster, D.J., Stansly, P.A., 2015. Biorational Insecticides for Integrated Pest Management in Tomatoes. http://edis.ifas.ufl.edu/in481.

Simmonds, M.S.J., Blaney, W.M., 1985. Some neurophysiological effects of azadirachtins on lepidopterous larvae and their feeding response. In: Schmutterer, H., Ascher, K.R.S. (Eds.), Natural Pesticides from the Neem Tree (*Azadirachta indica* A. Juss) and Other Tropical Plants, pp. 163–179.

Singh, R.P., Saxena, R.C. (Eds.), 1999. *Azadirachta indica* A. Juss. Oxford and IBH Publishing Co. Pvt. Ltd, New Delhi, p. 322.

Singh, R.P., Chari, M.S., Raheja, A.K., Kraus, W., 1996a. Neem and Environment, vol. 1. (WNC 1993) Oxford & IBH Publishing Co. Pvt. Ltd, New Delhi, p. 617.

Singh, R.P., Chari, M.S., Raheja, A.K., Kraus, W., 1996b. Neem and Environment, vol. 2. (WNC 1993). Oxford & IBH Publishing Co. Pvt. Ltd, New Delhi, pp. 621–1225.

Singh, R.P., Doharey, K.L., Saxena, P., 1999. Evaluation of neem dust formulation against Khapra beetle *Trogoderma granarium* everts and cigarette beetle *Lasioderma serricorne* F. In: Singh, R.P., Saxena, R.C. (Eds.), *Azadirachta indica* A. Juss. Oxford & IBH Co. Pvt. Ltd, New Delhi, pp. 67–77.

Sridhar, S., Arumugasamy, S., Vijaylakshmi, K., Balasubrmanian, A.V., 2001. Vrkshayurveda: Ayurveda of Plants. Centre for Indian Knowledge Systems, Chennai, p. 47.

Subramaniam, B., 1993. Potential pesticidal plants of India. In: Parmar, B.S., Devkumar, D. (Eds.), Botanical and Biopesticides. SPS Publication No.4 Society of Pesticide Science, India and Westville Publishing House, New Delhi, pp. 74–100.

Sulochna, C.B., 1977. Plant viruses and vector specificity. In: Ananthakrishnsn, T.N. (Ed.), Insects and Host Specificity. Macmillan Co. of India Ltd, New Delhi, pp. 63–68.

Tewari, D.N., 1992. Monograph on Neem (*Azadirachta indica* A. Juss). International Book Distribution, Dehradun, India, p. 279.

Toscano, N.C., Yoshida, H.A., Henneberry, T.J., 1997. Responses to azadirachtin and pyrethrum by two species of *Bemisia* (Homoptera: Aleyrodidae). Journal of Economic Entomology 90 (2), 583–589.

Trivedi, A., Sharma, S.K., Ameta, O.P., Sharma, S.K., 2014. Management of viral and leaf spot disease complexes in organic farming of blackgram. Indian Phytopathology 67 (1), 97–101.

Vyas, B.N., Ganesan, S., Raman, K., Godrej, N.B., Mistry, K.B., 1999. Effects of three plant extracts and Achook, a commercial neem formulation on growth and development of three noctuid pests. In: Singh, R.P., Saxena, R.C. (Eds.), *Azadirachta indica* A. Juss. Oxford & IBH Co. Pvt. Ltd, New Delhi, pp. 103–109.

Walter, J.F., 1999. Adjuvants, activators, and synergists. Do additives improve the activity of azadirachtin? In: Singh, R.P., Saxena, R.C. (Eds.), *Azadirachta indica* A. Juss. Oxford & IBH Co. Pvt. Ltd, New Delhi, pp. 47–56.

WNC, 1993. In: World Neem Conference 24–28 February, 1993, Bangalore, India. Indian Society of Tobacco Science, CTRI, Rajahmundry, A.P., India, p. 90.

Semiochemicals

N. Bakthavatsalam
ICAR-National Bureau of Agricultural Insect Resources, Bangalore, India

1. INTRODUCTION

The term *semiochemicals* was first proposed by Law and Regnier (1971) (*simeon*-marker or signal) and broadly describes the chemicals employed for both intraspecies and interspecies communication. Chemicals used for the communication between individuals of the same species are termed pheromones (*pherein*–to transfer, *hormone*–to excite), while those used between individuals of different species are called *allelochemicals*, a term coined by Whittaker (1970). A pheromone is defined as a chemical or a mixture of chemicals that is released to the exterior by an organism that causes one or more specific reactions in a receiving individual of the same species (Shorey, 1977). Allelochemicals on the other hand are defined as nonnutrient substances originating from an organism that affect the behavior, physiological condition, or ecological welfare of an organism of a different species (Scriber, 1984). Allelochemicals are further divided into several classes of chemicals, such as allomones, kairomones, synomones, and apneumones, depending on their action. Of all the semiochemicals, pheromones have been widely used for pest management, while allelochemicals are scarcely used. In this review, semiochemicals that have been widely used in monitoring, mass trapping, and mating disruption of insect pests and for enhancing the performance of entomophages, mainly based on field studies, are described.

2. HISTORICAL BACKGROUND

Pheromones were studied in a large number of animals. In fact, the number of studies conducted on insect semiochemicals, more specially on pheromones, constitute more than 80% of all the species studied, probably due to increased demand for their potential use in insect pest management.

The earliest literature on the probable role of pheromones was by the French naturalist Jean Henri Fabre who in 1870 observed that the female peacock moth, *Saturnia pyri* (Denis and Schiffermüller), was able to attract hundreds of male moths from miles away (Schwarcz, 2003). Joseph Linter in 1870 noticed that females of the spice bush silk moth attracted 50 males and suggested that chemists should come forward to help entomologists in the identification of pheromones (Schwarcz, 2003). During the 1930s, several groups of chemists started working on the identification of pheromones (Schwarcz, 2003). However, not much progress was achieved until 1957 when Schneider recorded small voltage fluctuations between the tip and base of the insect antenna poststimulation with pheromones, resulting in the development of the electroantennogram (EAG) (Schneider, 1957; Schneider et al., 1967), an essential tool for the study of the chemical ecology of insects.

Bethe (1932) proposed the term *ectohormone* to cover many chemical interactions, including common attraction to a food smell. However, Karlson and Lusher (1959) proposed the term *pheromone*, a chemical used to communicate between individuals of the same species. It was much later during 1960–1970 that the functional and applied aspects of pheromones were explored, especially in terms of pest management, by utilizing these chemicals for monitoring, mating disruption, and mass trapping of pests (Beroza et al., 1961; Shoray and Gaston, 1967).

With the advancement of sensitive electrophysiological instruments, such as the EAG, gas chromatography coupled electroantennogram detector (GCEAD), gas chromatography coupled single sensillum recorder (GCSSR), gas chromatography coupled mass selective detector (GCMSD), Fourier transformed infrared spectroscopy (FTIR), and nuclear magnetic resonance spectroscopy (NMR) coupled with various sample preparation techniques, such as solid-phase microextraction techniques (SPME), the identification of pheromones and other semiochemicals was facilitated for various insect species. Since then, several review articles, books, and bulletins were published on pheromones (Campion and Nesbitt, 1982; Burkholder and Ma, 1985; Mayer and McLaughlin, 1991; Howse, 1998; Wyatt, 2003, 2014; Blomquist and Vogt, 2003; Ryan, 2007; Beroza, 2012; Carde and Minks, 2012; Howse et al., 2013).

In this chapter, the uses of pheromones and other allelochemicals have been described for various insects. Earlier studies reviewed the use of pheromones for different purposes such as mass trapping, mating disruption, and monitoring (Molinari et al., 2000; Lecocq, 2000).

3. SEMIOCHEMICALS: RESEARCH TECHNIQUES

Development of semiochemical products (such as pheromones) involves various steps including isolation of the pheromone, identification, synthesis, identification of a suitable dispenser, loading of optimum dose, use of stabilizers for sustained release at optimum dose, fabrication of suitable traps, and standardizing the number of traps with dispensers per unit area for mass trapping, monitoring, mating disruption, and male annihilation techniques (MAT) (Cork, 2004; Heuskin et al., 2011).

Isolation of the semiochemicals is generally done using methods such as solvent extraction and volatile entrapment. Polar or nonpolar solvents, depending on the nature of the solubility of the semiochemical, are used in the solvent extraction methods to effectively trap the volatiles, which are again used for analytical chemistry. Nevertheless, the presence of other nonvolatile and unimportant contaminants is a major problem in this method. In volatile entrapment, the volatiles are directly trapped using adsorbents such as Poropak or Tenax, and the volatiles thus trapped are desorbed using polar/nonpolar solvents and then analyzed (Peacock et al., 1975). Though this method is useful in avoiding contaminants and nonvolatile compounds, the quantity of volatiles trapped is sometimes very low and thus difficult to detect in the analytical equipment, necessitating longer periods of experimental setup, sometimes extending up to several days time.

With the advancement in analytical techniques, SPME has become an efficient and easy-to-use method for trapping volatiles. The SPME is a handy device fitted with adsorbents that adsorb volatiles that can easily be desorbed in gas chromatography (GC) for further analysis (Frerot et al., 1997).

The electroantennogram is a versatile electrophysiological tool that measures the difference in electrical potential between the tip and base of the antenna on exposure to the volatiles (Figure 1). The electrical response (amplitude) is measured in millivolts. Comparison of amplitude with reference to control enables the quantification of responses of insects to volatiles. Factors such as age and sex of the insects, their hunger status, photoperiodism, types of antenna, and preparation methods of volatiles greatly influence the electroantennal responses.

Coupling GC to an electroantennogram has further made the identification of semiochemicals easy (Figure 2). The simultaneous electrophysiological response of the antenna to different fractions (peaks) enables the selective identification of semiochemical (Schulz, 2005).

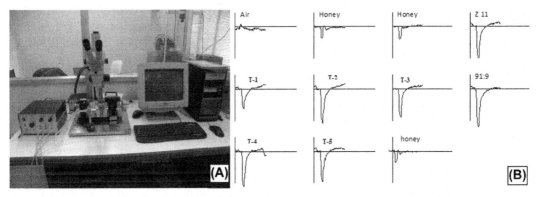

FIGURE 1   (A) Electroantennogram unit consisting of stimulus controller, microscope, and recording unit. (B) Sample electroantennogram traces showing the amplitudes in wave form different treatments (marked T-1 to T-5).

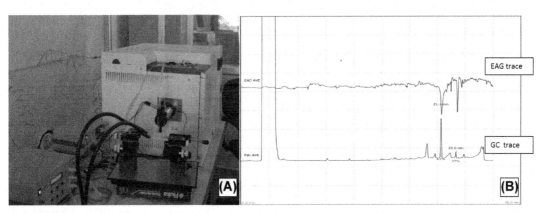

FIGURE 2   (A) Gas chromatograph coupled electroantennogram detector (GC-EAD) coupled to Agilent GC, a sophisticated tool to isolate electrophysiologically active volatiles from a matrix and identify them using other analytical equipment. (B) Sample of GC-EAD traces indicating electrophysiologically active GC peaks at various retention times.

GCMS is also an easy tool that enables the identification of compounds based on mass spectra created by the mass spectrometer and comparing with spectra available in libraries such as Wiley, Pesticide library, etc. (Pickett, 1991).

While, the above techniques enable the identification of the chemicals, identification of the functional group is done through the FTIR by comparing with the control functional group. NMR is most effectively used to identify the position of double bonds and describe pheromone structure (Hans and Miller, 2012).

Once identification of the compounds along with the structure is established, the synthesis becomes easy by identifying short and efficient schemes with good recovery. Schemes for synthesis of several pheromones are available on "Pherobase." The synthesized compounds are again tested in EAG and GCMS to characterize them both chemically and functionally.

4. PHEROMONE CHEMISTRY: SYNTHESIS AND TRAP DESIGNS

The chemistry of pheromones is an important aspect that has high relevance to intra- and interspecific communication. The number of carbon atoms forming the skeleton range from one carbon (formic acid; *Formica rufa*; Löfqvist, 1976) to 37 carbons (heptatriacontadiene–*Pikonema alaskensis* (Rohwer)). The functional groups in the compound, such as aldehyde (*Helicoverpa armigera*; Nesbitt et al., 1979), acetate esters (*Spodoptera littoralis*; Shani and Klug, 1980), amines (*Chrysoteuchia topiaria*; Ujvary et al., 1993), carboxylic acids (*Bombus hypnorum*; Cahliková et al., 2004), ketones (*Xylotrechus quadripes*; Hall et al., 2006), phenols (*Crematogaster difformis*; Jones et al., 2005), and alcohols (*Rhynchophorus ferrugineus*; Oehlschlager et al., 1992) are important aspects determining the pheromone activity. The position of double bonds is also very crucial as it determines the isomerism, the isomers such as "E," e.g., *Scirpophaga excerptalis* (Wu et al., 1990) and "Z," e.g., *H. armigera* (Nesbitt et al., 1979) are of common occurrence.

Pheromones are usually blends of multiple compounds differing in structure and amount. It is the blend of these major and minor compounds that causes the species specificity of pheromones. For example, *H. armigera* utilizes Z-11-hexadecenal and Z-9-hexadecenal at 97:2. 1, while *Helicoverpa assulta* (Guenee) uses the same compounds in an opposite blend (5:94. 5) (Wang et al., 2005). Geographical variations between populations in the blend ratio are also present, but still a large degree of similarity is still present (Tamhankar et al., 2003; Bakthavatsalam, unpublished data). The structural complexities of some pheromone compounds are presented in Figure 3.

Besides studies in lepidopterans, coleopterans, and dipterans, considerable studies have been made on semiochemicals of hemipterans, neuropterans, and hymenopterans (Hardie and Minks, 1999), however, their usage in the field for monitoring, mass trapping, or mating disruption is limited and hence not covered in this chapter. Significant contributions have been made on pheromones for homopterans, especially mealybug pheromones for monitoring the cryptic and sessile mealybugs.

Pheromones and kairomones have been identified and synthesized for several species of mealybugs, viz. *Crisicoccus matsumotoi*, *Maconellicoccus hirsutus*, *Phenacoccus madeirensis*, *Planococcus* spp., *Planococcus ficus*, *Pseudococcus comstocki*, *Pseudococcus* spp. (Bierl-Leronhardt et al., 1981; Bichina et al., 1982; McCullough et al., 1991; Tabata et al., 2012; Arai, 2002; Zada et al., 2003; Zhang and Nie, 2005; Millar et al., 2005, 2009; Kukovinets et al., 2006; Ho et al., 2007, 2009; Hashimoto et al., 2008; Unelius et al., 2011; Tabata et al., 2012).

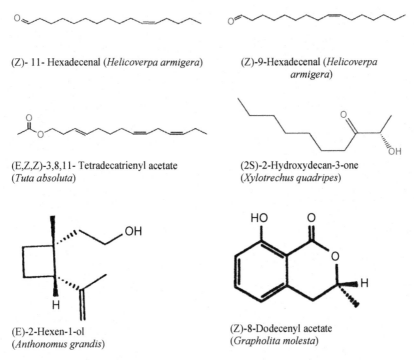

(Z)- 11- Hexadecenal (*Helicoverpa armigera*)

(Z)-9-Hexadecenal (*Helicoverpa armigera*)

(E,Z,Z)-3,8,11- Tetradecatrienyl acetate (*Tuta absoluta*)

(2S)-2-Hydroxydecan-3-one (*Xylotrechus quadripes*)

(E)-2-Hexen-1-ol (*Anthonomus grandis*)

(Z)-8-Dodecenyl acetate (*Grapholita molesta*)

FIGURE 3   Structural formulae of pheromone compounds of some insects.

Synthesis of pheromones or other semiochemicals is done through a variety of schemes. The challenges posed in synthesis are in assembling the number of carbon atoms, adding correct functional groups, ensuring correct double-bond geometry, and mixing of stereoisomers at proper proportions. The detailed synthesis schemes for straight-chain compounds, branched-chain alkenes, cyclic compounds, and chiral compounds with a few examples have been summarized (Howse, 1998); in addition, individual specific research articles and *Pherobase* offer schemes for synthesis of pheromone compounds.

Semiochemicals synthesized and isolated have been tested using wind tunnels, cage studies, and field studies to prove their activity. It becomes necessary to optimize the release of volatiles simulating quantities emitted naturally in a given period of time as well as for sustained release for longer periods. The dispensers (vials, septa, discs, rubber tubes, or plywood pieces) serve this purpose, and the loading of semiochemicals in dispensers is quite crucial for optimized and sustained release. Addition of antioxidants such as butylated hydroxy aniline (BHA) or butylated hydroxy toluene (BHT) and acetic acid function as stabilizers for pheromone dispensers (Cork, 2004). Nevertheless, the problem persists in using plant volatiles that are highly unstable.

Trap design is crucial for the efficient trapping of insects. For lepidopterans, the common trap is a sleeve trap, while for small insects yellow sticky traps or delta traps are used (Bakthavatsalam and Bhagat, 2008). Fruit fly traps are used considering their active flying habits and bucket traps for beetles and weevils. It is also a general practice that for beetles and weevils baits are also added in the traps along with some insecticide to restrict trapped insects from moving outside. Use of soap solution is favored for fruit flies when water traps are used, and addition of malathion or 2,2,dichlorovinyl dimethyl phosphate (DDVP) commonly known as dichlorvos, is advocated for

FIGURE 4 Various pheromone traps: (A) Cross-vane traps for coffee stem borer, (B) Bucket traps for coconut red palm weevil, (C) Delta trap for gall wasps, (D) Fruit fly trap with water, (E) Yellow sticky trap for smaller insects, and (F) Fruit fly trap with plywood piece impregnated with insecticide.

killing the fruit flies trapped in wooden dispensers. Yellow traps, tent-shaped delta traps, and multiseason plastic delta traps are often used for trapping mealybug males. A few examples of traps used for attracting various insect species are given in Figure 4.

5. MONITORING OF PESTS

Pheromones are excellent tools for quarantine monitoring, correlating pest population with field populations, prediction modeling, and temporospatial dispersal patterns of selective insect pests. Qualitative estimation (detection) and quantitative estimation (population dynamics, dispersal) are often necessary for the management of pest populations, and pheromones become handy in generating these data. These estimates are used to determine applications, viz. early warning, quarantine, schedule of management measures, population trends, dispersion, and risk assessment.

Effective monitoring of the pests is achieved through the use of appropriate dispensers containing the exact blend of the pheromone compounds, placed in optimum quantities of traps, which are spread one to two per acre at appropriate places in the fields (Witzgall et al., 2010; Witzgall, 2005; Shinoda, 2007; Colazza et al., 2007). Of course, the needs of the end user, depending on their skill level, have to be taken into consideration while designing the monitoring models. Monitoring of insects with pheromones has been consistently done in different species of insects (Table 1).

TABLE 1 Insects and Pheromones Used for Monitoring

Insect pest	Host	Pheromone	References
Agriotes brevis Candeze	Polyphagous	(E)-3,7-dimethyl-2,6,octadeinyl 3-methyl butanoate "Geranyl butyrate"	Toth et al. (2003)
Agriotes lineatus (L.)	Maize	(E)-3,7-dimethyl-2,6,ocatdienyl octanoate "Geranyl octanoate"	Veron and Toth (2007)
Agriotes litigious (Rossi)	Polyphagous	(E)-3,7-dimethyl-2,6-octadienyl 1,3 methyl butanoate "Geranyl isovalerate"	Toth et al. (2003)
Agriotes obscurus (Linnaeus)	Polyphagous	(E)-3,7-dimethyl-2,6,ocatdienyl octanoate "Geranyl octanoate"	Veron and Toth (2007)
Agriotes sordidus Illiger	Polyphagous	(E)3,7,dimethyl 2,6,octadienyl octanoate "Geranyl hexanoate"	Pozzati et al. (2006)
Anthonomus rubi Herbst	Strawberry	(Z)-2-(3,3-dimethyl)cyclohexylidene ethanol	Cross et al. (2006)
Aonidiella aurantii (Maskell)	Citrus	3,methyl-6-isopropenyl-9-decenyl acetate	Colazza et al. (2007)
Cameraria ohridella Deschka and Dimić	Horse chestnut	(E,Z)-8,10-tetradecadienal	Oltean et al. (2005)
C. ohridiella Deschka and Dimić	Chestnut	(E,Z)-8,10-tetradecadienal MK type trap	Grodner et al. (2005)
Carposina sasakii Matsumura	Apple	(Z)-7-eicosen-11-one	Feng et al. (2005)
Cydia molesta (Busck)	Peach	(Z)-8-dodecenyl acetate Ecodian Combi CM+Al dispensers	Kutinkova and Dozhuvinov (2012) and Rumine (2002)
Cydia pomonella Linnaeus	Apple	(E)-8-10-dodecadien-1-ol Codlemone (ethyl (E,Z) 2,4 decadienoate)	Knight et al. (2005)
Cylas formicarius (Fabricius)	Sweet potato	(Z) 2,3-dodecenyl-E-2 butenoate	Mondal et al. (2006)
Diabrotica virgifera LeConte	Maize	(E,Z)-24-methoxyphenyl-2-propenal "Raluca Ripan"	Muresanu and Popor (2005)
Diorhabda elongata (Brullé)		(E,Z)-2,4-heptadien-1-ol	Cosse et al. (2006)
Diprion jingyuanensis Xiao et Zhang	Chinese pine	(1S, 2R, 6R)-1,2,6-trimethyl dodecyl propionate	Zhang et al. (2005)
Elasmopalpus lignosellus (Zeller)	Corn	(Z)-9-tetradecenyl acetate "Pherocon 1C traps"	Loera et al. (1995)
Ephestia cautella (Walker)	Store house	(Z,E)-9,12-tetradecadienyl acetate, (Z),9-tetradecenyl acetate "Gachon Reg" "Biostop Reg"	Finkelman et al. (2007)

(Continued)

TABLE 1   Insects and Pheromones Used for Monitoring—cont'd

Insect pest	Host	Pheromone	References
Grapholita dimorpha Komai	Polyphagous	E−8-dodecenyl acetate and Z-8-dodecenyl acetate	Murakami et al. (2005)
Helicoverpa armigera Hubner	Polyphagous	Z-11-hexadecenal; Z-9-hexadecenal 97:3	Nesbitt et al. (1979) and Bruce and Cork (2001)
Lasioderma serricorne (F.)	Polyphagous	(2s,3R,1s)-2,3-dihydro-3,5,dimethyl-2-ethyl,6,(1-methyl-2-octo butyl) 4H-pyran-4-one "New Serrico Reg"	Finkelman et al. (2007) and Saeed et al. (2007)
Melittia cucurbitae (Harris)	Squash	(E,Z)2,13,octadecadienyl acetate Harstack trap	Jackson et al. (2005)
Metisa plana Walker	Oil palm	Heptacosane vane trap with pheromone	Kamarudin and Arshad (2006)
Neodiprion sertifer (Geoffroy)	Pine	(2, 3S, 7S)-3,7-dimethyl 2-pentadecanol (diprionol)	Lyytikainen-Saarenmaa et al. (2006)
Orgyia leucostigma (Smith)		(Z,Z)-6,9-heneicosadien-11-one ethylene ketal	Grant et al. (2006)
Ostrinia nubilalis (Hübner)	Corn	Z-11-tetradecenyl aetate E-11-tetradecenyl acetate tetradecenyl acetate Wire mesh corn trap	Pelozuelo et al. (2006)
Pectinophora gossypiella (Saunders)	Cotton	Gossyplure	Salem et al. (1990), Kabre and Dharne (2009), and Huber and Hoffmann (1979)
Scirpophaga incertulas Walker	Rice	Z-11-Hexadecenal: Z-9-Hexadecenal (3:1)	Patel and Desai (2004)
Sitodiplosis mosellana (Gehin)	Wheat	(2S-7R)- 2,7-nonadiyl dibutyrate	Bruce et al. (2007)
Stenotus rubrovittatus (Matsumura)	Rice	Hexyl butyrate, (E)-hexenyl butyrate, (E)-4-oxo-2-hexenal 7 m from edge	Yasuda and Higuchi (2012)
Thaumetopoea pityocampa Schif	Pine	(Z)-13-hexadecen-11-yn-1-ol-acetate Plate sticky trap Pityolure	Jactel et al. (2006)
Tortrix viridana Linnaeus	Oak	(Z)-11-tetradecenyl acetate Delta traps	Askary et al. (2005)
Trigonotylus caelestialum (Kirkaldy)	Rice	Hexyl hexanoate, (E)-2-hexenyl hexanoate, octyl butyrate	Yasuda and Higuchi (2012)
Tuta absoluta Meyrick	Tomato	(E,Z,Z)-3,8,11-tetradecatrienyl acetate Delta traps	Bavaresco et al. (2005)

Better utility of pheromones for monitoring can only be appreciated if the pheromone catches coincide with the field population. Trap populations are indicative of field populations and best correlated with populations in subsequent weeks as observed in *H. armigera* (Verma and Sankhyan, 1993). Computations based on pheromone catches have been done extensively in India in crops like chickpea, cotton, tomato, sunflower, and citrus (Sah et al., 1988; Srivastava et al., 1991; Prasad and Neupane, 1992; Duwedi et al., 1996; Loganthan and Uthamasamy, 1998; Visalakshi et al., 2000; Mohaptra et al., 2007).

In a field ecosystem, especially crops with multispecies pest environment, often the question posed is whether the pheromone of one species will affect the trap catches of other insect species. However, studies on combinations of pheromones of the forest tent caterpillar, *Malacosoma disstria* Hübner and large aspen tortrix, *Choristoneura conflictana* (Walker) reveal that they did not influence each other (Evenden, 2005).

Dispersal patterns of alien invasives and weed biocontrol agents need continuous monitoring, with pheromones being an effective tool for documenting their dispersal. Invasive pests of international significance continue to spread to nontraditional areas, and their presence can easily be monitored by using pheromone traps, e.g., *Diabrotica virgifera* LeConte was monitored in different quarantine houses (Hummel et al., 2005). Pheromones are also an indicator for assessing mating disruption technology, as observed in *Cydia molesta* (Busck) (Kovanc and Walgenbach, 2005).

The data obtained from pheromone traps have been used for developing population prediction models using several abiotic parameters, such as position, age of pheromone, number of days since monitoring, average monthly temperature, monthly precipitation, month and previous numbers as in bark beetles, *Ips typographus* (DeGreer) and *Pitogenes chalcographus* (Linnaeus) (Jurc et al., 2006). A threshold catch in pheromone traps helps determine the levels of egg and larvae present in the field (Lopez et al., 1990). Prediction modeling has been successfully developed for *Bactrocera oleae* (Gmelin), *Planococcus ficus* Signoret, San Jose scale *Quadraspidiotus perniciosus* Comst (Duca et al., 2005; Frantisek and Jittka, 2005; Ortu et al., 2006) and *Oryctes rhinoceros* (Linnaeus) (Kamarudin and Wahid, 2004). Management decisions, such as chemical control schedules and release of natural enemies are often made based on the pheromone data (Olivero et al., 2005).

The monitoring of mealybugs, *Pseudococcus comstocki*, *Planococcus citri* and *P. ficus* has been achieved through the use of pheromone 2, 6-dimethyl-1, 5-leptadien (at 200 μg per trap, 1R *cis*-3-isopropyl-2-2-dimethyl cyclobutyl methyl acetate (at 2000 μg, and (R) lavandulyl (S)-2-methyl butanoate, respectively (Smetnik and Rizinskya, 1988; Franco et al., 2003; Millar et al., 2002). These data on pheromone trap catches were used for scheduling releases of predator, *Cryptolaemus montrouzieri*, and parasitoid, *Leptomastix dactylopii*, in citrus orchards. Mass trapping and mating disruption using pheromones was attempted for *P. citri*, *Pseudococcus calceolariae* and *Planococcus kraunliae* (Branco et al., 2006; Rotunda et al., 1979; Teshiba et al., 2009; Walton et al., 2004).

Though monitoring using pheromones has been used for several pests, the common constraint is that in several instances there is no correlation between the number of insects trapped and field populations, mainly due to the fact that the trapped insects might have migrated from other fields. Further, the movement of males from other plants (Franco et al., 2003) or the interaction of natural enemies (Rotunda et al., 1979) and lack of sufficient quantity of synthesized pheromone materials to demonstrate in field trials may undermine the effectiveness of pheromone traps.

6. MASS TRAPPING

Mass trapping of insect pests is done through pheromone traps. Basically, the traps and dispensers remain the same as that of monitoring, but the number of traps per unit area is increased to effectively trap more insects. In essence, monitoring is used for estimating the pest population along with resultant management decisions, whereas mass trapping is for total eradication of insect populations. Pheromone dosage in some exceptional cases is reduced to half with a view to reduce the cost of the technology. However, the number of traps and their placement is vital for efficient trapping. For instance, in the management of *Anomala vitis* Fabricius and *Anomala dubia* Scopoli placing traps on one to two outside rows caught 92% of all beetles, with ultimate reduction in damage by the grubs (Voigt and Toth, 2004).

Cryptic habits of several coleopteran borers necessitate use of pheromones for mass trapping and species such as *Cosmopolites sordidus* (Germar) (Rhino et al., 2010), *Oryctes elegans* (Tabrizian et al., 2009), *Oryctes rhinoceros* (Linnaeus) (Ponnamma et al., 2002), *Ips acuminatus* Gyllenhal (Perez and Sierra, 2006), *R. ferrugineus* (Soroker et al., 2005; Mohammadpur and Faghih, 2008; Jayanth et al., 2007; Guarino et al., 2013), and *Xylotrechus quadripes* (Chevrolet) (Jayarama et al., 2007). Mass trapping of *Cylas formicarius* using Z-3-dodecenyl-E-2-butenoate at 0.001–1 mg/trap was successful, ultimately resulting in good crops with reduced damage (Talekar and Lee, 1989; Setokuchi et al., 1991; Teli and Salunkhe, 1993; Yasuda, 1995; Pillai et al., 1996; Smit et al., 2001; Chiranjeevi and Reddy, 2003); however, lack of availability of synthesized pheromone in greater quantities is still a constraint for its successful mass trapping. A list of insects for which mass trapping is done using pheromones is provided in Table 2.

Leaf feeding and boring habits of lepidopterans make mass trapping using pheromones one of the most effective tools. Exploitation of sex pheromones for the management of *Ephestia kuehniella* (Zeller), *Chilo suppressalis* (Walker) (Sheng et al., 2000, 2002; Jiaoet al., 2003, 2005; Su et al., 2003), *Chilo infuscatellus* Snell. and *Chilo sacchariphagus indicus* (Bojer) (David et al., 1985; Chelvi et al., 2010), *Plutella xylostella* (Linnaeus), (Chow et al., 1977; Maa et al., 1987; Chisholm et al., 1979; Reddy and Urs, 1997; Wang et al., 2004; Huangfu et al., 2005), and *Pectinophora gossypiella* (Saunders) is in vogue. The number of traps used, however, vary from 9 traps/ha (*P. gossypiella*; Huber and Hoffmann, 1979; Nassef et al., 1999; Campion and Nesbitt, 1983; Neto and Habib, 1996; Ahmad and Attique, 1993; Patil et al., 2008; Nandihalli et al., 1993) to 42/ha (*E. kuehniella*; Trematerra and Gentile, 2010). Long-term usage of mass trapping has been known to reduce major pests, such as *Cydia pomonella* (Linnaeus), *P. gossypiella*, bark beetles, palm weevils, corn rootworm *Anthonomus grandis* Boheman and *Lymantria dispar* (Linnaeus) (El Sayed et al., 2006). However, establishment of the effectiveness of this technology requires larger plot trials.

Integrated pest management with mass trapping as one of the modules has been successful for several pests. The use of pheromones for mass trapping has led to great reduction in the pest load of many crops in India, viz. *Scirpophaga incertulas* (Walker), where the use of pheromone traps has led to greatly reduced pest incidence and an appreciable reduction in insecticide usage in rice (Cork et al., 2005b). In the management of brinjal, installation of 25–30 traps/ac, containing a blend of E-11-hexadecenyl acetate and E-11-hexadecenol (100:1) along with insecticidal sprays such as neem, effectively managed *Leucinodes orbonalis* (Guenee) (Cork et al., 2001, 2003, 2005a; Rath and Dash, 2005; Dash et al., 2005). Although the management of pests using mass trapping is largely successful, in several instances it is unsuccessful as

TABLE 2 Insects and Pheromones Used for Mass Trapping

Species	Compound	References
Anomala vitis Fabricius *A. dubia* Scopoli	(E)-2-Nonen-1-ol	Voigt and Toth (2004)
Anthonomus rubi Herbst.	(Z)-2-(3,3-Dimethyl)-cyclohexylidene ethanol	Cross et al. (2006)
Bactrocera oleae (Gmelin)	Yellow triangular cardboard Ecotrap	Bjelis (2006), Latif et al. (2004), Rizzi et al. (2005), and Liaropoulos et al. (2005)
Cameraria ohridella Deschka	(E,Z)-8,10-Tetradecadienal	Haddad (2006), Hastings and Jullien (2006), and Colazza et al. (2007)
Ceratitis capitata (Wiedemann)	Ammonium acetate, putrescine and triethylamine in low rate polythene dispensers of borax and FA-3 Nulure plus propylene glycol Probodelt Tephry and Probodelt	Ros et al. (2002), Alemany et al. (2002), and Batllori et al. (2005)
Chilo sacchariphagus indicus (Bojer)	(Z)-13-Octadecenyl acetate, (Z)-13-Octadecen-1-ol	David et al. (1985) and Chelvi et al. (2010)
Chilo infuscatellus Snell.	(Z)-11-Hexadecen-1-ol	David et al. (1985) and Chelvi et al. (2010)
Chilo suppressalis (Walker)	41:5:4 Z-11-hexadecenal, Z-13-octadecenal, and Z9-hexadecenal	Jiao et al. (2003, 2005), Su et al. (2003), and Sheng et al. (2000, 2002)
Cosmopolites sordidus (Germar)	(1S,3R,5R,7S)-1-Ethyl-3,5,7-trimethyl-2,8-dioxabicyclo[3.2.1]octane	Rhino et al. (2010)
Cylas formicarius (Fabricius)	Z-3-dodecenyl-E-2-butenoate	Teli and Salunkhe (1993), Talekar and Lee (1989), Chiranjeevi and Reddy (2003), Yasuda (1995), Setokuchi et al. (1991), Smit et al. (2001), Pillai et al. (1996), and Mason and Jansson (1990)
Diabrotica virgifera LeConte	4-methoxy cinnamaldehyde in cone cup (Metcalf) type trap	Hummel et al. (2005)
Ephestia kuehniella (Zeller)	2 mg of pheromone (Z,E)-9,12, tetradecadienyl acetate	Trematerra and Gentile (2010)
Ips acuminatus Gyllenhal	ipsenol, ipsdienol and *cis*-verbenol	Perez and Sierra (2006)
Leucinodes orbonalis (Guenee)	blend of E-11-hexadecenyl acetate and E-11-hexadecenol (100:1)	Cork et al. (2001, 2003, 2005a), Rath and Dash (2005), and Dash et al. (2005)
Metisa plana Walker	sticky vane traps with four receptive females and placed in three trap insects	Norman et al. (2010)

(Continued)

TABLE 2   Insects and Pheromones Used for Mass Trapping—cont'd

Species	Compound	References
Oryctes elegans Prell. and *R. ferrugineus* Olivier	mixed pheromone dispensers	Mohammadpur and Faghih (2008)
O. elegans Prell	4 methyl octanoic acid	Tabrizian et al. (2009)
Oryctes rhinoceros (Linnaeus)	ethyl-4-methyloctanoate	Ponnamma et al. (2002)
Pectinophora gossypiella (Saunders)	Z-E-7,11-hexadecadienyl acetate and Z,Z,7,11,hexadecadienyl acetate	Huber and Hoffman (1979), Nassef et al. (1999), Neto and Habib (1996), Ahmad and Attique (1993), Patil et al. (2008), Nandihalli et al. (1993), and Campion and Nesbitt (1983)
Phyllopertha horticola (Linnaeus)	Z-3-hexanal-1-ol with membrane dispenser	Ruther and Mayer (2005)
Plutella xylostella (Linnaeus)	Z-11-hexadecenal, Z-11-hexadecenyl acetate at 3:7	Chow et al. (1977), Maa et al. (1987), Chisholm et al. (1979), Reddy and Urs (1997), Wang et al. (2004), and Huangfu et al. (2005)
Rhabdoscelus obscurus (Boisduval)	Pheromone + cut sugarcane	Sallam et al. (2007) and Reddy et al. (2005)
Rhynchophorus ferrugineus Olivier	Ferrugineol (2-tetradecanone) with ethyl acetate	Jayanth et al. (2007) and Soroker et al. (2005)
Scirpophaga incertulas (Walker)	Z-11-hexadecenal and Z-9-hexadecenal (3:1)	Varma et al. (2004) and Cork et al. (2005b)
Thaumetopoea pityocampa, Denis and Schiffermüller	(Z)-13-Hexadecen-11-yn-1 acetate	Baronio et al. (1992)
Xylotrechus quadripes (Chevrolet)	Cross-vane traps along with pheromone vials containing 72-hydroxy-3-decanone	Jayarama et al. (2007)

observed in the case of *Anthonomus rubi* Herbst. (Cross et al., 2006). To be effective as a management tool, the use of mass trapping is suggested for longer periods of usage. For instance, continuous use of *P. xylostella* pheromone reduced larval and adult populations of insects in China (Huangfu et al., 2005).

The efficacy of the trap type also differs with the kind of insect targeted. Infestation of grubs of *D. virgifera* was dramatically reduced in soil with the use of 4-methoxy cinnamaldehyde in a cone cup (Metcalf) type trap. Use of funnel traps baited with Z-3-hexanal-1-ol with membrane dispenser reduced the infestation of *Phyllopertha horticola* (Linnaeus) in apple (Ruther and Mayer, 2005).

In addition, the number of traps and their placement is crucial for efficient trapping. For instance, in the management of *A. vitis* Fabricius and *A. dubia* Scopoli, placing the traps on one to two outside rows caught 92% of all beetles, leading to ultimate reduction in damage by the grubs (Voigt and Toth, 2004). There is always a direct correlation between the plot size and success of mass trapping technology, for instance the mass trapping is highly successful where the treatments are imposed in bigger holdings, such as on 5 ha plots for the management of *S. incertulas* compared to 1 or 3 ha plots (Varma et al., 2004).

Mass trapping is especially effective in coleopterans owing to the availability of aggregation pheromone for a variety of species. Traps loaded with ferrugineol with ethyl acetate and fermenting mixture of dates and sugarcane molasses at 10/ha effectively reduced the infestation of *R. ferrugineus* and mixed pheromone dispensers for *O. elegans* and *R. ferrugineus* were effective against both the species (Mohammadpur and Faghih, 2008). The efficacy of pheromone in mass trapping *R. ferrugineus* is well documented (Jayanth et al., 2007). In India, coffee white stem borer *Xylotrechus quadripes* (Chevrolet) was effectively managed through the use of cross-vane traps at 10 traps/ac along with pheromone vials containing 75 mg of pheromone, 2-hydroxy-3-decanone (Jayarama et al., 2007). Bucket traps baited with pheromone and cut sugarcane attracted more *Rhabdoscelus obscurus* Boisduval in Guam (Reddy et al., 2005).

Although pheromones are species specific, it has often been observed that several nontarget insects, including honeybees, are trapped owing to faulty trap designs. Cross-vane traps used for coffee stem borer quite commonly trap other insects including beneficials, such as honeybees, neuropteran predators, and hymenopteran parasitoids. Appropriate technological interventions through change in trap design have significantly reduced the capture of nontarget insects (Weber et al., 2005).

In some instances, pheromones have been found not effective in mass trapping the pests. Lethmayer et al. (2004) and Cross et al. (2004) observed that population levels of *A. rubi* as well as damage caused did not reduce in spite of use of aggregation pheromone. Several factors that have been recognized as contributors to the failure of mass trapping are migration of populations from other regions, ineffective trap designs, insufficient number of traps, and nonadoption in wider or adjacent areas.

7. MATING DISRUPTION

7.1 Mating Disruption for Pests

Mating disruption is a term coined for the reduction in egg load or larval population through disruption of mating by inundating large quantities of pheromone in the field. The disruption is achieved through confusion (males are overtly confused), trail masking, and laying false trails (Campion et al., 1980). Successful mating disruption is indicated by a drop in the trap catches over time, reduced mating in fields as witnessed through visual observations, lesser incidence of pests or damage in comparison to untreated areas, and the absence of spermatophores in the females. Sometimes avoidance or delayed mating with calling females was also attributed to effectively reducing the fecundity of the females (Jones and Sasaki, 2001). Ecological factors such as emergence patterns of diapausing pupae, diurnal rhythms, and weather factors play a significant role in the effectiveness of mating disruption. Laboratory experiments are often necessary to optimize the dosage and efficacy of pheromones for mating disruption. In a study conducted with *Teia anartoides*, up to 50% of males located virgin females in control cages, while no mates were located in cages with pheromones (Suckling et al., 2002). A delay in mating results in the profound reduction in the fecundity of insects, for example, *Lobesia botrana* (Denis and Schiffermüller) laid significantly reduced number of eggs when mating was disrupted (Torres Vila et al., 2002).

Successful implementation of mating disruption in *C. suppressalis* on rice was achieved in several countries using various dispensers and doses as recommended by their respective developmental/research institutions (Yang et al., 2001a; Tanaka et al., 1987; Sudo et al., 1998; Ueno and Hayaska, 1997; Alfaro et al., 2009; Casagrande, 1993; Jiao et al., 2003). In India, mating disruption using a blend of Z-9-hexadecenal, Z-11-hexadecenal and Z-9-octadecenal (1:10:1) in a polyvinyl chloride (PVC) resin formulation was found to be effective (Cork and Basu, 1996). The sugarcane borer, *C. sacchariphagus indicus*, uses a pheromone blend of Z-1-octadecenal and Z-13-octadecenol (7:1) (David et al., 1985). Campion et al. (1987) and Rajabalee (1990) illustrated mating disruption using pheromone blends in sugarcane. A list of insect species for which mating disruption has been conducted is given in Table 3.

Irrespective of pheromone dispensers used, successful implementation of mating disruption has been achieved by using Gossyplure for *P. gossypiella* management (Carde et al., 1997; Sohi et al., 1999) either through the use of 250 dispensers/ha (Papa et al., 2000) or microencapsulated sprays, and rope dispensers (Muthukrishnan and Balasubramanian, 1999; Prasad et al., 2009; Radhika and Reddy, 2006). Similarly, successful mating disruption was realized in the management of *Spodoptera litura* (Srinivas and Rao, 1999; Krishnaiah, 1986; Wang et al., 2008; Katase et al., 2006) and *Spodoptera exigua* (Cheng and Zhang, 2008).

Highly effective, slow-releasing and cost-effective dispensers including sprayable formulations along with suitable traps substantially reduced the infestations of *Sparganothis sulfureana* in cranberries and *Sesamia nonagrioides* (Lefebvre) (Polavarapu et al., 2001). Multiple applications instead of single applications are effective in several instances (Polavarapu et al., 2001; Albajes et al., 2002). Commercial formulations such as PB-ROPEL Isomate M Plus, Isomate M 100, Isomate OFM, RAK1+2 dispensers, or RAK1+2R Isonet and Biodegradable dispensers are a few examples of pheromone dispensers used for successful mating disruption in *C. pomonella*, *Adoxphyes orana fasciata* Walsingham, *Grapholita lobarzewskii* (Nowicki), *Cydia moelsta*, *Eupoecilia ambiguella* (Huber), and *L. botrana* (Denis and Schiffermüller) (Angeli et al., 1999; Zuber, 1999; Molinari et al., 2000; Mansour and Mohamad, 2001; Yang et al., 2001b; Okazaki et al., 2001; Charmilllot and Pasquier, 2001; Maini and Accinelli, 2001; Charmillot et al., 2001; Waldner, 2001). Traditionally, the dosage needed for mating disruption was calculated based on the quantum per unit area, however, recent technologies describe release rates per unit area per unit time.

In a multiple species pest environment, chemical ecologists have been confronting the use of generic blends of pheromones. It has been seen that a generic blend for *Platynota idaeusalis* (Walker) suppresses captures of *Argyrotaenia velutinana* Brown & Cramer but not *A. flavedana* Clemens (Gronning et al., 2000). Use of multiple species mating disruptors (Confuzer-pReg) was effective against insects such as *Homona magnanima* Diakonoff, *Adoxophyes honmai* Yasuda and *Carposina niponensis* Walsingham (Izawa et al., 2000).

To be an effective management tool, mating disruption needs to be practiced for longer periods of time. For instance, release of pheromones at 23.7 to 26.4 mg/ha/h over 86–91 days reduced the capture in monitoring traps by 98%, suggesting a high level of disruption (Trimble et al., 2001). Use of pheromones for nine years in an area of 985 ha for *C. pomonella* and 255 ha in *C. molesta* recorded less than 1% damage, however, the cost of application of pheromones was more than the insecticide (Dallago, 2000). Wide area mating disruption

TABLE 3 Insects, Crop and Pheromones Used in Mating Disruption

Species	Crop	Pheromone component/formulation	References
Adoxophyes orana fasciata (Walsingham)	Fruit crops	(Z)-9-Tetradecenyl acetate	Okazaki et al. (2001)
Anarais lineatella (Zell)	Fruit crops	(E)-5-Decenyl acetate, (E)-5-Decen-1-ol	Trematerra et al. (2000)
Ascotis selenaria cretacea (Butler)	Tea	(Z,Z)-6,9-3,4-Epoxynonadecadiene, (Z,Z,Z)-3,6,9-Nonadecatriene3000-5000 tubes/ha	Ohtani et al. (2001)
Chilo sacchariphagus indicus	Sugarcane	Z-1-Octadecenal and Z-13-Octadecenol (7:1)	David et al. (1985), Campion et al. (1987), and Rajabalee (1990)
Chilo suppressalis (Walker)	rice	Z-9-hexadecenal, Z-11-hexadecenal and Z-9-octadecenal (1:10:1)	Cork and Basu (1996)
Cossus cossus Linnaeus	Fruit crops	Z,3-decenyl acetate + Z, 5-dodecenyl acetate at 4:1	Pasqualinin and Natale (1999)
Cydia funebrana (Treitschke) and *Anarsia lineatella* (Zeller)	Stone fruits	(E)-8-Dodecenyl acetate	Molinari (2001)
Cydia molesta (Busck)	Stone fruit, grape	(Z)-8,Dodecenyl acetate, Isomate OFM plus and Isomate OFM Rosso formulations 40 g/ha	Brown and Il'ichev (2000), Simon et al. (1999), and Sexton and Il'ichev (2001)
Cydia pomonella (Linnaeus)	Apple, peach & pear, walnut, stone fruits	(E,E)-8,10-Dodecadien-1-ol, RAK-3 & RAK-3 + carpovirus 300 dispensers/ha or isomate C+ dispensers 200/ac or Checkmate CM 500/ha or 427 dispensers/ha	Aalbers (2001), Ciglar et al. (2000), Novo et al. (2000), Quareles (2000), Villa-Gil et al. (1998), Knight et al. (2001), and Angeli et al. (2000)
Cylas brunneus Fabricius, *Cylas puncticollis* (Boheman)	Sweet potato	Dodecyl (E)-2 butenoate PVC resin	Downham et al. (2001)
Ephestia kuehniella (Zeller)	2.5 μg 3 TDA for 3 months	(Z,E)-9,12-Tetradecadienyl acetate	Suss et al. (1999)
Eucosma notanthes (Meyrick)	Carambola	53.0 g/ha for 5 months (Z)-8-Dodecenyl acetate	Hung et al. (2001)
Eupoecilia ambiguella (Hübner)	Grape	(Z)-9-Dodecenyl acetate	Zingg and Hohn (2000) and Vrabl and Matis (1999)
E. ambiguella (Hübner)	Grape	(Z)-9-Dodecenyl acetate	
Grapholita molesta (Busck) *Synanthedon exitiosa* (Say)	Peach	(Z)-8-Dodecenyl acetate, (Z,Z)-3,13-octadecadienyl acetate	Hull and Felland (1999)
Helicoverpa armigera (Hübner)	Cotton	Diamolure, (Z)-11-Hexadecenal: Z-9-Hexadecenal, Selibate Reg HA at 40 g/ha	Toyoshima et al. (2001) and Chamberlain et al. (2000)

(Continued)

TABLE 3 Insects, Crop and Pheromones Used in Mating Disruption—cont'd

Species	Crop	Pheromone component/ formulation	References
Laspeyresia pomonella (Hübner)	Apple	15–30 traps/ha (E,E)-8,10-Dodecadien-1-ol	Emel and Bul- gynskaya (1999)
Lobesia botrana (Denis and Schiffermüller)	Grape	(E)-7-Dodecenyl acetate	Charmillot and Pasquier (2000)
L. botrana (Denis and Schiffermüller)	Grape	(E)-7-Dodecenyl acetate	Varner et al. (2001)
Lymantria sp.	Apple	Disparlure with V mount sticky agent, (Z)-7,8-epoxy-2-methyloctadecane	Thorpe et al. (2000)
Neodiprion sertifer (Geoffroy)		(2,3,7S)-3,7-Dimethylpentadecan-2-ol	Ostrand et al. (1999)
Pectinophora gossypiella (Saunders)	Cotton	(Z,Z)-7,11-Hexadecadienyl acetate	Papa et al. (2000), Ogawa (2000), and Sohi et al. (1999)
Platynota idaeusalis (Walker)	Apple	(E)-11-Tetradecen 1-ol, (E)-11-Tetradecenyl acetate	Borchert and Walgenbach (2000)
Prays oleae	Fruit crops	Z-hexadecenal at 40 g/ha	Mazomenos et al. (1999)
Sparganothis pilleriana (Denis and Schiffermüller)	Grape wine	E–9-dodecenyl acetate, Z-9-dodecenyl acetate, E-11-tetradecenyl acetate, Z-11-tetradecenyl acetate, E-9-dodecen-1-ol, Z-9-dodecen-1-ol	Schmidt et al. (2000)
Sesamia nonaguiodes (Lefèbvre)	Polyphagous	(Z)-11-Hexadecenyl acetate, (Z)-11-Hexadecenyl 1-ol, (Z)-11-Hexadecenal, SesalineReg	Ameline and Frest (1999)
S. pilleriana (Denis and Schiffermüller)	Grape wine	E-9-dodecenyl acetate, Z-9-dodecenyl acetate, E-11-tetradecenyl acetate, Z-11-tetradecenyl acetate, E-9-dodecen-1-ol and Z dodecen-1-ol	Schmidt et al. (1999)
Spodoptera litura (Fabricius)	Polyphagous	9Z11E-9,11, tetradecadienyl acetate, 9Z12E-9,12, tetradecadienyl acetate (10:1) 50 g/ha	Srinivas and Rao (1999), Krishnaiah (1986), Wang et al. (2008), and Katase et al. (2006)
S. litura (Fabricius)	Yam	(Z,E)-9,11-Tetradecadienyl acetate, (Z,E)-9,12-tetradecadienyl acetate	Kakinuma et al. (2000)
Spodoptera exigua	Polyphagous	Microencapsulated pheromone formulation	Chen and Zhang (2008)
Zeuzera pyrina (Linnaeus)	Grape	E,2, Z, 3-decadecenyl acetate + E,13-Z,13 decadecenyl acetate at 95:5	Guario et al. (2001) and Pasqualinin and Natale (1999)

(WAMD) was effectively practiced in Australia for the management of *C. molesta* (Il' ichev et al., 1999). Dispensers with higher doses of *C. molesta* pheromone have a disorienting effect on males while dispensers with lower doses caught higher numbers (Maini and Accinelli, 2000). Split application of pheromone at lower rates may be more effective in maintaining pheromone levels than a single dose at higher rates, as observed in the case of *S. sulfureana* (Fabricius) in cranberries (Polavarapu et al., 2001). Metered semiochemical timed release system (MSTRSTM), a new approach, was used to effectively manage *Ostrinia nubilalis* (Hübner) with the pheromone blend of Z-11-tetradecenyl acetate and E-11-tetradecenyl acetate at 97:3 at grassy aggregation sites (Fadamiro and Baker, 1999). Application techniques, both effective and economical, have been identified for the management of various insects. An aerial application technique using a micron air atomizer with pheromone flakes and sticker and polyvinyl chloride spirals at 1481 effectively reduced the incidence of gypsy moth on apples, *P. gossypiella* and *P. idaeusalis* (Trent and Thistle, 1999; Kehat et al., 1999; Meissner et al., 2000). Paraffin emulsions consisting of 30% paraffin, 4% pheromone, 4% soybean oil, 1% vitamin E, emulsifier and remainder water has been suggested for effective control of *C. molesta* (Atterholt et al., 1999).

7.2 Mating Disruption as a Component of Integrated Pest Management

A combination of insecticides along with pheromone baits offer excellent results as observed in the case of *Rhyacionia buoliana* (Denis and Schiffermüller) and *Eucosma sonomana* Kearfott. Baits with 3.5–20.0 g/ha pheromone with rapid-acting toxic insecticides (7.92 g/ha) resulted in 86.6–100% reduction of trap catches in comparison to 500–800 g/ha of conventional sprays (Daterman et al., 2001).

In Switzerland, mating disruption is normally practiced on several crops (Zuber, 1999). Mating disruption with RAK1+2 dispensers or RAK1+2R and Isonet dispensers were effective for managing populations of grape berry moth, *E. ambiguella* (Hübner) and grapevine moth, *L. botrana* (Denis and Schiffermüller) (Charmillot et al., 2001). Integration of insecticide or biocontrol with mating disruption was effective in pest management, e.g., 120 mg/or 210 mg/ha pheromone along with tebufenozide 240 g/l and 1 l/ha controlled *L. botrana* (Denis and Schiffermüller) and *E. ambiguella* (Hübner), respectively, in grape yards (Emery and Schmid, 2001). A combination of half rate of Isomate C dispensers (500/ha) and insecticide sprays gave good control of *C. molesta* (Nunez, 2000). Apart from directly reducing the pest infestation, mating disruption also causes reduction of insecticide sprays and, thus, forms a major component of integrated pest management (IPM) as observed in the management of *Cydia* and *Anarsia* in peach orchards in Italy (Ceredi and Borghesi, 2002).

Since biocontrol programs also hold a major share in IPM, use of biocontrol agents such as *Bacillus thuringiensis*, *Trichogramma*, chrysopids, and granulosis virus along with pheromones for mating disruption is desirable. Such integration of management measures has been found to effectively control *C. pomonella* and *Adoxophyes orana* (Fischer von Roslerstamm) in apple, *P. gossypiella* on cotton, and *C. pomonella* in peaches (Albert and Wolff, 2000; Ahmad et al., 2001; Charmillot and Pasquier, 2002; Boselli, 1998).

Unlike mass trapping, mating disruption facilitates the abundance of natural enemies. The population of predatory coccinellids and syrphids increased rapidly in peach orchards implemented with mating disruption as one of the components for the management of pests

(Molinari et al., 1999). Neither the activity of pollinators nor pollination was affected when codling moth pheromones were used (Vaissiere et al., 2001). A fairly comprehensive study conducted at China for the management of *P. gossypiella* indicated that control could be achieved with the use of PB-ROPEL formulation along with protection to natural enemies, reduction in flower damage, reduction in pesticide use to three to seven rounds with the subsequent reduction in cost by 13.46–48.08% along with increase in cotton yield by 2.6–6.84% (Yang et al., 2001b).

Several other benefits also accrue when mating disruption was practiced. Charmillot et al. (2000) observed that mating disruption was better than classical biocontrol for *L. botrana* (Denis and Schiffermüller) and *E. ambiguella* (Hübner) in vineyards. Further, the insecticide-resistant populations of pests responded same manner to the pheromone blend as the nonresistant ones (Freort et al., 1999), indicating the prospective use of pheromones for their management.

7.3 Undesirable Effects of Mating Disruption

However, all is not beneficial with mating disruption treatments. Undesirable higher incidence of other pests had been reported in plots treated with pheromones for mating disruption. For instance, the apple leaf roller, *Argyrotaenia citrana* (Fernald) and *Pandemis pyrusana* Kearfott were higher in orchards treated with codling moth pheromone compared to untreated plots (Walker and Welter, 2001). Similarly, arrival of *Scaphoideus titanus* Ball was noticed in areas where large-scale adoption of mating disruption was used for *C. ambiguella* Hübner (Mattedi and Mescalchin, 2002). Migration of *C. molesta* in Australia from pear to peach blocks was observed when mating disruption studies were carried out (Il'ichev et al., 1999).

Adaptation to the continuous exposure to pheromones also results in resistance to mating disruption or low trap shutdown as observed in smaller tea tortrix, *Adoxophyes honmai* (Yasuda) and *Homona magnanima* Diakonoff, for their pheromone Z-11-tetradecnyl acetate in some orchards. However, by using blends instead of individual compounds, the resistance was broken down resulting in 99% disruption (Mochizuki et al., 2002, 2001). Resistance to mating disruption was observed to be 31.0% compared to 18.1% in control *Trichoplusia ni* (Hübner) after the fourth generation (Evenden and Haynes, 2001).

7.4 Failures of Mating Disruption Techniques

Migration of moths from other areas, insufficient pheromone load or delivery, and availability of noncrop hosts for several pests are some of the reasons contributing to the failure of mating disruption as observed in *Campylomma verbasci* (Meyer) (Helsen and Blommers, 2000), *P. xylostella*, *Laspeyresia splendana* (Hübner), and *Pammene fasciana* (Linnaeus) (Schroeder et al., 2000; Mansilla Vazquez et al., 1999). High population levels of pests such as *Spodoptera exigua* (Hübner) and *C. pomonella* also make mating disruption ineffective (Kerns, 2000; Mansour and Mohamad, 2001). The mating disruption in case of use of Isomate C+ at 635 dispensers/ha was not effective in controlling *C. pomonella* on walnuts (Gonzalez and Cepeda, 1999). Migration of moths from other areas was considered as a factor for failure of *C. molesta* mating disruption control in peaches (Sexton and Il'ichev, 2000). Other constraints in adoption of mating disruption include time-consuming technologies and incompetence with insecticides (Verhaeghe et al., 2001; Duran et al., 2000; d'-Arcier et al., 2001).

8. PARAPHEROMONES

Parapheromones are analogues of pheromones that are synthesized and used for trapping the insects, however, they are not conspecific in origin. Renou and Guerrero (2000) define parapheromones as compounds with very different origins. In Diptera, the male attractants are of plant origin, such as methyl eugenol, cuelure and trimedlure, which are extensively used for trapping several species of fruit flies.

Extracts of *Ocimum sanctum* and *Zieria smithii*, which were observed to be natural attractants for males of fruit flies (*Bactrocera dorsalis*) contain methyl eugenol as the major component (Shah and Patel, 1976; Tan and Lee, 1982; Fletcher et al., 1975). Howlett (1877–1920) was the first to discover methyl eugenol as an attractant for mango fruit flies (Verghese et al., 2013). Similarly, other compounds such as cuelure and trimedlure were identified from plants which act as attractants for other species of *Bactrocera*.

Generally, methyl eugenol, cuelure and trimedlure have been used for biodiversity studies (Fadlelmula and Ali, 2014), population studies, and quarantine monitoring. Fruit fly diversity in a given geographical area is mapped based on the catches in the parapheromone trap, with a foresightedness to estimate the economic losses. Drew (1974) observed that 56 species responded to cuelure while 23 species responded to methyl eugenol. The species of fruit flies that respond to parapheromones are given in Table 4.

Bactrocera dorsalis, often termed mango fruit fly, is distributed throughout the world except in some island nations, while *B. cucurbitae*, or melon fly, is another pest on cucurbitaceous plants throughout the world. In India, trap catches with methyl eugenol indicated the presence of *B. dorsalis*, *B. zonata*, *B. correcta*, and *B. verbascifoliae* with *B. dorsalis* the most predominant species, followed by *B. zonata* (Nandre and Shukla, 2012; Ukey et al., 2013). Mapping studies in larger areas in China using methyl eugenol traps documented the presence of *Bactrocera tau*, *Bactrocera scutellata*, *Bactrocera cucurbitae*, *Bactrocera minax*, and *B. dorsalis*, and *B. minax* which occurred in 27 villages and towns in 2011, and its occurrence area reached 1666.7 ha in the whole county with a loss of production above 400 tons (Xiao et al., 2013).

Trap catches along with the larval infestation enables us to identify the probable damage to be caused in fruit orchards of mango (*Mangifera indica*), guava (*Psidium guajava*), grapefruit (*Citrus paradise*), banana (*Musa* spp.), papaya (*Carica papaya*), cantaloupe (*Cucumis melo*), brazilia (*Terminalia brasiliensis*), usher (*Calotropis procera*), and wild strawberry (*Fragaria vesca*) by *Bactrocera* spp. (Fadelmula and Ali, 2014). Varietal variation of mango often influences the diversity of species in trap catches, e.g., populations of *B. dorsalis* were found more in the variety Mallika followed by variety Banganapalli, and lowest in variety Alphonso (Nagaraj et al., 2014). Methyl eugenol can also be used for the quarantine monitoring of highly invasive species, such as *Bactrocera invadens* by employing traps one per km or 100 ha and mapping the traps along with periodical collection of data (Anonymous, 2012). Methyl eugenol also attracts other species of insects belonging to other orders, viz. *Platycheirus*, *Melanostoma*, *Meliscaeva*, *Ferdinanda*, *Hadromyia*, *Blera* and *Melangyna* (Symphora), *Empis* (Diptera), *Orchesia* (Coleoptera) (Dowell, 2015), and many species of chrysopids (Umeya and Hirao, 1975).

Methyl eugenol and cuelure have also been realized in population monitoring of several crops (Liu et al., 1982; Bagle and Prasad, 1983; Shukla and Prasad, 1985; Tan and Jaal, 1986; Gupta et al., 1990; Khattak et al., 1990; Thakur et al., 2013; Darwish et al., 2015). Population dynamics studies using methyl eugenol traps indicated that major fruiting season and rainfall have a positive

TABLE 4 Parapheromones Used for Monitoring Diversity of Fruit Flies

Compound	Species responsive	References
Methyl eugenol	*Bactrocera (Dacus) dorsalis* (Hendel)	Ibrahim et al. (1979), Venkatachalam et al. (2014), and Vargas et al. (2014)
	Bactrocera (Dacus) musae (Tryon)	Smith (1977)
	Bactrocera (Dacus) cacuminata (Her.)	Stegeman et al. (1978)
	Bactrocera (Dacus) umbrosus	Ibrahim and Hashim (1980)
	Bactrocera (Dacus) opillae Drew and Hardy	Fitt (1981)
	Bactrocera zonata (Saunders)	Verghese and Jayanthi (2001)
	Bactrocera (Dacus) caudatus F.	Tan and Lee (1982)
	Bactrocera (Dacus) correcta (Bezzi)	Weems (1987)
	Acanthiophilus helianthi Rossi	
	Bactrocera arecae (Hardy and Adachi)	Win et al. (2014)
	Bactrocera carambolae Drew and Hancock	Win et al. (2014)
	Bactrocera kandiensi Drew and Hancock	Win et al. (2014)
	Bactrocera latilineola Drew and Hancock	Win et al. (2014)
	Bactrocera malaysiensis Drew and Hancock	Win et al. (2014)
	Bactrocera neocognata Drew and Hancock	Win et al. (2014)
	Bactrocera raiensis Drew and Hancock	Win et al. (2014)
	Bactrocera verbascifoliae Drew and Hancock	Win et al. (2014)
	Bactrocera umbrosa Fab.	Pramudi et al. (2013)
	Bactrocera. occipitalis Bazzi	Pramudi et al. (2013)
	Bactrocera latifrons Hendel	Pramudi et al. (2013)
	B. carambolae Drew and Hancock	Pramudi et al. (2013)
	Bactrocera papayae Drew and Hancock	Pramudi et al. (2013)
	Bactrocera albistrigata (Meijere)	Pramudi et al. (2013)
	Bactrocera tau (Walker)	Xiao et al. (2013)
	Bactrocera invadens Drew, Tsuruta and White	Asawalam and Nwachukwu (2012)
	Carpomyia vesuviana Costa	Win et al. (2014)
Cuelure	*Bactrocera (Dacus) tryoni* (Frogg.)	Hopper (1978)
	Bactrocera (Dacus) neohumeralis Hardy	
	Bactrocera (Dacus) cucurbitae	Tan and Lee (1982)
	Bactrocera (Dacus) caudatus	
	Bactrocera (Dacus) frauenfeldi (Schiner)	
	Bactrocera (Dacus) coccipitalis (Bez.)	
	Bactrocera (Dacus) tau (Walker)	

TABLE 4 Parapheromones Used for Monitoring Diversity of Fruit Flies—cont'd

Compound	Species responsive	References
	Bactrocera pernigra Ito	Sasaki et al. (1985)
	Bactrocera (Dacus) bivitatus	
Trimedlure	*Ceratitis capitata* (Wiedemann)	Shelly et al. (2014)
Cera lure	*Ceratitis capitata*	Leonhardt et al. (1996)
Mixed lure	*Ceratatis rosa* Karsch	Grové et al. (2014)
α-copaene	*Ceratitis capitata*	Ortu (1995)
Ethyl-2-methyl butyrate from durian flesh	*Bactrocera dorsalis*	Mo et al. (2014)

correlation with adult emergence (Agrawal and Deepa, 2013; Munj et al., 2013; Mayamba et al., 2014). Boopathi et al. (2013) developed a model using the capture data from methyl eugenol traps for the prediction of fruit fly incidence on chili, based on ambient temperature.

The attraction of fruit flies to parapheromones has been also exploited for the management of their populations. Mixture of methyl eugenol and cuelure acts synergistically on captures of *B. cucurbitae* and *Bactrocera d'emmerezi* Bezzi (Ramsamy et al., 1989; Liu and Lin, 1993). Methyl eugenol and terpinyl acetate, basil oil, and methyl eugenol, and extracts of *Melaleuca bracteata* are other parapheromone combinations used for monitoring of fruit flies, *B. invadens*, and *C. cosyra* (Vayssières et al., 2014; Chang et al., 2013; Kardinan and Hidayat, 2013).

Several types of dispensers, viz. fiberboard pieces (4 × 4 × 0.5 cm) (Chu et al., 1985), gel formulation (Bhagat et al., 2013), Min-U-Gel (Vargas et al., 2008, 2014), and cotton wick (Qureshi et al., 1992; Singh et al., 2015), placed inside traps such as Steiner trap (Hooper and Drew, 1978), McPhail trap (Xiao et al., 2013), Lynfield trap (Ali et al., 2013), green delta trap, white delta trap, open yellow trap, carton trap (Othman and Aulaqi, 2011), solid mallet TMR wafers, Mallet CMR wafers with DDVP, "Rakshak," ammonia trap or cera trap bottle fly trap, collapsible trap, liquid formulations (Suk Ling et al., 2013), and Rakshak (a commercial formulation) (Munj et al., 2013) have been developed for field capture of fruit flies and are used with variable results (Grewal and Kapoor, 1987; Vargas et al., 2012; Dominiak and Nicol, 2012; Munj et al., 2013; Royer et al., 2014; Nagaraj et al., 2014; Singh et al., 2015).

Insecticides, such as malathion, sumithion, and naled, have also been mixed along with the parapheromones for effective killing of trapped pests (Shankar et al., 2011; Manrakhan et al., 2011; Shelly et al., 2011; Jang et al., 2011; Gazia et al., 2013). The color of the trap and their location from the trap source (up to 300 m) and trap height is not known to influence the catches (Hooper and Drew, 1979; Chu et al., 1985).

Factors such as half-life of the parapheromone, age of the males (15 days old highly attracted to methyl eugenol), increasing concentration of methyl eugenol, diurnal cycle (with more catches in dawn and less in dusk), etc. affect the fruit fly trap catches and trap density (Masuda and Endo, 1981; Iga, 1982; Karunaratne and Karunaratne, 2012).

Parapheromones and the behavioral response of the targeted insects have led to the development of two new techniques, MAT and aroma therapy. MAT involves the use of a high density of bait stations consisting of a male lure combined with an insecticide to reduce the population of male fruit flies to such a low level that mating does not occur (www.spc.int/Ind/project/male-annihilation-technique). In this method, methyl eugenol or cuelure traps are used in sufficient numbers (6–20 per acre) from the start of fruiting season, with frequent change of dispensers resulting in the reduction in the availability of males for mating.

Total eradication of fruit flies has been attempted in several countries using MATs (Ushio et al., 1982; Sonda and Ichinohe, 1984; Koyama et al., 1984). The number of traps, generally 9 (Cunningham and Suda, 1986; Marwat et al., 1992; Liquido, 1993) to 16 per ha (Singh and Sharma, 2013) and sometime up to 42 (Ousmane et al., 2014) need to be standardized depending on the lure. Area of the plot and population load in the experimental area are crucial for the success in MAT. Specialized commercial formulations like STATIC Spinosad ME, a reduced-risk MAT formulation consisting of an amorphous polymer matrix in combination with methyl eugenol and spinosad, performs equally well with other treatments of Min-U-Gel mixed with methyl eugenol and naled (Dibrom) (Vargas et al., 2014). Mango is the crop where male annihilation MAT is most successful in reducing infestation substantially (Verghese et al., 2005, 2007; Reji et al., 2012; Singh et al., 2013; Shelly et al., 2013).

Due to several practical and technical reasons, the success of MAT is not satisfactory, either due to the fact that the emerging males are always found foraging outside the trap resulting in a female: male ratio of 1:171 (Rizk et al., 2014) or due to mating incidence of females prior to laying of traps or the tendency to mate multiply (Barclay and Hendrichs, 2014). These aspects necessitate thorough understanding of behavioral ecology of the pest prior to the use of such management tools.

Aroma therapy (Haq et al., 2014) is based on the observation that parapheromones, methyl eugenol, or cuelure, consumed by males (a common occurrence in most insect species), are accumulated in the rectal glands, which in turn increases the mating advantage for those males (Orankanok et al., 2013; Obra et al., 2013; Kumaran et al., 2013, 2014; Haq et al., 2014). Even feeding for 30 s enables the males to have mating advantage for 35 days (Shelly and Dewire, 1994; Shelley, 1995). This behavioral attribute is often exploited in using sterile male techniques, wherein the sterilized males fed with the methyl eugenol or cuelure are released in the field after three days of exposure to ensure mating with females resulting in no progeny (Haq et al., 2014).

9. PLANT VOLATILES AS ATTRACTANTS OR REPELLENTS

Insects often utilize a particular plant volatile or a set of volatiles for their various behavioral attributes such as aggregation, orientation, oviposition, etc. These chemicals, either alone or in combination, have been exploited for trapping the insects at field levels. Metcalf and Metcalf (1992) in their book *Plant Kairomones in Insect Ecology and Control* enumerated the various volatiles used as attractants. Phenyl acetaldehyde, terpineol acetate, α farnesene, amyl benzoate, amyl salicylate, carvone, anethole, anisaldehyde, anisic acid, estragole indole. β-caryophyllene, β-caryophyllene oxide, ethyl acetate, eugenol, isoeugenol, cinnamyl alcohol, α-phellandrene, α-copaene, α-cubebene, coumarin, geraniol, phenylethanol, caproic acid, allyl isothiocyanate 1, 8, cineole, anisaldehyde, δ-cadinene, γ-cadinene, β-elemene, hexyl propanoate, hexyl butanoate, β-elemene, methyl eugenol, α-humulene and α-bulnesene are some of the volatiles used for attracting several insect species (Metcalf and Metcalf, 1992; Leal, 2001; Fassotte et al., 2014). A list of plant volatiles that are used for the attraction of insects appears in Table 5.

TABLE 5 Plant Volatisles as Kairomones for Various Insects

Volatile	Species	References
Eugenol	*Megalopta amoena* (Spinola), *Megalopta aegis* (Vachal)	Knoll and Santos (20sssss12)
Methyl salicylate	*M. amoena* (Spinola), *M. aegis* (Vachal)	Knoll and Santos (2012)
Eucalyptol	*M. amoena* (Spinola), *M. aegis*(Vachal)	Knoll and Santos (2012)
β-ocimene along with pheromone blend	*Hyphantria cunea* (Drury)	Tang et al. (2012)
Limonene	*Ceratitis capitata* (Wiedemann)	Ioannou et al. (2012)
Linalool (repellent)	*C. capitata* (Wiedemann)	Ioannou et al. (2012)
E,E-alpha farnesene, R,E nerolidol, R,E,E-alpha farnesene 10,11 oxide	*Amblypelta lutescens* (Distant)	Khrimian et.al. (2012)
Blend of β-caryophyllene, n-decanal, n-tridecene, methyl salicylate, Geranyl acetate and methyl benzoate (attractant)	*Trigonotylus caelestialium* (Kirkaldy)	Hori and Enya (2013)
Terpinolene (repellent)	*Sitophilus granarius* (Linnaeus)	Benelli et al. (2012)
β-pinene (repellent)	*Sitophilus granarius* (Linnaeus)	Benelli et al. (2012)
Sabinene (repellent)	*Sitophilus granarius* (Linnaeus)	Benelli et al. (2012)
Blend of α-pinene and β-caryophyllene (attractant)	*Dendroctonus armandi* (Tsai and Li)	Xie and Lu (2012)
Phenylacetaldehyde	*Spodoptera albula* (Walker), *Autographa californica* (Speyer), *Cetonia aurata* (Linnaeus)	Landolt et al. (2013)
Phenyl ethyl alcohol with methyl eugenol + (E)-anethol	*Potesia cuprea* (Fabricius)	Vuts et al. (2007)
Ethanol	*Ambrosiodmus tachygraphus* (Zimmermann), *Anisandrus sayi* Hopkins, *Dryoxylon onoharaensum*, *Monarthrum mali* (Fitch), *Xyleborinus saxesenii* (Ratzeburg), *Xyleborus affinis* Eichhoff, *Xyleborus ferrugineus* F., *Xylosandrus compactus* (Eichhoff), *Xylosandrus crassiusculus* (Motschulsky), *Xylosandrus germanus* (Blandford)	Miller and Rabaglia (2009)
Alpha pinene	*Acanthocinus nodosus* F., *Acanthocinus obsoletus* Olivier, *Alaus myops* (Fabr.), *Arhopalus rusticus nubilus* (LeConte), *Asemum striatum* Goggi, *Buprestis lineata* (F.), *Hylobius pales* (Herbst), *Monochamus titillator* L., *Pachylobius picivorus* (Germar), *Prionus pocularis* Dalman, *Xylotrechus integer* (Haldeman), *Xylotrechus sagittatus* (Germar), *Dendroctonus terebrans* (Olivier), *Hylastes proculus* Erichson, *Hylastes salebrosus* Eichhoff, *Hylastes tenuis* Eichhoff, *Ips grandicollis* (Eichhoff), *Myloplatypus flavicornis* F.	Miller (2006)

(Continued)

TABLE 5 Plant Volatisles as Kairomones for Various Insects—cont'd

Volatile	Species	References
Alpha pinene + ethanol	*Acanthocinus nodosus* F., *Acanthocinus obsoletus* (Olivier), *Arhopalus rusticus nubilus* (LeConte), *M. titillator* (F.), *Xylotrechus sagittatus sagittatus*(Germar), *Alaus myop* (F.), *Hylobius pales* Herbst, *Pachylobius picivorus* (Germar), *Xyloborus pubescens* Zimmermann, *Hylastes porculus* Edmundston, *Hylastes tenuis* J. R. Baker and S. B. Bambara, *Pityophthorus cariniceps* LeConte and Horn	Miller (2006) and Miller and Rabaglia (2009)
Ipsenol + ipsdienol + ethanol + α-pinene	*Acanthocinus nodosus, Acanthocinus obsoletus, Astylopsis arcuata* (LeConte), *Astylopsis sexguttata* (Say), *Monochamus scutellatus* (Say), *M. titillator* F., *Rhagium inquisitor* (L.), *Buprestis lineata* F., *Ips avulsus* (Eichhoff), *Ips calligraphus* (Germar), *Ips gradicollis* (Eichhoff), *Orthotomicus caelatus* (Eichhoff), *Gnathotrichus materiarus* (Fitch)	Miller et al. (2011)
A-pinene, β-pinene, and ethanol	*Hylurgus ligniperda* (F.)	Petrice et al. (2004)

Ethanol is known as an attractant for several coleopteran insects. Funnel traps baited with ethanol attracted 10 species of ambrosia beetles and two species of bark beetles (Miller and Rabaglia, 2009). In India, a combination of methanol and ethanol has been successfully utilized for trapping the coffee berry borer, *Hypothenemus hampei*. Alpha pinene, a terpenoid from the plants, ipsenol, and methyl butanol are some of the volatiles that attract several coleopterans, viz. *Hylastes* spp. *Monochamus galloprovincialis* (Olivier), *Xylotrechus quadripes*, etc. (Miller, 2006; Miller and Rabaglia, 2009; Ibeas et al., 2007; Bakthavatsalam, unpublished data).

Application of KLP+traps baited with 1, 4-dimethoxybenzene in PE bag dispensers are recommended for *Diabrotica speciosa*, while KLP+traps with dual (pheromone and floral) lures for *Diabrotica barberi* (Toth et al., 2014). A patented product with the combination of cineole, D-limonene, α-pinene, anisyl alcohol, butyl salicylate, and phenyl acetaldehyde named MAGNET applied at one strip for 144 m or 72 m (depending on the pest intensity) was able to control *H. armigera* on cotton (Magnet, Insect attractant technology, Cotton, Technical manual, AgBiotech, Australia). Adding glacial acetic acid to pear ester or to (E)-β-ocimene significantly increased the catches of only codling moth or oriental fruit moth, respectively (Knight et al., 2014). Uninfested plants often produce chemicals such as caryophyllene oxide and caryophyllene alcohol, which act as attractants for chrysopids and coccinellids (Flint et al., 1979, 1981).

Weed-infesting insects need constant monitoring to document their dispersal and population. A blend of four green leaf volatiles (GLV) compounds, mimicking a natural blend ratio was highly attractive to male and female *Diorhabda elongata* Brullé, a weed biocontrol agent for salt cedar *Tamarix* spp. (Cosse et al., 2006), *Arhopalus ferus* (Mulsant), *Hylurgus ligniperda* Fabricius, and *Hylastes ater* Paykul in New Zealand (Brockerhoff et al., 2006).

Food supplements, and not parapheromones, mixed with insecticide poisons offer total eradication of fruit flies through mass trapping. Hydrolyzed protein with deltamethrin resulted in reduction in *B. oleae* (Bjelis, 2006). Integration of releases of *Opius concolor* (Szepligeti) at 120/tree along with ecotrap (commercial bait) resulted in the reduction in fruit damage by *B. oleae* (Rizzi et al., 2005; Liaropoulos et al., 2005). Nulure with borax and FA-3 plus propylene glycol, Probodelt, a combination of ammonium acetate, putrescine, and trimethylamine in polythene are some of the food baits used for mass trapping *Ceratitis capitata* (Ros et al., 2002; Alemany et al., 2002; Batllori et al., 2005).

Repellent activity of several plant extracts is also known, however, most of them remain as basic studies and not many of these products are used as repellents. Chemicals such as α terpineol, nerolidol, δ-cadinene, β-eudesmol, terpinolene, cedraol, L-menthol, phenylpropanoids, allyl phenol, limonene, eight cineole, osajin, pomiferin, and cucurbitacin are progressive for use as repellents against storage pests and several coleopteran pests (Rao et al., 2000; Peterson et al., 2000; Aggarwal et al., 2001; Yatagai et al., 2002; Agrawal et al., 2002).

However, the main constraint in using plant volatiles is their highly oxidative nature, which requires addition of antioxidants. Also, lack of specificity makes the plant volatiles a generalist trap, sometimes trapping nontarget insects such as honeybees and predators—a major criticism voiced by conservation biologists.

10. ANTIAGGREGATION PHEROMONE

Verbenone, derived from female beetles or auto-oxidation of α-pinene is a bark beetle antiaggregation pheromone first reported from *Dendroctonus ponderosae* (Pitman et al., 1969). An ovipositing female marks the site with verbenone, which is perceived by conspecific females enabling them to avoid oviposition on the same tree. This has been commercially exploited to develop verbenone, which was effectively used at a release rate of 5 mg/capsule per 24 h at 25 °C (Ammam et al., 1988; Fettig, 2013). Verbenone had direct effect on natural enemies with catches of specialist predators, viz. *Thanasimus undulatus* (Say), *Enoclerus sphegens* (F.), *E. lecotei* (Wolcott) and *Lasconotus* complex decreasing with increasing rate of verbenone, whereas generalist predators like *Lasconotus subcostulatus* and *Corticalis preeter* did not respond (Lindgreen and Miller, 2002). Frontalin at higher doses and methylcyclohexanone (MCH) (3-methylcyclohex-2-en-1-one) also act as antiaggregation pheromones and are used for species such as *Ips pini* and *Pseudotsuga menziesii* (Miller, 2001; Ross et al., 1996).

11. AUTOCONFUSION

This is a new and modern technology where the males are used to attract and confuse the conspecific males by using a novel delivery system based on electrostatically chargeable powder that adheres to the cuticle of an insect, presumably males. Males on being attracted to the dispenser having the "electro stat" TM powder containing the appropriate pheromone will be coated with the pheromone, and they get camouflaged as females, confusing other males. The advantage of this technique is that the point sources of pheromone deliveries are reduced, resulting in reduced costs and labor; in addition, quantities of pheromone are less than or equivalent to the amount released in by the population. Field trials in apple, rice, storage godowns, and forestry using 1000 times less pheromone have shown promising results

(Trematerra et al., 2013). However, this technology needs further refinement before it becomes viable for other insect species.

12. SEMIOCHEMICALS FOR NATURAL ENEMIES

The interest in the studies on semiochemicals for natural enemies has obviously evolved due to the compelling demand to improve the efficacy of natural enemies in the field to achieve success in biocontrol. A comprehensive account of various techniques in manipulating natural enemy populations using habitat and semiochemicals was published recently (Rodriguez et al., 2013). Parasitoids and predators utilize several volatile cues emanating from cuticle, egg chorion, glandular secretions, moth scales, frass, fecal matter of larvae, and honey dew secretions of the soft-bodied insects as cues for locating prey. Several important publications have been published on the chemical ecology of entomophages, synomones, kairomones, and related aspects (Bakthavatsalam et al., 2013; Barbosa and Castellanos, 2005; Brown et al., 1970; Colazza and Wajnberg, 2013; Dickens, 1999; Gurr and Wratten, 1999).

12.1 Host Insect Kairomones

Various semiochemicals from hosts, including kairomones found in the scales of lepidopteran hosts or plants that they feed on, have been shown to intensify the searching behavior of the *Trichogramma* spp. females, which subsequently results in increased rate of parasitism (Jones et al., 1973; Lewis et al., 1972, 1975a,b; Lewis and Martin, 1990; Nordlund et al., 1976, 1977; Shen et al., 1991; Nordlund and Greenberg, 2002). Compounds such as octacosane, pentacosane, hexacosane, tricosane, hexatriacontane, and docosane were considered as favorable hydrocarbons, which increased the activity of *Trichogramma* spp. (Padmavathi and Paul, 1998; Singh et al., 2002; Paul et al., 2002; Paramasivan et al., 2004; Yadav et al., 2005, 2008; Rani et al., 2007). A dust formulation or rubber septa impregnated with – n-tricosane increased parasitization by *T. chilonis* in cotton but not by *T. japonicum* in rice (Paramasivan et al., 2004; Bakthavatsalam and Tandon, 2006).

Mass priming is a technique in which exposure of bioagents to the kairomonal substances prior to release increases parasitizing efficiency of the parasitoids. Aphid parasitoid, *Aphytis ervi* Haliday, on exposure to O-coffenyl tyrosine showed higher parasitism against California red scale in field conditions (Hare and Morgan, 1997). The adult *T. chilonis* exposed to nonacosane also recorded higher parasitism in the net house studies (Bakthavatsalam, unpublished). L-tryptophan, on hydrolysis either alone or in combination with the larval kairomone, tricosane, was found to increase the abundance of chrysopids (Ben Saad and Bishop, 1976a,b; Bakthavatsalam and Singh, 1996; Bakthavatsalam et al., 2007). Commercial products such as "lady bug attractant" are available for attracting coccinellids, lacewings, and hover flies (www.plantnatural.com).

12.2 Host Pheromones as Kairomones

Pheromones are often recorded as kairomones for several entomophages. Sex pheromones of hosts often facilitate the host searching behavior of parasitoids such as *Bracon brevicornis*

Wesmael, *Anisopteromalus calandrae* Howard, *Venturia canescens* Gravenhorst, *Apanteles rufricus* Hal., *Anagyrus pseudococci*, and *Microplitis demolitor* Wilkinson (Pereira, 1998; Zaki, 1996; Zaki et al., 1998; Mansour et al., 2010). Aphid alarm pheromone (β-farnesene) and sex pheromone [(+)– 4aS, 7S, 7aR) nepetalactone] also are involved in cross-talk with aphid predators or parasitoids (Micha and Wyss, 1996; Powell et al., 1993; Hardie et al., 1991). Even predators such as *Thanasimus formicarius* Linnaeus are attracted to ipsdienol or *cis*-verbenol (Bakke and Kvamme, 1981).

12.3 Food Supplements

Providing suitable food enhances the local natural enemy population and encourages immigration of natural enemies to pest-containing areas, discourages emigration from pest containing areas, increases birth rates, and reduces death rates. Use of sucrose either alone or in combination with yeast improved the density of natural enemies, viz. *Coccinella* spp., *Hippodamia* sp., *Chrysoperla* spp., and *Orius tristicolor* (Ferguson). Products like *Amino Feed*® (Nemec, 2001), *Eliminate*® (Mathews and Stephen, 1999), *Envirofeast*® (Mensah, 1997), *Feed wheast*® (Hagen et al., 1976) and *Prey-feed*® (Welty and Alcaraz, 1995) are available to increase the efficiency of natural enemies and are suggested for use in the conservation biocontrol. However, the use of these compounds is very limited and only *Amino feed* and *Prey feed* are commercially available.

12.4 Synomones

The defense of plants to pest attack may be direct (chemicals that retard the growth or cause death of herbivores) or indirect (through the production of herbivore-induced plant volatiles), which acts as recruiting agent for the natural enemies. In field experiments, several entomophages such as *Chrysopa nigricornis* Burmeister, *Stethorus picipes* Cacey, *O. tristicolor*, *Geocoris pallens* Stal, *Deraeocoris brevis* (Uhler), and *Hemerobius* sp. responded to the methyl salicylate (James, 2003; James and Price, 2004). *Cis*-hexenol, (Z)-3-hexenyl acetate, E–2-hexen-al, indole, *cis*-jasmine, methyl jasmonate, geraniol, nonanol, octyl aldehyde, *cis*-3-hexen-1-ol, 4-hydroxy-4-methyl-2-pentaonane, and benzaldehyde were also found to attract various groups of entomophages (James, 2005; James and Price, 2004; James et al., 2008; Khan et al., 2008; Huang et al., 2009; Xu et al., 2008; Wijk et al., 2008; Logli et al., 1996).

The secondary metabolites of plants often regulate the attraction of the parasitoids; the larval parasitoid, *Cotesia rubecula* (Marshall) was attracted to *Pieris rapae* (L.) reared on nitrile-releasing plants than the larvae feeding on isothiocyanate-releasing plants, probably nitriles acting as a cue for *C. rubecula* (Geervliet et al., 1994).

13. CONCLUSION

Identification of semiochemicals has been made for a variety of insects in different countries. However, several insect species belonging to orders Thysanoptera, Hemiptera, Hymenoptera, Neuroptera, Trichoptera, Odonata, and Dermaptera need to be further investigated to meaningfully contribute to the science of chemical ecology as well as for field applications.

Rapid strides have been made on the isolation, identification, and synthesis of semiochemicals. However, new improved and sensitive analytical instruments with the sensitivity at par with the insect antenna need to be fabricated for the identification of volatiles. Synthesis schemes for some pheromones, especially for mealybug pheromones, is complex, resulting in low quantity of pheromones produced, which ultimately restricts its use for mass trapping and mating disruption. The cost of pheromones, especially mealybug pheromones, is prohibitive (Dunkelblum, 1999), and there is need for cheaper pheromones. Generic pheromones (Hanks et al., 2007) or combinations of pheromones for various pests in a crop will be a boon for farmers, simplifying the application of pheromone technology.

Dispensers need to be modernized and nanogels and nanomaterials are being attempted to be used as dispensers having sustained release in fields (Bhagat et al., 2013). However, there is still immense scope for discovering new products of high-quality adsorbents using modern polymer technology.

Monitoring of insects using pheromones has been attempted for several insect pests. However, there is a need to identify more insect species that can be considered for monitoring and population modeling. With climate change, pheromones can play a much better role in the monitoring of pests, natural enemies, and pollinators. Monitoring population and distribution patterns of weed biocontrol agents using pheromones forms an interesting chemoecological study unfolding the complexities involved in the establishment of introduced weed killers.

Quarantine monitoring is an international issue. Insects such as *Phyllotreta sinuata* (Steph.), *Anoplophora chinensis* (Forster), *Helopeltis schoutedeni* Reuter, *Bactrocera* (Dacus) *ferrugineus* Cotes, *Tessaratoma papillosa*, *Leguminivora glycinivorella* (Matsumara) *Cryptophlebia leucotreta* (Meyr), *Adoxophyes privatana* Walker, *Epiphyas postvittana* Walker, *Brontispa longissima* Gestro, *Coryphoderma tristis* Drury, *Gonipterus scutellatus* Gyllanhal, *Eucalyptolyma maideni* Froggatt, *Glycaspis brimblecombei* Moore, and *Thaumastocoris peregrines* Carpintero and Dellape are considered as potential invasive pests for India (Bakthavatsalam and Shylesha, 2011) and several other countries and prior availability of pheromone will ensure the task easier for quarantine monitoring. The international community needs to entrust this task to a competent institution so as to ensure the availability of pheromones for quarantine monitoring.

In several symposia conducted after 1960, the need for the use of pheromones has been stressed time and again (Wyatt, 2003). Pheromone technology needs to be extended to the pests infesting all plants including forestry. Mating disruption has been proven as an effective tool under wide area pest management programs (Sexton and Il'ichev, 2000; Ioriatti, 2008). However, the technology remains a distant dream for several pest species due to prohibitive cost involved in using pheromones for mating disruption. Efforts are needed to develop cost-effective pheromones and dispensers that will enable us to use the pheromone judiciously and optimally. Escalating labor wages need to be taken into consideration while developing suitable dispensers.

A simple tool using pheromones as a sensor will be of great help for the farmers. Trained dogs are often used to detect infestation of mealybugs using pheromones (https://www.wineinstitute.org/files/animalsandpestmanagement.pdf) and such innovations will be highly rewarding for pest management. Development of a highly sensitive pheromone sensor will enable us to detect the cryptic pests at initial stage of infestation. An integrated approach involving chemists, entomologists, and electronic engineers will pave way for the development of such pheromone sensors.

Farmers' awareness about the semiochemical technology should help us to disseminate the technology to wide area with ease. Field demonstrations, training, workshops, publication of pamphlets, bulletins and use of media will enable us to usher in the uptake of pheromone technology.

Chemical ecology and semiochemical education need to be a part of curricula in major universities to strengthen the semiochemical research. Pheromone technology including mode of application, action, and dissipation is in no way comparable to the insecticides, yet the registration under the Insecticide Regulations Act is considered mandatory for usage in mating disruption, necessitating generation of expensive toxicological data. Regulatory authorities should consider deleting pheromones from the Insecticides Act along with suitable amendment and guidelines for use of pheromones and other semiochemicals after due consultations with a panel of experts.

Acknowledgments

The author is thankful to the Director, ICAR- National Bureau of Agricultural Insect Resources (NBAIR), Bengaluru for the facilities extended. The author expresses his gratitude to Dr K. Srinivasa Moorthy and Dr K. Subaharan, principal scientists, NBAIR, for suggestions in improving the manuscript. The technical help received from Dr C. B. Soumya, Ms Thilagavathy, Ms Khushboo Sinha and Mr M. Murali, Senior Research Fellows, NBAIR is also gratefully acknowledged.

References

Aalbers, P., 2001. Four years of pheromone disruption in Zeeland. Fruitteelt Den Haag 91 (8), 24–25.

Aggarwal, K.K., Tripathi, A.K., Ahmad, A., Prajapati, V., Verma, N., Kumar, S., 2001. Toxicity of L-menthol and its derivatives against four storage insects. Insect Science and its Application 21 (3), 229–235.

Agrawal, N., Deepa, M., 2013. Population dynamics of fruit fly species caught through methyl eugenol traps at different locations of Kanpur, Central U.P. Journal of Entomological Research 37 (1), 87–90.

Agrawal, A.A., Janssen, A., Bruin, J., Posthumus, M.A., Sabelis, M.W., 2002. An ecological cost of plant defence: attractiveness of bitter cucumber plants to natural enemies of herbivores. Ecology Letters 5 (3), 377–385.

Ahmad, Z., Attique, M.R., 1993. Control of pink bollworm with gossyplure in Punjab, Pakistan. Bulletin OILB/SROP 16 (10), 141–148.

Ahmad, N., Ashraf, M., Fatima, B., 2001. Integration of mating disruption technique and parasitoids for the management of cotton bollworms. Pakistan Journal of Zoology 33 (1), 57–60.

Albajes, R., Konstantopoulou, M., Etchepare, O., Eizaguirre, M., Frerot, B., Sans, A., Krokos, F., Ameline, A., Mazomenos, B., 2002. Mating disruption of the corn borer *Sesamia nonagrioides* (Lepidoptera: Noctuidae) using sprayable formulations of pheromone. Crop Protection 21 (3), 217–225.

Albert, R., Wolff, R., 2000. New measures against leaf rollers in small orchards. Anzeiger fur Schadlingskunde 73 (3), 79–82.

Alemany, A., Alonso, D., Miranda, M.A., 2002. Evaluation of improved Mediterranean fruit fly attractants and retention systems in the Balearic Islands (Spain). In: Proceedings of the 6th International Symposium on Fruit Flies of Economic Importance, Stellenbosch, South Africa, 6–10 May 2002, pp. 355–359.

Alfaro, C., Navarro Llopis, V., Primo, J., 2009. Optimization of pheromone dispenser density for managing the rice striped stem borer, *Chilo suppressalis* (Walker), by mating disruption. Crop Protection 28 (7), 567–572.

Ali, S.A.I., Mahmoud, M.E.E., ManQun, W., Mandiana, D.M., 2013. Survey and monitoring of some tephritidae of fruit trees and their host range in river Nile State, Sudan. Persian Gulf Crop Protection 2 (3), 32–39.

Ameline, A., Frest, B., 1999. Mechanisms involved in mating disruption in *Sesamia nonagrioides* Lefebvre by Sesaline reg. Annals De La Socieltie Entomologique De, France 35 (Suppl.), 447–452.

Amman, G.D., Thier, R.W., McGregor, M.D., Schmitz, R.F., 1988. Efficacy of verbenone in reducing logepole pine infestation by mountain pine beetles in Idaho. Canadian Journal of Forest Research 19, 60–65.

Angeli, G., Rama, F., Forti, D., Monta, L.D., Bellinazzo, S., 1999. Control of *Cydia pomonella* in walnuts by mating disruption. Bulletin OILB/SROP 22 (9), 83–89.

Angeli, G., Bellinazzo, S., Monta, L.D., Rizzi, C., Rama, F., 2000. Control of *Cydia pomonella* L. in walnuts (*Juglans regia* L.) with mating disruption technique. In: GF 2000 Atti, Giornate Fitopatologiche, Perugia, 16–20 April 2000, Volume Primo, pp. 361–366.

Anonymous, 2012. Trapping guidelines for surveillance of *Bactrocera invadens* in fruit production areas. Department of Agriculture, Foresty and Fisheries, South Africa.

Arai, T., 2002. Attractiveness of sex pheromone of *Pseudococcus cryptus* Hempel (Homoptera: Pseudococcidae) to adult males in a citrus orchard. Applied Entomology and Zoology 37, 69–72.

Asawalam, E.F., Nwachukwu, J.C., 2012. Incidence of *Bactrocera invadens* (Drew, Tsuruta & White) [Diptera: Tephritidae] infestations on mango (*Mangifera indica*) trees in Abia state Nigeria. Journal of Sustainable Agriculture and the Environment 13 (1), 61–64.

Askary, H., Varandi, H.B., Vatandoust, A., Tarbizian, M., 2005. Monitoring of *Tortrix viridana* distribution by sex pheromone in Mazandaran Province of Iran. Iranian Journal of Forest & Range Protection Research 2 (2), 133–142.

Atterholt, C.A., Delwiche, M.J., Rice, R.E., Krochta, J.M., 1999. Controlled release of insect sex pheromones from paraffin wax and emulsions. Journal of Controlled Release 57 (3), 233–247.

Bagle, B.G., Prasad, V.G., 1983. Effect of weather parameters on population dynamics of oriental fruit fly, *Dacus dorsalis* Hendel. Journal of Entomological Research 7 (2), 95–98.

Bakke, A., Kvamme, T., 1981. Kairomone response of *Thanasimus* predators to pheromone components of *Ips typographus*. Journal of Chemical Ecology 7, 305–312.

Bakthavatsalam, N., Unpublished. Polymorphism in pheromone reception in Helicoverpa armigera. Final Research Project Report. RPP-III. National Burau of Agriculturally Important Insects, Bangalore, 31 pp.

Bakthavatsalam, N., Shylesha, A.N., 2011. Invasive pests of agricultural importance in India and their management through classical biological control. In: Sabu Thomas, K. (Ed.), Selected Beneficial and Harmful Insects of Indian Subcontinent. Lap Lambert Academic Publishing, pp. 3–20.

Bakthavatsalam, N., Singh, S.P., 1996. L-tryptophan as an ovipositional attractant for *Chrysoperla carnea* (Stephens) (Neuroptera: Chrysopidae). Journal of Biological Control 10, 21–27.

Bakthavatsalam, N., Tandon, P.L., 2006. Kairomones for increasing the efficiency of chrysopids and trichogrammatids. In: Semiochemicals in Crop Protection: Ongoing Technologies. VII Annual Discussion Meeting in Entomology, Chennai, 2 December 2006.

Bakthavatsalam, N., Tandon, P.L., Patil, S.B., Hugar, B., Hosamani, A., 2007. Kairomone formulations as reinforcing agents for increasing abundance of *Chrysoperla carnea* (Stephens) in cotton ecosystem. Journal of Biological Control 21, 1–8.

Bakthavatsalam, N., Tandon, P.L., Bhagat, D., 2013. Trichogrammatids: behavioural ecology. In: Sithantham, S., Jalali, S.K., Ballal, C.R., Bakthavatsalam, N. (Eds.), Biological Control of Insect Pests Using Egg Parasitoids. Springer India, pp. 77–102.

Bakthavatsalm, N., Bhagat, D., 2008. Pheromone technology: an ecofriendly method of pest management. Vatika Autumn 2008 issue 3, 9–14.

Barbosa, P., Castellanos, I., 2005. Ecology of Predator-Prey Interactions. Oxford Univ. Press, p. 416.

Barclay, H.J., Hendrichs, J., 2014. Models for assessing the male annihilation of *Bactrocera* spp. with methyl eugenol baits. Annals of the Entomological Society of America 107 (1), 81–96.

Baronio, P., Baldassari, N., Scaravelli, D., 1992. Quantitative development of a population of *Thaumetopoea pityocampa* (Den. & Schiff.) (Lepidoptera, Thaumetopoeidae) assessed by the mass trapping technique. Frustula Entomologica 15, 1–9.

Batllori, J.L., Vilajeliu, M., Creixell, A., Carbo, M., Garcia, N., Esteba, G., Raset, F., Vayreda, F., Gine, M., Curos, D., Cornella, J., 2005. Reduction of insecticide sprayings by using alternative methods in commercial apple orchards. Bulletin OILB/SROP 28 (7), 83–88.

Bavaresco, A., Torres, A.N.L., Pilati, G., 2005. Use of synthetic sexual pheromone for monitoring the seasonal fluctuation of *Tuta absoluta* in Planalto Norte of Santa Catarina State. Agropecuaria Catarinense 18 (2), 83–86.

Ben Saad, A.A., Bishop, G.W., 1976a. Attraction of insects to potato plants through the use of aritificial honey dews and aphid juice. Entomophaga 21, 49–57.

Ben Saad, A.A., Bishop, G.W., 1976b. Effect of artificial honeydews on insect communities in potato fields. Environmental Entomology 5, 453–457.

Benelli, G., Flamini, G., Canale, A., Molfetta, I., Cioni, P.L., Conti, B., 2012. Repellence of *Hyptis suaveolens* whole essential oil and major constituents against adults of the granary weevil *Sitophilus granarius*. Bulletin of Insectology 65 (2), 177–183.

Beroza, C.G., Callow, R.W., Johnston, N.C., 1961. The isolation and synthesis of queen substance, 9-oxodec-trans-2-enoic acid, a honeybee pheromone. Proceedings of Royal Society Series B 155, 417.

Beroza, M., 2012. Chemicals Controlling Insect Behaviour. Elsevier Technology & Engineering, p. 184.

Bethe, A., 1932. Neglected hormones. Naturwissenschaften 20, 177.

Bhagat, D., Samanta, S.K., Bhattacharya, S., 2013. Efficient management of fruit pests by pheromone nanogels. Scientific Reports 3, 1294.

Bichina, T.I., Kovalev, B.G., Smetnik, A.I., Vaganova, L.D., 1982. A test of the pheromone of the Comstock mealybug [Russian]. Zashchita Rastenii 10, 46.

Bierl-Leonhardt, B.A., Moreno, D.S., Schwarz, M., Fargerlund, J., Plimmer, J.R., 1981. Isolation, identification and synthesis of the sex pheromone of the citrus mealybug, *Planococcus citri* (Risso). Tetrahedron Letters 22, 389–392.

Bjelis, M., 2006. Control of the olive fruit fly (*Bactrocera oleae* Gmel., Diptera: Tephritidae) by mass trapping method. Fragmenta Phytomedicaet Herbologica 29 (1/2), 35–48.

Blomquist, G.J., Vogt, R.G., 2003. Insect Pheromone Biochemistry and Molecular Biology: The Biosynthesis and Detection of Pheromone and Plant Volatiles. Academic Press, p. 768.

Boopathi, T., Singh, S.B., Ngachan, S.V., Manju, T., Ramakrishna, Y., 2013. Influence of weather factors on the incidence of fruit flies in chilli (*Capsicum annuum* L.) and their prediction model. Pest Management in Horticultural Ecosystems 19 (2), 194–198.

Borchert, D.M., Walgenbach, J.F., 2000. Comparison of pheromone-mediated mating disruption and conventional insecticides for the management of tufted apple bud moth (Lepidoptera: Tortricidae). Journal of Economic Entomology 3 (3), 769–776.

Boselli, M., 1998. Mating disruption in the control of codling moth: two years of experience in Emilia-Romagna region. Atti, Giornate fitopatologiche, Scicli e Ragusa, maggio 3 (7), 251–256.

Branco, M., Jactel, H., Fanco, J.C., Mendel, J.C., 2006. Modelling response to insect trap captures to pheromone dose. Ecological Modelling 197, 247–257.

Brockerhoff, E.G., Jones, D.C., Kimberley, M.O., Suckling, D.M., Donaldson, T., 2006. Nationwide survey for invasive wood-boring and bark beetles (Coleoptera) using traps baited with pheromones and kairomones. Forest Ecology and Management 228 (1/3), 234–240.

Brown, D.J., Il'ichev, A.L., 2000. The potential for the removal of organophosphate insecticdes from stone fruit production in the Goulburn valley, Australia. Acta - Horticulture 525, 85–91.

Brown Jr., W.L., Eisner, T., Whittaker, R.W., 1970. Allomones and kairomones: transpecific chemical messengers. Biological Sciences 20, 21–22.

Bruce, T.J., Cork, A., 2001. Electrophysiological and behavioral responses of female *Helicoverpa armigera* to compounds identified in flowers of African marigold, *Tagetes erecta*. Journal of Chemical Ecology 27, 1119–1131.

Bruce, T.J.A., Hooper, A.M., Ireland, L., Jones, O.T., Martin, J.L., Smart, L.E., Oakley, J., Wadhams, L.J., 2007. Development of a pheromone trap monitoring system for orange wheat blossom midge, *Sitodiplosis mosellana* in the UK. Pest Management Science 63 (1), 49–56.

Burkholder, W.E., Ma, M., 1985. Pheromones for monitoring and control of stored product insects. Annual Review of Entomology 30, 257–272.

Cahlíková, L., Hovorka, O., Ptáček, V., Valterová, I., 2004. Exocrine gland secretions of virgin queens of five bumblebee species (Hymenoptera: Apidae, Bombini). Zeitschrift für Naturforschung C 59 (7–8), 582–589.

Campion, D.G., Nesbitt, F., 1982. Recent advances in the use of pheromones in developing countries with particular reference to mass-trapping for the control of the Egyptian cotton leaf worm *Spodoptera littoralis* and mating disruption for the control of pink bollworm, *Pectinophora gossypiella*. In: Les mediateurs chimiques agissant sur le comportement des insects Symposium International Versailles, 16–20 November 1981.

Campion, D.G., Nesbitt, B.F., 1983. The utilisation of sex pheromones for the control of stem borers. Insect Science and its Application 4 (1/2), 191–197.

Campion, D.G., Hunter-Jones, P., McVeigh, L.J., Hall, D.R., Lester, R., Nesbitt, B.F., 1980. Modifications of the attractiveness of the primary pheromone component of the Egyptian cotton leafworm. *Spodoptera littoralis* (Boisduval) (Lepidoptera: Noctuidae), by secondary pheromone components and related chemicals. Bulletin of Entomological Research 70, 417–434.

Campion, D.G., Hall, D.R., Prevett, P.F., 1987. Use of pheromones in crop and stored products pest management: control and monitoring. Insect Science and its Applications 8 (4–6), 737–741.

Carde, R.T., Minks, A.K., 2012. Insect Pheromone Research: New Directions. Springer Science & Business Media, p. 684.

Carde, R.T., Mafra Neto, A., Staten, R.T., Kuenen, L.P.S., 1997. Understanding mating disruption in the pink boll-worm moth. Bulletin OILB/SROP 20 (1), 191–201.

Casagrande, E., 1993. The commercial implementation of mating disruption for the control of the rice stemborer, *Chilo suppressalis*, in rice in Spain. Bulletin OILB/SROP 16 (10), 82–89.

Ceredi, G., Borghesi, L., 2002. Sexual confusion of *Cydia* and *Anarsia* in peach orchards in Emilia Romagna. Informa-tore Agrario 58 (3), 65–68.

Chamberlain, D.J., Brown, N.J., Jones, O.T., Casagrande, E., 2000. Field evaluation of a slow release pheromone for-mulation to control the American bollworm, *Helicoverpa armigera* (Lepidoptera: Noctuidae) in Pakistan. Bulletin of Entomological Research 90 (3), 183–190.

Chang, C.L., Cho, I.K., Li, Q.X., Manoukis, N.C., Vargas, R.I., Elsevier B.V., 2013. A potential field suppression system for *Bactrocera dorsalis* Hendel. Journal of Asia-Pacific Entomology 16 (4), 513–519.

Charmillot, P.J., Pasquier, D., 2001. The small fruit tortix *Grapholita lobarzewskii*: mating disruption and population dynamics. Revue Suisse de Viticulture, Arboriculture et Horticulture 33 (3), 119–124.

Charmillot, P.J., Pasquier, D., 2000. Mating disruption to control grape moths: success and failure. Bulletin OILB/SROP 23 (4), 145–147.

Charmillot, P.J., Pasquier, D., 2002. Combination of mating disruption (MD) technique and granulosis virus to control resistant strains of codling moth *Cydia pomonella*. Revue Suisse de Viticulture, Arboricultureet Horticulture 34 (2), 103–108.

Charmillot, P.J., Pasquier, D., Bolay, J.M., Jeanrenaud, M., Zingg, D., Zufferey, E., 2000. Mating disruption and clas-sical control of grape moths in canton Vaud's vineyards in 1999. Revue Suisse de Viticulture, d' Arboriculture et d' Horticulture 32 (2), 83–88.

Charmillot, P.J., Pasquier, D., Zufferey, E., Bovard, A., 2001. Trial to control grape moths by mating disruption in Dezaley vineyard in 1999 and 2000. Revue Suisse de Viticulture, Arboriculture et Horticulture 33 (5), 247–251.

Chelvi, C.T., Kandasamy, R., Singh, J.P., 2010. Case study on the use of pheromone technology for the control of shoot borer and internode borer in sugarcane. Cooperative Sugar 41 (7), 61–64.

Chen, Z.L., Zhang, Z.N., 2008. Development of microencapsulated sex pheromone formulations for insects. Chinese Bulletin of Entomology 3, 362–367.

Chiranjeevi, C., Reddy, D.D.R., 2003. Evaluation of an integrated pest management practice for sweet potato weevil *Cylas formicarius* (F) in Andhra Pradesh. Journal of Research ANGRAU 31 (2), 22–28.

Chisholm, M.D., Underhill, E.W., Steck, W.F., 1979. Field trapping of diamondback moth *Plutella xylostella* using synthetic sex attractants. Environmental Entomology 8, 516–518.

Chow, Y.S., Lin, Y.M., Hsu, C.L., 1977. Sex pheromone of the diamondback moth (Lepidoptera: Plutellidae). Bulletin of the Institute of Zoology, Academia Sinica 16 (2), 99–105.

Chu, Y.I., Yeh, W.I., Chen, Y.S., 1985. Estimation of the active space of methyl eugenol and its economic evaluation on the attraction. Plant Protection Bulletin 27 (4), 401–411.

Ciglar, I., Baric, B., Tomsic, T., Subic, M., 2000. Control of codling moth (*Cydia pomonella*) by mating disruption tech-nique. Agronomski Glasmik 63 (1/2), 85–93.

Colazza, S., Wajnberg, E., 2013. Chemical ecology of insect parasitoids: towards a New Era. In: Wajnberg, E., Colazza, S. (Eds.), Chemical Ecology of Insect Parasitoids. Wiley-Blackwell, pp. 1–8.

Colazza, S., Peri, E., Bue, P., 2007. Insect pheromones as management tool to control citrus insect pests: present situ-ations and outlooks. Informatore Fitopatologico 57 (1), 35–41.

Cork, A., Basu, S.K., 1996. Control of the yellow stem borer, *Scirpophaga incertulas* by mating disruption with a PVC resin formulation of the sex pheromone of *Chilo suppressalis* (Lepidoptera: Pyralidae) in India. Bulletin of Ento-mological Research 86 (1), 1–9.

Cork, A., Alam, S.N., Das, A., Das, C.S., Ghosh, G.C., Farman, D.I., Hall, D.R., Maslen, N.R., Vedham, K., Phythian, S.J., Rouf, F.M.A., Srinivasan, K., 2001. Female sex pheromone of brinjal fruit and shoot borer, *Leucinodes orbonalis* blend optimisation. Journal of Chemical Ecology 27 (9), 1867–1877.

Cork, A., Alam, S.N., Rouf, F.M.A., Talekar, N.S., 2003. Female sex pheromone of brinjal fruit and shoot borer, *Leu-cinodes orbonalis* (Lepidoptera: Pyralidae): trap optimization and application in IPM trials. Bulletin of Entomologi-cal Research 93 (2), 107–113.

Cork, A., Alam, S.N., Rouf, F.M.A., Talekar, N.S., 2005a. Development of mass trapping technique for control of brinjal shoot and fruit borer, *Leucinodes orbonalis* (Lepidoptera: Pyralidae). Bulletin of Entomological Research 95 (6), 589–596.

Cork, A., Iles, M.J., Kamal, N.Q., Choudhury, J.C.S., Rahman, M.M., Islam, M., 2005b. An old pest, a new solution: commercializing rice stem-borer pheromones in Bangladesh. Outlook on Agriculture 34 (3), 181–187.

Cork, A., 2004. A Pheromone Manual. Natural Resources Institute, UK, p. 743.

Cosse, A.A., Bartelt, R.J., Zilkowski, B.W., Dean, D.W., Andress, E.R., 2006. Behaviourally active green leaf volatiles for monitoring the leaf beetle *Diorhabda elongata*, a biocontrol agent of salcedar, *Tamarix* spp. Journal of Chemical Ecology 32 (12), 2695–2708.

Cross, J.V., Hall, D.R., Innocenzi, P.J., Burgress, C.M., 2004. Exploiting sex aggregation pheromone of strawberry blossom weevil *Anthonomus rubi*. Bulletin OILB/SROP 27 (4), 125–132.

Cross, J.V., Hall, D.R., Innocenzi, P.J., Hesketh, H., Jay, C.N., Burgess, C.M., 2006. Exploiting the aggregation pheromone of strawberry blossom weevil *Anthonomus rubi* (Coleoptera: Curculionidae): Part 2. Pest monitoring and control. Crop Protection 25 (2), 155–166.

Cunningham, R.T., Suda, D.Y., 1986. Male annihilation through mass-trapping of male flies with methyl eugenol to reduce infestation of Oriental fruit fly (Diptera: Tephritidae) larvae in papaya. Journal of Economic Entomology 79 (6), 1580–1582.

Dallago, G., 2000. Mating disruption on the orchard in Trentino: experiences of 9 years (1991–1999). In: GF 2000 Atti, Giornate Fitopatologiche, Perugia, 16–20 April 2000, Volume Primo, pp. 327–332.

d'-Arcier, F.F., Blnc, M., Vidal, C., Speich, P., Bues, R., 2001. Chemical control of *Metcalf purinosa*: can be integrated with control of other pests makes integrated crop protection difficult. Phytoma 537, 15–17.

Darwish, D.Y.A., Rizk, M.M.A., Abdel-Galil, F.A., Temerak, S.A.H., 2015. Analysis of factors influencing population density of the peach fruit fly (PFF), *Bactrocera zonata* (Saunders) (Diptera: Tephritidae) in Assiut, Northern Upper Egypt. Archives of Phytopathology and Plant Protection 48 (1), 62–72.

Dash, A.N., Mukherjee, S.M., Mishra, P.R., Sontakke, B.K., 2005. Evaluation of integrated insect pest management (IPM) modules on irrigated rice. Journal of Plant Protection and Environment 2 (2), 55–59.

Daterman, G., Eglitis, A., Czokajlo, D., Sack, C., Kirsch, P., 2001. Attract and kill technology for management of European pine shoot moth (*Rhyacionia buoliana*) and Western pine shoot borer (*Eucosma sonomana*). Journal of Forest Science 47 (Special Issue 2), 66–69.

David, H., Nesbitt, B.F., Easwaramoorthy, S., Nandagopal, V., 1985. Application of sex pheromones in sugarcane pest management. Proceedings of the Indian Academy of Sciences, Animal Sciences 94 (3), 333–339.

Dickens, J.C., 1999. Predator-prey interactions: olfactory adaptations of generalist and specialist predators. Agricultural and Forest Entomology 1, 47–54.

Dominiak, B.C., Nicol, H.I., 2012. Chemical analysis of male annihilation blocks used in the control of Queensland fruit fly, *Bactrocera tryoni* (Froggatt) in New South Wales. Plant Protection 27 (1), 31–35.

Dowell, R.V., 2015. Attraction of non-target insects to three male fruit fly lures in California. Pan-Pacific Entomologist 91 (1), 1–19 22.

Downham, M.C.A., Smit, N.E.J.M., Laboke, P.O., Hall, D.R., Odongo, B., 2001. Reduction of pre-harvest infestations of African sweetpotato weevils *Cylas brunneus* and *C. puncticollis* (Coleoptera: Apionidae) using a pheromone mating-disruption technique. Crop Protection 20 (2), 163–166.

Drew, R.A.I., 1974. The responses of fruit fly species (Diptera: Tephritidae) in the South Pacific area to male attractants. Journal of the Australian Entomological Society 13 (4), 267–270.

Duca, P.L., Spanedda, A.F., Terrosi, A., Pucci, C., 2005. A forecasting model of the live fruit fly infestation based on monitoring of males. Bulletin of OILB/SROP 28 (9), 59–65.

Dunkelblum, E., 1999. Scale insects. In: Hardi, J., Minks, A.K. (Eds.), Pheromones of Non-Lepidopteran Insects Associated with Agricultural Plants, pp. 251–276.

Duran, J.M., Alvarado, M., Ortiz, E., Rosa, A.de la, Sanchez, A., Serrano, A., 2000. Flight curves of *Pectinophora gossypiella* (Saunders, 1843) (Lepidoptera, Gelechiidae), cotton pink bollworm, in western Andalucia. Boletin de Sanidad Vegetal, Plagas 26 (2), 229–238.

Duwadi, V.R., Paneru, R.B., Chand, S.P., 1996. Monitoring of chickpea pod borer (*Helicoverpa armigera* Hubner) by using male attractant pheromone trap at Pakhribas Agricultural Centre. PAC Technical Paper Pakhribas Agricultural Centre 169, 9.

El Sayed, A.M., Suckling, D.M., Wearing, C.H., Byers, J.A., 2006. Potential of mass trapping for long-term pest management and eradication of invasive species. Journal of Economic Entomology 99 (5), 1550–1564.

Emel, Y.Y.A., Bul- gynskaya, M.A., 1999. Use of pheromones for control of the codling moth, *Laspeyresia pomonella* L. (Lepidoptera: Tortricidae) by elimination and disorientation of males. Entomologicheskoe obozrenie 78 (3), 555–564.

Emery, S., Schmid, A., 2001. Control of grape moth in areas with a high initial population: mating disruption combined with a growth regulator treatment. Revue Suisse de Viticulture, Arboriculture et Horticulture 33 (2), 101–105.

Evenden, M.L., Haynes, K.F., 2001. Potential for the evolution of resistance to pheromone-based mating disruption tested using two pheromone strains of the cabbage looper. *Trichoplusia ni*. Entomologia Experimentalis et Applicata 100 (1), 131–134.

Evenden, M.L., 2005. Potential for combining sex pheromones for the forest tent caterpillar (Lepidoptera: Lasiocampidae) and the large aspen tortrix (Lepidoptera: Tortricidae) with monitroing traps targeting both species. Canadian Entomologist 137 (5), 615–619.

Fadamiro, H.Y., Baker, T.C., 1999. Reproductive performance and longevity of female European corn borer, *Ostrinia nubilalis*: effects of multiple mating, delay in mating, and adult feeding. Journal of Insect Physiology 45 (4), 385–392.

Fadlelmula, A.A., Ali, E.B.M., 2014. Fruit fly species, their distribution, host range and seasonal abundance in Blue Nile State, Sudan. Persian Gulf. Crop Protection 3 (3), 17–24.

Fassotte, B., Fischer, C., Durieux, D., Lognay, G., Haubruge, E., Francis, F., Verheggen, F.J., 2014. First evidence of a volatile sex pheromone in lady beetles. PLoS One 9 (12), e115011.

Feng, M.X., Jiang, D.D., Wang, J.Q., Li, Q., 2005. Experiment using sex pheromone for monitoring *Carposina sasakii*. China Fruits 6, 27–28.

Fettig, C.J., 2013. Chemical ecology and management of bark beetles in western coniferous forests. Journal of Biofertilizers and Biopesticides 4 (1). http://dx.doi.org/10.4172/2155-6202.1000e111.

Finkelman, S., Navarro, S., Rindner, M., Dias, R., 2007. First evaluation of the efficiency of a pheromone trap as a monitoring tool. Bulletin OILB/SROP 30 (2), 71.

Fitt, G.P., 1981. The influence of age, nutrition and time of day on the responsiveness of male *Dacus opiliae* to the synthetic lure, methyl eugenol. Entomologia Experimentalis et Applicata 30 (1), 83–90.

Fletcher, B.S., Bateman, M.A., Hart, N.K., Lamberton, J.A., 1975. Identification of a fruit fly attractant in an Australian plant, *Zieria smithii*, as O-methyl eugenol. Journal of Economic Entomology 68 (6), 815–816.

Flint, H.M., Salter, S.S., Walters, S.S., 1979. Caryophyllene: an attractant for the green lacewing. Environmental Entomology 8, 1123–1125.

Flint, H.M., Merkle, J.R., Sledge, M., 1981. Attraction of male *Collops vittatus* in the field by caryophyllene alcohol. Environmental Entomology 10, 301–304.

Franco, J.C., Gross, S., Silva, E.B., Suma, P., Russo, A., Mendel, Z., 2003. Is mass-trapping a feasible management tactic of the citrus mealybug in citrus orchards? Anais-do-Instituto-Superior-de-Agronomia 49, 353–367.

Frantisek, K., Jittka, S., 2005. Monitoring of San Jose scale, *Quadraspidiotus pernicisus* Comst. by pheromone traps and timing of control of crawlers. Bulletin OILB/SROP 28 (7), 183–187.

Frerot, B., Malosse, C., Cain, A., 1997. Solid-phase microextraction (SPME): a new tool in pheromone identification in lepidoptera. Journal of High Resolution Chromatography 20 (6), 340–342.

Frerot, B., Beslay, D., Malosse, C., Renou, M., Coupard, H., Bouvier, J.C., Sauphanor, B., 1999. Rationalising chemical control. Analysis of a recent loss of effectiveness in sexual trapping of the codling moth: what connection with insecticide resistance? Phytoma 522, 34–37.

Gazia, E.F., El-Hosary, R.A., El-Zaher, T.R.A., 2013. Studies on the control of *Bactrocera zonata* (Saunders) by male annihilation techniques in two locations. American-Eurasian Journal of Toxicological Sciences 5 (2), 48–54.

Geervliet, J.B.F., Vet, L.E.M., Dicke, M., 1994. Volatiles from damaged plants as major cues in long range host searching by the specialist parasitoid *Cotesia rubecula*. Entomologia Experimentalis et Applicata 73, 289–297.

Gonzalez, R.H., Cepeda, D., 1999. The carob-bean moth, *Spectrobates ceratoniae* (Zeller) (Pyralidae), a barrier to the implementation of mating disruption for control of codling moth on walnuts. Revista Fruticola 20 (2), 57–67.

Grant, G.G., Liu, W., Slessor, K.N., Abou-Zaid, M.M., 2006. Sustained production of the labile pheromone component, (Z, Z,)-6-9-heneicosadien-11-one form a stable precursor for monitoring the white marked tussock moth. Journal of Chemical Ecology 32 (8), 1731–1741.

Grewal, J.S., Kapoor, V.C., 1987. A new collapsable fruit fly trap. Journal of Entomological Research 11 (2), 203–206.

Grodner, J., Przybyz, E., Kolk, A., Slusarski, S., Jabonski, T., Bichta, P., 2005. Application of synthetic pheromone for population monitoring of horse chestnut leaf miner, *Cameraria ohridiella* Deschka et Dimic. Pestycydy 4, 7–16.

Gronning, E.K., Borchert, D.M., Pfeiffer, D.G., Felland, C.M., Walgenbach, J.F., Hull, L.A., Killian, J.C., 2000. Effect of specific and generic sex attractant blends on pheromone trap captures of four leafroller species in mid-Atlantic apple orchards. Journal of Economic Entomology 93 (1), 157–164.

Grové, T., De Beer, M.S., Cronje, R., 2014. Monitoring fruit flies in litchi orchards in South Africa and determining the presence of alien invasive *Bactrocera* species. Acta Horticulturae 1029, 425–432.

Guarino, S., Peri, E., Bue, P.lo, Germana, M.P., Colazza, S., Anshelevich, L., Ravid, U., Soroker, V., 2013. Assessment of synthetic chemicals for disruption of *Rhynchophorus ferrugineus* response to attractant-baited traps in an urban environment. Phytoparasitica 41 (1), 79–88.

Guario, A., Bari, G., Marinuzzi, V., Milella, G., Alfarano, L., Falco, R., Papa, G.K., Nasole, C., 2001. Biology of *Zeuria pyrina* and control through sexual confusion. Informatore Agrario 57 (44), 57–61.

Gupta, D., Verma, A.K., Bhalla, O.P., 1990. Population of fruit-flies (*Dacus zonatus* and *D. dorsalis*) infesting fruit crops in north-western Himalayan region. Indian Journal of Agricultural Sciences 60 (7), 471–474.

Gurr, G.M., Wratten, S.D., 1999. 'Integrated biological control': a proposal for enhancing success in biological control. International Journal of Pest Management 45, 81–84.

Haddad, Y., 2006. Utilization of pheromones in green spaces. PH Revue Horticole (486), 21–26.

Hagen, K.S., Greany, P., Sawall Jr., E.F., Tassan, R.L., 1976. The use of food sprays to increase the effectiveness of entomophagous insects. In: Proceedings of Tall Timbers Conference on Ecological Animal Control by Habitat Management. Tall Timbers Research Station, Florida, pp. 59–81. Number 3.

Hall, D.R., Cork, A., Phythian, S.J., Chittamuru, S., Jayarama, B.K., Venkatesha, M.G., Sreedharan, K., Vinod Kumar, P.C., Sertharame, H.G., Naidu, R., 2006. Identification of components of male-produced pheromone of coffee white stemborer, *Xylotrechus quadripes*. Journal of Chemical Ecology 32 (1), 195–219.

Hanks, L.M., Millar, J.G., Moreina, A.J.A., Barbour, J.D., Lacye, E.S., McElfresh, J.S., Reuter, F.R., Ray, A.M., 2007. Using generic pheromone lures to expedite identification of aggregation pheromones for the cerambycid beetles *Xylotrechus nauticus, Phymatodes lecontei* and *Neoclytus modestus modestus*. Journal of Chemical Ecology 33 (5), 889–907.

Hans, E.H., Miller, T.A., 2012. Techniques in Pheromone Research. Springer Science & Business Media Science, p. 464.

Haq, I., Vreysen, M.J.B., Cacéres, C., Shelly, T.E., Hendrichs, J., 2014. Methyl eugenol aromatherapy enhances the mating competitiveness of male *Bactrocera carambolae* drew & Hancock (Diptera: Tephritidae). Journal of Insect Physiology 68, 1–6.

Hardie, J., Minks, A.K., 1999. Pheromones of Non Lepidopteran Insects Associated with Agricultural Plants. CABI Publishing, Wallingford, UK, p. 466.

Hardie, J., Nottinghma, S.F., Powell, W., Wadhams, C.J., 1991. Synthetic aphid sex pheromone lures female parasitoids. Entomologia Experimentalis et Applicata 61, 97–99.

Hare, J.D., Morgan, J.W.D., 1997. Mass priming *Aphytis*: behavioural improvement of insectary-reared biological control agents. Biological Control 10, 207–214.

Hashimoto, K., Morita, A., Kuwahara, S., 2008. Enantioselective synthesis of a mealybug pheromone with an irregular monoterpenoid skeleton. Journal of Organic Chemistry 73, 6913–6915.

Hastings, C., Jullien, J., 2006. Utilization of pheromones in vegetable and ornamental crops. PHM Revue Horticole (486), 11–20.

Helsen, H., Blommers, L., 2000. Mating disruption of *Campylomma verbasci* is difficult. Fruitteelt Den Haag 90 (22), 14–15.

Heuskin, S., Verheggen, F.J., Haubruge, E., Wathelet, J.P., Lognay, G., 2011. The use of semiochemical slow-release devices in integrated pest management strategies. Biotechnologie, Agronomie, Société et Environnement 15 (3), 459–470.

Ho, H.-Y., Hung, C.C., Chuang, T.-H., Wang, W.L., 2007. Identification and synthesis of the sex pheromone of the passionvine mealybug, *Planococcus minor* (Maskell) Journal of Chemical Ecology 33, 1986–1996.

Ho, H.Y., Su, Y.T., Ko, C.H., Tsai, M.Y., 2009. Identification and synthesis of the sex pheromone of the Madeira mealybug, *Phenacoccus madeirensis* Green. Journal of Chemical Ecology 35, 724–732.

Hooper, G.H.S., Drew, R.A.I., 1978. Comparison of the efficiency of two traps for male tephritid fruit flies. Journal of the Australian Entomological Society 17 (1), 95–97.

Hooper, G.H.S., Drew, R.A.I., 1979. Effect of height of trap on capture of tephritid fruit flies with cuelure and methyl eugenol in different environments. Environmental Entomology 8 (5), 786–788.

Hooper, G.H.S., 1978. Effect of combining methyl eugenol and cuelure on the capture of male tephritid fruit flies. Journal of the Australian Entomological Society 17 (2), 189–190.

Hori, M., Enya, S., 2013. Attractiveness of synthetic volatile blends of flowering rice panicles to *Trigonotylus caelestialium* (Kirkaldy) (Heteroptera: Miridae). Journal of Applied Entomology 137 (1/2), 97–103.

Howse, P., Stevens, J.M., Jones, G.A.D., 2013. Insect Pheromones and Their Use in Pest Managment. Springer Science & Business Meida, p. 369.

Howse, P., 1998. Insect Pheromone and Their Use in Pest Management. Springer, p. 369.

Huang, Y., Han-Ban, Y., Tang, Q., Xu, H., Wang, YuG., 2009. EAG and behavioural responses of *Apanteles* sp. (Hymenoptera: Braconidae) parasitizing tea geometrids by volatiles from tea shoots. Acta Entomologia Sinica 52, 1191–1198.

Huangfu, W.G., Chen, R.X., Wang, Y.J., Lu, D.Q., Hu, H.F., Yao, H.Y., Sun, Y.T., Zhou, J.B., 2005. Mass trapping and control efficacy of different synthetic sex pheromones on diamondback moth, *Plutella xylostella*. Acta Agriculturae Zhejiangensis 17 (6), 395–397.

Huber, R.T., Hoffmann, M.P., 1979. Development and evaluation of an oil trap for use in pink bollworm pheromone mass trapping and monitoring programs. Journal of Economic Entomology 72 (5), 695–697.

Hull, L.A., Felland, C.M., 1999. Mating disruption of peach insects – will it work? Pennsylvania Fruit News 79 (4), 50–54.

Hummel, H.E., Urek, G., Modic, S., Hein, D.F., 2005. Monitoring presence and advance of the alien invasive western corn rootworm beetle in eastern Slovenia with highly sensitive Metcalf traps. Communications in Agricultural and Applied Biological Sciences 70 (4), 687–692.

Hung, C.C., Hou, R.F., Hwang, J.S., 2001. Assessment of the effects of using sex pheromone for control of the carambola fruit borer, *Eucosoma notanthes* Mayrick. Plant Protection Bulletin, Taipei 43 (2), 57–68.

Ibeas, F., Gallego, D., Diez, J.J., Pajares, J.A., 2007. An operative kairomonal lure for managing pine sawyer beetle *Monochamus galloprovincialis* (Coleoptera: Cerymbycidae). Journal of Applied Entomology 131 (1), 13–20.

Ibrahim, A.G., Hashim, A.G., 1980. Efficacy of methyl-eugenol as male attractant for *Dacus dorsalis* Hendel (Diptera: Tephritidae). Pertanika 3 (2), 108–112.

Ibrahim, A.G., Singh, G., King, H.S., 1979. Trapping of the fruit-flies, *Dacus* spp. (Diptera: Tephritidae) with methyl eugenol in orchards. Pertanika 2 (1), 58–61.

Iga, M., 1982. Density of methyl eugenol-traps for efficient trapping of adult males of the oriental fruit fly, *Dacus dorsalis* Hendel (Diptera: Tephritidae), on Ogasawara Islands. Japanese Journal of Applied Entomology and Zoology 26 (3), 172–176.

Il' ichev, A.L., Hossain, M.S., Jerie, P.H., 1999. Migration of oriental fruit moth *Grapholita molesta* Busk. (Lepidoptera: Tortricidae) under wide area mating disruption in Victoria, Australia. Bulletin OILB/SROP 22 (11), 53–62.

Ioannou, C.S., Papadopoulos, N.T., Kouloussis, N.A., Tananaki, C.I., Katsoyannos, B.I., 2012. Essential oils of citrus fruit stimulate oviposition in the Mediterranean fruit fly *Ceratitis capitata* (Diptera: Tephritidae). Physiological Entomology 37 (4), 330–339.

Ioriatti, C., 2008. New techniques to control harmful insect pests through the use of semiochemicals. Gorgolili 5 (2), 495–507.

Izawa, H., Fuju, K., Matoba, T., 2000. Control of multiple species of lepidopterous insect pests using a mating disruptor and reduced pesticide applications in Japanese pear orchards. Japanese Journal of Applied Entomology and Zoology 44 (3), 165–171.

Jackson, D.M., Canhilal, R., Carner, G.R., 2005. Trap monitoring squash vine borers in cucurbits. Journal of Agricultural and Urban Entomology 22 (1), 27–39.

Jactel, H., Menassieu, P., Vetillar, F., Barthelemy, B., Piou, D., Frerot, B., Roussellet, J., Goussand, F., Branco, M., Battish, A., 2006. Population monitoring of the pine processionary moth (Lepidoptera: Thaumetopoeidae) with pheromone-baited traps. Forest Ecology and Management 235 (1/3), 96–106.

James, D.G., Price, T.S., 2004. Field testing of methyl salicylate for recruitment and retention of beneficial insects in grapes and hops. Journal of Chemical Ecology 30, 1613–1628.

James, D.G., 2003. Field evaluation of herbivore-induced plant volatiles as attractant for beneficial insects: methyl salicylate and the green lacewing, *Chrysopa nigricornis*. Journal of Chemical Ecology 29, 1601–1609.

James, D.G., 2005. Further evaluation of synthetic herbivore-induced plant volatiles as attractants for beneficial insects. Journal of Chemical Ecology 31, 481–495.

James, S.P., Babu, A., Perumalsamy, K., Muraleedharan, N., Selvasundaram, R., 2008. Red spider mite, *Oligonychus coffeae* (Nietner) induced plant volatiles from tea leaves, an invitation to predators. Journal of Plantation Crops 36, 425–429.

Jang, E.B., Khrimian, A., Siderhurst, M.S., 2011. Di- and tri-fluorinated analogs of methyl eugenol: attraction to and metabolism in the oriental fruit fly, *Bactrocera dorsalis* (Hendel). Journal of Chemical Ecology 37 (6), 553–564.

Jayanth, K.P., Mathew, M.T., Narabenchi, G.B., Bhanu, K.R.M., 2007. Impact of large scale mass trapping of red palm weevil *Rhynchophorus ferrugineus* Olivier in coconut plantations in Kerala using indigenously synthesized aggregation pheromone lures. Indian Coconut Journal 38 (2), 2–9.

Jayarama, D'Souza, M.V., Hall, D.R., 2007. (S)-2-hydroxy-3-decanone as sex pheromone for monitoring and control of coffee white stem borer, (*Xylotrechus quadripes*). In: 21st International Conference on Coffee Science. Montpellier, France, 11–15 September 2006, pp. 1291–1300.

Jiao, X.G., Xuan, W.J., Wang, H.T., Su, J.W., Sheng, C.F., 2003. Advances in the use of sex pheromone to control *Chilo suppressalis*. Entomological Knowledge 40 (3), 193–199.

Jiao, X.G., Xuan, W.J., Sheng, C.F., 2005. Mass trapping of the overwintering generation stripped stem borer, *Chilo suppressalis* (Walker) (Lepidoptera: Pyralidae) with the synthetic sex pheromone in northeastern China. Acta Entomologica Sinica 48 (3), 370–374.

Jones, V.P., Sasaki, A.M., 2001. Demographic analysis of delayed mating in mating disruption: a case study with *Cryptophelbia illepida* (Lepidoptera: Tortricidae). Journal of Economic Entomology 94 (4), 785–792.

Jones, R.L., Lewis, W.J., Beroza, M., Bierl, B.A., Sparks, A.N., 1973. Host-seeking stimulants (kairomones) for the egg parasite, *Trichogramma evanescens*. Environmental Entomology 2, 593–596.

Jones, T.H., Brunner, S.R., Edwards, A.A., Davidson, D.W., Snelling, R.R., 2005. 6-alkylsalicylic acids and 6-alkyl-resorcylic acids from ants in the genus *Crematogaster* from Brunei. Journal of Chemical Ecology 31 (2), 407–417.

Jurc, M., Perkom, M., Dzeroski, S., Demsar, D., Hrasovec, B., 2006. Spruce bark beetles (*Ips typographus, Pityogenes chalcographus*, Col.: Scolytidae) in the Dinaric mountain forests of Slovenia: monitoring and modelling. Ecological Modelling 194 (1/2), 213–226.

Kabre, G.B., Dharne, P.K., 2009. Mass trapping of *Earias vittella* (Fab) and *Pectinophora gossypiella* Saund. using indigenously synthesized sex pheromone in cotton ecosystem. Karnataka Journal of Agricultural Sciences 22 (3), 664–665.

Kakinuma, K., Sudo, K., Senbongi, I., Takayama, Y., Imuta, T., 2000. Control of the common cutworm, *Spodoptera litura* (Fabricius), by synthetic sex pheromone in Chinese yam fields. Annual Report of the Kanto Tosan Plant Protection Society 47, 125–127.

Kamarudin, N., Wahid, B.M., 2004. Immigration and activity of *Oryctes rhinoceros* within a small oil palm replanting area. Journal of Oil Palm Research 16 (2), 64–77.

Kamarudin, N., Arshad, O., 2006. Potentials of using the pheromone trap for monitoring and controlling the bagworm, *Metisa plana* Wlk (Lepidotera: Pscyhidae) on young oil palm in a small holder plantation. Journal of Asia-Pacific Entomology 9 (3), 281–285.

Kardinan, A.K., Hidayat, P., 2013. Potency of *Melaleuca bracteata* and *Ocimum* sp. leaf extracts as fruit fly (*Bactrocera dorsalis* complex) attractants in guava and star fruit orchards in Bogor, West Java, Indonesia. Journal of Developments in Sustainable Agriculture 8 (2), 79–84 17.

Karlson, R., Luscher, M., 1959. "Pheromones": a new term for a class of biologically active substances. Nature 183, 55–56.

Karunaratne, M.M.S.C., Karunaratne, U.K.P.R., 2012. Factors influencing the responsiveness of male oriental fruit fly, *Bactrocera dorsalis*, to methyl eugenol (3, 4 dimethoxyalyl benzene). Tropical Agricultural Research and Extension 15 (4) (Unpaginated).

Katase, M., Shimizu, K., Oki, H., Uchida, S., Nagata, K., Naito, T., 2006. Control effects of synthetic sex pheromone as a communication disruption agent against diamondback moth, *Plutella xylostella* (Linnaeus) and common cutworm, *Spodoptera litura* (Fabricius) in herbs cultivated in large scale greenhouses. Annual Report of the Kanto Tosan Plant Protection Society 53, 115–118.

Kehat, M., Anshelevich, L., Gordon, D., Harel, M., Zilberg, L., Dunkelblum, E., 1999. Effect of density of pheromone sources, pheromone dosage and population pressure on mating of pink bollworm, *Pectinophora gossypiella* (Lepidoptera: Gelechiidae). Bulletin of Entomological Research 89 (4), 339–345.

Kerns, D.L., 2000. Mating disruption of beet armyworm (Lepidoptera: Noctuidae) in vegetables by a synthetic pheromone. Crop Protection 19 (5), 327–334.

Khan, Z.R., James, D.G., Midega, C.A.O., Pickett, J.A., 2008. Chemical ecology and conservation biological control. Biological Control 45, 210–214.

Khattak, S.U., Afsar, K., Hussain, N., Khalil, S.K., Alamzeb, 1990. Annual population incidence of oriental fruit fly (*Dacus dorsalis*) Hendel in a fruit orchard at Peshawar, Pakistan. Bangladesh Journal of Zoology 8 (2), 131–138.

Khrimian, A., Fay, H.-A.-C., Guzman, F., Chauhan, K., Moore, C., Aldrich, J.-R., 2012. Pheromone of the banana-spotting bug, *Amblypelta lutescens lutescens* Distant (Heteroptera: Coreidae): identification, synthesis, and field bioassay. Psyche: A Journal of Entomology 2012 Article ID 536149.

Knight, A., Christianson, B., Cockfield, S., Dunley, J., 2001. Areawide management of codling moth and leafrollers. Good Fruit Grower 52 (11), 31–33.

Knight, A.L., Hilton, R., Light, D.M., 2005. Monitoring codling moth (Lepidoptera: Tortricidae) in apple with blends of Ethyl (E,Z)-2,4-decadienoate and codlemone. Environmental Entomology 34 (3), 598–603.

Knight, A.L., Hilton, R., Basoalto, E., Stelinski, L.L., 2014. Use of glacial acetic acid to enhance bisexual monitoring of tortricid pests with kairomone lures in pome fruits. Environmental Entomology 43 (6), 1628–1640.

Knoll, F.R.N., Santos, L.M., 2012. Orchid bee baits attracting bees of the genus *Megalopta* (Hymenoptera, Halictidae) in Bauru region, Sao Paulo, Brazil: abundance, seasonality, and the importance of odors for dim-light bees. Revista Brasileira de Entomologia 56 (4), 481–488.

Kovanc, O.B., Walgenbach, J.F., 2005. Monitoring the oriental fruit moth with pheromone and bait traps in apple orchards under different management regimes. International Journal of Pest Management 51 (4), 273–279.

Koyama, J., Teruya, T., Tanaka, K., 1984. Eradication of the oriental fruit fly (Diptera: Tephritidae) from the Okinawa Islands by a male annihilation method. Journal of Economic Entomology 77 (2), 468–472.

Krishnaiah, K., 1986. Studies on the use of pheromones for the control of *Spodoptera litura* Fab. on blackgram grown in rice fallows. Indian Journal of Plant Protection 14 (43), 46.

Kukovinets, O.S., Zvereva, T.I., Kesradze, V.G., Galin, F.Z., Frolova, L.L., Kuchin, A.V., Sprikhin, L.V., Abdullin, M.I., 2006. Novel synthesis of *Planococcus citri* pheromone. Chemistry of Natural Compounds 42 (2), 216–218.

Kumaran, N., Balagawi, S., Schutze, M.K., Clarke, A.R., 2013. Evolution of lure response in tephritid fruit flies: phytochemicals as drivers of sexual selection. Animal Behaviour 85 (4), 781–789.

Kumaran, N., Hayes, R.A., Clarke, A.R., 2014. Cuelure but not zingerone make the sex pheromone of male *Bactrocera tryoni* (Tephritidae: Diptera) more attractive to females. Journal of Insect Physiology 68, 36–43 46.

Kutinkova, H., Dzhuvinov, V., 2012. Biological control of oriental fruit moth on peach in Bulgaria. Acta Horticulturae 962, 449–454.

Landolt, P.J., Toth, M., Meagher, R.L., Szarukan, I., 2013. Interaction of acetic acid and phenylacetaldehyde as attractants for trapping pest species of moths (Lepidoptera: Noctuidae). Pest Management Science 69 (2), 245–249.

Latif, A., Abdullah, K., Akram, M., 2004. Enhancing trap catches of melon fruit flies *Bactrocera* spp. (Diptera: Tephritidae) by integration of visual and olfactory cues. Proceedings of Pakistan Congress of Zoology 24, 205–211.

Law, J.H., Regnier, F.E., 1971. Pheromones. Annual Review of Biochemistry 40, 533–548.

Leal, W.S., 2001. Molecules and macromolecules involved in chemical communication of scarab beetles. Pure and Applied Chemistry 73 (3), 613–616.

Lecocq, R., 2000. Insects and mites, alternate method: waiting for mating disruption. Arboriculture Fruitere 539, 41–44.

Leonhardt, B.A., Cunningham, R.J., Avery, J.W., DeMil, A.B., Warthen Jr., J.D., 1996. Comparison of ceralure and trimedlure attractants for the male Mediterranean fruit fly (Diptera: Tephritidae). Journal of Entomological Science 31 (2), 183–190.

Lethmayer, C., Hausdorf, H., Blumel, S., 2004. The first field experiences with sex aggregation pheromones of the strawberry blossom weevil, *Anthonomus rubi*, in Austria. Bulletin OILB/SROP 27 (4), 133–139.

Lewis, W.J., Martin, W.R., 1990. Semiochemicals for use with parasitoids: status and future. Journal of Chemical Ecology 16, 3067–3089.

Lewis, W.J., Jones, R.L., Sparks, A.N., 1972. A host seeking stimulant for the egg parasite *Trichogramma evanescens*; its source and a demonstration of its laboratory and field activity. Annals of Entomological Society of America 65, 1087–1089.

Lewis, W.J., Jones, R.L., Nordlund, D.A., Gross, H.R., 1975a. Kairomones and their use for management of entomophagous insects: II. Principles causing increases in parasitization rates by *Trichogramma* spp. Journal of Chemical Ecology 1, 349–360.

Lewis, W.J., Jones, R.L., Nordlund, D.A., Sparks, A.N., 1975b. Kairomones and their use for management of entomophagous insects: I. Utilization for increasing rate of parasitism by *Trichogramma* spp. Journal of Chemical Ecology 1, 343–347.

Liaropoulos, C., Mavraganis, V.G., Broumas, T., Ragoussis, N., 2005. Field tests on the combination of mass trapping with the release parasite *Opius concolor* (Hymenoptera: Braconidae), for the control of the olive fruit fly *Bactrocera oleae* (Diptera: Tephritidae). Bulletin OILB/SROP 28 (9), 77–81.

Lindgreen, B.S., Miller, D.R., 2002. Effect of verbenone on attraction of predatory and wood boring beetles (Coleoptera) to kairomone in lodgepole pine forests. Environmental Entomology 31, 766–773.

Liquido, N.J., 1993. Reduction of oriental fruit fly (Diptera: Tephritidae) populations in papaya orchards by field sanitation. Journal of Agricultural Entomology 10 (3), 163–170.

Liu, Y.C., Lin, J.S., 1993. The response of melon fly *Dacus cucurbitae* Coquillet to the attraction of 10% MC. Plant Protection Bulletin (Taipei) 35 (1), 79–88.

Liu, Y.C., Cavalloro, R., Balkema, A.A., 1982. Population studies on the oriental fruit fly, *Dacus dorsalis* Hendel, in central Taiwan. In: Proceedings of the CEC/IOBC International Symposium, Athens, Greece, 16–19 November 1982, pp. 62–67.

Loera, J., Lynch, R., Rodriguex, R., 1995. Lesser cornstalk borer monitoring with the use of pheromone traps *Elasmopalpus lignosellus* (Zeller) (Lepidoptera: Pyralidae). Agronomia Mesoamericana 6, 75–79.

Löfqvist, J., 1976. Formic acid and saturated hydrocarbons as alarm pheromone for the ant *Formica rufa*. Journal of Insect Physiology 22 (10), 1331–1346.

Loganathan, M., Uthamasamy, S., 1998. Efficacy of a sex pheromone formulation for monitoring *Heliothis armigera* Hubner moths on cotton. Journal of Entomological Research 22 (1), 35–38.

Logli, F.V., Bangers, A.G., Clement, J.L., 1996. Role of volatiles in the search for a host by parasitoid, *Diglyphus isaeass.* Journal of Chemical Ecology 22, 541–558.

Lopez, J.D., Shaver, T.N., Dickerson, W.A., 1990. Population monitoring of *Heliothis* spp. using pheromones. In: Ridgeway, R., Silverstein, R.M., Inscoe, A. (Eds.), Behaviour-Modifying Chemicals for Insect Management. Marcel Dekker, New York, pp. 473–496.

Lyytikainen-Saarenmaa, P., Varama, M., Anderbrant, O., Kukkola, M., Kokkonen, A.M., Hedenstrom, E., Hogberg, H.E., 2006. Monitoring the European pine sawfly with pheromone traps in maturing Scots pine stands. Agricultural and Forest Entomology 8 (1), 7–15.

Maa, C.J.W., Chen, Y.W., Ying, Y.J., Chow, Y.S., 1987. Effect of environmental factors to male adult catch by synthetic female sex pheromone trap of the diamondback moth, *Plutella xylostella* L. in Taiwan. Bulletin of the Institute of Zoology, Academia Sinica 26 (4), 257–269.

Maini, S., Accinelli, G., 2000. Mating disruption confusion method and sexual distraction: comparison among different dispenser types for *Cydia molesta* (Busck) (Lepidoptera Tortricidae). Bollettino dell' Istituto di Entomologia 'Guido-Grandi'della Universita degli Studi di Bologna 54, 113–122.

Maini, S., Accinelli, G., 2001. Mating disruption confusion method and sexual distraction: comparison among different dispenser types for *Cydia molesta* (Busck). Informatore Fitopatologico 51 (10), 36–40.

Manrakhan, A., Brown, L., Venter, J.H., Stones, W., Daneel, J.H., 2011. The *Bactrocera invadens* surveillance programme in South Africa. SA Fruit Journal 10 (6), 78–80.

Mansilla Vazquez, P., Perez Otero, R., Salinero Corral, M.C., Vela Fernandez, P., 1999. Integrated control of chestnut pests in the Verin area (Orense): results of three years' experience. Boletin de Sanidad Vegetal, Plagas 25 (3), 297–310.

Mansour, M., Mohamad, F., 2001. Mating disruption for codling moth, *Cydia pomonella* (L.) (Lepidoptera: Tortricidae), control in Syrian apple orchards. Polskie Pismo Entomologiczne 70 (2), 151–163.

Mansour, R., Suma, P., Mazzeo, G., Buonocore, E., Lebdi, K.G., Russo, A., 2010. Using a kairomone-based attracting system to enhance biological control of mealybugs (Hemiptera: Pseudococcidae) by *Anagyrus* sp. near *pseudococci* (Hymenoptera: Encyrtidae) in Sicilian vineyards. Journal of Entomological and Acarological Research 42 (3), 161–170.

Marwat, N.K., Hussain, N., Khan, A., 1992. Suppression of population and infestation of *Dacus* spp. by male annihilation in guava orchard. Pakistan Journal of Zoology 24 (1), 82–84.

Mason, L.J., Jansson, R.K., 1990. Potential of mass trapping and mating disruption for sweet potato weevil management. Proceedings of the Interamerican Society for Tropical Horticulture 34, 89–95.

Masuda, T., Endo, S., 1981. Disappearance of fruit fly lures and insecticides on cotton rope impregnated with these mixtures in the field. Pesticide Science 6 (2), 227–230.

Mathews, P.L., Stephen, F.M., 1999. Effect of artificial honey diet and varied environmental conditions on longevity of *Coeloides pissoidis* (Hymenoptera: Braconidae), a parasitoid of *Dendroctonus frontalis* (Coleoptera: Scolytidae). Environmental Entomology 28, 729–734.

Mattedi, L., Mescalchin, E., 2002. Side effects of [sexual] confusion against *Clysia ambiguella* in large areas. Informatore Agrario 58 (15), 101–102.

Mayamba, A., Nankinga, C.K., Isabirye, B., Akol, A.M., 2014. Seasonal population fluctuations of *Bactrocera invadens* (Diptera: Tephritidae) in relation to mango phenology in the Lake Victoria Crescent, Uganda. Fruits 69 (6), 473–480.

Mayer, M.S., McLaughlin, J.R., 1991. Handbook of Insect Pheromones and Sex Attractants. CRC Press, p. 1083.

Mazomenos, B.E., Stefanou, D., Pantazi Mazomenou, A., 1999. Small plot mating disruption trials: effect of various pheromone doses dispensed per hectare on male *Prays oleae* trap catches and fruit damage. Bulletin OILB/SROP 22 (9), 91–94.

McCullough, D.W., Bhupathy, M., Piccolino, E., Cohen, T., 1991. Highly efficient terpenoid pheromone synthesis via regio- and stereo controlled processing of allylthiums generated by reductive lithiation of allyl phenyl thioethers. Tetrahedron 47 (47), 9727–9736.

Meissner, H.E., Atterholt, C.A., Walgenbach, J.F., Kennedy, G.G., 2000. Comparison of pheromone application rates, point source density and dispensing methods for mating disruption of tufted apple bud moth (Lepidoptera: Tortricidae). Journal of Economic Entomology 93 (3), 820–827.

Mensah, R.K., 1997. Local density responses of predatory insects of *Helicoverpa* spp. to a newly developed food supplement 'Envirofeast' in commercial cotton in Australia. International Journal of Pest Management 43, 221–225.

Metcalf, L., Metcalf, R., 1992. Plant Kairomones in Insect Ecology and Control. Chapman and Hall, Newyor and London, p. 168.

Micha, S.G., Wyss, U., 1996. Aphid alarm pheromone (E)-β-farnesene: a host finding kairomone for the aphid primary parasitoid, *Aphidius uzbekistanicus* (Hymenoptea: Aphidiinae). Chemoecology 7, 132–139.

Millar, J.G., Daane, K.M., McElfresh, J.S., Moreira, J.A., Malakar-Kuenen, R., Guillen, M., Bentley, W.J., 2002. Development and optimization of methods for using sex pheromone for monitoring the mealybug *Planococcus ficus* (Homoptera: Pseudococcidae) in California vineyards. Journal of Economic Entomology 95, 706–714.

Millar, J.G., Daane, K.M., McElfresh, J.S., Moreira, J.A., Bentley, W.J., 2005. Chemistry and applications of mealybug sex pheromones. In: Petroski, R.J., Tellez, M.R., Behle, R.W. (Eds.), Semiochemicals in Pest and Weed Control. American Chemical Society, Washington, DC, pp. 11–27.

Millar, J.G., Moreira, J.A., McElfresh, J.S., Daane, K.M., Freund, A.S., 2009. Sex pheromone of the longtailed mealybug: a new class of monoterpene structure. Organic Letters 11, 2683–2685.

Miller, D.R., Rabaglia, R.J., 2009. Ethanol and (-) alpha-pinene: attractant kairomones for bark and ambrosia beetles in the southeastern US. Journal of Chemical Ecology 35 (4), 435–438.

Miller, D.R., Asaro, C., Crowe, C.M., Duerr, D.A., 2011. Bark beetle pheromones and pine volatiles: attractant kairomone lure blend for longhorn beetles (Cerambycidae) in pine stands of the Southeastern United states. Journal of Economic Entomology 104 (4), 1245–1257.

Miller, D.R., 2001. Frontalin interrupts attraction of *Ips pini* (Coleoptera: Scolytidae) to ipsdienol. Candaian Entomologist 133 (3), 407–408.

Miller, D.R., 2006. Ethanol and (-)-α-pinene: attractant kairomones for large wood boring beetles in Southeastern USA. Journal of Chemical Ecology 32, 779–794.

Mo, R.J., Ouyang, Q., Zhong, B.E., Wu, W.J., 2014. Attractiveness of durian fruit volatiles to *Bactrocera dorsalis* (Hendel). Chinese Journal of Applied Entomology 51 (5), 1336–1342.

Mochizuki, F., Fukumoto, T., Noguchi, H., Sugie, H., 2001. Resistance to a communication disruptant with a sex pheromone in two tea pests. Aroma Research 2 (2), 185–189.

Mochizuki, F., Fukumoto, T., Noguchi, H., Sugie, H., Morimoto, T., Ohtani, K., 2002. Resistance to a mating disruptant composed of (Z)-11-tetradecenyl acetate in the smaller tea tortrix, *Adoxophyes honmai* (Yasuda) (Lepidoptera: Tortricidae). Applied Entomology and Zoology 37 (2), 299–304.

Mohammadpur, K., Faghih, A., 2008. Investigation on the possibility of co mass trapping of the populations of red palm weevil, *Rhynchophorus ferrugineus* and date palm fruit stalk borer, *Oryctes elegans* using pheromone traps. Applied Entomology and Phytopathology 75 (2), 39–55.

Mohapatra, S.D., Aswal, J.S., Mishra, P.N., 2007. Monitoring population dynamics of tomato fruit borer, *Helicoverpa armigera* Hubner moths through pheromone traps in Uttaranchal Hills. Indian Journal of Entomology 69 (2), 172–173.

Molinari, F., Baviera, C., Melandri, M., 1999. Activity of aphid antagonists in peach orchards in northern Italy. Bulletin OILB/SROP 22 (11), 19–26.

Molinari, F., Cravedi, P., Rama, F., Reggiori, F., Pane, M.D., Boselli, M., 2000. The control of *Cydia molesta* in pome fruit orchards using sex pheromones through the method of "Disorientation". In: GF 2000 Atti, Giornate Fitopatologiche, Perugia, 16–20 April 2000, Volume Primo, pp. 333–339.

Molinari, F., 2001. The use of pheromones against stone fruit pests. Informatore Fitopatologico 51 (10), 16–20.

Mondal, S., Tarafdar, J., Sen, H., 2006. Use of sex pheromone trap for monitoring and management of sweet potato weevil (*Cylas formicarius* Fabricius). Environment and Ecology 245 (Special 3), 525–528.

Munj, A.Y., Salvi, B.R., Jalgaonkar, V.N., Damodhar, V.P., Narangalkar, A.L., 2013. Rakshak methyl eugenol traps: an ideal tools developed by Dr. B.S. Konkan Krishi Vidyapeeth, dapoli to minimize the mango fruit fly incidence (*Bactrocera dorsalis* Hendel) during unseasonal rains. Annals of Plant Physiology 27 (1/2), 54–57.

Murakami, Y., Sugie, H., Fukumoto, T., Mochizuki, F., 2005. Sex pheromone of *Grapholita dimorpha* Komai (Lepidoptera: Tortricidae), and its utilization for monitoring. Applied Entomology and Zoology 40 (3), 521–527.

Muresanu, F., Popov, C., 2005. Monitoring of western corn rootworm (*Diabrotica virgifera virgifera*) in central area of Romania. Analale Institutului National-de-Cercetare Dezvoltare Agricola Fundulea 72, 173–178.

Muthukrishnan, N., Balasubramanian, M., 1999. .Studies on the effect of new gossyplure pheromones in cotton ecosystem. Annals of Plant Protection Sciences 7 (1), 1–7.

Nagaraj, K.S., Jaganath, S., Srikanth, L.G., Husainnaik, M., 2014. Attraction of fruit fly to different quantities of methyl eugenol in mango orchard. Trends in Biosciences 7 (11), 1113–1115.

Nandihalli, B.S., Patil, B.V., Hugar, P., 1993. Optimisation of pheromone traps for cotton pink bollworm, *Pectinophora gossypiella* Saunders. Karnataka Journal of Agricultural Sciences 6 (3), 300–301.

Nandre, A.S., Shukla, A., 2012. Composition of fruit fly, *Bactrocera* species in Navsari district in sapota. Uttar Pradesh Journal of Zoology 32 (2), 253–255.

Nassef, M.A., Hamid, A.M., Walson, W.M., 1999. Mass trapping of pink bollworm with gossyplure. Alexandria Journal of Agricultural Research 44 (1), 327–334.

Nemec, S., 2001. Field Trial for the Assessment of Attractant Qualities of Aminofeed®, Envirofeast® and Predfed® on Ingard® Cotton. Agri-Chem Research Report: Agrichem Manufacturing Industries Pty Ltd., Loganholme, Queensland, Australia.

Nesbitt, B.F., Beevor, P.S., Hall, D.R., Lester, R., 1979. Female sex pheromone components of the cotton bollworm, *Heliothis armigera*. Journal of Insect Physiology 25 (6), 535–541.

Neto, M.A., Habib, M., 1996. Evidence that mass trapping suppresses pink bollworm populations in cotton fields. Entomologia Experimentalis et Applicata 81 (3), 315–323.

Nordlund, D.A., Greenberg, S.M., 2002. Comparison of the influence of external treatments on oviposition by *Trichogramma pretiosum* and *Trichogramma minutum* (Hymenoptera: Trichogrammatidae) into stretched plastic artificial eggs. Subtropical Plant Science 54, 49–53.

Nordlund, D.A., Lewis, W.J., Jones, R.L., Gross Jr., H.R., 1976. Kairomones and their use for management of entomophagous insects. IV. Effect of kairomones on productivity and longevity of *Trichogramma pretiosum* Riley (Hymenoptera: Trichogrammatidae). Journal of Chemical Ecology 2, 67–72.

Nordlund, D.A., Lewis, W.J., Todd, J.W., Chalfan, R.B., 1977. Kairomones and their use for management of entomophagous insects VII. The involvement of various stimuli in the differential response of *Trichogramma pretiosum* Riley to two suitable hosts. Journal of Chemical Ecology 3, 513–518.

Norman, K., Siti Nurulhidayah, A., Othman, A., Mohd Basri, W., 2010. Pheromone mass trapping of bagworm moths, *Metisa plana* Walker (Lepidoptera: Psychidae), for its control in mature oil palms in Perak, Malaysia. Journal of Asia-Pacific Entomology 13 (2), 101–106.

Novo, R.J., Igarzabal, D., Vigilianco, A., Ruosi, G., Bracocmonte, E., Penaloza, C., 2000. Control of *Cydia molesta* (Busck.) (Lepidoptera: Olethreutidae) in Cordoba (Argenitne) by mating disrupion technique. AgriScientia 17, 29–34.

Nunez, S., 2000. Entomological outlook for IFP implementation in Uruguay. Acta Horticulturae 525, 363–365.

Obra, G.B., Resilva, S.S., Hendrichs, J., Pereira, R., 2013. Influence of adult diet and exposure to methyl eugenol in the mating performance of *Bactrocera philippinensis*. Journal of Applied Entomology 137 (1), 210–216.

Oehlschlager, A.C., Pierce, H.D., Morgan, B., Wimalaratne, P.D.C., Slessor, K.N., King, G.G.S., Gries, R., Bordan, J.H., Jirum, L.F., Chinchilla, C.M., Mexzan, R.G., 1992. Chirality and field activity of rhynchophorol, the aggregation pheromone of the American palm weevil. Naturwissenschaften 79 (3), 134–135.

Ogawa, K., 2000. Pest control by pheromone mating disruption and the role of natural enemies. Journal of Pesticide Science 25 (4), 456–461.

Ohtani, K., Witjaksono, Fukumoto, T., Mochizuki, F., Yamamoto, M., Ando, T., 2001. Mating disruption of the Japanese giant looper in tea gardens permeated with synthetic pheromone and related compounds. Entomologia Experimentalis et Applicata 100 (2), 203–209.

Okazaki, K., Arakarva, A., Noguchi, H., Mochizuki, F., 2001. Further studies on mating disruptants for the summer fruit tortrix moth, *Adoxophyes orana fasciata*. Japanese Journal of Applied Entomology and Zoology 45 (3), 137–141.

Olivero, J., Garcia, E.J., Ortiz, A., Questada, A., Sanchz, A., 2005. Essay of chemical control strategies against *Euzophera pinguis* (Haworth) (Lepidoptera: Pyralidae) based on the flight monitoring with sexual pheromones. Boletin de Sanidad Vegetal, Plagas 31 (3), 459–469.

Oltean, I., Ghizdavu, I., Perju, T., Bunescu, H., Bodis, I., Porca, M.M., Dinuta, A., Oprean, I., Gansca, L., Maxim, S., Ciotlaus, I., 2005. The horse chestnut leaf-miner, *Cameraria ohridella* Deschka-Dimic species monitoring with the aid of sexual attractants. Buletinul-Universitatii-de-Stiinte-Agricole-si-Medicina-Veterinara-Cluj-Napoca-Seria-Agriculture 61, 90–95.

Orankanok, W., Chinvinijkul, S., Sawatwangkhoung, A., Pinkaew, S., Orankanok, S., Hendrichs, J., Pereira, R., 2013. Methyl eugenol and pre-release diet improve mating performance of young *Bactrocera dorsalis* and *Bactrocera correcta* males. Journal of Applied Entomology 137 (1), 200–209.

Ortu, S., Cocco, A., Lentini, A., 2006. Utilisation of the sexual pheromones of *Planococus ficus* and *Planococcus citri* in vineyeards. Bulletin OILB/SROP 29 (11), 207–208.

Ortu, S., 1995. The protection of citrus fruit against Mediterranean fruit fly (*Ceratits capitata* Wieldemann). Informatore Fitorpathologica 45 (1), 10–16.

Ostrand, F., Wedding, R., Jirle, E., Anderbrant, O., 1999. Effect of mating disruption on reproductive behavior in the European pine sawfly, *Neodiprion sertifer* (Hymenoptera: Diprionidae). Journal of Insect Behavior 12 (2), 233–243.

Othman, A.K.M., Aulaqi, W.A., 2011. Efficiency of some colored pheromone traps for monitoring fruit fly, *Bactrocera zonata* (Saunders) (Tephritidae: Diptera) on mango crop in Abyan Governorate. Journal of Natural and Applied Sciences 15 (2), 299–307.

Ousmane, Z.M., Aboubacar, K., Correia, Z.A., da, C., Kadi, H.A.K., Abdourahamane, T.D., 2014. Agro-ecological management of mango fruit flies in the northern part of Guinea-Bissau. Journal of Applied Biosciences 75, 6250–6258.

Padmavathi, C., Paul, A.V.N., 1998. Saturated hydrocarbons as kairomonal source for the egg parasitoid, *Trichogramma chilonis* Ishii (Hym.: Trichogrammatidae). Journal of Applied Entomology 122, 29–32.

Papa, G., Silva, R.B., Almeida, F.J., 2000. Efficacy and total release interval of mating disruption pheromone on the control of pink bollworm *Pectinophora gossypiella* in cotton under field conditions in Brazil. In: Proceedings Beltwide Cotton Conferences, San Antonio, USA, 4–8 January 2000, vol. 2, pp. 1022–1024.

Paramasivan, A., Paul, A.V.N., Dureja, P., 2004. Kairomones of *Chilo partellus* (Swinhoe) and their impact on the egg parasitoid *Trichogramma chilonis* Ishii. Indian Journal of Entomology 66, 78–84.

Pasqualini, E., Natale, D., 1999. *Zeuzera pyrina* and *Cossus cossus* (Lepidoptera, Cossidae) control by pheromones: four years advances in Italy. Bulletin-OILB/SROP 22 (9), 115–124.

Patel, K.G., Desai, H.R., 2004. Monitoring rice yellow stem borer, *Scirpophaga incertulas* (Walker), using sex pheromone/light traps. Insect Environment 10 (2), 51–52.

Patil, S.B., Udikeri, S.S., Hirekurubar, R.B., Guruprasad, G.S., 2008. Management of pink bollworm, *Pectinophora gossypiella* (Saunders) through mass trapping in cotton. Pesticide Research Journal 20 (2), 210–213.

Paul, A.V.N., Singh, S., Singh, A.K., 2002. Kairomonal effect of some saturated hydrocarbons on the egg parasitoids *Trichogramma brasiliensis* (Ashmead) and *Trichogramma exiguum*, Pinto, Platner and Oatman (Hym., Trichogrammatidae). Journal of Applied Entomology 126, 409–416.

Peacock, J.W., Cuthbert, R.A., Gore, W.E., Lanier, G.N., Pearce, G.T., Silverstein, R.M., 1975. Collection on Porapak Q of the aggregation pheromone of *Scolytus multistriatus* (Coleoptera: Scolytidae). Journal of Chemical Ecology 1 (1), 149–160.

Pelozuelo, L., Fagih, A., Espahbodi, A., Genestier, G., Guenego, H., Malosse, C., Frerot, B., 2006. Efficiency of pheromone baited traps for monitoring of the European corn borer *Ostrinia nubilalis* (Lep.:Crambidae) in Mazandaran Province. Applied Entomology and Phytopathology 73 (2), 19–31.

Pereira, P., 1998. The use of pheromone traps for monitoring *Ephestia kuehniella* Zeller (Lepidoptera: Pyralidae) and detection of parasitoids in flour mills. Bulletin OILB-SROP 21, 111–117.

Perez, G., Sierra, J.M., 2006. Pheromone baits and traps effectiveness in mass trapping of *Ips acuminatus* Gyllenhal (Coleoptera: Scolytidae). Boletin de Sanidad Vegetal, Plagas 32 (2), 259–266.

Peterson, C., Fristad, A., Tsao, R., Coats, J.R., 2000. Osajin and pomiferin, two isoflavones purified from osage orange fruits, tested for repellency to the maize weevil Coleoptera: Curculionidae. Environmental Entomology 29 (6), 1133–1137.

Petrice, T.R., Haack, R.A., Poland, T.M., 2004. Evaluation of three trap types and five lures for monitoring *Hylurgus ligniperda* (Coleoptera: Scolytidae) and other local scolytids in New York. Great-Lakes-Entomologist 37 (1/2), 1–9.

Pickett, J.A., 1991. Gas chromatography-mass spectrometry in insect pheromone identification: three extreme case histories. In: Chromatography and Isolation of Insect Hormones and Pheromones. Chromatographic Society Symposium Series, pp. 299–309.

Pillai, K.S., Palaniswami, M.S., Rajamma, P., Ravindran, C.S., Premkumar, T., 1996. An IPM approach for sweet potato weevil. Tropical Tuber Crops: Problems, Prospects and Future Strategies 1996, 329–339.

Pitman, G.B., Vite, J.P., Kinzer, G.W., Fentiman Jr., A.F., 1969. Specificity of population aggregating pheromone of *Dendroctonus*. Journal of Insect Physiology 15, 363–366.

Polavarapu, S., Lonergan, G., Peng, H., Neilsen, K., 2001. Potential for mating disruption of *Sparganothis sulfureana*. Journal of Economic Entomology 94 (3), 658–665.

Ponnamma, K.N., Lalitha, N., Khan, A.S., 2002. Bioefficacy of the pheromone rhinolure in the integrated pest management for rhinoceros beetle, *Oryctes rhinoceros* L. a major pest of oil palm. In: Proceedings of the 15th Plantation Crops Symposium Placrosym XV, Mysore, India, 10–13 December 2002, pp. 525–530.

Powell, W., Hardie, J., Hick, A.J., Holler, C., Mann, J., Merritt, L., Nottingham, S.F., Wadhams, L.J., Witthinrich, J., Wright, A.F., 1993. Responses of the parasitoid *Praon volucre* (Hymenoptera: Braconidae) to aphid sex pheromone lures in cereal fields in autumn: Implications for parasitoid manipulation. European Journal of Entomology 90, 435–438.

Pozzati, M., Reggiani, A., Ferrari, R., Zucchi, L., Burzio, G., Furlan, L., 2006. Monitoring Elateridae with pheromone traps. Informatore Agrario 62 (3), 56–59.

Pramudi, M.I., Puspitarini, R.D., Rahardjo, B.T., 2013. Classification of fruit flies in South Kalimantan based on morphology and molecular characters. Journal of Tropical Life Science 3, 3.

Prasad, L.N., Neupane, F.P., 1992. Monitoring the chickpea borer, *Helicoverpa armigera* Hubner over period 1987/88–1989/90 by pheromone trap at Rampur, Chitwan, Nepal. ACIAR Food Legume Newsletter 17, 9.

Prasad, N., Rao, N.H.P., Mahalakshmi, M.S., 2009. Efficacy of PB Rope L as mating disruptant for management of pink bollworm, *Pectinophora gossypiella* Saunders in cotton. Journal of Plant Protection and Environment 6 (1), 6–10.

Quareles, W., 2000. Mating disruption success in codling moth IPM. IPM Practioner 22 (5/6), 1–12.

Qureshi, Z.A., Siddiqui, Q.H., Hussain, T., 1992. Field evaluation of various dispensers for methyl eugenol, an attractant of *Dacus zonatus* (Saund.) (Dipt., Tephritidae). Journal of Applied Entomology 113 (4), 365–367.

Radhika, P., Reddy, B.S., 2006. Management of pinkbollworm, *Pectinophora gossypiella* (Saunders) with PB ropel and IPM approach. Asian Journal of Bio Science 1 (2), 68–69.

Rajabalee, A., 1990. Management of *Chilo* spp. on sugar cane with notes on mating disruption studies with the synthetic sex pheromone of *C. sacchariphagus* in Mauritius. Insect Science and Its Application 11 (4/5), 825–836.

Ramsamy, M.P., Joomaye, A., Tawananshah, T., 1989. The field response of Mauritian tephritids to parapheromones. Revue Agricole et Sucrière de l'Île Maurice 68 (1–3), 70–75.

Rani, P.U., Kumari, S.I., Sriramakrishna, T., Sudhakar, T.R., 2007. Kairomones extracted from rice yellow stem borer and their influence on egg parasitization by *Trichogramma japonicum* Ashmead. Journal of Chemical Ecology 33, 59–73.

Rao, G.P., Sharma, S.R., Sing, P.K., 2000. Fungitoxic and insect repellent efficiency of limonene against sugarcane pests. Journal of Essential Oil Bearing Plants 3 (3), 157–163.

Rath, L.K., Dash, B., 2005. Evaluation of a nonchemical IPM module for the management of brinjal shoot and fruit borer. Vegetable Science 32 (2), 207–209.

Reddy, G.V.P., Urs, K.C.D., 1997. Mass trapping of diamondback moth *Plutella xylostella* in cabbage fields using synthetic sex pheromones. International Pest Control 39 (4), 125–126.

Reddy, G.V.P., Cruz, Z.T., Bamba, J., Muniappan, R., 2005. Development of a semiochemical based trapping method for the New Guinea sugarcane weevil, *Rhabdoscelus obscurus* in Guam. Journal of Applied Entomology 129 (2), 65–69.

Reji, R.O.P., Paul, A., Jiji, T., 2012. Field evaluation of methyl eugenol trap for the management of Oriental fruit fly, *Bactrocera dorsalis* Hendel (Diptera: Tephritidae) infesting mango. Pest Management in Horticultural Ecosystems 18 (1), 19–23.

Renou, M., Guerrero, A., 2000. Insect parapheromones in olfaction research and semiochemical-based pest control strategies. Annual Review of Entomology 48, 605–630.

Rhino, B., Dorel, M., Tixier, P., Risede, J.M., 2010. Effect of fallows on population dynamics of *Cosmopolites sordidus*: toward integrated management of banana fields with pheromone mass trapping. Agricultural and Forest Entomology 12 (2), 195–202.

Rizk, M.M.A., Abdel-Galil, F.A., Temerak, S.A.H., Darwish, D.Y.A., 2014. Factors affecting the efficacy of trapping system to the peach fruit fly (PFF) males, *Bactrocera zonata* (Saunders) (Diptera: Tephritidae). Archives of Phytopathology and Plant Protection 47 (4), 490–498 24.

Rizzi, I., Petacchi, R., Guidotti, D., 2005. Mass trapping technique in *Bactrocera oleae* control in Tuscany Region: results obtained at different territorial scale. Bulletin OILB/SROP 28 (9), 8390.

Rodriguez, S.C., Isaccs, R., Bleaw, B., 2013. Manipulation of natural enemies in aqueous agroecosystem: habitat and semiochemicals for sustainable insect pest control. In: Solanki, S., Laruamendy, M.L. (Eds.), Integrated Pest Management and Pest Control: Current and Future Tactics. InTechopen.com, pp. 89–126.

Ros, J.P., Wong, E., Olivero, J., Castillo, E., 2002. Improvements of traps, attractants and killing agents against *Ceratitis capitata* Wied. How to do the mass trapping technique a good way to control this pest? Boletin de Sanidad Vegetal, Plagas 28 (4), 591–597.

Ross, D.W., Daterman, G.E., Munson, A.S., Thier, R.W., 1996. Optimal dose of an antiaggregation pheromone (3-methycyclohex-2-en-1-one) for protecting live Douglas-fir from attack by *Dendroctonus pseudotsugae* (Coleoptera: Scolytidae). Journal of Economic Entomology 89 (5), 1204–1207.

Rotundo, G., Tremblay, E., Giacometti, R., 1979. Final results of mass captures of the citrophilous mealybug males (*Pseudococcus calceolariae* Mask.) (Homoptera Coccoidea) in a citrus grove. Bollettino del Laboratorio di Entomologia Agraria "Filippo Silvestri", Portici 36, 266–274.

Royer, J.E., De Faveri, S.G., Lowe, G.E., Wright, C.L., 2014. Cucumber volatile blend, a promising female-biased lure for *Bactrocera cucumis* (French 1907) (Diptera: Tephritidae: Dacinae), a pest fruit fly that does not respond to male attractants. Australian Entomology 53 (3), 347–352.

Rumine, P., 2002. *Cydia molesta* monitoring using sex pheromone traps. Italus-Hortus 9 (4), 34–38.

Ruther, J., Mayer, C.J., 2005. Response of garden chafer, *Phyllopertha horticola*, to plant volatiles: from screening to application. Entomologia Experimentalis et Applicata 115 (1), 51–59.

Ryan, M., 2007. Insect Chemoreception: Fundamental and Applied. Springer Science & Business Media, p. 330.

Saeed, M., Khan, S.M., Shahid, M., 2007. Effective monitoring of cigarette beetle, *Lasioderma serricorne* (F.) Coleoptera: Anobiidae in tobacco warehouses of NWFP- Pakistan. Sarhad Journal of Agriculture 23 (1), 123–128.

Sah, L.N., Sahu, R., Neupane, F.P., 1988. Monitoring the chickpea pod-borer, *Heliothis armigera* Hubner by a pheromone trap. Journal of the Institute of Agriculture and Animal Science, Nepal 9, 107–109.

Salem, S.A., Radwan, S.M.E., Hamaky, M.A., 1990. Prospects of using sex pheromone for the control of cotton boll-worms, *Earias insulana* Boisd. and *Pectinophora gossypiella* Saund. in cotton fields. Annals of Agricultural Science, Moshtohor 28 (3), 1743–1752.

Sallam, M.N., Peck, D.R., Mc Avoy, C.A., Donald, D.A., 2007. Pheromone trapping of the sugarcane weevil borer, *Rhabdoscelus obscurus* (Boisduval) (Coleoptera: Curculionidae): an evaluation of trap design and placement in the field. Australian Journal of Entomology 46 (3), 217–223.

Sasaki, F., Fuke, S., Masuda, H., 1985. Notes on some fruit flies collected by lure trapping and host plant survey. Research Bulletin of the Plant Protection Service 21, 1–16.

Schmidt, T.A., Louis, F., Fukumoto, T., Zebity, C.P.W., Arn, H., 1999. Mating disruption tests to control *Sparganothis pilleriana* (Schiff.) (Lepidoptera: Tortricidae). Bulletin OKB/SROP 22 (9), 77–82.

Schmidt, T.A., Louis, F., Zebitz, C.P.W., Arn, H., 2000. First steps to control *Sparganothis pilleriana* (Schiff.) by the use of pheromone. Bulletin-OILB/SRO 23 (4), 163–166.

Schneider, D., Block, B., Boeckh, J., Priesner, E., 1967. Die Reaktionder männlichen Seidenspinner auf Bombykol und seine Isomeren: Elektroantennogramm und Verhalten. Zeitschrift für vergleichende Physiologie 54, 192–209.

Schneider, D., 1957. Elektrophysiologische Untersuchungen von Chemo- und Mechanoreceptoren de Antenne des Seidenspinners *Bombyx mori* L. Zeitschrift für vergleichende Physiologie 40, 8–41.

Schroeder, P.C., Shelton, A.M., Ferguson, C.S., Hoffmann, M.P., Petzoldt, C.H., 2000. Application of synthetic sex pheromone for management of diamondback moth, *Plutella xylostella*, in cabbage. Entomologia Experimentalis et Applicata 94 (3), 243–248.

Schulz, S., 2005. The Chemistry of Pheromones and Other Semiochemicals II. Springer Science & Business Media, p. 333.

Schwarcz, J., 2003. Why Moths Are Confused Bachelors? , p. 6 The Gazette December 14.

Scriber, J.M., 1984. Host plant suitability. In: Bell, W.J., Carde, R.T. (Eds.), Chemical Ecology of Insects. Chapman & Hall, London, pp. 3–36.

Setokuchi, O., Nakamura, Y., Kubo, Y., 1991. A fibreboard formulation of the sweet-potato weevil, *Cylas formicarius* (Fabricius), sex pheromone and an insecticide for mass-trapping. Japanese Journal of Applied Entomology and Zoology 35 (3), 251–253.

Sexton, S.B., Il'ichev, A.L., 2000. Pheromone mating disruption with reference to oriental fruit moth *Grapholita molesta* (Busck). (Lepidoptera: Tortricidae) literature review. General and Applied Entomology 29, 63–68.

Sexton, S.B., Il'ichev, A.L., 2001. Release rate characteristics of two pheromone formulations for control of oriental fruit moth *Grapholita molesta* (Busck) (Lepidoptera: Tortricidae) by mating disruption. General and Applied Entomology 30, 31–34.

Shah, A.H., Patel, R.C., 1976. Role of tulsi plant (*Ocimum sanctum*) in control of mango fruitfly, *Dacus correctus* Bezzi (Tephritidae: Diptera). Current Science 45 (8), 313–314.

Shani, A., Klug, J.T., 1980. Sex pheromone of Egyptian cotton leafworm (*Spodoptera littoralis*): Its chemical transformations under field conditions. Journal of Chemical Ecology 6 (5), 875–881.

Shankar, M., Rao, S.R.K., Umamaheswari, T., Reddy, K.D., 2011. Influence of methyl eugenol soaked wood blocks on the capturing efficiency of mango fruit flies, *Bactrocera* spp. Journal of Plant Protection and Environment 8 (2), 106–107.

Shelly, T.E., Dewire, F., 1994. Chemical mediated mating success in male oriental fruit flies (Diptera: Tephritidae). Annals of Entomological Society of America 87 (3), 375–382.

Shelly, T.E., Kurashima, R., Nishimoto, J., Diaz, A., Leathers, J., War, M., Joseph, D., Elsevier B.V., 2011. Capture of *Bactrocera* fruit flies (Diptera: Tephritidae) in traps baited with liquid versus solid formulations of male lures. Journal of Asia-Pacific Entomology 14 (4), 463–467.

Shelly, T., Renshaw, J., Dunivin, R., Morris, T., Giles, T., Andress, E., Diaz, A., War, M., Nishimoto, J., Kurashima, R., 2013. Release-recapture of *Bactrocera* fruit flies (Diptera: Tephritidae): comparing the efficacy of liquid and solid formulations of male lures in Florida, California and Hawaii. Florida Entomologist 96 (2), 305–317.

Shelly, T., Nishimoto, J., Kurashima, R., 2014. Distance-dependent capture probability of male Mediterranean fruit flies in trimedlure-baited traps in Hawaii. Journal of Asia-Pacific Entomology 17 (3), 525–530.

Shelly, T.E., 1995. Methyl eugenol and the mating competitiveness of irradiated male *Bactrocera dorsalis* (Diptera: Tephritidae). Annals of Entomological Society of America 88 (6), 883–886.

Shen, B.J., Qio, H.G., Qio, Z.T., Fu, W.J., 1991. Behavioural responses of *Trichogramma dendrolimi* Matsumura to different plants. Kunchong Zhi Shi 28, 359–361.

Sheng, C.F., FuAn, Y., YongBao, W., ChunQiang, Z., YanWen, X., 2000. Field trials for mass trapping of rice stem borer *Chilo suppressalis* by sex pheromone. Plant Protection 26 (5), 4–5.

Sheng, C.F., Wang, W.D., Jiao, X.G., Song, F.B., Xuan, W.J., 2002. A preliminary study on effect of applying sex pheromone into mass trapping of rice stem borer, *Chilo suppressalis*. Journal of Jilin Agricultural University 24 (5), 58–61, 65.

Shinoda, K., 2007. Pheromone traps for monitoring stored product insects: their present situation and future issues. Aroma Research 8 (1), 22–26.

Shoray, H.H., Gaston, L.K., 1967. Pheromones. In: Kilgore, W.W., Pant, R.C. (Eds.), Pest Control: Biological, Physical and Selected Chemical Methods. Academic Press, New York, pp. 241–265.

Shorey, H.H., 1977. Manipulation of insect pests of agricultural crops. In: Shorey, H., McKelvey, J.J. (Eds.), Chemical Control of Insect Behaviour: Theory and Application. Wiley, New York, pp. 353–367.

Shukla, R.P., Prasad, V.G., 1985. Population fluctuations of the oriental fruit fly, *Dacus dorsalis* Hendel in relation to hosts and abiotic factors. Tropical Pest Management 31 (4), 273–275.

Simon, S., Corroyer, N., Getti, F.X., Girard, T., Combe, F., Faurial, J., Bussi, C., 1999. Organic farming optimization of techniques. Arboriculture-Fruitiere 533, 27–32.

Singh, S., Sharma, D.R., 2013. Management of fruit flies in rainy season guava through male annihilation technique using methyl eugenol based traps. Indian Journal of Horticulture 70 (4), 512–518.

Singh, S., Paul, A.V.N., Dureja, P., Singh, A.K., 2002. Kairomones of two host insects and their impact on the egg parasitoids *Trichogramma brasiliensis* (Ashmead) and *T. exiguum* Pinto, Platner and Oatman. Indian Journal of Entomology 64, 96–106.

Singh, M., Gupta, D., Gupta, P.R., 2013. Population suppression of fruit flies (*Bactrocera* spp.) in mango (*Mangifera indica*) orchards. Indian Journal of Agricultural Sciences 83 (10), 1064–1068.

Singh, S., Sharma, D.R., Kular, J.S., 2015. Eco-friendly management of fruit flies in peach with methyl eugenol based traps in Punjab. Agricultural Research Journal 52 (1), 47–49.

Smetnik, A.I., Rizinskya, E.M., 1988. A new method is introduced. Zashchita Rastenii (Moskya) 3, 42–43.

Smit, N.E.J.M., Downham, M.C.A., Laboke, P.O., Hall, D.R., Odongo, B., 2001. Mass-trapping male *Cylas* spp. with sex pheromones: a potential IPM component in sweet potato production in Uganda. Crop Protection 20 (8), 643–651.

Smith, E.S.C., 1977. Studies on the biology and commodity control of the banana fruit fly, *Dacus musae* (Tryon), in Papua New Guinea. Papua New Guinea Agricultural Journal 28 (2/4), 47–56.

Sohi, A.S., Singh, J., Brar, D.S., Russell, D., 1999. Further studies on mating disruption in pink bollworm *Pectinophora gossypiella* (Saunders) using sex pheromone as a component of IPM program in irrigated cotton fields in Punjab. Pest management and Economic Zoology 7 (1), 31–38.

Sonda, M., Ichinohe, F., 1984. Eradication of the oriental fruit fly from Okinawa island and its adjacent islands. Japan Pesticide Information 44, 1–6.

Soroker, V., Blumberg, D., Haberman, A., Hamburger Rishard, M., Reneh, S., Talebaev, S., Anshelevich, L., Harari, A.R., 2005. Current status of red palm weevil infestation in date palm plantations in Israel. Phytoparasitica 33 (1), 97–106.

Srinivas, K., Rao, P.A., 1999. Management of *Spodoptera litura* (F.) infesting groundnut by mating disruption technique with synthetic sex pheromone. Journal of Entomological Research 23 (2), 115–119.

Srivastava, C.P., Pimbert, M.P., Ahmad, K., 1991. Monitoring of *Helicoverpa armigera* (Hubner) in Pakistan with pheromone traps. International Chickpea Newsletter 25, 20–22.

Stegeman, B.R., Rice, M.J., Hooper, G.H.S., 1978. Daily periodicity in attraction of male tephritid fruit flies to synthetic chemical lures. Journal of the Australian Entomological Society 17 (4), 341–346.

Su, J.W., Xuan, W.J., Sheng, C.F., Ge, F., 2003. The sex pheromone of rice stem borer, *Chilo suppressalis* in paddy fields: suppressing effect of mass trapping with synthetic sex pheromone. Chinese Journal of Rice Science 17 (2), 171–174.

Suckling, D.M., Charles, J.G., Allen, D., Stevens, P.S., 2002. Possibility of control of painted apple moth (*Teia anartoides*) using single component mating disruption. New Zealand Plant Protection 55, 1–6.

Sudo, K., Senbongi, I., Takahashi, Y., Maehara, H., Ogura, Y., Nomura, Y., 1998. Control of rice stem borer, *Chilo suppressalis* (Walker) by communication disruption using its sex pheromone in large fields [in Japan]. Proceedings of the Kanto Tosan Plant Protection Society 45, 159–161.

Suk Ling, W., Shelly, T., Elsevier B.V., 2013. Capture of *Bactrocera* fruit flies in traps baited with liquid versus solid formulations of male lures in Malaysia. Journal of Asia-Pacific Entomology 16 (1), 37–42.

Suss, L., Locatelle, D.P., Marrone, R.V., 1999. Mating suppression of the Mediterranean flour moth (*Ephestia kuehniella* Zeller) (Lepidoptera: Pyralidae) in a food industry. Bollettino Di Zoologia Agraria E- Di- Bach 31 (1), 59–66.

Tabata, J., Narai, Y., Sawamura, N., Hiradate, S., Sugie, H., 2012. A new class of mealybug pheromones: a hemiterpene ester in the sex pheromone of *Crisicoccus matsumotoi*. Naturwissenschaften 99 (7), 567–574.

Tabrizian, M., Mohammadpoor, K., Tabak, S.N., 2009. Synthesis and field evaluation of aggregation pheromone of date palm fruit stalk borer, *Oryctes elegans*. Applied Entomology and Phytopathology (Special), 51–60.

Talekar, N.S., Lee, S.T., 1989. Studies on the utilization of a female sex pheromone for the management of sweet potato weevil, *Cylas formicarius* (F.) (Coleoptera: Curculionidae). Bulletin of the Institute of Zoology, Academia Sinica 28 (4), 281–288.

Tamhankar, A.J., Rajendran, T.P., Rao, N.H., Lavekar, R.C., Jeyakumar, P., Monga, D., Bambawale, O.M., 2003. Variability in response of *Helicoverpa armigera* males from different locations in India to varying blends of female sex pheromone suggests male sex pheromone response polymorphism. Current Science 84, 448–450.

Tan, K.H., Jaal, Z., 1986. Comparison of male adult population densities of the oriental and *Artocarpus* fruit-flies, *Dacus* spp. (Diptera: Tephritidae), in two nearby villages in Penang, Malaysia. Researches on Population Ecology 28 (1), 85–89.

Tan, K.H., Lee, S.L., 1982. Species diversity and abundance of *Dacus* (Diptera: Tephritidae) in five ecosystems of Penang, West Malaysia. Bulletin of Entomological Research 72 (4), 709–716.

Tanaka, F., Yabuki, S., Tatsuki, S., Tsumuki, H., Kanno, H., Hattori, M., Usui, K., Kurihara, M., Uchiumi, K., Fukami, J., 1987. Control effect of communication disruption with synthetic pheromones in paddy fields in the rice stem borer, *Chilo suppressalis* (Walker) (Lepidoptera: Pyralidae). Japanese Journal of Applied Entomology and Zoology 31 (2), 125–133.

Tang, R., Zhang, J.P., Zhang, Z.N., 2012. Electrophysiological and behavioral responses of male fall webworm moths (*Hyphantria cunea*) to herbivory-induced mulberry (*Morus alba*) leaf volatiles. PLoS One 7 (11), e49256.

Teli, V.S., Salunkhe, G.N., 1993. Monitoring of adults of sweet potato weevil, m, *Cylas formicarius* with sex pheromone. Journal of Insect Science 6 (2), 283–284.

Teshiba, M., Shimizu, N., Sawamura, N., Narai, Y., Sugie, H., Sasaki, R., Tabata, J., Tsutsumi, T., 2009. Use of a sex pheromone to disrupt the mating of *Planococcus kraunhiae* (Kuwana) (Hemiptera: Pseudococcidae). Japanese Journal of Applied Entomology and Zoology 53 (4), 173–180.

Thakur, M., Gupta, D., Singh, M., 2013. Monitoring of fruit fly populations using parapheromone in peach orchards. Journal of Insect Science 26 (1), 34–41.

Thorpe, K.W., Leonard, D.S., Mastro, V.C., Mc Lane, W., Reardon, R.C., Seller, P., Webb, R.E., Talley, S.E., 2000. Effectiveness of gypsy moth mating disruption from aerial applications of plastic laminate flakes with and without a sticking agent. Agricultural and Forest Entomology 2 (3), 225–231.

Torres Vila, L.M., Rodriguez Molina, M.C., Stockel, J., 2002. Delayed mating reduces reproductive output of female European grapevine moth, *Lobesia botrana* (Lepidoptera: Tortricidae). Bulletin of Entomological Research 92 (3), 241–249.

Toth, M., Bakcsa, F., Csonka, E., Szarukan, I., Benedek, P., 2003. Species spectrum of flea beetles (*Phyllotreta* spp., Coleoptera, Chrysomelidae) attracted to alkyl isothiocyanate-baited traps in Hungary. In: 3rd International Plant Protection Symposium at Debrecen University and 8th Trans-Tisza Plant Protection Forum, 15–16 October 2003, Debrecen, Hungary, pp. 154–156.

Tóth, M., Viana, P.A., Vilela, E., Domingue, M.J., Baker, T.C., Vuts, J., 2014. KLP+ ("hat") trap with semiochemical lures suitable for trapping two *Diabrotica* spp. exotic to Europe. Acta Phytopathologica et Entomologica Hungarica 49 (2), 211–221.

Toyoshima, G., Kobayashi, S., Yoshihama, T., 2001. Control of *Helicoverpa armigera* (hubner) by mating disruption using diamolure in lettuce fields. Japanese Journal of Applied Entomology and Zoology 45 (4), 183–188.

Trematerra, P., Gentile, P., 2010. Five years of mass trapping of *Ephestia kuehniella* Zeller: a component of IPM in a flour mill. Journal of Applied Entomology 134 (2), 149–156.

Trematerra, P., Sciarretta, A., Gentile, P., 2000. Trials on combined mating disruption in *Anarsia lineatella* (Zeller) and *Cydia molesta* (Busck) using CheckMateReg. SF Dispenser. In: GF 2000 Atti, Giornate Fitopatologiche, Perugia, 16–20 April 2000, Volume Primo, pp. 349–354.

Trematterra, P., Athanassion, C.G., Sciarretta, A., Kavallieratos, N.G.M., Buchelos, T.C., 2013. Efficacy of the autoconfusion system for mating disruption of *Ephestia kuehniella* (Zeller) and *Plodia interpunctella* (Hubner). Journal of Stored Products Research 55, 90–98.

Trent, A., Thistle, H., 1999. Aerial application of gypsy moth pheromone flakes and sticker. In: ASAE/CSAE SCGR Annual International Meeting, Toronto, Ontario, Canada, 18–21 July 1999, p. 9.

Trimble, R.M., Pree, D.J., Carter, N.J., 2001. Integrated control of oriental fruit moth (Lepidoptera: Tortricidae) in peach orchards using insecticide and mating disruption. Journal of Economic Entomology 94 (2), 476–485.

Ueno, K., Hayasaka, T., 1997. Control effect on mating disruption of rice stem borer, *Chilo suppressalis* (Walker) (Lepidoptera: Pyralidae) with synthetic pheromones in paddy fields [in Japan]. Annual Report of the Society of Plant Protection of North Japan 48, 159–163.

Ujvary, I., Dickens, J.C., Kamm, J.A., McDonough, L.M., 1993. Natural product analogs: stable mimics of aldehyde pheromones. Archives of Insect Biochemistry and Physiology 22, 393–411.

Ukey, N.S., Chandele, A.G., Wagh, S.S., Bansode, G.M., 2013. Species composition of fruit flies, *Bactrocera* spp. (Diptera: Tephritidae) infesting guava in Maharashtra. Pest Management in Horticultural Ecosystems 19 (2), 242–244.

Umeya, K., Hirao, J., 1975. Attraction of the jackfruit fly, *Dacus umbrosus* F. (Diptera: Tephritidae) and lace wing, *Chrysopa* sp. (Neuroptera: Chrysopidae) by lure traps baited with methyl eugenol and cue-lure in the Philippines. Applied Entomology and Zoology 10 (1), 60–62.

Unelius, C.R., El-Sayed, A.M., Twidle, A., Bunn, B., Zaviezo, T., Flores, M.F., Bell, V., Bergmann, J., 2011. The absolute configuration of the sex pheromone of the citrophilous mealybug, *Pseudococcus calceolariae*. Journal of Chemical Ecology 37 (2), 166–172.

Ushio, S., Yoshioka, K., Nakasu, K., Waki, K., 1982. Eradication of the oriental fruit fly from Amami islands by male annihilation (Diptera: Tephritidae). Japanese Journal of Applied Entomology and Zoology 26 (1), 1–9.

Vaissiere, B., Matti, M., Morison, N., Sauphanor, B., 2001. Mating disruption and pollinating insects. Proof of the harmlessness of this codling moth control method. Phytoma 534, 44–48.

Vargas, R.I., Stark, J.D., Hertlein, M., Neto, A.M., Coler, R., Piñero, J.C., 2008. Evaluation of SPLAT with spinosad and methyl eugenol or cue-lure for "attract-and-kill" of oriental and melon fruit flies (Diptera: Tephritidae) in Hawaii. Journal of Economic Entomology 101 (3), 759–768.

Vargas, R.I., Souder, S.K., Mackey, B., Cook, P., Morse, J.G., Stark, J.D., 2012. Field trials of solid triple lure (trimedlure, methyl eugenol, raspberry ketone, and DDVP) dispensers for detection and male annihilation of *Ceratitis capitata*, *Bactrocera dorsalis*, and *Bactrocera cucurbitae* (Diptera: Tephritidae) in Hawaii. Journal of Economic Entomology 105 (5), 1557–1565.

Vargas, R.I., Souder, S.K., Hoffman, K., Mercogliano, J., Smith, T.R., Hammond, J., Davis, B.J., Brodie, M., Dripps, J.E., 2014. Attraction and mortality of *Bactrocera dorsalis* (Diptera: Tephritidae) to STATIC Spinosad ME weathered under operational conditions in California and Florida: a reduced-risk male annihilation treatment. Journal of Economic Entomology 107 (4), 1362–1369.

Varma, N.R.G., Krishnaiah, K., Pasalu, I.C., Rao, P.R.M., 2004. Influence of field size in management of yellow stem borer (YSB), *Scirpophaga incertulas* Walker through pheromone mediated mass trapping in rice. Indian Journal of Plant Protection 32 (1), 39–41.

Varner, M., Mattedi, L., Rizzi, C., Mescalchin, E., 2001. Mating disruption in viticulture: experiences in Trentino. Informatore Fitopatologico 51 (10), 23–29.

Vayssières, J.F., Sinzogan, A., Adandonon, A., Rey, J.Y., Dieng, E.O., Camara, K., Sangaré, M., Ouedraogo, S., Hala, N., Sidibé, A., Keita, Y., Gogovor, G., Korie, S., Coulibaly, O., Kikissagbé, C., Tossou, A., Billah, M., Biney, K., Nobime, O., Diatta, P., N'Dépo, R., Noussourou, M., Traoré, L., Saizonou, S., Tamo, M., 2014. Annual population dynamics of mango fruit flies (Diptera: Tephritidae) in West Africa: socio-economic aspects, host phenology and implications for management. Les Ulis Cedex A, France, Fruits (Paris) 69 (3), 207–222.

Venkatachalam, A., Sithanantham, S., Kumar, S.S., Mathivanan, N., Ramkumar, D., Janarthanan, S., Marimuthu, T., 2014. Seasonal catches of fruit flies in traps with two lure sources in noni. Indian Journal of Entomology 76 (4), 317–320.

Verghese, A., Jayanthi, P.D.K., 2001. A convenient polythene sachet trap for fruit flies, *Bactorcera* spp. Insect Environment 6 (4), 191.

Verghese, A., Sreedevi, K., Kalleshwaraswamy, C.M., 2005. Pest management in horticultural ecosystems. Special Issue on Fruit Fly 11 (2), 81–180.

Verghese, A., Sreedevi, K., Nagaraju, D.K., 2007. Pre and post harvest IPM for the mango fruit fly *Bactrocera dorsalis* (Hendel) - fruit flies of economic importance: from basics to applied knowledge. In: Proc 7th International Symposium of Fruit Flies of Economic Importance, 10–15 September 2006, Salvadar, Brazil, pp. 179–182.

Verghese, A., Shivananda, T.N., Jayanthi, P.D.K., Sreedevi, K., 2013. Frank Milburn Howlett (1877–1920): discoverer of Pied Pipers' lure for the fruit flies. Current Science 105 (2), 260–262.

Verhaeghe, A., Penet, C., Garcin, A., 2001. Alternative methods of controlling the codling moth. Acta Horticulturae 544, 399–404.

Verma, A.K., Sankhyan, S., 1993. Pheromone monitoring of *Heliothis armigera* (Hubner) and relationship of moth activity with larval infestation on important cash crops in mid-hills of Himachal Pradesh. Pest Management and Economic Zoology 1 (1), 43–49.

Vernon, R.S., Toth, M., 2007. Evaluation of pheromones and a new trap for monitoring *Agriotes lineatus* and *Agriotes obscurus* in the Fraser valley of British Columbia. Journal of Chemical Ecology 33 (2), 345–351.

Villa-Gil, F., Sasot-Bayona, J.A., Balduque-Martin, R., Casanova, G.J., Valencia-Sancho, G., 1998. Study on the control of codling moth, *Cydia pomonella*, in organic farming of pear and apple in Aragon. Georgica 6, 143–151.

Visalakshmi, V., Rao, P.A., Krishnayya, P.V., 2000. Utility of sex pheromone for monitoring *Heliothis armigera* (Hub.) infesting sunflower. Journal of Entomological Research 24 (3), 255–258.

Voigt, E., Toth, M., 2004. Protection of peach fruits against the scarabs *Anomala vitis* Fabr. and *A. dubia* Scop. (Coleoptera, Scarabaeidae, Melolonthinae) through mass trapping: a three year study. Novenyvedelem 40 (7), 329–332.

Vrabl, S., Matis, G., 1999. A contribution to the investigation of rape berry moth mating disruption. In: Zbornik-predavanj-in-referatov-4-Slovenskega-Posvetovanja-o-Varstvu-Rastlin-v-Portorozu-od-3-do-4-Marca-1999, pp. 295–300.

Vuts, J., Imrei, Z., Toth, M., 2007. Improving the field activity of the synthetic floral bait in *Cetonia a. aurata* and *Potosia cuprea* (Coleoptera: Scarabaeidae: Cetoniinae). In: 12 Tiszantuli Novenyvedelmi Forum, 17–18 October 2007, Debrecen, Hungary, pp. 176–183.

Waldner, W., 2001. Pheromones in pome tree control. Informatore Fitopatologico 51 (10), 10–14.

Walker, K.R., Welter, S.C., 2001. Potential for outbreaks of leafrollers (Lepidoptera: Tortricidae) in California apple orchards using mating disruption for codling moth suppression. Journal of Economic Entomology 94 (2), 373–380.

Walton, V.M., Daane, K.M., Pringle, K.L., 2004. Monitoring *Planococcus ficus* in South African vineyards with sex pheromone-baited traps. Crop Protection 23, 1089–1096.

Wang, X.P., Le, V.T., Fang, Y.L., Zhang, Z.N., 2004. Trap effect on the capture of *Plutella xylostella* (Lepidoptera: Plutellidae) with sex pheromone lures in cabbage fields in Vietnam. Applied Entomology and Zoology 39 (2), 303–309.

Wang, H.L., Zhao, C.H., Wang, C.Z., 2005. Comparative study of sex pheromone composition and biosynthesis in *Helicoverpa armigera*, *H. assulta* and their hybrid. Insect Biochemistry and Molecular Biology 35 (6), 575–583.

Wang, Y., Wu, H., Liu, C., Ren, C., Zou, Y., 2008. Mass trapping of the *Spodoptera litura* moth with synthetic sex pheromone lures and control efficacy in the fields. Acta Phytophylacica Sinica 5, 475–476.

Weber, D.C., Robbins, P.S., Averill, A.L., 2005. *Hoplia equina* (Coleoptera: Scarabaeidae) and nontarget capture using 2 tetra decanone baited traps. Environmental Entomology 34 (1), 158–163.

Weems, H.V., 1987. Guava fruit fly, *Dacus* (Strumeta) *correctus* (Bezzi) (Diptera: Tephritidae). Florida Department of Agriculture and Consumer Services 291, 4.

Welty, C., Alcaraz, S., 1995. Integrating Biological and Chemical Control of Sweet Corn Pests. http://ipm.osu.edu/mini/95m-5.htm.

Whitaker, R.H., 1970. The biochemical ecology of higher plants. In: Sondheimer, E., Simeone, J.B. (Eds.), Chemical Ecology. Academic Press, New York, pp. 43–70.

Wijk, M., van Bruijin, P.J.O., Sabelis, M.W., 2008. Predatory mite attraction to herbivore-induced plant odours is not a consequence of attraction to individual herbivore-induced plant volatiles. Journal of Chemical Ecology 34, 791–803.

Win, N., Mi, K.M., Oo, T.T., Win, K.K., Park, J.Y., Park, J.K., 2014. Occurrence of fruit flies (Diptera: Tephritidae) in fruit orchards from Myanmar. Korean Journal of Applied Entomology 53 (4), 323–329.

Witzgall, P., Ktisch, P., Cork, A., 2010. Sex pheromones and their impact on pest management. Journal of Chemical Ecology 36 (1), 80–100.

Witzgall, P., 2005. Use of pheromones and other semiochemicals in integrated pest control. Annals of Warsaw Agricultural University, Horticulture and Landscape Architecture 26 (5), 5–11.

Wu, D.M., Yan, Y.H., Cui, J.R., Liu, M.Y., Ren, D.F., Chen, A., Zhang, Q., 1990. Sex pheromone of sugarcane tip borer *Scirphophaga excerptalis* Walker. Chinese Science Bulletin 35, 1469–1473.

Wyatt, T.D., 2003. Pheromones and Animal Beahaviour: Communication by Smell and Taste. Cambridge University Press, UK, p. 391.

Wyatt, T.D., 2014. Pheromones and Animal Behaviour. Chemical Signals and Signatures. Cambridge University Press, p. 405.

Xiao, X.H., Liu, C., Wu, H.H., Yang, C.H., Li, H., 2013. Study on species and occurrence regulation of fruit flies in Xuwen County of Chongqing, China. In: The Proceedings of Chinese Society of Plant Protection in 2013, Shandong, China, 22–25 October, pp. 128–134.

Xie, A.S., Lv, J.S., 2012. An improved lure for trapping the bark beetle *Dendroctonus armandi* (Coleoptera: Scolytinae). European Journal of Entomology 109 (4), 569–577.

Xu, Y.X., Sun, X.G., He, Z., Liu, X.H., Ge, F., 2008. Electroantennogram responses of *Carcelia matsukarehae* to the volatiles of *Pinus massoniana* damaged by *Dendrolimus punctatus*. Chinese Bulletin of Entomology 43, 319–322.

Yadav, B., Paul, A.V.N., Gautam, R.K., 2008. Kairomonal effect of different combinations of synthetic kairomones on the egg parasitoids *Trichogramma exiguum*, Pinto and Platner and *Trichogramma japonicum* (Ashmead). Indian Journal of Entomology 70, 68–70.

Yadav, B., Paul, A.V.N., Gopal, M., Gautam, R.K., 2005. Kairomonal effect of some saturated fatty acids on the egg parasitoids *Trichogramma exiguum*, Pinto and Platner and *Trichogramma japonicum* (Ashmead). Indian Journal of Entomology 67, 7–11.

Yang, F., Sheng, C.F., Wei, Y.B., Zhu, C.Q., Xiong, Y.W., 2001a. Study on mating disruption caused by synthetic sex pheromones for controlling rice stem borer. Plant Protection 27 (3), 4–6.

Yang, K.S., Cheng, F.R., Pan, Z.Y., Ling, J.Y., Song, J.X., Zhu, J.B., 2001b. Experiments on the control of pink bollworm with the sex pheromone PB ROPEL. China Cottons 28 (5), 10–12.

Yasuda, T., Higuchi, H., 2012. Sex pheromone of *Stenotus rubrovittatus* and *Trigonotylus caelestialium*, two mirid bugs causing pecky rice, and their application in insect monitoring in Japan. Psyche: A Journal of Entomology 2012 Article id 435640.

Yasuda, K., 1995. Mass trapping of the sweet potato weevil *Cylas formicarius* (Fabricius) (Coleoptera: Brentidae) with a synthetic sex pheromone. Applied Entomology and Zoology 30 (1), 31–36.

Yatagai, M., Makihara, H., Oba, K., 2002. Volatile components of Japanese cedar cultivars as repellents related to resistance to *cryptomeria* bark borer. Journal of Wood Science 48 (1), 51–55.

Zada, A., Dunkelblum, E., Assael, F., Harel, M., Cojocaru, M., Mendel, Z., 2003. Sex pheromone of the vine mealybug, *Planococcus ficus* in Israel: occurrence of a second component in a mass-reared population. Journal of Chemical Ecology 29, 977–988.

Zaki, F.N., El-Saadany, G., Gomma, A., Saleh, M., 1998. Increasing rates of parasitism of the larval parasitoid, *Bracon brevicornis* (Hymenoptera: Braconidae) by using kairomones, pheromones and a supplementary food. Journal of Applied Entomology 122, 565–567.

Zaki, F.N., 1996. Effect of some kairomones and pheromones on two hymenopterous parasitoids, *Apanteles ruficrus* and *Mircoplitis demolitor* (Hymenoptera: Braconidae). Journal of Applied Entomology 120, 555–557.

Zhang, A.J., Nie, J.Y., 2005. Enantioselective synthesis of the female sex pheromone of the pink hibiscus mealybug, *Maconellicoccus hirsutus*. Journal of Agriculture and Food Chemistry 53 (7), 2451–2455.

Zhang, Z., Wang, H., Chen, G., Anderbrant, O., Zhang, Y., Zhou, S., Hedenstorm, E., Hogberg, H.E., 2005. Sex pheromone for monitoring flight periods and population densities of the saw fly, *Diprion jingyanensis* Xiao et Zhang (Hym.: Diprionidae). Journal of Applied Entomology 129 (7), 368–374.

Zingg, D., Hohn, H., 2000. Confusion method in viticulture. Obst-und-wenbau 13B (26), 668–671.

Zuber, M., 1999. On the increase of mating disruption in arboriculture and viticulture in Switzerland, 1996–1998. Bulletin OILB/SROP 22 (9), 125–127.

Further Reading

Beevor, P.S., David, H., Jones, O.T., 1990. Female sex pheromones of *Chilo* spp. (Lepidoptera: Pyralidae) and their development in pest control applications. Insect Science and its applications 11, 787–794.

Bhanu, K.R.M., Hall, D.R., Mathew, T., Malvika, C., Prabhakara, M.S., Awalekar, R.V., Jayanth, K.P., 2011. Monitoring of *Opisina arenosella* by using female sex pheromones. APACE 39.

Mochida, O., Arida, G.S., Tatsuki, S., Fukami, J., 1984. A field test on a 3rd component of the female sex pheromone of the rice stem borer moth, *Chilo suppressalis*, in the Philippines. Entomologia Experimetnalis et Applicata 36, 295–296.

Sakagami, A., 2000. Recent advances in fruit tree pest management with pheromones in Japan. Bulletin of the National Institute of Fruit Tree Science 34, 17–42.

CHAPTER

20

Insect Hormones (as Pesticides)

S. Subramanian, K. Shankarganesh

Division of Entomology, Indian Agricultural Research Institute, New Delhi, India

1. INTRODUCTION

Growth and development of insects is punctuated by stages or instars. The development of the insect from one stage to another is regulated by hormones, including prothoracicotropic hormones (PTTHs), ecdysteroids, and juvenile hormones (JHs). The JH inhibits the genes that promote development of adult characteristics causing the insect to remain as nymph or larva. JH has a unique terpenoid structure and is the methyl ester of epoxy farnesoic acid. This sesquiterpene exists in at least six different forms. JH III is the most common type and is present in most insects, whereas JH I and II are the principal ones in Lepidoptera. The insect molting hormone, 20-hydroxyecdysone (20E), is secreted as ecdysone by a pair of endocrine glands that are located in the prothorax of Lepidoptera and other insects (prothoracic glands) or in the ventroposterior part of the head (ventral glands). Ecdysone is converted into 20E, which manifests its action via molt-inducing effects through interaction with amino acid residues in the ligand-binding domain of the ecdysone receptor (EcR) protein.

Chitin is a major constituent of the cuticle, the outermost layer of insects, which serves as an exoskeleton and protects insects against dehydration, microbial infection, and physical injury. Insects consistently synthesize and degrade chitin in a highly controlled manner to allow ecdysis and regeneration of the peritrophic matrices. JHs and ecdysones, chitin synthesis and cuticle formation have been the prime targets for JH analogues (JHA), chitin synthesis inhibitors (CSIs), and ecdysone agonists (EAs).

The discovery of juvabiones strengthened the vision of developing JHA as nontoxic and selectively acting "third-generation insecticides" (Williams, 1967). The search for synthetic juvenoids led to the development of JHAs, such as methoprene, hydroprene, and kinoprene, during the 1970s followed by the innovations of more potent JHAs, like fenoxycarb and pyriproxyfen, with broad-spectrum action and photostability.

Ecofriendly Pest Management for Food Security
http://dx.doi.org/10.1016/B978-0-12-803265-7.00020-8

The term IGR was introduced by Schneiderman (1972) for substances that are analogues or antagonists of these hormones and interfere with insect development. Since then, the compounds that adversely interfere with the growth and development of insects have been collectively referred to as *insect growth regulators* (IGRs). JH analogues, ecdysone agonists, and chitin synthesis inhibitors have been referred to as IGRs. Pener and Dhadialla (2012) considered the term *insect growth disruptors* (IGDs) as more appropriate for chemicals that disrupt growth and development in insects.

Serendipitous discovery of benzoylphenyl urea (BPU) derivative, diflubenzuron, evinced research interest in the development of a number of BPU derivatives, like novaluron, lufenuron, hexaflumuron, and non-BPU compounds, like buprofezin (thiadiazine) and etoxazole (oxazoline). Development of the first nonsteroidal ecdysone agonist (EA) belonging to the chemistry class of bisacylhydrazines stimulated the search for potent ecdysone agonists, such as tebufenozide, methoxyfenozide, halofenozide, and chromafenozide, for pest management with high degree of selectivity against specific groups of insect pests.

Presently, a number of commercial IGRs are available, as standalone compounds, combination products with other IGRs and insecticides, and unique bait formulations and custom-designed products to cater to the management of crop and stored product pests, termites, pests of public health importance, and veterinary and urban pests.

2. PHYSIOLOGY OF INSECT MOLTING

Insects undergo molts from the embryonic stage, and the molting process continues as larval, pupal, and adult stages in holometabolous insects, and nymphal instars and adults in hemimetabolous insects. Molting occurs in insects to accommodate the growth of the insects under the influence of 20E with the regulation by timing and titers of JH. These hormones change their roles to regulating reproductive processes in the adult stages of the insects (Chapman et al., 2012).

The exoskeleton becomes too small for the growing body tissues of an immature insect, and it must molt its cuticle to accommodate the growth of the internal organs and tissues. Stretch receptors in the insect body are stimulated by the increasing growth of body tissues, and when it attains a critical size, the brain secretes PTTH into the circulating hemolymph. PTTH binds with specific receptor proteins on prothoracic glands (PGLs). Multiple biochemical reactions are initiated that result in synthesis of ecdysone by PGL. Ecdysone is converted to 20-hydroxyecdysone by epidermal cells. Epidermal cells enlarge and change shape to become columnar. They begin to divide by mitosis, and with the more numerous cells per unit area, there is an increase in surface tension leading to separation of epidermal cells from old cuticle.

Apolysis, the separation of epidermal cells from old cuticle, marks the beginning of a molt and of a new instar. Ecdysial space, a minute space created by the separation of epidermal cells from the old cuticle, is filled with molting fluid containing inactive chitinolytic enzymes, proteinases for digesting the cross-linking proteins and chitin of the old endocuticle. The fall in the 20E titer triggers the activation of enzymes in the molting fluid for digestion of the procuticle underlying the old cuticle. The preparation for ecdysis starts with the resorption of the molting fluid (Locke, 1998). Finally, the clearance of 20E titer completely from the system triggers the release of eclosion hormone leading to the ecdysis of the larva leaving behind the remnants of the old cuticle (Zitnanova et al., 2001).

2.1 Juvenile Hormone

Juvenile hormone is secreted by the corpora allata prior to each molt. JH secretion by the corpus allatum is regulated by two neurohormones, the allatotropins that stimulate secretion and the allatostatins that inhibit synthesis. The main role of JH in immature insects is to inhibit the genes that promote development of adult characteristics, causing the insect to remain as nymph or larva. During the last larval or nymphal instar, the corpora allatum becomes atrophied and stops producing juvenile hormone. This releases inhibition on development of adult structures and causes metamorphosis into an adult (hemimetabolous) or a pupa (holometabolous).

In adult insects, the neurosecretory cells of the brain release a brain hormone that reactivates the corpora allatum, stimulating the production of juvenile hormone. Stimulation of yolk production in adult females and stimulation of the male accessory glands to produce the proteins of seminal fluid and spermatophore cases are under the influence of JH.

Juvenile hormone has a unique terpenoid structure and is the methyl ester of epoxy farnesoic acid. This sesquiterpene exists in at least six different forms. JH III is the most common type and is present in most insects, whereas JH I and II are the principal ones in Lepidoptera.

2.2 Ecdysone

The insect molting hormone, 20-hydroxyecdysone (20E; also referred to as ecdysterone) is secreted as ecdysone by a pair of endocrine glands that are located in the prothorax of Lepidoptera and other insects (prothoracic glands) or in the ventroposterior part of the head (ventral glands).

Molting hormone affects many cells throughout the body, but its prime function is to stimulate a series of physiological events that lead to synthesis of a new exoskeleton. As long as ecdysteroid levels remain above a critical threshold in the hemolymph, other endocrine structures remain inactive (inhibited). This hormone triggers ecdysis, the physical process of shedding the old exoskeleton. The insect molting hormone, 20E, manifests its molt-inducing effects by interaction with amino acid residues in the ligand-binding domain of the ecdysone receptor protein.

Toward the end of apolysis, ecdysteroid concentration falls, and neurosecretory cells in the ventral ganglia begin secreting eclosion hormone. The rise in concentration of eclosion hormone stimulates other neurosecretory cells in the ventral ganglia to secrete bursicon, a hormone that causes hardening and darkening of the integument (tanning) due to the formation of quinone cross-linkages in the exocuticle (sclerotization).

2.3 Chitin Synthesis

Chitin is a major constituent of the cuticle, the outermost layer of insects, which serves as an exoskeleton and protects insect against dehydration, microbial infection, and physical injury. Chitin is also an integral part of insect peritrophic matrices, which act as a permeability barrier between the food bolus and the midgut epithelium and protect the gut from mechanical disruption, toxins, and pathogens. Insects consistently synthesize and degrade chitin in a highly controlled manner to allow ecdysis and regeneration of the peritrophic matrices.

3. CONCEPT OF INSECT GROWTH REGULATORS

Carrol Williams proposed the term "third-generation pesticide" in 1967 envisioning the potential use of the insect juvenile hormone as an insecticide. With the development of juvabiones and boom of synthetic juvenoids, the term *IGRs* was introduced by Schneiderman (1972) for substances that are analogues or antagonists of these hormones and interfere with insect development. Since then, the compounds that adversely interfered with the growth and development of insects have been collectively referred to as insect growth regulators.

Dhadialla et al. (2010) used the term, "insect growth and development disrupting insecticides" to refer JHAs, CSIs and EA as these compounds rather deregulate the insects' growth and normal development. Pener and Dhadialla (2012) considered the term "insect growth disruptors (IGD)" as more appropriate for chemicals that disrupt growth and development in insects and hope that the term IGD will replace IGR.

3.1 History and Evolution of IGRs

3.1.1 *Juvenile Hormone Analogues*

The history of juvenoids began with the preparation of the crude extract of the male Cecropia moths, *Hyalophora cecropia*. The extracts when injected into pupae or when applied on the surface of pupal body caused retention of pupal characters. Subsequent investigations revealed that lipid extracts prepared from different invertebrates also had positive JH responses (Schneiderman and Gilbert, 1959).

The discovery of "paper factor" having JH activity factor from the balsam fir tree (*Abies balsamea*) was a landmark event in the search for JHA from plants (Sláma and Williams, 1965, 1966; Williams and Sláma, 1966). Many naturally occurring JHAs, such as juvabione, from Canadian balsam fir (Bowers et al., 1966), and dehydrojuvabione from European balsam fir (Cerný et al., 1967), led Sláma and Williams (1966) to speculate that JHA is an evolutionary adaptation of plants to develop resistance against herbivores. The discovery of juvabiones strengthened the vision of developing JHA as a nontoxic and selectively acting third-generation insecticide (Williams, 1967).

In the initial stage of juvenoid research, special mention must be made about the preparation called "Law and Williams mixture" as it was used for dose response juvenoid

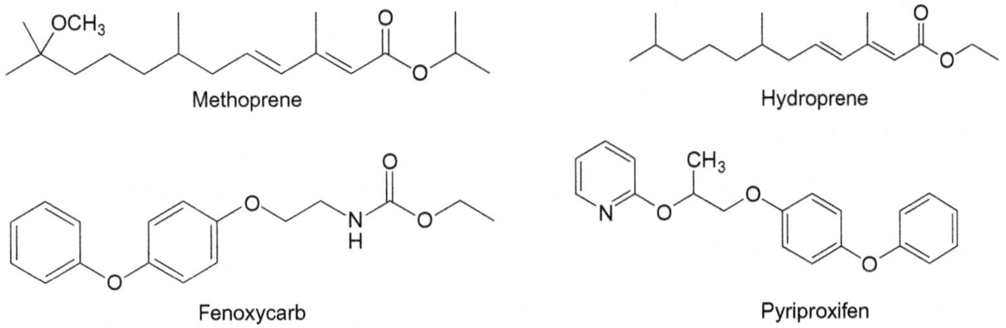

FIGURE 1   Juvenile hormone analogues widely used for pest control.

investigations (Law et al., 1966). A large number of sesquiterpenoid juvenoids were prepared and screened for their JH activity against insects (Bowers, 1971). In addition to these compounds, a particular type of the acyclic isoprenoid juvenoid, methoprene (developed by Zoecon Corp. in 1972) showed outstanding JH activity against insects and was the first commercial IGR (registered under the trademark Altosid™) for insect control (Henrick et al., 1973). An important milestone in the JHA was the development of fenoxycarb with broad-spectrum activity and higher photostability than the earlier JHAs (Grenier and Grenier, 1993). The next notable discovery was of pyriproxyfen, regarded probably as the most potent JHA available today (Hatakoshi, 2012) with its broad-spectrum activity against a wide range of pests (see Figure 1 for JHA).

STORY OF PAPER FACTOR

Karel Sláma brought the fire bug, *Pyrrhocoris apterus* Linnaeus, from (the then) Czechoslovakia to the late Carroll Williams laboratory at Harvard University in the United States (US). There was difficulty in maintenance of this insect culture in the US although it had been easily cultured in Czechoslovakia. It was observed that instead of developing into normal adults, the nymphs developed into nymphal-adult intermediates termed *adultoids*, or extra instar nymphs were obtained. While analyzing the causes, it was inferred that the molting deformities were due to the "paper factor," the newspaper strips used in rearing jars. The American newspapers made from the balsam fir (*A. balsamea*) contained the juvabione, whereas the trees in Czechoslovakia did not.

Several years later, William S. Bowers isolated and characterized the active factor, which he named *juvabione*.

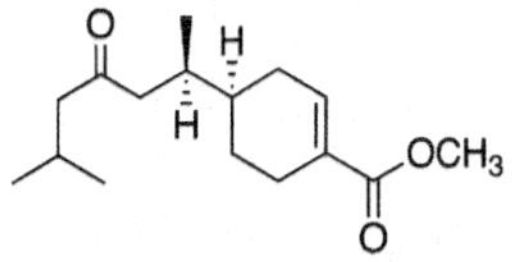

Juvabione – "The Paper Factor"

3.1.2 Chitin Synthesis Inhibitors

The first chitin synthesis inhibitor, diflubenzuron, belonging to the benzoylphenyl urea class of chemistry, was discovered serendipitously by Philips-Duphar Company in the 1970s.

This discovery of diflubenzuron sparked the research interest in benzoylphenyl ureas resulting in the development of a number of BPU derivatives such as triflumuron (Hamman and Sirrenberg, 1980), chlorfluazuron (Haga et al., 1982), teflubenzuron (Becher et al., 1983), hexaflumuron (Sbragia et al., 1983), flufenoxuron (Anderson et al., 1986), novaluron (Ishaaya et al., 1996), and lufenuron (developed by Ciba-Geigy in 1998).

A number of CSIs not related to the BPU category of compounds, such as etoxazole (Ishida et al., 1994), buprofezin, cyromazine, and dicyclanil have been recently developed. These products do not bear any resemblance to BPUs nor share any structural similarity among them (see Figure 2 for BPU and non-BPU derivatives).

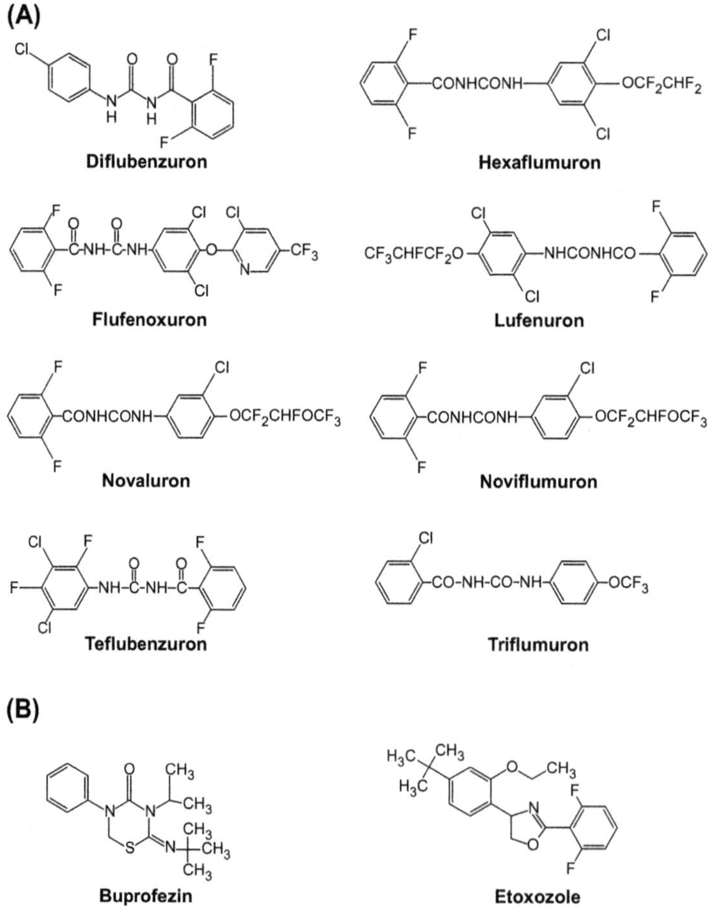

FIGURE 2 Chitin synthesis inhibitors in use for pest control: (A) benzoylphenyl urea compounds; (B) non-benzoylphenyl urea compounds.

SERENDIPITOUS DISCOVERY OF CSI

Benzoylphenyl urea compounds were discovered, mostly by chance, in the early 1970s by Philips-Duphar Company. The first compound, coded as DU-19111, was obtained by combination of two herbicides, dichlobenil and diuron (see). It was routinely tested for herbicidal and insecticidal activity. When tested on insect larvae, there was no immediate knockdown effect, but treated individuals showed various degrees of molt deformities. The newly synthesized compound had a difluorobenzoyl group on one end of the urea bridge and a chlorophenyl group on the other side and came to be known as diflubenzuron and marketed under the name Dimilin (Retnakaran et al., 1985).

Diflubenzuron: The First BPU Derivative

3.1.3 Ecdysone Agonists

Ecdysone, 20E, and many other related steroids are collectively termed *ecdysteroids*. Despite many attempts, there has been no commercial success in developing natural or synthetic ecdysteroids. Rohm and Haas Co. first reported the development of a nonsteroidal compound (under the code name RH-5849) belonging to the chemistry class of bisacylhydrazines (Aller and Ramsay, 1988).

Although, the first bisacylhydrazine was not commercialized, further synthesis and screening of about 4000 bisacylhydrazine structural analogues resulted in selection of 22 candidate compounds. Rigorous field testing led to the development of three successful analogues, viz., RH-5992 (tebufenozide), RH-2485 methoxyfenozide, and RH-0345 halofenozide (Heller et al., 1992; Le et al., 1996; RohMid, 1996), potent ecdysone agonists for insect control (see Figure 3 for structures of ecdysone agonists).

SERENDIPITOUS DISCOVERY OF BISACYLHYDRAZINES

In 1984, a researcher with Rohm and Hass Company, Hsu, during the process of synthesis of benzoyl hydrazide (which was to be used for synthesis of another class of compounds), obtained an undesired product, 1,2-di (4-chlorobenzoyl)-1-t-butyl hydrazine as contaminant. Instead of discarding this contaminant (coded as RH-5849), Hsu tested it for biological activity and observed ecdysteroid activity in insects. This compound was not commercialized because of its low potency, but its novel mode of action stimulated the search for bisacylhydrazines and culminated in developing some commercial EAs such as tebufenozide, methoxyfenozide, and halofenozide.

RH-5849- The First Bisacylhydrazin

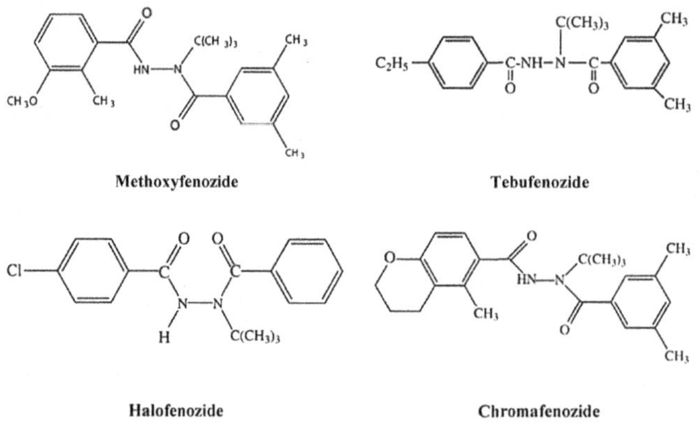

FIGURE 3   Ecdysone agonists in use for pest control.

4. HORMONE INSECTICIDES

Progress has gone a long way since the initial suggestion of using JH analogues as third-generation insecticides as an alternative to insecticides. Over the years, a number of structural modifications have been incorporated in the JH analogues to increase the spectrum of their action. A number of combination products involving hormone analogues and other IGRs and insecticides have also been developed to enhance their utility in diverse pest management programs.

4.1 Juvenile Hormone Analogues

Since the early 1970s, persistent efforts by the agrochemical industry have led to the discovery of a large number of new acyclic or alicyclic sesquiterpenoid juvenoids and a particular acyclic type juvenoid, viz., methoprene and hydroprene, found to have exemplary laboratory and field efficacy. Methoprene, under the trade name Altosid™ (by Zoecon Corporation, USA) was the first commercial hormone insecticide made available for pest control. Subsequently, a number of methoprene formulations solely or in combination with non-IGR molecules have been developed and marketed. Some of the commercial methoprene-based formulations are: Altosid™, Apex™, Biopren™, Duplex™, Extinguish™, Inhibitor™, Juvenon™, Protect™, and Viodat™. Methoprene is combined with other insecticides or IGRs for synergistic effect against stored product insects. Hydroprene, commercially available as, viz., Altozar™, Gentrol™, Gencor™, Entocone™, and Pointsource™, is effective on several orders of insects (Tomlin, 2009). As hydroprene is slightly more volatile, it is advantageous as an aerosol formulation in urban pest management (Phillips and Throne, 2010).

As most of the early synthetic isoprenoid compounds were slow acting and less potent, structural modifications on these analogues have improved their knock-down action. The most important structural innovation was incorporation of a 4-phenoxyphenyl group into the juvenoid molecule. The first juvenoid of this 4-phenoxyphenyl series registered for practical use was fenoxycarb. Structurally, it shows similarity with the juvenoid peptides and was found to have enormous juvenoid activity (Dorn et al., 1981). Discovery of fenoxycarb marked a milestone in the development of JHA for pest control as it is more potent and photostable than the earlier JHAs (see Figure 1).

Fenoxycarb under the trade names Precision™, Award™, Logic™, Polyon™, Torus™, Pictyl™, Varikill™, Dicare™, and Helgar™ are used against a wide range of crop pests, stored product pests, mites, household pests, termites, and ants (Grenier and Grenier, 1993).

Investigations on isoprenoid juvenoids on the effects of chain length (Mori et al., 1975), pyridyl structures (Kramer et al., 1979; Ichikawa et al., 1980), phenoxy propyl ether structures (Niwa et al., 1989), and 4-alkylphenyl aralkyl ethers (Hyashi et al., 1990) on juvenoid activity culminated in the development of pyriproxyfen, a 4-phenoxyphenoxy-type compound with a pyridyl structure in the side chain. Pyriproxyfen is regarded as the most potent juvenoid for practical use (Hatakoshi, 2012). It has been commercialized for pest control under the trade names Sumilarv™, Admiral™, Knack™, Distance™, Esteem™, Seize™, Nylar™, and Archer™. The effects of pyriproxyfen have been assayed on a wide range of insect species including mosquitoes (Okazawa et al., 1991), ants (Banks and Lofgren, 1991), and cockroaches (Koehler and Patterson, 1991).

A number of combination products involving pyriproxyfen and other insecticidal compounds have been developed and registered under different trade names for control of crop pests, phytophagous mites, vectors, and veterinary pests (Amelotti et al., 2009; Fourie et al., 2010; Juan-Blasco et al., 2011).

4.2 Chitin Synthesis Inhibitors

There are two groups of CSIs, benzoylphenyl ureas and compounds not related to benzoylphenyl ureas (see Figure 2). Some of the compounds that are not inhibiting chitin synthesis but only interfering with normal cuticle deposition have also been included under CSIs.

4.2.1 Benzoylphenyl Urea Compounds

Diflubenzuron (developed by Philips-Duphar Company in 1970) belong to the BPU group of compounds. This discovery sparked the research on benzoylphenyl urea as a new class of insecticides leading to the development of a number of BPU derivatives as successful chitin synthesis inhibitors.

Benzoylphenyl ureas have a central urea moiety, with most of the complex substitutions occurring on the phenyl end leaving the benzoyl part to remain relatively simple. It was hypothesized that this end was the one that attached itself to the unidentified receptor resulting in inhibition of chitin synthesis (Nakagawa et al., 1991). The extensive substitutions at the phenyl or the anilide end probably account for the differential effects on pest populations. Chlorfluazuron with an unoccupied orthoposition in the phenyl end and a dichloropyridyloxy group in the paraposition made it highly hydrophobic and electron withdrawing, making it very active. Similarly, teflubenzuron has four halogen substituents in the phenyl end, making it both hydrophobic and electron withdrawing. Benzoylphenyl urea compounds may have a common mode of action and block a postcatalytic step of chitin synthesis (Nauen and Smagghe, 2006; Van Leeuwen et al., 2012).

Diflubenzuron is effective against phytophagous insect pests including beetles, moths, jumping lices, scale insects and mites (Ishaaya, 1990). Flucycloxuron has a wider spectrum of activity against mites, while lufenuron has been particularly effective against many lepidopteran pests due to its classic larvicidal along with transovarial–ovicidal and ovicidal actions. Fluazuron has been shown to be effective against ticks and mites. Hexaflumuron has been effective against the larval stages of Lepidoptera, Coleoptera, and Diptera. Another major use has been in bait incorporation against the subterranean termites. Noviflumuron has been found to be relatively fast acting especially against cockroaches and termites.

4.2.2 *Non-benzoylphenyl Urea Compounds as CSI*

Among the CSIs not related to the benzoylphenyl urea class of compounds, buprofezin, etoxazole, cyromazine, and dicyclanil have been widely used for pest control in agricultural crops and public health management systems.

Buprofezin belonging to the group of thiadiazines is found efficacious against the homopterans, such as mealybugs, scale insects, and whiteflies. Its actions on insects include inhibition of cuticle deposition and chitin biosynthesis in the brown planthopper, *Nilaparvata lugens* (Stål) (Uchida et al., 1985); prevention of formation of a lamellate cuticle in nymphs of greenhouse whitefly, *Trialeurodes vaporariorum* Westwood (De Cock and Degheele, 1991); and inhibition of cholinesterase activity in insects (Cottage and Gunning, 2006).

Etoxazole belonging to the chemical class of oxazolines was synthesized by Ishida et al. (1994). Experimental findings suggest that benzoylurea and etoxazole have similar modes of action by blocking a postcatalytic step of chitin synthesis (Nauen and Smagghe, 2006). Etoxazole has been found especially effective against plant-feeding spider mites (Nauen et al., 2001) and is reported to have insecticidal action as well against crop pests such as rice leafhoppers, diamondback moth of crucifers, and peach aphids (Suzuki et al., 2002).

Cyromazine and dicyclanil, which do not inhibit chitin synthesis but interfere with cuticle formation, are also considered as molt inhibitors. Cyromazine is an aminotriazine and a cyclopropyl derivative of melamine marketed under the trademarks Neoprex™, Trigard™, and Vetrazin™. It is found to be giving good control of stable flies in winter hay (Taylor et al., 2012). Dicyclanil under the commercial trade name Clik™ has been found to be efficacious against blowfly on sheep and lamb.

4.3 Ecdysone Agonists

There are many natural phytoecdysteroids (Dinan et al., 2009) and zooecdysteroids present in plants and animals (Lafont and Koolman, 2009). Despite many attempts to develop insecticides based on the structures of natural or synthetic ecdysteroids, none has been developed commercially. The first nonsteroidal EA belonging to the bisacylhydrazine (BAH) class of chemistry was developed by Rohm and Haas Co. (Aller and Ramsay, 1988). The compound RH-5849 induced a premature but unsuccessful molt by interfering with normal cuticle formation leading to death of insects. Although RH-5849 was not commercialized due to low potency, the improved derivatives of BAH, such as tebufenozide (Heller et al., 1992), methoxyfenozide (Le et al., 1996), and halofenozide (RohMid, 1996), were released for commercial use against insect control. Chromafenozide was registered for use in Japan (Yanagi et al., 2006) and furan tebufenozide or fufenozide was registered for use in China (Zhang, 2005).

Bisacylhydrazines are true ecdysone agonists and their activity is manifested via interaction with the ecdysone receptor complex (EcR); via the ecdysone receptor they activate genes that are dependent upon increasing titers of 20E. Their presence in the hemolymph inhibits the release of eclosion hormone, resulting in an unsuccessful lethal molt. Thus, due to the action of BAH, the molting process is completely derailed at the physiological and molecular levels leading to precocious lethal molt in susceptible larvae.

Bisacylhydrazines, viz., methoxyfenozide, tebufenozide, and chromafenozide, have selectivity in their action against lepidopteran pests, and *in vitro* studies suggest that this selectivity is due to the specific binding of these EAs with the EcR from lepidopteran insects. These three EAs have been registered for commercial use against lepidopteran larvae. Halofenozide has broad-spectrum activity against lepidopteran and coleopteran pests. Both tebufenozide and methoxyfenozide exhibit selective contact and ovicidal activity (Trisyono and Chippendale, 1997; Sun and Barrett, 1999; Sun et al., 2000).

5. IGR IN PEST MANAGEMENT

Inspired by the concept of third-generation pesticides proposed by Carrol Williams, there was a boom in synthesis of isoprenoid juvenoids in the 1970s to develop "biorational" compounds as an alternative to chemical insecticides. Stimulated by the serendipitous discoveries of a benzoylphenyl urea analogue and an ecdysone agonist, a number of CSIs and ecdysone agonists have been developed over the years. In the past three decades, there has been a quantum increase in the availability of IGRs with broad-spectrum action to suit diverse pest management programs. There has been a paradigm shift in the use of IGRs as a safe alternative to chemical insecticides to current use of novel combination products involving IGRs and insecticides for pest control (see Tables 1 and 2 for an overview of IGRs with details of registered combination products).

5.1 Management of Crop Pests

JHAs can be broadly categorized into two groups: the terpenoid compounds, such as methoprene, hydroprene and kinoprene; and phenoxy JHAs, like fenoxycarb and pyriproxyfen. JHAs have been found effective especially against adult stages of dipteran and homopteran insects but not against lepidopteran insect pests. As JHAs are especially effective against adult stages of the insects, their poor efficacy against lepidopteran crop pests may be because most of these pests cause the damage as larvae (Retnakaran et al., 1985).

Terpenoid JHAs, such as methoprene and kinoprene, have been used successfully against mosquito larvae, scales, and mealybugs. Phenoxy JHAs cause morphogenetic effects, ovicidal action, and sterility on treated insects. Fenoxycarb offers effective control against fire ants, fleas, and sucking insect pests of crops. It is highly effective against sucking pests, such as apple wooly aphid, *Eriosoma lanigerum* (Nicholas et al., 2005); mealybugs, *Phenacoccus pergandei* (Zhou et al., 2008), and scales, *Diaspidiotus perniciosus* (Sazo et al., 2008). Pyriproxyfen has been used for controlling aphids, whiteflies, and psyllids causing damage to crops. It has been used for control of locusts, *Locusta migratoria* Linnaeus (Pener et al., 1997), California red scale, *Aonidiella aurantii* Maskell, citrus psyllid, *Diaphorina citri* Kuwayama (Boina et al., 2010), and greenhouse whitefly, *T. vaporariorum* (Oouchi and Langley, 2005).

TABLE 1 A Consolidated Table on Commercial IGR Products

S.No	IGR compound	Trade names	Target pests
I	**JH Analogue**		
1	Methoprene	Altosid®, Aquaprene®, Precor®, Kabat®, Pharorid®, Dianex®, Apex®, Minex® Fleatrol®, Ovitrol®, Extinguish® and Diacon®, Biopren®, Dianex®, Duplex®, Inhibitor®, Juvenon®, Lafarex®, MoorMan's®, Protect®, Viodat®	*Rhyzopertha dominica* (Chanbang et al., 2008); Sandfly, *Phlebotomus papatasi* (Mascari et al., 2011)
2	Hydroprene	Altozar®, Gencor®, Mator®, Raid® Gentrol® Entocone® and Pointsource®	*Plodia interpunctella* (Mohandass et al., 2006a,b)
3	Kinoprene	Enstar® AQ, Enstar® II	*Bemisia argentifolii* (Rothwangl et al., 2004)
4	Fenoxycarb	Insegar®, Logic®, Pictyl®, Torus®, Varikill®, Preclude®, Precision®, Award®, Polyon®, Dicare®, Helgar® and Comply®	German cockroach, *Blattella germanica* (Kaakeh et al., 1997; Fathpour et al., 2009); aphid, *Eriosoma lanigerum* (Nicholas et al., 2005). Scale insects, *Phenacoccus pergandei* (Zhou et al., 2008)
5	Pyriproxyfen	Aciprox®, Adeal®, Admiral®, Archer®, Ardito®, Atominal®, Brai®, Distance®, Duss®, Editor®, Epingle®, Esteem®, Farewell®, Juvinal®, Knack®, Lano®, Maraka®, MIO Pyriproxyfen®, Nemesis®, Pitch®, Pyrilate®, Pyriphar®, Pyriprem®, Scalex®, Seize®, Terva®, Wopro-pyriproxyfen®, Sumilarv® and Nylar®	*Locusta migratoria migratorioides*, and the desert locust, *Schistocerca gregaria* (Pener et al., 1997); California red scale, *Aonidiella aurantii* (Eliahu et al., 2007; Grafton-Cardwell et al., 2006); Pharaoh ant, *Monomorium pharaonis* (Lim and Lee, 2005)
II	**Chitin Synthesis Inhibitor**		
	A. BPU derivatives		
6	Diflubenzuron	Dimilin, Adept, Bi-Larv, Hekmilin, Hilmilin, icromite, Device, Diflorate, Diflox, Dimisun, Du-Dim, Forester, Patron and Pestanal	*Helicoverpa armigera* (Kumar et al., 1996); *Prostephanus truncates* and *R. dominica* Kavallieratos et al. (2012)
7	Bistrifluron	Xterm®	*Coptotermes formosanus* (Kubota et al., 2007)
8	Chlorfluazuron	Aim®, Atabron®, Fertabron® and Helix®	*Spodoptera litura* (Perveen, 2006); *Spodoptera littoralis* (Korrat et al., 2012); striped stem borer, *Chilo suppressalis* (He et al., 2008); *Coptotermes acinaciformis* (Sukartana et al., 2009)
9	Chlorbenzuron	Chlorbenzuron®	Migratory locust, *L. migratoria manilensis* (Lei et al., 2002); Soft scale, *Eulecanium gigantean* (Yue et al., 2011)
10	Flufenoxuron	Cascade® and Floxate®	*S. litura* (Bakr et al., 2010)
11	Hexaflumuron	Consult®, Recruit II®, SentriTech® and Shatter®	*H. armigera* (Rafiee et al., 2008)

TABLE 1 A Consolidated Table on Commercial IGR Products—cont'd

S.No	IGR compound	Trade names	Target pests
12	Lufenuron	Match®, Luster®,Manyi®, Sorba®, Lufenox®, Axor®	*H. armigera* (Gogi et al., 2006); *Musca domestica* (Khalil et al., 2010)
13	Novaluron	Rimon®, Diamond®, Pedestal®	Codling moth, *Cydia pomonella*, and the oriental fruit moth, *Grapholita molesta* (Magalhaes and Walgenbach, 2011)
14	Noviflumuron	Recruit III® and Recruit IV®	Termite
15	Teflubenzuron	Nomolt®, Dart®, Nemolt®, Diaract®, Nobelroc®, Teflurate®	Diamondback moth, *Plutella xylostella*
16	Triflumuron	Starycide®, Alsystin®, Certero®, Poseidon®, Baycidal®	Tomato leafminer, *Tuta absoluta* (Silva et al., 2011)
17	Fluazuron	Acatak®	*Boophilus microplus* (Bull et al., 1996)
18	Flucycloxuron	Andalin®	Oriental armyworm, *Mythimna separata*
	B. Non-BPU derivatives		
19	Buprofezin	Applaud®,Courier®, Maestro®, Podium®, Profezon®, Sunprofezin®, Viappla® and Dadeci®	Brown planthopper, *Nilaparvata Lugens*, Kanno et al. (1981) & Uchida et al. (1985) Californian red scale, *A. aurantii* (Yarom et al., 1988); Persimmon bud mite, *Aceria diospyri* (Ashihara et al., 2004); *N. lugens* (Hegde and Nidagundi, 2009); *Phenacoccus solenopsis* (Patel et al., 2010); *Amrasca biguttula biguttula* (Patel et al., 2012);
20	Etoxazole	Barok®, Baroque®, Borneo®, Zeal®; in combination with fenpropathrin [a pyrethroid]; Biruku®	Red spider mite, *Tetranychus urticae* (Kim and Yoo, 2002).
21	Cyromazine	Trigard®,Larvadex®, Neporex®, Cliper®, Cyromate®, Garland®, Genialroc®,Manga®, Sun-Larwin®, Trivap®, Vetrazin®	Stable flies, *Stomoxys calcitrans* (Taylor et al., 2012)
22	Dicyclanil	Clik®	Blowfly on sheep and lamb (Cohen, 2010).
III	**Ecdysone Agonists**		
23	Tebufenozide	Confirm®, Mimic®, Conidan® (+imidacloprid)	*S. exempta, Spodoptera exigua, S. littoralis, M. brassicae* and *G. mellonella* (Smagghe and Degheele, 1994)
24	Methoxyfenozide	Falcon®, Intrepid®, Integro®, Pacer® and Prodigy®	*H. armigera* (Moosa Saber et al., 2013).
25	Halofenozide	Match 2®	Lepidopteran and coleopteran larvae
26	Chromafenozide	ANS-118, Virtu®, Phares®, Killat®, Podex®, Cyclone ® and Kanpei®	Common cutworm, *S. litura*, the beet armyworm, *S. exigua* (Yanagi et al., 2006)

TABLE 2 List of Combination Products Involving IGR Compounds

S.No.		Combination products	Mixture	Targets pest
I	Juvenile Hormone Analogues	(S)-Methoprene + Fipronil + Amitraz	(Certifect®)	Ticks
		Pyriproxyfen + Cyphenothrin	Sergeant's Gold®	African yellow dog tick, *Haemaphysalis elliptica* (Acari: Ixodidae), and against the cat flea, C. felis (Fourie et al., 2010)
		Pyriproxyfen + Permethrin + Benzoyl benzoate	175 CS®, Allergoff ®200 EC and Fedorex-Profi®	Storage mites (Stará et al., 2011a,b)
		Pyriproxyfen + chlorpyriphos	Inesfly IGR FITO®	Ant (Juan-Blasco et al., 2011).
		Pyriproxyfen + Chlorpyrifos and Diazinon	Inesfly® 5A IGR	Pyrethroid resistant populations of *Anopheles gambiae* and *Culex quinquefasciatus* (Mosqueira et al., 2010)
		Pyriproxyfen 5% EC + Fenpropathrin 15% EC		Cotton, brinjal, and okra: whitefly, shoot and fruit borer
II	Chitin synthesis Inhibitors	Diflubenzuron 20% + Deltamethrin 2% SC		*Spodoptera littoralis*
		Lufenuron + Spinosad + Indoxacarb	Match® 5%	Grapes tortricid, *Lobesia botrana*
		Lufenuron + Fenoxycarb	Lufox®	*Rhyzopertha dominica* (Kavallieratos et al., 2012)
		Novaluron + Bifenthrin	RimOn Fast is a 50:50	Lygus bugs in cotton
		Etoxazole + Fenpropathrin	Biruku®	*Tetranychus urticae* (Tomlin, 2009)
		Novaluron 5.25% + Indoxacarb 4.5% SC	Plethora	Tomato: fruit borer and leaf eating caterpillar
		Buprofezin + Deltamethrin	Dadeci®	Homopteran, Coleopteran and Acarina pests (Kanno et al., 1981)
		Buprofezin 15% + Acephate 35% w/w WP		BPH of rice
		Buprofezin 5.65% w/w EC + Deltamethrin 0.72% w/w		BPH, leaf folders of rice
		Buprofezin 20% w/w SC + Flubendiamide 4%		Yellow stem borer, leaf folder, BPH
		Hexaconazole 5% w/w WG + Flubendiamide 3.5%		Stem borer, leaf folder

Most of the BPU compounds have been found very effective against lepidopteran insect pests. Diflubenzuron has been found effective against a wide range of lepidopteran pests and locusts (Latchininsky and Gapparov, 2007; Tail et al., 2010). Chlorfluazuron and hexaflumuron are very effective against the striped stem borer, *Chilo suppressalis* Walker in rice (He et al., 2008). Flucycloxuron is effectively used as an acaricide against spider mites. Flufenoxuron and novaluron are particularly effective against apple codling moth, *Cydia pomonella* L. (Charmillot et al., 2001; Magalhaes and Walgenbach, 2011). Lufenuron has been used effectively to control fruit flies, such as *Ceratitis capitata* (Weidemann), *Bactrocera dorsalis* (Handel) *and Bactrocera latifrons* (Handel) (Casanã-Giner et al., 1999). Non-BPU compounds, such as buprofezin, have shown efficacy against sucking insect pests including whitefly, *Bemisia tabaci* (Gennadius), mealybugs, *Planococcus citri* Risso and *Phenacoccus solenopsis* Tinsley, California red scale, and *A. aurantii* (Kanno et al., 1981; Yarom et al., 1988; Patel et al., 2010), while etoxazole is especially toxic to mite pests (Nauen et al., 2001).

5.2 Management of Stored Product Pests

IGRs have been in use for the management of stored product pests in different ways, such as surface treatments on warehouses, grain protectants and as application in bulk storage of grains. Absorptive dusts and inert materials, such as diatomaceous earth, have been used against stored product insects as physical control measures because they cause abrasion of cuticular lipids from the exoskeleton of the insects. Recent studies have shown that addition of methoprene to diatomaceous earth or a contact insecticide would offer better control of stored product pests (Athanassiou et al., 2011).

Alongside insecticidal chemicals, JHAs have been evaluated for their role in control of stored product pests. Some of the JHAs including methoprene, hydroprene, diflubenzuron, fenoxycarb, and pyriproxyfen have been found effective as surface treatments against the stored product pests.

Methoprene is found effective against some stored product insects, like lesser grain borer, *Rhyzopertha dominica* Fabricius (Chanbang et al., 2008). It is often combined with other insecticides or with CSIs like diflubenzuron (Daglish and Wallbank, 2005) or with pyrethrin (Sutton et al., 2011) for synergistic effect against stored product insects. Hydroprene is found effective against eggs and late instars of Indian meal moth, *Plodia interpunctella* Hübner and adults of red flour beetles, *Tribolium castaneum* (Herbst) and *T. confusum* Jacquelin du Val (Mohandass et al., 2006b).

Fenoxycarb was found to be effective against storage mites (Collins, 2006). Kavallieratos et al. (2012) reported that sprays of fenoxycarb, as well as a combination of fenoxycarb + lufenuron, caused slight or moderate direct mortality of adults of the stored product pests, the larger grain borer, *Prostephanus truncatus* (Horn) and *R. dominica*.

Efficacy of pyriproxyfen in controlling Indian meal moth, *P. interpunctella*, was reported by Ghasemi et al. (2010). It caused direct mortality of adults of *P. truncatus* and *R. dominica* and reduced the progeny production by 90% and 100%, respectively, in these two species (Kavallieratos et al., 2012). It was found to be relatively less effective against adults and eggs than immature stages of mold mite, *Tyrophagus putrescentiae* Schrank (Sánchez-Ramos & Castañera, 2003). Stará et al. (2011b) found that storage mites were more or less susceptible to the multicombinatorial formulation of acaricide + permethrin + pyriproxyfen + benzoyl benzoate (Allergoff™ 200 EC and Fedorex-Profi™).

Diflubenzuron is effective against a wide range of stored product pests (Oberlander et al., 1997). Daglish and Wallbank (2005) reported that a mixture of IGRs such as diflubenzuron (at 1 mg/kg) + methoprene (1 mg/kg) was effective against rice weevil, *Sitophilus oryzae* L. and *R. dominica*. They opined that diflubenzuron was the more effective component of this IGR mixture as this mixture was effective against a methoprene-resistant strain of *R. dominica* in Australia. Collins (2006) reported that diflubenzuron administered in wheat diet caused 57–96% reduction in populations of storage mites, *Acarus siro* L. Sobotnik et al. (2008) found that ingestion of diflubenzuron affects the peritrophic matrix in *A. siro* presumably by inhibiting chitin synthesis. Kavallieratos et al. (2012) observed that diflubenzuron spray at 5 ppm suppressed the multiplication of stored product pests, viz., *P. truncatus* and *R. dominica*, and the efficacy of CSIs like flufenoxuron, lufenuron, and triflumuron was slightly inferior to diflubenzuron against *P. truncatus*.

Investigations also revealed the effects of other CSIs, such as novaluron, lufenuron, flufenoxuron, and hexaflumuron on stored product pests. Elek and Longstaff (1994) observed that chlorfluazuron was more effective than teflubenzuron in four Coleopteran pests of stored grains. Teflubenzuron was the least effective against the rice weevil, *S. oryzae*. Topical application of hexaflumuron was more effective (at LC_{50} values of 7.5 mg/kg) than the LC_{50} obtained for teflubenzuron against the eggs of the pulse pod borer, *Callosobruchus maculatus* (F.) (Abo-Elghar et al., 2004). Immersing seeds into high concentrations of cyromazine (up to 5%) reduced adult emergence of *C. maculatus* (Al-Mekhlafi et al., 2012). Salokhe et al. (2003) reported that the neonates of *T. castaneum* exposed to sublethal (LC_{20} and LC_{30}) concentrations of flufenoxuron for 24 h often result in adults producing mostly nonviable eggs with reduced egg hatchability and death of first instar larvae.

There have been some reports on disadvantages of using JHA for stored product pest management. JHA application may induce a viable giant extra larval instars with increased longevity resulting in increased damage for a longer period as in *T. confusum* (Smet et al., 1989). Diet administration of pyriproxyfen at 0.1 ppm was found to completely inhibit the occurrence of adults of the next generation in two strains of *T. castaneum*, but the longevity of the larvae was greatly extended, resulting in giant larvae (Kostyukovsky et al., 2000).

5.3 Management of Public Health and Urban Pests

IGR compounds play a major role in managing the insect pests associated with human health. Because of specificity in their mode of action and safety to nontarget organisms and the environment, these IGR-based insecticides are more preferred than other synthetic insecticides. Most of the IGRs have high potency against mosquitoes, other pests and vector species. The IGRs such as diflubenzuron, lufenuron, triflumuron, novaluron, and methoprene are being used to manage insect vectors of human diseases.

Methoprene was the first commercial IGR used for control of mosquito larvae, and it has been used worldwide against a number of dipteran insects including mosquito larvae, cattle horn fly, *Haematobia irritans* (L.), stable fly, *Stomoxys calcitrans* L., and sandfly, *Phlebotomus papatasi* Loew (Mascari et al., 2011). Hydroprene because of its high volatile nature is a preferred aerosol formulation used for controlling household insect pests, like cockroaches. Fenoxycarb is found more toxic to German cockroach, *Blattella germanica* (Kaakeh et al., 1997) and is effective against cat flea, *Ctenocephalides felis* (Bouché) (Rajapakse et al.,

2002). Application of pyriproxyfen to adult females produced no viable offspring in tsetse fly, *Glossina morsitans morsitans* (Langley et al., 1990); it inhibited the adult emergence of *Aedes albopictus* (Skuse), *Aedes aegypti* (L.) (Suárez et al., 2011), *Anopheles culicifacies* Giles and *Anopheles subpictus* Grassi (Yapabandara et al., 2001).

Diflubenzuron inhibited the adult emergence of 16 species of mosquitoes (Grosscurt, 1978). Novaluron was highly effective against second and fourth instar larvae of *A. aegypti* (Mulla et al., 2003). Triflumuron was found to inhibit adult emergence in the susceptible Rockefeller strain of *A. aegypti* (Martins et al., 2008). It was found more effective than diflubenzuron or lufenuron against *Culex quinquefasciatus* Say (Suman et al., 2010).

Five species of mosquitoes, viz., *Anopheles albimanus* Weidemann, *Anopheles pseudopunctipennis* Theobald, *A. aegypti*, *A. albopictus*, and *C. quinquefasciatus*, were found to be highly susceptible to novaluron (Arredondo-Jiménez & Valdez-Delago, 2006). Novaluron effectively controlled sandfly, *P. papatasi*, by feed-through application to rodent, *Mesocricetus auratus* (Mascari and Foil, 2010). It greatly reduced adult emergence of the housefly, *Musca domestica* L., through feeding and dipping methods (Cetin et al., 2006). Jambulingam et al. (2009) recommended novaluron for control of *C. quinquefasciatus*, a major vector of lymphatic filariasis in Tamil Nadu State, India. Novaluron and Buprofezin were highly effective against *C. quinquefasciatus* larvae and pupae in pools, drains, and tanks at the dosages of 0.05 and 0.5 mg/l (Rajasekar and Jebanesan, 2012).

Cyromazine effectively controlled the maggots of housefly, *M. domestica*, and the face fly, *M. autumnalis* (Hall and Foehse, 1980; Williams and Berry, 1980); it was found effective against the sheep blowflies, *Lucilia cuprina* Wiedemann and *Lucilia sericata* Meigen, and other blowflies (Bisdorff and Wall, 2008); it reduced the emergence of stable flies up to 97% when it was sprayed on feeding sites (Taylor et al., 2012).

5.4 Termite Management

The utility of CSIs, especially diflubenzuron, in controlling termites such as subterranean termite, *Heterotermes indicola* (Wasmann) and Eastern subterranean termite *Reticulitermes flavipes* Kollar, was first demonstrated by Doppelreiter and Korioth (1981). However, subsequent laboratory and field testing with CSIs gave mixed responses against different termite species. Faragalla et al. (1985) could not confirm the efficacy of diflubenzuron against wheat termite, *Microcerotermes* spp., while subsequent laboratory investigations have proved that diflubenzuron (Su and Scheffrahn, 1993) and lufenuron (Su and Scheffrahn, 1996a) inhibited the ecdysis of *R. flavipes* albeit with no effect on Formosan subterranean termite, *Coptotermes formosanus* Shiraki. Investigations by several researchers confirmed that hexaflumuron significantly inhibits ecdysis in several subterranean termites including *Reticulitermes*, *Coptotermes*, and *Heterotermes* species (Su, 1994; DeMark et al., 1995; Su and Scheffrahn, 1996b; Forschler and Ryder, 1996; Grace et al., 1996; Su et al., 1997).

5.4.1 Termite Baits with CSIs

Termites have a symbiotic association with a diverse community of microorganisms for their survival, growth and development. The symbionts are lost at each molt and therefore, they have to be replenished after molts. Termites feed each other (proctodaeal trophallaxis) for transferring symbiotic microorganisms and as some castes are unable to feed themselves they also need stomodeal trophallaxis.

Use of baits with CSIs allows bait toxicants to be horizontally transferred from termite to termite in the colony by proctodaeal and stomodeal trophallaxis without causing immediate death of workers (Verma et al., 2009).

King et al. (2005) found in a no-choice and choice experiments with *R. flavipes* that noviflumuron caused quicker and higher levels of mortality than diflubenzuron. Su (2005) on evaluation of baits against *C. formosanus* reported that noviflumuron baits (Recruit III™) could achieve 100% mortality compared to 25–35% mortality with fipronil and thiamethoxam baits.

Efficacy of hexaflumuron in termite baits has been widely investigated. Saran and Rust (2008) reported that addition of other carbohydrates, especially 3% xylose improved uptake of hexaflumuron containing baits by workers of western subterranean termite, *R. hesperus*. Vahabzadeh et al. (2007) found that lufenuron as a bait component was superior to hexaflumuron against *R. flavipes*. Karr et al. (2004) observed that noviflumuron showed greater potency and faster action than hexaflumuron on *R. flavipes*. Studies by Osbrink et al. (2011) revealed that chlorfluazuron bait reduced populations of colonies of subterranean termites, *C. formosanus*, and *R. flavipes* in about three years in New Orleans, USA. It was also observed that chlorfluazuron baits were less effective than hexaflumuron. Use of Requiem™ bait matrix containing 0.1% (w/w) chlorfluazuron eliminated colonies of the subterranean termite, *Coptotermes curvignathus* in Jakarta, Indonesia within about 6–8 weeks of applying the baits (Sukartana et al., 2009).

Baiting with bistrifluron was proven effective against *C. formosanus* (Kubota et al., 2007), *Coptotermes acinaciformis* (Froggatt) (Evans, 2010), *R. speratus* (Kubota et al., 2007), and *Globitermes sulphureus* (Neoh et al., 2011). Bistrifluron was especially effective in solid pelletized alpha cellulose bait (Xterm™). Comparative studies revealed that bistrifluron bait was slightly more efficacious than hexaflumuron against subterranean termites (Kubota, 2011).

5.4.2 Monitoring cum Baiting Systems with Chitin Synthesis Inhibitors

Many studies were done on the use of CSI baits against natural field colonies of subterranean termites. The main difficulty in evaluation of the CSIs against subterranean termites is detecting their presence, and many times the pest goes unnoticed until they inflict considerable damage. Su et al. (1995) described a monitoring/baiting station device to detect and eliminate subterranean termites with CSI bait, and this technology was commercialized as the Sentricon™ termite colony elimination system. Subsequently, a number of other termite baiting systems were commercialized and their efficacy was evaluated.

Noviflumuron-baited Sentricon™ system eliminated within a year, 15 colonies of *C. formosanus* in New Orleans, USA (Hussender et al., 2007). Getty et al. (2007) by using Sentricon™ with baits of 0.5% hexaflumuron (Recruit II™) and with noviflumuron (Recruit III™) in a large-scale field trial controlled *Reticulitermes hesperus* and an unidentified species of *Reticulitermes*. Use of an improved monitoring cum bait system involving Sentricon™ stations with improved bait containing cellulose and 0.5% hexaflumuron resulted in elimination of eight field colonies of *Coptotermes gestroi* and subterranean termite, *Schedorhinotermes* sp. in Malaysia, respectively, in 42 and 77 days (Sajap et al., 2009). Comparative evaluation of three baiting systems, viz., Sentricon™ with hexaflumuron bait (Recruit II™), Terminate™, and FirstLine bait system revealed little difference in efficacy between these three baiting systems against *R. flavipes* (Glenn et al., 2008).

5.5 Insect Growth Regulators and Entomopathogens

A criticism of the entomopathogenic fungi is that they act too slowly (Lomer et al., 2001). The combined use of chemical insecticides, ideally at a low, sublethal rate along with entomopathogenic fungi has been an attractive approach to counter this criticism. Hassan and Charnley (1987) first observed that the larval integument of tobacco hornworm, *Manduca sexta* (L.) was weakened by diflubenzuron so that the fungus could more easily penetrate through the cuticle and its infectivity was quickened. This observation on synergizing effect of diflubenzuron in enhancing the pathogenicity of entomopathogenic fungi stimulated the research interest on combinations of CSIs and entomopathogens.

The effect of combinations of diflubenzuron on infectivity of entomopathogenic fungi, *Beauveria bassiana* Vuill, was explored by several researchers. Olson and Oetting (1999) showed that diflubenzuron interfered with the infectivity of *B. bassiana* in aphids. Nishimatsu and Jackson (1998) observed varied effects for the combinations of *B. bassiana* and diflubenzuron ranging from no interaction to additive to synergism depending on the concentration of the fungal conidia and dosage of the CSI used.

Irigaray et al. (2003) noted that an important consideration while determining the effects of a chemical on the fungus *in vitro* should be the type of formulations used. The role of additives in the formulations of IGR or type of formulations or application method in determining the infectivity of entomopathogens was elaborated by several researchers. It was stressed that the choice of carrier should be done in such a way that the carrier would not harm the activity of either fungus or the CSI. Anderson and Roberts (1983) observed the critical role of additives rather than the active ingredient of IGR on conidial germination of *B. bassiana* in determining the growth of the fungus. The wettable powder and flowable formulations caused no inhibition and often increased colony counts, whereas, the emulsifiable concentrate formulation frequently inhibited *B. bassiana* germination (Anderson et al., 1986). Olson and Oetting (1999) recommended that tank mixtures of fungus and diflubenzuron should not be done and diflubenzuron applied on foliage must dry before *B. bassiana* application.

Combinations of sublethal concentrations of CSI or insecticides and entomopathogenic fungi have been found to increase the mortality of pest insects. Quintela and McCoy (1997) observed that combinations of *B. bassiana* and *Metarhizium anisopliae* Metchnikoff with sublethal doses of imidacloprid synergistically increased the mortality of citrus root weevil, *Diaprepes abbreviatus* L. In another study, Purwar and Sachan (2006) found similar results with mustard aphid *Lipaphis erysimi* (Kaltenbach). Bitsadze et al. (2012) observed that sublethal concentrations of *B. bassiana* and CSIs (diflubenzuron and novaluron) increased the mortality when both the IGR and fungi were applied simultaneously.

However, there are some reports on the negative effects of IGRs on entomopathogens. The conidial viability of *M. anisopliae* was reduced by teflubenzuron (Nomolt™) or lufenuron (Luke and Bateman, 2006; de Almeida Alves et al., 2011).

5.6 Novel Utilities of Insect Growth Regulators in Insect Pest Management

Studies have demonstrated some novel utilities of IGRs in pest management programs, such as improved harvest of virus occlusion bodies in JHA-treated insects, horizontal transmission of JHA from exposed adults to oviposition sites and from exposed males to virgin females in vector insects, transovarial ovicidal effect on crop and stored product pests, extraordinary long-term persistence of a combination product involving IGR, and improved annihilation of fruit flies in a combination system involving sterile insect and chemosterilant bait containing an IGR.

Elvira et al. (2010) brought out an interesting aspect of utilizing fenoxycarb in mass production of a nuclear polyhedrosis virus (NPV). They reported that harvest of NPV was markedly enhanced in supernumerary sixth-instar larvae obtained by application of fenoxycarb in the beet armyworm, *Spodoptera exigua* (Hübner). They suggested that fenoxycarb could be used for mass production of viral occlusion bodies in insects. Kwon and Kim (2007) found that pyriproxyfen enhanced pathogenic effect of *Bacillus thuringiensis kurstaki* on the diamondback moth, *Plutella xylostella* (L.) by probably exerting an immunosuppressive activity.

Itoh et al. (1994) exposed the blood-fed adult females of *A. aegypti* females to a surface treated with pyriproxyfen; when the exposed adults were allowed to lay eggs into a cup of water containing fourth-instar larvae, the development of larvae into adults was highly inhibited. Such horizontal transmission of pyriproxyfen from exposed adults to larvae in water was confirmed by Chism and Apperson (2003) and Sihuincha et al. (2005). Field application of this technology for vector control of *A. aegypti* was described by Devine et al. (2009). Gaugler et al. (2012) observed the horizontal transfer of pyriproxyfen in *A. albopictus* and reported venereal transfer of the compound from contaminated males to virgin females and subsequent transmission by the females to oviposition sites.

Transovarial effect of application of IGR was recorded in several crop and stored product pests. Wise et al. (2007) observed that application of novaluron (Rimona™) strongly reduced the larval survival of plum weevil, *Conotrachelus nenuphar* Herbst, after mated females were exposed to treated substrate. Transovarial effect of pyriproxyfen was observed in *B. tabaci* on cotton (Ishaaya and Horowitz, 1992); of novaluron on apple codling moth, *C. pomonella* (Kim et al., 2011); of flucycloxuron on mites *Tetranychus urticae* Koch and *Panonychus ulmi* Koch (Grosscurt, 1993).

Salokhe et al. (2003) reported that transovarial effects were manifested in production of abnormal nonviable eggs in *T. castaneum* when adult beetles were exposed to diet containing flufenoxuron. Transovarial effect of novaluron on *T. castaneum* was demonstrated by Trostanetsky and Kostyukovsky (2008).

Mosqueira et al. (2010) observed the long-term residual effects of IGR–insecticide combination product against vector insects, such as *Anopheles gambiae* and *C. quinquefasciatus*. They observed in a field trial in Benin, West Africa, that insecticide paints of Inesfly™ 5A IGR (containing chlorpyrifos + diazinon + pyriproxyfen) could result in up to 93% and 55% mortality for *A. gambiae* and *C. quinquefasciatus* at nine months after application.

A study conducted on comparative evaluation of sterile insect technique and combination of sterile insect + chemosterilant bait system containing lufenuron reported that plots exposed to the combined method resulted in significantly higher reduction in populations of Mediterranean fruit fly, *C. capitata* (Navarro-Llopis et al., 2011).

6. EFFECT ON NATURAL ENEMIES

IGRs are considered as ecofriendly control measures in comparison to synthetic chemical insecticides. Compared to conventional insecticides, they do not cause harm to natural enemies of crop pests, like parasites and predators. Studies have demonstrated the beneficial impact of

IGRs on natural enemy population in agroecosystems. Van Driesche and Lyon (2003) showed that combining kinoprene application with release of an endoparasite, *Eretmocerus eremicus* Rose effectively maintained a low level of population density of the whitefly, *B. tabaci* on poinsettia. Flufenoxuron was found safe to hymenopteran parasitoids such as, *Diadegma semiclausum* Hellen and *Oomyzus sokolowskii* (Kurdjimov) (Haseeb et al., 2005), *Telenomus remus* Nixon (Carmo et al., 2010a), *Trichogramma pretiosum* Riley (Carmo et al., 2010b), *Eretmocerus mundus* Mercet, *E. eremicus*, and *Encarsia formosa* Gahan (Sugiyama et al., 2011).

Despite these apparent advantages, there are concerns about their safety to natural enemies as they may exert lethal or harmful sublethal effects on beneficial fauna such as predators (Hattingh, 1996; Nakahira et al., 2010; Carvalho et al., 2011) and parasitoids (Hattingh, 1996; Schneider et al., 2004) of agroecosystem.

6.1 Effect on Parasitoids

Kanzaki and Tanaka (2010) observed that endoparasitoid wasp *Meteorus pulchricornis* Wesmael was more susceptible to pyriproxyfen than the host larvae of oriental armyworm, *Mythimna separata* (Walker). Rothwangl et al. (2004) reported that pyriproxyfen was comparatively less toxic than kinoprene to *Leptomastix dactylopii* (Howard), parasitoids of the citrus mealybug, *P. citri*. Some studies have recorded the negative effects of pyriproxyfen on endoparasitoids, *Pseudacteon* spp. (Farnum and Loftin, 2010) and on natural enemies of whitefly (Simmons and Abd-Rabou, 2011). Negative effects of lufenuron on egg parasitism by *T. pretiosum* were recorded by Bastos et al. (2006) and Vianna et al. (2009). However, Carvalho et al. (2010) reported that lufenuron was generally harmless to *T. pretiosum*.

Studies with topical and feeding assays by mixing chlorfluazuron with honey (as adult feed) against parasitoid of diamondback moth, *D. semiclausum* and *O. sokolowskii*, parasitoids of *P. xylostella* revealed that contact toxicity was negligible to both the parasitoids, but feeding the adults with chlorfluazuron + honey reduced parasitism by 84% and 77.5% in *D. semiclausum* and *O. sokolowskii*, respectively (Haseeb et al., 2005).

6.2 Effect on Predators

Lufenuron was found harmful to some predatory insects like the lacewing, *Ceraeochrysa cubana* Hagen (Carvalho et al., 2011) and the bugs, *Orius laevigatus* (Say) (Angeli et al., 2005) and *Pilophorus typicus* (Nakahira et al., 2010). Hexaflumuron was toxic by contact or by ingestion to the predatory bug, *O. laevigatus* (Angeli et al., 2005). Zhao et al. (2012) found that hexaflumuron and chlorfluazuron were the least toxic compounds among 30 insecticides tested on the wasp, *T. japonicum*, an egg parasitoid of crop pests. Cutler et al. (2006) showed that Novaluron was toxic to the predatory bugs, *Podisus maculiventris* (Say) (a natural enemy of the Colorado potato beetle) and mirid bugs, *P. typicus* (Nakahira et al., 2010). Toxicity of pyriproxyfen to predatory beetles of scale insects was recorded by Mendel et al. (1994) and Hattingh (1996).

Etoxazole was toxic to the eggs of predatory mites, *Amblyseius womersleyi* Schicha (Kim and Seo, 2001), *Phytoseiulus persimilis* (Athias-Henroit) (Kim and Yoo, 2002), *Neoseiulus californicus* and *Neoseiulus womersleyi* Schicha (Amano et al., 2004). Etoxazole was found toxic against predatory mites, *Orius insidiosus* (Say) and *Orius strigicollis* (Poppius), natural enemies of phytophagous mites and thrips, respectively (Ashley et al., 2006).

7. EFFECT ON BEES AND POLLINATORS

IGRs have generally been found to exert no direct lethal effect on honeybees and non-Apis bees. However, investigations on some of the most common pollinators, viz., European honeybee, *Apis mellifera* L. and Asian honeybee, *Apis cerana* L. and the bumblebee, *Bombus terrestris* L. through topical application of IGRs on bees or honeycomb and oral feeding of bees by mixing the IGRs with sucrose syrup or pollen indicated the indirect effects exerted by IGRs on pollinators. Review of available literature revealed distinct differences in susceptibility between *Apis* and *Bombus* species to IGRs (Tasei, 2001).

7.1 Effect on Bumblebees

Evaluation of CSIs on bumblebees revealed the detrimental effect of some of the CSIs tested on reproduction of bumblebees when applied through pollen. Diflubenzuron was found to be more toxic to bumblebee brood. Diflubenzuron and teflubenzuron caused complete inhibition of egg hatching irrespective of route of application. Flufenoxuron caused 50% reduction in reproduction as contact application; but it completely inhibited reproduction when the same dose of the compound was given in syrup to the bees. Novaluron through syrup and with pollen was found to completely block reproduction; buprofezin and cyromazine in syrup moderately or completely inhibited reproduction of bumblebees. Both buprofezin and cyromazine were found to induce strong mortality by the pollen route. Ecdysone agonists, methoxyfenozide and tebufenozide, tested by oral and topical application showed no adverse effect on the bumblebees (Mommaerts et al., 2006). However, studies by Malone et al. (2007) revealed that novaluron did not show any lethal or sublethal toxicity on adults and offspring of *B. terrestris*, but it affected the quality of drones.

7.2 Effect on Bees

JHAs were found to accelerate behavioral aging of honeybee workers. Fenoxycarb strongly affected honeybee brood (Aupinel et al., 2007). When it was mixed in syrup, it affected the colony viability and overwintering survival of *A. mellifera*. It also affected the colony to requeen itself as none of the treated queens mated (Thompson et al., 2005). Aupinel et al. (2007) observed that although fenoxycarb did not cause any lethal effect on larvae of *A. mellifera*, at doses ≥ 25 ng larva^{-1}, it completely inhibited the emergence from pupae. Thompson et al. (2005) found no apparent long-term effects of tebufenozide on colonies and queen development in *A. mellifera*.

7.3 Effect on Other Pollinators

Scott-Dupree et al. (2009) found that direct contact exposure to technical grade novaluron was nontoxic to eastern bumblebees, *Bombus impatiens* Cresson, orchard mason bees, *Osmia lignaria* Say, leafcutting bees on alfalfa, *Megachile rotundata* Fabricius. In contrast, Hodgson et al. (2011) reported that a novaluron formulation (Rimon™ 0.83 EC) exerted a negative effect on eggs and larvae of *M. rotundata* when sprayed on the alfalfa fields.

8. RESISTANCE TO INSECT GROWTH REGULATORS

Insect resistance to IGRs, especially JHAs, was considered to be unlikely because of their unique mode of action in structurally mimicking an essential hormone in the endogenous system (Williams, 1967). However, several studies have shown the development of resistance in insects to all IGRs, such as JHAs, CSIs, and EAs. Some of the recent records (indicative but not exhaustive) of insect pests showing resistance to the IGRs are furnished in Table 3.

TABLE 3   Insects Showing Resistance to IGRs

S.No	Name of the compound	Name of the species	Country	Year	Author
Juvenile Hormone Analogues					
1	Methoprene	*Musca domestica*	Denmark	2003	Kristensen and Jespersen (2003)
		Culex pipiens pipiens	Cyprus		Vasquez et al. (2009)
2	Fenoxycarb	*Cydia pomonella*	Czechoslovakia	2007	Stara and Kocourek (2007)
3	Pyriproxyfen	*Bemisia tabaci*	Israel	1994	Horowitz and Ishaaya (1994)
			Israel-Carmel Coast	1995	Horowitz et al. (2005)
			Israel-Carmel Coast	2001	Horowitz et al. (2005)
			Spain-Murcia-Campo de Cartagena	2005	Fernandez et al. (2009)
			Pakistan-Multan	2009	Basit et al. (2011)
		Trialeurodes vaporariorum	Germany	2011	Karatolos et al. (2012)
		M. domestica	Pakistan	2014	Shah et al. (2015)
Chitin Synthesis Inhibitors					
4	Diflubenzuron	*C. pomonella*	USA	1988	Moffitt et al. (1988)
			China	2001	Jia et al. (2009), Che et al. (2012), Su and Sun (2014), and Gao et al. (2014)
			Italy	1994	Cravedi (2000)
		M. domestica	Denamrk	2001	Kristenson and Jesperson (2003)
		Plutella xylostella	Thailand	1988	Vattanatangum (1988)
		Spodoptera exigua	Netherlands	1995	Van Laecke et al. (1995)
		Spodoptera littoralis	Egypt	1997	Smagghe and Degheele (1997)
		Tuta absoluta	Brazil	2005	Silva et al. (2011)

(Continued)

TABLE 3   Insects Showing Resistance to IGRs—cont'd

S.No	Name of the compound	Name of the species	Country	Year	Author
5	Flufenoxuron	C. pomonella	Spain	2007	Rodriguez et al. (2011)
6	Triflumuron	C. pomonella	France	1990	Sauphanor and Bouvier (1995).
		P. xylostella	Thailand	1988	Vattanatangum (1988)
		T. absoluta	Brazil	2005	Silva et al. (2011)
			France	1990 and 2001	Sauphanor and Bouvier (1995) and Bouvier et al. (2002)
7	Teflubenzuron	S. exigua	France		
		S. exigua	Netherlands	1995	Van Laecke et al. (1995)
		T. absoluta	Brazil	2005	Silva et al. (2011)
		S. littoralis	Israel and Egypt, Netherlands	1990 and 1997	Ishaaya and Klein (1990), Smagghe and Degheele (1997) and Van Laecke e al. (1995)
		P. xylostella	Malaysia; China and Thailand		Vattanatangum (1988)
		C. pomonella	France	1990–2001	Sauphanor and Bouvier (1995)
8	Chlorfluazuron	P. xylostella	Thailand	1988	Vattanatangum (1988)
		P. xylostella	Nicaragua	2000	Perez et al. (2000)
		S. exigua	China	2009	Che et al. (2012)
		S. exigua	China	2011	Su and Sun (2014)
9	Lufenuron	S. exigua	Pakistan	2010	Ishtiaq and Saleem (2011)
10	Flucycloxuron	Tetranychus urticae	Belgium		Van Leeuwen et al. (2005)
11	Buprofezin	B. tabaci	Israel		Horowitz and Ishaaya (1994)
			Spain-		Fernandez et al. (2009)
			Pakistan-Multan		Basit et al. (2011)
			Pakistan-Punjab-Vehari		Basit et al. (2013)
		Nilaparvata lugens	China	2005, 2006 and 2007	Wang et al. (2008)
		Sogatella furcifera	China	2010–2013	Su et al. (2013) and Zhang et al. (2014)
		T. vaporariorum	Belgium	1995	De Cock et al. (1995)

TABLE 3 Insects Showing Resistance to IGRs—cont'd

S.No	Name of the compound	Name of the species	Country	Year	Author
Ecdysone Agonists					
12	Tebufenozide	*C. pomonella*	France	1995	Sauphanor and Bouvier (1995)
			China-Anhui-Hexian County	2003	Fuentes (2003)
		Planotortrix octo	New Zealand	1988	Wearing (1998)
		P. xylostella	China	2006	Cao and Zhaojun Han (2006)
		S. exigua	Thailand	2000	Moulton et al. (2000)
			China	2001–2011	Jia et al. (2009), Che et al. (2012), Su and Sun (2014), and Gao et al. (2014)
13	Methoxyfenozide	*S. littoralis*	Egypt	1997	Smagghe and Degheele (1997)
		S. exigua	USA	2005	Osorio et al. (2008)
			Pakistan	2010	Ishtiaq et al. (2012)
			China	2011	Gao et al. (2014)
		Choristoneura rosaceana	Canada	2001	Smirle et al. (2002)

The first report of development of resistance in insects to JHA was in a strain of *T. castaneum* showing resistance to JH1 (Dyte, 1972). Resistance to JHAs has been extensively reported since then in several insect pests. Resistance to methoprene was recorded in *M. domestica* (Cerf and Georghiou, 1972), *C. pipiens* (Georghiou et al., 1975), lesser grain borer, *R. dominica* and red flour beetle, *T. castaneum* (Daglish, 2008). Resistance to fenoxycarb was recorded in apple codling moth *C. pomonella*. It was observed that a diflubenzuron-resistant population of apple codling moth was exhibiting cross-resistance to fenoxycarb (Sauphanor and Bouvier, 1995); 45 out of the 47 field populations of *C. pomonella* drawn from France, Italy, Switzerland, Spain, and Armenia showed various degrees of resistance to fenoxycarb (Reyes et al., 2007).

Whitefly, *B. tabaci*, is resistant to several groups of neurotoxins and is found to show widespread resistance to pyriproxyfen in several countries (Horowitz et al., 2008; Crowder et al., 2007; Dennehy et al., 2010; Fernández et al., 2009; Luo et al., 2010). The Q biotype of *B. tabaci* is considered to exhibit greater resistance to pyriproxyfen than the B biotype (Horowitz et al., 2005). Greenhouse whitefly, *T. vaporariorum* (Karatolos et al., 2012) and housefly, *M. domestica* (Shah et al., 2015), were also found to show resistance to pyriproxyfen.

Ever since the first report of resistance in housefly, *M. domestica* against diflubenzuron (Pimprikar and Georghiou, 1979), resistance to CSIs has been recorded by several research groups across the world. Resistance to CSIs such as diflubenzuron, flufenoxuron, chlorfluazuron, and buprofezin have been recorded from insect pests including diamondback moth,

P. xylostella (Vattanatangum, 1988); housefly, *M. domestica* (Keiding et al., 1992); beet army-worm, *S. exigua* (Van Laecke et al., 1995); tomato leaf miner, *Tuta absoluta* (Silva et al., 2011); rice brown planthopper, *N. lugens* (Wang et al., 2008); cotton leafworm, *S. littoralis* (Van Laecke et al., 1995); and *B. tabaci* (Horowitz and Ishaaya, 1994).

Resistance to tebufenozide was recorded first in apple codling moth (Sauphanor and Bouvier, 1995). Several species of insects including armyworm, *S. exigua* (Moulton et al., 2002); greenheaded leafroller, *Planotortrix octo* Dugdale (Wearing, 1998); and diamondback moth, *P. xylostella* (Cao and Zhaojun Han, 2006) have shown resistance to ecdysone ago-nists, such as tebufenozide and methoxyfenozide.

9. CONCLUSIONS

Insect hormones have been the target site for IGRs, viz., JH analogues, chitin synthesis inhibitors, and ecdysone agonists. Rigorous efforts by chemists have shaped the develop-ment of improved analogues of JHA, CSIs, and ecdysone agonists to cater to diverse pest management programs. The efficacy and photostability of early isoprenoid juvenoids have been improved through structural modifications in the basic juvenoid structure to develop newer analogues such as fenoxycarb and pyriproxyfen. Besides the BPU derivatives of chitin synthesis inhibitors, newer classes of CSIs including thiadiazines (buprofezin) and oxazolines (etoxazole) have been developed in the recent past with improved efficacy against sucking pests and mites. Understanding the molecular basis of hormones and elucidation of the bio-chemical pathway of chitin synthesis will help to identify new target sites for novel IGR com-pounds with improved potency and spectrum of action for sustainable pest management. There is a scope for developing chemicals with anti-JH, anti-ecdysteroid activities.

Acknowledgments

The authors gratefully acknowledge the constant support and encouragement from Drs N. Ramakrishnan, V. V. Ramamurthy, and B. Subrahmanyam. The authors also thank the support and help from Drs. G.T.Gujar, Chitra Srivastava, and Sachin S. Suroshe.

References

Abo-Elghar, G.E., El-Sheikh, A.E., El-Sayed, F.M., El-Maghraby, H.M., El-Zun, H.M., 2004. Persistence and residual activity of an organophosphate, pirimiphosmethyl, and three IGRs, hexaflumuron, teflubenzuron and pyriproxyfen, against the cowpea weevil, *Callosobruchus maculatus* (Coleoptera: Bruchidae). Pest Management Science 60, 95–102.

Aller, H.E., Ramsay, J.R., 1988. RH-5849—a novel insect growth regulator with a new mode of action. Brighton Crop Protection Conference: Pests and Diseases 2, 511–518.

de Almeida Alves, M.M.T., Orlandelli, R.C., Lourenço, D.A.L., Pamphil, J.A., 2011. Toxicity of the insect growth regulator lufenuron on the entomopathogenic fungus *Metarhizium anisopliae* (Metschnikoff) Sorokin assessed by conidia germination speed parameter. African Journal of Biotechnology 10, 9661–9667.

Al-Mekhlafi, F.A., Mashaly, A.M.A., Wadaan, M.A., Al-Mallah, N.M., 2012. Effect of different applicable conditions of the insect growth regulator (cyromazine) on the southern cowpea weevils, *Callosobruchus maculatus* reared on peas. Pakistan Journal Of Zoology 44, 481–488.

Amano, H., Ishii, Y., Kobori, Y., 2004. Pesticide susceptibility of two dominant phytoseiid mites, *Neoseiulus californicus* and *N. womersleyi*, in conventional Japanese fruit orchards (Gamasina: Phytoseiidae). Journal of the Acarological Society of Japan 13, 65–70.

Amelotti, I., Catalá, S.S., Gorla, D.E., 2009. Experimental evaluation of insecticidal paints against *Triatoma infestans* (Hemiptera; Reduviidae), under natural climatic conditions. Parasites &Vectors 2, 30.

Anderson, T.E., Roberts, D.W., 1983. Compatibility of *Beauveria bassiana* strain with formulations used in Colorado potato beetle (Coleoptera: Chrysomelidae) control. Journal of Economic Entomology 76, 1437–1441.

Anderson, M., Fisher, J.P., Robinson, J., 1986. Flufenoxuron–an acylurea acaricide/insecticide with novel properties. In: Proceedings of British Crop Protection Conference on Pests and Diseases. Brighton, UK, pp. 89–96.

Angeli, G., Baldessari, M., Maines, R., Duso, C., 2005. Side-effects of pesticides on the predatory bug *Orius laevigatus* (Heteroptera: Anthocoridae) in the laboratory. Biocontrol Science and Technology 15, 745–754.

Arredondo-Jiménez, J.I., Valdez-Delgago, K.M., 2006. Effect of novaluron (Rimon™ 10EC) on the mosquitoes *Anopheles albimanus, Anopheles pseudopunctipennis, Aedes aegypti, Aedes albopictus* and *Culex quinquefasciatus* from Chiapas, Mexico. Medical and Veterinary Entomology 20, 377–387.

Ashihara, W., Kondo, A., Shibao, M., Tanaka, H., Hiehata, K., Izumi, K., 2004. Ecology and control of eriophyid mites injurious to fruit trees in Japan. Japan Agricultural Research Quarterly 38, 31–41.

Ashley, J.L., Herbert, D.A., Lewis, E.E., Brewster, C.C., Huckaba, R., 2006. Toxicity of three acaricides to *Tetranychus urticae* (Tetranychidae: Acari) and *Orius insidiosus* (Anthocoridae: Hemiptera). Journal of Economic Entomology 99, 54–59.

Athanassiou, C.G., Kavallieratos, N.G., Vayias, B.J., Tomanović, Z., Petrović, A., Rozman, V., Adler, C., Korunic, Z., Milovanović, D., 2011. Laboratory evaluation of diatomaceous earth deposits mined from several locations in central and southeastern Europe as potential protectants against coleopteran grain pests. Crop Protection 30, 329–339.

Aupinel, P., Fortini, D., Michaud, B., Marolleau, F., Tasei, J.-N., Odoux, J.F., 2007. Toxicity of dimethoate and fenoxycarb to honey bee brood (*Apis mellifera*), using a new in vitro standardized feeding method. Pest Management Science 63, 1090–1094.

Bakr, R.F.A., El-Monairy, O.M., El-Barky, N.M., El-Shourbagy, N.M.B., 2010. Toxicological and behavioral effects of chlorfluazuron on pheromone production and perception of *Tribolium castaneum* (Coleoptera: Tenebrionidae). Egyptian Academic Journal of Biological Sciences C, Physiology and Molecular Biology 2, 61–72.

Banks, W.A., Lofgren, C.S., 1991. Effectiveness of the insect growth regulator pyriproxyfen against red imported fire ant (Hymenoptera: Formicidae). Journal of Entomological Science 26, 331–338.

Basit, M., Saeed, S., Saleem, M.A., Denholm, I., Shah, M., 2013. Detection of resistance, cross-resistance, and stability of resistance to new chemistry insecticides in *Bemisia tabaci* (Homoptera: Aleyrodidae). Entomological Society of America 106, 1414–1422.

Basit, M., Sayyed, A., Saleem, M., Saeed, S., 2011. Cross-resistance, inheritance and stability of resistance to acetamiprid in cotton whitefly, *Bemisia tabaci* Genn (Hemiptera: Aleyrodidae). Crop Protection 30, 705–712.

Bastos, C.S., de Almeida, R.P., Suinaga, F.A., 2006. Selectivity of pesticides used on cotton (*Gossypium hirsutum*) to *Trichogramma pretiosum* reared on two laboratory-reared hosts. Pest Management Science 62, 91–98.

Becher, H.M., Becker, P., Prokic-Immel, R., Wirtz, W., 1983. CME-134, a new chitin synthesis inhibiting insecticide. In: Proceeding of 10th International Congress of Plant Protection, vol. 1, pp. 408–415.

Bisdorff, B., Wall, R., 2008. Sheep blowfly strike risk and management in Great Britain: a survey of current practice. Medical and Veterinary Entomology 22, 303–308.

Bitsadze, N., Jaronski, S., Khasdan, V., Abashidze, E., Abashidze, M., Latchininsky, A., Samadashvili, D., Sokhadze, I., Rippa, M., Ishaaya, I., Horowitz, A.R., 2012. Joint action of *Beauveria bassiana* and the insect growth regulators diflubenzuron and novaluron, on the migratory locust, *Locusta migratoria*. Journal of Pesticide Science 86 (2), 293–300. http://dx.doi.org/10.1007/s10340-012-0476-4.

Boina, D.R., Rogers, M.E., Wang, N., Stelinski, L.L., 2010. Effect of pyriproxyfen, a juvenile hormone mimic, on egg hatch, nymph development, adult emergence and reproduction of the Asian citrus psyllid, *Diaphorina citri* Kuwayama. Pest Management Science 66, 349–357.

Bouvier, J.C., Boivin, T., Besley, D., Sauphanor, B., 2002. Age-dependent response to insecticides and Enzymatic Variation in susceptible and resistant codling moth larvae. Archives of Insect Biochemistry and Physiology 51, 55–66.

Bowers, W.S., 1971. Chemistry and biological activity of morphogenetic agents. Mitteilungen der Schweizerischen Entomologischen Gesellschaft 44, 115–130.

Bowers, W.S., Fales, H.M., Thompson, M.J., Uebel, E.C., 1966. Juvenile hormone: identification of an active compound from balsam fir. Science 154, 1020–1021.

Bull, M.S., Swindale, S., Overend, D., Hess, E.A., 1996. Suppression of *Boophilus microplus* population with fluazuron–an acarine growth regulator. Australian Veterinary Journal 74, 468–470.

Cao, G., Han, Z., 2006. Tebufenozide resistance selected in *Plutella xylostella* and its cross-resistance and fitness cost. Pest Management Science 62, 746–751.

Casaná-Giner, V., Gandía-Balaguer, A., Mengod-Puerta, C., Primo-Millo, J., Primo-Yú fera, E., 1999. Insect growth regulators as chemosterilants for *Ceratitis capitata* (Diptera: Tephritidae). Journal of Economic Entomology 92, 303–308.

Carmo, E.L., Bueno, A.F., Bueno, R.C.O.F., 2010a. Pesticide selectivity for the insect egg parasitoid *Telenomus remus*. BioControl 55, 455–464.

Carmo, E.L.do, Bueno, A.F., Bueno, R.C.O.F., Vieira, S.S., Goulart, M.M.P., Carneiro, T.R., 2010b. Selectivity of pesticides used in soybean crops to *Trichogramma pretiosum* Riley, 1879 (Hymenoptera: Trichogrammatidae) pupae. Arquivos do Instituto Biológico, São Paulo 77, 283–290 (in Portuguese with English abstract).

Carvalho, G.A., Carvalho, C.F., Ferreira, M. do N., 2011. Toxicity of acaricides to eggs and adults of *Ceraeochrysa cubana* (Hagen, 1861) (Neuroptera: Chrysopidae). Ciência e Agrotecnologia 35, 165–171 (in Portuguese with English abstract).

Carvalho, G.A., Godoy, M.S., Parreira, D.S., Rezende, D.T., 2010. Effect of chemical insecticides used in tomato crops on immature *Trichogramma pretiosum* (Hymenoptera: Trichogrammatidae). Revista Colombiana de Entomologia 36, 10–15.

Cerf, D.C., Georghiou, G.P., 1972. Evidence of cross-resistance to a juvenile hormone analogue in some insecticide-resistant house flies. Nature 239, 401–402.

Cerný, V., Dolejš, L., Lábler, L., Šorm, F., Sláma, K., 1967. Dehydrojuvabione - a novel compound with juvenile hormone activity from balsam fir. Tetrahedron Letters 1967, 1053–1057.

Cetin, H., Erler, F., Yanikoglu, A., 2006. Larvicidal activity of novaluron, a chitin synthesis inhibitor, against the house fly, *Musca domestica*. Journal of Insect Science 6 (50).

Chanbang, Y., Arthur, F.H., Wilde, G.E., Throne, J.E., Subramanyam, B., 2008. Susceptibility of eggs and adult fecundity of the lesser grain borer, *Rhyzopertha dominica*, exposed to methoprene. Journal of Insect Science 8, 48.

Chapman, R.F., 2012. In: Simpson, S.J., Douglas, A.E. (Eds.), The Insects Structure and Function. Cambridge University Press, p. 956.

Charmillot, P.J., Gourmelon, A., Fabre, A.L., Pasquier, D., 2001. Ovicidal and larvicidal effectiveness of several insect growth inhibitors and regulators on the codling moth *Cydia pomonella* L. (Lep., Tortricidae). Journal of Applied Entomology 125, 147–153.

Che, W., Shi, T., Wu, Y., Yang, Y., 2012. Insecticide resistance status of field populations of *Spodoptera exigua* (Lepidoptera: Noctuidae) from China. Journal of Economic Entomology 106 (4), 1855–1862.

Chism, B.D., Apperson, C.S., 2003. Horizontal transfer of the insect growth regulator pyriproxyfen to larval microcosms by gravid *Aedes albopictus* and *Ochlerotatus triseriatus* mosquitoes in the laboratory. Medical and Veterinary Entomology 17, 211–220.

Cohen, E., 2010. Chitin biochemistry: synthesis, hydrolysis and inhibition. Advances in Insect Physiology 38, 5–74.

Collins, D.A., 2006. A review of alternatives to organophosphorus compounds for the control of storage mites. Journal of Stored Products Research 42, 395–426.

Cottage, E.L., Gunning, R.V., 2006. Buprofezin inhibits acetylcholinesterase activity in B-biotype *Bemisia tabaci*. Journal of Molecular Neuroscience 30, 39–40.

Cravedi, P., 2000. Resistance to pesticides. In: Insects and Acari in Italy. http://www.iacr.bbsrc.ac.uk/enmaria/database/italy/cravedi1.html.

Crowder, D.W., Dennehy, T.J., Ellers-Kirk, C., Yafuso, C.M., Ellsworth, P.C., Tabashnik, B.E., Carrière, Y., 2007. Field evaluation of resistance to pyriproxyfen in *Bemisia tabaci* (B biotype). Journal of Economic Entomology 100, 1650–1656.

Cutler, G.C., Scott-Dupree, C.D., Tolman, J.H., Harris, C.R., 2006. Toxicity of the insect growth regulator novaluron to the non-target predatory bug *Podisus maculiventris* (Heteroptera: Pentatomidae). Biological Control 38, 196–204.

De Cock, A., Degheele, D., 1991. Effects of buprofezin on the ultrastructure of the third instar cuticle if *Trialeurodes vaporariorum*. Tissue Cell 23, 755–762.

De Cock, A., Ishaaya, I., Van De Veire, M., Degheele, D., 1995. Response of Buprofezin-Susceptible and -Resistant strains of *Trialeurodes vaporariorum* (Homoptera: Aleyrodidae) to Pyriproxyfen and Diafenthiuron. Journal of Economic Entomology 88, 763–767.

Daglish, G.J., Wallbank, B.E., 2005. Efficacy of diflubenzuron plus methoprene against *Sitophilus oryzae* and *Rhyzopertha dominica* in stored sorghum. Journal of Stored Products Research 41, 353–360.

Daglish, G.J., 2008. Impact of resistance on the efficacy of binary combinations of spinosad, chlorpyrifos-methyl and s-methoprene against five stored-grain beetles. Journal of Stored Products Research 44, 71–76.

DeMark, J.J., Benson, E.P., Zungoli, P.A., Kard, B.M., 1995. Evaluation of hexaflumuron for termite control in the southeast U.S. Down to Earth 50, 20–26.

Dennehy, T.J., Degain, B.A., Harpold, V.S., Zaborac, M., Morin, S., Fabrick, J.A., Nichols, R.L., Brown, J.K., Byrne, F.J., Li, X., 2010. Extraordinary resistance to insecticides reveals exotic Q biotype of *Bemisia tabaci* in the New World. Journal of Economic Entomology 103, 2174–2186.

Devine, G.J., Perea, E.Z., Killeen, G.F., Stancil, J.D., Clark, S.J., Morrison, A.C., 2009. Using adult mosquitoes to transfer insecticides to *Aedes aegypti* larval habitats. Proceedings of the National Academy of Sciences of the United States of America 106, 11530–11534.

Dhadialla, T.S., Retnakaran, A., Smagghe, G., 2010. Insect growth- and development disrupting insecticides. In: Gilbert, L.I., Gill, S.S. (Eds.), Insect Control. Elsevier, New York, pp. 121–184.

Dinan, L., Harmatha, J., Volodin, V., Lafont, R., 2009. Phytoecdysteroids: diversity, biosynthesis and distribution. In: Smagghe, G. (Ed.), Ecdysone: Structures and Functions. Springer Science + Business Media, Dordrecht, The Netherlands, pp. 3–45.

Doppelreiter, V.H., Korioth, M., 1981. Entwicklungshemmung durch Diflubenzuron bei den Bodentermiten Heterotermes indicola und Reticulitermes flavipes. Zeitschrift für Angewandte Entomologie 91, 131–137.

Dorn, S., Frischknecht, M.L., Martinez, V., Zurflüh, R., Fischer, U., 1981. A novel non-neurotoxic insecticide with a broad activity spectrum. Zeitschrift fur Pflanzenkrankheiten und Pflanzenschutz 88, 269–275.

Dyte, C.E., 1972. Resistance to synthetic juvenile hormone in a population of the flour beetle, *Tribolium castaneum*. Nature 238, 48–49.

Elek, J.A., Longstaff, B.C., 1994. Effect of chitin-synthesis inhibitors on stored-product beetles. Pesticide Science 40, 225–230.

Eliahu, M., Blumberg, D., Horowitz, A.R., Ishaaya, I., 2007. Effect of pyriproxyfen on developing stages and embryogenesis of California red scale (CRS), *Aonidiella aurantii*. Pest Management Science 63, 743–746.

Elvira, S., Williams, T., Caballero, P., 2010. Juvenile hormone analog technology: effects on larval cannibalism and the production of *Spodoptera exigua* (Lepidoptera: Noctuidae) nucleopolyhedrovirus. Journal of Economic Entomology 103, 577–582.

Evans, T.A., 2010. Rapid elimination of field colonies of subterranean termites (Isoptera: Rhinotermitidae) using bistrifluron solid bait pellets. Journal of Economic Entomology 103, 423–432.

Faragalla, A.A., Badawi, A.I., Dabbour, A.I., 1985. Field evaluation of the effects of the juvenile hormone analogs (JHA's) and diflubenzuron (Dimilin) on termites of the genus *Microcerotermes* (Isoptera: Termitidae) in the central region of Saudi Arabia. Sociobiology 11, 29–37.

Farnum, J.M., Loftin, K.M., 2010. Impact of methoprene and pyriproxyfen on *Pseudacteon tricuspis* (Diptera: Phoridae), a parasitoid of the red imported fire ant, *Solenopsis invicta* (Hymenoptera: Formicidae). Florida Entomologist 93, 584–589.

Fathpour, H., Meftahi, G.H., Afrouz, T., 2009. Contact effect of the juvenoid, fenoxycarb, on growth and reproduction of German cockroach *Blattella germanica* (Dictyoptera: Blattellidae). Iranian Journal of Biology 22, 343–353.

Fernandez, E., Gravalos, C., Javier Haro, P., Cifuentes, D., Bielza, P., 2009. Insecticide resistance status of *Bemisia tabaci* Q-biotype in south-eastern Spain. Pest Management Science 65, 885–891.

Forschler, B.T., Ryder, J.C., 1996. Subterranean termite, *Reticulitermes* spp. (Isoptera: Rhinotermitidae), colony response to baiting with hexaflumuron using a prototype commercial termite baiting system. Journal of Entomological Science 31, 143–151.

Fourie, J.J., Fourie, L.J., Horak, I.G., Snyman, M.G., 2010. The efficacy of a topically applied combination of cyphenothrin and pyriproxyfen against the southern African yellow dog tick, *Haemaphysalis elliptica*, and the cat flea, *Ctenocephalides felis*, on dogs. Journal of the South African Veterinary Association 81, 33–36.

Fuentes, E., 2003. Polilla de la Manzana. Pomaceas Boletin Tecnico 3 (6), 1–3.

Gao, M., Li, X., Mu, W., Zhang, P., Zhou, C., 2014. Resistant levels of *Spodoptera exigua* to eight various insecticides in Shandong, China. Journal of Pesticide Science 39, 7–13.

Gaugler, R., Suman, D., Wang, Y., 2012. An autodissemination station for the transfer of an insect growth regulator to mosquito oviposition sites. Medical and Veterinary Entomology 26, 37–45.

Getty, G.M., Solek, C.W., Sbragia, R.J., Haverty, M.I., Lewis, V.R., 2007. Large-scale suppression of a subterranean termite community using Sentricon™ Termite Colony Elimination System: a case study in Chatsworth, California, USA. Sociobiology 50, 1041–1050.

Georghiou, G.P., Ariaratnam, V., Pasternak, M.E., Lin, C.S., 1975. Organophosphorus multiresistance in *Culex pipiens quinquefasciatus* in California. Journal of Economic Entomology 68, 461–467.

Ghasemi, A., Sendi, J.J., Ghadamyari, M., 2010. Physiological and biochemical effect of pyriproxyfen on Indian meal moth *Plodia interpunctella* (Hübner) (Lepidoptera: Pyralidae). Journal of Plant Protection Research 50, 416–422.

Glenn, G.J., Austin, J.W., Gold, R.E., 2008. Efficacy of commercial termite baiting systems for management of subterranean termites (Isoptera: Rhinotermitidae) in Texas. Sociobiology 51, 333–362.

Gogi, M.D., Sarfraz, R.M., Dosdall, L.M., Arif, M.J., Keddie, A.B., Ashfaq, M., 2006. Effectiveness of two insect growth regulators against *Bemisia tabaci* (Gennadius) (Homoptera: Aleyrodidae) and *Helicoverpa armigera* (Hübner) (Lepidoptera: Noctuidae) and their impact on population densities of arthropod predators in cotton in Pakistan. Pest Management Science 62, 982–990.

Grace, J.K., Yates, J.R., Tome, C.H.M., Oshiro, R.J., 1996. Termite-resistant construction: use of a stainless steel mesh to exclude *Coptotermes formosanus* (Isoptera: Rhinotermitidae). Sociobiology 28, 365–372.

Grafton-Cardwell, E.E., Lee, J.E., Stewart, J.R., Olsen, K.D., 2006. Role of two insect growth regulators in integrated pest management of citrus scales. Journal of Economic Entomology 99, 733–744.

Grenier, S., Grenier, A.-M., 1993. Fenoxycarb, a fairly new insect growth regulator: a review of its effects on insects. Annals of Applied Biology 122, 369–403.

Grosscurt, A.C., 1978. Diflubenzuron: some aspects of its ovicidal and larvicidal mode of action and an evaluation of its practical possibilities. Pesticide Science 9, 373–386.

Grosscurt, A.C., 1993. Factors influencing the acaricidal activity of flucycloxuron. Entomologia Experimentalis et Applicata 69, 201–208.

Haga, T., Tobi, T., Koyanagi, T., Nishiyama, R., 1982. Structure activity relationships of a series of benzoylpyridyloxyphenyl-urea derivatives. In: Fifth International Congress on Pesticide Chemistry. IUPAC, Kyoto, Japan, p. 7.

Hall, R.D., Foehse, M.C., 1980. Laboratory and field tests of CGA-72662 for control of the house fly and face fly in poultry, bovine and swine manure. Journal of Economic Entomology 73, 564–569.

Hamman, I., Sirrenberg, W., 1980. Laboratory evaluation of SIR 8514, a new chitin synthesis inhibitor of the benzoylated urea class. Pflanzenschutz-Nachrichten Bayer 33, 1–34.

Haseeb, M., Amano, H., Liu, T.X., 2005. Effects of selected insecticides on *Diadegma semiclausum* (Hymenoptera: Ichneumonidae) and *Oomyzus sokolowskii* (Hymenoptera: Eulophidae), parasitoids of *Plutella xylostella* (Lepidoptera: Plutellidae). Insect Science 12, 163–170.

Hassan, A.E.M., Charnley, A.K., 1987. The effect of Dimilin on the ultrastructure of the integument of *Manduca sexta*. Journal of Insect Physiology 33, 669–676.

Hatakoshi, M., 2012. Pyriproxyfen: a new juvenoid. In: Kramer, W., Schirmer, U., Jeschke, P., Witschel, M. (Eds.), Modern Crop Compound, second ed. Wiley-VCH Verlag GmbH & Co. KGaA, Weinheim, pp. 963–998.

Hattingh, V., 1996. The use of insect growth regulators—implications for IPM with citrus in southern Africa as an example. Entomophaga 41, 513–518.

He, Y.-P., Shao, Z.-R., Chen, W.-M., Liang, G.-M., Li, Y.-P., Zhou, W.-J., Shen, J.L., 2008. Laboratory screening of alternative insecticides for highly toxic insecticides against the striped stem borer (*Chilo suppressalis*) on rice. Zhongguo Shuidao Kexue (Chinese Journal of Rice Science) 22, 313–320 (in Chinese and English abstract).

Hegde, M., Nidagundi, J., 2009. Effect of newer chemicals on planthoppers and their mirid predator in rice. Karnataka Journal of Agricultural Sciences 22, 511–513.

Heller, J.J., Mattioda, H., Klein, E., Sagenmüller, A., 1992. Field evaluation of RH 5992 on lepidopterous pests in Europe. Brighton Crop Protection Conference Pests and Diseases 1, 59–65.

Henrick, C.A., Staal, G.B., Siddall, J.B., 1973. Alkyl 3,7,11-trimethyl-2,4-dodecadienoates, a new class of potent insect growth regulators with juvenile hormone activity. Journal of Agricultural and Food Chemistry 21, 354–359.

Hodgson, E.W., Pitts-Singer, T.L., Barbour, J.D., 2011. Effects of the insect growth regulator, novaluron on immature alfalfa leaf cutting bees, *Megachile rotundata*. Journal of Insect Science 11 (43).

Horowitz, A.R., Ishaaya, I., 1994. Managing resistance to insect growth regulators in the sweetpotato whitefly (Homoptera: Aleyrodidae). Journal of Economic Entomology 87 (4), 866–871.

Horowitz, A.R., Kontsedalov, S., Khasdan, V., Ishaaya, I., 2005. Biotypes B and Q of *Bemisia tabaci* and their relevance to neonicotinoid and pyriproxyfen resistance. Archives of Insect Biochemistry and Physiology 58, 216–225.

Horowitz, R., Kontsedalov, S., Khasdan, V., Breslauer, H., Ishaaya, I., 2008. The biotypes B and Q of *Bemisia tabaci* in Israel distribution, resistance to insecticides and implications for pest management. Journal of Insect Science 8, 23–24.

Hussender, C., Simms, D.M., Riegel, C., 2007. Evaluation of treatment success and patterns of reinfestation of the Formosan subterranean termite (Isoptera: Rhinotermitidae). Journal of Economic Entomology 100, 1370–1380.

Hyashi, T., Iwamura, H., Fujita, T., 1990. Electrostatic and stereochemical aspects of insect juvenile hormone active compounds: a clue for high activity. Journal of Agricultural and Food Chemistry 38, 1972–1977.

Ichikawa, Y., Komatsu, M., Takigawa, T., Mori, K., Kramer, K.J., 1980. Synthesis of pyridyl terpenoid ether analogs of juvenile hormone. Agricultural and Biological Chemistry 44, 2709–2715.

Irigaray, F.J.S.C., Marco-Mancebón, V., Pérez-Moreno, I., 2003. The entomopathogenic fungus *Beauveria bassiana* and its compatibility with triflumuron: effects on the twospotted spider mite *Tetranychus urticae*. Biological Control 26, 168–173.

Ishaaya, I., 1990. Benzoylphenyl ureas and other selective control agents–mechanism and application. In: Casida, J.E. (Ed.), Pesticides and Alternatives. Elsevier, Amsterdam, pp. 365–376.

Ishaaya, I., Klein, M., 1990. Response of susceptible laboratory and resistant field strains of *Spodoptera littoralis* (Lepidoptera: Noctuidae) to teflubenzuron. Journal of Economic Entomology 83, 59–62.

Ishaaya, I., Horowitz, A.R., 1992. Novel phenoxy juvenile hormone analog (pyriproxyfen) suppresses embryogenesis and adult emergence of sweetpotato whitefly (Homoptera: Aleyrodidae). Journal of Economic Entomology 85, 2113–2117.

Ishaaya, I., Yablonski, S., Mendelson, Z., Mansour, Y., Horowitz, A.R., 1996. Novaluron (MCW-275), a novel benzoylphenyl urea, suppressing developing stages of lepidopteran, whitefly and leafminer pests. In: Proceedings of British Crop Protection Conference on Pests and Diseases, vol. 3. Brighton, UK, pp. 1013–1020.

Ishida, T., Suzuki, J., Tsukidate, Y., 1994. YI-5301, a novel oxazoline acaricide. In: Proceedings of British Crop Protection Conference on Pests and Diseases, vol. 1. Brighton, UK, pp. 37–44.

Ishtiaq, M., Saleem, M.A., 2011. Generating susceptible strain and resistance status of field populations of *Spodoptera exigua* (Lepidoptera: Noctuidae) against some conventional and new chemistry insecticides in Pakistan. Journal of Economic Entomology 104, 1343–1348.

Ishtiaq, M., Saleem, M.A., Razaq, M., 2012. Monitoring of resistance in *Spodoptera exigua* (Lepidoptera: Noctuidae) from four districts of the Southern Punjab, Pakistan to four conventional and six new chemistry insecticides. Crop Proctection 33, 13–20.

Itoh, T., Kawada, H., Abe, A., Eshita, Y., Rongsriyam, Y., Igarashi, A., 1994. Utilization of bloodfed females of *Aedes aegypti* as a vehicle for the transfer of the insect growth regulator pyriproxyfen to larval habitats. Journal of the American Mosquito Control Association 10, 344–347.

Jambulingam, P., Sadanandane, C., Nithiyananthan, N., Subramanian, S., Zaim, M., 2009. Efficacy of novaluron against *Culex quinquefasciatus* in small- and medium-scale trials, India. Journal of the American Mosquito Control Association 25, 315–322.

Jia, B., Liu, Y., Zhu, Y.C., Liu, X., Gao, C., Shen, J., 2009. Inheritance, fitness cost, and mechanism of resistance to tebufenozide in *Spodoptera exigua* (Hubner) (Lepidoptera: Noctuidae). Pest Management Science 65, 996–1002.

Juan-Blasco, M., Tena, A., Vanaclocha, P., Cambra, M., Urbaneja, A., Monzo, C., 2011. Efficacy of a micro-encapsulated formulation compared with a sticky barrier for excluding ants from citrus canopies. Journal of Applied Entomology 135, 467–472.

Kaakeh, W., Scharf, M.E., Bennett, G.W., 1997. Comparative contact activity and residual life of juvenile hormone analogs used for German cockroach (Dictyoptera: Blattellidae) control. Journal of Economic Entomology 90, 1247–1253.

Kanno, H., Ikeda, K., Asai, T., Maekawa, S., 1981. 2-tert-Butylimino-3-isopropyl-5- phenyl perhydro-1,3,5-thiadiazin-4-one (NNI 750), a new insecticide. Proceedings of British Crop Protection Conference Pests and Diseases 1, 59–66.

Kanzaki, S., Tanaka, T., 2010. Different responses of a solitary (*Meteorus pulchricornis*) and a gregarious (*Cotesia kariyai*) endoparasitoid to four insecticides in the host *Pseudaletia separata*. Journal of Pesticide Science 35, 1–9.

Karatolos, N., Williamson, M.S., Denholm, I., Gorman, K., ffrench-Constant, R., Nauen, R., 2012. Resistance to spiromesifen in *Trialeurodes vaporariorum* is associated with a single amino acid replacement in its target enzyme acetyl-coenzyme A carboxylase. Insect Molecular Biology 21 (3), 327–334.

Karr, L.L., Sheets, J.J., King, J.E., Dripps, J.E., 2004. Laboratory performance and pharmacokinetics of the benzoylphenylurea noviflumuron in eastern subterranean termites (Isoptera: Rhinotermitidae). Journal of Economic Entomology 97, 593–600.

Kavallieratos, N.G., Athanassiou, C.G., Vayias, B.J., Tomanović, Z., 2012. Efficacy of insect growth regulators as grain protectants against two stored-product pests in wheat and maize. Journal of Food Protection 75, 942–950.

Keiding, J., Elkhodary, A.S., Jespersen, J.B., 1992. Resistance risk assessment of 2 insect development inhibitors, diflubenzuron and cyromazine, for control of the housefly, *Musca domestica* L. 2. Effect of selection pressure in laboratory and field populations. Pesticide Science 35, 27–37.

Khalil, S.I.Y., Raslan, S.A., Hegab, O.I., Abd El-Sattar, O.S.G., 2010. Toxicity of Spinosad, Methoxyfenozide and Chlo-ropyrifos Used to Control Cotton Bollworms to Honeybee Foragers (Ph.D. thesis). Fac. Agric., Zagazig Univ., Egypt, 196 p.

Kim, S.-H.S., Wise, J.C., Gökçe, A., Whalon, M.E., 2011. Novaluron causes reduced egg hatch after treating adult codling moths, *Cydia pomonella*: support for transovarial transfer. Journal of Insect Science 11, 126.

Kim, S.S., Seo, S.G., 2001. Relative toxicity of some acaricides to the predatory mite, *Amblyseius womersleyi* and the twospotted spider mite, *Tetranychus urticae* (Acari: Phytoseiidae, Tetranychidae). Applied Entomology and Zoology 36, 509–514.

Kim, S.S., Yoo, S.S., 2002. Comparative toxicity of some acaricides to the predatory mite, *Phytoseiulus persimilis* and the two spotted spider mite, *Tetranychus urticae*. BioControl 47, 563–573.

King, J.E., DeMark, J.J., Griffin, A.J., 2005. Comparative laboratory efficacy of noviflumuron and diflubenzuron on *Reticulitermes flavipes* (Isoptera: Rhinotermitidae). Sociobiology 45, 779–785.

Koehler, P.G., Patterson, R.S., 1991. Incorporation of pyriproxyfen in a German cockroach (Dictyoptera: Blatellidae) management programme. Journal of Economic Entomology 84, 917–921.

Korrat, E.E.E., Abdelmonem, A.E., Helalia, A.A.R., Khalifa, H.M.S., 2012. Toxicological study of some conventional and nonconventional insecticides and their mixtures against cotton leaf worm, *Spodoptera littoralis* (Boisd.) (Lepi-doptera: Noctuidae). Annals of Agricultural Sciences 57 (2), 145–152.

Kostyukovsky, M., Chen, B., Atsmi, S., Shaaya, E., 2000. Biological activity of two juvenoids and two ecdysteroids against three stored product insects. Insect Biochemistry and Molecular Biology 30, 891–897.

Kramer, K.J., McGregor, H.E., Mori, K., 1979. Susceptibility of stored-product insects to pyridyl ether analogues of juvenile hormone. Journal of Agricultural and Food Chemistry 27, 1215–1217.

Kristensen, M., Jespersen, J.B., 2003. Larvicide resistance in *Musca domestica* (Diptera: Muscidae) populations in Denmark and establishment of resistant laboratory strains. Journal of Economic Entomology 96 (4), 1300–1306.

Kubota, S., 2011. Colony elimination of subterranean termites by bait application using benzoylphenylurea com-pounds, with special reference to bistrifluron. In: Stoytcheva, M. (Ed.), Pesticides in the Modern World—Risks and Benefits. InTech Open Access Publishers, pp. 347–362.

Kubota, S., Shono, Y., Matsunaga, T., Tsunoda, K., 2007. Termiticidal efficacy of bistrifluron as a bait toxicant against the Japanese subterranean termites *Coptotermes formosanus* and *Reticulitermes speratus* (Isoptera: Rhinotermitidae). Sociobiology 50, 623–631.

Kumar, S., Dahiya, B., Chauhan, R., 1996. Bioefficacy of diflubenzuron against *Helicoverpa armigera*. Pest Management and Economic Zoology 4, 59–63.

Kwon, S., Kim, Y., 2007. Immunosuppressive action of pyriproxyfen, a juvenile hormone analog, enhances patho-genicity of *Bacillus thuringiensis* subsp. *kurstaki* against diamondback moth, *Plutella xylostella* (Lepidoptera: Ypo-nomeutidae). Biological Control 42, 72–76.

Lafont, R., Koolman, J., 2009. Diversity of ecdysteroids in animal species. In: Smagghe, G. (Ed.), Ecdysone: Structures and Functions. Springer Science + Business Media, Dordrecht, The Netherlands, pp. 47–71.

Langley, P.A., Felton, T., Stafford, K., Oouchi, H., 1990. Formulation of pyriproxyfen, a juvenile hormone mimic, for tsetse control. Medical and Veterinary Entomology 4, 127–133.

Latchininsky, A.V., Gapparov, F., 2007. Locust control in Central Asia MiGs versus micronairs. Outlooks on Pest Management 18, 100–104.

Law, J.H., Yuan, C., Williams, C.M., 1966. Synthesis of a material with juvenile hormone activity. Proceedings of the National Academy of Sciences (American) 55, 576–578.

Le, D.P., Thirugnanam, M., Lidert, Z., Carlson, G.R., Bryan, J.B., 1996. RH-2485: a new selective insecticide for cater-pillar control. Brighton Crop Protection Conference Pests and Diseases 2, 481–486.

Lei, Z.-R., Wen, J.-Z., Lu, Y.-G., Wang, Y., Huang, H., 2002. Evaluations on the toxicity and applicability of several insect growth regulators for controlling oriental migratory locust, *Locusta migratoria manilensis* (Meyen). Plant Protection (China) 28, 5–7 (in Chinese with English abstract).

Lim, S.P., Lee, C.Y., 2005. Effects of juvenile hormone analogs on new reproductives and colony growth of pharaoh ant (Hymenoptera: Formicidae). Journal of Economic Entomology 98, 2169–2175.

Locke, M., 1998. Epidermis. In: Harrison, F.W., Locke, M. (Eds.), Microscopic Anatomy of Invertebrates. Insecta, vol. 11A. Wiley-Liss, New York, pp. 75–138.

Lomer, C.J., Bateman, R.P., Johnson, D.L., Langewald, J., Thomas, M., 2001. Biological control of locusts and grass-hoppers. Annual Review of Entomology 46, 667–702.

Luke, B.M., Bateman, R.P., 2006. Effects of chemical and botanical insecticides used for locust control on *Metarhizium anisopliae* var. *acridum* conidia after short- to medium-term storage at 30°C. Biocontrol Science and Technology 16, 761–766.

Luo, C., Jones, C.M., Devine, G., Zhang, F., Denholm, I., Gorman, K., 2010. Insecticide resistance in *Bemisia tabaci* biotype Q (Hemiptera: Aleyrodidae) from China. Crop Protection 29, 429–434.

Magalhaes, L.C., Walgenbach, J.F., 2011. Life stage toxicity and residual activity of insecticides to codling moth and oriental fruit moth (Lepidoptera: Tortricidae). Journal of Economic Entomology 104, 1950–1959.

Malone, L.A., Scott-Dupree, C.D., Todd, J.H., Ramankutty, P., 2007. No sub-lethal toxicity to bumblebees, *Bombus terrestris*, exposed to Bt-corn pollen, captan and novaluron. New Zealand Journal of Crop and Horticultural Science 35, 435–439.

Martins, A.J., Belinato, T.A., Lima, J.B.P., Valle, D., 2008. Chitin synthesis inhibitor effect on *Aedes aegypti* populations susceptible and resistant to organophosphate temephos. Pest Management Science 64, 676–680.

Mascari, T.M., Foil, L.D., 2010. Laboratory evaluation of novaluron as a rodent feedthrough insecticide against sand fly larvae (Diptera: Psychodidae). Journal of Medical Entomology 47, 205–209.

Mascari, T.M., Mitchell, M.A., Rowton, E.D., Foil, L.D., 2011. Evaluation of juvenile hormone analogues as rodent feed-through insecticides for control of immature phlebotomine sandflies. Medical and Veterinary Entomology 25, 227–231.

Mendel, Z., Blumberg, D., Ishaaya, I., 1994. Effects of some insect growth regulators on natural enemies of scale insects. Entomophaga 39, 199–209.

Moffitt, H.R., Westigard, P.H., Montey, K.D., van de Baan, 1988. Resistance to deflubenzuron in the codling moth (Lepidoptera: Tortricidae). Journal of Economic Entomology 81, 1511–1515.

Mohandass, S., Arthur, F.H., Zhu, K.Y., Throne, J.E., 2006a. Hydroprene: mode of action, current status in stored-product pest management, insect resistance, and future prospects. Crop Protection 25, 902–909 Corrigendum, (2007). Crop Protection 26, 173.

Mohandass, S., Arthur, F.H., Zhu, K.Y., Throne, J.E., 2006b. Hydroprene prolongs developmental time and increases mortality of the Indianmeal moth (Lepidoptera: Pyralidae) eggs. Journal of Economic Entomology 99, 1007–1016 47, 205–209.

Mommaerts, V., Sterk, G., Smagghe, G., 2006. Bumblebees can be used in combination with juvenile hormone analogues and ecdysone agonists. Ecotoxicology 15, 513–521.

Mori, K., Takigawa, T., Manabe, Y., Tominaga, M., Matsui, M., Kiguchi, K., Akai, H., Ohtaki, T., 1975. Effect of the molecular chain length on biological activity of juvenile hormone analogs. Agricultural and Biological Chemistry 39, 259–265.

Mosqueira, B., Chabi, J., Chandre, F., Akogbeto, M., Hougard, J.-M., Carnevale, P., Mas-Coma, S., 2010. Efficacy of an insecticide paint against malaria vectors and nuisance in West Africa—part 2: field evaluation. Malaria Journal 9, 341.

Moulton, J.K., Pepper, D.A., Dennehy, T.J., 2000. Pro-active Management of Beet Armyworm (*Spodoptera exigua*) Resistance to the IGR's. Tebufenozide and Methoxyfenozide. The University College of Agriculture 2000 Vegetable Report, Index at http://ag.arizona.edu/pubs/crops/az1177/.

Moulton, J.K., Pepper, D.A., Jansson, R.K., Dennehy, T.J., 2002. Pro-active management of beet armyworm (Lepidoptera: Noctuidae) resistance to tebufenozide and methoxyfenozide: baseline monitoring, risk assessment, and isolation of resistance. Journal of Economic Entomology 95, 414–424.

Mulla, M.S., Thavara, U., Tawatsin, A., Chompoosri, J., Zaim, M., Su, T., 2003. Laboratory and field examination of novaluron, a new acylurea insect growth regulator, against *Aedes aegypti* (Diptera: Culicidae). Journal of Vector Ecology 28, 241–254.

Nakagawa, Y., Izumi, K., Oikawa, N., Kurozumi, A., Iwamura, H., Fujita, T., 1991. Quantitative structure-activity relationships of benzoylphenylurea larvicides. VII. Separation of effects of substituents in the multisubstituted anilide moiety on the larvicidal activity against *Chilo suppressalis*. Pesticide Biochemistry and Physiology 40, 12–26.

Nakahira, K., Kashitani, R., Tomoda, M., Kodama, R., Ito, K., Yamanaka, S., Momoshita, M., Arakawa, R., 2010. Side effects of vegetable pesticides on a predatory mirid bug, *Pilophorus typicus* Distant (Heteroptera: Miridae). Applied Entomology and Zoology 45, 239–243.

Nauen, R., Smagghe, G., 2006. Mode of action of etoxazole. Pest Management Science 62, 379–382.

Nauen, R., Stumpf, N., Elbert, A., Zebitz, C.P.W., Kraus, W., 2001. Acaricide toxicity and resistance in larvae of different strains of *Tetranychus urticae* and *Panonychus ulmi* (Acari: Tetranychidae). Pest Management Science 57, 253–261.

Navarro-Llopis, V., Vacas, S., Sanchis, J., Primo, J., Alfaro, C., 2011. Chemosterilant bait stations coupled with sterile insect technique: an integrated strategy to control the Mediterranean fruit fly (Diptera: Tephritidae). Journal of Economic Entomology 104, 1647–1655.

Neoh, K.-B., Jalaludin, N.A., Lee, C.T., 2011. Elimination of field colonies of the mound building termite *Globitermes sulphureus* (Isoptera: Termitidae) by bistrifluron bait. Journal of Economic Entomology 104, 607–613.

Nicholas, A.H., Spooner-Hart, R.N., Vickers, R.A., 2005. Abundance and natural control of the woolly aphid *Eriosoma lanigerum* in an Australian apple orchard IPM program. BioControl 50, 271–291.

Nishimatsu, T., Jackson, J.J., 1998. Interaction of insecticides, entomopathogenic nematodes, and larvae of the western corn rootworm (Coleoptera: Chrysomelidae). Journal of Economic Entomology 91 (2), 410–418.

Niwa, A., Iwamura, H., Nakagawa, Y., Fujita, T., 1989. Development of (phenoxyphenoxy)- and (benzylphenoxy)- propyl ethers as potent insect juvenile hormone mimetics. Journal of Agricultural and Food Chemistry 37, 462–467.

Oberlander, H., Silhacek, D.L., Shaaya, E., Ishaaya, I., 1997. Current status and future perspectives of the use of insect growth regulators for the control of stored product insects. Journal of Stored Products Research 33, 1–6.

Okazawa, T., Bakotee, B., Suzuki, H., Kawada, H., Kere, N., 1991. Field evaluation of an insect growth regulator, pyriproxyfen, against *Anopheles punctulatus* on North Guadalcanal, Solomon Islands. Journal of the American Mosquito Control Association 7, 604–607.

Olson, D.L., Oetting, R.D., 1999. Compatibility of insect growth regulators and *Beauveria bassiana* (Balsamo) Vuillemin in controlling green peach aphids (Homoptera: Aphididae) on greenhouse chrysanthemums. Journal of Entomological Science 34, 286–294.

Oouchi, H., Langley, P., 2005. Control of greenhouse whitefly (*Trialeurodes vaporariorum*) using visually attractive targets impregnated with pyriproxyfen. Journal of Pesticide Science 30, 50–52.

Osbrink, W.L.A., Cornelius, M.L., Lax, A.R., 2011. Areawide field study on effect of three chitin synthesis inhibitor baits on populations of *Coptotermes formosanus* and *Reticulitermes flavipes* (Isoptera: Rhinotermitidae). Journal of Economic Entomology 104, 1009–1017.

Osorio, A., Martinez, A.M., Schneider, M.I., Diaz, O., Corrales, J., Aviles, M.C., Smagghe, G., Pineda, A., 2008. Monitoring of beet armyworm resistance to spinosad and methoxyfenozide in Mexico. Pest Management Science 64, 1001–1007.

Patel, J.J., Patel, P.B., Patel, H.C., 2012. Bio-Efficacy of Buprofezin 70% DF against jassid infesting okra. AGRES – An International e-Journal 1 (3), 395–399.

Patel, M.G., Jhala, R.C., Vaghela, N.M., Chauhan, N.R., 2010. Bio-efficacy of buprofezin against mealy bug, *Phenacoccus solenopsis* Tinsley (Hemiptera: Pseudococcidae) an invasive pest of cotton. Karnataka Journal of Agricultural Sciences 23, 14–18.

Pener, M.P., Dhadialla, T.S., 2012. An overview of insect growth disruptors; applied aspects. In: Advances in Insect Physiology, vol. 43. p. 162.

Pener, M.P., Ayali, A., Kelmer, G., Bennettová, B., Němec, V., Rejzek, M., Wimmer, Z., 1997. Comparative testing of several juvenile hormone analogues in two species of locusts, *Locusta migratoria* migratorioides and *Schistocerca gregaria*. Pesticide Science 51, 443–449.

Perez, C.J., Alvarado, P., Narvaez, C., Miranda, F., Hernandez, L., Vanegas, H., Hruska, A., Shelton, A.M., 2000. Assessment of insecticide resistance in five insect pests attacking field and vegetable crops in Nicaragua. Journal of Economic Entomology 93 (6), 1779–1787.

Perveen, F., 2006. Reduction in egg hatch after a sublethal dose of chlorfluazuron to larvae of the common cutworm, *Spodoptera litura*. Physiological Entomology 31, 39–45.

Phillips, T.W., Throne, J.E., 2010. Biorational approaches to managing store-product insects. Annual Review of Entomology 55, 375–397.

Pimprikar, G.D., Georghiou, G.P., 1979. Mechanisms of resistance to diflubenzuron in the house fly, *Musca domestica* L. Pesticide Biochemistry and Physiology 12, 10–22.

Purwar, J.P., Sachan, G.C., 2006. Synergistic effect of entomogenous fungi on some insecticides against Bihar hairy caterpillar *Spilarctia obliqua* (Lepidoptera: Arctiidae). Microbiological Research 161 (1), 38–42.

Quintela, E.D., McCoy, C.W., 1997. Pathogenicity enhancement of *Metarhizium anisopliae* and *Beauveria bassiana* to first instars of *Diaprepes abbreviatus* (Coleoptera: Curculionidae) with sublethal doses of imidacloprid. Environmental Entomology 26, 1173–1182.

Rafiee, D.H., Hejazi, M.J., Nouri Ganbalani, G., Saber, M., 2008. Toxicity of some bio-rational and conventional insecticides to cotton bollworm, *Helicoverpa armigera* (Lepidoptera: Noctuidae) and its ectoparasitoid, *Habrobracon hebetor* (Hymenoptera: Braconidae). Journal of Entomological Society of Iran 28 (1), 27–37.

Rajapakse, C.N.K., Meola, R., Readio, J., 2002. Comparative evaluation of juvenoids for control of cat fleas (Siphonaptera: Pulicidae) in topsoil. Journal of Medical Entomology 39, 889–894.

Rajasekar, P., Jebanesan, A., 2012. Efficacy of IGRs compound novaluron and buprofezin against *Culex quinquefasciatus* mosquito larvae and pupal control in pools, drains and tanks. International Journal of Research in Biological Sciences 2 (1), 45–47.

Retnakaran, A., Granett, J., Ennis, T., 1985. Insect growth regulators. In: Kerkut, J.A., Gilbert, L.I. (Eds.), Comprehensive Insect Physiology, Biochemistry and Pharmacology, vol. 12. Pergamon Press, Oxford, pp. 529–601.

Reyes, M., Franck, P., Charmillot, P.J., Ioriatti, C., Olivares, J., Pasqualini, E., Sauphanor, B., 2007. Diversity of insecticide resistance mechanisms and spectrum in European populations of the codling moth, *Cydia pomonella*. Pest Management Science 63, 890–902.

Rodriguez, M., Marques, T., Bosch, D., Avilla, J., 2011. Assessment of insectide resistance in eggs and neonate larvae of *Cydia pomonella* (Lepidoptera: Tortricidae). Pesticide Biochemistry and Physiology 100, 151–159.

RohMid, L.L.C., 1996. RH-0345, turf and ornamental insecticide. Technical Information Bulletin 1, 8.

Rothwangl, K.B., Cloyd, R.A., Wiedenmann, R.N., 2004. Effects of insect growth regulators on citrus mealybug parasitoid *Leptomastix dactylopii* (Hymenoptera: Encyrtidae). Journal of Economic Entomology 97, 1239–1244.

Saber, M., Parsaeyan, E., Vojoudi, S., Bagheri, M., Mehrvar, A., Kamita, S.G., 2013. Acute toxicity and sublethal effects of methoxyfenozide and thiodicarb on survival, development and reproduction of *Helicoverpa armigera* (Lepidoptera: Noctuidae). Crop Protection 43, 14–17.

Sánchez-Ramos, I., Castañera, P., 2003. Laboratory evaluation of selective pesticides against the storage mite *Tyrophagus putrescentiae* (Acari: Acaridae). Journal of Medical Entomology 40, 475–481.

Sajap, A.S., Lee, L.C., Shah, Z.M., 2009. Elimination of subterranean termite colonies with hexaflumuron in an improved bait matrix, Preferred Textured Cellulose (PTC). Sociobiology 53, 891–901.

Salokhe, S.G., Pal, J.K., Mukherjee, S.N., 2003. Effect of sublethal concentrations of flufenoxuron on growth, development and reproductive performance of *Tribolium castaneum* (Herbst) (Coleoptera: Tenebrionidae). Invertebrate Reproduction & Development 43, 141–150.

Saran, R.K., Rust, M.K., 2008. Phagostimulatory sugars enhance uptake and horizontal transfer of hexaflumuron in the western subterranean termite (Isoptera: Rhinotermitidae). Journal of Economic Entomology 101, 873–879.

Sauphanor, B., Bouvier, J.C., 1995. Cross-resistance between benzoylureas and benzoylhydrazines in codling moth, *Cydia pomonella* L. Pesticide Science 45, 369–375.

Sazo, L., Araya, J.E., Esparza, S., 2008. Control of San Jose scale nymphs, *Diaspidiotus perniciosus* (Comstock), on almond and apple orchards with pyriproxyfen, fenoxycarb, chlorpyriphos, and mineral oil. Chilean Journal of Agricultural Research 68, 284–289.

Sbragia, R.J., Bishabri-Ershadi, B., Risterink, R.H., Clilfford, D.P., Dutton, R., 1983. XRD-473, a new acylurea insecticide effective against *Heliothis*. In: Proceedings of British Crop Protection Conference on Pests and Diseases, vol. 1. Brighton, UK, pp. 417–424.

Schneider, M.I., Smagghe, G., Pineda, S., Viñuela, E., 2004. Action of insect growth regulator insecticides and spinosad on life history parameters and absorption in third-instar larvae of the endoparasitoid *Hyposoter didymator*. Biological Control 31, 189–198.

Schneiderman, H.A., Gilbert, L.I., 1959. The chemistry and physiology of insect growth hormones. In: Rudnick, D. (Ed.), Cell, Organism and Milieu. Ronal Press Comp., pp. 157–187.

Schneiderman, H.A., 1972. Insect hormones and insect control. In: Menn, J.J., Beroza, M. (Eds.), Insect Juvenile Hormones. Academic Press, New York, pp. 3–27.

Scott-Dupree, C.D., Conroy, L., Harris, C.R., 2009. Impact of currently used or potentially useful insecticides for canola agroecosystems on *Bombus impatiens* (Hymenoptera: Apidae), *Megachile rotundata* (Hymenoptera: Megachilidae), and *Osmia lignaria* (Hymenoptera: Megachilidae). Journal of Economic Entomology 102, 177–182.

Shah, R.M., Abbas, N., Shad, S.A., Sial, A.A., 2015. Selection, resistance risk assessment, and reversion toward susceptibility of pyriproxyfen in *Musca domestica* L. Parasitology Research 114, 487–494.

Sihuincha, M., Zamora-Perea, E., Orellana-Rios, W., Stancil, J.D., López-Sifuentes, V., Vidal-Oré, C., Devine, G.J., 2005. Potential use of pyriproxyfen for control of *Aedes aegypti* (Diptera: Culicidae) in Iquitos, Perú. Journal of Medical Entomology 42, 620–630.

Silva, G., Picano, M., Bacci, L., Crespo, A., Rosado, J., Guedes, R., 2011. Control failure likelihood and spatial dependence of insecticide resistance in the tomato pinworm, *Tuta absoluta*. Pest Management Science 67, 913–920.

Simmons, A.M., Abd-Rabou, S., 2011. Populations of predators and parasitoids of *Bemisia tabaci* (Hemiptera: Aleyrodidae) after the application of eight biorational insecticides. Pest Management Science 67, 1023–1028.

Sláma, K., Williams, C.M., 1965. Juvenile hormone activity for the bug *Pyrrhocoris apterus*. Proceedings of the National Academy of Sciences of the United States of America 54, 411–414.

Sláma, K., Williams, C.M., 1966. "Paper factor" as an inhibitor of the embryonic development of the European bug, *Pyrrhocoris apterus*. Nature 210, 329–330.

Smagghe, G., Degheele, D., 1997. Comparative toxicity and tolerance for the ecdysteroid mimic tebufenozide in a laboratory and field strain of cotton leafworm (Lepidoptera: Noctuidae). Journal of Economic Entomology 90 (2), 278–282.

Smagghe, G., Degheele, D., 1994. Action of the nonsteroidal ecydysteroid mimic RH-5849 on larval development and adult reproduction of insects of different orders. Invertebrate Reproduction & Development 25, 227–236.

Smet, H., Rans, M., De Loof, A., 1989. Activity of new juvenile hormone analogues on a stored food insect, *Tribolium confusum* (J. Du Val) (Coleoptera: Tenebrionidae). Journal of Stored Products Research 25, 165–169.

Smirle, M.J., Lowery, D.T., Zurowski, C.L., 2002. Resistance and cross-resistance to four insecticides in populations of oblique banded leafroller (Lepidoptera: Tortricidae). Journal of Economic Entomology 95 (4), 820–825.

Sobotnik, J., Kudlikova-Krizkova, I., Vancova, M., Munzbergova, Z., Hubert, J., 2008. Chitin in the peritrophic membrane of *Acarus siro* (Acari: Acaridae) as a target for novel acaricides. Journal of Economic Entomology 101, 1028–1033.

Stara, J., Kocourek, F., 2007. Insecticidal resistance and cross-resistance in populations of *Cydia pomonella* (Lepidoptera: Tortricidae) in Central Europe. Journal of Economic Entomology 100 (5), 1587–1595.

Stará, J., Nesvorná, M., Hubert, J., 2011a. The toxicity of selected acaricides against five store product mites under laboratory assay. Journal of Pesticide Science 84, 387–391.

Stará, J., Stejskal, V., Nesvorná, M., Plachý, J., Hubert, J., 2011b. Efficacy of selected pesticides against synanthropic mites under laboratory assay. Pest Management Science 67, 446–457.

Su, J., Sun, X., 2014. High level of metaflumizone resistance and multiple insecticide resistance in field populations of *Spodoptera exigua* (Lepidoptera: Noctuidae) in Guangdong Province, China. Crop Protection 61, 58–63.

Su, J., Wang, Z., Zhang, K., Tian, X., Yin, Y., Zhao, X., Shen, A., Gao, C., 2013. Status of insecticide resistance of the Whitebacked planthopper, *Sogatella furcifera* (Hemiptera: Delphacidae). Florida Entomologist 96.

Su, N.-Y., 1994. Field evaluation of hexaflumuron bait for population suppression of subterranean termites (Isoptera: Rhinotermitidae). Journal of Economic Entomology 87, 389–397.

Su, N.-Y., Scheffrahn, R.H., 1993. Laboratory evaluation of two chitin synthesis inhibitors, hexaflumuron and diflubenzuron, as bait toxicants against the Formosan subterranean termite and eastern subterranean termite (Isoptera: Rhinotermitidae). Journal of Economic Entomology 86, 1453–1457.

Su, N.-Y., Scheffrahn, R.H., 1996a. Comparative effects of two chitin synthesis inhibitors, hexaflumuron and lufenuron, in a bait matrix against subterranean termites (Isoptera: Rhinotermitidae). Journal of Economic Entomology 89, 1156–1160.

Su, N.-Y., Scheffrahn, R.H., 1996b. Fate of subterranean termite colonies (isoptera) after bait applications ± an update and review. Sociobiology 27, 253–275.

Su, N.-Y., 2005. Response of the Formosan subterranean termites (Isoptera: Rhinotermitidae) to baits or nonrepellent termiticides in extended foraging arenas. Journal of Economic Entomology 98, 2143–2152.

Su, N.-Y., Ban, P.M., Scheffrahn, R.H., 1997. Remedial baiting with hexaflumuron in above-ground stations to control structure infesting populations of the Formosan subterranean termite (Isoptera: Rhinotermitidae). Journal of Economic Entomology 90, 809–817.

Su, N.-Y., Thoms, E.M., Ban, P.M., Scheffrahn, R.H., 1995. Monitoring baiting station to detect and eliminate foraging populations of subterranean termites (Isoptera: Rhinotermitidae) near structures. Journal of Economic Entomology 88, 932–936.

Suárez, J., Oviedo, M., Álvarez, L., González, A., Lenhart, A., 2011. Evaluation of the insect growth regulator pyriproxyfen against populations of *Aedes aegypti* from Trujillo, Venezuela. Revista Colombiana de Entomología 37, 91–94 (in Spanish with English abstract).

Sugiyama, K., Katayama, H., Saito, T., 2011. Effect of insecticides on the mortalities of three whitefly parsitoid species, *Eretmocerus mundus, Eretmocerus eremicus* and *Encarsia Formosa* (Hymenoptera: Aphelinidae). Applied Entomology and Zoology 46, 311–317.

Sukartana, P., Sumarni, G., Broadbent, S., 2009. Evaluation of chlorfluazuron in controlling the subterranean termite *Coptotermes curvignathus* (Isoptera: Rhinotermitidae) in Indonesia. Journal of Tropical Forest Science 21, 13–18.

Suman, D.S., Parashar, B.D., Prakash, S., 2010. Efficacy of various insect growth regulators on organophosphate resistant immatures of *Culex quinquefasciatus* (Diptera: Culicidae) from different geographical areas of India. Journal of Entomology 7, 33–43.

Sun, X., Barrett, B.A., 1999. Fecundity and fertility changes in adult codling moth (Lepidoptera: Tortricidae) exposed to surfaces treated with tebufenozide and methoxyfenozide. Journal of Economic Entomology 92, 1039–1044.

Sun, X., Barrett, B.A., Biddinger, D.J., 2000. Fecundity and fertility changes in adult redbanded leafroller and oblique banded leafroller (Lepidoptera: Tortricidae) exposed to surfaces treated with tebufenozide and methoxyfenozide. Entomologia Experimentalis et Applicata 94, 75–83.

Sutton, A.E., Arthur, F.H., Zhu, K.Y., Campbell, J.F., Murray, L.W., 2011. Residual efficacy of synergized pyrethrin + methoprene aerosol against larvae of *Tribolium castaneum* and *Tribolium confusum* (Coleoptera: Tenebrionidae). Journal of Stored Products Research 47, 399–406.

Suzuki, J., Ishida, T., Kikuchi, Y., Ito, Y., Morikawa, C., Tsukidate, Y., Tanji, I., Ota, Y., Toda, K., 2002. Synthesis and activity of novel acaricidal/insecticidal 2,4-diphenyl-1,3-oxazolines. Journal of Pesticide Science 27, 1–8.

Tail, G., Porcheron, P., Doumandji-Mitiche, B., 2010. Diflubenzuron et évolution des taux des ecdystéroïdes dans les ovaries et dans les oeufs du criquet pèlerin, *Schistocerca gregaria*, (Orthoptera: Acrididae). Journal of Orthoptera Research 19, 363–370.

Tasei, J.N., 2001. Effects of insect growth regulators on honey bees and non-Apis bees. A review. Apidologie 32, 527–545.

Taylor, D.B., Friesen, K., Zhu, J.J., Sievert, K., 2012. Efficacy of cyromazine to control immature stable flies (Diptera: Muscidae) developing in winter hay feeding sites. Journal of Economic Entomology 105, 726–731.

Thompson, H.M., Wilkins, S., Battersby, A.H., Waite, R.J., Wilkinson, D., 2005. The effects of four insect growth-regulating (IGR) insecticides on honeybee (*Apis mellifera* L.) colony development, queen rearing and drone sperm production. Ecotoxicology 14, 757–769.

Tomlin, C.D.S. (Ed.), 2009. The Pesticide Manual, fifteenth ed. British Crop Protection Council, Alton, Hampshire.

Trisyono, A., Chippendale, M., 1997. Effect of the non – steroidal ecdysone agonists, methoxyfenozide and tebufenozide, on the European corn borer (Lepidoptera: Pyralidae). Journal of Economic Entomology 90 (6), 1486–1492.

Trostanetsky, A., Kostyukovsky, M., 2008. Transovarial activity of the chitin synthesis inhibitor novaluron on egg hatch and subsequent development of larvae of *Tribolium castaneum*. Phytoparasitica 36, 38–41.

Uchida, M., Asai, T., Sugimoto, T., 1985. Inhibition of cuticle deposition and chitin biosynthesis by a new insect growth regulator, buprofezin, in *Nilaparvata lugens* Stal. Agricultural and Biological Chemistry 49, 1233–1234.

Vahabzadeh, R.D., Gold, R.E., Austin, J.W., 2007. Effects of four chitin synthesis inhibitors on feeding and mortality of the eastern subterranean termite, *Reticulitermes flavipes* (Isoptera: Rhinotermitidae). Sociobiology 50, 833–859.

Van Driesche, R.G., Lyon, S., 2003. Commercial adoption of biological control-based IPM for whiteflies in poinsettia. Florida Entomologist 86, 481–483.

Van Laecke, K., Smagghe, G., Degheele, D., 1995. Detoxifying enzymes in greenhouse and laboratory strain of beet armyworm (Lepidoptera: Noctuidae). Journal of Economic Entomology 88 (4), 777–781.

Van Leeuwen, T., Van Pottelberge, S., Tirry, L., 2005. Comparative acaricide susceptibility and detoxifying enzyme activities in field-collected resistant and susceptible strains of *Tetranychus urticae*. Pest Management Science 61, 499–507.

Van Leeuwen, T., Demaeght, P., Osborne, E.J., Dermauw, W., Gohlke, S., Nauen, R., Grbic, M., Tirry, L., Merzendorfer, H., Clark, R.M., 2012. Population bulk segregant mapping uncovers resistance mutations and the mode of action of a chitin synthesis inhibitor in arthropods. Proceedings of the National Academy of Sciences of the United States of America 109, 4407–4412.

Vasquez, M.I., Violaris, M., Hadjivassilis, A., Wirth, M.C., 2009. Susceptibility of Culex pipiens (Diptera: Culicidae) Field Populations in Cyprus to conventional Organic Insecticides, Bacillus thuringiensis subsp. israelensis, and Methopren. Journal of Medical Entomology 46 (4), 881–887.

Vattanatangum, A., 1988. Recent Problems on Chemical Control of Thailand Agricultural Insect Pests Related to Insecticide Resistance of Diamondback Moth and Other Major Species. Report meeting of the joint research project "Insect toxicological studies on resistance to insecticides and integrated control of the diamondback moth." March 1988. Bangkok, Thailand.

Verma, M., Sharma, S., Prasad, R., 2009. Biological alternatives for termite control: a review. International Biodeterioration & Biodegradation 63, 959–972.

Vianna, U.R., Pratissoli, D., Zanuncio, J.C., Lima, E.R., Brunner, J., Pereira, F.F., Serraõ, J.E., 2009. Insecticide toxicity to *Trichogramma pretiosum* (Hymenoptera: Trichogrammatidae) females and effect on descendant generation. Ecotoxicology 18, 180–186.

Wang, Y., Gao, C., Xu, Z., Zhu, Y.C., Zhang, J., Li, W., Dai, D., Lin, Y., Zhou, W., Shen, J., 2008. Buprofezin susceptibility survey, resistance selection and preliminary determination of the resistance mechanism in *Nilaparvata lugens* (Homoptera: Delphacidae). Pest Management Science 64, 1050–1056.

Wearing, C.H., 1998. Cross-resistance between azinphos methyl and tebufenozide in the green headed leafroller, *Planotortrix octo*. Pesticide Science 54, 203–211.

Williams, C.M., Sláma, K., 1966. The juvenile hormone. VI. Effects of the "Paper factor" on the growth and metamorphosis of the bug, *Pyrrhocoris apterus*. Biological Bulletin 130, 247–253.

Williams, C.M., 1967. Third-generation pesticides. Scientific American 217, 13–17.

Williams, R.E., Berry, J.G., 1980. Evaluation of CGA72662 as a topical spray and feed additive for controlling house flies breeding in chicken manure. Poultry Science 59, 2207–2212.

Wise, J.C., Kim, K., Hoffmann, E.J., Vandervoort, C., Gökçe, A., Whalon, M.E., 2007. Novel life stage targets against plum curculio, *Conotrachelus nenuphar* (Herbst), in apple integrated pest management. Pest Management Science 63, 737–742.

Yanagi, M., Tsukamoto, Y., Watanabe, T., Kawagishi, A., 2006. Development of a novel lepidopteran insect control agent, chromafenozide. Journal of Pesticide Science 31, 163–164.

Yapabandara, A.M.G.M., Curtis, C.F., Wickramasinghe, M.B., Fernando, W.P., 2001. Control of malaria vectors with the insect growth regulator pyriproxyfen in a gem-mining area in Sri Lanka. Acta Tropica 80, 265–276.

Yarom, I., Blumberg, D., Ishaaya, I., 1988. Effects of buprofezin on California red scale (Homoptera: Diaspididae) and Mediterranean black scale (Homoptera: Coccidae). Journal of Economic Entomology 81, 1581–1585.

Yue, C., Zhang, J., Zhang, X., 2011. Damage regularity and control techniques of and *Eulecanium gigantean* around Tarim Basin in Xinjiang. Procedia Engineering 18, 133–138.

Zhang, X.N., 2005. Novel insect growth regulator fufenozide (JS118). World Pestic. (¼Shijie Nongyao) 27, 48–49 (in Chinese).

Zhang, K., Zhang, W., Zhang, S., Wu, S., Ban, L., Su, J., Gao, C., 2014. Susceptibility of *Sogatella furcifera* and *Laodelphax striatellus* (Hemiptera: Delphacidae) to six insecticides in China. Journal of Economic Entomology 107, 1916–1922.

Zhao, X., Wu, C., Wang, Y., Cang, T., Chen, L., Yu, R., Wang, Q., 2012. Assessment of toxicity risk of insecticides used in rice ecosystem on *Trichogramma japonicum*, an egg parasitoid of rice lepidopterans. Journal of Economic Entomology 105, 92–101.

Zhou, L., Ma, J.-Q., Wang, J.-C., 2008. Bionomics and control efficacy of the scale insect *Phenacoccus pergandei*. Chinese Bulletin of Entomology 45, 808–810.

Zitnanova, I., Adams, M.E., Zitnan, D., 2001. Dual ecdysteroid action on the epitracheal glands and central nervous system preceding ecdysis of *Manduca sexta*. Journal of Experimental Biology 204, 3483–3495.

Integrated Pest Management

P. Karuppuchamy, Sheela Venugopal

Agricultural Research Station, Tamil Nadu Agricultural University, Bhavanisagar,
Tamil Nadu, India

1. INTRODUCTION

A green revolution is one of the most significant achievements during the post-independent period in India that resulted in self-sufficiency in food production. Maintaining the status of self-sufficiency is the current focus because of the exploding human population, shrinking land holdings, and depleting water resources. Obviously, it is vital to achieve sustainable crop production without compromising the productivity levels of cultivated crops. One of the main limiting factors for this is undoubtedly the biotic stress. It is estimated that herbivorous insects eat about 20% of the crops grown for human consumption. The losses are likely to mount with increasing monocropping, fertilization, irrigation, and other important features of intensive agriculture that may intensify further in coming years to produce more and more food and fiber to meet the growing demands of the increasing population. These losses threaten global food security and are a serious economic and nutritional burden to farmers and consumers around the world.

2. LOSSES DUE TO PESTS AND EMERGING PEST PROBLEMS

The major constraints limiting higher productivity are the onslaught of insect pests and diseases, which are favored by intensive agriculture. The development of resistance, secondary pest outbreaks, and emergence of new pest problems have increased monetary losses. Singh et al. (2003) reported that the annual crop losses due to insect pests and diseases in India vary up to 38% (insect pests and diseases 26%, weeds 10%, and birds 1–2%). Reports indicate that the losses caused by the specific major pests may be higher. *Helicoverpa armigera* (Hubner) in cotton causes yield losses of up to 20–25%. Raheja and Tiwari (1997) had reported that losses due to American bollworm alone may be around rupees 10,000 million annually while the losses due to insect pests and diseases in rice (18.6%) amounted to rupees 55,120 million (US$1102.4 million). The overall losses due to insect pests were estimated to be rupees 60 billion (US$1.2 billion) in

Ecofriendly Pest Management for Food Security
http://dx.doi.org/10.1016/B978-0-12-803265-7.00021-X

1983 (Krishnamurthy Rao and Murthy, 1983), rupees 200 billion (US$4 billion) in 1993 (Jayaraj and Regupathi, 1993), and rupees 292.4 billion (US$8.6 billion) in 1996 (Dhaliwal and Arora, 1996). In recent years, the leaf curl disease of cotton transmitted by *Bemisia tabaci* has been reported in the states of Rajasthan, Punjab, and Haryana in virulent form and more than 100,000 ha of cotton have been infected with this disease in Rajasthan alone. Insect pests inflict direct losses and also act as vectors.

The national governments, private industries, universities, non-governmental organizations, and international centers have all been working to manage pest problems to improve the agricultural productivity on a sustainable basis to feed the growing population. Until recently, use of synthetic chemical pesticides had been the widely used approach for reducing crop loss due to pests and diseases. More and more quantities of chemicals were used for agricultural intensification to feed an ever-growing population. Regular and indiscriminate use of broad-spectrum synthetic insecticides has caused turbulence in the environment because farmers began to place excessive reliance on pesticide use. Instead of need-based spraying, farmers have reverted to time-based spraying, and consequently this has led to many undesirable problems like buildup of insecticide resistance in insects, upset in the balance of life in nature due to weakening of biotic pressure, pest resurgence, destruction of bioagents, secondary pest outbreaks, etc., besides health hazards. Often insect outbreaks occurred as a result of excessive use of nitrogenous fertilizers, use of high-yielding varieties, and introduction of new crops, destruction of forests and bringing forest area under cultivation, monoculture, accidental and intentional introduction of a new pest in a new area, destruction of natural enemies and large scale storage of food grains, besides indiscriminate use of pesticides. Overuse, misuse, and improper use of pesticides endanger the health of farm workers and consumers of agricultural products worldwide (Goodell, 1984). Many examples exist that document the environmental and health risks from the indiscriminate use of pesticides. Insecticide resistance (Atwal and Dhaliwal, 1997), resurgence of crop pests (Chelliah, 1987), poisoning of bees (Dhaliwal and Arora, 1996), poisoning of birds (Muralidharan et al., 1992) and fish (Wadhwani and Lal, 1972), suppression of soil microbes (Dhaliwal and Arora, 1996), and human toxicity and health hazards (Atwal and Dhaliwal, 1997; Nag, 1983) are some of the ill effects of insecticides. The global community has expressed a willingness to reduce its reliance on chemical pesticides and to move toward a more balanced approach to pest management that relies on cultural, biological, and biorational control measures. The shift is driven by the high cost of pesticides, increased pest resistance to pesticides, and the negative impacts of pesticides on biodiversity, food and water quality, human and animal health (Rola and Pingali, 1993), and the environment. Rachel Carson's book *Silent Spring*, published in 1962, alerted the public to problems with pesticides. Also, as the global economy moves toward a free market economy and free trade, strict pesticide residue regulations in European and North American markets are forcing many exporting countries to redesign their pest management strategies to remain globally competitive (Schillhorn van Veen et al., 1997; Henson and Loader, 2001). This has forced a shift from a chemical-based pest management paradigm to ecologically/biologically based pest management paradigms. The importance of integrated approaches involving various eco-friendly tactics to manage pests efficiently was then recognized, which necessitated the use of target-specific compounds with low persistence.

3. INTEGRATED PEST MANAGEMENT: DEFINITION AND SCOPE

The modern concept of pest management is based on ecological principles and integration of different control tactics into a pest management system. Bartlett (1956) and Stern et al. (1959) defined integrated control as applied pest control by blending of biocontrol agents with chemical control interventions. Since then, the concept of pest management has gained importance. Yet from a historical view, the concept of integrating chemical control with other tactics was proposed much earlier (Hoskins et al., 1939). Geier and Clark (1961) introduced the word *management* for control and called this concept of pest control the protective management of noxious species or pest management. In 1967, Smith and Van Den Borsch stated that the determination of the insect numbers is broadly under the influence of the total agroecosystem, and the role of the principal elements is essential for integrated pest management. The backbone of management of pests in an agricultural ecosystem is the concept of economic injury level. According to the Food and Agriculture Organisation (FAO) (1967), IPM was defined as a pest management system in the context of associated environment and population dynamics in pest species. It utilizes all suitable techniques and methods in as compatible a manner as possible and maintains the pest population at levels below those that cause economic injury. All the available techniques are evaluated and consolidated into a unified program to manage pest populations so that economic damage is avoided and adverse side effects on the environment minimized (National Academy of Sciences, 1969). In 1972, the term *integrated pest management* was accepted by the Council of Environmental Quality where "I" stands for "Integration," which is harmonious use of multiple methods to control the impact of a single pest as well as multiple pests; "P" for "Pest," which refers to any organism that is detrimental to humans including vertebrates and invertebrates or weeds or pathogens; and "M" for "Management," which refers to a set of decisions or rules based on ecological principles, economic, and social considerations.

Apple et al. (1979) defined IPM as the optimization of the pest control in an economically and ecologically sound manner, i.e., IPM should be ecologically sound and economically viable. Pedigo (1989) defined pest management as a comprehensive technology that uses combined means to reduce the status of the pest to tolerable levels while maintaining a quality environment. Luckmann and Metcalf (1994) defined IPM as the intelligent selection and use of pest control tactics that will ensure favorable economic, ecologic, and sociologic consequences. IPM is a decision support system for the selection and use of pest control tactics, singly or harmoniously coordinated into a management strategy, based on cost–benefit analyses that take into account the interests of and impacts on producers, society, and the environment (Kogan, 1998). The United States (US) federal definition of IPM underwent a recent paradigm shift toward "risk reduction" as a way of placing focus on the consequences and impacts of pests and pest management tactics rather than just pesticides: "IPM is a long-standing, science-based, decision-making process that identifies and reduces risks from pests and pest management related strategies. It coordinates the use of pest biology, environmental information, and available technology to prevent unacceptable levels of pest damage by the most economical means, while posing the least possible risk to people, property, resources, and the environment. IPM provides an effective strategy for managing pests in all arenas from agricultural, residential, and public areas to wild lands. IPM serves as an umbrella to provide an effective, all encompassing, low-risk approach to protect resources and people from pests" (Anonymous, 2004).

Some of the underlying principles enumerated from the above definitions are:

- Framing of reliable monitoring techniques.
- Identifying the pest that might cause economic loss.
- Understanding the ecology and biology of the pest.
- Developing the pest management strategy by integrating the biological knowledge with the technical application.
- Integrating several techniques instead of relying on a single one.
- Reducing the pest infestation and not wiping out the pest population.
- Maintaining the quality of the ecosystem.

IPM misinterpreted

- **IPM is not new**: In one form or another it has been around since the advent of agriculture. Scientifically based programs specifically focused in this area, however, are only a few decades old.
- **IPM is not implemented overnight**: The development of an IPM program may take years of research and involve participants such as university researchers, extension workers, pest control advisors, industry scientists, and, most importantly, farmers.
- **IPM is not organic farming**: Organic farming is a philosophical approach to crop production that relies on no synthetic inputs for either pest control or plant nutrition. Organic farmers are prevented from using some of the low-risk techniques and technologies available to growers practicing IPM, simply because they are synthetic.
- **IPM is not a formula to eliminate or reduce pesticide use**: Well-developed, science-based IPM programs have consistently resulted in reduced pesticide use, as they employ a wider array of pest management techniques. IPM programs, by design, result in safer and more judicious use of pesticides. IPM is not a rigid program of management techniques. IPM is a balance of all suitable techniques, providing the grower with options to manage pests within a given crop production system.
- **IPM programs are not universal**: Depending upon the pest complex and the geography, programs may differ dramatically for the same crop in different geographies.

4. CONCEPTS OF IPM

IPM seeks to minimize the disadvantages associated with use of pesticides and maximize socioeconomic and ecological advantages. To achieve the desired outcome out of IPM, one should bear in mind the following:

4.1 Understanding the Agricultural Ecosystem

An agroecosystem contains a lesser diversity of animal and plant species than a natural ecosystem like a forest. A typical agroecosystem contains only one to four major crop species and six to ten major pest species. An agroecosystem is intensively manipulated by man and subjected to sudden alterations, such as plowing, intercultivation, and treatment with pesticides. These practices are critical in pest management as pest populations are greatly influenced by these practices. Agroecosystems can be more susceptible to pest damage and catastrophic outbreaks owing to lack of diversity in species of plants and insects and sudden alternations imposed by weather and man. However, an agroecosystem is a complex of food chains and food webs that interact together to produce a stable unit.

4.2 Planning of Agricultural Ecosystem

In an IPM program the agricultural system can be planned in terms of anticipating pest problems and also the ways to reduce them by integrating crop protection with the crop production system. Growing of susceptible varieties should be avoided and related crops should not be grown. Okra followed by cotton increases incidence of the spotted borer. Groundnut followed by soybean increases the incidence of the leaf miner.

4.3 Cost–Benefit Ratio

Based on the possibility of pest damage by predicting the pest problem and by defining the economic threshold level, emphasis should be given to the cost–benefit ratio. The crop life table has to be determined to provide solid information analysis of pest damage as well as the cost–benefit ratio in pest management. Benefit–risk analysis comes when a chemical pesticide is applied in an agroecosystem for considering its impact on society as well as environment relevant to its benefits.

4.4 Tolerance of Pest Damage

The pest-free crop is neither necessary in most cases for high yields nor appropriate for insect pest management. Any pest exists at some tolerance level. Castor crop can tolerate up to 25% defoliation by defoliators. Exceptions occur in case of plant disease transmission by vectors. The relationship between density of pest population and profitability of control measures is expressed through threshold values such as economic threshold and economic injury levels.

4.5 Leaving a Pest Residue

Natural enemy populations are gradually eliminated not only in the absence of their respective insect hosts but also because of the indiscriminate use of broad-spectrum insecticides, which in turn also eliminate natural enemies. Therefore, it is an important concept of pest management to leave a permanent pest residue below the economic threshold level so that natural enemies will survive.

4.6 Timing of Treatments

Treatment in terms of pesticide spray should be need based, with minimum number of sprays, timely scheduling, combined with improved techniques of pest monitoring, e.g., use of pheromone traps for monitoring of pest populations.

4.7 Public Understanding and Acceptance

In order to deal with various pest problems, special effort should be made for effective communication to the community for better understanding and acceptance of pest management practices. The IPM practices followed should be economical and sustainable.

5. TOOLS AND INPUTS OF IPM

Different inputs of IPM

1. Pest surveillance and monitoring
2. ETL
3. Knowledge of the ecology of pests

Different tools of IPM

4. Cultural methods
5. Host plant resistance/genetic/biotechnological methods
6. Mechanical methods
7. Physical methods
8. Biological methods
9. Behavioral methods
10. Regulatory/legal methods
11. Chemical methods

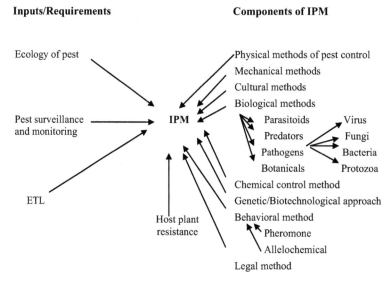

Tools or components of integrated pest management.

5.1 Pest Surveillance, Monitoring, and Forecasting

Pest surveillance forms the backbone of an effective pest management system and refers to the systematic or constant watch on the population dynamics of pests, its incidence, and damage on each crop at fixed intervals in order to predict the pest outbreak to forewarn the farmers to take up timely crop protection measures. Monitoring phytophagous insects and their natural enemies is a fundamental tool in IPM for making management decisions. Monitoring is the estimation of changes in insect distribution and abundance, information about insects' life history, and influence of abiotic and biotic factors on pest populations.

With the help of pest surveillance programs, the population dynamics and the key natural mortality factors operating under field conditions can be known, which in turn helps in devising the appropriate management strategies. Three basic components of pest surveillance are determination of the level of incidence of the pest species, the loss caused by the incidence, and the economic benefits that the control will provide. The two types of pest surveillance are roving survey and fixed plot survey. Roving survey refers to the quick survey method where fields are visited once during the cropping period. A minimum of 10 places are randomly selected in a village for this survey. Fixed plot survey is conducted by researchers in fixed plots of 50 cents each selected in research stations or in farmers' fields, and the fields are surveyed at regular intervals for pest population/damage throughout the growing season of a crop until harvest. Sampling and counting of pests may be done by in situ counting, handpicking, net sweeping, suction trap, light trap, sticky trap, pheromone trap, etc. Based on the surveillance and monitoring, a forecasting system can be developed integrating a computer database about weather predictions, costs and efficiency of various management tactics, prices of inputs, farm produce, etc.

5.1.1 Advantages

1. One can know how a pest is multiplying in an area and when it is expected.
2. Minimize the cost of plant protection by reducing the amount of pesticides used and in turn reduce environmental pollution.
3. Pest control measures can be initiated in time due to advance forecasting.
4. Useful for pest forecasting.
5. To find out natural enemy population.
6. To study the influence of weather parameters on pests.
7. Mark endemic areas.
8. Maintain the stability of the agro ecosystem.

5.1.2 Forecasting and Forewarning

Forecasting is predicting how something (e.g., the weather) will develop or predicting in advance or judging to be probable, while forewarning is warning in advance or beforehand. Weather forecasting and its relevance in the field of agriculture is traditionally known. The Indian Meteorological Department has served the farming community since 1875. Prasada Rao (2005) elaborated the importance of weather forecasting and the influence of weather factors on pest incidence. Recurring crop losses can be minimized if reliable forecasts on the incidence of pests and diseases are given in a timely manner based on weather variables.

5.1.3 Weather and Pest Incidence

Geographical distribution of pests is mainly based on climatic factors. Population dynamics of pests are influenced by biotic and abiotic factors. Seasonal periodical incidence and outbreaks are ascribed to congenial weather conditions either directly or indirectly through quantity and quality of food crops. The impact of various weather components on pests is location and crop specific. If the occurrence of pests in time and space can be predicted in advance with reasonable accuracy on the basis of relevant weather parameters, appropriate and timely control measures can be programmed and adopted, thus minimizing crop losses. Hence, insect pest forewarning systems play a major role in IPM and the sustenance of agricultural production at desired levels.

5.1.4 Pest Prediction Model

Several regression models have been developed for the prediction of crop pest outbreaks based on weather variables. A relationship could be established between weather variables and incidence of such pests through correlation analysis and fitting regression equations. These are more location specific and often fail to explain the incidence of the same species in other agroclimatic zones as the crop environments are totally different. It is important to assess the threshold variables of weather parameters affecting the different stages of insect life cycles for useful developing models. A thorough knowledge of the interaction between host (H) and pest (P) should be acquired before attempting any model as the physical environment (E) greatly influences both the host and the pest. Forecasting pest incidence based on weather variables is difficult due to complex interactions with numerous biotic factors involved. To predict growth and development of insect pests, degree-day (DD) simulation models are used extensively. The DD represents the accumulation of thermal units above a particular minimum (threshold) temperature for a 24h period. The threshold temperature represents the limit below which the development process is arrested. To predict the stage of development from DD values, the thermal constant (sum of DD) for an event must be established through experimentation (Pedigo, 2002).

5.1.5 Indian Locust Warning Organization

An early warning system for the desert locust in India is being used because of the importance of locusts. An account of the locust waring organization (LWO) was compiled by Ram Asre (2004). The Indian LWO operates a centralized forewarning system for the desert locust to keep the state authorities informed about the current desert locust situation on a regular and timely basis as reported through the surveys conducted in the scheduled desert area (SDA) by LWO staff. Survey data and field reports received from different circle offices of LWO are analyzed at LWO field headquarters, Jodhpur, together with data on ecological conditions and weather, etc. These are compared with the historical data of analogous situations in order to provide meaningful forewarning. The outcome of these surveys is sent to central headquarters, Directorate of Plant Protection, Quarantine and Storage, Faridabad, where the data are compiled and collated, and Locust Situation Bulletins are issued every two weeks throughout the year. These bulletins are circulated among all national and international agencies related to the locust control work to appraise them of the latest developments in Indian Thar Desert. The state authorities are alerted/warned immediately when any significant development of locusts is noticed.

5.2 Economic Threshold Level Concept

The terms used to express the levels of pest population are:

1. **General equilibrium position (GEP)**: It is the average population density of insects over a long period of time unaffected by temporary interventions of pest control. However, the economic injury level may be at any level well above or below the general equilibrium.
2. **Economic injury level (EIL)**: This is the lowest population at which the pest will cause economic damage or the pest level at which the damage can no longer be tolerated; therefore, it is the level at or before which the control measures are initiated. The amount of injury that will justify the artificial control measures is called the economic damage. EIL is usually expressed as the number of insects per unit area.
3. **Economic threshold level (ETL)**: It is the index for making pest management decisions. ETL is defined as the population density at which control measures should be applied to prevent increasing pest population from reaching the economic injury level. The relationship between EIL and ETL can be expressed as when no action is taken at ETL the population reaches or exceeds EIL. The ETL indicates the number of insects (density or intensity) that should trigger management action. For this reason, it is sometimes called the action threshold. Although expressed in insect numbers, the ETL is really a time parameter, with pest numbers used as an index for when to implement management (see Figure 1). The relationships between the EIL and the ETL are shown in the figure, which demonstrates the action taken when a population level exceeded the ETL and forced down the population before it could reach the EIL. No action is needed at levels below ETL. Economic injury level is defined as the lowest population density that will cause economic damage (Stern et al., 1959; Pedigo, 1996). It is also defined as a critical density where the loss caused by the pest is equal to or higher than the cost of the control measure.

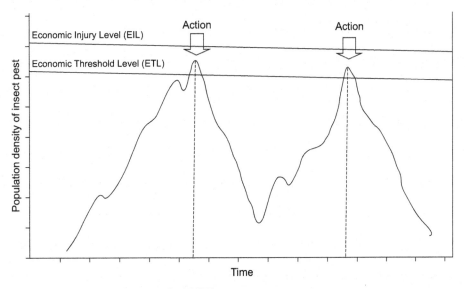

FIGURE 1   ETL and EIL.

EIL can be calculated using the following formula:

$$\text{EIL} = \frac{C}{V \times I \times D \times K} \text{ (or) } \frac{C}{\text{VIDK}}$$

where EIL = Economic injury level in insects/production (or) insects/ha
C = Cost of management activity per unit of production (Rs/ha)
V = Market value per unit of yield or product (Rs/ton)
I = Crop injury per insect (% defoliation/insect)
D = Damage or yield loss per unit of injury (Ton loss/% defoliation)
K = Proportionate reduction in injury from pesticide use.

5.2.1 Economic Threshold Level for Important Pests of Rice

Stem borer—two egg masses/m^2 or 10% dead hearts in vegetative stage or 2% white ear head in flowering stage.
Leaf folder—10% leaf damage at vegetative phase and 5% of flag leaf damage at flowering.
Gall midge—10% silver shoots; whorl maggot—25% damaged leaves.
Thrips—60 numbers in 12 wet hand sweeps in nursery or rolling of the first and second leaves in 10% of seedlings.
Brown plant hopper—1 hopper/tiller in the absence of predatory spider and two hoppers/tiller when spider is present at 1/hill.
Green leaf hopper—60/25 net sweeps or 5/hill at vegetative stage or 10/hill at flowering or 2/hill in tungro endemic area.
Ear head bug—five bugs/100 ear heads at flowering and 16 bugs/100 ear heads from milky stage to grain maturity.

5.2.2 Economic Threshold Level for Important Pests of Cotton

Thrips—50 nymphs or adults/50 leaves; Aphids—15% of infested plant.
Leaf hopper—50 nymphs or adults/50 leaves; Spider mite—10 mites/cm^2 leaf area.
Spotted bollworm or spiny bollworm—10% infested shoots/squares/bolls.
Pink bollworm—10% infested fruiting parts.
American bollworm *H. armigera*—One egg or one larva/plant.
Whiteflies—5–10/leaf; Stem weevil—10% infestation.
Tobacco cutworm—eight egg masses/100 m row.

5.3 Ecology of the Pest

Pest ecology is the study of interactions between insects and their environment. Understanding pest ecology has both basic and applied goals. The basic goals are to improve our understanding and ability to model interactions and feedback, in order to predict changes in the ecosystem. The applied goals are to evaluate and manage the extent to which the insect responds to environmental changes. Understanding the ecological factors that contribute to pest outbreaks in agroecosystems is important when implementing an IPM program. Many of the preventative methods of IPM aim to reduce the influence of the above factors on pest outbreaks through careful management of the agroecosystem. An IPM program should be based on sound ecological principles involving the knowledge of interaction between the fauna (pests and their natural enemies) and its environment.

5.4 Cultural Methods of Pest Control

Over thousands of years our forefathers have shared their crops with pests. Through trial and error they learned the best crops to grow, the best sowing times, and the best crop combinations and spacing that would reduce the insect pests' share of the harvest. They selected seeds and plants that were tolerant to the pests in the fields. Any farmers who did not know the good farming practices did not survive. The diversity of a cropping system can be increased by intercropping, trap cropping, and presence of weeds or by crops grown in the adjacent fields. When interplanted crops or weeds in the crop are also suitable host plants for a particular pest they may reduce feeding damage to the main crop by diverting the pest. On the other hand, these may also serve as an essential source of food or shelter at some point in the crop life cycle or during some part of the season, enabling the pest to maintain or build up its numbers in the field and so attack the main crop more severely.

Cropping systems like intercropping, mixed cropping, alley cropping, strip cropping, etc., when carefully selected, can reduce pest incidence by the following:

1. Acting as a barrier, a hazard or camouflage.
2. Acting as alternate hosts, diverting the pest away from the crop at risk.
3. Benefiting the pest's natural enemies.
4. Minimizing the risks involved in monocultures. Raising lab, cowpea, or other short-statured pulses as a mixed crop with sorghum minimizes infestation by stem borer.

Intercropping contributes to:

1. The diversity of agro ecosystem.
2. Change in the microclimate.
3. The buildup of population of insect pests.

Pest outbreaks are fewer in mixed strands due to crop diversity than in sole strands. Regulation of pests in intercropping systems is due to:

1. Physical means (protection from wind, hiding, shading, alteration of color or shape of the stand, etc.).
2. Biological interference—production of adverse chemical stimuli, presence of predators, parasitoids, etc.

Cultural interventions are often labor intensive and are largely preventative. They can affect pest populations in three ways. First, they can make the crop plant or agroecosystem unacceptable to the pest, and the pest will avoid the crop. Second, they can displace the crop plant in time or space, causing it to be unavailable to the pest during the period when it normally feeds. Third, they can make the agroecosystem a dangerous place for the pest by increasing the beneficial population. Some cultural interventions are listed below.

These cultural interventions or agronomic practices can be categorized into the following four subdivisions:

5.4.1 Spatial Methods

The way that crops are arranged on a farm can affect their susceptibility to pest outbreaks.

1. **Cropping pattern**: Many pests have multiple alternative hosts. By avoiding planting alternative hosts near each other, pest buildup can be avoided. Alternatively, planting a crop that attracts or encourages beneficials of neighboring crops should be considered.
2. **Plant spacing**: The space between plants affects air flow and sunlight penetration and therefore moisture and humidity levels. Wider spacing often results in reduced pest incidence. Sathpathi et al. (2012) reported a comparatively higher brown plant hopper (BPH) *Nilaparvata lugens* (Stål) population in rice at closer spacing than wider spacing. Teetes (1981) reported that compact seedling transplantation changed paddy plant canopy growth, which in turn generated a microclimatic environment that was conducive for BPH multiplication. Prasad Ram et al. (2003) also concluded that dense seedling plantation influenced BPH incidence positively.
3. **Intercropping**: Growing two or more crops in the same, alternate, or double rows often results in reduced pest problems. Carefully chosen intercrops can disrupt visual and odor cues that pests use to locate crop plants, physically limiting dispersion and enhancing natural enemies. Intercropping chickpea with barley or wheat controlled *H. armigera* to a level below its economic threshold (Ghosh et al., 1989). Tomato intercropped with cabbage in 1:1 ratio has been reported to afford maximum reduction of diamond back moth, *Plutella xylostella* Linnaeus, and leaf webber, *Crocidolomia binotalis* Zeller, larvae on cabbage. Reduction in insect incidence was attributed to possible release of volatile substances from late crop growth stages of tomato, which inhibited oviposition by incoming moths. Intercropping pulses in cotton reduced the population of leaf hopper on cotton (Rabindra, 1985), and lablab bean in sorghum reduced the sorghum stem borer incidence. Intercropping with seven rows of green gram or black gram in *kharif* red gram and two rows of cumbu in *rabi* red gram were found to encourage and conserve parasitoids and predators. Growing red gram or cowpea as an intercrop attracts the adult moths of red hairy caterpillar, *Amsacta* spp., to lay more eggs. Intercropping of groundnut with pearl millet reduced the incidence of thrips, leaf hoppers, and leaf miners, whereas the same with sunflower and castor increased the incidence of thrips and leaf hoppers, respectively. When pearl millet was grown as an intercrop in groundnut, the parasitic activity of *Goniozus* sp. was considerably enhanced. The pollen grains of the pearl millet were preferably used as food by the adult parasitoids.
4. **Strip cropping**: Growing two or more crops in broad, alternating strips has many of the benefits of intercropping but allows a broader range of mechanized field operations.
5. **Border crops**: Crops sown at the periphery of the main crop that specifically prevent pests from entering a crop field or harbor beneficials and can consist of a single crop species, a mixture of species, or a mixture of wild species. Growing castor along the border of cotton or chilies fields and irrigation channels act as an indicator or trap crop for *Spodoptera litura* (Fabricius). *Spodoptera litura* prefers to lay eggs in the border crop castor, thereby preventing the pest entering into the main crop, viz., cotton or chilies. Growing trap crops like marigold *Tagetes* sp. attracts pests like American bollworm to lay

eggs; barrier crops like maize/sorghum prevent migration of sucking pests like aphids; and guard crops like castor attract *S. litura* in cotton fields as reported by Murthy and Venkateshwarulu (1998).

6. **Trap crops**: Sacrificial crops that are more attractive than the main crop to a particular pest. A trap crop is grown specifically to lure pests away from the main crop. Once the trap crop is infested, it is often sprayed or destroyed. This protects the main crop and reduces the need for pest control measures. Often a trap crop is the same species as the main crop but is a different cultivar or grown in a different way. For example, a small sowing of maize preceding the main crop will effectively trap aerial pests. This is because the earlier sown, taller plants are more attractive to the pests. Trap crops of different species can be employed as well. Eggplants are more attractive to many potato pests than potatoes are, and therefore small, sacrificial sowings can be used to protect potato crops from various insect pests. Plant strands are also grown to attract other organisms, like nematodes, to protect target crops from pest attack. Mustard can be grown as a trap crop at two rows per 25 cabbage rows for the management *P. xylostella*. Mustard crop is sown 15 days prior to cabbage planting or 20-day-old mustard seedlings are planted. The mustard crop that acts as trap crop should be periodically sprayed to prevent dispersal of the larvae. Planting 40-day-old African tall marigold in 25-day-old tomato seedlings (1:16 rows) reduces *H. armigera* damage.

5.4.2 Sequence-Related Methods

Previous crops can have an effect on pest problems in the present crop. Planting crops in a particular sequence can reduce pest problems in general and soil-borne pest problems in particular.

1. **Crop rotation**: Growing crops that are hosts for an insect pest in a successive manner will provide a continuous supply of food and uninterrupted breeding facilities, which will result in an increase of the pest population. Crop rotation can be used to interfere with the pests of one crop by culturing another. Crop rotations have long been used by farmers both as a fertility and pest management tool. Planting a succession of different crops in a field over several years prevents the buildup of pests that can occur when the same crop is grown repeatedly. Crop rotations are particularly effective in controlling soil-borne diseases, as well as soil-dwelling organisms that do not disperse well. Crop rotations work best when the crops in the rotation have few shared pests. Some crops directly reduce pest problems for the succeeding crop. Crop rotation with less favorable crops like sorghum, gingelly, black gram, horse gram, and paddy can be effective in managing *H. armigera* in red gram. African marigolds release an antinematode compound called thiophene, which can reduce nematode loads for subsequent horticultural crops. Cucurbits or sweet potatoes are often included in a rotation because they tend to choke out persistent weeds. Okra following a cotton crop will suffer from increased pest infestation because most of the pests of cotton are also pests of okra.

2. **Multiple cropping**: Multiple cropping, or multicropping, is a cropping system in which several crops are grown in succession within a season. Often the same multicropping scheme is followed year after year. The crops grown in a multicropping scheme are often limited by climate or pests. In many tropical countries, wet rice is grown during the rainy season, followed by cool-season vegetables such as cabbage. If sufficient irrigation

or soil water is available, a third, drought-tolerant crop may be grown (e.g., legumes). Many farmers grow a cereal crop immediately before higher-value crops, such as tobacco, cotton, or vegetables. Multicropping a monocotyledonous crop with a dicotyledonous crop effectively disrupts soil-borne pests. It also improves weed management, as the persistent monocot weeds of cereals can be easily identified in a dicot crop and vice versa.

5.4.3 Tillage Operations

Summer plowing is a specific tillage operation aimed at destruction of overwintering insects. Summer plowing after rain showers destroys the pupae of red hairy caterpillar *Amsacta* spp. in groundnut. Plowing and hoeing help to bury the stages of insects or expose stages of soil-inhabiting insects to be picked up by birds. Trimming and plastering of field bunds in rice fields expose eggs of grasshoppers and eliminate the bugs breeding in grasses. In temperate climates, autumn plowing and disking exposes overwintered forms of many insects to predation, freezing, and so on.

5.4.4 Planting Materials and Inputs

Both the genetic makeup and the health of planting materials can affect crop susceptibility to pest problems.

1. **Disease-free plants and seeds**: Many pest problems are borne on planting stock or seeds. Virus and disease loads on vegetatively propagated stock can reduce yields by as much as 50%. Farmers can greatly improve plant health by starting with clean planting materials. Many countries have certification programs for seed and planting stock. Certified seed is often more expensive than locally available seed, but the extra cost should be balanced against the potential risks of using uncertified seed. Vegetatively propagated crops are often grown in tissue culture to obtain disease-free strains.
2. **Altered planting/harvest dates**: Planting and harvesting out of synchrony with pest populations can reduce pest damage. If a particular pest responds to a particular environmental cue, then this pest can be avoided by adjusting sowing or harvesting dates. Sowing may be delayed until the pest has emerged and dispersed, or the crop can be sown earlier so that it will reach a resistant growth stage by the time the pest emerges. Changes in sowing and planting dates can help in the avoidance of egg-laying periods by certain pests, the establishment of tolerant plants before pest occurrence, the maturation of crops before abundance of pests, and the synchronization of pests with natural enemies. Adjusting time of planting to asynchronize the vulnerable stages of the crop and peak pest incidence is essential in pest management as reported by Krishnaiah et al. (1983). Early sowing and increased seed rate of 12.5 kg/ha is effective against sorghum shoot fly, *Atherigona soccata* Rondani. Rice leaf folder, *Cnaphalocrocis medinalis* Guenee, incidence was more in September-planted rice, and green leaf hopper and brown plant hopper were maximum on crops planted during August 16 (Karuppuchamy and Uthamasamy, 1986). Karuppuchamy et al. (1982) reported lesser incidence of gall midge *Orseolia oryzae* (Wood-Mason) in October planted rice.
3. **Mulches**: Mulches can effectively control pests, in particular weeds and some ground-dwelling insects. The main usefulness of mulches for pest management is that they can

effectively control weeds. Mulches also help retain soil moisture, preventing aphids and thrips problems but potentially promoting fungal pathogens. Mulch provides habitat for a diversity of beneficial. Research has shown that straw mulch can suppress early season Colorado potato beetle activity by creating a microenvironment that increases the number of predators like ground beetles, lady beetles, and lacewings. (http://www.extension.org/pages/18574/managing-the-soil-to-reduce-insect-pests#.VWLyttKeDGc).

4. **Irrigation**: The amount of water available in an agroecosystem can have a large effect on pest populations. Withholding water and flooding a field are the two ways in which irrigation can be used to manage pests. First, applying an optimum amount of water to a crop results in quick-growing, healthy, and pest-resistant plants. Overwatering can cause plants to become lush and prone to fungal diseases and insects, such as thrips and aphids. On the contrary, plants stressed from underwatering will be susceptible to pest attacks. Flooding a field can drown nonaquatic pests, and newly irrigated soil particles may swell enough to crush soil-dwelling pests. Raising and lowering water levels in a flooded field can augment predation of above-water pests by aquatic predators. Adjusting irrigation can speed or slow crop maturity, preventing a crop from ripening at the same time that a pest population peaks. Alternate flooding and draining of water significantly reduced green leaf hopper and brown plant hopper populations and whorl maggot, *Hydrellia sasakii* Yuasa and Isitani damage in rice (Karuppuchamy and Uthamasamy, 1984).

5. **Fertilization/manuring**: Soil fertility affects the growth and composition of a crop, which in turn affects its susceptibility to various pests. Soils with adequate, balanced fertility result in quick-growing, healthy plants that resist pests. Fast-growing plants reduce opportunities for pests that attack certain growth stages, such as stem borers, by shortening their period of vulnerability. Fast-growing plants can also compensate for damage that does occur. Fast-growing plants quickly form a canopy, discouraging pests that disperse aerially or seek bare ground. Although high fertility can increase potential yield, balanced fertility that is appropriate to the intended crop is the key to pest prevention. Many diseases and insect pests are associated with too much nitrogen fertilizer in proportion to phosphorous, potassium, and other elements. Building balanced soils through the addition of compost and organic matter can help reduce nutrient imbalances. Relying on nonchemical sources for at least part of a crop's fertility requirements can indirectly reduce pest problems. Rice brown plant hopper *N. lugens* population increased with increased N (nitrogen) application (Karuppuchamy and Uthamasamy, 1984).

6. **Pasturage**: Grazing animals in a crop field before, during, or after harvest can reduce pest problems. Domestic cattle, buffalo, pigs, ducks, poultry, and fish can drastically change the nature of a crop field through their grazing activities.

7. **Efficient harvest and storage**: A clean, well-timed efficient harvest can break pest cycles. Crop materials that can harbor pests until the next season should be removed. Crops should be harvested at the proper growth stage. Harvesting early or late can cause pest problems in storage. Properly managing storage pests maintains value of the stored crop and prevents it from becoming a pest vector for future crops. Drying, curing, and other primary processing activities should be planned and carried out with IPM in mind. Stored crop IPM follows the same procedures as field IPM. First, the crop is stored in a way that prevents pest problems. Second, pests are monitored, identified, evaluated for crossing ETL, and intervened if necessary.

5.5 Host Plant Resistance/Genetic Control

Crop plants that are bred to resist certain pests form the central pivot of IPM and is compatible with all other methods of pest control. Resistance is the inheritable character that enables the plant to inhibit the growth of an insect population or to recover from injury caused by populations that were not inhibited, or the ability of a strain to tolerate or avoid factors that would prove lethal or reproductively degrading to the majority of strains in a normal population.

Many diseases and insects have been managed effectively by breeding genetic resistance into modern varieties. The mechanisms of host–plant resistance are diverse, and varieties with resistance to multiple pests have been developed. Biotechnology, in particular, the controversial approach of genetic engineering, presents many interesting possibilities for future resistant varieties as well as concerns about its safety. The mechanisms of resistance are diverse. Crops resist pests through color, palatability, hairiness, waxy coatings, gross morphology, gumminess, necrosis, hardness, phenological shifts, toxin production, nutrition, integration with biocontrol, and compensation. The above mechanism can be broadly categorized into the following: (1) Nonpreference, where the crop plants are avoided by the pest because they do not "like" it; (2) Antibiosis, where plants cause a reduction in the biological performance of the pest; and (3) Tolerance, where plants can tolerate the pest and still provide an adequate yield. Many resistant varieties have lost their resistance through mass plantings and adaptation of pest populations. Resistance management plans are designed to reduce or prevent the development of resistance by planting resistant varieties in a particular way. For example, many maize growers grow a strip of an older, susceptible variety around a new resistant variety so that the pest has something preferable to eat and maintain its existing resistance status.

Therefore, varieties having different bases of resistance should simultaneously be in the field to reduce the chances of development of new insect biotypes. Multiple pest resistance has the potential to play a more important role in crop pest management. Development of TKM 6, possessing multiple resistance to several rice pests, is a notable achievement, and this variety is used as a donor in the development of many national and international varieties currently in cultivation (Chelliah and Bharathi, 1985). Genotypes favoring colonization by the natural enemies due to the presence of allelochemicals should preferably be utilized in IPM systems.

Wide/wild hybridization enriches the gene pool of domesticated crops and has helped in improving productivity, adaptability to stress environments, resistance to pests and diseases, as well as improving quality of food, feed, and fiber. Hybrids were produced between *Oryza sativa* and *Oryza ridleyi* by adopting embryo rescue technique to transfer yellow stem borer resistance to cultivated rice (Maheswaran, 1998). **Sterilization** of pests is another genetic approach where a large number of sterile male insects are released into an agroecosystem aiming to quickly reduce the local population. **Genetic displacement** is also a tactic where a wild pest population is replaced with a population that has been bred to be less of a pest problem. **Genetic improvement of beneficials** is a recent tool where improved beneficials with higher level of adaptation, survival, and reproduction are bred and released into the agroecosystem. It is also possible to isolate strains of entomopathogens, particularly virus, with increased virulence and genetically improve them further (Rabindra, 2000).

5.6 Mechanical Methods of Pest Control

Mechanical methods of pest control are the reduction or suppression of insect pest population by means of manual devices or labor. Mechanical interventions involve the direct removal or destruction of the pest.

1. **Handpicking and weeding**: Hand picking and destruction of pests is widely practiced wherever labor and time permits. Egg masses, larvae, or nymphs and sluggish adults can be handpicked and destroyed (e.g., egg masses of paddy stem borer and groundnut hairy caterpillar). Early stages of *S. litura* and hairy caterpillars are easily located by their typical damage symptoms. Most of the insects can be collected with hand nets and destroyed. Collection and destruction of fallen fruits is effective against fruit flies and fruit borers. Manual collection and destruction of pink bollworm attacked rosette flowers, withered and drooped terminals infested by spotted bollworm can reduce the incidence of these pests in cotton.

2. **Traps**: Traps are often used for monitoring pest populations, but trapping can also be used for control. Many kinds of traps exist. Banding with sticky bands or bands impregnated with repellent can exclude pests from tree crops. Traps using mechanical force such as suction, pitfall, and emergence traps are also used in pest management.

3. **Shaking**: Shaking crop plants can dislodge pests to groundsheets where they can be collected and destroyed.

4. **Pruning**: Pruning to remove egg masses of pests can be an effective way of controlling orchard and ornamental pests.

5. **Barriers**: Barriers, such as greenhouses, screens, or mulches, prevent pests from reaching a crop for feeding or laying eggs. Several types of barriers are described here.

   a. **Screens**: Screens are commonly used in greenhouses to allow circulation of air but prevent entry of pests. Screen houses are similar to greenhouses, except that their outer covering consists of screening that is appropriately sized for the target insect. Screens can also be used in fields to protect valuable plants.

   b. **Greenhouses and other structures**: Built structures that can be closed off from the outside environment can very effectively exclude pest organisms. Greenhouses provide protection not only from extreme weather and temperatures but also from dispersing pests. Many greenhouses can be sealed for fumigation, which is often required if pests establish a population inside.

   c. **Row covers and mulches**: Row covers and mulches are placed on the ground around the stem of the crop plants. They are effective against soil-borne pests such as cabbage maggots.

   d. **Trenching**: Some insect pests are unable to escape from trenches, and lining a crop field with a trench can provide substantial protection. For example, Colorado potato beetles are trapped by a "V"-shaped trench with a sharp slope, resulting in substantial reductions in adults and egg deposition within potato crops. Provision of preventive barriers like digging of 30–60-cm-wide and 60-cm-deep trenches or erecting 30-cm-high tin sheet barriers around the fields is useful against pests like hairy caterpillars. The groundnut red hairy caterpillar and rice army worm are prevented from migrating from one field to another by digging trenches around the field.

e. **Bags**: Valuable fruits are sometimes covered with individual bags. This method is time consuming and labor intensive, but can lead to premium prices for the resulting unblemished fruits. Bagging/wrapping of pomegranate and mango fruits in paper bags avoids infestation of the pomegranate butterfly, *Deudorix isocrates* Fabricius (Karuppuchamy, 1995) and mango fruit fly *Bactrocera dorsalis* (Hendel). Entire bunches of banana can also be protected by using large bags.

f. **Packaging**: Many materials have been developed that physically exclude pests from stored crop products. These include metal foils, plastics such as cellophane and polypropylene, and paper.

g. **Fences**: A well-built fence can protect crops from many pests as long as it is properly installed and maintained. In particular, mammals and low-flying insects can be excluded by good fencing.

h. **Banding/Swabbing**: Tin bands are fixed over coconut palms to prevent damage by rats. Putting a band of grease around the tree trunk will prevent the migration of mealybugs from soil to mango trees. This is called mechanical exclusion. Screens and other barriers like mosquito nets and fly proof cages are used to prevent flies, mosquitoes, and other insects. Ant pans exclude ants. In the case of mango mealybug, *Drosicha mangiferae* (Green), banding of trees with 20-cm-wide alkaline/polythene (400 gauge) during the middle of December prevents the climbing of newly hatched nymphs of mealy bugs from the ground. Banding should be done two to three feet above the ground level and just below the junction of branching on all the trees in the orchard, including other fruit trees like guava, jack, and citrus.

5.7 Physical Methods of Pest Control

Physical interventions exploit a physical characteristic of the environment in order to manipulate pest populations. Different temperatures, humidity levels, and even atmosphere can be used to manage pests, as can mechanical intervention. In situations where the farmer has a large degree of control over the physical environment, such as greenhouses, physical interventions can be the most important methods of IPM. Even in field situations, physical manipulations such as compaction, flooding, or mulching can adversely affect potential pests.

1. **Temperature**: Heat can be used to manage pests. Examples include soil solarization, flaming, burning, etc.

2. **Water**: Moisture level can be manipulated to manage pests. Flooding is an example.

3. **Air**: Atmospheric manipulation can be used to manage a pest. This method is usually associated with storage of harvested product in high-nitrogen or high-carbon dioxide environments.

4. **Light**: The behavior of certain species of insects being attracted to light could be advantageously used in their management. The light traps could be used both for monitoring and as a means of control. Mohan and Janarthanan (1985) reported that the rice stem borer and the brown plant hopper responded more toward a yellow light source, while the rice leaf folder and green leaf hoppers responded to a green light source.

5. **Physical agents**: Physical forces minimize certain pests. A material called drie-die consists of highly porous, finely divided silica gel that when applied abrades the insect cuticle thus encouraging loss of moisture resulting in death. It is mainly used against

stored product pests. Kaolinic clay after successive activation with acid and heat can be mixed with stored grain. The clay minerals absorb the lipoid layer of the insect cuticle by which the insects lose their body moisture and die due to desiccation. Artificial heating and cooling of stored products prevent insect damage. Usually high temperatures are more effective than low temperatures. Stored products can be exposed to 55 °C for 3h to avoid stored product pests. Steam sterilization of soil kills soil insects and is done by vapor heat treatment: Heated air is saturated with water (>RH 90%) for a specified period of 6–8h for raising pulp temperature to 43–44.5 °C for mango against fruit flies. Activated clay or vegetable oil at 1% effectively reduced the damage by pulse beetle, *Callosobruchus chinensis* (Linnaeus). Solar heat treatment of sorghum seeds for 60s using solar drier resulted in 100% mortality of rice weevil, *Sitophilus oryzae* (L.) and red flour beetle, *Tribolium castaneum* Herbst. There was no reduction in germination percent of seeds (Mohan et al., 1987). Biogas fumigation for a period of five days resulted in complete mortality of eggs, grubs, and adults of pulse beetle (Mohan et al., 1989).

5.8 Biocontrol

Biocontrol is the use of an organism to reduce the population density of another organism. It is the most successful, most cost-effective, and environmentally safest method of pest management. It is nature's own way to keep numbers of pest organisms at low levels. Biocontrol is present in all ecosystems, both natural and man-made, and is always active. The result of natural biocontrol is that the earth is green and that plants can produce sufficient biomass to sustain other forms of life. Without biocontrol, the production of energy by plants would be a tiny fraction of what is produced currently.

- Purposeful use of an organism to reduce a plant/animal population that is inimical to man.
- For ecologists, biocontrol is a stabilizing process of ecosystem services.
- For crop production specialists, biocontrol forms population management.
- At low pest densities, a generalist works well.
- At high pest densities, specialist parasitoids and pathogens work well.
- Pests in agricultural fields are not permanent residents and will colonize fields each season.

In pest management, biocontrol usually refers to the action of parasitoids, predators, or pathogens on a pest population, which reduces its numbers below a level causing economic injury. Insects and pathogens that attack insect pests, noninsect pests, and weeds are considered biocontrol agents. Biocontrol is a part of natural control and can apply to any type of organism, pest or not, and regardless of whether the biocontrol agent occurs naturally, is introduced by humans, or manipulated in any way. Biocontrol differs from chemical, cultural, and mechanical controls in that it requires maintenance of some level of food supply (e.g., pest) in order for the biocontrol agent to survive and flourish. Therefore, biocontrol alone is not a means by which to do the pest management.

Important Characteristics

- Often relatively inexpensive and can be "permanent" for those biocontrol agents that can survive multiple years and become self-perpetuating.
- Effectiveness can be from low to high.

- Can be disrupted by other pest management tactics, especially broad-spectrum pesticides.
- Suppressive effects are density dependent; it will have its greatest impact when pest densities are high.
- Often pest specific, not broad spectrum.
- Often has a lag time between buildup of the pest population and buildup of the biocontrol agent; generally not fast acting.
- Good tactic to include in a multifaceted approach (IPM); fits in well with cultural, mechanical, and some chemical controls.
- Most successes have been in perennial crops (orchards, vineyards), pasture and field or forage crops that can withstand a moderate level of pest injury.

Biological interventions are usually highly selective, there are rarely negative side effects (except in the case of classical biological control), and released organisms are self-perpetuating. Pest resistance is rare because predators and parasitoids tend to coevolve with pests, although there are cases of pests developing resistance to frequently applied pathogens. Rearing and release of bioagents is often simple and inexpensive. Disadvantages include slow action, unpredictability, and incompatibility with pesticides. Using biological intervention effectively requires good observations and a sound understanding of the biology of pests and their natural enemies.

There are three types of biocontrol agents (see Figure 2):

1. **Predators** catch and consume pest prey. Major groups of predators include the insect orders Hemiptera, Neuroptera, Diptera, Coleoptera, and Hymenoptera, the Arachnida, and vertebrates such as snakes, birds, and fish. Predators are often fairly generalist, preying on a wide range of prey according to abundance and ease of capture. The green lace wing, *Chrysoperla* spp., is a general predator feeding on soft-bodied sucking pests, like mealybugs, aphids, thrips, scales, and mites, and eggs and young larvae of lepidopteran pests.
2. **Parasitoids** lay eggs on or in a pest host, and the emerging parasitoid larvae consume and ultimately kill the pest. Most parasitoids are far smaller than their prey, and therefore mass rearing and release of parasitoids can be relatively convenient. Parasitoids are found in the insect order Hymenoptera and the Tachinid family of flies (Diptera). Parasitoids usually exhibit high host specificity. The egg parasitoid *Trichogramma* species are the most widely used natural enemy in the world, partly because they are easy to mass rear and they attack many crop pests.
3. **Pathogens** infect pests with debilitating diseases and include fungi, bacteria, viruses, nematodes, and other microbes. Fungi, particularly Deuteromycetes, can infect pests externally under favorable conditions, but other pathogens must be ingested in order to be effective as control agents. Pathogens are very specific to their hosts. They are often referred to as biopesticides because they can be applied in ways similar to chemical interventions. The usefulness of nuclear polyhedrosis viruses in the management of polyphagous pests like *H. armigera* and *S. litura* on several crops has been demonstrated (Rabindra, 1998).

Categories of biological interventions are listed below.

5.8.1 Natural Biocontrol (Conservation)

Conservation biocontrol involves any practice that increases colonization, establishment, reproduction, and survival of native or previously established natural enemies in the

FIGURE 2 **Example of three types of biocontrol agents.** (A) Pathogen: NPV-infected larva of *Helicoverpa armigera*. (B) Egg parasitoid: release of *Trichogramma* in the field. (C) Predator: chrysopa release as eggs in cotton field for the control of sucking pests.

agroecosystem (Landis et al., 2000). This definition can be expanded to include any method that encourages biodiversity, but usually this category encompasses measures that encourage a specific beneficial. Natural biocontrol takes advantage of the beneficial already present in the agroecosystem by conserving and promoting self-sustaining populations. Birds, bats, snakes, insects, nematodes, fungi, bacteria, viruses, and other groups of organisms can contribute to pest management. Well-managed natural biocontrol can handle the majority of pest problems without requiring additional interventions. Increasing the biodiversity of the agroecosystem will increase natural biocontrol, as will devoting a larger percentage of the total area to noncrop or perennial plantings. Flowering herbs and aromatic plants tend to attract beneficial insects, while compost and organic matter will improve the soil habitat for natural biocontrol. Conservation is defined as the actions to preserve and release of natural enemies by environmental manipulations or alter production practices to protect natural enemies that are already present in an area or nonuse of those pest control measures that destroy natural enemies.

Important conservation measures are:

5.8.2 Use Selective/Modified Insecticide Molecules Safer to Natural Enemies

Modifications to pesticide regimens include reducing or eliminating insecticide use, using pest-specific insecticides when needed, making applications when beneficial arthropods are not active, and making treatment decisions based on monitoring and the presence of vulnerable life stages.

While total independence from chemical control is not feasible for most situations, reductions in insecticide use are possible through IPM programs based on rigorous pest monitoring, established treatment thresholds, and/or insect population models (Horn, 1988; Pimental, 1997). Thus, insecticides are used only when needed to prevent crop damage that results in economic loss. When insecticide use is warranted, adverse effects on natural enemies can be minimized by using selective, pest-specific products that are only effective against the target pest and its close relatives. Selective chemistries include microbial insecticides, insect growth regulators, botanicals, and novel insecticides with specific modes of action against target insects. Alternatively, insecticide applications can be timed so they do not coincide with natural enemy activity; dormant or inactive predators and parasitoids are not exposed to broad-spectrum insecticides applied when they are dormant or inactive (van Driesche and Bellows, 1996). This strategy requires a thorough understanding of the crop, agroecosystem, and the biology and life cycle of important natural enemies in the system.

5.8.3 Conserving Natural Enemies via Habitat Manipulation

Natural enemies are attracted to habitats rich in food, shelter, and nesting sites (Landis et al., 2000; Rabb et al., 1976). Avoidance of cultural practices that are harmful to natural enemies and usage of favorable cultural practices, cultivation of varieties that favor colonization of natural enemies, providing alternate hosts for natural enemies, preservation of inactive stages of natural enemies, and providing pollen and nectar for adult natural enemies are some of the habitat manipulation strategies to conserve natural enemies.

5.8.4 Augmentation

The aim of augmentative biocontrol is to improve the numerical ratio between pest and natural enemy to increase pest mortality. It involves the release of natural enemies, typically mass reared in an insectary, either to inoculate or inundate the target area of impact (Obrycki et al., 1997; Parrella et al., 1992; Ridgway, 1998). Inoculative releases involve relatively low numbers of natural enemies, typically when pest populations are low or at the beginning of a growth cycle or season. Inundation involves relatively high numbers of natural enemies released repeatedly throughout the growth cycle or season. Thus, inundative release of natural enemies is similar to insecticide use in that releases are made when pests achieve high density enough to cause economic harm to the crop. In both types of release, the objective is to inflict high mortality by synchronizing the life cycles of the pest and natural enemy. Hence, an effective monitoring program of pest population is essential to the success of augmentative biocontrol. Augmentation is often used to "tip the balance" in favor of the beneficial population, thereby preventing a destructive peak in the pest population. Augmentation is a short-term solution, and biocontrol agents that are released in an augmentation release will disperse quickly if the agroecosystem cannot provide them with food and shelter. Conservation and creation of habitat for the released beneficial species should accompany augmentation in order to make the releases more sustainable. In recent years, sugarcane woolly aphid, *Ceratovacuna lanigera*, has emerged as one of the serious pests threatening sugarcane cultivation in many states of India including Tamil Nadu. Studies carried out against the aphid and its natural enemies revealed that the mass production and inoculative release of predators, namely, *Dipha aphidivora* (1000/ha) or *Micromus igorotus* (2500/ha) effectively controlled the sugarcane woolly aphid population when released two to three times depending on the incidence of the pest (PDBC, 2008; Kalayanasundaram et al., 2010) (Figure 3).

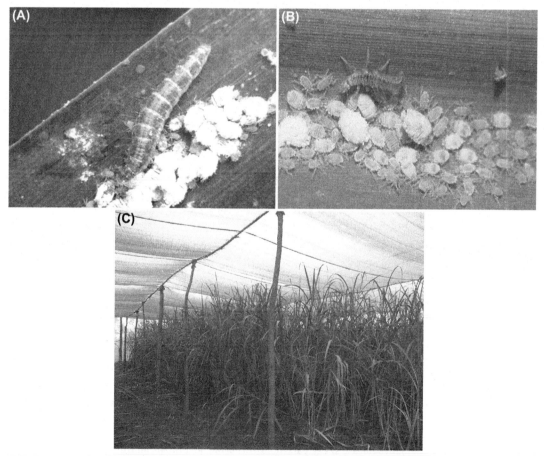

FIGURE 3 Mass production of *Dipha* under shade net for the control of sugarcane woolly aphid (SWA) *Ceratovacuna lanigera*. (A) *Dipha* larva; (B) *Micromus* grub (C) Inside view of shade net.

5.8.5 Inundative Release

Releasing large number of beneficials into an agroecosystem in order to manage a pest problem is inundative release. Inundation is the biological equivalent of a chemical pesticide. Large numbers of biocontrol agents are released with the intention of reducing or eliminating the pest population in a short period of time.

Normally, the released agent disperses or dies soon afterward due to lack of food or hosts. Inundation is effective against pests that only experience a single generation per season or that occur in rare outbreaks. For example, *Trichogramma* (a parasitoid) is mass released against lepidopteran borers in rice, cotton, sugarcane, fruits, and vegetable crops. Karuppuchamy (1995) reported more than 50% reduction in fruit damage caused by anar butterfly *D. isocrates* in pomegranate orchards by inundative release of *Trichogramma chilonis* at 2.5 lakh parasitoids/ha. four times at 10 days interval (see Figure 4).

5.8.6 Seasonal Inoculative Release

Seasonal inoculative release is releasing a small number of natural enemies into an agroecosystem early in the season in order to give the bioagents a "head start" on pest population. Seasonal inoculations are used to establish natural enemy populations in areas where they previously existed in small numbers or did not exist at all. The release of bio-agents into a recently planted greenhouse is a good example of a seasonal inoculation. Seasonal inoculations are a good way to establish a stable pest–beneficial relationship early in the season. By releasing a significant number of bioagents early in the season, pest and bioagent numbers will not fluctuate widely, preventing pest population peaks that would require additional intervention.

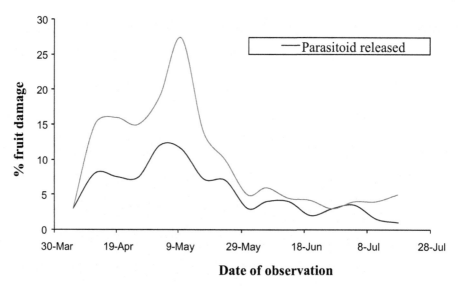

FIGURE 4 Effect of inundative release of egg parasitoid *Trichogramma chilonis* on fruit damage to pomegranate by anar butterfly *Deudorix isocrates*. Blue (dark gray in print versions) line—parasitoid released plot. Pink (light gray in print versions) line—untreated check (unreleased plot).

5.8.7 *Classical Biocontrol*

Classical biocontrol aims at introducing the exotic natural enemies of inadvertently introduced alien organisms (which have become pests in the absence of natural checks in the new environment) in order to reestablish the balance between the pests and natural enemies. Introduction of host-specific organisms from the country of origin of the pests offers some highly effective and environmentally friendly solutions to the problem of invading alien pests. The crops that we are familiar with today in many countries are not indigenous to those countries. Rather, they have been introduced, either historically or recently, often along with their pests. Where a pest organism is introduced into an exotic environment, the indigenous natural enemies that had controlled its population do not exist. There are many documented instances of pests that seemed innocuous in their native lands becoming huge pests in their new countries. An important example in the Asia–Pacific region is the introduction and spread of the apple snail, *Pomacea canaliculata* Lamarck, which has had a devastating effect on rice cultivation in several countries. In Brazil and Paraguay, where it is native, natural enemies maintain the population of apple snails at less than one or two per square meter of rice. In Asian rice, however, densities of 300 or 400 per square meter have been recorded. The obvious reaction to a problem like this is to research what natural enemies exist in a pest's native region and introduce them to the new area where the pest is a problem. While this is a good idea in theory, often the factors that allow particular natural enemies of a pest to control it back home are overlooked or misunderstood. Thus classical biocontrol has seen mixed success, with some introduced enemies doing their job, some being ineffectual, and still others causing even more problems. According to Waage and Greathead (1988), of the 563 attempts to establish classical biocontrol against insect pests, 40% were successful and 31% of releases against weed pests (126 releases) were deemed successful.

The invasive papaya mealybug, *Paracoccus marginatus* (Williams and Granara de Willink) was first noticed in India in June 2008. Three exotic parasitoids, viz., *Acerophagus papayae* (Noyes and Schauff), *Anagyrus loecki* (Noyes and Menezes), and *Pseudleptomastix mexicana* (Noyes and Schauff) were introduced through National Bureau of Agricultural Insect Resources (NBAIR) (Shylesha et al., 2011) of which *A. papayae* was the most promising parasitoid. Mass multiplication of the exotic parasitoid *A. papayae* was taken up for large-scale field release (see Figure 5).

About 5.7 million parasitoids of *A. papayae* were mass multiplied and released in farmers' fields at 100 parasitoids/field/village or block in more than 2500 locations throughout Tamil Nadu from 2009 to 2011, especially on papaya, tapioca, and mulberry severely attacked by *P. marginatus*. A spectacular control of papaya mealybug within a period of 90 days after release of the imported parasitoid *A. papayae* is an example of successful classical biocontrol programs carried out by Tamil Nadu Agricultural University (Kalyanasundaram et al., 2010, 2011; Jonathan et al., 2011).

5.9 Chemical Interventions

Chemical interventions introduce organic and inorganic substances into the agroecosystem to manage pest problems. They can be collected and derived from organisms (botanicals, pheromones, allelochemicals, insect growth regulators) or man-made (synthetic), collected from other natural sources (inorganics).

FIGURE 5   Biocontrol of papaya mealybug utilizing the parasitoid *Acerophagus papayae*. (A) Severe incidence of papaya mealybug. (B) *Acerophagus papayae* parasitoid laying egg on papaya mealybug. (C) Mass culturing of papaya mealybug parasitoid *A. papayae* in the biocontrol laboratory.

5.9.1 Botanicals

Much before the advent of synthetic insecticides, botanical pesticides were used to protect agricultural crops from insect ravages. Botanicals are pesticidal substances derived or refined from plants. Grainge and Ahmed (1988) listed about 2400 plant species having insecticidal and acaricidal properties distributed in 189 plant families. Neem (*Azadirachta indica*) and pyrethrum (*Chrysanthemum cinerariaefolium*) are commonly used commercial botanicals.

All parts of the neem tree are biologically active, however, maximum insecticidal activity is found in the seed kernels. To date, neem products have been evaluated against 450–500 species of insects in different countries worldwide, and 413 of these are reportedly susceptible at different concentrations (Schmutterer and Singh, 1995). Concentrations ranging from 0.5% to 5% of various neem seed kernel extracts (NSKE) have generally been found to deter the feeding of most of the insects evaluated so far (Singh, 2000). The seed kernel extracts or azadirachtin when fed or applied to juvenile stages arrest their growth. Depending on the dose, the insects are either killed before reaching adult stage or produce malformed and miniature adults. The other physiological effects recorded are prolongation of larval period (up to two months), production of larval-pupal and pupal-adult intermediates. Azadirachtin affects vitellogenesis, insect vigor, longevity, and fecundity in insects. Both neem seed oil and extract of neem seed kernel have been reported to deter oviposition by adult insects (Karuppuchamy, 1995).

Even today some innovative farmers puddle green leaves and twigs of neem in rice nursery beds to produce robust seedlings and simultaneously ward off attack by early pests like leaf hoppers, plant hoppers, and maggots. The repellent and antifeedant effects of neem have been reported against a wide range of rice insect pests including *N. lugens,* white backed plant hopper, *Sogatella furcifera* (Horvath), *C. medinalis,* and ear-cutting caterpillar, *Mythimna separata* (Walker) (Saxena, 1989). NSKE (5%) and neem oil (0.5%) were very effective in controlling cotton whitefly (Jayaraj et al., 1988) and anar butterfly *D. isocrates* (Karuppuchamy and Balasubramanian, 2001).

Garlic and hot peppers are used on a noncommercial basis by many farmers and gardeners. While most biopesticides are produced on a small or local scale, there is an increasing demand for commercially produced biopesticide formulations where the amount and strength of the active ingredient is known. Botanicals are known only at a local level and are produced and used mainly by small farmers. Rather than adopt a widespread botanical such as neem into an IPM program, IPM practitioners should evaluate indigenous botanicals for inclusion in an IPM program.

5.9.2 Semiochemicals

Semiochemicals are produced by organisms that modify the behavior of animals. The most important types of semiochemicals for IPM are pheromones and allomones. Pheromones are emitted by members of a species to modify the behavior of other members of same species. Allomones are like pheromones, except they are emitted by one species in order to modify the behavior of another species.

The most commonly used pheromones in agriculture are sex attractants. These chemicals are produced by females to attract males for mating and are used by IPM practitioners to attract males into traps. Pheromone traps are often used to determine population density by sampling the number of males caught in a trap in a certain amount of time. Sex pheromones also help in mass trapping and mating disruption of insects. The synthetic sex pheromone "gossyplure" is used for trapping cotton pink bollworm *Pectinophora gossypiella* Saunders (Karuppuchamy and Balasubramanian, 1990). Large-scale trials in mating disruption in India on the pink bollworm demonstrated an increase of 35% in yield (Sharma and Singh, 1982) in Gujarat area using 50–75 g of gossyplure pheromone per hectare with seven applications at 8–22 days interval. Alarm pheromones can be used to repel certain species from crops. A traditional practice in Mexico takes advantage of the alarm pheromones released by burning beetles. Placing burned beetle pests in bean fields overnight effectively repels living beetle pests by the next morning.

Allomones are produced by many plants to repel herbivores and prevent them from feeding. Many secondary products of plants, such as tannins, cyanins, etc. are in fact antiherbivore allomones. Some act by repelling, others directly by affecting the growth and development of the pest organism. Allomones have not been used very much as an applied intervention, but they are often the basis for companion planting and other cultural interventions.

5.9.3 Hormones or Insect Growth Regulators

Chemicals that interfere with the normal growth and development of insect pests are insect growth regulators (IGRs). Hormones are chemicals produced in one part of a pest's body that affect the growth and behavior of other parts of the body. The most successful hormones used in IPM are insect growth regulators. These hormones affect the development of juvenile insects, either causing death or abnormality in newly hatched insects, or preventing sexual maturity. Insect growth regulators are usually synthetic versions of naturally occurring hormones. They are highly selective and have extremely low toxicity to other organisms (including humans). An IGR does not necessarily have to be toxic to its target but may lead instead to various abnormalities that impair insect survival (Siddall, 1976). Interestingly, most of the IGRs that have shown effectiveness against insect pests cause the rapid death of the insect through failure of a key regulatory process to operate or function. IGRs generally control insects either through regulation of metamorphosis or interference with reproduction (Riddiford and Truman, 1978). Compounds developed to disrupt metamorphosis ensure that no reproductive adults are formed. Those that specifically interfere with reproduction may include the development of adults with certain morphogenetic abnormalities that reduce their reproductive potential. Adults may be sterile or possess abnormally developed genitalia, which hinders the mating process or the capacity to produce fertile offspring (Tunaz and Uygun, 2004). Chitin synthesis inhibitors, juvenile hormone analogues. and anti-juvenile hormone analogues are examples of insect growth regulators.

5.9.4 Pesticides

The term *pesticide* is used to describe chemicals that kill pests including insects, other animals, mites, diseases, or even weeds. Inorganic pesticides are substances derived or refined from nonliving natural sources. They are termed *inorganic* because they do not contain carbon compounds. Many of them contain heavy metals that are persistent and toxic to humans. Inorganic pesticides that were used in the past but are seldom used today contained arsenic, cyanide, and mercury. In general, inorganic compounds are little used in modern agriculture.

The exceptions to this rule are copper and sulfur-based compounds used as fungicides. Sulfur is a widely used and safe way to control fungal diseases and mites. Bordeaux mixture, a combination of sulfur, copper, and lime, is an important fungicide used in orchards.

Synthetic pesticides are made from carbon-containing compounds and are widely used and available throughout most of the world. Synthetic pesticides are often classified according to their chemical makeup, their intended target pests (insecticides, fungicides, etc.), or their mode of action. In the later stages, various pesticides belonging to organochlorinated, organophosphorus, carbamate, pyrethroids, neonicotinoids, and other new molecules were developed for use in pest management. Of them, neonicotinoids and other new molecules are quite effective against target insects and relatively safer to human beings.

Insecticides are the most powerful tool available for use in pest management. They are highly effective, rapid in curative action, adoptable to most situations, flexible in meeting changing agronomic and ecological conditions, and economical. Insecticides are the most reliable means of reducing crop damage when the pest population exceeds the ETLs. When used properly based on ETLs, insecticides provide a dependable tool to protect the crop from insect pests (Sharma, 2009). Chemical pesticides will continue to be one of the essential components in IPM programs.

There are 256 pesticides registered in India. The pesticides' regulations in India are governed by two different bodies: the Central Insecticides Board and Registration Committee (CIBRC) and the Food Safety and Standards Authority of India (FSSAI). CIBRC was established in 1968 under the Department of Agriculture and Co-operation of Ministry of Agriculture. It is responsible for advising central and state governments on technical issues related to manufacture, use, and safety issues related to pesticides. Its responsibilities also include recommending uses of various types of the pesticides depending on their toxicity and suitability, determining the shelf life of pesticides, and recommending a minimum gap between the pesticide applications and harvesting of crop produce (waiting period) (http://cibrc.nic.in/cibrc.htm). The other part of the CIBRC, the registration committee, is responsible for registering pesticides after verifying the claims of the manufacturer or importer related to the efficacy and safety of the pesticides (http://cibrc.nic.in/registration_committee.htm).

Pesticide residues occur in edible substances as a result of intentional use of pesticides. To protect human well-being and unintentional exposure, residue analysis is performed to estimate tolerance limit of maximum residue level and to know the level of metabolites and degraded products. Pesticide residues in food over and above the prescribed tolerance limit are considered harmful (Handa et al., 1999). FFSAI is responsible for recommending tolerance limits of various pesticides in food commodities. It was established under the Food Safety and Standards Act, 2006 (http://www.fssai.gov.in/AboutFSSAI/introduction.aspx).

Many IPM programs are based largely on the use of synthetic pesticides. IPM practitioners need to calculate economic injury levels and make recommendations, and using synthetic pesticides is often simpler and more predictable than other IPM interventions. However, every effort should be made to integrate other interventions into IPM programs in order to reduce potential harmful effects from synthetic pesticide use. A program that makes limited, intelligent use of synthetic pesticides is often more effective than one that relies entirely on synthetic pesticides for crop protection, or one that abandons synthetic pesticides suddenly and completely. Synthetic pesticides can be used effectively to complement other IPM methods, such as biocontrol. Using a synthetic pesticide that selectively kills pests will increase the proportion of beneficials to pests, possibly increasing the efficiency of biocontrol. Understanding the biology and life cycles of both pests and beneficials is crucial to using synthetic pesticides effectively in IPM.

6. LEGISLATIVE/LEGAL/REGULATORY METHODS OF PEST CONTROL

Plant quarantine is defined as the legal enforcement of the measures aimed to prevent pests from spreading or to prevent them from multiplying further in case they have already gained entry and have established in new restricted areas. The importance of imposing restrictions on the movement of pest-infested plants or plant materials from one country to another was realized when the grapevine phylloxera was introduced into France from America around 1860 and the San Jose scale spread into the US in the latter part of the eighteenth century and caused severe damage. The first Quarantine Act in the US came into was in 1905, while India passed an act in 1914 entitled "Destructive Insect and Pests Act of 1914" to prevent the introduction of any insect, fungus, or other pests into the country. This was later supplemented by a more comprehensive act in 1917. The legislative measures in force now in different countries can be grouped into five classes:

1. Legislation to prevent the introduction of new pests and weeds from foreign countries (international quarantine).
2. Legislation to prevent the spread of already established pests, diseases, and weeds from one part of the country to another (domestic quarantine).
3. Legislation to enforce farmers to apply effective control measures to prevent damage by already established pests.
4. Legislation to prevent the adulteration and misbranding of insecticides and determine their permissible residue tolerance levels in food stuffs.
5. Legislation to regulate the activities of people engaged in pest control operations and application of hazardous insecticides.

Today, IPM has attained many successes but fallen short on some issues. Due to the awareness and biological understanding of how insecticide resistance develops, and because insecticides are so expensive to develop, in 1984 the insecticide manufacturers created the Insecticide Resistance Action Committee (IRAC) to encourage the responsible use of their products in a manner that minimizes the risks of insecticides in target pest populations (IRAC, 2010). Frisbie and Smith (1989) coined the term *biologically intensive* IPM, which involves reliance on ecological methods of control based on knowledge of a pest's biology. Horn (1988) outlined how principles of insect ecology could be incorporated into insect pest management strategies. More recently, Koul and Cuperus (2007) published "Ecologically Based Integrated Pest Management," essentially capturing the breadth and depth of the evolution that IPM has undergone over the past 60 years.

7. SUMMARY AND CONCLUSIONS

India has devised and implemented many IPM programs encompassing research, extension, and education with the objective to reduce the use of chemical pesticides, improve farm profitability, conserve environment, and reduce adverse effects of pesticides on human health. Under the ambit of the IPM program, the government of India has established 31 central IPM centers in 28 states and 1 union territory. Overall consumption of chemical pesticides in the country was reduced from 75,033 MT (Tech. grade) during 1990–1991 to 41,822 MT (Tech. grade) during 2009–2010. One of the reasons for reduction of pesticides is adoption of IPM in several field and horticultural crops in India. The use of biopesticides/neem-based pesticides increased from 123 MT during 1994–1995 to 1262 MT during 2009–2010.

Although IPM is accepted in principle as the most attractive option for the protection of agricultural crops from the ravages of pests, its implementation at the farmers' level is rather limited due to lack of their proven economic feasibility, short shelf life, slow effect, and incompatibility with chemical pesticides. Pesticides still remain as the means of intervention and as an essential component of IPM strategies. This suggests that the research should target overcoming these technological problems. The success of IPM would be determined by the extent to which these constraints are alleviated. With the growing demand for organically cultivated agricultural products, these eco-friendly IPM strategies, if streamlined, will find greater application in the coming century. The National Research Council officially introduced the concept of "ecologically based pest management," calling for a new paradigm for IPM in the twenty-first century (National Research Council, 1996). Benbrook et al. (1996) promoted the idea of moving IPM along a continuum from simple to complex, or "biointensive." Hence, ecologically based or biointensive IPM that involves sensible and need-based use of pesticides can significantly reduce their harmful effects to the environment.

References

Anonymous, 2004. National Road Map for Integrated Pest Management. USDA-CSREES, Washington, DC, p. 7. http://www.ipmcenters.org/Docs/IPMRoadMap.pdf.

Apple, J.L., Benepal, P.S., Berger, R., Bird, G.W., Ruesink, W.G., Maxwell, F.G., Santlemann, P., White, G.B., 1979. Integrated Pest Management, a Program of Research for the State Agricultural Experimental Stations and Colleges of 1890. p. 190.

Atwal, A.S., Daliwal, G.S., 1997. Agricultural Pests of South Asia and Their Management. Kalyani Publishers, Ludhiana, p. 486.

Bartlett, B.R., 1956. Natural predators. Can selective insecticides help to preserve biotic control? Agricultural Chemicals 11 (2) 42–44, 107–109.

Benbrook, C.M., Groth III, E., Halloran, J.M., Hansen, M.K., Marquardt, S., 1996. Pest Management at the Crossroads. Consumers Union of the United States, New York, USA. ISBN: 0890439001.

Chelliah, S., 1987. Insecticide induced resurgence of rice brown plant hopper, *Nilaparvata lugens* (Stål). In: Jayaraj, S. (Ed.), Resurgence of Sucking Pests, pp. 1–14.

Chelliah, S., Bharathi, M., 1985. Rice brown plant hopper *Nilaparvatha lugens* – status and strategies in management. In: Jayaraj, S. (Ed.), Integrated Pest and Disease Management. TNAU, pp. 34–53.

Council on Environmental Quality (CEQ), 1972. Integrated Pest Management. U.S. Govt. Printing Office, Washington, DC, p. 41.

Dhaliwal, G.S., Arora, R., 1996. Principles of Insect Pest Management. National Agricultural Technology Information Centre, Ludhiana, India, p. 374.

FAO, 1967. Food and Agriculture Organisation. In: Papers Presented at the FAO Symposium on Crop Losses, Rome, Italy.

Frisbie, R.E., Smith Jr., J.W., 1989. Biologically intensive integrated pest management: the future. In: Menn, J.J., Stienhauer, A.L. (Eds.), Progress and Perspectives for the 21st Century. Entomological Society of America, Lanham, Maryland, USA, pp. 156–184. ISBN: 0938522361.

Geier, P.W., Clark, L.R., 1961. An ecological approach to pest control. In: Proceedings: Technical Meeting of the International Union for Conservation of Nature and Natural Resources (8th, 1960: Warsaw, Poland), pp. 10–18.

Ghosh, P.K., Devendra, P., Prasad, R., Shaw, S.P., 1989. Management of pod borer in chick pea. In: Maximising Crop Production in Rainfed and Problem Areas. IARI, New Delhi, India, pp. 122–126.

Goodell, G., 1984. Challenges to international pest management research and extension in the Third World: do we really want IPM to work? Bulletin of the Entomological Society of America 30 (3), 18–26.

Grainge, M., Ahmed, S., 1988. Hand Book of Plants with Pest Control Properties. John Wiley & Sons, New York.

Handa, S.K., Agnihotri, N.P., Kulshrestha, G., 1999. Pesticide Residues, Significance, Management and Analysis. Research Periodicals and Book Publishing House, Texas, USA, p. 226.

Henson, S., Loader, R., 2001. Barriers to agricultural exports from developing countries: the role of sanitary and phytosanitary requirements. World Development 29, 85–102.

Horn, D.J., 1988. Ecological Approach to Pest Management. The Guilford Press, New York, USA. ISBN: 0898623022.

Hoskins, W.M., Borden, A.D., Michelbacher, A.E., 1939. Recommendations for a more discriminating use of insecticides. In: Proceedings of the 6th Pacific Science Congress, vol. 5, pp. 119–123.

Insecticide Resistance Action Committee (IRAC), 2010. Downloaded on 03.08.11. Available from: http://www.irac-online.org/about/irac/.

Jayaraj, S., Rangarajan, A.V., Murugesan, S., Santharam, G., Vijayaraghavan, S., Thangaraj, D., 1988. Studies on the outbreak of whitefly, *Bemicia tabaci* (Genn.) on cotton in Tamil Nadu. In: Jayaraj, S. (Ed.), Resurgence of Sucking Pests. Proc. Natl. Symp. TNAU, Coimbatore, pp. 103–115.

Jayaraj, S., Regupathi, A., 1993. Novel concept and future of pesticides in third world. In: Dhaliwal, G.S., Singh, B. (Eds.), Pesticides: Their Ecological Impact in Developing Countries. Commonwealth Publishers, New Delhi, pp. 335–363.

Jonathan, E.I., Karuppuchamy, P., Kalyanasundaram, M., Suresh, S., Mahalingam, C.A., 2011. Status of papaya mealybug in Tamil Nadu and its management. In: Proceedings of the National Consultation Meeting on Strategies for Deployment and Impact of the Imported Parasitoids of Papaya Mealybug, Classical Biological Control of Papaya Mealybug (*Paracoccus marginatus*) in India, pp. 24–33.

Kalyanasundaram, M., Karuppuchamy, P., Kannan, M., Suganthi, M., 2010. *Dipha aphidivora* (Meyrick), a potential bioagent for the management of sugarcane woolly aphid. In: Ignacimuthu, S., David, B.V. (Eds.), Non Chemical Insect Pest Management. Elite Publishing House Pvt. Ltd, New Delhi, pp. 164–168.

Kalyanasundaram, M., Karuppuchamy, P., Divya, S., Sakthivel, P., Rabindra, R.J., Shylesha, A.N., 2011. Impact on release of the imported parasitoid Acerophagus papayae for the management of papaya mealybug Paracoccus marginatus in Tamil nadu. In: Proceedings of the National Consulation Meeting on Strategies for Deployment and Impact of the Imported Parasitoids of Papaya Mealybug, Classical Biological Control of Papaya Mealybug (Paracoccus marginatus) in India, pp. 68–72.

Karuppuchamy, P., 1995. Studies on the Management of Pests of Pomegranate with Special Reference to Fruit Borer *Virachola isocrates* (Fabr.) (Ph.D. thesis). Submitted to Tamil Nadu Agricultural University, Coimbatore, p. 180.

Karuppuchamy, P., Balasubramanian, M., 1990. Field evaluation of gossyplure, the synthetic sex pheromone of *Pectinophora gossypiella* in Tamil Nadu. Indian Journal of Entomology 52 (2), 170–179.

Karuppuchamy, P., Balasubramanian, G., 2001. IPM for pomegranate fruit borer Deudorix isocrates (Fabr.). In: Verghese, A., Parvatha Reddy, P. (Eds.), IPM in Horticultural Crops: Emerging Trends in the New Millenium. IIHR, Bangalore, pp. 41–42.

Karuppuchamy, P., Uthamasamy, S., Chakkaravarthy, G., 1982. Influence of planting time on gall midge incidence at Aduthurai. IRRN 7 (4), 11.

Karuppuchamy, P., Uthamasamy, S., 1984. Influence of flooding, fertilizer and plant spacing on insect pest incidence. IRRN 9 (6), 17.

Karuppuchamy, P., Uthamasamy, S., 1986. Influence of time of planting on the incidence of rice pests. Madras Agricultural Journal 73 (11), 606–609.

Koul, O., Cuperus, G.W., 2007. Ecologically Based Integrated Pest Management. CAB International, Wallingford, UK. ISBN:1845930649.

Krishnaiah, K., Quayum, M.A., Rao, C.S., Reddy, P.C., Charyulu, A.M.R.K., 1983. Integrated pest management in rice in Warangal district. In: Chelliah, S., Balasubramanian, M. (Eds.), Pest Management in Rice. TNAU, Coimbatore, pp. 326–342.

Krishnamurthy Rao, B.H., Murthy, K.S.R.K., 1983. In: Proceedings of National Seminar on crop losses due to insect pests. Indian Journal of Entomology (Special Issue), Vol I-II. Entomological Society of India, Hyderabad, India.

Kogan, M., 1998. Integrated pest management: historical perspectives and contemporary development. Annual Review of Entomology 43, 243–270.

Landis, D.A., Wratten, S.D., Gurr, G.M., 2000. Habitat management to conserve natural enemies of arthropod pests in agriculture. Annual Review of Entomology 45, 175–201. ISSN 0066-4170.

Luckman, W.H., Metcalf, R.L., 1994. The pest management concept. In: Metcalf, R.L., Luckman, W.H. (Eds.), Introduction to Insect Pest Management. John Wiley & Sons, New York, pp. 1–34.

Maheswaran, M., 1998. Exploitation of wild species in selected crops for the development of insect resistance crops. In: Asaf Ali, K., et al. (Eds.), Training Course on Host Plant Resistance to Insects and Mites in Crop Plants - Manual. TNAU, Coimbatore, pp. 74–90.

Mohan, S., Janarthanan, R., 1985. On certain behavioural response of major pests of rice to different light sources. In: Proceedings of National Seminar on Behavioural and Physiological Approaches in the Management of Crop Pests. TNAU, Coimbatore, pp. 94–99.

Mohan, S., Balasubramanian, G., Gopalan, M., Jayaraj, S., 1987. Solar heat treatment. A novel method to check rice weevil and red flour beetle infestation in sorghum during storage. Madras Agricultural Journal 74, 235–236.

Mohan, S., Devadoss, C.T., Jayaraj, S., Mohanasundaram, M., 1989. Biogas fumigation to control pulse beetle, *Callosobruchus chinensis*. Bulletin of Grain Technology 27, 196–198.

Muralidharan, S., Ragupathy, A., Sundaramoorthy, T., 1992. Organochlorine residues in the eggs of selected colonial water birds breeding at Keoladeo National Park, Bharatpur, India. In: N.H.R.I. Symposium Series, vol. 7. Environment Canada, Saskatoon, pp. 189–195.

Murthy, R.L.N., Venateswarulu, P., 1998. Introducing ecofriendly farming techniques and inputs in cotton. In: Proceedings of the Workshop on 'Ecofriendly Cotton Held at Agricultural College and Research Institute, Madurai, Tamil Nadu, October 27–28, 1995, p. 110.

Nag, D.D., 1983. Human safety in manufacture, handling and usage of pesticides. In: Course on Toxicology of Pesticides. Regional Research Laboratory, Hyderabad, pp. 220–229.

National Academy of Science, 1969. Insect-pest management and control. Principles of plant and animal pest control. National Academy of Sciences Publication 1695 (3), 448–449.

National Research Council (NRC), 1996. Ecologically Based Pest Management: New Solutions for a New Century. National Academy Press, Washington, DC, USA. ISBN: 0309053307.

Obrycki, J.J., Lewis, L.C., Orr, D.B., 1997. Augmentative releases of entomophagous species in annual systems. Biological Control 10 (1), 30–36. (September 1997) ISSN 1049-9644.

Parrella, M.P., Heinz, K.M., Nunney, L., 1992. Biological control through augmentative releases of natural enemies: a strategy whose time has come. American Entomologist 38 (3), 172–179. ISSN 1046-2821.

PDBC, 2008. In: Bhummanavar, B.S., Rabindra, R.J. (Eds.), Annual Report 2007–08. Project Directorate of Biological Control, Bangalore, p. 132.

Pedigo, L.P., 1989. Entomology and Pest Management. Macmillan Publishing Co., Inc., NY.

Pedigo, L.P., 1996. Entomology and Pest Management, second ed. Prentice-Hall Pub., Englewood Cliffs, NJ, p. 679.

Pedigo, L.P., 2002. Entomology and Pest Management, fourth ed. Prentic Hall Inc., New Jersey, USA, pp. 200–206.

Pimentel, D. (Ed.), 1997. Techniques for Reducing Pesticide Use: Economic and Environmental Benefits. John Wiley & Sons, New York, p. 444.

Prasad Ram, B., Pasalu, I.C., Thammi, N.B., Ram Gopal Verma, N., 2003. Influence of nitrogen and rice varieties on population build up of brown plant hopper *Nilaparvata lugens* (Stål). Journal of Entomology Research 27 (2), 167–170.

Prasada Rao, G.S.L.H.V., 2005. Agricultural Meteorology, second ed. Kerala Agricultural University, Thrissur, Kerala, India, p. 326.

Rabb, R.L., Stinner, R.E., van den Bosch, R., 1976. Conservation and augmentation of natural enemies. In: Huffaker, C.B., Messenger, P.S. (Eds.), Theory and Practice of Biological Control. Academic Press, New York, NY, USA, pp. 233–254. ISBN: 0123603501.

Rabindra, R.J., 1985. Transfer of plant protection technology in dry crops. In: Jayaraj, S. (Ed.), Integrated Pest and Disease Management. Proceedings of National Seminar. Tamil Nadu Agricultural University, Coimbatore, pp. 337–383.

Rabindra, R.J., 1998. Development of microbial pesticides: present and future prospects. In: Abstract of the National Symposium on Development of Microbial Pesticides and Insect Pest Management. Research and Development Centre, Hindustan Antibiotics Ltd, Pune, p. 10. 12–13 November, 1998.

Rabindra, R.J., 2000. Genetic improvement of baculoviruses for insect pest management. In: Proceedings of National Symposium on Microbials in Insect Pest Management. February 24–25. Entomology Research Institute, Loyola College, Chennai.

Raheja, A.K., Tiwari, C.G., 1997. Awareness Programme on Pesticides and Sustainable Agriculture. ICAR Publication, p. 69.

Ram Asre, 2004. Early warning system for desert locust in India. Plant Protection Bulletin 56 (3+4), 24–26.

Riddiford, L.M., Truman, J.W., 1978. Biochemistry of insect hormones and insect growth regulators. In: Rockstein, M. (Ed.), Biochemistry of Insects. Acad. Press, New york, pp. 307–357.

Ridgway, R.L., 1998. Mass-reared natural enemies: application, regulation, and needs. Thomas Say Publications in Entomology: Proceedings. Entomological Society of America, Lanham, Maryland, USA. ISBN: 0938522663.

Rola, A.C., Pingali, P.A., 1993. Pesticides, Rice Productivity, and Farmers' Health: An Economic Assessment. International Rice Research Institute, Manila, Philippines, p. 100.

Satpathi, C.R., Chakraborty, K., Acharjee, P., 2012. Impact of seedling spacing and fertilizer on brown plant hopper, *Nilaparvata lugens* Stål. incidence in rice field. Journal of Biological and Chemical Research 29 (1), 26–36.

Saxena, R.C., 1989. Insecticides from neem. In: Arnason, J.T., Philogene, B.J.R., Morand, P. (Eds.), Insecticides of Plant Origin. American Chemical Society, Washington, DC, pp. 110–135.

Schillhorn van Veen, T.W., Forno, D., Joffe, S., Umali-Deininger, D.L., Cooke, S., 1997. Integrated pest management: strategies and policies for effective implementation. In: Environmentally Sustainable Development Studies and Monograph Series No. 13. The World Bank, Washington, DC.

Schmutterer, H., Singh, R.P., 1995. List of insect pests susceptible to neem products. In: Schmutterer, H. (Ed.), The Neem Tree, *Azadirachta indica* A. Juss. and Other Meliaceous Plants: Sources of Unique Natural Products for Integrated Pest Management, Medicine, Industry and Other Purposes. VCH, Weinheim, Germany, pp. 325–326.

Sharma, H.C., Singh, O.P., 1982. Sex pheromones for the control of pink bollworm, *Pectinophora gossypiella*, their feasibility under Indian conditions. Cotton Development 11 (4), 35–38.

Sharma, H.C., 2009. Biotechnological Approaches for Pest Management and Ecological Sustainability. CRC Press.

Shylesha, A.N., Dhanyavathy, Shivaraj, 2011. Mass production of parasitoids for the classical biological control of papaya mealybug, Paracoccus marginatus. In: Proceedings of the National Consultation Meeting on Strategies for Deployment and Impact of the Imported Parasitoids of Papaya Mealybug, Classical Biological Control of Papaya Mealybug (Paracoccus marginatus) in India, pp. 63–67.

Siddall, J.B., 1976. Insect growth regulators and insect control: a critical appraisal. Environmental Health Perspectives 14, 119–126.

Singh, R.P., 2000. Botanicals in pest management: an ecological perspective. In: Dhaliwal, G.S., Singh, B. (Eds.), Pesticides and Environment. Commonwealth Publishers, New Delhi, pp. 279–343.

Singh, A., Singh, S., Rao, S.N., 2003. Integrated pest management in India. In: Maredia, K.M., Dakouo, D., Mota-Sanchez, D. (Eds.), Integrated Pest Management in the Global Arena. CABI Publishing, pp. 209–222.

Smith, R.F., van den Bosch, R., 1967. Integrated control. In: Kilgore, W.W., Doutt, R.L. (Eds.), Pest Control: Biological, Physical and Selected Chemical Methods. Academic Press, New York, pp. 295–340.

Stern, V.M., Smith, R.F., van den Bosch, R., Hagen, K.S., 1959. The integrated control concept. Hilgardia 29, 81–101.

Teetes, G.L., 1981. The environmental control of insects using planting time and plant spacing. In: Prementel, D. (Ed.), Hand book of Pest management, vol. 1. CRC Press, Florida, USA, pp. 209–221.

Tunaz, H., Uygun, N., 2004. Insect growth regulators for insect pest control. Turkish Journal of Agriculture and Forestry 28, 377–387.

van Driesche, R.G., Bellows, T.S., 1996. Biological Control. Chapman & Hall, New York, NY, USA. ISBN: 0412028611.

Waage, J.K., Greathead, D.J., 1988. Biological control. Philosophical Transactions of the Royal Society of London. Series B 318, 111–126.

Wadhwani, A.M., Lal, I.J., 1972. Harmful Effects of Pesticides. Indian Council of Agricultural Research, New Delhi, pp. 77–79.

Further Reading

Koul, O., Dhaliwal, G.S., Cuperus, G.W., 2004. Integrated Pest Management: Potential, constraints and challenges. CABI publishing, p. 343.

Metcalf, R.L., Luckmann, W.H. (Eds.), 1994. Introduction to Insect Pest Management, third ed. John Wiley & sons inc, p. 661.

Rabindra, R.J., Palaniswamy, S., Karuppuchamy, P., Ramaraju, K., Philip Sridhar, R. (Eds.), 1999. Proceedings of the National Training on Ecology Based Pest Management. TNAU, Coimbatore. p. 277. 1–21 September 1999.

Biotechnological Approaches

Arun K. Tripathi[1], Shikha Mishra[2]

[1]CSIR-Central Institute of Medicinal and Aromatic Plants, Lucknow, India;
[2]CSIR-Central Drug Research Institute, Lucknow, India

1. INTRODUCTION

Plants produce secondary metabolites as defense chemicals against insect herbivores. Molecular marker–assisted plant breeding comprised of genes responsible for the production of defensive chemicals toward insect resistance (Smith, 2005) may facilitate development of crops with desirable genetic traits. Only a few crops have been developed using molecular techniques and further work is required to develop insect resistant crops using modern biotechnological tools and techniques. Biotechnological approaches to develop insect resistant plants began long ago, and commercial application started in 1996 with success of transgenic *Bacillus thuringiensis* (Bt) crops (James, 2008).

This chapter covers the issues ranging from insect–host plant resistance to the application of molecular approaches for pest management. The authors have also attempted to evaluate how genetic engineering can be used to create and deploy safe and cost-effective biotechnology-derived products for pest management. Further, some of the major developments in the areas of molecular biology, genetics, and biotechnology and the potential impacts that they could have on integrated pest management worldwide, are also discussed.

2. GENETIC ENGINEERING APPROACHES

Novel crop genotypes may be developed through genetic engineering approaches utilizing tools and techniques of molecular biology. Plants have a range of insecticidal proteins and molecules that are effective against insect pests of economic importance. Some of them are elaborated below.

2.1 Transgenes

A transgene is genetic material transferred from one organism to another through an appropriate molecular mechanism. The introduction of a transgene has the potential to change the phenotype of an organism and forms the basis of the most revolutionary crop improvement technology. The crop transgenes thus far commercialized were designed to aid in crop protection against insects and weeds. The first commercial introduction of transgenes into field crops occurred in 1996. The gene encoding the cowpea trypsin inhibitor was subsequently transferred into rice (Xu et al., 1996) and potato (Gatehouse et al., 1997) but did not provide sustainable insect protection and was thus not commercially viable. However, cowpea trypsin inhibitor produced by transgenic tobacco plant showed enhanced protection against the lepidopteran pest *Heliothis virescens* (Harsulkar et al., 1999). Commercial development of insecticidal genes has focused on the soil microorganism Bt toxins.

2.1.1 Bt Endotoxins

Toxicity of Bt toxins to a range of insects and easy isolation of the gene from the bacterial plasmid facilitated the development of the first insect-resistant transgenic plants. Each Bt strain plasmid encodes a specific toxin with specific action against insects. More than 150 different Cry toxins have been cloned and tested for their toxicity on various insect species.

At the sporulation stage of Bt, δ-endotoxins are accumulated as protoxins in crystalline form. After solubilization, these protoxins are liberated in the insect midgut and are cleaved off at C-terminal part to release ~66-kDa active N-terminal toxic molecule. The crystalline form of protoxin took place as a result of intermolecular disulfide bonds, which helps in bridging the protoxin molecules containing well-conserved cysteine residues (Cry1Ac). Among the δ-endotoxins, two are crystals forming (Cry) proteins, e.g., Lepidoptera specific, Cry1Aa (Grochulski et al., 1995), and Coleoptera specific, Cry3A (Li et al., 1991).

One of the major breakthroughs in the development of Cry1Ac transgenics has been the designing of synthetic versions of the gene with codon modification to remove the putative polyadenylation sequences and to bring in the codon usage favorable for high expression in dicotyledonous plants (Perlak et al., 1990). Bt toxins are used in a variety of transgenic crops, e.g., cotton for protection against various lepidopteran pests, maize for protection against the European corn borer, *Ostrinia nubilalis*, and potatoes for protection against the Colorado potato beetle, *Leptinotarsa decemlineata*. So far, most of the genetically modified crops that exhibit insecticidal activity consist mainly of the toxin *Cry1Ac* in transgenic cotton and Cry1Ab in transgenic corn (Tabashnik and Carriere, 2010).

However, a particular toxin is generally effective against only a limited range of closely related species. Some insect populations (e.g., diamondback moth) have become resistant to the Bt toxin after prolonged exposure. Resistance in a particular insect pest population means that Bt would no longer be effective in controlling that pest. It is generally accepted that the intensive use of Bt in genetically engineered plants will accelerate the selection pressure on insect populations to develop resistance. Other bacteria also provide a resource for identification of insect-specific toxin genes such as those derived from *Bacillus cereus* and the entomopathogenic nematode-associated bacterium *Photorhabdus luminescens*.

Monsanto, Rohm and Haas, Ciba–Geigy, Agracetus, Agrigenetics Advanced Sciences, Calgene, the US Department of Agriculture (USDA), Sandoz Crop Protection, Crop Genetics

International, and the University of California are some of the major players who have developed and field-tested tomato, tobacco, cotton, walnut, and potato plants genetically engineered to contain the insect-killing Bt toxin from *Bacillus thuringiensis*.

2.2 Vegetative Insecticidal Proteins

Vegetative insecticidal proteins (Vip), the second generation of insecticides, are produced during the vegetative growth stage of Bt (Estruch et al., 1996). This type of protein includes Vip1, Vip2, and Vip3. Vip1 and Vip2 are binary toxins that have coleopteran specificity, whereas Vip3 toxins have lepidopteran specificity (Estruch et al., 1996). To date, there are approximately 82 kinds of vegetative insecticidal protein genes that have been identified and cloned. These genes can be classified into three groups, eight subgroups, 25 classes, and 82 subclasses according to the encoded amino acid sequence similarity. The Vip proteins, bearing no similarity to δ-endotoxins (Lee et al., 2003), have become an important new class of insecticidal proteins.

Vip induces insect gut paralysis and complete lysis of gut epithelium. Vip3A induces lethal toxicity toward larvae of *Agrotis ipsilon* and *Spodoptera frugiperda* (Yu et al., 1997) and Vip3Aa14 toward larvae of *Spodoptera litura* and *Plutella xylostella* (Bhalla et al., 2005).

2.3 Pyramiding of Genes

Pyramiding of genes is a combination of genes to get multimechanistic resistance in plants, e.g., cloning of cowpea trypsin inhibitor (CpTI) gene and pea lectin gene to produce a transgenic tobacco and transgenic potato expressing lectin and bean chitinase. Gene pyramiding may yield medium-to-long-term control (Kelly et al., 1995) because it is the combination of several resistance genes in a single cultivar conferring resistance to a specific pest. A successful example of this approach in rice breeding is the pyramiding of three genes (*xa5*, *xa13*, and *Xa21*) that confer resistance to bacterial blight caused by *Xanthomonas oryzae pv. Oryzae* (Xoo) (Sanchez et al., 2000). This strategy can be aided by molecular markers that allow simultaneous monitoring of several resistance genes (Kelly et al., 1995). This method has become one of the effective ways in breeding rice varieties for resistance.

2.4 Bifunctional Inhibitors

Bifunctional inhibitors may be single inhibitor with dual targets or at least two inhibitors that have different targets. Some examples of bifunctional inhibitors are alpha-amylase/trypsin inhibitors (Haq et al., 2004) and trypsin/carboxypeptidase A inhibitors (Chiche et al., 1993). Similarly, expression of a fusion protein composed of a cystatin and a serine PI has been used to control certain nematode pathogens in transgenic plants (Urwin et al., 1998). Oppert et al. (2005) demonstrated synergism between soybean Kunitz trypsin inhibitor and the cysteine protease inhibitor L-trans-epoxy succinyl leucylamide (4-guanidino) butane (E64) in artificial diet bioassays with the coleopteran beetle *Tribolium castaneum*. Transgenic tobacco plants expressing both a Bt toxin and a CpTI were more protected from *Heliothis armigera* damage compared to transgenic tobacco expressing the Bt toxin alone (Fan et al., 1999).

3. PROTEIN ENGINEERING APPROACHES

Improvement of the crop performance along with management of insect pests can be effectively done through genetic engineering and breeding plant varieties resistant to insects. Plants have evolved a range of adaptations in nature against insect attack for their survival, which is described as host–plant resistance.

3.1 Host–Plant Resistance

Plants adopt various morphological, biochemical, and molecular mechanisms for their protection against insect herbivores. Among them, biochemical mechanisms are mediated by both direct and indirect defenses against insects. Some of the defensive compounds are produced in response to plant damage and others are produced constitutively. These defensive compounds adversely affect insect growth, development, and reproduction. Thus plants exert both constitutive and induced type of resistance (Zhang et al., 2008).

3.1.1 *Constitutive Resistance*

Constitutive resistance is always present in plants. Composition and concentration of constitutive defenses exhibit a range of variability from mechanical defenses to digestibility reducers and toxic compounds (Figure 1). Constitutive defenses include quantitative

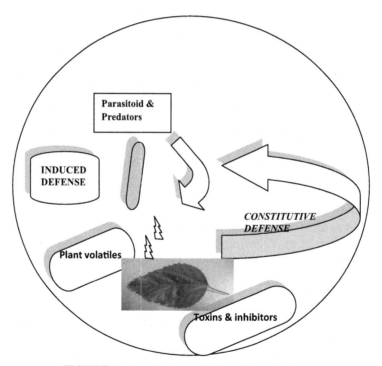

FIGURE 1   Constitutive and induced plant defenses.

defenses as well as mechanical defenses. Quantitative chemicals are present in bulk in plants (5–40% dry weight) and their mobilization is difficult within the plant (Nina and Lerdau, 2003). Examples of such chemicals are digestibility reducers like sylvatiins produced in the petals of the plant *Geranium sylvaticum* (wood cranesbill) (Tuominen et al., 2015) to protect itself from Japanese beetles. Sylvatiin is a group of hydrolyzable tannins that contain additional acetylglucose moiety attached to the galloyl groups of galloyl glucoses and chebulinic acid. Ingestion of sylvatiin paralyzes the beetle within 30 min and the compound usually wears off within a few hours. Meanwhile, the beetle is often consumed by its own predators.

3.1.2 Induced Resistance

Mechanism of induced resistance helps in understanding the types of insect herbivores likely to be affected by induced responses, and accordingly elicitors of induced responses may be sprayed on the plants to build up the natural defense system against insect damage. Induced resistance can be best applied in crop plants that readily produce the inducible responses upon mild infestation by insect herbivores. Such type of resistance may be a part of integrated pest management. Inducible defenses (see Figure 1) are less costly and may be produced when needed (Harvell and Tollrian, 1998). Induced resistance has been reported in more than 100 species of plants, such as Arabidopsis (De Vos et al., 2006), tobacco (Wu et al., 2007), tomato (Egusa et al., 2008), soybean (Bi et al., 1994), etc.

3.1.2.1 INDUCED RESISTANCE MECHANISM OF PLANTS

Regulation of signaling networks involved in induced defense (Reymond and Farmer, 1998; Ton et al., 2002) are mediated by the plant hormones salicylic acid (SA), jasmonic acid (JA), and ethylene (ET) (Figure 2). Insects and necrotrophic pathogens are resisted by JA/ET-dependent defenses, whereas other pathogens are more sensitive to SA-dependent responses (Glazebrook, 2005). However, JA is the main signaling molecule that is involved in the plant's defense system against insect herbivores (Reymond and Farmer, 1998). Apart from this, induced defense response also produces defensive compounds like proteinase inhibitors affecting insect feeding and digestion (Zavala et al., 2004), volatile compounds attracting parasitoids and predators of the insects feeding on the plant (Kessler and Baldwin, 2001), and extrafloral nectar as arrestants to carnivorous arthropods present on infested plants (Heil and Silva, 2007). For example, leaves of tomato (*Solanum lycopersicum*) accumulate JA after attack by *Manduca sexta* larvae, which results in the activation of genes encoding proteinase inhibitor proteins that inhibit digestive serine proteinases of herbivorous insects and reduce further insect feeding (Howe, 2004).

To understand the mechanism of constitutive and induced plant defense responses against herbivory, a variety of molecular and biochemical approaches has been adopted (Walling, 2000; Wu and Baldwin, 2009). As a result, several classes of defensive chemicals have been identified and are elaborated below.

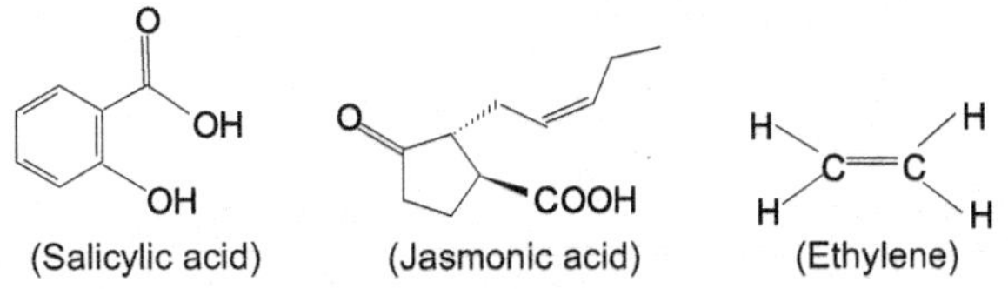

(Salicylic acid)   (Jasmonic acid)   (Ethylene)

FIGURE 2   Chemical structures of some plant hormones involved in induced resistance.

3.2 Single Gene Defensive Molecules

Plant protease inhibitors (PIs) are a part of the natural plant defense proteins that inhibit insect digestive enzymes (Zhu-Salzman and Zeng, 2015) by interfering with insect protein digestion. They bind to digestive proteases resulting in an amino acid deficiency, and as a result, insect growth and development, fecundity, and survival are adversely affected (Lawrence and Koundal, 2002).

Protease inhibitors along with α-amylases inhibitors constitute major tools for improving the resistance of plants to insects. PIs are the products of single genes, therefore they have practical advantages over genes encoding for complex pathways and they are effective against a wide range of insect pests, e.g., transferring trypsin inhibitor gene from *Vigna unguiculata* to tobacco conferred resistance against lepidopteran insect species such as *H. armigera* and *S. litura*, and coleopteran species such as *Diabrotica undecimpunctata* and *Anthonomus grandis* (Hilder et al., 1987). Expression of serine and cysteine proteinase inhibitors in transgenic plants has shown resistance to some insect pest species including Lepidoptera and Coleoptera (Alfonso-Rubi et al., 2003).

PIs are classified based on the type of enzyme they inhibit: serine protease inhibitors, cysteine protease inhibitors, aspartic protease inhibitors, or metallocarboxy-protease inhibitors (Bode and Huber, 2000). Plant serine proteinase inhibitors (Volpicella et al., 2011) are further subclassified into a number of subfamilies based on their amino acid sequences and structural properties, e.g., Kunitz type, Bowman–Birk type, Potato I type, and Potato II type inhibitors (Bode and Huber, 1992). The proteins in Kunitz-like family generally inhibit serine proteinases, but they also include inhibitors of cysteine and aspartate proteases (Heibges et al., 2003).

An arthropod's ability to acquire amino acids involves multiple proteins working in concert. The multiple action potential of arthropod-inducible proteins (AIPs) may be utilized to target various nutritional vulnerabilities in arthropods (Kempema et al., 2007).

3.3 Enzymatic Defense Molecules

Imbalance in utilization of plant proteins by the insects results in drastic effects on insect physiology. The enzymes that impair the nutrient uptake by insects through formation of electrophiles include polyphenol oxidases, lipoxygenases, cholesterol oxidases, etc. Apart from these, alteration of gene expression under stress including insect attack leads to qualitative and quantitative changes in proteins, which in turn play an important role in signal transduction and oxidative defense.

3.3.1 Threonine Deaminase

Essential amino acids are required by insects for plant protein digestion. Higher plants use threonine deaminase (TD) to catalyze the dehydration of threonine (Thr) to α-ketobutyrate and ammonia as the committed step in the biosynthesis of isoleucine (Ile). Cultivated tomato (*S. lycopersicum*) and related *Solanum* species contain a duplicated TD paralog (pTD2) that is coexpressed with a suite of genes involved in herbivore resistance. Tomato uses different TD isozymes to perform these functions.

The constitutively expressed TD1 has a housekeeping role in Ile biosynthesis, whereas expression of TD2 in leaves is activated by the jasmonate signaling pathway in response to herbivore attack. Ingestion of tomato foliage by specialist (*M. sexta*) and generalist (*Trichoplusia ni*) insect herbivores triggers proteolytic removal of TD2's C-terminal regulatory domain, resulting in an enzyme that degrades Thr without being inhibited through feedback by Ile. Thus, the processed form (pTD2) of TD2 accumulates to high levels in the insect midgut and feces (frass). Purified pTD2 showed biochemical properties indicating a postingestive role in defense.

TD2 has a defensive function related to Thr catabolism in the gut of lepidopteran herbivores (Gonzales-Vigil et al., 2011). This proteolytic activation step occurs in the gut of lepidopteran but not coleopteran herbivores. Enzymatic defense molecules such as TD2 are effective at lower concentrations, whereas PIs must be in higher concentration to exert the defensive function. Other enzymes involved in amino acid metabolism (e.g., Arg decarboxylase, Try decarboxylase, etc.) are frequently induced by insect feeding and require further examination for their potential in restricting amino acid utilization.

3.3.2 Polyphenol Oxidases

Polyphenol oxidases (PPOs and some peroxidases) are other group of enzymes that may impair nutrition through forming electrophiles and oxidize mono- or dihydroxyphenolics, e.g., the oxidation of *o*-diphenols forms reactive *o*-quinones, which are potent electrophiles capable of polymerizing or forming covalent adducts with the nucleophilic groups of proteins (e.g., -SH or ε-NH$_2$ of Lys) (Felton et al., 1992). PPOs are widespread in plants and are inducible by wounding (herbivory) and JA (Thaler et al., 2001). They provide resistance potential in transgenic plants against caterpillars (Wang and Constabel, 2004).

3.3.3 Cholesterol Oxidases

Cholesterol oxidases were first discovered in *Streptomyces* (Purcell et al., 1993). Ingestion of the enzyme cholesterol oxidase by boll weevil (*A. grandis*) larvae induces morphological changes in the midgut tissues. Cholesterol oxidase disrupts the midgut epithelium at low doses and lyses the midgut cells at higher doses (Purcell et al., 1993). Corbin et al. (2001) transformed tobacco (*Nicotiana tabacum* L.) plants with the cholesterol oxidase *choM* gene and expressed cytosolic and chloroplast-targeted versions of the ChoM protein. Transgenic leaf tissues expressing cholesterol oxidase exerted insecticidal activity against boll weevil larvae. When produced cytosolically, cholesterol oxidase could metabolize phytosterols in vivo, so the transgenic plants exhibiting cytosolic expression of *choM* gene accumulated low levels of saturated sterols known as sitostanols (Figure 3) and displayed severe developmental aberrations.

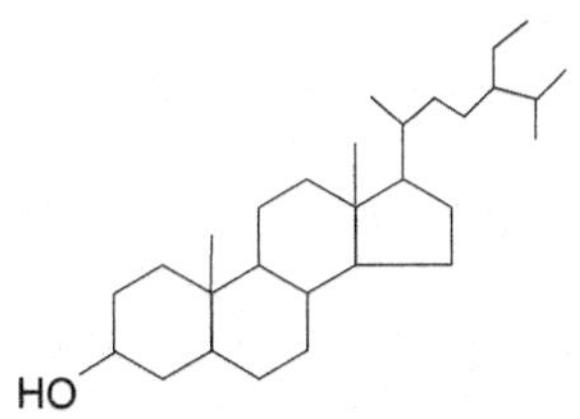

FIGURE 3   Chemical structure of sitostanol.

3.3.4 *Lipoxygenases*

Lipoxygenases (LOXs) are widely distributed in plants and animals and known as non-heme iron-containing dioxygenases. LOX in plants comprised of a single polypeptide chain with a molecular mass of 94–104 kDA (kilo Dalton). Catalytic site of the enzyme is at the carboxy-terminal domain, which contains a nonheme iron atom that is coordinated with five amino acids: three histidines, one asparagine, and the carboxyl group of the carboxy-terminal isoleucine. The amino-terminal domain may be involved in membrane or substrate binding.

LOX are categorized as "type 1" (mainly extra-plastidial) and "type 2" (mainly plastidial), but many different isoenzymes also exist depending on the particular species, e.g., soybean lipoxygenase exists in eight different isoforms. They include both soluble cytoplasmic and membrane-bound enzymes. In Arabidopsis, lipoxygenase activity is located mainly in the plastidial envelope and stroma of leaf chloroplasts. Isoforms in different subcellular regions may provide different pools of hydroperoxy fatty acids, which serve as substrates for alternative metabolic pathways and physiological functions.

Lipoxygenases catalyze the addition of molecular oxygen to polyunsaturated fatty acids containing a (*cis,cis*)-1,4-pentadiene system to yield an unsaturated fatty acid hydroperoxide. They function by damaging midgut membranes, e.g., lipoxygenase from soybean retards the growth of *M. sexta* when incorporated into artificial diet (Shukle and Murdock, 1983).

3.4 Redox Potential Disruptors

Various metabolic functions of the cell are regulated by oxidation–reduction status (redox). Any perturbations in the redox status of cells either by external or internal stimuli elicit distinct responses, resulting in alteration of cell function (Das and Carl, 2002). Some enzymes can disrupt the arthropod redox status. Disturbances in gut redox state may cause proliferation of oxyradicals that damage proteins, lipids, and DNA. Enzymes that produce a superoxide radical (e.g., NADH oxidase) or hydrogen peroxide (e.g., oxalate oxidases, polyamine oxidases, peroxidases, etc.) could function as defensive proteins in the insect herbivore gut. Depending upon redox conditions, enzymes could impose either oxidative or reductive stress. Insects rely upon ascorbate (Asc), reduced glutathione (GSH), and NADH/NADPH as reductants, and enzymes that deplete any of these reductants could disrupt the normal redox state (Felton and Summers, 1993).

3.5 Vegetative Storage Proteins

Vegetative storage proteins (VSPs) have been identified in many plants, such as soybean (*Glycine max*; Wittenbach, 1983), potato (*Solanum tuberosum*; Mignery et al., 1988), sweet potato (*Ipomoea batatas*; Maeshima et al., 1985), white clover (*Trifolium repens*; Goulas et al., 2003), alfalfa (*Medicago sativa*; Meuriot et al., 2004), etc. These proteins are stored in various vegetative storage organs at higher concentrations and act as reservoirs for amino acids that facilitate source–sink interactions in a number of plants (Staswick, 1994).

In plants, some proteins have dual or multiple roles and often possess enzymatic as well as other activities. Examples of such proteins are the legume storage protein vicilins conferring insect resistance (Yunes et al., 1998); patatin, the storage protein from potato tubers as a lipid acyl hydrolase (Andrews et al., 1988); sporamin belonging to the superfamily of trypsin inhibitors (Yeh et al., 1997); the 32-kD VSP from alfalfa as a chitinase (Meuriot et al., 2004); and jacalin-like lectins in the bark of the black mulberry tree (*Morus nigra*) (Van Damme et al., 2002). In Arabidopsis (*Arabidopsis thaliana*) VSPs (AtVSPs) are induced by JA application, insect feeding, and other environmental stresses (Reymond et al., 2004), and a positive correlation has been found between AtVSP expression and insect resistance (Ellis and Turner, 2001). Anti-insect activity of many VSPs is due to enzymatic functions (phosphatase activity) (Liu et al., 2005).

Identification of such interlinked counterdefense protein genes that facilitate insect adaptation to dietary challenges is possible through functional genomic and proteomic studies. Therefore, targeting transcription factors that interact with common *cis*-elements of these counterdefense-related proteins could be an attractive approach in biotechnology-based insect control.

3.6 R-Gene-Mediated Plant Resistance

Plant resistance genes are commonly known as R genes encoding proteins with nucleotide binding site (NBS) and leucine rich repeat (LRR) domains (NBS-LRR proteins). They confer resistance to insect herbivores. Three types of R genes have been cloned: Mi-1.2, Vat, and Bph14. Mi-1.2 confers resistance in tomato to certain clones of *Macrosiphum euphorbiae* (potato aphid), two whitefly biotypes, a psyllid, and three nematode species (Rossi et al., 1998; Casteel et al., 2006; Walling, 2008); *Vat* confers resistance to one biotype of the melon-cotton aphid *Aphis gossypii* (Dogimont et al., 2010); while Bph14 confers resistance to the rice brown plant hopper *Nilaparvata lugens* (Du et al., 2009).

However, R-gene-mediated resistance is often limited to one clone of an insect species, e.g., specific aphid biotypes may evade or suppress plant defenses agreeing with the gene-for-gene model in plant–pathogen interactions (Dangl and Jones, 2001). For example, herbivore-induced expression of *terpene synthase 23* (*TPS23*) in a maize variety resulted in production of the volatile compound (E)-β-caryophyllene that caused a stronger attraction of the natural enemies of the insect herbivore compared with a maize variety that did not induce *TPS23* expression (Kollner et al., 2008). Apart from these, conservation of cuticle integrity has been shown as an important component of wheat resistance against the Hessian fly (Kosma et al., 2010).

The mechanism of resistance against the brown plant hopper in rice seems to involve the deposition of callose in sieve elements of the phloem in order to prevent the insect from taking up phloem sap (Du et al., 2009), and the Vat gene in melon seems to confer resistance via enhanced sieve element wound healing (Martin et al., 2003). The future challenge will be to further elucidate the resistance mechanisms acting downstream of insect R-genes and also to identify the insect-derived compounds (effector proteins) that are triggering the resistance reaction, which might be present in the insect saliva (Bonaventure et al., 2011).

Plant breeding companies prefer to exploit R-gene-mediated resistance in view of backcross breeding success. However, insects may rapidly generate new virulent biotypes to break down this type of resistance. For example, some *histocompatibility* genes that have been introgressed in wheat cultivars to control Hessian fly populations have been defeated within 10 years after being first deployed (Cambron et al., 2010). Russian wheat aphid biotypes virulent to wheat varieties containing a *dominant* resistance gene have also been found (Haley et al., 2004).

Pyramiding of R-genes may be a good approach to increase the effectiveness of R-genes in insect–pest management (Yencho et al., 2000). Next-generation sequencing and tools developed to study plant–pathogen interactions, such as NBS profiling (Jacobs et al., 2010) and effector genomics (Vleeshouwers et al., 2008), can facilitate the quick discovery of novel genes of the NBS-LRR family.

3.7 RNA Interference Technology in Host Plant Resistance

Ribonucleic acid interference (RNAi) is also known as posttranscriptional gene silencing (PTGS). It is a method of blocking gene function by inserting short sequences of RNA that match part of the target gene's sequence, thus no proteins are produced. RNAi has provided a way to control pests and diseases, introduce novel plant traits, and increase crop yield. Using RNAi, scientists have developed novel crops such as nicotine-free tobacco, nonallergenic peanuts, decaffeinated coffee, and nutrient-fortified maize. Sequence-specific gene silencing has practical applications in functional genomics, therapeutic intervention, agriculture, and other areas.

RNAi study facilitates directly observing the effects of the loss of function of specific genes in living systems. It is one of the most important technological breakthroughs in modern biology and provides an efficient means for blocking expression of a specific gene and evaluating its response to chemical compounds or changes in signaling pathways.

RNA is normally single stranded and performs the main functions in protein synthesis, which involve several steps. A single strand of messenger RNA (mRNA) is made of the template provided by DNA. The mRNA then causes amino acids to form chains in the exact order to produce a certain protein. RNAi interferes between these two processes by interfering with or cutting up the target mRNA. The result is that the proteins are not formed and unmodified genes are only interfered with or silenced.

Without certain proteins, pests may not develop, certain processes cannot continue, and so the process or the organism fails to function or reproduce. There are many potential applications of this technology such as controlling insects, diseases and nematodes, insect-transmitted diseases, and reversing pesticide resistance. Just like one's genetic makeup, this technology is highly specific. One of the newer approaches to RNAi-based technology has been delivery of dsRNA (double-stranded RNA) through feeding insects. RNAi is also referred to as dsRNA technology since the interference is actually initiated due to the presence of dsRNA.

The use of RNAi for insect control is less developed. Insect genes can be downregulated by injection of dsRNA or by oral administration of high concentrations of exogenously supplied dsRNA as part of an artificial diet, but a much more efficient method of delivering dsRNA is needed before RNAi technology can be used to control pests in the field (Huvenne and Smagghe, 2010). Overall health impacts from insect or plant disease organism-targeted dsRNA is of less concern in humans due to RNA specificity and enzymes in our blood and stomach that rapidly degrade RNA.

An important question arises of whether the plant-induced RNAi amplifies and spreads within the insect? It is plausible that a small amount of dsRNA ingested by the insect could be processed, amplified, and mobilized throughout its body. Insect genomics indicates that cotton bollworm and *Diabrotica virgifera* (western corn rootworm) lack the RNA-dependent RNA polymerase needed to drive this RNAi amplification. Perhaps unamplified RNAi is effective against the P450 and V-ATPase (*vacuolar-type proton ATPase*) genes because they are mainly expressed in the midgut and the gut cells of the feeding insect receive a continuous supply of hairpin RNA. The hairpin RNA, an artificial RNA molecule with a tight hairpin turn that can be used to silence target gene expression via RNA interference (RNAi) would be diced into siRNAs (*small interfering RNA*) in these cells and may also spread into surrounding cells and tissues via the intercellular dsRNA transport SID proteins (*Systemic RNA interference defective protein*), whose genes are present in almost all animals, including arthropods.

3.8 Recombinant Protease Inhibitors

In order to enhance the inhibitory potency of protease inhibitors against insect pests, protein engineering approaches have been adopted (Goulet et al., 2008), e.g., fusion protein engineering that integrates complete or partial inhibitor sequences (Benchabane et al., 2008). Such protein engineering strategies, together with transgene stacking (or gene pyramiding) in plants involving protease inhibitor combinations (Abdeen et al., 2005) or protease inhibitors combined with other pesticidal proteins (Han et al., 2007), have clearly confirmed the practical potential of these proteins in plant protection.

4. ADVANCEMENT IN PROTEOMIC TECHNIQUES

4.1 Shotgun Proteomics

Shotgun proteomics refers to the use of bottom-up proteomics techniques in identifying proteins in complex mixtures using an automated method named multidimensional protein identification technology (MudPIT), which combines multidimensional chromatography with electrospray ionization tandem mass spectrometry (Wolters et al., 2001). Shotgun proteomics identifies proteins from tandem mass spectra of their proteolytic peptides. Proteomics has emerged as an enormously powerful method for gaining insight to different physiological changes at cellular level (Natera et al., 2000), but relatively no attempts have been made to apply this technique to study insect toxicity, i.e., possibilities to study insecticide toxicity at the molecular level using proteomics technology. The study of partial amino acid sequence analysis will greatly contribute to the molecular biology for the identification of target genes and predicting their functions. Recently, shotgun proteomics after proteolytic enzyme digestion of samples has been used as an alternative technology capable of identifying hundreds of proteins from single samples.

4.2 Next-Generation Sequencing Technology

Next-generation sequencing technology is an efficient tool to understand how insects adapt to plant defenses. Results obtained from next-generation sequencing experiments provided whole-genome insight into the midgut transcriptome of several lepidopteran herbivores.

These studies also advanced knowledge of the mechanisms by which lepidopteran insects actively modify their digestive physiology to adapt to host plant defenses, which is key to understanding the process of host plant selection by insect pests. Further, high-throughput transcriptome sequencing may be quite useful in delineating mechanisms of metabolic adaptation and chemical coevolution in any plant–insect interaction. Microarray data indicate that a large set of genes are upregulated in response to herbivory, but thus far very few gene products have been shown to play a direct role in plant defense.

Natural genetic and transcriptomic variation in crop species with large and complex genomes can be best studied by next-generation sequencing technology (Morozova and Marra, 2008; Edwards et al., 2013), as a first step to identify potential genes and alleles that may be valuable for resistance to insects.

5. ENVIRONMENTAL IMPACT OF BIOTECHNOLOGY

The greatest concern about the commercial transgenic cultivation is threat to agricultural biodiversity and the environment. Therefore, it is necessary to focus future research toward developing transgenic varieties that may contribute to preserve and enhance the biodiversity.

5.1 Regulatory Aspects of Products Derived from Biotechnology

In order to meet the continuously increasing demand for food and feed, genetically modified (GM) crops were first introduced to farmers in 1995 in the expectation to have better crop yield. Soybean and maize are the primary GM crops along with cotton and canola, but other crops with combinations of herbicide tolerance, insect resistance, and nutritional improvements are being developed. There is growing demand for biotechnology-based crops in the global market (Ronald et al., 2015). However, toxicology, allergy, and environmental safety of the crops and their transgenic traits are the point of consideration for their registration. In these contexts, protein and molecular assessments of the genetic traits provide a thorough description that helps in identifying safety concerns, if any.

5.2 Molecular and Protein Characterization of Genetically Modified Crops

GM crops have evolved to include a thorough safety evaluation for their use in human food and animal feed. For the safety considerations of GM crops, GM DNA inserted into the crop genome is evaluated for its content, position, and stability in order to ensure that the transgenic novel proteins are safe from a toxicity, allergy, and environmental perspective. The inserted DNA of a transgenic crop consists of a gene that expresses a protein with specific trait as well as supporting DNA, such as promoters and terminators. A vector DNA construct containing the DNA of interest is verified as accurate prior to insertion into the host plant genome. Apart from these, the grain that provides the processed food or animal feed is also tested to evaluate its nutritional content and identify unintended effects to the plant composition when warranted. To provide a platform for the safety assessment, composition equivalence testing is done, under which the GM crop is compared to non-GM comparators.

New technologies, such as mass spectrometry and well-designed antibody-based methods, allow better analytical measurements of crop composition, including endogenous allergens. Many of the analytical methods and their intended uses are based on regulatory guidance documents, some of which are outlined in globally recognized documents such as Codex Alimentarius. The quality and standardization of testing methods can be supported, in some cases, by employing good laboratory practices and is recognized in some countries to ensure quality data.

6. FUTURE PERSPECTIVES

Understanding plant defense mechanisms and prediction of interacting traits in the form of ecological output are the challenges for the future. Another challenge in the era of systems biology is to use the massive amounts of quantitative data that will be acquired from both plant and herbivore data sets to construct and validate predictive models for defense against specific herbivores.

Apart from these, combination of transcriptomics and proteomics represents a powerful approach for protein discovery and enables confident identifications to be made in nonmodel insects. Efforts are required to focus research priorities in the following areas of importance:

- High-throughput screening protocols for field resistance to herbivores
- Identification of new sources of resistance to herbivores
- Characterization of insect biotypes and genes with resistance and their efficient deployment
- Evaluation of existing varieties and elite inbreds/hybrids for field resistance to insects
- Understanding the genetic basis of field resistance for the development of the next-generation of resistant cultivars.
- Developing durable insect-resistant varieties/hybrids through pyramiding of mapped genes
- Marker-assisted selection
- Establishing functional SNPs for selection for insect resistance.

The development of biotechnology should be shaped within the context of sustainable agricultural systems, ecologically based systems that reflect the goals of long-term economic viability, productivity, and natural resource stability. Therefore, emphasis should be on adopting ecologically based, sustainable farming systems. Biotechnology techniques should only be used within ecological research to search for innovative and sustainable solutions to agriculture's economic, social, and environmental problems.

References

Abdeen, A., Virgos, A., Olivella, E., Villanueva, J., Aviles, X., Gabarra, R., Prat, S., 2005. Multiple insect resistance in transgenic tomato plants over-expressing two families of plant proteinase inhibitors. Plant Molecular Biology 57, 189–202.

Alfonso-Rubi, J., Ortego, F., Castariera, P., Carbonero, P., Diaz, I., 2003. Transgenic expression of trypsin inhibitor CMe from barley in *indica* and *japonica* rice, confers resistance to the rice weevil *Sitophilus oryzae*. Transgenic Research 12, 23–31.

Andrews, D.L., Beames, B., Summers, M.D., Park, W.D., 1988. Characterization of the lipid acyl hydrolase activity of the major potato (*Solanum tuberosum*) tuber protein, patatin, by cloning and abundant expression in a baculovirus vector. Biochemical Journal 252, 199–206.

Benchabane, M., Goulet, M.C., Dallaire, C., Cote, P.L., Michaud, D., 2008. Hybrid protease inhibitors for pest and pathogen control – a functional cost for the fusion partners? Plant Physiology and Biochemistry 46, 701–708.

Bhalla, R., Dalal, M., Panguluri, S.K., Jagadish, B., Mandaokar, A.D., Singh, A.K., Kumar, P.A., 2005. Isolation, characterization and expression of a novel vegetative insecticidal protein gene of *Bacillus thuringiensis*. FEMS Microbiology Letters 243, 467–472.

Bi, J.L., Felton, G.W., Mueller, A.J., 1994. Induced resistance in soybean to *Helicoverpa zea*: role of plant protein quality. Chemical Ecology 20, 183–198.

Bode, W., Huber, R., 1992. Natural protein proteinase inhibitors and their interactions with proteinases. European Journal of Biochemistry 204, 433–451.

Bode, W., Huber, R., 2000. Structural basis of the endoproteinase-protein inhibitor interaction. Biochimica et Biophysica Acta 1477, 241–252.

Bonaventure, G., Vandoorn, A., Baldwin, I.T., 2011. Herbivore-associated elicitors: FAC signaling and metabolism. Trends in Plant Science 16, 294–299.

Cambron, S.E., Buntin, G.D., Weisz, R., Holland, J.D., Flanders, K.L., Schemerhorn, B.J., Shukle, R.H., 2010. Virulence in Hessian fly (Diptera: Cecidomyiidae) field collections from the southeastern United states to 21 resistance genes in wheat. Journal of Economic Entomology 103, 2229–2235.

Casteel, C.L., Walling, L.L., Paine, T.D., 2006. Behavior and biology of the tomato psyllid, *Bactericerca cockerelli*, in response to the *Mi-1.2* gene. Entomologia Experimentalis et Applicata 121, 67–72.

Chiche, L., Heitz, A., Padilla, A., Lenguyen, D., Castro, B., 1993. Solution conformation of a synthetic bis-headed inhibitor of trypsin and carboxypeptidase A: new structural alignment between the squash inhibitors and the potato carboxypeptidase inhibitor. Protein Engineering 6, 675–682.

Corbin, D.R., Grebenok, R.J., Ohnmeiss, T.E., Greenplate, J.T., Purcell, J.P., 2001. Expression and chloroplast targeting of cholesterol oxidase in transgenic tobacco plants. Plant Physiology 126, 1116–1128.

Dangl, J.L., Jones, J.D.G., 2001. Plant pathogens and integrated defense responses to infection. Nature 411, 826–833.

Das, K.C., Carl, W.W., 2002. Redox systems of the cell: possible links and implications. Proceedings of the National Academy of Sciences of the United States of America 99, 9617–9618.

De Vos, M., Van Zaanen, W., Koornneef, A., Korzelius, J.P., Dicke, M., Van Loon, L.C., Pieterse, C.M.J., 2006. Herbivore-induced resistance against microbial pathogens in Arabidopsis. Plant Physiology 142, 352–363.

Dogimont, C., Bendahmane, A., Chovelon, V., Boissot, N., 2010. Host plant resistance to aphids in cultivated crops: genetic and molecular bases, and interactions with aphid populations. Comptes Rendus Biologies 333, 566–573.

Du, B., Zhang, W., Liu, B., Hu, J., Wei, Z., Shi, Z., He, R., Zhu, L., Chen, R., Han, B., He, G.C., 2009. Identification and characterization of *Bph14*, a gene conferring resistance to brown planthopper in rice. Proceedings of the National Academy of Sciences of the United States of America 106, 22163–22168.

Edwards, D., Batley, J., Snowdon, R.J., 2013. Accessing complex crop genomes with next-generation sequencing. Theoretical and Applied Genetics 126, 1–11.

Ellis, C., Turner, J.G., 2001. The *Arabidopsis* mutant *cev1* has constitutively active jasmonate and ethylene signal pathways and enhanced resistance to pathogens. Plant Cell 13, 1025–1033.

Egusa, M., Akamatsu, H., Tsuge, T., Otani, H., Kodama, M., 2008. Induced resistance in tomato plants to the toxin-dependent necrotrophic pathogen *Alternaria alternata*. Physiological and Molecular Plant Pathology 73, 67–77.

Estruch, J.J., Warren, G.W., Mullins, M.A., Nye, G.J., Craig, J.A., Koziel, M.G., 1996. Vip3A, a novel *Bacillus thuringiensis* vegetative insecticidal protein with a wide spectrum of activities against lepidopteran insects. Proceedings of the National Academy of Sciences of the United States of America 93, 5389–5394.

Fan, X., Shi, X., Zhao, J., Zhao, R., Fan, Y., 1999. Insecticidal activity of transgenic tobacco plants expressing both Bt and CpTI genes on cotton bollworm (*Helicoverpa armigera*). Chinese Journal of Biotechnology 15, 1.

Felton, G.W., Donato, K.K., Broadway, R.M., Duffey, S.S., 1992. Impact of oxidized plant phenolics on the nutritional quality of dietary protein to a noctuid herbivore, *Spodoptera exigua*. Journal of Insect Physiology 38, 277–285.

Felton, G.W., Summers, C.B., 1993. Potential role of ascorbate oxidase as a plant defense protein against insect herbivory. Journal of Chemical Ecology 19, 1553–1568.

Gatehouse, A.M.R., Davidson, G.M., Newell, C.A., Merryweather, A., Hamilton, W.D.O., Burgess, E.P.J., Gilbert, R.J.C., Gatehouse, J.A., 1997. Transgenic potato plants with enhanced resistance to the tomato moth, *Lacanobia oleracea*: growth room trials. Molecular Breeding 3, 49–63.

Glazebrook, J., 2005. Contrasting mechanisms of defense against biotrophic and necrotrophic pathogens. Annual Review of Phytopathology 43, 205–227.

Gonzales-Vigil, E., Bianchetti, C.M., Phillips Jr., G.N., Howe, G.A., 2011. Adaptive evolution of threonine deaminase in plant defense against insect herbivores. Proceedings of the National Academy of Sciences of the United States of America 108, 5897–5902.

Goulet, M.C., Dallaire, C., Vaillancourt, L.P., Khalf, M., Badri, M.A., Preradov, A., Duceppe, M.O., Goulet, C., Cloutier, C., Michaud, D., 2008. Tailoring the specificity of a plant cystatin toward herbivorous insect digestive cysteine proteases by single mutations at positively selected amino acid sites. Plant Physiology 146, 1010–1019.

Goulas, E., Le, D.F., Ozouf, J., Ourry, A., 2003. Effects of a cold treatment of the root system on white clover (*Trifolium repens* L.) morphogenesis and nitrogen reserve accumulation. Journal of Plant Physiology 160, 893–902.

Grochulski, P., Masson, L., Borisova, S., Pusztai-Carey, M., Schwartz, J.L., Brousseu, R., Cyzler, M., 1995. *Bacillus thuringiensis* CryIA(a) insecticidal toxin: crystal structure and channel formation. Journal of Molecular Biology 254, 447–464.

Haley, S.D., Peairs, F.B., Walker, C.B., Rudolph, J.B., Randolph, T.L., 2004. Occurrence of a new Russian wheat aphid biotype in Colorado. Crop Science 44, 1589–1592.

Han, L., Wu, K., Peng, Y., Wang, F., Guo, Y., 2007. Efficacy of transgenic rice expressing *Cry1Ac* and *CpTI* against the rice leaf folder, *Cnaphalocrocis medinalis* (Guenee). Journal of Invertebrate Pathology 96, 71–79.

Haq, S.K., Atif, S.M., Khan, R.H., 2004. Protein proteinase inhibitor genes in combat against insects, pests, and pathogens: natural and engineered phytoprotection. Archives of Biochemistry & Biophysics 431, 145–159.

Harsulkar, A.M., Giri, A.P., Patankar, A.G., Gupta, V.S., Sainani, M.N., Ranjekar, P.K., Deshpande, V.V., 1999. Successive use of non-host plant proteinase inhibitors required for effective inhibition of *Helicoverpa armigera* gut proteinases and larval growth. Plant Physiology 121, 497–506.

Harvell, C.D., Tollrian, R., 1998. Why inducible defenses? In: Tollrian, R., Harvell, C.D. (Eds.), The Ecology and Evolution of Inducible Defenses. Princeton University Press, Princeton, New Jersey, pp. 3–9.

Heibges, A., Salamini, F., Gebhardt, C., 2003. Functional comparison of homologous membranes of three groups of Kunitz-type enzyme inhibitors from potato tubers (*Solanum tuberosum* L.). Molecular Genetics & Genomics 269, 535–541.

Heil, M., Silva, B.J.C., 2007. Within-plant signaling by volatiles leads to induction and priming of an indirect plant defense in nature. Proceedings of the National Academy of Sciences of the United States of America 104, 5467–5472.

Hilder, V.A., Gatehouse, A.M.R., Sheerman, S.E., Barker, R.F., Boulter, D., 1987. A novel mechanism of insect resistance engineered into tobacco. Nature 330, 160–163.

Howe, G.A., 2004. Jasmonates as signals in the wound response. Plant Growth Regulation 23, 223–237.

Huvenne, H., Smagghe, G., 2010. Mechanisms of dsRNA uptake in insects and potential of RNAi for pest control: a review. Journal of Insect Physiology 56, 227–235.

James, C., 2008. Global Status of Commercialized Biotech/GM Crops. ISAAA Briefs 39 (International Service for the Acquisition of Agri-biotech Applications), Ithaca, NY.

Jacobs, M.M.J., Vosman, B., Vleeshouwers, V.G.A.A., Visser, R.G.F., Henken, B., van den Berg, R.G., 2010. A novel approach to locate *Phytophthora infestans* resistance genes on the potato genetic map. Theoritical Applied Genetics 120, 785–796.

Kelly, J.D., Afanador, L., Haley, S.S., 1995. Pyramiding genes for resistance to bean common mosaic virus. Euphytica 82, 207–212.

Kempema, L.A., Cui, X., Holzer, F.M., Walling, L.L., 2007. Arabidopsis transcriptome changes in response to phloem-feeding silver leaf whitefly nymphs. Similarities and distinctions in responses to aphids. Plant Physiology 143, 849–865.

Kessler, A., Baldwin, I.T., 2001. Defensive function of herbivore-induced plant volatile emissions in nature. Science 291, 2141–2144.

Kollner, T.G., Held, M., Lenk, C., Hiltpold, I., Turlings, T.C.J., Gershenzon, J., Degenhardt, J., 2008. A maize (E)-beta-caryophyllene synthase implicated in indirect defense responses against herbivores is not expressed in most American maize varieties. Plant Cell 20, 482–494.

Kosma, D.K., Nemacheck, J.A., Jenks, M.A., Williams, C.E., 2010. Changes in properties of wheat leaf cuticle during interactions with Hessian fly. Plant Journal 63, 31–43.

Lawrence, K., Koundal, K.P., 2002. Plant protease inhibitors in control of phytophagous insects. EJB Electronic Journal of Biotechnology 5, 93–109.

Lee, M.K., Walters, F.S., Hart, H., Palekar, N., Chen, J.S., 2003. The mode of action of the *Bacillus thuringiensis* vegetative insecticidal protein Vip3A differs from that of Cry1Ab δ-endotoxin. Applied and Environmental Microbiology 69, 4648–4657.

Li, J., Carroll, J., Ellar, D.J., 1991. Crystal structure of insecticidal delta-endotoxin from *Bacillus thuringiensis* at 2.5 Å resolutions. Nature 353, 815–821.

Liu, Y.L., Ahn, J.E., Datta, S., Salzman, R.A., Moon, J., Huyghues-Despointes, B., Pittendrigh, B., Murdock, L.L., Koiwa, H., Zhy-Salzman, K., 2005. *Arabidopsis* vegetative storage protein is an anti-insect acid phosphatase. Plant Physiology 139, 1545–1556.

Maeshima, M., Sasaki, T., Asahi, T., 1985. Characterization of major proteins in sweet-potato tuberous roots. Phytochemistry 24, 1899–1902.

Martin, B., Rahbe, Y., Fereres, A., 2003. Blockage of stylet tips as the mechanism of resistance to virus transmission by *Aphis gossypii* in melon lines bearing the Vat gene. Annals of Applied Biology 142, 245–250.

Meuriot, F., Noquet, C., Avice, J.C., Volenec, J.J., Cunningham, S.M., Sors, T.G., Caillot, S., Ourry, A., 2004. Methyl jasmonate alters N partitioning, N reserves accumulation and induces gene expression of a 32-kDa vegetative storage protein that possesses chitinase activity in *Medicago sativa* taproots. Physiologia Plantarum 120, 113–123.

Mignery, G.A., Pikaard, C.S., Park, W.D., 1988. Molecular characterization of the patatin multigene family of potato. Gene 62, 27–44.

Morozova, O., Marra, M.A., 2008. Applications of next-generation sequencing technologies in functional genomics. Genomics 92, 255–264.

Natera, S.H., Guerreiro, N., Djordjevic, M.A., 2000. Proteome analysis of differentially displayed proteins as a tool for investigation of symbiosis. Molecular Plant Microbe Interaction 3, 995–1009.

Nina, T., Lerdau, M., 2003. The evolution of function in plant secondary metabolites. International Journal of Plant Sciences 164 (3 Suppl.), S93–S102.

Oppert, B., Morgan, T.D., Hartzer, K., Kramer, K.J., 2005. Compensatory proteolytic responses to dietary proteinase inhibitors in the red flour beetle, *Tribolium castaneum* (Coleoptera : Tenebrionidae). Comparative Biochemistry and Physiology, C: Toxicology & Pharmacology 140, 53–58.

Perlak, F.J., Deaton, R.W., Armstrong, T.A., Fuchs, R.L., Sims, S.R., Greenplate, J.T., Fischhoff, D.A., 1990. Insect resistant cotton plants. Bio/Technology 8, 939–943.

Purcell, J.P., Greenplate, J.T., Jennings, M.G., 1993. Cholesterol oxidase: a potent insecticidal protein active against boll weevil larvae. Biochemical Biophysical Research Communications 196, 1406–1413.

Reymond, P., Farmer, E.E., 1998. Jasmonate and salicylate as global signals for defense gene expression. Current Opinion in Plant Biology 1, 404–411.

Reymond, P., Bodenhausen, N., Van-Poecke, R.M.P., Krishnamurthy, V., Dicke, M., Farmer, E.E., 2004. A conserved transcript pattern in response to a specialist and a generalist herbivore. Plant Cell 16, 3132–3147.

Ronald, V.R., Poulsen, L.K., Wong, G.W.K., Ballmer-Weber, B., Gao Zhongshan, B.K., Jia Xudong, J., 2015. Food allergy: definitions, prevalence, diagnosis and therapy. Chinese Journal of Preventive Medicine 49, 1–17.

Rossi, M., Goggin, F.L., Milligan, S.B., Kaloshian, I., Ullman, D.E., Williamson, V.M., 1998. The nematode resistance gene *Mi* of tomato confers resistance against the potato aphid. Proceedings of the National Academy of Sciences of the United States of America 95, 9750–9754.

Sanchez, A.C., Brar, D.S., Huang, N., Li, Z., Khush, G.S., 2000. Sequence tagged site marker-assisted selection for three bacterial blight resistance genes in rice. Crop Science 40, 792–797.

Shukle, R.H., Murdock, L.L., 1983. Lipoxygenase, trypsin inhibitor and lectin from soybeans: effects on larval growth of *Manduca sexta* (Lepidoptera: Sphingidae). Environmental Entomology 12, 787–791.

Smith, C.M., 2005. Plant Resistance to Arthropods: Molecular and Conventional Approaches. Springer, Dordrecht, The Netherlands. 423 pp.

Staswick, P.E., 1994. Storage proteins of vegetative plant-tissue. Annual Review of Plant Physiology & Plant Molecular Biology 45, 303–322.

Tabashnik, B.E., Carriere, Y., 2010. Field-evolved resistance to Bt cotton: *Helicoverpa zea* in the US and pink bollworm in India. Southwest Entomologist 35, 417–424.

Thaler, J.S., Stout, M.J., Karban, R., Duffey, S.S., 2001. Jasmonate-mediated induced plant resistance affects a community of herbivores. Ecological Entomology 26, 312–324.

Ton, J., Van Pelt, J.A., Van Loon, L.C., Pieterse, C.M.J., 2002. Differential effectiveness of salicylate-dependent and jasmonate/ethylene-dependent induced resistance in *Arabidopsis*. Molecular Plant Microbe Interaction 15, 27–34.

Tuominen, A., Jari, S., Maarit, K., Juha-Pekka, S., 2015. Sylvatiins, acetylglucosylated hydrolysable tannins from the petals of *Geranium sylvaticum* show co-pigment effect. Phytochemistry 115, 239–251. http://dx.doi.org/10.1016/j.phytochem.2015.01.005.

Urwin, P.E., McPherson, M.J., Atkinson, H.J., 1998. Enhanced transgenic plant resistance to nematodes by dual proteinase inhibitor constructs. Planta 204, 472–479.

Van Damme, E.J.M., Hause, B., Hu, J.L., Barre, A., Rouge, P., Proost, P., Peumans, W.J., 2002. Two distinct jacalin-related lectins with a different specificity and subcellular location are major vegetative storage proteins in the bark of the black mulberry tree. Plant Physiology 130, 757–769.

Volpicella, M., Leoni, C., Costanza, A., de Leo, F., Gallerani, R., Ceci, L.R., 2011. Cystatins, serpins and other families of protease inhibitors in plants. Current Protein and Peptide Science 12, 386–398.

Vleeshouwers, V.G.A.A., Rietman, H., Krenek, P., Champouret, N., Young, C., Oh, S.K., Wang, M., Bouwmeester, K., Vosman, B., Visser, R.G.F., Jacobsen, E., Govers, F., Kamoun, S., van der Vossen, E.A.G., 2008. Effector genomics accelerates discovery and functional profiling of potato disease resistance and *Phytophthora infestans* avirulence genes. PLoS One 3, e2875.

Walling, L.L., 2000. The myriad plant responses to herbivores. Journal of Plant Growth Regulation 19, 195–216.

Walling, L.L., 2008. Avoiding effective defenses: strategies employed by phloem-feeding insects. Plant Physiology 146, 859–866.

Wang, J.H., Constabel, C.P., 2004. Polyphenol oxidase overexpression in transgenic *Populus* enhances resistance to herbivory by forest tent caterpillar (*Malacosoma disstria*). Planta 220, 87–96.

Wittenbach, V.A., 1983. Purification and characterization of a soybean leaf storage glycoprotein. Plant Physiology 73, 125–129.

Wolters, D.A., Washburn, M.P., Yates, J.R., 2001. An automated multidimensional protein identification technology for shotgun proteomics. Analytical Chemistry 73, 5683–5690.

Wu, J., Baldwin, I.T., 2009. Herbivory-induced signalling in plants: perception and action. Plant Cell Environment 32, 1161–1174.

Wu, J., Hettenhausen, C., Meldau, S., Baldwin, I.T., 2007. Herbivory rapidly activates MAPK signaling in attacked and unattacked leaf regions but not between leaves of *Nicotiana attenuata*. Plant Cell 19, 1096–1122.

Xu, D., Xue, Q., McElroy, D., Mawal, Y., Hilder, V.A., Wu, R., 1996. Constitutive expression of a cowpea trypsin inhibitor gene, *CpTi*, in transgenic rice plants confers resistance to two major rice insect pests. Molecular Breeding 2, 167–173.

Yeh, K.W., Chen, J.C., Lin, M.I., Chen, Y.M., Lin, C.Y., 1997. Functional activity of sporamin from sweet potato (*Ipomoea batatas* Lam.): a tuber storage protein with trypsin inhibitory activity. Plant Molecular Biology 33, 565–570.

Yencho, G.C., Cohen, M.B., Byrne, P.F., 2000. Applications of tagging and mapping insect resistance loci in plants. Annual Review of Entomology 45, 393–422.

Yu, C., Mullins, M., Warren, G., Koziel, M., Estruch, J., 1997. The *Bacillus thuringiensis* vegetative insecticidal protein Vip3A lyses midgut epithelium cells of susceptible insects. Applied and Environmental Microbiology 63, 532–536.

Yunes, A.N.A., de Andrade, M.T., Sales, M.P., Morais, R.A., Fernandes, K.V.S., Gomes, V.M., Xavier-Filho, J., 1998. Legume seed vicilins (7S storage proteins) interfere with the development of the cowpea weevil (*Callosobruchus maculatus* (F). Journal of the Science of Food and Agriculture 76, 111–116.

Zavala, J.A., Patankar, A.G., Gase, K., Hui, D., 2004. Baldwin IT: manipulation of endogenous trypsin proteinase inhibitor production in *Nicotiana attenuata* demonstrates their function as antiherbivore defenses. Plant Physiology 134, 1181–1190.

Zhang, P.J., Shu, J.P., Fu, C.X., Zhou, Y., Hu, Y., Zalucki, M.P., Liu, S.S., 2008. Trade-offs between constitutive and induced resistance in wild crucifers shown by a natural, but not an artificial, elicitor. Oecologia 157, 83–92.

Zhu-Salzman, K., Zeng, R., 2015. Insect response to plant defensive protease inhibitors. Annual Review of Entomology 60, 233–252.

GMO and Food Security

Mala Trivedi[1], Rachana Singh[1], Manish Shukla[2],
Rajesh K. Tiwari[1]

[1]Amity Institute of Biotechnology, Amity University Uttar Pradesh, Lucknow, India;
[2]Plant Production Research Centre, Directorate General of Agriculture and Livestock
Research, Muscat, The Sultanate of Oman

1. INTRODUCTION

Climate change leading to natural calamities, like flood, drought, dwindling natural resources for agricultural practices, food crisis, and gaps in demand and supply of energy are some of the issues concerning scientists and policy makers worldwide. All these problems are more complicated in developing countries in comparison to developed ones. Food crisis is growing at an alarming rate as it is getting difficult to keep pace in agriculture production with the rate of population growth (Table 1). The problem is further worsened due to scarcity of funding in agricultural research. Earlier conventional breeding was the only method for production of plants with desirable characteristics. Here desirable traits are selected and combined by repeated crossings for several generations. Self-fertilization is followed for at least seven generations after crossing for desirable traits. This is very time consuming and a minimum of 7–15 years are required to produce new varieties (Southgate et al., 1995). With the ever-increasing population size, impact of a green revolution is also diminishing (Khush, 2001). Therefore, scientists are looking for new technologies to provide food security.

2. GENETICALLY MODIFIED TECHNOLOGY IN AGRICULTURE

Genetic engineering is the technology that allows introduction of desirable genes in a highly targeted and accelerated manner. It overcomes the barrier of sexual incompatibility between species and vastly increases the size of the available gene pool (Southgate et al., 1995). Transgenic technology is extensively used for plants and microbes, and to some extent also for animals. Recombinant DNA technology is the basic tool for the development of transgenics. In this technique, a gene of interest that may or may not be native is

Ecofriendly Pest Management for Food Security
http://dx.doi.org/10.1016/B978-0-12-803265-7.00023-3

TABLE 1 Undernourishment Around the World, 1990–1992 to 2012–2014: Number of Undernourished (Millions) and Prevalence (%) of Undernourishment

	1990–1992		2000–2002		2005–2007		2008–2010		2012–2014	
	No.	%	No.	%	No.	%	No.	%	No.	%
World	1014.5	18.7	929.9	14.9	946.2	14.3	840.5	12.1	805.3	11.3
Developed regions	20.4	<5	21.1	<5	15.4	<5	15.7	<5	15.6	<5
Developing regions	994.1	23.4	908.7	18.2	930.8	17.3	824.9	14.5	790.7	13.5

inserted into the plant/animal or endogenous genes are modified. There are several important applications of this technology, e.g., development of resistance to abiotic (drought, extreme temperature, or salinity) and biotic (insects and pathogens) stresses. This is vital for plant growth and/or survival. Other important applications are improvement of the nutritional content of the plant, an application that could be of particular use in the developing world. Some important industrial products, like monoclonal antibodies, vaccines, plastics, and biofuels, are being developed using transgenic technology. Other applications include the production of nonprotein (bioplastic) or nonindustrial (ornamental plant) products. A number of animals have also been genetically engineered to increase growth and maturation and to decrease susceptibility to diseases. For example, salmon have been engineered to grow larger and mature faster and cattle have been manipulated to exhibit resistance to mad cow disease (Takeda and Matsuoka, 2008; Sticklen, 2005; Conard, 2005; Ma et al., 2003).

Crop plants are one of the most frequently cited examples of genetically modified organisms (GMOs). Some benefits of genetic engineering are increased crop yields, reduced costs for food or drug production, reduced need for pesticides, enhanced nutrient composition and food quality, resistance to pests and diseases, greater food security, and medical benefits to the world's growing population. Advances have also been made in developing crops that mature faster and tolerate aluminum, boron, salt, drought, frost, and other environmental stressors, allowing plants to grow under conditions where they might not otherwise flourish (Takeda and Matsuoka, 2008).

Several crop modifications are done using these methods that are now in widespread use. The best known of these are crop plants into which a gene from the soil bacterium *Bacillus thuringiensis* has been introduced. *Bacillus thuringiensis* has long been used as a biopesticide because it produces proteins that are toxic to the larvae of certain group of insects that feed on crop plants but is quite safe to animals and humans. The gene coding for the toxin is often called "the Bt gene" for its bacterial origin. The Bt toxin genes, which constitute a family of closely related proteins, have been introduced into several different crops, primarily corn and cotton. In the United States (US) and Europe, pest-protected crop varieties have been produced almost exclusively by companies such as, Monsanto, DuPont, and Syngenta. In other parts of the world, including China and India, such crop modifications are being done by both public and private research establishments (Fedoroff, 2010).

Another important genetic modification that is helpful in food security is "herbicide resistance." Herbicides or weedicides are the chemicals that kill unwanted plants that grow along with the crop plants. Most of these chemicals block some important aromatic acid production pathway of the weed plants, leading to their death. Therefore, a gene is introduced into the crop plants to make them tolerant to these chemicals. Examples of some genetically modified (GM) herbicide tolerant crops are soybean and canola (Tan et al., 2006).

Papaya ring spot disease was devastating to papaya farming in Hawaii. It is an insect-transmitted viral disease. Development of transgenic papaya resistant to ring spot virus saved papaya farmers from great losses every year and also saved the industry (Gonsalves et al., 2004). Here studies have shown that the resistance is due to posttranscriptional gene silencing. It is a process in which a small RNA derived from a double-stranded version of the viral gene transcript initiates the cleavage of invading viral RNA (Tennant et al., 2001). This remarkable method of crop protection enhances a mechanism present in plants and protects the plant from further infection by the same and closely related viruses, much as the development of immunity protects people and animals from reinfection by pathogens.

2.1 Genetically Modified Food Crops

More than 900 million people of the developing world are chronically undernourished and are surviving on less than 2000 Kcal/day. Many of them (approximately 1.3 billion) are living on less than US$1/day and do not have secure access to food (Pinstrup-Anderson et al., 1999; FAO, 2001; Smith et al., 2000). Farmers of such countries depend entirely on small-scale agriculture for their subsistence. Because of their financial condition and mindset, they cannot afford to buy herbicides and other pesticides, which leads to a poor crop growth, pest susceptibility, and reduced yields, with little monetary gain (Christou and Twyman, 2004). According to a recent estimate, there will be a huge increase in population by 2040 with over 95% increases in developing countries. It is estimated that food production will need to be increased 40% in order to feed the human population that will be coping with decreasing fertile lands and water resources (Byerlee et al., 2000).

The genetically modified plant technologies are one of the most important approaches to meet these upcoming challenges. Major laboratories are working on different aspects: (1) to genetically modify plants to increase crop yields, and/or (2) to improve the nutritional content. Genetic engineering is one of the methods being used to meet breeding goals and to develop robust, high yielding cultivars custom tailored to specific regional conditions. It offers possibilities of resistance to viral, bacterial, and fungal pathogens, including tolerance to drought and salinity. According to 2010 data, transgenic crops are grown on more than 300 million acres spreading across 25 countries. Major transgenic crops grown by small landholder, poor farmers are: cotton, maize, soybean, and canola. Other important transgenics being grown are papaya, squash, tomato, poplar, petunia, sweet pepper, alfalfa, and sugar beet.

2.1.1 Golden Rice and Other Varieties

The developed world has access to all types of food to meet their nutritional requirements. In contrast, the people in developing countries live on only one staple food to meet their nutritional requirements, i.e., cereals. GM technology improves nutritional content in the staple food to meet nutrition requirements. According to reports, approximately more than 2 million people suffer from various defects, particularly loss of vision. Vitamin A deficiency is a considerable public health problem in many developing countries. It affects 140 million preschool children and 7 million pregnant women worldwide. Of these, up to 3 million children die every year (UN SCN report, 2005). The deficiency is most common in poverty households where diets are mainly dependent on staple foods with comparatively low nutritional value (Allen, 2003). Widespread consumption of golden rice promises to improve the situation in rice-eating populations (Qaim, 2010). The "Golden Rice Project" was started by scientists to combat vitamin A deficiency in the children of developing world. Cereals are deficient in Vitamin A, however, other plants can synthesize beta-carotene, which is a precursor of vitamin A. Initially a rice plant with a moderate level of beta-carotene was developed (Ye et al., 2000), and later a more advanced variety, "Golden Rice-2," was developed keeping in mind the farmers of poor countries of the world.

An application for the approval of food and feed use of LL62 rice has been submitted to the European Union (EU). It is still undergoing safety evaluations. This GM rice cultivar has been engineered for herbicide resistance. Rice breeding is in progress in many countries and projects have been launched at international agricultural research centers.

2.1.2 Potatoes

Potatoes are an important food and also as a raw material for the starch industry. The starch produced by potatoes is a mixture of two types of starch: amylose and amylopectin. Amylopectin is 80% of the starch content in potatoes. It is very useful in the food, paper, and chemical industries as paste, glue, and lubricant. Amylose is made up of long chain-like molecules and is used predominantly in the production of films and foils. Both of these starches are useful in human nutrition. But for the processing industry, a mixture of different starches is a problem. Industries need to separate starches using expensive processes. Classical breeding methods have not yet been able to provide an amylose-free potato that has acceptable yield and resistance to pests and diseases. However, genetic engineering offers a targeted approach to suppress the production of amylose (GMO Compass, 2015). Genetically modified amylopectin potatoes have been tested for several years. In the meantime, applications have been presented to European regulatory authorities for approving the cultivation of these potatoes as a raw material for starch production. Because the postprocessing residues would be fed to livestock, a request for approval of the potatoes as feed has also been submitted. Starch-modified GM potatoes could soon be grown in European fields.

Several GM potato cultivars with improved resistance to viruses and to the potato beetle have been approved in the US and Canada. These GM potatoes were planted on approximately 25,000 ha in 1999. But the cultivation of GM potatoes ceased in subsequent years because of limited economic benefits, therefore, major US companies refused to use GM potatoes for processing. Attempts to confer pest and disease resistance to potatoes using genetic engineering have not been so successful.

Research is underway on potatoes using genetic engineering to confer resistance to *Phytophthora infestans*, late blight of potato. This plant disease is most dangerous because it spreads rapidly under warm and moist conditions, leading to devastating losses. Genetic engineers have given a promising new strategy and the first field trials with fungus-resistant GM potatoes have started. The next few years will confirm if this new concept is effective (GMO Compass, 2015).

2.1.3 Brinjal

The eggplant (*Solanum melongena*), commonly known as brinjal, is the fourth most widely grown vegetable in India, and a significant export crop (Government of India, 2009). Production of brinjal is significantly affected by heavy infestation of a fruit borer pest, *Leucinodes orbonalis*. It bores initially into young shoots and later into the fruits. Farmers apply copious amounts of insecticides and the pest has developed resistance against them; breeders have not had success developing insect-resistant plants. However, use of Bt technology proved to be effective in control of *L. orbonalis* in brinjal.

Field trials on Bt brinjal began in 2006 but the Supreme Court of India halted them in September 2006. After this stop order, the Genetic Engineering Advisory Committee (GEAC) formed the Expert Committee I (EC-I) to address public concerns and allowed large-scale trials on the brinjal to run at 10 research institutions across the country in 2007 (11 in 2008) including the Indian Institute of Vegetable Research (IIVR). After the successful results of large-scale trials in 2009, GEAC voted to approve the release of Bt brinjal. But there was a great deal of controversy about this decision, mainly in southern India including Kerala. The key issue was that wild relatives of brinjal are important ingredients in medical formulations in classical and popular traditions (Chithprabhu and Glenn, 2013). People were apprehensive that any genetic modification in brinjal might lead to undesirable changes in the gene pool of wild varieties. However, outcrossing studies were established between wild-type and Bt brinjal to address this issue. The Ministry of Environment and Forest, India, and Department of Biotechnology (DBT), New Delhi, India, issued a report stating that " as such there is no natural crossing among cultivated and wild species of brinjal" (Government of India, 2009).

The analysis of large-scale field trial results revealed that use of Bt had resulted in a significant reduction in insecticide use. Overall the quantities of insecticides used against fruit and shoot borer (FSB) were reduced by 77.2% (IIVR, 2009; AICVIP/ICAR, 2007). The data showed that yield of Bt brinjal hybrids was consistently higher than that of non-Bt hybrids and the gain by Bt hybrids was 37.3% over non-Bt hybrids and 54.9% over popular hybrids. Farmers benefited at multiple levels—they could save on quantity of insecticide used, the cost of insecticides, and labor used in insecticide spraying. Brinjal production rose from 30,000 tons to 119,000 tons. Thus, Bt technology is expected to generate high yield and huge savings (Rs. 47 crore to Rs. 187 crore) (Kumar et al., 2011).

2.1.4 Soybeans

Worldwide soybean production was 216 million tons in 2007. The world's leading soybean producers are the United States (33%), Brazil (27), Argentina (21%), and China (7%). India also produces a significant amount of soybean. Soybeans as a legume have not been used directly in India. The bulk of soybeans are crushed and used as feed for animals, which are the main source of meat and fat in India. Soybean oil is also used as cooking oil (Manuwa, 2011).

The first genetically modified soybean was planted in the United States in 1996. It possesses a gene that confers herbicide resistance. A GM soybean is cultivated in nine countries covering more than 60 million hectares. Argentina (98%) and the US (85%) produce the bulk of soybeans. GM soybeans are approved without any restriction in either of these two countries. In Brazil, GM soybeans were approved in 2007, and now 64% of the country's soybean crop is genetically modified. Paraguay, Canada, Uruguay, and South Africa are also large-scale producers of GM soybeans (GMO Compass, 2015).

GM soybeans may have unpredictable side effects on human health. There are chemical compounds in GM soybean that cannot be detected easily but can harm humans. The GM soybeans cause liver damage and toxemia, harm beneficial gut bacteria and shift the balance of gut microflora, and ultimately affect human health (Herman, 2003). However, a thorough scientific investigation is required to validate these findings.

2.1.5 Tomato

Tomato (*Solanum lycopersicum* L.) is an important vegetable crop. It is a rich source of vitamins, minerals, and fiber, and a dietary source of antioxidants. Tomato is also a model system for studies on fruit development and functional genomics (Klee and Giovannoni, 2011). A wide range of microbial pathogens and insect pests attack different parts of the tomato plant, causing severe losses to the crop productivity. Tomato leaf curl virus resistant transgenic tomatoes have been developed (Raj et al., 2005). An efficient agrobacterium-mediated transgenic tomato line Ab25 E was developed for pest resistance. Transgenic tomatoes showed overexpression of modified *cry1Ab* gene and efficient binding of Cry1Ab toxin with different receptors and other possible target proteins located in the midgut of the insects. This overexpressing homozygous transgenic line can be deployed in tomato breeding programs for the introgression of insect-resistant trait in commercially important varieties of tomatoes (Koul et al., 2014).

2.1.6 Alfalfa

Alfalfa is engineered to produce novel compounds for industrial and diagnostic purposes. The main focus includes abiotic stress tolerance, biotic stress tolerance, herbicide resistance, and improvement in forge quality. Abiotic stresses, such as drought, freezing, flooding, and biotic stresses, viz., diseases, are associated with increased production of oxygen free radicals. A class of metalloproteins called superoxide dismutases (SODs) has the ability to detoxify oxygen free radicals, converting them to hydrogen peroxide and molecular oxygen. Alfalfa engineered to overexpress SOD has exhibited improved winter survival (McKersie et al., 1999) and reduced secondary injury symptoms with enhanced winter stress (McKersie et al., 2000).

2.1.7 Canola (Rapeseed)

Rapeseed, *Brassica napus*, was initially not considered important for knocking out of undesirable traits. Improved cultivars were free from erucic acid and glucosinolates. Erucic acid gives a bitter taste and prevents canola from being used as food, and glucosinolates present in oil cakes are toxic as animal feed. A new cultivar, known as double zero rapeseed, is named canola (Canadian oil) to differentiate from nonedible wild varieties. GM canola for herbicide resistance (glyphosate and glufosinate) was produced by Canada in 1995. Canada is amongst the top five countries producing GM crops (James, 2009). Herbicide-resistant canola allows farmers to manage some of the most difficult weeds (Devine and Buth, 2001). The yield of GM canola is high when sprayed with glyphosate and glufosinate in comparison to nontransformed canola (Harker et al., 2000).

2.1.8 Maize

Maize is one of the most important cultivated cereals and major profitable food crops grown in many countries around the world. Major maize-producing countries are the US, China, Mexico, France, Argentina, and India. The US alone grows 332 million metric tons of maize annually. This crop is attacked by pests and diseases leading to yield loss. Recently, some potyviruses have also caused great losses to maize cultivation. Genetically modified maize constituted 85% of the maize planted in the United States in 2009. GM maize has been developed for herbicide resistance as well as virus tolerance (Ishida et al., 1996). Bt corn was initially engineered to target only the European corn borer (*Ostrinia nubilalis*); in 2003, another variety of Bt corn that is resistant to corn rootworm (*Diabrotica* spp.) was introduced (Lundmark, 2007). According to a survey done by Animal and Plant Health Inspection Service (APHIS), Bt corn is the second-most common GM crop (after soybean), with 12.4 million hectares planted in 2002. GM corn starch and soybean lecithin are two ingredients found in

70% of the processed food supply (Gewin, 2003). A recent development in GM maize is the commercial release of drought-tolerant GM maize under the name of Drought Gard maize (Castiglioni et al., 2008).

A number of GM crops carrying novel traits have been developed and released for commercial agriculture production with the rapid advancements in biotechnology. Scientists have attempted various approaches for developing GM crops in last two decades. The first GM fruit crop marketed commercially was the well–known "Flavr Savr" tomato (Bruening and Lyons, 2000), which had been modified to contain reduced levels of the cell wall softening enzyme polygalacturonase (Smith et al., 1988). Since then, there has been a massive expansion in the research and commercial cultivation of transgenic crops worldwide. These include pest-resistant cotton, maize, canola (mainly Bt), herbicide glyphosate-resistant soybean, cotton and viral disease-resistant potatoes, papaya, and squash. In addition, various GM crops are under development and are expected to be commercially released shortly with traits for biofortification, phytoremediation, and production of pharmaceuticals, such as rice with a high level of carotenoid for production of vitamin A. Commercial cultivation of GM crops started in the early 1990s. Insect resistance and herbicide tolerance are the main GM traits that are under commercial cultivation, and the main crops are: soybean, maize, canola, and cotton. GM crops are currently commercially cultivated on about 100 million hectares in some 22 developed and developing countries. Argentina, Brazil, China, and India are the largest developing country producers of GM crops. The choice of GM crops varies among the developing countries, with insect-resistant cotton being the most important commercially produced transgenic crop in Asian and African countries.

2.2 Genetically Modified Nonfood Crops

Under this category are crops that are not used as food but are economically important, such as cotton, medicinal plants, and ornamental plants.

2.2.1 Cotton

Bt cotton contains genes from *B. thuringiensis* that make the plant resistant to the cotton bollworm complex. This inbuilt insect resistance confers with savings from use of pesticides and higher yields to farmers (Qaim and Zilberman, 2003). Bt cotton adoption is associated with significant benefits to farmers in various countries (Carpenter, 2010; Qaim, 2009; Pray et al., 2002; Subramanian and Qaim, 2010; Ali and Abdulai, 2010). In addition to productivity gains (Bennett et al., 2004, 2006; Crost et al., 2007; Morse et al., 2007; Sadashivappa and Qaim, 2009), it entails reduced acute pesticide poisoning among smallholder farmers (Kouser and Qaim, 2011).

In India, cotton is primarily grown by smallholder farmers with farm sizes of less than 15 acres and cotton holdings of 3–4 acres on average. The first Bt cotton was commercially released in India in 2002. By 2011, 7 million farmers had adopted Bt cotton on 26 million acres, around 90% of the total cotton area in India (James, 2011). A survey of Indian cotton farmers between 2002 and 2008, in four major cotton-producing states, revealed that Bt cotton share increased to 93% in 2005 and reached 99% in 2008 (Kathage and Qaim, 2012).

2.2.1.1 IMPACT ON COTTON PROFITS

GM cotton influences profits from farming mainly through three channels: (1) changes in yield, (2) reduction in pesticide cost, and (3) changes in seed cost (Qaim et al., 2006). The Bt impact per acre does not change significantly over time. However, total cotton profits per farm rose because of increased Bt adoption. Currently, 26 million acres are under Bt cotton farming. This implies annual net gain of almost 50 billion INR (1 billion US $) in cotton profits in India (Kathage and Qaim, 2012). Sadashivappa and Qaim (2009) published the data that show comparative advantage of Bt over conventional cotton in India (Table 2).

2.2.2 Medicinal Plants

Genetic transformation of medicinal plants includes transformation of plants for the production of medicinally important active components without damaging them. Various methods are adopted for the transfer of a gene of interest, and of them agrobacterium-mediated transformation is the most important and popular method adopted in most of the studies (Gelvin, 2003; Trivedi et al., 2013)

Another GM technology known as "hairy roots" has been recognized as a potential alternative source of many secondary compounds in medicinally important plants as the genetic transformation does not affect the natural root synthetic capacities (Flores and Medina-Bolivar, 1995).

Several medicinal plants were genetically modified for their root system for production of medicinally important compounds in the last several years. In an independent study, it was reported that by using hairy root technology, *Gentiana macrophylla* hairy roots grew 33 times faster, *Scutellaria baicalensis* grew 37 times faster, and *Glycyrrhiza uralensis* grew 38 times faster than normal root in standardized liquid culture condition and accumulated the higher amount of pharmaceutically important compounds (Tiwari et al., 2007, 2008). This higher alkaloid production, fast-growing hairy root, offers exciting possibilities for large-scale production of larger biomass by bioreactors and stable high production of metabolites for pharmaceutical use. With hairy root cultures the year-round production of pharmaceuticals is possible, the extraction of active component is also less expensive, less laborious, and ecofriendly. The use of hairy root cultures as a raw material will not only reduce the dependence of pharmaceutical industry on the natural habitat of plants but ensure the availability of ecologically

TABLE 2　Comparative Advantage of Bt over Conventional Cotton in India

	2002	2004	2006	Average
Insecticide use	−50%	−51%	−21%	−41%
Yield	+34%	+35%	+43%	+37%
Seed cost	+221%	+208%	+68%	+166%
Total cost	+17%	+11%	+24%	+17%
Gross revenue	+33%	+37%	+40%	+37%
Profit	+69%	+129%	+70%	+89%
Profit gain in US $/ha	+111$	+142$	+152$	+135$

unpolluted, microbial spore-free raw material without being affected by seasonal variation of useful compounds. Moreover, hairy root regenerants showed more excessive adventitious root system, potentially capable of producing a large mass of root in per unit area. This possibility would further be investigated and refined for agricultural purposes.

3. GM TECHNOLOGY AND ANIMALS

Not only plants but animals are also being genetically modified by introducing foreign genes (i.e., genes of interest). These animals have good quality traits with enhanced protein and high meat content.

Food animals are being engineered to grow more quickly, require less feed, or leave behind less environmental damaging waste. The University of Guelph, Canada, has developed a GM pig that efficiently digests phosphate from feed. As a result, the pig excretes less phosphate in its feces, which reduces its environmental impact. This must be seen against the background of the growing environmental problems in areas of intensive agricultural production caused by the accumulation of phosphate and nitrogen in soil and water from the application of manure and fertilizers. The GM pig, however, is not so much to do with environmental or food safety but the role it could play in encouraging further strengthening of livestock farming. As the GM pig puts less of a burden on the environment, this may actually reduce the stress to find other solutions (COGEM topic report, 2011).

3.1 Healthier Food

Food animals, such as pigs, are under development to contain increased levels of omega-3 fatty acids, providing a more healthful product. Farm animals can also be engineered to provide leaner meat or more milk (source: US FDA report). GM sheep have been developed with an inserted transgenic growth hormone to make them grow faster and produce more milk (CSIRO, 2002).

Gene technology has already found some application in food production through bacterial expressed bovine somatotropin (BST), which is used in the United States and some countries in eastern Europe to increase the milk production of cows.

Food-producing farm animals, such as cattle, are important to human nutrition. Genetic engineering offers the potential to increase this food supply. Australia started selling clones of top breeding bulls, whose offspring produce more meat (Taipei Times, 2001).

3.2 Fish Production

Fish are considered essential food resources in some parts of India and world. Fish consumption is on the increase around the world. According to the National Marine Fisheries Services, the amount of fish consumed per person in the US increased from ~5.08 kg in 1910 to ~7.53 kg in 2004. Salmon consumption in the US increased ninefold between 1987 and 1999; during that time, total European salmon consumption increased four times. Salmon consumption in Japan doubled between 1992 and 2002. Atlantic salmon is a good source of protein and an excellent source of omega-3 fatty acids, which are thought to aid in cardiovascular health (US FDA, 2015).

In 2006, the Food and Agriculture Organization (FAO) of the United Nations predicted that there will be a 25% increase in global demand for seafood by 2030, based on the current rate of increase in consumption of fish, and by 2030, the world population will be 8 billion. The increase in demand for fish can only be met by aquaculture, which accounts for about one-third of the fish caught annually. Recent studies indicate that half the fish eaten around the world today is farmed (Naylor et al., 2009). It contributes to the food supply, economy, and health of the nation (West, 2006). For example, salmon has been engineered to grow larger and mature faster (Devlin et al., 1994).

Genetic engineering in the fish industry could greatly enhance the food supply. Transgenic fish, in particular salmon and trout species, with additional copies of growth-promoting genes (Devlin et al., 1995) were the first commercialized animals. In early 1992, transgenic salmons carrying additional growth hormone genes were shown to grow significantly faster than control salmons (Du et al., 1992).

In addition, research is exploring the possibilities of raising the quality of fish meat for human consumption. Using salmon genes to alter the ratio between omega-3 and omega-6 fatty acids is one of them. Worldwide, 50 fish species have been genetically modified, creating more than 400 different fish/trait combinations. A fast-growing GM river carp has been developed in China and a fast-growing GM tilapia in Cuba. Transgenic trout and shellfish are also being developed, but there has been no application for marketing authorization (CONGEM topic report, 2011).

The amount of fish available in oceans and lakes is declining, while the global population and fish consumption are increasing. This is why fish are being produced by aquaculture. Atlantic salmon for consumption are now almost entirely produced in fish farms. The fast-growing GM salmon can help boost fish production (Aqua Bounty Technologies, 2010).

4. GM MICROORGANISMS IN FOOD PRODUCTION

Microorganisms play important roles in different sectors of agriculture, food processing, pharmaceutical industry, and environmental management. Having said that microbial processes are under the control of genes, there is urgent need to continually improve and optimize their specific processes through genetic improvement. Recombinant DNA technology presents a number of benefits in this area, since specific metabolic pathways can be manipulated with more precision and completely new functions can now be engineered into microbes. Many microbes are being manipulated with the objectives of improving process control, yields, and efficiency as well as the quality, safety, and consistency of bioprocessed products.

4.1 Lactic Acid Bacteria

Lactic acid bacteria have a long history of use in fermented food products. With advancements in technology they were modified by introducing new genes or by modifying their metabolic functions. These modifications made the bacteria better fitted to technological processes, leading to improved organoleptic properties or to new applications, such as bacteria-producing therapeutic molecules that can be taken orally. Probiotics have many potential therapeutic uses but have not been universally accepted because of lack of understanding of their action. Lactic acid bacteria (LAB) have been modified by traditional and genetic engineering methods to produce new varieties. Modern techniques of molecular biology have facilitated the identification of probiotic LAB strains, but only a few have been modified by recombinant DNA technology because of consumer resistance, especially in Europe (Ahmed, 2003).

New strains of LAB, *Lactobacillus* sp., are very much in demand in industry as they have been produced to provide much improved starter cultures for cheese production. The potential for more rapid production and maturation of cheese and a whole range of other fermented products is being explored. Moreover, LAB are excellent producers of peptidases, which are widely used in food technology. This characteristic is thought to have enormous potential, associated with the production of traditional foods (http://www.biotopics.co.uk/edexcel/biotechnol/gmmfoo.html).

4.2 Enzymes

Enzymes are very important in food processing. Apart from enhancing nutritional value, enzymes can be used to influence flavor, aroma, texture, appearance, and speed of production. Enzymes are extracted from edible plants and tissues of food animals. Microorganisms are the most important sources of these enzymes and have been used for centuries in food manufacturing. Microorganisms, such as *Rhizomucor pusillus*, *Rhizomucor miehei*, *Endothia parasitica*, *Aspergillus oryzae*, and *Irpex lactis*, are used extensively for rennet production in cheese manufacturing. Rennet has been used in cheese making since its beginning. Rennet contains a protease enzyme that coagulates milk, causing it to separate into solids (curds) and liquids (whey). Traditionally, the enzyme was obtained from the "stomach" of calves, but this source became limited and expensive. Some other issues also arise with the animal use, and vegetarians find this source unacceptable. Considering the economic value of cheese rennet, a gene for calf chymosin was one of the first genes for mammalian enzymes that was cloned and expressed in microorganisms (Nishimori et al., 1982). Many laboratories have cloned the gene for calf prochymosin in *Escherichia coli*, *Saccharomyces cerevisiae*, *Kluyveromyces lactis*, *Aspergillus nidulans*, *Aspergillus niger*, and *Trichoderma reesei* (Foltman, 1993; Mellor et al., 1983; Goff et al., 1984; Christensen et al., 1988). The chymosin produced by these modified organisms is identical to the bovine product. This was achieved some time ago for chymosin (rennin), used in cheese manufacture.

4.3 Yeasts

Much attention is currently being focused on the possible uses of yeasts. Chymosin from transgenic yeast was the first enzyme from GMO to gain regulatory approval for food use. In 1994, the Brewing Research Foundation International gained approval for the use of a GM yeast to make low-carbohydrate beer. The yeast *Saccharomyces cerevisiae* var *diastaticus* produces the enzyme amylase, which is capable of hydrolyzing starch residues that normally remain in the brew (Renault, 2002).

4.4 Single-Cell Protein

The increasing demand for protein-rich food worldwide prompted researchers to look for alternative protein sources as a supplement to the conventional protein sources.

Single-cell proteins (SCP) are dried cells of microorganisms such as algae, bacteria, yeasts, molds, and higher fungi that are grown in large-scale culture systems for use as protein rich sources of food for human and animal consumption. SCPs are an alternative and an innovative way to solve the global food problem (Anupama and Ravindra, 2000). The term SCP refers to total proteins extracted from pure microbial cell culture or by using a number of different microorganisms including bacteria, fungi, and algae (Anupama and Ravindra, 2000). Besides high protein content (about 60–82% of dry cell weight), SCP also contains fats, carbohydrates, nucleic acids, vitamins, and minerals (Asad et al., 2000; Jamel et al., 2008). SCP is rich in certain essential amino acids like lysine and methionine, which are limiting in most plant and animal foods. Microorganisms are grown on waste materials, and then they are harvested and purified or processed to produce protein-rich sources of food for humans and livestock. Methanol, cane, molasses, wood acid hydrolysates, CO_2, sulfite waste liquors, and whey have been used as substrates for growing specific strains of bacteria, yeasts, algae, and filamentous fungi (Goldberg, 1988).

Many microorganisms are suitable for this process, including algae, yeasts, fungi, and bacteria. Advantages of using microbes as SCP include: (1) their rapid growth rates, (2) great varieties of substrates can be exploited, (3) easy manipulation of genetic characteristics, (4) fairly high protein content, and (5) they grow independently of climate. *Spirulina* (blue green algae) is an important example of algae used in SCP production. Similarly, biomass obtained from *Chlorella* and *Scenedesmus* is harvested and used as a source of food by tribal communities in certain parts of the world. Alga is used as a food in many different ways and its advantages include simple cultivation, effective utilization of solar energy, faster growth, and rich in protein content.

Many species used as a source of protein rich food are *Candida, Hansenula, Pichia, Torulopsis, Aspergillus,* and *Saccharomyces* (Sadiq et al., 2014). They all are yeasts. Modern technology for producing yeast single-cell protein has largely developed since World War II. Today, yeast products for human or animal consumption are produced on a commercial scale in many places, viz., US and Russia, etc. In addition, baker's yeast, which is grown on molasses, is sold as a food flavoring and nutritive ingredient in addition to being used as a leavening agent. Among bacteria, *Cellulomonas* and *Alcaligenes* are the most commonly used in SCP production, and *Methylophilus methylotrophus* is used in animal feed (Windass et al., 1980). The large-scale application of genetic engineering has been reported by Windass et al. (1980) for single-cell protein production. Imperial Chemical Industries (ICI, UK) announced a major process improvement obtained through the genetic engineering of *M. methylotrophus*, a methanol-consuming microorganism, with a plant capacity of approximately 60,000 tons of single-cell protein product per year (Sherwood, 1974).

The SCP product contains a concentration of nucleic acids (6–10%) that has no nutritive value and its use elevates serum uric acid levels and results in kidney stone formation (Nasseri et al., 2011). Therefore, the product is treated with acid alkanes and enzymes to remove nucleic acid and is then supplemented with essential amino acids and flavors. *Yarrowia lipolytica* isolated from the marine fish gut was found to contain 47.6 g of crude protein per 100 gm of cell dry weight and had potential use as single cell protein. Wang et al. (2009) have genetically modified the marine-derived yeast *Y. lipolytica* with high protein content.

5. STUDIES ON FOOD SAFETY

There are many GM products available in the market for consumption as food. Therefore, it is imperative to check quality of GM food. Feeding experiments conducted with Bt cotton seed meal on fish, chickens, cows, and buffalo indicated that Bt cotton seed meal is nutritionally equivalent, wholesome, and as safe as the non-Bt cotton seed meal (Senthil and Singhal, 2004).

6. SAFETY ISSUE WITH GMOs AS FOOD

The term *biosecurity* refers to a set of preventive measures designed to reduce the risk of transmission of infectious diseases in crops and livestock, quarantined pests, invasive *alien* species, and living modified organisms (Koblentz, 2010). GM plants or crops are under strict regulation of the respective governments. In the US, the Food and Drug Administration (FDA), the Environmental Protection Agency, and the US Department of Agriculture, Animal and Plant Health Inspection Service provide approval for the marketing and consumption (EFSAJ, 2004). In Europe, the European Food Safety Authority is the regulatory authority along with participation of states for risk assessments. In India, the Government of India Agricultural Biosecurity Authority does safety testing and monitors GMOs before commercialization.

6.1 Biosafety Policy: India

Biosafety in the context of GMO plants is basically related to the protection of the environment along with human and animal health from the possible adverse effects of the GMOs and the products developed by the use of modern biotechnology. Issues related to biosafety need strict attention at all stages of development and before release of GM crops for field cultivation.

During the erstwhile Indian government, Agricultural Biosecurity Bill, 2013, was tabled in the Lok Sabha with the aim to repeal the Destructive Insects and Pests Act, 1914, and the Livestock Importation Act, 1898, and establish the Agricultural Biosecurity Authority of India (Authority). The chief objective of the bill was to establish an integrated National Biosecurity System covering plant, animal, and marine issues. The second objective was to combat threats of bioterrorism arising from pests and weeds by using the system.

The Agricultural Biosecurity Authority of India is expected to be responsible for: (1) regulating the import and export of plants, animals, and related products; (2) preventing the introduction of quarantine pests from outside India; and (3) implementing postentry quarantine measures. The Indian Government first issued rules and procedures for handling GMOs in December 1989. The Department of Biotechnology, under the Ministry of Science and Technology, published these rules and procedures in January 1990 (DBT, 1990). These "Recombinant DNA Safety Guidelines" were revised and republished as "Revised Guidelines for Safety in Biotechnology" in 1994 (DBT, 1994). They described the biosafety measures that must be undertaken in India both for research activities under containment and also for large-scale open environmental release of GM agricultural and pharmaceutical materials. A subsequent 1998 revision of the guidelines further elaborated procedures for screening transgenic plants and seeds for toxicity and allergies (DBT, 1998).

Under the Environment Protection Act (1986), the Ministry of Environment & Forests, India, has developed the rules for the manufacture, use, import, export, and storage of hazardous microorganisms/genetically engineered organisms or cells. These rules and regulations cover the areas of research as well as large-scale applications of GMOs and products made therefrom throughout the country. The rules also cover the application of hazardous microorganisms that may not be genetically modified. Hazardous microorganisms include those that are pathogenic to humans, animals as well as plants. The rules cover activities involving manufacture, use, import, export, storage, and research on hazardous natural microorganisms and GM organisms including microorganisms, plants, and animals.

7. ECOLOGICAL CONCERNS

With the advent of recombinant DNA technology a large number of GM crops are being developed and released for field-testing and commercialization. Therefore, it is imperative to check potential risks associated with them concerning the environment and human health.

The risks of transgenic crops to the environment and human health have been assessed worldwide, which depends upon crop physiology, ecology, crop management, and its regulation. To date there are no reports on health hazards of the commercialization of GM plants. However, a number of issues are being debated because the potential long-term cumulative effects are difficult to predict. Various environmental issues of GM crops including horizontal gene transfer, gene flow, outcrossing, evolution of virulent strains of pests, and impact of soil microbes need to be addressed (Randhawa and Durga, 2001; Bhalla et al., 2007).

7.1 GM Crops and Biodiversity

Biodiversity is the variability in the living organisms at species level residing in a particular area. Local cultivars and many times their wild relatives have often been donors of many useful traits including resistance to pests and diseases. There are chances of contamination of gene pool if large-scale commercialization of a GM crop is done. For example, development of insect-resistant crops will reduce the use of broad-spectrum insecticides, preserve the biodiversity, and in turn reduce pesticide residues in water and soil. In 2004, the Ministry of Agriculture, India, issued the Seed Bill that requires mandatory registration of all types of seed including GM seeds. To maintain biodiversity, new strategies are practiced while commercialization of crops, for example, refugee seeds in transgenic (Bt) cotton that have both transgenic and nontransgenic seeds grown in the same field to maintain resistance and biodiversity. There is little evidence of secondary pest outbreaks in Bt crops, requiring substantial use of insecticides (Whitehouse et al., 2005). In India, the Ministry of Environment issued the Biological Diversity Act in 2002 and Biodiversity Rules in 2004, which regulate the use of biological resources including genes used for improving crops and livestock through genetic intervention.

A study was conducted in the United Kingdom to check the impact of GM crops on the surrounding flora and fauna, particularly their long-term effects, (www.defra.gov.uk/environment/gm/fse). In this study, a total of four crops were tested, and in three, wildlife was reduced in the land where GM crops were grown; however, in the fourth crop, the opposite results were found. Further study proved that the differences were not because of genetic modification of the crop but because of some new agricultural practices by the farmers.

This led the government to take appropriate steps to check the effect of GM crops on non-GM crops, before providing any approval for commercialization of GM crops(www.defra.gov.uk/environment/gm/fse).

7.2 Effects on Pests

Continuous exposure of insects and pests to pest-resistant varieties may lead to evolution of super pests (virulent strains/biotypes/races). Such changes will help pests to overcome the adverse conditions (e.g., crops expressing genes for pest resistance) for their survival.

7.3 Outcrossing Prevention Strategies

There are two types of gene transfer: (1) vertical gene flow, which is flow of genes in the same species, and (2) horizontal gene transfer, where genes are transferred to unrelated species. To avoid gene flow between GM and conventional crops, the best strategy should be cultivation of both the crops either in isolation or at a great distance. Therefore, transformation techniques that reduce outcrossing should be used, and also the potential for gene flow to other crop plants and wild relatives must be considered on a case-by-case basis.

For the prevention of gene flow in the environment, some foolproof strategies should be followed for GM crops. Song et al. (2002a) reported a high percentage of outcrossing in rice *Oryza rufipogon* (wild rice) by cultivated rice *Oryza sativa*. To check, gene flow in transgenic and nontransgenic crops proper isolation strategies should be formulated. In another study on GM rice for herbicide resistance gene "bar gene," Song et al. (2002b) revealed that gene flow between transgenic rice and wild-type species was not possible because of reproductive barriers between these species.

Mascia and Flovell (2004) reported some strategies to prevent outcrossing of GM crops with wild species. Physical isolation is one of the methods to prevent gene flow, where a crop must be grown in physical isolation and protected environment. Another strategy proposed by Mascia and Flovell (2004) is "land fallowing," which is land in which a GM crop is grown should not be used the next year. This will help in identifying seed growth in the next year. One more strategy to check outcrossing is by targeting a gene of interest to cytoplasmic organelle, i.e., maternal inheritance, e.g., gene transfer through chloroplast (pollen grains lack chloroplasts). Cultivation of GM food crops along with nonfood GM crops, which are important for industry, may lead to health hazards as reported in Canada where farmers were growing two varieties of rapeseed with high and low erucic acid varieties. Erucic acid in high concentration is toxic to humans and is used as a lubricant in industry. Rapeseed varieties producing low erucic acid are nontoxic and used for consumption by humans as cooking oil (canola). Farmers growing both these varieties of rapeseed have developed standard procedures for growth and processing to avoid outcrossing or gene flow between these two varieties. Canadian farmers have developed systems to routinely keep the two apart during growth and processing (Mascia and Flovell, 2004).

7.4 Impacts on Other Species

There are apprehensions about the invasiveness of GM crops. However, there is no evidence of change in some important traits in GM cotton because of *cry1Ac* gene (Percival et al., 1999). The negative impact of GM crops on nontarget organisms is also a serious threat. Nontarget organisms are the ones living in or around the agricultural fields that are not intended to be targeted. There are concerns that the genes for biotic stresses may affect the organisms other than the target pests. Herbicide-resistant GM crops are considered to have no direct toxic effect on nontarget organisms because the enzymes conferring herbicide tolerance are normally expressed in the plants and are not known to have toxic properties (Carpenter, 2001). GM plants that produce insecticidal substances should continue, subject to careful testing to ensure safety and minimize environmental risks. A targeted approach to an ecological risk assessment is recommended particularly in a regulatory context (Alonso et al., 2006). There is also concern with respect to insect-resistant transgenic crops expressing Cry proteins from *B. thuringiensis* that they could harm organisms other than the pests. However, no adverse effects on nontarget organisms resulting from direct toxicity of the expressed Bt toxins have so far been recorded in laboratory and field studies. The Bt crops are more target specific and have fewer side effects on nontarget organisms than most of current insecticides used.

There are two viewpoints about the exposure of organisms to Bt toxins: (1) constitutive expression of Bt toxin in all plant tissues throughout the season leads to continuous exposure of nontarget organisms to insecticidal protein and thus increases the risk of nontarget effects, and (2) the other view is constitutive expression of Bt toxin as a potential advantage as it would reduce the environmental exposure of nontarget organisms to the toxin (Poppy and Sutherland, 2004). The insects consuming the GM plant part only may express toxins throughout Bt strains are species specific whereas with conventional insecticide spraying many more insects are exposed. This has been confirmed by the recent large-scale studies conducted in commercially managed *Bt* and non-*Bt* cotton fields in the US (Head et al., 2005; Torres and Ruberson, 2005).

8. TRANSGENICS AND SOIL MICROORGANISMS

After harvesting of GM crops, the same field would be utilized for another crop after regular agronomical practices. This may pose concern to those plants in which recombinant product is found in root or stem, as after harvesting and plowing recombinant product present in root and stem may persist and in turn affect soil microorganisms.

The main concerns are: (1) whether the commercial cultivation of GM crops is inducing changes in the biodiversity of soil and soil processes, and (2) whether these changes exceed natural variations caused by a multitude of environmental factors or the variation found in the conventional crop systems caused by different crops, cultivars, and crop rotation as well as impacts of agricultural operations. It is important to evaluate the ecological significance of changes induced by GM crops in the context of these variations and to assess whether changes such as the accumulation of toxins or altered community structure of soil organisms prove to be reversible. The concern is that ultra-high doses of chemicals used for the management of resistance may lead to soil pollution with high doses of the toxin and may affect the soil biota. The studies from GM crop residues suggest that there will be no acute effects on soil health. However, monitoring will be necessary to evaluate the long-term effects.

Degradation of Bt toxin varies, depending on parameters, such as temperature and soil type. Initial degradation of toxin is rapid with a low percentage that may remain in the soil ecosystem following one growing season. It has been shown that Bt toxin may bind to the clay and humic acid compounds. However studies reveal no accumulation of Bt toxin after several years of cultivation. Neither laboratory nor field studies have shown lethal/sublethal effects of Bt toxins on nontarget soil organisms (nematodes, earthworms, Collembola, mites, etc.). Hence, systematic development, exchange, and commercialization of transgenic crops after addressing the biosafety issues pertaining to the environment followed by regular postrelease monitoring to evaluate the long-term impacts would be required to sustain biodiversity and to harness the benefits of GM crops (Bhalla et al., 2007).

9. CONCLUSIONS

Genetically modified organisms are organisms that have been modified by using recombinant DNA technology tools. The main objective of GM is to create a diversity of plants serving human needs and is a boon for both producers and consumers. GMOs are now an integral part of many agriculturally based commodities, adding billions of dollars per year to the global economy, and are major sources of income and food for developed and developing countries including India.

At present, about 170 million hectares of world crop acreage is planted with GM crops with an annual growth rate of 6% (James, 2012a,b). GM crops have been commercialized in the past 15 years, and approximately 28 countries are growing them (James, 2012a,b). Rice forms the main food source for almost half of the world's population. Herbicide-resistant genetically modified rice and tomatoes are growing in fields of several countries around the world. Virus-resistant papayas are commercially grown in Hawaii. Cultivation and production of Bt cotton has grown exponentially since then, and India has become the second largest producer of cotton and leading cotton exporter in the world (Choudhary and Gaur, 2010).

The future of genetically modified crops look promising. Many developing countries are expected to plant GM crops in next two to three years, especially in Asia. Some African countries may also contribute to GM crop hectarage in the near future, with the first drought-tolerant maize planned for release in Africa in 2017. The first stacked soybean (with herbicide tolerance and insect resistance traits) will be planted in Brazil; vitamin A–enriched golden rice could be released in the Philippines; drought-tolerant sugarcane in Indonesia; and biotech maize in China. Not only crops but animals are also modified genetically to exploit their milk, meat, and wool. GM foods have the potential to solve the world's hunger and malnutrition problems, and help protect and preserve the environment by increasing yield and reducing reliance on pesticides and herbicides. Genetic modification of microorganisms is also done for development of various industrially important products.

Genetic engineering is the inevitable wave of the future and we cannot afford to ignore a technology that has enormous potential. However, we must proceed with caution to avoid causing unintended harm to human health and the environment as a result of our enthusiasm for this powerful technology.

Acknowledgments

The authors are grateful to Dr A.K. Chauhan, Founder President, and Mr Aseem Chauhan, Chancellor Amity University Haryana & Chairperson AMITY Lucknow, for providing necessary facilities and support. We also extend our gratitude to Maj. Gen. K.K Ohri, AVSM (Retd.), Pro Vice Chancellor, Amity University, Uttar Pradesh Lucknow Campus, for constant support and encouragement.

References

Ahmed, F.E., 2003. Genetically modified probiotics in foods. Trends in Biotechnology 21 (11), 491–497.

AICVIP/ICAR, 2007. Performance of Multi-location Research Trials of Bt Brinjal Hybrids, Indian Institute of Vegetable Research (Indian Council of Agricultural Research), Varanasi.

Alonso, G.M., Jacobs, E., Raybould, A., Nickson, T.E., Sowig, P., Willekens, H., Van der, K.P., Layton, R., Amijee, F., Fuentes, A.M., Tencalla, F., 2006. A tiered system for assessing the risk of genetically modified plants to non-target organisms. Environmental Biosafety Research 5 (2), 57–65.

Ali, A., Abdulai, A., 2010. The adoption of genetically modified cotton and poverty reduction in Pakistan. Journal of Agriculture and Economy 61, 175–192.

Allen, L.H., 2003. Interventions for micronutrient deficiency control in developing countries: past, present and future. Journal of Nutrition 133, 38755–38785.

Anupama, Ravindra, P., 2000. Value added Food: single cell protein. Biotechnology advances 18, 459–479.

Aqua Bounty Technologies, Inc, 2010. An Atlantic Salmon (*Salmo salar* L.0 Bearing a Single Copy of the Stably Integrated Alpha-locus in the EO-1X Line. Submitted to the Center for Veterinary medicine USA Food and Drug Administration, Maryland, USA. Available at: http://www.fda.gov/downloads/AdvisoryCommittees/CommitteesMeetingMaterials/VeterinaryMedicineAdvisoryCommittee/UCM224760.pdf.

Asad, M.J., Asghan, M., Yaqub, M., Shahzad, K., 2000. Production of single cell protein delignified corn cob by *Arachniotus* species. Pakistan Journal of Agricultural Sciences 37, 3–4.

Bennett, R., Ismael, Y., Kambhampati, U., Morse, U., 2004. Economic impact of genetically modified cotton in India. AgBioForum 7, 96–100.

Bennett, R., Kambhampati, U., Morse, S., Ismael, Y., 2006. Farm-level economic performance of genetically modified cotton in Maharashtra, India. Review of Agricultural Economics 28, 59–71.

Bhalla, S., Chalam, V.C., Khetarpal, R.K., 2007. Biosafety concerns of *Bt* crops. In: Randhawa, G.J., Bhalla, S., Chalam, V.C., Sharma, S.K. (Eds.), Cartagena Protocol on Biosafety: Decisions to Diagnostics. National Bureau of Plant Genetic Resources, New Delhi.

Bruening, G., Lyons, J.M., 2000. The case of the FLAVR SAVR tomato. California Agriculture 54, 6–7.

Byerlee, D., Helsey, P., Pingali, P.L., 2000. Realising yield gains for food staples in developing countries in the early 21st century: prospects and challenges. In: Chang, B.M., Colombo, M., Soronolo, M. (Eds.), Food Needs of the Developing World in the 21st Century. Vatican City: Political Academy of Sciences, pp. 207–250.

Carpenter, J.E., 2001. Case Studies in Benefits and Risks of Agricultural Biotechnology: Roundup Ready Soybean and *Bt* Field Corn. National Centre for food and Agricultural Policy, Washington, DC.

Carpenter, J.E., 2010. Peer-reviewed surveys indicate positive impact of commercialized GM crops. Nature Biotechnology 28, 319–321.

Castiglioni, P., Warner, D., Bensen, R.J., Anston, D.C., Harrison, J., Stoecker, M., Abad, M., Kumar, G., Salvador, S., D'Ordine, R., 2008. Bacterial RNA chaperons confer abiotic stress tolerance in plants and improved grain yield in maize under water-limited conditions. Plant Physiology 147 (2), 446–455.

Chithprabhu, K., Glenn, D.S., 2013. The trials of Genetically modified food: BT eggplant and ayurvedic medicine in India. Food Culture Society 16 (1), 21–42.

Choudhary, B., Gaur, K., 2010. *Bt* Cotton in India: A Country Profile. ISAAA Series of Biotech Crop Profiles. ISAAA, Ithaca, NY.

Christensen, T., Woeldike, H., Boel, E., Mortensen, S.B., Hjortshoej, K., Thim, L., Hansen, M.T., 1988. High level expression of recombinant genes in *Aspergillus oryzae*. Biotechnology 6, 1419–1422.

Christou, P., Twyman, R.M., 2004. The potential of genetically enhanced plants to address food insecurity. Nutrition Research Reviews 17, 23–42.

COGEM topic report, 2011. Genetically Modified Animals: A Wanted and Unwanted Reality.

Conrad, U., 2005. Polymers from plants to develop biodegradable plastics. Trends in Plant Science 10, 511–512.

Crost, B., Shankar, B., Bennett, R., Morse, S., 2007. Bias from farmer self-selection in genetically modified crop productivity estimates: evidence from Indian data. Journal of Agricultural Economics 58, 24–36.

CSIRO, 2002. GM Sheep Grow Bigger, Produce More Milk and Wool (news item November 22, 2002, Ref 2002/234).

DBT (Department of Biotechnology, Government of India), 1990. http://envfor.nic.in/divisions/csurv/geac/annex-5.pdf.

DBT (Department of Biotechnology, Government of India), 1994. http://shodhganga.inflibnet.ac.in/bitstream/10603/18185/12/12_chapter%206.pdf.

DBT (Department of Biotechnology, Government of India), 1998. http://shodhganga.inflibnet.ac.in/bitstream/10603/18185/12/12_chapter%206.pdf.

Devine, M.D., Buth, J.L., 2001. Advantages of Genetically Modified Canola:a Canadian Perspective, Proceedings, Brighten Crop Protection Conference- Weeds. British Crop Protection Council, Farnham, Surrey, UK, pp. 267–272.

Devlin, R.H., Yesaki, T.Y., Blagi, C.A., Donaldson, E.M., 1994. Extraordinary salmon growth. Nature 371, 209–210.

Devlin, R.H., Yesaki, T.Y., Donaldson, E.M., Hew, C.L., 1995. Transmission and phenotypic effects of an antifreeze/GH gene construct in coho salmon (*Oncorhynchus kisutch*). Aquaculture 137, 161–170.

Du, S.J., Gong, Z., Fletcher, G.I., Shears, M.A., King, M.J., Idler, D.R., Hew, C.I., 1992. Growth enhancement in transgenic Atlantic salmon by use of "all fish" chimeric growth hormone gene construct. Biotechnology 10, 176–181.

EFSAJ, 2004. Guidance document of the genetically modified organisms for the risk assessment of genetically modified plants and derived food and feed. European Food safety Authority Journal 99, 1–94.

Food and Agriculture Organisation, 2001. The State of Food Insecurity in the World. FAO, Rome.

Fedoroff, N.V., 2010. The Past, present and future of crop genetic modification. New Biotechnology 27 (5), 461–465.

Flores, H.E., Medina-Bolivar, F., 1995. Root culture and plant natural products: "unearthing" the hidden half of plant metabolism. Plant Tissue Culture and Biotechnology 1, 59–74.

Foltman, B., 1993. In: Fox, P.F. (Ed.), Cheese – Chemistry, Physics and Microbiology –General Aspects, vol. 1. second ed. Chapman & Hall, London, pp. 37–68.

Gelvin, S.B., 2003. Agrobacterium-mediated plant transformation: the biology behind the "gene-jockeying" tool. Microbiology and Molecular Biology Review 67, 16–37.

GMO Compass, 2015. http://www.gmo-compass.org/eng/grocery_shopping/crops/.

Goff, C.G., Moi, D.T., Kohno, T., Gravius, T.C., Smith, R.A., Yamasaki, E., Taunton-Rigby, A., 1984. Expression of calf prochymosin in *Saccharomyces cerevisiae*. Gene 27 (1), 35–46.

Gonsalves, D., Gonsalves, C., Ferreira, S., Pitz, M., Manshardt, R., Slightom, J., 2004. Transgenic virus resistant papaya: from hope to reality for controlling papaya ringspot virus in Hawaii. APSnet Features. http://www.apsnet.org/online/feature.

Goldberg, I., 1988. Future prospects of genetically engineered single cell protein. Trends in Biotechnology 6 (2), 32–34.

GoI (Government of India), 2009. Report of the Expert Committee (EC-II) on Bt Brinjal Event EE-I, submitted to the Genetic Engineering Approval Committee (GEAC). Ministry of Environment and Forests, New Delhi.

Gwein, V., 2003. Genetically modified corn-environmental benefits and risks. PLoS Biology 1 (1), e8. http://dx.doi.org/10.1371/journal.pbio.0000008.

Harker, K.N., Blackshaw, R.E., Kirkland, K.J., Derksen, D.A., Wall, D., 2000. Herbicide tolerance in Canola: weed control and yield comparison in western Canada. Canadian Journal of Plant Science 80, 647–654.

Head, G., Moar, M., Eubanks, M., Freeman, B., Ruberson, J., Hagerty, A., Turnipseed, S., 2005. A multi year, large scale comparison of arthropod populations on commercially managed *Bt* and non-*Bt* cotton fields. Environmental Entomology 34, 1257–1266.

Herman, E.M., 2003. Genetically modified soybeans and food allergies. Journal of Experimental Botany 54, 1317–1319.

Ishida, V., Saito, H., Ohta, S., Hiei, Y., Komari, T., Kumashiro, T., 1996. High efficiency transformation of maize (*Zea mays* L.) mediated by *Agrobacterium tumefaciens*. Nature Biotechnology 6, 745–750.

IIVR (Indian Institute of Vegetable Research), 2009. Performance of Bt Brinjal Hybrids Containing cry 1AC Gene During Large Scale Trials, 2007–08, and 2008–09, IIVR (July 2008 and April 2009), Varanasi.

Jamel, P., Alam, M.Z., Umi, N., 2008. Media optimization for bio proteins production from cheaper carbon source. Journal of Engineering Science and Technology 3 (2), 124–130.

James, C., 2009. Global Status of Commercialized Biotech/GM Crops. 2009. ISAAA, Brief 41, Ithaca, NY.

James, C., 2011. Global Status of Commercialized Biotech/GM Crops. ISAAA Brief 43. ISAAA, Ithaca, NY.

James, C., 2012a. Executive Summary on Global Status of Commercialized Biotech/GM Crops. ISAAA, Ithaca, NY.

James, C., 2012b. Global Review of Commercialized Transgenic Crops: 2012. ISAAA Briefs No.44, Executive Summary. ISAAA, Ithaca, NY.

Kathage, J., Qaim, M., 2012. Economic impacts and impact dynamics of Bt (Bacillus thuringiensis) cotton in India. Proceedings of the National Academy of Sciences of the United States of America 109 (29), 11652–11656.

Khush, G.S., 2001. Green Revolution: the way forward. Nature Review of Genetics 2, 815–822.

Klee, H.J., Giovannoni, J.J., 2011. Genetics and control of tomato fruit ripening and quality attributes. Annual Review of Genetics 45, 41–59.

Koblentz, G.D., 2010. Biosecurity reconsidered: calibrating biological threats and responses. International Security 34 (4), 96–132.

Koul, B., Srivastava, S., Sanyal, I., Tripathi, B., Sharma, V., Amla, D.V., 2014. Transgenic tomato line expressing modified Bacillus thuringiensis cry1Ab gene showing complete resistance to two lepidopteran pests. Springer Plus 3, 84.

Kouser, S., Qaim, M., 2011. Impact of Bt cotton on pesticide poisoning in smallholder agriculture: a panel data analysis. Ecological Economics 70, 2105–2113.

Kumar, S., Prasanna, L.P.A., Wankhede, S., 2011. Potential benefits of Bt brinjal in India- an economic assessment. Agricultural Economics Research Review 24, 83–90.

Lundmark, C., 2007. Genetically modified maize. Bioscience 57 (11), 996.

Ma, J.K.C., Drake, P.M.W., Christou, P., 2003. The production of recombinant pharmaceutical proteins in plants. Nature 4, 794–805.

Manuwa, S.I., 2011. Properties of soybean for best postharvest options. In: El Shemy, H.A. (Ed.), Agricultural and Biological Sciences. Soybean Physiology and Biochemistry. www.intechopen.com.

Mascia, P.N., Flovell, R.B., 2004. Safe and acceptable strategies for producing foreign materials in plants. Current Opinion in Plant Biology 2004 (7), 189–195.

McKersie, B.D., Murnaghan, J., Jones, K.S., Bowley, S.R., 2000. Iron-superoxide dismutase expression in transgenic alfalfa increases winter survival without a detectable increase in photosynthetic oxidative stress tolerance. Plant Physiology 122, 1427–1437.

McKersie, B.D., Bowley, S.R., Jones, K.S., 1999. Winter survival of transgenic alfalfa overexpressing superoxide dismutase. Plant Physiology 119, 839–847.

Mellor, J., Dobason, M.J., Roberts, N.A., Tuite, M.F., Emtage, J.S., White, S., Lowe, P.A., Patel, T., Kingsman, A.J., Kingsman, S.M., 1983. Efficient synthesis of enzymatically active calf chymosin in Saccharomyces cerevisiae. Gene 24 (1), 1–14.

Morse, S., Bennett, R., Ismael, Y., 2007. Inequality and GM crops: a case-study of Bt cotton in India. AgBioForum 10, 44–50.

Nasseri, A.T., Rasoul- Amini, S., Moromvat, M.H., Ghasemi, Y., 2011. Single cell protein: production and process. American Journal of Food Technology 6 (2), 103–116.

Naylor, R., Hardy, R.W., Bureau, D.P., Chiu, A., Elliott, M., Farrell, A.P., Forster, I., Gatlin, D.M., Goldburg, R.J., Hua, K., Nichols, P.D., 2009. Feeding aquaculture in an era of finite resources. Proceedings of the National Academy of Sciences of the United States of America 106 (36), 15103–15110.

Nishimori, K., Kawaguchi, Y., Hidaka, M., Beppu, T., 1982. Expression of cloned prochymosin gene sequence in Escherichia coli. Gene 19, 337–344.

Percival, A.E., Wendeland, J.F., Stewart, J.M., 1999. Taxonomy and germplasm resources. In: Smith, W.C. (Ed.), Cotton: Origin, History, Technology and Production. Wiley, NewYork, pp. 33–63.

Pinstrup-Anderson, P., Pandra-Lorch, R., Rosegrant, M.W., 1999. World Food Prospects: Critical Issues for the Early Twenty-first Century Food policy report. International Food Policy Research Institute, Washington, DC.

Poppy, G.M., Sutherland, J.P., 2004. Can biological control benefit from genetically modified crops? tritrophic interactions on insect-resistant transgenic plants. Physiological Entomology 29, 257–268.

Pray, C.E., Huang, J., Hu, R., Rozelle, S., 2002. Five years of Bt cotton in China—The benefits continue. The Plant Journal 31, 423–430.

Qaim, M., Zilberman, D., 2003. Yield effects of genetically modified crops in developing countries. Science 299, 900–902.

Qaim, M., Subramanian, A., Naik, G., Zilberman, D., 2006. Adoption of Bt cotton and impact variability: insights from India. Review of Agricultural Economics 28, 48–58.

Qaim, M., 2009. The economics of genetically modified crops. Annual Review of Resource Economics 1, 665–693.

Qaim, M., 2010. Benefits of genetically modified crops for the poor: household income, nutrition and health. New Biotechnology 27 (5), 552–557.

Randhawa, G.J., Durga, P.C., 2001. Potential impact of transgenic crops on biodiversity. In: Randhawa, G.J., Khetarpal, R.K., Tyagi, R.K., Dhillon, B.S. (Eds.), Transgenic Crops and Biosafety Concerns. National Bureau of Plant Genetic Resources, New Delhi, pp. 58–64.

Raj, S.K., Singh, R., Pandey, S.K., Singh, B.P., 2005. Agrobacterium mediated tomato transformation and regeneration of transgenic lines expressing tomato leaf curl virus coat protein gene for resistance against TLCV infection. Current Science 88 (10), 1674–1679.

Renault, P., 2002. Genetically modified lactic acid bacteria: applications to food or health and risk assessment. Biochimie 84 (11), 1073–1087.

Sadashivappa, P., Qaim, M., 2009. *Bt* cotton in India: development of benefits and the role of government seed price interventions. AgBioForum 12, 172–183.

Sadiq, A., Khan, Z., Ahmad, B., Khan, I., Ali, J., 2014. Production of single cell protein from orange peels using *Aspergillus niger* and *Saccharomyces cerevisiae*. Global Journal of Biotechnology and Biochemistry 9 (1), 14–18.

Senthil, K., Singhal, K.K., 2004. Chemical composition and nutritional evaluation of transgenic cottonseed for ruminants. The Indian Journal of Animal Sciences 74 (5), 868–871.

Sherwood, M., 1974. Single cell protein comes of age. New Scientist 64, 634–639.

Smith, C.J.S., Watson, C.F., Ray, J., Bird, C.R., Morris, P.C., Schuch, W., Grierson, D., 1988. Antisense RNA inhibition of polygalacturonase gene expression in transgenic tomatoes. Nature 334, 724–726.

Smith, L.C., El Obeid, A.E., Jensen, H.H., 2000. The geography and causes of food insecurity in developing countries. Agriculture Economics 22, 199–215.

Song, X.L., Qiang, S., Liu, L.L., Xu, Y.H., 2002a. Assessment on gene flow through detection of sexual compatibility between transgenic rice with bar gene and *Echinochloa crusgalli* var. mitis. Agricultural Sciences in China 35, 1228–1231.

Song, Z.P., Lu, B.R., Zhu, Y.G., Chen, J.K., 2002b. Pollen competition between cultivated and wild rice species (*Oryza sativa and O. rufipogon*). New Phytologist 153, 289–296.

Southgate, E.M., Davey, M.R., Power, J.B., Merchant, R., 1995. Factors affecting the genetic engineering of plants by microprojectile bombardment. Biotechnology Advances 13, 631–657.

Sticklen, M., 2005. Plant genetic engineering to improve biomass characteristics for biofuels. Current Opinion in Biotechnology 17, 315–319.

Subramanian, A., Qaim, M., 2010. The impact of *Bt* cotton on poor households in rural India. Journal of Development Studies 46, 295–311.

Taipei Times, 2001. Cloned Cattle Set to Revolutionize Beef, Milk Markets. http://www.taipeitimes.com/News/biz/archives/2001/08/25/100125.

Takeda, S., Matsuoka, M., 2008. Genetic approaches to crop improvement: responding to environmental and population changes. Nature Reviews 444–457.

Tan, S., Evans, R., Singh, B., 2006. Herbicidal inhibitors of amino acid biosynthesis and herbicide tolerant crops. Amino Acids 30, 195–204.

Tennant, P., Fermin, G., Fitch, M.M., Manshardt, R.M., 2001. Papaya ringspot virus resistance of transgenic rainbow and sun up is affected by gene dosage, plant development, and coat protein homology. European Journal of Plant Pathology 107, 645–653.

Tiwari, R.K., Trivedi, M., Guang, Z.C., Guo, G.-Q., Zheng, G.-C., 2007. Genetic transformation of *Gentiana macrophylla* with *Agrobacterium rhizogenes*: growth and production of secoiridoid glucoside gentiopicroside in transformed hairy root cultures. Plant Cell Reports 26, 199–210.

Tiwari, R.K., Trivedi, M., Guang, Z.-C., Guo, G.-Q., Zheng, G.-C., 2008. *Agrobacterium rhizogenes* mediated transformation of *Scutellaria baicalensis* and production of flavonoids in hairy roots. Biologia Plantarum 52 (1), 26–35.

Torres, J.B., Ruberson, J.R., 2005. Canopy and ground dwelling predatory arthropods in commercial *Bt* and non-*Bt* cotton fields: patterns and mechanisms. Environmental Entomology 34, 1242–1256.

Trivedi, M., Sharma, A., Yadav, S.K., 2013. Plants as source of lead Molecules. In: Kumar, et al. (Ed.), Introduction to Drug Designing and Development. Nova Science Publisher, USA, pp. 49–80.

United States Food and drug administration, 2015. An Overview of Atlantic Salmon, Its Natural History, Aquaculture and Genetic Engineering.

UN SCN Fifth Report on the World Nutrition Situation for Improved Development Outcomes, 2005. United Nations System, Standing Committee on Nutrition, Geneva.

Wang, F., Yue, L., Wang, L., Li, J., Wang, X., Chi, Z., 2009. Genetic modification of the marine derived yeast *Yarrowia lipolytica* with high protein content using a GPI- anchor fusion expression system. Biotechnology Progress 25 (5), 1297–1303.

West, C., 2006. Economics and ethics in the genetic engineering of animals. Harvard Journal of Law and technology 19 (2), 414–442.

Whitehouse, M.E.A., Wilson, L.J., Fitt, G.P., 2005. A comparison of arthropod communities in transgenic Bt and conventional cotton in Australia. Environmental Entomology 34, 1224–1241.

Windass, J.D., Worsey, M.J., Pioli, E.M., Pioli, D., Barth, P.T., Atherton, K.T., Dart, E.C., Byrom, D., Powell, K., Senior, P.J., 1980. Improved conversion of methanol to single-cell protein by *Methylophilus methylotrophus*. Nature 287 (5781), 396–401.

Ye, X.D., Al-Babili, S., Kloti, A., 2000. Engineering the provitamin A (beta-carotene) biosynthetic pathway into (carotenoid-free) rice endosperm. Science 287, 303–305.

Further Reading

AVRDC, 2003. Development of an Integrated Pest Management Strategy for Eggplant Fruit and Shoot Borer in South Asia. Technical Bulletin 28. Asian Vegetable research and Development Center, Tainan.

Choudhary, B., Gheysen, G., Buysse, J., Meer, P., Burssens, S., 2014. Regulatory options for genetically modified crops in India. Plant Biotechnology Journal 12, 135–146.

D'Souza, A.L., Rajkumar, C., Cooke, J., Bulpitt, C.J., 2002. Probiotics in prevention of antibiotic associated diarrhoea: meta analysis. British Medical Journal 324, 1361–1364.

Daniell, H., 1999. GM Crops: public perception and scientific solutions. Trends in Plant Science 4, 467–469.

Echelard, Y., Ziomek, C.A., Meade, H.M., 2006. Production of recombinant therapeutic proteins in the milk of transgenic animals. Biopharm International 19 (8), 36–46.

Eenennaam, A.V., Davis, U.C., 2008. Genetically Engineered Animals: An Overview. http://animalscience.ucdavis.edu/.

FAO, WFP, IFAD, 2012. The State of Food Insecurity in the World 2012. Economic Growth Is Necessary but Not Sufficient to Accelerate Reduction of Hunger and Malnutrition. FAO, Rome.

FAO, 2009. How to Feed the World in 2050. Discussion Paper, High-level Expert Forum. The Food and Agricultural Organization, Rome, Italy.

James, C., 2010. Global Status of Commercialized Biotech and GM Crops: 2010. International Service for the Acquisition of Agribiotech Applications (ISAAA), NY.

Kouwe, P.V., Layton, R., Amijee, F., Fuentes, A.M., Tencalla, F., 2006. A tiered system for assessing the risk of genetically modified plants to non-target organisms. Environmental Biosafety Research 5 (02), 57–65.

Kuroiwa, Y., Kasinathan, P., Choi, Y.J., Naeem, R., Tomizuka, K., Sullivan, E.J., Knott, J.G., Duteau, A., Goldsby, R.A., Osborne, B.A., Ishida, I., Robl, J.M., 2002. Cloned transchromosomic calves producing human immunoglobulin. Nature Biotechnology 20, 889–894.

Kudlu, C., Stone, G.D., 2013. The trials of genetically modified food: *Bt* eggplant and ayurvedic medicine in India. Food, Culture and Society 16 (1), 21–42.

McFarland, L.V., Elmer, G.W., 1995. Biotherapeutic agents: past, present and future. Micro Ecology and Therapy 23, 46–73.

Paine, J.A., Shipton, C.A., Chaggar, S., 2005. Improving the nutritional content of Golden Rice through increased provitamin A content. Nature Biotechnology 23, 482–487.

Patel, T.B., Pequignot, E., Parker, S.H., Leavitt, M.C., Greenberg, H.E., Kraft, W.K., 2007. Transgenic avian-derived recombinant human interferon-alpha 2b (AVI-005) in healthy subjects: an open-label, single dose, controlled study. International Journal of Clinical Pharmacology and Therapeutics 45, 161–168.

Phillips, T., 2008. Genetically modified organisms (GMOs): transgenic crops and recombinant DNA technology. Nature Education 1 (1), 213.

Potrykus, I., 2001. Golden Rice and beyond. Plant Physiology 125, 1157–1161.

Rosegrant, M.,W., Paisner, M.S., Mejer, S., Wit cover, J., 2001. Global Food Outlook Trends, Alternatives and Choices. A 2020 Vision for Food Agriculture and the Environment Initiative. IFPRI, Washington, DC. 20,21.

Royan Institute, 2009. Iran's First Transgenic Goats Produced in Royan Institute. (press release January 30, 2009).

Singhal, K.K., Tyagi, A.K., Rajput, Y.S., 2001. Effect of Feeding Cottonseed Produced from *Bt* Cotton on Feed Intake, Milk Production and Composition in Lactating Cows. Report submitted to Mahyco, Mumbai.

Teitelbaum, J.E., Walker, W.A., 2002. Nutritional impact of pre- and probiotics as protective gastrointestinal organisms. Annual Review of Nutrition 22, 107–138.

Union of Concerned Scientists, 2009. Failure to Yield: Evaluating the Performance of Genetically Engineered Crops. UCS Publications, Cambridge, MA.

United States Department of Energy, 2007. Office of Biological and Environmental Research, Human Genome Program. Human Genome Project Information: Genetically modified foods and organisms.

Van Berkel, P.H., Welling, M.M., Geerts, M., van Veen, H.A., Ravensbergen, B., Salaheddine, M., Pauwels, E.K., Pieper, F., Nuijens, J.H., Nibbering, P.H., 2002. Large scale production of recombinant human lactoferrin in the milk of transgenic cows. Nature Biotechnology 20, 484–487.

Van Eenennaam, A., 2008. Genetically Engineered Animals: An Overview. http://animalscience.ucdavis.edu/.

World Bank, 2000. World Development Report 2000/2001. Attacking Poverty. World Bank, Washington, DC. www.defra.gov.uk/environment/gm/fse.

Websites

http://www.biotopics.co.uk/edexcel/biotechnol/gmmfoo.html

http://www.fda.gov/ForConsumers/ConsumerUpdates/ucm143980.htm

http://www.fda.gov/AdvisoryCommittees/CommitteesMeetingMaterials/VeterinaryMedicineAdvisoryCommittee/ucm222635.html

http://www.isaaa.org/kc

http://www.gmo-compass.org/eng/grocery_shopping/crops/

http://www.gmo-compass.org/eng/grocery_shopping/crops/23.genetically_modified_potato.html

http://www.isaaa.org/resources/publications/briefs/41/ececutivesummary/default.asap

http://www.gmo-compass.org/eng/grocery_shopping/crops/19.genetically_modified_soybean.html

http://www.thehindu.com/news/national/geac-clears-field-trials-for-gm-crops/article6225697

FAO Biotechnology www.fao.org/biotech/en/

Index

Note: Page numbers followed by "f" indicate figures, "t" indicate tables, and "b" indicate boxes.

Printed in the United States
By Bookmasters